HANDBOOK OF POLYMER SYNTHESIS

PLASTICS ENGINEERING

Founding Editor
Donald E. Hudgin
Professor
Clemson University
Clemson, South Carolina

1. Plastics Waste: Recovery of Economic Value, *Jacob Leidner*
2. Polyester Molding Compounds, *Robert Burns*
3. Carbon Black-Polymer Composites: The Physics of Electrically Conducting Composites, *edited by Enid Keil Sichel*
4. The Strength and Stiffness of Polymers, *edited by Anagnostis E. Zachariades and Roger S. Porter*
5. Selecting Thermoplastics for Engineering Applications, *Charles P. MacDermott*
6. Engineering with Rigid PVC: Processability and Applications, *edited by I. Luis Gomez*
7. Computer-Aided Design of Polymers and Composites, *D. H. Kaelble*
8. Engineering Thermoplastics: Properties and Applications, *edited by James M. Margolis*
9. Structural Foam: A Purchasing and Design Guide, *Bruce C. Wendle*
10. Plastics in Architecture: A Guide to Acrylic and Polycarbonate, *Ralph Montella*
11. Metal-Filled Polymers: Properties and Applications, *edited by Swapan K. Bhattacharya*
12. Plastics Technology Handbook, *Manas Chanda and Salil K. Roy*
13. Reaction Injection Molding Machinery and Processes, *F. Melvin Sweeney*
14. Practical Thermoforming: Principles and Applications, *John Florian*
15. Injection and Compression Molding Fundamentals, *edited by Avraam I. Isayev*
16. Polymer Mixing and Extrusion Technology, *Nicholas P. Cheremisinoff*
17. High Modulus Polymers: Approaches to Design and Development, *edited by Anagnostis E. Zachariades and Roger S. Porter*
18. Corrosion-Resistant Plastic Composites in Chemical Plant Design, *John H. Mallinson*
19. Handbook of Elastomers: New Developments and Technology, *edited by Anil K. Bhowmick and Howard L. Stephens*
20. Rubber Compounding: Principles, Materials, and Techniques, *Fred W. Barlow*
21. Thermoplastic Polymer Additives: Theory and Practice, *edited by John T. Lutz, Jr.*
22. Emulsion Polymer Technology, *Robert D. Athey, Jr.*
23. Mixing in Polymer Processing, *edited by Chris Rauwendaal*
24. Handbook of Polymer Synthesis, Parts A and B, *edited by Hans R. Kricheldorf*
25. Computational Modeling of Polymers, *edited by Jozef Bicerano*
26. Plastics Technology Handbook: Second Edition, Revised and Expanded, *Manas Chanda and Salil K. Roy*

HANDBOOK OF POLYMER SYNTHESIS

Second Edition

Hans R. Kricheldorf
Universität Hamburg
Hamburg, Germany

Oskar Nuyken
Technical University
München, Germany

Graham Swift
GS Polymer Consultants
Chapel Hill, North Carolina, U.S.A.

CRC Press
Taylor & Francis Group
Boca Raton London New York

CRC Press is an imprint of the
Taylor & Francis Group, an **informa** business

First published 2005 by Marcel Dekker

Published 2019 by CRC Press
Taylor & Francis Group
6000 Broken Sound Parkway NW, Suite 300
Boca Raton, FL 33487-2742

First issued in paperback 2020

ISBN 13: 978-0-367-57822-0 (pbk)
ISBN 13: 978-0-8247-5473-0 (hbk)

Library of Congress Cataloging-in-Publication Data
A catalog record for this book is available from the Library of Congress.

Visit the Taylor & Francis Web site at
http://www.taylorandfrancis.com

and the CRC Press Web site at
http://www.crcpress.com

Preface

The purpose of the 1st edition of this handbook was to present a condensed but comprehensive review of the methods used for syntheses and modifications of the most important classes of polymers. The good acceptance of this handbook by the international scientific community has prompted the publisher to launch a second edition updating the literature up to the year 2000 for the most widely studied groups of polymers. The editors hope that this 2nd edition will provide the chemists with an useful first hand information on new preparative methods in the field of polymer science.

Contents

List of Contributors

E. Bacher, Technische Universität München, Garching, Germany

Harald Braun, Technische Universität München, Garching, Germany

James Crivello, Rensselaer Polytechnic Institute, Troy, New York

Javier de Abajo, Institute of Polymer Science and Technology, Madrid, Spain

José G. de la Campa, Institute of Polymer Science and Technology, Madrid, Spain

Ulrich Frenzel, Technische Universität München, Garching, Germany

J. V. Grazulevicius, Kaunas University of Technology, Kaunas, Lithuania

Manfred L. Hallensleben, Institut für Makromolekulare Chemie, Universität Hannover, Hannover, Germany

B. Hinrichs, University of Hamburg, Hamburg, Germany

Samuel J. Huang, Institute of Materials Science, University of Connecticut, Storrs, Connecticut

Walter Kaminsky, Institute of Technical and Macromolecular Chemistry, University of Hamburg, Hamburg, Germany

Hans R. Kricheldorf, Institute of Technical and Macromolecular Chemistry, University of Hamburg, Hamburg, Germany

Krzysztof Matyjaszewski, Center for Macromolecular Engineering, Carnegie Mellon University, Pittsburgh, Pennsylvania

Bettina K. M. Müller, Technische Universität München, Garching, Germany

Herbert Naarmann, (emerit) BASF AG Ludwigshafen

Oskar Nuyken, Technische Universität München, Garching, Germany

Zoran S. Petrović, Pittsburg State University, Kansas Polymer Research Center, Pittsburg, Kansas

M. Rojahn, Technische Universität München, Garching, Germany

James Spanswick, Center for Macromolecular Engineering, Carnegie Mellon University, Pittsburgh, Pennsylvania

P. Strohriegl, Universität Bayreuth, Makromolekulare Chemie I, and Bayreuther Institut für Makromolekülforschung (BIMF), Bayreuth, Germany

R. Weberskirch, Technische Universität München, Garching, Germany

V. Wiederhirn, Technische Universität München, Garching, Germany

Dieter Wöhrle, University of Bremen, Bremen, Germany

1

Polyolefins

Walter Kaminsky
University of Hamburg, Hamburg, Germany

I. INTRODUCTION

The polyolefins production has increased rapidly in the 40 years to make polyolefins the major tonnage plastics material worldwide. In 2003, 55 million tons of polyethene and 38 million t/a polypropene were produced [1]. These products are used for packing material, receptacles, pipes, domestic articles, foils, and fibers. Polyolefins consist of carbon and hydrogen atoms only and the monomers are easily available. Considering environmental aspects, clean disposal can be achieved by burning or by pyrolysis, for instance. Burning involves conversion to CO_2 and H_2O, exclusively.

By copolymerization of ethene and propene with higher n-olefins, cyclic olefins, or polar monomers, product properties can be varied considerably, thus extending the field of possible applications. For this reason terpolymers of the ethene/propene n-olefin type are the polymers with the greatest potential. Ethene can be polymerized radically or by means of organometallic catalysts. In the case of polyisobutylene a cationic polymerization mechanism takes place. All other olefins (propene, 1-butene, 4-methylpentene) are polymerized with organometallic catalysts. The existence of several types of polyethene as well as blends of these polymers provides the designer with an unusual versatility in resin specifications. Thus polyethene technology has progressed from its dependence on one low-density polymer to numerous linear polymers, copolymers, and blends that will extend the use of polyethene to many previously unacceptable applications.

Polypropene also shows versatility and unusual growth potential. The main advantage is improved susceptibility to degradation by outdoor exposure. The increase in the mass of polypropene used for the production of fibers and filaments is inive of the versatility of this polymer.

Synthetic polyolefins were first synthetisized by decomposition of diazomethane [2]. With the exception of polyisobutylene, these polymers were essentially laboratory curiosities. They could not be produced economically. The situation changed with the discovery of the high pressure process by Fawcett and Gibson (ICI) in 1930: ethene was polymerized by radical compounds [3]. To achieve a sufficient polymerization rate, a pressure of more than 100 MPa is necessary. First produced in 1931, the low density polyethene (LDPE) was used as isolation material in cables.

Due to its low melting point of less than 100 °C LDPE could not be applied to the production of domestic articles that would be used in contact with hot water.

1

Important progress for a broader application was made when Hogan and Banks [4] (Phillips Petroleum) and Ziegler et al. [5] found that ethene can be polymerized by means of activated transition metal catalyst systems. In this case the high density polyethene (HDPE), a product consisting of highly linear polymer chains, softens above 100 °C. Hogan polymerized ethene using a nickel oxide catalyst and later a chromium salt on an alumina-silica support. Zletz [6] used molybdenum oxide on alumina in 1951 (Standard Oil); Fischer [7] used aluminum chloride along with titanium tetrafluoride (BASF 1953) for the production of high-density polyethene. The latter catalyst has poor activity and was never used commercially. Zieglers [5] use of transition metal halogenides and aluminum organic compounds and the work of Natta [8] in applying this catalyst system for the synthesis of stereoregular polyolefins were probably the two most important achievements in the area of catalysis and polymer chemistry in the last 50 years. They led to the development of a new branch of the chemical industry and to a large production volume of such crystalline polyolefins as HDPE, isotactic polypropene, ethane-propene rubbers, and isotactic poly(l-butene). For their works, Ziegler and Natta were awarded the Nobel Prize in 1963. The initial research of Ziegler and Natta was followed by an explosion of scientific papers and patents covering most aspects of olefin polymerization, catalyst synthesis, and polymerization kinetics as well as the structural, chemical, physical, and technological characteristics of stereoregular polyolefins and olefin copolymers. Since that first publication, more than 20 000 papers and patents have been published on subjects related to that field. Several books and reviews giving detailed information on the subjects of these papers have been published [9–19].

The first generation of Ziegler–Natta catalysts, based on $TiCl_3/AlEt_2Cl$, was characterized by low polymerization activity. Thus a large amount of catalyst was needed, which contaminated the raw polymer. A washing step that increased production costs was necessary. A second generation of Ziegler–Natta catalysts followed, in which the transition metal compound is attached to a support ($MgCl_2$, SiO_2, Al_2O_3). These supported catalysts are of high activity. The product contains only traces of residues, which may remain in the polymer. Most Ziegler–Natta catalysts are heterogeneous. More recent developments show that homogeneous catalyst systems based on metallocene-alumoxane and other single-site catalysts can also be applied to olefin polymerization [20–23]. These systems are easy to handle by laboratory standards, and show highest activities and an extended range of polymer products.

The mechanism of Ziegler–Natta catalysis is not known in detail. A two-step mechanism is commonly accepted: First, the monomer is adsorbed (π-complex bonded) at the transition metal. During this step the monomer may be activated by the configuration established in the active complex. Second, the activated monomer is inserted into the metal–carbon bond. In this sequence the metal-organic polymerization resembles what nature accomplishes with enzymes.

Ziegler–Natta catalysts are highly sensitive, to oxygen, moisture, and a large number of chemical compounds. Therefore, very stringent requirements of reagent purity and utmost care in all manipulations of catalysts and polymerization reactions themselves are mandatory for achieving experimental reproducibility and reliability. Special care must be taken to ensure that solvents and monomers are extremely pure. Alkanes and aromatic compounds have no substantial effect on the polymerization and can therefore be used as solvents. Secondary alkenes usually have a negative effect on polymerization rates, and alkynes, allenes (1,2-butadiene), and conjugated dienes are known to act as catalyst poisons, as they tend to form stable complexes.

Almost all polar substances exert a strong negative influence on the polymerization. COS and hydrogen sulfide, particularly, are considered to be strong catalyst poisons, of which traces of more than 0.2 vol ppm affect a catalyst's activity. Neither the solvent nor the gaseous monomer should contain water, carbon dioxide, alcohols, or other polar substances in excess of 5 ppm. Purification may be carried out by means of molecular sieves.

The termination of the polymerization reaction by the addition of carbon monoxide is used to determine the active centers (sites) of the catalyst. Hydrogen is known to slightly reduce the catalyst's activity. Yet it is commonly used as an important regulator to lower the molecular weights of the polyethene or polypropene produced.

II. POLYETHENE

The polymerization of ethene can be released by radical initiators at high pressures as well as by organometallic coordination catalysts. The polymerization can be carried out either in solution or in bulk. For pressures above 100 MPa, ethene itself acts as a solvent. Both low- and high-molecular-weight polymers up to 10^6 g/mol can be synthesized by either organometallic coordination or high pressure radical polymerization. The structure of the polyethene differs with the two methods. Radical initiators give more-or-less branched polymer chains, whereas organometallic coordination catalysts synthesize linear molecules.

A. Radical Polymerization

Since the polymerization of ethene develops excess heat, radical polymerization on a laboratory scale is best carried out in a discontinuous, stirred batch reactor. On a technical scale, however, column reactors are widely used. The necessary pressure is generally kept around 180 to 350 MPa and the temperature ranges from 180 to 350 °C [24–29]. Solvent polymerization can be performed at substantial lower pressures and at temperatures below 100 °C. The high-pressure polymerization of ethene proceeds via a radical chain mechanism. In this case chain propagation is regulated by disproportionation or recombination.

$$R\cdot + H_2C = CH_2 \longrightarrow R-CH_2-CH_2\cdot \tag{1}$$

$$R-CH_2-CH_2\cdot + nH_2C = CH_2 \longrightarrow R-(CH_2-CH_2)_n-CH_2-CH_2\cdot \tag{2}$$

$$R-(CH_2-CH_2)_n-CH_2-CH_2\cdot + R-(CH_2-CH_2)_m-CH_2-CH_2\cdot$$

recombination | disproportionation

$$R-(CH_2-CH_2)_{n+m+2}-R$$

$$R-(CH_2-CH_2)_n-CH_2-CH_3$$
$$+ R-(CH_2-CH_2)_m-CH=CH_2$$

$$\tag{3}$$

The rate constants for chain propagation and chain termination at 130° and 180 MPa can be specified as follows [30]:

$$M_p = 5.93 \times 103\,L \times mol^{-1}\,s^{-1}$$
$$M_t = 2 \times 108\,L \times mol^{-1}\,s^{-1}$$

Intermolecular and intramolecular chain transfer take place simultaneously. This determines the structure of the polyethene. Intermolecular chain transfer results in long flexible side chains but is not as frequent as intramolecular chain transfer, from which short side chains mainly of the butyl type arise [31,32].

Intermolecular chain transfer:

$$P_1—CH_2—CH_2 \cdot + P_2—CH_2—CH_2—P_3 \longrightarrow \tag{4}$$

$$P_1—CH_2—CH_3 + P_2—CH—CH_2—P_3 + nH_2C{=}CH_2 \longrightarrow P_2—CH_2—CH_2—P_3$$
$$\underset{(CH_2—CH_2)_{n-1}—CH_2—CH_2\,\cdot}{|} \tag{5}$$

Intramolecular chain transfer:

$$P—CH_2 \overset{CH_2}{\diagup} \underset{CH_2 \diagup}{\overset{CH_2}{|}} CH_2 \longrightarrow P—CH \overset{CH_2}{\diagup} \underset{CH_3 \diagup}{\overset{CH_2}{|}} CH_2 \tag{6}$$

$$P—CH—C_4H_9 + nH_2C{=}CH_2 \longrightarrow P—CH—(CH_2—CH_2)_{n-1}—CH_2—CH_2 \cdot$$
$$\underset{C_4H_9}{|} \tag{7}$$

Radically created polyethene typically contains a total number of 10 to 50 branches per 1000 C atoms. Of these, 10% are ethyl, 50% are butyl, and 40% are longer side chains. With the simplified formulars (6) and (7), not all branches observed could be explained [33,34]. A high-pressure stainless steal autoclave (0.1 to 0.51 MPa) equipped with an inlet and outlet valve, temperature conductor, stirrer, and bursting disk is used for the synthesis. Best performance is obtained with an electrically heated autoclave [35–41].

To prevent self-degeneration, the temperature should not exceed 350 °C. Ethene and intitiator are introduced by a piston or membrane compressor. An in-built sapphire window makes it possible to observe the phase relation. After the polymerization is finished, the reaction mixture is released in two steps. Temperature increases are due to a negative Joule–Thompson effect. At 26 MPa, ethene separates from the 250 °C hot polymer melt. After further decompression down to normal pressure, the residual ethene is removed [42–46]. Reaction pressure and temperature are of great importance for the molecular weight average, molecular weight distribution, and structure of the polymer. Generally, one can say that with increasing reaction pressure the weight average increases, the distribution becomes narrower, and short- and long-chain branching both decrease [47].

Table 1 Peroxides as initiators for the high-pressure polymerization of ethene.

Peroxide	Molecular weight	Half-time period of 1 min by a polymerization temperature (°C)
$(H_3C)_3\text{-COOC}(CH_3)_3$	146.2	190
$(H_3C)_3C\text{—}\overset{\displaystyle O}{\overset{\|}{C}}=C(CH_3)_3$	174.2	110
$[H_5C_2\text{—}\overset{\displaystyle O}{\overset{\|}{C}}\text{—O}]_2$	146	115
$H_9C_4\text{—}\underset{\displaystyle C_2H_5}{\overset{}{CH}}\text{—}\overset{\displaystyle O}{\overset{\|}{C}}\text{—O—OC}(CH_3)_3$	216.3	130
$[H_3C\text{—}(CH_2)_6\text{—}\overset{\displaystyle O}{\overset{\|}{C}}\text{—O}]_2$	286.4	120
$(H_3C)_3C\text{—}CH_2\text{—}\underset{}{\overset{\displaystyle CH_3}{CH}}\text{—}CH_2\text{—}\overset{\displaystyle O}{\overset{\|}{C}}\text{—O—O—C}(CH_3)_3$	230.3	160
$(H_3C)_3C\text{—}CH_2\text{—}\underset{\displaystyle CH_3}{\overset{\displaystyle CH_3}{C}}\text{—}CH_2\text{—}\overset{\displaystyle O}{\overset{\|}{C}}\text{—OO—C}(CH_3)_3$	246.4	100
$[H_3C\text{—}(CH_2)_{10}\text{—}\overset{\displaystyle O}{\overset{\|}{O}}\text{—O}]_2$	194.2	120
$H_5C_6\text{—}\overset{\displaystyle O}{\overset{\|}{C}}\text{—O—O—C}(CH_3)_3$	194.2	170
$[H_9C_4\text{—O—}\overset{\displaystyle O}{\overset{\|}{C}}\text{—O}]_2$	234.3	90

Oxygen or peroxides are used as the initiators. Initiation is very similar to that in many other free-radical polymerizations at different temperatures according to their half-live times (Table 1). The pressure dependence is low. Ethene polymerization can also be started by ion radiation [48–51]. The desired molecular weight is best adjusted by the use of chain transfer reagents. In this case hydrocarbons, alcohols, aldehydes, ketones, and esters are suitable [52,53].

Table 2 shows polymerization conditions for the high-pressure process and density, molecular weight, and weight distribution of the polyethene (LDPE). Bunn [54] was the first to study the structure of polyethene by x-ray. At a time when there was still considerable debate about the character of macromolecules, the demonstration that wholly synthetic and crystalline polyethene has a simple close-packed structure in which the bond angles and bond lengths are identical to those found in small molecules such

Table 2 Polymerization conditions and product properties of high-pressure polyethene (LDPE).

Pressure (MPa)	Temp. (°C)	Regulator (propane) (wt%)	Density (g/cm³)	Molecular weight MFI	Distribution
165	235	1.6	0.919	1.3	20
205	290	1.0	0.915	17.0	10
300	250	3.9	0.925	2.0	10

Source: Ref. 29.

as $C_{36}H_{74}$ [55–57], strengthened the strictly logical view that macromolecules are a multiplication of smaller elements joined by covalent bonds. LDPE crystallizes in single lamellae with a thickness of 5.0 to 5.5 nm and a distance between lamellae of 7.0 nm which is filled by an amorphous phase. The crystallinity ranges from 58 to 62%.

Recently, transition metals and organometallics have gained great interest as catalysts for the polymerization of olefins [58,59] under high pressure. High pressure changes the properties of polyethene in a wide range and increases the productivity of the catalysts. Catalyst activity at temperatures higher than 150 °C is controlled primarily by polymerization and deactivation. This fact can be expressed by the practical notion of catalyst life time, which is quite similar to that used with free-radical initiators. The deactivation reaction at an aluminum alkyl concentration below 5×10^{-5} mol/l seems to be first order reaction [60]. Thus for various catalyst-activator systems, the approximate polymerization times needed in a continuous reactor to ensure the best use of catalyst between 150 to 300 °C are between several seconds and a few minutes. Several studies have been conducted to obtain Ziegler–Natta catalysts with good thermal stability. The major problem to be solved is the reduction of the transition metal (e.g., $TiCl_3$) by the cocatalyst, which may be aluminum dialkyl halide, alkylsiloxyalanes [60], or aluminoxane [59].

Luft and colleagues [61,62] investigated high-pressure polymerization in the presence of heterogeneous catalysts consisting of titanium supported on magnesium dichloride or with homogeneous metallocene catalysts. With homogeneous catalysts, a pressure of 150 MPa (80 to 210 °C) results in a productivity of 700 to 1800 kg PE/cat, molecular weights up to 110 000 g/mol, and a polydispersity of 5 to 10, with heterogeneous catalysts, whereas the productivity is 3000 to 7000 kg PE/cat, molecular weight up to 70 000 g/mol, and the polydispersity 2.

B. Coordination Catalysts

Ethene polymerization by the use of catalysts based on transition metals gives a polymer exhibiting a greater density and crystallinity than the polymer obtained via radical polymerization. Coordination catalysts for the polymerization of ethene can be of very different nature. They all contain a transition metal that is soluble or insoluble in hydrocarbons, supported by silica, alumina, or magnesium chloride [5,63]. In most cases cocatalysts are used as activators. These are organometallic or hydride compounds of group I to III elements; for example, $AlEt_3$, $AlEt_2Cl$, $Al(i\text{-}Bu)_3$, $GaEt_3$, $ZnEt_2$, n-BuLi, amyl Na [64]. Three groups are used for catalysis:

1. Catalysts based on titanium or zirconium halogenides or hydrides in connection with aluminum organic compound (Ziegler catalysts)

2. Catalysts based on chromium compounds supported by silica or alumina without a coactivator (Phillips catalysts)
3. Homogeneous catalysts based on metallocenes in connection with aluminoxane or other single site catalysts such as nickel ylid, nickel diimine, palladium, iron or cobalt complexes.

Currently, mainly Ziegler and Phillips catalysts as well as some metallocene catalysts [63] are generally used technically.

Three different processes are possible: the slurry process, the gas phase process, and the solvent process [65–68]:

1. ***Slurry process.*** For the slurry process hydrocarbons such as isobutane, hexane, n-alkane are used in which the polyethene is insoluble. The polymerization temperature ranges from 70 to 90 °C, with ethene pressure varying between 0.7 and 3 MPa. The polymerization time is 1 to 3 h and the yield is 95 to 98%. The polyethene produced is obtained in the form of fine particles in the diluent and can be separated by filtration. The molecular weight can be controlled by hydrogen; the molecular weight distribution is regulated by variation of the catalyst design or by polymerization in several steps under varying conditions [69–73]. The best preparation takes place in stirred vessels or loop reactors.

   In some processes the polymerization is carried out in a series of cascade reactors to allow the variation of hydrogen concentration through the operating steps in order to control the distribution of the molecular weights. The slurry contains about 40% by weight polymer. In some processes the diluent is recovered after centrifugation and recycled without purification.

2. ***Gas phase polymerization.*** Compared to the slurry process, polymerization in the gas phase has the advantage that no diluent is used which simplifies the process [74–76]. A fluidized bed that can be stirred is used with supported catalysts. The polymerization is carried out at 2 to 2.5 MPa and 85 to 100 °C. The ethene monomer circulates, thus removing the heat of polymerization and fluidizing the bed. To keep the temperature at values below 100 °C, gas conversion is maintained at 2 to 3 per pass. The polymer is withdrawn periodically from the reactor.

3. ***Solvent polymerization.*** For the synthesis of low-molecular-weight polyethene, the solvent process can be used [77,78]. Cyclohexane or another appropriate solvent is heated to 140 to 150 °C. After addition of the catalyst, very rapid polymerization starts. The vessel must be cooled indirectly by water. Temperature control is also achieved via the ethene pressure, which can be varied between 0.7 and 7 MPa.

In contrast to high-pressure polyethene with long-chain branches, the polyethene produced with coordination catalysts has a more or less linear structure (Figure 1) [79]. A good characterization of high-molecular-weight-polyethenes gives the melt rheological behaviour [80] (shear viscosity, shear compliance). The density of the homopolyethenes is higher but it can be lowered by copolymerization. Polymers produced with unmodified Ziegler catalysts showed extremely high molecular weight and broad distribution [81]. In fact, there is no reason for any termination step, except for consecutive reaction. Equations (8) to (11) show simplified chain propagation and chain termination steps [11].

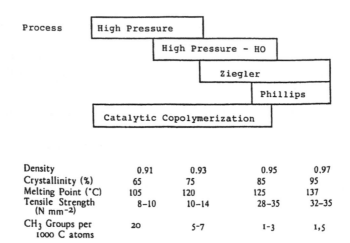

Process	High Pressure			
	High Pressure - HO			
	Ziegler			
	Phillips			
Catalytic Copolymerization				

Density	0.91	0.93	0.95	0.97
Crystallinity (%)	65	75	85	95
Melting Point (°C)	105	120	125	137
Tensile Strength (N mm^{-2})	8–10	10–14	28–35	32–35
CH$_3$ Groups per 1000 C atoms	20	5-7	1-3	1,5

Figure 1   Comparison of various polyethenes.

Chain propagation:

$$\text{(8)}$$

Chain termination:

(a)   By β elimination with H transfer to monomer

$$\text{(9)}$$

(b)   By hydrogenation

$$\text{(10)}$$

(c)   By β elimination forming hydride

$$CH_2 = CH - R$$

$$+$$

$$M - H \qquad (11)$$

(M=transition metal)

Termination via hydrogenation gives saturated polymer and metal hydride. The termination of a growing molecule by an α-elimination step forms a polymer with an olefinic end group and a metal hydride. In addition, an exchange reaction with ethene forming a polymer with an olefinic end group and an ethyl metal is observed.

1.   Titanium Chloride-Based Catalysts

The first catalyst used by Ziegler et al. [5,82] for the polymerization of ethene was a mixture of $TiCl_4$ and $Al(C_2H_5)_3$, each of which is soluble in hydrocarbons. In combination they form an olive-colored insoluble complex that is very unstable. Its behavior is very sensitive to a number of experimental parameters, such as Al/Ti ratio, temperature and time of mixing of all components, and absolute and relative concentrations of reactants [83]. After complexation, $TiCl_4$ is reduced by a very specific reduction process. This reduction involves alkylation of $TiCl_4$ with aluminum alkyl molecules followed by a dealkylation reduction to a trivalent state:

Complexation: $TiCl_4 + AlEt_3 \rightleftharpoons TiCl_4 \cdot AlEt_3$ (12)

Alkylation: $TiCl_4.AlEt_3 \rightleftharpoons EtTiCl_3 \cdot AlEt_2Cl$ (13)

Reduction: $2EtTiCl_3 \rightleftharpoons 2TiCl_3 + Et_2$ (14)

Under drastic conditions, $TiCl_3$ can be reduced to $TiCl_2$ in a similar way. The actual $TiCl_3$ product is a compound alloyed with small amounts of $AlCl_3$ and probably some chemisorbed $AlEt_2Cl$. The mechanistic process is very complex and not well understood.

Instead of $Al(C_2H_5)_3$, also $Al(C_2H_5)_2Cl$, $Al_2(C_2H_5)_3Cl_3$, or $Al(i\text{-}Bu)_3$ could be used. These systems, called first-generation catalysts, are used for the classic process of olefin polymerization. In practice, however, the low activity made it necessary to deactivate the catalyst after polymerization, remove the diluent, and then remove the residues of catalyst with HCl and alcohols. This treatment is followed by washing the polyethene with water and drying it with steam. Purification of the diluent recovered and feedback of the monomer after a purification step involved further complications. The costs of these steps reduced the advantage of the low-pressure polymerization process. Therefore, it was one of the main tasks of polyolefin research to develop new catalysts (second generation catalysts) that are more active, and can therefore remain in the polymer without any disadvantage to the properties (Table 3) [84]. The process is just as sensitive to perturbation, it is cheaper, and energy consumption as well as environmental loading are lower. It is also possible to return to the polymerization vessel diluent containing a high amount of the aluminum alkyl. The second generation is based on $TiCl_3$ compounds or supported catalysts $MgCl_2/TiCl_4/Al(C_2H_5)_3$ or $CrO_3(SiO_2)$ (Phillips).

Table 3 Comparison of various catalyst processes for ethene polymerization.

First generation	Second generation
Catalyst preparation	Catalyst preparation
Polymerization	Polymerization
Limited influence to molecular weight and weight distribution	Great variation of molecular weight and weight distribution
Catalyst deactivation with alcohol	
Filtration	Filtration
Washing with water (HCl), wastewater treatment, purification, and drying of diluent	Feedback of diluent
Drying of PE	Drying of PE
Finishing	Finishing
Thermal degradation of molecular weight, blending	
Stabilization	Stabilization

Source: Ref. 84.

2. Unsupported Titanium Catalysts

There is a very large number of different combinations of aluminum alkyls and titanium salts to make high mileage catalysts for ethene polymerization, such as α-TiCl$_3$ + AlEt$_3$, AlEt$_2$Cl, Al(i-Bu)$_3$, and Ti(III)alkanolate-chloride + Al(i-$_{hexyl}$)$_3$ [85]. TiCl$_3$ exists in four crystalline modifications, the a, b, g, and d forms [86]. The composition of these TiCl$_3$s can be as simple as one Ti for as many as three Cl, or they can have a more complex structure whereby a second metal is cocrystallized as an alloy in the TiCl$_3$. The particular method of reduction determines both composition and crystalline modification. α-TiCl$_3$ can be synthesized by reduction of TiCl$_4$ with H$_2$ at elevated temperatures (500 to 800 °C) or with aluminum powder at lower temperatures (about 250 °C); in this case the α-TiCl$_3$ contains Al cations [87]. More active are γ- and δ-TiCl$_3$ modifications. They are formed by heating the α-TiCl$_3$ to 100 or 200 °C. The preferred α-TiCl$_3$ contains Al and is synthesized by reducing TiCl$_4$ with about 1/3 part AlEt$_3$ or 1 part AlEt$_2$Cl. A modem TiCl$_3$ catalyst has a density of 2.065 g/cm^3, a bulk density of 0.82, a specific surface area (BET) of 29 m^2/g, and a particle size of 10 to 100 μm. The polymerization activity is in the vicinity of 500 L mol^{-1} × s^{-1} [88].

3. Supported Catalysts

MgCl$_2$/TiCl$_4$ catalysts. Good progress in increasing the polymerization activity was made with the discovery of the MgCl$_2$/TiCl$_4$-based catalysts [89]. Instead of MgCl$_2$, Mg(OH)Cl, MgRCl, or MgR$_2$ [90–94] can be used. The polymerization activity goes up to 10 000 L mol^{-1} s^{-1}. At this high activity the catalyst can remain in the polyethene. For example, the specific volume (BET) of the catalystis 60 m^2/g [95]. The high activity is accomplished by increasing the ethene pressure. The dependence is not linear as it was for first-generation catalysts, and the morphology is also different. The polyethene has a cobweb-like structure, whereas first generation catalysts produced a worm-like structure [90,91]. The cobweb structure is caused by the fact that polymerization begins at the surface of the catalyst particle. The particle is held together by the polymer. While polymerization is in progress, the particle grows rapidly and parts of it break. Cobweb structures are formed by this fast stretching process of the polyethene.

It is known that in the case of these supported catalysts the higher activity is linked to a higher concentration of active titanium. In contrast to first-generation catalysts in which only 0.1 to 1% of all titanium atoms form active sites, in supported catalysts 20 to 80% of them are involved in the formation of active sites [97,98].

Solvay workers [99] have investigated extensively the supported $Mg(OH)Cl/TiCl_4/AlEt_3$ catalyst and related systems including $MgSO_4$, $MgOSiO_2$, and MgO. It is not clear whether all of the Ti centers in the supported catalysts are isolated. The high activity suggests the incorporation of small $TiCl_3$ crystallites into the $Mg(OH)Cl$. Fink and Kinkelin [100] prepared a high-activity catalyst by combination of MgH_2 and $TiCl_4$. The MgH_2 has a much greater surface area ($90\,m^2/g$). It reacts with the $TiCl_4$ under the evolution of hydrogene. By 30 °C and 2 bar ethene pressure, 110 kg of PE per gram of Ti could be obtained.

4. Phillips Catalyst

The widely investigated Phillips catalyst, which is alkyl free, can be prepared by impregnating a silica-alumina (87:13 composition [101–103] or a silica support with an aqueous solution of CrO_3). High surface supports with about 400 to $600\,g/m^2$ are used [104]. After the water is removed, the powdery catalyst is fluidized and activated by a stream of dry air at temperatures of 400 to 800 °C to remove the bound water. The impregnated catalysts contain 1 to 5 wt% chromium oxides. When this catalyst is heated in the presence of carbon monoxide, a more active catalyst is obtained [105]. The Phillips catalyst specifically catalyzes the polymerization of ethene to high-density polyethene. To obtain polyethene of lower crystallinity, copolymers with known amounts of an α-olefin, usually several percent of 1-butene can be synthesized. The polymerization can be carried out by a solution, slurry, or gas-phase (vapor phase) process.

The chromium oxide-silica is inactive for polymerizing ethylene at low temperatures but becomes active as the temperature is increased from 196 °C (the melting point for CrO_3) to 400 °C. Interactions of chromium oxide with SiO_2 and Al_2O_3 take place.

Hogan [103] calculated that for a silica support of $600\,m^2/g$ and about 5% Cr(VI), the average distance between adjacent Cr atoms is 10 Å. This corresponds to the accepted population of silanol groups on silica after calcination. The structures (15) and (16) are proposed:

$$
\begin{array}{cc}
\overset{\displaystyle OH \quad OH}{\underset{|\qquad |}{-Si-O-Si-}} + CrO_3 & \longrightarrow \qquad \overset{O\diagdown \qquad \diagup O}{\underset{\displaystyle \underset{|\qquad |}{-Si-O-Si-}}{\overset{Cr}{\underset{O \qquad O}{\diagup \quad \diagdown}}}}
\end{array} \tag{15}
$$

$$
\begin{array}{cc}
\overset{\displaystyle OH \quad OH}{\underset{|\qquad |}{-Si-O-Si-}} + 2CrO_3 & \longrightarrow \qquad \overset{\displaystyle \overset{O}{\parallel} \quad \overset{O}{\parallel}}{O=Cr-O-Cr=O} \\
& \qquad \qquad \underset{\displaystyle \underset{|\qquad |}{-Si-O-Si-}}{\overset{|\qquad |}{O \qquad O}}
\end{array} \tag{16}
$$

It has been calculated that between 0.1 and 0.4wt% of the total chromium forms active centers [105]. A difficult question relates to the valences of chromium in the active sites. Valences of II, III, IV, V, and VI have been established [106]. Because of the small number of total chromium atoms that are active centers, it has not been possible to unequivocally assign the active valence [107,108]. Krauss and Hums [109] concluded that the reduction of hexavalent chromium centers linked to support produced coordinately unsaturated Cr(II) surface compounds. A speciality of the Phillips catalyst is that there is no influence of hydrogen to control the molecular weight of the polyethylene. Only by higher activation temperatures can the molecular weight be lowered.

5. Homogeneous (Single Site) Catalysts

Among the great number of Ziegler catalysts, homogeneous systems have been preferentially studied in order to understand the elementary steps of the polymerization which is simpler in soluble systems than in heterogeneous systems. The situation has changed since in recent years homogeneous catalyst based on metallocene and aluminoxane [12,110], nickel and palladium diimin complexes [111], and iron and cobalt compounds were discovered which are also very interesting for industrial and laboratory synthesis. Some special polymers can only be synthesized with these catalysts.

In comparison to Ziegler systems, metallocene catalysts represent a great development: they are soluble in hydrocarbons, show only one type of active site and their chemical structure can be easily changed. These properties allow one to predict accurately the properties of the resulting polyolefins by knowing the structure of the catalyst used during their manufacture and to control the resulting molecular weight and distribution, comonomer content and tacticity by careful selection of the appropriate reactor conditions. In addition, their catalytic activity is 10–100 times higher than that of the classical Ziegler–Natta systems.

Metallocenes, in combination with the conventional aluminum alkyl cocatalysts used in Ziegler systems, are indeed capable of polymerising ethene, but only at a very low activity. Only with the discovery and application of methylaluminoxane (MAO) it was possible to enhance the activity, surprisingly, by a factor of 10 000 [113]. Therefore, MAO plays a crucial part in the catalysis with metallocenes.

Kinetic studies and the application of various methods have helped to define the nature of the active centers, to explain the aging effects of Ziegler catalysts, to establish the mechanism of interaction with olefins, and to obtain quantitative evidence of some elementary steps [9,112–115]. It is necessary to differentiate between the soluble catalyst system itself and the polymerization system. Unfortunately, the well-defined bis(cyclopentadienyl)titanium system is soluble, but it becomes heterogeneous when polyethylene is formed [116].

The polymerization of olefins, promoted by homogeneous Ziegler catalysts based on biscyclopentadienyltitanium(IV) or analogous compounds and aluminum alkyls, is accompanied by a series of other reactions that greatly complicate the kinetic interpretation of the polymerization process:

Insertion

$$mt \overset{\uparrow}{\underset{CH_2=CH_2}{\rule{3cm}{0.4pt}}} R \longrightarrow mt-CH_2-CH_2-R \tag{17}$$

Alkyl exchange

$$mt\diagup\!\!\!\!^{R}_{R'} + mt' \longrightarrow mt\diagdown^{R}_{R'} + {}^{\diagdown}mt'$$ (18)

H exchange

$$mt \diagup\!\!\!\!^{CH_2-CH_2}_{CH_3-CH_2} \begin{bmatrix} \\ H \\ \end{bmatrix} mt' \longrightarrow mt \diagup^{CH_2-CH_2}\!\!\diagdown mt' \\ + \\ CH_3-CH_3$$ (19)

Reduction

$$mt \diagup\!\!\!\!^{CH_2-CH_2}\!\!\diagdown mt \longrightarrow CH_2{=}CH_2 \quad mt \quad mt$$ (20)

Concomitant with continued olefin insertion into the metal–carbon bond of the transition metal aluminum complex, alkyl exchange and hydrogen-transfer reactions are observed. Whereas the normal reduction mechanism for transition metal organic complexes is initiated by release of olefins with formation of a hydride followed by hydride transfer to an alkyl group, a reverse reaction takes place in the case of some titanium and zirconium acompounds. A dimetalloalkane is formed by the release of ethane. In second step, ethene is evolved from the dimetalloalkane:

$$Ti(IV){-}CH_2{-}CH_2{-}Ti(IV) \rightarrow CH_2{=}CH_2 + 2Ti(III)$$ (21)

leaving two reduced metal atoms. Some of the aging processes occurring with homogeneous and heterogeneous Ziegler catalysts can be explained with the aid of these side reactions.

Table 4 summarizes important homogeneous Ziegler catalysts. The best known systems are based on bis(cyclopentadienyl)titanium(IV), bis(cyclopentadienyl)zirconium(IV), terabenzyltitanium, vanadium chloride, allyl metal, or chromium acetylacetonate with trialkylaluminum, alkylaluminum halides, or aluminoxanes. Breslow [126] discovered that bis(cyclopentadienyl)titanium(IV) compounds, which are easily soluble in aromatic hydrocarbons, could be used instead of titanium tetrachloride as the transition metal compound together with aluminum alkyls for ethene polymerization. Subsequent research on this and other systems with various alkyl groups has been conducted by Natta [127], Belov et al. [128,129], Patat and Sinn [130], Shilov [131], Henrici-Olive and Olive [132], Reichert and Schoetter [133], and Fink et al. [134,135]. With respect to the kinetics of polymerization and side reactions, this soluble system is probably the one that is best understood. It is found that the polymerization takes place primarily if the titanium exists as titanium(IV) [136,137]. According to Henrici-Olive and Olive [138], the speed of polymerization decreases with increasing intensity of ESR signals of the developing titanium(III) compound.

The increase in length of the polymer chain occurs by insertion of the monomer in to a metal–carbon bond of the active complex. Dyachkovskii et al. [139] and Eisch et al. [140] were the first to believe, based on kinetic measurements and synthesis, that the insertion takes place on a titanium cation. An ion of the type $(C_5H_5)_2Ti^+$-R, derived from

Table 4 Homogeneous catalysts for ethene polymerization.

System	Transition metal (M) compound	Polymerization temperature (°C)	Normalized activity	Catalyst yield	Refs
$Cp_2TiCl_2/AlMe_2Cl^a$	1:2.5–1:6	30	40–200		117
$Cp_2TiCl_2/AlMe_2Cl/H_2O$	1:6:3	30	2000		117
$Cp_2TiCl_2/AlEt_2Cl$	1:2	15	7–45		118
Cp_2TiMe_2/MAO	$1:10^5.5 \times 10^2$	20	35 000	>15 000	110
Cp_2TiMe_2/MAO	1:100	20	200	>5 000	119
Cp_2ZrCl_2/MAO	1:1000	70	400 000	>10 000	120
$VO(acac)_2/Et_2AlCl/activator$	1:50	20	180		121
Cp_2VCl_2/Me_2AlCl	1:5	50	13		122
$Zr(allyl)_4$		80	2.0		
$Hf(allyl)_4$		160	0.6		
$Cr(ally)_3$		80	0.3		123
$Cr(acac)_3/EtAlCl$	1:300	20	150		121
$Ti(benzyl)_4$		20(80)	8×10^{-3} (0.2)		124,125
$Ti(benzyl)_3Cl$		20	0.4		124,125
$Ti(benzyl)_4$					

complexing and dissociation,

$$(C_5H_5)_2TiRCl + AlRCl_2 \rightleftharpoons (C_5H_5)_2TiRCl \cdot AlRCl_2 \tag{22}$$

$$(C_5H_5)_2TiRCl \cdot AlRCl_2 \rightleftharpoons [(C_5H_5)_2TiRCl_3]^+ + [AlRCl_3]^- \tag{23}$$

could be the active species of polymerization. Sinn and Patat [137] drew attention to the electron-deficient character of those main-group alkyls that afford complexes with the titanium compound. Fink and co-workers [141] showed by ^{13}C-NMR spectroscopy with ^{13}C-enriched ethene at low temperatures (where no alkyl exchange was observed) that in higher halogenated systems, insertion of the ethene takes place only into a titanium–carbon bond.

At low polymerization temperatures with benzene as a solvent, Hocker and Saeki [142] could prepare polyethene with a molecular weight distribution $M_W/M_n = 1.07$ using the bis(cyclopentadienyl)titanium dichloride/diethylaluminum chloride system. The molecular weight could be varied in a wide range by changing the polymerization temperature. Using ally$_4$Zr(allylZrBr$_3$) at a polymerization temperature of 160 °C (80 °C) yields polyethene with a density of 0.966 g/cm, M_n of 10,500, (700), 3.0 CH$_3$ groups per 1000 °C and 0.4 vinyl groups. The benzene- and allyl-containing transition metals are working without any cocatalyst and therefore are alkyl free. If transition metal organometallic compounds such as $Cr(allyl)_3$, $Zr(allyl)_4$, $Zr(benzyl)_4$, $Ti(benzyl)_4$, and Cr(cyclopentadienyl)$_2$ are supported on Al_2O_3 Or SiO_2, the activity increases by a factor of more than 100 [124,143].

Apparently, soluble catalysts are obtained by reaction of $Ti(OR)_4$ with AlR_3 [144]. High-molecular-weight polyethene is obtained in variable amounts, with Al/Ti ratios ranging between 10 and 50. Similar results are attained by replacing titanium alkoxide by $Ti(NR_2)_4$ [145]. Soluble catalytic systems are also obtained by reaction of Ti(acac)$_3$ [146] and Cr(acac)$_3$ [147] with AlEt$_3$ as well as by reaction of Cr(acac)$_3$ and VO(acac)$_2$ with AlEt$_2$Cl in the presence of triethyl phosphite [121]. With vanadium catalysts the activity reaches its maximum at Al/V ratio = 50. Under these conditions up to 67% vanadium is in the bivalent oxidation state. Bivalent and trivalent compounds will be active.

6. Aluminoxane as Cocatalysts

The use of metallocenes and alumoxane as cocatalyst results in extremely high polymerization activities (see Tables 4 and 5). This system can easily be used on a laboratory scale. The methylalumoxane (MAO) is prepared by careful treatment of trimethylaluminum with water [148]:

$$nH_2O + (n+1)Al(CH_3)_3 \rightarrow (H_3C)_2Al - [O - \overset{\overset{\displaystyle CH_3}{|}}{Al}]_n - CH_3 + 2nCH_4 \qquad (24)$$

MAO is a compound in which aluminum and oxygen atoms are arranged alternately and free valences are saturated by methyl substituents. It is gained by careful partial hydrolysis of trimethylaluminum and, according to investigations by Sinn [149] and Barron [150], it consists mainly of units of the basic structure $[Al_4O_3Me_6]$, which contains four aluminum, three oxygen atoms and six methyl groups. As the aluminum atoms in this structure are co-ordinatively unsaturated, the basic units (mostly four) join together forming clusters and cages. These have molecular weights from 1200 to 1600 and are soluble in hydrocarbons.

If metallocenes, especially zirconocenes but also titanocenes, hafnocenes and other transition metal compounds (Figure 2) are treated with MAO, then catalysts are acquired that allow the polymerization of up to 100 tons of ethene per g of zirconium [151–153]. At such high activities the catalyst can remain in the product. The insertion time (for the insertion of one molecule of ethene into the growing chain) amounts to some 10^{-5} s only (Table 6). A comparison with enzymes is not far-fetched.

As shown by Tait under these conditions every zirconium atom forms an active complex and produces about 20 000 polymer chains per hour. At temperatures above 50 °C, the zirconium catalyst is more active than the hafnium or titanium system; the latter is decomposed by such temperatures. Transition metal compounds containing some halogene show a higher activity than systems that are totally free of halogen. Of the cocatalysts, methylalumoxane is much more effective than the ethylaluminoxane or isobutylalumoxane.

It is generally assumed that the function of MAO is firstly to undergo a fast ligand exchange reaction with the metallocene dichloride, thus rendering the metallocene methyl

Table 5 Ethene polymerization[a] with metallocene/methylaluminoxane catalysts.

Metallocene[b]	Structure	Activity [kg PE/(mol Zr.h.c_{mon}]	Molecular weight (g/mol)
Cp_2ZrCl_2	6	60 900	62 000
$[Me_2C(Ind)(Cp)]ZrCl_2$	8	3330	18 000
$[En(IndH_4)_2]ZrCl_2$	9	22 200	1 000 000
$[Em(Ind)_2]ZrCl_2$	11	12 000	350 000
$[En(Ind)_2]HfCl_2$	12	2900	480 000
$[Me_2Si(Ind)_2]ZrCl_2$	13	36 900	260 000
$[Me_2Si(2,4,7-Me_3Ind)_2]ZrCl_2$	15	111 900	250 000
$[Me_2C(Flu)(Cp)]ZrCl_2$	18	2000	500 000

[a]Ethene pressure = 2.5 bar. temp. = 30 °C. [metallocene] = 6.25×10^{-6} M. Metallocene/MAO = 250. Solvent = toluene; [b]Cp = cyclopentadienyl; Ind = indenyl; En = C_2H_4; Flu = fluorenyl.

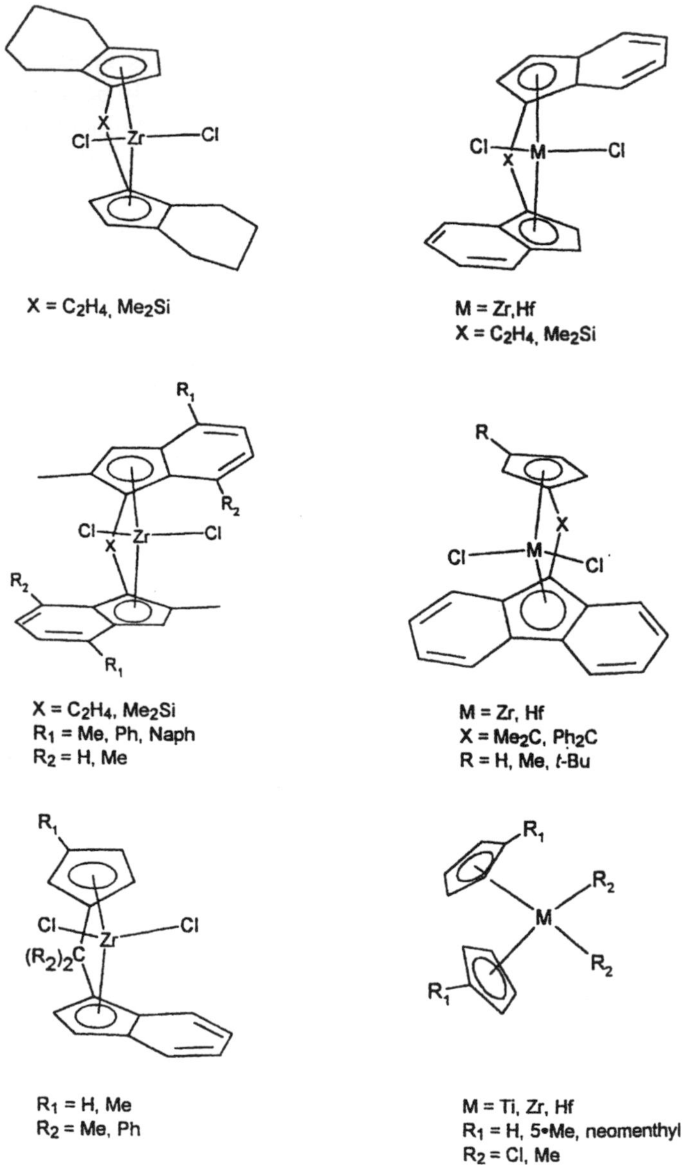
Cl
X
Zr
Cl
X = C2H4, Me2Si
Cl
X
M
Cl
M = Zr,Hf
X = C2H4, Me2Si
R1
R2
Cl
Zr
Cl
X
R2
R1
X = C2H4, Me2Si
R1 = Me, Ph, Naph
R2 = H, Me
R
X
Cl
M
Cl
M = Zr, Hf
X = Me2C, Ph2C
R = H, Me, t-Bu
R1
Cl
Zr
Cl
(R2)2C
R1 = H, Me
R2 = Me, Ph
R1
R2
M
R2
R1
M = Ti, Zr, Hf
R1 = H, 5•Me, neomenthyl
R2 = Cl, Me

Figure 2 Some classes of metallocene catalysts used for olefin polymerization.

and dimethyl compounds (Figure 3). In the further step, either Cl^- or CH_3^- is abstracted from the metallocene compound by al Al-center in MAO, thus forming a metallocene cation and a MAO anion [156,157]. The alkylated metallocene cation represents the active center (Figure 4). Meanwhile, other weakly coordinating cocatalysts, such as tetra(perfluorophenyl)borate anions $[(C_6F_5)_4B]^-$, have been successfully applied to the activation of metallocenes [158–161].

Polyethenes synthesized by metallocene-alumoxane have a molecular weight distribution of $M_w/M_n = 2$, 0.9 to 1.2 methyl groups per 1000 C atoms, 0.11 to 0.18 vinyl groups, and 0.02 trans vinyl group per 100 C atoms. The molecular weight can easily be lowered by increasing the temperature, increasing the metallocene concentration, or

Table 6 Polymerization activity of bis(cyclopentadienyl)zirconium dichloride/
methylalumoxane catalyst applied to ethene in 330 ml of toluene.

Activity (95 °C), 8 bar	39.8×10^6 g PE/g Zr·h
[Zirconocene]	6.2×10^8 mol/l
[Alumoxane] (M = 1200)	7.1×10^{-4} mol/l
Molecular weight of the polyethene obtained	78 000
Degree of polymerization	2800
Macromolecules per Zr atom per hour	46 000
Rate of growth of one macromolecule	0.087 s
Turnover time	3.1×10^{-5} s

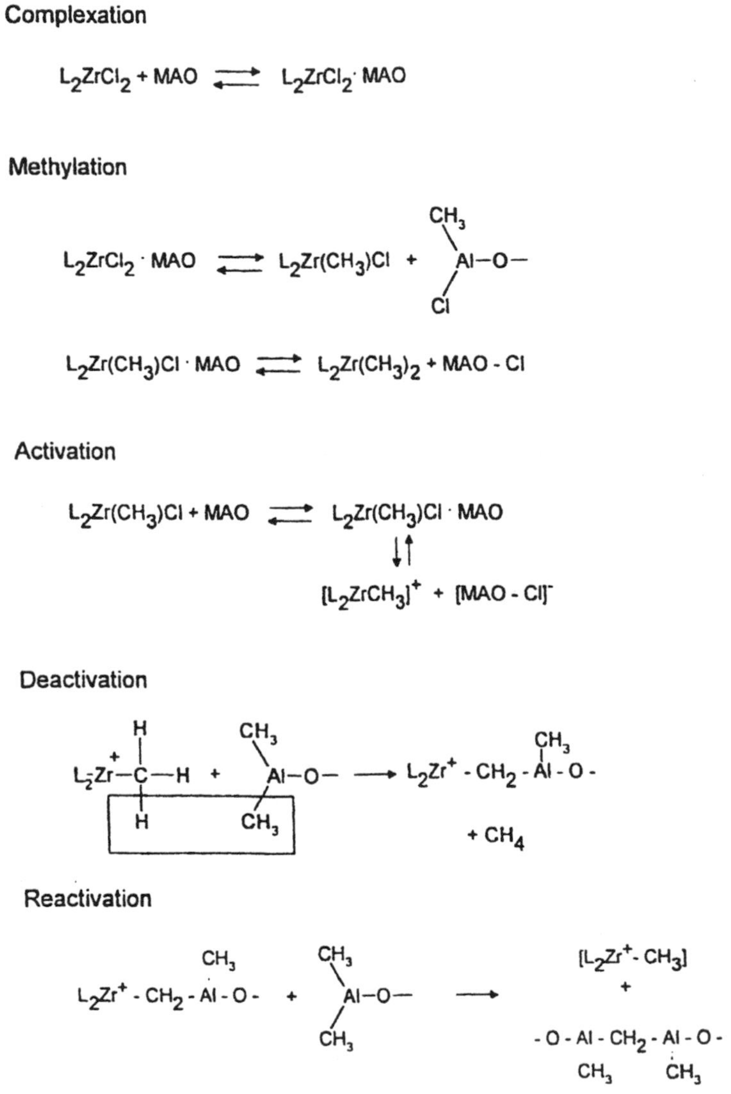

Figure 3 Reactions of zirconocenes with MAO.

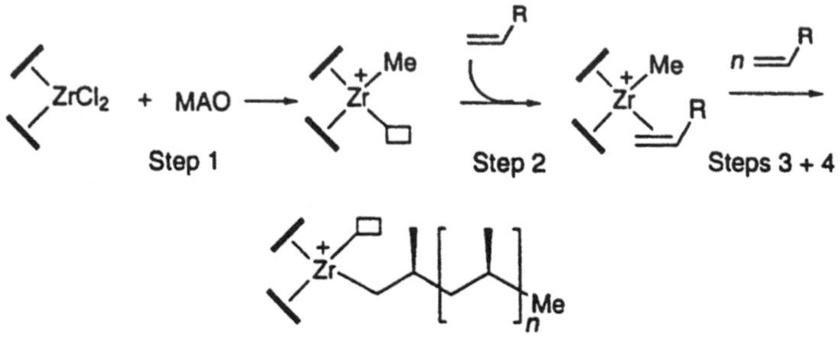

Figure 4 Mechanism of the polymerization of olefins by zirconocenes. Step 1: The cocatalyst (MAO: methylalumoxane) converst the catalyst after complexation into the active species that has a free coordination position for the monomer and stabilizes the latter. Step 2: The monomer (alkene) is allocated to the complex. Step 3: Insertion of the alkene into the zirconium alkyl bond and provision of a new free coordination position. Step 4: Repetition of Step 3 in a very short period of time (about 2000 propene molecules per catalyst molecule per second), thus rendering a polymer chain.

decreasing the ethene concentration. The molecular weight distribution can be decreased up to 1.1 (living polymerization) by bis(phenoxy-imine)titanium complexes [161]. Molecular weights of 170 000 were obtained. The molecular weight is also lowered by the addition of small amounts) (0.1 to 2 mol%) of hydrogen (e.g., without H_2, $M_w = 170\,000$; adding 0.5 mol% H_2, $M_w = 42\,000$) [155].

7. Late Transition Metal Catalyst

Brookhart et al. [57,58] described square planar nickel and palladium-diimine systems which are capable of polymerizing ethene to high molecular weight polymers with activities comparable to the metallocene catalyst systems when activated with methylaluminoxane.

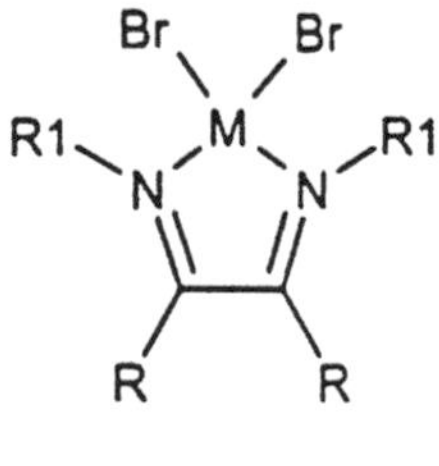

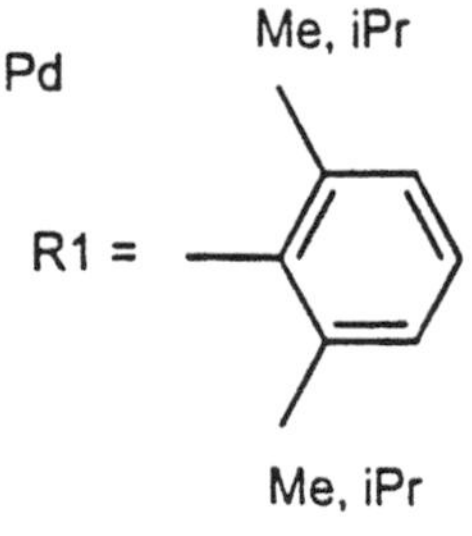

(25)   (26)

Important for the polymerization activity is the substituent 1 which has to be a bulky aryl group. The task of this substituent is to fill up the coordination spheres below and above the square plane of the complex and thus enable the growing polymer chain to stay coordinated to the metal center. This is one of the main differences to the well-known SHOP catalysts invented by Keim et al. [164] and Ostoja-Starzewski and Witte [165] which produces mainly ethene oligomers.

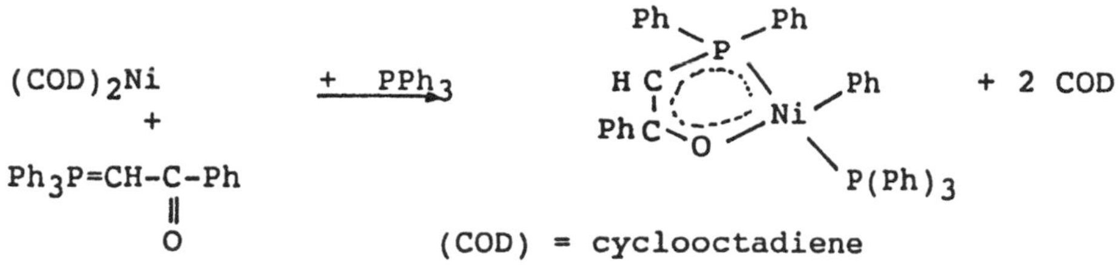

$$(COD)_2Ni$$
$$+$$
$$Ph_3P=CH-C-Ph$$
$$\overset{\parallel}{O}$$

$$+ \quad PPh_3 \quad \longrightarrow$$

(structure)

$$+ \quad 2 \ COD$$

(COD) = cyclooctadiene

(27)

The use bis(ylid)nickel catalysts by reaction of nickel oxygen complexes and phosphines [166].

For the one-component catalyst, it is possible to use solvents of various polarities. Even in THF or acetone there is good activity. The best solvents are methylene chloride or hexane. If the hydrogen next to the oxygen in the ylid is replaced by larger groups, the activity increases and reaches at 10-bar ethene pressure and 100 °C about 50 000 mol of reacted ethene per mole of nickel [167].

A very interesting feature of this new catalyst generation is that chain isomerization processes can take place during the polymerization cycles. This results in more or less branched polymers with varying product properties depending on polymerization conditions and catalyst type. The number of isomerization cycles which are carried out directly one after another determines the nature of the branching formed. Branches ranging from methyl to hexyl and longer can be formed.

The extent of branching can be tailored precisely by tuning the polymerization conditions and products, from highly crystalline HDPE to completely amorphous polymers with glass transition temperatures of about −50 °C. These products are different to all known conventionally produced copolymers due to their content and distribution pattern of short chain branching [168].

Another new catalyst generation based on iron and cobalt. The direct iron analogs of the nickel-diimine catalysts derived from structures (25) and (26) did not seem to be very active in olefin polymerization at all. The electronic and steric structure analysis shows why: the nickel d^8-system favors a square planar coordination sphere but the iron d^6-system favors a tetrahedral one. It is very likely that these tetrahedral coordination sites are not available for olefin insertion, and hence no polymerization can take place.

The next logical step was the employment of another electron donating atom in the ligand structure in order to obtain a trigonal-bipyramidal coordination sphere. Gibson and Brookhart both succeeded with a catalyst system based on an iron–bisiminopyridyl complex. The structures (28)–(30) illustrate the three types of catalysts [169,170].

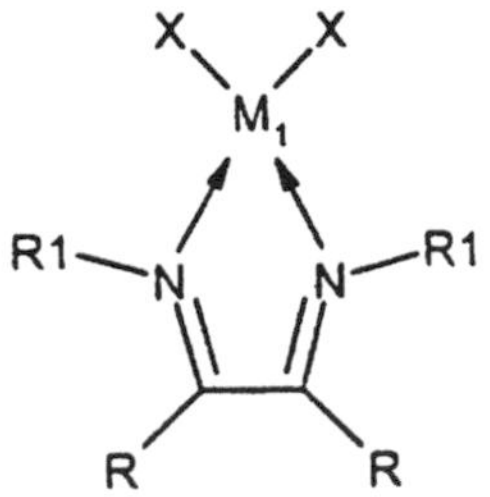

(28)

(structure)

Square planar

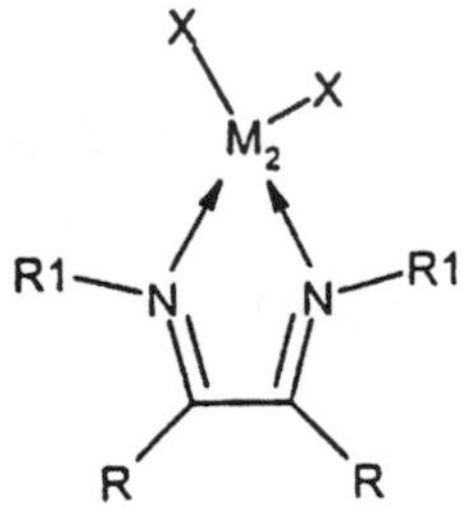

(29)

Tetrahedral

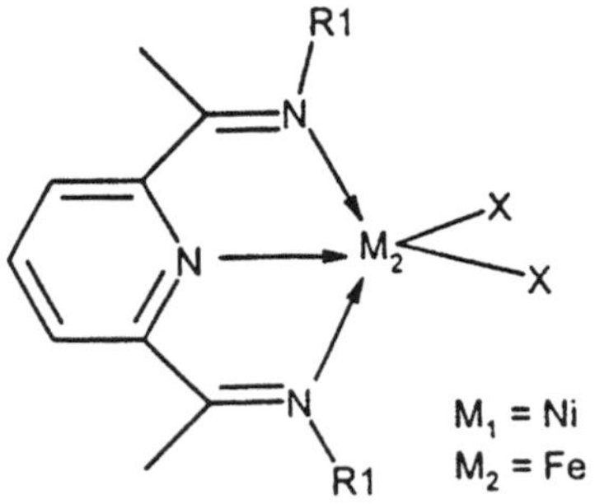

(30)

M_1 = Ni
M_2 = Fe

Trigonal-bipyramidal

The ethene polymerization activity of these new family of catalysts is comparable with the one obtained with the most productive metallocenes under similar conditions if activated with methylaluminoxane. Again, the nature of the aryl substituents R1 plays a major role in controlling the molecular weight of the polymers.

In contrast to nickel-diimine catalysts no chain isomerization takes place and thus only linear HDPE is formed.

In 1998, Grubbs [171,172] reported on a new type of neutral nickelII-complexes with salicylaldimin ligands (structure (31)). With these catalysts low branched polyethylenes were obtained with a narrow molecular weight distribution. The copolymerization of ethene and norbornene is possible.

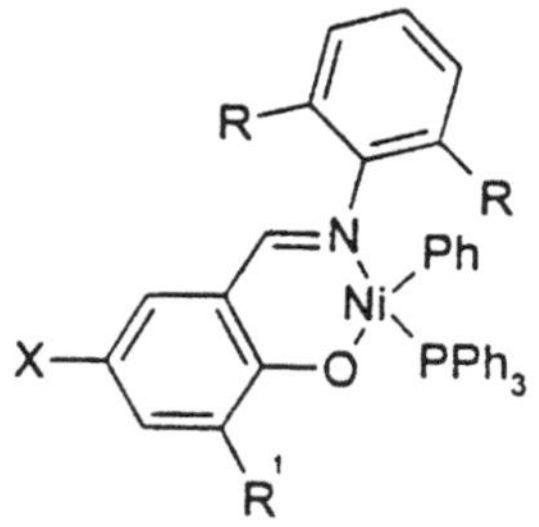

(31)

C. Copolymers of Ethene

The properties of polyethene could be varied in a wide range by copolymerization of ethene with other comonomers. Most commercial products contain at least small amounts of other monomers. In general, adding comonomers to the polymerization reduces the polyethenes crystallinity, thereby reducing the melting point, the freezing point, and in many cases the tensile strength and modulus. At the same time, optical properties are improved and polarity is increased. The architecture of the copolymer can be controlled experimentally by the following factors: operating conditions, chemical composition and physical state of used catalyst, physical state of the copolymer being formed, and structure of the comonomers.

The practically most important copolymer is made from ethene and propene. Titanium- and vanadium-based catalysts have been used to synthesize copolymers that have a prevailingly random, block, or alternating structure. Only with Ziegler or single site catalyst, longer-chain α-olefins can be used as comonomer (e.g., propene, 1-butene, 1-hexene, 1-octene). In contrast to this, by radical high-pressure polymerization it is also possible to incorporate functional monomers (e.g., carbon monoxide, vinyl acetate). The polymerization could be carried out in solution, slurry, or gas phase. It is generally accepted [173] that the best way to compare monomer reactivities in a particular polymerization reaction is by comparison of their reactivity ratios in copolymerization reactions.

The simplest kinetic scheme of binary copolymerization in the case of olefin insertion reaction is

$$\text{Cat} - M_1 - \text{polymer} + M_1 \xrightarrow{k_{11}} \text{Cat} - M_1 - M_1 - \text{polymer} \tag{32}$$

$$\text{Cat} - M_1 - \text{polymer} + M_2 \xrightarrow{k_{12}} \text{Cat} - M_2 - M_1 - \text{polymer} \tag{33}$$

$$\text{Cat} - M_2 - \text{polymer} + M_1 \xrightarrow{k_{21}} \text{Cat} - M_1 - M_2 - \text{polymer} \tag{34}$$

$$\text{Cat} - M_2 - \text{polymer} + M_2 \xrightarrow{k_{22}} \text{Cat} - M_2 - M_2 - \text{polymer} \tag{35}$$

$$r_1 = \frac{k_{11}}{k_{12}} \qquad r_2 = \frac{k_{22}}{k_{21}} \tag{36}$$

where k_{11} and k_{22} are the homopolymerization propagation rates for monomers M_1 and M_2 and k_{12} and k_{21} are cross-polymerization rate constants. The definition of reactivity ratios is

$$\frac{d[M_1]}{d[M_2]} = \frac{[M_1]r_1[M_1] + [M_2]}{[M_2][M_1] + r_2[M_2]} \tag{37}$$

The product $r_1 \times r_2$ usually ranges from zero to 1. When $r_1 \times r_2 = 1$, the copolymerization is random. As $r_1 \times r_2$ approaches zero, there is an increasing tendency toward alternation.

1. Radical Copolymerization

At elevated temperatures, ethene can be copolymerized with a number of unsaturated compounds by radical polymerization [174–180] (Table 7). The commercially most important comonomers are vinyl acetate [181], acrylic acid, and methacrylic acid as well as their esters. Next to these carbon monoxide is employed as a comonomer, as it promotes the polymer's degradability in the presence of light [182].

As a consequence of the diversified nature of the comonomers, a large number of variants of copolymer composition can be realized, thus achieving a broad variation of properties. The copolymerization can be carried out in the liquid monomer, in a solvent, or in aqueous emulsion. When high molecular mass is desired, solvents with low chain transfer constants (e.g., *tert*-butanol, benzene, 1,4-dioxane) are preferred. Solution

Table 7 Copolymerization of ethene (M_1) with various comonomers (M_2).

Comonomer	r_1	r_2	Pressure (MPa)	Temp. (°C)
Propene	3.2	0.62	102–170	120–220
1-Butene	3.2	0.64	102–170	130–220
Isobutylene	2.1	0.49	102–170	130–220
Styrene	0.7	1	150–250	100–280
Vinyl acetate	1	1	110–190	200–240
Vinyl chloride	0.16	1.85	30	70
Acrylic acid	0.09		196–204	140–226
Acrylic acid methylester	0.12	13	82	150
Acrylnitrile	0.018	4	265	150
Methacrylic acid	0.1		204	160–200
Methacrylic acid methylester	0.2	17	82	150

polymerization permits the use of low polymerization temperatures and pressures. Poly(ethylene-co-vinyl acetate, for instance, is produced at 100 °C and 14 to 40 MPa [183].

For the polymerization of ethene with vinyl acetate and vinyl chloride, emulsion polymerization in water is particularly suitable. The polymerizates have gained some importance as adhesives, binding materials for pigments, and coating materials [184,185].

2. Linear Low-Density Polyethene (LLDPE)

In contrast to LDPE produced with the high-pressure process, the tensile strength in LLDPE is much higher. Therefore, there has been a considerable boost in the production of LLDPE [186]. All Ziegler catalysts listed earlier are suitable for the copolymerization of ethene with other monomers. Monomers that decrease the melting point and crystallinity of a polymer at low concentrations are of great interest. Portions of 2 to 5 mol% are used. Longer-chained monomers such as 1-hexene are more effective at the same weight concentration than smaller units such as propene. It results in a branched polyethene with methyl branching (R) if propene is used, ethyl if butene is used, and so on.

$$
\begin{array}{ccccccccccccc}
H & H & H & H & H & H & & H & H & H & H & H & H \\
| & | & | & | & | & | & & | & | & | & | & | & | \\
C & = & C & + & C & = & C & + & C & = & C & \rightarrow & C & - & C & - & C & - & C & - & C & - & C \\
| & | & | & | & | & | & & | & | & | & | & | & | \\
H & H & H & R & H & H & & H & H & H & R & H & H
\end{array}
\tag{38}
$$

Important for the copolymerization are the different ractivities of the olefins. The principal order of monomer reactivities is well known [187]; ethene > propene >1-butene > linear α-olefins > branched α-olefins. Normally propene reacts 5 to 100 times slower than ethene, and 1-butene 3 to 10 times slower than propene. Table 8 shows the reactivity ratios for the copolymerization of ethene with other olefins. The data imply that the reactivity of the polymerization center is not constant for a given transition metal compound but depends on the structure of the innermost monomer unit of the growing polymer chain and on the cocatalyst.

On a laboratory scale, single site catalysts based on metallocene/MAO are highly useful for the copolymerization of ethene with other olefins. Propene, 1-butene, 1-pentene, 1-hexene, and 1-octene have been studied in their use as comonomers, forming linear low-density polyethene (LLDPE) [188,189]. These copolymers have a great industrial potential and show a higher growth rate than the homopolymer. Due to thee short branching from

Table 8 Reactivity ratios of ethene with various comonomers and heterogeneous TiCl$_3$ catalyst by 70 °C.

Comonomer	Cocatalyst	r_1	r_2	Ref.
Propene	Al(C$_6$H$_{13}$)$_3$	15.7	0.11	174
Propene	AlEt$_3$	9.0	0.10	174
1-Butene	AlEt$_3$	60	0.025	178
4-Methyl-1-pentene	AlEt$_2$Cl	195	0.0025	177
Styrene	AlEt$_3$	81	0.012	179

Table 9 Results of ethene reactivity ratio determinations with soluble catalysts[a].

Metallocene	Temp. (°C)	α-Olefin	r_1	r_2	$r_1 \cdot r_2$
Cp$_2$ZrMe$_2$	20	Propene	31	0.005	0.25
[En(Ind)$_2$]ZrCl$_2$	50	Propene	6.61	0.06	0.40
[En(Ind)$_2$]ZrCl$_2$	25	Propene	1.3	0.20	0.26
Cp$_2$ZrCl$_2$	40	Butene	55	0.017	0.93
Cp$_2$ZrCl$_2$	60	Butene	65	0.013	0.85
Cp$_2$ZrCl$_2$	80	Butene	85	0.010	0.85
[En(Ind)$_2$]ZrCl$_2$	30	Butene	8.5	0.07	0.59
[En(Ind)$_2$]ZrCl$_2$	50	Butene	23.6	0.03	0.71
Cp$_2$ZrMe$_2$	60	Hexene	69	0.02	1.38
[Me$_2$Si(Ind)$_2$]ZrCl$_2$	60	Hexene	25	0.016	0.40

the incorporated α-olefin, the copolymers show lower melting points, lower crystallinities, and lower densities, making films formed from these materials more flexible and better processable. Applications of the copolymers can be found in packaging, in shrink films with a low steam permeation, in elastic films, which incorporate a high comonomer concentration, in cable coatings in the medical field because of the low part of extractables, and in foams, elastic fibers, adhesives, etc. The main part of the comonomers is randomly distributed over the polymer chain. The amount of extractables is much lower than in polymers synthesized with Ziegler catalysts.

The copolymerization parameter r_1, which says how much faster an ethene unit is incorporated into the growing polymer chain than an α-olefin, if the last inserted monomer was an ethene unit, lies between 1 and 60 depending on the kind of comonomer and catalyst. The product $r_1 \cdot r_2$ is important for the distribution of the comonomer and is close to one when using C$_2$-symmetric catalysts [190] (Table 9).

Under the same conditions, syndiospecific (C$_s$-symmetric) metallocenes are more effective in inserting α-olefins into an ethene copolymer than isospecific working (C$_2$-symmetric) metallocenes or unbridged metallocenes. In this particular case, hafnocenes are more efficient than zirconocenes, too.

An interesting effect is observed for the polymerization with ethylene(bisindenyl)-zirconium dichloride and some other metallocenes. Although the activity of the homopolymerization of ethene is very high, it increases when copolymerizing with propene [191].

The copolymerization of ethene with other olefins is effected by the variation of the Al/Zr ratio, temperature and catalyst concentration. These variations change the molecular weight and the ethene content. Higher temperatures increase the ethene content and lower the molecular weight.

Studies of ethene copolymerization with 1-butene using the Cp_2ZrCl_2/MAO catalyst indicated a decrease in the rate of polymerization with increasing comonomer concentration.

3. Ethene-Propene Copolymers

The copolymers of ethene and propene, with a molar ratio of 1:0.5 up to 1:2, are of great industrial interest. These EP-polymers show elastic properties and, together with 2–5 wt% of dienes as third monomers, they are used as elastomers (EPDM). Since there are no double bonds in the backbone of the polymer, it is less sensitive to oxidation reaction. Ethylidenenorbornene, 1,4-hexadiene and dicyclopentadiene are used as dienes. In most technical processes for the production of EP and EPDM rubber, soluble or highly disposed vanadium components have been used in the past (Table 10) [192–195]. Similar elastomers which are less coloured can be obtained with metallocene/MAO catalyst at a much higher activity [196]. The regiospecificity of the metallocene catalysts towards propene leads exclusively to the formation of head-to-tail enchainments. Ethylidenenorbornene polymerizes via vinyl polymerization of the cyclic double bond and the tendency of branching is low. The molecular weight distribution of about 2 is narrow [197].

At low temperatures the polymerization time to form one polymer chain is long enough to consume one monomer and then to add another one. So, it becomes possible to synthesize block copolymers if the polymerization, catalyzed especially by hafnocenes, starts with propene and, after the propene is nearly consumed, continues with ethene.

High branching, which is caused by the incorporation of long chain olefins into the growing polymer chain, is obtained with silyl bridged amidocyclopentadienyltitanium compounds (structure (39)) [198–200].

(39)

Table 10 Results of ethene reactivity ratio determinations with soluble catalysts[a].

Catalyst	Cocatalyst	Temp. (°C)	$r_1(M_1)$	$r_2(M_2)$	$r_1 \cdot r_2$	Ref.
VCl_4	$AlEt_2Cl$	21	3.0	0.073	0.23	192
VCl_4	Al-i-Bu_2Cl		20.0	0.023	0.46	193
$VOCl_3$	Al-i-Bu_2Cl	30	16.8	0.052	0.87	192
$V(acac)_3$	Al-i-Bu_2Cl	20	16.0	0.04	0.64	193
$VOCl_2(OEt)$	Al-i-Bu_2Cl	30	16.8	0.055	0.93	194
$VOCl_2$	Al-i-Bu_2Cl	30	18.9	0.069	1.06	194
$VO(OBu)_3$	Al-i-Bu_2Cl	30	22.0	0.046	1.01	194
$VO(OEt)_3$	Al-i-Bu_2Cl	30	15.0	0.070	1.04	194
$VO(OEt)_3$	$AlEt_2Cl$	30	26.0	0.039	1.02	195

[a]Monomer 1 = ethene, monomer 2 = propene.

These catalysts, in combination with MAO or borates, incorporate oligomers with vinyl endgroups which are formed during polymerization by β-hydrogen transfer resulting in long chain abranched polyolefins. In contrast, structurally linear polymers are obtained when catalysed by other metallocenes. Copolymers of ethylene with 1-octene are very flexible materials as long as the comonomer content is less than 10%.

With higher 1-octene content they show that elastic properties polyolefin elastomers (POE) are formed [201]. EPDM is a commercially important synthetic rubber. The dienes as terpolymers are curable with sulfur. This rubber shows a higher growth rate than the other synthetic rubbers [202]. The outstanding property of ethene-propene rubber is its weather resistance since it has no double bonds in the backbone of the polymer chain and thus is less sensitive to oxygen and ozone. Other excellent properties of this rubber are its resistance to acids and alkalis, its electrical properties, and its low-temperature performance [203].

EPDM rubber is used in the automotive industry for gaskets, wipers, bumpers, and belts. In the tire industry, EPM and EPDM play a role as a blending component, especially for sidewalls. Furthermore, EPDM is used for cable insulation and in the housing industry, for roofing as well as for many other purposes, replacing special rubbers [204].

For technical uses, the molecular weight (M_w) is in the range 100 000 to 200 000. EPDM rubber, synthesized with vanadium catalyst, show a molecular weight distribution between 3 and 10, indicating that two and more active centers are present.

The properties of the copolymers depend to a great extent on several structural features of the copolymer chains as the relative content of comonomer units, the way the comonomer units are distributed in the chain, the molecular weight and molecular weight distribution, and the relative content of normal head-to-tail addition or head-to-head/tail-to-tail addition.

4. Ethene-Cycloolefin Copolymers

Metallocene/methylaluminoxane (MAO) catalysts can be used to polymerize and copolymerize strained cyclic olefins such as cyclobutene, cyclopentene, norbornene, DMON and other sterically hindered olefins [205–210]. While polymerization of cyclic olefins by Ziegler–Natta catalysts is accompanied by ring opening [10], homogeneous metallocene [211], nickel [212,213], or palladium [214,215], catalysts achieve exclusive double bond opening polymerization.

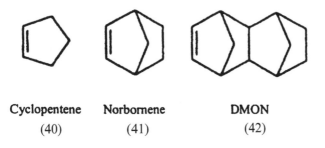

Cyclopentene   Norbornene   DMON
     (40)   (41)   (42)

Copolymerization of these cyclic olefins with ethylene or α-olefins cycloolefin copolymers (COC) can be produced, representing a new class of thermoplastic amorphous materials [217–220]. Early attempts to produce such copolymers were made using heterogeneous $TiCl_4$/$VAlEt_2Cl$ or vanadium catalysts, but first significant progress was made by utilizing metallocene catalysts for this purpose. They are about ten times more active than vanadium systems and by careful choice of the metallocene, the comonomer distribution may be varied over a wide range by selection of the appropriate cycloolefin and its degree of incorporation into the polymer chain. Statistical copolymers become amorphous at comonomer incorporations beyond 10–15 mol% cycloolefin.

COCs are characterized by excellent transparency and very high, long-life service temperatures. They are soluble, chemically resistant and can be melt-processed. Due to their high carbon/hydrogen ratio, these polymers feature a high refractive index, e.g. 1.53 for ethene-norbornene copolymer at 50 mol% norbornene incorporation. Their stability against hydrolysis and chemical degradation, in combination with their stiffness lets them become desirable materials for optical applications, e.g. for compact disks, lenses, optical fibers and films. The first commercial COC plant run by Ticona GmbH with a capacity of 30 000 tons a year commerced production in September 2000 and is located in Oberhausen, Germany.

The first metallocene-based COC material was synthesized from ethene and cyclopentene [218]. While homopolymerization of cyclopentene results in 1,3-enchainment of the monomer units [219], isolated cyclopentene units are incorporated into the ethene-cyclopentene copolymer chain by 1,2-insertion. Ethylene is able to compensate the steric hindrance at the α-carbon of the growing chain after and before the insertion of cyclopentene [220].

Ethene-norbornene copolymers are most interesting for technical applications as they can be made from easily available monomers and provide glass transition temperatures up to 200 °C. Table 11 presents the activities and comonomer ratios for the several applied catalysts of C_2- and C_s-symmetry. C_s-symmetric zirconocenes are more active in the copolymerization than for the homopolymerization of ethene. Under the chosen conditions, $[En(Ind)_2]ZrCl_2$ develops the highest activity while the highest comonomer incorporation is achieved by $[Ph_2C(Ind)(Cp)]ZrCl_2$.

Due to different incorporation ratios of the cyclic olefin into the copolymer, the glass transition temperature can vary over a wide range which is basically independent of the applied catalyst. A copolymer containing 50 mol% of norbornene yields a material with a glass transition point of 145 °C. Considering COCs of different comonomers with equal comonomer ratios, increased T_g values can be observed for the bulkier comonomer, for instance 72 °C for ethene-norbornene and 105 °C for ethene-DMON at comonomer mole ratio $X_{Co} = 0.30$ each.

The copolymerization parameters r_1 and r_2 were calculated from the rates of incorporation, determined by ^{13}C NMR spectroscopy, dependent on the reaction temperature.

Table 12 shows the temperature dependence of the copolymerization parameters r_1 and r_2 and of the influence of the catalyst systems. Metallocene catalysts show low r_1 values, which increases with the temperature and allows the easy incorporation of bulky cycloolefins into the growing polymer chain. Surprisingly, the copolymerization parameter $r_1 = 1.8$–3.1 for cyclopentene and norbornene is surprisingly low. The r_1 value of 2 means that ethylene is inserted only twice as fast as norbornene.

The product $r_1 \cdot r_2$ shows whether statistical insertion ($r_1 \cdot r_2$) or alternating one ($r_1 \cdot r_2 = 0$) has occurred. The different catalysts produce copolymers with structures that are between statistical and alternating.

Due to different incorporation values of the cyclic olefin in the copolymer, the glass transition temperature can vary over a wide range that is independent of most of the used catalysts (Figure 5). A copolymer with 50 mol% of norbornene yields a material with a glass transition point of 145 °C. A T_g of 205 °C can be reached by higher incorporation rates.

Table 11 Copolymerization of norbornene (N) and ethene (E) by different metallocene/MAO catalysts at 30 °C. Conditions: MAO/Zr = 200, c(Zr) = 5 × 10⁻⁶ mol/l; p(E) = 2.00 bar, c(N) = 0.05 mol/l.

Catalyst	t [min]	Activity [kg/mol h]	Incorp. of norbornene [weight %]
Cp₂ZrCl₂	30	1200	21.4
[En(Ind)₂]ZrCl₂	10	9120	26.1
[Me₂Si(Ind)₂]ZrCl₂	15	2320	28.4
[En(IndH₄)₂]ZrCl2	40	480	28.1
[Me₂C(Flu)(Cp)]ZrCl₂	10	7200	28.9
[Ph₂C(Flu)(Cp)]ZrCl₂	10	6000	27.3
[Ph₂C(Ind)(Cp)]ZrCl₂	15	2950	33.3

Table 12 Copolymerization parameters r_1 and r_2 of ethene/cycloolefin copolymerization with different metallocene/MAO catalysts.

Cycloolefin	Catalyst	Temp. in °C	r_1	r_2	$r_1 \cdot r_2$
Cyclopentene	[En(IndH₄)₂]ZrCl₂	0	1.9	<1	~1
Cyclopentene	[En(IndH₄)₂]ZrCl₂	25	2.2	<1	~1
Norbornene	[Me₂Si(Ind)₂]ZrCl₂	30	2.6	<2	~1
Norbornene	[Me₂C(FIu)(Cp)]ZrCl₂	30	3.4	0.06	0.2
Norbornene	[Ph₂C(Flu)(Cp)]ZrCl₂	0	2.0	0.05	0.1
Norbornene	[Ph₂C(Flu)(Cp)]ZrCl₂	30	3.0	0.05	0.15
Norbornene	[Me₂C(Flu)(t-BuCp)]ZrCl₂	30	3.1	0	0
DMON	[Ph₂C(Flu)(Cp)]ZrCl₂	50	7.0	0.02	0.14
DMON	[Ph₂C(Ind)(Cp)]ZrCl₂	50	6.4	0.10	0.64
DMON	[Ph₂C(Flu)(Cp)]HfCl₂	50	7.1	0.04	0.28

The T_g values are raised with a bulkier cycloolefin, regarding the same incorporation rate of 30% (norbornene: $T_g = 72$ °C; DMON: $T_g = 105$ °C). The highest glass transition temperature with 229 °C was reached by a copolymer of ethene and 5-phenylnorbornene.

Copolymerization of ethene and norbornene with [Me₂C(Flu)(*tert*-BuCp)]ZrCl₂ leads to a strong alternating structure [221]. This copolymer is crystalline and shows a melting point of 295 °C, good heat resistance and resistance against unpolar solvents.

5. Ethene-Copolymerization by Styrene or Polar Monomers

The copolymerization of ethene and styrene is possible by single site catalysts such as metallocenes and amido (see structure (33)) [222,223]. Amounts of more than 50 mol% of styrene could be incorporated into the copolymer. In dependence of the styrene content the copolymers show elastic to stiff properties. The polymerization happens by both 1,2- and 2,1-insertion of the styrene unit; the regioselectivity is low.

While it is difficult to copolymerize ethene and polar monomers by Ziegler- or single-site catalysts because of the great reactivity of the active sites to polar groups, it is commercialized to use free radical polymerization by high ethene pressure.

Vinyl acetate and acrylate esters used as comonomers containing sufficient stabilizer to prevent the homopolymerization. The effect of the copolymerization with polar

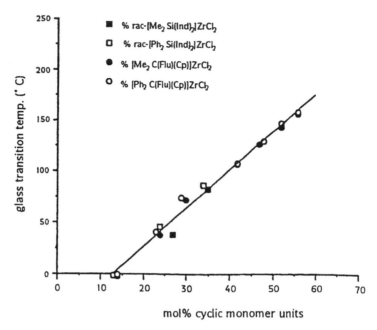

Figure 5 Glass transition temperatures of norbornene/ethene copolymers catalyzed with different zirconocenes.

monomers is to reduce the crystallinity and to receive materials for blending. Acrylate esters such as methyl acrylate, ethyl acrylate, butyl acrylate, methyl methacrylate form flexible copolymers. They provide enhanced adhesion, particularly in coextruded films or laminates.

Late transition metal complexes are more efficient in the copolymerization of ethene and polar monomers. Nickel or palladium complexes (see strtuctures (30)–(32) are functional-group tolerant allowing the copolymerization of ethene and methyl methacrylate or CO [224–227].

III. POLYPROPENE

A. Homopolymerization

In contrast to the polymerization of ethene, only coordination catalysts are successful in polymerizing propene to a crystalline polymer. The cationic polymerization of propene with concentrated sulfuric acid leads to oily or waxy amorphous polymers of low molecular weight [228]. Next to strong acids, catalysts such as complex Lewis acids may serve as initiators in the cationic polymerization of propene. The polymerization is conducted at temperatures between −100 and 80 °C. Chlorinated hydrocarbons are commonly used as solvents.

Under cationic conditions, migration of the C–C double bond is observed. Like all other α-olefins, propene cannot be polymerized via an anionic route. The same applies to free-radical polymerization. In polymerization with Ziegler–Natta catalysts, propene or longer-chained α-olefins are inserted into the growing chain in a head-to-tail fashion with high selectivity. Every CH_2-group (head) is followed by a CH(R)-group (tail) with a tertiary carbon atom bearing a methyl or even larger alkyl group:

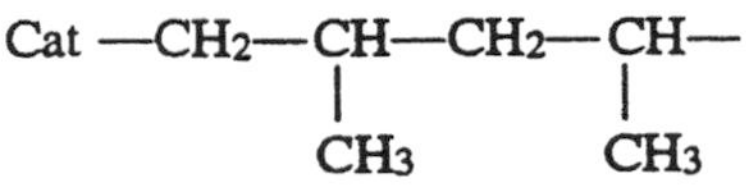

$$\text{Cat} \quad -CH_2-\underset{\underset{CH_3}{|}}{CH}-CH_2-\underset{\underset{CH_3}{|}}{CH}- \tag{43}$$

This construction principle is mandatory for the stereoregular structure of the polypropene molecule. In addition, head to head

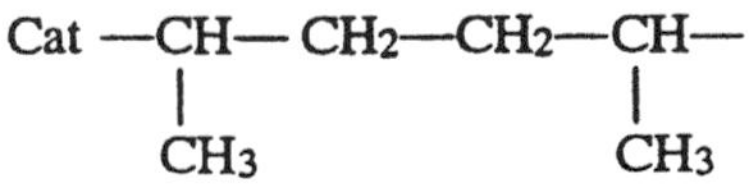

$$\text{Cat} \quad -\underset{\underset{CH_3}{|}}{CH}-CH_2-CH_2-\underset{\underset{CH_3}{|}}{CH}- \tag{44}$$

and tail-to-tail

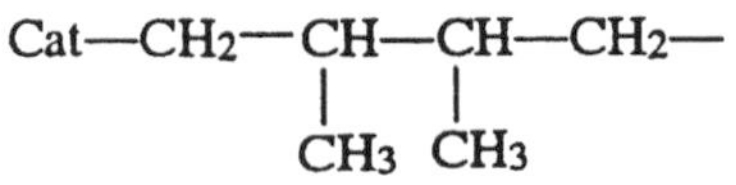

$$\text{Cat}-CH_2-\underset{\underset{CH_3}{|}}{CH}-\underset{\underset{CH_3}{|}}{CH}-CH_2- \tag{45}$$

arrangements occur. These links can be detected by IR and ^{13}C NMR spectroscopy. Exclusive head-to-tail bonding is a mandatory but not a sufficient condition for stereo-regularity. Another important detail is the sterical orientation of the pendant methyl groups with respect to the main C–C axis of the polymer molecule.

Natta formulated three different structures [229]

Isotactic structure

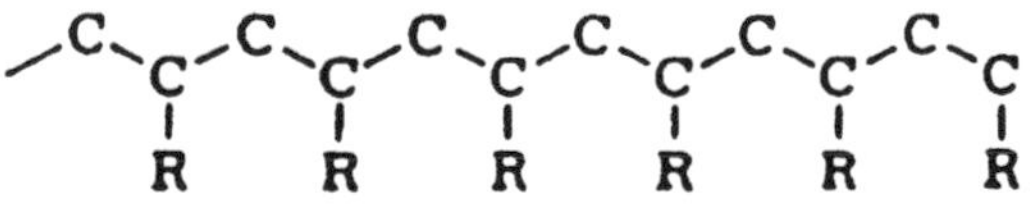

$$\tag{46}$$

Syndiotactic structure

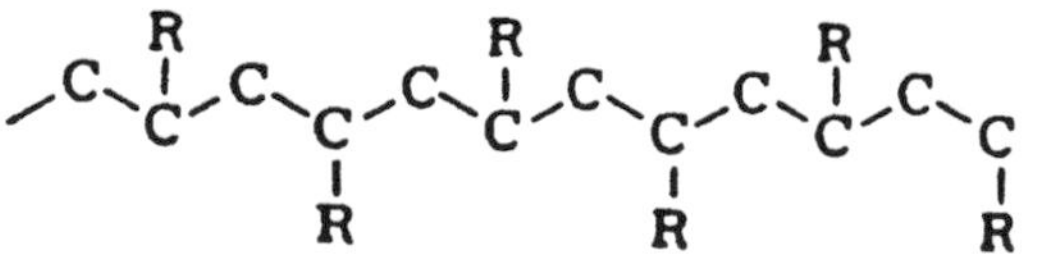

$$\tag{47}$$

Atactic structure

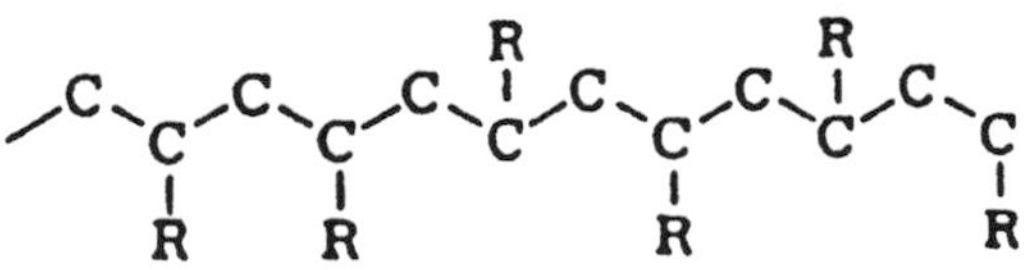

$$\tag{48}$$

1. In the structure all pendant methyl groups are located on one side of the zigzag plane; these polymers are called isotactic

2. For polymers in which the position of the pendant methyl groups is alternatingly above and below the backbone plane, the term syndiotactic is used.
3. When the pendant methyl groups are randomly positioned above and below the plane, the polymer is said to be atactic.

While the structural description of low molecular weight compounds with asymmetric carbon atoms is explicit, there are no similarly accurate rules for the description of polymers. Tertiary carbon atoms in polyolefin chains are not asymmetric in a general chemical sense. Even with one of the substituents bearing a double bond at its end and the other terminated by an ethyl group, they are very similar. Therefore, these carbon atoms are often called pseudo asymmetric. The differences between the three forms of polypropene with identical molecular weight distribution and branching percentage are considerable (Table 13).

B. Isotactic Polypropene

In view of the stereochemistry, Natta managed to synthesize isotactic crystalline polypropene with the combination catalysts that have previously been discovered by Ziegler [230]. He thus achieved a breakthrough for a technical application of polypropene. The most widely used catalyst for the stereospecific polymerization of propene still consists of titanium halogenides and alkylaluminum compounds. In addition to this catalyst, a large number of other systems have been tested. Table 14 lists important heterogeneous systems.

Table 13 Some characteristics of polypropene.

Characteristic	Isotactic	Syndiotactic	Atactic
Melting point (°C)	160–171	130–160	–
Crystallinity (%)	55–65	50–75	0
Tensile strength (kP/cm^2)	320–350		0

Table 14 Heterogeneous catalysts for the propene polymerization.

Catalyst[a]	Activity (g PP/g Ti · h · atm)	Part of isotactic PP (%)	Refs
$TiCl_4/Al(C_2H_5)_3(1:3)$	30	27	231
α-$TiCl_3/Al(C_2H_5)_2Cl$	25	87	232
α-$TiCl_3/AlCl_3Al/(C_2H_5)_2Cl$	120	80	233
β-$TiCl_3/Al(C_2H_5)_3/H_2$		15	234
$TiCl_3/LiAlH_4$			235
$TiCl_3/LiAl_2H_7/NaF$	70	90	236
β-$TiCl_3/Al(C_2H_5)_2Cl/LB^1$	99	95	237
β-$TiCl_3/AlCl_3/Al(C_2H_5)_2Cl/LB^2$	520	98	238
$TiCl_3/TiCl_3CH_3$	Low	Low	239
Ti/I_2	Low	Low	240

[a]LB^1 = Lewis base 1, methylmethacrylate; LB^2 = Lewis base 2, diisoanyl ether.

The nature of the ligands and the valency of the transition metal atoms essentially govern activity, productivity, and stereospecifity. Another strong influence is exerted by the nature of the cocatalyst. It consists of organometallic compounds of the main groups 1 and 3 of the periodic table. For the propene polymerization, alkyls of lithium [64], beryllium [240], magnesium [240], zinc [241], aluminum [64], and gallium [10] have been used. Aluminum alkyls have been proven to be particularly suitable. Nowadays they are used exclusively as cocatalysts since they are superior to all other organometallic compounds as far as activity, stereospecifity, accessibility, and availability are concerned. Only lithium alanate is an exception to this. It possesses higher thermostability and is therefore preferred for solution polymerization at temperatures between 150 and 200 °C.

Heterogeneous catalysts are suspended in the solvent. The Ziegler catalyst $TiCl_4/Al(C_2H_5)_3$ affords polypropene with very low stereospecifity (compare Table 14). One criterion for the determination of stereospecifity is the isotacticity index, which is defined as the percentage of polymer that is insoluble in boiling heptane [10]. Natta achieved a substantial increase in stereoselectivity by using $TiCl_3$ instead of $TiCl_4$ [240].

$$3TiCl_4 + Al \rightarrow 3TiCl_3 \cdot AlCl_3 \qquad (49)$$

The aluminum halogenide content of $TiCl_3$ leads to the formation of defects in the crystal lattice, thereby effecting an increase in activity. At temperatures up to 100 °C, β-$TiCl_3$ is formed, which upon tempering assumes the layered structure of γ-$TiCl_3$. Above 200 °C, α-$TiCl_3$ is formed. Today, $TiCl_3$ in combination with $Al(C_2H_5)_2Cl$ is still used as a catalyst for the polymerization of propene. It is referred to as a first-generation catalyst. The use of $Al(C_2H_5)_3$ decreases the stereospecifity and $Al(C_2H_5)Cl_2$ drastically lowers the catalytic activity [242]. Table 15 gives the influence of various ligands on stereoregularity for the system $TiCl_3/Al(C_2H_5)_2X$ [243].

The preferred metal alkyls possess ethyl and isobutyl ligands. Typical examples are $AlEt_3$, $AliBu_3$, $AlEt_2Cl$, and Al-$(i$-$Bu)_2Cl$. The stereoregularity of the polypropene decreases with increasing size of R in AlR_3 [244]. Vanadium salts attracted much attention because they led predominantly to statistical copolymers, as opposed to block copolymers produced with titanium salts.

Depending on reaction temperatures, first-generation catalysts produce increasing amounts of atactic polypropene (8 to 20%) next to the isotactic main product. By modification with electron donors (Lewis bases; see also Table 14) of the desired complexation tendency, the atactic polymerization sites can be largely deactivated, thus raising the isotaxy index to 94 to 98% [245]. It is obvious that atactic polymerization

Table 15 Varying X in $Al(C_2H_5)_2X/TiCl_3$ catalysts polymerizing.

X	Rate of polymerization (relative to $X = C_2H_5$)	Stereoregularity I.I. (%)
C_2H_5	100	85
F	30	83
Cl	33	93
Br	33	95
I	9	98

Source: Ref. 243.

centers have a greater tendency towards complexation than do isotactic ones. Catalysts that are modified in this manner are also known as second-generation catalysts.

The partial blocking of active sites leads to a decrease of catalytic activity. Due to a tremendous increase in surface area of the $TiCl_3$ the activity of the modified catalyst can be increased by a factor of 2 to 5.

1. Kinetic Aspects

To date, numerous papers dealing with the kinetics of propene polymerizations have been published [246–263]. Since in the course of the polymerization the $TiCl_3$ crystallites break into smaller pieces, thereby exposing new active centers, the kinetical investigation of the reaction is made more difficult. For the majority of systems, however, it was found that the polymerization rates are proportional to the concentrations of catalyst and monomer but do not depend on the aluminum organic component as long as a threshold concentration is maintained.

$$r = k_p \cdot [TiCl_3]^1 \cdot [C_3H_6]^1 \cdot [Al(C_2H_5)_2Cl]^\circ \tag{50}$$

This means that there is practically no dependence of the propene polymerization rate on the $Al(C_2H_5)_3/TiCl_3$ ratio over a wide range. However, a dependence of the reaction rate on the metal component ratio was observed by Tait [264] and Zakharov et al. [265] in the presence of AlR_3/VCl_3 and $Al(C_2H_5)_3/AlCl_3/TiCl_3$ systems. It must be remarked that extremely high aluminum alkyl concentrations of 0.3 mol/l were used, whereas these are normally 0.005 mol/l. The authors introduced kinetic models of the Langmuir–Hinshelwood type with reversible adsorption of aluminum alkyl on the transition metal halogenide surface.

Other differences in behaviour between the investigated catalytic systems concern the dependence of the polymerization rate on time. In the first minutes the activity increases until it reaches a maximum value, which keeps constant for several hours [266,267].

2. Active Sites

To measure the activity of the catalyst, it is necessary to know something about the portion of titanium atoms that form active sites. A lot of studies have been carried out to evaluate the concentration of active sites and their location [268]. Such studies are facilitated by the fact that the first polymer chains forming on active sites preserve only a trace of their origin as well as by the fact that the variation of their molecular weight with polymerization time depends on the number of active sites. On the other hand, a correct determination of the active sites is made complex by the nature of the phenomena occurring during the polymerization.

There are two parameters linked to the concentration of active sites, the polymerization rate (propagation rate) and the growing time (average lifetime) of the polymer chains. The various methods are summarized as follows:

1.   Variation of the molecular weight as a function of polymerization time [268–278] (kinetic method). This could only be obtained at low temperatures and low monomer conversions.
2.   Determination of the number of labeled alkyl groups bound to the polymer chains (end groups) of polymers obtained with catalysts prepared in the presence of ^{14}C-labeled aluminum alkyls [268,279].

Table 16 Percentage of active sites (C*) and propagation constants (k_p) for propene polymerization with TiCl₃ catalysts.

Catalyst system	Temp. (°C)	k_p (L mol^{-1}s^{-1})	C* (% Ti active/Ti total)	Method[a]	Refs
α-TiCl₃/AlEt₃	70		0.7–1.7	MW,¹⁴C	268
α-TiCl₃/AlEt₂Cl	70		0.3–0.6	MW,¹⁴C	268
α-TiCl₃/Al(I-Bu)₃	80	2.9	0.54	K	273
α-TiCl₃/AlEt₃	70		0.54	I	269
α-TiCl₃/AlEt₂Cl	70	1.1	3.6	¹⁴CO	274
δ-TiCl₃/0.3AlCl₃/AlEt₃	70	100	0.58	¹⁴CO	275
δ-TiCl₃/AlEt₂Cl	70	124	2.8	T	276
δ-TiCl₃/0.3AlCl₃/Al(i-Bu)₃	70	90	0.8	¹⁴CO	275
γ-TiCl₃/AlEt₂Cl	50	80	1.5	MW,K	277

[a]MW, molecular weight variation method; K, kinetic method; I, inhibitor method; ¹⁴C, ratoactive alkyl method; T, tritiated quenching agent method; ¹⁴CO, radioactive carbon monoxide method.

3. Inhibition of the active sites with compounds such as methanol, iodine, or allenes [280]. Since it is most unlikely for these compounds to react with active sites only, the method gives too high values.
4. Reaction between the transition metal–carbon bonds present in the polymerization system and a labeled quenching agent such as ^{131}I$_2$, tritiated alcohols or water [276], deuterated methanol or water, and ^{14}CO or ^{14}CO$_2$ [281,282].

Table 16 summarizes some results of the propagation rate k_p and the number of active sites for various TiCl₃ catalysts. It can be seen that 0.3–3.6% of the total titanium atoms form active sites.

Polymerization does not seem to involve proper kinetic chain termination phenomena, but polymer chain transfer processes, making the active site available to initiate a new polymer chain. These can be formulated as a chain transfer process with the monomer M:

$$\text{Cat}-(\text{M})_n-\text{H} + \text{M} \rightarrow \text{Cat}-\text{MH} + (\text{M})_n \tag{51}$$

or as a transfer process with the organometallic compound:

$$\text{Cat}-(\text{M})_n-\text{H} + \text{AlR}_3 \rightarrow \text{Cat}-\text{R} + \text{AlR}_2-(\text{M})_n-\text{H} \tag{52}$$

There is evidence that many polymer chains are bound to aluminum at the end of polymerization [283]. Very important is a spontaneous termination process by β-hydride extraction:

$$\text{Cat}-\text{CH}_2-\text{CH}(\text{CH}_3)-(\text{M})_n-\text{H} \rightarrow \text{Cat}-\text{H} + \text{CH}_2{=}\text{C}(\text{CH}_3)-(\text{M})_n-\text{H} \tag{53}$$

When operating in the presence of H₂ as molecular weight regulator, saturated polymer chains are formed:

$$\text{Cat}-(\text{M})_n-\text{H} + \text{H}_2 \rightarrow \text{Cat}-\text{H} + \text{H}-(\text{M})_n-\text{H} \tag{54}$$

Dividing the chain propagation rate constant k_p by the average transfer rate constant k_t gives the average molecular weight M_n of the polymer chain. M_w/M_n was assumed to be in the range 5 to 10 [284].

3. Mechanism

Various models of catalytic centers and of monomeric unit addition mechanism have been proposed to interpret the isospecific polymerization of α-olefins with Ziegler–Natta catalytic systems [285–293]. For the α-olefins the combination of x-ray diffraction and IR analysis showed very early that the polymers obtained with the Ziegler catalytic system [294] are substantially linear polymers with head-to-tail enchainments. The regioselectivity of the amorphous product is slightly lower than that of the crystalline polymer. α-Olefin polymerization is shown to occur through a cis-insertion reaction by using deuterated propene. The cis addition to the double bond was proven when Miyazawa and Ideguchi [295] and Natta et al. [296] established that the polymer of cis-ld1-propene is erythro-diisotactic, whereas the polymer from trans-ld1-propene is threo-diisotactic. The metal atom of the catalyst bearing the growing chain and the growing chain end are added simultaneously to the double bond of the incoming monomer:

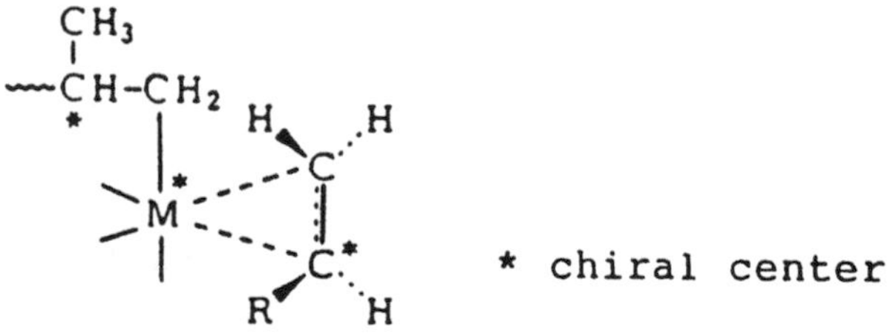

(55)

To synthesize isotactic polypropene, the catalytic center must sharply discriminate between the two prochiral faces of the α-olefin. To do this, the catalytic system must possess one or more chirality centers.

    Considering the simplest model of a monometallic catalytic center (55), there certainly is a chiral carbon atom in the growing chain in α-position with respect to the metal atom; furthermore, the metal atom itself can be a center of chirality [297], which being bound to a solid surface could maintain its absolute configuration during the insertion reaction. Therefore, stereoselectivity is caused by the chirality of the catalytically active center and not by chiral atoms in the growing chain.

    One model for the active center, proposed by Arlmann and Cossee [298,299], is based on monometallic catalytic centers (56) with hexacoordinated transition metal:

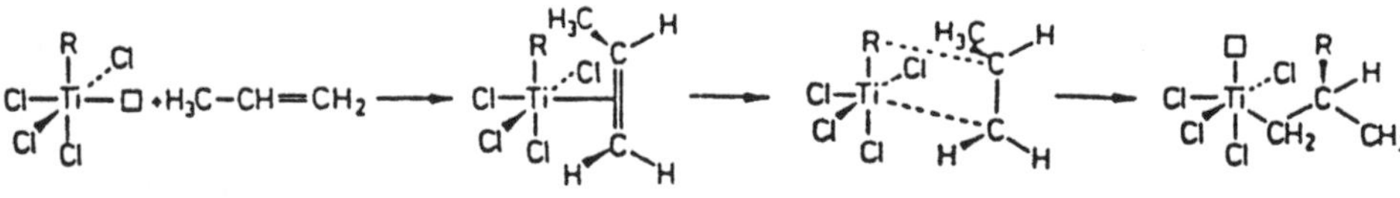

(56)

The Ti atoms close to the TiCl$_3$ surface have a vacant octahedral site, one chlorine ligand singly bonded and four chlorine ligands bridge bonded with neighboring Ti atoms. By reaction with the aluminum alkyl, the singly bonded chlorine atom is substituted by an alkyl group with the formation of a Ti–C bond. The olefin is complexed on the vacant site with the double bond parallel to an octahedral axis. Two orientations are therefore possible, giving rise to the stereospecifity. After monomeric unit insertion, the Ti–R bond enters the vacant site and another olefin molecule is coordinated.

This model is modified by Pino [300,301], Corradini [302], Kissin [303], Keii [304], Terano [305], Cecchin [306] to other titanium complexes. Bimetallic models between the titanium compound and the cocatalyst were discussed by Sinn and Patat [137], Pino [301], and Zakharov [307]. Others suggest that the growing polymer chain is bound to the transition metal through a double bond (carbene complex) and that the insertion reaction occurs through formation of a metal-cyclobutane intermediate [308,309].

4. Supported Catalysts

The traditional Ziegler–Natta catalyst, based on $TiCl_3$ and aluminum alkyls (first generation), is not active enough to do without the removal of catalyst residues from the polymer. This is why only a small part of the titanium present on the side surface of $TiCl_3$ crystallites is deemed to be active in propene polymerization. Researchers have endeavored to obtain better utilization of the titanium halogenide by trying to attach it to the surface of proper supports. Great industrial interest is evidenced by the numerous patent applications following the initial Shell patent [310–319]. Commonly used supports are $MgCl_2$, $CoCl_2$, SiO_2, $Mg(OH)_2$, $Mg(OH)Cl$, $MgR(Cl)$, MgO, $MgCO_3$, SiO_2, and SiO_2/Al_2O_3 [320–333]. The preferred halides are those having the same layered lattice structure as δ-$TiCl_3$. The dimensions of $MgCl_2$ and $COCl_2$ (ionic radii of Mg^{2+} and Co^{2+} are 0.066 and 0.072 nm) make them particularly suitable carriers for $TiCl_4$ (ionic radius of Ti^{4+}, 0.068 nm). These catalysts were demonstrated to substantially increase the activity in propene polymerizations using $AlEt_3$ or $AlEt_2Cl$ as cocatalysts (second generation). Furthermore, the use of electron donors, notably esters of carbocylic acids such as ethylbenzoate, was demonstrated to increase stereoselectivity (third generation) [334–347]. With $TiCl_4$ supported on SiO_2, the activity is low but the crystallinity of the resulting polypropylene is high [348]. The addition of $NaCl$, $CaCl_2$, or $BaCl_2$ increases the activity by a factor of up to 5.

Soga studied propene polymerizations with catalytic systems based on $Mg(OH)_2$, $Mg(OH)Cl$, or $MgCl_2/TiCl_4/AlEt_3$ [94,322]. Unlike the $TiCl_4/AlEt_3$ system, these catalysts exhibit an almost constant overall rate of polymerization (4.1 g PP/g Cat h atm) for at least 2 h. Catalysts obtained by reaction of Ti benzyl and cyclopentadienyl derivates with $Mg(OH)Cl$ have been investigated as well as Grignard reagents together with $TiCl_4$ at varying ratios have been investigated [317].

The most important catalysts are obtained by supporting titanium halides on activated $MgCl_2$. By combination with the cocatalyst AlR_3, a very high activity is given, although the stereospecifity is low (Table 17) [321]. The discovery of catalysts supported on activated $MgCl_2$ and modified Lewis bases has solved the problems of low stereospecifity.

5. Role of Lewis Base Esters

The catalyst can be prepared on different routes such as ball milling, vibration milling, or chemical conversions [349]. First, commercially available anhydrous $MgCl_2$ is ball milled with ethyl bonzoate over 20 h to afford active $MgCl_2$. By this process the dimensions of the agglomerated primary $MgCl_2$ crystallites (60 × 30 nm) are broken (3 × 2 nm) and stabilized by ethyl benzoate. The support develops a surface area of 50 to 300 m^2/g [334]. Second, the ball milled support is mixed with $TiCl_4$ by further ball milling of the catalyst support in

Table 17 Polymerization of propene with supported $MgCl_2/TiCl_4$ catalysts by 70 °C.

Catalyst	Ti (%)	Cocatalyst	Activity (kg PP/mol Ti h)	I.I. (%)
$MgCl_2/TiCl_4$	3.6	$AlEt_3$	870	49
$MgCl_2/TiCl_4$	3.6	$AlEt_2Cl$	45	33
$MgCl_2/TiCl_4/EB$	2	$AlEt_3$	650	98
$TiCl_3 \cdot 0.3\ AlCl_3$	24	$AlEt_3$	23	76

the presence of $TiCl_4$ or by suspending the $MgCl_2/EB$ in hot undiluted $TiCl_4$. The resulting solid is washed to remove soluble titanium complexes. The catalyst contains 1 to 5 wt% Ti and 5 to 20 wt% ethyl benzoate or diisobutylphthalate. The Lewis base used in this procedure is called an *internal Lewis base*.

Therefore the function of the internal donor in $MgCl_2$-supported catalysts is twofold. One function is to stabilize small primary crystallites of magnesium chloride; the other is to control the amount and distribution of $TiCl_4$ in the final catalyst. Activated magnesium chloride has a disordered structure comprising very small lamellae.

An essential part of every Ziegler catalyst is the cocatalyst. Supported $MgCl_2/TiCl_4$ or $MgCl_2/EB/TiCl_4$ are combined with $AlEt_3$ or $AlEt_3/EB/$ to give high polymerization activities. The donor used for this procedure is called an *external Lewis base*. Carboxylic acid esters or aromatic silanes, preferably alkoxisilane or derivatives such as *para*-ethyl anisate, are described as external Lewis bases [338,345]. Silyl ethers $R_nSi(OR')_{4-n}$ such as Ph_3SiOCH_3, $Ph_2Si(OCH_3)_2$, $PhSi(OCH_3)_3$, and $(C_2H_5)Si(OCH_3)$ have also been found to be highly active promoters in stereospecific olefin polymerization [350,351].

Both internal and external Lewis bases react with aluminum alkyls forming a 1:1 complex in the first step (57) [352,353]. The second step is an alkylation reaction affording a new alkoxyaluminum species (58).

$$AlEt_3 + \quad (57) \quad \rightleftharpoons \quad (58)$$

Kashiwa found that suitable amounts of ethenebenzoate increased the yield of iso-tactic polymers; at the same time, the production of atactic polymer is strongly decreased. So the aromatic ester simultaneously acts as a poison of the aspecific sites and as an activator of the isospecific sites: It is reasonable to assume that a highly active and stereospecific catalyst can be obtained by selectively using only the stereospecifically active centers in the $MgCl_2/TiCl_4/AlEt_3$ catalyst system. The Lewis base increase M_n for isotactic polymers and slightly decrease M_n for atactic ones. Such catalyst yields 100 to 2000 kg/g Ti of polypropene containing 95 to 98% isotactic polymer in about 3 h. The high activity makes it possible to leave the titanium (0.5 to 2 ppm) in the polymer. The concentration of active sites reaches values of 5 to 20% measured by the ^{14}CO method [354,355].

A new type of catalysts contain a diether such as 2,2-disubstituted-1,3-dimethoxy-propane and have high stereospecficity even in the absence of an external donor [356]. The polypropene yield obtained under typical polymerization conditions (liquid monomer, 70 °C, 1–2 h) has increased from 30–80 kg PP/g cat for the third generation to 80–160 kg PP/g cat [342,357].

Giannini [358] has indicated that, on preferential lateral cleavage surfaces, the magnesium atoms are coordinated with 4 or 5 chlorine atoms, as opposed to 6 chlorine atoms in the bulk of the crystal. These lateral cuts correspond to (110) and (100) faces of $MgCl_2$. It has been proposed that bridged dinuclear Ti_2Cl_8 species can coordinate to the (100) face of $MgCl_2$ and on contact with an alkylaluminum cocatalyst these species are reduced to Ti_2Cl_6 units in which the environment of the Ti atoms is chiral [358], a necessary condition for isospecific polymerization. In the absence of a Lewis base, $TiCl_4$ will coordinate to both the (100) and the (110) faces of $MgCl_2$. In the presence of an internal donor, however, there will be a competition between the donor and $TiCl_4$ for the available coordination sites. One possible function of the internal donor is that, due to the higher acidity of the coordination sites on the (110) face, preferential coordination of the donor on these sites will avoid the formation if Ti species having poor selectivity.

The requirement for an external donor when using catalysts containing an ester as internal donor is due to the fact that, when the catalyst is brought into contact with the cocatalyst (most commonly $AlEt_3$), a large proportion of the internal donor is lost as a result of alkylation and/or complexation reactions. In the absence of an external donor, this leads to poor stereoselectivity due to increased mobility of the titanium species on the catalyst surface [360].

In contrast to ester internal donors, the diethers, having greater affinity towards $MgCl_2$ than towards AlR_3, are not displaced from the catalyst surface on contact with the cocatalyst [361]. Consequently, highly isotactic poly(propene) can be obtained even in the absence of an external donor.

Studies by Busico et al. [362] have indicated that active species in $MgCl_2$-supported catalysts can isomerize very rapidly (during the growth time of a single polymer chain) between three different propagating species. The chain can therefore contain, in addition to highly isotactic segments, sequences that can be attributed to weakly isotactic (isotactoid) and to syndiotactoid segments.

For the production of isotactic polypropene, different processes can be used (see production of polyethene). In the laboratory, the catalyst is suspended in dry pure heptane (or other hydrocarbons) in inert gas atmosphere and then bubble in propane gas at 30 to 60 °C. Polymers form a while solid permeating the catalyst particles.

In industry scale bulk polymerization is carried out in liquid propene, solution process, or in gas phase using a stirred bed as well as a fluidised bed process [357,363–365]. Mostly used is the sheripal process using a spherioidal catalyst and liquid propene [357,363]. The polymerization takes place at 70 °C and 4 MPa circulating liquid propene round one or more loop reactors. A single axial flow agitator in each loop maintains high flow rates to ensure good heat transfer to the water-cooled jackets, whilst also preventing any polymer particles settling from the slurry. Continuously metered catalyst, triethylaluminum and a Lewis base stereoregulator are fed into the reactor to maintain polymerization and stereo control.

6. Homogeneous Catalysts

With ansa(chiral) titanocenes, zirconocenes, and hafnocenes in combination with methylalumoxane (MAO) it is possible to obtain highly isotactic polypropene [366–374]. When changing the symmetry of the complex, different structures of the polypropene are yielded. The activity of these hydrocarbon soluble catalysts are extremely high.

The microstructure of polypropene in terms of the enchainment of the monomer units and their configuration is determined by the regio- and stereospecificity of the insertion of the monomer. Depending on the orientation of the monomer during insertion into the transition-metal-polymeryl bond, primary (1,2-) and secondary (2,1-) insertions are possible (Figure 6). Consecutive regiospecific insertion results in regioregular head to tail enchainment (1,3-branching) of monomer units while regioirregularities cause the formation of head to head (1,2-branching) and tail to tail (1,4-branching) structures.

Generally, metallocenes favor consecutive primary insertions due to their bent sandwich structure. Secondary insertion also occurs to an extent determined by the structure of the metallocene used and the experimental setup (especially temperature and monomer concentration). Secondary insertions cause an increased steric hindrance to the next primary insertion. The active center is blocked and therefore is regarded as a resting state of the catalysts [375]. The kinetic hindrance of chain propagation by another insertion favors chain termination and isomerization processes. One of the isomerization processes observed in metallocene catalysed polymerization of propene leads to the formation of 1,3-enchained monomer units [376–379]. The mechanism is discussed to involve transition metal mediated hydride shifts [380,381].

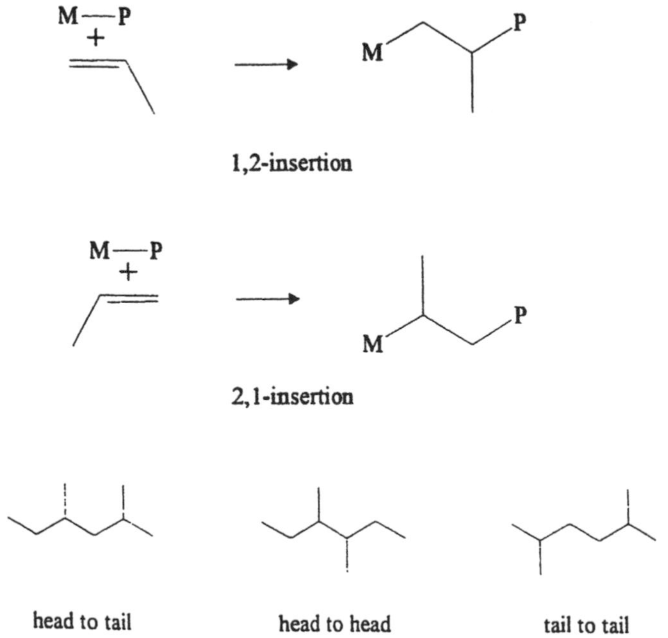

Figure 6   Primary (1,2) and secondary (2,1) insertion in propene polymerization.

Another type of steric isomerism observed in polypropene is related to the facts that propene is prochiral and polymers have pseudochiral centers at every tertiary carbon of the chain. The regularity of the configuration of successive pseudochiral centers determines the tacticity of the polymer. If the configuration of two neighbored pseudochiral centers is the same this 'diad' is said to have a *meso* arrangement of the methyl groups (Figure 7). If the pseudochiral centers are enantiomeric, the diad is called *racemic*. A polymer containing only *meso* diads is called isotactic, while a polymer consisting of *racemic* diads only is named syndiotactic. Polypropene in which *meso* and *racemic* diads are randomly distributed is atactic (see structures (46)–(48)).

A single step of the polymerization is analogous to a diastereoselective synthesis. Thus for achieving a certain level of chemical stereocontrol, chirality of the catalytically active species is necessary. In metallocene catalysis, chirality may be located at the transition metal itself, the ligand, or the growing polymer chain, e.g. the terminal monomer unit. Therefore two basic mechanisms of stereocontrol are possible [382]: (a) catalytic site control (also referred to as enantiomorphic site control), which is connected to chirality at the transition metal or the ligand, and (b) chain end control which is caused by the chirality of the last inserted monomer unit. These two mechanisms cause microstructures (Figure 8) which may be described by different statistics, while in the case of catalytic site control, errors are corrected due to the regime of the catalytic site (Bernoullian statistics), chain end controlled propagation is not capable of doing so (Markovian statistics).

  a.   Isotactic Polypropene.   In the mid-1980s, the first metallocene/MAO catalysts for the isotactic polymerization of propene were described. Ewen [369] found Cp_2TiPh_2/ MAO to produce isotactic polypropene at low temperatures by chain end control mechanism (stereoblock structure). When using a mixture of *racemic* and *meso* $[En(Ind)_2]TiCl_2$ in combination with MAO, he obtained a mixture of isotactic and atactic polypropene, the isotactic polymer having a microstructure in accordance with catalytic site control (isoblock structure). The use of pure racemic $[En(Ind)_2]ZrCl_2$ yielded for the first time pure isotactic polypropene formed by metallocene/MAO catalysts [366,382]. These investigations were the beginning of rapid development in the area of metallocene catalyzed polymerization of propene which resulted in the invention of tailor-made

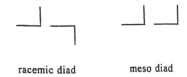

racemic diad   meso diad

Figure 7   Schematic drawing of the *racemic* and *meso* diad of poly(α-olefins).

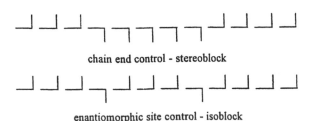

chain end control - stereoblock

enantiomorphic site control - isoblock

Figure 8   Microstructures of polypropenes resulting from different types of stereocontrol during insertion.

metallocenes for different microstructures based on the mechanistic understanding of stereocontrol.

According to the structure of the metallocene, different microstructures are realized. Generally, among the rigid metallocenes, some different structures may be distinguished, and there are also metallocenes which will be referred to later, having fluctuating structures.

b. C_{2v} Symmetric Metallocenes. C_{2v} symmetric metallocenes like Cp_2MCl_2 or $[Me_2Si(Flu)_2]ZrCl_2$ are achiral. The only stereocontrol observed is both chain end type and low, due to the fact that the chiral center of the terminal monomer unit of the growing chain is in β-position due to 1,2-insertion of the monomers. A significant influence on the tacticity is only observed at low temperatures (Table 18) and much more pronounced in case of titanocenes and hafnocenes due to their shorter M–$C\alpha$ bonds bringing the chiral β-carbon closer to the active center [385–387]. Therefore mainly atactic polypropene is produced [388].

c. C_2 Symmetric Metallocenes. rac-Ethenebis(indenyl)zirconium dichloride and rac-ethenebis(4,5,7,8-terahydroindenyl)zirconium dichloride were the first chiral metallocenes investigated and found to produce isotactic polypropene (see Figure 2). During the last ten years a lot of variations of these metallocenes have been published and patented, aiming for higher activities, molecular weights, tacticities and thereby higher melting points. Table 19 summarizes some of the developments leading to catalysts which

Table 18 Chain end control by $Cp_2M(2methylbutyl)_2$/MAO at low temperatures determined as isotacticity index I.I. and sequence length of *meso* and *racemic* blocks.

Temp (°C)	Zr/30	Zr/7	Zr/−20	Zr/−60	Ti/−3
mmmm%	0.052	0.085	0.106	0.140	0.430
n_{iso}	2.02	2.28	2.36	2.53	4.09
n_m	1.50	1.69	1.74	1.88	3.47
n_r	1.48	1.32	1.29	1.23	1.13

Table 19 Comparison of the productivity, molecular weight, melting point and isotacticity obtained in polymerization experiments with various metallocene/MAO catalysts (bulk polymerization in liquid propene at 70 °C, Al/Zr ratio 15 000) showing the broad range of product properties [389].

Metallocene	Productivity [kg PP/(mmol Zr h)]	$M_w \times 10^{-3}$ [g/mol]	m.p. [°C]	Isotacticity [% mmmm]
$[En(Ind)_2ZrCl_2$	188	24	132	78.5
$[Me_2Si(Ind)_2]ZrCl_2$	190	36	137	81.7
$[Me_2Si(IndH_4)_2]ZrCl_2$	48	24	141	84.5
$[Me_2Si(2Me-Ind)_2]ZrCl_2$	99	195	145	88.5
$[Me_2Si(2-Me-4iPr-Ind)_2]ZrCl_2$	245	213	150	88.6
$[Me_2Si(2,4Me_2-Cp)_2]ZrCl_2$	97	31	149	89.2
$[Me_2Si(2Me-4tBu-Cp)_2]ZrCl_2$	10	19	155	94.3
$[Me_2(2Me-4,5Benzlnd)_2]ZrCl_2$	403	330	146	88.7
$[Me_2Si(2Me-4Ph-Ind)_2]ZrCl_2$	755	729	157	95.2
$[Me_2Ge(2Me-4Ph-Ind)_2]ZrCl_2$	750	1135	158	–
$[Me_2Si(2Me-4Naph-Ind)_2]ZrCl_2$	875	920	161	99.1

can produce polypropenes with properties comparable to the ones reached by using supported $TiCl_4$ catalysts [389–391].

Systematic investigation of bis(indenyl)zirconocenes showed that the main chain termination reaction is β-hydrogen transfer with the monomer [392,393]. This reaction is very effectively suppressed by substituents (Me, Et) in position 2 of the indenyl-ring [394,395]. Substituents in position 4 also cause an enhancement in molecular weight by reducing 2,1-misinsertions which preferably result in chain termination by β-hydrogen elimination. Due to the fact that primary insertion is sterically hindered after a regioerror occurs and therefore the catalyst is in a resting state after a 2,1-insertion, suppression of this type of misinsertion also leads to enhanced activities. Using aromatic substituents in position 4 results in additional electronic effects. Thus the most active catalysts feature a methyl or ethyl group in position 2 and an aromatic group in position 4 of the indenyl rings.

Besides the bis(indenyl) ansa compounds, C_2 symmetric bridged bis(cyclopenta-dienyl) metallocenes of zirconium and hafnium (structure (59)) were found to be able to produce isotactic

(59)

polypropene (Table 20) [396]. The key for high isotacticity are substituents in positions 2,4,3′ and 5′ generating a surrounding of the transition metal similar to the one in bis(indenyl) metallocenes.

In this type of metallocenes the chirality is due to the chirality of the ligand and the two chlorines (e.g. the position of the growing chain and the coordinating monomer) are

Table 20 Polymerization behavior of metallocenes based on bridged biscyclopentadienyl compounds. All polymerizations were performed at 30°C in 500 ml toluene at 3 bar. Al/M = 10 000, [M] = 0.002 mmol, $t = 2$ h.

Metallocene (structure 5)	Productivity [kg PP/(mmol Zr h)]	$M_W \times 10^{-3}$ [g/mol]	m.p. [°C]	Isotacticity [% mmmm]
$[Me_2Si(2,3,5Me_3Cp)_2]ZrCl_2$	1.6	134	162	97.7
$[Me_2Si(2,4Me_2Cp)_2]ZrCl_2$	11.1	87	160	97.1
$[Me_2Si(3tBuCp)_2]ZrCl_2$	0.3	10	149	93.4
$[Me_2Si(3MeCp)_2]ZrCl_2$	16.3	14	148	92.5
$[Me_2Si(2,3,5Me_3Cp)_2]HfCl_2$	0.30	256	163	98.7
$[Me_2Si(2,4Me_2Cp)_2]HfCl_2$	0.10	139	162	98.5
$[Me_2Si(3tBuCp)_2]HfCl_2$	0.03	17	157	–
$[Me_2Si(3MeCp)_2]HfCl_2$	1.61	67	148	–

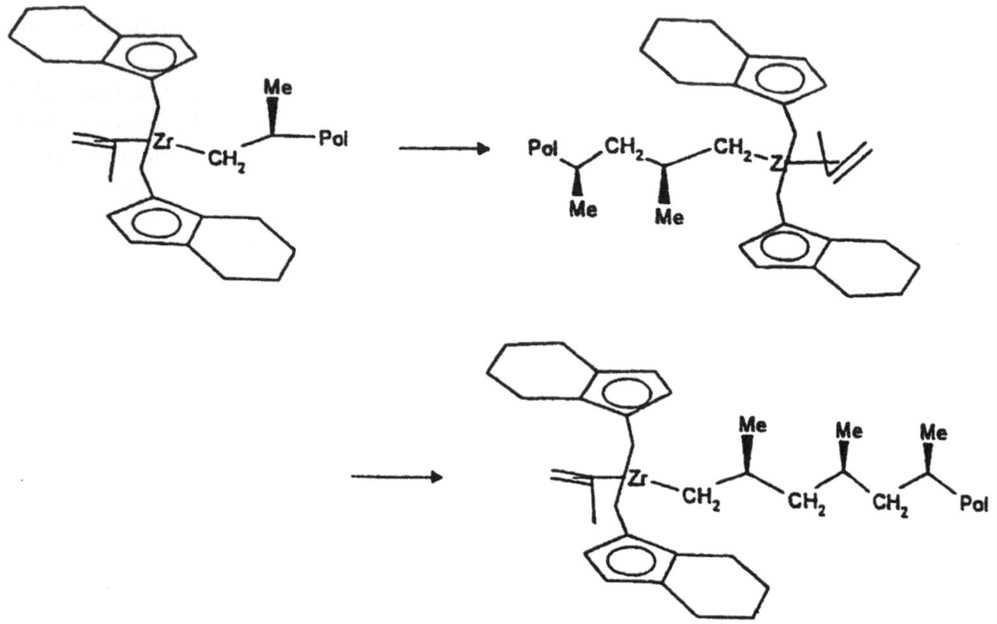

Figure 9 Mechanism of the isotactic polymerization of propene using an alkylzirconocenium ion generated from a C_2–symmetric bis(indenyl)zirconocene. The orientation shown for the first inserted propene with the methyl group down is favored over the other possibility with the methyl group up.

homotopic. According to a model of Pino et al. [397–400] the conformation of the growing polymer chain is determined by the structure of the incoming monomer and is forced into a distinct orientation by steric interactions of its side chain with the polymer chain (Figure 9) [401–406].

The C_2-symmetric metallocenes give polypropylenes with a high melting point (162 °C) and ^{13}C-NMR spectroscopically measured tacticities (mmmm pentades) of 97 to 99%. The properties and melting point of isotactic polypropenes prepared by metallocene catalysts are determined by the amount of irregularities (stereo- and regioerrors) randomly distributed along the polymer chain. Thus the term stereospecificity does not refer to extractable aPP as for conventional PP always having a melting point of 160–165 °C. Metallocene catalysts depending on their substitution pattern can give a wide range of homopolymers having melting points between 125 and 165 °C (Table 21).

The molecular weight distribution of these iPPs ($M_w/M_n = 2$–2.5) is lower than that of conventional produced PP ($M_w/M_n = 5$–20). For applications demanding broader molecular weight distributions, two or more metallocenes may be combined to give a tailor-made molecular weight distribution. Compared with conventional iPP grades, metallocene products show enhanced mechanical strength which can be improved by tailoring the molecular weight distribution.

The low melting points obtained with some metallocene catalysts, even at high pentad isotacticities, are caused by 2,1- and 1,3-misinsertions [408,409]. Low melting point polymers with conventional catalysts are obtained by copolymerization with small amounts of ethene.

The excellent performance of metallocenes in copolymerizations also offer improvements in impact copolymers. In the wide variety of properties of impact copolymers, the stiffness of the material is determined by the matrix material, while the impact resistance

Table 21 Comparison of isotactic polypropenes prepared by different metallocene/MAO catalysts [En(IndH$_4$)$_2$]ZrCl$_2$ (I), [Me$_2$Si(4,5Benzlnd)$_2$]ZrCl$_2$ (II), [Me$_2$Si(4,6iPrlnd)$_2$]ZrCl$_2$ (III) at 70 °C in a bulk polymerization at Al/Zr = 15 000 to conventional isotactic PP prepared by a TiCl$_4$/MgCl$_2$ catalyst (IV) [407].

	(I)	(II)	(III)	(IV)
Melting point [°C]	139	151	160	162
M_W/M_n	2.2	2.3	2.5	5.8
Modulus [N/mm^2]	1060	1440	1620	1190
Hardness [N/mm^2]	59	78	86	76
Impact resistance Izod [mJ/mm^2]	128	86	100	103
Light transmission rate [% 1-mm plate]	56	44	35	34
Melt flow rate [°/min]	2	2	2	2

largely depends on the elastomeric phase. While conventional catalysts show some inhomogeneities in the ethene/propene rubber phase due to crystalline ethene rich sequences, the more homogeneous comonomer distribution obtained with metallocene catalysts results in a totally amorphous phase [410].

Using highly stereoselective metallocenes, highly crystalline, stiff polypropene types are produced. These polymers exhibit a stiffness 25–30% above that of conventional polypropenes, resembling that of polypropenes filled with talcum or other minerals [411]. Packages made from these polypropenes may have reduced thickness of the walls, are easier recycled, show enhanced impact strength, heat resistance, lower density and less aging.

Metallocenes are also interesting for the production of new iPP waxes for use as pigment dispersants, toner or lacquer surfaces [412]. The molecular weight of about 10 000 to 70 000 g/mol combined with melting points between 140 and 160 °C is easily obtained by the choice of the metallocene

The homogeneous system is capable of producing isoblock polypropene (see Figure 8). More stereoerrors shown as lower isotacticity are formed. [En(Ind)$_2$]ZrCl$_2$ (see Table 19) produces a polypropene with an isotacticity of only 78.5%. This means that about 15 propene units are incorporated in the same stereospecific structure, one is reversed and the following one has the same orientation as the previous block. The block of 15 same oriented units is called 'isotactic sequence length'.

A different structure is obtained when 4 to 12 propene units have the same stereospecific structure; then there is a change and the following block of propene units has the opposite stereospecific structure. This results in a stereoblock polypropene (see Figure 8). Such structures of polypropene can be synthesized with a catalyst in which the chirality is further or the rotation of the ligands is hindered away from the transition metal [411]. Bis(neomenthyl)zirconium dichloride is an example of this type. The catalyst possesses three chiral carbons at the neomethyl group bonded to the cyclopentadienyl ring. The stereoblock length depends on the polymerization temperature. With increasing temperature, the stereoblock length increases. Products with smaller isotactic block length have a lower melting point and are more flexible.

The methylaluminoxane as cocatalyst can be replaced by a mixture of trimethyl-aluminum and dimethylfluoroaluminum, or by N,N-dimethylanilineumtetrakis(penta-fluorophenyl)boron, showing similar polymerization activities [412,413]. Cocatalyst-free propene polymerization systems are found by Watson [414] like (C$_5$Me$_5$)$_2$LuCH$_3$, but the activity and the molecular weights are low and there is no stereotacticity. Homogeneous vanadium catalysts are not able to produce isotactic polypropene but yield a syndiotactic polymer.

C. Syndiotactic Polypropylene

Syndiotactic polypropene was first isolated by Natta [415] through separation of polymer obtained with α- or γ-TiCl$_3$ and Al(C$_2$H$_5$)$_2$F or LiC$_4$H$_9$ and on TiCl$_4$ and LiC$_4$H$_9$. The yield of boiling hexane-soluble polypropene was only 1 to 10%. Higher yields are obtained with homogeneous catalyst based on vanadium compounds (Table 22). For highly syndiotactic polypropene the Al/V ratio is between 2 and 10; when a weak Lewis base such as anisole is added, it is 1:1 [416–421].

The syndiospecific increases as the polymerization temperature is lowered and as the steric hindrance of the alkyl groups of the cocatalyst increases. Higher syndiospecificity is attainable when operating in n-heptane instead of toluene. The viscosimetric molecular weight of polymers obtained in the presence of anisole increases almost linearly with polymerization time and is in the range of 10 000 units (intrinsic viscosity in tetralin at 35 °C = 0.5 to 1). The polymerization rate is low; the resulting polymer amounts to only a few percent of the converted monomer [417].

Experiments with various deuterated propenes have shown that syndiospecific polymerization with vanadium catalysts takes place via cis addition of the monomeric units to the growing chain [423]. The polymerization is not thoroughly regiospecific; that is, 1–2 insertions are possible [424,425]. Next to syndiotactic stereoblocks, the polymer also contains irregular stereoblocks [426–428].

In 1988, Ewen and Razavi developed a catalyst for the syndiotactic polymerization of propene based on C_s-symmetric metallocenes (Table 23) [429–431].

From these prochiral metallocenes, chiral metallocenium ions can be produced in which chirality is centred at the transition metal itself. Due to the flipping of the polymer chain, the metallocene alternates between the two enantiomeric configurations and produces a syndiotactic polymer [432–437].

Syndiotactic polypropene produced by metallocene catalysts shows a higher level of irregularities than isotactic ones. Comparing samples of the same degree of tacticity, the syndiotactic polymer exhibits a lower melting point, lower density (strongly) depending on the tacticity, ranging from 0.87 to 0.89 g/cm^3, lower crystallinity, and a lower crystallization rate [438]. The small crystal size in syndiotactic polypropene causes a higher clarity of the material but is also responsible for its inferior gas barrier properties

Table 22 Syndiotactic polypropene produced with vanadium catalysts[a].

Vanadium compound	Cocatalyst	Al:V:An (Mot)	Solvent	Temp. (°C)	I.S.
VCl$_4$	Al(C$_2$H$_5$)Cl	5:1:1	Toluene	−78	1
VCl$_4$	Al(C$_2$H$_5$)Cl	10:1:1	Toluene	−78	0.9
VCl$_4$	Al(i-C$_4$H$_9$)$_2$Cl	5:1:1	Toluene	−78	1.3
VCl$_4$	Al(i-C$_4$H$_9$)$_2$Cl	5:1:1	n-Heptane	−78	2.4
VCl$_4$	Al(C$_2$H$_5$)Cl	5:1:1	n-Heptane	−78	1.9
VCl$_4$	Al(C$_2$H$_5$)Cl	2:1:1	n-Heptane	−78	1.3

Source: Ref. 422.
[a]Conditions: time 18–20 h, vanadium concentration 1–2.6 mmol in 100 mL of solvent, temperature −78 °C, I.S. = crystallinity index due to syndiotactic polymer, determined by IR methods.

Table 23 Syndiotactic polypropenes prepared by different metallocene catalysts (see Figure 2). Polymerizations were carried out at 60 °C in 1 l of liquid propene.

Metallocene	Productivity [kg PP/(gM h)]	M_w [kg/mol]	m.p. [°C]	Syndiotacticity [rrrr %]
[Me$_2$C(Flu)(Cp)]ZrCl$_2$	180	90		0.82
[Me$_2$C(Flu)(Cp)]HfCl$_2$	3	778		0.73
[En(Flu)(Cp)]ZrCl$_2$	50	171	111	0.71
[Ph$_2$C(Flu)(Cp)]ZrCl$_2$	3138	478	133	0.87
[Ph$_2$C(Flu)(Cp)]HfCl$_2$	28	1950	102	0.74

preventing applications in food packaging. However, the resistance against radiation allows medical applications. Other advantages of sPP are the higher viscous and elastic moduli at higher shear rates and its outstanding impact strength which disappears at low temperatures due to the independence of the glass transition temperature on the tacticity.

Commercial product of syndiotactic polypropene utilizes a silica supported metallocene in a bulk suspension process at 50–70 °C and a pressure of 30 kg/cm^2 [439].

Variation of C_s-symmetric metallocenes leads to C_1-symmetric ones (structure (60)). If a methyl group is introduced at position 3 of the cyclopentadienyl ring, stereospecificity is disturbed at one of the reaction sites so every second insertion is random. A hemiisotactic polymer is produced [440–442]. If steric hindrance is bigger (for example a *tert*-butyl group is introduced instead of the methyl group), stereo selectivity is inverted and the metallocene catalyses the production of isotactic polymers [432,443–450].

(60)

1. Elastomeric Polypropene

Elastomeric polypropenes (Table 24) of two different types may be prepared using metallocene catalysts: (1) polypropenes being elastomeric due to a high content of 1,3-enchainments; and (2) polypropenes having a stereoblock structure prepared using oscillating or C_1-symmetric metallocenes.

Oscillating metallocenes are obtained if unbridged substitutet metallocenes have a significant rotational isomerization barrier (Figure 10). Early attempts concentrated on substituted cyclopentadienyl and indenyl compounds [411,451–453].

Most recent efforts by Coates and Waymouth have shown that 2-phenylindenyl-groups are well suited for this purpose [454]. They oscillated between the enantiomeric and meso arrangements giving rise to a stereoblock polypropene containing atactic (produced

Table 24 Properties of enastomeric polypropenes prepared by [MeHC(Ind)(C$_5$Me$_4$)]TiCl$_2$ (1), [Me$_2$C(Ind)(Cp)]HfCl$_2$ (2), [Me$_2$C(Ind)(Cp)]ZrCl$_2$ (3), and (2PhInd)$_2$ZrCl$_2$ (4).

Catalysts	1	2	3	3	4
M_w (kg/mol)	127	30	50	380	889
Isotacticity (mmmm %)	40	38	54	52	28
Melting point (°C)		47/61	54/93	53/84	125–145
Crystallinity %	6.7	7.2	19.1	16.7	0.2
Elastic recovery					
after 100% strain	93		92	95	
after 200% strain	91		90	93	
after break	86		86	84	

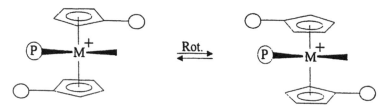

Figure 10 Oscillating metallocene: by rotation of the cyclopentadienyl rings the metallocenes epimerizes.

by the meso rotamer) and isotactic (produced by the chiral rotamer) sequences. The block length was strongly dependent on the temperature.

    C_1-symmetric metallocenes are able to produce also elastomeric polypropene if consecutive insertions take place on the same active site in addition to chain migratory insertion. Polypropenes containing blocks of atactic and isotactic sequences are produced, the block lengths depending on the rate of chain stationary insertion or site isomerization vs. chain migratory insertion [447,455–458]. Rieger [455] used bridged indenyl-fluorenyl metallocenes (structure (61)) to catalyze polypropenes with tailored isotacticities between 20 and 80%.

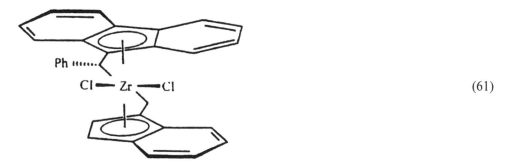

(61)

D. Atactic Polypropene

Atactic polypropene (aPP) is a head-to-tail polymer that is incapable of crystallizing, due to the statistically irregular sequence of sterical orientations of methyl groups connected to the tertiary carbon atoms [459]. It is amorphous to x-rays. The IR spectrum features strong absorption bands at 1155 and 970 wavenumbers, respectively [460].

The glass transition temperature below which atactic polypropene becomes solid is in the vicinity of $-12\,^{\circ}C$ [461]. Atactic polypropene is employed in polymer blends for adhesives, and carpet coatings [462]. It can be produced with the aid of Ziegler–Natta catalysts. Without the use of stereoselectivity regulators, the synthesis yields 20 to 70% of atactic polypropene, which can be isolated through extraction with n-hexane or n-heptane [463].

In contrast to atactic polypropene produced with Ziegler–Natta catalysts, the polymer synthesized by a cationic mechanism does not possess an exclusive head-to-tail sequence but is of irregular structure with head-to-head and tail-to-tail enchainments accounting for up to 10% of the polymer [464]. However, there are still syndiotactic or isotactic portions in atactic polymers produced with heterogeneous systems (using vanadium catalysts). Very pure atactic polypropene is obtained with metallocene (see Table 18) [465–467]. The atactic polypropene extracted from Ziegler–Natta catalysis shows a broad molecular weight distribution and low molecular weight and is used as additive to oil and bitumen.

Using metallocene catalysts, aPPs convering the whole range of molecular weights of technical interest with narrow molecular weight distributions of $M_w/M_n = 2$ can be produced [468,469]. The main characteristics of high molecular weight aPP produced by $[Me_2Si(Flu)_2]ZrCl_2$ are low density, high transparency, softness, low modulus, and high elongation caused by the totally amorphous state of the polymer. High molecular weight aPP has potential applications in blends with other polyolefin, upgrading transparency, softness, elastic recovery and elongation.

1. Supported Metallocene Catalysts

Metallocene catalysts which are to be used as drop-in catalysts in existing plants for polyolefin production have to be heterogenized due to the fact that current technology is based on gas phase and slurry processes. Thus the metallocenes are to be fixed on a carrier. Carriers may be divided into three groups: (1) metals have been used as fillers; (2) inorganics like silica, aluminia, zeoliths or $MgCl_2$ [470,476]; and (3) organic materials like cyclodextrins [477], starch (as a filler) [478] and polymers (polystyrenes, polyamides) have been used to support either the metallocene or the cocatalyst.

Looking at the preparation of supported metallocenes, synthesis of the metallocene on the carrier is found as well as fixing a metallocene either via functionality at the ligand or by direct reaction with the carrier, in both cases followed by activation with MAO or trialkylaluminum, but more common is heterogenization of the cocatalyst prior to mixing the modified carrier with the metallocene and activation by trialkylaluminum.

Soga et al. have reported the synthesis of ansa zirconocenes on silica by synthesizing a precursor of the bridge anchored using SiO_2–OH groups on the surface of silica [479].

Activation of these catalysts is done using either MAO or triisobutylaluminum. In all cases the polymers obtained are isotactic despite the fact that synthesis of fixed bis(indenyl) metallocenes may result in inseparable meso and racemic diastereomers and bridged bis(fluorenyl) metallocenes are not chiral.

A similar procedure starts with a metallocene carrying an additional functionality at the ligand which can be used for bonding on the carrier. These metallocenes may be fixed on silica (probably after modification of the SiO_2–OH groups) or other inorganic carriers as well as on polymeric materials.

Table 25 Comparison of metallocenes in homogeneous phase and supported on silica fixed MAO at 40 °C. (I) [En(IndH$_4$)$_2$], (II) [Me$_2$C(Flu)Cp)]ZrCl$_2$, (III) Cp$_2$ZrCl$_2$.

Catalyst	Cocatalysts	Activity (kg/molZr h)	Tm (°C)	M_n (kg/mol)	M_w/M_n	Tacticity
(I)/homogeneous	MAO 3 mmol	2070	111	3.3	1.9	71%mmmm
(I)IMAO/SiO$_2$	TEA 1 mmol	77	140	5.3	2.5	90%mmmm
(I)/MAO/SiO$_2$	TIBA 2 mmol	382	105	6.6	1.8	69%mmmm
(II)/homogeneous	MAO 13 mmol	758	123	39.3	1.8	77% rrrr
(II)/MAO/SiO$_2$	TIBA 2 mmol	141	133	45.2	1.9	83% rrrr
(III)/homogeneous	MAO 10 mmol	132	–	0.3	–	–
(III)/MAO/SiO$_2$	TIBA	99	–	1.8	–	–

Metallocenes may also be fixed on silica or alumina by direct reaction of the two components. Marks has shown that reaction of dimethyl metallocenes with alumina results in the formation of an active catalyst [480]. Others have investigated the direct reaction of metallocene dichlorides with silica (and alumina) and faced the problem of metallocene decomposition. Nevertheless an active species is formed which produces isotactic polypropene but with rather low activities.

Two approaches have been followed to generate a supported methylaluminoxane: (1) the reaction of a carrier containing hydroxyl groups (starch, silica) with trimethylaluminum; and (2) fixing MAO itself by reaction with OH groups of the carrier. In both cases heterogenization of the cocatalyst is followed by reaction of the heterogeneous MAO with a metallocene dichloride to generate a metallocene bond to the supported MAO. These catalysts are usually activated by trialkylaluminums or additional small amounts of MAO. Metallocenes fixed on supported MAO exhibit similar behavior in polymerization as do their homogeneous analogues (Table 25). Transfer of knowledge about catalysts design gathered in homogeneous processes is possible. Therefore these techniques are most widely used to fix metallocenes onto a carrier.

E. Copolymers

The most important copolymer of propene with ethene is described under ethene polymerization. There are copolymers of propene with longer-chained α-olefins, too. Copolymers with 1-butene [484], 1-hexene [485], 4-methyl-1-pentene [486], styrene [487], and vinylcyclohexane [488] are amorphous over a wide range of composition and show elastic properties. Table 26 gives the copolymerization parameters $r_1 = k_{11}/k_{12}$ and k_{22}/k_{21}. Conventional Ziegler–Natta titanium-based catalysts produce block copolymers of butadiene and propene while vanadium-based catalysts give highly alternating copolymers [489,490]. Metallocene catalysts are highly active for the copolymerization of propene with other 1-olefins and cycloalkenes [491–498].

The copolymerization of butadiene and propene is possible by zirconocene/MAO catalysts.

Table 26 Copolymerization of propene (1) with various α-olefins (2).

Comonomer (2)	Catalyst	r_1	r_2	Ref.
1-Butene	TiCl$_3$/AlEt$_3$	4.6	0.5	484
1-Hexene	TiCl$_3$/AlEt$_2$Cl	4.18	0.16	485
1-Methyl-1-pentene	TiCl$_3$/AlEt$_2$Cl	6.4	0.31	486
Styrene	TiCl$_3$/AlEt$_3$	20.5	0.3	487
Vinyl cyclohexane	TiCl$_3$/AlEt$_2$Cl	80	0.049	488

With the system ethylenebis(indenyl)zirconium dichloride/methylaluminoxane, 6% of 1,5-hexadiene could be incorporated [499,500]. The copolymerization parameter r_1 for this reaction is in the range of 14 to 19. The polymerization rate decreases with increasing 1,5-hexadiene concentration in the reaction mixture. The block copolymers are mostly mixtures of the copolymer and both homopolymers [501].

1. Polymer Characterization and Compounding

Isotactic and syndiotactic polypropene are partly crystalline polyolefins. Therefore, their properties largely depend on their degree of crystallinity and crystal structure. These parameters can be influenced by crystallization temperature, cooling rate, tensile stress, tempering, molecular weight, and molecular weight distribution. Isotactic polypropene forms a 3.1 helix with identical configuration at every tertiary carbon atom, whereas in syndiotactic polypropene the configuration of the tertiary carbon atoms is alternating. The helix conformations are states of particularly low energy and are therefore favored by the system. In the helix the pendant methyl groups have a well-defined tilt pointing up and down, but they are always oriented toward the outside (away from the center) [502]. The helical phase is a racemic mixture consisting of pairs of dextrogyric and levogyric helices. This is responsible for the high melting point of 162–175 °C.

In contrast to this, atactic polypropene is amorphous even at room temperature.

The degree of crystallinity of iso- or syndiotactic polypropene is normally on the order of 60 to 70%. There are several modifications of isotactic polypropene. The most important is the monoclinic a type. Its unit cell, with dimensions of 0.67, 2.10, and 0.65 nm, contains three monomeric units of four polymer chains, respectively [503]. Next to this, the trigonal a and triclinic g-modifications occur. They are best characterized by spectroscopical methods IR [502,503], x-ray diffraction [504], ^{1}H-NMR, and ^{13}C-NMR [505,506].

Stabilizers and additives are blended into the crude product to prevent the polymer chains from degrading. The tertiary carbon atoms bearing the pendant methyl groups are fairly sensitive to oxidation. That is why the processing must proceed under exclusion of air with antioxidants present in the polymer.

Substituted phenols that act as scavengers for radicals formed in the course of the degradation reaction as well as thio compounds and organophosphites are added as antioxidants. Benzophenone and carbon black act as UV stabilizers. Calcium stearate or other carboxylic acids are added to bind chloride originating from the catalyst. Further additives can be acceptors for coloring, anti-electrostatics, nucleating agents to promote crystallization, lubricants, or flame retardants [507,508]. Polypropene can also be blended with other polymers as rubbers (EPDM) or polyethene [509–511]. In these products the mixed phase is embedded in a continuous polypropene matrix. No genuine solution is obtained. Additional new properties can be achieved by compounding the polypropene with solid inorganic or organic filling materials [512]. Rigidity and hardness, especially, can be improved in this way. Possible filling materials are glass fibers, glass balls, talcum, chalk, aluminum oxide, metal powder, wood flour, or carbon fibers [513]. Up to 50 wt% of these filling materials can be directly worked in as a master batch.

IV. POLYMERS OF HIGHER α-OLEFINS

Higher α-olefins are less easily polymerized, as they pose more steric hindrance. Propene is about three times more active than 1-butene, depending on the catalyst system. The polymerization rate is progressively decreased with increasing size of the linear alkyl group. On the other hand, 2-olefins cannot be homopolymerized. On the other hand, 2 olefins cannot be homopolymerized by Ziegler catalysts. Nickel-diimine/MAO or borate systems are able to homopolymerize trans-2-butene to polymers with low glass-transition temperatures until 53 °C [514]. The industrial interest in polymers of higher α-olefins is much lower than that for polyethene and polypropene. Only poly-l-butene and poly (4-methyl-l-pentene) have some commercial use. The scientific importance of higher 1-olefns is explained by their ability to form chains with variable isotacticities. Because of the longer alkyl group, higher α-olefins show only 1,2-insertion. A sufficiently large number of different linear, branched, cyclic, and aromatic olefins have been investigated.

A. Poly(1-Butene)

Isotactic poly(1-butene) crystallizes from the melt into an unstable modification which is slowly converted into a thermodynamically more stable form. The two modifications differ in helical structure and density; therefore, the product undergoes deformation in the course of time [515,516]. For the polymerization of butene the same catalysts are used as for propene.

The reactivity is lower. In contrast to isotactic polypropene isotactic poly(1-butene) is soluble in boiling heptane despite the ethyl side group. Poly(1-butene) is produced in suspension, solution, or in liquid-1-butene. Because of the lower activity of the catalyst, the polymer usually has to be purified by a number of washing steps from catalyst residues.

Mainly $TiCl_3$/$AlEt_2Cl$ or $MgCl_2$/$TiCl_4$/EB/$AlEt_3$ catalysts are used. $TiCl_4$ (0.2 wt%) on MgO and $AlEt_3$ as cocatalysts yields 60 kg of poly(1-butene) per hour and gram of Ti [517]. With $TiCl_4$ and $(C_6H_{13})_2$ Mg atactic polybutene is formed [518].

Also soluble catalyst based on ethene bis(indenyl)zirconium dichloride/methylaluminoxane can be used [519]. The ^{13}C-NMR spectroscopically measured isotacticity is in excess of 97%, the molecular weight low (44 000), and the crystallinity 66.9%. Similar to polypropene poly(1-butene) crystallizes in four different modifications [520]. To influence the crystallinity, 1-butene was copolymerized with ethene or propene [521,522] or compounded [523,524]. Poly(1-butene) shows very good stability against stress, corrosion, cracking and is therefore used for pressure tubes.

B. Polymers of Other α-olefins

The other α-olefins were investigated considering the polymerization rate, helix conformation, and melting point of the polymer (Table 27) [525]. It was found that

Table 27 α-Olefin relative reactivity (propene = 1) in polymerization with heterogeneous catalysts and melting points of the isotactic polymers.

Monomer	Reactivity	Melting point (°C)
Propene	1	170
1-Butene	0.22–0.62	140
1-Pentene	0.2–0.45	80
1-Hexene	0.16–0.36	−20
1-Heptene	0.2–0.45	18
1-Octene	0.1–0.3	–
1-Nonene	0.1–0.3	19
1-Decene	0.12–0.28	34
1-Dodecene	0.1–0.2	49
1-Tetradecene	0.1–0.2	57
1-Hexadecene	0.1–0.2	68
1-Octadecene	0.1–0.15	71

poly(1-pentene) gives two monoclinic modifications (4.1 and 3.1 helix); poly(1-hexene) forms a 7.2 helix. The polymerization activity and the melting points of the isotactic polymers decrease with growing alkyl groups [526]. The melting points of the isotactic polymers pass through a minimum at −20 °C for poly(1-hexene).

Campbell observed a decrease in activity for α-olefins with branched alkyl groups [527]. A significant decrease in olefine activity takes place when the branch moves closer to the double bond. 3-Methylpentene is three times less active than 4-methylpentene. When the branch is moved away from the double bond to carbon 5 or 6, however, the olefin activities increase. The olefin activity is drastically decreased when both carbons in position 3 or 4 are substituted (the activity of 3,4-dimethyl-1-pentene is five times less than that of 3-methyl-1-pentene). Lower activities are also found with growing size of the substituent at the branch. The poly(4-methyl-1-pentene) has drawn most interest in this group [528,529]. It has a higher melting point of 235 °C and higher transparency than polypropene and forms a 7.2 helix. $MgCl_2/TiCl_4/AlEt_3$ can be used as catalyst and phthalic acid diisobutyl ester as Lewis acid. The isotacticity index of the polymer is 98.2%.

When the 3-alkyl is a ring structure as in vinylcyclohexane, its contribution to the steric repulsion becomes lower and the polymerization activity increases [530–535].

C. Polycycloolefins

Many but not all cycloolefins from cyclobutene to cyclododecene have been polymerized successfully. A wide range of Ziegler catalysts were explored [536,537]. The main interest is directed to the polymerization of cyclopentene. There are two pathways for the polymerization of cycloalkenes [538–543]. One is the double-bond opening:

(62)

and the other is the ring opening.

$$n \; \bigpentagon \longrightarrow \; +CH = CH-CH_2-CH_2-CH_2 +_{\overline{n}} \tag{63}$$

As of the steric hindrance and the possibility to form allylic intermediates, cyclopentene inserts by metallocene catalysts with a 1,3-structure while polynorbornene shows a 1,2-structure (structure (64)) [544].

$$(64)$$

When polymerization occurs by double-bond opening, the product can be syndiotactic, isotactic, or atactic.

The ring opening polymerization can give two structures, the trans-poly(cyclopentenamer) or the cis-poly(cyclopentenamer). Both pathways were found to be followed with Ziegler catalysts. Table 28 gives results of cycloolefin polymerizations. Cyclopropene polymerized spontaneously above $-80\,°C$ [545]. Cyclobutene is very active and polymerized easily with various catalysts [546]. Cyclopentene has been polymerized by metathesis reaction. With heterogeneous catalysts both ring opening and double-bond opening are found at the same time [547]. Cyclohexene has not been polymerized because of the high ring stability of the normal twisted-chair conformation that it takes [537].

Cyclic alkenes such as cyclobutene, cyclopentene, and norbornene can be polymerized by double-bond opening with metallocene/methylalumoxane or late transition catalysts [548–551]. The activities for the polymerization of cyclobutene and cyclopentene are high, whereas the activity of norbornene is significantly lower. The melting points are

Table 28 Polymerization of cycloolefins.

Monomer and catalyst system	Double-bond opening	Ring opening		Ref.
		trans	*cis*	
Cyclobutene				
$V_{(acac)3}/AlEt_2Cl$	100	0	0	542
$TiCl_4/AlEt_3$	5	65	30	541
$WCl_6/AlEt_3$	30	40	30	542
Cyclopentene				
$MoCl_5/AlEt_3$	0	0	100	543
$WCl_6/AlEt_3$	0	100	0	543
$VCl_4/AlMe_2Cl$	50–80	30–50		540
$[En(Ind)_2]ZrCl_2/MAO$	100	0	0	539
Cycloheptene				
$MoCl_5/AlEt_5$	0	93	7	543
Cyclooctene (cis)				
$WCl_6/AlEt_3$	0	85	15	543
Cyclodexene (cis)	0	85	15	538

Table 29 Polymerization of norbornene with different catalysts at 20 °C.

Catalyst	[M] (mol/l)	Polymerization time (°C)	Al:M molar ratio	Activity (kg Pol/mol M)	Molecular weight (g/mol)
Ni(acac)$_2$	5×10^{-6}	0.5	9 400	11 000	2.4×10^{-6}
Pd(acac)$_2$	2×10^{-6}	0.5	25 000	27 000	Insoluble
Ni(cod)$_2$	5×10^{-6}	0.5	9 400	3 800	1.7×10^{-6}
(Pd-Allyl-Cl)$_2$	5×10^{-6}	0.3	10 000	44 000	Insoluble
Cp$_2$ZrCl$_2$	7×10^{-4}	72	5	1.6	Insoluble
[En(Ind)]$_2$ZrCl$_2$	10^{-5}	48	5 000	40	Insoluble
CpTiCl$_3$	6×10^{-4}	2	603	400	Insoluble
Co(acac)$_2$	2×10^{-5}	1	2 600	220	Insoluble

surprisingly high. Under vacuum (to have avoid oxidation) they were found to be 485 °C for polycyclobutene, 395 °C for polycyclopentene.

Table 29 compares the activities for the homopolymerization with different transition metal [M] catalysts using MAO as cocatalyst [551].

Nickel-O- and palladium-O-complexes are very active catalysts for the polymerization of norbornene and also for cyclopentene [552–554]. Nickel catalysts produce soluble polymers with a molecular weight of over one million while polymers obtained with palladium or metallocene complexes are insoluble. The soluble polymers have an atactic structure. The microstructure of the polynorbornene depends on the catalyst used and is isotactic by synthesis with chiral metallocenes.

The processing of homopolynorbornene would be extremely difficult, due to a melting point higher than the decomposition, especially if the polymer is insoluble. Dimethanooctahydronaphthalene [DMON] is more rigid than norbornene. The copolymerization product of norbornene or DMON with ethene is amorphous, featuring high T_g values of 160 °C. The copolymers are insoluble in hydrocarbons and have an excellent transparency, thermal stability, and chemical resistance [555].

D. Polyisobutene

Isoolefins with branchings adjacent to the double bond cannot be homopolymerized by active centers of the Ziegler–Natta type. Corresponding olefins with vinylidene double bonds, however, are classical monomers for cationic polymerization [556]. In this context Ziegler catalysts can act as cationic catalysts [557]. In the group of vinylidene olefins, isobutene is the one of greatest importance. It is obtained as the major component (45 to 50 vol%) of the C$_4$ fraction in the naphtha cracking process [558]. The first polymerization was carried out by Lebedew [559].

The main distinguishing feature of technical poly(isobutenes) is their molecular weight. Molecular weights range from about 3000 for viscous oily liquids to 40 000 to 120 000 for sticky products and 300 000 to 2 500 000 for elastic rubbery materials [560,561]. Poly(isobutenes) have low glass transition temperatures and thermal conductivity as well as high electrical resistivity and chemical resistance. They are soluble in hydrocarbons but insoluble in alcohols [562,563]. The cationic polymerization of isobutylene can be carried out as a precipitation reaction at temperatures below 0 °C with Friedel–Crafts catalysts (e.g., AlCl$_3$, BF$_3$) in chloromethane or other solvents [564–573]. The Lewis acid as initiator is activated incombination with a proton sosurce which is usually present as a trace impurity (e.g. water, hydrogen chloride). The mechanism includes the following reactions [558,574]:

$$HCl \; + \; AlCl_3 \; \rightleftharpoons \; H^{\oplus} \, [AlCl_4]^{\ominus} \tag{65}$$

Activation

$$H^{\oplus} \, [AlCl_4]^{\ominus} \; + \; H_2C{=}\underset{\underset{CH_3}{|}}{\overset{\overset{CH_3}{|}}{C}} \; \longrightarrow \; H_3C{-}\underset{\underset{CH_3}{|}}{\overset{\overset{CH_3}{|}}{C}}{}^{\oplus} \, [AlCl_4]^{\ominus} \tag{66}$$

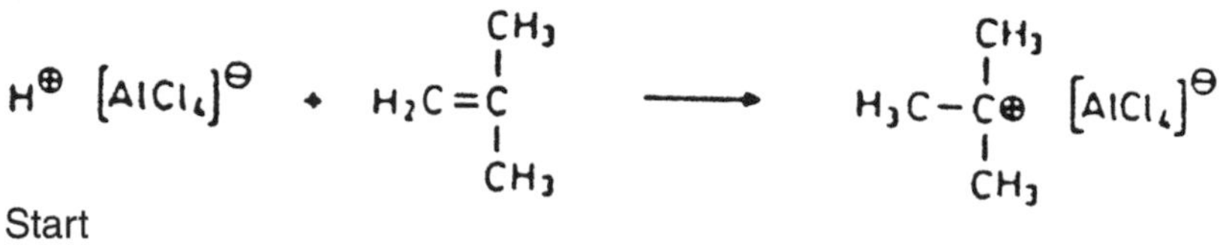

Start

$$H_3C{-}\underset{\underset{CH_3}{|}}{\overset{\overset{CH_3}{|}}{C}}{}^{\oplus} \, [AlCl_4]^{\ominus} \; + \; n \; H_2C{=}\underset{\underset{CH_3}{|}}{\overset{\overset{CH_3}{|}}{C}} \; \longrightarrow \; H{-}\!\left[CH_2{-}\underset{\underset{CH_3}{|}}{\overset{\overset{CH_3}{|}}{C}}\right]_n\!{-}CH_2{-}\underset{\underset{CH_3}{|}}{\overset{\overset{CH_3}{|}}{C}}{}^{\oplus} \, [AlCl_4]^{\ominus}$$

Propagation

$$\tag{67}$$

$$H{-}\!\left[CH_2{-}\underset{\underset{CH_3}{|}}{\overset{\overset{CH_3}{|}}{C}}\right]_n\!{-}CH_2{-}\underset{\underset{CH_3}{|}}{\overset{\overset{CH_3}{|}}{C}}{}^{\oplus} \, [AlCl_4]^{\ominus} \; \longrightarrow \; H{-}\!\left[CH_2{-}\underset{\underset{CH_3}{|}}{\overset{\overset{CH_3}{|}}{C}}\right]_n\!{-}CH_2{-}\underset{\underset{CH_3}{|}}{\overset{\overset{CH_2}{||}}{C}} \; + \; H^{\oplus} \, [AlCl_4]^{\ominus}$$

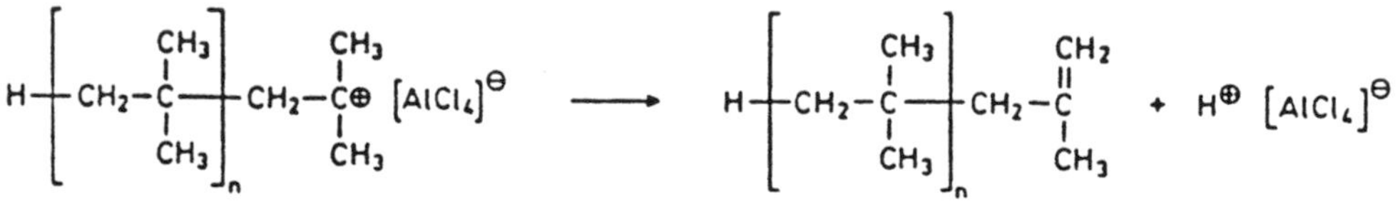

Deactivation

$$\tag{68}$$

To obtain high molecular weights, the polymerization temperature and with it the rate of the transfer reactions must be lowered. At $-100\,°C$ poly(isobutylene) of molecular weight $300\,000\,g/mol$ is obtained [575].

The reaction is strongly exothermic with a reaction enthalpy of $356\,kJ/mol$ and is usually finished within seconds or a few minutes, even at low temperatures. The molecular mass is kept low by adding α-olefins and diisobutene. Addition of 0.25% of diisobutene, for instance, reduces the molecular weight from $260\,000$ to $45\,000$. In contrast, the addition

Table 30 Copolymerization of isobutene (M_1) with different comonomers (M_2); copolymerization parameters r_1 and r_2.

Comonomer	r_1	r_2	Refs.
Styrene	1.8–2.8	1.4–5.5	577
4-Chlorostyrene	1.0–1.7	1.0–14.7	578
α-Methylstyrene	0.20–1.2	1.4–5.5	579,580
Indene	1.1	2.2	580
1,3-Butadiene	43	0	581
trans-1,3-Pentadiene	2.3	–	582
cis-1,3-Pentadiene	5.0	–	582
Cyclopentadiene	0.2–0.8	1.5–6.3	583

of organometallic amides [e.g., zinc bis(di-ethylamide), titanium tetra(diethylamide)] has the opposite effect, raising the molecular weight to more than 1 million [576].

1. Copolymers of Isobutene

Isobutene can be copolymerized with numerous unsaturated compounds via a cationic route [577–586]. Table 30 lists various comonomers and copolymerization parameters. The isobutene portion in the copolymers usually exceeds 90%. The use of aluminum organic compounds (e.g., $AlEt_2Cl$) as opposed to aluminum trichloride permits better control of the copolymerization, as they are weaker Lewis acids. Hydrogen chloride or halogens must be added as cocatalysts that are capable of regenerating the carbocations. The organoaluminum catalysts are produced at $-78\,°C$ with boron trifluoride.

Isobutene/isoprene copolymers (butyl rubbers) are the technically most important copolymers of isobutene. The polymerization is carried out as a continuous suspension polymerization in an agitator vessel with 0.8 to 3 mol% isoprene in chloromethane [558]. Further solvents, such as hydrogen sulfide or heptane, can be added to prevent partial cross-linking of the isoprene at higher isoprene contents [587]. The isoprene is incorporated with a trans-1,4 linkage [588]. Further catalysts can be zinc bisdiisobutyl-amide/boron trifluoride or titanium tetrachloride/cumylic chloride. When the monomer mixture is added to the latter, the molecular weight of the copolymer increases with reaction time (i.e., the reaction proceeds quasi-living).

Isobutene can also be copolymerized with 1,3-butadiene, cyclopentadiene, indene, or α-pinene [589–593].

In copolymerization with styrene, isobutene is the monomer with the highest reactivity [594–596]. In analogy with this it is possible to produce copolymers with a-methylstyrene or 4-chlorostyrene.

2. Block Copolymers and Graft Polymers

Block copolymers of isobutene and styrene can be produced by homopolymerizing styrene in dichloromethane with titanium(IV)chloride and 2-chloro-2-phenylpropane as initiator system to a desired chain length followed by addition of isobutene [597]. At a reaction temperature of $-50\,°C$ the molecular weight of the block copolymer is $M_n = 45\,000$ and that of the styrene block is $M_n = 29\,000$. Homopolystyrene and homopoly(isobutene) are removed by extraction with pentane and butanone [598].

With 2-bromo-6-chloro-2,6-dimethylheptane as a starting molecule it is possible to specifically produce diblocks that are not contaminated by homopolymerizates [599]. The two terminal halogen groups differ in reactivity. Addition of aluminum alkyls successively generates carbocations that can be used for a stepwise initiation. The more reactive chlorine group is converted first, starting the growth of a polystyrene block. After this the bromine group is activated, which then leads to the growth of the poly(isobutene) block. Products of this type feature two glass transition temperatures [$-74\,°C$ for the poly(isobutene) block and $-96\,°C$ for the polystyrene block].

The development of a synthesis for telechelic poly(isobutenes) with boron trichloride catalysts (Inifer technique) has led to an extension of the spectrum of isobutene block copolymers [600,601].

In addition to isobutene-styrene, di- and triblock copolymers can be synthesized by isobutene/a-methylstyrene. Further, block copolymers of isobutene can be made by polyaddition reactions with hydroxy- and dihydroxytelechelics of poly(isobutene).

Poly(isobutene)/polycarbonate blocks are produced by conversion of poly (bisphenol-A carbonate) that contains isocyanate groups and dihydroxytelechelics of poly(isobutene).

It is possible to graft isobutene onto activated polymer molecules when they can be alkylated in an electrophilic reaction [602–604]. Polystyrene, for instance, is alkylated with Friedel–Crafts catalysts. The grafting is simplified by the presence of halogen atoms in the main polymer chain. Addition of diethylaluminumchloride or triethylaluminum leads to the formation of cations. The graft polymers usually contain between 10 and 60% of isobutene units and feature outstanding elasticity and dimensional stability at higher temperatures. Also, copolymers of isobutene and isoprene, butadiene, styrene, and so on, can be grafted in this manner [605–607].

A controlled synthesis of functionalized poly(isobutenes) was not possible until Kennedy and Smith [596,608] introduced the Inifer concept to cationic polymerization. Inifer systems are mono-, bi-, or trifunctional initiators and transfer reagents (structures (69)–(71)) that lead to cationic polymerization reactions. Terminal halogen groups are introduced by controlled chain transfer reactions while terminating reactions are supressed. In this way one can synthesize linear, triaxial, or star-shaped poly(isobutenes) with reactive terminal halogen groups [609–612]. These groups can subsequently be converted to hydroxyl, phenyl, carboxy, silyl, and other functional groups.

Minifer	**Binifer**	**Trinifer**
(69)	(70)	(71)

3. Polymers of 1,1- and 1,2-Disubstituted Olefins

Next to isobutene, other 1,1-dialkyl substituted ethenes can also be polymerized cationically. Suitable monomers are 2-methyl-1-butene, 2-methyl-1-pentene, and 2,3-dimethyl-1-butene [613] Polymers with very high molecular weights of $M_n > 300\,000$ are obtained by catalysis with aluminum alkyl halogenides. Also, cyclic hydrocarbons with a methylene group (methylenecyclopropane, methylene cyclobutane, methylene cyclohexane, α-pinene) are suitable monomers [614–619]. 1,1-Disubstituted ethenes with stronger steric hindrance as camphene or 2-methylene-bicyclo-[2.2.1] heptane, however, could not be polymerized cationically [574].

With Friedel–Crafts catalysts or mineral acids, 2-butene can only be oligomerized while nickel II-diimin complexes give high molecular weight polymers [514]. Copolymers of 2-butene with ethane by metallocene catalysts are possible [620]. Also, cyclopentene, cyclohexene,hexene, and norbornene can only be converted to products of low molecular weights with boron trifluoride/hydrogen fluoride or AlCl$_3$ [621–623]. In these cases the steric shielding effect of long aliphatic chains or rings prevents an effective cationic attack.

REFERENCES

1. Ko, H. S. (2004). *JUPAC Congress – Macro 2004*, Paris, 4–9 July.
2. Meerwein, H., and Bierneleit, W. (1928). *Ber.*, *616*: 1845.
3. Fawcett, E. W., Gibson, R. U., Perrin, M. W., Paton, J. G., and Williams, E. G. (1936). GB Pat. 471590 to ICI., C.A. (1938), *32*: 1362.
4. Hogan, J. P., and Banks, R. L. (1958). US Pat. 2825721 assigned to Phillips Petroleum Co., CA. (1958), *52*: 8621I.
5. Ziegler, K., Holzkamp, E., Martin, H., and Breil, H. (1955). *Angew. Chem.*, *67*: 541.
6. Zletz, A. (1954). US Pat. 2692.257, assigned to Standard Oil of Indiana (Amoco), C.A. (1955) *49*: 2777 D.
7. Fischer, M. (1953). Germ. Pat. 874215 to BASF A.-G., C.A. (1957) *51*: 10124 G.
8. Natta, G. (1955). *J. Am. Chem. Soc.*, *77*: 1708.
9. Keii, T. (1972). *Kinetics of Ziegler-Natta Polymerization*, Kodanska Ltd. Tokyo, and Chapman & Hall Ltd. London.
10. Boor, J. (1979). *Ziegler-Natta Catalysts and Polymerizations*, Academic Press, New York.
11. Kissin, Y. V. (1985). *Isospecific Polymerization of Olefins*, Springer-Verlag, Berlin.
12. Sinn, H., and Kaminsky, W. (1980). *Advances in Organometallic Chemistry*, Vol. 18., Academic Press, New York, p. 99.
13. Quirk, R. P., ed. (1988). *Transition Metal Catalyzed Polymerizations Ziegler-Natta and Metathesis Polymerizations*, Cambridge University Press, New York.
14. Seymour, R. B., and Cheng, T., eds. (1985). *History of Polyolefins*, D. Reidel Publ. Co., Dordrecht.
15. Seymour, R. B., and Cheng, T., eds. (1987). *Advances in Polyolefins—The World's Most Widely Used Polymers*, Plenum Press, New York.
16. Fink, G., Mülhaupt, R., and Brintzinger, H. H., eds. (1995). *Ziegler Catalysts*, Springer-Verlag, Berlin.
17. Kaminsky, W., and Sinn, H., eds. (*1988*). *Transition Metals and Organometallics as Catalysts for Olefin Polymerization*, Springer Verlag, Berlin.
18. Blom, R., Follestad, A., Rytter, E., Tilsel, M., and Ystenes, M., eds. (2001). *Organometallic Catalyst and Olefin Polymerization*, Springer Verlag, Berlin.
19. Kaminsky, W., ed. (1999). *Metalorganic Catalysts for Synthesis and Polymerization*, Springer Verlag, Berlin.
20. Sinn, H., Kaminsky, W., Vollmer, H.-J., and Woldt, R. (1980). *Angew. Chem. Int. Ed. Engl.*, *19*: 390.
21. Kaminsky, W. (1996). *Macromol. Chem. Phys.*, *197*: 3907.
22. Brintzinger, H. H., Fischer, D., Mülhaupt, R., Rieger, B., and Waymouth, R. (1995). *Angew. Chem.*, *107*: 1255.
23. Scheirs, J., and Kaminsky, W., eds. (2000), *Metallocene-Based Polyolefins*, Vol. I + II, Wiley, Chichester.
24. Fawcett, E. W., and Gibson, R. O. (1934). *J. Chem. Soc.*, *136*: 386.
25. Evans, M. G., and Polany, M. (1936). *Trans. Faraday Soc.*, *32*: 133.
26. Bailey, A. I., and Daniels, H. (1973). *J. Physical Chemistry*, *77*: 501.
27. Nicholson, A. E., and Norrish Disc, R. G. W. (1956). *Faraday Soc.*, *22*: 97.
28. Constantin, D., Hert, M., and Machon, J. P. (1981). *European Polym. J.*, *17*: 115.
29. Payer, W. (1987). *Houben Weyl Methoden der Organischen Chemie*, Vol. E 20/2, Makromolekulare Stoffe, G. Thieme Verlag, New York, p. 689.
30. Ehrlich, P., and Takahashi, T. (1982). *Macromolecules*, *15*: 714.
31. Roedel, M. J. (1953). *J. Am. Chem. Soc.*, *75*: 6110.
32. Wickham, W. T. (1962). *J. Polym. Sci.*, *60*: 68.
33. Vile, J., Hendra, P. J., Willis, H. A., Cudby, M. E. A., and Bunn, A. (1984). *Polymer*, *25*: 1173.
34. Tatsukami, Y., Takahashi, T., and Yoshioka, H. (1980). *Makromol. Chem.*, *181*: 1107.
35. Luft, F., Lim, P., and Yokawa, M. (1983). *Makromol. Chem.*, *184*: 207.

36. Luft, G., and Seidl, H. (1985). *Angew. Makromol. Chem.*, *129*: 61.
37. Tapavicza, St. von, Buback, M., and Frank, E. U. (1975). *High Temp.-High Press.*, *7*: 535.
38. Tobita, H. (1995). *J. Polym. Sci. Polym. Phys.* *33*: 841–853.
39. Rätzch, M., and Baumann, H. J. (1976). *Plaste Kautsch.*, *23*: 890.
40. Franko-Filipasie, B. R., and Michaelson, R. C. (1984). *Chem. Eng. Progr.*, *80*: 65.
41. Livingston, E. H. (1984). *Chem. Eng. Progr.*, *80*: 70.
42. Ehrlich, P. (1965). *J. Polym. Sci.*, *Part A*, *3*: 131.
43. Steiner, R., and Horle, K. (1972). *Chem.-Ing.-Techn.*, *44*: 1010.
44. Luft, G., and Lindner, A. (1976). *Angew. Makromol. Chem.*, *56*: 99.
45. Lin, D. D., and Prausnitz, J. M. (1980). *Ind. Eng. Chem. Process Res. Dev.*, *19*: 205.
46. Spahl, R., and Luft, G. (1982). *Ber. Bunsenges. Phys. Chem.*, *86*: 621.
47. Luft, G., and Steiner, R. (1971). *Chemiker Z.*, *95*: 11.
48. Wiley, R. H., Lipscomb, N. T., Johnston, F. J., and Guillet, J. E. (1962). *J. Polym. Sci.*, *57*: 867.
49. Machi, S., Hagiwara, M., Gotoda, and M., Kagiya, T. (1965). *J. Polym. Sci.*, *Part A*, *3*: 2931.
50. Munari, S., and Russa, S. (1966). *J. Polym. Sci.*, *4*: 773.
51. Buback, M., and Vögele, H.-P. (1985). *Makromol. Chem. Rapid Commun.*, *6*: 481.
52. Luft, G., Bitsch, and H., Seidl, H. (1977). *J. Macromol. Sci. Chem.*, *11*: 1089.
53. Tidwell, P. W., and Mortimer, G. A. (1970). *J. Polym. Sci, Part A-1*, *8*: 1549.
54. Bunn, C. W. (1939). *Trans Faraday Soc.*, *35*: 482.
55. Ballard, D. G. H., Schelten, J., Cheshire, P., Janke, E., and Nevin, A. (1982). *Polymer*, *23*: 1875.
56. Crist, B., Graessley, W. W., and Wignall, G. D. (1982). *Polymer*, *23*: 1963.
57. Breitmeir, U., and Bailey, A. I. (1979). *Surface Science*, *89*: 191.
58. Gouarderes, R. (1981). Eur. Pat. 30171 ATO-Chimie S.A., C.A. (1981), *95*: 98633.
59. Kolwert, A., and Herwig, J. (1983). Ger. Pat. 3150270 to EC Erdölchemie GmbH, C.A. (1983), *99*: 88743.
60. Machon, J. P. (1988). *Transition Metal Catalyzed Polymerizations, Ziegler-Natta* and *Metathesis Polymerizations* (Quirk, R., ed.), Cambridge Univ. Press, New York, p. 344.
61. Grönig, H., and Luft, G. (1986). *Angew. Makromol. Chem.*, *142*: 161.
62. Luft, G., Rau, A., Dyroff, A., Götz, C., Schmiz, S., Wieczorek, T., Klimech, R., and Gonioukh, A. (1999). In [19], p. 651.
63. Choi, K., and Ray, W. H. (1985). *Rev. Macromol. Chem. Phys.*, *C 25*: 18.
64. Natta, G., Pasquon, I., Zimbelli, A., and Gatti, G. (1961). *J. Polymer Sci.*, *51*: 387.
65. Ray, W. H., and Laurence, R. L. (1977). *Chemical Reactor Theory* (Lapidus, L., and Amudson, N.R., eds.). Englewood Cliffs, Prentice Hall.
66. Reichert, K.-H., and Geisler, H. (1983). *Polymer Reaction Engineering*, Hanser Verlag, Munich.
67. Sebastian, D. H., and Biesenberger, J. A. (1983). *Polymerization Engineering*, Wiley, New York.
68. Soga, K., and Terano, M., eds. (1994). *Catalyst Design for Tailor-made Polyolefins*, Kodansha-Elsevier, Tokyo
69. Böhm, L. (1980). *Angew. Makromol. Chem.*, *89*: 1.
70. Stevens, J. (1970). *Hydrocarbon Processing*, *4*: 179.
71. Böhm, L. L. (1986). *Catalytic Polymerization of Olefins* (Keii, T., and Soga, K., eds.). Kodansha Ltd., Tokyo.
72. Scheirs, J., Böhm, L. L., Boot, J. C., and Leevers, P. S. (1996). *Trends Polym. Sci.*, *4*: 408–415
73. Böhm, L. L., and Passing (1976). *Makromol. Chem.*, *177*: 1097.
74. Rasmussen, D. M. (1972). *Chem. Eng.*, *79/21*: 104.
75. Wisseroth, K. (1977). US Pat. 4012573 to BASF A.-G., C.A., *77*: 49149.
76. Ray, W. H. (1983). Chapter 5 in *ACS Symposium Series*, 226.
77. Clark, A., and Bailey, G. C. (1963). *J. Catal.*, *2*: 241.
78. De Bree, S. D. (1974). *Hydrocarbon Processing*, *53*: 115.
79. Rudolph, H., Trautvetter, W., and Weirauch, K. (1972). *Chemische Technologie* (Winnacker, K., and Kiichler, L., eds.), Vol. 5., Carl Hanser Verlag, Munich, p. 60.
80. Fleissner, M. (1981). *Angew. Makromol. Chem.*, *94*: 197.
81. Wesslau, H. (1956). *Makromol. Chem.*, *20*: 111.

82. Ziegler, K., Martin, H., and Stedefeder, J. (1959). *Tetrahedron Lett.*, *20*: 12.
83. Schindler, A. (1965). *Crystalline Olefin Polymers* (Raff, R.A., and Doak, K.W., eds.), Wiley, New York, p. 153.
84. Böhm, L. L. (1984). *Chem.-Ing. Tech.*, *56*: 674.
85. Diedrich, B. (1975). *Appl. Polym. Symp.*, *26*: 1.
86. Wilchinsky, Z. W., Looney, R. W., and Törnqvist, E. G. M. (1973). *J. Catal.*, *28*: 351.
87. Natta, G., Pasquon, I., and Giachetti, E. (1957). *Angew. Chem.*, *69*: 213.
88. Reichert, K. H. (1981). *Angew. Makromol. Chem.*, *94*: 1.
89. Kashiwa, N. (1972). US Pat. 3647772 to Mitsui Petrochemical Ind. Ltd., C.A. (1970), *73*: 67051.
90. Weissermel, K., Cherdron, H., Berthold, J., Diedrich, B., Keil, K.-D., Rust, K., Strametz, H., and Toth, T. (1975). *J. Polym. Sci. Polym. Symp.*, *51*: 187.
91. Greco, A., Bertolini, G., and Cesca, S. (1980). *J. Appl. Polym. Sci.*, *25*: 2045.
92. Bart, I. C. J., Bazzi, I. W., Calcaterra, M., Albizzati, E., Giannini, U., and Parodi, S. (1981). *Z. Anorg. All. Chem.*, *482*: 121.
93. Sobota, P., Utko, J., and Lis, T. (1984). *J. Chem. Soc., Dalton Trans*, 2077.
94. Soga, K., Katano, S., Akimoto, Y., and Kagiya, T. (1973). *Polymer J.*, *5*: 128.
95. Böhm, L. L. (1984). *J. Appl. Polym. Sci.*, *29*: 279.
96. Graff, R. J. L., Kortleve, G., and Vonk, C. G. (1970). *J. Polym. Sci., Part B*, *8*: 735.
97. Doi, Y., Murata, M., and Yano, K. (1982). *Ind. Eng. Chem. Prod. Res. Dev.*, *21*: 580.
98. Zakharov, I. I., and Zakharov, V. A. (1982). *J. Mol. Catal.*, *14*: 171.
99. Delbouille, A., and Derroitte, J. L. (1972). US Pat. 3658722 to Solvay et Cie., C.A. (1970), *72*: 79626.
100. Fink, G., and Kinkelin, E. (1988). *Transition Metal Catalyzed Polymerizations, Ziegler-Natta and Metathesis Polymerizations* (Quirk, R., ed.), Cambridge Univ. Press, New York, p. 161.
101. Hogan, J. P., and Banks, R. L. (1958). US Pat. 2825721 to Phillips Petroleum Co., C.A. (1958), *52*: 8621 H.
102. Clark, A. (1967). *Ind. Eng. Chem.*, *59*: 29.
103. Hogan, J. P. (1970). *J. Polym. Sci., Part A-1*, *8*: 2637.
104. Baker, L. M., and Carrick, W. L. (1970). *J. Org. Chem.*, *35*: 774.
105. Clark, A. (1969). *Rev. 3*: 145.
106. Miesserov, K. G. (1966). *J. Polym. Sci., Part A 1*, *4*: 3047.
107. Yermakov, Y7.I., and Zakmarov, V. A. (1968). *Proc. Intern. Congr.*, 4th Reprint, p. 16.
108. Krauss, H. L., Hagen, K., and Hums, E. (1985). *J. Mol. Catal.*, *28*: 233.
109. Hums, E., and Krauss, H. C. (1986). *Z. Anorg. Allgem. Chem.*, *537*: 154.
110. Andersen, A., Cordes, H. G., Herwig, J., Kaminsky, W., Merck, A., Mottweiler, R., Pein, J., Sinn, H., and Vollmers, H. J. (1976). *Angew. Chem.*, *88*: 689; *Angew. Chem. Int. Ed. Engl.*, *15*: 630.
111. Ittel, S. D., Johnson, L. K., and Brookhart, M. (2000). *Chem. Rev.*, *100*: 1169.
112. Britovsek, G. J. P., Bruce, M. I., Gibson, Y. C., Kimbereley, B. S., Maddox, J. P., Mastroianni, S., McTavish, S. J., Redshaw, C., Solan, G. A., Strömberg, S., White, A. J. P., and Williams, D. J. (1999). *J. Am. Chem. Soc.*, *121*: 8728.
113. Tritto, I., Donetti, R, Sacchi, M. C., Locatelli, P., and Zannoni, G. (1997). *Macromolecules*, *30*: 1247.
114. Waters, J. A., and Mortimer, G. A. (1972). *J. Polym. Sci., Part A 1*, *10*: 895.
115. Breslow, D. S., and Newburg, N. R. (1957). *J. Am. Chem. Soc.*, *79*: 5072.
116. Reichert, K. H., and Meyer, K. R. (1969). *Kolloid Z.*, *232*: 711.
117. Long, W. P., and Breslow, D. S. (1975). *Justus Liebigs Ann. Chem.*, *1975*: 463.
118. Breslow, d. S., and Newburg, N. R. (1959). *J. Am. Chem. Soc.*, *81*: 81.
119. Herwig. J., and Kaminsky, W. (1983). *Polym. Bull.*, *9*: 464.
120. Kaminsky, W., Bark, A., Spiehl, R., Möller-Lindenhof, N., and Niedoba, S. (1988). In [17], p. 29.
121. Henrici-Olivé, S. (1971). *Angew. Chem.*, *83*: 782.
122. Karapinka, G. L., and Carrick, W. L. (1961). *J. Polym. Sci.*, *55*: 145.

123. Wilke, G., Boddanovic, B., Hardt, P., Heimbach, P., Keim, W., Kröner, M., Oberkirch, W., Tanaka, K., Steinrücke, E., Walter, D., and Zimmermann, H. (1966). *Angew. Chem.*, *78*: 157.
124. Ballard, D. G. H. (1973). *Adv. Catal.*, *23*: 263.
125. Giannini, U., Zuchini, U., and Albizzati, E. (1970). *Polym. Lett.*, *8*: 405.
126. Breslow, D. S. (1958). US Pat. 2, 827,446 to Hercules.
127. Natta, G. (1955). *J. Nucl. Inorg. Chem.*, *49*: 1885.
128. Belov, G. P., Kuznetsov, V. I., Solovyeva, T. I., Chirkov, N. M., and Ivanchev, S. S. (1970). *Makromol. Chem.*, *140*: 213.
129. Agasaryan, A. B., Belov, G. P., Dartyan, S. P., and Erisyan, M. L. (1975). *Eur. Polym. J.*, *11*: 549.
130. Patat, E., and Sinn, H. (1958). *Angew. Chem.*, *70*: 496.
131. Shilov, A. E. (1960). *Dokl. Akad. Nauk SSSR*, *132*: 599.
132. Henrici-Olivé, G., and Olivé, S. (1969). *J. Organomet. Chem.*, *16*: 339.
133. Reichert, K. H., and Schoetter, E. (1968). *Z. Phys. Chem.*, *57*: 74.
134. Fink, G., Rottler, R., Schnell, D., and Zoller, W. (1976). *J. Appl. Polym. Sci.*, *20*: 2779.
135. Fink, G., and Rottler, R. (1981). *Angew. Makromol. Chem.*, *94*: 25.
136. Clauss, K., and Bestian, H. (1962). *Justus Liebigs Ann. Chem.*, *654*: 8.
137. Sinn, H., and Patat, F. (1963). *Angew. Chem.*, *75*: 805.
138. Henrici-Olivé, G., and Olivé, S. (1967). *Angew. Chem.*, *79*: 764.
139. Dyachkovskii, F. S., Shilova, A. K., and Shilov, A. E. (1967). *J. Polym. Sci., Part C.*, *16*: 2333.
140. Eisch, J. J., Piotrowski, A. M., and Brownstein, S. K. (1985). *J. Am. Chem. Soc.*, *107*: 7219.
141. Fink, G., and Schnell, D. (1982). *Angew. Makromol. Chem.*, *105*: 39.
142. Höcker, H., and Saeki, K. (1971). *Makromol. Chem. Soc.*, *148*: 107.
143. Zakharov, V. A., and Yermakov, Y. I. (1979). *Catal. Rev. Sci. Eng.*, *19*: 67.
144. Farina, M., and Ragazzini, M. (1958). *Chim. Ind. (Milan)*, *40*: 816.
145. Tajima, Y., and Kuniola, E. (1968). *J. Polym. Sci. PartA 1*, *6*: 241.
146. Hiraki, K., Ivrone, T., and Hirai, H. (1970). *J. Polym. Sci. Part A 1*, *8*: 2543.
147. Bawn, C. E. H., North, A. M., and Wolker, J. S. (1964). *Polymer*, *5*: 419.
148. Kaminsky, W. (1986). *Polymerisation and Copolymerisation; Houben Weyl: Methoden der organischen Chemie*, Vol. E18/T2, *Organo-π-metall-Verbindungen als Hilfsmittel in der organischen Chemie*, 4th ed. (Bartl, H., and Falbe, J., eds.), Georg Thieme, New York, p. 907.
149. Sinn, H. (1995). *Macromol. Symp.*, *97*: 27.
150. Koide, Y., Bott, S. G., and Barron, A. R. (1996). *Organometallics*, *15*: 2213.
151. Kaminsky, W., Miri, M., Sinn, H., and Woldt, R. (1983). *Makromol. Chem. Rapid Commun.*, *4*: 417.
152. Sinn, H., Kaminsky, W., Vollmer, H. J., and Woldt, R. (1981). US Pat. 4,404,344, DEP 3,007,725 to BASF; C.A. (1981), *95*: 187927.
153. Kaminsky, W., Hähnsen, H., Kulper, K., and Woldt, R. (1983). US Pat. 4,942,199, DEP 3,127,133 to Hoechst AG; C.A. (1983), *98*: 126804.
154. Tait, P. (1988). In: *Transition Metals and Organometallics as Catalysts for Olefin Polymerization* (Kaminsky, W., and Sinn, H., eds.), Springer, Berlin.
155. Kaminsky, W., and Lüker, H. (1989). *Makromol. Chem. Rapid Commun*, *5*: 225.
156. Eisch, J. J., Bombrick, S. I., and Zheng, G. X. (1993). *Organometallics*, *12*: 3856.
157. Jordan, R. F., Dasher, W. E., and Echols, S. F. (1986). *J. Am. Chem. Soc.*, *108*:1718.
158. Sishta, C., Hathorn, R. M., and Marks, T. J. (1992). *J. Am. Chem. Soc.*, *114*: 114.
159. Powell, J., Lough, A., and Saeed, T. (1997). *J. Chem. Soc., Dalton Trans.*, 4137.
160. Duchatean, R., Lancaster, S. J., Thornton-Pelt, M., and Bochmann, M. (1997). *Organometallics*, *16*: 4995.
161. Erker, J. K. G., and Fröhlich, R. (1997). *J Am. Chem. Soc.*, *119*: 11165.
162. Johnson, L. K., Killian, C. M., Arthur, S. D., Feldman, J., McCord, E. F., Mclain, S. J., Kreutzer, K. A., Bennett, M. A., Coughlin, E. B., Ittel, S. D., Parthasarathy, A., Dempel, D. J., and Brookhart, M. S. (1996). *DuPont*, WO 96/23010.

163. Johnson, L. K., Killian, C. M., and Brookhart, M. (1995). *J. Am. Chem. Soc.*, *117*: 6414–6415.
164. Keim, W., Appel, R., Gruppe, S., and Knoch, F. (1987). *Angew. Chem.*, *99*: 1042.
165. Ostoja-Starzewski, K. A., and Witte, J. (1985). *Angew. Chem.*, *97*: 610.
166. Ostoja-Starzewski, K. A., and Witte, J. (1987). *Angew. Chem.*, *99*: 76.
167. Ostoja-Starzewski, K. A., and Witte, J. (1988). *Transition Metal Catalyzed Polymerizations, Ziegler-Natta and Metathesis Polymerizations* (Quirk, R. P., ed.), Cambridge University Press, New York.
168. Pyrlik, O., and Arndt, M. (1999). In: *Proceedings of the International Symposium of Organometallic Catalysts for Synthesis and Polymerisation*, Hamburg.
169. Britovsek, G. J. P., Gibson, V. C., Kimberley, B. S., Maddow, P. J., McTavish, S. J., Solan, G. A., White, A. J. P., and Williams, D. J. (1988). *Chem. Commun.*, 849–850.
170. Small, B. L., Brookhart, M., and Bennett, M. A. (1998). *J. Am. Chem. Soc.*, *120*: 4049.
171. Wang, C., Friedrich, S., Younkin, T. R., Li, T. R., Grubbs, R. H., Bansleben, T. R., and Day, M. W. (1998). *Organometallics*, *17*: 3149.
172. Younkin, T. R., Connor, E. F., Henderson, J. I., Friedrich, S. K., Grubbs, R. H., and Bansleben, D. A. (2000). *Science*, *287*: 460.
173. Natta, G., Panusso, F., and Siamesi, D. (1959). *Makromol. Chem.*, *30*: 238.
174. Ehrlich, P., and Mortimer, G. A. (1970). *Fortschr. Hochpolym. Forsch.*, *7*: 386.
175. Mortimer, G. A. (1965). *J. Polym. Sci.*, *3*: 343.
176. Taniguchi, J., Yoshikawa, K., and Tatsugami, Y. (1963). D.E. Pat. 1,520,825 to Sumitomo Chemical Co.; C.A. (1964), *61*: 16274.
177. Luft, G., and Bitsch, H. (1973). *Angew. Makromol. Chem.*, *32*: 17.
178. Erussalimsky, B., Tumarkin, N., Duntoff, F., Lyobetzky, S., and Goldenberg, A. (1967). *Makromol. Chem.*, *104*: 288.
179. Brown, F. E., and Ham, G. E. (1964). *J. Polym. Sci. Part A*, *2*: 3623.
180. Uschold, R. E., and Finlag, J. B. (1974). *Appl. Polym. Symp.*, *25*: 205.
181. Aliani, G., Smits, J. A. F., and Lechat, J. B. (1982). E.E. Pat. 0,078,122 to Exxon; C.A. (1983), *99*: 39553.
182. Hudgin, D. E. (1977). DE OS 2,725,244 to Atlantic Richfield Co.; C.A. (1977), *84*: 40222.
183. Bartl, H. (1957). D.E. Pet. 1,126,613 to Bayer; C.A. (1960), *57*: 2427.
184. Lindemann, M. K. (1968). *Paint Manuf.*, *70*: 4623.
185. Edser, M. H. (1982). *Paint Resin*, *51*: 17.
186. Glenz, W. (1986). *Kunststoffe*, *76*: 834.
187. Kissin, Y. V. (1974). *Adv. Polym. Sci.*, *15*: 91.
188. Kaminsky, W., and Schlobohm, M. (1986). *Makromol. Chem., Macromol. Symp.*, *4*: 103.
189. Rossi, A., Zhang, J., and Odian, G. (1996). *Macromolecules*, *29*: 2331.
190. Herfert, N., and Fink, G. (1992). *Makromol. Chem.*, *193*: 1359.
191. Kaminsky, W., and Drögemüller, H. (1989). In: *Polymer Reaction Engineering*, (Reichert, K. H., and Geiseler, W., eds.), VCH Berlin, p. 372.
192. Zambelli, A., Giongo, G. M., and Natta, G. (1968). *Makromol. Chem.*, *112*: 183.
193. Pajaro, G. (1959). *Energia Nucl. (Milan)*, *6*: 273.
194. Natta, G., Mozzanti, G., Valvassori, A., Sartori, G., and Fiumani, O. (1960). *J. Polym. Sci.*, *51*: 911.
195. Natta, G., Mozzanti, G., Valvassori, A., Sartori, G., and Morero, D. (1960). *Chim. Ind. (Milan)*, *42*: 125.
196. Kaminsky, W., and Miri, M. (1985). *J. Polym. Sci., Polym. Chem. Ed.*, *23*: 2151.
197. Chien, J. C. W., and He, D. (1991). *J. Polym. Sci., Part A*, *29*: 1585.
198. Stevens, J. (1993). *Proc. MetCon*, Houston, May 26–28, p. 157.
199. Shapiro, P. J., Bunel, E., Schaefer, W. P., and Bercaw, J. W. (1990). *Organometallics*, *9*: 867.
200. Okuda, J., du Plooy, K. E., Massa, W., Kang, H.-C., and Rose, U. (1996). *Chem. Ber.*, *129*: 275.
201. Chum, S. P., Kao, C. L, and Knight, G. W. (2000). In [23], p. 261.
202. (1984). *Chem. Eng. News*, *62*: 35.

203. Borg, E. L. (1981). *Rubber Technology*(Morton, M., ed.), 2nd ed., R.E. Krieger, Melbourne, Bla., p. 220.
204. Allen, R. D. (1983). *J. Elastomers Plast.*, *15*: 19.
205. Kaminsky, W. (1996). *Macromol. Chem. Phys.*, *197*: 3907.
206. Grassi, A., Maffei, G., Milione, S., and Jordan, R. F. (2001). *Macromol. Chem. Phys.*, *202*: 1239.
207. Kaminsky, W., Bark, A., and Steiger, R. (1992). *J. Mol. Catal.*, *74*: 109.
208. Kaminsky, W., Beulich, I., and Arndt-Rosenau, M. (2001). *Macromol. Symp.*, *172*: 211.
209. Kaminsky, W., Bark, M., and Arndt, M. (1991). *Makromol. Chem.*, *Macromol. Symp.*, *47*: 83.
210. Herfert, N., Montag, P., and Fink, G. (1993). *Makromol. Chem.*, *94*: 3167.
211. Kaminsky, W., and Arndt, M. (1997). *Adv. Polym. Sci.*, *127*: 143.
212. Demming, T. J., and Novak, B. M. (1993). *Macromolecules*, *26*: 7089.
213. Goodall, B. L., Barnes, D. A., Benedict, G. M., McIntosh, L. H., and Rhodes, L. F. (1997). *Polym. Mater. Sci. Eng.*, *76*: 56.
214. Mehler, C., and Risse, W. (1991). *Makromol. Chem. Rapid Commun.*, *12*: 255.
215. Heitz, W., and Haselwander, T. F. A. (1997). *Macromol. Rapid Commun.*, *18*: 689.
216. Cherdron, H., Brekner, M.-J., and Osan, F. (1994). *Angew. Makromol. Chem.*, *223*: 121.
217. Kaminsky, W., and Arndt-Rosenau, M. (2000). In: *Metallocene-based Polyolefins* (Scheirs, J., and Kaminsky, W. eds.), Vol. 2, Wiley Series in Polymer Science, Chichester, p. 91.
218. Kaminsky, W., and Spiehl, R. (1989). *Makromol. Chem.*, *190*: 515.
219. Collins, S., and Kelly, W. M. (1992). *Macromolecules*, *25*: 233.
220. Kaminsky, W., Engehausen, R., and Kopf, J. (1995). *Angew. Chem.*, *107*: 2469; (1995). *Angew. Chem., Int. Ed. Engl.*, *34*: 2273.
221. Arndt, M., and Beulich, I. (1998). *Macromol. Chem. Phys.*, *199*: 1221.
222. Sernetz, F. G., Mülhaupt, R., Amor, F., Eberle, T., and Okuda, J. (1997). *J. Polymer. Sci.: Part A*, *35*: 1571.
223. Stevens, J. (1993). *Proc. MetCon*, Houston, p. 157.
224. Laine, T. V., Klinga, M., Maaninen, A., Aitola, E., and Leshela, M. (1999). *Act. Chem. Scand.*, *53*: 968.
225. Drent, E., and Budzelaar, P. H. M. (1996). *Chem. Rev.*, *96*: 663.
226. Shiono, T., Yoshida, K., and Soga, K. (1990). *Makromol. Chem. Rapid Commun.*, *11*: 169.
227. Drent, E., van Broeckhoren, J. A. M., and Budzelaar, P. H. M. (1996). *Rec. Trav. Chim. Pays-Bas*, *115*: 263.
228. Kennedy, J. P., and Squires, R. G. (1967). *J. Macromol. Sci. Chem.*, *1*: 805.
229. Natta, G. (1955). *J. Polym. Sci.*, *16*: 143.
230. Natta, G. (1956). *Angew. Chem.*, *68*: 393.
231. Natta, G., Pino, P., Mazzanti, G., and Longi, P. (1957). *Gazz. Chim. Ital.*, *87*: 549.
232. Natta, G., Pino, P., Mazzanti, G., and Longi, P. (1957). *Gazz Chim. Ital.*, *87*: 570.
233. Tomquist, E., Seelbach, W., and Langer, A.W. (1964). US Pat. 3128252 to Esso Research.
234. Brit. Pat. 807204 (1959). To Hercules Powder, C.A. (1959), *53*: 13660 G.
235. Edwards, M. B., and Hagemeyer, H. J. (1963). Brit. Pat. 930 633 to Eastman Kodak, C.A. (1964), *61*: 4508 F.
236. Edwards, M. B., Park, K. V., Carter, E. H. (1966). Fr. Pat. 1450785 to Eastman Kodak, C.A. (1964), *66*: 95603.
237. Brit. Pat. 1092390 (1966). To Mitsubishi Petrochem. Corp., C.A. (1968), *68*: 13557.
238. Chien, J. C. W. (1980). *Preparation and Properties of Stereoregular Polymers* (Lenz, R. W., and Ciardelli, F., eds.), D. Reidel, Dordrecht, p. 115.
239. Beermann, C., and Bastian, H. (1959). *Angew. Chem.*, *71*: 618.
240. Natta, G. (1959). *J. Polym. Sci.*, *34*: 21.
241. Boor, J. (1963). *J. Polym. Sci*, *Part C*, *1*: 237.
242. Ingberman, A. K. (1966). *J. Polym. Sci.*, *Part A*: 2781.
243. Natta, G., Porri, G., and Fiore, L. (1959). *Gazz. Chim. Ital.*, *89*: 761.
244. Dawans, F., and Teyssie, Ph. (1963). *Bull. Soc. Chim. Fr.*, *10*: 2376.

245. Natta, G., Corradini, P., and Allegra, E. (1959). *Atti Accad. Naz. Lin Cei, Cl. Sci. Fis. Mat. Nat. Rend. 26*: 159.
246. Keii, T., Terano, M., Kimura, K., and Ishi, K. (1988). *Transition Metals and Organometallics as Catalysts for Olefin Polymerization* (Kaminsky, W., and Sinn, H., eds.), Springer Press, p. 1.
247. Kashiwa, N., and Yoshitake, J. (1984). *Polymer Bull., 11*: 479.
248. Kashiwa, N., and Yoshitake, J. (1984). *Polymer Bull., 12*: 99.
249. Spitz, R., Lacombe, J. L., and Guyot, A. (1984). *J. Polym. Sci. Polym. Chem. Ed., 22*: 2625.
250. Amendola, P., and Zambelli, A. (1989). *Makromol. Chem., 185*: 2451.
251. Chien, J. C. W., and Kuo, C.-I. (1985). *J. Polym. Sci. Polym. Chem. Ed., 23*: 761.
252. Kissin, Y. V., and Beach, D. L. (1984). *J. Polym. Sci. Polym. Chem. Ed., 22*: 333.
253. Anderson, I. H., Burnett, G. M., and Tait, P. J. (1962). *J. Polym. Sci., 56*: 361.
254. Seidov, N. M., Guseinov, F. O. P., Ibragimov, K. D., Abasov, A. I., and Etandiev, M. A. (1977). *Vysokomol. Soedin A, 19*: 1523.
255. Galli, P., Luciani, L., and Cecchin, G. (1981). *Angew. Makromol. Chem., 94*: 63.
256. Keii, T., Suzuki, E., Tamura, M., Murata, M., and Doi, Y. (1982). *Makromol. Chem., 183*: 2285.
257. Guastalla, G., and Giannini, G. (1983). *Makromol. Chem., Rapid Comm., 4*: 519.
258. Soga, K., Terano, M., and Ikeda, S. (1980). *Polymer Bull., 3*: 179.
259. Caunt, A. (1981). *Br. Polymer J., 13*: 22
260. Giannini, G. (1981). *Makromol. Chem. Suppl., 5*: 216.
261. Doi, Y., Ueki, S., and Keii, T. (1980). *Polymer, 21*: 1352.
262. Chien, J. C. W. (1984). *Catal. Rev.-Sci. Eng., 26*: 613.
263. Nielsen, R. P. (1983). *Transition Metal Catalyzed Polymerization* (Quirk, ed.), Vol. 4A, Harwood Acad. Publ., New York, p. 47.
264. Tait, P. J. T. (1975). *Coordination Polymerization* (Chien, J. C. W., ed.), Academic Press, New York, p. 155.
265. Zakharov, V. A., Bukatov, G. D., Chumaevskii, N. B., and Yermakov, Y. I. (1977). *Makromol. Chem., 178*: 967.
266. Natta, G., Mazzanti, G., Giannini, U., De Luca, D., and Bandini, F. (1964). *Makromol. Chem., 76*: 54.
267. Keii, T., Soga, K., and Saiki, N. (1967). *J. Polym. Sci., Part C., 16*: 1507.
268. Natta, G., and Pasquon, I. (1959). *Advances in Catalysis*, Vol. XI, Academic Press, New York, p. 1.
269. Kissin, Y. V., Mezhikovsky, S. M., and Chirkov, N. M. (1970). *Europ. Polym. J., 6*: 267.
270. Tait, P. J. T. (1979). *Developments in Polymerization* (Howard, R. N., ed.), Applied Science Publ., London, p. 81.
271. Mejzlik, J., Lesna, M., and Kratochvila, J. (1986). *Adv. Polym. Sci., 81*: 83.
272. Bukatov, G. D., Zakharov, V. A., and Yermakov, Yu. I. (1978). *Makromol. Chem., 179*: 2097.
273. Chirkov, N. M. (1969). *IUPAC Intern. Macromol. Symposium*, Budapest, p. 297.
274. Cooper, H. W., Guillet, J. E., Combs, R. L., and Joyner, F. G. (1966). *Polym. Sci., Part A., 4*: 2583.
275. Zakharov, V. A., Bukatov, G. D., and Yermakov, Y. I. (1983). *Polym. Sci. Technol., 19*: 267.
276. Kohn, E., Shurmans, H. J. L., Cavender, J. Va., and Mendelson, R. A. (1962). *J. Polym. Sci., 58*: 681.
277. Grierson, B. M. (1965). *Makromol. Chem., 84*: 93.
278. Schnecko, H., Dost, W., and Kern, W. (1969). *Makromol. Chem., 121*: 159.
279. Burfield, D. R. (1978). *J. Polym. Sci. Polym. Chem. Ed., 16*: 3301.
280. Petts, R. W., and Waugh, K. C. (1982). *Polymer, 23*: 897.
281. Mejzlik, J., and Lesna, M. (1977). *Makromol. Chem., 178*: 261.
282. Warzelhan, V., Burger, T. F., and Stein, D. J. (1982). *Makromol. Chem., 183*: 489.
283. Natta, G., Pasquon, I., and Giuffr, L. (1961). *Chem. Ind. (Milan), 43*: 871.

284. Tsvetkova, V. I., Plusnin, A. N., Bolshakova, R. F., Uvarov, B. A., and Chirkov, N. M. (1969). *Vysokomol. Soedin, A11*: 1817.
285. Allegra, G. (1971). *Makromol. Chem.*, *145*: 235.
286. Zambelli, A., Locatelli, P., and Rigomonti. E. (1979). *Macromolecules*, *12*: 156.
287. Corradini, P., Barone, V., Fusco, R., and Guerra, G. (1979). *Eur. Polym. J.*, *15*: 1133.
288. Bozik, J. E., Vogel, R. F., Kissin, Y. V., and Beach, D. L. (1984). *J. Appl. Polym. Sci.*, *29*: 3491.
289. Galli, P., Barbe, P., Guidetti, G., Zannetti, R., Martorana, A., Marigo, A., Bergozza, M., and Fishera, A. (1983). *Eur. Polym. J.*, *19*: 19.
290. Kashiwa, N., and Yoshitake, J. (1983). *Makromol. Chem. Rapid Commun.*, *4*: 41.
291. Muñoz-Escalona, A., and Parada, A. (1979). *Polymer*, *20*: 474.
292. Boucher, D. G., Parsons, I. W., and Harward, R. N. (1974). *Makromol. Chem.*, *175*: 3461.
293. Bier, G. (1982). *Polym. Bulletin*, *7*: 177.
294. Zambelli, A., Sacchi, M. C., and Locatelli, P. (1979). *Macromolecules*, *12*: 1051.
295. Miyazawa, T., and Ideguchi, T. (1963). *J. Polym. Sci.*, *Part B*, *1*: 389.
296. Natta, G., Farina, M., and Peraldo, M. (1960). *Chim. Ind.* (*Milan*), *42*: 255.
297. Natta, G. (1958). *J. Inorg. Nucl. Chem.*, *8*: 589.
298. Arlmann, E. J., and Cossee, P. (1964). *J. Catal.*, *3*: 99.
299. Cossee, P. (1967). *The Stereochemistry of Macromolecules* (Ketley, A. D., ed.), Dekker Press, New York, p. 145.
300. Pino, P., and Mülhaupt, R. (1983). *Transition Metal Catalyzed Polymerizations* (Quirk, R. P., ed.), MMI Press, Symposium Series No. 4, p. 1.
301. Pino, P. (1965). *Adv. Polym. Sci.*, *4*: 393.
302. Corradini, P., Guerra, G., Fusco, R., and Barone, V. (1980). *Eur. Polym. J.*, *16*: 835.
303. Kissin, Yu. V., and Chirkov, N. M. (1970). *Eur. Polym. J.*, *6*: 525.
304. Keii, T. (1995). *Macromol. Theory. Simul.*, *4*: 947.
305. Liu, B., Matsuoka, H., and Terano, M. (2001). *Macromol. Rapid Commun,*. *22*: 1.
306. Cecchin, G., Marchetti, E., and Baruzzi, G. (2001). *Macromol. Chem. Phys.*, *202*: 1987.
307. Bukatov, G. D., and Zakharov, V. (2001). *Macromol. Chem. Phys.*, *202*: 2003.
308. Ivin, K. I. J., Rooney, J. J., Stewart, C. D., Green, M. L. H., and Mahtab, R. (1978). *J. Chem. Soc. Chem. Commun.*, 607.
309. Zambelli, A., Locatelli, P., Sacchi, M. G., and Rigamonti, E. (1980). *Macromolecules*, *13*: 798.
310. Hewett, W. A., and Shokal, E. C., Brit. Pat. 904.510 (1960) to Shell, C.A.
311. Hewett, W. A. (1965). *J. Polym. Sci.*, *B5*: 855.
312. Luciani, L., and Barbe, P. C. (1975). Belg. Pat. 845593 to Montedison, C.A., *86*: 140706.
313. Toyota, A., Yoshitugu, K., and Kashiwa, N. (1977). Ger. Pat. 2809318 to Mitsui Petrochem., C.A. (1978), *89*: 216056.
314. Langer, A. W. (1977). Ger. Pat. 2804838 to Exxon, C.A. (1978), *89*: 198265.
315. Miyoshi, M., Tajioma, Y., Matsuura, K., Kuroda, N., and Matsuno, M. (1978). Ger. Pat. 2914812 to Nippon Oil, C.A. (1980), *92*: 7209.
316. Ho, A., Sasahi, H., Osawa, M., Iwao, T., and Iwata, K. (1976). Ger. Pat. 2734652 to Mitsui Toatsu, C.A. (1978), *88*: 137207.
317. Yamaguchi, K., Kahogawa, G., Masuo, M., Suga, Y., and Kitadu, H. (1977). Jpn. Pat. 54062289 to Mitsubishi Chem., C.A. (1979), *91*: 108486.
318. Staiger, A. (1978). Ger. Pat. 2831830 to BASF, C.A. (1980), *92*: 164525.
319. Wagner, K. P. (1978). Brit. Pat. 2018789 to Hercules, C.A. (1980), *92*: 147511.
320. McKenna, T. F., Cokljat, D., Spitz, R., and Schweich, D. (1999). *Catalysis Today*, *48*: 101.
321. Duck, E. W., Grant, D., and Kronfli, E. (1979). *Eur. Polym. J.*, *15*: 625.
322. Soga, K., Izumi, K., Terano, M., and Ikeda, S. (1980). *Makromol. Chem.*, *181*: 657.
323. Ivanchev, S. S., Baulin, A. A., and Rodionov, A. G. (1980). *J. Polym. Sci. Polym. Chem. Ed.*, *18*: 2045.
324. Kashiwa, N. (1980). *Polym. J.*, *12*: 603.
325. Hsieh, H. L. (1980). *Polym. J.*, *12*: 596.

326. Kashiwa, N., and Yoshitake, J. (1982). *Makromol. Chem. Rapid Commun.*, *3*: 211.
327. Matsuoka, H., Liu, B., Nakatani, H., and Terano, M. (2001). *Macromol. Rapid Commun.*, *22*: 326.
328. Oleshko, V. P., Crozier, P. A., Cantrell, R. D., and Westwood, A. D. (2001). *Macromol. Rapid Commun.*, *22*: 34.
329. Bukatov, G. D., Shepelev, S. H., Zakharov, V. A., Sergeev, S. A., and Yermakov, Yu. I. (1982). *Macromol. Chem.*, *183*: 2657.
330. Chien, J. C. W., Wu, J. C., and Kuo, I. J. (1982). *J. Polym. Sci., Part A-1*, *20*: 2091.
331. Caunt, A. D., Licchelli, J. A., Parsons, I. V., Haward, R. N., and Al-Hillo, M. R. Y. (1983). *Polymer*, *24*: 121.
332. Barbe, P. C., Noristi, L., Baruzzi, G., and Marchetti, E. (1983). *Makromol. Chem. Rapid Commun.*, *4*: 249.
333. Soga, K., Kozumi, T., and Yanagihara, H. (1989). *Pro Makromol. Chem.*, *190*: 31.
334. Giannini, U., Cassata, A., Longi, P., and Mazzocchi, R. (1976). DE.P. 2643143 to Montedison S.P.A. and Mitsui Petrochem., C.A. (1973), *87*: 68893.
335. Luciani, L., Kashiwa, N., Barbe, P. C., and Toyota, A., (1976). Ger.Pat. 2643143 to Montedison SPA and Mitsui Petrochem., C.A. (1973), *87*: 68893.
336. Kioka, M., Kitani, H., and Kashiwa, N., (1980). Ger. Pat. 3022738 to Mitsui Petrochem., C.A. (1981), *94*: 122303.
337. Muñoz-Escalona, A., Martin, A., Hidalgo (1981). *Eur. Polym. J.*, *17*: 367.
338. Kashiwa, N., Kawasaki, M., and Yoshitake, J. (1986). *Catalytic Polymerization of Olefins* (Keii, T., and Soga, K., eds.), Elsevier, Tokyo, p. 43.
339. Pino, P., Fochi, G., Piccolo, O., and Giannini, U. (1982). *J. Am. Chem. Soc.*, *104*: 7381.
340. Kissin, Y. V., and Sivak, A. J. (1984). *J. Polym. Sci.*, *22*: 3747.
341. Chien, J. C. W., Kuo, C. I., and Aug, T. (1985). *J. Polym. Sci.*, *23*: 723.
342. Chadwick, J. C., Morini, G., Balbontin, G., Camurati, I., Heere, J. J. R., Mingozzi, I., and Testoni, F. (2001). *Macromol. Chem. Phys.*, *202*: 1995.
343. Barbe, P. C., Cecchin, G., and Noristi, L. (1987). *Adv. Polym. Sci.*, *81*: 1.
344. Barino, L., and Scordamaglia, R. (1995). *Macromol. Symp.*, *89*: 101.
345. Chien, J. C. W., and Bres, P. L. (1986). *J. Polym. Sci. Polym. Chem. Ed.*, *24*: 2483.
346. Terano, M., Kataoka, T., and Keii, T. (1987). *Macromol. Chem.*, *188*: 1477.
347. Chien, J. C. W., Wu, J. C., and Kuo, C. I. (1983). *J. Polym. Sci. Polym. Chem. Edc.*, *21*: 725.
348. Muñoz-Escalona, A. (1983). *Transition Metal Catalyzed Polymerizations* (Quirk, R. P., ed.), MMI Press Symposium Series, Vol. 4, p. 323.
349. Goodall, B. L. (1983). *Transition Metal Catalyzed Polymerizations* (Quirk, R. P., ed.), MMI Press Symposium Series, Vol. 4, p. 323.
350. Sormunen, P., Iiskola, E., Vähäsarja, E., Pakkanen, T. T., and Pakkanen, T. A. (1987). *J. Organomet. Chem.*, *319*: 327.
351. Iiskola, E., Sormunen, P., and Garoff, T. (1988). *Olefin Polymerization* (Kaminsky, W., and Sinn, H., eds.), Springer Verlag, Heidelberg, p. 113.
352. Soga, K., Shiono, T., and Doi, Y. (1989). *Macromol. Chem.*, *189*: 1531.
353. Soga, K., Park, J. R., Uchino, H., Kozumi, T., and Shiono, T. (1989). *Macromolecules*, *22*: 3824.
354. Soga, K. (1987). *Sekiyu Gakkaishi*, *30*: 359.
355. Matsko, M. A., Bukatov, G. D., Mikenas, T. B., and Zakharov, V. A. (2001). *Macromol. Chem. Phys.*, *202*: 1435.
356. Albizzati, E., Giannini, U., Morini, G., Galimberti, M., Barino, L., and Scordamaglia, R. (1995). *Macromol. Symp.*, *89*: 73.
357. Galli, P., Cecchin, G., Chadwick, J. C., Del Duca, D., and Vecellio (1999). In [19] p. 14.
358. Corradini, P., Busico, V., and Guerra, G. (1988). In [18] p. 337.
359. Farina, M., and Puppi, C. (1993). *J. Mol. Catal.*, *82*: 3.
360. Noristi, L., Barbe, P. C., and Baruzzi, G. (1991). *Makromol. Chem.*, *192*: 1115.
361. Albizzati, E., Giannini, U., Morini, G., Smith, C. A., and Zeigler, R. C. (1995). In [16] p. 413.

362. Busico, V., Cipullo, R., Monaco, G., Palarico, G., Vacatello, M., Chadwick, J. C., Segre, A. L., and Sudmeijer, O. (1999). *Macromolecules*, *32*: 4173.
363. Ferrero, M. A., and Chiavetta, M. G. (1990). *Polym. Plast. Technol. Eng.*, *29*: 263.
364. Hattori, N. (1986). *Chem. Econ. Eng. Rev.*, *18*: 21.
365. Ficker, H. K., Goeke, G. L., and Powers, G. W. (1987). *Plast. Eng.*, *43*: 29.
366. Kaminsky, W., Külper, K., Brintzinger, H. H., Wild, F. R. W. P. (1985). *Angew. Chem. Int. Ed. Engl.*, *24*: 507.
367. Kaminsky, W., Külper, K., Buschermöhle, M., and Luker, H. (1984). Ger.Pat. 3443087 to Hoechst AG.
368. Wild, F. R. W. P., Wasiucionek, M., Huttner, G., and Brintzinger, H. H. (1985). *J. Organomet. Chem.*, *288*: 63.
369. Ewen, J. A. (1984). *J. Am. Chem. Soc.*, *106*: 6355.
370. Drögemüller, H., Niedoba, S., and Kaminsky, W. (1986). *Polymer Reaction Engineering* (Reichert, K. H., and Geiseler, W., eds.), Hüthig & Wepf, Heidelberg, p. 299.
371. Kaminsky, W. (1986). *Catalytic Polymerization of Olefins* (Keii, T., and Saga, K., eds.), Elsevier Kodansha, Tokyo, p. 293.
372. Kaminsky, W. (1986). *History of Polyolefins* (Seymour, R. B., and Cheng, T., eds.), Reidel Publ. Co., New York.
373. Spaleck, W. (2000). In [23], p. 425.
374. Leclerc, M. K., and Brintzinger, H. H. (1995). *J. Am. Chem. Soc.*, *117*: 1651.
375. Busico, V., Cipullo, R., and Corradini, P. (1993). *Makromol. Chem. Rapid Commun.*, *117*: 1651.
376. Grassi, A., Zambelli, A., Resconi, L., Albizzati, E., and Mazzocchi, R. (1989). *Macromolecules*, *21*: 617.
377. Soga, K., Shiono, T., Takemura, S., and Kaminsky, W. (1987). *Makromol. Chem. Rapid Commun.*, *8*: 305.
378. Cheng, H. N., and Ewen, J. A. (1989). *Makromol. Chem.*, *190*: 1931.
379. Rieger, B., Mu, X., Mallin, D. T., Rausch, M. D., and Chien, J. C. W. (1990). *Macromolecules*, *23*: 3559.
380. Schupfner, G., and Kaminsky, W. (1995). *J. Mol. Catal. A. Chem.*, *102*: 59.
381. Busico, V., and Cipullo, R. (1994). *J. Am. Chem. Soc.*, *117*: 1652.
382. Sheldon, R. A., Fueno, T., Tsuntsugo, T., and Kurukawa, J. (1965). *J. Polym. Sci. Part B, 3*: 23.
383. Kaminsky, W., Külper, K., and Niedoba, S. (1986). *Makromol. Chem. Macromol. Symp.*, *3*: 377.
384. Farina, M., Disilvestro, G., and Terragni, A. (1995). *Makromol. Chem. Phys.*, *196*: 353.
385. Resconi, L., Piemontesi, F., Franciscono, G., Abis, L., and Fiorani, T. (1992). *J. Am. Chem. Soc.*, *114*: 1025.
386. Resconi, L., Abis, L., and Franciscono, G. (1992). *Macromolecules*, *25*: 6814.
387. Kaminsky, W., and Arndt, M. (1994). In: *Catalysts Design for Tailor-made Polyolefins* (Soga, K., and Terano, M. eds.), Kodansha Ltd., Tokyo, p. 179.
388. Resconi, L. (2000). In [23], p. 467.
389. Spaleck, W., Antberg, M., Aulbach, M., Bachmann, B., Dolle, V., Haftka, S., Küber, F., Rohrmann, J., and Winter, A. (1994). In: *Ziegler Catalysts* (Fink, G., Mülhaupt, R., and Brintzinger, H. H., eds.), Springer Verlag, Berlin, p. 83.
390. Spaleck, W., Küber, F., Winter, A., Rohrmann, J., Bachmann, B., Kiprof, P., Behn, J., and Herrmann, W. A. (1994). *Organometallics*, *13*: 954.
391. Spaleck, W., Aulbach, M., Bachmann, B., Küber, F., and Winter, A. (1995). *Macromol. Symp.*, *89*: 221.
392. Jüngling, S., Mülhaupt, R., Stehling, U., Brintzinger, H. H., Fischer, D., and Langhauser, F. (1995). *Macromol. Symp.*, *97*: 205.
393. Resconi, L., Fait, A., Piemontesi, F., Colonnesi, M., Rychlicki, H., and Zeigler, R. (1995). *Macromolecules*, *28*: 6667.

394. Spaleck, W., Antberg, A., Rohrmann, J., Winter, A., Bachmann, B., Kiprof, P., Behn, J., and Herrmann, W. A. (1992). *Angew. Chem.*, *104*: 1373; *Angew. Chem. Int. Ed. Engl.*, *31*: 1347.
395. Stehling, U., Diebold, J., Kirsten, R., Röll, W., Brintzinger, H. H., Jünglich, S., Mülhaupt, R., and Langhauser, F. (1994). *Orgonometallics*, *13*: 964.
396. Mise, T., Miya, S., and Yamazaki, H. (1989). *Chem. Letters*, 1853.
397. Pino, P., Cioni, P., Wei, J., Rotzinger, B., and Arizzi, S. (1988). In: *Transition Metal Catalyzed Polymerizations* (Quirk, R. P., ed.), Cambridge University Press, Cambridge, p. 1.
398. Pino, P., Cioni, P., and Wei, J. (1987). *J. Am. Chem. Soc.*, *109*: 6189.
399. Waymouth, R., and Pino, P. (1990). *J. Am. Chem. Soc.*, *112*: 4911.
400. Pino, P., and Galimberti, M. (1989). *J. Organomet. Chem.*, *370*: 1.
401. Corradini, P., and Guerra, G. (1988). In: *Transition Metal Catalyzed Polymerizations* (Quirk, R. P., ed.), Cambridge University Press, Cambridge, p. 553.
402. Corradini, P., Busico, V., and Guerra, G. (1987). In: *Transition Metals and Organometallics as Catalysts for Olefin Polymerisation* (Kaminsky, W., and Sinn, H. eds.), Springer Verlag, Berlin, p. 337.
403. Corradini, P., Guerra, G., Vacatello, M., and Villani, V. (1988). *Gazz. Chim. Ital.*, *118*: 173.
404. Venditto, V., Guerra, G., Corradini, P., and Fusco, R. (1991). *Polymer*, *31*: 530.
405. Corradini, P., and Guerra, G. (1991). *Progr. Polym. Sci.*, *16*: 239.
406. Corradini, P. (1993). *Macromol. Chem. Macromol. Symp.*, *66*: 11.
407. Antberg, M., Dolle, V., Haftka, S., Rohrmann, J., Spaleck, W., Winter, A., and Zimmermann, H. J. (1991). *Macromol. Chem. Macromol. Symp.*, *48/49*: 333.
408. Rieger, B., Mu, X., Mallin, D. T., Rausch, M. D., and Chien, J. C. W. (1990). *Macromolecules*, *23*: 3559.
409. Soga, K., Shiono, T., Takemura, S., and Kaminsky, W. (1987). *Macromol. Chem. Rapid Commun.*, *8*: 305.
410. Langhauser, F., Fischer, D., and Seelert, S. (1995). *Proceedings of the Intern. Congress on Metallocene Polymers Metallocenes '95*, Schotland Business Research Inc., Brussels p. 243.
411. Kaminsky, W., and Buschermöhle, M. (1987). *Recent Advances in Mechanistic and Synthetic Aspects of Polymerization* (Fontanille, M., and Guyot, A., eds.), Reidel Publishing Co., Dordrecht, p. 503.
412. Zambelli, A., Longo, P., and Grassi, A. (1989). *Macromolecules*, *21*: 2186.
413. Hlatky, G. G., Turner, H. W., and Eckmann, R. R. (1989). *J. Am. Chem. Soc.*, *111*: 2728.
414. Watson, P. L. (1982). *J. Am. Chem. Soc.*, *104*: 337.
415. Natta, G., Pasquon, I., and Zambelli, A. (1962). *J. Am. Chem. Soc.*, *84*: 1488.
416. Zambelli, A., Natta, G., Pasquon, I., and Signorini, R. (1967). *J. Polym. Sci., Part C*, *16*: 2485.
417. Zambelli, A., Natta, G., and Pasquon, I. (1963). *J. Polym. Sci., Part C*, *4*: 411.
418. Zambelli, A., Pasquon, I., Signorini, R., and Natta, G. (1968). *Macromol. Chem.*, *112*: 160.
419. Natta, G., Mazzanti, G., Crespi, G., and Moraglio, G. (1957). *Chim. Ind. (Milan)*, *39*: 275.
420. Boor jr., I., and Youngan, E. A. (1966). *J. Polym. Sci., Part A-1*, *4*: 1861.
421. Youngman, E. A., and Boor jr, I. (1967). *Makromol. Rev.*, *2*: 33.
422. Pasquon, I. (1967). *J. Pure Appl. Chem.*, *15d*: 465.
423. Zambelli, A., Giongo, G. M., and Natta, G. (1968). ¯*Makromol. Chem.*, *112*: 183.
424. Takegami, Y., and Suzuki, T. (1969). *Bull. Chem. Soc. Jap.*, *42*: 1060.
425. Suzuki, T., and Takegami, Y. (1970). *Bull. Chem. Soc. Jap.*, *43*: 1484.
426. Zambelli, A., and Allegra, G. (1980). *Macromolecules*, *13*: 42.
427. Zambelli, A., and Tosi, C. (1974). *Adv. Polym. Sci.*, *15*: 31.
428. Doi, Y. (1979). *Macromolecules*, *12*: 248.
429. Ewen, J. A., Jones, R. L., Razavi, A., and Ferrara, J. (1988). *J. Am. Chem. Soc.*, *110*: 6255.
430. Ewen, J. A., Alder, M. J., Jones, R. L., Curtis, S., and Cheng, H. N. (1991). In: *Catalytic Olefin Polymerization* (Keii, T., and Soga, K., eds.), Kodansha Ltd., Tokyo, p. 439.

431. Ewen, J. A., Elder, M. J., Jones, R. L., Haspeslagh, L., Atwood, J. L., Bott, S. G., and Robinson, K. (1991). *Polym. Prepr. Am. Chem. Soc.*, *32*: 469.
432. Razavi, A., Nafpliotis, L., Vereecke, D., DenDauw, K., Atwood, J. L., and Thewald, U. (1995). *Macromol. Symp.*, *89*: 345.
433. D'Aniello, C. D., Guadagno, L., Naddeo, C., and Vittoria, V. (2001). *Macromol. Rapid Commun.*, *22*: 104.
434. Razavi, A., Peters, L., Nafpliotis, L., and Atwood, J. L. (1994). In: *Catalysts Design for Tailor-made Polyolefins* (Soga, K., and Terano, M., eds.), Kodansha Ltd., Tokyo.
435. Zambelli, A., Sessa, I., Grisi, F., Fusco, R., and Accomazzi, P. (2001). *Macromol. Rapid Commun.*, *22*: 297.
436. Ewen, J. A. (1999). *J. Mol. Catal. A. Chem.*, *140*: 225.
437. Milano, G., Guerra, G., Pellecchia, C., and Cavallo, L. (2000). *Organometallics*, *19*: 1343.
438. Chowdhury, J., and Moore, S. (1993). *Chem. Eng.*, 36.
439. Kaminaka, M., and Soga, K. (1991). *Makromol. Chem. Rapid Commun.*, *12*: 367.
440. Farina, M., DiSilvestro, G., and Sozzani, P. (1991). *Progr. Polym. Sci.*, *16*: 219.
441. Farina, M., DiSilvestro, G., and Sozzani, P. (1993). *Macromolecules*, *26*: 946.
442. Herfert, N., and Fink, G. (1993). *Makromol. Chem. Macromol. Symp.*, *66*: 157.
443. Ewen, J. A. (1995). *Macromol. Symp.*, *89*: 181.
444. Spaleck, W., Aulbach, M., Bachmann, B., Küper, F., and Winter, A. (1995). *Macromol. Symp.*, *89*: 237.
445. Ewen, J. A., and Elder, M. J. (1995). In: *Ziegler Catalysts* (Fink, G., Mülhaupt, R., and Brintzinger, H. H., eds.), Springer Verlag, Berlin, p. 99.
446. Razavi, A., Vereecke, D., Peters, L., DenDauw, K., Nafpliotis, L., and Atwood, J. L. (1995). In: *Ziegler Catalysts* (Fink, G., Mülhaupt, R., and Brintzinger, H. H., eds.), Springer Verlag, Berlin, p. 99.
447. Mallin, D. T., Rausch, M. D., Lin, G. Y., Dong, S., and Chien, J. C. W. (1990). *J. Am. Chem. Soc.*, *112*: 2030.
448. Chien, J. C. W., Llinas, G. H., Rausch, M. D., Lin, G. Y., and Winter, H. H. (1991). *J. Am. Chem. Soc.*, *113*: 8569.
449. Llinas, G. H., Dong, S. H., Mallin, D. T., Rausch, M. D., Lin, G. Y., Winter, H. H., and Chien, J. C. W. (1992). *Macromolecules*, *25*: 1242.
450. Gauthier, W. J., Corrigan, J. F., Taylor, N. J., and Collins, S. (1995). *Macromolecules*, *28*: 3771.
451. Erker, G., Nolte, R., Tsay, Y. H., and Krüger, K. (1989). *Angew. Chem.*, *101*: 642.
452. Erker, G., Aulbach, M., Knickmeier, M., Wingbergmühle, D., Krüger, K., Nolte, M., and Werner, S. (1993). *J. Am. Chem. Soc.*, *115*: 4590.
453. Erker, G., Aulbach, M., Krüger, C., and Werner, S. (1993). *J. Organomet. Chem.*, *450*: 1.
454. Coates, G. W., and Waymouth, R. M. (1995). *Science*, *267*: 217.
455. Voegele, I., Dietrich, U., Hackmann, M., and Rieger, B. (2000). In [23], p. 485.
456. Rieger, B., Jany, G., Fawzi, R., and Steinmann (1997). *Organometallics*, *16*: 544.
457. Alt, A., Milius, W., and Palackal, S. (1994). *J. Organomet. Chem.*, *472*: 113.
458. Rieger, B., Repo, T., and Jany, G. (1995). *Polym. Bull.*, *35*: 87.
459. Natta, G., and Danusso, F. (1959). *J. Polym. Sci.*, *34*: 3.
460. Folt, V. L., Shipman, J. J., and Krimm, S. (1962). *J. Polym. Sci.*, *61*: 17.
461. Wilkingson, R. W., and Dole, M. (1962). *J. Polym. Sci.*, *58*: 1089.
462. Dürrscheidt, W., Hahmann, O., Kehr, H., Nissing, W., and Potthoff, P. (1976). *Kunststoffe*, *66*: 572.
463. Marconi, W., Cesca, S., and Della Fortuna, G. (1964). *Chim. Ind.* (*Milan*), *46*: 1131.
464. Henrici, Oliv, G., and Oliv, S. (1970). *Polym. Lett.*, *8*: 205.
465. Kaminsky, W. (1983). *Transition Metal Catalyzed Polymerizations* (Quirk, R. P., ed.), MMI Press Sympos. Series, Vol. 4. Part A., Harwood Academic Publ., New York, p. 225.
466. Kaminsky, W. (1986). *Angew. Makromol. Chem.*, *145/146*: 149.
467. Kaminsky, W., and Niedoba, S. (1987). *CLB Chemie für Labor and Betrieb*, *38*: 398.

468. Silvestri, R., Resconi, L., and Pelliconi, A. (1995). *Proceedings of the International Congress on Metallocene Polymers Metallocene '95*, Schotland Business Res.Inc., Brussels, p. 207.
469. Kaminsky, W. (1995). *Macromol. Symp.*, *89*: 203.
470. Soga, K., and Kaminaka, M. (1992). *Makromol. Chem. Rapid Commun.*, *13*: 221.
471. Soga, K., and Kaminaka, M. (1993). *Makromol. Chem.*, *194*: 1745.
472. Soga, K., Kaminaka, M., and Shiono, T. (1993). *Proceedings of the MetCon'93*, Catalysts Consultants Inc., Houston.
473. Soga, K., Kim, H. J., and Shiono, T. (1994). *Makromol. Chem. Rapid Commun.*, *15*: 139.
474. Chien, J. C. W., and He, D. (1991). *J. Polym. Sci. Part A. Polym. Chem.*, *29*: 1603.
475. Collings, S., Kelly, W. M., and Holden, D. A. (1992). *Macromolecules*, *25*: 1780.
476. Janiak, C., and Rieger, B. (1994). *Angew. Makromol. Chem.*, *215*: 47.
477. Lee, D., Yoon, K (1994). *Macromol. Rapid Commun.*, *15*: 841.
478. Kaminsky, W. (1983). In: *Transition Metal Catalyzed Polymerization, Alkenes, and Dienes* (Quirk, R. P., ed.), Academic Publishers, New York, p. 225
479. Soga, K, Arai, T., Nozawa, H., and Uozumi, T. (1995). *Macromol. Symp.*, *97*: 53.
480. Marks, T. J. (1992). *Acc. Chem. Res.*, *25*: 57.
481. Kaminsky, W., and Renner, F. (1993). *Makromol. Chem. Rapid Commun.*, *14*: 239.
482. Kaminsky, W., Renner, F., and Winkelbach, H. (1994). *Proceedings of the MetCon'94*, Houston TX, p. 323.
483. Sacchi, M. C., Zucchi, D., Tritto, I., and Locatelli, P. (1995). *Macromol. Rapid Commun.*, *16*: 581.
484. Hayashi, I., and Ohno, K. (1965). *Chem. High. Polym.* (*Japan*), *22*: 446.
485. Piloz, A., Pham, W. T., Deiroix, J. Y., and Gullot, J. (1975). *J. Macromol. Sci. Chem.*, *A9*: 517.
486. Shteinbak, V. S., Amerik, V. V., Yakobson, F. I., Kissin, Y. V., Ivanyukov, D. V., and Krentsel, B. A. (1975). *Eur. Polym. J.*, *11*: 457.
487. Ashikari, N., Kanemitsu, T., Yanagisawa, K., Nakagawa, K., Okamoto, H., Kobayashi, S., and Nishioka, A. (1964). *J. Polym. Sci.*, *Part A2*: 3009.
488. Yakobson, F. I., Amerik, V. V., Ivanyukov, D. V., Petrova, V. F., Kissin, Y. V., and Krentsel, B. A. (1971). *Vysokomol Soedin, A* 13: 2699.
489. Wohlfarth, L., Frank, W., and Arnold, M. (1991). *Plaste u. Kautschuk*, *38*: 369.
490. Furukawa, J., Kobayashi, E., and Haga, K. (1973). *J. Polym. Sci., Part A-1*, *11*: 629.
491. Brüll, R., Pasch, H., Raubenheimer, H. G., Sanderson, R., v. Reenen, A. J., and Wahner, U. M. (2001). *Macromol. Chem. Phys.*, *202*: 1281.
492. Kaminsky, W. (2001). *Polymeric Materials, Sci. A. Eng.*, *84*: 31.
493. Kono, H., Mori, H., and Terano, M. (2001). *Macromol. Chem. Phys.*, *202*: 1319.
494. Arndt, M., Kaminsky, W., Schauwienold, A.-M., and Weingarten, U. (1998). *Macromol. Chem. Phys.*, *199*: 1135.
495. Waymouth, R. M. (1998). *Angew. Chem.*, *110*: 964.
496. Coates, G. W., and Waymouth, R. M. (1995). *Science*, *257*: 217.
497. Arnold, M., Henschke, O., and Köller, E. (1996). *Macromol. Reports*, *A 33*: 219.
498. Henschke, O., Knorr, I., and Arnold, M. (1998). *J. Macromol. Sci., Pure Appl. Chem.*, *A 35*: 473.
499. Kaminsky, W., and Dögemüller, H. (1989). In: *Polymer Reaction Engineering* (Reichert, K.-H., and Geiseler, W., eds.), VCH-Press, Weinheim.
500. Kaminsky, W., and Drögemüller, H. (1990). *Makromol. Chem. Rapid Commun.*, *11*: 89.
501. Heggs, T. G. (1973). In: *Block Copolymers* (Allport, D. C., and Janes, W. H., eds.), Wiley, New York, p. 513.
502. Luongo, J. P. (1960). *J. Appl. Polym. Sci.*, *3*: 302.
503. Burfield, D. R., and Loi, P. S. T. (1988). *J. Appl. Polym. Sci.*, *36*: 279.
504. Ruland, W. (1987). *Macromolecules*, *20*: 87.
505. Zambelli, A. (1975). *Macromolecules*, *8*: 687.
506. Randall, J. C. (1974). *J. Polym. Sci., Part A-2*, *12*: 703.

507. Rohe, D. (1986). *Chem. Ind.*, *10*: 873.
508. Fink, H. W. (1974). *Kunststoffe*, *64*: 64.
509. Lovinger, A. J., and Williams, M. L. (1980). *J. Appl. Polym. Sci.*, *25*: 1703.
510. Kojima, M., and Satake, H. (1984). *J. Polym. Sci., Polym. Phys. Ed.*, *22*: 285.
511. Lohse, D. J. (1986). *Polym. Eng. Sci.*, *26*: 1500.
512. Döring, E. (1981). *Plastverarbeiter*, *32*: 1629.
513. Theberge, J. E. (1981). *Polym. Plast. Technol. Eng.*, *16*: 41.
514. Leatherman, M. D., and Brookhart, M. (2001). *Macromolecules*, *34*: 2748.
515. Davis, J. C. (1973). *Chem. Eng.*, 32.
516. Foglia, A. J. (1969). *Appl. Polym. Symp.*, *11*: 1.
517. Rodionov, A. G., Baulin, A. A., and Ivanchev, S. S. (1981). *Intern. Polym. Sci. Technol.*, *8*: 38.
518. Soga, K., Ohtake, M., and Doi, Y. (1985). *Makromol. Chem.*, *186*: 1129.
519. Kaminsky, W., and Schlobohm, M. (1986). *Makromol. Chem. Symp.*, *4*: 103.
520. Goldbach, G., and Peitscher, G. (1968). *Polymer Letters*, *6*: 783.
521. Korcz, W. H. (1984). *Adhesives Age, Nov 84*: 19.
522. Rangwala, H. A., Dalla Lana, I. G., Szymura, J. A., Fiederow (2000). *J. Polym. Sci., Polym. Chem.*, *34*: 3379.
523. Cortazar, M., and Guzman, G. M. (1981). *Polym. Bull.*, *5*: 635.
524. Heiland, K., and Kaminsky, W. (1992). *Makromol. Chem.*, *193*: 601.
525. Turner-Jones, A., and Aizlewood, J. M. (1963). *Polymer Letters*, *1*: 471.
526. Fujita, T. (1978). *Pure Appl. Chem.*, *50*: 987.
527. Campbell, T. W., and Haven, A. C. (1959). *J. Appl. Polym. Sci.*, *1*: 73.
528. Raine, H. C. (1969). *Appl. Polym. Sym.*, *11*: 39.
529. Westall, J. (1970). *Mod. Plast. Ency.*, *47, No. 10A*: 152.
530. Danusso, F., and Sianesi, D. (1962). *Chim. Ind. (Milan)*, *44, No. 5*: 474.
531. Vandenberg, E. J. (1977). *Macromol. Synth.*, *6*: 39.
532. Soga, K., and Monoi, T. (1989). *Macromolecules*, *22*: 3823.
533. Soga, K., Yu, C. H., and Shiono, T. (1988). *Makromol. Chem. Rapid Commun.*, *9*: 351.
534. Grassi, A., Longo, P., Proto, A., and Zambelli, A. (1989). *Macromolecules*, *22*: 104.
535. Ishihara, N., Kuramoto, M., and Uoi, M. (1988). *Macromolecules*, *21*: 3356.
536. Dall'Asta, G., and Matroni, G. (1971). *Makromol. Chem.*, *16/17*: 51.
537. Calderon, N. (1972). *J. Makromol. Sci., Rev. Macromol. Chem.*, *C 7*: 105.
538. Dall'Asta, G., Manetti (1968). *Eur. Polym. J.*, *4*: 145.
539. Arndt, M., and Kaminsky, W. (1995). *Macromol. Symp.*, *95*: 167.
540. Boor, J., Youngman, E. H., and Dimbat, M. (1966). *Makromol. Chem.*, *90*: 26.
541. Dall'Asta, G., Mazzanti, G., Natta, G., and Porri, L. (1963). *Makromol. Chem.*, *56*: 224.
542. Natta, G., Dall'Asta, G., Mazzanti, G., and Matroni, G. (1963). *Makromol. Chem.*, *69*: 163.
543. Natta, G., Dall'Asta, G., Bassi, I., and Carella, G. (1966). *Makromol. Chem.*, *91*: 87.
544. Collins, S., and Kelly, W. M. (1992). *Macromolecules*, *25*: 233.
545. Wiberg, K. B., and Bartley, W. J. (1960). *J. Am. Chem. Soc.*, *82*: 6375.
546. Natta, G., Dall'Asta, G., and Porri, L. (1965). *Makromol. Chem.*, *81*: 253.
547. Dall'Asta, G. (1964). *Chim. Ind. (Milan)*, *46*: 1525.
548. Kaminsky, W., Bark, A., and Däke, I. (1989). *Intern. Symp. on 'Recent Development in Olefin Polymerization Catalysts'*, Oct. 23–26, Tokyo.
549. Breuning, S., and Risse, W. (1992). *Makromol. Chem.*, *193*: 2915.
550. Goodall, B. L., McIntosh, L., and Rhodes, L. F. (1995). *Makromol. Symp.*, *89*: 421.
551. Gosmann, M. (2000). Thesis, University of Hamburg.
552. Heitz, W., Haselwander, T. F. A., and Krügel, S. A. (1996). *Makromol. Chem. Phys.*, *197*: 3435.
553. Goodall, B. L., Barnes, D. A., Benedikt, G. M., McIntosh, L. H., and Rhodes, L. F. (1997). *Polym. Mat. Sci. Eng.*, *76*: 56.
554. Arndt, M., and Gosmann, M. (1998). *Polymer Bulletin*, *41*: 433.

555. Tsutsui, T., Kioka, M., and Kashiwa, T. (1986). Jap. Pat. 61-221206 to Mitsui, C.A. (1987) 106: 85227.
556. Ullmann (1980), *Enzyklopädie der technischen Chemie* (Bartholomä, E., Biekert, E., Hellmann, H., Lei, H., Weigert, W. M., and Wiese, E., eds.), Vol. 19, Verlag Chemie, Weinheim, p. 216.
557. Bacskai, R., and Lapporte, S. J. (1963). *J. Polym. Sci., Part A, 1*: 2225.
558. Meurer, P. (1987). *Makromolekulare Stoffe* (Houbel-Weyl), Vol. 20.2, Georg Thieme Press, Stuttgart.
559. Lebedew, S. W., and Filonenko, E. P. (1925). *Ber. Dtsch. Chem. Ges., 58*: 163.
560. Ferraris, E. (1965). *Mater. Plast. Elastomeri, 31*: 1325.
561. Glaser, R., and Schmidt, E. (1973). *Kunststoffe, 62*: 610.
562. Roundet, J., and Gandini, A. (1989). *Makromol. Chem. Rapid Commun., 10*: 277.
563. Kennedy, J. P., and Marechal, E. (1982). *Carbocationic Polymerization,* J. Wiley & Sons, New York.
564. Mayr, H., Schneider, R., and Pork, R. (1986). *Makromol. Chem. Makromol. Symp., 3*: 19.
565. Carr, A. G., Dawson, D. M., and Bochmann, M. (1998). *Macromolecules, 31*: 2035.
566. Hasenbein, N., and Bandermann, F. (1987). *Makromol. Chem., 188*: 83.
567. Koroskenyi, B., Wang, L., and Faust, R. (1997). *Macromolecules, 30*: 7667.
568. Faust, R, and Kennedy, J. P. (1987). *J. Polym. Sci, Polym. Chem. Ed., 25*: 1847.
569. Kuntz, I., Cheng, M. D., Dekmezian, A. H., and Chang, S. (1987). *J. Polym. Sci., Part A, Polym. Chem., 25*: 3127.
570. Zsuga, M., Kennedy, J. P., and Kelen, T. (1988). *Polym. Bull., 19*: 427.
571. Nesmelov, A. I., Muracher, V. B., Eshova, E. A., and Byrikhin, V. S. (1988). *Vysokomol. Soedin Ser. A, 30*: 1957.
572. Barsan, F., and Baird, M. C. (1995). *J. Chem. Soc. Chem. Commun.,* 1065.
573. Marek, M., Pecka, J., and Halaska, V. (1988). *Makromol. Chem. Macromol. Symp., 13,14*: 443.
574. Kennedy, J. P. (1975). *Cationic Polymerization of Olefin. A Critical Inventory,* J. Wiley and Sons, New York.
575. Kennedy, J. P., and Thomas, R. M. (1962). *Adv. Chem. Ser., 34*: 111.
576. Imanishi, Y., Yamamoto, R., Higashimura, T., Kennedy, J. P., and Okamura, S. (1967). *J. Macromol. Sci. Chem., A 1*: 877.
577. Okamura, S., Higashimura, T., and Takeda, T. (1961). *Kobunshi Kagaku, 18*: 389.
578. Overberger, C. G., and Kamath, V. G. (1963). *J. Am. Chem. Soc., 85*: 446.
579. Okamura, S., Higashimura, T., Imanishi, Y., Yamamoto, R., and Kumura, K. (1967). *J. Polym. Sci., Part C, 16*: 2365.
580. Marechal, E., Menissez, J.-P., Richard, J. P., and Zaffran, C. (1968). *C.R. Acad Sci., Ser. C 266*: 1427.
581. Kennedy, J. P., and Canter, N. H. (1967). *J. Polym. Sci., Part A-1, 5*: 2712.
582. Priola, A., Cesca, S., Ferraris, G., and Bruzzone, M. (1975). *Makromol. Chem., 176*: 1969.
583. Imanishi, Y., Yamane, T., Momiyama, Z., and Higashimura, T. (1966). *Kobunshi Kagaku, 23*: 152.
584. Roundet, J., and Gandini, A. (1989). *Makromol. Chem. Rapid Commun., 10*: 277.
585. Kuntz, I., and Rose, K. D. (1989). *J. Polym. Sci., Polym. Chem. Ed., 27*: 107.
586. Sawamoto, M., and Higashimura, T. (1986). *Makromol. Chem. Macromol. Symp., 3*: 83.
587. Kennedy, J. P., and Gillham, J. K. (1972). *Adv. Polym. Sci., 10*: 1.
588. Priola, A., Ferraris, G., Maina, M. D., and Giusti, P. (1975). *Makromol. Chem., 176*: 2271.
589. Cesca, S., Priola, A., Ferraris, G., Busetto, Co., and Bruzzone, M. (1976). *J. Polym. Sci., Polym. Symp., 56*: 159.
590. Kennedy, J. P., and Squires, R. G. (1967). *J. Makromol. Sci. Chem., 1*: 861.
591. Kaszas, G., Györ, M., Kennedy, J. P., and Tüdös, F. (1982), *J. Makromol. Sci. Chem., 18*: 1367.
592. Heublein, G., Knöppel, G., and Winnefeld, I. (1987). *Acta Polym., 38*: 10.
593. Kennedy, J. P., and Chou, T. (1976). *Adv. Polym. Sci., 21*: 1.

594. Iino, M., and Tokura, N. (1964). *Bull. Chem. Soc. Jpn.*, *37*: 23.
595. Overberger, C. G., and Kamath, V. G. (1959). *J. Am. Chem. Soc.*, *81*: 2910.
596. Kennedy, J. P., and Smith, R. A. (1980). *J. Polym. Sci., Polym. Chem. Ed.*, *18*: 1523.
597. Heublein, G., Rottmayer, H. H., Stadermann, D., and Vogel, J. (1987). *Acta Polym.*, *38*: 559.
598. Fodor, Z., Kennedy, J. P., Kelen, T., and Tüdös, F. (1987). *J. Macromol. Sci. Chem.*, A 24: 735.
599. Kennedy, J. P., and Melby, E. G. (1974). *Prepr. Am. Chem. Soc. Div. Polym. Chem.*, *15*: 180.
600. Kennedy, J. P., Huang, S. Y., and Smith, R. A. (1980). *J. Macromol. Sci. Chem.*, *14*: 1085.
601. Speckhard, T. A., Verstrate, G., Gibson, P. E., and Cooper, S. L. (1983). *Polym. Eng. Sci.*, *23*: 337.
602. Percec, V., Guhaniyogi, S. C., Kennedy, J. P., and Ivan, B. (1982). *Polym. Bull.*, *8*: 25.
603. Kennedy, J. P. (1977). *J. Appl. Polym. Sci., Appl. Polym. Symp.*, *30*: 1.
604. Jiang, Y., and Jean, M. J. (1989). *Polym. Prepr. Am. Chem. Soc. Div. Polym. Chem.*, *30*: 127.
605. Sigwalt, P., Polton, A., and Miskovic, M. (1976). *J. Polym. Sci., Polym. Symp.*, *56*: 13.
606. Chapiro, A. (1978). *J. Polym. Sci., Polym. Symp.*, *56*: 431.
607. Nuyken, O., and Pask. S. D. (1989). In: *Comprehensive Polymer Science* (Allen, G., and Bevington, J. C., eds.), Pergamon, New York, p. 619.
608. Kennedy, J. P. (1984). *J. Appl. Polym. Sci., Appl. Polym. Symp.*, *39*: 21.
609. Zsuga, M., Faust, R., and Kennedy, J. P. (1989). *Polym. Bull.*, *21*: 273.
610. Zsuga, N., Kennedy, J. P., and Kelen, T. (1989). *J. Macromol. Sci. Chem.*, A 25: 1305.
611. Dittmer, T., Gruber, F., and Nuyken, O. (1989). *Macromol. Chem.*, *190*: 1755.
612. Dittmer, T., Nuyken, O., and Pask, S. D. (1988). *J. Chem. Soc. Perkin Trans. II, 1988*: 151.
613. Van Lohnizen, O. E., and De Vries, K. S. (1968). *J. Polym. Sci., Part C, 16*: 3943.
614. Pinazzi, C., and Brossas, J. (1965). *C.R. Acad. Sci. Compt. Rend.*: 261; *C.A.* (1966) *64*: 6765.
615. Ketley, A. D., Berlin, A. J., and Fisher, L. P. (1967). *J. Polym. Sci., Part A-1, 5*: 227.
616. Kennedy, J. I., Elliot, J. J., and Butler, P. E. (1968). *J. Macromol. Chem.*, *2*: 1415.
617. Roberts, W. J., and Day, A. R. (1951). *J. Am. Chem. Soc.*, *72*: 1226.
618. Zlamal, Z., Katzka, A., and Ambroz, L. (1966). *J. Polym. Sci., Part A 1, 4*: 367.
619. Marvel, C. S., Hassley, J. R., and Longone, D. T. (1959). *J. Polym. Sci.*, *40*: 551.
620. Ahn, C.-H., Tahara, M., Uozumi, T., Jin, J., Tsubaki, S., Sano, T., and Soga, K. (2000). *Macromol. Rapid Commun.*, *21*: 385.
621. Hofmann, F. C. (1933). *Chem. Ztg.*, *57*: 5.
622. Kennedy, J. P., and Makowski, H. S. (1967). *J. Macromol. Sci. Chem.*, *1*: 345.
623. Saback, M. B., and Farona, M. F. (1987). *Polym. Bull.*, *18*: 441.
624. Yoshida. Y., Mohri, J., and Fujita, T. (2004). *J. Am. Chem. Soc.*, *126*: 12023.

2
Polystyrenes and Other Aromatic Poly(vinyl compound)s

Oskar Nuyken
Technische Universität München, Garching, Germany

I. STYRENE

Discovery of the styrene monomer is credited to Newman [1] who isolated it by steam distillation from liquid amber, which contains cinnamic acid, yielding styrene via decarboxylation. The first polymerization was described by Simon [2]. Commercial styrene polymerization was begun about 1925 [3]. Cracking of ethylbenzene became the major manufacturing route for the monomer. The first commercialization was based on bulk polymerization using the can process [4,5]. Polystyrene production has grown by an average of 2.8% per year over the last 10 years, reaching 8 500 000 metric tons worldwide in 1995 [6]. The most general-purpose polystyrene is produced by solution polymerization in a continuous process with the aid of peroxide initiation. Suspension polymerization is used for products for which a small spherical form is desirable. Emulsion polymerization is the method of choice for ABS resins.

Polystyrene is a glasslike solid below 100 °C. Below this temperature it shows considerable mechanical strength. Rubber-modified polystyrene is a two-phase system, rubber dispersed in polystyrene being the continuous phase. Advantage is taken of the complex interaction of those systems in many applications in which high stress-crack resistance is needed. Polystyrene is nonpolar, chemically inert, resistant to water, and easy to process. It is the material of choice for many food-packing, optical, electronic, medical, and automotive applications. Tensile strength can be increased by controlled orientation of polystyrene.

A. Synthesis [5,7]

All current styrene production starts with ethylbenzene followed by dehydrogenation over ferrum oxide catalysts yielding crude styrene:

$$\text{C}_6\text{H}_5-\text{CH}_2-\text{CH}_3 \xrightleftharpoons{600\text{–}650\ ^{\circ}\text{C}} \text{C}_6\text{H}_5-\text{CH}=\text{CH}_2 \;+\; \text{H}_2 \tag{1}$$

Side reactions are reduced by keeping the conversion low or by adding water as a diluent. It is also possible to synthesize styrene by the oxidation of ethylbenzene:

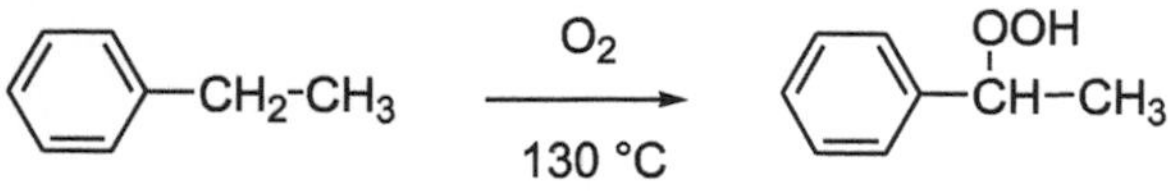

$$\text{Ph-CH}_2\text{-CH}_3 \xrightarrow[130\ °C]{O_2} \text{Ph-}\underset{\text{OOH}}{\text{CH}}\text{-CH}_3 \tag{2}$$

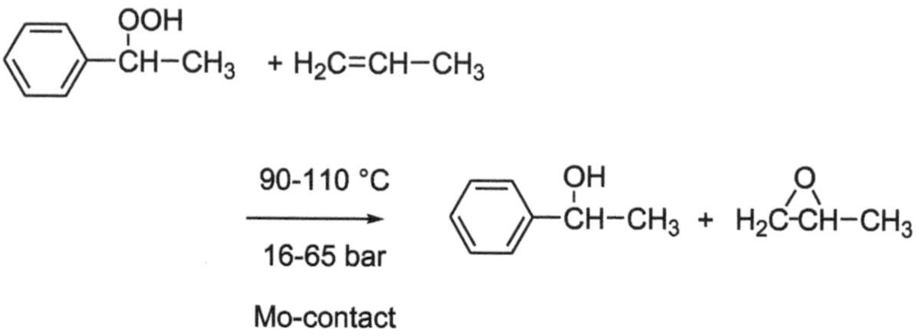

$$\text{Ph-}\underset{\text{OOH}}{\text{CH}}\text{-CH}_3 \ +\ H_2C=CH-CH_3$$

$$\xrightarrow[\substack{16\text{-}65\ \text{bar} \\ \text{Mo-contact}}]{90\text{-}110\ °C} \text{Ph-}\underset{\text{OH}}{\text{CH}}\text{-CH}_3 \ +\ H_2\text{C-CH-CH}_3 \tag{3}$$

$$\text{Ph-}\underset{\text{OH}}{\text{CH}}\text{-CH}_3 \longrightarrow \text{Ph-CH=CH}_2 \ +\ H_2O \tag{4}$$

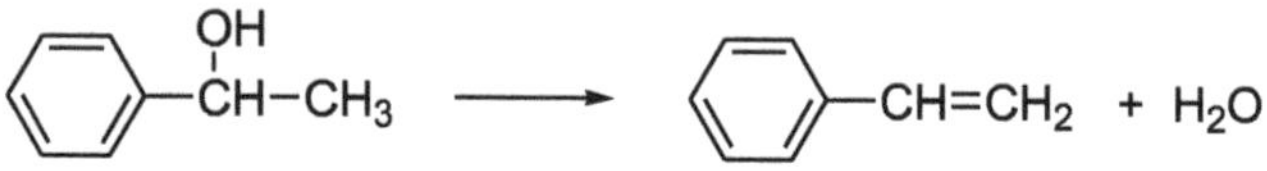

An alternative pathway starts from toluene and ethene:

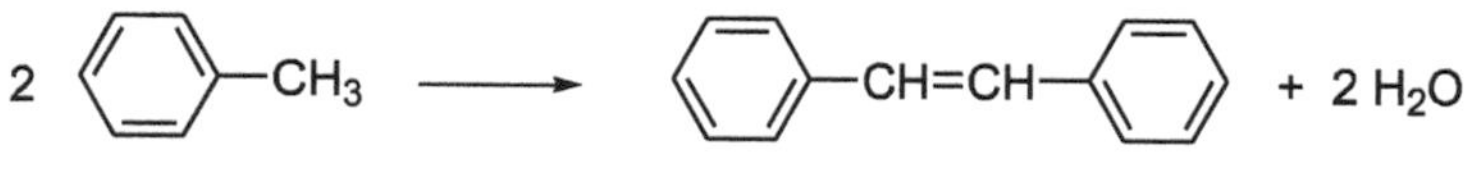

$$2\ \text{Ph-CH}_3 \longrightarrow \text{Ph-CH=CH-Ph} \ +\ 2\ H_2O \tag{5}$$

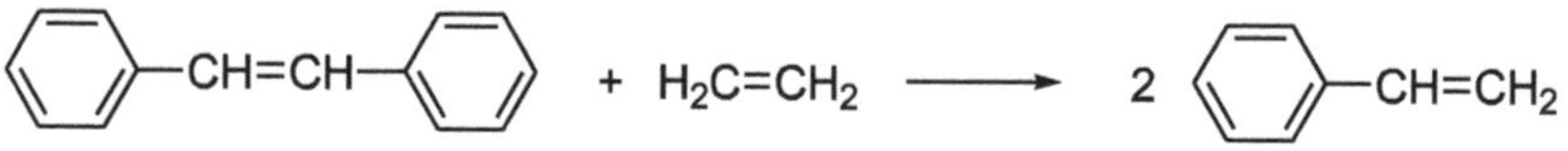

$$\text{Ph-CH=CH-Ph} \ +\ H_2C=CH_2 \longrightarrow 2\ \text{Ph-CH=CH}_2 \tag{6}$$

It is also possible to synthesize styrene from butadiene:

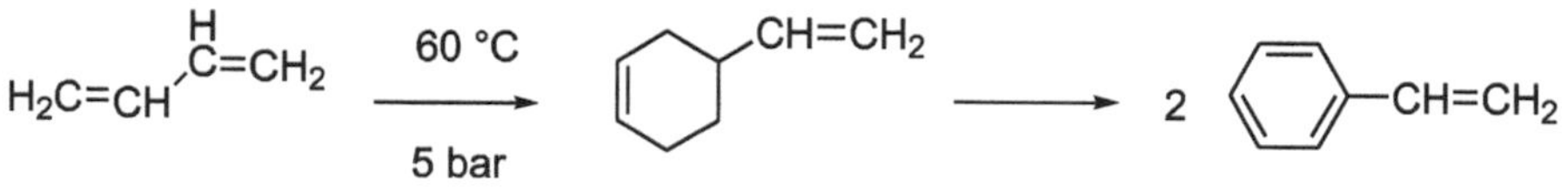

$$H_2C=CH-\underset{H}{\overset{}{C}}=CH_2 \xrightarrow[5\ \text{bar}]{60\ °C} \text{(cyclohexene-CH=CH}_2) \longrightarrow 2\ \text{Ph-CH=CH}_2 \tag{7}$$

Physical properties and health and safety information for styrene are given in the literature [8].

B. Homopolymerization

The C–C double bond of styrene can act either as electron-donating or as electron-withdrawing center. Therefore, it cannot only be polymerized by radicals but also anionically or cationically or by coordination initiators.

1. Radical Polymerization

Styrene can be polymerized without a chemical initiator simply by heating (spontanous polymerization). The first step is a Diels–Alder reaction.

(8)

axial-1-phenyltetraline
equatorial-1-phenyltetraline

Only the *axial* isomer can react with a further styrene atom, yielding two radicals, which can start a radical polymerization:

(9)

Some of the possible dimers, including 1-phenyltetraline, and a trimer have been identified [9–11]:

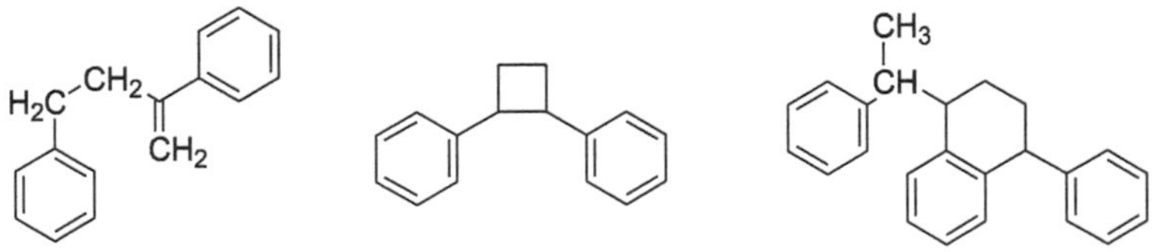

(10)

This type of polymerization requires third-order initiation kinetics [12,13]:

$$R_i = k_d[M]^3$$

Consequently, the rate of propagation should be of 5/2 order with respect to styrene [13,14]:

$$R_p = k_p \left(\frac{k_d}{2k_t} \right)^{1/2} [M]^{5/2}$$

However, the reaction does not follow a simple kinetic order with respect to monomer over the entire range of conversion and temperature. The order is negative below 50 °C. At 75 °C the polymerization is zero order in monomer for the first 65% conversion. At 127 °C the polymerization is first order for the first 85%. Only above 200 °C does the polymerization follow theory [15].

   a. Initiators. The list of initiators available for radical polymerization of styrene is very long [16–18], including azo compounds, peroxides, redox systems and many more. An interesting development is the application of initiators like

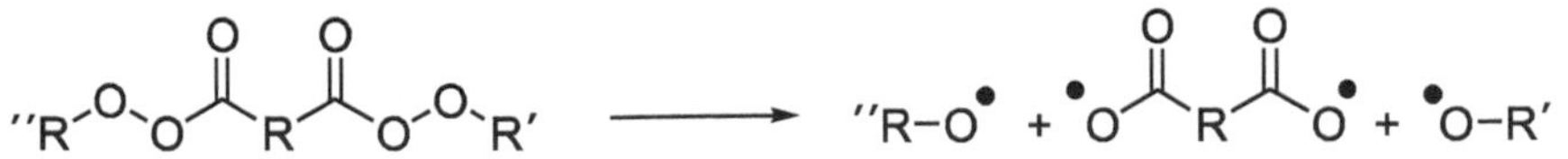

(11)

which decompose to form four radicals. It is even more interesting to have a different half-life for both peroxide groups. This presents novel opportunities for changing the molecular weight and its distribution [19,20].

Table 1 Chain transfer constants in styrene polymerization [28].

Compound	C_{tr}	$T/°C$
Benzene	2×10^{-5}	100
Toluene	5×10^{-5}	100
Ethylbenzene	1.4×10^{-4}	100
Isopropylbenzene	2×10^{-4}	100
Styrene	5×10^{-5}	100
Dichloromethane	1.5×10^{-5}	60
Chloroform	5×10^{-6}	60
Tetrachloromethane	1.8×10^{-2}	100
1-Dodecanthiol	1.3×10^{1}	130
1-Hexanthiol	1.5×10^{1}	100
Ethylthioglyconate	5.8×10^{1}	60

b. Inhibitors. During shipping and storage styrene needs an inhibitor. The most efficient inhibitors—quinones, hindered phenols, and amines [21]—require traces of oxygen to function. *t*-Butyl-catechol at 15 to 50 ppm is the most common inhibitor for commercial styrene [22]. It is also possible to use nitrophenol, hydroxylamine, and nitrogen oxide compounds [23]. The inhibitors have to be removed before polymerization, in order to avoid an induction period.

Traces of metal such as iron or copper [24] and sulfur compounds [25] are the cause of retardation effects in styrene polymerization.

c. Chain Transfer. In styrene polymerization the chain transfer agent can be the solvent, monomer, initiator, polymer, or an added chemical agent. As $C_{tr} = k_p/k_{tr}$ increases, the chain transfer agent becomes more effective. Some examples are given in Table 1. The most important property affected by chain transfer is the molecular weight of the polymer. The transfer to monomer has a value of 10^{-5} which can be neglected. However, since the transfer constant to the Diels–Alder dimer *axial*-1-phenyltetralin is about 113 at 80 °C [26], this may cause experimental error. Any transfer to polymer would lead to branched structures in the final product. Although this reaction has been investigated to some extent, there is no conclusive evidence that it is an important reaction [27]. The most important aspect of chain transfer is the control of molecular weight by the adequate use of added transfer agent. Mercaptanes are by far the most widely used chemicals for this purpose.

d. Termination Reactions. The free-radical polymerization of styrene is terminated almost exclusively by the combination of two growing chains [29,30]:

$$(12)$$

Termination is diffusion controlled at all temperatures below 150 °C [31,32]. Increasing viscosity leads to a reduction in the termination rate [33]. However, the resulting Trommsdorff effect is comparably small for polystyrene [22].

e. Processing [34]. Free-radical polystyrene can be synthesized either by bulk, solution, suspension, or emulsion techniques. Techniques for preparing polystyrene on a laboratory scale are described in detail in Refs. [35–37]. The bulk process needs pure styrene; it is very simple and yields polymers with high clarity. Due to its poor control, this process is not used commercially. In solution polymerization styrene is diluted with solvents, which makes temperature control easier. However, solvents normally reduce the molecular weight and polymerization rate. Both processes can be carried out either in batch or continuously. The advantages are more uniform products and low volatile levels. The main disadvantage is the transportation of highly viscous finished product.

Suspension polymerization is still an important mode of polystyrene production, although it has lost ground to continuous solution polymerization. The polymerization system contains monomer suspended in water, stabilizing agents, and initiators to speed polymerization. The easy heat control and removal of the finished polymer count as advantages. Contamination with stabilizing agents is considered a disadvantage.

Emulsion polymerization requires water as a carrier with emulsifying agents. It yields extremely small particles. Advantages are rapid reactions and excellent heat control. Disadvantages are the contamination of polymer with the emulsifier, water, its deficit in clarity, and the limitation to batch processing. However, this type of processing is important for ABS polymers.

2. Controlled Radical Polymerization

The controlled radical polymerization combines the advantages of living ionic systems, as there are narrow molecular weight distributions, linear increase of the DP with the reaction time and the possibility of the formation of block copolymers, with the main advantage of the radical polymerization, the low sensitivity against impurities. The general idea of controlled radical polymerization is to avoid the bimolecular, irreversible termination reactions, typically obtained in a free radical polymerization (combination, disproportionation etc.) by decreasing the number of growing radical chains. Thus, although the reaction itself becomes comparably slow, the molecular mass can be very well controlled and very narrow molecular weight distributions can be obtained.

Early attempts to realize the controlled radical polymerization of styrene involved the concept of reversible termination of growing polymer chains by iniferters (*ini*tiation, trans*fer*, *ter*mination) [38]. These iniferters based on dithiocarbamates were the first species with photochemically labile C–S bonds.

(13)

Another way of reversible termination was introduced by the same group [39,40]. They showed, that at the decomposition of phenylazotriphenylmethane both a phenyl and a trityl radical are generated. The phenyl radical initiates polymerization, while the trityl radical does not, due to its mesomeric stabilization.

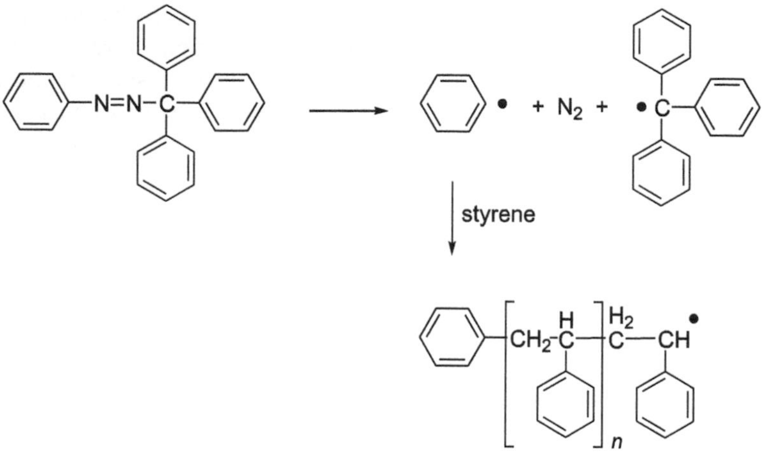

(14)

Instead, the trityl radical acts as a radical trap and efficiently terminates polymerization by primary radical coupling. As a result of steric crowding between the pendant groups on the polymer chain and the phenyl groups of the trityl moiety, as much as a result of the stability of the triphenylmethylradical, the C–C bond can redissociate at elevated temperature and add more monomer.

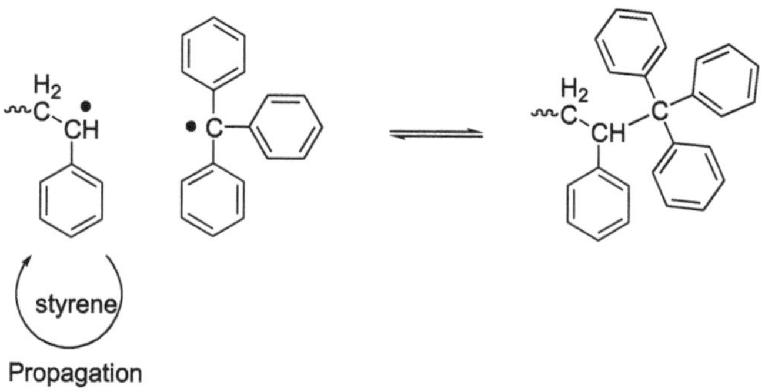

(15)

Propagation

Following this approach, Rizzardo et al. and Georges et al. introduced the use of stable nitroxide free radicals, such as 2,2,6,6-tetramethylpiperidinyloxy (TEMPO), as reversible terminating agents to cap the growing polymer chain [41,42].

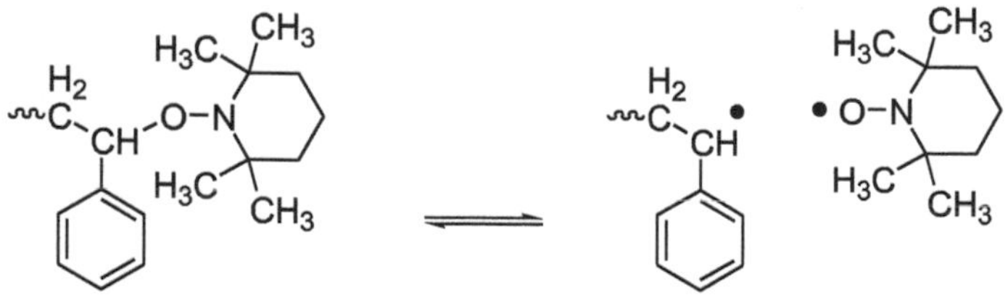

(16)

It has been demonstrated, that at elevated temperatures narrow molecular weight distribution polystyrene (PDI = 1.1–1.3) could be prepared using bulk polymerization conditions. In the polymerization of styrene, temperatures around 120 °C are required in order to obtain a sufficient rate of monomer insertion, because of the stability of the C–O bond.

A very similar approach is the use of triazolinyl counter radicals as an alternative to the nitroxides [43,44]. The electron spin density is not localized, as in the case of TEMPO, but delocalized in a extended π system.

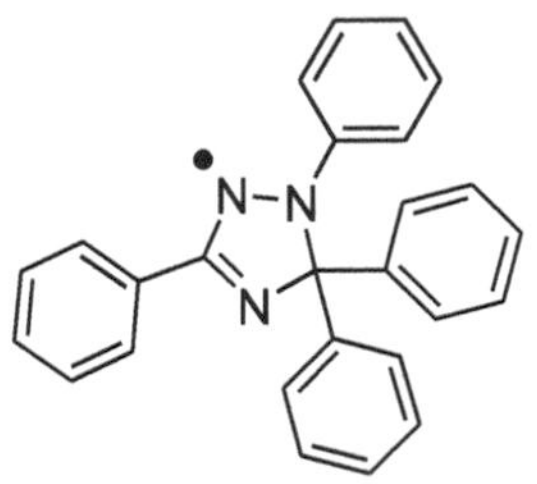

(17)

With a polymerization temperature of 120 °C, a three-fold higher polymerization rate in comparison to TEMPO could be obtained in styrene polymerization. Furthermore, in contrast to TEMPO mediated polymerization, polymers up to a molecular weight of 100 000 g/mol can be obtained in good yields. The mechanism of the polymerization process is not identical to the TEMPO mediated polymerization, but the control is introduced by a self regulation process [45]. In 1995, a further approach to controlled radical polymerization, the Atom Transfer Radical Polymerization (ATRP) was independently reported by Matyjaszewski [46] and Sawamoto [47]. These systems are based on the dynamic equilibrium of a reversible redox reaction between halogen endgroups of the polymers and transition metal catalysts. The catalysts are mainly Cu(I), Fe(II) or Ru(II) complexes with different ligands. The copper based systems usually contain nitrogen ligands like bipyridines, multidentate amines and Schiff bases. A general review over copper mediated ATRP is given in Ref. [48]. The ruthenium based catalysts show a wide variety of structures with arenes, phosphines and halogens as ligands [49]. Iron based catalysts are also applied, most of them are containing bipyridines, trialkylamines, phosphines or phosphites [50].

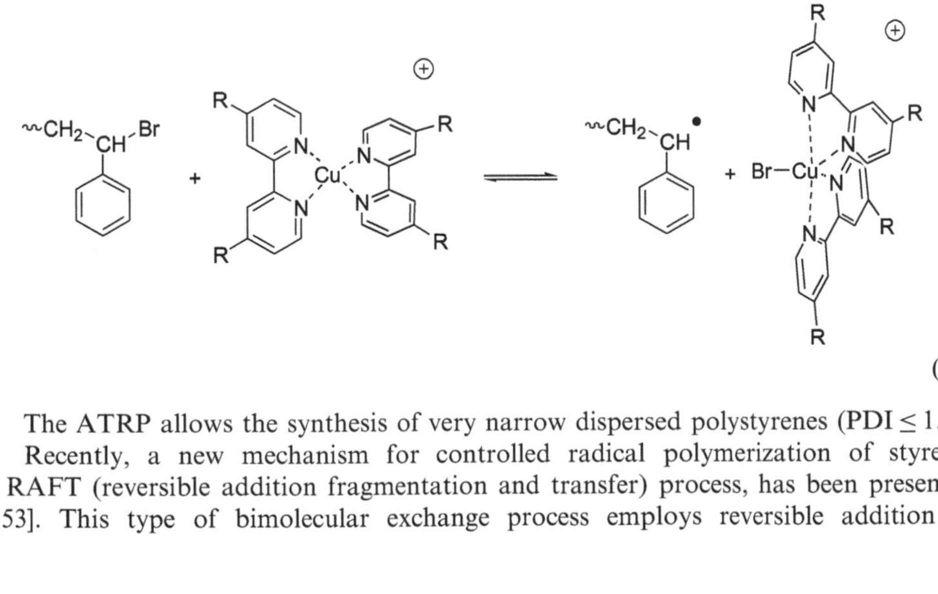

(18)

The ATRP allows the synthesis of very narrow dispersed polystyrenes (PDI ≤ 1.1).

Recently, a new mechanism for controlled radical polymerization of styrene, the RAFT (reversible addition fragmentation and transfer) process, has been presented [51,53]. This type of bimolecular exchange process employs reversible addition of the radicals to a nonpolymerizable double bond. The RAFT process is best represented by the use of several dithioesters as transfer reagents and the mechanism can be divided in three main steps. The addition of a growing polymer chain to the transfer reagent with subsequently homolytic fragmentation of the S–R bond (transfer) is the first step:

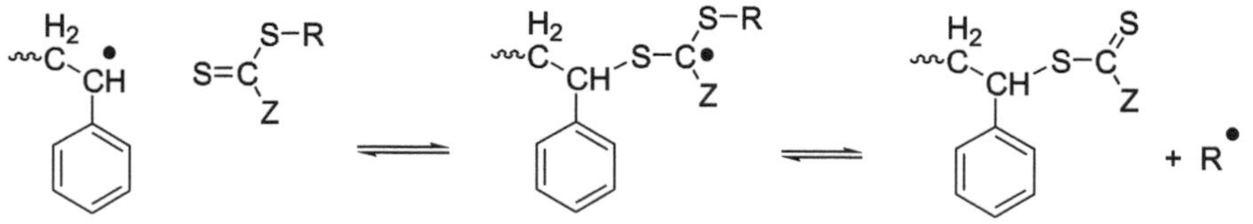

(19)

The reinitiation of a new active polymer chain with the resulting radical R• follows.

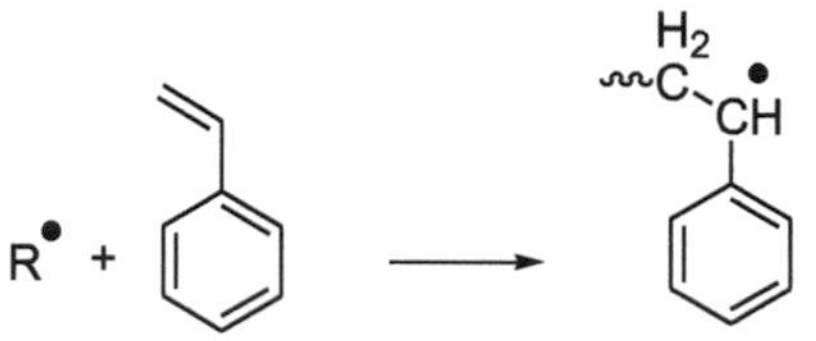

(20)

Subsequent addition-fragmentation steps set up an equilibrium between the two propagating radicals and the dormant polymeric dithiocarbonylthio compound by way of an intermediate stabilized radical. Throughout the polymerization (and at the end) the vast majority of the polymer chains are end capped by a thiocarbonylthio group (dormant chains).

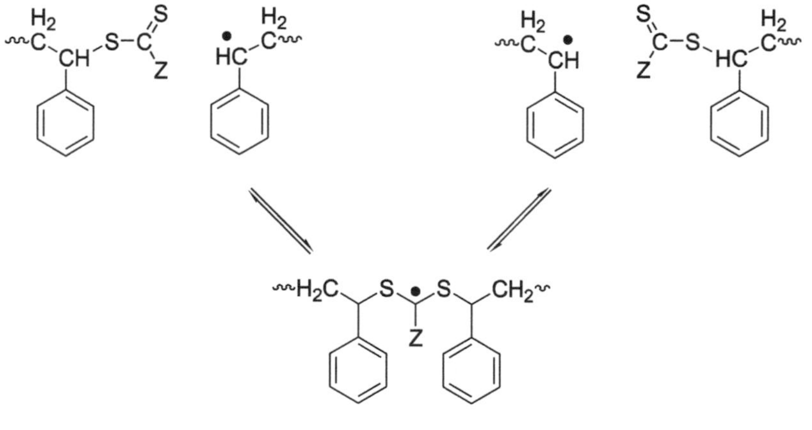

(21)

In general, every kind of monomer needs a different dithioester for best results, whereby the most suitable compound for styrene contains phenyl for Z and 2-phenyl-propyl for R. Polystyrene with a PDI down to 1.07 can be obtained by use of this compound [52].

3. Anionic Polymerization

The phenyl group of styrene is able to act as an electron-donating or an electron-withdrawing center. This situation allows the growing end of the polymer to be either a carbeniumion or a carbanion, as shown in more detail in this chapter.

   a.   Initiation.   A highly purified monomer is reacted with a strong base. Although several initiators are known, organolithium compounds are the most studied and probably the best understood initiators [54,55].

$$\tag{22}$$

   This initiation is much faster than the propagation step. All styryl anions are therefore formed almost instantaneously. Since no termination occurs, the degree of polymerization can be calculated easily on the basis of the following equation:

$$\overline{DP}_n = \frac{[M]}{[I]}$$

   Furthermore, one can observe that

$$\overline{DP}_n \approx \overline{DP}_w \approx \overline{DP}_z$$

   This has been discussed in detail in several reviews [56–58].
   b.   Propagation.   The ideal polymerization of this type (the cationic polymerization follows the same kinetics) obeys the following equation:

$$R_p = k_p \cdot [C^{\pm}] \cdot [M]$$

where $C^{\pm}$ represents the molar concentration of active ionic chain ends. The rate constant is strongly affected by the solvent [59]: for example k_p is $2\,L\,mol^{-1}s^{-1}$ in benzene compared to $3800\,L\,mol^{-1}s^{-1}$ in 1,2-dimethoxyethane at 25 °C. In addition to solvent, the counterion affects the rate of polymerization. The effect of the counterions is often explained on the basis of their sizes (e.g., increasing solvation with decreasing size yields a greater concentration of free ions and higher polymerization rates) [60]. For the growing end of poly(styryllithium), an association of two growing chains has been discussed [61,62]. This complex dissociates if polar solvents such as THF and diethylether are added, resulting in an increase in the polymerization rate. Instead of using a monofunctional initiator, it is possible to use a bifunctional anionic initiator.
   One of the best described systems involves the reaction between sodium and naphthalene, forming a radical anion that transfers this character to the monomer. The two radical anions combine quickly to form a dianion [57].

$$\tag{23}$$

$$(24)$$

$$(25)$$

$$(26)$$

Another bifunctional initiator is formed by the reaction of 1,3-bis(1-phenylvinyl)-benzene with organolithium, yielding a dianion with good solubility in hydrocarbons [63–65]:

$$(27)$$

A similar system was described by Guyot et al. [66]:

$$(28)$$

The living nature of polystyryl anion has been applied for the system of diblock, triblock, and multiblock copolymers [67–70]. Examples for commercial products are Kraton rubber (Shell Oil Co.), a styrene–butadiene–styrene triblockcopolymer, and Styrolux (BASF AG), a styrene–butadiene–styrene starblockcopolymer. First, a block polystyrene can be prepared that remains active; then a new monomer can be added. Termination with dimethylsilicium dichloride yields a triblock copolymer:

$$(29)$$

Another advantage of the living nature of the chain end is that it can be converted into interesting functional end groups [71–74]. The reactions of the anionic end group can be divided into three categories: (1) coupling for chain extensions, yielding special structures such as star-branched polymers; (2) coupling with other polymers, yielding block or graft copolymers; and (3) coupling with polymerizable groups for the reaction in further polymerizations as monomer (macromonomer). Reaction of poly(styryllithium) first with oxirane, then with methacryloyl chloride, is an example for the transformation of the living chain end into a polymeric monomer [75,76].

$$(30)$$

Copolymerization of this macromonomer with common monomers such as methyl methacrylate yields graft copolymers that are not contaminated with homopolymers. Many examples of this reaction type are described in the literature [77–81]. More details are given later in the chapter.

4. Cationic Polymerization

The commercial use of cationic polymerization of styrene is practically nonexistent at this time because of the low temperature needed, the uncontrollable molecular weight, and residual acidic initiator. However, since numerous basic papers are published on this topic it is discussed here briefly. The initiators of a cationic polymerization of styrene can be carried out in the presence of strong acids like: protonic acids like perchloric, hydrochloric or sulfuric acid or Lewis acids such as BF_3, BCl_3, and $AlCl_3$ [82–86]. Additionally, alumina, silica, and molecular sieves were used to initiate cationic polymerization. A list of studies of the cationic polymerization of styrene is given in Ref. [86]. One of the most interesting developments in this field is the polymerization of styrene initiated by $HClO_4$ [87]. The reaction involves three stages: stage 1: fast polymerization, showing electrical conductivity and all orange–red color; stage 2: characterized by the absence of conductivity and color; and stage 3: when the conductivity and color reappear. These observations are explained by intermediate perchlorate esters [88].

Cationic polymerization of styrene in the presence of salts such as $(n\text{-Bu})_4N^+ClO_4^-$ has been shown to accelerate the polymerization rate compared to the rate for salt-free polymerization [89]. This result is explained on the basis of the following reaction:

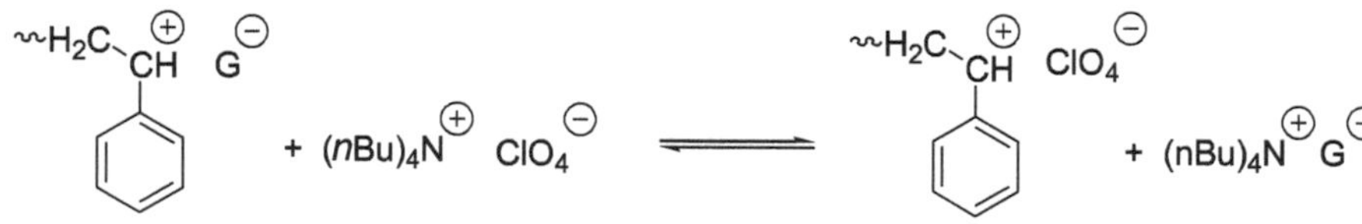

$$\text{(31)}$$

Polymerization of styrene with $BF_3 \cdot OEt_2$ in the presence of electron acceptors (EA) such as tetracyanoethylene (TCE) and chloranil (CA) strongly increased its rate if chlorinated solvents were used. This was not observed in the 'donor solvent' benzene and also not in the acceptor solvent nitrobenzene [90]. This is explained by complexation of the counterion, enhancing the ion separation and therefore increasing the rate.

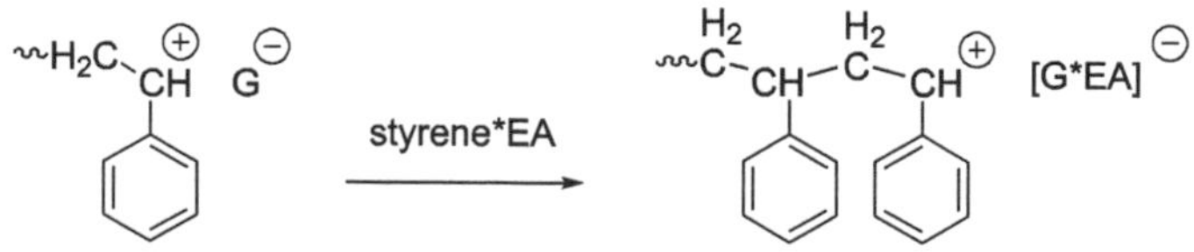

$$\text{(32)}$$

Interesting effects were observed if cationic polymerization was carried out under an electric field [91,92]. Depending on the solvent, the degree of ion separation is decreased under the influence of the electrical field. However, if toluene is applied, the effect is small due to the low ε value of the solvent. In the range of intermediate values of ε (dichloroethane) the highest changes in rate are observed. In nitrobenzene, a solvent with high ε values, the ion separation is almost complete. Therefore, application of an external field does not affect the free ion concentration and the rate. Some research was carried out in the field of electro-initiated polymerization of styrene [93].

When the system $Et_4N^+BF_4^-$/styrene/nitrobenzene was electrolyzed with 0.35 mA for 60 min, some polystyrene was observed around the anode. Another field in which cations are assumed to be the intermediate is styrene polymerization by γ-rays [94]. Careful work showed that the rate of polymerization gradually changes from 0.5 order, indicating a radical process, to a first-order process, suggesting an unimolecular termination characteristic for the cationic mechanism [95–99].

Normally, molecular weight is difficult to control in cationic polymerization of styrene. This is not only because of transfer to polymer and solvent but also of transfer to monomer. Friedel–Crafts reactions during growth with aromatic solvents significantly decrease the molecular weight [100]. A living carbocationic polymerization of styrene has been described [101]:

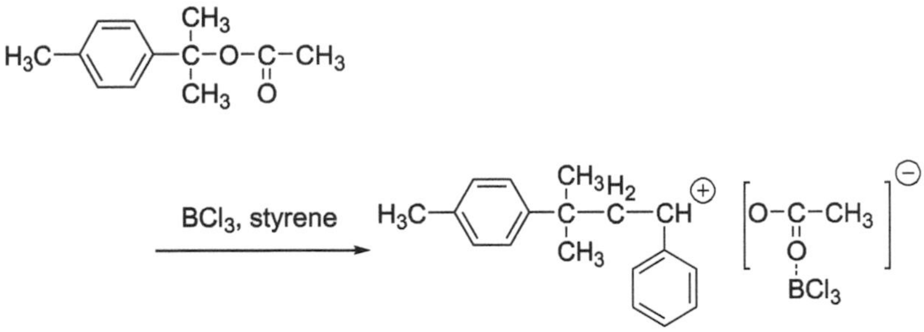

(33)

Addition of further monomer results in an increase in molecular weight. The danger of termination by indane formation seems to be reduced by adding the monomer in small portions [102,103]:

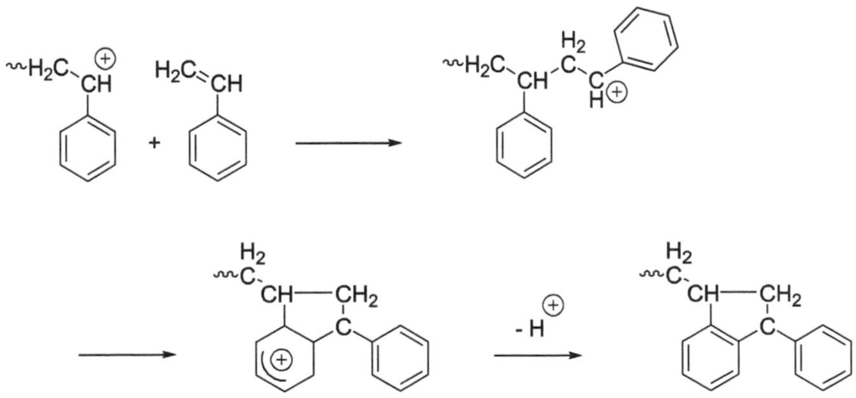

(34)

5. Coordination Polymerization

Styrene can be polymerized to stereoregular structures by coordination catalysts. Highly isotactic polystyrene is prepared using Ziegler–Natta-type catalysts obtained from the reaction between $TiCl_4$ and $AlEt_3$ [104,105] and of a $TiCl_3$/Al(i-Bu)$_3$ mixture [106] in a temperature range of 0 °C to 10 °C. The Al/Ti ratio has to be 3:1 for the formation of isotactic polystyrene [107,108]. A detailed description of preparations for isotactic polystyrene is given in Ref. [109].

Syndiotactic polystyrene has also been described [110–113]. A mixture of methylaluminoxane (MAO) and cyclopentadienyltitanium(III)chloride was used as catalyst, whereby the active species was postulated to be a cationic complex [CpTi(III) (Polymer)Sty]$^+$ [114]. The stereocontrol in this catalyst is induced by the phenylgroups of the growing polymer chain and not by the symmetry of the catalyst as in most type of coordination catalysts.

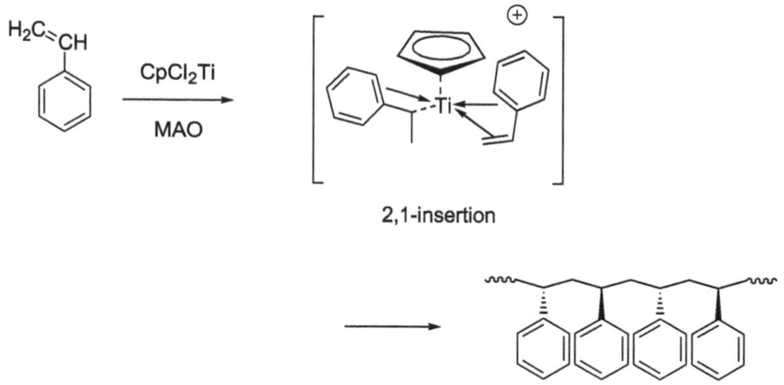

2,1-insertion

(35)

Syndiotactic polystyrene (sPS) is a hard, stiff material with high temperature stability and excellent isolator properties. The E-module of ca. 10^9 Mpa is similar to that of poly-amide 66, and therefore much higher than in amorphous polystyrene. These properties lead to new, very interesting applications, especially if sPS is blended with polyamides [6].

C. Copolymerization

A large number of copolymers can be prepared with styrene. However, only the most interesting combinations of monomers are described here. From the reactivity ratios r_1 and r_2, one can calculate the composition of the instantaneously produced copolymer with the following equation:

$$\frac{[M_1]}{[M_2]} = \frac{1 - r_2}{1 - r_1}$$

The reactivity ratios for selected comonomers are listed in Table 2.

1. Poly(styrene-co-methyl methacrylate) (PSMMA)

Due to the reactivity ratios, the copolymer shows less composition shift than other systems do [121]. The copolymer should contain more than 30 wt % styrene to avoid degradation at temperatures above 250 °C [122]. The SMMA copolymers are interesting because of improved light, outdoor and weather stability, and higher clarity [123]. These copoly-mers are produced by either bulk, solution, or suspension polymerization. Alternating structures are derived in the presence of ZnCl$_2$ [124] or EtAlCl$_2$ [125].

2. Poly(styrene-co-maleic anhydride) (PSMA)

Copolymerization of styrene with maleic anhydride yields alternating structures, probably due to the formation of charge transfer complexes [126,127]. Statistical copolymers

Table 2 Styrene (M_1) comonomer reactivity ratio; a comprehensive list of reactivity ratios is given in Ref. [120].

Monomer 2 (M_2)	r_1 ($= k_{11}/k_{12}$)	r_2 ($= k_{22}/k_{21}$)	$T/°C$	Ref.
Methyl methacrylate	0.52	0.46	60	[115]
Maleic anhydride	0.02	0.00	80	[116]
Acrylonitrile	0.4	0.04	60	[117]
Butadiene	0.5	1.40	50	[118]
1,3-Divinylbenzene	0.65	0.60	60	[119]

are produced if the process is carried out in a continuous batch reactor [128]. Copolymers with small amounts of maleic anhydride are the basis for several commercial products, whereby the primary benefit of maleic anhydride is a greatly increased heat resistance.

3. Poly(styrene-co-acrylonitrile) (PSAN)

PSAN is probably the most important copolymer of styrene because of its improved chemical resistance, improved mechanical properties, and better heat stability. The disadvantage is that a higher portion of acrylonitrile often yields yellow products [129,130].

For high-quality PSAN, formation of homopolystyrene has to be avoided. A small amount of polystyrene will produce hazy PSAN because of phase separation. The copolymerization is initiated by radicals and is carried out in bulk, solution, suspension, and emulsion [131–133]. Alternating copolymers result if $ZnCl_2$ or $EtAlCl_2$ was added to the monomer mixture [134].

4. Poly(styrol-co-acrylic ester) and Poly(styrol-co-acrylic acid salts)

This copolymerization is initiated by radicals in bulk, solution, or suspension. Emulsion polymerized copolymers of styrene and acrylic esters are important basic materials for coating resins. Copolymerization of styrene with acrylic acid salts (Zn^{2+}, Co^{2+}, Ni^{2+}, and Cu^{2+}) in methanol as solvent yields copolymers that form ionomers with properties of reversible networks [135].

5. Poly(styrene-co-butadiene) (SB) and Poly(styrene-co-acrylonitrile-co-butadiene) (ABS)

SB polymers are prepared by emulsion polymerization [136]. Molecular weight is controlled by the addition of a chain transfer agent. ABS rubber is produced in emulsion [137,138] or solution [139,140]. The polybutadiene is generally prepared separately, then the SAN copolymerization is started in presence of a certain amount of polybutadiene. Under certain conditions the rubber is grafted by the growing chain. Both products show very high stress resistance in contrast to pure PS.

6. Copolymerization with Divinylbenzene

Copolymerization of styrene with small amounts of bifunctional monomers such as divinylbenzene is used for the synthesis of networks. The polymerization technique of choice is bead polymerization. Polymer porosity can be controlled by the addition of polystyrene, which can be extracted after polymerization has been completed. Sulfonation of such networks yields cation-exchange resins; anion-exchange resins can be synthesized by chloromethylation followed by nucleophilic substitution of the chlorine by amine groups and quaternization of the amino groups [141].

D. Block Copolymers

Block copolymers containing two different monomers are industrially synthesized by means of anionic techniques exclusively. Three pathways are possible:

1. Adding a monofunctional initiator to a *mixture* of monomers, if the reactivity ratios are different enough; otherwise, it is more convenient to polymerize one monomer to complete conversion and then add the other monomer. Following the first line, A–B-diblocks can be produced. The stepwise addition of different monomers leads to tri- and multiblocks.
2. Bifunctional initiators are used for the synthesis of triblock copolymers
3. Triblock and star-shaped polymers are available if the polymerization of one monomer is started with a monofunctional initiator, then the second monomer is added. Finally, the polymerization is terminated by adding a two- or multifunctional terminator.

Normally, blockcopolymers from styrene and butadiene are synthesized by methods 1 and 3. These reactions are reviewed in Ref. [142]. From a mixture of styrene and 1,3-butadiene in a nonpolar solvent, first the 1,3-butadiene is polymerized. With increasing relative concentration, more and more styrene will be incorporated into the polymer chain until only styrene is left forming the styrene block [143]. The sequence is affected by the addition of polar substances favoring the incorporation of styrene from the beginning [144]. A complete description for the synthesis of triblock poly(styrene-isoprene-styrene) is given in Ref. [145].

Pathway 2 is generally chosen in the academic literature. In addition to sodium naphthalene, dilithium compounds are often used [146–150]. The following terminators are described for the third pathway: silicon tetrachloride and tintetrachloride dicarbonic acid ester [151], divinylbenzene [152], and polymers formed from divinylbenzene [153] containing numerous vinyl groups yielding star-shaped polymers. Block copolymers of styrene and butadiene or isoprene are synthesized commercially in large ranges.

For the synthesis of block copolymers containing styrene and methyl methacrylate, a living polystyrene is formed in the first step, then MMA is added at low temperature ($-78\,^{\circ}C$) to avoid reactions of the living polyanion with the ester group [154–156]. Other combinations of monomers are possible if a living polymerization is terminated with a dichloroazo initiator, yielding a polymeric initiator useful for further polymerizations [157,158]:

(36)

A possibility to synthesize block copolymers by conventional radical polymerization is given by the application of polymeric initiators [162]. Partial decomposition of the polymeric initiators in the presence of styrene yields block copolymers containing polystyrene and part of the polymeric initiator, still containing some initiator functions. These functions can be decomposed in a following step in the presence of an additional monomer yielding block copolymers [157,159–161].

Block copolymers can also be synthesized by controlled radical polymerization [162]. This technique is very interesting for probable industrial applications, because of its low sensitivity against impurities and the mild reaction conditions. In principle, all methods of the controlled radical polymerization, as they are described in the chapter above, can be utilized more or less successfully for the formation of block copolymers.

Georges et al. showed the the formation of block copolymers of styrene with butadiene, isoprene, acrylate and methycrylate by use of TEMPO or Proxyl(2,2,5,5-tetramethyl-1-pyrrolidinyloxy) as stable counter radicals [163,164].

(37)

If nitroxyl radicals are used in the presence of organoaluminium complexes in association with various ligands, the homolytic cleavage of the counter radical is strongly activated [165–167]. If such systems are applied to vinyl acetate (VAc) or MMA, the produced polymers undergo further stepwise polymerization of styrene, leading to PVAc-*b*-PSt or PMMA-*b*-PSt.

ATRP is also a method to obtain blockcopolymers from different monomers with styrene [168,169]. If the first block is terminated with a halogen end group, it works as a macroinitiator in the presence of transition metal catalysts under a reversible redox mechanism, following the same scheme as in homopolymerization. The formation of tri- and multiblockstructures is also practicable.

The RAFT process also allows the formation of block copolymers from styrene [170]. After polymers have been synthesized in the presence of dithioesters as transfer reagents, they are terminated with a dithioester group. This group can be easily activated with small amounts of a conventional radical initiator, so that the polymerization of the second block proceeds. With this method, block copolymers of styrene with *N,N*-dimethylacrylamide, methylstyrene or methyl methacrylate have been synthesized.

E. Graft Copolymerization

1. Polystyrene Backbone

Graft copolymers containing styrene in the main chain and other monomers in their side chains are available by numerous methods, including conventional radical [171], controlled radical, anionic [174], and cationic [175] polymerization and by copolymerization of macromonomers [80,174]. Grafting methods via conventional radical polymerization are reviewed by Nuyken and Weidner [171]. The following reaction scheme demonstrates the principles and universality of the methods applying polymeric initiators:

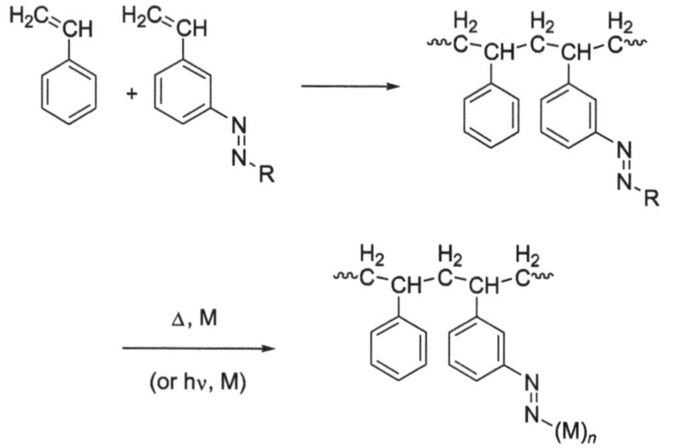

(38)

Grafting via controlled radical polymerization follows the same principle. Copolymerization of styrene with styrene derivatives, containing a nitroxyl group [172] or a halogen group [173] results in a macroinitiator for graft copolymerization via the grafting from method.

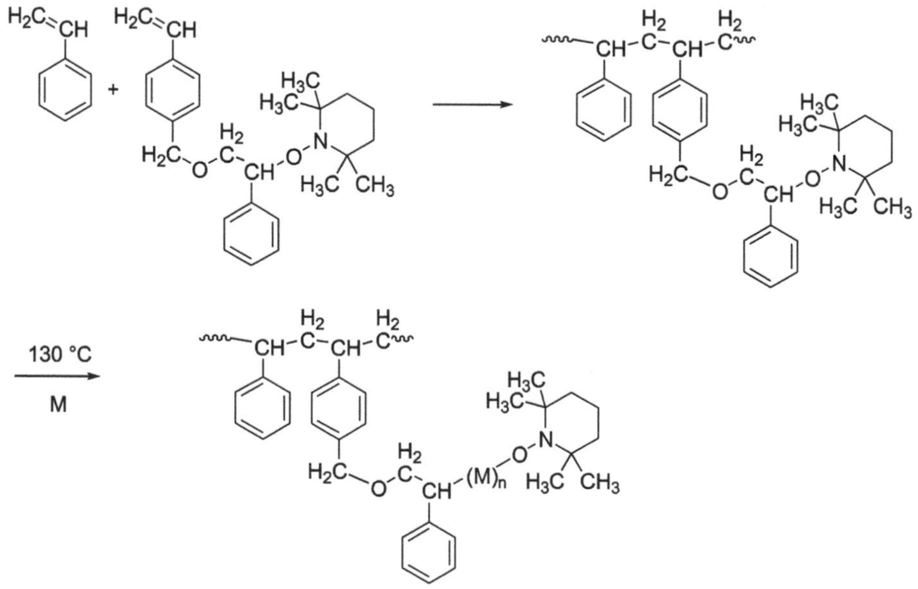

(39)

Another way is to brominate poly(styrene) with NBS to obtain a macroinitiator for graft copolymerization via ATRP [176].

Furthermore, copolymers of styrene and suitable monomers can be grafted via anionic mechanism [177,178]:

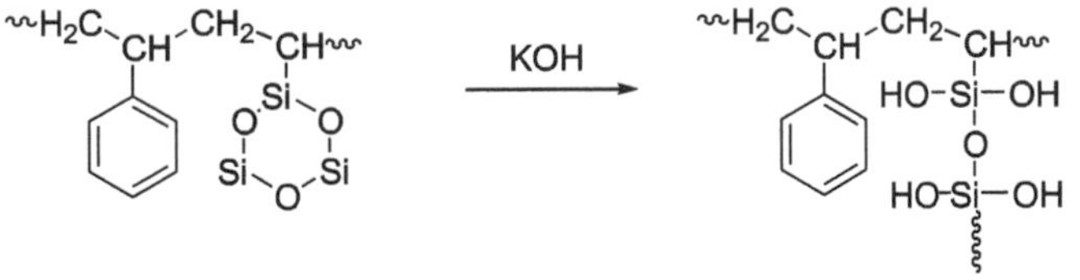

(40)

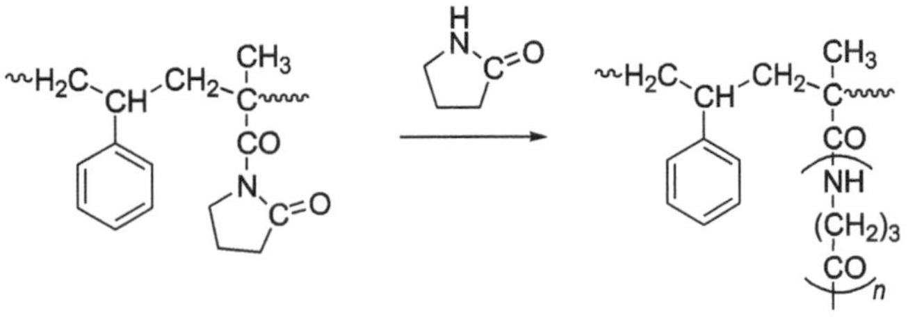

(41)

Examples of cationic grafting are given by Kennedy et al. [175]:

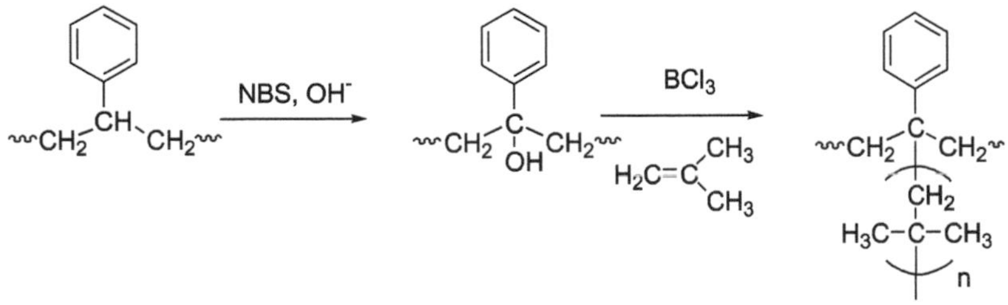

(42)

In another example, a copolymer of styrene and acryloyl chloride is reacted with silver hexafluoroantimonate and then grafted with THF [179,180]:

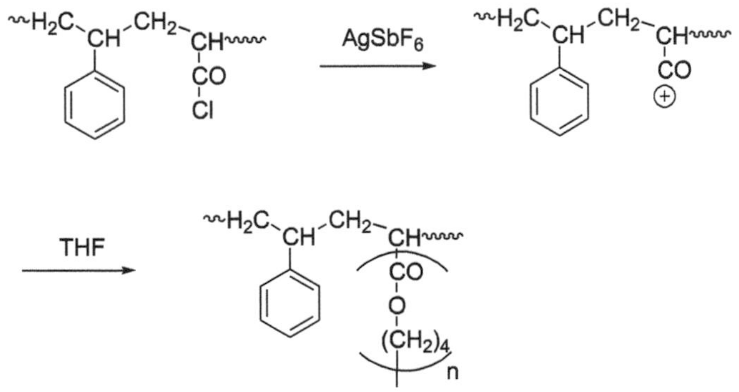

(43)

The macromonomer method is probably the most promising development at present concerning the synthesis of well-defined graft copolymers. The developments in this field up to 1999 have been reviewed by Ito and Kawaguchi [181]. Any macromonomer having a head group that is copolymerizable with styrene can be utilized for this purpose.

2. Polystyrene Sidearm

The most important example of graft copolymers having polystyrene sidearms is high-impact polystyrene (HIPS), in which polystyrene is the continuous phase and poly-butadiene grafted with polystyrene forms the separated phase. Grafting occurs when some of the radicals react with the double bonds in polybutadiene [182–186]. Grafting is also possible onto poly(ethene-co-propene-co-butadiene) [187] and polyacrylic ester [188]. High-impact polystyrene is reviewed in detail in Ref. [189].

Graft copolymers can also be synthesized by a macromonomer method [80,174, 181,190]. The advantages of this method are its variability and the fact that homo-polymerization can be avoided. The following example can be considered as representative for other possibilities: Radical polymerization of styrene in the presence of iodine-acetic acid yields polystyrene having carboxylic end groups. Reactions of this functionalized polymer with glycidyl-methacrylate yield a macro-monomer having a methacrylic end group [191].

Another method to develop polystyrene terminated by a methacrylic unit is described by Schulz and Milkovich [78]. There, a living polystyrene is converted into an alcoholate function by addition of oxirane and then esterificated with methacrylic chloride.

3. Branched and Hyperbranched Polystyrene

Frechet et al. first presented a way to obtain branched and hyperbranched polystyrene via cationic polymerization [192], which is called the self condensing vinyl polymerization. In this method, 3-(1-chloroethyl)ethenyl benzene was the monomer and has been polymerized in the presence of $SnCl_4$.

Matyjaszewski et al. presented a route to branched and hyperbranched polystyrene via ATRP in presence of Cu(I) [193]. In this case the monomer was p-(chloromethyl)-styrene (CMS). CMS acts as both initiator and monomer. The degree of branching can be varied by adding different amounts of styrene.

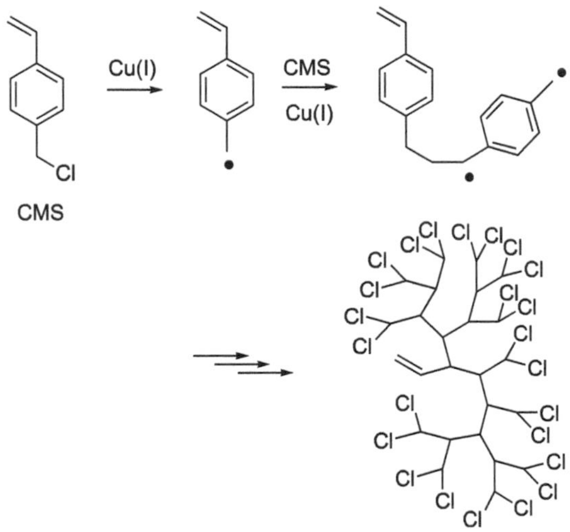

(49)

The TEMPO mediated radical polymerization also has been successfully used for the synthesis of hyperbranched polystyrene, if 4-[2(phenyl)-2-(1-2,2,6,-tetramethylpiperidinyl-oxy)ethyloxy] methylstyrene was used as monomer, following the same approach [194].

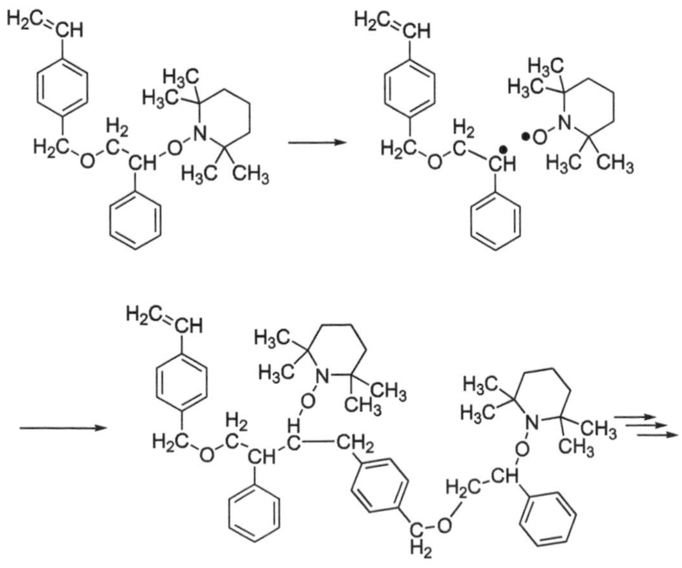

(50)

Hyperbranched polystyrene with a polar shell can be prepared by thermal induced radical polymerization of 3-vinylphenylazo-methylmalonodinitrile [195].

$$(51)$$

II. SUBSTITUTED STYRENES

(This section was prepared by O. Nuyken, M. Lux and M. Heller.)

A. α-Methylstyrene

$$(52)$$

Probably the most intensively studied derivative of styrene with regard to its polymerization behavior is α-methylstyrene. It is produced commercially by the dehydrogenation of isopropyl-benzene (cumene) and also as a by-product in the production of phenol and acetone by the cumene oxidation process. The polymerization characteristics of α-methylstyrene are considerably different from those of styrene. Whereas radical polymerization of the pure monomer proceeds very slowly and is therefore not a practical technique [196], both ionic and coordination-type polymerization can be used to prepare poly(α-methylstyrene) (PMS).

1. Cationic Polymerization

The scientific literature concerning cationic polymerization of α-methylstyrene has been completely reviewed by Bywater [197] up to about 1962 and Kennedy [198] gives an essentially complete list of publications in this field (excluding irradiation studies) up to about 1972. It was found that the cationic polymerization of α-methylstyrene, especially at low temperatures, yielded highly stereoregular PMS, but for a long time the stereochemical structure was discussed controversially. Brownstein et al. [199] studied the structure of PMS synthesized with the cationic initiators BF_3, $AlCl_3$, and $TiCl_4$ in toluene at $-70\,°C$ and $SnCl_4$ in nitromethane and ethylene chloride at -30 and $-35\,°C$, respectively. They proposed a predominantly (80 to 90%) syndiotactic configuration by assigning the split signals of the α-methyl group in ^{1}H-NMR spectra (60 MHz) to the syndiotactic, heterotactic, and isotactic triad with decreasing field strength. This proposition was based mainly on inspection of Hirschfelder–Taylor atomic models. Sakurada et al. [200] showed that polymerization in toluene/n-hexane mixtures with BF_3–OEt_2 at -78, -60, and $0\,°C$, with $TiCl_4$ and $AlEt_2Cl$ at $-78\,°C$ and with $AlEtCl_2$ at -78, -30, and $0\,°C$ yielded more stereoregular polymers than anionic (K, Na, BuLi) and coordinative ($AlEt_3$/$TiCl_4$) polymerization. Ohsumi and co-workers [201] investigated the effect of various reaction parameters on the stereoregularity of the PMS obtained. They found that the nature of the solvent decisively influences the steric course of the polymerization. While good solvents for the polymer (e.g., toluene, methylene chloride) resulted in highly stereoregular polymers, the nonsolvent n-hexane yielded largely atactic polymers. Interestingly, the nature of the cationic initiator (BF_3–OEt_2, $SnCl_4$ CCl_3COOH, $AlBr_3$ CCl_3COOH, $TiCl_4$) had only a slight influence on stereoregularity, but polymerization temperature was also an important parameter. Only below about $-60\,°C$, highly stereoregular polymer was formed. Both, Sakurada and Ohsumi believed the polymers to have mostly an isotactic configuration. Ramey and Statton investigated the spectra using 100- and 220-MHz ^{1}H-NMR spectra at elevated temperatures [202] and investigations using high-magnetic-field instruments, ^{13}C-NMR, and partially deuterated polymers [203–205] seem to confirm the original assignment, and therefore it can be concluded that cationic polymerization of α-methylstyrene usually yields highly syndiotactic polymers. Using BF_3–OEt_2, $TiCl_4$, and $AlCl_3$, Kunitake and Aso [206] were able to obtain 100% syndiotactic PMS in toluene at $-75\,°C$. Both increasing temperature and addition of methylcyclohexane to the toluene drastically decreased the percentage of syndiotactic triads. Matsuguma and Kunitake [207] later extended these studies and used other conventional Lewis acids and triphenylmethyl salts $Ph_3C^+X^-$, where $X^- = AlCl_4^-$, $SnCl_5^-$, $AlBr_4^-$, BF_4^-, and $SbCl_6^-$ at $-78\,°C$ in solvent mixtures of different polarity. While in the least polar solvent (methylcyclohexane/methylene chloride 4:1) stereoregularity of the polymer obtained varied widely with the initiator, in the most polar solvent (methylene chloride/acetonitrile 7/3) all initiators yielded highly syndiotactic polymer. Other authors used more exotic catalyst systems such as m-chlorobenzoic acid in liquid sulfur dioxide [208], $tert$-butyl chloride/Et_2AlCl in methylene chloride [209], or 9-anthranylmethyl hexafluorophosphate [210]. Kennedy and co-workers published a series of papers [211–213] in which they investigated the polymerization ability of the cationic initiation systems H_2O/$SnCl_4$, H_2O/BCl_3, and pentamethylbenzyl chloride/$SnCl_4$ with and without the use of the proton trap 2,6-di-tert-butylpyridine (DtBP). In the systems H_2O/BCl_3 in methylene chloride at -60 to $-20\,°C$, H_2O/$SnCl_4$ in ethyl chloride at -122 to $-40\,°C$, and pentamethylbenzyl chloride/$SnCl_4$ in methylene chloride at -80 to $-30\,°C$ addition of DtBP was found to have the same effects on the polymerization of α-methylstyrene. With increasing amounts of DtBP, the conversion of the polymerization drastically decreased, but molecular weights increased and the molecular weight distributions narrowed. The initiator system H_2O/$SnCl_4$ in methylene chloride at $-60\,°C$ behaved slightly differently. In this case increasing amounts of DtBP first increased the molecular weight, but from concentrations of about $10^{-4}\,mol/L$, molecular weights decreased drastically. The influence of DtBP was explained in terms of acting as a scavenger to trap the protons, which usually emerge during chain transfer to the monomer, while it does not influence the other elementary events of the polymerization reaction. Kennedy and co-workers [214] also reported about the discovery of the 'quasi-living' carbocationic polymerization systems H_2O/BCl_3 and cumyl chloride/BCl_3 in methylene chloride/methylcyclohexane mixtures at $-50\,°C$.

The first examples of living carbocationic polymerization have been reported by Miyamoto et al. [215] and Faust and Kennedy [216]. Higashimura and coworkers [217] showed the first living polymerization of α-methylstyrene (α-MeSt), with the HCl-adduct of 2-chloroethylvinylether/SnBr$_4$ initiating system at −78 °C in CH$_2$Cl$_2$. Fodor and Faust [218] reported the living polymerization of α-MeSt using the cumylchloride, (CH$_3$)$_3$C–CH$_2$–C(CH$_3$)$_2$–CH$_2$–C(Ph)$_2$–OCH$_3$ (TMPDPEOMe) or the HCl adduct of α-MeSt dimer (DiαMeSt) as initiators and BCl$_3$ as coinitiator.

Because of the tertiary benzylic cation, poly(α-MeSt$^+$) is more reactive than poly(styrene$^+$) and readily undergoes side reactions such as β-proton elimination and chain transfer to monomer (indanyl ring formation). In order to obtain living polymerization, these reactions could be eliminated at low temperature (−80 °C and −60 °C) and by non-polar solvent mixtures (CH$_2$Cl$_2$:hexane or CH$_2$Cl$_2$:cyclohexane). BCl$_3$ was chosen because termination is absent using this Lewis acid [214]. Cumylchloride was found to be an inefficient initiator.

TMPDPEOMe and DiαMeStHCl are very efficient initiators. The polymerization was much faster in CH$_2$Cl$_2$:hexane mixture than in CH$_2$Cl$_2$:cyclohexane. The polymerization rate was higher at −80 °C than at −60 °C but found to be living at both temperatures. Poly(α-MeSt) was yielded with controlled molecular weight and $M_w/M_n \sim 1.1$–1.2. The chain ends remained living up to 40 min (−80 °C).

Kwon et al. [219] later investigated SnCl$_4$ as coinitiator for the living polymerization of α-MeSt using Diα-MeStHCl as initiator (proton trap DTBP, methylcyclohexane/CH$_2$Cl$_2$ 60/40). They found a little deviation from linearity which may indicate termination but they could demonstrate the absence of chain transfer.

2. Coordinative Polymerization

Few data have been reported on the coordinative Ziegler–Natta-type polymerization of α-MeSt. Some papers of Sakurada [200,220–222] describe successful preparation of PMS with organometallic systems. The system AlEt$_3$/TiCl$_4$ was examined below −70 °C and the influence of the solvent (toluene/n-hexane mixtures), Al/Ti ratio, and catalyst aging conditions on yield and molecular weight was investigated. It was found that an Al/Ti ratio of 1.0 to 1.2 and a solvent mixture containing about 70% toluene were the best conditions and high-molecular-weight PMS with DP between 1000 and 4500 (determined by viscosity measurements) could be obtained. Although ^{1}H-NMR showed a smaller degree of stereoregularity in these polymers than in PMS prepared cationically, Sakurada claims them to be crystallizable after heat stretching under 'certain' conditions, while attempts to crystallize the more stereoregular cationic samples failed.

3. Anionic Polymerization

Living anionic polymerization of α-MeSt has been investigated by various research groups, especially with the intention to obtain polymers with narrow molecular weight distribution close to the theoretical Poisson distribution. Due to the low ceiling temperature (61 °C) of PMS [221], it is possible to purify and initiate the polymerization system at elevated temperatures (e.g., room temperature), where no high polymers can be formed, but upon rapid cooling to low temperatures (e.g., −78 °C) the growth of all chains is started simultaneously. By this procedure the limiting effects of mixing monomer solution and initiator as in usual living anionic polymerization can be minimized. McCormick [221] used naphthalene-sodium as an initiator for the polymerization in THF at dry-ice temperature after initiation at elevated temperature. Examination of the polymers obtained showed bimodal molecular weight distributions with one peak having double the molecular weight of the other, especially when the initiating species (monomer mixed with naphthalene-sodium) was added to the monomer solution instead of initiation in the monomer solution. M_w/M_n values were varying from 1.42 to 1.04 with increasing molecular weight (values determined by sedimentation and viscosity measurements). The fact of the bimodal distribution was attributed to termination of one side of the initiating oligo(α-MeSt) dianion by impurities in the first stage of the polymerization reaction.

Similar results were obtained by Wenger [222,223], who used oligo(α-MeSt) dianions formed by the reaction of sodium with the monomer as the initiating species for the polymerization in THF. Even after several attempts to improve the polymerization procedure, bimodal molecular weight distributions were still obtained. The best M_w/M_n value reported was 1.03 (determined by light scattering and osmotic pressure measurements). PMS with extremely narrow molecular weight distribution was synthesized by Fujimoto et al. [224] using LiBr- and LiOH-free n-BuLi as the initiator. This monofunctional initiator was reacted with the monomer at 40 °C for 30 min and then the mixture was cooled quickly to −78 °C, where the polymerization was allowed to proceed for several hours. The polymers had extremely narrow molecular weight distributions ($M_w/M_n < 1.01$ by GPC) and were also characterized by light scattering, osmotic pressure and viscosity measurements, and sedimentation). In this paper also the influence of such additives as LiOH, LiBr, LiOBu, and LiNEt$_2$ to the initiator was studied. Several workers have been working with the equilibrium anionic polymerization under different conditions. Worsfold and Bywater [225] in THF between −40 and 0 °C, Wyman and Song [226] in bulk between −20 and 50 °C, and Leonard and Malkotra [227] in p-dioxane between 5 and 40 °C determined the conversion of monomer to polymer as a function of temperature and from that the thermodynamic values. The paper of Wyman and Song [226] gives M_w/M_n values in the range 1.5 to 1.8 and molecular weights in the order of 10^5 g/mol.

Concerning the design of positive electron-beam resists [228,229], anionic polymerization was used to introduce 2-phenylallylgroups at the end of poly(α-MeSt) chains.

4. Living Radical Polymerization

p-Br or p-nitroxide-α-methylstyrenes have been polymerized with a solid supported 2,2,6,6-tetramethyl-1-piperidin-N-oxyl (TEMPO)-initiator [230].

B. *cis*- and *trans*-β-Methylstyrene

(53)

It is well known that α,β-disubstituted olefins usually cannot be polymerized to high polymers, especially by radical initiators. This is due to the steric hindrance of the β substituent in the transition state of the propagating species [231]. For β-methylstyrene it was found that conditions for radical polymerization yielded only dimers [232]. Only copolymers of β-MeSt have been reported using radical polymerization [233], yielding phenoxy-phenyl maleimide-β-MeSt copolymers under participation of CT-complex. β-MeSt is also a good transfer agent of propene polymerization but becomes partially incorporated into the polymer [234]. Anionic polymerization of the trans isomer with BuLi and other catalysts [235] yielded no polymer in apolar solvents and only oligomers in THF. There are also only few reports about coordination-type polymerization [236,237]. However, the introduction of the methyl group in the β position of the vinyl double bond increases the electron density enough to allow cationic polymerization despite the steric hindrance. Poly-(β-MeSt) was also obtained by electroinitiated anodic polymerization in CH_2Cl_2 [238].

1. Cationic Polymerization

The polymerization of β-methylstyrene (isomer distribution not specified) was first investigated by Staudinger and Dreher [239]. Its polymerization initiated by BF_3 in toluene between −80 and −60 °C yielded products with molecular weight 1000 to 3000 g/mol, while $SnCl_4$ did not yield methanol-insoluble polymer. The authors proposed their polymer to consist of 1,3 units based on pyrolysis experiments and viscosity data.

Later this work was reinvestigated by Kennedy and Langer [212,240]. $AlCl_3$ in methylene chloride at −60 °C yielded 77% polymer, which was only partially soluble in toluene. IR investigations established the presence of methyl groups in the polymer chain and it was thus concluded that the product consisted mainly of conventional 1,2 units. Conclusions in a publication by Murahashi et al. [241] on the 1,3 polymerization of β-methylstyrene with various cationic catalysts have been corrected by the same authors in a subsequent paper [241], where they admit that substantial indene impurities in the monomer led to erroneous results. Polymerization of an isomer mixture (87% trans, 13% cis) by Shimizu et al. [238] using $BF_3 \cdot OEt_2$ in bulk at 0 to 15 °C yielded 5 to 9% oligomers ($M_n = 720$ g/mol) after several days. IR spectra showed that the product consisted mainly of conventional 1,2 units. Also, copolymerization studies of Mizote et al. [242,243] confirm the formation of 1,2 enchainments by cationic polymerization of β-methylstyrene.

C. *ar*-Methylstyrene

(54)

As styrene *o*-, *m*-, and *p*-methylstyrene are polymerizable by all of the conventional mechanisms. A copolymer containing 60% *m*- and 40% *p*-MeSt is commercially produced by thermal polymerization.

1. Radical Polymerization

Polymerization of all three *ar*-methylstyrenes using 1 mol% AIBN in toluene at 80 °C as the initiator was carried out by Kawamura et al. [244]. The stereoregularity of the polymers was characterized in terms of probability of racemic addition to the propagating end (Pr) by ^{13}C-NMR at 25 MHz using the splitting pattern of the aromatic

Table 3 Propagation coefficients (k_p) and activation energies (E_a) of different *para*-substituted (X) styrenes.

X	k_p 40 °C	k_p 30 °C	k_p 20 °C	E_a/kJ mol^{-1}
MeO	94	60	38	34.9
Me	112	75	48	32.4
Cl	197	133	85	32.1
H	160	110	70	31.5

C carbon atoms. Pr values of 0.83, 0.75, and 0.72 were found for *o*-, *m*-, and *p*-MeSt polymers, respectively, indicating increasing amounts of syndiotactic structures with the substituent located nearer the polymer backbone. *p*-MeSt has been polymerized by Mutschler et al. [245] using AIBN as the initiator in cyclohexane in the temperature range 50 to 70 °C. Molecular weights were determined by light scattering and GPC and varied from $M_w \sim 5000$ to $270\,000$ g/mol with polymerization conditions ($M_w/M_n = 1.55$ to 3.00). Coote and Davis [246] investigated the propagation kinetics of *m*-MeSt and other *para*-substituted styrenes (X = OCH_3, F, Cl, Br). Pulsed laser polymerization measurements of the homopropagation rate coefficients (k_p) at different temperatures are reported. Further the activation energies (E_a) have been calculated (Table 3). The authors tested the applicability of the Hammett relationship and found only resonable qualitative description of the trend in the data.

2. Cationic Polymerization

Kanoh et al. [247] studied the cationic polymerization of *p*-MeSt by iodine in ethylene chloride at 30 °C. They found the propagation rate constant to be 5.7l/mol min, which was about 25 times higher than for styrene under the same conditions. Kanoh et al. [247] also studied the solvent effect on the polymerization of *p*-MeSt with iodine at 30 °C. They used ethylene chloride, chloroform, carbon tetrachloride, and mixtures thereof as the solvent. The rate constant was highly dependent on the solvent and was several orders of magnitude larger for polymerizations carried out in ethylene chloride than in carbon tetrachloride. Kennedy et al. [248] studied the polymerization of *o*-MeSt by H_2SO_4, $AlBr_3$, or BF_3–OEt_2 in chlorinated solvents at low temperatures and compared it to thermal polymerization in bulk. The polymers showed identical IR spectra and it was thus concluded that also with this monomer exclusively, conventional 1,2 polymerization had occurred as it did with *p*-MeSt [212]. Heublein and Dawczynski [249] used the $SnCl_4/H_2O$ system to polymerize *p*-MeSt at 0 °C in different solvents. Using optimized conditions for initiator and monomer concentrations, they obtained overall rate constants in five different solvents, ranging from 119 min^{-1} in hexane to 5910 min^{-1} in nitromethane. Molecular weights were in the range 2000 ± 500 g/mol (VPO). Recent publications have dealt with 'living like' or even living cationic polymerization of *p*-methylstyrene. Tanizaka et al. [250] described the polymerization of *p*-methylstyrene initiated by acetyl perchlorate at −78 °C in CH_2Cl_2 containing Bu_4NClO_4 or in CH_2Cl_2/toluene (1:4 v/v), which led to long-lived polymers with a relatively narrow molecular weight distribution ($M_w/M_n = 1.1$ to 1.4). Faust and Kennedy [251] found the catalyst system cumyl acetate/BCl_3 in CH_3Cl and C_2H_5Cl at −30 and −50 °C to be living in terms of having a linear yield M_w/M_n plot, but the molecular weight distributions were unusual broad ($M_w/M_n = 2$ to 5). Other esters were also used and were found to behave like cumyl acetate in the initiating system. Kojima et al. [252] reported the living cationic polymerization of p-MeSt with use of HI-ZnX$_2$ (X = Cl, I) intiating systems in toluene and CH$_2$Cl$_2$. Later, Fodor and Faust [253] used another initiator system TiCl$_4$/CH$_2$Cl$_2$:methylcyclohexane 40:60/$-80\,°$C; initiator TMPCl (2,4,4-trimethyl-1-pentylchloride); protontrap DTBP for the living carbocationic polymerization of p-MeSt and for block copolymerization with isobutylene. They found linear ln([M]$_0$/[M])/vs time plots but higher molecular weights than calculated (slow initiation) and $M_w/M_n \sim 2$. Cationic photopolymerization of p-MeSt initiated by phosphonium and arsonium salts is reported by Abu-Abdoun et al. [254]. The effects of photolysis time, light intensity and salt structure on the rate of polymerization are presented.

3. Coordinative Polymerization

The coordinative polymerization of o-, m-, and p-methylstyrene has already been reported in the fundamental studies of Natta et al. [255]. With TiCl$_4$/AlEt$_3$ (1:3) in benzene at 70 °C they could polymerize all three derivatives in 5 to 50% yield. The o- and m-methylstyrene polymers were found to be crystallizable, while poly(p-methylstyrene) was amorphous. The authors believed their crystallizable polymers to be highly isotactic. Later investigations on tacticity by ^{13}C-NMR [244] showed that the o- and m-derivatives were really highly isotactic, while the p-derivative was rather atactic, and no isotactic-rich polymer could be separated by extraction with ethyl methyl ketone [polymerization conditions: TiCl$_4$/AlEt$_3$ (1:3) in hexane at 60 °C]. Despite the fact that p-methylstyrene did not form a crystallizable polymer, Hodges and Drucker [256] were able to obtain a crystallizable copolymer of ar-methylstyrenes. Using a commercial monomer mixture (33% o, 65% p, 2% m isomer), they obtained a polymer with TiCl$_3$/AlEt$_3$ in benzene at 60 to 70 °C, which contained 80 to 90% methyl ethyl ketone-insoluble stereospecific polymer, which could be crystallized by annealing. The resulting polymer contained 15 to 20% o-, 80 to 85% p-, and traces of m-methylstyrene units. Zambelli et al. [257] reported homo- and copolymerisation of p-MeSt, p-ClSt and styrene, comparing η^5- and none-η^5 homogeneous titanium catalysts. They report a syndiotactic polymerization of p-MeSt and random copolymerisation with styrene. But they found no stereoregularity polymerizing p-ClSt (see also [290,313]) with systems such as Ti(OC$_4$H$_9$)$_4$–MAO or TiBz$_4$–MAO. With η^5-catalyst CpTiCl$_3$–MAO, syndiotactic-specific polymerization of both, p-MeSt and p-ClSt is possible. The polymerization might be defined 'electrophilic' since it becomes faster turning from p-ClSt to p-MeSt (increasing electron density of the double bond). η^5-catalysts seem to be stronger 'electrophiles' than non-η^5-catalysts.

Most of the newer publications concerning coordinative metallocene catalysis and methylstyrene polymerization investigate poly(ethylene-co-p-methylstyrene)-elastomers. The important breakthrough for controlled copolymerization came with the development of the metallocene catalysts with constrained ligand geometry which provide the spatially opened catalytic site for monomer insertion including relatively large monomers. Chung and Lu [258,259] made ar-MeSt-ethylene copolymers using [(C$_5$Me$_4$SiMe$_2$N$^+$Bu)]TiCl$_2$ and Et(Ind)$_2$ZrCl$_2$ catalysts ore Ziegler–Natta catalysts, namely MgCl$_2$/TiCl$_4$/electron donor/AlEt$_3$ and TiCl$_3$·AA/Et$_2$AlCl [260]. Kotani et al. [261] reported on polymerization of p-substituted styrenes (p-ClSt, p-MeSt) with rhenium and iron complexe catalysts. With a Re(V)-oxide system in toluene at 60 °C the polymerization of p-MeSt proceeded slower than that of styrene to reach 90% conversion. With an Fe(II)–CpI system in dioxane at 80 °C p-MeSt leveled off around 60% conversion and p-ClSt polymerized faster than styrene.

4. Anionic Polymerization

Higashi et al. [262] investigated the anionic polymerization of p-MeSt with n-amylsodium in n-hexane at 0 °C. The polymer obtained could be fractionated in benzene-soluble, acetone-soluble, and insoluble fractions. The IR spectra of these fractions showed distinct differences, but no further explanation was given. Hirohara et al. [263,264] reported on the kinetics of anionic polymerization of o-, m-, and p-methylstyrene in methyltetrahydrofuran at 25 °C using several organometallic initiators.

5. Living Radical Polymerization

Devenport et al. [265] showed that the autopolymerization of styrenic derivatives like p-MeSt in the presence of 2,2,6,6-tetramethyl-1-piperidin-N-oxyl (TEMPO) is a 'living' process. Molecular weight can be controlled by varying the ratio of vinyl monomer to TEMPO (varied from 100 to 400; $M_n \sim 9500$–36 500; PD ~ 1.24–1.32). They found a correspondence to Mayo mechanism for autopolymerization of styrene.

Schmidt-Naake et al. [266] did dynamic DSC-measurements concerning TEMPO polymerization of p-MeSt and p-ClSt and could show the ability to polymerize in the presence of TEMPO. They also found the exothermal peak of the living polymerization in the same temperature range as of the thermal polymerization.

RAFT processing with p-MeSt has been reported by Chong et al. [267] who synthesized MeSt-p-MeSt-AB diblock copolymers with $M_n = 20\,300/25\,460$ and PD $\sim 1.15/1.19$.

A series of substituted styrenes (p- and m-MeSt among others) were polymerized 'living' by atom transfer radical polymerization (ATRP) by Qiu and Matyjaszewski [268]. The effect of substituents is discussed with regard to the Hammett equation. m-MeSt could be polymerized up to 90% conversion with $M_n \sim 110\,000$, PD ~ 1.2; the conversion of p-MeSt was 50% ($M_n \sim 4000$, PD ~ 1.5). Monomers with electron withdrawing substituents result in better control and polymerize faster than those bearing electron donating substituents (3-CF$_3$, 4-CF$_3$ > 4-Br, 4-Cl > 4-F, 4-H > 3-CH$_3$ > 4-OCH$_3$ > 4-CH$_3$ > 4-C(CH$_3$)$_3$). This is because the stabilities of different substituted polystyryl radicals are similar to the change of bond dissociation energy (BDE) of C-3 Hal in the dormant polystyryl halides, caused by substitution. It is the decrease in the BDE of C-3 Hal by electron withdrawing substituents that accounts for the larger equilibrium constant k_{eq} for atom transfer.

D. ar-Methoxystyrene

(55)

As other styrene derivatives with donor substituents, o- and p-MeOSt (MeOSt) are readily polymerizable by cationic and radical mechanisms, whereas these monomers poison conventional Ziegler–Natta catalysts, and side reactions with anionic initiators can occur. There are only a few data available on the polymerization of the m isomer.

1. Radical Polymerization

The thermal polymerization of p-MeOSt has already been mentioned by Staudinger and Dreher [239]. Heating a bulk sample to 90 °C for several days yielded a polymer with a DP = 390 (by viscosity measurements). Later, Russian authors [269] polymerized thermally all three isomers at 100 to 125 °C and found the p and m isomers to polymerize less rapidly than styrene, but the o isomer more rapidly. The stereoregularity of poly(p-MeOSt) prepared by thermal polymerization in bulk at 60 °C was examined by Yuki et al. [270]. 100-MHz ^{1}H-NMR spectra showed a rather split signal for the methoxy group and was interpreted in terms of pentad sequences. The analysis of the thermally polymerized sample showed a rather low content of syndiotactic triads. Kawamura et al. [244] studied the ^{13}C-NMR spectra of o- and p-MeOSt polymers prepared with BPO in toluene at 80 °C. They found both polymers to be rich in syndiotactic sequences [o derivative, $P_r = 0.80$; p derivative, $P_r = 0.71$ (P_r = probability of racemic addition of monomer to the growing chain)].

Actual kinetic investigations on free radical polymerization of methoxystyrenes have been made by Coote and Davis [246] and are reported in the ar-MeSt chapter.

2. Cationic Polymerization

Staudinger and Dreher [239] polymerized p-MeOSt using SnCl$_4$ in benzene at 0 °C. The polymer obtained was fractionated in five fractions with molecular weight 300 to 13 000 g/mol (cryoscopic measurements). Matsushita et al. [271] polymerized p-MeOSt using BF$_3$–OEt$_2$ in toluene, methylene chloride, and mixtures of these solvents in the temperature range −78 to −20 °C. The authors found that the molecular weight increased with decreasing temperature in toluene, while in methylene chloride no significant temperature dependence was observed. The samples polymerized in the more polar solvent generally had the higher molecular weight. Okamura et al. [272] studied the polymerization kinetics of p-MeOSt initiated by iodine in methylene chloride and carbon tetrachloride and found the polymerization reaction to be much faster in the more polar methylene chloride. Imanishi et al. [273] were using BF$_3$–OEt$_2$ in several chlorinated solvents at −20, 0, and 30 °C to polymerize both o- and p-MeOSt. It was found that both the polarity of the solvent and the temperature did not affect the ratio of monomer transfer to propagation in the case of the *ortho* derivative, but that this rate decreased with decreasing temperature and with increasing dielectric constant of the solvent in the case of the *para* derivative. p-MeOSt was polymerized by Heublein and Dawczynski [249] with an optimized SnCl$_4$/H$_2$O system in methylene chloride at 0 °C. They found the polymerization to proceed 14 times faster than with styrene and the polymer obtained had $M_n = $ 19 000 g/mol (VPO). Several initiating systems have been investigated with the aim to obtain living cationic polymerization of p-MeOSt. Higashimura et al. [205] used iodine in methylene chloride and carbon tetrachloride at −15 °C and 0 °C and found this system to yield long-lived but not really living polymers, with best results in carbon tetrachloride at −15 °C. Under these conditions, the M_w/M_n value was 1.3 to 1.4 at any conversion in monomer addition experiments. Heublein et al. [249] polymerized p-methylstyrene in 1,2-dichloroethane at −15 °C with the initiating systems Ph$_3$CBr/I$_2$ and Ph$_3$CSCN/I$_2$. The propagating species was found to be long-lived and M_w/M_n values of about 1.5 were obtained, which increased to about 2.1 in a monomer addition experiment after two additions. The same authors also used picric acid [249] and triphenylmethylium picrate/picric acid [249] in 1,2-dichloroethane at room temperature. Recently, Higashimura et al. [205] found that HI/ZnI$_2$ in toluene is a really living system not only at −15 °C, but unusually also at 25°C. The polymers obtained had molecular weights of 7000 to 12 000 g/mol and the molecular weight distribution was very narrow ($M_w/M_n = 1.04$ for both temperatures). Even in monomer addition experiments the M_w/M_n values were below 1.1. Sawamoto and Higashimura [274] investigated the living cationic polymerization system HI/ZnI_2 for p-MeOSt in polar and non polar solvents. p-MeOSt can be polymerized in toluene to yield living polymers (PD > 1.1) even at room temperature. When polymerized in a polar solvent (CH_2Cl_2) p-MeOSt results in polymers with bimodal MWDs and are not living. On addition of $nBu_4N^+I^-$ the higher mass polymer is completely eliminated and the resultant polymer fraction had very narrow MWD (PD > 1.1) again. This is due to $nBu_4N^+I^-$ is shifting the equilibrium of the activated species from *not* living dissociated to the living non dissociated form in CH_2Cl_2 as solvent. Hall and co-workers [275,276] used the rather unusual initiators trialkylsilyl triflate in methylene chloride at −78 °C and bis(trifluoromethanesulfonyl)methane in nitroethane and nitromethane/methylene chloride mixtures at 0 °C to obtain poly(p-MeOSt) in good yields, but usually with rather broad molecular weight distribution.

A series of end functionalized poly(p-MeOSt)s has been synthesized by Shohi et al. [277] using functional iodine based initiators with ZnI_2 (α-end functionality) or alcohol quenching reagents (ω-end functionality). Satoh et al. [278,279] published controlled cationic polymerization methods in aqueous media (organic/aqueous phase = 5/3) with $Yb(Otf)_3$—which is known as a unique Lewis acid characterized by its tolerance toward water—at −30 °C. Monomer conversion reached 98% (200 h), PD was ~1.4, molecular weight increased with conversion. The same authors also investigated cationic polymerization of p-MeOSt with BF_3OEt_2/ROH systems at 0 °C in the presence of excess water (PD ~ 1.3) [280]. At high conversions (>90%) a high weight GPC shoulder was observed which suggests a Friedel–Crafts-chain-coupling reaction.

3. Coordination Polymerization

Natta et al. [255] investigated the polymerization of MeOSts with 'modified Friedel–Crafts catalysts' with the aim to obtain stereoregular polymers, which could not be obtained with conventional Ziegler–Natta catalysts due to poisoning of the active centers. While the *m* isomer could not be polymerized at all, both the *o* and *p* isomers gave varying yields of polymers with $AlCl_2Et$ (27%/92%), $AlClEt_2$ (4%/82%), $TiCl_2(OAc)_2$ (trace/7%), and $TiCl_2(OBu)_2$ (0%/12%) in toluene at −78 °C for 6 h. The poly(p-MeOSts) had a much higher molecular weight ([η] = 2) than the poly(o-MeOSts) ([η] = 0.1). Both the *o*- and *p*-MeOSt polymers could not be crystallized, but by catalytic hydrogenation the *ortho* derivative could be converted to crystallizable poly(2-methoxyvinylcyclohexane), which established that the starting polymer was also stereoregular. The *para* derivative could not be hydrogenated despite great efforts and attempts with varying conditions, the reasons being unknown. Therefore, it could not be determined whether poly(p-MeOSt) was also of stereoregular architecture.

4. Anionic Polymerization

Few papers about the anionic polymerization of MeOSts, mostly kinetic studies, have been published. Bumet and Young [281] studied the initiation reaction of n-BuLi with p-MeOSt in hexane. They found the rate of initiation to be proportional to added small amounts of THF and stated the occurrence of side reactions by the appearance of an absorption at 500 nm in the UV spectra. Geerts et al. [282] polymerized both *o*- and *p*-MeOSt with *n*-BuLi in toluene at 20 °C and found severe differences in the polymerization behavior of both monomers. While initiation of the *ortho* isomer was instantaneous and no appreciable termination could be observed, the *para* isomer showed a slow initiation with an induction period and a relatively fast termination reaction. The kinetics of propagation of living poly(*p*-MeOSt) has been investigated by Takaya et al. [283] with cumyl cesium and sodium and potassium (α-methylstyrene tetramers, respectively, as initiators in THF. Kawamura et al. [244] mention the polymerization of *o*-methylstyrene with *n*-BuLi in toluene at −25 °C, resulting in highly isotactic polymer according to ^{13}C-NMR analysis.

5. Living Radical Polymerization

p-MeOSt has been 'living' free radical polymerized in the presence of TEMPO by Devenport et al. [265]. The molecular masses could be controlled ($M_n = 11\,000$–$34\,000$; PD = 1.19–1.34). Qiu and Matyjaszewski [268] found *p*-MeOSt to be the only exception in ATRP among a series of substituted styrenes (see also chapter *ar*-methylstyrene). In this case no polymer was formed by ATRP. The resulting products consist of oligomers, dominantly dimers. The reason could be that the electron donating methoxy group may direct the reaction toward the heterolysis of the C–Br complex bond to generate a cation. Another reason could be the oxidation of the *p*-MeOSt radical by an electron transfer process from a Cu(II) species.

E. *ar*-Chlorostyrene

(56)

Both *o*- and *p*-ClSt are readily polymerizable by radical and cationic mechanisms. Few data have been published about coordination polymerization of *ar*-ClSts, and anionic polymerization is not practicable, due to side reactions of the initiator with the chloro substituent [281].

1. Radical Polymerization

Rubens [284] describes the radical polymerization of *o*- and *p*-ClSt with BPO, γ-irradiation, and thermally without initiator. Thermal polymerization at 70 to 90 °C in bulk showed that the initial rates of polymerization for styrene, *p*-ClSt, and *o*-ClSt are in the ratio 1:4:10.7. Both initiation with BPO and γ-irradiation yielded polymers with respectable molecular weight (DP$_n$ = 1000 to 6000), which was similar to the values obtained by thermal polymerization. Breitenbach et al. [285] published a short note on the emulsion polymerization of *o*-ClSt with and without $K_2S_2O_8$ initiator at 50 °C.

Both systems yielded polymers with molecular weights of about 35 000 g/mol, with the initiated polymerization being three to four times faster than the thermal one. Olaj [286] reports on the kinetics of thermally and AIBN-started radical polymerization of o-ClSt in bulk at 30 °C. He found the rate of polymerization to be about 15 times higher than that of styrene. Propagation kinetics of para substituted styrenes have been investigated by Coote and Davis [246] (see also Section II.C on ar-methylstyrenes).

2. Cationic Polymerization

Polymerization of p-ClSt with CCl_3COOH has been mentioned by Brown and Matheson [287]. Kanoh et al. [247] polymerized the para isomer with iodine in 1,2-dichloroethane at 30 °C and found this monomer to be about one-third as reactive as styrene under the same conditions.

Imanishi et al. [273] studied the reactivity of ClSt monomers toward propagation with $SnCl_4$ CCl_3COOH at 30 °C and found it to increase in the order p-ClSt, o-ClSt, styrene. Brown and Pepper [288] used $HClO_4$ in 1,2-dichloroethane at 25 °C to obtain poly(p-ClSt) with relatively low molecular weight ($M_n = 1380$ g/mol). The rate constant for propagation was found to be 3.5 times lower than that of styrene. Heublein and Dawczynski [249] polymerized p-ClSt with an optimized $SnCl_4/H_2O$ system in methylene chloride at 0 °C. They received a polymer with molecular weight 14 700 g/mol and found the rate of propagation to be about one-fifth that of styrene.

3. Coordination Polymerization

Natta et al. [255] report on the polymerization of all three ar-ClSt isomers. With $TiCl_4/AlEt_3$ (1:3) in benzene at 70 °C for 7 h, they found the m isomer to yield 23% ([η] ≈ 3.5) and the p isomer to yield 28% polymer ([η] ≈ 2.1), while the o isomer failed to polymerize at all. The polymers derived from the m and p isomers were both described to be amorphous. In a later study, Nagai et al. [289] polymerized p-ClSt with $TiCl_4/AlEt_3$ (1:2) in n-heptane at 70 °C and received a polymer, which could be separated in a xylene-soluble and a xylene-insoluble fraction with distinct differences in their IR spectra. Annealing of the insoluble fraction at 160 °C for 1 h resulted in a moderately crystalline material which was characterized by X-ray measurements.

p-ClSt could be polymerized syndiotactic specific with $CpTiCl_3$–MAO catalyst but not with non $η^5$-catalysts like $Ti(OC_4H_9)$-MAO, because these catalytic systems are not 'electrophilic' enough for insertion of the double bond with relatively low electron density [257]. Poly(p-ClSt) was also synthesized with rhenium(V)oxo complexes and half metallocene carbonyl complexes of iron(II) and iron(I) [261]. Using $[FeCpI:FeCpI(CO)_2]$, p-ClSt polymerized much faster than styrene (90% conversion, 25 h) to give polymers with $M_n = 9400$, PD = 1.29. Like allready shown above (ar-methylstyrenes), in the newer metallocene literature chlorostyrene is mainly used for copolymerizations like the copolymerization of p-ClSt and m-ClSt with styrene [290], using $Ti(O-menthol)_4$-MAO-system to yield atactic copolymers. The mechanism was found to be coordinated cationic. Proto and Senatore [291] reported on the synthesis of an alternating ethylene-p-ClSt copolymer, prepared in the presence of a homogeneous zirconium-Ziegler–Natta catalytic system ('Arai-type' catalyst). They found and isotactic p-ClSt arrangement and surprisingly a comparable reactivity of p-ClSt and styrene in the presence of the catalyst. With the catalytic system $CpTiMe_3$-$B(C_6F_5)_3$ the copolymerization of p-ClSt with ethylene only afforded random oligomer fractions [292].

F. Divinylbenzene

(57)

A commercial mixture of *m*- and *p*-divinylbenzene, ethylvinylbenzenes, and diethylbenzenes is produced by dehydrogenation of an isomeric mixture of diethylbenzenes and is used as a cross-linking agent in a large number of different polymer materials. Thermal polymerization of this mixture is easily possible but results in a brittle, highly cross-linked, unsatisfactory polymer [293]. Under certain conditions, however, it is also possible to receive soluble homopolymers of divinylbenzenes by radical, cationic, or anionic mechanisms.

1. Radical Polymerization

Aso et al. [294,295] studied the radical polymerization of pure *o*-divinylbenzene by AIBN in benzene solution at 20 to 90 °C. The products obtained were either totally or at least partially soluble in organic solvents such as aromatic hydrocarbons or chloroform. A conversion of 70% to a totally soluble product could be reached with $[M_0] = 0.6\,mol/L$ at 70 °C. The amount of pendant double bonds in the polymers was determined by use of IR spectroscopy and bromination and found to be 30 to 90% of the maximum value calculated for one double bond per monomer unit. The authors suggested that cyclization had occurred in addition to conventional 1,2-polymerization.

It was found that the amount of pendant double bonds decreased with both increasing monomer concentration and increasing temperature. A kinetic study on the polymerization of *m*- and *p*-divinylbenzene was published by Wiley et al. [296]. Using toluene solutions at 70 °C and BPO as initiator they found the *meta* isomer to polymerize nearly twice as fast as the *para* isomer. The possibility of the preparation of microgels by emulsion polymerization of *p*-divinylbenzene was investigated by Obrecht et al. [297]. They found that in the case $K_2S_2O_8$ was used as the initiator many sulfate radical anions reacted with remaining double bonds of the microgels produced, which resulted in an increased solubility of these products in methanol.

2. Cationic Polymerization

Aso and Kita [294] described the polymerization of *o*-divinylbenzene by cationic initiators in the temperature range −78 to 20 °C in various solvents. They obtained soluble polymers in 10 to 70% yield with moderate molecular weight (3000 to 14 000 g/mol). As with radical initiation they found less than one pendant double bond per monomer unit and suggested a mechanism of intramolecular cyclization polymerization competing with conventional 1,2-polymerization of only one double bond. The ease of cyclization was dependent on the initiator used ($SnCl_4$ $CCl_3COOH > TiCl_4$ $CCl_3COOH > BF_3 \cdot OEt_2$), but no significant influence of the solvent could be observed. In later investigations [294] the effect of various solvents was investigated more closely and it was found that cyclization was favored in less polar solvents, such as CCl_4 or toluene, while very little cyclization was found in polar solvents such as nitrobenzene or acetonitrile. The number of initiator systems was also extended and the tendency for cyclization was found to decrease in the following order: $AlCl_3 > AlBr_3 > SnCl_4 > TiCl_4 > FeCl_3 > BF_3\text{-}OEt_2 > ZnCl_2$. Hasegawa and Higashimura [298] obtained soluble polymers from both pure p-divinylbenzene and a mixture of m and p isomers (70/30) through a proton-transfer polyaddition reaction catalyzed by acetyl perchlorate in benzene and 1,2-dichloroethane at 5 and 70 °C. The polymers consisted of two different structural units: an unsaturated unit (58a), which is produced by the proton-catalyzed reaction of two vinyl groups, as in the cationic dimerization of styrene, and unit (58b), which is the result of conventional 1,2 polymerization.

(58a) (58) (58b)

(58a): unsaturated unit produced by proton-catalyzed reaction of two vinyl groups, (58b): unit as the result of conventional 1,2 polymerization.

It was found that the polymerization yielded exclusively structural unit (58a) at low monomer concentration (0.1 mol/L) and high temperature (70 °C). These polymers were terminated by vinyl groups on both ends (determined by ^{1}H-NMR of oligomers) and were therefore thought to be suitable as telechelics. Both increasing of the monomer concentration and lowering the temperature increased the amount of structural unit (58b) and the tendency toward cross-linking. Soluble polymers with molecular weights up to 25 000 g/mol (GPC maximum) could be obtained by sequential monomer addition, which was needed to keep the monomer concentration low at all times during the polymerization.

3. Anionic Polymerization

Anionic polymerization of o-divinylbenzene was examined by Aso et al. [294]. The authors used n-BuLi, phenyllithium, and naphthalene/alkali metal in THF, ether, dioxane, and toluene at temperatures between -78 and 20 °C. Generally, it was found that as with radical and cationic initiators, a competition between cyclopolymerization and conventional 1,2-polymerization occurs, with the tendency for cyclization to be lower than with the other mechanisms. The polymerization initiated with the lithium organic compounds resulted in polymers with up to 92% double bonds per monomer unit (THF, 20 °C). Polymerization with lithium, potassium, and sodium naphthalene also showed a rather weak tendency for cyclization. In THF at 0 °C and 20 °C the cyclization tendency increased with decreasing ionic radii of the counter cation, while in dioxane the reverse effect was observed, and in ether still another dependence was found (K > Li > Na). Nitadori and Tsuruta [299] used lithium diisopropyl amide in THF at 20 °C to polymerize m- and p-divinylbenzene. The authors obtained soluble products with molecular weight up to 100 000 g/mol (GPC) and showed the polymers to contain pendant double bonds by IR and NMR spectra. It seemed to be important that a rather large excess of free amine (the initiator was formed by reaction of n-BuLi with excess diisopropylamine) was present in the polymerization mixture. In later studies [300,301] a closer view was taken on polymerization kinetics and the steric course of the polymerization reaction.

An interesting application of anionic polymerization is the 'living dispersion polymerization' (LDP) which was reported by Kim et al. [302]. LDP is one of the best methodologies to prepare µm sized polymer particles. The authors did LDP-copolymerization of styrene and divinylbenzene using poly(*t*-butylstyryl)lithium as macromolecular initiator/stabilizer.

4. Living Radical Polymerization

As in the case of anionic living polymerization, the ATRP polymerization allows the synthesis of polymer networks by the end linking process [303]. A difunctional initiator (bis(2-bromopropionyloxy)ethane) allowed the preparation of difunctional polymer precursors that can be used to prepare polymer networks with divinylbenzene-end linking. Divinylbenzene also gives access to a self condensing TEMPO functionalized AB* monomer [304].

G. *p*-Diisopropenylbenzene

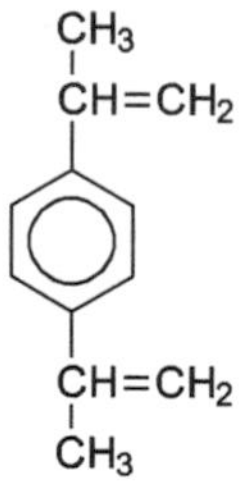

(59)

1. Cationic Polymerization

The cationic polymerization of *p*-diisopropenylbenzene (DIPB) has been studied by Brunner et al. [305] and independently at the same time by Mitin and Glukhov [306]. Under conditions that favor dimerization of (α-methylstyrene, Brunner et al. [305] could prepare soluble polymers, which contained predominantly polyindane structures, with catalysts such as BF_3, $SnCl_4$, H_2SO_4, and H_3PO_4. They visualized the polymerization proceeding according to the following scheme:

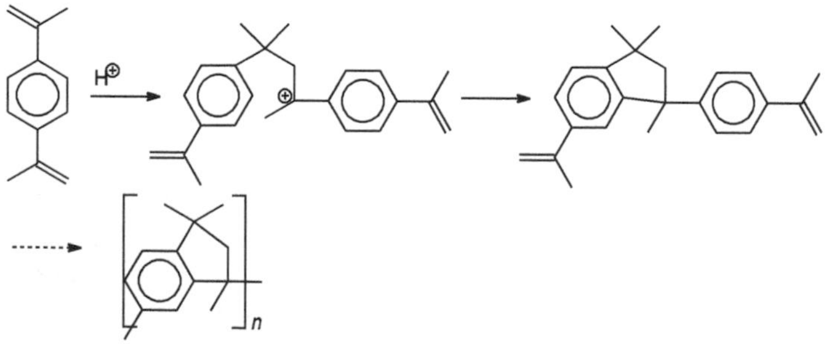

(60)

(60): polymerization mechanism of *p*-diisopropenylbenzene.

Mitin and Glukhov [306] used $SnCl_4/HCl$ in toluene as the catalyst. They obtained a soluble polymer with molecular weight 7000 g/mol. The structure elucidation was based on infrared spectroscopy. Higher-molecular-weight polyindanes were prepared by D'Onofrio [307] using the heterogeneous catalyst system BuLi/TiCl₄/HCl in toluene at 25°C and 100°C. Up to 93% of the product was soluble in toluene and the reduced viscosity at a concentration of 0.2 g per 100 mL was in the range 0.3 to 0.8. A closer view of this polymerization reaction was taken by Dittmer et al. [308]. Using $AlCl_3$, aqueous H_2SO_4, and CF_3COOH as the initiator in 1,2-dichloroethane at 85°C, they obtained polyindanes of moderate molecular weight (up to 5000 g/mol by GPC). Careful structure elucidation by ^{1}H-NMR, ^{13}C-NMR, IR, UV, and EA led to the conclusion that polymers with more than 99% indane structure were formed.

Nuyken et al. [309] took a closer look at the many different structures which may result by cationic polymerization of diisopropenylbenzene. This lead to a strategy to produce telechelic poly(indane)s.

(61)

(61): telechelic poly(indane)s.

Substitution of an alkyl side chain onto the isopropenyl groups of 1,4-diiso-propenylbenzene leads to substituted polyindanes with $T_g = 26$°C and a thermal stability up to 340°C (weight loss 2%).

2. Anionic Polymerization

DIPB can be polymerized with organometallic initiators to form soluble, linear polymers with a pendant isopropenyl group [310–312]. This is because the reactivity of the pendant double bond in the polymer is about three to four orders of magnitude lower than that of the double bond in the monomer [312]. Lutz et al. [310] used 1-phenylethylpotassium as catalyst for living anionic polymerization of DIPB in TBF at −30°C. As long as reaction times were not too long, soluble polymers with moderately sharp molecular weight distributions ($M_w/M_n = 1.09$ to 1.25) were obtained. ^{1}H-NMR showed almost exactly one remaining double bond per monomer unit. Very long reaction times led to very broad molecular weight distributions and eventually to cross-linked polymers. Okamoto and Mita [312] also polymerized DIPB in THF at −30°C using naphthalene sodium as the initiator and showed that not only the conventional polymer bonds but also the cross-links are thermodynamically reversible.

III. VINYL ARENES

(This section was prepared by O. Nuyken.)

This section deals with polymers derived from monomers bearing a vinyl group on an aromatic ring, excluding styrenes and vinyl arenes containing heteroatoms in the ring.

Most interest of vinyl arene containing polymers focused on vinylnaphthalenes, vinylpyrenes, vinylanthracenes, vinylphenanthrenes and vinylbiphenyls. The research work until 1968 was summerized by Heller and Anyos [314].

vinylphenanthrene   vinylpyrene   vinylbiphenyl

vinylnaphthalene   vinylanthracene

The significance of poly(vinyl arene)s bases on their photochemical and photophysical properties, which are employed in a variety of investigations and applications: Substituted poly(1-vinylpyrene)s provide fluorescent materials, whose absorption and emission properties can be tailored by means of the substituents [315]. Vinyl arenes (mostly vinylnaphthalenes) incorporated by copolymerization in polymeric materials are widely employed as fluorescence probes to elucidate the microstructure of polymers [316–320]. A very interesting aspect of poly(vinyl arene)s represents their use as photocatalyst [321], for instance in the photochemical dechlorination of polychlorinated biphenyls [322] or in the synthesis of previtamine D_3 [323]. In this context, the investigation of energy transfer within poly(vinyl arene)s and their copolymers has attracted great interest [324–330]. Other applications employing the photon-harvesting effect represent the copolymerization of 1-vinylnaphthalene and ethylene-propylene-diene terpolymer (EPDM) [331] or styrene-acrylonitrile copolymers [332], which provides materials with improved light resistance. Furthermore, applications of poly(vinyl arene)s were described, which do not rely on the photochemical and photophysical properties of these materials: For instance, vinylnaphthalene or vinylbiphenyl containing copolymers support lithiation reactions [333], as well as the selectivity of polymeric receptors could improved by the incorporation of vinylbiphenyl [334].

Vinyl arene monomers generally can be prepared by dehydration of the corresponding carbinol, which is usually obtained by acetylation of the corresponding arene and reduction of the ketone. The carbinol can also be obtained by the reaction of the arene Grignard reagent with a carbonyl compound. A further method preparing vinyl arenes includes the Wittig reaction of the corresponding aromatic aldehyde.

A. Cationic Polymerization

Cationic polymerizations of vinyl arene monomers have so far failed to produce high molecular weight polymers. Most research work focused on investigations of transfer and termination reactions, as well as isomerizations, which obviously cause the low molecular weights of the products. Bunel et al. investigated the cationic polymerization and copolymerization of 1-vinylnaphthalene, 2-vinylnaphthalene, 1-vinyl-4-methoxynaphthalene, 9-vinylanthracene and 4-vinylbiphenyl [335]. The 1- and 2-vinylnaphthalenes and 1-vinyl-4-methoxynaphthalene could be polymerized cationically under various conditions. Although the yields were quantitative in most cases, the molecular weights of the obtained products were very low. The reactivity ratios for the various vinyl arenes and styrene were determined in methylene chloride with titanium tetrachloride as the initiator. They were found to be in good agreement with those calculated by quantum chemical means [335].

In 1958, Bergmann and Katz reported the rapid cationic polymerization of 9-vinyl-anthracene [336]. Michel showed in 1964 that this reaction does not lead to the expected 1,2-addition products [337]. Instead, a predominantly across-the-ring polymerization takes place. This agrees with investigations about the isomerization of the carbocation of 9-vinylanthracene [335].

In a series of studies Coudane et al. investigated the cationic polymerization of different vinyl arenes [338]. The behavior of the various monomers was described as follows: 1- and 2-vinylnaphthalenes, 1-, 2- and 9-vinylanthracenes and 2- and 3-vinylphenanthrenes polymerize to low molecular weight products; 2-propenyl-2-naphthalene, 2-propenyl-2-anthracene and 2-propenyl-2-phenanthrene give low molecular weight polymers or oligomers; 2-propenyl-9-anthracene does not polymerize cationically and 2-propenyl-1-anthracene and 2-propenyl-9-phenanthrene give only dimers. Schulz et al. described the quantitative oligomerization of 2-propenyl-2-naphthalene by means of different initiators [339]. According to Coudane et al. [338], steric hindrance caused by hydrogen atoms in the 'peri' position to the isopropenyl group is responsible for the differences in the polymerization of these monomers. When there is at least one hydrogen in this position, the isopropenyl group turns out of the aromatic ring plane. This causes deconjugation and destabilization of the carbocation occurs.

Blin et al. determined the transfer and termination constants for the cationic polymerization of 1- and 2-vinylnaphthalenes and 3-vinylphenanthrene in methylene chloride and titanium tetrachloride as initiator [340]. For each monomer these constants increased with increasing temperature. Their values were significantly larger than those of styrene at the same conditions. This is supposed to be a consequence of the existence of highly reactive aromatic sites, which permit Friedel–Crafts attacks by the growing chain.

Stolka et al. studied the polymerization of 2-vinylanthracene and 2-propenyl-2-anthracene [341]. They found that if highly purified these monomers do not polymerize cationically. However, 2-vinylanthracene, which was repeatedly crystallized for purification, apparently contained some cocatalytic species, because in this case the monomer could be polymerized in methylene chloride by boronium trifluoride. The maximum molecular weight of this product was $M_w = 17\,200\,\mathrm{g\,mol^{-1}}$ determined by GPC and membrane osmometry.

The cationic polymerization of 9-vinylanthracene, 9-vinyl-10-methylanthracene, 1-vinylanthracene and 2-propenyl-1-anthracene was also investigated by Pearsons group [342]. The latter two monomers failed to polymerize cationically, while the first gave polymers of low molecular weight at different conditions. Several groups investigated the polymerization of acenaphthalene by means of various initiators and solvents [343–347]. The reaction takes place in two stages: a fast initial reaction followed by a propagation reaction at a lower rate [345,346]. The propagation rate is first order with respect to the monomer. The degree of polymerization was found to be independent of both the monomer and the initiator concentration [346,348]. The polymerization proceeds up to complete conversion without any loss of active centers. Monomer addition experiments showed the same propagation rate as for the first reaction. They showed neither an initial jump nor an induction period, indicating that the active centers were already created and the initiation was quantitative [346]. It is noteworthy that the molecular weight of the polymers were unaffected by the second addition of monomer [345,346]. According to Anasagasti et al. the constancy of the molecular weights may be a consequence of competition between transfer to monomer processes and propagation processes [346].

Electro-initiated polymerizations of various vinyl arenes led to low molecular weight products. The propagation reaction of these polymerization appears to proceed via a cationic mechanism [349–352].

B.   Anionic Polymerization

Pearson et al. investigated the anionic polymerization of various vinyl arene monomers [341,342,353,354]. They found that highly purified 1-vinylanthracene [342], 2-vinylan-thracene [341], 2-propenyl-1-anthracene [342] and 2-propenyl-2-anthracene [341] can be polymerized by living polystyrene and α-methylstyrene tetramer dianion as the initiator. The polymerization were effective at temperatures below −40 °C and optimum results were obtained at −78 °C. With increasing temperature both the yield and molecular weight of the polymers decrease and the molecular weight distribution becomes broader. The low solubility of the monomers in typical solvents for anionic polymerizations at low temperatures results in extremely low polymerization rates. The reactions proceed in a heterogeneous system with the soluble portion of the monomer, which is consumed in the polymerization, being replenished from nondissolved particles. High yields and high molecular weights can only be obtained by polymerization over a period of several days. This results in broad molecular weight distributions of the polymers (often $M_w/M_n > 2$).

The anionic polymerization of 9-vinylanthracene gives only low molecular weight products [342], which agrees with Rembaum's and Eisenberg's results [355]. Stolka et al. [342] found no proof of the proposed [337,355] across-the-ring addition; instead, the IR and UV spectra of their polymers indicated the conventional 1,2-addition pattern. 2-Propenyl-1-anthracene could not be polymerized anionically [342]. Attempts to initiate polymerizations by means of electron-transfer-type initiators (e.g., sodium naphthalene and sodium biphenyl) were unsuccessful [341,342,353,354]. The polymerization of 1-vinylpyrene initiated by electron-transfer initiators showed the characteristics of a 'living' polymer system [356,357]. Block copolymers of the AB and ABA type were synthesized with ethylene oxide, styrene and isopropene [357].

Schulz et al. investigated the homo- and copolymerization of 2-propenyl-2-naphthalene by n-butyllithium in tetrahydrofuran at −78 °C [358,359]. Homopolymeriza-tion as well as copolymerization led to quantitative yields and narrow molecular weight distributions. Molecular weights of poly(2-propenyl-2-naphthalene), determined by light scattering measurements, yielded values up to 270 000 g mol^{-1} [358]. Diblock and triblock copolymers of 2-propenyl-2-naphthalene and 1,3-butadiene were synthesized and characterized by DTA, DSC, GPC, IR, electron microscopy and stretch-strain measurements [359]. The molecular weights of these polymers range between 66 000 and 230 000 g mol^{-1}, the M_w/M_n values between 1.08 and 1.25. 2-Vinylnaphthalene and 2-propylene-2-naphthalene can also be polymerized by means of *sec*-butyl lithium as initiator and toluene as the solvent [360]. The anionic polymerization of acenaphthalene by different initiators led to colored, low molecular weight products [361].

C. Ziegler-Natta Polymerization

In 1958, Natta et al. published a study on Ziegler–Natta polymerizations of various vinyl monomers [362]. Natta and his co-workers showed that these polymerizations are very sensitive to steric hindrance at the double bound. In addition, Ziegler–Natta catalysts can generate a series of Lewis acids, which lead to complex side reactions. Therefore, only few attempts have been made to polymerize vinyl arenes by Ziegler–Natta catalysts. Heller and Miller obtained stereoregular polymers consisting of 1-vinylnaphthalene, 2-vinylnaphthalene and 4-vinylbiphenyl [363]. Polymerization by triethyl aluminum/titanium tetrachloride gave polymers in 75% to 90% conversion, which were characterized by IR and ^{1}H-NMR spectroscopy and found to be at least 90% isotactic. Only 1-vinylnaphthalene produced a crystallizable polymer [362,363].

Stolka et al. investigated the polymerization of 2-vinylanthracene and 2-propenyl-2-anthracene by means of triethyl aluminum/titanium tetrachloride [341]. The latter monomer gave soluble polymers with no stereoregularity, while the first gave completely amorphous and insoluble products. 3-Vinylpyrene can be polymerized up to 90% conversion by triethyl aluminum/titanium tetrachloride. The reaction led to poorly crystalline polymers. It is suggested that a cationic process is involved in this polymerization [364]. In 1990, Benito et al. described the polymerization of 2-vinylnaphthalene by means of triethyl aluminum/vanadium trichloride, which led to amorphous, low molecular weight polymers [365].

D. Free-Radical Polymerization

The AIBN-initiated bulk polymerization of 1-vinylnaphthalene led to low molecular weight polymers. The molecular weights are limited by a chain transfer reaction with the monomer; the chain transfer constant is about 300 times larger than that of styrene [366]. The bulk polymerization of 2-vinylnaphthalene led to products with molecular weights of 66 000 g mol^{-1}. Molecular weights of 115 000 g mol^{-1} for poly(2-vinylnaphthalene) and 250 000 g mol^{-1} for poly(1-vinylnaphthalene) were obtained by means of emulsion polymerization techniques [367]. Enzyme-mediated free radical polymerization of 2-vinylnaphthalene at ambient temperature gave polymers with $M_n = 116 000$ g mol^{-1} and a M_w/M_n value of 2.28 in 91% yield [368]. The free radical copolymerization of 1-vinylnaphthalene with styrene, methylmethacrylate and acrylonitrile was carried out to elucidate the relative reactivity ratios of the different monomers [369]. Poly(sodium styrenesulfonate-*block*-2-vinylnapthalene) could be obtained by means of nitroxide-mediated 'living' radical polymerization [370]. The nitroxide controlled radical polymerization was also employed on investigations about the monomer reactivity of styrene and 4,4′-divinylbiphenyl in copolymerization experiments [371]. The copolymerization of 2-propenyl-2-naphthalene and maleic acid anhydride by AIBN at 60 °C in toluene led to alternating copolymers [372,373]. Engel and Schulz showed that in these system the copolymerization of the two monomers competes with the formation of the Diels-Alder adduct [373]. 2-Propenyl-2-naphthalene did not homopolymerize radically.

The free radical polymerization of several vinylanthracenes was investigated by Stolka et al. [341,342]. 2-Vinylanthracene can be polymerized quantitatively by di-*tert*-butyl peroxide in solution, but high molecular weights could not be obtained [341]. Chain transfer reactions were favored by the high temperatures required for solubilization of the monomer. 2-Propenyl-2-anthracene could not be polymerized radically [341]. Bulk polymerizations of 9-vinylanthracene and 9-vinyl-10-methylanthracene gave only low molecular weight polymers [342,355]. Polymerization of 1-vinylanthracene by AIBN in benzene at 55 °C led to high yields (>70%) of low molecular weight polymers [342]. A series of studies about copolymerizations of various vinyl arene monomers with styrene, methyl methacrylate and methyl acrylate showed that all vinyl arenes except 9-vinyl-anthracene are more reactive in copolymerization than the respective comonomer [374–376]. The addition of styrene or methyl methacrylate to vinylnaphthalene generally increases the molecular weight of the copolymers [367,375]. As in homopolymerization, emulsion copolymerizations produce copolymers having higher molecular weights relative to those prepared by bulk polymerization [367]. The copolymerization of vinylanthra-cenes and vinylphenantrenes with styrene showed the same order of decreasing acitivity (9-vinylphenantrene > 1-vinylanthracene > 9-vinylanthracene) as the homopolymerization [376,377].

High molecular weight polyacenaphthalene could be obtained by polymerization with free radical initiators (e.g., benzoyl peroxide) [378] or by thermal initiation [343,379]. Functionalized poly(acenaphthalene)s of low molecular weight were synthesized by Springer and Win, using 4,4′-dicyano-4,4′-azodivaleric acid as initiator [361]. The copolymerization behaviour of acenaphthalene and styrene, methyl methacrylate, maleic acid anhydride and some other comonomers were examined by Ballesteros et al. [380]. The ring strain in acenaphthalene is responsible for its good polymerizability. Acenaphthalene is preferentially combined into each copolymer synthesized in this work. The authors showed that the Q, e scheme is not useful in describing the copolymerization of such steric hindered monomer [380].

IV. N-VINYLCARBAZOLE

(This section was prepared by O. Nuyken.)

Poly(N-vinylcarbazole) (PVK) is a vinyl aromatic polymer produced from the monomer N-vinyl-carbazole (NVK):

(62)

It was first prepared by Reppe and co-workers in 1934 [381]. The polymerization reaction as well as the structure and properties of NVK and PVK are largely controlled by the electronic and steric influence of the carbazole group. The nitrogen atom donates its unshared electron pair, thus creating an electron-rich double bond. NVK can be poly-merized by radical initiators and by cationic initiators but fails to polymerize anionically.

Due to the bulkiness of the carbazole groups the polymer chains are stiff in nature and the polymer has a glass transition temperature of 227 °C [382], which is among the highest known for vinyl polymers. PVK exhibits excellent thermal stability up to at least 300 °C [383–385]. Unfortunately, this property has never been fully utilized because of the extreme brittleness of the material. PVK is soluble in common organic solvents such as benzene, toluene, chloroform, and tetrahydrofuran.

In its early days PVK was mainly used as a high-temperature dielectric material in the electric industry. Although it served as a replacement for mica in the 1940s, its extreme brittleness and poor mechanical and processing characteristics severely limited its applications, and in time it was replaced by materials with improved properties. The discovery of photoconductivity in PVK by Hoegl [386,387] in 1959 created a strong industrial interest in photoelectrically active organic materials. In IBM's Copier I series introduced in 1970, an organic photoconductor, a charge transfer complex of PVK with 2,4,7-trinitrofluorenone, was utilized commercially for the first time. Today, organic photoconductors are frequently used in photocopiers, laser printers, and electrophotographic printing plates. Additionally copolymers of NVC [388,389] and doped homopolymers [390,391] are interesting materials for organic light emitting diodes (OLEDs) as hole transport materials.

In recent years several papers have been published that deal with the electrochemical polymerization of NVK [392–394]. This type of polymerization yields polymers that are already doped and exhibit electrical conductivities up to ca. 10^{-4} S/cm. The polymers can be discharged and recharged several times without decomposition, and their use as an electrode material for a lithium battery has been proposed [395]. The properties of PVK have been compiled in a number of review articles. A very detailed description of the chemical and physical properties of the polymer is given in a book by Pearson and Stolka [396] and in two review articles [397,398]. The methods of polymerization have been reviewed by Börner [399] and by Sandler and Wolf [400]. In his review Rooney focuses on the cationic polymerization of NVK [401].

A. Monomer Synthesis

NVK is a white crystalline material that melts at 65 °C. It is soluble in most aromatic, chlorinated, and polar organic solvents. The material should be handled with care because it may cause severe skin irritations. Like other aromatic amines, NVK is suspected to cause cancer [402]. Most published synthetic routes use carbazole, which is readily available from coal-tar distillation, as the starting material. The industrial process is the vinylation reaction of carbazole and alkali metal hydroxide or preformed alkali metal salts of carbazole with acetylene [381,403,404]. The reaction is normally carried out in high-boiling solvents at slight pressures.

Another procedure involves the dehydration of N(2-hydroxyethyl)carbazole [405,406], which is obtained by the reaction of potassium carbazole with ethylene oxide or 2-chloroethanol. Dehydrochlorination of N(2-chloroethyl)carbazole was reported to be a convenient laboratory scale method [407,408]. Several other methods, such as direct vinylation with ethylene catalyzed by PdCl$_2$ [409] or with vinyl chloride [410] and some transvinylation reactions, have been reported. Although a variety of purification techniques have been developed [404], complete elimination of impurities from NVK prepared from coal tar seems to be impossible. NVK from coal tar contains substantial amounts of sulfur compounds as well as a variety of condensed aromatic impurities such as anthracene. For scientific purposes, where high-purity NVK and PVK are required, synthetic carbazole prepared from cyclohexanone and phenylhydrazine [408,411] should be used as a starting material.

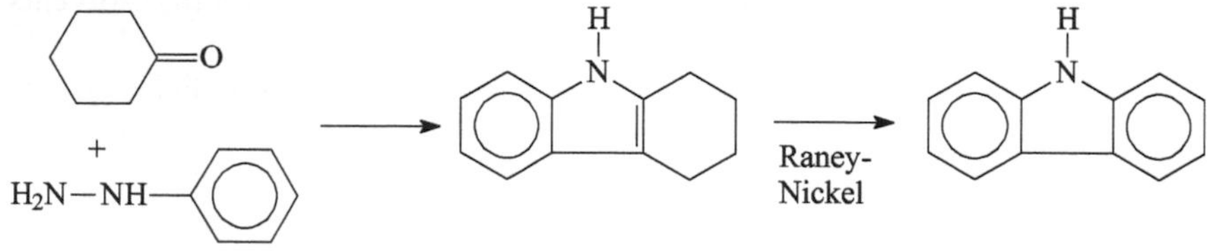

(63)

B. Radical Polymerization

Free-radical initiators such as azobisisobutyronitrile (AIBN) or peroxides readily polymerize NVK. The reaction rate of free-radical polymerization in cyclohexanone at $70\,^{\circ}C$ using 0.01 mol% AIBN as initiator has been determined [412]. NVK polymerizes 10 times faster than styrene.

The termination rate constant for NVK does not satisfy the Arrhenius relationship and becomes markedly reduced when the temperature is lowered to $-30\,^{\circ}C$. This result has been attributed to a hindered rotation of the stiff PVK macroradicals, which markedly retards the diffusion-controlled termination reaction.

The heat of polymerization was determined by Rooney [413], Bowyer and Ledwith [414], and Rodriguez and Leon [415] to be $92\pm4\,kJ/mol$. Almost colorless polymer with high molecular weight can be achieved by bulk polymerization of NVK initiated with AIBN [416]. Besides bulk polymerization, which is carried out as a technical process, NVK can be polymerized in suspension [417] and by precipitation polymerization [418].

The electron-releasing carbazole group facilitates electron transfer even to weak electron acceptors by stabilizing the resulting carbocation. The large amount of low-molecular-weight polymer in the polymerization of NVK formed by a cationic process initiated by benzoyl peroxide [419,420] or AIBN in the presence of CCl_4 [421] and CBr_4 [422,423] in aprotic solvents can be explained by this type of reaction. Both the reaction kinetics and the polymer product differ markedly from the AIBN-initiated reaction [424,425].

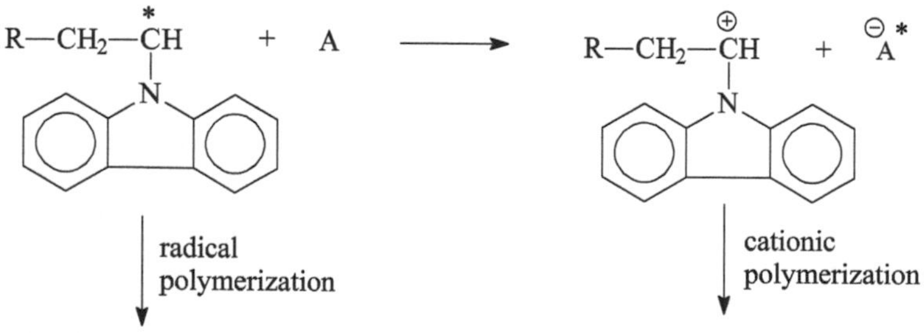

(64)

Poly(styrene-co-NVK) copolymers and NVK-homopolymers with controlled molecular weight and narrow polydispersities can be synthesized by N-oxyl-controlled free radical copolymerization using benzoyl peroxide as initiator and 2,2,6,6-0tetramethyl-piperidine-N-oxyl as terminating agent. The copolymerization behaves in a living fashion and allows the synthesis of blockcopolymers [426,427].

C. Cationic Polymerization

As a strong basic monomer, NVK undergoes cationic polymerization with almost all cationic initiators, such as protonic acids, Lewis acids, metal salts, and organic cation salts. Since many of these compounds are also electron acceptors toward electron-rich monomers such as NVK, both addition and electron-transfer modes of initiation have to be considered. Because of the high reactivity of NVK, it is often difficult to distinguish between these two initiation processes.

1. Protonic and Lewis Acids

Probably the most obvious method for the cationic polymerization of NVK involves treatment of the monomer with acids. So NVK can be polymerized by hydrochloric acid [428], perchloric acid [429,430], ammonium perchlorate [431] and several other Bronsted acids [432]. Due to the nucleophilicity of NVK, even relatively weak carboxylic acids will cause polymerization [433]. In dichloroethane, a solvent of medium polarity, oligomers bearing a terminal acetate ester were isolated from the reaction of NVK with an excess of acetic acid [434]. This suggests that proton addition to the β carbon of the monomer is followed by the addition of the counterion to the α carbon, forming a polarized ester. The lability of the ester linkage was demonstrated by the rapid hydrolysis of the monomeric adduct in the presence of atmospheric moisture.

The cationic polymerization of NVK was reported as early as 1937 by Reppe and co-workers [435], who used boron trifluoride as initiator. The polymerization is normally carried out at low temperatures ($-78\,^\circ$C) in dichloromethane. With highly purified monomer and solvent and an optimum amount of initiator, high-molecular-weight polymer is obtained in quantitative yield [399,429]. The amount of catalyst required for complete conversion is very small (2×10^{-5} mol of BF$_3$ etherate per mol of monomer). With more than 10^{-4} mol of BF$_3$ etherate per mol of NVK, a crosslinked polymer is obtained. Several other compounds, such as arsenic trichloride [436], metal nitrates [437,438], and sodium tetrachloroaurate [439] were used as initiators for the cationic polymerization of NVK. Despite the fact that diagnostic tests indicate that most of the polymerizations discussed above are cationic in nature, only a small amount of kinetic and mechanistic information is derived from these experiments. The principal difficulty is that the concentration, and sometimes even the identity, of the initiating species is unknown. Furthermore, the nature and the extent of dissociation of the counterion is not well understood. This lack of knowledge prevents the determination of the fundamental kinetic parameters and therewith a detailed understanding of the polymerization reaction. One approach to overcoming these difficulties involves the use of organic cation salts as initiators.

2. Organic Cation Salts

Not all organic salts are able to initiate the polymerization of NVK. While stable ammonium salts such as benzyltrimethylammonium hexafluoroantimonate do not initiate the polymerization, stable ammonium and tropylium salts will. A kinetic study of NVK polymerization initiated by stable tropylium hexachloroantimonate and perchlorate salts provided the first propagation rate constants for that monomer [440]. The authors showed that initiation involves direct addition of the tropylium ion to NVK via formation of a charge transfer complex [441]. This mechanism was proved by isolation of the corresponding methanol adduct. In this particular system the initiation reaction was almost instantaneous and quantitative, and no chain termination was observed. Under these conditions the propagation rate constants can be determined from conversion-time plots. The propagation rate coefficient for the free cation species was $4.6 \times 10^5\,\text{L/(mol s)}$ for the tropylium hexachloroantimonate salt in dichloromethane at $0\,°C$. This value is about five orders of magnitude larger than the rate coefficient for free-radical propagation.

(65)

Even optically active PVKs can be prepared by using chiral organic salts [442].

3. Iodine and Hydrogen Iodide

Not long ago criteria for the generation of living cationic polymerizations have been developed [443–445]. These criteria include an almost instantaneous and quantitative initiation and the absence of chain transfer and termination reactions. These requirements were satisfied when iodine was used to polymerize NVK in dichloromethane solution containing tetrabutylammonium iodide at low temperatures [446,447]. Polymer molecular weights increased in a linear fashion with monomer consumption during the reaction and upon the addition of successive aliquots of monomer.

Sawamoto et al. [448] used hydrogen iodide as initiator for the living cationic polymerization of NVK in toluene and dichloromethane solution. In dichloromethane with added tetrabutylammonium iodide at $-78\,°C$, the polymerization is in a living fashion, demonstrated by the linear dependence of the polymer molecular weight from the monomer/initiator ratio. The polymers had a molecular weight distribution $(M_w/M_n = 1.2)$ markedly narrower than that for iodine as initiator $(M_w/M_n = 1.5)$. S. Oh [449] used the initiator system 1-iodo-1-(2-methylpropyloxy)-ethane/tetrabutyl-ammoniumperchlorate/tetrabutylammonium iodide at $-25\,°C$ in dichloromethane and got PVK with very narrow molecular weight distribution $(M_w/M_n \leq 1.15)$ as well.

D. Charge Transfer Polymerization

Polymerization reactions involving charge transfer interactions have been of considerable interest since Scott et al. [450] and Ellinger [451] reported independently in 1963 that donor monomers such as NVK could be polymerized spontaneously in the presence of electron-accepting species such as tetracyanoethylene and chloranil. Since then, a large number of publications on the charge transfer polymerization of NVK have appeared which are summarized in several review articles [452–455]. Unfortunately, many of these investigations have contributed to the mechanistic confusion, since purity considerations and possible side reactions were ignored. Studies dealing with the polymerization of NVK with chloranil and related compounds must be interpreted with caution, since these compounds are difficult to free from traces of acidic impurities such as 2-hydroxy-3,5,6-trichloro-1,4-benzochinone [456,457]. A different problem arises when NVK is polymerized by tetranitromethane [458]. This compound was shown to undergo a sequence of reactions leading to nitroalkanes and trinitromethane, which acts as cationic initiator [459].

Charge transfer polymerizations are frequently accompanied by other, non-polymer-producing processes. The NVK/tetracyanoethylene system has been studied in detail [460,461] and provides an excellent example of the complexities encountered in charge transfer polymerization. The complicated reaction pattern that ensures on mixing NVK with electrophilic alkenes has been elucidated in 1986/87 [462,463].

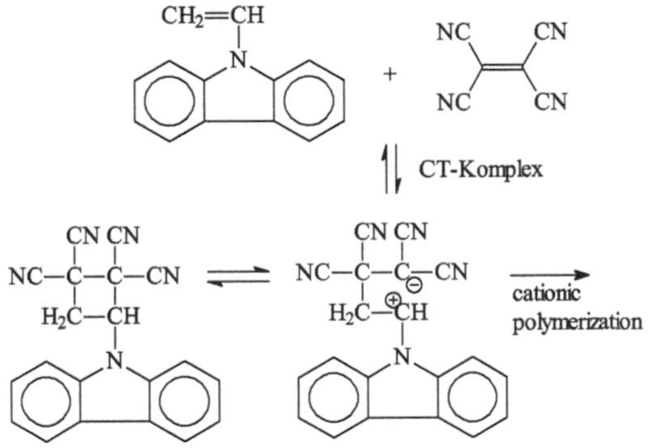

(66)

The zwitterionic tetramethylene intermediate has been trapped with methanol as a linear 1-methoxybutane, indicating that the addition of tetracyanoethylene to NVK occurs in a stepwise manner. Once the zwitterions are formed, the course of the reaction is determined largely by experimental conditions. With excess NVK, cationic polymerization occurs. If the concentrations of NVK and tetracyanoethylene are comparable, collapse of the intermediate to the corresponding cyclobutane is favored. The cyclobutane formation is reversible, and the isolated cyclobutane alone is capable of initiating the cationic polymerization of NVK. The polymerization of NVK by the stable radical cation salts phenothiazin hexafluoroantimonate [464] and tris-p-bromophenylamminium hexachloroantimonate [465] were shown to occur via charge transfer initiation too.

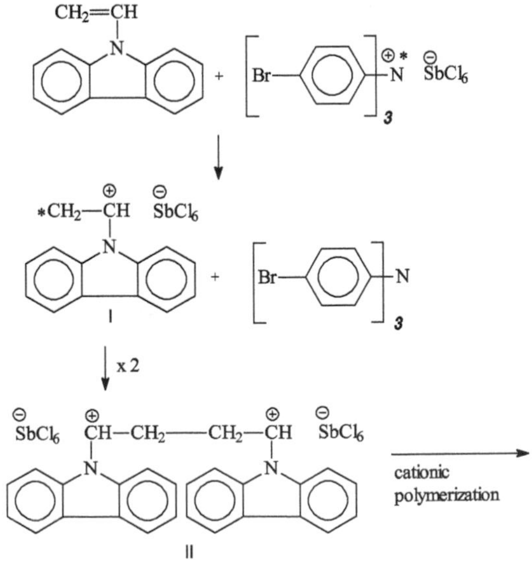

(67)

In the case of the polymerization with tris-*p*-bromophenylamminium hexachloroantimo-nate the existence of the dication (II), which is probably formed by dimerization of the radical cation (I), was confirmed by a trapping experiment with methanol from which the 1,4-bismethoxy derivative could be isolated. So the initiation by stable radical cation salts can occur by both addition of the cation to the double bond of NVK as in the case of tropylium salts and by electron transfer, such as with tris(*p*-bromophenyl)amminium salts. As described throughout this section, NVK is readily polymerized by strong electron acceptors such as tetracyanoethylene. With many weak acceptors, however, thermal activation is required to induce polymerization, and usually no appreciable reaction occurs at room temperature in the dark. Many of these systems, including charge transfer complexes of NVK and maleic anhydride, pyromellitic dianhydride, fumaronitrile, and dimethyl terephthalate, can readily be polymerized by irradiation [453,466]. These photo-induced reactions of NVK with weak electron acceptors show multireaction courses, including cationic homopolymerization of NVK, cyclodimerization, radical homopoly-merization, and copolymerization of NVK with an electron-accepting monomer. These competitive reaction pathways have been investigated by several groups, and a generalized reaction scheme based on solvent basicity has been presented in a review article by Shirota and Mikawa [453].

Specially designed allyl-onium salts which initiate via addition-fragmentation reac-tions are shown by Yagci [467]. In these systems, the polymerization conditions can be tuned not only to the desired wavelengths but also to temperature ranges by choosing appropriate radical initiators. In the first step photochemically or thermally generated free radicals add to the olefinic double bond of allyl thiophenium, allyl pyridinium, allyl alkoxy-pyridinium and allyl phosphonium salts. The intermediate thus produced is relatively unstable and prone to fragmentation. In the second step a new radical is generated which initiates the polymerization.

(68)

E. Solid-State Polymerization

Solid-state polymerization of a monomer, if it can be achieved topochemically, is an interesting approach to polymer single crystals. Several publications describing the poly-merization of NVK in the crystalline state initiated by high-energy radiation [468–471], cationic catalysts [472], hologen vapors, [473,474], and redox systems [475] have been published. However, the crystal structure of NVK monomer [476] provides strong evidence that these polymerizations are not topochemical since the arrangement of the monomer units in the crystal lattice does not align the vinyl groups in a position favorable for polymerization. This would suggest that all solid-state polymerizations of NVK occur by a mechanism involving monomer reorientation due to local melting of the crystal. The fact that the PVK produced in all these polymerizations is amorphous is also consistent with a nontopochemical polymerization.

V. N-VINYLPYRROLIDONE

(This section was prepared by O. Nuyken, W. Billig-Peters and T. Frasch.)

Poly(N-vinylpyrrolidone) (PVP) has become quite an important polymer as it can be used as a blood plasma extender or substitute, due to its low toxicity. PVP is obtained mainly by radical polymerization in solution with hydrogen peroxide or azobisisobutyronitrile (AIBN) as initiator. Cationic, charge-transfer, and radiation initiation are also possible. Polymerization can be carried out in bulk (including solid state), in solution, or in suspension. Molecular weights between 2500 and 10^6 g/mol are described. The glass transition temperature depends on the molecular weight; T_g values up to 175 °C can be observed. PVP is able to form complexes with various compounds (e.g., with iodine), which leads to effective disinfectants of very low toxicity. Although it is no longer used as a blood plasma extender, it is still of importance for applications such as adhesives, textile auxiliaries, and dispersing agents and is applied in pharmaceutical material, and detergents.

In most publications, particularly in commercial data sheets, the term N-vinylpyrrolidone with the abbreviation NVP is used for the monomer and PVP is used for the corresponding polymer; therefore, these abbreviations are used here. However, frequently other names are used, such as N-vinyl-2-pyrrolidone; N-vinyl-2-pyrrolidinone; 1-vinyl-2-pyrrolidinone; 1-vinyl-2-pyrrolidone; 1-vinylpyrrolidin-2-on; 1-ethenyl-2-pyrrolidinone; and 2-pyrrolidinone-1-ethenyl.

$$\text{PVP} \qquad \text{NVP} \qquad (69)$$

A.   Monomer Synthesis

NVP is one of the various products of the acetylene chemistry [477] discovered by Reppe [478–481]. The reaction between acetylene and formaldehyde yields 1,4-butinediol, which is hydrogenated to 1,4-butanediol. After the oxidative cyclization to γ-butyrolactone on Cu contact (700 °C), the reaction with ammonia leads to γ-hydroxycarbonamide, which is cyclosized to yield α-pyrrolidone. At high temperatures and with alkaline as catalyst, α-pyrrolidone and acetylene react and form NVP.

$$HC\equiv CH + 2\ H_2C=O \longrightarrow HO-CH_2-C\equiv C-CH_2-OH$$

$$\downarrow +2\,H_2$$

$$HO-CH_2\text{-}CH_2\text{-}CH_2\text{-}CH_2-OH \xleftarrow{\ -2\,H_2\ }$$

(70)

$$HC\equiv CH \longrightarrow$$

B. Radical Polymerization

1. Bulk Polymerization

In one of the first bulk polymerizations of NVP [482–486], in which H_2O_2 was used as initiator, the resulting products of PVP were yellowish brown, nonuniform, and of low molecular weight. Initiation with benzoyl peroxide did not give satisfactory results either [482,483,485,486]. The initiator of choice is AIBN [487]. Breitenbach et al. [488,489] have polymerized NVP by heating it over 100 °C in the presence of AIBN. The same authors demonstrated that initiator-free polymerization is also possible. AIBN as initiator fulfilled the conditions of radical-induced vinyl polymerization (e.g., the rate of polymerization was found to be proportional to the square root of the initiator concentration). However, the average degree of polymerization is nearly independent of the AIBN concentration [482,488,489], indicating that transfer may play an important role in this process. The energy of activation of the transfer reaction between the growing chain and the monomer is not very much higher than that of the growing reaction [482]. Bulk polymerization processes have also been used to study the copolymerization of NVP with a variety of monomers, such as methacrylates and vinyl chloride, and reactivity ratios of several copolymer systems have been determined [482]. A slow polymerization reaction is observed in the presence of daylight and by heating over 140 °C without initiators [482,483,488,489]. The resulting polymer is a transparent, glassy material.

Luberoff et al. [490] employed as initiators metal salts such as chloride, bromide, iodide, sulfate, and nitrate of mercury(II); the chloride, bromide, and iodide of bismuth; and the chloride of antimony. The mechanism of polymerization with these initiator systems is still unknown, but the experimental results show that the systems are neither Friedel–Crafts nor free-radical catalysts.

2. Aqueous Solution Polymerization

Radical initiation in solution is the most important method for the synthesis of linear poly(N-vinylpyrrolidone) with weight-average molar masses from 2500 to 1 million. The relevant techniques are described in detail by Bömer [491], Sandler and Karo [492], and Kern and Cherdron [482]. Polymerization of NVP is carried out in neutral or basic media to avoid decomposition into pyrrolidone and acetaldehyde as it occurs in acid media if hydrogen peroxide or other peroxides are applied as initiators [484]. In early patents [493–495] for the polymerization of NVP, sodium (or potassium) sulfite is claimed to function as polymerization initiator [492]. The most important initiators for technical processes are redox systems from hydrogen peroxide as oxidizing agents and ammonia or organic amines as reduction agents [496–498]. The amines also function as buffering agents [496–499]. Furthermore, AIBN is found to be an efficient initiator for the polymerization of NVP [497,498] without the need of a buffering agent [487,499–501].

The kinetics of NVP polymerization with redox initiators in aqueous solution have been studied in detail, especially for the system hydrogen peroxide-ammonia [482,491, 492,496,498]. According to these studies, the rate of polymerization increases with the amount of H_2O_2, whereas the average molecular weight decreases synchronously. The rate of polymerization also increases with the square root of the ammonia concentration. In contrast to this, the average molecular weight is independent of the ammonia concentration. The influence of the monomer concentration is almost peculiar. It has some effect on the rate of polymerization and the induction period. With increasing monomer concentration up to approximately 30% (per weight) monomer in the water solution, the rate of polymerization increases. From 30% to approximately 60% (per weight), the rate remains constant. Above this concentration it declines sharply. The average molecular weight is independent of the monomer concentration and temperature. All polymers, which are obtained in aqueous solution using H_2O_2 as initiator, are glossy and transparent. They have an hydroxy and a formyl end group, which are formed during chain termination by splitting off pyrrolidone. Therefore, PVP produced in aqueous solution contains small amounts of pyrrolidone. There is no evidence for other chain termination reactions. Breitenbach and Schmidt [488,489], Bond and Lee [499], and Senogles and Thomas [487,500] and Lizravi [501] have studied in detail the AIBN initiation of the polymerization of NVP in aqueous solution. This reaction is autocatalytic from the start, even in fairly diluted systems.

Initiation (aqueous solution):

$$H_2O_2 \longrightarrow 2\,HO*$$

Propagation (aqueous solution):

(71)

Termination (aqueous solution):

An obvious difference between bulk and aqueous solution polymerization is the higher molecular weight of polymers synthesized in water. In general, it has to be noted that the rate of polymerization is proportional to the square root of the AIBN concentration. Using a viscometric technique for kinetic studies of the polymerization of NVP, Bond and Lee [499] have confirmed the observation of Breitenbach and Schmidt [488,489] that the degree of polymerization of PVP is not influenced significantly by the concentration of initiator or by the reaction temperature. Beside AIBN, dialkylperoxides such as tert-butyl(2,2-dimethylpropanoyl)peroxide [502] or peroxodiphosphate-Ag$^+$ [503] are employed as initiators for the polymerization of NVP in aqueous or aqueous/alcohol solutions.

3. Solution Polymerization in Organic Solvents

One of the earliest examples of solution polymerization process is found in the patent of Schuster et al. [485]. There, NVP is heated in ethanol solutions with H_2O_2 and benzoylperoxide. Breitenbach and Schmidt [488,489] have studied the polymerization of NVP in benzene initiated with AIBN. They have noted that both the rate of conversion and the intrinsic viscosity of the polymer are independent of the monomer concentration but depend on the square root of the AIBN concentration. Hayashi et al. [492,504] used acetone as solvent and AIBN as an initiator for NVP polymerization. In patents [497,498] it was stated that batch polymerization of NVP in aqueous solution with H_2O_2 often gives rise to gel formation. This difficulty can be overcome by replacing at least part of the water with substances such as isopropyl alcohol, trichloracetic acid, α-mercaptoethanol, methyl ethyl ketone, DMF, ethanolamine, and thioglycolic acid. A continuous polymerization procedure is claimed to be particularly effective [492,497,498]. The transfer properties of H_2O_2 have been used to control the molecular weight when the polymerization of NVP was carried out in ethanol, initiated by AIBN but in the presence of varying amounts of H_2O_2 [505]. A retarding effect was observed if the polymerization was carried out in dimethylformamide and ferric chloride was added [506]. The mechanism suggested for the polymerization of NVP in organic solvents is shown below. It is characteristic of this type of polymerization that the chain end group is formed by a solvent molecule.

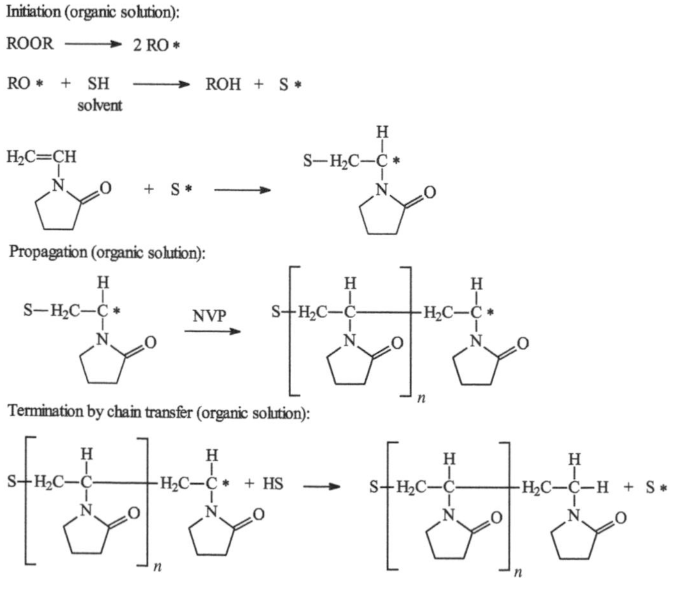

(72)

4. Suspension Polymerization

Since NVP is quite soluble in pure water suspension, polymerization of this monomer is obviously likely only if specialized techniques are used. Monagle [507] claimed that a nontacky, hard, yet water-soluble bead of a copolymer of 80% acrylamide and 20% NVP can be produced in a mixture containing at least 40% tertiary butanol (a nonsolvent for the polymer) in water if electrolytes such as KCl, NH_4CI, Na_2SO_4, NaCl, NH_4OH, and so on, were added. Ammonium persulfate, potassium persulfate, H_2O_2, and AIBN were used as initiators. It is also possible to produce cross-linked PVP bead polymer in an aqueous system in the presence of inorganic salts [508,509].

C. Cationic Polymerization

An early cationic polymerization experiment is reported by Schildknecht et al. [510], who used BF_3-etherate as initiator. After a short reaction time, the polymer was recovered as a sticky solid at room temperature. Combinations of triethylboron with peroxides and amines are claimed to be efficient initiators for the polymerization and copolymerization of NVP [511]. Aluminum trialkyls and aluminum alkylchlorides ($AlEt_{1.5}Cl_{1.5}$) have also been utilized for the polymerization of NVP [512,513].

Another group of initiator systems are metal halides (e.g., $ZrCl_4$, $TiCl_4$, etc.) and metallic oxyhalides (e.g., vanadium oxychloride) prepared on silica gel such as Cab-O-Sil (with a surface OH content of ca. 1.5 mEq/g) or on a carbon black such as channel black or furnace black [492,513]. NVP and i-Bu_3Al are added to a toluene slurry of this material. The reaction mixture is heated up to 80 °C to yield 25% of PVP [513]. The polymerization initiated by carbon black at room temperature was totally inhibited by pyridine and DMF, which indicates the cationic nature of the reaction [514]. The polymerization rate is first order in carbon black [492]. The activation energy of the polymerization was determined to be 85 kJ/mol from an Arrhenius plot [514].

It is suggested that the cationic polymerization of NVP is initiated by surface carboxyl groups and that PVP is grafted onto the surface of carbon black. The amount of PVP grafted onto the surface was determined by a semimicrokjeldahl: 1.1% nitrogen, which is equivalent to 0.08 mEq/g of PVP [514]. $POCl_3$ serves as initiator for the polymerization of NVP in CH_2Cl_2 (30 °C). However, PVP shows an extremely low molecular weight, corresponding to the low degree of polymerization ($P_n = 8$) [515]. From the retarding effect of basic additives and of water, the authors have concluded that the chain carriers must be cations. We can conclude that cationic polymerization leads to oligomers only and has no technical importance [510–516].

D. Radiation and Solid-State Polymerization

Radiation and solid-state polymerization are combined in this section because it is common to induce polymerization of liquid monomers by radiation below their melting points. In a patent published in 1955 [517], a simple procedure for UV-induced polymerization of the monomer is described. The resulting polymer has a k value close to 110. A number of sensitizers are used in this field, such as $ZnCl_2$ and air [518], anthrachinone [519], and various metal perfluoroalkane sulfonates [492,520]. The polymerization of liquid NVP initiated by γ-radiation is much faster than that of the solid NVP. However, the polymerization of the solid NVP becomes faster during irradiation and can reach rates comparable with those observed for liquid NVP.

This autoacceleration has been related to the crystal structure, temperature, and amount of polymer added to the monomer before irradiation was begun [521]. The activation energy for the polymerization of liquid NVP was found to be 37 kJ/mol, while that for solid-state polymerization is 73 kJ/mol [522]. With irradiation, cross-linked-PVP can also be produced, to give photo resins [523].

E. Charge Transfer Polymerization

Photopolymerization of NVP with MMA in the presence of $ZnCl_2$ and the thermal copolymerization of these two monomers takes place naturally in the presence of oxygen after an induction period. Oxygen seems to be essential and to participate in the formation of the active species. Tamura and Tanaka [518] have stated that a charge transfer polymerization process is involved here. The UV-induced terpolymerization of maleic anhydride (MA), NVP, and MMA exhibits charge transfer polymerization characteristics without $ZnCl_2$ being present [523]. The UV spectrum indicates the formation of an one-to-one complex between MA and NVP. Polymerization proceeds in the presence of oxygen to a significant extent only if all three monomers (MA, NVP, and MMA) are present in the system [524]. The same authors have concluded from the reactivity ratios of NVP/MMA that charge transfer complexes are formed during polymerization [525].

F. Properties and Applications

1. Properties of Poly(*N*-Vinylpyrrolidone)

Radical polymerization of NVP leads to a Schulz–Flory molecular weight distribution that is, in contrast to normal findings, broader for polymers of higher molecular weight than for those of lower molecular weight. This is explained on the basis of intensive branching during the formation of high-molecular-weight polymers, due to grafting of the polymer chain [478].

In solution, PVP probably exists as a random coil; suggestions concerning a possible helix structure have not yet been proven [526]. The size of the coil of PVP is important for applications in the medical field, in particular for the ability of the human body to excrete the polymer [527]. Depending on the molecular weight, the coil dimensions in an aqueous solution of sodium chloride have an average end-to-end distance of between 1 and 100 pm [478,526].

A special and unusual property of PVP is its solubility in water as well as in various organic solvents. The reason for this phenomenon is that PVP has hydrophilic as well as hydrophobic functional groups, and therefore interactions with various solvents are possible [478]. It is difficult to produce a PVP completely free of water because of its high hygroscopicity. Aqueous solutions of PVP show that the viscosity is not affected by electrolytes [478]. Polymers with molecular weight (M_w) of more than 1 million are generally used as thickening agents [528].

PVP cannot be processed in the melt because of its low decomposition temperature and the extremely poor flow properties of the melt. Films coated from solution are very brittle, clear, and glassy. The glass transition temperature (T_g) increases with molecular weight and reaches a plateau at about 175 °C [529], which corresponds to a molecular weight of 1 million. At low molecular weight T_g falls to values below 100 °C. The addition of small amounts of water also leads to lower T_g values [529].

2. Applications of *N*-Vinylpyrrolidone Polymers

A survey of the most important applications of NVP polymers is given in the reviews of Börner [491], Wood [530], Haaf et al. [478], and Lorenz [531]. So PVP is used in pharmaceuticals (e.g., PVP-iodine complexes as local antiseptics having bactericidal, fungicidal, and virucidal properties), cosmetics, detergents, foods (e.g., stabilization of beverages and the extremely high interaction constant of gallic acid with PVP is utilized for clearing fruit juices and beer containing polyphenols of similar structure as gallic acid), adhesives, textiles, and last but not least, in polymers as a suspending agent in two-phase polymerization systems. In most cases PVP is present as an essential auxiliary and not as the active substance itself.

Cross-linked PVP is applied as a semipermeable membrane [492]. Furthermore, PVP can be used for the encapsulation of hydrophobic material [492]. Poly(*N*-vinylpyrrolidone-co-styrene) was found to be an extraordinary catalyst in two-phase reactions [532]. For special applications one can design copolymers of NVP with different olefins and diolefins, or it is possible to graft NVP onto cellulose, polyolefins, keratin, collagen, and others, or to graft PVP with olefins [492,533–542].

VI. VINYLPYRIDINES

(This section was prepared by O. Nuyken and N. Stoeckel.)

Although a large number of derivatives of vinylpyridine have been synthesized, only three monomers [(a), (b), and (c)] have become commercially important [543–545] (Table 4). Polyvinylpyridines are of particular interest because they can act as polyelectrolytes, and furthermore, because they can form complexes at the basic nitrogen moiety.

Vinylpyridines are basic; they have an intensive smell and are soluble in common organic solvents. Since they have a high tendency to polymerize, they have to be stabilized for storing and transportation. Typical stabilizers are amines, phenols, thiols, and polysulfides. Their purity is determined by means of double-bond analysis [546], polarography [547], UV spectroscopy [548], and gas chromatography [549].

A. Monomers

A general route to vinylpyridines is given in Figure 2. The most important methods industrially are dehydration (X = OH) and dehydrogenation (X = H). On a laboratory scale dehydrohalogenation (X = halogen) is often used for the synthesis of these monomers.

The dehydrogenation is carried out in the gas phase at 500 to 800°C using catalysts similar to those applied in styrene production starting from ethyl benzene.

Table 4 Properties of the monomers (a), (b), and (c)[a]

	(a)	(b)	(c)
Density at 20°C (g/cm^3)	0.977	0.988	0.958
Boiling point	70°C/4 kPa	65°C/2 kPa	75°C/2 kPa
Solubility in water (g/L)	4.92	5.62	5.67

[a]A comprehensive list of vinylpyridines and their properties is given in Ref. 543.

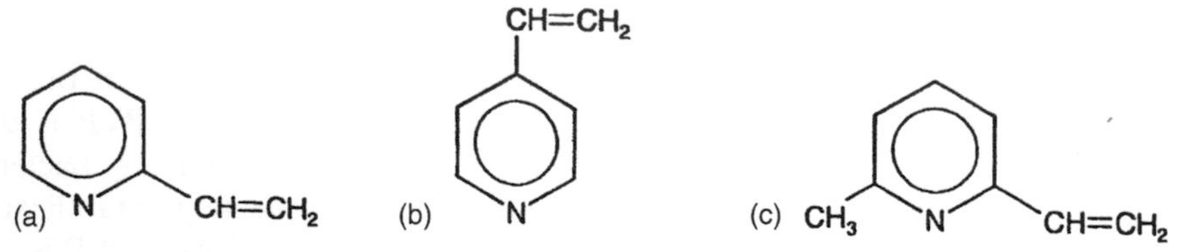

Figure 1   2-vinylpyridine (a), 4-vinylpyridine (b) and 2-vinyl-6-methyl-pyridine (c).

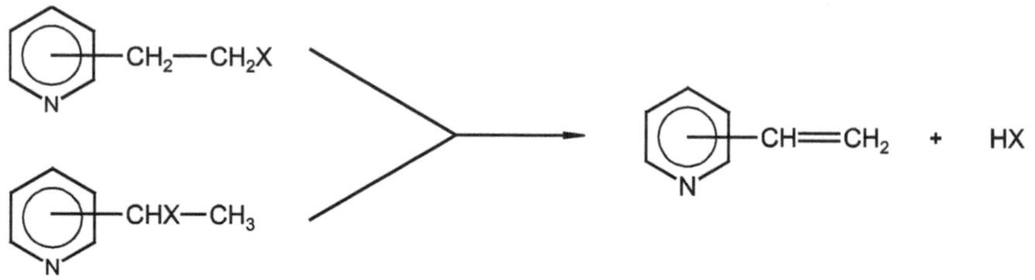

Figure 2   Dehydration (X = OH), dehydrogenation (X = H), dehydrohalogenation (X = halogen).

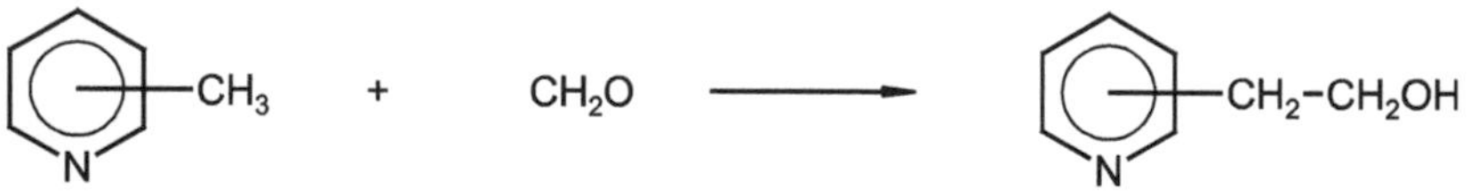

Figure 3   Formation of β-(2-pyridyl)-ethyl alcohol from 2- or 4-picoline with formaldehyde.

Catalysts such as Cr_2O_3/Mo_2O_3 [549], $Fe_2O_3/Cr_2O_3/KOH$ [550], $ZnO/Al_2O_3/CaO$ [551,552] and several others [543,545] have been described. This process is utilized primarily for the production of (c). As this synthetic route suffers from numerous by-products, purification methods have been published in patents [550–553].

The dehydration route is applied for the synthesis of (a) and (b). The starting materials are synthesized by reaction of 2- or 4-picoline with formaldehyde [554–556].

The syntheses of other hydroxyalkyl pyridines and their dehydration reactions are described in Refs. [557–559]. As catalysts, alkali hydroxide, H_2SO_4, $KHSO_4$, and P_2O_5 are used for liquid-phase reactions. For gas-phase dehydration, Al_2O_3 is applied.

B. Homopolymerization

Vinylpyridines undergo both radical and anionic polymerization. The presence of basic nitrogen in vinylpyridine introduces unique and complicating factors into the polymerization. Cationic polymerizations have not yet been reported.

1. Radical Polymerization

Vinylpyridine shows a polymerization tendency similar to that observed for styrene. The kinetics of the polymerization initiated by peroxides or azo compounds does not fulfill the classical equation

$$-\frac{\mathrm{d}M}{\mathrm{d}t} = k[T]^{1/2}[M]^n \tag{73}$$

as the monomer coefficient is $n = 1.3$ to 1.5 [560] instead of 1; in some cases $n = 2$ is observed [561]; in certain investigations perfect agreement with equation (73) for the radical polymerization of (a) has been shown [562]. Investigations with initiators marked with ^{14}C show clearly that termination by radical combination is favored and that transfer to monomer can be neglected [563]. In addition to conventional radical initiators cupric salts can be used as initiators [564–566]. A cation radical formed from monomer is claimed to be the initiating species [566] resulting from complex formation between monomer and cupric salt leading to an electron transfer from monomer to copper.

The free-radical polymerization of vinylpyridine in emulsion has also been described [567–571]. In contrast to classical emulsion polymerizations in which the micelles are the sole propagating sites, for monomers with high hydrophilicity such as vinylpyridines, the aqueous phase is apparently important as the locus of polymerization.

Recently, it has been reported that 4-vinylpyridine undergoes controlled radical polymerization in the presence of 2,2,6,6-tetramethylpiperidin-N-oxyl (TEMPO). In calorimetric experiments a polymerization behavior similar to that observed for the controlled living polymerization of styrene was found. Furthermore, the resulting polymers showed a small polydispersity compared to polymers obtained by free-radical polymerization [572].

2. Anionic Polymerization

The high electronegativity of the nitrogen atom activates the monomer toward attack by carbanions. The polymerization can be initiated by carbanions [571]; by electron-transfer initiators [573]; electrochemically [574]; by sodium in liquid ammonia [575]; by a Grignard reagent [576]; by magnesium, beryllium, and lithium compounds [577,578]; or by transition metal allyl compounds [579].

a. Polymerization Initiated by Alkali Metal Compounds. Polymerization of (b) with n-butyl lithium in heptane or toluene at −30 °C yields a polymer having a deep orange–red color, which is stable for months if moisture or other polar impurities are avoided [571]. Polymerization of (a) initiated by diphenyl sodium yields polymers with small polydispersity [580]. The type of active center, possible side reactions, and the stereochemistry of the anionic polymerization of (a) and (b) are still subject to discussion [581–584].

b. Polymerization Initiated by Alkali Earth Metal Compounds. This initiator class is important for the synthesis of isotactic, crystalline poly(2-vinylpyridine) [578] with a degree of isotacticity of 0.9 [585,586]. Similar attempts with (b) were not successful. Although the initiators are insoluble in toluene they can be dissolved by addition of monomer, indicating the formation of complexes with the monomer and the polymer, respectively. The tacticity of poly(2-vinylpyridine) depends on the type of gegenion [587,588]; the following order was observed: Ca > Sr > Ba.

c. Polymerization Initiated by Transition Metal Allyl Compounds. Allyl derivatives of chromium, molybdenum and tungsten have been used as initiators for the polymerization of 2- and 4-vinylpyridines. Variation of reaction parameters such as temperature and polarity of the solvent gave rise to a variety of different polymer architectures. Synthesized polymers contained block and graft sequences as well as macrocyclic fragments [579].

   d. Polymerization of Activated Monomers. Acidic molecules can initiate the polymerization of vinylpyridine.

   RX can be an alkyl halide [589,591], dimethyl sulfate [590,592], or protonic acid [593]. The active ends are not active to other monomers, such as styrene and acrylonitrile [589,591]. This polymerization technique has been extended to the use of polyacids as activators [594,595]. Polymerization of (b) in water with polyacrylic acid [596–598], polystyrene sulfuric acid [599], and polyphosphoric acid [600] has been utilized for this type of polymerization (matrix polymerization).

   4-Vinylpyridinium salts (nitrate, perchlorate, hydrogen sulfate) polymerize in solution spontaneously without the addition of initiator. In diluted systems the following reaction was observed [601].

   However, monomer concentrations of $c > 1\,mol/L$ yield polymers with normal structure, as described in equation (4) [602].

   In contrast to the other 4-vinylpyridinium salts, 4-vinylpyridinium triflates can only be polymerized by means of an initiator or spontaneously at by heating to 120–125 °C. This opens up a synthetic route to poly(4-vinylpyridinium) block copolymers [603–605].

C. Copolymerization

1. Statistical Copolymers

Vinylpyridines can be copolymerized radically with numerous common monomers. In many copolymerizations vinylpyridines are used in small quantities only. The most important technical product is a terpolymer from 1,3-butadiene (70 to 80% by weight), styrene (10 to 15%, and 10 to 15% of 2-vinylpyridine) [606]. Copolymers of methacrylic esters of long-chain alcohols and vinylpyridine have been applied as an oil additive. This and other applications for vinylpyridine copolymers are reviewed in detail in Ref. [607]. Copolymerization parameters are listed in Refs. [543,545,608,609]; some of the newer results are given in Tables 5 and 6.

2. Alternating Copolymers

Alternating copolymers are the result of the copolymerization of 2-vinylpyridine with butylvinyl ether in the presence of acetic acid [(a): acetic acid = 1:1.3] [610]. Other examples for alternating copolymers are (1a)/vinyl acetate copolymerized in the presence of acetic acid [611] and (c)/acrylonitrile as well as (c)/acrylic amide, using $K_2S_2O_8$ as complexing initiator [612].

3. Block and Graft Copolymers

Living anionic polymerization can be used for the synthesis of AB block copolymers in which (a) or (b) is added to a living chain of the comonomer. The alternative procedure is limited due to the low reactivity of a chain end formed from (a) or (b). Block copolymers from butadiene and (a) or (b) show remarkable reduction of the 'cold flow' [613,614]. ABA block copolymers are the result of the polymerization of butadiene (B) with a bifunctional initiator followed by the addition of (a) or (b) (A) [615]. AB block copolymers from

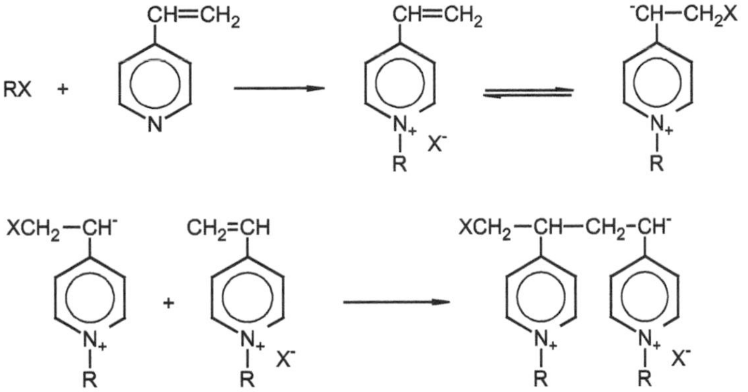

Figure 4 Polymerzation of activated monomers (RX = alkyl halide, dimethyl sulfate or protonic acid).

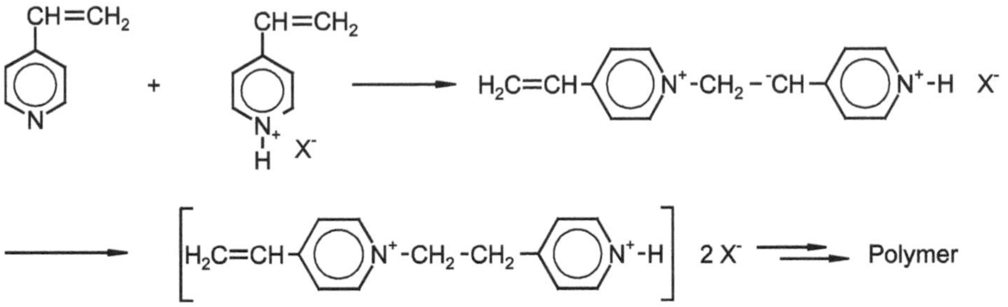

Figure 5 Polymerization of 4-vinylpyridinium salts in diluted systems (X = nitrate, perchlorate, hydrogen sulfate).

Table 5 Selected reactivity ratios for copolymerization of 2-vinylpyridine (M_1).

M_2	r_1	r_2	Remarks	Ref.
Acrylic acid butyl ester	2.5	0.097	60 °C, bulk	630
	2.59	0.11		544
Acrylic acid methyl ester	1.56	0.168	60 °C, bulk, AIBN	631
	2.14	0.21	60 °C, benzene, AIBN	632
Acrylonitrile	4.00	0.083		633
Butadiene	0.90	0.94	50 °C, $K_2S_2O_8$, in emulsion	634
1,3-Butadiene 1,4-dicarboxylic acid, diethyl ester	0.40	0.80	50 °C, DMSO, AIBN	635
Cyclopropene ketal[a]	2.08	0.071	60 °C, BPO, benzene	636
Methacrylic acid glycidyl ester	0.62	0.51	60 °C, THF, AIBN	637
Methacrylic acid 2-hydroxy ethyl ester	0.69	0.58	60 °C, bulk, AIBN	638
Methacrylic acid methyl ester	1.13	0.30	60 °C, benzene, AIBN	632
	0.76	0.35		544
Styrene	1.1	0.55		638
	0.75	0.46		544
	1.26	0.44		639

[a] 6,6-Dimethyl-4,8-dioxaspiro(2,5)oct-l-ene.

Table 6 Selected reactivity ratios for copolymerization of 4-vinylpyridine (M_i).

M_2	r_1	r_2	Remarks	Ref.
Acrylic acid ethyl ester	2.58	0.29		640
Butyl vinyl ether	2.38	0	601°C, benzene, AIBN	641
Methacrylic acid 2-hydroxyethyl ester	0.95	0.66	60 °C, bulk, AIBN	638
Methacrylic acid methyl ester	1.77	0.18	60 °C, benzene, AIBN	642
	1.05	0.54	65 °C, toluene, AIBN	643
	0.77	0.58	60 °C, bulk, AIBN	
Styrene	0.75	0.57	50 °C, benzene, AIBN	644
	0.69	0.375		639
n-Vinyl caprolactame	4.43	0.096		645
Vinyl imidazole	1.07	0.21	60 °C, bulk	646

styrene and (a) are synthesized by different procedures [616–619]. ABA block coplymers of (a) and styrene can be obtained by using the dipotassium salt of α-methylstyrene tetramer as a bifunctional initiator [620]. Triblock copolymers from styrene/1,3-butadiene and (b) are available by sequential polymerization [621] as well as triblock copolymers from isoprene/styrene/(a) [622].

A comprehensive list of the grafting reactions of (a) and (b) onto different polymer backbones is given in Ref. 543. The methods summarized there include radiation techniques [623], grafting by radical transfer [624], and grafting initiated by functional groups in backbone polymers [625,626]. Macromonomers of (a) have been synthesized by means of anionic polymerization techniques and have been copolymerized with styrene [627,628]. Only a few examples are known in which polymers from (a) and (b) were used as backbone [629]. Star shaped block copolymers with four arms were prepared by coupling living styrene/(a) block copolymers with 1,2,4,5-tetrakis-bromomethyl-benzene [626].

REFERENCES

1. Warner, A. J. (1952). *Styrene: Its Polymers, Copolymers and Derivatives* (Bounty, R. H., and Boyer, R. F., eds.), Reinhold, New York, p. 3.
2. Warner, A. J. (1952). *Styrene: Its Polymers, Copolymers and Derivatives* (Bounty, R. H., and Boyer, R. F., eds.), Reinhold, New York, p. 1.
3. Moore, E. R. (1989). *Encyclopedia of Polymer Science and Engineering*, Vol. 16 (Mark, H. F., Bikales, N. M., Overberger, C. G., and Menges, G., eds.), Wiley, New York, p. 2.
4. Amos, J. L. (1974). *Polym. Eng. Sci.*, *14*: 1.
5. James, D. H., and Castor, W. M. (1994). *Ullmann's Encyclopedia of Industrial Chemistry*, Vol. A25 (Gerhartz, W., and Elvers, B., eds.) Verlag Chemie, Weinheim, p. 329.
6. Gausepohl, H., and Warzelhan, V. (1997). *Angew. Makromol. Chem.*, *244*: 17
7. Weissermel, H., and Arpe, H.-J. (1994). *Industrielle Organische Chemie*, 4. Aufl., VCH-Weinheim, New York, Basel, Cambridge, Tokyo.
8. James, D. H., Gardner, J. B., and Mueller, E. C. (1989). *Encyclopedia of Polymer Science and Engineering*, 2nd ed., Vol. 16 (Mark, H. F., Bikales, N. M., Overberger, C. G., and Menges, G., eds.), Wiley, New York, p. 5.
9. Olaj, O. F., Kauffman, H. F., and Breitenbach, J. W. (1977). *Makromol Chem.*, *178*: 2707.
10. Mayo, F. R. (1968). *J. Am. Chem. Soc.*, *90*: 1289.

11. Stein, D. F., and Mosthaf, H. (1968). *Angew. Makromol. Chem.*, *2*: 39.
12. Hiatt, R. R., and Bartlett, P. D. (1959). *J. Am. Chem. Soc.*, *81*: 1149.
13. Pryor, W. A., and Coco, J. H. (1970). *Macromolecules*, *3*: 500.
14. Mayo, F. R. (1953). *J. Am. Chem. Soc.*, *75*: 6133.
15. Meister, B. J., and Malanga, M. T. (1989). *Encyclopedia of Polymer Science and Engineering*, 2nd ed., Vol. 16 (Mark, H. F., Bikales, N. M., Overberger, C. G., and Menges, G., eds.), Wiley, New York, p. 24.
16. Masson, J. C. (1989). *Polymer Handbook* (Brandrup, J., and Immergut, E. H., eds.), Wiley, New York, p. II-1.
17. Sheppard, C. S. (1985). *Encyclopedia of Polymer Science and Engineering*, 2nd ed., Vol. 2 (Mark, H. F., Bikales, N. M., Overberger, C. G., and Menges, G., eds.), Wiley, New York, p. 143.
18. Sheppard, C. S. (1988). *Encyclopedia of Polymer Science and Engineering*, 2nd ed., Vol. I (Mark, H. F., Bikales, N. M., Overberger, C. G., and Menges, G. eds.), Wiley, New York, p. 1.
19. Kamath, V. R. (1981). *Mod. Plast.*, *58*: 106.
20. Kamath, V. R., and Harpell, G. A. (1978). DOS 2,449,519 to Pennwalt Corp.; C.A. (1978), *89*: 90475y.
21. Miller, R. F., and Nicholson, M. P. (1984). U.S. Pat. 4,465,881 to Atlantic Richfield Co.; C.A. (1984), *101*: 192661q.
22. Dow Family of styrenic Monomers, Tech. Bull. 11 5-608-85, The Dow Chemical Company, Midland, Mich. (1985).
23. Watson, J. M. (1978). U.S. Pat. 4,086,147 to Cosden Techn. Inc.; C.A. (1978), *89*: 90437n.
24. Watson, J. M., and Bracke, W. J. I. (1983). U.S. Pat. 4,396,462 to Cosden Techn. Inc.; C.A. (1983), *99*: 140574j.
25. George, M. H. (1967). *Vinyl Polymerization* (Ham, G. H., ed.), Marcel Dekker, New York, pp. 186–188.
26. Olaj, O. F., Kauffmann, H. F., and Breitenbach, J. W. (1977). *Makromol. Chem.*, *178*: 2707.
27. Henrici-Olivé, G., and Olivé, S. (1960). *J. Polym. Sci.*, *48*: 329.
28. Berger, K. C., and Brandrup, G. (1989). *Polymer Handbook* (Brandrup, J., and Immergut, E. H., eds.), Wiley, New York, p. II-81.
29. Bevington, J. C., Melville, H. W., and Taylor, R. P. (1954). *J. Polym. Sci.*, *14*: 463.
30. Ayrey, G., Levitt, F. G., and Mazza, R. J. (1965). *Polymer*, *6*: 157.
31. Weickert, G., and Thiele, R. (1983). *Plaste Kautschuk*, *8*: 432.
32. Mita, I., and Horie, K. (1987). *Rev. Macromol. Chem. Phys. Part C*, *27*: 91.
33. Tulig, T. J., and Tirrell, M. (1981). *Macromolecules*, *14*: 1501.
34. Meister, B. J., and Malanga, M. T. (1989). *Encyclopedia of Polymer Science and Engineering*, 2nd ed., Vol. 16 (Mark, H. F., Bikales, N. M., Overberger, C. G., and Menges, G., eds.), Wiley, New York, p. 46.
35. Overberger, C. G. (1963). *Macromolecular Synthesis*, Vol. I (Overberger, C. G., ed.), Wiley, New York, p. 4.
36. Dezelic, N., Petres, J. J., and Dezelic, G. J. (1977). *Macromolecular Synthesis*, Vol. 6 (Mulvaney, J., ed.), Wiley, New York, p. 85.
37. Sherrington, D. C. (1982). *Macromolecular Synthesis*, Vol. 8 (Pearce, E. M., ed.), Wiley, New York, p. 69.
38. Otsu, T., and Yoshida, M. (1982). *Makromol. Chem. Rapid Commun.*, *3*: 127.
39. Otsu, T., Yoshida, M., and Tazaki, T. (1982). *Makromol. Chem. Rapid Commun.*, *3*: 133
40. Otsu T., Yoshida, M., and Kuriyama, A. (1982). *Polym. Bull.*, *7*: 45
41. Solomon, D.H., Rizzardo, E., and Cacioli, P. *Eur. Pat. Appl.*, EP135280; C. A. (1985) *102*: 221335q).
42. Georges, M. K., Veregin, R. P. N., Kazmaier, P. M., and Hamer, G. K. (1993). *Macromolecules*, *26*: 2987.
43. Colombani, D., Steenbock, M., Klapper, M., and Müllen, K. (1997). *Macromol. Rapid Comm.*, *18*: 243.

44. Steenbock, M., Klapper, M., Müllen, K., Bauer, C., and Hubrich, M. (1998). *Macromolecules*, *31*: 5223.
45. Klapper, M., Brand, T., Steenbock, M., and Müllen, K. (2000). *ACS Symp. Ser.*, *768*: 152.
46. Wang, J. S., and Matyjaszewski, K. (1995). *Macromolecules*, *28*: 7901.
47. Kato, M., Kamigaito, M., Sawamoto, M., and Higashimura, T. (1995). *Macromolecules*, *28*: 1721.
48. Matyjaszewski, K. (1999). *Chem. Eur. J.*, *5*(11): 3095.
49. Simal, F., Jan, D., Demonceau, A., and Noels, A. F. (2000). *ACS Symp. Ser.*, *768*: 223.
50. Matyjaszewski, K., Wei, M., Xia, J., and McDermott, N. E. (1997). *Macromolecules*, *30*: 8161.
51. Chiefari, J., Chong, Y. K., Ercole, F., Kristina, J., Jeffery, J., Le, T. P. T., Maydunne, R. T. A., Meijs, G. F., Moad, C. L., Moad, G., and Rizzardo, E. (1999).*Macromol. Symp. Ser.*, *143*: 291.
52. Rizzardo, E., Chiefari, J., Maydunne, R. T. A., and Moad, G. (2000). *ACS Symp. Ser.*, *768*: 278.
53. Chong, Y. K., Le, T. P. T., Moad, G., Rizzardo, E., and Thang, S. H. (1999). *Macromolecules*, *32*: 2071.
54. Hsieh, H. (1965). *J. Polym. Sci. Part A*, *3*: 163.
55. Roovers, J. E. L., and Bywater, S. (1975). *Macromolecules*, *8*: 251.
56. Szwarc, M. (1956). *Nature*, *178*: 1168.
57. Szwarc, M., Levy, M., and Milkovich, R. (1956). *J. Am. Chem. Soc.*, *78*: 2656.
58. Szwarc, M. (1968). *Carbanions: Living Polymers and Electron Transfer Processes*, Wiley-Interscience, New York, pp. 9–22.
59. Szwarc, M.. and Smid, J. (1964). *Progress in Reaction Kinetics*, Vol. 2 (Porter, G., ed.), Pergamon Press, Oxford, p. 250.
60. Bhattacharyya, D. N., Smid, J.. and Szwarc, M. (1965). *J. Phys. Chem.*, *69*: 624.
61. Worsfold, D. J.. and Bywater, S. (1972). *Macromolecules*, *5*: 393.
62. Kaspar, M., and Trekoval, J. (1980). *Coll. Czech. Chem. Commun.*, *45*: 1047.
63. Tung, L. H., Lo, G. Y. S., Rakshys, J. W., and Beyer, D. E. (1977). DOS 2,634,391 to Dow Chemical Comp.; C.A. (1977), *86*: 190663q.
64. Tung, L. H., Lo, G. Y. S., and Beyer, D. E. (1979). JP 54,063,186 to Dow Chemical Comp.; C.A. (1979), *91*: 158857y.
65. Tung, L. H., Lo, G. Y. S., and Beyer, D. E. (1978). *Macromolecules*, *11*: 616.
66. Guyot, P., Favier, J. C., Uytterhoeven, H., Fontanille, M., and Sigwalt, P. (1981). *Polymer*, *22*: 1724.
67. Noshay, A., and McGrath, J. (1977). *Block Copolymers: Overview and Critical Survey*, Academic Press, Orlando, Fla.
68. Folkes, M. J. (1985). *Processing, Structure and Properties of Block Copolymers*, Applied Science Publishers, Barking, Essex, England.
69. Morton, M. (1983). *Anionic Polymerization: Principles and Practice*, Academic Press, Orlando, Fla.
70. Fetters, L. J. (1969). *J. Polym. Sci. Polym. Symp.*, *26: 1*.
71. Morton, M., and Fetters, L. J. (1967). *Macromol. Rev.*, *2*: 7 1.
72. Bywater, S. (1974). *Progr. Polym. Sci.*, *4*: 54.
73. Fetters, L. J. (1969). *J. Polym. Sci. Polym. Symp.*, *26*: 22.
74. Richards, D. H., Eastmond, G. C., and Stewart, M. J. (1989). *Telechelic Polymers: Synthesis and application* (Goethals, E. J., ed.), CRC Press, Boca Raton, Fla., p. 33ff.
75. Milkovich, R., and Chiang, M. T. (1972). DOS 2,208,340 to CPC International Incorp.; C.A. (1973), *78*: 4789x.
76. Milkovich, R., and Chiang, M. T. (1972). DOS 2,208,340; C.A. (1973), *78*: 4789x.
77. Masson, P., Franta, E., and Rempp, P. (1982). *Macromol Chem. Rapid Commun.*, *3*: 499.
78. Schulz, G. O., and Milkovich, R. (1982). *J. Appl Polym. Sci.*, *27*: 4773.
79. Chujo, Y., and Yamashita, Y. (1989). *Telechelic Polymers: Synthesis and Applications* (Goethals, E. J., ed.), CRC Press, Boca Raton, Fla., P. 163.

80. Percec, V., Pugh, C., Nuyken, O., and Pask, S. D. (1989). *Comprehensive Polymer Science*, Vol. 6 (Ellen, G., and Bevington, J. C., eds.), Pergamon Press, Oxford, p. 281.

81. Nuyken, O., and Pask, S. D. (1989). *Encyclopedia of Polymer Science and Engineering*, 2nd ed., Vol. 16 (Mark, H. F., Bikales, N. M., Overberger, C. G., and Menges, G. eds.), Wiley, New York, p. 494.

82. Mathieson, A. R. (1963). *The Chemistry of Cationic Polymerization* (Plesch, P. H., ed.), Macmillan, New York, p. 235.

83. Dunn, D. J. (1979). *Developments in Polymerization*, Vol. 1 (Haward, R. N., ed.), Applied Science Publishers, Barking, Essex, England, p. 46.

84. Ledwith, A., and Sherrington, D. C. (1974). *Reactivity, Mechanism and Structure in Polymer Chemistry* (Jenkins, A. D., and Ledwith, A., eds.), Wiley-Interscience, New York, p. 252.

85. Gandini, A., and Cheradame, H. (1980). *Adv. Polym. Sci., 34/35:* 202.

86. Kennedy, J. P. (1975). *Cationic Polymerization of Olefins: A Critical Inventory*, Wiley, New York, p. 228.

87. Gandini, A., and Plesch, P. H. (1964). *Proc. Chem. Soc., 240.*

88. Matyjaszewski, K. (1988). *Macromol. Chem. Macromol. Symp., 13/14:* 433.

89. Matsuda, T., and Higashimura, T. (1971). *J. Polym. Sci. Part A-1, 9:* 1563.

90. Panayotov, I. M., Dimitrov, I. K., and Bakerdjiev, I. E. (1969). *J. Polym. Sci. Part A-1, 7:* 2421.

91. Ise, N. (1969). *Adv. Polym. Sci., 6:* 347.

92. Sakurada, I., Ise, N., and Hayashi, Y. (1967). *J. Macromol. Sci. Chem. Part A, 1:* 1039.

93. Breitenbach, J. W., and Srna, C. (1962). *Pure Appl. Chem., 4:* 245.

94. Pinner, S. H. (1963). *The Chemistry of Cationic Polymerization* (Plesch, P. H., ed.), Macmillan, New York, p. 611.

95. Westlake, J. F., and Huang, R. Y. (1972). *J. Polym. Sci. Polym. Chem. Ed., 10:* 3053.

96. Williams, F. (1968). *Fundamental Process in Radiation Chemistry* (Ausloos, P., ed.), Wiley, New York, p. 515.

97. Potter, R. C., Schneider, C., Ryska, M., and Hummel, D. O. (1968). *Angew. Chem. Intern. Ed., 7:* 845.

98. Williams, F. (1963). *Quart. Rev., 17:* 101.

99. Chapiro, A., and Stannet, V. (1960). *J. Phys. Chem., 57:* 55.

100. Kamardi, M., and Miyama, H. (1968). *J. Polym. Sci. Polym. Chem. Ed.,* Part A 1, *6:* 1537.

101. Kennedy, J. P., and Mishra, M. K. (1988). EP 265,053 A2; C.A. (1988), *110:* 155017z.

102. Puskas, J., Kaszas, G., Kennedy, J. P., Kelen, T., and Tüdös, F. (1982/1983). *J. Macromol. Sci. Chem., 18:* 1315.

103. Faust, R., Fehervari, A., and Kennedy, J. P. (1982/1983). *J. Macromol. Sci. Chem., 18:* 1209.

104. Natta, G., Danusso, F., and Sianesi, D. (1957). Ital. Pat. 558,314 to Montecatini; C.A. (1959), *53:* 1835g.

105. Natta, G., Danusso, F., and Sianesi, D. (1958). *Macromol Chem., 28:* 253.

106. Overberger, C. G., Ang, F., and Mark, H. (1959). *J. Polym. Sci., 35:* 38 1.

107. Natta, G., Danusso, F., and Pasquon, I. (1957). *Collect. Czech. Chem. Commun., 22:* 191.

108. Danusso, F., and Sianesi, D. (1958). *Chim. Ind. (Milan), 40:* 450.

109. Kem, R. J., Hurst, H. G., and Richards, W. R. (1960). *J. Polym. Sci., 45:* 195.

110. Ishihara, N., Seimiya, T., Kuramoto, M., and Uoi, M. (1986). *Macromolecules, 19:* 2465.

111. Affimendola, P., Pellecehia, C., Longo, P., and Zambelli, A. (1987). *Gazz. Chim. Ital., 117–65.*

112. Pellecehia, C., Longo, P., Grassi, A., Ammendola, P., and Zambelli, A. (1987). *Macromol. Chem. Rapid Commun., 8:* 227.

113. Zambelli, A., Longo, P., Pellecchia, C., and Grassi, A. (1987). *Macromolecules, 20:* 2035.

114. Zambelli, A., Pellecchia, C., Oliva, L., Longo, P., and Grassi, A. (1991). *Macromol. Chem., 192:* 223.

115. Vanderhoff, J. W. (1954). *Monomeric Acrylic Ester* (Riddle, E. H., ed.), Reinhold, New York, p. 94.

116. Chapman, C. B., and Valentine, L. (1959). *J. Polym. Sci., 34:* 319.

117. Lewis, F. M., Mayo, F. R., and Hulse, W. F. (1945). *J. Am. Chem. Soc.*, *67*: 1701.
118. Meehan, E. J. (1946). *J. Polym. Sci.*, *1*: 318.
119. Wiley, R. H., and Sale, E. E. (1960). *J. Polym. Sci.*, *42*: 491.
120. Greenley, R. Z. (1989). *Polymer Handbook* (Brandrup, J., and Immergut, E. H., eds.), Wiley, New York, p. II-214.
121. Walling, C. (1949). *J. Am. Chem. Soc.*, *71*: 1930.
122. Meister, B. J., and Malanga, M. T. (1989). *Encyclopedia of Polymer Science and Engineering* 2nd ed., Vol. 16 (Moore, E. R., Bikales, N. M., Overberger, C. G., and Menges, G., eds.), reprinted from 2nd ed. (Mark, H. F., ed.), Wiley, New York, p. 39.
123. Echte, A. (1987). *Houben-Weyl: Methoden der organischen Chemie*, 4th ed., Vol. E20/II (Bartl, H. and Falbe, J., eds.), Georg Thieme, Stuttgart, p. 962.
124. Bataille, P. F., and Granger, F. (1983). *Colloid Polym. Sci.*, *261*: 914.
125. Patnaik, B. K., and Gaylord, N. G. (1977). *Macromolecular Syntheses*, Coll. Vol. 1 (Moore, J. A., ed.), Wiley, New York, p. 517.
126. Trivedi, B. C., and Culbertson, B. M. (1982). *Maleic Anhydride*, Plenum Press, New York, p. 359.
127. Tsuchida, E., Tomono, T., and Sano, H. (1972). *Macromol. Chem.*, *151*: 245.
128. Hanson, A. W., and Zimmermann, R. L. (1957). *Ind. Eng. Chem.*, *49*: 1803.
129. Kent, R. W. (1981). U.S. Pat. 4,243,781 to Dow Chemical; C.A. (1981), *94*: 209496b.
130. Kent, R. W. (1981). U.S. Pat. 4,268,652 to Dow Chemical; C.A. (1981), *95*: 62981r.
131. Fisler, T. M. (1967). Brit. Pat. 1,093,349 to Ministry Chem. hid., Romania; C.A. (1967), *68*: 30414j.
132. Kohlpoth, G., Komischke, P., Mixich, J., and Kilger, H. J. (1972). DOS 2,057,250 to Knapsack; C.A. (1972), *77*: 75805q.
133. Murray, A. G. (1969). DOS 1,802,089 to Uniroyal Inc.; C.A. (1969), *71*: 39877g.
134. Barton, J., and Wemer, P. (1979). *Makromol. Chem.*, *180*: 989.
135. Wojtczak, Z., and Grenowski, A. (1985). *Makromol. Chem.*, *186*: 139.
136. Tate, D. P., and Bethea, T. W. (1985). *Encyclopedia of Polymer Science and Engineering*, 2nd ed., Vol. 2 (Mark, H. F., Bikales, N. M., Overberger, C. G., Menges, G. eds.), Wiley, New York, p. 553.
137. Childers, C. W., and Fisk, C. F. (1958). U.S. Pat. 2,820,773 to U.S. Rubber Co.; C.A. (1958), *52*: 5871b.
138. Calvert, W. C. (1966). U.S. Pat. 3,238,275 to Borg-Warner Corp.; C.A. (1966), *64*: 16071f.
139. Bredeweg, C. J. (1980). U.S. Pat. 4,239,863 to Dow Chemical Co.; C.A. (1981), *94*: 84863k.
140. Burk, R. D. (1981). Brit. Pat. Appl. GB 2,076,412 to Monsanto Co.; C.A. (1982), *96*: 86440k.
141. Arden, T. V., and DeDardel, F. (1994). *Ullmann's Encyclopedia of Industrial Chemistry*, Vol. A14 (Gerhartz, W., and Elvers, B., eds.), Verlag Chemie, Weinheim, p. 393.
142. Morton, M. (1983). *Anionic Polymerization Principles and Practice*, Academic Press, Orlando, Fla., p. 179.
143. Young, R. N., Quirk, R. P., and Fetters, L. J. (1984). *Adv. Polym. Sci.*, *56*: 1.
144. Sitola, A. (1977). *Acta Polytech.*, *134*: 1935.
145. Falk, J.C., Benedetto, M. A., VanFleet, J., and Ciaglia, L. (1982). *Macromolecular Syntheses*, Vol. 8 (Pearce, E. M., ed.), Wiley, New York, p. 61.
146. Bandermann, F., Speikamp, H. D., and Weigel, L. (1985). *Makromol. Chem.*, *186*: 2017.
147. Tung, L. H. (1979). *J. Appl Sci.*, *24*: 953.
148. Bastelberger, T. (1983). *Ph.D. thesis*, University of Bayreuth, West Germany.
149. Bastelberger, T., and Höcker, H. (1984). *Angew. Makromol. Chem.*, *125*: 53.
150. Morton, M., Helminak, T. E., Gadkary, S. D., and Bueeke, F. (1962). *J. Polym. Sci.*, *57*: 471.
151. Finarz, G., Gallot, Y., and Rempp, P. (1962). *J. Polym. Sci.*, *58*: 1363.
152. Sheridan, M. M., Hoover, J. M., Ward, T. C., and McGrath, J. E. (1984). *ACS Polym. Prepr.*, *25*(2): 102.
153. Eschwey, H., Hallensleben, M. L., and Buchard, W. (1973). *Makromol. Chem.*, *173*: 235.
154. Freyss, D., Leng, M., and Rempp, P. (1964). *Bull Soc. Chim. France*, 221.

155. Baer, M. (1964). *J. Polym. Sci. Part A, 2*: 417.
156. Kotaka, T., Tanaka, T., and Inagaki, H. (1972). *Polym. J., 3*: 327.
157. Qui, X. Y., Ruland, W., and Heitz, W. (1984). *Angew. Makromol Chem., 125*: 69.
158. Echte, A., Stein, D., Fahrbach, G., Adler, H. S., and Gerberding, K. (1972). DOS 2,546,377 to BASF AG; C.A. (1977), 87.6652g.
159. Woodword, A. E., and Smets, G. (1955). *J. Polym. Sci., 17:* 51.
160. Oppenheimer, C., and Heitz, W. (1981). *Angew. Makromol Chem., 98*: 167.
161. Piirma, I., and Chou, L. P. H. (1979). *J. Appl Polym. Sci., 24*: 2051.
162. Améduri, B., Boutevin, B., and Gramain, Ph. (1997). *Adv. Polym. Sci., 127*: 87.
163. Georges, M. K., Veregin, R. P. N., Kazmaier, P. M., and Hamer, G. K. (1993). *Macromolecules, 26*: 2987.
164. Georges, M. K., Veregin, R. P. N., Kazmaier, P. M., and Hamer, G. K. (1995). *Macromolecules, 28*: 4391.
165. Matyjaszewski, K., Gaynor, S., Greszta, D., Mardare, D., and Shigemoto, T. (1995). *Macromol. Symp., 88*: 89.
166. Mardare, D., and Matyjaszewski, K. (1993). *ACS Polym. Prepr., 34*: 566.
167. Mardare, D., and Matyjaszewski, K. (1994). *Macromolecules, 27*: 645.
168. Gaynor, S. and Matyjaszewski, K. (1997). *ACS Polym. Prepr., 38*(1): 758.
169. Matyjaszewski, K. (1997). *ACS Polym. Prepr., 38*(2): 383.
170. Chong, Y. K., Tam, P. T. L., Moad, G., Rizzardo, E., and Thang, S. H. (1999). *Macromolecules, 32*: 2071.
171. Nuyken, O. and Weidner, R. (1986). *Adv. Polym. Sci., 73174*: 145.
172. Hawker, C. J. (1995). *Angew. Chem. Int. Ed. Engl., 34*: 1456.
173. Matyjaszewski, K., Beers, K. L., Coca, S., Gaynor, S., Miller, P., Paik, H.-J., and Teodorescu, M. (1999). *ACS Polym. Prepr., 40*(2): 95.
174. Rempp, P. F., and Lutz, P. J. (1989). *Comprehensive Polymer Science*, Vol. 6 (Allen, G., and Bevington, J. C., eds.), Pergamon Press, Oxford, p. 403.
175. Kennedy, J. P., and Vidal, A. (1975). *J. Polym. Sci. Polym. Chem. Ed., 13*: 1765.
176. Liu, S., and Sen, A. (2000). *Macromolecules, 33*: 5106.
177. Zhdanov, A. A., Zavin, B. G., and Blokhina, G. O. (1986). *Vysokomol Soedin. Part A*, 28.
178. Nitatori, Y., Rempp, P., and Franta, E. (1978). *Makromol. Chem., 179*: 941.
179. Franta, E., Reibel, L., Lehmann, J., and Penczek, S. (1976). *J. Polym. Sci. Polym. Symp., 56*: 139.
180. Franta, E., Afshar-Taromi, F., and Rempp, P. (1976). *Makromol Chem., 177*: 2191.
181. Ito, K., and Kawaguchi, S. (1999). *Adv. Polym.Sci., 142*: 129.
182. Brydon, A., Bumett, G. M., and Cameron, G. G. (1973). *J. Polym. Sci. Polym. Chem. Ed., 11*: 3255.
183. Brydon, A., Bumett, G. M., and Cameron, G. G. (1974). *J. Polym. Sci. Polym. Chem. Ed., 12*: 1011.
184. Riess, C., and Locatelli, J. L. (1975). *Adv. Chem. Ser., 142*: 186.
185. Echte, A., Haaf, F., and Hambrecht, J. (1981). *J. Angew. Chem. Intern. Ed. Engl., 20*: 344.
186. Jiang, M., Huang, X., and Yu, T. (1983). *Polymer, 24*: 1259.
187. Freund, G., Lederer, M., and Strobel, W. (1977). *Angew. Chem., 58159*: 199.
188. Platzer, N. (1977). *Chemtech, 7*: 634.
189. Bubeck, R. A. (1989). *Encyclopedia of Polymer Science and Engineering*, 2nd ed., Vol. 16 (Mark, H. F., Bikales, N. M., Overberger, C. G., and Menges, G. eds.), Wiley, New York, p. 88.
190. Rempp, P., and Franta, E. (1984). *Adv. Polym. Sci., 58*: 1.
191. Yamashita, Y., Ito, K., Mizuno, H., and Okada, H. (1982). *Polym. J., 14*: 255.
192. Frechet, J. M. J., Henmi, M., Gitsov, I., Aoshima, S., Leduc, M. R., and Grubbs, B. (1995). *Science 269*: 1080.
193. Gaynor, S. G., Edelman, S., and Matyjaszewski, K. (1996). *Macromolecules, 29*: 1079.

194. Hawker, C. J., Frechet, J. M. J., Grubbs, R. B., and Dao, J. (1995). *J. Am. Chem. Soc.*, *117*: 763.
195. Wieland, P. C., Nuyken, O., Schmidt, M., and Fischer, K. (2001). *Submitted.*
196. Coulter, K. E., and Kehde, H. (1970). *Encyclopedia of Polymer Science and Technology*, Vol. 13 (Mark, H. F., Gaylord, N. G., and Bikales, N., eds.), Wiley-Interscience, New York, p. 151.
197. Bywater, S. (1963). *The Chemistry of Cationic Polymerization* (Plesch, P. H., ed.), Macmillan, New York, p. 305.
198. Kennedy, J. P. (1975). *Cationic Polymerization of Olefins: A Critical Inventory*, Wiley-Interscience, New York, p. 270.
199. Brownstein, S., Bywater, S., and Worsfold, D. J. (1961). *Makromol Chem.*, *48*: 127.
200. Sakurada, Y., Matsumoto, M., Itnai, K., Nishioka, A., and Kato, Y. (1963). *Polym. Lett.*, *1*: 633.
201. Ohsumi, Y., Higashimura, T., and Okamura, S. (1965). *J. Polym. Sci. Part A*, *3*: 3729.
202. Ramey, K. C., Statton, G. L., and Jankowski, W. C. (1969). *J. Polym. Sci. Part B*, *7*: 693.
203. Elgert, K.-F., Seiler, E., Puschendorf, G., Ziemann, W., and Cantow, H.-J. (1971). *Makromol Chem.*, *144*: 73.
204. Inoue, Y., Nishioka, A., and Chujo, R. (1972). *Makromol. Chem.*, *156*: 207.
205. Higashimura, T., Takeda, T., Sawamoto, M., and Urya, T. (1978). *J. Polymer Sci. Polym. Chem. Ed.*, *16*: 503.
206. Kunitake, T., and Aso, C. (1970). *J. Polym. Sci. Part A-1*, *8*: 665.
207. Matsuguma, Y., and Kunitake, T. (1971). *Polym. J.*, *2*: 353.
208. Akcelrud, L., and Gomes, A. S. (1978). *J. Polym. Sci. Polym. Chem. Ed.*, *16*: 2423.
209. Trivedi, P. D., Acharya, H. K., and Bhardwaj, L. S. (1981). *Polym. Bull. (Berlin)*, *5*: 393.
210. Skorokhodov, S. S., Stepanov, V. V., Toncheva, V. D., Velichkova, R. S., and Panayotov, I. M. (1986). *J. Polym. Sci. Part A*, *24*: 327 1.
211. Gubariiyogi, S. C., and Kennedy, J. P. (1981). *Polym. Bull. (Berlin)*, *4*: 267.
212. Kennedy, J. P., and Chou, R. T. (1982). *J. Macromol Sci. Chem. Part A*, *18*: 3.
213. Guhaniyogi, S. C., Kennedy, J. P., and Kelen, T. (1982). *J. Macromol Sci. Chem. Part A*, *18*: 77.
214. Faust, R. F., A., and Kennedy, J. P. (1982). *J. Macromol Sci. Chem. Part A*, *18*: 1209.
215. Miyamoto, M., Sawamoto, M., and Higashimura, T. (1984). *Macromolecules*, *17*: 265.
216. Faust, R., and Kennedy, J. P. (1986). *Polymer Bulletin*, *15*: 317.
217. Higashimura, T., Kamigaito, M., Kato, M., Hasebe, T., and Sawamoto, M. (1993). *Macromolecules*, *26*: 2670.
218. Fodor, Z., and Faust, R. (1998). *J. Macromol. Sci., Pure Appl. Chem.*, *A35*: 375.
219. Kwon, Y., Cao, X., and Faust, R. (1999). *Macromolecules*, *32*: 6963.
220. Sakurada, Y. (1963). *J. Polym. Sci. Part A*, *1*: 2407.
221. McCormick, H. W. (1957). *J. Polym. Sci.*, *25*: 488.
222. Wenger, F. (1960). *Preprints of the Polymer Division, 137th Meeting of the ACS*, Cleveland, Ohio, Apr.
223. Wenger, F. (1960). *Makromol. Chem.*, *37*: 143.
224. Fujimoto, T., Ozaki, N., and Nagasawa, M. (1965). *J. Polym. Sci. Part A*, *3*: 2259.
225. Worsfold, D. J., and Bywater, S. (1957). *J. Polym. Sci.*, *26*: 299.
226. Wyman, D. P., and Song, I. H. (1968). *Makromol Chem.*, *115*: 64.
227. Leonard, J., and Malhottal, S. L. (1971). *J. Polym. Sci. Part A-1*, *9*: 1983.
228. Nagasaki, Y., Yamazaki, N., Kato, M. (1997). *J. Photopolym. Sci. Technol.*, *10*: 321.
229. Nagasaki, Y. (1998). *ACS Symp. Ser.*, *706*: 276.
230. Hodges, J. C., Harikrishnan, L. S. and Ault-Justus, S. (2000). *J. Comb. Chem.*, *2*: 80.
231. Alfrey, T., Bohrer, J. J., and Mark, H. (1952). *Copolymerization*, Interscience, New York, p. 49.
232. Nakatsuka, K. (1951). *Kobunshi Kagaku*, *8*: 133.
233. Vukovic, R., Fles, D., and Erceg, A. (1997). *J. Macromol. Sci., Pure Appl. Chem.*, *A 34*: 1619.

234. Barson, C. A., Bevington, J. C., and Hunt, B. J. (1998). *Polymer, 39*: 1345.
235. Okamoto, Y., Yuki, H., and Murahashi, S. (1968). *Bull. Chem. Soc. Jpn., 41*: 197.
236. Endo, K., and Otsu, T. (1986). *J. Polym. Sci. Part A Polym. Chem. Ed., 24*: 1615.
237. Shimizu, A., Otsu, T., and Imoto, M. (1 965). *Polym. Lett., 3*: 103 1.
238. Calafate, B. A. L., Musco, I. M. P., and Mano, E. B. (1990). *Bull. Electrochem., 6*: 769.
239. Staudinger, H., and Dreher, E. (1935). *Justus Liebigs Ann. Chem., 517*: 73.
240. Kennedy, J. P. (1964). *J. Polym. Sci. Part A, 2*: 517.
241. Murahashi, S., Nozakura, S., Tsuboshima, K., and Kotake, Y. (1964). *Bull Chem. Soc. Jpn., 37*: 706.
242. Mizote, A., Higashimura, T., and Okamura, S. (1968). *J. Polym. Sci. Part A-1, 6*: 1825.
243. Mizote, A., Higashimura, T., and Okamura, S. (1968). *Pure Appl Chem., 16*: 457.
244. Kawamura, T., Uryu, T., and Matsuzaki, K. (1982). *Makromol Chem., 183*: 125.
245. Mutschler, H., Schröder, U., Fabner, E., Ebert, K. H., and Hamielec, A. E. (1985). *Polymer, 26*: 935.
246. Coote, M. L., and Davis, T. P. (1999). *Macromolecules, 32*: 4290.
247. Kanoh, N., Gotoh, A., Higashimura, T., and Okamura, S. (1963). *Makromol. Chem., 63*: 106.
248. Magagnini, P. L., Plesch, P. H., and Kennedy, J. P. (1971). *Eur. Polym. J., 7*: 1161.
249. Heublein, G., and Dawczynski, H. (1972). *J. Prakt. Chem., 314*: 557.
250. Tanizaki, A., Sawamoto, M., and Higashimura, T. (1986). *J. Polym. Sci. Polym. Chem. Ed., 24*: 87.
251. Faust, R., and Kennedy, J. P. (1988). *Polym. Bull. (Berlin), 19*: 29.
252. Kojima, K., Sawamoto, M., and Higashimura, T. (1990). *J. Polym. Sci., Part A: Polym. Chem., 28*: 3007.
253. Fodor, Z., and Faust, R. (1994). *J. Macromol. Sci., Pure Appl. Chem., A31*: 1985.
254. Abu-Abdoun, I. I., and Aale, A. (1993). *Eur. Polym. J., 29*: 1439.
255. Natta, G., Danusso, F., and Sianesi, D. (1958). *Makromol Chem., 28*: 253.
256. Hodges, W., and Drucker, A. (1959). *J. Polym. Sci., 39*: 549.
257. Zambelli, A., Pellecchia, C., Oliva, L., Longo, P., and Grassi, A. (1991). *Makromol. Chem., 192*: 223.
258. Chung, T. C., and Lu, H. L. (1997). *J. Polym. Sci., Part A: Polym. Chem., 35*: 575.
259. Chung, T. C., and Lu, H. L. (1998). *J. Polym. Sci., Part A: Polym. Chem., 36*: 1017.
260. Lu, H. L., Hong, S., and Chung, T. C. (1999). *J. Polym. Sci., Part A: Polym. Chem., 37*: 2795.
261. Kotani, Y., Kamigaito, M., and Sawamoto, M. (2000). *Macromolecules, 33*: 6746.
262. Higashi, H., Tanaka, T., Tanimura, M., and Egashira, T. (1961). *Makromol. Chem., 43*: 245.
263. Nakayama, M., Hirohara, H., Takaya, T., and Ise, N. (1970). *J Polym. Sci., Part A-1, 8*: 3653.
264. Hirohara, H., Nakayama, M., Kawabata, R., and Ise, N. (1970). *J. Chem. Soc. Faraday Trans. 1, 68*(1): 51.
265. Devenport, W., Michalak, L., Malmström, E., Mate, M., Kurdi, B., and Hawker, C. J. (1997). *Macromolecules, 30*: 1929.
266. Schmidt-Naake, G., and Stenzel, M. (1998). *Angew. Makromol. Chem., 254*: 55.
267. Chong, Y. K., Le, T. P. T., Moad, G., Rizzardo, E., and Thang, S. H. (1999). *Macromolecules, 32*: 2071.
268. Qiu, J., and Matyjaszewski, K. (1997). *Macromolecules, 30*: 5643.
269. Zapelova, N. P., and Koton, M. M. (1957). *Zhur. Obsc. Khim., 27*: 2138; C.A., 52: 6247c; see also C.A., 52: 8064e, 52: 12804a, 53: 19942b.
270. Yuhi, H., Okamoto, Y., Kuwae, Y., and Hatada, K. (1969). *J. Polym. Sci. Part A-1, 7*: 1933.
271. Matsushita, S., Higashimura, T., and Okamura, S. (1960). *Kobunshi Kagaku, 17*: 456.
272. Okamura, S., Kanoh, N., and Higashimura, T. (1961). *Makromol. Chem., 47*: 19.
273. Imanishi, Y., Matsushita, S., Higashimura, T., and Okamura, S. (1964). *Makromol. Chem., 70*: 68.
274. Sawamoto, M., and Higashimura, T. (1992). *Makromol. Chem., Macromol. Symp., 60*: 47.
275. Gong, M. S., and Hall, H. K., Jr. (1986). *Macromolecules, 19*: 301 1.

276. Hall, H. K., Jr., and Atsumi, M. (1988). *Polym. Bull. (Berlin)*, *19*: 319.
277. Shohi, H., Sawamoto, M., and Higashimura, T. (1992). *Macromolecules*, *25*: 53.
278. Satoh, K., Kamigaito, M., and Sawamoto, M. (1999). *Macromolecules*, *32*: 3827.
279. Satoh, K., Kamigaito, M., and Sawamoto, M. (2000). *J. Polym. Sci., Part A: Polym. Chem.*, *38*: 2728.
280. Satoh, K., Kamigaito, M., and Sawamoto, M. (2000). *Macromolecules*, *33*: 5830.
281. Burnett, G. M., and Young, R. N. (1966). *Eur. Polym. J.*, *2*: 329.
282. Geerts, J., van Beylen, M., and Smets, G. (1969). *J. Polym. Sci. Part A-1*, *7*: 2859.
283. Takaya, K., Hirohara, H., Nakayama, M., and Ise, N. (1971). *Trans. Faraday Soc.*, *67*: 119.
284. Rubens, L. C. (1965). *J. Appl. Polym. Sci.*, *9*: 1473.
285. Breitenbach, J. W., Edelhauser, H., and Hochrainer, R. (1966). *Monatsh. Chem.*, *97*: 217.
286. Olaj, O. F. (1966). *Monatsh. Chem.*, *97*: 1437.
287. Brown, C. P., and Mathieson, A. R. (1957). *J. Chem. Soc., Part III*: 3608.
288. Brown, G. R., and Pepper, D. C. (1965). *Polymer*, *6*: 497.
289. Nagai, K., Kobayashi, M., and Nagai, E. (1959). *Bull. Chem. Soc. Jpn.*, *32*: 771.
290. Soga, K., Nakatani, H., and Monoi, T. (1990). *Macromolecules*, *23*: 953.
291. Proto, A., and Senatore, D. (1999). *Macromol. Chem. Phys.*, *200*: 1961.
292. Pellecchia, C., Pappalardo, D., Oliva, L., Mazzeo, M., and Gruter, G.-J. (2000). *Macromolecules*, *33*: 2807.
293. Coulter, K. E., Kehde, H., and Hiscock, B. F.(1971). *Vinyl and Diene Monomers*, Part 2 (Leonard, E. C., ed.), Wiley-Interscience, New York, p. 540.
294. Aso, C., Nawata, T., and Hisashi, K. (1963). *Makromol. Chem.*, *68*: 1.
295. Aso, C. (1970). *Pure Appl. Chem.*, *23*: 287.
296. Wiley, R. H., Jin, J.-I., and Ahn, T.-O. (1969). *J. Macromol. Sci. Chem. Part A*, *8*: 1543.
297. Obrecht, W., Seitz, U., and Funke, W. (I 975). *Makromol Chem.*, *176*: 277 1.
298. Hasegawa, H., and Higashimura, T. (1980). *Macromolecules*, *13*: 1350.
299. Nitadoti, Y., and Tsuruta, T. (1978). *Makromol. Chem.*, *179*: 2069.
300. Tsuruta, T. (1985). *Makromol Chem. Suppl*, *13*: 33.
301. Nagasaki, Y., Ito, H., and Tsuruta, T. (1986). *Makromol. Chem.*, *187*: 23.
302. Kim, K. U., Kim, J., Ahn, Y. H., and Im, S. S. (1997). *Macromol. Symp.*, *118*: 143.
303. Asgarzadeh, F., Ourdouillie, P., Beyou, E., and Chaumont, P. (1999). *Macromolecules*, *32*: 6996.
304. Dao, J., Benoit, D., and Hawker, C. J. (1998). *J. Polym. Sci., Part A: Polym. Chem.*, *36*: 2161.
305. Brunner, H., Palluel, A. L. L., and Walbridge, D. J. (1958). *J. Polym. Sci.*, *28*: 629.
306. Mitin, Y. V., and Glukhov, N. A. (1957). *Dokl. Akad. Nauk SSSR.*, *115*: 97.
307. D'Onofrio, A. A. (1964). *J. Appl Polymer Sci.*, *8*: 521.
308. Dittmer, T., Gruber, F., and Nuyken, O. (1989) *Makromol. Chem.*, *190*: 1755.
309. Nuyken, O., Maier, G., Yang, D., and Leitner, M. B. (1992). *Makromol. Chem., Macromol. Symp.*, *60*: 57.
310. Lutz, P., Beinert, G., and Rempp, P. (1982). *Makromol. Chem.*, *183*: 2787.
311. Nasitowa, R. M., Murav'eva, L. S., Mushina, E. A., and Krentsel, B. A. (1979). *Russ. Chem. Rev.*, *48*: 692, and literature cited there.
312. Okamoto, A., and Mita, 1. (1978). *J Polym. Sci. Polym. Chem. Ed.*, *16*: 1187.
313. Grassi, A., Longo, P., Proto, A., and Zambelli, A. (1989). *Macromolecules*, *22*: 104.
314. Heller, J., and Anyos, T. (1971). *Encyclopedia of Polymer Science and Technology*, Vol. 14 (Mark, H. F., Gaylord, N. G., and Bikales, N., eds.), Wiley, New York, pp. 257–272.
315. Katritzky, A. R., Zhu, D.-W., and Schanze, K. S. (1993). *J. Polym. Sci., Polym. Chem. Ed.*, *31*: 2187.
316. Todesco, R. V., Basheer, R. A., and Kamat, P. V. (1986). *Macromolecules*, *19*: 2390.
317. Tsujii, Y., Tsuchida, A., Ito, S., and Yamamoto, M. (1991). *Macromolecules*, *24*: 4061.
318. Tsuchida, A., Ikawa, T., and Yamamoto, M. (1995). *Polymer*, *36*: 3103.
319. Hussey, D. M., and Fayer, M. D. (1999). *Macromolecules*, *32*: 6638.
320. Kobayashi, T., Fukaya, T., and Fujii, N. (2000). *J. Membr. Sci.*, *164*: 157.

321. Guillet, J. E. (1991). *Pure Appl. Chem., 63*: 917.
322. Sustar, E., Nowakowska, M., and Guillet, J. E. (1992). *J. Photochem. Photobiol. A: Chem., 63*: 357.
323. Nowakowska, M., Foyle, V. P., and Guillet J. E. (1993). *J. Am. Chem. Soc., 115*: 5975.
324. Bai, F., Chang, C.-H., and Webber, S. E. (1986). *Macromolecules, 19*: 2484.
325. Itaya, A., and Okamoto, K. (1987). *Bull. Chem. Soc. Jpn., 60*: 83.
326. Bigger, S. W., Ghiggino, K. P., and Ng, S. K. (1989). *Macromolecules, 22*: 800.
327. Morishma, Y., Lim, H. S., Nozakura, S., and Sturtevant, J. L. (1989). *Macromolecules, 22*: 1158.
328. Lin, J., and Fox, M. A. (1994). *Macromolecules, 27*: 902.
329. Porter, C. F. C., and Philipps, D. (1994). *Eur. Polym. J., 30*: 189.
330. Moriyama, T., Monobe, K., Miyasaka, H., and Itaya, A. (1997). *Chem. Phys. Lett., 275*: 291.
331. Park, D.-J., Ha, C.-S., and Cho, W.-J. (1993). *J. Macromol. Sci., Pure Appl. Chem., A30*: 949.
332. Bratschkov, C., and Braun, D. (1989). *Angew. Makromol. Chem., 168*: 135.
333. Gómez, C., Ruiz, S., and Yus, M. (1999). *Tetrahedron, 55*: 7017.
334. Asanuma, H., Hishiya, T., Ban, T., Gotoh, S., and Komiyama, M. (1998). *J. Chem. Soc., Perkin Trans., 2*: 1915.
335. Bunel, C., Cohen, S., LaGuerre, J. P., and Marechal, E. (1975). *Polym. J., 7*: 320.
336. Bergmann, E. D., and Katz, D. (1958). *J. Chem. Soc.*: 3816.
337. Michel, R. H. (1964). *J. Polym. Sci., Polym. Chem. Ed., 2*: 2533.
338. Coudane, J., Brigodiot, M., and Marechal, E. (1979). *Polym. Bull., 1*: 465.
339. Engel, D., and Schulz, R. C. (1979). *Makromol. Chem., 180*: 2991.
340. Blin, P., Bunel, C., and Marechal, E. (1981). *J. Polym. Sci., Polym. Chem. Ed., 19*: 891.
341. Stolka, M., Yanus, J. F., and Pearson, J. M. (1976). *Macromolecules, 9*: 710.
342. Stolka, M., Yanus, J. F., and Pearson, J. M. (1976). *Macromolecules, 9*: 715.
343. Barrales-Rienda, J. M., and Pepper, D. C. (1967). *Polymer, 8*: 337.
344. Giusti, P., Cerrai, P., Andruzzi, F., and Magagnini, P. L. (1971). *Eur. Polym. J., 7*: 165.
345. Prosser, H. J., and Young, R. N. (1975). *Eur. Polym. J., 11*: 403.
346. Anasagasti, S., and Leon, L. M. (1984). *Polym. Bull., 11*: 13.
347. Pask, S. D., Plesch, P. H., and Kingston, S. B. (1981). *Makromol. Chem., 182*: 3031.
348. Holdecroft, G. E., and Plesch, P. H. (1984). *Makromol. Chem., 185*: 27.
349. Smith, J. D. B., Phillips, D. C., and Davies, D. H. (1979). *J. Polym. Sci., Polym. Chem. Ed., 17*: 1411.
350. Phillips, D. C., Davies, D. H., Smith, J. D. B., and Spewock, S. (1977). *J. Polym. Sci., Polym. Chem. Ed., 15*: 1563.
351. Kamat, P. V., Basheer, R., and Fox, M. A. (1985). *Macromolecules, 18*: 1366.
352. Kamat, P. V., and Gupta, S. K. (1988). *Polymer, 29*: 1329.
353. Stolka, M., Yanus, J. F., and Pearson, J. M. (1976). *Macromolecules, 9*: 374.
354. Stolka, M., Yanus, J. F., and Pearson, J. M. (1976). *Macromolecules, 9*: 719.
355. Rembaum, A., and Eisenberg, A. (1966). *Macromol. Rev., 1*: 57.
356. Tanikawa, K., Hirata, H., Kusabayashi, S., and Mikawa, H. (1969). *Bull. Chem. Soc. Jpn., 42*: 2406.
357. O'Malley, J. J., Yanus, J. F., and Pearson, J. M. (1972). *Macromolecules, 5*: 158.
358. Engel, D., and Schulz, R. C. (1981). *Makromol. Chem., 182*: 3279.
359. Engel, D., and Schulz, R. C. (1983). *Eur. Polym. J., 19*: 967.
360. Cunningham, R. E., and Colvin, H. A. (1992). *Polymer, 33*: 5073.
361. Springer, J., and Win, T. (1978). *Makromol. Chem., 179*: 29.
362. Natta, G., Danusso, F., and Sianesi, D. (1958). *Makromol. Chem., 28*: 253.
363. Heller, J., and Miller, D. B. (1967). *J. Polym. Sci., Polym. Chem. Ed., 5*: 2323.
364. Tanikawa, K., Kusabayashi, S., Hirata, H., and Mikawa, H. (1968). *Polym. Lett., 6*: 275.
365. Benito, Y., Rodriguez, J. G., Baeza, J. G., Fernandez-Sanchez, C., and Gomez-Anton, M. R. (1990). *Eur. Polym. J., 26*: 689.
366. Lashaek, S., Broderick, E., and Bernstein, P. (1959). *J. Polym. Sci., 39*: 223.

367. Golubeva, A. V., Usmanova, N. F., and Vansheidt, A. A. (1961). *J. Polym. Sci.*, *52*: 63.
368. Singh, A., Ma, D., and Kaplan, D. L. (2000). *Biomacromolecules*, *1*: 592.
369. Braun, D., Essmail-Pour, A., and Czerwinski, W. K. (1989). *Colloid Polym. Sci.*, *267*: 1096.
370. Nowakowska, M., Zapotoczny, S., and Karewicz, A. (2000). *Macromolecules*, *33*: 7345.
371. Ide, N., and Fukuda, T. (1997). *Macromolecules*, *30*: 4268.
372. Hallensleben, M. L., and Lumme, I. (1971). *Makromol. Chem.*, *144*: 261.
373. Engel, D., and Schulz, R. C. (1979). *Makromol. Chem.*, *180*: 2987.
374. Price, C. C., Halpern, B. D., and Voong, S. T. (1953). *J. Polym. Sci.*, *11*: 575.
375. Loshaek, S., and Broderick, E. (1959). *J. Polym. Sci.*, *39*: 241.
376. Katz, D. (1963). *J. Polym. Sci., Polym. Chem. Ed.*, *1*: 1635.
377. Bergmann, E. D., and Katz, D. (1958). *J. Chem. Soc.*: 3216.
378. Moacanin, J., Rembaum, A., Laudenslager, R. K., and Adler, R. (1967). *J. Macromol. Sci.*, *A1*: 1497.
379. Stelter, T., and Springer, J. (1984). *Makromol. Chem.*, *185*: 1719.
380. Ballesteros, J., Howerd, G. J., and Teasdale, L. (1977). *J. Macromol. Sci.*, *A11*: 29.
381. Reppe, W., and Keyssner, E. (1935). Ger. 618,120 to IG Farbenindustrie; C.A. (1936), *30*: 110.
382. Bergfjord, J. A., Penwell, R. C., and Stolka, M. (1979). *J. Polym. Sci. Polym. Phys. Ed.*, *17*: 711.
383. Davidge, H. (1959). *J. Appl. Chem.*, *9*: 553.
384. Pielichowski, J. (1972). *Therm. Anal.*, *4*: 339.
385. Barrales-Rienda, J. M., Gonzales-Ramos, J., and Dabrio, M. V. (1975). *Angew. Makromol. Chem.*, *43*: 105.
386. Hoegl, H., Süs, O., and Neugebauer, W. (1959). Ger. 1,068,115 to Kalle AG (1961); C.A. (1961), *55*: 20742a.
387. Hoegl, H. (1965). *J. Phys. Chem.*, *69*: 755.
388. Heischkel, Y., and Schmidt, H.-W. (1991). *Macromol. Chem. Phys.*, *85*: 119.
389. Kim, H. K., Hong, S. I., Cho, H. N, Kim, D. Y., and Kim, C. Y. (1997). *Polym. Bull.*, *38*: 169.
390. Chen, B., Hou, J., Xue, S., and Liu, S. (1996). *Semicond. Photonics Technol.*, *2*: 293.
391. Jiang, X., Register, R. A., Killeen, K. A., Thompson, M. E., Peschenitzka, and Sturm, J. C. (2000). *Chem. Mater.*, *12*: 2542.
392. Kanega, H., Shirota, Y., and Mikawa, H. (1984). *J. Chem. Soc. Chem. Commun.*, 158.
393. Shirota, Y., Noma, N., Kanega, H., and Mikawa, H. (1984). *J. Chem. Soc. Chem. Commun.*, *470*.
394. Biswas, M., and Roy, A. (1993). *Polymer. 34*: 2903.
395. Kakuta, T., Shirota, Y., and Mikawa, H. (1985). *J. Chem. Soc. Chem. Commun.*, 553.
396. Pearson, J. M., and Stolka, M. (1981). *Polymer Monographs*, Vol. 6, Poly(N-Vinyl-Carbazole), Gordon and Breach, New York.
397. Penwell, R. C., Ganguly, B. N., and Smith, T. W. (1978). *J. Polym. Sci. Macrom. Rev.*, *13*: 63.
398. Tazuke, S., and Okamura, S. (1971). *Encyclopedia of Polymer Science and Technology*, Vol. 14 (Mark, H. F., Gaylord, N. G., and Bikales, N., eds.), Wiley, New York, P. 281.
399. Bömer, B. (1987). *Houben-Weyl: Methoden der Organischen Chemie,* Vol. E 20 (Bartl, H., and Falbe, J., eds.), Georg Thieme, Stuttgart, p. 1277.
400. Sandler, S. R., and Wolf, C. (1980). *Organic Chemistry: A Series of monographs*, Vol. 29-III (Wasserman, H. H., ed.), Academic Press, Orlando, Fla., p. 162.
401. Rooney, J. M. (1989). *Comprehensive Polymer Science*, Vol. 3 (Allen, G., ed.), Pergamon Press, Oxford, p. 697.
402. Klimisch, H. J., and Kieczka, H. (1983). *Ullmanns Enzyklopädie der technischen Chemie,* 4th ed., Vol. 23 (Bartholomé, E., Biekert, E., Hellmann, H., Lei, H., Weigert, W. M., and Wiese, E., eds.), Verlag Chemie, Weinheim, p. 597.
403. Reppe, W., and coworkers (1956). *Justus Liebigs Ann. Chem.*, *601*: 81.
404. Davidge, H. (1959). *J. Appl. Chem.*, *9*: 241.
405. Flowers, R. G., Muller, H. G., and Flowers, L. W. (1948). *J. Am. Chem. Soc.*, *70*: 3019.
406. Polaczek, J., Fraczek, K., Tecza, T., and Lisicki, J. (1968). Pol. 54,803; C.A. (1969), *70*: 28822.

407. Clemo, G. R., and Perkins, W. H. (1924). *J. Chem. Soc., 125*: 1804.
408. Lopatinskii, V. P., and Zherebtsov, L. P. (1965). *Izv. Tomsk. Politekhn. Inst., 136*: 23; C.A. (1966), *65*: 18550.
409. Stern, E. W., and Spector, M. L. (1971). U.S. 3,564,007 to Grace Comp.; C.A. (1971), *75*: 35732r.
410. Reppe, W., Keyssner, E., and Nikolai, F. (1937). Ger. 646,995 to IG Farbenindustrie; C.A. (1937), *31*: 6258.
411. Rogers, C. U., and Corson, B. B. (1947). *J. Am. Chem. Soc., 69*: 2910.
412. Chernobai, A. V., Tirakyants, Z. S., and Delyatitskaya, R. Y. (1967). *Vysokomol. Soedin.,* 9: 664; C.A. (1967), *67*: 22260.
413. Rooney, J. M. (1976). *J. Polym. Sci. Polym. Symp., 56*: 47.
414. Bowyer, P. M., Ledwith, A., and Sherrington, D. C. (1971). *Polymer, 12*: 509.
415. Rodriguez, M., and arid Leon, L. M. (1983). *J. Polym. Sci. Polym. Lett., 21*: 881.
416. Ellinger, L. P. (1965). *J. Appl. Polym. Sci., 9*: 3939.
417. Davidge, H. (1960). Brit. 831,913 to British Oxygen Comp.; C.A. (1961), *54*: 16925.
418. Hort, E. V. (1971). Ger. 2,111,293 to GAF Corp.; C.A. (1972), *76*: 46676.
419. Bevington, J. C., and Dyball, C. J. (1976). *J. Polym. Sci. Polym. Chem. Ed., 14*: 1819.
420. Bevington, J. C., Dyball, C. J., and Leech, J. (1979). *Makromol Chem., 180*: 657.
421. Breitenbach, J. W., and Srna, C. (1963). *J. Polym. Sci. Polym. Lett., 1*: 263.
422. Olaj, O. F., Breitenbach, J. W., and Kauffmann, H. F. (1971). *J. Polym. Sci. Polym. Lett.,* 9: 877.
423. Breitenbach, J. W., Kauffmann, H. F., and Olaj, O. F. (1971). *Monatsh. Chem., 102*: 385.
424. Bevington, J. C., and Dyball, C. J. (1975). *J. Chem. Soc. Faraday Trans. 1, 71*: 2226.
425. Jones, R. G., Catterall, E., Bilson, R. T., and Booth, R. G. (1972). *J. Chem. Soc. Chem. Commun.,* 22.
426. Baethge, H., Butz, S., and Schmidt-Naake, G. (1997). *Macromol. Rapid Commun., 18*: 911.
427. Baethge, H., Butz, S., Han, C.-H., and Schmidt-Naake, G. (1999). *Angew. Makromol. Chem., 267*: 52.
428. Solomon, O. F., Ciuta, I. Z., and Cobiann, N. (1964). *J. Polym. Sci. Polym. Lett., 2*: 311.
429. Sargent, D. E. (1945). U.S. 2,560,251 to GAF Corp.; C.A. (1951), *45*: 9306.
430. Tsubokawa, M., Omichi, H., and Okui, N. (1994). *Polym. Bull., 33*: 187.
431. Gal, Y. S., Ryoo, M. S., Jung, B., Cho, H. N., and Choi, S. K. (1993). *Korean Polym. J., 1*: 100.
432. Biswas, M. (1976). *J. Macromol. Sci. Rev. Macromol. Chem., 14*: 1.
433. Gandini, A., and Plesch, P. H. (1966). *J. Chem. Soc. Part B,* 7.
434. Gandini, A., and Prieto, S. (1977). *J. Polym. Sci. Polym. Lett., 15*: 337.
435. Reppe, W., Keyssner, E., and Dorrer, E. (1937). U.S. 2,072,465 to IG Farbenindustrie; C.A. (1937), *31*: 2717.
436. Biswas, M., and Chakravarty, D. (1973). *J. Polym. Sci. Polym. Chem. Ed., 11*: 7.
437. Tazuke, S., Tjoa, T. B., and Okamura, S. (1967). *J. Polym. Sci. Part A-1, 5*: 1911.
438. Kawamura, T., and Matsuzaki, K. (1981). *Makromol Chem., 182*: 3003.
439. Tazuke, S., Asai, M., and Okamura, S. (1968). *J. Polym. Sci. Part A-1, 6*: 1809.
440. Ledwith, A., and Sambhi, M. (1965). *J. Chem. Soc. Chem. Commun.,* 64.
441. Bowyer, P. M., Ledwith, A., and Sherrington, D. C. (1971). *Polymer, 12*: 509.
442. Chen, C., Su, Q., Liu, W., Chen, Y., and Xi, F. (1996). *J. Macromol. Sci., Pure Appl. Chem., A33*: 1017.
443. Higashimura, T., and Sawamoto, M. (1984). *Adv. Polym. Sci., 62*: 49.
444. Higashimura, T., and Sawamoto, M. (1989). *Comprehensive Polymer Science*, Vol. 3 (Allen, G., ed.), Pergamon Press, Oxford, p. 673.
445. Tsuchida, A., Umeda, A., Nagata, A. Yamamoto, M., Fukui, H., Sawamoto, M., and Higashimura, T. (1994). *Macromol. Rapid. Commun., 15*: 475.
446. Higashimura, T., Teranishi, H., and Sawamoto, M. (1980). *Polym. J. (Tokyo), 12*: 393.
447. Higashimura, T., Deng, Y. X., and Sawamoto, M. (1983). *Polym. J. (Tokyo), 15*: 385.

448. Sawamoto, M., Fujimori, J., and Higashimura, T. (1987). *Macromolecules, 20*: 916.
449. Saehoon, O. (1995), *Ph.D. thesis*, TU München.
450. Scott, H., Miller, G. A., and Labes, M. M. (1963). *Tetrahedron Lett.*, 1073.
451. Ellinger, L. P. (1963). *Chem. Ind. (London)*, 1982.
452. Hyde, P., and Ledwith, A. (1974). *Molecular Complexes,* Vol. 2 (Foster, R., ed.), Elek Science, London.
453. Shirota, Y., and Mikawa, H. (1978). *J. Macromol. Sci. Rev. Macromol. Chem., 16*: 129.
454. Ellinger, L. P. (1968). *Adv. Macromol Chem., 1*: 169.
455. Hall, H. K., Jr. (1983). *Angew. Chem. Intern. Ed. Engl., 22*: 440.
456. Natsume, T., Akana, Y., Tanabe, K., Fujimatsu, M., Shimitsu, M., Shiroto, Y., Hirata, H., Kusabayashi, S., and Mikawa, H. (1969). *J. Chem. Soc. Chem. Commun.*, 189.
457. Aoki, S., and Stille, J. K. (1970). *Macromolecules, 3*: 472.
458. Pac, J., and Plesch, P. H. (1967). *Polymer, 8*: 237.
459. Gumbs, R., Penczek, S., Jagur-Grodzinski, J., and Szwarc, M. (1969). *Macromolecules, 2*: 77.
460. Nakamura, T., Soma, M., Ohnishi, T., and Tamura, K. (1970). *Makromol Chem., 135*: 241.
461. Bawn, C. E. H., Ledwith, A., and Sambhi, M. (1971). *Polymer, 12*: 209.
462. Gotoh, T., Padias, A. B., and Hall, H. K., Jr. (1986). *J. Am. Chem. Soc., 108*: 4920.
463. Abdelkader, M., Padias, A. B., and Hall, H. K., Jr. (1987). *Macromolecules, 20*: 944.
464. Abu-Abdoun, I. I., and Ledwith. A. (1997). *Eur. Polym. J., 33*: 1671.
465. Ledwith, A., and Sherrington, D. C. (1972). *Macromol. Synth.* (Moore, J. A., ed.), *4*: 183.
466. Tazuke, S. (1973). *Pure Appl. Chem., 34*: 329.
467. Atmaca, L., Kayihan, I., and Yagci, Y. (2000). *Polymer, 41*: 6035.
468. Chapiro, A., and Hardy, G. (1962). *J. Chim. Phys., 59*: 993.
469. Kroh, J., and Pekala, W. (1966). *Bull. Acad. Pol. Sci., 14*: 55.
470. Galdecki, Z., Karolak, J., Pekala, W., and Kroh, J. (1967). *Bull. Acad. Pol. Sci., 15*: 209.
471. Ayscough, P. B., Roy, A. K., Croce, R. G., and Munari, S. (1968). *J. Polym. Sci. Part A-1, 6*: 1307.
472. Higashimura, T., Matsuda, T., and Okamura, S. (1970). *J. Polym. Sci. Part A-1*, 8: 483.
473. Myers, R. A., and Christman, E. M. (1968). *J. Polym. Sci. Part A-1, 6*: 945.
474. Tsuji, K., Takahura, K., Niski, M., Hayaski, K., and Okamura, S. (1966). *J. Polym. Sci. Part A-1, 4*: 2028.
475. Matsuda, T., Higashimura, T., and Okamura, S. (1968). *J. Macromol. Sci. Chem. Part A-2,* 43.
476. Tsutsui, K., Hirotsu, K., Umesaki, M., Kurakashi, M., Shimado, A., and Higuchi, T. (1975). *Acta Crystallogr. Part B, 32*: 3049.
477. Bauer, L. N., Healy, R. B., and Stringer, H. R. (1959). Ger. to Rohm & Haas Co.; C.A. (1961), *55*: 14997i.
478. Haaf, F., Scanner, A., and Straub, F. (1985). *Polym. J., 17*: 143.
479. Reppe, W. (1954). *Polyvinylpyrrolidon,* Verlag Chemie, Weinheim; see (1953), *Angew. Chem., 65*: 577.
480. Reppe, W., and coworkers, (1956). *Justus Liebigs Ann. Chem., 601*: 81, 135.
481. Luckenbach, R. (1978). *Beilstein's Handbuch der organischen Chemie* (Luckenbach, R., et al., eds.), 3/4 Ergänzungswerk, Vol. 21/4, Springer-Verlag, Berlin, p. 3147.
482. Kern, W., and Cherdron, H. (1962). *Houben-Weyl: Methoden der organischen Chemie,* Vol. XIV/1 (Müller, E., ed.), Georg Thieme, Stuttgart, p. 1106.
483. Fikentscher, H., and Herle, K. (1945). *Mod. Plas., 23*: 157; C.A. (1946), *40*: 20332.
484. Schuster, C., Sauerbier, R., and Fikentscher, H. (1939). Ger. 757,355 to IG Farbenindustrie; C.A. (1954), *48*: 4959.
485. Schuster, C., Sauerbier, R., and Fikentscher, H. (1943). U.S. 2,335,454; C.A. (1944), *38*: 27701.
486. Sidel'kovskaya, F. P., Askarov, M. A., and Ibragimov, F. (1964). *Vysokomol. Soedin, 6*:1810; (1964), *Polym. Sci. USSR (Engl. Transl.), 6*: 2005; C.A. (1965), *62*: 6563c.
487. Senogles, E., and Thomas, R. A. (1975). *J. Polym. Sci. Polym. Symp., 49*: 203.
488. Breitenbach, W., and Schmidt, A. (1952). *Monatsh. Chem., 83*: 833, 1288.

489. Breitenbach, W. (1957). *J. Polym. Sci.*, *23*: 949.
490. Luberoff, B. J., and Gersumky, W. D. (1964). U.S. 3,162,625 to American Cyanamid Co.; C.A. (1965), *62*: 6593b.
491. Bömer, B. (1987). *Houben-Weyl: Methoden der organischen Chemie*, 4th ed., Vol. E20/2 (Bartl, H., and Falbe, J., eds.), Georg Thieme, Stuttgart, p. 1267.
492. Sandler, S. R., and Karo, W. (1977). *Progress in Organic Chemistry: Polymer Syntheses*, Vol. 29, (Wasserman, H. H., ed.) Academic Press, Orlando, p. 232.
493. Schildknecht, C. E. (1953). U.S. 2,658,045 to General Aniline & Film Corp.; C.A. (1954), *48*: 2413f.
494. Reppe, W., Schuster, C., and Hartmann, A. (1943). Ger. 737,663 to IG Farbenindustrie; C.A. (1944), *38*: 37572.
495. Reppe, W., Schuster, C., and Hartmann, A. (1941). U.S. 2,265,450 to IG Farbenindustrie; C.A. (1942), *36*: 20524.
496. Reppe, W., Herle, K., and Fikentscher, H. (1943). Ger. 922,378 to BASF; C.A. (1955), *49*: 5914.
497. Traylor, T. G., and Voeks, J. F. (1962). U.S. 3,048,555 to Dow Chemical Co.; C.A. (1962), *57*: 15372i.
498. Voeks, J. F., and Traylor, T. G. (1963). Ger. 1,157,397 to Dow Chemical Co.; C.A. (1964), *60*: 4227c.
499. Bond, J., and Lee, P. I. (1969). *J. Polym. Sci.*, *7*(1): 379; (1971). *J. Polym. Sci.*, *9*(6): 1775; (1971), *J. Polym. Sci.*, *9*(6): 1777.
500. Senogles, E., and Thomas, R. A. (1978). *J. Polym. Sci. Polym. Lett. Ed.*, *16*: 555.
501. Cizravi, J. C., Tay, T. Y., and Pon, E. C. (2000). *J. Appl. Polym. Sci.*, *75*: 239.
502. Barabas, E. S., and Cho, J. R. (1984). Eur. 104,042 to General Aniline & Film Corp.; C.A. (1984), *101*: 24133a.
503. Gupta, K. C. (1994). *J. Appl. Polym. Sci.*, *53*: 71.
504. Hayashi, K., and Smets, G. (1958). *J. Polym. Sci.*, *27*: 275.
505. Brit. 102112 (1966) to General Aniline & Film Corp.; C.A. (1966), *64*: 19821h.
506. Chernobai, A. V. (1968). *Vysokomol. Soedin. Ser. A*, *10*: 1716 (1968), *Polym. Sci. USSR (Engl. Transl)*, *10*: 1986.
507. Monagle, D. J. (1967). U.S. 3,336,270 to Hercules Inc.; C.A. (1967), *67*: 91245f.
508. Field, N. D., and Williams, E. P. (1970). Ger. 1,929,501 to General Aniline & Film Corp.; C.A. (1970), *72*: 67774k.
509. Field, N. D., and Williams, E. P. (1970). Fr. 2,010,746 to General Aniline & Film Corp.; C.A. (1970), *73*: 46270y.
510. Schildknecht, C. E., Zoss, A. O., and Grosser, F. (1949). *Ind. Eng. Chem.*, *41*: 2891; C.A. (1950), *44*: 3739b.
511. Mottus, E. H., and Fields, J. E. (1966). U.S. 3,275,611 to Monsanto Co.; C.A. (1967), *66*: 11288p.
512. Nakaguchi, K., Kawasumi, S., Hirooka, M., Yabuuchi, H., and Takao, H. (1970). Jpn. 7,009,952 to Sumitomo Chemical Co., Ltd.; C.A. (1970), *73*: 35965c.
513. MacKenzie, J. C., and Orzechowski, A. (1966). U.S. 3,285,892 to Cabot Corp.; C.A. (1967), *66*: 19020r.
514. Tsubokawa, N., Takeda, N., and Kanamaru, A. (1980). *J. Polym. Sci. Polym. Lett. Ed.*, *18*: 625.
515. Biswas, M., and Mishra, P. K. (1975). *Polymer*, *16*(8): 621.
516. Breitenbach, J. W., Galinovski, F., Nesvadba, H., and Wolf, E. (1955). *Naturwissenschaften*, *42*: 155, 440.
517. Brit. 725,674 (1955) to General Aniline & Film Corp.; C.A. (1955), *49*: 12037a.
518. Tamura, H., and Tanaka, M. (1969). *Bull. Chem. Soc. Jpn.*, *42*: 3042; C.A. (1970), *72*: 13096x.
519. Frunze, N. K. (1969). *Rev. Roum. Chim.*, *14*: 1309; C.A. (1970), *72*: 112201j.
520. Kropp, J. E. (1970). Ger. 2,106,018 to Minnesota Mining and Manufacturing Co.; C.A. (1971), *74*: 42836k.
521. Munari, S., Tealdo, F. V., Vigo, F., and Bonta, G. (1966). *Proc. Tihany Symp. Radiat. Chem.*, *2*: 573; C.A. (1968), *68*: 30142u.

522. Hardy, G., and Morsi, M. A. (1969). *Kinet. Mech. Polyreactions Int. Symp. Macromol. Chem. Prepr.*, *4*: 75; C.A. (1971), *75*: 64348a.
523. Wainer, E., Lewis, J. M., and Shirey, J. E. (1971). Ger. 2,112,416 to Horizons Research Inc.; C.A. (1972), *76*: 8964g.
524. Tamura, H., and Tanaka, M. (1969). *Bull. Chem. Soc. Jpn.*, *42*: 3041; C.A. (1970), *72*: 13094v.
525. Tamura, H., and Tanaka, M. (1970). Kobunshi Kagaku, 27: 652; C.A. (1971), *75*: 6461 w; (1970), *Kobunshi Kagaku*, *27*: 736; C.A. (1971), *75*: 6459b.
526. Tonelli, A. E. (1982). *Polymer*, *23*: 676.
527. Handler, J. S., and Orloff, J. (1973). *Handbook of Physiology* (Orloff, J., and Berliner, R. W., eds.), American Physiological Society, Washington, D.C., Sec. 8, p. 791.
528. Davidson, R. L., and Sittig, M. (1968). *Water-Soluble Resins* (Davidson, R. L., and Sittig, M., eds.), Reinhold, New York, p. 137; C.A. (1968), *69*: 107337s.
529. Tan, Y. Y., and Challa, G. (1976). *Polymer*, *17*(8): 739.
530. Wood, A. S. (1970). *Kirk-Othmer Encyclopedia of Chemical Technology*, 2nd ed., Vol. 21 (Mark, H. F., McKetta, J. J., and Othmer, D. F., eds.), Wiley, New York, p. 427ff.
531. Lorenz, D. H. (1971). *Encyclopedia of Polymer Science and Technology*, Vol. 14 (Mark, H. F., Gaylord, N. G., and Bikales, N. M., eds.), Wiley, New York, p. 239.
532. Kondo, S., Ozeki, M., Nakashima, N., Suzuki, K., and Tsuda, K. (1988). *Angew. Makromol. Chem.*, *163*: 139.
533. Koetsier, D. W., Tan, Y. Y., and Challa, G. (1980). *J. Polym. Sci. Polym. Chem. Ed.*, *18*: 1933.
534. Alberda van Ekenstein, G. O. R., Koetsier, D. W., and Tan, Y. Y. (1981). *Eur. Polym. J.*, *17*: 839, 845.
535. Rajan, V. S., and Ferguson, J. (1982). *Eur. Polym. J.*, *18*: 633.
536. Osada, Y., Takase, M., and Iriyama, Y. (1983). *Polym. J.*, *15*: 81.
537. Quella, F., Czerwinski, W. K., and Braun, D. (1987). *Makromol. Chem.*, *188*: 2403.
538. Braun, D., and Czerwinski, W. K. (1987). *Makromol. Chem.*, *188*: 2371.
539. Braun, D., and Czerwinski, W. K. (1987). *Makromol. Chem.*, *188*: 2389.
540. Manickam, S. P. P., and Subbaratnam, N. R. (1980). *Makromol Chem.*, *181*: 2637.
541. Georgiev, G. S., and Zubov, V. P. (1980). *Polym. Bull.*, *2*: 325.
542. Gao; J.-P., Li, Z.-C., and Huang, M.-Z. (1995). *J. Appl. Polym. Sci.*, *55*: 1291.
543. Tazuke, S., and Okamura, S. (1971). *Encyclopedia of Polymer Science and Technology*, Vol. 14 (Mark, H. F., Gaylord, N. G., and Bikales, N., eds.), Wiley, New York, p. 637.
544. Khan, I. M. (1989). *Encyclopedia of Polymer Science and Engineering*, 2nd ed., Vol. 17 (Mark, H. F., ed.), Wiley, New York, p. 567.
545. Bömer, B. (1985). *Houben-Weyl: Methoden der organischen Chemie*, Vol. 20 (Bartl, H., and Falbe, J., eds.), Georg Thieme, Stuttgart, p. 1287.
546. Hays, J. T. (1952). U.S. Pat. 2,611,769 to Hercules Powder Co.; C.A. (1953), *47*: 9367i.
547. Yoshida, M. (1960). *Kogyo Kagaku Zasshi*, *63*: 893; C.A. (1962), *57*: 10533g.
548. Takayama, Y. (1959). *Kogyo Kagaku Zasshi*, *62*: 658; C.A. (1962), *57*:79136.
549. Wagner, C. R. (1956). U.S. Pat. 2,732,376 to Phillips Petroleum Co.; C.A. (1956), *50*: 9450g.
550. Mahan, J. E. (1956). U.S. Pat. 2,769,811 to Phillips Petroleum Co.; C.A. (1957), *51*: 7433d.
551. Runge, F., Naumann, G., and Morgner, M. (1957). Ger. (East) Pat. 13,099 to Farbenfabrik Wolfen; C.A. (1959), *53*: 5292g.
552. Runge, F., Naumann, G., and Morgner, M. (1960). Brit. Pat. 828,205 to Farbenfabrik Wolfen; C.A. (1960), *54*: 13149b.
553. Yoshida, M. (1958). *Yuki Gosei Kagaku Kyokai Shi*, *16*: 571; C.A. (1959), *53*: 1342c.
554. Mahan, J. E. (1950). U.S. Pat. 2,513,600 to Phillips Petroleum Co.; C.A. (1950), 44: 9987a.
555. Winterfeld, K., and Heinen, C. (1951). *Justus Liebigs Ann. Chem.*, *573*: 85.
556. Profft, E. (1955). *Chem. Tech. (Berlin)*, *7*: 511; C.A. (1957), *51*: 6629e.
557. Mahan, J. E. (1950). U.S. Pat. 2,534,285 to Phillips Petroleum Co.; C.A. (1951), *45*: 3425i.
558. Salisbury, L. F. (1949). Brit. Pat. 632,611 to E.I. du Pont de Nemours & Co.; C.A. (1950), *44*: 4513.

559. Dimond, H. L., Fleckenstein, L. J., and Shrader, M. O. (1958). U.S. Pat. 2,848,456 to Pittsburgh Coke & Chemical Co.; C.A. (1959), *53*: 1384d.

560. Ostroverkhov, V. G., Vakarchuk, I. S., and Sinyavskii, V. G. (1961). *Vysokomol. Soedin.*, *3*: 1197; C.A. (1962), *56*: 8921d.

561. Karkozov, V. G., Gavurina, R. K., Polonskii, V. S., and Smirnova, A. I. (1968). *Vysokomol. Soedin. Ser. A, 10*: 1343.

562. Korshak, V. V., Zubakova, L. B., and Zhovnirovskaya, A. B. (1985). *Vysokomol. Soedin. Ser. B, 27*: 7; C.A. (1985), *102*: 96150x.

563. Bengough, W. I., and Henderson, W. (1965). *Trans. Faraday Soc., 61*: 141.

564. Potts, M. F., and Hudson, P. F. (1956). U.S. Pat. 2,767,159 to Phillips Petroleum Co.; C.A. (1957), *51*: 4054c.

565. Tazuke, S., and Okamura, S. (1965). *Polym. Lett., 3*: 135.

566. Tazuke, S., and Okamura, S. (1966). *J. Polym. Sci. Part A-1, 4*: 14 1.

567. Pritchard, J. E., Opheim, M. H., and Moyer, P. H. (1955). *Ind. Eng. Chem., 47*: 863; C.A. (1955), *49*: 9320f.

568. Fitzgerald, E. B., and Fuoss, R. M. (1950). *Ind. Eng. Chem., 42*: 1603; C.A. (1950), *44*: 10366g.

569. Katchalsky, A., Rosenheck, K., and Altman, B. (1957). *J. Polym. Sci., 23*: 955.

570. Crescentini, L., Gechele, G. B., and Pizzoli, M. (1965). *Eur. Polym. J., 1*: 293.

571. Spiegelman, P. P., and Parravano, G. (1964). *J. Polym. Sci. Part A, 2*: 2245.

572. Schmidt-Naake, G., and Stenzel, M. (1998). *Angew. Makroml. Chem., 254*: 55.

573. Fontanille, M., and Sigwalt, P. (1967). *Bull. Soc. Chim. France, 4083.*

574. Bhadani, S. N., and Parravano, G. (1970). *J. Polym. Sci. Part A-I, 8*: 225.

575. Laurin, D., and Parravano, G. (1968). *J. Polymer. Sci. Part A-1, 6*: 1047.

576. Fontanille, M., and Sigwalt, P. (1966). *Compt. Rend. Ser. C, 263*: 624.

577. Natta, G., Mazzanti, G., Dall'Asta, G., and Longi, P. (1960). *Makromol. Chem., 37*: 160.

578. Natta, G., Mazzanti, G., Longi, P., Dall'Asta, G., and Bernatdini, F. (1961). *J. Polym. Sci., 51*: 487.

579. Erussalimsky, B. L., Fedorova, L. A., Klenin, S. I., and Shibaev, L. A. (1998). *Macromol. Symp., 128*: 221.

580. Smid, J., and Szwarc, M. (1962). *J. Polym. Sci., 61*: 31.

581. Khan, I. M., Meverden, C. C., and Hogen-Esch, T. E. (1983). *ACS Polym. Prepr., 24*: 151.

582. Meverden, C. C., and Hogen-Esch, T. E. (1984). *Makromol. Chem. Rapid Commun., 5*: 749.

583. Soum, A. H., and Hogen-Esch, T. E. (1985). *Macromolecules, 18*: 690.

584. Meverden, C. C., and Hogen-Esch, T. E. (1983). *Makromol. Chem. Rapid Commun., 4*: 563.

585. Soum, A., and Fontanille, M. (1980). *Macromol. Chem., 181*: 799.

586. Soum, A. H., Tien, C.-F., and Hogen-Esch, T. E. (1983). *Macromol. Chem. Rapid Commun., 4*: 243.

587. Lindsell, W. E., Robertson, F. C., and Soutar, I. (1983). *Eur. Polym. J., 19*: 115.

588. Tang, L. C., and Francois, B. (1983). *Eur. Polym. J., 19*: 715.

589. Kargin, V. A., Kabanov, V. A., Aliev, K. V., and Razvodovskii, E. F. (1965). *Dokl. Akad. Nauk SSSR, 160*: 604; C.A. (1965), *62*: 14837g.

590. Kabanov, V. A., Patrikeeva, T. I., and Kargin, V. A. (1966). *Dokl. Akad. Nauk SSSR, 168*: 1350; C.A. (1966), *65*: 17050e.

591. Kabanov, V. A., Aliev, K. V., and Kargin, V. A. (1968). *Vysokomol. Soedin. Ser. A, 10*: 1618.

592. Kabanov, V. A., Patrikeeva, T. I., Kargina, 0. V., and Kargin, V. A. (1966). *Intern. Symp. Macromol. Chem. Tokyo*, preprint 2-2-09.

593. Kabanov, V. A., and Petrovskaya, V. A. (1968). *Vysokomol. Soedin. Ser. B, 10*: 797; C.A. (1969), *70*: 78444n.

594. Kargin, V. A., Kabanov, V. A., and Kargina, 0. V. (1965). *Dokl. Akad. Nauk SSSR, 161*: 1131; C.A. (1965), *63*: 3068h.

595. Kargina, 0. V., Kabanov, V. A., and Kargin, V. A. (1968). *J. Polym. Sci. Part C, 22*: 339.

596. Salamone, J. C., Snider, B., and Fitch, W. L. (1971). *J. Polym. Sci. Part A-1, 9*: 1493.

597. Kabanov, V. A., Kargina, 0. V., and Ulyanova, M. V. (1976). *J. Polym. Sci. Polym. Chem. Ed*, *14*: 2351.
598. Kabanov, V. A. (1979). *Makromol. Chem. Suppl.*, *3*: 41.
599. Kargina, 0. V., Mishustina, L. A., Svergun, V. I., Lukovkin, G. M., Yevdakov, V. P., and Kabanov, V. A. (1974). *Vysokomol. Soedin. Ser. A*, *16*: 1755.
600. Gvozdetskii, A. N., Kim, V. O., Smetanyuk, V. I., Kabanov, V. A., and Kargin, V. A. (1971). *Vysokomol. Soedin. Ser. A*, *13*: 2409.
601. Mielke, I., and Ringsdorf, H. (1972). *Macromol. Chem.*, *153*: 307.
602. Salamone, J. C., Ellis, E. J., and Mahmud, M. U. (1985). *Macromol. Synth.*, *9*: 85.
603. Fife, W. K., Ranganathan, P., and Zeldin, M. (1989). *Polym. Prepr. Am. Chem. Soc. Div. Polym. Chem.*, *30*(2): 123.
604. Ranganathan, P., Fife, W. K., and Zeldin, M. (1990). *J. Polym. Sci. Part A: Polym. Chem.*, *28*: 2711.
605. Fife, W. K., Zeldin, M., Xin, Y., and Parish, C. (1991). *Polym. Prepr. Am. Chem. Soc. Div. Polym. Chem.*, *32*(1): 579.
606. Jpn. Kokai Tokkyo Koho JP 60 13,869 (1985) to Nippon Zeon Co.; C.A. (1985), *102*: 205280n.
607. Luskin, L. S. (1974). *Functional Monomers*, Vol. 2 (Yocum, R. H., and Nyquist, E. B., eds.), Marcel Dekker, New York.
608. Brandrup, J., and Immergut, E. H., eds. (1975). *Polymer Handbook*, Wiley, New York, p. II-295.
609. Greenley, R. Z. (1980). *J. Macromol. Sci. Chem. Part A*, *14*: 445.
610. Semchikov, Y. D., Ryabov, A. V., and Kashaeva, V. N. (1969). *Vysokomol. Soedin. Ser. B*, *11*: 726; C.A. (1970), 72; 32302c.
611. Semchikov, Y. D., Ryabov, A. V., and Kashaeva, V. N. (1970). *Vysokomol. Soedin. Ser. B*, *12*: 567; C.A. (1970), *73*: 110165x.
612. Trubitsyna, S. N. (1984). *Vysokomol. Soedin. Ser. B*, *26*: 719; C.A. (1985), *102*: 25141j.
613. Brit. Pat. 895,980 (1962) to Phillips Petroleum Co.; C.A. (1963), *58*: 14140a.
614. Strobel, C. W. (1964). U.S. Pat. 3,402,162 to Phillips Petroleum Co.; C.A. (1968), *69*: 107495s.
615. Strobel, C. W. (1960). U.S. Pat. 3,144,429 to Phillips Petroleum Co.; C.A. (1964), *61*: 10854b.
616. Grosius, P., Gallot, Y., and Skoulios, A. (1969). *Macromol. Chem.*, *127*: 94.
617. Grosius, P., Gallot, Y., and Skoulios, A. (1970). *Macromol. Chem.*, *132*: 35.
618. Ishizu, K., Kashi, Y., Fukutomi, T., and Kakurai, T. (1982). *Makromol. Chem.*, *183*: 3099.
619. Matsushita, Y., Nakao, Y., Saguchi, R., Choshi, H., and Nagasawa, M. (1986). *Polym. J.*, *18*: 493.
620. Matsushita, Y., Nomura, M., Watanabe, J., Mogi, Y., Noda, I., and Imai, M. (1995). *Macromolecules*, *28*: 6007.
621. Arai, K., Kotaka, T., Kitano, Y., and Yoshimura, K. (1980). *Macromolecules*, *13*: 1670.
622. Mogi, Y., Kotsuji, H., Kaneko, Y., Mori, K., Matsushita, Y., and Noda, I. (1992). *Macromolecules*, *25*: 5408.
623. Stannett, V., Williams, J. L., Gosnell, A. B., and Gervasi, J. A. (1968). *Polym. Lett.*, *6*: 185.
624. Kolesnikov, G. S., Tevlina, A. S., Vasyukov, S. E., and Smimov, V. S. (1968). *Plast. Massy*, *4*: 22; C.A. (1968), *69*: 19914x.
625. Cumberbirch, R. J. E., and Holker, J. R. (1966). *J. Soc. Dyer Colourists*, *82*: 59; C.A. (1966), *64*: 12861c.
626. Matsushita, Y., Noda, I., and Torikai, N. (1997). *Macromol. Symp.*, *124*: 121.
627. Ragunath Rao, P., Masson, P., Lutz, P., Beinert, G., and Rempp, P. (1984). *Polym. Bull.*, *11*: 115.
628. Ragunath Rao, P., Lutz, P., Lamps, J. P., Masson, P., and Rempp, P. (1986). *Polym. Bull.*, *15*: 69.
629. Kryazhev, Y. G., and Rogovin, Z. A. (1964). *VysokomoL Soedin.*, *6*: 672.
630. Funt, B. L., and Ogryzlo, E. A. (1957). *J. Polym. Sci.*, *25*: 279.
631. Tamikado, T. (1960). *J. Polym. Sci.*, *43*: 489.

632. Natansohn, A., Maxim, S., and Feldman, D. (1979). *Polymer, 20*: 629.
633. Dmitrieva, S. I., Roskin, E. S., and Ezrielev, A. I. (1978). *Vysokomol. Soedin. Ser. B, 20*: 208; C.A. (1978), *88*: 191576s.
634. Dasgupta, S., and Thomas, W. W. (1979). *J. Polym. Sci. Polym. Chem. Ed., 17*: 923.
635. Bando, Y., Dodou, T., and Minoura, Y. (1977). *J. Polym. Sci. Polym. Chem. Ed., 15*: 1917.
636. Cook, G. A., and Butler, G. B. (1985). *J. Macromol. Sci. Chem. Part A, 22*: 1049.
637. Tanaka, Y., Atsukawa, M., Shimura, Y., Okada, A., Sakuraba, H. and Sakata, T. (1975). *J. Polym. Sci. Polym. Chem. Ed., 13*: 1017.
638. Mikes, F., Strop, P., Seycek, O., Roda, J., and Katal, J. (1974). *Eur. Polym. J., 10*: 1029.
639. Jenkins, A. D., and Jenkins, J. (1996). *Macromol. Symp., 111*: 159.
640. Niwa, M., Matsumoto, T., Kagami, M., and Kajiyama, K. (1984). *Sci. Eng. Rev. Doshisha Univ., 25*: 192; C.A. (1985), *102*: 149848q.
641. Shaikhutdinov, E. M., Nurkeeva, Z. S., and Chebeiko, S. V. (1978). *Izv. Akad. Nauk Kaz. SSR, Ser. Khim., 28*: 53; C.A. (1979), *90*: 122130w.
642. Natansohn, A., Maxim, S., and Feldman, D. (1978). *Eur. Polym. J., 14*: 729.
643. Orbay, M., Laible, R., and Dulog, L. (1982). *Makromol Chem., 183*: 47.
644. Petit, A., Cung, M. T., and Neel, J. (1987). *Eur. Polym. J., 23*: 507.
645. Cobianu, N., Boghina, C., Marculescu, B., Cincu, C., and Amalinei, I. (1974). *Rev. Roum. Chim., 19*: 1251; C.A. (1975), *82*: 58272z.
646. Skvortsova, G. G., Skushnikova, A. I., Domnina, Y. S., and Brodskaya, E. L. (1977). *Vysokomol Soedin. Ser. A, 19*: 2091.

3

Poly(vinyl ether)s, Poly(vinyl ester)s, and Poly(vinyl halogenide)s

Oskar Nuyken and Harald Braun
Technische Universität München, Garching, Germany

James Crivello
Rensselaer Polytechnic Institute, Troy, New York

I. POLY(VINYL ETHER)S

A. Introduction

1. Definition and Historical Background

Vinyl ethers comprise that class of olefinic monomers which possess a double bond situated adjacent to an ether oxygen. These monomers include those compounds which have various substituents attached to the carbon atoms of the double bond as well as the unsubstituted compounds. Due to the presence of the neighboring oxygen atom, the double bond possesses a highly electronegative character, a feature that dominates both the organic and polymer chemistry of these compounds. The analogous vinyl thioethers are also known [1] and their chemistry closely parallels that of their corresponding vinyl ether counterparts. Beginning with the accidental discovery by Wislicenus [2] that elemental iodine catalyzes the violent exothermic polymerization of ethyl vinyl ether, the polymerization of these monomers has been the subject of many investigations over the years and continues to occupy the attention of investigators today. In particular, the field of the cationic polymerization of vinyl ethers is a very lively field engaging the efforts of academic as well as industrial workers. Apart from the interesting chemistry of these compounds, the chief incentive for these efforts is their versatility in a wide variety of technical applications. Among the many uses of poly(vinyl ethers) and their copolymers are applications such as adhesives, surface coatings, lubricants, greases, elastomers, molding compounds, films, thickeners, anticorrosion agents, fiber and textile finishes, and numerous others.

Vinyl ether monomers, and their polymerization and copolymerization, have been the subjects of several excellent past reviews [3–7] and some more recent one [8–10]. These reviews have provided a rich source of background material for the present chapter and the reader is referred to them for specific details concerning such topics as manufacturing methods, economics, toxicity, and special applications of poly(vinyl ether) homopolymers and copolymers.

2. Synthesis of Vinyl Ether Monomers

Vinyl ether monomers are accessible by a number of synthetic methods. A comprehensive listing of these monomers, their physical characteristics, and their commercial suppliers may be found in the review article by Lorenz [5]. Given below are brief descriptions of the major synthetic methods for the preparation of these compounds, with special emphasis on those developed in the past few years.

The oldest, most versatile, and major commercial method for the synthesis of vinyl ethers is by the base-catalyzed condensation of acetylene with alcohols first described by Reppe and co-workers [11–13].

$$R-OH + HC \equiv CH \xrightarrow[120-180\,°C]{MOH} RO-CH=CH_2 \tag{1}$$

Presumably this reaction proceeds by formation of the metal alcoholate, which undergoes nucleophilic addition to the acetylenic double bond. The resulting adduct then regenerates the alcoholate by proton exchange. Sodium and potassium hydroxides are the most common catalysts employed for this reaction.

The oxidative vinylation reaction of ethylene with alcohols in the presence of oxygen has been reported [14,15] to give vinyl ethers in high yields. Like many Wacker-type reactions, this reaction is typically catalyzed by heterogeneous and homogeneous catalysts containing palladium.

$$R-OH + H_2C=CH_2 + 1/2\,O_2 \xrightarrow{PdCl_2, CuCl, HCl} RO-CH=CH_2 + H_2O \tag{2}$$

While the direct oxidative vinylation reaction shown above has many advantages over the acetylene route to the preparation of vinyl ethers, it has yet to be commercialized.

Acetals can be thermally cracked at temperatures between 250 and 400 °C over heterogeneous catalysts such as palladium on asbestos [16], thoria [17], or metal sulfates on alumina [18], as shown in the following equation.

$$H_2C-CH \begin{smallmatrix} OR \\ \\ OR \end{smallmatrix} \xrightarrow[250 - 400°C]{cat.} H_2C=CH-OR \quad + \quad R-OH \tag{3}$$

It is also possible to prepare vinyl ethers by a transvinylation reaction between an alcohol and a vinyl ether as shown in equation (4). The reaction can be catalyzed by palladium(II) complexes [19] and by mercury salts [20–23].

$$RO-CH+CH_2+R'-OH \xrightarrow{cat.} R'O-CH=CH_2+R-OH \tag{4}$$

This method is especially recommended for the preparation of vinyl ether monomers bearing functional groups that are sensitive to the basic conditions of the vinylation reaction using acetylene. Transvinylation reactions can also be carried out between alcohols and vinyl acetate, a reaction that has been described by Adelman [24] is also being catalyzed by salts of mercury such as mercuric sulfate.

$$H_3C-CO_2-CH=CH_2+R-OH \xrightarrow{cat.} RO-CH=CH_2+H_3C-COOH \tag{5}$$

The dehydrochlorination of 1- and 2-chloroalkyl ethers with sodium or potassium hydroxide provides a simple and direct route to the synthesis of the corresponding vinyl ethers [25,26]. It is the method of choice for the preparation of 2-chloroethyl vinyl ether from 2-dichlorodiethyl ether [27,28].

$$(Cl-CH_2CH_2)_2O + NaOH \longrightarrow Cl-CH_2CH_2-O-CH=CH_2 + NaCl + H_2O \quad (6)$$

The compound shown above, 2-chloroethyl vinyl ether, undergoes facile nucleophilic displacement reactions and can thus be used as a valuable synthon for a variety of specialized vinyl ether monomers [29–31].

$$Cl-CH_2CH_2-O-CH=CH_2 + ROH \xrightarrow[\text{NaOH}]{R'_4NBr} RO-CH_2CH_2-O-CH=CH_2 + NaCl \quad (7)$$

Allylic ethers can be conveniently isomerized to the corresponding vinyl ethers in the presence of potassium t-butoxide [32,33] or such transition metal catalysts as tris(triphenylphosphine)-ruthenium dichloride [34].

$$RO-CH_2-CH=CH_2 \xrightarrow[(Ph_3P)_3RuCl_2]{t-BuOK\ or} RO-CH=CH-CH_3 \quad (8)$$

Finally, there are cyclic vinyl ethers, such as 2,3-dihydrofuran and 3,4-dihydro-2H-pyran and their derivatives. This types of cyclic vinyl ethers are in the present of academic interest only. 2,3-Dihydrofuran can be synthesized by different ways [35–38].

The first synthesis starts with 1,4-butandiol and give the ring by the release of water and hydrogen [35].

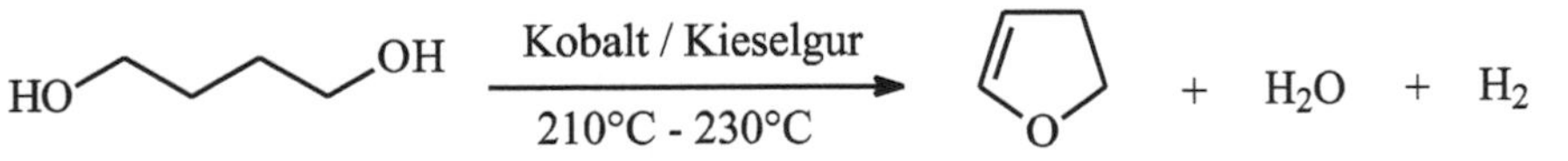

$$(9)$$

The other possibility is to start from butadiene, to epoxidize one doublebond, using a new type of silver catalyst [36]. After heating of 1-epoxy-3-butene, 2,5-dihydrofuran is built by changing the ring size. The other possibility to get 2,5-dihydrofuran is, to treat 1,4-Dichloro-but-2-ene with strong bases at high temperatures [37], which yield 2,5-dihydrofuran which can easily be converted into 2,3-dihydrofuran in the presence of isomerization catalysts, such as Fe(CO)$_5$, KOC(CH$_3$)$_3$ or Ru(PPh$_3$)$_2$Cl$_2$ [38].

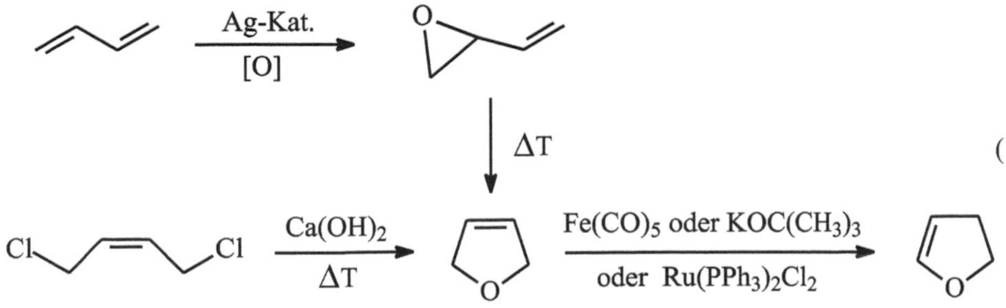

$$(10)$$

3. Polymer Synthetic Methods

The methods used for the synthesis of poly(vinyl ethers) fall into three major classifications; cationic, coordination-cationic, and free-radical polymerizations. Typical examples of cationic agents are Lewis and Bronsted acids and iodine. Ziegler-Natta catalysts comprise agents that initiate coordination-cationic polymerization, while azo compounds and peroxides are initiators for free-radical polymerization. In Table 1 are listed various examples of the above types of initiators, together with the conditions under which their polymerizations were carried out. Insofar as was obtainable from the references cited, the table also includes conversion, molecular weight, and tacticity data. Since the literature for the polymerization of vinyl ethers is particularly extensive, no attempt was made to cite every reference available for each initiator. Rather, typical and usually the best and most complete example in the authors' judgment was selected for inclusion in this table. In the following sections the state of present knowledge about the mechanism and utility of the various methods and initiator types are summarized. For an in-depth discussion of the mechanisms of individual catalyst systems, the reader is referred to the publication of Lal [39] and the review by Gandini and Cheradame [40].

B. Cationic Methods

The reactivity of vinyl ethers in cationic polymerization depends not only on the initiator used but also on the structure of the vinyl ether itself. To generalize, it may be said that vinyl ethers possessing highly branched alkyl groups are more reactive than those bearing straight-chain alkyl groups and have a greater tendency toward stereoregularity in the final polymer. Substitution by alkyl groups at either the α or β positions on the vinyl group increases the electron density of the vinyl group and hence, its tendency to polymerize. *cis*-Propenyl ethers are more reactive than *trans*-propenyl ethers in nonpolar solvents, whereas in polar solvents their reactivity is comparable [41]. Under cationic conditions, aryl vinyl ethers tend to undergo side reactions leading to rearrangements instead of polymerization [1]. Using simple cationic initiators at elevated temperatures there is, in most cases, no stereochemical control, and atactic polymers result. In contrast, using BF_3 etherate at low temperatures, Schildknecht and his co-workers [42,43] were able to prepare crystalline, isotactic poly(isobutyl vinyl ether) as the first recorded example of a stereoregular polymer. Later studies by Blake and Carlson [44] demonstrated that when these polymerizations are carried out in nonpolar solvents, they proceed from a homogeneous phase to a gel-like phase. Stereoregular polymers are produced from both phases by a mechanism of slow chain propagation. Since that time, stereoregular polymers have been prepared using a wide variety of catalysts, including metal halides, organometallic halides, metal oxyhalides, metal oxides, metal sulfates, stable carbenium ion salts, and Ziegler–Natta coordination catalysts. Cationic polymerizations of vinyl ethers are subject to the usual chain transfer and termination processes in the presence of hydroxyl-, aldehyde-, and basic-containing impurities that inhibit polymerization and limit and stop chain growth. Due to the propensity for vinyl ethers to hydrolyze in aqueous acidic media, water is usually to be avoided as a solvent in these types of polymerizations [45,46].

1. Bronsted and Lewis Acids

Due to the highly electron-rich character of their double bonds, vinyl ethers are susceptible to cationic polymerization using a variety of Bronsted and Lewis acids as initiators. Bronsted acids as weak as H_2SO_3 ($SO_2 + H_2O$) and H_3PO_4 effect the cationic polymerization of these monomers. Reppe and co-workers [47,48] and later Favorskii and Shostokovskii [49] were among the first to employ both protonic and Lewis acids as initiators for the cationic polymerization of vinyl ethers. In the case of protonic acids, direct protonation of vinyl ether may occur as shown in equation (11) to give a carbocation species stabilized by the neighboring ether oxygen.

$$
HX + CH_2=CH \atop \underset{R}{\overset{|}{O}} \longrightarrow
\left\{
H_3C-\underset{\underset{R}{\overset{|}{O}}}{\overset{\overset{H}{|}}{C}}{}^{\oplus}
\rightleftharpoons
H_3C-\underset{\underset{R}{\overset{|}{O^{\oplus}}}}{\overset{\overset{H}{|}}{C}}
\right\} X^-
$$

$$
\Bigg\downarrow \; n \; CH_2=CH \atop \underset{R}{\overset{|}{O}}
\tag{11}
$$

$$
H_3C-\underset{\underset{R}{\overset{|}{O}}}{\overset{\overset{H}{|}}{C}}
\left(CH_2-\underset{\underset{R}{\overset{|}{O}}}{CH}\right)_n
CH_2-\underset{\underset{R}{\overset{|}{O}}}{\overset{\overset{H}{|}}{C}}{}^{\oplus} \; X^-
$$

Much work has been done using boron trifluoride complexes as initiators for vinyl ether polymerization, due principally to the use of these catalysts in the industrial production of poly(vinyl ether)s. The nature of the complexing agent has been found to influence the rate of polymerization in the following order: anisole > diisopropyl ether > diethyl ether > n-butyl ether > tetrahydrofuran [39,50]. BF_3 requires a protogen (water, alcohol, etc.) as a coinitiator to initiate polymerization. Protogens may be deliberately added or may be present in the polymerization mixtures due to adventitious moisture or other hydroxylic impurities. Similarly, many other but not all Lewis acids require protogens to initiate polymerization efficiently, and their mechanisms are similar to that given above. Commercial catalysts typically consist of BF_3 complexed with water [51] or diethyl ether [52] and are especially active for the polymerization of lower alkyl vinyl ethers. Polymerizations conducted using these initiator systems have come to be known as flash polymerizations because they are typically carried out at -40 to $-79\,^{\circ}C$ in the presence of a low-boiling hydrocarbon. Solvent such as ethane or propane used to control the exotherm of the polymerization by evaporative cooling (flashing off). Conversions are commonly very high, approaching 100%, although the molecular weights tend to be rather low. Flash polymerization is the current method of choice for the preparation of poly(vinyl ethers) on an industrial scale.

A variety of other Lewis acids, including $AlCl_3$, $SnCl_4$, $FeCl_3$, $MgCl_2$, TiF_4, $ZnCl_2$, $EtAlCl_2$, Et_2AlCl, and aluminum and titanium alkoxides have also been used to initiate the cationic polymerization of alkyl vinyl ethers. Like polymerizations using BF_3, these polymerizations are highly exothermic, requiring low temperatures and dilution with solvents to avoid violent runaway polymerizations. Among the most active Lewis acid catalysts, as well as those giving the best stereochemical control, are $EtAlCl_2$ and Et_2AlCl [53–59].

Table 1

Initiator	Monomer(s) $RO\text{-}CH\text{=}CH_2$ R=	Solvent	Temp. (°C)	Mol. Weight (viscosity)	Conversion (%)	Physical state tacticity	Ref.
Protonic acids							
Al(OR)$_3$-HF	CH_3	CH_2Cl_2/n-heptane	0–25	η_{red} = 7.2 dL/g	93	crystalline	173
	various	hydrocarbons	10–120				174
Al$_2$(SO$_4$)$_3$-3H$_2$SO$_4$	i-C$_4$H$_9$	CS$_2$	5	η_{inh} = 1.9 dL/g	14–100	isotactic	175
	CH$_3$, i-C$_3$H$_7$, C$_4$H$_9$, i-C$_4$H$_9$					isotactic	102, 176
Cr$_2$(SO$_4$)$_3$-H$_2$SO$_4$	i-C$_4$H$_9$, C$_4$H$_9$, CH$_3$	n-hexane	40	$[\eta]$ = 0.7–1.4 dL/g	80–93	crystalline	177
NH$_4$ClO$_4$	various	DMF		oligomers			178
SO$_2$	C$_4$H$_9$, vinyl thioethers	bulk	−10				31, 179
Lewis acids							
BF$_3$·Et$_2$O	i-C$_4$H$_9$	bulk	−78	$[\eta]$ = 0.4–0.8 dL/g		isotactic	180
	i-C$_4$H$_9$	CH$_2$Cl$_2$/n-haxane	−78			isotactic	181
	C$_4$H$_9$	propane	−45–80			atactic	182
	C$_4$H$_9$	propane	−50	low		atactic	183
	i-C$_4$H$_9$	propane	−78	high		atactic	184
	various	propane/butane	−25	variable	15–100	atactic	185
	t-C$_4$H$_9$	CH$_2$Cl$_2$/H$_3$CNO$_2$	−78			syndiotactic	186
BF$_3$·2H$_2$O	various	bulk	3–5		9.2–93	atactic	51
SnCl$_2$	i-C$_4$H$_9$, i-C$_5$H$_{11}$	benzene	12			atactic	187
SnCl$_4$	i-C$_4$H$_9$	CH$_2$Cl$_2$/n-hexane	−78		88	isotactic	181
Et$_2$AlCl	i-C$_3$H$_7$, i-C$_4$H$_9$, neo-C$_5$H$_{11}$	toluene/propylene	−78	$[\eta]$ = 0.9–1.29 dL/g		isotactic	56
EtAlCl$_2$	C$_4$H$_9$	toluene/heptane	−80	$[\eta]$ = 2.9 dL/g	98	isotactic	55
	i-C$_4$H$_9$	toluene	−80	$[\eta]$ = 2.2 dL/g	75–85	crystalline	53
TiF$_3$	i-C$_4$H$_9$	CH$_2$Cl$_2$/heptane	60	$[\eta]$ = 2.2 dL/g	79	crystalline	188
SbCl$_5$	i-C$_4$H$_9$	toluene	−10	$[\eta]$ = 0.2 dL/g	100	isotactic	189
MgCl$_2$	Allyl	toluene		22–60000 g/mol			
Iodine	i-C$_4$H$_9$	bulk	25	η_{sp}/c = 0.55 dL/g	97	atactic	100

I_2	C_4H_9, c-C_6H_{11}	diethyl ether	25			atactic	79
	i-C_4H_9, i-C_4H_9, 2-Cl-C_2H_4	ethylene chloride	30			atactic	80
	1,2-Divinyloxyethane	CH_2Cl_2	−5		low		139
	1,2-Divinyloxyethane	$CHCl_3$	0				190
Carbonium salts and cation radicals							
[Tropylium]$^+$SbCl$_6^-$	i-C_4H_9	CH_2Cl_2	0	100	$[\eta] = 0.7$–0.8 dL/g		87,91,186
[Trityl]$^+$ SbCl$_6^-$	i-C_4H_9	CH_2Cl_2	0	100	$[\eta] = 0.8$ dL/g		87
[Trityl]$^+$ X$^-$	CH_3, C_2H_5, i-C_4H_9	CH_2Cl_2	−40–0				191
[9,10-Diphenyl anthracene]$^+$ClO$_4^-$	t-C_4H_9	CH_2Cl_2/toluene	−76	64–90	$M_n = 24{,}500$ g/mol	isotactic	92
	n-C_4H_9, i-C_4H_9	CH_3CN, $C_6H_5NO_2$	−20–10				93–95
[Pyrene]$^+$ ClO$_4^-$	i-C_4H_9	CH_3NO_2	5		oligomers		96
[Rubene]$^+$ ClO$_4^-$	i-C_4H_9	CH_2Cl_2	10–20		100		94,95
[Triphenylene]$^+$ ClO$_4^-$	i-C_4H_9	CH_2Cl_2			10–20		94
Grignard reagents							
C_4H_9MgBr	C_4H_9, i-C_4H_9, i-C_3H_7	bulk	25	74–90	high	crystalline	192
	i-C_4H_9	n-hexane	60–70	43–49	$M_w = 280.000$–900000 g/mol	crystalline	99
	CH_3, C_2H_5, C_4H_9, i-C_4H_9	cyclohexane	80	47	$[\eta] = 3.9$ dL/g	crystalline	98
Metal oxyhalides							
AlOCl, AlOBr, AlOI	i-C_4H_9	CH_2Cl_2	−78	79–100	$\eta_{red} = 0.14$–0.31 dL/g	isotactic	193
CrO_2Cl_2	i-C_4H_9	pet. ether	0–20	99	$\eta_{red} = 2.62$ dL/g	isotactic	100
$VOCl_2$	i-C_4H_9	pet. ether	0–20	30	$\eta_{red} = 1.26$ dL/g	atactic	193
WO_2Cl_2	i-C_4H_9	pet. ether	0–20	25	$\eta_{red} = 0.54$ dL/g	atactic	193
Metal sulfates							
$Fe_2(SO_4)_3$ H_2SO_4 3-4 H_2O	C_2H_5, i-C_3H_7	pentane	−20	14–95	$\eta_{inh} = 1.7$–3.5 dL/g	crystalline	102
$Fe_2(SO_4)_3$, VOSO$_4$, V(SO$_4$)$_2$	i-C_4H_9	bulk	25–30			crystalline	194
Al(O-i-Pr)$_3$ + H_2SO_4	i-C_3H_7	CH_2Cl_2	25	70	$\eta_{sp} = 4.1$ dL/g	crystalline	195
TiO(SO$_4$), VO(SO$_4$), Cr(SO$_4$)$_2$	CH_3, i-C_3H_7 C_4H_9, t-C_4H_9	CH_2Cl_2	25				196

(continued)

Table 1 Continued

Initiator	Monomer(s) RO–CH=CH$_2$ R=	solvent	Temp. (°C)	Mol. Weight (viscosity)	Conversion (%)	Physical state tacticity	Ref.
Fe$_2$(SO$_4$)$_3$, NiSO$_4$, Al(O-i-Pr)$_3$, Al$_2$(SO$_4$)$_3$ H$_2$SO$_4$, MgSO$_4$ H$_2$SO$_4$, Cr$_2$(SO$_4$)$_3$ H$_2$SO$_4$	i-C$_4$H$_9$	hexane	25	[η] = 1–2 dL/g	60–80	isotactic	197
Metal oxides							
Fe$_2$O$_3$	i-C$_4$H$_9$	toluene	25	η$_{sp}$ = 0.21–0.41 dL/g	43	isotactic	104
Cr$_2$O$_3$	i-C$_4$H$_9$	toluene	80	η$_{sp}$ = 2.14 dL/g	32	isotactic	103
MoO$_2$	i-C$_4$H$_9$	bulk	25	η$_{sp}$ = 0.56 dL/g	62	atactic	100
V$_2$O$_3$	i-C$_4$H$_9$	bulk	25	η$_{sp}$ = 0.22 dL/g	22	atactic	100
NiO$_2$	i-C$_4$H$_9$	bulk	25	η$_{sp}$ = 0.14 dL/g	36	atactic	100
SiO$_2$	i-C$_4$H$_9$	bulk	25	η$_{sp}$ = 0.26 dL/g	71	atactic	100
Ziegler–Natta (coordination catalysts)							
VCl$_3$ + Al(i-C$_4$H$_9$)$_3$ + THF	CH$_3$	ether	30	η$_{red}$ = 2.5 dL/	79	crystalline	130
TiCl$_3$ + Al(i-C$_4$H$_9$)$_3$	CH$_3$	ether/n-heptene	30		10	amorphous	130
TiCl$_4$ + Al(i-C$_4$H$_9$)$_3$	i-C$_4$H$_9$	bulk	–78	[η] = 2–7 dL/g		cristalline	198
	allyl	bulk	–78	low			198
TiCl$_4$ + Al(C$_2$H$_5$)$_3$	C$_2$H$_5$	benzene	–40–100	low		amorphous	199
(C$_6$H$_5$)$_2$Cr + TiCl$_4$	i-C$_4$H$_9$	toluene	25		10	crystalline	200
Photochemical (UV) Initiators							
(C$_6$H$_5$)$_2$I$^+$BF$_4^-$, (C$_6$H$_5$)$_2$I$^+$PF$_4^-$, (C$_6$H$_5$)$_2$I$^+$AsF$_6^-$, (C$_6$H$_5$)$_2$I$^+$SbF$_6^-$	2-Cl-C$_2$H$_4$	CH$_2$Cl$_2$	25 (hv 5 s)	[η] = 0.15 dL/g	74	atactic	201

$(4\text{-}t\text{-}C_4H_9\text{-}C_6H_4)(C_6H_5)_2S^+$ $AsPF_6^-$	$2\text{-}Cl\text{-}C_2H_4$	CH_2Cl_2	25		92	atactic	202
Electrochemical Initiation							
	$i\text{-}C_4H_9$	$CH_2Cl_2/(Bu)_4N^+BF_4^-$	25		80–90		203
	C_4H_9	$CH_3CN/NaClO_4$	0	low	45–72		93
	C_4H_9	$CH_3CN/(Bu)_4N^+ClO_4^-$	0	3500–7000 g/mol	70–80		93
	$i\text{-}C_4H_9$	1,2-dichloroethane/$(Bu)_4N^+I_3^-$	25		40–50		128
	$i\text{-}C_4H_9$	CH_3CN/di-isopropyl ether/$(Bu)_4N^+ClO_4^-$	25	6400 g/mol			129
Free-radical initiators							
Di-t-butyl peroxide	C_2H_5, $i\text{-}C_4H_9$	cyclohexane	159	low	78		131
Cumene hydroperoxide							
t-Butyl hydroperoxide							
2,2'-Azobisisobutyronitrile	aryl vinyl ethers	bulk	75	530–743 g/mol	22		134, 204
Azo compounds	various	bulk, DMF	75	low	94		205
$NaHSO_3/(NH_4)_2S_2O_8$	$2\text{-}Cl\text{-}C_2H_4$	bulk	50				133
Miscellaneous initiators							
Sulfur	various	bulk	20–80	high			206
Molecular sieves	$i\text{-}C_4H_9$	benzene	30	$[\eta] = 0.02\text{--}0.07$ dL/g	90		207
$ZnCl_2$ + t-BuCl	CH_3	bulk	−20–30	$[\eta] = 0.2\text{--}1.3$ dL/g			208
$Al(Et)_3$ + $POCl_3$ + $SOCl_2$ + V_2O_5 + t-BuCl	$i\text{-}C_4H_9$	bulk	30		10–91		101
Carbon (channel back)	$i\text{-}C_4H_9$	CCl_4	20		84		180

aCoupled means that both processes occur in the same space; decoupled means that both processes occur in separate spaces.

2. Hydrogen Iodide-Iodine (Living/Controlled Cationic Polymerization)

The synthesis of polymers with controlled end groups, molecular weight distribution, and the preparation of well-characterized block polymers requires polymerization methods in which the growing chain end is well defined and undergoes chain growth in the absence of termination and chain transfer. Until recently, these conditions have been observed only in certain anionic polymerizations and were unknown although highly sought after in cationic polymerizations. In 1984, workers at Kyoto University [60] described the first example of a living cationic polymerization consisting of vinyl ethers employing the initiator system HI/I_2. Since that time, a number of additional papers have appeared by this same group of researchers which describe in some details the characteristics of this particular initiator system [61]. Various well-characterized functional polymers and block polymers were prepared using this new initiator [62]. The absence of termination and transfer using the HI/I_2 initiator system was attributed to a tight association of the stabilization of the growing carbocationic end group by the counterion.

$$ \tag{12} $$

In the first step, HI adds to the vinyl ether monomer to give a 1:1 adduct. Next, the carbon-iodide bond of the adduct is activated by iodine, allowing insertion of the incoming monomer at the end of the chain. In this mechanism, I_2 behaves as a weak electrophile that activates the C–I bond of the vinyl ether-HI adduct by association. Accordingly, the Highashimura group has termed HI the *initiator* and I_2 the *activator*.

Another mechanism that perhaps better explains the living character of the HI/I_2– vinyl ether system has been put forth by Matyjaszewski [63] and involves the polymerization occurring through a six-membered transition state involving the C–I chain end, I_2, and the incoming vinyl ether monomer.

$$ \tag{13} $$

In addition to the HI/I_2 initiator system described above, the Kyoto group [64–66] have described several new initiators that also display living character in the polymerization of vinyl ether monomers. They report that isobutyl vinyl ether may be polymerized in the presence of HI in combination with tin and zinc halides or trimethylsilyliodide and zinc iodide to give polymers with very narrow molecular weight distributions and predetermined chain lengths. Furthermore, they also reported that $EtAlCl_2$ which has been complexed with acetic acid, dioxane, ethyl acetate, or water similarly gives living poly(isobutyl vinyl ether) polymers. What is particularly remarkable is the observation that low temperatures are not necessary to obtain living polymers; in some cases temperatures as high as 70 °C were used. Again, stabilization of the growing carbocation was invoked as a rationale for the living cationic polymerizations, although it should be pointed out that in every case, six-membered transition states similar to that proposed by Matyjaszewski could also satisfactorily explain the observations. Nuyken and Kröner [67] showed that tetraalkylammonium salts could be used as activators together with HI to initiate the living polymerization of *iso*-butyl vinyl ether. In particular, tetra-*n*-butylammonium perchlorate was found to be especially effective in giving high conversions of polymers with narrow molecular weight distributions. Polar solvents such as dichloromethane, in which the tetraalkylammonium salt is most soluble, give the highest polymerization rates, propably due to the greater interaction of the growing chain end and the ammonium salt. This field of the living cationic polymerization of vinyl ethers is currently under rapid and intense development. The goals of the synthesis of well-characterized terminal functional and block polymers appears to be in hand. More detailed information about new living systems are given in some reviews [68–78].

3. Polymerization of Vinyl Ethers with Iodine

Historically, iodine was the first initiator used for the polymerization of vinyl ether monomers. It is therefore paradoxical that the mechanism of its initiation reaction has been elucidated only recently. Originally, Eley and Saunders [79] proposed that iodine undergoes self-ionization of the type shown below to generate the I^+ cation, which then attacks the double bond of the monomer.

$$2I_2 \longrightarrow I^+ + I_3^- \tag{14}$$

Somewhat later, Okamura et al. [80] suggested that iodine may form a π complex with the vinyl ether and considered the possibility that more than one initiation mechanism could be involved, depending on the polarity of the solvent used. Unfortunately, the kinetics of the polymerization fail to justify a purely ionic mechanism [81].

Parnell and Johnson [82] proposed the following mechanism for the initiation of cationic polymerization by iodine:

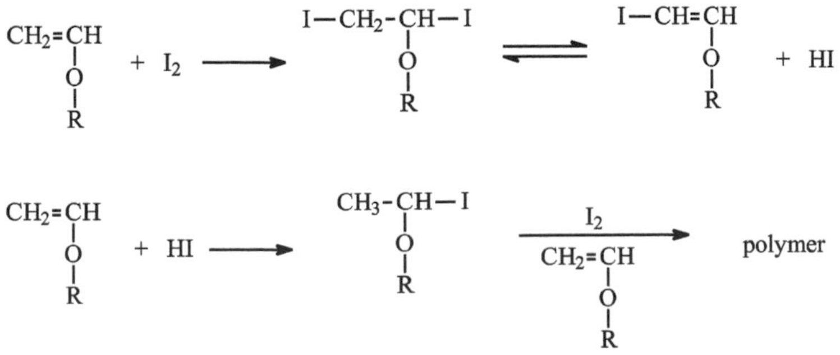

$$(15)$$

Subsequent work by Ledwith and Sherrington [83] confirmed this proposal by showing that the initial products of the reaction of iodine with vinyl ethers were the corresponding 1,2-diiodoethane adducts. Plesch [84] demonstrated that the usual inhibition period observed in olefin polymerizations could be reduced by the addition of HI and suggested that HI may be generated by an in situ elimination reaction involving the diiodide adducts. In 1973, Janjua and Johnson [85] reported that this in fact takes place in vinyl ether polymerizations. Finally, Johnson and Young [86] demonstrated unequivocally that vinyl ether polymers produced using iodine as an initiator contain iodine bound as the end groups and further, that on introduction of additional monomer, the chains can continue to grow. They recognized that these polymerizations had some of the characteristics of living polymerizations. It may thus be seen that the polymerization of vinyl ethers by the HI/I$_2$ initiator system and by iodine alone share some of the same elements and may, indeed, be proceeding by much the same mechanism. The use of iodine to prepare block polymers or well-controlled terminal functional polymers has not yet been carried out. If these prospects are realized, the use of iodine as an initiator of the cationic polymerization of vinyl ethers would receive considerably more attention than in the past.

4. Initiation by Stable Carbenium and Carbenium Ion-Radical Salts

Stable carbenium ion salts such as the tropylium (a) and trityl ion (b) salts are especially convenient and facile initiators of vinyl ether polymerizations, and although not commercially feasible, have been studied extensively in many academic laboratories.

$$(16)$$

(a)   (b)

The first systematic study using both trityl and tropylium salts was reported by Bawn and his coworkers in two major papers [87,88]. Further investigations were carried out by Ledwith et al. [89] and by Chung et al. [90]. All the available evidence appears to confirm that initiation takes place by a direct electrophilic addition:

$$(17)$$

Evidence for this mechanism consists of the observation of aryl groups at the chain ends when trityl salts were used as well as by the discovery that the number of active polymerizing species corresponds to the number of initiating molecules of salt used. These initiators provide good control over the polymerization of vinyl ether monomers with high conversions provided that polymerizations are carried out in the presence of pure, good solvents that give homogeneous reaction mixtures. Other than the observation by Okamura *et al.* [91] that the tropylium hexachloroantimonate-initiated polymerization of isobutyl vinyl ether gives isotactic polymers, there appears to be no information concerning the tacticity of the polymers obtained using carbenium ion salt initiators [92].

The relatively recent development of electrochemical methods for the synthesis of stable cation-radical salts, such as the perylene (18a) and 9,10-diphenylanthracene (18b) cation-radicals has permitted their use as initiators for vinyl ether polymerizations.

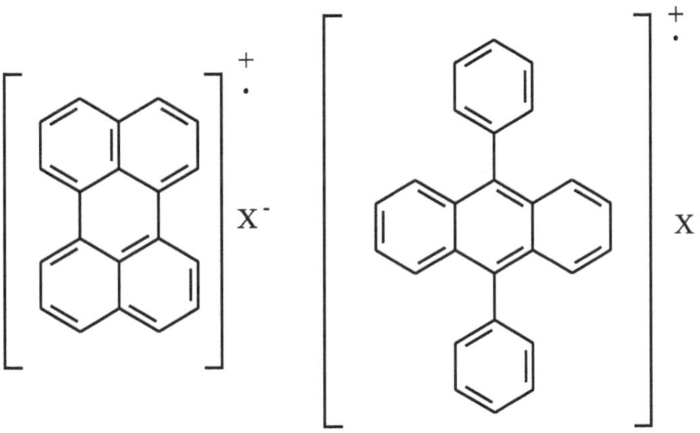

$$(18)$$

(a)   (b)

Although these initiators tend to be somewhat air and moisture sensitive, they are stable and can be stored for reasonably long times under dry-box conditions. The first work in this area was done by Mengoli and Vidotto [93], who prepared the stable 9,10-diphenylanthracene cation radical and used it to study the polymerization of n-butyl vinyl ether. Similarly, Funt and co-workers [94,95] carried out the in situ electrochemical generation of both the 9,10-diphenylanthracene and rubrene cation radicals in dichloromethane and acetonitrile solutions of iso-butylvinyl ether and studied the polymerization kinetics. Finally, Oberrauch et al. [96] prepared the perylene cation radical, isolated it, and investigated its use in the polymerization of isobutyl vinyl ether. While appearing structurally deceptively similar to simple carbenium salts, the chemistry of initiation and propagation reactions using stable carbenium ion-radical salts is considerably more complex. To account for the observations that first, there is some incorporation of residues derived from the initiator, and second, that at least 50% of the initiating cation radicals are isolated as the parent hydrocarbons, the two initiation mechanisms shown below have been proposed by Glasel et al. [94] involving simple addition and electron transfer, respectively:

$$R\overset{+}{\cdot}X^- + M \longrightarrow RM\overset{+}{\cdot} X^- \tag{19}$$

$$R\overset{+}{\cdot}X^- + M \longrightarrow R + M\overset{+}{\cdot}X^- \tag{20}$$

In addition to the mechanisms above, Oberrauch et al. [96] proposed that disproportion reactions could occur between the initiating cation radicals and their monomer adducts, leading not only to the parent hydrocarbons but also to dicationic species, the latter of which can undergo propagation from two cationic sites. Considerably more work is required not only to clarify the mechanisms involved but also to provide more details about the structure, conversions, and molecular weights of the vinyl ether polymers that are obtained using these new types of initiators.

5. Grignard Reagents

Organomagnesium halides (Grignard reagents) are active catalysts for the polymerization of vinyl ethers. Both alkyl and aryl Grignard reagents may be used; however, there is some difference in their reactivity. For example, phenylmagnesium bromide is more active than n-butylmagnesium bromide. Although the precise mechanism of initiation is not known, based on the observation that poly(vinylcarbazole) produced using Grignard reagents contains magnesium bound to the chain, Biswas and John [97] suggested that the initiating species possibly involves the RMg^+ ion. Kray [98] was the first to use Grignard reagents for the polymerization of vinyl ethers. Bruce and Farrow [99] demonstrated conclusively that pure Grignard reagents are not by themselves active initiators. However, when these reagents are exposed to a trace of oxygen or particularly carbon dioxide, they have been shown to give high-molecular-weight, highly crystalline, isotactic poly(isobutyl vinyl ether). An alkoxymagnesium halide intermediate was proposed as the active catalyst

6. Inorganic Halides and Oxyhalides

Inorganic halides and oxyhalides such as $POCl_3$, SO_2Cl_2, and CrO_2Cl_2 are powerful catalysts for the polymerization of vinyl ether monomers. It has been suggested by Gandini and Cheradame [40] that $POCl_3$ hydrolyzes in the presence of traces of water to give HCl and that this acid is responsible for the observed polymerizations. CrO_2Cl_2 gives crystalline, stereoregular poly(isobutyl vinyl ether) [100]. The other catalysts cited above generally react with vinyl ethers to give ill-defined chars rather than simple polymers. In contrast, reaction of these catalysts with a vinyl ether monomer in the presence of triethylaluminum produces well-controlled stereoregular polymerizations [101].

7. Metal Sulfates

Metal sulfates complexed with sulfuric acid are easily prepared, especially efficient heterogeneous catalysts for the preparation of stereoregular (isotactic), high-molecular-weight poly(vinyl ethers). What is most remarkable are the high rates of polymerization that can be achieved and the ability of these catalysts to produce polymers of high stereoregularity at temperatures above $0\,°C$. Lal and McGrath [102] found that vinyl ethers having linear alkyl groups polymerize faster than those having branched groups and that among the straight-chain alkyl vinyl ethers the following order was found: ethyl > n-butyl > n-hexyl = n-octyl. They also observed that with these catalysts, the nature of the solvent that is employed is quite important. In aromatic solvents such as benzene or toluene, the polymerization is considerably slower than in heptane. However, the degree of stereoregularity in heptane is higher.

The exact mechanism for the stereospecific polymerization of vinyl ethers by metal sulfate sulfuric acid catalysts has not been elucidated, although many theories have been advanced. However, it is generally assumed that the first step must consist of initiation by protonation of the monomer by acidic sites bound to the catalyst lattice. It has been further suggested [102] that steric considerations associated with the heterogeneous nature of the catalyst play a major role in determining the mode of insertion of the monomer between the anion situated at the surface of the heterogeneous catalyst and the growing polymer chain. This may involve coordination of the ether oxygen of the end group of the growing chain and the monomer to the same or adjacent coordination sites located on the surface of the heterogeneous catalyst.

8. Metal Oxides

Metal oxides, especially those of the transition metals, can catalyze the polymerization of vinyl ethers. Especially noteworthy as a catalyst is Cr_2O_3 [103], which gives high yields of high-molecular-weight isotactic poly(isobutyl vinyl ether). Fe_2O_3 gives atactic vinyl ether polymers, while the same material, which has been produced by calcination from $Fe_2(SO)_3$, gives isotactic polymers [104]. The mechanism of catalysis by these materials is not known but probably involves the presence of metal cations on the surface of the heterogeneous catalyst particles which serve as Lewis acid sites responsible for electrophilic attack on the monomer. The active catalysts are prepared by calcining the metal oxide at a high temperature followed by crushing the product to produce powdered catalyst with the correct particle size.

9. Photochemical Initiation

In recent years there has been a great deal of activity in the design and synthesis of photo-initiators for cationic polymerization. The main motivation for this work has been to use these photo-polymerizations as new ultrahigh-speed methods of making nonpolluting coatings. Although aimed primarily at the polymerization of epoxides, cationic photo-polymerization has also been applied to vinyl ether monomers. As photo-initiators a considerable number of photosensitive onium salts have been investigated and found to be active. These include diazonium [105], diaryliodonium [106], diaryliodosonium [107], triarylsulfonium [108], triarylsulfoxonium [109], dialkylphenacylsulfonium [110], dialkyl-4-hydroxyphenylsulfonium [111], tetraarylphosphonium [112], certain N-substituted phenacyl ammonium [112], ferrocenium salts [113] and phenacylsulfonium salts [114,115].

Of these, the diaryliodonium, triarylsulfonium, and ferrocenium salts are most practical and most often employed. Shown in (21a) to (21c) are the structures of typical members of these classes of compounds.

(a)   (b)   (c)

(21)

In the case of photo-initiator types (a) and (b), it has been clearly shown that mode of initiation involves the formation on photolysis of a Brønsted acid, HX, which corresponds to the anion associated with the starting salt. Equations (22) and (23) give the mechanism of the photolysis proposed for diaryliodonium salts.

$$Ar_2I^+X^- \xrightarrow{h\nu} ArI^+X^- + Ar \tag{22}$$

$$ArI^+X^- + \text{solvent-H} \longrightarrow ArI + \text{solvent} + HX \tag{23}$$

This Brønsted acid, which is generated by interaction of the aryliodonium cation radical with the solvent or monomer, is the true initiator of cationic polymerization of the vinyl ether monomers.

This is also probably the case with the other classes of photoinitiators, with the exception of the diazonium and ferrocenium salts. The photolysis of diazonium salts is well known to generate Lewis acids which initiate polymerization either by themselves or in combination with a protogen. Similarly, the photolysis of ferrocenium salts proportedly proceeds by the mechanism shown in equation (24), which involves the formation of the Lewis acid shown. Apparently, this iron-containing Lewis acid is strong enough to initiate many types of cationic polymerization, including those of vinyl ethers.

There are several advantages of carrying out the photo-initiated cationic polymerization of vinyl ether monomers. Whereas it is difficult to achieve homogeneous polymerizations by the addition of strong Lewis or Brønsted acids to the very highly reactive vinyl ether monomers, photo-initiators (21a) to (21c) dissolve in vinyl ethers to give homogeneous, stable solutions. The desired acid is then generated in situ by photolysis of the photo-initiator. Controlled polymerization of those monomers can then be carried out by adjusting the light intensity. Because of their high rates of polymerization, multifunctional vinyl ether monomers are ideal for thin, cross-linked coating applications which must be applied at high rates of speed. Of course, such applications are limited to rather thin layers and to those planer or curved substrate topographies to which light can be directed. For more detailed investigations, there are several review articles, that will give a good overview of this very interesting topic [116–120].

10. Radiation Techniques

Ionizing radiation is capable of initiating the polymerization of vinyl ethers. The primary process [121] appears to consist of the electrolytic removal of electrons from monomer molecules with consequent formation of the corresponding cation radicals. Polymerization then proceeds by a cationic mechanism by further interaction of the cation radicals with themselves and the monomer. Typical radiation techniques consist of ^{60}Co γ-rays [121–123] and pulse radiolysis (high-energy electrons) [124]. Since ionizing radiation produces cations free of negative counterions, the so-called bare or free cations that are generated are exceptionally reactive and in highly purified systems give high yields of highmolecular-weight polymers. The rates of propagation observed in radiation-induced cationic polymerizations of vinyl ethers is reportedly substantially higher than those same polymerizations carried out using chemical initiators [125]. It has been noted that methyl vinyl ether is resistant to radiation-induced cationic polymerization and undergoes only slow polymerization, which appears to be of a free-radical nature [126]. So far, there have been no reports regarding the tacticity of poly(vinyl ether)s produced by ionizing radiation.

A related method for inducing the cationic polymerization of vinyl ether monomers is by field ionization [127]. This technique involves introducing a monomer solution between two electrodes at high electrical potential. At the positively charged electrode, which is sharpened to a fine point, the local electric field strips electrons from the monomer to generate cation radicals according to the following equation:

$$M \longrightarrow M^+ + e^- \tag{25}$$

Radiation-induced polymerizations of vinyl ether monomers must be regarded as special techniques and are not generally applicable to laboratory or commercial production of poly(vinyl ethers).

11. Electrochemical Initiation

The electrolysis of vinyl ethers in the presence of a supporting electrolyte either a tetraalkylammonium salt, an inorganic salt such as sodium perchlorate, or sodium tetraphenylborate readily leads to polymerization. In all cases, the mechanism of polymerization appears to be cationic, although different workers differ with respect to the precise steps involved. For example, Cerai and coworkers [128] have proposed that when tetra-n-butylammonium triiodide is used as the supporting electrolyte, the triiodide anion undergoes oxidation by the following anodic process, which generates elemental iodine:

$$2I_3^- \xrightarrow[-2e]{\text{anode}} 3I_2 \tag{26}$$

The actual polymerization is thus, in fact, initiated by iodine. On the other hand, Breitenbach et al. [129] have suggested that direct anodic oxidation of the monomer occurs when tetra-n-butylammonium perchlorate is used as a supporting electrolyte to give cation-radical species which dimerize to give dications that are the active species responsible for polymerization.

Whatever the precise mechanism is, it has been noted that generally polymerizations of vinyl ethers occur rapidly under electrolytic conditions to give high yields of polymer per Faraday of current passed. However, in most cases, only low-molecular-weight polymers are obtained. Until major breakthroughs are made, electrochemical initiation must be regarded as a rather special, nonroutine technique for the polymerization of vinyl ethers.

C. Coordination Cationic Polymerizations

Coordination cationic catalysis of the polymerization of vinyl ethers is probably involved to some extent in several of the heterogeneous catalysts already cited above. However, the best characterized examples of coordination catalysts are the modified Ziegler–Natta catalysts termed PSV catalysts (pretreated stoichiometric vanadium) discovered by Vandenberg [130]. Although catalysts containing vanadium are most generally used, analogous catalysts containing titanium are also effective; those containing nickel, chromium, and molybdenum are significantly less active. These catalysts are characterized by their ability to yield highly crystalline poly(vinyl ethers) at room temperature. The catalysts are typically prepared by adding a trialkyl aluminum compound to a solution of VCl_4 in heptane. After aging the catalyst for 2 h and then heating for 16 h at 90 °C, this catalyst is further treated with iso-Bu_3Al-tetrahydrofuran complex and further heated. Under the best conditions, for example, the PSV catalysts give 20 to 41% conversions of crystalline isotactic poly(methylvinyl ether) together with 50% conversion to amorphous polymer. Apart from methyl and ethyl vinyl ether, no other straight-chain alkyl vinyl ethers give stereoregular polymers. In contrast, branched alkyl vinyl ethers polymerized in the presence of PSV catalysts to give highly crystalline polymers but with only low conversions. To account for the stereoregularity, Vandenberg [130] put forth the following schematic representation of the stereospecific propagation step:

(27)

Here Cl′ is the chloride counterion, Cl is a chlorine atom, M′ is one type of metal ion (usually vanadium), M″ is another metal center, X is a bridging group (Cl or OR), and A is a coordinate bond that is broken and replaced by bond B. M″ is thus freed to coordinate with another monomer molecule. Both metal centers are located at the surface of the insoluble component of the catalyst.

The coordination-catalyzed polymerization of vinyl ethers, particularly with the PSV catalysts, give the most highly stereoregular polymers that have yet been obtained. Such polymers are characterized by their high crystallinities, high melting points, and high molecular weights. PSVcatalyzed polymerizations appear also to proceed in a more controlled fashion than Lewis acid- or Brønsted acid-initiated polymerizations.

D. Free-Radical Polymerizations

The free-radical homopolymerization of vinyl ether monomers can be accomplished using various peroxide [131], azo [132], and redox initiators [133]. Polymerization under free-radical conditions gives only low-molecular-weight oligomers which have reported uses as lubricating oils [132]. Aryl vinyl ethers are polymerized by AIBN [134] and also give oligomeric materials. Because of the high temperatures and long reaction times required for carrying out free-radical polymerizations of vinyl ethers and the low-molecular-weight of the polymers obtained, these types of polymerizations are rarely carried out either in industry or in academia. Some fundamental studies are, however, worth noting. Matsumoto et al. [135,136] carried out a detailed investigation of the polymerization of *n*-butyl vinyl ether using various radical initiators and compared the structure of the oligomers that were formed to polymers produced by typical cationic initiators. While the structures of the two polymers were identical, they concluded that extensive chain-transfer processes were occurring and that the free-radical polymerization behavior of vinyl ethers was similar to that of allylic monomers.

Divinyl ether monomers undergo cyclopolymerization under free-radical as well as cationic conditions. If the polymerizations are carried to high conversion (>30 to 35%), gelation occurs. However, the soluble polymers that are produced at high dilution and low conversion often have rather complex backbone structures. For example, the polymerization of divinyl ether proceeds to give a polymer that incorporates tetrahydrofuran, vinyloxy, and dioxabicyclo[3.3.0]octane units [137,138]:

$$(28)$$

Work by Nishikubo et al. [139] showed that the polymerization of divinyl ethers derived from aliphatic diols gave polymers with different structures, depending on whether cationic or free-radical initiators were used. Shown in equation (29) are the structures of the polymers obtained from polymerization of ethylene glycol divinyl ether using AIBN and iodine.

$$(29)$$

The related monomers divinyl formal (30a), acetal (30b), and dimethylketal (30c) also undergo facile free-radical polymerization to give mainly soluble polymers [140–145].

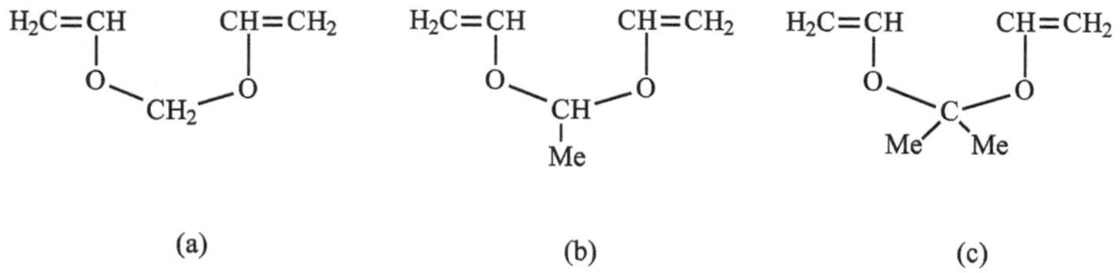

(a)   (b)   (c)

(30)

Detailed NMR analysis of the polymers produced by the polymerization of these compounds showed that the main backbone structures consist of *cis*-4,5-disubstituted 1,3-dioxolane units with some *trans*-disubstituted isomeric segmers and pendant 1,3-dioxolane segmers present as minor structural units [146]. Equation (27) depicts the polymerization of (30a). Similar compounds, dimethyldivinyloxysilane and dimethldivinyloxygermane, undergo analogous free-radical-induced cyclopolymerizations [147]. The structures of the polymers contain, in addition to 1,3-dioxa-2-silanole segmers, the corresponding six-membered rings and pendant vinyloxysilane groups.

In the last years many attemps are made to polymerize vinyl ethers with radical initiators, with only little success. Only the combination of vinylethers with other monomers can bring the success [148,149].

E. Copolymerization

1. Cationic Copolymerization

Copolymerization between two different vinyl ether monomers proceeds well in the presence of typical cationic initiators. These copolymerizations result, in most instances, in random copolymers being formed. In many cases, however, the polymers obtained display some blockiness, due to the differences in reactivity between the two monomers. For example, block polymers are obtained between isobutyl vinyl ether and 4-methoxystyrene (a phenylogous vinyl ether) using iodine as an initiator [150]. Vinyl ethers also catonically copolymerize with 1-alkoxybutadienes to give rubbery polymers having segments with pendant double bonds as shown in equation (32), which can be used as cross-linking sites for vulcanization [151,152].

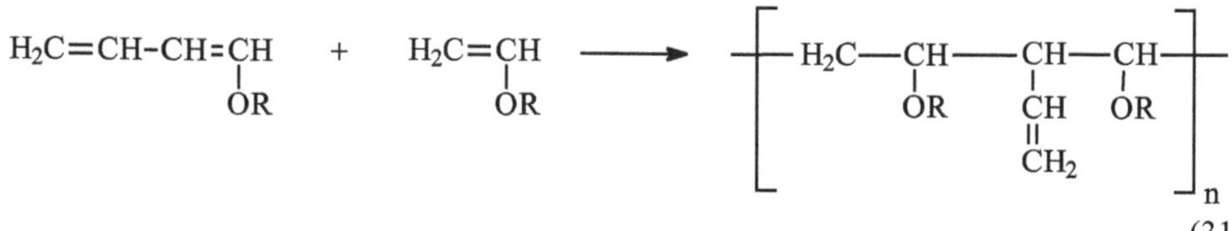

(31)

It is interesting that cationic initiators can be used to produce copolymers between vinyl ether and acrylate monomers. For example, the polymerization of *n*-butyl vinyl ether with methyl methacrylate gives an alternating copolymer when carried out in toluene at $0\,°C$ using butyl chlorotriethyldialuminum [153,154]. Copolymers produced by cationic copolymerization have found some commercial uses, among which are as elastomers (isobutylene with 2-chloroethyl vinyl ether [155] and allyl vinyl ether with methyl or isobutyl vinyl ether [156,157] and thickeners (copolymers of methyl and octadecyl vinyl ethers [158]).

There are also great possibilities to create new polymers by the combination of different vinyl ethers. There are several articles where tailor made polymers were synthesized by the copolymerization of different vinyl ethers [159–163].

2. Free-Radical Copolymerization

Although the free-radical homopolymerization of vinyl ether monomers proceeds rather poorly, the copolymerization of these compounds with especially vinyl monomers containing electron-poor double bonds is very facile [1]. Copolymers produced by free-radical techniques are of considerable commercial importance. In general, bulk, solution, emulsion, or suspension techniques can be used. Since the product of the reactivity ratios for vinyl ether monomers with electron-poor vinyl monomers are always near zero, there is a strong tendency toward alternation in the copolymers that are formed [3]. Of particular importance are the 1:1 alternating copolymers of various vinyl ethers with maleic anhydride, which find commercial uses as adhesives, floculants, lubricants, lacquers, greases, and processing aids, among many others. These polymers are amazingly adaptable materials whose range of properties can readily be modified by lengthening of the chain of the alkyl group, partial hydrolytic ring opening of the anhydride groups, as well as salt formation of the carboxyl groups that are formed and copolymerization with other comonomers. An excellent indepth review of this topic may be found in an article by Hort and Gasman of the GAF Corporation [7].

The 1:2 stoichiometric copolymerization of divinyl ethers with maleic anhydride gives interesting results. Using divinyl ethers with long alkylene groups such as 1,4-tetramethylene divinyl ether (1,4-butanediyl divinyl ether) gives cross-linked gels as expected [164]. At the same time, Butlerand co-workers [165,166] observed that divinyl ether itself forms a soluble, high-molecular weight 1:2 copolymer. The structure of the copolymer has been elucidated by a number of authors [167–170] and appears to consist of the combination of (32a) and (32b), due to cyclopolymerization.

(32)

(a)   (b)

This copolymers have a variety of biological and physiological properties, ranging from antifungal, bacteriostatic, and most important, antiviral and antitumor effects [171,172].

II. POLY(VINYL ACETATE)

(This section was prepared by O. Nuyken, J. Crivello and C. Lautner)

A. Introduction

1. Definition and Historical Background

Soon after the first preparation of vinyl acetate by the reaction of acetic acid with acetylene and its polymerization by Klatte [209] in 1912, methods for its industrial-scale synthesis were developed first in Germany, then in Canada [210]. At the same time, the chemistry was extended to the preparation and polymerization of vinyl esters of other aliphatic and aromatic carboxylic acids. The new polymers found immediate uses in paints, lacquers, and adhesives. Steady improvements in the industrial-scale monomer synthesis, particularly in the discovery of new catalysts for the acetic acid-acetylene condensation and development of a low-cost synthesis route based on ethylene have made vinyl acetate a comparatively inexpensive monomer. Besides the original applications, which still dominate the major uses of poly(vinyl acetate), this polymer finds additional utility as thickeners, plasticizers, textile finishes, plastic and cement additives, paper binders and chewing gum bases, among many others. At the same time, the uses and production of polymers of the higher vinyl esters have not kept pace with that of poly(vinyl acetate), primarily due to their higher cost. Consequently, the current world-wide production of these materials remains low.

The chemistry of vinyl acetate and its higher vinyl ester homologs has been the subject of several reviews [211–215]. These reviews have provided a rich source of background material for the present article, and the reader is referred to them for specific details concerning such topics as an in-depth discussion of plant design, manufacturing details, economics, toxicology, sample formulations and special applications of poly(vinyl acetate) and its homologous poly(vinyl esters). In this section we deal exclusively with various aspects of chemistry relating to the polymerization of poly(vinyl acetate). Due to the chemical similarity of the higher homologs, a direct analogy may be drawn to these materials as well.

The published literature, particularly the patent literature, of vinyl ester polymerization is very extensive. No attempt will be made here to cover this field comprehensively. Rather, selected examples will be drawn from various sources which represent state-of-the-art methods for the preparation of these polymers from both a commercial and a laboratory point of view.

2. Synthesis of Vinyl Ester Monomers

In the following discussion the major methods that have been developed for the synthesis of vinyl acetate in particular and vinyl ester monomers in general are described. The oldest process for making vinyl acetate and some of the more volatile vinyl ester monomers is the condensation of acetylene with a carboxylic acid:

$$HC \equiv CH + CH_3-COOH \xrightleftharpoons{\text{catalyst}} H_3C-COO-CH=CH_2 \qquad (33)$$

This reaction may be carried out either in the liquid state or by a vapor-phase reaction. The older liquid-phase process based on the passing of acetylene through the liquid carboxylic acid at 40 to 50 °C is catalyzed by mercury salts, typically mercuric sulfate, in the presence or absence of promoters [209,210,216]. In point of fact, reaction (33) proceeds with considerable reversibility; it is therefore necessary to work at high pressures and/or to remove the product vinyl ester in order to obtain a good yield. The more recently developed vapor-phase process for vinyl acetate synthesis is carried out at 180 to 210 °C using a zinc acetate catalyst [214]. Much effort has been expended on both of these processes in optimization of the yield of vinyl acetate and minimization of the byproducts (mainly ethylidine diacetate) through manipulation of the reaction conditions and improvements in the catalyst technology.

The modern commercial process for making vinyl acetate is based on palladium-catalyzed oxidative coupling of ethylene and acetic acid [217]. This process has largely supplanted the older acetylene based method of preparing vinyl acetate. Again, this reaction can be carried out by either a liquid- or a gas-phase process. The basic chemistry of the liquid-phase reaction is shown in the following equations (34)–(36).

$$H_2C=CH_2+2CH_3COOLi + PdCl_2$$

$$\longrightarrow CH_3COO-CH=CH_2 + 2LiCl + Pd + CH_3COOH \tag{34}$$

$$Pd + 2Cu^{2+}+4Cl^- \longrightarrow PdCl_2+2CuCl \tag{35}$$

$$2CuCl + \frac{1}{2}O_2+2HCl\longrightarrow 2CuCl_2+H_2O \tag{36}$$

Since, as shown in equation (34), palladium metal is precipitated as a byproduct of the reaction, it is necessary to reoxidize it back to the Pd^{2+} state. This is accomplished with a palladium-copper couple, as depicted in equations (35) and (36), which is driven by oxygen. The reaction is carried out by contacting a mixture of ethylene and oxygen with a mixture of acetic acid, lithium acetate, and the palladium-copper couple at temperatures of 80 to 150 °C. The vapor-phase process is carried out under pressure at high temperatures (120 to 150 °C) using a fixed-bed palladium catalyst [218]. The oxidative acylation of ethylene can also be used for the preparation of the higher vinyl esters, although it is not currently used for that purpose, due to the low demand for those materials.

Depending on the reaction conditions, ethylidine diacetate can be the major product of the metal-catalyzed reaction of acetylene with acetic acid and is also a byproduct of the oxidative acylation of ethylene. In addition, ethylidine diacetate is readily prepared by the reaction of acetaldehyde with acetic anhydride (37). A commercial-scale synthesis of vinyl acetate developed and piloted by the Celenese Corporation involved the pyrolysis of ethylidine diacetate obtained from acetaldehyde (38) [219,220].

$$CH_3CHO + (CH_3CO)_2O \longrightarrow CH_3CH(OOCCH_3)_2 \tag{37}$$

$$CH_3CHO(OOCCH_3)_2 \longrightarrow CH_3COOH+H_2C=CH-OOCCH_3 \tag{38}$$

Vinyl esters of carboxylic acids, which are not amenable to preparation by other synthetic techniques, are readily prepared by transvinylation. As depicted in equation (39), a carboxylic acid can undergo a vinyl exchange reaction with vinyl acetate in the presence of mercuric acetate as a catalyst [215,221]:

$$RCOOH + CH_2=CH-OOCCH_3 \xrightarrow[H_2SO_4]{Hg(OAc)_2} CH_2=CH-OOCR + CH_3COOH \quad (39)$$

Both the starting materials and byproducts of the reaction are low-boiling liquids that are removed by volatilization after the reaction, leaving the desired vinyl ester. This method is especially advantageous for the synthesis of high-molecular-weight vinyl esters which cannot be prepared by alternative methods that involve volatilization of the product during synthesis or purification.

A related specialized method consists of the reaction of divinylmercury with carboxylic acids (40,41). The reaction proceeds through a vinyl acyloxymercury intermediate [222]:

$$RCOOH + (CH_2=CH)_2Hg \longrightarrow CH_2=CH_2 + CH_2=CH-Hg-OOCR \quad (40)$$

$$CH_2=CH-Hg-OOCR \longrightarrow CH_2=CH-OOCR + Hg \quad (41)$$

This synthesis gives high yields of the desired vinyl esters but is rather cumbersome due to the necessity of preparing the starting divinyl mercury compound.

Acid chlorides react with acetaldehyde in the presence of tertiary amines to give high yields of vinyl esters according to the following reaction (42) [223]:

$$RCOCl + CH_3CHO \longrightarrow CH_2=CH-OOCR + HCl \quad (42)$$

The reaction of vinyl chloroformate with the sodium salt of a carboxylic acid generates the corresponding vinyl ester in good yields in most cases (43) [224]. This method constitutes a very good laboratory synthesis of vinyl esters.

$$CH_2=CH-OCOCl + RCOONa \longrightarrow CH_2=CH-OOCR + CO_2 + NaCl \quad (43)$$

Finally, glycol diesters can be thermolyzed to give vinyl esters (44) [225].

$$RCOO-CH_2-CH_2-OOCR \longrightarrow CH_2=CH-OOCR \quad (44)$$

Only the first three methods have had or continue to have commercial importance. The other methods are suitable for laboratory-scale syntheses and for the preparation of specific vinyl esters.

B. Monomer Reactivity and Polymer Structure

1. General Reactivity Considerations

The semiempirical Alfrey–Price Q and e values for vinyl acetate are, respectively, 0.026 and -0.22 [226]. With some exceptions, the reactivity of the higher vinyl esters is similar to that of vinyl acetate and is reflected in similarity of their Q and e values. From these values one can qualitatively conclude that compared to styrene, the vinyl acetate double bond is slightly more electron rich and that there is comparatively little resonance interaction between the double bond and the acetate group. In terms of its reactivity, vinyl acetate more closely resembles ethylene and other saturated olefins than styrene. Consequently, vinyl acetate and the related higher vinyl esters are reluctant to undergo either anionic or cationic polymerization. An additional complication is the presence of the ester carbonyl, which presents a competing site for attack by both anions and cations. For these reasons, the known polymerization chemistry of the vinyl esters almost exclusively proceeds by a free-radical mechanism. Compared to styrene, the ability of vinyl esters to react with a radical and to stabilize it through resonance is less. Once it is formed, the radical is very reactive toward further addition of monomers or other side reactions. This reactivity gives rise to a higher rate constant for propagation for vinyl acetate than for styrene.

The polymerization of vinyl acetate and other vinyl esters is effectively initiated using virtually and free-radical source. Thus a wide range of azo, peroxide, hydroperoxide, and redox initiator systems, as well as light and high-energy radiation, can be used. Polymerizations are inhibited or retarded in the presence of oxygen, phenols, quinones, nitro aromatic compounds, acetylenes, anilines, and copper compounds. Thus the monomer purity of vinyl esters is critical for their successful polymerization and for good molecular weight control. Vinyl esters of long-chain-saturated carboxylic acids tend to be less reactive than vinyl acetate, and the rate of polymerization decreases as the length of the chain increases [227,228]. Vinyl esters derived from unsaturated carboxylic acids, such as vinyl oleate, vinyl linoleate, and vinyl 10,12-octadecadienoate do not homopolymerize by themselves [229] and act as retarders in most copolymerizations [230].

2. Structure of Poly(vinyl acetate)

The structure of poly(vinyl acetate) produced by free-radical methods is complex. First, both head-to-head and head-to-tail addition can take place (45), resulting in the incorporation of the two types of repeating units shown in the backbone of the polymer [231].

$$\text{\small{wwCH}_2-\text{CH}-\text{CH}_2-\text{CH}ww} \qquad \text{\small{wwCH}_2-\text{CH}-\text{CH}-\text{CH}_2ww} \qquad (45)$$
$$\qquad\qquad |\qquad\quad | \qquad\qquad\qquad\qquad |\qquad\quad |$$
$$\qquad\quad \text{OAc}\quad\text{OAc} \qquad\qquad\qquad \text{OAc}\quad\text{OAc}$$

The proportion of head-to-tail and head-to-head repeating groups in the polymers is dependent on the temperature at which the polymerization is carried out. Higher head-to-head enchainment is obtained as the temperature is increased. These two types of repeating groups can be detected in the polymer by first removing the acetoxy groups by hydrolysis. The 1,2-glycols, which are formed by head-to-head enchainmnent, are readily cleaved by oxidants such as lead tetraacetate, which results in a lowering of the molecular weight. During the polymerization of vinyl acetate, extensive chain transfer takes place and gives rise to considerable branching in the final polymer. Branching is a particularly important process in the latter stages of emulsion and suspension polymerizations, which are carried to very high conversion. Chain transfer to monomer occurs predominantly at the acetyl methyl groups with a reported chain transfer constant of $C_m = 2.4 \times 10^{-4}$. Chain transfer to polymer is also facile and a chain transfer constant C_p of 2.36×10^{-4} has been recorded [232]. Hydrogen abstraction at the tertiary positions along the chain as well as at the pendant acetoxy groups appears to take place and leads to extensive branching at these sites [233]. When vinyl esters of long-chain fatty acids are polymerized, branching is even more facile than with poly(vinyl acetate), due to the presence of more easily abstractable methylene hydrogens in the hydrocarbon groups situated on both the monomers and the polymers [234–236]. In poly(vinyl acetate) there is also some evidence for unsaturation in the polymer chain which arises due to hydrogen abstraction at the tertiary positions along the backbone of the polymer followed by chain transfer to monomer [237].

$$\text{~CH}_2-\overset{\bullet}{\text{C}}\text{~} + \text{CH}_2=\text{CH} \longrightarrow \text{~CH}=\text{C~} + \text{CH}_3-\overset{\bullet}{\text{CH}}$$
$$\text{OAc} \quad\quad \text{OAc} \quad\quad\quad\quad \text{OAc} \quad\quad \text{OAc}$$

$$(46)$$

Poly(vinyl acetate) prepared by conventional free-radical techniques is completely atactic, and noncrystalline as determined by NMR and x-ray studies [214,238]. A stereoregulation method for free-radical polymerization of vinyl esters using fluoro-alcohols as solvents is described [239].

C. Free-Radical Polymerization Methods

1. Emulsion Polymerization

The chief large-scale commercial method employed for the polymerization of vinyl acetate and its higher vinyl ester homologs is emulsion polymerization. Poly(vinyl acetate) emulsions are stable but dry rapidly to give coherent films and coatings. In vinyl acetate emulsion polymerizations, typically, the polymers are not isolated, but rather the polymerization reaction mixtures are used directly in the various applications themselves. Because of their relative simplicity, emulsion polymerizations are also recommended for laboratory-scale preparation of poly(vinyl acetate).

Besides vinyl acetate monomer, three other components are necessary to carry out an emulsion polymerization: water, an emulsifier and/or a protective colloid, and a water-soluble initiator. Most commonly, anionic long-chain alkyl sulfonates are used as surfactants in amounts up to 6%. Studies have shown that the rate of polymerization is dependent on the amount of emulsifier present, with the rates increasing as the amount of emulsifier is increased up to a certain point and then falling off as free-radical chain transfer to the surfactant becomes a serious competing side reaction [240]. In general, surfactants are used in combination with a protective colloid. Especially useful as protective colloids are poly(vinyl alcohol), hydroxyethyl cellulose, alkyl vinyl ether-maleic anhydride and styrene-allyl alcohol copolymers, and gum arabic. Water-soluble initiators, particularly potassium persulfate, alkali peroxydisulfates, hydrogen peroxide, and various redox systems, are most commonly used.

Additional additives are also often included for various purposes. For example, buffers are added primarily to control the pH of the solution between pH 4 and 6 to prevent hydrolysis of the poly(vinyl acetate). An additional purpose of a buffer is to eliminate variations in radical generation of those initiators whose decomposition rates are pH dependent. Chain transfer agents are also commonly added to control the molecular weight within certain tolerances. A variety of thiols, aldehydes, or halogen compounds have been employed for this purpose.

Emulsion polymerizations have been subject to much technical optimization over the past several decades because of their commercial importance. The polymers produced using this type of polymerization are in the form of a creamlike latex in which the polymer particle sizes are on the order of 0.1 to 0.2 μm, which can be used directly in paint and adhesive formulations. An excellent general description of the emulsion technology of poly(vinyl acetate) may be found in the review article by Lindemann [213].

2. Suspension Polymerization

The suspension or bead polymerizations of vinyl acetate are also carried out in water. Typically, the polymerizations are run with an initiator that is soluble in vinyl acetate monomer and insoluble in water. A suspending agent, such as poly(vinyl alcohol), gelatin, and various water-soluble cellulose derivatives, have been used as well as water-insoluble inorganic materials such as calcium carbonate, barium carbonate, and barium sulfate [213]. Depending on such factors as concentration of monomer present, agitation rate, reactor vessel configuration, polymerization temperature, and type and amount of suspending agent, the particle size can vary widely [241]. Usually, spherical particle or beads with diameters in the range 0.01 to 3 mm are desired. Conversions are typically high; however, as with all bulk-type polymerizations, some residual monomer remains. Care must be taken to reduce the monomer level in the polymer so that the beads which are formed are not tacky and do not adhere to one another. Suspension polymerization is used as the method of choice for the commercial production of poly(vinyl acetate) intended for conversion to poly(vinyl alcohol). In this case, the traces of residual monomer is of little concern. Laboratory-scale syntheses using suspension polymerization are also successful provided that efficient, high-speed stirring is employed.

3. Bulk Polymerization

Bulk polymerization of vinyl acetate can be carried out simply by dissolving any one of a variety of common organic free-radical initiators in the monomer and heating to dissociate the initiator. Reasonable care should be taken to eliminate oxygen and other impurities that retard or inhibit the polymerization. As with most other types of vinyl polymerization, this type is generally suitable only for the polymerization of small masses of monomer. Because of the rapidity of vinyl acetate polymerizations, its high heat of polymerization, and the poor heat-transfer characteristics of the polymer, bulk polymerization of large masses of monomer can result in runaway conditions leading to partial decomposition of the polymer that is formed. For these reasons, bulk polymerizations of vinyl acetate are not commonly practiced on a commercial scale, although some recent innovative reactor designs have been published that preport to enable the continuous bulk polymerization of vinyl acetate [242].

4. Solution Polymerization

The polymerization of vinyl acetate can be carried out in a wide variety of solvents in which both the monomer and polymer are soluble. Azo, peroxide, and hydroperoxide initiators as well as many other organic-soluble initiators can be used. Solvents with low chain-transfer constants, such as benzene, toluene, acetic acid, acetic anhydride, acetone, and cyclohexanone, are required to obtain reasonably high molecular weights. Solution techniques are especially convenient for the laboratory preparation of poly(vinyl esters) and are used in certain commercial applications in which the polymers are sold directly as solutions.

5. Photopolymerization

Direct UV irradiation of vinyl acetate at 255 nm, or more advantageously, irradiation in the presence of photo initiators, induces facile free-radical polymerization [243,244]. Benzoin, benzoin alkyl ethers, biacetyl, and alkoxyacetophenones are particularly efficient photo initiators [245]. Polymerization is typically carried out by irradiating a solution of vinyl acetate and a photo initiator in a quartz reaction vessel. The polymerizations are generally run under nitrogen using a medium-pressure mercury arc lamp or a mercury-doped xenon arc lamp as the UV irradiation source. Photochemical polymerizations have been carried out on a laboratory scale but have not been found useful for the commercial preparation of poly(vinyl acetate). High-speed UV-initiated coating processes that have many commercial applications have not been applied to vinyl acetate, due to its high volatility.

6. High-Energy Radiation Polymerization

Cobalt-60 γ-ray irradiation induces the facile polymerization of vinyl acetate and produces highmolecular-weight polymers [246]. Polymerization can be carried out in solution, bulk, and emulsion. The emulsion γ-ray irradiation polymerization of vinyl acetate has been of particular interest, and considerable labor has been expended on studies designed to explore the effects of dose, irradiation intensity, type of emulsifier, monomer concentration, and so on, on the course of the polymerization [247–249]. Particularly attractive aspects of this technique of radiation polymerization are the excellent efficiency of polymerization together with the high rates of polymerization attained with γ-irradiation. Further, radiation-polymerized poly(vinyl esters) do not contain initiator derived end groups, which can be the source of polymer thermal oxidative and photo degradation. A proposal for a commercial process based on radiation polymerization of vinyl acetate has been published [250,251]. In general, radiochemical initiation must be regarded as a special, nonroutine process for the preparation of poly(vinyl acetate).

7. Miscellaneous Methods

Vinyl acetate has been polymerized by a wide variety of nonconventional initiator systems which are documented primarily in the patent literature. Most of these polymerizations must be regarded as laboratory curiosities. Free-radical mechanisms are clearly involved in most instances; however, examples of cationic and anionic types of polymerization are also known. Typical of initiators whose mechanisms may be, respectively, free-radical, cationic or anionic are organometallic compounds [252–258]. Organoboron [252–254], organo-aluminum [254,255], organolithium [253–256], organomagnesium [257], and organotitanium [258] compounds have been used as well as their combinations with oxygen [259], peroxides [260], metal halides [259–263], and alcohols [264]. Electropolymerization has also been briefly explored as a route to the preparation of poly(vinyl acetate) [265].

D. Controlled Radical Polymerization Methods

Reversible Addition-Fragmentation Chain Transfer (RAFT) polymerization using xanthanes and dithiocarbamates is described [266]. Narrow polydispersities and good control of molecular weight for polymers of $M_n < 30\,000$ are achieved for these polymers. The living nature of RAFT polymerization allows the synthesis of block copolymers, star polymers and gradient copolymers [266].

The homopolymerization of vinyl acetate with the Atom Transfer Radical Polymerization Method (ATRP) has not yet been successful [267].

E. Modification of Poly(vinyl acetate)

The most important chemical modification of poly(vinyl acetate) is its conversion to poly(vinyl alcohol) (47) [268].

$$\left[CH_2-\underset{\underset{OAc}{|}}{CH}\right]_n \quad \xrightarrow[CH_3OH]{NaOH} \quad \left[CH_2-\underset{\underset{OH}{|}}{CH}\right]_n + nCH_3COOCH_3 \qquad (47)$$

Usually, basic conditions are used, although acid-catalyzed hydrolysis has also been employed. Hydrolysis can be carried out using NaOH and water or, more advantageously, methanolysis employing NaOH and methanol. By controlling the degree of hydrolysis, one can obtain polymers with different properties. Complete hydrolysis (<1.5 mol% remaining acetate groups) of high-molecular-weight poly(vinyl acetate) to poly(vinyl alcohol) results in a polymer that is soluble in hot water but insoluble in cold water and is crystalline, hard, and friable. In contrast, poly(vinyl acetate) that has been 88% hydrolyzed (12 mol% remaining acetate groups) is soluble in water at room temperature, while a polymer containing 30 mol% acetate groups is soluble only in a mixture of alcohol and water. Poly(vinyl alcohol) finds many uses, such as adhesives, thickeners, emulsifiers, and paper and textile treatments.

Another important modification of poly(vinyl acetate) is its derivatization with aldehydes to poly(vinyl acetal)s [269]. This can be accomplished by first hydrolyzing the poly(vinyl acetate) to poly(vinyl alcohol) and then carrying out a subsequent acetalation reaction with an aldehyde and a strong mineral acid in water. Alternatively, poly(vinyl acetate) can be converted in a single one-pot reaction with acetic acid as a solvent directly to the poly(vinyl acetal) by reaction with water, an aldehyde, and a mineral acid catalyst.

$$\tag{48}$$

Reaction (48) is a considerable simplification of the actual overall chemistry and structure of poly(vinyl acetals). Of course, some intermolecular condensation also takes place between two poly(vinyl alcohol) chains and intramolecularly between the 1,2-glycols derived from head-to-head enchainmnent to give dioxolane units in the chain. There are also present branches and some residual acetate groups. Typically, formaldehyde and, especially, butyraldehyde have been used in reaction (48) to make the respective poly(vinyl formal) and poly(vinyl butyral). The latter material finds considerable use in interlayers for safety glass and in adhesives.

F. Copolymers of Vinyl Acetate

As mentioned previously, the Alfrey–Price Q and e values for vinyl acetate are 0.026 and −0.22, respectively [226]. Thus vinyl acetate is rather sluggish in its free-radical copolymerization, with most monomers, particularly olefinic monomers, bearing electron-donating subtitutents. The copolymerization reactivity ratios reflect the reluctance of vinyl acetate to enter into copolymerization with other monomers [270]. Nevertheless, vinyl acetate copolymers with a great many electron-rich as well as electron-poor olefins have been prepared. Especially significant from a commercial point of view are copolymers with ethylene, vinyl chloride, acrylates, methacrylates, fumarates, and maleates. Often, mixtures of three and more comonomers are used in these copolymerizations.

Free-radical copolymerizations can be carried out by any one of the usual methods used for the preparation of the homopolymers themselves. Particularly advantageous for the synthesis of these copolymers are emulsion copolymerizations [271]. Separate emulsions each containing one, two, or more monomers can be equilibrated with one another and then polymerized to give the copolymer latex. Random ethylenevinyl acetate copolymers, which have found considerable commercial use in adhesives, coatings, and molding compounds are prepared by various techniques. Those polymers that have high ethylene contents are normally prepared by bulk or solution methods while high-pressure emulsion techniques are employed for copolymers rich in vinyl acetate [272]. Vinyl acetate is often polymerized in minor amounts with vinyl chloride and with acrylic monomers to modify their polymers and to impart special properties, such as plasticization and dyeability.

While most copolymers of vinyl acetate are random copolymers, alternating copolymers are formed when the reactivity ratios for the two monomers are suitable. This occurs spontaneously when vinyl acetate is polymerized with electron-poor monomers such as maleic anhydride [273]. Alternatively, it has been reported that acrylonitrile which has been precomplexed with zinc chloride gives alternating polymers with vinyl acetate [274]. Block polymers of vinyl acetate with methyl methacrylate, acrylonitrile, acrylic acid, and n-vinyl pyrrolidone have been prepared by the strategy of preparing poly(vinyl acetate) macroradicals in poor solvents in which the macroradicals are occluded. Addition of a second monomer swells the polymer coils, and polymerization continues with the addition of the new monomer [275].

III. POLY(VINYL CHLORIDE)

(This section was prepared by O. Nuyken, J. V. Crivello and J. P. Robert)

A. Introduction

Poly(vinyl chloride) (PVC) is one of the most important thermoplastics produced by the chemical industry [276,277]. For the year 1999, the worldwide total production of PVC was estimated at about 24.3 million tons [278]. Well-known PVC materials, for example, are tubes, valves, flexible pipes, or floor coverings. As can be seen in this short enumeration, it is possible to produce rubberlike up to hard PVC. This variety of applications is attributed to the polar structure of the macromolecule, which permits the use of specific plasticizers [279–281] and specific additives useful for manufacturing [282–285]. Another important factor in PVC technology is grain morphology [286–289].

Grain morphology determines the variety of possible manufacturing processes. On the commercial preparation of PVC, therefore, not only polymerization prescription but also the size of reaction vessel and the shape of the stirrer determine the applications. Theoretical as well as experimental investigations have been performed by Kiparissides et al. [290] and others [291]. Altogether, the PVC production exhibits a considerable complicated area.

Vinyl chloride (VC), the monomer, was first obtained in 1835 by Regnault [292] when he treated 1,2-dichloroethane with potassium hydroxide:

$$\underset{\underset{Cl}{|}}{\overset{\overset{Cl}{|}}{H_2C-CH_2}} + KOH \longrightarrow \underset{\underset{Cl}{|}}{HC=CH_2} + KCl + H_2O \tag{49}$$

After keeping this reaction mixture in sunlight for 4 days, a white powder, PVC, was formed. Although chemists [293–296] continued investigations on the syntheses and properties of the monomer and polymer, full-scale commercial production of PVC started not before 1930. Today, the technical preparation of VC [297–299] is mainly based on the thermic fission of 1,2-dichloroethane (EDC):

$$\underset{\underset{Cl}{|}}{\overset{\overset{Cl}{|}}{H_2C-CH_2}} \xrightarrow{\Delta} \underset{}{\overset{\overset{Cl}{|}}{H_2C=CH_2}} + HCl \quad \Delta H = +67 \text{ kJ/mol} \tag{50}$$

The synthesis of VC by hydrochlorination of ethine in the presence of mercury(II) chloride plays nowadays a subordinated role:

$$HC\equiv CH + HCl \xrightarrow{[HgCl_2]} \underset{\underset{Cl}{|}}{H_2C=CH_2} \quad \Delta H = \text{-96 kJ/mol} \tag{51}$$

1,2-Dichloroethane again is produced from ethene and chlorine:

$$H_2C=CH_2 + Cl_2 \longrightarrow \underset{\underset{Cl}{|}}{\overset{\overset{Cl}{|}}{H_2C-CH_2}} \quad \Delta H = \text{-183 kJ/mol} \tag{52}$$

or in the presence of ethene, oxygen and hydrogen chloride:

$$H_2C=CH_2 + HCl + 0,5 O_2 \longrightarrow \underset{\underset{Cl}{|}}{H_2C=CH} + H_2O \tag{53}$$

the manufacturing process of VC is often a combination of (50), (52) and (53). The hydrogen chloride in (50) is used for the oxychloration (53).

Vinyl chloride is a colorless, combustible gas at atmospheric pressure and room temperature. Therefore, polymerization is carried out under pressure or temperature below $-13.4\,°C$, the boiling point [300]. Besides its inclination to polymerize, halogenation or hydrohalogenation also takes place smoothly. In contact with oxygen it reacts readily to form chloride peroxide [301], which decomposes into formaldehyde, carbon-monoxide, and hydrogen chloride [302]. Without oxygen, light, or high temperatures, however, pure and dry VC is very stable and storage for a longer period is possible. Its most unpleasant property is to cause cancer of the liver, lung, and brain [303,304]. Therefore, working with VC demands some safety precautions. For technical polymerization the autoclave engineering is the method of choice. It should also be noted that if PVC is produced, only a partial conversion is achieved. Unreacted monomer has to be eliminated by vacuum technology in an effective hood.

B. Radical Polymerization: General Aspects

In this section we present a short overview of the polymerization of VC. The most common method is polymerization by free radicals [305]. According to the ease of homolytic splitting of the π bond in the monomer, radical polymerization takes place in the presence of suitable initiation systems. In general, there are three methods for producing radicals available for the polymerization of VC: (A) thermal cleavage of azo or peroxo compounds, (B) oxidation-reduction processes, and (C) metal alkyls in connection with oxygen. After the initiation step, chain growth takes place rapidly:

$$R-CH_2-\overset{\bullet}{C}H + H_2C=CH \longrightarrow R-CH_2-CH-CH_2-\overset{\bullet}{C}H \quad \Delta H = -96 \text{ kJ/mol}$$
$$\quad\quad\quad \mid \quad\quad\quad\quad \mid \quad\quad\quad\quad\quad\quad\quad\quad \mid \quad\quad\quad \mid$$
$$\quad\quad\quad Cl \quad\quad\quad\quad Cl \quad\quad\quad\quad\quad\quad\quad\quad Cl \quad\quad\; Cl$$

$$(54)$$

The final step is termination of chain growth mostly by radical transfer reaction to monomer [306], whereas combination or disproportionation are observed only to a small extent. The monomer radical is able to start a new chain. The most widely used procedures for preparation of commercially PVC resins are, in order of their importance, suspension, emulsion, bulk, and solution polymerization. A common feature of the first three methods is that PVC precipitates in liquid VC at conversions below 1%. The free polymerization of VC in a precipitating medium exhibits an accelerating rate from the beginning of reaction up to high conversion [307]. This behavior is called autoacceleration and is typical for heterogeneous polymerization of halogenated vinyls and acrylonitrile [308].

Detailed studies of polymerization mechanism and analysis of microstructure have been carried out [291,309]. Primary structure of PVC is demonstrated in equation (54). During VC polymerization, however, the monomer has still another possibility for reaction with the macroradical, as follows:

$$R-CH_2-\overset{\bullet}{C}H + H_2C=CH \longrightarrow R-CH_2-CH-CH-\overset{\bullet}{C}H_2$$
$$\quad\quad\quad \mid \quad\quad\quad\quad \mid \quad\quad\quad\quad\quad\quad\quad\quad \mid \quad\; \mid$$
$$\quad\quad\quad Cl \quad\quad\quad\quad Cl \quad\quad\quad\quad\quad\quad\quad\quad Cl \;\; Cl$$

$$(55)$$

Reaction (54) is a head-to-tail addition, whereas reaction (55) leads to a head-to-head structure [305]. Head-to-head addition is hampered by steric and resonance stabilization reasons. This irregularity in primary structure yields short-chain branching, which was studied by Talamini [310] by the use of IR-spectroscopy. It was found that samples prepared at 50 °C show structure units as described in equation (55) of approximately 1.5%. Furthermore, branching is independent of conversion but dependent on temperature. If the polymerization temperature is lowered, Talamini reports that branching decreased.

Examinations on ^{13}C-NMR spectroscopy by Starnes et al. and Hjertberg et al. [311–315] verify the results of Talamini for short-chain branching. The carbon atoms at branching points (methine groups) show chemical shifts between 35 and 40 ppm, methyl groups at 10 to 20 ppm, and methylene groups at 25 to 35 ppm. Furthermore, the spectral data reveal the occurrence of longer-chain branches.

Many authors [310–315] have proposed a mechanism on branching involving chain transfer. Short-chain branches were attributed to an intramolecular chain transfer mechanism. Long-chain branching might be the result of transfer involving a PVC macromolecule and a polymer radical. Butyl branches can be expected from an intramolecular back-biting mechanism. Presumably, chain transfer involving monomeric VC may also lead to branching. The full detailed mechanism for chain transfer, both intramolecular and intermolecular (monomer or polymer), is still not completely understood.

By means of ^{1}H-NMR spectroscopy of unreduced PVC samples, Hjertberg et al. [316], and others [317] have demonstrated that the most frequent unsaturated end group is R–CH$_2$–CH=CH–CH$_2$Cl and the most frequent saturated end group is R–CH$_2$–CHCl–CH$_2$Cl. To explain the formation of both structures, it has been suggested that the loss of a chlorine radical leads to an olefinic group that can be transformed into the allylic structure by isomerization. The chlorine radical is able to start a new chain, which leads to the saturated head group mentioned above.

From calorimetric measurements (differential scanning calorimetry) of PVC samples it was shown that considerable variations in the glass transition temperature (T_g) can be obtained [318]. In comparison to ordinary PVC, where T_g is approximately 80 °C, PVC synthesized at low temperature reaches values up to 100 °C. Mijangos et al. [319] studied the dependence of T_g on tacticity of PVC. Their results show that the differences in T_g may be attributed to changes in the stereospecifity and crystallinity of resins produced at various polymerization temperatures. It was shown that the content of syndiotactic segments is responsible for the ability to crystallize and for higher T_g. In contrast, isotactic and heterotactic sequences appear to affect T_g in a negative manner. By evaluation of ^{13}C-NMR spectra, Martinez et al. [320] calculated the concentration of syndiotactic, heterotactic, and isotactic segments, respectively, on the correlation of the areas of corresponding signals at 57.5, 56.5, and 55.5 ppm with a compensating polar planimeter as well as by means of the built-in electronic integrator. In a typical commercial polymer [319], the probability of syndio (Pss), hetero (Psi + Pis), and isotactic (Pii) triads is Pss = 0.297, Psi + Pi = 0.495, and Pii = 0.208. In comparison to a product prepared at −50 °C, Pss = 0.426, Psi + Pis = 0.442, and Pii = 0.132. In an earlier publication, Millan et al. [321] measured the tacticity by determining the ratio between the absorbances of the infrared bands at 615 and 690 cm^{-1}, which refer to the syndiotactic and isotactic structures of the polymer chain, respectively. Similar results on the dependence of tacticity on the polymerization temperature were obtained by Abdel-Amin [322], Talamini and Vidotto [323], and Hassan [324] when they determined stereoregularity by means of ^{1}H-NMR spectroscopy. Furthermore, not only tacticity, but also molecular weight, influences T_g. It was found that T_g increases rapidly with the number-average molecular weight over the range 500 to 10 000 and to level off at an almost constant value of 80 °C [325].

Hjertberg et al. [326] investigated the thermal stability of PVC products prepared at different temperatures. They found the low-temperature samples show higher stability than that of ordinary resins. This behavior was attributed to the smaller content of tertiary chlorine atoms in the low-temperature PVC. It is assumed that the tertiary chlorine is the most important labile structure in the backbone because dehydrochlorination is favored there [327–331]. Tertiary carbon atoms are met at points of branching. Further dehydrochlorination leads to conjugated double bonds in the macromolecule. The maximum length of these polyene sequences is 20 to 30 units.

The degradation rate was followed by measuring evolved hydrochlorine conducto-metrically at 190 °C. UV-visible spectroscopy was used to determine the polyene units. Abbas and Sörvik [332] investigations on the thermal degradation of PVC samples with different amount of polyene sequences show that these internal double bonds seem to be the units that are responsible for chain scission. Another important factor that influences the stability of polymers is the presence of oxygen during polymerization. Garton and George [333] reported that in an oxygen atmosphere, the resulting resin has lower molecular weight and possesses a poor thermal stability. The influence of oxygen is also reported by Zilberman in 1992 [301].

As could be seen in this short overview, the PVC field exhibits considerable complexity. The most important factor that controls chain branching, stereospecifty, and crystallinity is the polymerization temperature.

C. Radical Polymerization: Procedures

1. Polymerization in Bulk

The bulk polymerization of VC is the third important manufacturing process for PVC [291,334–336]. The advantage of bulk polymerization is, in contrast to the more common suspension or emulsion polymerization, that products are free of protective colloids, suspending agents, surfactants, buffers, water, additives, or solvents. There is, however, a great problem for technical application. This problem is to remove the heat generated during polymerization and, related to that, to control the rate of reaction. The industrial-scale bulk polymerization based on the Pechiney-Saint-Gobain process avoids the heat problem by using a two-stage process [337,338]. In a first step VC is prepolymerized to approximately 10% conversion. Then the reacting mass is dropped into a second autoclave. This reactor is specially designed to stir powdery material and is equipped with a condenser. The products exhibit desirably high porosity and high bulk density, coupled with good transparency upon plasticization. To avoid agglomeration of the beads, it is very important to control the rate of agitation in industrial plants [339,340]. Experimental procedures for bulk polymerization in laboratory scale are relatively simple [311,341]. Normally, the monomer is heated in the presence of a small amount of a monomer-soluble initiator under a suitable condensing or pressure system until the desired conversion of monomer into polymer has been achieved. The residual VC has to be recovered by distillation in an effective hood. Some of the common initiators suitable for bulk and for suspension polymerization are listed in Table 2.

Further monomer-soluble initiators mentioned in patents are di(2-ethylhexanoyl) peroxide [345], 3,5,5-trimethylhexanoyl peroxide [346], di(t-butyl) peroxyoxalate [347], di(carballyloxy isopropyl peroxydicarbonate) [348], di-2-butoxyethyl peroxydicarbonate [349], and di-4-chlorobutyl peroxydicarbonate [350]. An elegant initiation technique for polymerization of VC is described by Ravey and Waterman [351]. They investigated the in situ formation of the initiator during reaction from stable precursors. This method avoids the problem of handling unstable initiators. Furthermore, the in situ mode proved to be more efficient at lower temperatures than the conventional systems [352].

Table 2 Suitable initiators for the radical polymerization of vinylchloride in bulk

Initiator	T [°C]	Refs.
Acetyl cyclohexylsulfonyl peroxide	40–55	[342–344]
Acetyl heptylsulfonyl peroxide	40–55	[342]
Dicyclohexyl peroxydicarbonate	45–60	[342]
Diethyl peroxydicarbonate	45–60	[342]
Diisopropyl peroxydicarbonate	45–60	[342,343]
Di-*s*-butyl peroxydicarbonate	45–60	[342,343]
Di-2-ethylhexyl peroxydicarbonate	45–60	[342–344]
Ditetradecyl peroxydicarbonate	45–60	[342]
Diacetyl peroxydicarbonate	45–60	[342,343]
Distearyl peroxydicarbonate	50–60	[342]
Di-3-methoxybutyl peroxydicarbonate	45–60	[343]
Dibenzyl peroxydicarbonate	45–65	[342]
Di-4-*t*-butylcyclohexyl peroxydicarbonate	50–80	[343,344]
t-Butyl peroctoate	50–65	[342]
t-Butyl pemeodecanoate	50–65	[343,344]
t-Butyl perpivalate	50–65	[342–344]
2,4,4-Trimethylpentyl 2-phenoxyperacetate	40–80	[344]
Bis(2,4-dichlorobenzoyl) peroxide	40–80	[344]
Bis(*o*-methylbenzoyl) peroxide	40–80	[343,344]
Dilauroyl peroxide	50–70	[342–344]
Dicapryloyl peroxide	50–70	[344]
Dibenzoyl peroxide	70–90	[342]
Azobis(2,4,4-trimethyl valeronitrile)	45–60	[342]
Azobis(2,4-dimethyl 4-methoxyvaleronitrile)	45–60	[342]
Azobis(2,4-dimethyl valeronitrile)	50–70	[342–344]
Azobis(isobutyronitrile)	55–80	[342,343]
Azobis(cyclohexylcarbonitrile)	80–100	[311]
t-Butyl azoisobutyronitrile	80–100	[342,343]

Many authors [307,309,340,353–355] have investigated the dependence of the molecular weight of polymers produced in bulk on the initiator concentration, type of initiator, conversion, and temperature. Their results show that M_n and M_w are apparent independent from the initiator concentration, type of initiator, and conversion. This behavior is the result of the strong chain transfer to monomer. The temperature, however, is suitable for regulation of the degree of polymerization, because the monomer transfer constant depends more strongly on temperature than does the rate constant of propagation. The intrinsic viscosity of PVC as a function of polymerization temperature shows that with increasing temperature, molecular weight decreases. Furthermore, at temperatures above 50 °C the rate of polymer formation initiated by monomer-radicals is significantly higher than that started by initiators. This is the reason for the limitation of M_n between 50 000 and 100 000. Most of the molecular weights given in the literature are determined by gel permeation chromatography and viscosity. For determination by means of gel permeation chromatography, a universal calibration curve was first obtained using polystyrene standards; then a PVC calibration curve was performed from it, using the hydrodynamic concept. The resulting equation for weight-average molecular weight is [306]:

$$[\eta] = 1.50 \times 10^{-4} \overline{M}_w^{0.77} \qquad (\text{THF}, 30°\text{C})$$

The number-average molecular weight was determined from viscosity measurements in cyclohexanone at 25 °C, frequently by the use of the empirical equation of Danusso et al. [356]:

$$[\eta] = 2.40 \times 10^{-4} \overline{M}_w^{0.77} \qquad \text{(Cyclohexanone, 25°C)}$$

As mentioned in the introduction, if PVC with higher thermal stability is desired, polymerization has to be carried out at low temperatures. For preparing PVC at low temperatures, specific initiators are necessary. For example, redox-type catalysts such as organic hydrogen peroxides with sulfur dioxide or sulfur trioxide [357,358], organic hydrogen peroxides with sulfinic acid or its derivatives [359,360], and organic hydrogen peroxides with hydroxy ketones [361] are described in the patent literature. These systems are applicable at temperatures between -30 and $+20$ °C.

Another type of initiator that is also useful in VC polymerization at low temperature is tri-n-butylborane in connection with oxygen [362]. The presence of a small amount of oxygen is necessary to produce alkyl peroxyborane in a first step. Further information about generation of the initiating species are still not available in full detail. Detailed results and discussions about low temperature PVC prepared by means of tri-n-butylborane are published by Talamini and Vidotto [363] and Braun et al. [364]. Talamini discussed the abnormal change of molecular weight with temperature. From -80 to $+25$ °C molecular weight goes through a maximum at -30 °C and decreases again with higher temperature. This feature should be attributed to the different viscosity arising from the different degree of swelling of the polymer particles by the monomer at different temperatures. With diminishing temperature and consequently increasing viscosity, the termination rate decreases until a critical value of viscosity is reached. At about -30 °C the viscosity reaches such a high value that the propagation rate decreases abruptly, too. This behavior is characteristic of a heterogeneous polymerization [365]. Braun investigated the dependence of conversion and molecular weight on initiator concentration. His results show that it is possible to get high-molecular-weight samples at low temperatures. For example, he received at -15 °C a polymer with $M_n = 300\,000$ and $M_w = 380\,000$. The influence of initiator concentration on conversion and molecular weight is very complex. It depends not only on the amount of borane but also on the amount of oxygen. Best results for high conversion and high molecular weight (see above) were achieved when the VC/borane/oxygen ratio was 1×10^3 mol:0.5 mol:0.14 mol. The average reaction time was 6 h. Braun also studied the structure of low-temperature PVC by means of ^{1}H-NMR spectroscopy. The signal at 3.78 ppm which is assigned to branching points was correlated to the signal at 4.5 ppm, the chloromethylene group of the backbone. His results support the view that branching decreases if temperature is decreased. The improved thermal stability was demonstrated by measuring dehydrochlorination conductometrically at 190 °C. Besides, it was found that stability increases when the molecular weight of samples rises.

Instead of a reducing agent for decomposing peroxides or azo compounds, UV irradiation can be applied for splitting initiator. Millan et al. [319,321,366] describe the use of an ultraviolet beam to activate initiator decomposition at low temperatures. In a series of experimental procedures it was established that at 0 °C a molecular weight of $M_n = 320\,000$ can be obtained. Results similar to those discussed for borane-initiated polymerization in reference to tacticity, thermal stability, and branching were obtained. This technique has very little practical value because special equipment is required.

2. Polymerization in Suspension

More than 80% of PVC resins are produced by suspension polymerization. In contrast to bulk polymerization for the preparation of a typical suspension charge, vinyl chloride is added to a suitable amount of water (weight ratio to VC is approximately 2:1) and one or more protective colloids (usually under 1% in reference to VC) [367–372]. In analogy to the bulk polymerization, monomersoluble initiators start reactions. Therefore, initiation systems described earlier are also useful in suspension polymerization [342]. Suspending agents are partially hydrolyzed poly(vinyl acetate), cellulose ethers, vinyl acetate/maleic anhydride copolymers, acrylic acid copolymers with vinyl esters, acrylic esters, or vinyl pyrrolidone, vinyl ether/maleic anhydride hydrolyzed vinyl acetate copolymers, poly(vinyl acetamide), poly(oxazoline), gelatine, lithium stearate, sodium lauroylsulfate, and combinations of two or more of these additives [303,305,373,374]. The presence of a suspending agent is necessary for stabilizing the monomer droplets to avoid coagulation and to control the dimension of the particles [373]. After polymerization, excess monomer is vented in an effective hood. The polymer is recovered by filtration, washed repeatedly with distilled water, and dried to constant weight under reduced pressure at about 45 °C [375].

Since kinetic equivalence of bulk and suspension polymerization has been demonstrated by Crosato-Amaldi et al. [376], it is not surprising that the average molecular weight of samples produced in bulk or suspension at the same conditions shows only slight differences in the degree of polymerization. Also, branching and tacticity yields similar results as discussed in bulk polymerization. This behavior seems to be the result of the very poor solvent capacity of VC for its own polymer and chain transfer onto monomer. Thus the suspension polymerization of VC can be considered as a micro bulk polymerization. However, since not all protective colloids can be removed, PVC has lower heat stability and clarity than bulk polymer. In many respects the polymer produced in bulk is similar to those prepared in suspension, but there are important morphological differences [377]. By means of scanning electron micrographs, particle size was investigated [378]. It was found that the grain size increased with higher temperature, but it was unaffected by the amount of initiator, type of protective colloid, and rate of agitation. These factors, however, influence the porosity and the morphology of the grains in a complex manner [379], but these topics are mostly a proprietary secret of PVC producers. Since the property of a PVC resin will be decided mostly from porosity and morphology, it is very important to control the rate of agitation and type of suspending agent if a reproducible product should be obtained. Low-temperature polymerization in suspension succeed with ordinary initiators in the presence of titane(III) chloride and sodium carboxylates [380], NN'-dimethylaniline [381], α-chlorolauroyl peroxide [382], and ferrous(II) hydroxide [383].

3. Polymerization in Emulsion

Another important way for producing PVC is the emulsion polymerization [384–388]. In contrast to both systems discussed before, the initiator here is water soluble. Free radicals are formed by potassium persulfate [344] or ammonium persulfate [305], sodium perborate [389], sodium percarbonate, sodium perphosphate [390], peracetic acid, water-soluble organic hydrogen peroxide (cumolhydroperoxide) [391], hydrogen peroxide [392], and water-soluble azo compounds [393,394]. In general, only a small amount of initiator (under 0.5 wt%) starts a reaction in the aqueous phase. The ratio between water and VC is approximately 2:1. Sometimes it is advantageous to add a reducing agent: for example, sulfites [395], sodium formaldehyde sulfoxylate [396], thiosulfates, dithionites [397], or sulfur dioxide [398]. Both in suspension and by applying emulsion polymerization, the presence of protective colloids is necessary for stabilizing lattices. Such an emulsifer can be fatty acid salts (sodium stearate), alkylsulfates (sodium laurylsulfate), alkylsulfonates (sodium di-n-butylsulfosuccinate) [395], alkylphosphates [399], ammonium alkylcarboxylates [400], salts of styrene/maleic acid copolymers [401], and fatty alcohol/poly(glycolcarboxylates) [402]. The concentration of these stabilizers ranges between 0.5 and 2 wt% in referring to VC. In analogy to suspension polymerization, the rate of agitation is very important for the latter application of emulsion resins [403,404]. Generally, the speed of agitation is more moderate than in suspension. Sometimes it is desirable to keep the pH value constant during reaction [405]; therefore, a buffer (e.g., phosphate, borate, and acetate) is added to the reaction mixture. The effect of pH on the rate of polymerization has been investigated for the persulfate system [406]. It was found that in the pH range 13 to 9, the rate of decomposition of persulfate increases; the rate of initiation of the polymerization also increases. Since the pH does not interfere with polymerization, it is possible to choose that medium in which optimal rates of decomposition for respective initiators can be obtained. Although the heat is controlled easily and safety hazards were lower than in bulk, there is one great disadvantage in emulsion polymerization—the costs of purification are high because spray drying is required [303].

The kinetics of the classical emulsion polymerization follows the theory of Smith and Ewart [407]. According to this theory, the degree of polymerization should be a function of the number of polymer lattices and initiator concentration. However, for the polymerization of VC in emulsion, the degree of polymerization and the molecular weight, respectively, are independent of these factors [403,408,409]. Presumably, this behavior can be attributed to the partial solubility of VC in water [410] and the chain transfer to monomer. A comparison of products prepared in bulk, suspension, and emulsion shows that the average molecular weight essentially is controlled by temperature, therefore, at similar polymerization temperatures similar values in molecular weight were obtained [411,412]. Several papers on the determination of molecular weight distribution have been published [334,412]. The results obtained for VC polymerization by free radicals at temperatures of 40 to 80 °C indicate that polydispersity, the quotient M_w/M_n, for the three methods discussed, is 2 or slightly higher. Therefore, similar distribution curves were measured by gel permeation chromatography. Due to the absence of additives for PVC resins produced in bulk, heat stability, clarity, and dynamic stability are better than for those polymers synthesized by emulsion and suspension. The emulsion products have the poorest heat stability because excessive emulsifying agents were used and the degree of branching is higher than in suspension polymers [413]. The desired properties of resulting resins determine which type of polymerization is applied. Stereoregular emulsion polymerization of VC at −30 to 20 °C was studied by Dimov and Slavtcheva [414]. The redox catalyst system consists of hydrogen peroxide/ferrous(II) sulfate/oxalic acid; for stabilizing lattices, sodium alkylsulfonate (alkyl groups range from C12 to C18) was used. Remarkable results were obtained if the molar ratio of ferrous(II) sulfate to hydrogen peroxide is 0.093 : 1 and the temperature is 0 °C. In conforming these conditions, the glass temperature increases up to 110 °C; stereoregularity is also increased. Participation of oxalic acid is intended to reduce Fe(III) to Fe(II), which gives an opportunity for more complete use of the redox components. Furthermore, oxalic acid establishes a pH of 3 and does not decrease thermostability of the formed polymer. Since similar properties for ascorbic acid are known, it is also useful for low-temperature polymerization [415,416].

Another initiation system described in the patent literature consists of ketone/hydrogen peroxide/dithionite/ferrous(II) salts/sulfuric acid [417].

4. Polymerization in Solution

As already mentioned, PVC precipitates in its own monomer. Therefore, it is necessary for solution polymerization to find a solvent for both. Such systems are tetrahydrofurane, acetone, cyclohexyl ketone, alkyl acetates, chlorinated alkyls, and diethyl oxalate [303,305]. Since the reaction is carried out in an organic medium, azo and organic peroxo compounds similar to those listed in Table 2 are suitable for initiation [305]. If desired conversion is achieved, the polymer may precipitate by means of aliphatic hydrocarbons, cyclohexane, benzene, methanol, and water. It has to be mentioned that the solution capacity of the solvent is an important factor. Kinetic studies show that the homogeneous VC polymerization is very complex. In solvents that are not high chain transfer agents (e.g., 1,2-dichloroethane), the molecular weight varies with the monomer concentration in the same manner as the polymerization rate [418]. When the monomer concentration increases, molecular weight also increases. Results in tetrahydrofurane are interpreted as an indication that the solvent functions as a retarder, forming relatively unreactive radicals [354]. In general, the degree of polymerization is lower than in heterogeneous polymerization. The described method is without commercial use because the cost of solvents and their recovery make the process unattractive.

5. Other Methods of Polymerization

PVC produced at higher temperatures includes short- and long-chain branches and lower crystallinity, respectively. These factors influence thermostability negatively. Therefore, systems were investigated for obtaining polymers of enhanced stereoregular structure. One possibility discussed before is to reduce the reaction temperature. Since anionic polymerization of vinyl monomers has been widely used for the preparation of macro-molecules with a high degree of linearity, Wesslen and Wirsen [419] studied the behavior of VC in the presence of an anionic starting system, which seems to be effective in vinyl polymerization. But t-butyllithium is an attractive initiator only in bulk or aliphatic hydrocarbon media. The polymers formed are less branched than a conventional polymer. The structure was determined by ^{1}H- and ^{13}C-NMR spectroscopy. Differential scanning calorimetry evaluation shows that the temperature of decomposition rises up to 300 °C. A reason for the broad molecular weight distribution found by gel permeation chromatography seems to be that termination of living ends takes place. This view is also supported by the observation that the ionic end group vanished during the reaction. Furthermore, only 5 to 10% of initiator (starting concentration is 1 wt% in reference to VC) is effective, the remainder obviously being deactivated through side reactions, presumably metallation. The termination step probably occurs by transfer to monomer and by formation of a complex between the growing end and lithium chloride formed in the metallation reaction during the initiation step [420]. The effect of temperature, initiator concentration, and monomer concentration on the conversion of the PVC samples prepared with t-butyllithium was studied. It was found that the conversion is directly proportional to the initiator and monomer concentration, respectively, whereas the average molecular weight is inversely proportional to the initiator concentration but directly proportional to the monomer concentration. Polymers with M_n on the order of 10 000 to 140 000 were obtained. Under improved conditions Kudrna et al. [421] were able to synthesize PVC with a M_w value up to 500 000. Low-molecular-weight oily polymer ($M_n = 4000$) was obtained if VC is polymerized with t-butylmagnesium chloride in tetrahydrofurane [422]; however, its conversion is low. Ziegler–Natta catalysts are another possibility for producing stereoregular polyvinyl compounds. Ordinary initiation systems were usually inactive for VC polymerization. If modified catalysts were employed (e.g., tetrabutoxytitanium or vanadium oxytrichloride/diethylaluminum chloride or ethylaluminum dichloride), a polymer could be isolated [423,424]. In all cases the conversion depends on the molar ratio of aluminum to titane or vanadium, respectively. Furthermore, when a complexing agent such as tetrahydrofurane or triethylamine was added to the reaction mixture, the yield of polymer increased remarkably. The catalytic systems were deactivated if alcohol was added. For the determination of the reaction mechanism, Yamazaki et al. [423] added tetrachloromethane to a reaction mixture. For a radical mechanism its addition should reduce the molecular weight by transfer and consequently diminish the viscosity of the polymer solution. Since this was not observed and diphenylpicrylhydrazyl did not influence the yield and the molar mass, a coordinated mechanism for the polymerization of VC is proposed in the presence of such catalysts. The infrared spectra of samples produced by a modified Ziegler–Natta system (vanadium oxytrichloride/tri-isobutylaluminium/tetrahydrofurane) and those received from conventional initiators show differences at 638 cm^{-1} (isotactic segments) and at 615 cm^{-1} (syndiotactic structure units) [424]. From these differences the authors concluded that crystallinity is increased in Ziegler–Natta polymers. Differential scanning calorimetry indicates that samples prepared by these systems have decomposition temperatures up to 335 °C compared to 250 to 295 °C for ordinary resins. Other investigators explain these results on the basis of a radical mechanism. Ulbricht et al. [425] reported the VC polymerization with titanium tetrachloride/diethyl ethyloxyaluminum/dioxane in methylcyclohexane. From their kinetic analysis they concluded that the mechanism is similar to a typical radical polymerization. In addition, the molecular weights determined by viscosity, and the tacticity measured by infrared spectroscopy shows no differences to samples produced by free radicals. In contrast, Guzman et al. [426] studied VC polymerization in the presence of tetrabenzyltitanium and claimed that this polymerization is true coordinated.

The polymerization of VC has also been initiated by radiation [427–429]. By using this technique it is possible to polymerize VC at −78 °C. The irradiation was performed in a gamma pool facility. Ellinghorst and Hummel [430,431] studied features of PVC prepared in bulk between −55 and +30 °C. For determining the tacticity and crystallinity by means of ^{1}H-NMR and infrared spectroscopy, results similar to those discussed at low-temperature polymerization by free radicals could be obtained. Polymerization temperature determines regularity. Polymerization at low temperature influences average molecular weight as follows: Lower conversion results in lower molecular weights. For example, at 20 °C and 10% conversion, $M_n = 90\,000$. At the same temperature but 50% conversion, a molecular weight of $M_n = 148\,000$ was found. For the determination of M_n, by viscosity, Ellinghorst and Hummel have used the following relation:

$$[\eta] = 49.8 \times 10^{-5} \overline{M}_w^{0.69} \qquad \text{(THF, 25 °C)}$$

When the polymerization was carried out at about 40 °C, it was found that kinetic results in bulk on the polymerization rate and the molecular weights are identical to the corresponding chemically initiated reaction [430,432]. Therefore, the main role of monomer chain transfer on the degree of polymerization has been confirmed because molecular weight remains constant over a large range of conversion. Plasma polymerization, which is mentioned only briefly [433,434], yielded a tough film with excellent adhesion to the glass. Polymerization in an electrodeless glow discharge is discussed in Yasuda and Lamaze [435].

6. Miscellaneous

Additionally to the procedures described earlier, improvements for thermostabilization is copolymerisation of vinyl chloride with suitable monomers. A great number of monomers were investigated to optimize the properties of resins. But only vinyl acetate, vinylidene chloride, ethylene, propylene, acrylonitrile, acrylic acid esters, and maleic acid esters, respectively, are of interest commercially [305,436,437]. The copolymerization was carried out in emulsion, suspension, and solution in connection with water- or oil-soluble initiators, as mentioned elsewhere. Another possibility for modifying PVC is grafting of VC on suitable polymers [305,438], blends of PVC with butadiene/styrene and butadiene/ methacryl acid esters copolymers [433], and polymer-analogous reactions on the macromolecule [439,440] (e.g., chlorination of PVC).

D. Summary

Poly(vinyl chloride) is one of the most widely used thermoplasts; therefore, intensive research in polymerization, mechanism, and properties were made, particularly by the chemical industry. But due to the complex nature of PVC, a clear understanding of macromolecular structure and stability has still not emerged. Chain branching and lower stereoregularity probably influence heat stability in a negative manner. Preparation at a lower temperature produces PVC with greater crystallinity. Radical-initiated polymerization in suspension, emulsion, and bulk are well-known methods for preparing PVC commercially, whereas in irradiated, anionic, and modified Ziegler–Natta catalysts, polymerization is usually of importance only for studying mechanism and properties. Copolymerization and grafting are further possibilities for new PVC products.

Due to controversial discussion about the environmental problems concerning PVC and its production, intensive research is focused on the recycling [441–445] of PVC and on the reduction of additives [446]. For example the viscosity of PVC pastes can be reduced by blending a fine particle resin (0.2–2 μm) and a resin of 15 μm monodisperse PVC particles in different ratios [447,448]. Thus the use of plasticizers can be decreased.

IV. POLY(VINYL FLUORIDE)

(This Section was prepared by O. Nuyken and B. Voit.)

A. Introduction

Poly(fluoroalkenes) have some interesting properties that explain their commercial importance: (1) large range of application temperature, (2) high resistance to solvent and chemical attack, (3) good mechanical and electrochemical properties, (4) long lifetime, and (5) high resistance to flammability. The most important fluoroalkenes are tetrafluoro-ethylene, hexafluoropropene, 1,1-difluoroethene (vinylidene fluoride), chlorotrifluoro-ethene, and 1-fluoroethene (vinyl fluoride). [449,450]

The monomer vinyl fluoride (VF) was first prepared by Swartz [451] in 1901, but its polymerization was reported as difficult or impossible due to the properties of the monomer (bp $= -2.2\,°C$, $T_c = 54.7\,°C$), the lack of purity of the early samples, and the limited knowledge of the techniques of free-radical polymerization. Starkweather, however, observed some polymerization in a toluene solution saturated with vinyl fluoride at $-35\,°C$ and maintained at 6000 atm and $67\,°C$ for 16 h [452]. But it was only after World War II, when the interest in fluoride-containing compounds increased, that developments in the preparation of poly(vinyl fluoride) of high molecular weights were patented, reported, and commercialized [453–455].

Today PVF is produced from Du Pont de Nemours & Co into films and coatings called Tedlar (earlier called Teslar). A dispersion-grade resin is sold by Diamond Shamrock. But compared with poly(tetrafluoroethylene), PVF is used only as speciality product in small quantities.

There are several good reviews on the polymerization of PVF [456–461]. Kalb et al. [456] were the first to study the effect of various reaction parameters on the polymerization process and the polymer properties in the case of chemical initiation. Cohen and Brasure et al. [457] also give a good overview of the techniques of the polymerization and of the properties of the products. Other reviews have been written by Sadler and Karo [458], Sianesi and Caporiccio [459], Usmanov et al. [460], and Reiher [461].

At atmospheric pressure vinyl fluoride (VF) is a colorless gas with an etheral odor. Its boiling point is $-72.2\,°C$, its critical temperature $T_c = 54.7\,°C$, and its critical pressure $P_c = 51.7$ atm. It is flammable at air between the limits of 2.6 and 21.7% VF by volume and an ignition temperature of $400\,°C$. A toxic effect is found at 20% VF by volume and above [456,457]. The first reported preparation was by Swartz [451] in 1901. VF was found after the reaction between 1,1-difluoro-2-bromoethane and zinc. Later the pyrolysis of 1,1-difluoroethane at $725\,°C$ over a chromium fluoride catalyst [462] or at $400\,°C$ in the presence of oxygen [463] are described.

$$CF_2H-CH_2Br \xrightarrow{\text{cat.}} CHF{=}CH_2 + HBr \qquad (56)$$

$$CF_2H-CH_3 \xrightarrow{\text{cat.}} CHF{=}CH_2 + H_2 \qquad (57)$$

The addition of hydrogen fluoride to acetylene is the most common commercial approach to the preparation of vinyl fluoride monomer. As catalysts, mercury oxide, carbon pellets, or charcoal-containing mercuric chloride or fluoride were mentioned [464].

$$CH{\equiv}CH + HF \xrightarrow{\text{cat.}} CHF{=}CH_2 \qquad (58)$$

As by-products, difluoroethane, vinylidene fluoride, acetylene, and oxygen are found. The commercial monomer is stabilized with terpenes. The purification of the monomer is evidently extremely important. Kalb et al. [456] studied the effect of impurities in the VF on the polymerization. He noted that 500 ppm oxygen in the polymerization vessel slightly inhibited polymerization, but 135 ppm promoted it. The presence of 2% acetylene strongly inhibited polymerization and low-molecular-weight brittle polymer was isolated. 1000 ppm acetylene accelerated the reaction and the polymer obtained was cross-linked and insoluble. Therefore, he used only VF with less than 5 ppm acetylene and less than 20 ppm oxygen for polymerization using the following purification steps [456,457]:

1. Distillation of VF to separate 1,1-difluoroethane, HF, and other impurities
2. Removal of last traces of HF by percolation through soda-lime towers

3. Separation of acetylene by scrubbing with ammonia cuprous chloride
4. Separation of oxygen by another fractional destination between -5 and $-25\,°C$ at 2.7 to 6.8 atm.

It was reported also that it is possible to purify vinyl fluoride by passing it through a special silica gel and freeing it from air by conventional degassing procedures.

It is known that vinyl fluoride resembles ethylene rather than vinyl halides and some other halo-substituted ethylenes, because it is very difficult to polymerize [452,453]. The most common free-radical initiators are generally employable at temperatures above the critical temperature of VF, and this implies that extremely high monomer partial pressures must be adopted during polymerization to assure formation of high polymers. This makes it necessary to apply stainless steel reactors (autoclaves) which are flushed thoroughly with nitrogen to remove oxygen. Furthermore, equipment and methods for handling gases under pressure and careful engineering is absolutely essential.

Generally, the mechanism of the polymerization is described to be a free-radical mechanism and follows theoretical kinetics. In the presence of solvents and additives the chain transfer constants have to be considered. The polymerization methods can be classified in two broad categories [458]:

1. *Chemically initiated processes.* This section covers all processes in which free radicals are generated chemically or thermally from typical free-radical initiators (azobisisobutyronitrile, organic peroxides) as well as Ziegler–Natta catalysts, organometallic systems, and donor–acceptor complexes.
2. *Radiation-initiated processes.* These include the use of ultraviolet radiation, γ-rays, glow discharges, and so on. Photopolymerization has been used to graft onto PVF.

B. Chemically Initiated Processes

1. Bulk Polymerization

True bulk polymerizations of vinyl fluoride are rare since PVF is insoluble at room temperature in all common solvents as well as in its liquefied monomer. Thus, the bulk polymerization is greatly influenced by the precipitation of the polymer from the monomer.

In the bulk polymerization described by Newkirk [453] a low conversion to polymer (4%) has been achieved. He used a steel bomb filled with liquid monomer containing 1% dibenzoyl peroxide. The polymerization temperature was $62\,°C$ and it was maintained for 112 h. The conversion did not increase with a longer reaction time. However, Newkirk reported that the presence of acetone substantially increased the yield of polymer up to 30%, because acetone enhanced the solubility of the initiator in the monomer.

In the case of bulk polymerization, the reaction kinetics seem to depend on the nature of the initiator. Different reaction orders and activation energies have been found for the monoperoxide of tributylborane and diisopropyl peroxydicarbonate as initiators. With the latter initiator a conversion of about 90% at a reaction temperature of $40\,°C$ is reported [465,466].

2. Solution Polymerization

In contrast to the polymer, vinyl fluoride is soluble in a variety of organic solvents (e.g., lower alcohols, ethers, dimethylformamide, and heptane). Therefore, during polymerization in solution one has to expect that the polymer precipitates as it forms. But a 'solvent' often increases the molar mass and the conversion and decreases the polymerization pressure. The molar mass of PVF and thus the effects of solvents on the properties of the resulting polymers is often determined by measuring the intrinsic viscosity of the products in dimethylformamide (DMF) at temperatures above 100 °C. Medium molar mass of PVF shows an intrinsic viscosity of 3 to 4 dL/g (M_n, 100 000 to 250 000 g/mol).

Of some interest is the fact that many solvents act as chain transfer agents and are used to control the molar mass of the product. Also Kalb et al. [456] mentioned the possibility of adding telogens (chain-terminating agents) to reduce the intrinsic viscosity of the product. He studied the effect using methanol or other alcohols. Polymerization with solvents such as toluene [452] and benzene [456] yielded a low-molar-mass polymer, and the conversion did not increase over 17% even when the pressure was raised to 9×10^7 Pa.

Polymerization in t-butyl alcohol and in aqueous methanol is also reported [467,468]. In t-butyl alcohol, azobisisobutyronitrile (AIBN) was used as the initiator. After 16 h at 50 °C, 76% conversion of monomer to polymer was found. When the alcohol is replaced by water, the conversion dropped but the intrinsic viscosity rose to 5.34 dL/g, a value that is considered to indicate a molar mass higher than 500 000 g/mol. In aqueous methanol with diisopropyl peroxydicarbonate as initiator a conversion of 89.7% was yielded after 12 h at 45 °C with a pressure between 2.63×10^7 and 4.41×10^7 Pa. Methanol can be replaced by DMF, but that decreased the molar mass of the polymer to a value of less than 50 000 g/mol.

3. Suspension Polymerization

The major part of commercially produced PVF is manufactured by suspension polymerization. The first commercial polymerization in the late 1940s was carried out in an aqueous suspension using dibenzoyl peroxide at 80 °C under high pressure of 9.12×10^7 Pa for 6.5 h, referred mainly in patents of Du Pont [462,469–471]. In the early 1950s a lower-pressure polymerization technique was described. Azo initiators were employed instead of dibenzoyl peroxide [472]. The reaction with azo initiators was conducted at 25 to 100 °C and between 0.25×10^7 and 1.01×10^7 Pa over 18 to 19 h.

Kalb et al. [456] compared the two polymerization types. First VF was polymerized with dibenzoyl peroxide at 85 °C for 10 h under 3×10^7 Pa. The conversion to polymer was 52%. Substitution of the initiator by AIBN allows the pressure to be reduced to 0.7×10^7 Pa and the temperature to 70 °C. After 19 h a 90% conversion to high-molar-mass polymer was obtained. When oil-soluble initiators were used, polymerization took place on top of the water surface. The polymer found was described as 'webs' [456]. As these webs grew, they collapsed and floated on the water.

Polymer properties, such as thermal stability and wettability, are greatly affected by the solubility of the initiator in the medium and the nature of the initiator fragments that are incorporated into the polymer chain. A list of some useful initiators and their influence on the polymer properties is given in Ref. [456]. AIBN was the most effective initiator. In a patent held by Kureha Chemical Industry Co., an example of the use of a suspending agent is given [473]. They used dipropyl peroxydicarbonate as initiator, methyl cellulose as suspending agent, and sodium phosphate as a buffer. The polymerization procedure is described as a two-stage heating cycle at two different temperatures and pressures to reduce the processing time [458].

Most suspension polymerization procedures are described in patents. Sadler and Karo have given a short overview of some procedures in a table containing initiators, polymerization temperatures, pressures, and additives (e.g., alcohols) which have been used as chain transfer reagent to control the molar mass. An almost complete list of procedures and patents has been given by Usmanov et al. [460].

4. Emulsion Polymerization

The emulsion polymerization of vinyl fluoride was generally found to be more difficult than that of other monomers [457]. The surfactants seem to affect the polymerization as extensive chain transfer agents. However, emulsion polymerization is the only method to obtain PVF as a dispersion and to reduce the polymerization pressure. Kalb et al. [456] used α,α-azobisisobutyroamidine hydrochloride and ammonium persulfate as water-soluble initiators. In contrast to suspension polymerization, polymerization occurred in the water phase and caused a change of color in the solution from initially clear to red, orange, and finally, to opaque. The molar mass was 'high' ($[\eta] > 4$) in the case of azo compound and 'medium' ($[\eta] \sim 3$) when ammonium persulfate was used. The changes in color are thought to be associated with the change in the particle size of the colloidal system. Conventional surfactants seem to be ineffective. Salts of higher perfluorinated carboxylic acids are more suitable emulsifiers and result in high conversion of vinyl fluoride to polymer.

In a procedure described in Ref. [474], sodium orthosilicate was added to the reaction medium to raise the polymer yield to 95%. The initiator was ammonium persulfate with a special emulsifier that is believed to be perfluorinated carboxylated emulsifier from 3M Corp [458]. Other initiators used are redox systems such as potassium persulfate-sodium metabisulfite or ammonium persulfate-sodium sulfite [475]. Different initiators are also of some interest because they affect the particle diameter of the emulsion. Polymer particles formed with a diameter between 0.36 and 18 μm, depending on the initiator system and the polymerization procedures, are reported [458]. In a patent of Dynamit Nobel [476], iodine-containing compounds such as ammonium iodide or isopropyl iodide are used. These compounds give rise to polymers of improved thermal stability and resistance to color deterioration. The procedure did not require the use of an emulsifier. After 150 min the internal pressure of the autoclave dropped from 2×10^6 to 3×10^5 Pa and a conversion to polymer of 86% was found. More recently Uschold [477] describes also an emulsion polymerization of PVF in high yields and having excellent color.

5. Polymerizations Initiated by Organometallic Compounds

Many catalyst and initiator systems have been studied to find a way of reducing the polymerization temperature. Vinyl fluoride, regardless of applied pressure, cannot be liquefied above 54.7 °C, its critical temperature. Therefore, at higher temperatures only relatively low concentrations of monomer are able to react even at high pressure. Consequently, low reaction temperatures are desired. The use of Ziegler–Natta catalysts or boron trialkyls was explored since it was known that these materials are frequently effective at moderate temperatures and low pressures [459]. There are some publications in which the author essentially studied organometallic systems with regard to their efficiency in polymerization of vinyl fluoride at lower temperatures [459,478,479]. Sianesi and Caporiccio [459] described the polymerization of VF in the temperature range 0 to 50 °C with the aid of Ziegler–Natta initiator systems based on vanadyl acetylacetonate and AlR(OR)Cl compounds.

Enhanced reaction rates and higher polymerization degrees were achieved with boron alkyls (and, to a lesser degree, Cd, Zn, and Be alkyls) activated by oxygen. For all these systems, the polymerization is considered to be of the free-radical type, although some properties (crystallinity, melting temperature) of the polymers obtained are shown to be remarkably improved over those described for ordinary high-pressure PVF. This may be attributed to an increased degree of chemical regularity of the chain [459]. The alkyl borons had to be activated by oxygen with the ratio O_2/BR_3 of 0.5 being near the optimum. The polymerization was carried out in bulk or in the presence of different solvents, such as dimethyl sulfoxide (DMSO), ethyl acetate, methylene chloride, and cyclohexane. The solvents did not influence the polymerization noticeably. Maximum conversion achieved was of the order of 70%. Higher conversion are impossible due to catalyst depletion, corresponding to the very fast initiation, and due to the increasing heterogeneity of the system brought about by the insolubility of the polymer. Instead of oxygen, hydrogen peroxides may be used to form the active oxygen trialkylborane complex [480].

Haszeldine et al. [478] studied the effect of vanadium compounds together with zinc and aluminum alkyls in tetrahydrofuran (THF). The most satisfactory modified Ziegler–Natta catalyst system is obtained when vanadium oxytrichloride and triisobutyl aluminum react together in the presence of an excess THF. The initiation mechanism and the structure of the active complexes is studied in a further publication of Haszeldine. The results demonstrate that the active centers in the catalyst system $VOCl_3/Al(iBu)_3/THF$ are formed by interaction of vanadium oxydichloride tetrahydrofuranate and triisobutylaluminum in the presence of THF:

$$VOCl_2 \cdot 2THF + AlR_3 \cdot THF \longleftrightarrow AlR_2Cl \cdot VOClR \cdot THF + THF$$
$$\text{complex I} \tag{59}$$

In addition, the following equilibria may be established in the catalyst system:

$$AlR_2Cl \cdot VOClR \cdot 2THF \longleftrightarrow VOClR \cdot 2THF + AlR_2Cl$$
$$\text{complex II} \tag{60}$$

$$AlR_2Cl \cdot VOClR \cdot 2THF + AlR_3 \longleftrightarrow VOClR \cdot 2THF + Al_2R_5Cl$$
$$\text{complex II} \tag{61}$$

The resulting complex (complex II), octahedral and with a vacant coordination position, is considered to be a potentially active site. The kinetic behavior has been shown to be consistent with a simple Ziegler–Natta kinetic scheme involving a rapid bimolecular decomposition of the active centers [481]. In general, all these reactions were found to have polymerization rates that are first order with regard to the monomer and 0.5 order with regard to the catalyst or initiator and oxygen [458].

Raucher et al. [479] reported a study of polymerization of VF at low temperatures and pressures with silver nitrate and tetraethyl lead as the catalytic system. DMSO was found to have a specific effect on the half-life time of the initiating system, which leads to long reaction times but high conversions. At 0 °C the reaction reached 80% conversion after 48 h. This effect was explained by the complexing power of DMSO on silver ions, which slows down the rate of formation of silver ethyl and permits the polymerization to proceed for longer periods than in other solvents with no complexation property. Similar results were obtained with t-butanol. Tetraethyl lead acts as a powerful chain transfer agent and determines the resulting polymer and the assumed kinetics [479].

C. Radiation-Induced Polymerization

The first photopolymerization of VF was reported by Newkirk in 1946 [453]. By irradiation with UV light of a wavelength of 253.7 nm in a quartz capillary at 27 °C for 2 days, a yield of 36% polymer was obtained. Later [482], the photopolymerization of vinyl fluoride, in the presence of peroxide, was studied in bulk as well as in DMSO or in t-butanol solution (50% by volume of the total liquid monomer composition) within a temperature range of −20 to +30 °C. The monomer mixture was irradiated with a mercury UV lamp. It was found that high reaction rates could be obtained despite the fact that the polymer precipitated at an early stage. In DMSO, the precipitate was a gel that was partially transparent to light, and by continuous supply of monomer to the polymerization, a transparent bulk polymer was formed.

VF gave only traces of polymer when irradiated under the described experimental conditions in absence of any catalyst. By addition of 1 mol% ditertiary butyl peroxide, 50% conversion was observed after 2 h. Tetraethyl lead was also quite effective. However, benzoyl peroxide and AIBN were much less efficient. DMSO and t-butanol accelerated the reaction while other solvents (ethyl acetate, hexane, DMF, acetone, etc.) had a retarding effect because they forced chain transfer reactions. At low conversions, the reaction kinetics were treated as pseudohomogeneous [482]. In an UV-initiated bulk polymerization at 25 °C, the reaction proceeded at a normal rate and conversion over 90% was achieved in 8 h although after 10% conversion, the whole volume of the glass ampoules was filled with a white nontransparent substance. The polymer obtained was a soft, highly porous solid with a density approximately 0.5 g/cm^3 from a monomer with a density of 0.6 g/cm^3. The polymer structure revealed open pores [482]. Varying the reaction conditions allowed to obtain a uniform block of a perfectly transparent, slightly yellowish solid [479].

The photochemically initiated polymerization and copolymerization of VF were also carried out in a continuous flow cylindrical reactor at room temperature and under pressures of up to 30×10^6 Pa in order to achieve higher yields and to control the polymerization conditions [483]. From ^{19}F-NMR analysis it turned out that the homopolymers prepared with the help of UV light were of higher regularity and contained less head-to-head addition than commercial PVF prepared at high temperatures.

The initiation of the polymerization with γ-radiation from a ^{60}Co source was studied by Usmanov et al. [460,484,485]. Polymerization was carried out by irradiation of the monomer, in both liquid and gaseous phase, with the use of γ-rays at 38 °C. The dose rates were 10 rad/s and 0.5 Mrad/s. Impurities such as acetylene greatly inhibited polymerization. Oxygen influenced the kinetics, a factor that confirms a free-radical mechanism. Liquids such as difluoroethane, benzene, and carbon tetrachloride reduced the polymerization rate and caused low-molar-mass polymer. On studying the thermal behavior of PVF it was found that the polymers obtained by γ-ray initiation in bulk were the most crystallized and had the lowest degree of irregularity in the polymer chain. Nearly no branches were found in contrast to the chemically initiated polymers produced in suspension [482,484].

Kobayashi et al. [486] described the use of a radio-frequency electric glow discharge ('plasma polymerization') to induce the polymerization of VF. Westwood [487] studied the polymerization of VF in a capacitively coupled radio-frequency glow discharge operating at 3.14 MHz and in the pressure range 13 to 700 Pa. He observed an increase in the polymer deposition rate with decreasing electrode temperature. The polymers examined had a cross-linked aliphatic structure. They were partially degraded with a remarkable halogen-deficiency, and had considerable C=C unsaturation. On exposure to air all glow discharge polymers take up remarkable amounts of oxygen. Nakamura et al. [488] studied the emulsion polymerization of fluoro olefins at high rates by ionizing radiation at dose rates 1×10^3 to 5×10^5 rad/h in the presence of a water-soluble fluorinated aliphatic carboxylic or a phosphate salt dispersing agent.

D. Technical Production and Properties

The polymerization of vinyl fluoride in an aqueous medium with water- or oil-soluble initiators is the most interesting process for industrial synthesis. A continuous process for commercial polymerization has been described [457] in which a mixture of monomer, water, and free-radical initiator is stirred at 100 °C and 265 atm. A small amount of low-molar-mass olefin (C3 to C7) as a chain regulator is added. These olefins help to avoid too-high-molar-mass poly(vinyl fluoride) being formed. When it is to be used as a coating for laminating structures designed for outdoor exposure, the polymerization is carried out by continuously passing a water-soluble initiator, water, and vinyl fluoride into a first reaction zone maintained at 60 to 90 °C and 1 500 to 9 000 psi; further polymerization subsequently takes place in a second reaction zone maintained at 90 to 140 °C and 4000 to 14 000 psi.

Commercially manufactured PVF resin is generally converted into films and protective coatings. There is no solvent known that dissolves PVF at room temperature; however, some 'latent' solvents do exist which dissolve the resin at higher temperatures. These solvents are used to prepare dispersions of small resin particles, similar to organosol of PVC. They disrupt the polymer order and lower the minimum temperature for film forming. Usually, the melting point of PVF is near the decomposition temperature, which inhibits the production of PVF films by melting the polymer without additives.

1. Effects of Polymerization Variables

Kalb et al. [456] have extensively studied the effects of various polymerization variables on the chemicially initiated (without organometallic initiators) polymerization of vinyl fluoride and on the physical and chemical properties of the resulting polymer. Polymers prepared within the temperature range 35 to 165 °C and under pressures from 10^5 to 10^8 Pa with a variety of initiating systems have been compared. Increasing initiator concentrations lower the initiator efficiency and at the same time polymers with decreasing molar masses are produced. The initiator efficiency increases with temperature to a maximum and then rapidly decreases.

The polymerization temperature has a pronounced effect on the degree of polymerization and on the structure of the products. Polymers prepared at high temperatures are reported to be branched and of lower molar mass. Increasing pressure increases the rate of polymerization and favors the formation of high-molar-mass polymer. If organic solvents, especially alcohols, are used as a polymerization medium instead of water, a high tendency of telomerization of vinyl fluoride accompanied by a drastic reduction of molar mass is observed [457].

Polymers initiated by organometallic or related initiator systems at low temperatures ($T < 50$ °C) and under low pressure ($P < 3 \times 10^6$ Pa) have a higher crystallinity and a higher melting point than that of poly(vinyl fluoride) prepared by high-pressure processes [459,478]. However, there is no evidence of increased stereoregularity in the polymer formed with these special initiator systems [459] since their IR spectra and their x-ray diffraction pattern are very similar [459]. Presumably the higher degree of crystallinity is due to an increase in chain regularity and a lower content of branches. High thermal stability and high melting points are also reported for the products of the UV-light-induced polymerizations carried out at low temperatures in bulk [484,489].

2. Properties of PVF

a. Solubility. Poly(vinylfluoride) is difficult to dissolve in ordinary solvents, also in supercritical CO_2 [490], a property that creates some difficulties in characterization. High-molar-mass PVF dissolves above 100 °C in N-substituted amides, dinitriles, ketones, tetramethylene sulfone, and tetramethylurea. For characterization in solution, DMF, dimethylacetamide, and γ-butyrolactone are mostly used. The solution characteristics in DMF have been studied intensively by Wallach and Kabayama [491], measuring the sedimentation velocity and the characteristics of osmometry and viscosimetry. Chapiro et al. [492] have described the unusual swelling behavior of films of PVF in various solvents, such as acetone, THF, acetonitrile, DMSO, DMF, and ethanol.

b. Molar Mass. The molar mass of PVF can be determined by light-scattering measurements, viscosimetry, osmometry, and sedimentation. For light scattering, solvent pairs are used; the intrinsic viscosity and the osmotic number-average molar mass are measured in DMF at temperatures of 90 to 130 °C. Low-molar-mass PVF, obtained from polymerization media other than water, at high temperature or with the use of telogens, has an intrinsic viscosity of 1 to 2 dL/g [456], which corresponds to a number-average molar mass from 20 000 to 90 000 g/mol. It is better soluble in DMF than the high-viscosity PVF ($[\eta] \sim 3$ to 4 dL/G, $M_n \sim 100\,000$ to 250 000 g/mol), which dissolves to an extent of 10 to 15% at 110 °C in DMF. The commercially important products show molar masses from 50 000 to 200 000 g/mol [461].

c. Thermal Properties. Usually, PVF has a melting point of 197 °C and decomposes above 170 °C in air and above 355 °C in an inert medium. Using special polymerization methods (organometallic initiators, γ-ray initiation) the chain regularity can be increased. For example, melting points of 235 °C and degrees of crystallinity of 50% have been found for polymerization with boron alkyl initiators [459]. Raucher and Levy [489] studied the thermal stability of PVF polymerized by photoinitiation. They found a two-step decomposition for low-molar-mass polymer, one step at 350 °C and the other at 450 °C. The high-molar-mass products showed a one-step decomposition thermogram at temperatures above 420 °C.

The thermal degradation of PVF in a nitrogen or oxygen atmosphere has been investigated in detail [493–495]. The initial stages of degradation start at 420 °C with the evolution of HF. With increasing temperature, benzene is found as a decomposition product. In measurements of dielectric properties, two loss procedures were observed. The high temperature loss at 41 °C was assigned to the glass transition temperature (T_g) [457].

d. Structure. The structure of poly(vinyl fluoride) was studied intensively by IR and Raman spectroscopy [496–498]. These spectroscopic investigations indicate an atactic polymer structure independent of the polymerization method. X-ray studies have shown a high degree of crystallinity of all PVF samples. This was deducted from the similarity in size of the fluorine atom compared to the hydrogen atom. Therefore, the crystal lattice can form without differentiation between the two atoms, and strong hydrogen bonding is found [499]. Dimensions of the unit cell of PVF were found by x-ray methods to be $a = b = 0.493$ nm and $c = 0.253$ nm. Golike [500] proposed a hexagonal unit cell, but from another study it was concluded that the true unit cell was an orthorhombic space group $Cm2m$ [501].

^{19}Fluorine-NMR spectroscopy is a powerful method for studying the microstructure of PVF, since ^{19}F shifts are highly sensitive to structural variables and can be supported by ^{1}H and ^{13}C investigations [502–504]. These studies show that in the polymer chain of commercial PVF, usually one monomer unit in six is added in the reverse direction (i.e., head-to-head). A decrease in the amount of head-to-head addition corresponds to a decrease in the temperature of polymerization. Chain branching is found to vary between 1 branch every 80 to 200 units depending on the reaction conditions [504]. At a reaction temperature of 20 °C, for example, the head-to-head addition was found to be only one unit in eight. Other properties of poly(vinyl fluoride) are good resistance to chemicals and solvents, water repellence, good mechanical properties, and an exceptional transparency to light of a wavelength greater than 230 nm.

3. Copolymers

In general, the conditions employed for homopolymerization apply to copolymerization as well. Comonomers mentioned are, for example, vinylidene fluoride, vinylidene carbonate, vinyl formate, and vinyl chloride [457], and others [505]. Copolymers of vinyl fluoride, vinylidene fluoride and other fluorinated comonomers have been described in a recent patent [506] as being e.g. suitable material as polymer electrolyte membranes for fuel cells, as ion exchange membranes or as electrolytes in lithium batteries. The polymerization of these copolymers has been carried out in an emulsion polymerization process with preemulsified monomers and the use of fluorinated emulsifiers. Recent publications described the radiation-induced grafting on PVF [507–511]. Grafting comonomers were N-vinylpyrrolidone, acrylamides, and acrylic acid. The author used commercial Tedlar (PVF) films for the experiments. Usmanov et al. [484] grafted vinyl fluoride to some natural and synthetic polymers (cellulose, poly(styrene), poly(tetrafluoroethylene), poly(methyl methacrylate), poly(vinyl chloride)) using γ-rays for initiation. The copolymers have good weather, solvent, and abrasion resistance if the fluorine content is sufficiently high.

4. Application

Due to its properties, poly(vinyl fluoride) is used in protective films and coatings to prevent abrasion, corrosion and moisture penetration and to increase the outdoor life of the substrate. It can be bonded to galvanized steel, plywood, timber hardboard, insulation board, aluminum, reinforced plastics, decorative laminates, roofing, pipe insulation, and paper. The relatively high price is compensated for special applications by the superior abrasion resistance and the exceptional toughness of poly(vinyl fluoride) as coating. Usually, the PVF film has much greater flexibility and formability than the substrates to which it is bonded. PVF is also considered as a protective coating for solar stills due to its high transparency, also to UV light, and its good weather resistance [496,512].

Due to the high chemical stability, more recently application as membrane material has been discussed for PVF and copolymers containing VF units. The use as ion exchange membranes as well as porous membranes for separation techniques and application in medicine are reported [510,511,513–516].

V. POLY(VINYLIDENE FLUORIDE)

(This section was prepared by O. Nuyken and B. Voit.)

A. Introduction

Vinylidene fluoride (VF$_2$) as well as vinyl fluoride (VF) resemble more ethylene than other vinyl monomers in regard to the properties and can also be polymerized in a manner analogous to that of ethylene. The first successful polymerization of vinylidene fluoride was reported by Du Pont in 1948 [517]. Poly(vinylidene fluoride) (PVF$_2$ or PVDF) is commercially available from various companies (Kureha Chemical Co., Pennalt Chemicals Corp., Solvay & Co., Süddeutsche Kalkstickstoff-Werke AG now Degussa, Diamond Shamrock, Du Pont) since 1965 under the trade names KF, Kynar, Foraflon, Vidar, and Dalvor, respectively. VF$_2$ can be polymerized by free-radical initiators to give high-molar-mass, crystalline polymers.

Compared with commodity polymers as well as with poly(tetrafluoroethylene), the commercial importance of PVF$_2$ is low. However, the properties that caused an increased interest in this polymer were its extraordinary piezoelectricity and pyroelectricity, first reported by Kawai [518], Bergmann et al. [519], and Nakamura and Wada [520]. This behavior combined with the mechanical and thermal properties render PVF$_2$ an unique transducer. Since the electric properties are mainly due to the chemical microstructure of the polymer, crystallization in relation to the polymerization processes has been studied extensively. The results have been reviewed by Lovinger [521]. A general review on the preparation and the properties of PVF$_2$ is given in the *Encyclopedia of Polymer Science and Technology* [541], in *Polymeric Materials Encyclopedia* [522,523], and in *Encyclopedia of Chemical Technology* [524].

The monomer vinylidene fluoride, $CH_2=CF_2$, is like vinyl fluoride, a colorless, nearly odorless gas. It boils at $-82\,°C$ with a critical temperature of $30.1\,°C$ and a critical pressure of $43.75\,atm$. It is commonly synthesized from acetylene (62) or vinylidene chloride via 1-chloro-1,1-difluoroethane [525,526], (63) and (64).

$$CH\equiv CH + 2HF \longrightarrow CH_3-CHF_2 + Cl_2 \longrightarrow CH_3-CCIF_2 + HCl \qquad (62)$$

$$CH_2= CCl_2 + 2HF \longrightarrow CH_3-CClF_2 + HCl \qquad (63)$$

$$CH_3-CCIF_2 \xrightarrow{800°C} CH_2=CF_2 + HCl \qquad (64)$$

Recently, also the synthesis of VF$_2$ from 1,1-difluorethane on contact with Cr_2O_3/Al_2O_3 based catalysts in the presence of oxygen has been reported [527]. Vinylidene fluoride is usually stored and shipped without polymerization inhibitors. If desired, terpenes or quinones can be added to inhibit polymerization. Before use the monomer has to be distilled and degassed several times to remove impurities [528]. Since vinylidene fluoride is a gas under normal conditions, most of the polymerization processes are carried out under pressure. The advantage of poly(vinylidene fluoride) is its good solubility in solvents such as dimethylacetamide (DMA), dimethylformamide (DMF), or dimethyl sulfoxide (DMSO). This fact allows its commercial processing without any problems. Industrially, polymerization is usually accomplished in suspension or emulsion as described in several patents [529–533]. Vinylidene fluoride can also be polymerized with coordination catalysts of the Ziegler–Natta type to achieve products with low content of chain defects [534,535]. While suspension and emulsion polymerizations are the only important commercial processes for the manufacture of PVF$_2$, radiation-induced polymerization in solution or in the gaseous state have been used extensively on laboratory scale [528,536,537].

B. Chemically Initiated Polymerization

1. Polymerization in Emulsion and Suspension

Water is commonly used as the polymerization medium, with peroxy compounds serving as the initiators for both suspension and emulsion polymerizations. In most cases reaction temperature is between 10 and 150 °C and pressure between 10 and 300 atm [541]. In the polymerization of vinylidene fluoride, proper selection of the radical initiator, the emulsifier, and the reaction medium is important.

For the emulsion polymerization a chemically stable, preferred a fluorinated surfactant must be employed. The use of sodium and ammonium perfluorooctanate [530,532,538], and perfluorochloro carboxylic acids or their water-soluble salts [529] have been described. Perfluoropolyethers can be used to prepare stable microemulsion [539]. The most common initiator is di-t-butyl peroxide [522,529,530,540,541]. Typical reaction conditions are $T = 122$ °C, $P_0 = 54.4$ atm, and a reaction duration of 18.5 h [529]. Under these conditions a conversion of 83% was observed and polymer with high molar mass and good molding characteristic was formed. Other initiators that have been used for the polymerization of vinylidene fluoride are diisopropylperoxydicarbonate [532], disuccinic acid monoperoxide [531], and other dialkyl percarbonates [533]. PVF$_2$ with good storage stability was prepared in emulsion using a bis-peroxy carbonate, added as 50% solution to methanol in water and methylcellulose (as stabilizer) at 25 °C [542]. One can carry out an surfactant free emulsion polymerization of VF$_2$ when suitable comonomers are used which result in a self-emulsifying system. This has been described for the copolymerization of chlorotrifuoroethylene and vinylidene fluoride using redoc initiating systems [543].

It is also possible to carry out the polymerization at 0 °C under the vapor pressure of the monomer in water with methylcellulose as surfactant [533]. High monomer conversion of 95 to 98% was observed. The intrinsic viscosity of 2.26 dL/g (in DMF) corresponds to a molar mass of about 10^6 g/mol. By modifying the reaction conditions (temperature and pressure) the molar mass could be controlled and η varied between 0.5 and 5 dL/g. Bis(t-butyl) hyponitrite and bis(α,α-dimethylbenzyl) hyponitrite can also act as initiators; however, the conversion to polymer is low [538]. Often higher melting point, and higher crystallinity was observed than for suspension-polymerized PVF$_2$ [544].

The procedures described for the suspension polymerization are similar to those described for the emulsion polymerization; indeed, these two types of polymerization are often not clearly distinguishable. Aqueous recipes with and without colloidal stabilizers are used [522,533,541]. N-Heptane is mentioned as solvent for the initiator [545]. Polymerization at high temperatures causes an increase of head-to-head or tail-to-tail units within the polymer chains. Görlitz et al. [499] found a linear increase of the polymer chain defects with the polymerization temperature. He studied PVF$_2$ samples produced in the temperature range from 50 to 130 °C in emulsion and suspension together with commercially available samples. The molar masses of the polymers range from 60 000 to 150 000 g/mol (M_n, GPC measurements in DMF). PVF$_2$ polymerized at 25 °C showed 4.1% head-to-head connection in the backbone; at 130 °C 5.4% chain defects were found by ^{19}F-NMR spectroscopy. Further analysis of the spectra showed that 3.5 to 6% of branches are realized in all the polymer samples. Furthermore, the chain defects influence the ability of the polymer to crystallize. Therefore, the melting points of the polymers decrease with the increase of the polymerization temperature (respectively, the increase of the chain defects).

2. Polymerization in Solution

Poly(vinylidene fluoride) is better soluble in high-polar solvents than poly(vinyl fluoride). Highly fluorinated or chlorofluorinated hydrocarbons ($CF_2Cl–CF_2Cl$, CCl_2F_2, CCl_3F, $CClF_2H$, $CCl_2F–CCl_2F$) [546,547] have been used as solvents for the solution polymerization of VF_2. Initiators such as di-t-butyl peroxide, dicumyl peroxide, t-butyl perbenzoate, benzoyl peroxide [546], and fluorinated initiators such as bis(perfluoropropenyl) peroxide and bis((ω-hydrooctafluorovaleryl) peroxide [547] have been applied successfully. The polymerizations carried out at about 70 °C and at pressures below 35 atm yielded polymer in lower conversion and with lower molar mass than normally found for emulsion polymerization [546,547]. PVF_2 in the range of 3000 to 5000 g/mol has been obtained in different solvents initiated by bis(1-t-butylcyclohexyl)peroxidicarbonate at 55 °C [548]. The ^{19}F NMR spectra indicate that the structure of PVF_2 prepared in solution consist predominantly of head-to-head sequences [549], but products with very low head-to-head sequences (2.8%) could also be prepared under special conditions [550].

3. Polymerization Initiated with Organometallic Compounds

The defects in the polymer chain and the resulting crystallization behavior of poly(vinylidene fluoride) is the topic of many publications [522,541] due to the special electric properties of the polymer which are influenced by its crystallinity. The aim is to produce defect-free PVF_2. Polymerizations initiated by organometallic initiator systems have been carried out at low temperatures to achieve low contents of head-to-head and tail-to-tail units in the polymer. Natta et al. [535] used triisobutyl borane activated with molecular oxygen and polymerized VF_2 at 25 °C in bulk. High-molar-mass polymers were observed showing a high degree of crystallinity and melting temperatures some 10 °C higher than those usually reported for PVF_2

In a study by Liepins et al. [534] a modified Ziegler–Natta catalytic system is described. Although a number of compounds were tested, only two of the catalytic systems produced an appreciable yield of polymer: $Et_3Al/TiCl_4/CH_2Cl_2$ and $Et_2AlOEt/Vacac/Et_2AlCl/THF$. The molar ratios of the different components in these systems were varied over a wide range. The best result was a 3% yield of PVF_2 with vanadyl acetylacetonate (V_{acac}) and 8% yield of a low molar mass product with the triethylaluminium/titanium tetrachloride/methylene chloride system. The same authors [534] have described a free-radical polymerization initiated by triisobutyl borane and oxygen. After 3 days at room temperature a polymer yield of 33% was obtained. Although the intrinsic viscosity of 0.234 dL/g in DMF was higher than in the case of the Ziegler–Natta polymerization, the molar mass was still low

For both types of polymerization the structure of the resulting polymers was studied by x-ray diffraction analysis, ^{19}F-NMR, and infrared analysis [534]. In the case of the trialkylborane systems the chain defects were given to be 3.2%, which is higher than the imperfections found for Ziegler–Natta polymers (2.7%). However, remarkable differences in the ^{19}F-NMR of the two polymers were observed. That of the polymer initiated with boronalkyl/oxygen was similar to the spectrum of conventional PVF_2. However, the spectrum of the polymer initiated by the Ziegler–Natta system contained five additional peaks. These lines were assigned to a second type of imperfection (the first type being head-to-head followed by tail-to-tail), which consists of a head-to-head followed by a tail-to-head sequence [534]. Despite the different structures a similar free-radical mechanism probably operates with both initiating systems.

Boron alkyls and Ziegler–Natta catalyst systems have also been mentioned in the patent literature [551,552]. The initiating systems have been modified by the addition of elemental titanium, solvents (lactones, amides), plasticizers (esters, carboxylic or phosphoric acids), and lead and cadmium compounds.

C. Radiation-Initiated Polymerization

The radiation-initiated polymerization of vinylidene fluoride is used on a laboratory scale only. The effect of polymerization conditions on the chain defects, content, and changes in the crystalline phases have been studied [528,536,537,544]. Doll and Lando [544] used a ^{60}Co source with an average dose rate of 0.33 Mrad/h. The polymerization was carried out between 0 and 400 °C at a pressure equal to the vapor pressure of the solvent–vinylidene fluoride mixture. Esters and ketones (acetone, methyl ethyl ketone, ethyl acetate, acetophenone), DMF, DMSO, and γ-butyrolactone were used as solvents. All these solvents are good chain-transfer agents for vinylidene fluoride. The molding characteristics of the resulting polymers were very poor and the intrinsic viscosity of the sample polymerized in acetone solution was low (0.183 dL/g) compared to that of a suspension-polymerized polymer (1.68 dL/g) [544].

No remarkable change was observed in the head-to-head content when this type of polymerization was utilized [544]. However, the polymers were highly crystalline (∼ 70%) and melting points 10 to 15 °C higher (170 to 175 °C) than those of the suspension-polymerized PVF_2 (160 °C) were detected. The same authors observed high-molar-mass polymer (intrinsic viscosity 1.76 to 2.1 dL/g) if the strongly hydrogen-bonding, fluorinated solvents, hexafluoroacetone sesquihydrate (HFAS), and trifluoroacetic acid (TFAA) were used for polymerization [544]. Polymers synthesized in these solvents had very good molding characteristics. Although lower radiation doses were used compared to polymerization in acetone or esters, higher conversions were observed. Solid-state polymerizations using γ-radiation as an initiator at liquid nitrogen temperatures resulted in a low yield of oily polymeric materials [544].

Weinhold et al. [528] describe the bulk polymerization at 0 °C initiated with γ-rays. The viscosity of the products was measured in N,N-dimethylacetamide (DMA) at 30 °C. The intrinsic viscosity ranged from 5 to 8 dL/g, indicating that the polymers had extremely high molar masses. The authors succeeded in preparing well-oriented phase III (or γ phase) PVF_2 by heating these polymer samples cast as films from DMA solution from 90 to 110 °C under vacuum.

The various crystalline structures of PVF_2 obtained by chemical and radiation-initiated polymerization are described by Gal'perin et al. [536,537]. PVF_2 samples of different molar masses could be prepared (0.3 to 10×10^{-5} g/mol, determined from viscosimetry in DMF). Radiation polymerization in the gaseous phase resulted in polymers with the highest molar masses. It was also shown that the conformation of PVF_2 chains was independent of the method of initiation but was influenced by the polarity of the medium in which polymerization took place. Radiation polymerization in polar solvents promoted formation of the β phase, while nonpolar solvents (or gaseous polymerization) yielded the α phase [521,537].

Other polymerization reactions have also been reported for VF_2, especially the glow-discharge polymerization in plasma [553–555]. The polymerization of VF_2 in a radio-frequency glow discharge with particular reference to the potential application as a gas-phase surface coating process has been studied by Westwood. [553,554] Using capacitive coupling of power into the monomer gas, rates of polymer deposition have been measured as a function of pressure, current density, and electrode surface temperature. An increase in deposition rate with decreasing electrode temperature has been observed. It was suggested that positive ions from the gas phase react on the cathode surface with absorbed monomer molecules and initiate the polymerization. The resulting polymers were characterized by IR spectroscopy and chemical analysis. It was concluded that the samples have a cross-linked structure. Although related to the conventional PVF_2 and containing some of the same structural features, glow-discharge polymers are partially degraded materials. The samples are marked by halogen deficiency and have considerable C=C unsaturation. They contained large amounts of oxygen in a variety of functional groups after exposure to air. Clark and Shuttlworth [555] prepared PVF_2 by radio-frequency (RF) and microwave glow discharges. The study showed that there was little difference in the structure and composition of the polymer films produced by these two methods. Radio-frequency plasma polymerization of VF can be used to change surface properties of non-fluorinated polymers like chlorobutyl rubber [556] or poly(ethylene) [557] and in general to obtain water repellent surfaces [558]. This surface modification leads to decreased friction and enhanced wettability.

D. Copolymers and Properties of Poly(vinylidene fluoride)

1. Copolymers

In the addition to homo-PVF_2, a large number of copolymers have also been synthesized which allow to optimize the mechanical properties of fluoropolymers. Most common are copolymers with vinyl fluoride, trifluoroethylene, tetrafluoroethylene, hexafluoropropylene, hexafluoroisobutylene, chlorotrifluoroethylene, and pentafluoro-propene [521,535, 559–562]. Copolymerization with nonfluorinated monomers is possible [563] in principle but has not yet found commercial use. Fluorocarbon monomers that can help to retain or enhance the desirable thermal, chemical, and mechanical properties of the vinylidene structure are more interesting comonomers. Copolymerization with hexafluoropropylene, pentafluoropropylene, and chlorotrifluoroethylene results in elastomeric copolymers [564]. The polymerization conditions are similar to those of homopoly(vinylidene fluoride) [564]. The copolymers have been well characterized by x-ray analysis [535], DSC measurements [565], and NMR spectroscopy [565,566].

Recently, also triblockcopolymers of polystyrene-b-PVF_2-b-polystyrene have been synthesized using atom transfer radical polymerization. A linear increase of molar mass with conersion and low polydispersities of the products could be achieved. These blockcopolymers can be used as compatibilizers. Blockcopolymer formation is also reported when, e.g., tetrafluorethylene and perfluoro(methyl vinyl ether) are copolymerized using a boron or iodine organic compound and then TFE and VF_2 are added [567].

Grafting of N-vinylpyrrolidone onto copolymers of VF and VF_2 by accelerated electrons has been investigated [568] to prepare membranes for separation processes. Conducting polymer films could be obtained by grafting pyrrole derivatives onto PVF_2 [569]. Sulfonated PVF_2, obtained by irradiation induced grafting with styrene and subsequent sulfonation, is of interest as proton exchange membranes in fuel cells since it exhibits similar behavior than *Nafion* [570]. Grafting with phosphonium compounds and also with N-isopropylacrylamide resulted in an enhanced performance of PVF_2 as porous membrane in purification processes [571,572]. Also MMA, acrylic acid, and other comonomers like styrene have been grafted onto PVF_2 mainly to change surface and membrane properties [573–575].

2. Properties

a. Solution Characteristics. Good solvents for poly(vinylidene fluoride) are N,N-dimethylacetamide (DMA), dimethylformamide (DMF), hexamethyl phosphoramide (HMPA), and dimethyl sulfoxide (DMSO). Acetone is strongly absorbed at room temperature and causes swelling and softening of the polymer. Generally, DMF or DMA are used to determine the solution characteristics. Welch [576] and Ali and Raina [577] found for the dependence of intrinsic viscosity on weight-average molar mass in DMA:

$$[\eta] = 2.02 \times 10^{-4} M_w^{0.0675} \text{at } 25\,^\circ C \quad \text{and} \quad [\eta] = 0.178 \times 10^{-4} M_w^{0.74} \text{at } 125\,^\circ C$$

based on light-scattering data. The commercial products show intrinsic viscosities from 0.86 to 1.63 dL/g at 125 °C corresponding to molar masses from 2.1 to 5.4×10^6 g/mol [521]. Gel permeation chromatography (GPC) is not suitable for these polymers, since PVF_2 of molar masses $>2.5 \times 10^3$ g/mol does not form molecularly dispersed solutions independent on the solvent. Chapiro et al. [578] studied the swelling behavior of PVF_2 films in cyclohexane, γ-butyrolactone, and DMSO, as well as in DMA, N-methylpyrrolidone, and propylene carbonate.

b. Bulk Characteristics. PVF_2 is a highly crystalline polymer with very good mechanical properties. Oriented poly(vinylidene fluoride) films and fibers exhibit exceptional mechanical strength. In addition, PVF_2 maintains superior resistance to elastic deformation under load, as well as exceptional ability to withstand repeated flexure or fatigue. Its resistance to abrasion according to a variety of tests is excellent [521,541].

The formation of some structure units in head-to-head or tail-to-tail position is an unavoidable phenomenon of the radical polymerization of vinyl monomers. In the case of poly(vinylidene fluoride) these chain defects are of particular importance because they affect the crystallization and thus the properties of the polymer [521]. In general, the percentage of monomer inversion in PVF_2 appears to be a function of the temperature of polymerization: The number of H–H and T–T units increase from 3.5% at 20 °C to 6.0% at 140 °C. A polymer with a very low content of chain defects has been prepared by Butler et al. [579]. The authors carried out the polymerization in bulk at 0 °C using trichloroacetyl peroxide as the initiator and obtained high-molar-mass PVF_2 with only 2.85% of reversed monomer units and low content of branches.

The extent of reversed monomeric units is routinely determined by high-resolution NMR methods, using predominately ^{19}F spectra and, less frequently, ^{13}C spectra [580–582]. ^{19}F-NMR can also be used to measure chain branching in head to-tail sequences. Plasma polymerized samples yield more complex spectra as a result of extensive cross-linking, branching, and unsaturation [554].

The degrees of crystallinity in PVF_2 are dependent on the thermal treatment. Values of 20 to 60%, 50% [521], and 68% [541] have been described. The polymorphism of PVF_2 has been the subject of numerous investigations [521,551,581,583–585]. Lovinger [521] has summarized the newest results on this topic. Four crystalline modifications are known: α, β, γ, δ. The α phase is the most common polymorph of PVF_2 and is normally obtained by crystallization from the melt at moderate or fast rates of cooling. The generally employed piezoelectric modification is well known as β form. It is routinely obtained by stretching of melt-crystallized films. The third phase was found in films crystallized from solution in DMA [528] and during solution crystallization at high pressure [559]. The δ phase has been found by poling the α form at high electric fields. Thus a polar version of the antipolar $αPVF_2$ could be obtained [521].

Different melting points for each crystalline phase have been cited [521]. They vary from 158 to 197 °C [541] (171 °C [586]). The glass transition temperature of the commercial poly(vinylidene fluoride) (α form) is −40 °C [541,586]. Poly(vinylidene fluoride) is a good heat insulator: It can dissipate intense thermal radiation through decomposition without flaming at temperatures higher than 340 °C. The decomposition reactions have been studied by Nguyen [583]. The predominant reaction appears to be the elimination of hydrogen fluoride; the second major product is trifluorobenzene produced by chain scission and cyclization. The latter becomes more predominant at temperatures above 530 °C. Unlike most vinylidene polymers, poly(vinylidene fluoride) undergoes cross-linking when exposed to ionizing radiation [585].

Its ferroelectric properties are one of the reasons for the interest in poly(vinylidene fluoride) and its copolymers. A material is piezoelectric if mechanical stress produces a change in electrical polarization (or vice versa), and pyroelectric if a change in polarization is induced by a change in temperature. PVF_2 is a ferroelectric material (i.e., it forms polar crystals that can change direction of polarization in an electric field). This property and its relation to the crystalline phases of PVF_2 have been studied extensively by Lovinger [521,587] and others [522,588]. It has been shown that temperature may affect only the rate [521] of polarization but not the ultimate value. Polarization has also found to depend on the relative amount of the α and β phases, increasing linearly with the β-form content [521]. Hysteresis phenomena in PVF_2 have been studied comprehensively by Furukawa et al. [589–591].

PVF_2 is among the very few crystalline polymers that are thermodynamically compatible with other polymers, such as poly(methyl methacrylate), poly(ethyl methacrylate), poly(methacrylate), and poly(vinyl methyl ketone) [521,592]. This topic is summarized in a review of Furukawa and Johnson [593]. The system PVF_2/PVF has also been studied [594].

3. Applications

PVF_2 has found applications as long-life finishes for exterior metal siding, in the electrical and electronic industries, and in a variety of chemical process equipment [541]. PVF_2 is, like PVF, very resistant to chemicals, to weathering, and to UV light. Its mechanical properties are good, the abrasion resistance is high, and thermal degradation does not occur below 315 °C [541]. It combines easy moldability, toughness, and flexibility. For coatings it is applied mainly as dispersion or powder coating. Furthermore, it finds also application in the CPI (chemical process industry) market, utilized as pipes, tubes, valves, fittings, filters and membranes due to processability and high chemical resistance [523].

The ferroelectricity, combined with excellent mechanical properties and the ease of fabrication, have rendered PVF_2 a versatile material for piezoelectric and pyroelectric applications (transducer, stems, ultrasonics, loudspeakers, microphones, finger-press switches, ultrared detectors). Furthermore, PVF_2 can be found in semiconductor applications and in the electrical-electronic market (plenum cables, aircraft wiring, computer back panel a.o.) [523]. Application in sensors and optical storage devices is also discussed [541,595]. PVF_2 electrets can be applied in lithium batteries [596].

As a special aspect, PFV_2 forms miscible blends with a large number of polymers, e.g., PMMA, PMA, poly(methyl ethyl ketone), poly(N-vinyl-2-pyrrolidone), poly(ε-caprolactone), and poly(vinyl acetate). Blends of $PMMA/PVF_2$ (under the trade name KYNAR 500® (Elf Atochem) are used for aluminum or steel coatings used for the facade of high-rise buildings a.s.o. In combination with polyolefines, PVF_2 is used in ink and toner formulations [523].

PVF_2 can be grafted with various monomers induced by irradiation, ozonolysis, or base induced dehydrofluorination [597]. This is used intensively to optimize the surface and membrane properties of PVF_2 [568–575]. Due to the high chemical resistance, PVF_2 is an interesting membrane material for separation and purification techniques, but even in fuel cell application.

E. Polymerization of Other Fluoroalkenes

While partially fluorinated alkenes such as vinyl fluoride and vinylidene fluoride polymerize with the same facility as tetrafluoroethylene, the latter is unique in the class of the perfluoroalkenes with the respect to the ease of polymerization as will be described later. Perfluoroalkenes such as hexafluoropropylene (HFP) and hexafluorobutadiene polymerize only with great difficulty as the result of the steric inhibition in the propagation step [598]. HFP can be converted to high-molar-mass polymer only at pressures above 1000 atm. The polymerizations are carried out most conveniently in a perfluorinated solvent using perfluorinated free-radical initiators [599]. For the polymerization of hexafluorobutadiene, similar conditions are reported [600]. Due to these very drastic polymerization conditions, these oligomers/polymers are not yet commercially applied.

Apart from the fluoro monomers vinyl fluoride (VF), vinylidene fluoride (VF_2), and tetrafluoroethylene (TFE), only chlorofluoroethylene has found commercial use as homopolymer. It is applied as thermoplastic resin based on its vapor-barrier properties, superior thermal stability ($T_{dec} > 350\,°C$), and resistance to strong oxidizing agents [601]. Chlorofluoroethylene is homo- and copolymerized by free-radical-initiated polymerization in bulk [602], suspension, or aqueous emulsion using organic and water-soluble initiators [603,604] or ionizing radiation [605], and in solution [606]. For bulk polymerization, trichloroacetyl peroxide [607] and other fluorochloro peroxides [608,609] have been used as initiators. Redox initiator systems are described for the aqueous suspension polymerization [603,604]. The emulsion polymerization needs fluorocarbon and chloro-fluorocarbon emulsifiers [610].

Oils and waxes with excellent chemical inertness are obtained from low-molar-mass poly(chlorotrifluoroethylene). They are prepared by polymerization of the monomer in the presence of suitable chain transfer agents [601]. Furthermore, chlorotrifluoroethylene has been copolymerized with vinylidene fluoride to elastomeric polymers by suspension and emulsion polymerization [601,611].

Most of the other fluorine-containing monomers such as trifluoroethylene, hexafluoropropylene, and pentafluoropropylene are used only for copolymerization with vinyl fluoride, vinylidene fluoride, and tetrafluoroethylene [506,521,535,559–562]. Those copolymers, after a convenient vulcanization procedure using peroxides, diisocyanates, or amines, can be applied as fluorocarbon elastomers [564]. Due to the fluorine content, they have high chemical resistance and often a broad temperature range for application [612]. Polymers of interest are the vinylidenefluoride/hexafluoropropylene copolymer and the vinylidene fluoride/hexafluoro-propylene-tetrafluoroethylene terpolymer. Also, a copoly-mer of vinylidene fluoride and 1-hydropentafluoro-propylene [612] or perfluoro(methyl vinyl ether) [613,614] is commercially available as elastomer. The co- or terpolymerization can be carried out in aqueous phase with an inorganic water-soluble persulfate as the initiator [611] and under conditions similar to those used for TFE and VF_2 alone [598].

Other fluorocarbon polymers result from the copolymerization of vinylidene fluoride with fluorine-containing acrylates [e.g., $F_2C=CH-COOR$, $F_2C=CF-COOCH_3$, $F_2C=C(CF_3)-COF$] and are mentioned mainly in the patent literature [615]. An interesting thermoplastic fluoropolymer from Allied Chemical Corp., called CM-1, is reported [616]. This copolymer from hexafluoroisobutylene and vinylidene fluoride can be radically polymerized in suspension or emulsion at a pressure of 10 to 20 bar and a temperature of 20 °C. The properties of the copolymer are comparable or even better than those of poly (tetrafluoroethylene).

VI. POLY(TETRAFLUOROETHENE) (PTFE)

(This section was prepared by O. Nuyken and R. Jordan.)

A. Introduction

Poly(tetrafluoroethene) (PTFE) was serendipitously discovered by Roy Plunkett at DuPont's Jackson Laboratory [617] in 1938, while attempting to prepare fluorocarbon derivatives. He described the formation of an inert, white, opaque solid by 'storing' (polymerizing) tetrafluoroethene (TFE) in a cylinder. Soon the unique properties of PTFE were discovered and various methods of the polymerization of TFE followed. PTFE is mainly manufactured by free-radical polymerization methods of TFE in aqueous media. Suitable initiators [618] include ammonium, sodium, and potassium persulfate, hydrogen peroxide, oxygen, and some organic peroxy compounds. Redox systems involving persulfate with either ferrous ion or bisulfite or the use of bisulfite with ferric ion are useful combinations [619]. Photo-initiated, radiation-initiated, glow-discharge and plasma polymerization are also applied.

PTFE with an estimated ultrahigh molar mass of $>10^7$ g/mol is in many respects a unique polymer due to its insolubility in any common solvent, chemical inertness even under extreme reaction conditions, high melting point, high melt viscosity of $>10^{11}$ Pa s, heat resistance, ultrahigh low surface free energy [620], dielectric properties [621], and a low coefficient of friction [622,623] in a wide temperature range [624]. Because of these properties conventional PTFE cannot be processed by common techniques such as melt-processing. To overcome this problem, several solutions were applied since the discovery of PTFE: (a) an extensive processing technology was developed [624–626], (b) main chain scission of PTFE by means of irradiation, resulting in molar masses around 10^4–10^5 g/mol, (c) the polymerization of TFA with various fluorinated comonomers to yield copolymers of reduced molar mass and viscosity

Indeed TFE can be copolymerized with numerous other monomers under conditions similar to those used for its homopolymerization. It was copolymerized with, e.g., hexafluoropropene (FEP) [627], perfluorinated ethers [628], isobutene [629], ethene [630] and propene [631]. In some cases it is used as a termonomer [632]. It is also used to prepare low molecular weight polyfluorocarbons [633] and carbonyl fluoride [634] as well as to form PTFE coatings in situ on metal surfaces [635].

However, recently Tervoort et al. reported on melt-processable PTFE blends of various commercially available fluoropolymers [636,637]. They identified a window of viscosities that permits standard melt-processing of PTFE.

Electrical applications such as hookup and hookup-type wire and coaxial cables, consume half of the PTFE produced. Thin and ultrathin coatings of PTFE are mainly used in advanced microelectronic applications, although novel applications for engineered interfaces are currently under discussion [638–641]. Mechanical and chemical applications, e.g., seals and piston rings, basic shapes, antistick bearings, mechanical tapes, coated glass fabrics and soft or hard packing. The PTFE micropowder (waxes) have significantly lower molecular masses than normal PTFE. They are commonly used as additives in plastics, inks and lubricants [624,642].

The first reliable report of the preparation of tetrafluoroethene (TFE) was given in 1933 by Ruff and Bretschneider [643], who decomposed tetrafluoromethane in an electric arc. Other syntheses are based on the dechlorination of *syn*-chlorodifluoromethane [644], the pyrolysis of chlorodifluoromethane [645], and the decarboxylation of sodium perfluoroproprionate [646]. Since then, a number of synthetic routes have been developed [644–646]. On a laboratory scale, depolymerization of PTFE by heating at about 600 °C is probably the preferred method to obtain a small amount of pure monomer (97%) [647], along with the highly toxic perfluoroisobutene as well as octafluorocyclobutane as a side product formed by the thermal $(2\pi + 2\pi)$ cyclodimerization of TFE [648,649]. The most common commercial approach for the preparation of TFE is the pyrolysis of chlorodifluoromethane [645,650]. The noncatalytic gas-phase reaction is carried out in a flow reactor at atmospheric pressure, yields over 95% TFE at 590 to 900 °C. The synthesis of TFE involves the following steps:

$$CaF_2 + H_2SO_4 \longrightarrow CaSO_4 + 2HF \tag{65}$$

$$CH_4 + 3Cl_2 \longrightarrow CHCl_3 + 3HCl \tag{66}$$

$$CHCl_3 + 2HF \longrightarrow CHClF_2 + 2HCl \tag{67}$$

$$2CHClF_2 \longrightarrow CF_2{=}CF_2 + 2HCl \tag{68}$$

Hydrogen fluoride is manufactured by the first step; chloroform and other chloromethanes are formed in the following steps. Then chloroform is partially fluorinated with hydrogen fluoride to chlorodifluoromethane (using antimony trifluoride as a catalyst) and pyrolyzed to give TFE. A large number of side products (hexafluoropropene, perfluorocyclobutane, 1-chloro-1,1,2,2-tetrafluoroethane, and 2-chloro-1,1,1,2,3,3-hexafluoropropane and highly toxic perfluoroisobutene) are formed in this process. Since TFE has to be very pure for the consecutive polymerization, the raw product of this pyrolysis has to be refined by a complex process [624]. Inhibitors such as *d*-limonene and terpene B are added to the purified monomer to prevent polymerization during storage [624]. Recently the storage of TFE in CO_2 was discussed, by this many associated dangers of working with TFE (autopolymerization, disproportionation) can be avoided [651].

Tetrafluoroethene (TFE) is a colorless, tasteless, odorless, nontoxic gas with a boiling point of −76.3 °C at 101.3 kPa, a critical temperature (T_c) of 33.3 °C and a critical pressure (P_c) of 3.94 MPa [652]. In the absence of air, TFE disproportionate violently into carbon and carbon tetrafluoride. The flammability limits of TFE in air are 14 to 43 vol%—forming explosive mixtures with air or oxygen.

Suspension and emulsion polymerization are the main commercial processes for the manufacture of PTFE. For industrial applications, these processes are described in several patents [652–676]. Suitable initiators for polymerization are peroxy compounds. The polymerization with coordination catalysts of the Ziegler–Natta type, the photo-initiation by metal carbonyls, combination of metal organyls with hydrogen peroxide, or the radiation-induced polymerization in solution or emulsion has been used extensively on a laboratory scale [677–685]. The polymerization of TFE is very exothermic (172 kJ/mol) [653].

B. Chemically Initiated Polymerization

1. Polymerization in Emulsion and Suspension [686]

For suspension and emulsion polymerization (also called granular polymerization [653]), gaseous TFE is added at 1.4 to 2.8×10^6 Pa to an aqueous solution of about 20 to 30 ppm of a water-soluble initiator, and 50 to 100 ppm of a dispersing agent [659]. Suitable initiators are ammonium, sodium, or potassium peroxodisulfate [654,655,659,660, 673–675], hydrogen peroxide [619], oxygen [663] or organic peroxy compounds (e.g., benzoyl peroxide or disuccinic acid peroxide) [654,663]. Later patents and process agreements are reporting that oxygen acids of VII elements and their salts (especially potassium permanganate) function as initiators [624,633,665,672]. Steininger and Fitz proposed redox systems for the polymerization containing conventional peroxidic oxidation components and water-soluble nitrogen compounds such as ammonium carbamate by in situ diimine formation under polymerization conditions [675]. Other redox systems are combinations of peroxydisulfate with iron(II) salts [654,655,675] or a combination of bisulfites and iron(III) salts [659]. As buffer systems, sodium phosphates, borax, or ammonia buffers are reported [660,664]. Dispersing agents are perfluoro- or ω-hydroperfluoroalkanoic acid salts. The use of ammonium perfluorooctanoate (1 to 1.5 wt%) and perfluorotributylamine has been described [672]. This has the advantages of compatibility with the polymer and, more important, it is complete resistance against perfluorocarbon radicals [653]. Polymerization temperature is maintained above the critical temperature for TFE at about 50 to 100 °C [674]. During the polymerization, TFE is added continuously to maintain a constant pressure. The TFE pressure is maintained below the critical pressure of TFE in the range of 2.0 to 3.6×10^6 Pa [672].

Dispersion polymerization of TFE is carried out under conditions quite different from those described for the granular system. The same initiator and dispersing agent can be used, however, in concentrations much higher than in the granular polymerization [655,676]. Nevertheless, the dispersing agent is still used at a concentration below the critical micelle concentration [687]. Bankhoff [654] obtained colloidal dispersion of PTFE in the presence of 0.1 to 12 wt% of a saturated hydrocarbon that has more than 12 C atoms per molecule. This wax of a saturated hydrocarbon should must be molten at the given reaction temperature and become solid during dispersion work-up. Its function is to reduce the coagulation tendency during the polymerization as well as to trap coagulated particles to retard the granular-type polymerization. In a similar way, cetane, mineral oil, or paraffin wax may be used as stabilizers [654,688,689]. Temperatures of 60 to 100 °C are employed, agitation is mild, and TFE pressure is maintained in the range of 2.0 to 3.6×10^6 Pa [672]. The product is a semistable dispersion of particles of a variety of shapes with dimensions of 0.1 to 0.3 μm [662,672], in contrast to granular particles with maximum particle diameters of 500 μm [653]. The aqueous colloidal dispersion contain about 10 wt% PTFE [662]. The body of the dispersions can be increased to 75% in the presence of polyethylene-glycolmono-*p*-octylphenyl ether as the dispersing agent [658].

Berry [657] reported on dispersions of PTFE that have been made in organic liquids, such as ketones and neutral esters of dicarboxylic acids with boiling points above 175 °C. These low viscosity dispersions containing 15 to 20% solid, and at this concentration they are practically nonthixotropic. The dispersions require no dispersing agents and do not coagulate. Brown [661] described the polymerization of TFE in perfluorinated saturated liquids, especially perfluoromethylcyclohexane, perfluorokerosine and perfluorodimethylcyclohexane, as solvents, initiated by AIBN or bis(heptafluoro-butyryl) peroxide. Under these conditions, higher yields and faster polymerization is reported. Anisotropic liquid crystalline PTFE dispersions can be prepared by radical polymerization of TFE in aqueous media containing anionic fluorocarbon surfactants $(C_8F_{17}SO_3K)$ [670,671].

2. Polymerization in Solution and Carbon Dioxide

Solution polymerization is of no practical value since the precipitated PTFE can not be further processed [653]. Hence, only few examples are reported. Highly fluorinated hydrocarbons such as hexafluoropropene [687] and chlorofluorohydrocarbons (CFCs) (e.g. 1,1,2-trichloro-1,2,2-trifluoroethane, dichlorodifluoro-methane, and chlorodifluoro-methane) [721] were used as solvents. The solution polymerization can be initiated by γ-rays [105] or by UV irradiation [687]. Technologically relevant is the copolymerization of TFE with various fluorinated monomers (e.g., perfluorovinyl ether), since the heterogeneous copolymerization in aqueous media results in polymers with a significant amount of carboxylic end groups that require complex post-polymerization treatments. However, the use of CFCs is cost intensive and environmentally more than problematic. In 1995 DeSimone demonstrated the homopolymerization of TFE in a CO_2/aqueous hybrid system which yielded granular and spherical high-molecular-weight PTFE resins [690]. Copolymerization of TFE with various fluorinated vinyl monomers in CO_2 were recently reviewed [691].

3. Polymerization Initiated with Organometalic Compounds

The polymerization of TFE with coordination catalysts of the organometalic type has been used on a laboratory scale only. Reiher et al. [666] described the TFE polymerization in liquid NH_3 initiated by alkali or alkali earth amides and acetylides. In addition, the polymerization in the presence of Et_3Al, Pr_3Al, $EtAlCl_2$, and Et_2Zn was studied. These compounds are dissolved in hydrocarbons, THF, dialiphatic ethers, dioxane, or mixtures of these.

Sianesi and Caporiccio [692] have shown that Ziegler–Natta catalysts can be used to polymerize TFE. With catalysts based on titanium tetraalcoholates and alkylaluminum derivatives. In halogenated solvents, slow polymerization of TFE to crystalline polymers was observed. The successful polymerization of TFE with $TiCl_4/Al(C_2H_5)_3$ or $TiCl_3/Al(C_2H_5)_3$ catalysts proved that the activity of the catalytic systems was suitable in the presence of TFE. It is well known that VII and VIII transition metal carbonyls are efficient initiators for the free-radical polymerization [77]. Bamford et al. [684,693,694] reported that manganese and rhenium carbonyls in the presence of a suitable chlorine or bromine derivatives (e.g. CCl_4) are efficient photoinitiators for the free-radical polymerization of liquid TFE. The following reactions were proposed:

$$Mn_2(CO)_{10} \longrightarrow Mn(CO)_4 + Mn(CO)_6 \tag{69}$$

$$Mn(CO)_4 + C_2F_4 \longrightarrow (CO)_4MnCF_2CF_2 \cdot \tag{70}$$

$$(CO)_4MnCF_2CF_2 \cdot + nM \longrightarrow (CO)_4MnCF_2M_{n-1}M \cdot \tag{71}$$

Another efficient photoinitiator for the polymerization of TFE is dimethyl(2,2'-bipyridyl)platinum (DMBP) [683]. The irradiation of DMBP + TFE is considered to yield an adduct that initially give rise to propagating radicals [Me_2(2,2-bipyridyl)$PtCF_2CF_2 \cdot$].

C. Radiation-Induced Polymerizations

1. Photo-Initiated Polymerization

Since the discovery of PTFE a considerable research efforts focused on the photo-initiated polymerization of TFE. Polymer has been produced in the mercury-sensitized photolysis of gaseous TFE. However, analysis showed perfluorocyclopropane to be the main product. At pressures less than 8 kPa it is reported to be the only product [695,696]. Photo-polymerization of gaseous TFE has also been reported in the presence of mercury bromide, phosgene, and nitrous oxide [697–699]. Atkinson [697] has observed that mercury bromide activated by light of 365 nm initiates the photochemical polymerization of TFE.

Marsh and Heicklen [698] have photolyzed phosgene with UV light of 254 nm in the presence of TFE. Under these conditions phosgene decomposes quantitatively to carbon monoxide and chlorine atoms that are scavenged by the olefin. The resulting radicals propagate by monomer addition or terminate by combination. No disproportionation products were found. The extent of polymerization was followed at temperatures from 23 to 300 °C by monitoring the pressure drop during the reaction. Number-average chain lengths from 2 to 200 were achieved. Rotating-sector experiments were performed at 150 °C and absolute rate constants were determined, assuming no activation energy for termination. Some polymer was found when TFE was reacted with oxygen atoms produced by the Hg-photosensitized decomposition of N_2O [699].

Haszeldine [700,701] obtained short-chain TFE polymers of the general formula $CF_3(CF_2CF_2)_nI$ ($n = 1$ to 10) by the reaction of trifluoroiodomethane with TFE, and some homologues have been isolated. The reaction was found to give short-chain polymers containing a terminal iodine atom, which by reaction with chlorine or a fluorinating agent could be exchanged and converted into inert oils and greases. Similar results were obtained when tribromofluoromethane [702] or trichloro-bromomethane [703] was used instead of trifluoroiodomethane.

Mungull et al. [704] reported that 1,2-dibromotetrafluoroethane was an efficient initiator for the photo-polymerization of TFE. However, no polymer formation was detected when a Pyrex filter was used instead of the quartz window. Since 1,2-dibromo-tetrafluoroethane absorbs strongly at $\lambda < 270$ nm photo-dissociation of the carbon-bromine bond was made responsible for the initiation. Other halocarbons were tested as initiators in this reaction: CCl_4, iodomethane, bromoethane, iodoethane, bromotrifluoro-methane, and pentafluoroiodoethane. All of these materials did act as initiators but yielded only oily products.

Direct photolysis of TFE results in surface photo-polymerization at lower monomer pressures [705] and solid polymer at monomer pressures above 1066 Pa. Wilkus and Wright [706] observed a white polymer in the gas phase during the direct photolysis of TFE vapor at pressures ranging from about 1330 to 10 013 Pa. A continuous, transparent deposit was formed at 330 °C. Unlike the surface photo-polymerization process, in which film deposition is restricted to the irradiated areas, the polymer was deposited on all internal surfaces of the reactor. Although most depositions were at substrate temperatures near 25 °C, variations in temperatures from 0 to 200 °C did not seem to have pronounced effects on the deposition rate. The direct photolysis requires absorption of light at or near the nonbonded continuum associated with dissociation of TFE at 215 nm.

It has also been reported that PTFE is one of the products received by a photosensitized reaction of perfluorocyclobutane at 147 nm. Davenport and Miller [707] have photolyzed perfluorocyclobutane mixed with small amounts of xenon using radiation from a xenon resonance lamp. Beside the main product, trans-2-C_4F_8, higher fluoro-carbons including PTFE were found. The influence of oxygen on the photolysis of TFE was studied by varying the amount of O_2 from few ppm to 50%. Addition of O_2 up to 20 ppm favors the formation of TFE during the photolysis of TFE at 185 nm and 254 nm [708]. At higher concentrations, oxygen appears to inhibit the polymerization and only small amounts of TFE oxide, carbonyl fluoride, and perfluorocyclopropane were found.

Bamford and Mullik [709] presented a kinetic study of TFE polymerization photo-initiated by $Mn_2(CO)_{10}$ in acetic acid at room temperature and close to atmospheric pressure. The rate of polymerization is proportional to the square root of the light intensity, and proportional to $[TFE]^{1.4}$, over the range 0.78 to 2.60 mol/L. Polymers prepared in this way show the infrared absorption near 2000 cm^{-1}, characteristic for $(CO)_5MnCF_2$–CF_2 end groups and confirming that the initiating radicals have the structure $(CO)_5MnCF_2CF_2 \cdot$. Although the group VI carbonyls are relatively ineffective, their efficiency can be increased by the presence of CCl_4, forming CCl_3 radicals as initiating species. In group VII, $Re_2(CO)_{10}$ and $Mn_2(CO)_{10}$ are very active, and from group VIII carbonyls, $Os_3(CO)_{12}$ is a better initiator than $Ru_3(CO)_{12}$. The efficiency of the latter increases in the presence of CCl_4.

2. Radiation-Initiated Polymerization

Bulk polymerization of tetrafluoroethene (TFE) by radiation was studied in the gas, liquid, and solid phase over a wide range of temperatures from −196 to 90 °C by a number of methods (e.g. NMR and FTIR spectroscopy). Volkova et al. [710] studied the radiation-induced polymerization in the gas phase from 12 to 90 °C. Different activation energies were found below and above 70 °C. Enslin et al. [711] reported that the rate of polymerization in the gas phase was a zero-order function of the monomer pressure. However, the rate of polymerization was profoundly influenced by the initial monomer pressure (4.6-order dependence) and on the radiation intensity (0.36-order dependence).

Bruk et al. studied [712] the radiation polymerization of TFE adsorbed on highly porous substrates. It has been shown that the rate of radiation polymerization of TFE on silica is controlled by the concentration of monomer in the adsorbed layer. The polymerization rate increased with increasing concentration of PTFE grafted to silica gel as a result of an increased number of active center formation caused by a more effective energy transfer from silica gel to grafted PTFE [713]. Bruk et al. [714] further reported that the rate of polymerization in the solid phase increases with temperature from $-196\,°C$ to $-131\,°C$.

Tabata et al. [715] carried out solid-state polymerization by initiating with γ-rays from ^{60}Co and electrons using a van de Graaff generator and they have observed that the polymerization rate changes only very little with temperature. The polytetrafluoroethene (PTFE) that was obtained by liquid- or solid-state polymerizations was examined by infrared and x-ray diffraction methods. From these experiments it was clearly demonstrated that the crystalline structure is quite different. The polymerization in the solid state is strongly influenced by the crystal lattice of the polymer.

Bruk et al. [716] described the low-temperature radiation polymerization of crystalline TFE in detail. It has been established that three solid-phase post-polymerization reactions can take place when irradiated specimens are heated above the melting point: low-temperature polymerization (in the interval 77 to 110 K), 'slow' polymerization close to the melting point (in the interval 128 to 138 K), and 'rapid' polymerization during melting of the crystal (142 K). Tabata et al. [717] have found that a significant post-polymerization takes place even in the liquid phase. Kinetic analysis has been made of the in-source and post-polymerizations [718,719]. Post-polymerization is explained by a long lifetime of polymer radicals in the liquid phase at $-78\,°C$ due to the slow combination rate of the polymer radicals caused by their rod-like shape.

The solution polymerization of TFE by radiation in various media was studied by Fujioka [720], Hisasue [721] and Tabata et al. [722]. Fujioka reported on the polymerization in fluorinated solvents such as $CHClF_2$, $CF_2ClCFCl_2$, and CCl_2F_2 in the low-temperature region from -78 to $-40\,°C$. The rate acceleration and post-polymerization observed in $CHClF_2$ indicate the presence of growing polymer radicals of longer life times. The maximum polymer molecular weight is 1.2×10^6, which is still lower than that of the commercial products prepared by chemical initiators. Tabata et al. [722] have also studied the solution polymerization between -40 and $0\,°C$ in $CHClF_2$. A maximum rate of the polymerization was observed at $-10\,°C$ yielding molecular weights lower than $1 \times 10^6\,g/mol$.

Kinetic studies on the radiation-induced polymerization and post-polymerization of TFE were carried out using chlorofluorohydrocarbons as solvents. The remarkable post-polymerization is again explained by the unusually slow rate of the bimolecular chain termination. A chain transfer reaction was also discussed by Hisasue et al. [721]. Suwa et al. [679] discussed the emulsion polymerization of TFE by radiation with ammonium perfluorooctanoate (FC-143) as the emulsifier. The polymerization rate is proportional to the 0.8 power of the dose rate and is almost independent of the emulsifier concentration (up to 2 wt% in water). Molecular weights between 10^4 and 10^6 were observed, which increases with reaction time but decreases with the emulsifier concentration. In general, the molecular weight of PTFE prepared by radiation-induced polymerization in solution and in emulsion is relatively low compared with commercial PTFE. However, it is also possible to produce molecular weight of up to 3×10^6 if an emulsifier-free polymerization are carried out [677,678,723].

An interesting discovery is that PTFE as a hydrophobic polymer forms a stable latex in the absence of emulsifier. Under certain conditions PTFE latex coagulates during polymerization and the polymerization rate decreases. Probably because polymerization proceeds mainly on the polymer particle surface. The observed rate acceleration and successive increase in polymer molecular mass may be due to a slow termination of propagating radicals in the rigid PTFE particles [723]. The size, distribution, and number of PTFE particles formed by radiation-induced emulsifier-free polymerization were measured by electron microscopy and automatic particle analyzer (centrifugation method). The polymer molecular mass ($>10^6$) is almost independent of the particle size [677].

To clarify this phenomenon, the effects of additives, in particular radical scavengers, on the molecular mass of PTFE and its polymerization behavior were studied. It was found that the molecular mass of PTFE increased by the addition of hydroquinone, benzoquinone, α-pinene, dl-limonene, and ethylenediamine but is decreased by oxygen and triethylamine. A PTFE latex with molecular mass higher than 2×10^7 was obtained in the presence of hydroquinone [724]. Tabata and Shimada [725] prepared a low molecular mass PTFE powder by exposing deoxygenated acetone solutions of TFE to radiation between $-130\,°C$ and room temperature. Thus irradiating an oxygen-free solution of TFE in acetone at $-80\,°C$ with 1.56 Mrad γ-rays gave powdered PTFE with molecular mass of 1.3×10^3 and a particle diameter of 0.3 μm. Radiation-induced copolymerizations of TFE with various monomers, including perfluoro-olefins [726,727] and with propylene [728] have been studied. In the case of TFE/propene, semibatch experiments at a constant pressure of $2.5 \times 10^5 \text{kg/m}^2$ and at $40\,°C$ at various dose rates were carried out. Copolymerization of TFE with styrene [729] was studied in bulk and in perfluorotoluene at $22\,°C$ at autogenous pressure.

3. Glow-Discharge and Plasma Polymerization

Glow-discharge and plasma polymerization has emerged as powerful techniques of thin-film deposition. Despite the numerous useful characteristics of the process, the mechanism of glow-discharge polymerization is still considered to be very complex and the reproducibility of the process is poor. A proper understanding of the polymerization process is lacking because of the large number of experimental parameters governing the plasma process, inaccessibility of the obtained polymers for standard analytical techniques and lack of tools that can monitor the plasma process without interfering the reaction [730–735].

Lippold et al. [736] investigated the polymerization of TFE on the anode of a hot glow-discharge. Rates of polymer deposition have been measured as a function of monomer pressure, discharge current density, and electrode surface temperature. A mechanism for the polymer formation is suggested. The results are different from those obtained by polymerization on the cathode of a glow-discharge. It is concluded that negative ions have a great impact on the polymer formation mechanism. Additionally, monomer molecules adsorbed at the electrode are polymerized by electron bombardment.

The plasma polymerization of TFE by itself and mixed with inert gases has been studied in the field-free zone inside a Faraday cage by Buzzard et al. [737]. The chemical structure of the products were analyzed by XPS, revealing linear and branched molecules. Linear chains are formed by less energetic plasma and at low monomer residence times.

Nakajima et al. carried out plasma polymerization in continuous-wave and pulsed radio-frequency discharges to establish the effects of reaction conditions on the kinetics of polymer deposition as well as on the polymer structure. Under conditions favoring low deposition rates, the dominant functional group is $-CF_2$. At higher deposition rates the of $-CF_2-$ group concentration is reduced and a cross-linked polymer was obtained [738].

A long tube with the coupling coil at the middle was used for the plasma polymerization of TFE in an inductively coupled radio-frequency glow discharge using a flow system. Deposition rates and the chemical nature of the polymer were detected as a function of location in the reactor tube relative to the coupling coil and of applied energy/unit TFE mass. A fluoro-poor polymer, containing many C–O bonds, was obtained at all locations at high wattage and high mass flow rate [739].

Morosoff et al. have studied the plasma polymerization of TFE in a capacitive coupled system with internal electrodes using radio-frequency power to identify the conditions, most compatible for continuous coating of plasma polymer on a substrate which was moved through the center of the inter-electrode gap. Without magnets the most active zone of the plasma was in the center of the inter-electrode gap, while the use of a magnetic field moved the active zone closer to the electrodes and resulted in a more efficient energy coupling [740].

Yanagihara and Yasuda applied the so-called double-probe method. The electrode temperature (T_e) and density of positive ions (n_p) were measured at various discharge wattages. Although the values of T_e and n_p may not be related directly to the polymer formation in a plasma, the method provides a direct measure of plasma energy density where plasma polymerization takes place, whereas it cannot be accurately estimated by the input energy of a discharge [741].

TFE polymerization in a cold plasma leads to polymers with considerable deficiencies of H or F atoms. With ESR, various kinds of radicals trapped in the macromolecular structure were detected [742]. A glow discharge of TFE gas under reduced pressure was applied to solid substrates with complex surface structures to deposit a thin film of plasma-polymerized PTFE [743].

Chen et al. reported on plasma polymerized TFE were IR spectroscopy, x-ray diffraction and elemental analysis showed that polymers formed in the glowing region of the plasma having a lower concentration of $-CF_2-$ groups and are cross-linked [744]. A structural interpretation of plasma polymerized TFE produced in a glow-discharge chamber is obtained by ^{19}F-NMR and IR spectroscopy, elemental analysis, and number-average mole mass detection. The number average molecular weight was 3600 for the majority of the polymer. The IR spectrum shows evidence of some C=C groups and ^{19}F-NMR spectroscopy indicated a highly branched structure [745]. PTFE films prepared by plasma polymerization using hydrogen as carrier gas were higher cross-linked and showed a higher dielectrical loss than the films prepared using argon. Both films contained radicals of polyene type [746].

D. Technical Production and Properties of PTFE

Polytetrafluoroethene is manufactured and sold in three forms: granular, fine powder and as an aqueous dispersion [624]. Granular polymerization of TFE was the first procedure used to make PTFE [747] and is, even after 50 years, of great industrial significance. The granular product can be molded in various forms. For ram extrusion of granular resin into long tubes and rods, a partially presintered resin is preferred. Granular PTFE resin is nonflammable. Dispersion polymerization of TFE yields a semistable dispersion of particles having a variety of shapes with dimensions of 0.1 to 0.3 µm [748]. In comparison, granular particles are of the order of 500 µm in their greatest dimension [748]. The raw dispersion is stabilized with a nonionic or anionic surfactant and concentrated to 60 to 65 wt% solids by electrodecantation, evaporation, or thermal concentration [749] and can be modified with chemical additives. The polymer produced by aqueous dispersion cannot be molded but is fabricated by dispersion coating or conversion to powder for paste extrusion.

PTFE is commonly known as Teflon®. Other Teflon types are copolymers of TFE (PFA, FEP, PDD) with altered properties for certain applications. For example, Teflon

AF, a copolymer of TFA and 2,2-bis-trifluoromethyl-4,5-difluoro-1,3dioxole, display the same chemical resistance as PTFE but is completely amorphous and with the lowest dielectric constant of 1.89 to 1.9 reported so far and unusually high gas permeation rates [750,751].

PTFE has a gas permeation rate of 420 cB (centi-Barrers) for oxygen and 1200 cB for carbon dioxide [750].

PTFE reacts with fluor under pressure to give CF_4. PTFE is thermoplastic with a density of 2.2 to 2.3 g/mL, the tensile strength is 13.8×10^6 to 31×10^6 Pa, and the elongation is 300 to 450%. PTFE has a long period stability up to 300 °C. Between 320 and 327 °C, a solid-phase transition is reported [652,752]. Prolonged heating at 390 °C causes degradation of the material and above 400 °C volatile losses were observed [652]. Due to the widespread use of PTFE in sensitive applications such as space aviation or nuclear power technology, its radiation degradation was extensively investigated [753,754].

PTFE cannot be dissolved in any common solvent below its melting point. However, PTFE and TFE containing copolymers are not strictly insoluble. PTFE shows small solubility in perfluorinated cerosin [755], perfluorinated oils [667] and in its oligomers (chain length above C_{22}) [756] at temperatures above 290 °C. Tuminello reported on the solubility of PTFE as well as an LCST behavior in cyclic perfluorocarbons [757]. However, the molecular weight of PTFE is difficult to determine by common methods. Its number average molecular weight has been estimated by end group analysis using radioactive labeling [619]. This technique has been applied to technical polymers of molecular weights between 4×10^5 and 9×10^6 g/mol. The molecular weight of PTFE was determined to range from 10^6 to 10^7 g/mol, although different values can be found in the literature stating that the average molecular weight should be at least $>10^7$ g/mol. In practice, relative determinations were obtained by means of specific gravity [758], crystallinity [758,759], zero strength time (ZST) [760,761] and melt viscosity measurements [762,763]. An additional method for the molecular weight determination of PTFE was applied by Suwa et al. [764], who studied the melting and crystallization behavior of virgin PTFE by using differential scanning calorimetry. The following quantitative relationship was found between number-average molecular weight of PTFE and the heat of crystallization in the molecular weight between 5.2×10^5 and 4.5×10^7:

$$M_n = 2.1 \times 10^{10} \times \Delta H_c^{-5.16} \tag{72}$$

where M_n is the number average molecular weight and ΔH_c is the heat of crystallization in J/g. At about 342 °C, virgin PTFE changes from a white crystalline material to an almost transparent amorphous gel. Differential thermal analysis indicates that the first melting of virgin polymer is irreversible (transition: extended chain structure/folded chain structure [765–769]) and that subsequent remeltings occur at 327 °C, which is generally regarded as the melting point. The melting process of PTFE begins near 300 °C [770]. Melting is accompanied by a volume increase of about 30%. Because the viscosity of the polymer at 380 °C is 10 GPa, the shape of the melt is stable. The melting point increases with increasing pressure at a rate of 1.52 °C/MPa [771]. Above 230 °C, decomposition rates become measurable (0.0001%/h). Small amounts of toxic perfluoroisobutylene were isolated at 400 °C and above; free fluorine was not found. Above 690 °C, the decomposition products burn but do not support combustion. Combustion products consist primarily of carbon dioxide, carbon tetrafluoride, and small quantities of toxic and corrosive hydrogen fluoride [624]. The molecular composition of PTFE, e.g., crystallinity, quantification of end groups, chain packing, morphology down to the molecular level etc. has been studied extensively by NMR [772–776] and IR spectroscopy [703,777,778,783], with x-rays [779,780] and scanning force microscopy [781].

Virgin PTFE has an exceptionally high crystallinity in the range 92 to 98%, which indicates an unbranched chain structure. The PTFE chains are rigid, since the fluorine atoms are too large to allow a planar zigzag structure and/or *trans-gauche* defects as in hydrocarbons [782]. Besides, at the melting point, the transition at 19 °C (triclinic to hexagonal unit cell causing a slight untwisting of the molecule from a twist of $180°/13\,CF_2$ groups to $15\,CF_2$ groups) is noteworthy because it significantly affects the product behavior. At the first-order transition at 30 °C, the hexagonal unit cell disappears and the rod-like hexagonal packing of the chain in the lateral direction is retained. Below 19 °C there is an almost perfect three-dimensional order; between 19 and 30 °C, the chain segments are disordered; and above 30 °C, the preferred crystallographic direction is lost— the molecular segments oscillate above their long axes with a random angular orientation in the lattice. As-polymerized PTFE features a number of microstructures such as highly developed crystals, fibrils and shish-kebab structures [783].

REFERENCES

1. Schildknecht, C. E. (1952). *Vinyl and Related Polymers*, Wiley, New York, p. 593.
2. Wislicenus, J. (1878). *Justus Liebigs. Ann. Chem.*, *192*: 106.
3. Field, N. D., and Lorenz, D. H. (1970). *High Polymers*, Vol. 24, *Vinyl and Diene Monomers* (Leonard, E. C., ed.), Wiley, New York, p. 365.
4. Sandler, S. R., and Karo, W. (1977). *Polymer Syntheses*, Vol. 2, Academic Press, New York, p. 214.
5. Lorenz, D. H. (1971). *Encyclopedia of Polymer Science and Technology*, Vol. 14 (Mark, H. F., Gaylord, N. G., and Bikales, N.M., eds.), Wiley, New York, p. 504.
6. Bikales, N. M. (1971). *Encyclopedia of Polymer Science and Technology*, Vol. 14 (Mark, H. F., Gaylord, N. G., and Bikales, N. M., eds.), Wiley, New York, p. 521.
7. Hort, E. V., and Gasman, R. C. (1983). *Kirk-Othmer Encyclopedia of Chemical Technology*, 3rd ed., Vol. 23 (Grayson, M., ed.), Wiley, New York, p. 937.
8. Sawamoto, M., and Higashimura, T. (1990). *Makromol. Chem., Macromol. Symp. (1990)*, *32 (Invited Lect. Int. Symp. Cationic Polym. Relat. Ionic Processes, 9th, 1989)*: 131.
9. Goethals, E. J., Reyntjens, W., and Lievens S., *Macromol. Symp. (1998)*, *132(132 international Symposium on Ionic Polymerization, 1997)*: 57.
10. Decker C., *Macromol. Symp. (1999)*, *143 World Polymer Congress, 37th International Symposium on Macromolecules, 1998)*: 45.
11. Reppe, W. (1934). U.S. 1,959,927; C.A. (1935) *28*: 4431.
12. Reppe, W. (1939). U.S. 2,157,348; U.S. 2,157,347, both to I.G. Farbenindustrie AG.
13. Reppe, W., and coworkers (1956). *Justus Liebigs Ann. Chem.*, *601*: 81.
14. Stern, E. W., and Spector, M. L. (1961). *Proc. Chem. Soc.*, 370; (1965). Neth. Pat. 6,411,879 to Imperial Chemical Industries; C.A. (1965), *63*: 9813.
15. Klass, D. L. (1977). U.S. 4,057,575 to Union Oil of California; C.A. (1978), 89: 6804; U.S. 4,161,610 to Union Oil of California; C.A. (1979), *91*: 123418.
16. Baur, K. (1934). U.S. 1,931,858 to I.G. Farbenindustrie AG; C.A. (1934), *28*: 485.
17. Calanac, M. (1930). *Compt. Rend.*, *190*: 881.
18. Kitabatake, M., Owari, S., and Kuno, K. (1967). *Soc. Synth. Org. Chem. Jpn.*, *25*: 70.
19. McKeon, J. E., Fritton, P., and Griswold, A. A. (1972). *Tetrahedron*, *28*: 227.
20. Baur, K. (1931). Ger. 525,836 to I.G. Farbenindustrie AG; C.A. (1931), *25*: 4556.
21. Watanabe, W. H., and Conlon, L. E. (1957). *J. Am. Chem. Soc.*, *79*: 2828.

22. Adelman, R. L. (1953). *J. Am. Chem. Soc.*, *75*: 2678.
23. Weeks, G. A., and Grant, W. J. (1954). Brit. 709,106 to British Oxygen; C.A. (1955), *49*: 10360.
24. Adelman, R. L. (1955). *J. Am. Chem. Soc.*, *77*: 1669.
25. Hurd, C. D., and Botteron, D. G. (1946). *J. Am. Chem. Soc.*, *68*: 1200.
26. Black, D. K., and Landor, S. R. (1965). *J. Chem. Soc.*, 5525.
27. Cretcher, L. H., Koch, J. A., and Pittenger, W. H. (1925). *J. Am. Chem. Soc.*, *47*: 1173.
28. Rehberg, C. E. (1949). *J. Am. Chem. Soc.*, *71*: 3247.
29. Butler, G. B., and Nash, J. L., Jr. (1958). *J. Am. Chem. Soc.*, *73*: 2538.
30. Gallucci, R. R., and Going, R. C. (1983). *J. Org. Chem.*, *48*: 342.
31. Crivello, J. V., and Conlon, D. A. (1983). *J. Polym. Sci. Polym. Chem. Ed*, *21*: 1785.
32. Prosser, T. J. (1961). *J. Am. Chem. Soc.*, *83*: 1701.
33. Price, C. C., and Snyder, W. H. (1961). *J. Am. Chem. Soc.*, *83*: 1772.
34. Crivello, J. V., and Conlon, D. A. (1984). *J. Polym. Sci. Polym. Chem. Ed*, 22: 2105.
35. Eliel, E. L., Nowak, B. E., Daignault, R. H., and Badding, V. G. J. (1965). *Org. Chem.*, *30*: 2441.
36. Havel, J. J., and Chan, K. H. (1976). *J. Org. Chem.*, *41*: 513.
37. Vogel, E., and Gunther, H. (1967). *Angew. Chem. 19, 79*: 429.
38. Eberbach, W., and Buchardt, B. (1978). *Chem. Ber., 111*: 3665.
39. Lal, J. (1968). *Polymer Chemistry of Synthetic Elastomers*, Part I (Kennedy, J. P., and Tornqvist, E., eds.), Interscience, New York, pp. 331–376.
40. Gandini, A., and Cheradame, H. (1980). *Adv. Polym. Sci.*, 34/35.
41. Higashimura, T., and Yamamoto, K. (1974). *Makromol Chem.*, *175*(4): 1139.
42. Schildknecht, C. E., Gross, S. T., and Zoss, A. O. (1949). *Ind. Eng. Chem.*, *41*: 1998.
43. Schildknecht, C. E., Zoss, A. O., and Grosser, F. E. (1949). *Ind Eng. Chem.*, *41*: 2891.
44. Blake, G. J., and Carlson, A. M. (1966). *J. Polym. Sci. Part A-1, 4*: 1813.
45. Mowers, W. A., and Crivello, J. V. (1999). *Polym. Mater. Sci. Eng.*, *81*: 479.
46. Reppe, W., and Kuehn, E. (1937). U.S. 2,098,108 to I.G. Farbenindustrie AG; C.A. (1938), *32*: 367; (1938), Ger. 625,017 to I.G. Farbenindustrie AG; C.A. (1938), *32*: 7615; (1940), U.S. 2,188,788 to I.G. Farbenindustrie AG; C.A. (1940), *34*: 4077.
47. Schildknecht, C. E. (1949). *Ind. Eng. Chem.*, *41*: 2894.
48. Reppe,W., and Schlichting, O. (1937). U.S. Pat. 2,104,000, 2,104,001, and 2,104,002 to I.G. Farbenindustrie AG; C.A. (1938), *32*: 1815.
49. Favorskii, A. E., and Shostokovskii, M. F. (1943). *J. Gen. Chem. USSR (Engl Transl)*, *13*: 1428; C.A. (1944), *38*: 330.
50. Eley, D. D., and Scabrooke, A. (1964). *J. Chem. Soc. Part A-1, 5*: 2226.
51. Copenhaver, J. W., and Bigelow, M. H. (1949). *Acetylene and Carbon Monoxide Chemistry*, Van Nostrand Rheinhold, New York, p. 49.
52. Ketley, A. D. (1962). *J. Polym. Sci.*, *62*: 281.
53. Ital. 597,550 (1959) to Montecatini; C.A. (1961). *55*: 17097.
54. Natta, G., Dall'Asta, G., and Oddo, N. (1960). Ger. 1,091,754 to Montecatini; C.A. (1961), *55*: 25361.
55. Dall'Asta, G., and Bassi, 1. W. (1961). *Chim. Ind. (Milan)*, *43*: 999; C.A. (1962), *56*: 8916.
56. Dall'Asta, G., and Oddo, N. (1960). *Chim. Ind. (Milan)*, *42*: 1234; C.A. (1961), *55*: 24532.
57. Aoshima, S., Onishi, H., Kamiya, M., Shachi, K., and Kobayashi, E. (1994). *J. Polym. Sci., Part A: Polym. Chem.*, *32*(5): 879.
58. Farooq, O. (1991). *Eur. Pat. Appl.*, 18.
59. Higashimura, T., Okamoto, S., Kishimoto, Y., and Aoshima, S. (1989). *Polym. J. (Tokyo) 21*(9): 725.
60. Miyamoto, M., Sawamoto, M., and Higashimura, T. (1984). *Macromolecules, 7*: 265.
61. Higashimura, T., Aoshima, S., and Sawamoto, M. (1986). *Makroinol. Chem. Macromol Symp. 3*: 99.
62. Sawamoto, M., Aoshima, S., and Higashimura, T. (1988). *Makromol. Chem. Macromol. Symp., 13/14*: 513.

63. Matyjaszewski, K. (1987). *J. Polym. Sci. Polym. Chem. Ed., 25*: 765.
64. Aoshima, S., Sawamoto, M., and Higashimura, T. (1988). *Makromol. Chem. Macromol. Symp., 13/14*: 457.
65. Sawamoto, M., Okamoto, C., and Higashimura, T. (1987). *Makromolecules, 20*(11): 2693.
66. Sawamoto, M., Kamigaito., M., Kojima, K., and Higashimura, T. (1988). *Polym. Bull, 19*: 359.
67. Nuyken, O., and Krbner, H. (1988). *Polym. Prepr., 29*(2): 87.
68. Ingrisch, S., Nuyken, O., and Mishra, M. K. (1999). *Plast. Eng. (N. Y.) 53 (Star and Hyperbranched Polymers)*: 77.
69. Sawamoto, M., Kanaoka, S., and Higashimura, T. (1999). *Hyper-Struct. Mol. I: [Int. Forum], 1st Meeting Date 1996*: 43.
70. Pugh, C., and Kiste, A. L. (1997). *Prog. Polym. Sci. 22*(4): 601.
71. Nuyken, O., Kroener, H., Riess, G., Oh, S., and Ingrisch, S. (1996), *Macromol. Symp. 101 (5th International Polymer Conference 'Challenges in Polymer Science and Technology', 1994)*: 29.
72. Aoshima, S., and Kobayashi, E. (1995). *Macromol. Symp. 95 (Synthesis of Controlled Polymeric Structures through Living Polymerizations and Related Processes)*: 91.
73. Sawamoto, M. (1994). *Macromol. Symp, 85*: 33.
74. Sawamoto, M., and Higashimura, T. (1992). *Makromol. Chem., Macromol. Symp., 60 (Int. Symp. Cationic Polym. Relat. Ionic Processes, 10th, 1991)*: 47.
74. Kennedy, J. P. (1990). *Chemtracts: Macromol. Chem., 1*(4): 260.
75. Sawamoto, M., and Higashimura, T. (1990). *Makromol. Chem., Macromol. Symp., 32 (Invited Lect. Int. Symp. Cationic Polym. Relat. Ionic Processes, 9th, 1989)*: 131.
76. Cho, C. G., and McGrath, J. E. (1992). *Pollimo, 16*(1): 29.
77. Kanaoka, S., Minoda, M., Sawamoto, M., and Higashimura, T. (1990). *J. Polym. Sci., Part A: Polym. Chem., 28*(5): 1127.
78. Choi, W., Sawamoto, M., and Higashimura, T. (1990). *Macromolecules, 23*(1): 48.
79. Eley, D. D., and Sanders, J. (1952). *J. Chem. Soc.*, 4167.
80. Okamura, S., Kano, N., and Higashimura, T. (1961). *Makromol. Chem., 47*: 35.
81. Higashimura, T., and Sawamoto, M. (1978). *Polym. Bull, 1*: 11.
82. Parnell, R. D., and Johnson, A. F. (1969). *IUPAC Symposium on Kinetics and Mechanisim of Polyreactions*, Paper 2/18.
83. Ledwith, A., and Sherrington, D. C. (1971). *Polymer, 12*: 344.
84. Plesch, P. H. (1966). *Pure Appl Chem., 12*: 117.
85. Janjua, K. M., and Johnson, A. F. (1973). *Preprints, International Symposium on Cationic Polymerization*, Rouen (France); Paper C15.
86. Johnson, A. F., and Young, R. N. (1976). *4th International Symposium on Cationic Polymerization*, Akron, Ohio; (*1976*), *J. Polym. Sci. Polym. Symp., 56*: 211.
87. Bawn, C. E. H., Fitzsimmons, C., and Ledwith, A. (1964). *Proc. Chem. Soc.*, 391.
88. Bawn, C. E. H., Fitzsimmons, C., Ledwith, A., Penfold, J., Sherrington, D. C., and Weightman, J. A. (1971). *Polymer, 12*: 119.
89. Ledwith, A., Lockett, E., and Sherrington, D. C. (1975). *Polymer, 16*: 31.
90. Chung, Y. J., Rooney, J. M., Squire, D. R., and Stannett, V. (1975). *Polymer, 16*: 527.
91. Okamura, S., Kodama, T., and Higashimura, T. (1962). *Makromol. Chem., 53*: 180.
92. Kunitake, T., and Takarabe, K. (1981). *Makromol. Chem., 182*: 817.
93. Mengoli, G., and Vidotto, G. (1970). *Makromol. Chem., 139*: 293; (1971), *Makromol. Chem., 139*: 293; (1971), *Makromol. Chem., 150*: 277.
94. Glasel, A., Murray, K., and Funt, B. L. (1976). *Makromol. Chem., 177*: 3345.
95. Funt, B. L., Severs, W., and Glasel, A. (1976). *J. Polym. Sci. Polym. Chem. Ed., 14*: 2763.
96. Obertauch, E., Salvatori, T., and Cesca, S. (1978). *J. Polym. Sci. Polym. Lett. Ed., 16*: 345.
97. Biswas, M., and John, K. J. (1978). *J. Polym. Sci. Polym. Chem. Ed., 16*: 971; 3025.
98. Kray, R. (1960). *J. Polym. Sci., 44*: 264.
99. Bruce, J. M., and Farrow, D. W. (1963). *Polymer, 4*: 407.
100. Nakano, S., Iwasaki, K., and Fukutani, H. (1963). *J. Polym. Sci. Part A-1*: 3277.

101. Biswas, M., and Kabir, G. M. A. (1978). *Polymer, 19*: 357.
102. Lal, J., and McGrath, J. (I 964). *J. Polym. Sci. Part A, 2*(8): 3369.
103. Iwasaki, K. (1962). *J. Polym. Sci., 56*: 27.
104. Hino, M., and Arata, K. (1980). *J. Polym. Sci. Polym. Chem. Ed., 18*: 235; (1980). *Chem. Lett., 8*: 963.
105. Licari, L. J., and Crepeau, P. C. (1965). U.S. Pat. 3,205,157; C.A. (1965), *63*: 17368a.
106. Crivello, J. V., and Lam, J. H. W. (1977). *Macromolecules, 10*: 1307.
107. Irving, E. (1985). Eur. Pats. 104,143 and 106,797.
108. Crivello, J. V., and Lam, J. H. W. (1979). *J. Polym. Sci. Polym. Chem. Ed., 17*: 977.
109. Irving, E., and Stark, B. P., (1983). *Brit. Polym. J., 15*: 24.
110. Crivello, J. V., and Lam, J. H. W. (1979). *J. Polym. Sci. Polym. Chem. Ed., 17*: 2877.
111. Crivello, J. V., and Lam, J. H. W. (1980). *J. Polym. Sci. Polym. Chem. Ed., 18*: 1021.
112. Crivello, J. V. (1978). U.S. Pat. 4,069,056 to General Electric; C.A. (1978), *88*: 138709f.
113. Meier, K., Bühler, N., Zweifel, H., Bemer, G., and Lohse, F. (1984). Eur. Pat. 094,915 to Ciba Geigy.
114. Kong, S., Crivello, J. V. (1999). *Polym. Prepr. (Am. Chem. Soc., Div. Polym. Chem), 40*(2): 569.
115. Crivello, J. V. (1999). *NATO Sci. Ser., Ser. E, 359 (Ionic Polymerizations and Rel. Processes)*: 45.
116. Yagci, Y., and Hizal, G. (1999). *Trends Photochem. Photobiol., 5*: 139.
117. Tsunooka, M. (1996). *Kobunshi, 45*(11): 786.
118. Decker, C., Elzaouk, B., and Biry, S. (1993). *Radcure Coat. Inks., 31*: 33.
119. Crivello, J. V., Bi, D., and Fan, M. (1994). *J. Macromol. Sci., Pure Appl. Chem., A31*(8): 1001.
120. Crivello, J. V. (1990). *Proc. Water-Borne Higher-Solids Coat. Symp., 17th*: 1.
121. Stannett, V., Kubota, H., Hsieh, W. C., and Squire, D. R. (1979). *Polym. Prepr., 20*(1): 383.
122. Fee, J. G., Port, W. S., and Witnauer, L. P. (1958). *J. Polym. Sci., 33*: 95.
123. Pinner, S. H., and Worrall, R. (1959). *J. AppL Polym. Sci., 2*(4): 122.
124. Hayashi, K. A., Yamazawa, Y., Takagaki, T., Williams, F., Hayashi, K., and Okamura, S. (1977). *Trans. Faraday Soc., 63*: 1489.
125. Goineau, A. M., Kohler, J., and Stannett, V. (1977). *J. Makromol. Sci. Chem. Part A, 11*: 99.
126. Desai, V. R., Suzuki, Y., and Stannett, V. (1977). *J. Macromol. Sci. Part A, 11*: 133.
127. Brendlé, M. C. C., and Ilovas, A. M. (1976). *Brit. Polym. J., 8*: 11.
128. Cerai, P., Giusti, P., Guerra, G., and Tricoli, M. (1974). *Eur. Polyin. J., 11*: 57.
129. Breitenbach, J. W., Sommer, F. F., and Unger, J. (1976). *Monatsch. Chem., 107*: 359.
130. Vandenberg, E. J. (1963). *J. Polym. Sci. Part C, 1*: 207.
131. Nelson, J. F., Banes, F. W., and Fitzgerald, W. P. (1961). U.S. Pat. 2,967,203 to Esso Co.; C.A. (1961), *55*: 8842.
132. Bortnick, N. M., and Melamed, S. (1956). US 2,734,890 to Röhm and Haas Co.; C.A. (1957), *51*: 3184.
133. Dreisbach, R. R., and Lang, J. L. (1958). U.S. Pat. 2,859,209 to Dow Chemical Co.; C.A. (1959), 53: 377a.
134. Shostakovskii, M. F., and Bogdanova, A. V. (1954). *Izv. Akad. Nauk SSSR Otd Khim. Nauk*, 919; (1957), *Izv. Akad. Nauk SSSR Otd. Khim. Nauk*, 387; C.A. (1955), *49*: 13941.
135. Matsumoto, A., Iwanami, K., and Oiwa, M. (1983). *Makromol. Chem. Rapid Commun., 4*: 277.
136. Matsumoto, A., Iwanami, K., and Oiwa, M. (1981). *J. Polym. Sci. Polym. Lett. Ed., 19*: 497.
137. Tsukino, M., and Kunitake, T. (1979). *Macromolecules, 12*: 387.
138. Tsukino, M., and Kunitake, T. (1979). *Polym. J., 11*: 437.
139. Nishikubo, T., Iizawa, T., and Yoshinaga, A. (1979). *Makromol. Chem., 180*: 2793.
140. Matsoyan, S. G. (1961). *J. Polym. Sci., 52*: 189.
141. Miyaki, T. (1961). *J Chem. Soc. Jpn. Ind Chem. Sec., 64*: 1272.
142. Minoura, Y., and Mitoh, M. (1965). *J. Polym. Sci. Part A, 3*: 2149.
143. Aso, C., Kunitake, T., and Ando, S. (1971). *J. Macromol. Sci. Part A, 5*: 167.

144. Dietrich, H. J., and Raymond, M. A. (1972). *J. Macromol. Sci. Part A*, 6: 191.
145. Raymond, M. A., and Dietrich, H. J. (1972). *J. Macromol. Sci. Part A*, 6: 206.
146. Tsukino, M., and Kunitake, T. (1985). *Polym. J.*, 17. 657, 943.
147. Kida, S., Nozakura, S., and Murahashi, S. (1972). *Polym. J.*, 3: 234.
148. Bevington, J. C.; Huckerby, T. N. and Jenkins, A. D. (1999) *J. Macromol. Sci., Pure Appl. Chem.*, *A36*(12): 1907.
149. Shaikhutdinov, E. M., Nurkeeva, Z. S., Mun, G. A., and Ermukhambetova, B. B. (1994). *Dokl. Nats. Akad. Nauk Resp. Kaz.*, *(3)*: 79.
150. Higashimura, T., Mitsuhashi, M., and Sawamoto, M. (1979). *Macromolecules*, *12*: 178.
151. Heck, R. F. (1962). U.S. Pat. 3,025,276 to Hercules; C.A. (1962), 57. 744 1.
152. Lal, J. (1962). U.S. Pat. 3,038,889 to Goodyear; C.A. (1962), 57.7441.
153. Hirooka, M., and Mashita, K. (1973). U.S. Pat. 3,752,788 to Sumitomo.
154. Hirooka, M., Takeya, K., Uno, Y., Yamane, A., and Maruyama, K. (1973). Ger. Offen. 2,065,345.
155. Ital. 632,714 to Monticatini; C.A. (1962), *58*: 4713.
156. Siebel, H. P., Herrle, L., and Fikentscher, H. (1956). Get. Pat. 946,849 to BASF AG; C.A. (1958), *52*: 17778.
157. Siebel, H. P. (1958). U.S. Pat. 2,830,032 to BASF AG; C.A. (1958), *52*: 13322.
158. Reppe, W., Hoelscher, F., and Schneevoight, A. (1938). U.S. Pat. 2,108,994 to GAF; C.A. (1938), *32*: 3167.
159. Zhang, X., Reyntjens, W., and Goethals, E. J. (2000). *Polym. Int.*, *49*(3): 277.
160. Aoshima, S., and Sugihara, S. (2000). *Nihon Yukagakkaishi*, *49*(10): 1061–1069.
161. Laus, M., Bignozzi, M. C., Fagnani, M., Angeloni, A. S., Galli, G., Chiellini, E., and Francescangeli, O. (1996). *Macromolecules*, *29*(15): 5111.
162. Neumann, M. G. and de Sena, G. L. (1999). *Polymer*, *40*(18): 5003.
163. Hadjikyriacou, S., and Faust, R. (1995). *Polym. Prepr. (Am. Chem. Soc., Div. Polym. Chem.) (1995), Volume Date 1995*, *36*(2): 174.
164. Pfitzner, K., Bruemmer, W., and Klockow, M. (1971). Get. Offen. 2,008,990 to Merck.
165. Butler, G. B. (1960). *J. Polym. Sci.*, *48*: 278.
166. Butler, G. B. (1971). *J. Macromol. Sci.*, *5*: 219.
167. Butler, G. B., and Chu, Y. C. (1979). *J. Polym. Sci. Polym. Chem. Ed.*, *17*: 859.
168. Aso, C., Ushio, S., and Sogabe, M. (1967). *Makromol. Chem.*, *100*: 100.
169. Samuels, R. J. (1977). *Polymer*, *18*: 452.
170. Kunitake, T., and Tsukino, M. (1979). *J. Polym. Sci. Polym. Chem. Ed.*, *17*: 877.
171. Breslow, D. S. (1976). *Pure Appl. Chem.*, *46*: 103.
172. Morahan, P. S., Regelson, W., Baird, L. G., and Kaplan, A. M. (1974). *Cancer Res.*, *34*: 506.
173. Chang, R. (1962). U.S. 3,025,281 to Hercules Powder Co.; C.A. (1962), 57: 4884.
174. Mosley, S. A. (1951). U.S. 2,549,921 to Union Carbide Corp.; C.A. (1951), 45: 5972.
175. Higashimura, T.,Watanabe, T., and Okatnura, S. (1963). *Kobunshi Kagaku*, 20: 680; C.A. (1964), *60*: 15989.
176. Lal, J. (1962). U.S. 3,062,789 to Goodyear Tire and Rubber Co.; C.A. (1963), *58*: 3524.
177. Okarnura, S., and Higashitnura, T. (1962). Jpn. 6,739 to Mitsubishi Chem. Ind. Co. Ltd.; C.A. (1963), 59: 11688; (1963), Jpn. 13,534 to Mitsubishi M Chem. Ind. Co. Ltd.; C.A. (1963), *59*: 14131.
178. Susorov, L. A., Sukhinin, V. S., Zheltukhina, A. N., and Romanova, E. G. (1985). *Otkrytiya Izobret.*, *34*: 81; C.A. (1986), *104*: 34492.
179. I.G. Farbenindustrie, A.G. (1936). Fr. 792,721; C.A. (1936), 30: 4176.
180. Tsubokawa, N., Takeda, N., and Kudoh, K. (1980). *Carbon*, *18*: 163.
181. Okamura, S., Higashimura, T., and Sakurada, I. (1959). *J. Polym. Sci.*, *39*: 507.
182. Zoss, A. O. (1952). U.S. 2,616,879 to General Aniline and Film Corp.; C.A. (1953), *47*: 4642; (1952), U.S. 2,609,364 to General Aniline and Film Corp.; C.A. (1953), *47*: 3026.
183. Schildknecht, C. E. (1950). U.S. 2,513,820 to General Aniline and Film Corp.; C.A. (1950), *44*: 9728.

184. Schildkniecht, C. E., and Dunn, P. H. (1956). *J. Polym. Sci., 20*: 597.
185. Fishbein, L., and Crowe, B. F. (1961). *Makromol. Chem., 48*: 221.
186. Higashimura, T., Suzuoki, K., and Okamura, S. (1965). *Makromol Chem., 86*: 259.
187. Losev, L. P., Fedotova, O. Y., and Shostakovskii, M. F. (1944). *J. Gen. Chem. (USSR), 14*: 889.
188. Natta, G., Dall'Asta, G., and Mazzanti, G. (1964). Belg. 636,435 to Montecatini; C.A. (1965), *61*: 16187.
189. Sikkema, D. J., and Angrad-Gaur, H. (1980). *Makromol Chem., 181*: 2259.
190. Seung, S. L. N., and Young, R. N. (1978). *J. Polym. Sci. Polym. Lett., 16*: 367.
191. Subira, F., Vairon, J.-P., and Sigwalt, P. (1988). *Macromolecules, 21*: 2339.
192. Shostakovskii, M. F., Khomertov, A. M., and Alimov, A. P. (1963). *Izv, Akad, Nauk, SSSR Ser. Khim., 10*: 1843; C.A. (1964). *60*: 4255.
193. Imai, H., Saegusa, T., Ohsugi, S., and Furukawa, J. (1965). *Makromol. Chem., 81*: 119.
194. Dunay, M., and Noether, H. D. (1964). U.S. 3,156,680 to Celanese Corp.; C.A. (1965), *62*: 2846.
195. Christman, D. L., and Vandenberg, E. J. (1962). U.S. 3,025,282 to Hercules Powder Corp.; C.A. (1962). *57*: 2415.
196. Heck, R. F., and Vandenberg, E. J. (1962). U.S. 3,025,283 to Hercules Powder Corp.; C.A. (1962). *57*: 2415.
197. Okamura, S., Higashimura, T., and Watanabe, T. (1961). *Makromol Chem., 50*: 137.
198. Lal, J. (1958). *J. Polym. Sci., 31*: 179; (1964). *J. Polym. Sci. Part A, 2*: 3369.
199. Gluesenkamp, E. W. (1962). U.S. 3,026,290 to Monsanto; C.A. (1962), *57*: 1080.
200. Roch, J. M., and Saunders, J. (1959). *J. Polym. Sci., 38*: 554.
201. Crivello, J. V., and Lam, J. H. W. (1977). *Macromolecules, 10*: 1307.
202. Crivello, J. V., and Lam, J. H. W. (1979). *J. Polym. Sci. Polym. Chem. Ed., 17*: 977, 2877.
203. Funt, B. L., and Blain, T. J. (1971). *J. Polym. Sci. Part A-1, 9*: 115.
204. Nishikubo, T., Iizawa, T., and Yoshinaga, A. (1979). *Makromol Chem., 180*: 2793.
205. Bortnik, N. M., and Melamid, S. (1956). U.S. 2,734,890; C.A. (1957), *51*: 3184.
206. Trofimov, B. A., Rappoport, L. Ya., Morozova, L. V., Minakova, T. T., Petrov, G. N., and Salnis, K. (1979). *Otkrytiya Izobret. Prom. Obraztsy Tovarnye Znaki, 30*: 97; C.A. (1979), *91*: 158342.
207. Biswas, M., and Maity, N. C. (1981). *J. Macromol Sci. Chem. Part A, 15*: 1553.
208. Shukys, J. E. (1960). Ger. 1,091,751; C.A. (1961), *55*: 25361.
209. Klatte, F. (1912). Ger 271,381; C.A. (1914), *8*: 991; (1914), US 1,084,581 to Chemische Fabriken Griesheim-Electron.
210. Schildknecht, C. A. (1952). *Vinyl and Related Polymers*, Wiley, New York, p. 323.
211. Daniels, W. (1983). *Kirk-Othmer Encyclopedia of Chemical Technology*, 3rd ed., Vol. 23, Wiley, New York, p. 817.
212. Lindemann, M. K. (1971). *Encyclopedia of Polymer Science and Technology*, Vol. 15 (Mark, H. F., Gaylord, N. G., and Bikales, N. M., eds.), Wiley, New York, p. 531.
213. Lindemann, M. K. (1971). *Encyclopedia of Polymer Science and Technology*, Vol. 15 (Mark, H. F., Gaylord, N. G., and Bikales, N. M., eds.), Wiley, New York, p. 577.
214. Leonard, E. C. (1970). *High Polymers*, Vol. 24, Part I, *Vinyl and Diene Polymers*, Wiley-Interscience, New York, p. 263.
215. Lindemann, M. K. (1970). *High Polymers*, Vol. 24, Part I, Vinyl and Diene Polymers, Wiley-Interscience, New York, p. 329.
216. Morrison, G. O., and Shaw, T. P. G. (1933). *Trans. Electrochem. Soc., 63*: 425.
217. Schaeffer, W. D. (1966). US 3,253,020; C.A. (1966), *65*: 3754; (1966), US 3,260,739; both to Union Oil of California; C.A. (1966), *65*: 8767.
218. Schwerdtel, W. (1968). *Chem. Eng. Technol., 40*: 781; (1968), *Hydrocarbon Proc., 47*(11): 187.
219. Oxley, H. F., Thomas, E. B., and Mills, W. G. B. (1947). US 2,425,389 to British Cellanese Corp.; C.A. (1947). *41*: 7410d; (1948). *42*: 1603.
220. Schnitzer, A. W. (1956). US 2,859,241 to Cellanese Corp.; C.A. (1959), *53*: 7991.

221. Adelman, R. L. (1949). *J. Org. Chem.*, *14*: 1057.
222. Foster, D. J., and Tobler, E. (1961). *J. Am. Chem. Soc.*, *83*: 851.
223. Sladkov, A. M., and Petrov, G. S. (1954). *Zh. Obshch. Khim.*, *24*: 450; C.A. (1955). *49*: 6093.
224. Pechukas, A. (1949). US 2,472,434 to Pittburgh Plate Glass; C.A. (1949), *43*: 7957.
225. Chitwood, H. C. (1941). US 2,251,983 to Carbide and Carbon Chemicals Corp.; C.A. (1941), *35*: 7416.
226. Young, L. J. (1961). *J. Polym. Sci.*, *54*: 411.
227. Kurian, C. J., and Muthana, M. S. (1959). *Makromol. Chem.*, *29*: 1 and 19.
228. Rostovshii, E. N., Ushakov, S. N., and Barinova, A. N. (1958). *Izv. Akad. Nauk SSSR, Otd. Khim Nauk*, 59; C. A. (1958), *52*: 11739.
229. Swern, D., Billen, G. N., and Knight, H. B. (1947). *J. Am. Chem. Soc.*, *69*: 2439.
230. Harrison, S. A., and Wheeler, D. H. (1951). *J. Am. Chem. Soc.*, *73*: 839.
231. Flory, P. J., and Leutner, F. S. (1948). *J. Polym. Sci.*, *3*: 880; (1950). *5*: 267.
232. Graessley, W. W., Uy, W. C., and Gandhi, A. (1969). *Ind. Eng. Chem. Fund*, *8*: 696.
233. Graessley, W. W., Hartung, R. D., and Uy, W. C. (1969). *J. Polym. Sci. Part A-2, 7*: 1919.
234. Buselli, A. J., Lindemann, M. K., and Blades, C. E. (1958). *J. Polym. Sci.*, *28*: 485.
235. Thompson, C. F., Port, W. S., and Witnauer, L. P. (1959). *J. Am. Chem. Soc.*, *81*: 2552.
236. Wheeler, O. L., Ernst, S. L., and Crozier, R. N. (1952). *J. Polym. Sci.*, *8*: 409.
237. Clarke, J. T., Howard, R. O., and Stockmayer, W. H. (1961). *Makromol. Chem.*, *44146*: 427.
238. Mayo, F. R., and Walling, C. (1950). *Chem. Rev.*, *46*: 191.
239. Okamoto, Y., Yamada, K., and Nakano, T. (2000). *ACS Symp. Ser. 768*: 57.
240. Okamura, S., and Motoyama, T. (1958). *J. Polym. Sci.*, *28*: 221.
241. Reichert, K. H., and Moritz, H. U. (1987). *Makromol. Chem. Macromol. Symp.*, *10/11*: 571.
242. Gabel, C., and Reichert, K. H. (1985). *Chem. Ing. Tech.*, *57*(7): 612.
243. Bumett, G. M., George, M. H., and Melville, H. W. (1955). *J. Polym. Sci.*, *16*: 31.
244. Bengough, W. L., and Melvile, H. W. (1958). *Proc. Roy. Soc. (London) Ser. A*, *225*: 330.
245. Encinas, M. V., Garrido, J., and Lissi, E. A. (1985). *J. Polym. Sci. Polym. Chem. Ed.*, *23*(9): 2481.
246. Chapiro, A. (1947). *J. Chem. Phys.*, *47*: 747; (1953). *50*: 689; (1957). *54*: 276.
247. Stannett, V. T., Gervasi, J. A., Kearney, J. J., and Araki, K. (1969). *J. Appl. Polym. Sci.*, *13*: 1175.
248. Hummel, D. O. (1963). *Angew. Chem.*, *75*: 330.
249. Woodward, D. G., and Ranshof, J. A. (1970). *J. Paint Technol.*, *42*: 467.
250. Stahel, E. B., and Stannett, V. T. (1969). *Large Radiation Sourcesfor Industrial Processes*, International Energy Agency, Vienna, p. 135.
251. Challa, R. R., Drew, J. H., Stannett, V. T., and Stahel, E. P. (1985). *J. Appl. Polym. Sci.*, *30*(11): 4261.
252. Furukawa, J., Tsuruta, T., and Inoue, S. (1957). *J. Polym. Sci.*, *26*: 234.
253. Inoue, S., Tsuruta, T., and Furukawa, J. (1961). *Makromol. Chem.*, *49*: 13.
254. Welch, F. J. (1964). US 3,127,380 and 3,127,383 to Union Carbide. C.A. (1964), *60*: 5661, 16003d.
255. Wexler, H., and Manson, J. A. (1965). *J. Polym. Sci. Part A-3*, 2903.
256. Bywater, S. (1965). *Fortschr. Hochpolymer. Forsch.*, *4*: 66.
257. Belogorodskaya, K. V., Nikolaev, A. F., Shibalovich, V. G., and Pushkareva, L. I. (1980). *Vysokomol. Soedin. Ser. B.*, *22*(6): 449; C.A. (1980). *93*(24): 221131r.
258. Blandy, C., Gervais, D., and Varion, J. P. (1984). *Eur. Polym. J.*, *20*(6): 619.
259. Mirable, F. A., and Werber, F. X. (1964). US 3,117,112 to W. R. Grace & Co.; C.A. (1964), *60*: 8155.
260. Florjanczyk, Z., and Siudakiewicz, M. (1986). *J. Polym. Sci. Polym. Chem. Ed.*, *24*(8): 1849.
261. Brachman, A. E. (1961). US 3,000,837 to E. I. du Pont de Nemours & Co.; C.A. (1962), *56*: 7516.
262. Gay, F. P. (1962). US 3,047,554 to E. I. du Pont de Nemours & Co.; C.A. (1962), *57*: 13999i.

263. Baker, W. P. (1960). *J. Polym. Sci.*, *42*: 578.
264. Prapas, A. G. (1962). US 3,069,403; C.A. (1963), *58*: 5798.
265. Chandrasekran, M., Pitchumani, S., and Krishan, V. (1986). *Bull. Electrochem.*, *2*(4): 395.
266. Rizzardo, E., Chiefari, J., Mayadunne, R. T. A., Moad, G., and Thang, S. H. (2000). *ACS Symp. Ser. 768*: 278.
267. Matyjaszewski, K. (2000). *ACS Symp. Ser. 768*: 2.
268. Lindemann, M. K. (1971). *Encyclopedia of Polymer Science and Technology*, Vol. 14 (Mark, H. F., Gaylord, N. G., and Bikales, N. M., eds.), Wiley, New York, p. 149.
269. Lindemann, M. K. (1971). *Encyclopedia of Polymer Science and Technology*, Vol. 14 (Mark, H. F., Gaylord, N. G., and Bikales, N. M., eds.), Wiley, New York, p. 208.
270. Brandrup, J., and Immergut, E. H. (1966). *Polymer Handbook, Interscience*, New York, p. II 141.
271. Lindemann, M. K., and Deacon, K. (1986). US 4,616,057 to Sun Chemical Corp.; C.A. (1986), *105*(26): 227574a.
272. Andrews, F., Jahns, R., and Weiher, R. (1970). *Plaste Kautschuk*, *17*: 153.
273. Imoto, E., and Horiuchi, H. (1953). *Chem. High Polym. (Jpn.)*, 8: 463; C.A., 1953, 47: 9663.
274. Chen, C. H. (1976). *J. Polym. Sci. Polym. Chem. Ed.*, *14*: 2109.
275. Seymour, R. B., and Stahl, G. A. (1977). *J. Macromol. Sci. Chem.*, *A11*: 53.
276. Semon, W. L., and Stahl, G. A. (1981). *J. Macromol. Sci. Chem.*, *A15*: 1263.
277. Starnes, W. H. (1996). *J. Vinyl Addit.*, 2: 277
278. Wirtschaftsdaten und Grafiken zu Kunststoffen, Verband Kunststofferzeugende Industrie e. V, www.vke.de (2000).
279. Krüger, R., Hoffmann, K., and Praetorius, W. (1972). *Kunststoffe*, *62*: 602.
280. Zeinalov, B. K., Radzhabov, D. T., and Mekhtieva, F. A. (1971). *J. Polym. Sci. Part C Polym. Symp.*, *33*: 353.
281. Rohlfing, W. (1995). *Kontakt Stud.*, *318*: 225.
282. Böhme, K. (1975). *Angew. Makromol Chem.*, *47*: 243.
283. Minsker, K. S. (1997). *Polym.-Plast. Technol. Eng.*, *36*: 513.
284. Catalina, F. (1997), *Rev. Plast. Mod.*, *73*: 542.
285. Klamann, J.-D. (1999). *Kunststoffe*, *89*: 56.
286. Davidson, J. A., and Witenhafer, D. E. (1980). *J. Polym. Sci. Polym. Phys. Ed.*, *18*: 51.
287. Zhao, J., Wang, X., and Fan, C. (1991). *Polymer*, *32*: 2674.
288. Faraday, A. G.; and Courtis, A. (1992). *Plast. Rubber Compos. Process. Appl.*, *18*: 9104.
289. Bortel, K., and Szewczyk, P. (1993). *Polimery*, *38*: 578.
290. Kiparissides, C., Daskalakis, G., Achilias, D. S., and Sidiropoulou, E. (1997). *Ind. Eng. Chem. Res.*, *36*: 1253.
291. Xie, T. Y., Hamielec, A. E., Wood, P. E., and Woods, D. R. (1991). *J. Vinyl Technol.*, *13*: 2.
292. Regnault, V. (1835). *Justus Liebigs Ann. Chem.*, *14*: 22.
293. Wurtz, A., and Frapolli. (1858). *Justus Liebigs Ann. Chem.*, *108*: 223.
294. Tollens, B. (1866). *Justus Liebigs Ann. Chem.*, *137*: 311.
295. Baumann, E. (1872). *Justus Liebigs Ann. Chem.*, *163*: 308.
296. Biltz, H. (1902). *Ber. Dtsch. Chem. Ges.*, *35*: 3524.
297. Bönnigshausen, K. H. (1986). *Kunststoff Handbuch*, Vol. 2/1 (Becker, G. W., Braun, D., and Felger, H. K., eds.), Carl Hanser, München, p. 39.
298. Bier, G. (1965). *Kunststoffe*, *55*: 228.
299. Patat, F., and Weidlich, P. (1949). *Helv. Chim. Acta*, *32*: 783.
300. Allsopp, M. W., and Giovanni, V. (2000). *Ullmann's Encyclopedia of Industrial Chemistry, Electronic Release.*, *6*: 1.
301. Zilberman, E. N. (1992). *J. Macromol. Sci–Rev. Macromol. Chem. Phys.*, *C32*: 235
302. Bauer, J., and Sabel, A. (1975). *Angew. Makromol. Chem.*, *47*: 15.
303. Sandler, R. S., and Karo, W. (1977). *Polymer Synthesis*, Vol. II (Blomquist, A. T., and Wasserman, H. H., eds.), Academic Press, Orlando, Fla., p. 307.
304. Neumüller, O. A. (1988). *Römpps Chemie-Lexikon*, Vol. 6, Franck'sche Verlagshandlung, Stuttgart, p. 4516.

305. Flatau, K., and Heidel, K. (1986). *Kunststoff Handbuch*, Vol. 2/1 (Becker, G. W., Braun, D., and Felger, H. K., eds.), Carl Hanser Verlag, München, p. 62.
306. Abdel-Alim, A. H., and Hamielec, A. E. (1972). *J. Appl. Polym. Sci.*, *16*: 783.
307. Mickley, H. S., Michaels, A. S., and Moore, A. L. (1962). *J. Polym. Sci.*, *60*: 121.
308. Ryska, M., Kolinsky, M., and Lim, D. (1967). *J. Polym. Sci. Part C*, *16*: 621.
309. Bier, G., and Krämer, H. (1956). *Kunststoffe*, *46*: 498.
310. Rigo, A., Palma, G., and Talamini, G. (1972). *Makromol. Chem.*, *153*: 219.
311. Starnes, W. H., Schilling, F. C., Plitz, I. M., Cais, R. E., Freed, D. J., Hartless, R. L., and Bovey, F. A. (1983). *Macromolecules*, *16*: 790.
312. Starnes, W. H., Schilling, F. C., Abbas, K. B., Cais, R. E., and Bovey, F. A. (1979). *Macromolecules*, *12*: 556.
313. Caraculacu, A., Buruiana, E. C., and Robila, G. (1978). *J. Polym. Sci. Polym. Chem. Ed*, *16*: 2741.
314. Hjertberg, T., and Sörvik, E. M. (1983). *Polymer*, *24*: 673.
315. Hjertberg, T., and Sörvik, E. M. (1983). *Polymer*, *24*: 685.
316. Hjertberg, T., and Sörvik, E. M. (1982). *J. Macromol. Sci. Chem. Part A, 17*: 983.
317. Jurriaan, C., van den Heuvel, M., and Weber, A. J. M. (1983). *Makromol. Chem.*, *184*: 2261.
318. Reding, F. P., Walter, E. R., and Welch, F. J. (1962). *J. Polym. Sci.*, *56*: 225.
319. Mijangos, C., Martinez, G., and Millan, J. L. (1988). *Makromol. Chem.*, *189*: 567.
320. Martinez, G., Mijangos, C., and Millan, J. (1985). *Eur. Polym. J.*, *21*: 387.
321. Millan, J., Carranza, M., and Guzman, J. (1973). *J. Polym. Sci. Part C Polym. Symp.*, *42*: 1411.
322. Abdel-Alim, A. H. (1975). *J. Polym. Appl. Sci.*, *19*: 1227.
323. Talamini, G., and Vidotto, G. (1967). *Makromol. Chem.*, *100*: 48.
324. Hassan, A. M. (1974). *J. Polym. Sci. Polym. Phys. Ed.*, *12*: 655.
325. Pezzin, G., Zilio-Grandi, F., and Sanmartin, P. (1970). *Eur. Polym. J.*, *6*: 1053.
326. Hjertberg, T., Martinsson, E., and Sörvik, E. (1988). *Macromolecules*, *21*: 603.
327. Troitskii, B. B., and Troitskaya, L. S. (1999), *Eur. Polym. J.*, *35*: 2215.
328. Minsker, K. S., Kulish, E. I., and Zaikov, G. E. (1994). *Int. J. Polym. Mater.*, *24*: 107.
329. Kulish, E. I., Kolesov, S. V., Minsker, K. S., and Zaikov, G. E. (1994). *Int. J. Polym. Mater.*, *24*: 123.
330. Starnes, W. H., Wallach, J. A., and Yao, H. (1996). *Macromolecules*, *29*: 7631.
331. Kolesov, S. V., Kulish, E. I., Zaikov, G. E., and Minsker, K. S. (2000), *Plast. Massy*, *1*: 17.
332. Abbas, K. B., and Sörvik, E. M. (1973). *J. Appl. Polym. Sci.*, *17*: 3577.
333. Garton, A., and George, M. H. (1974). *J. Polym. Sci. Polym. Chem. Ed.*, *12*: 2779.
334. Huber, H., and Mitterberger, W. D. (1973). *Kunststoffe*, *63*: 762.
335. Jentet, P. (1995). *Actual. Chim.*, *6*: 12.
336. Talamini, G., Vicentini, A., and Kerr, J. (1998). *Polymer*, *39*: 1879.
337. Chatelain, J. (1973). *Brit. Polym. J.*, *5*: 457.
338. Thomas, M. (1964). Fr. Pat. 1,357,736 to Produits Chimiques Pechiney, Saint-Gobain; C.A. (1964), *61*: 3231c.
339. Boissel, J., and Fischer, N. (1977). *J. Macromol. Sci. Chem. Part A*, *11*: 1249.
340. Cotman, J. D., Gonzalez, M. F., and Claver, G. C. (1967). *J Polym. Sci. Part A-1*, *5*: 1137.
341. Jenckel, E., Eckmans, H., and Rumbach, B. (1949). *Makromol. Chem.*, *4*: 15.
342. Rauer, K., and Groepper, J. (1972). *Kunststoffe*, *62*: 699.
343. Jaspers, H., and Strolenberg, O. K. (1976). *Kunststoffe*, *66*: 688.
344. Verhelst, W. F., Oosterwijk, H. H. J., and van der Bend, H. T. (1980). *Kunststoffe*, *70*: 224.
345. Winter, H., Meyer, H., Edl, W., Meister, W., and Schwarzer, H. (1970). Ger. Pat. 1,927,761 to Elektrochemische Werke München AG; C.A. (1971), *74*: 54344m.
346. Pajaczkowski, A. (1961). Ger. 1,116,408 to Imperial Chemical Industries Ltd.; C.A. (1962), *57*: 2509h.
347. Nakanishi, K., and Tadaza, T. (1970). Jpn. 70 36,514 to Japan Carbide Industries Co., Inc.; C.A. (1971), *74*: 64645z.

348. Kirillov, A. I., Molkov, A. D., and Ivanova, L. F. (1971). S.U. 308,00019; C.A. (1972), *76*: 100390q.
349. Neth. 6,500,231 (1965) to Kureha Chemical Industry Co., Ltd.; C.A. (1966), *64*: 527lb.
350. Amagi, Y., and Shiigi, Y. (1969). Jpn. 69 13,146 to Kureha Chemical Industry Co., Ltd.; C.A. (1969), *71*: 81889n.
351. Ravey, M., and Waterman, J. A. (1977). *J. Polym. Sci. Polym. Chem. Ed.*, *15*: 2521.
352. Barter, J. A., and Kellar, D. E. (1977). *J. Polym. Sci. Polym. Chem. Ed.*, *15*: 2545.
353. Kuchanov, S. I., and Bort, D. N. (1973). *Polym. Sci. USSR Part A*, *15*: 2712.
354. Vidotto, G., Crosato-Arnaldi, A., and Talamini, G. (1968). *Makromol. Chem.*, *114*: 217.
355. Ravey, M., and Waterman, J. A. (1975). *J. Polym. Sci. Polym. Chem. Ed.*, *13*: 1475.
356. Danusso, F., Moraglio, G., and Gazzera, S. (1954). *Chim. Ind.*, *(Milan)*, *36*: 883.
357. Monaco, S. L., Mazzolini, C., Patron, L., and Moretti, A. (1971). Ger. 2,046,143 to Chatillon Societa Anonima Italiana per le Fibre Tessili Artificiali SpA; C.A. (1971), *75*: 6590n.
358. Monaco, S. L., Mazzolini, C., Patron, L., and Moretti, A. (1967). Ital. 792,244 to Chatillon Societa Anonima Italiana per le Fibre Tessili Artificiali SpA; C.A. (1969), *70*: 20927f.
359. Blankenstein, G., and Rosendahl, F. K. (1973). Ger. 2,158,581 to Bayer AG; C.A. (1973), *79*: 5478le.
360. Chatelain, J., and Steinbach-van Gaver, G. (1968). Fr. 1,538,572 to Produits Chimiques Pechiney Saint-Gobain; C.A. (1968), *71*: 3909e.
361. Schrage, K. (1969). Fr. 2,000,081 to Dynamit Nobel AG; C.A. (1970), *72*: 44322x.
362. Fordham, J. W. L., Burleigh, P. H., and Sturm, C. L. (1959). *J. Polym. Sci.*, *41*: 73.
363. Talamini, G., and Vidotto, G. (1961). *Makromol. Chem.*, *50*: 129.
364. Braun, D., Bezdadeo, E., and Dethloff, M. (1986). *Angew. Makromol. Chem.*, *142*: 183.
365. Sobue, H., and Kubota, H. (1966). *Makromol. Chem.*, *90*: 276.
366. Martinez, G., Mijangos, C., and Millan, J. (1983). *J. Appl. Polym. Sci.*, *28*: 33.
367. Moore, J. A. (1977). *Macromolecular Synthesis*, Coll. Vol. I (Overberger, C. G., ed.), Wiley, New York, p. 195.
368. Hauss, A. F. (1971). *J. Polym. Sci. Part C Polym. Symp.*, *33*: 1.
369. Won, H. Y., and Kim, I. W. (1993). *Kobunja Kwahak Kwa Kisul.*, *4*: 263.
370. Fauvarque, J. (1998). *Caoutch. Plast.*, *75*: 33.
371. Dixon, K. W. (2000). *Annu. Tech. Conf. - Soc. Plast. Eng.*, *58*: 3414.
372. Longeway, G. D., and Witenhafer, D. E. (2000). *Annu. Tech. Conf. - Soc. Plast. Eng.*, *58*: 3417.
373. Bieringer, H., Flatau, K., and Reese, D. (1984). *Angew. Makromol. Chem.*, *123/124*: 307.
374. Shiraishi, M., and Toyoshima, K. (1973). *Brit. Polym. J.*, *5*: 419.
375. Macoveanu, M., and Reichert, K. H. (1975). *Angew. Makromol. Chem.*, *44*: 141.
376. Crosato-Arnaldi, A., Gasparini, P., and Talamini, G. (1968). *Makromol. Chem.*, *117*: 140.
377. Barclay, L. M. (1976). *Angew. Makromol. Chem.*, *52*: 1.
378. Smallwood, P. V. (1986). *Polymer*, *27*: 1609.
379. Eliassaf, J. (1974). *J. Macromol. Sci. Chem. Part A*, *8*: 459.
380. Lambling, C., and Percheron, J. C. (1968). Fr. 1,512,417 to Produits Chimiques Pechiney, Saint Gobain; C.A. (1969), *70*: 58402u.
381. Sanches, J. (1970). Ger. 2,028,363 to Pennwalt Corp.; C.A. (1971), *74*: 64587g.
382. Steinbach-van Gaver, G., and Fagnoni, Y. (1967). Fr. 1,482,879 to Produits Chimiques Pechiney, Saint Gobain; C.A. (1968), *68*: 13728x.
383. Flatau, K. (1986). *Kunststoff Handbuch*, Vol. 2/1 (Becker, G. W., Braun, D., and Felger, H. K., eds.), Carl Hanser, München, p. 152.
384. Karakas, G., and Orbey, N. (1989). *Br. Polym. J.*, *21*: 399.
385. Sæthre, B., and Pedersen, S. (1996). *Kjemi.*, *56*: 20.
386. Melis, S., Ghielmi, A., Storti, G., and Morbidelli, M. (1998). *Entropie*, *34*: 65.
387. Forcolin, S., Marconi, A. M., Ghielmi, A., Butte, A., Storti, G., and Morbidelli, M. (1999). *Plast. Rubber Compos.*, *28*: 109.
388. Suenaga, Y., and Akimoto, A. (1999). *Colloids Surf.*, *153*: 321.

389. Hohenstein, W. P., and Mark, H. (1946). *J. Polym. Sci.*, *1*: 127.
390. Castrantas, H. M., Mucenieks, P. R., Cohen, B., and Mackellar, D. G. (1970). Ger. 2,002,865 to FMC Corp.; C.A. (1970), *73*: 88658m.
391. Flatau, K. (1986). Kunststoff Handbuch, Vol. 2/1 (Becker, G. W., Braun, D., and Felger, H. K., eds.), Carl Hanser, München. p. 165.
392. Tauer, K., and Behnisch, J. (1985). *Angew. Makromol. Chem.*, *131*: 157.
393. Bartl, H., Roos, E., Schuster, K., and Schmidt, A. (1980). Ger. 2,841,046 to Bayer AG; C.A. (1980), *93*: 73935p.
394. Fredrickson, M. J., Sturt, A. C., and Williams, R. H. (1974). Ger. 2,352,846 to BP Chemicals International Ltd.; C.A. (1974), *81*: 106570y.
395. Hopff, H., and Fakla, I. (1965). *Makromol. Chem.*, *88*: 54.
396. Pierce, J. K. (1972). U.S. 3,642,740 to *Diamond Shamrock Corp.*; C.A. (1972), *77*: 6365j.
397. Goffart, J. L. (1965). Neth. 6,409,653 to Pittsburgh Plate Glass Co.; C.A. (1965), *63*: 4414a.
398. Visger, R. L., and Hahn, H. G. (1966). U.S. 3,255,164 to Dow Chemical Co.; C.A. (1966), *65*: 3992b.
399. Kraft, P. (1968). S. Afr. 6,705,994 to Stauffer Chemical Co.; C.A. (1969), *70*: 38534k.
400. O'Donnell, R. T. (1960). U.S. 2,957,858 to Rubber Corp. of America; C.A. (1961), *55*: 6934e.
401. Hoenig, W., and Freytag, J. (1959). Ger. 1,108,908 to Dynamit-Nobel AG; C.A. (1962), *56*: 6184f.
402. Egli, R. (1972). Ger. 2,151,703 to Lonza Ltd.; C.A. (1972), *77*: 49138e.
403. Ugelstad, J., Mork, P. C., Dahl, P., and Rangnes, P. (1969). *J. Polym. Sci. Part C Polym. Symp.*, *27*: 49.
404. Geil, P. H. (1977). *J. Macromol. Sci. Chem. Part A*, *11*: 1271.
405. Kalka, J., Winter, H., and Kania, A. (1977). Ger. 2,531,780 to Chemische Werke Hüls AG; C.A. (1977), *86*: 107422s.
406. Liegeois, J. M. (1971). *J. Polym. Sci. Part C Polym. Symp.*, *33*: 147.
407. Smith, W. V., and Ewart, R. H. (1948). *J. Chem. Phys.*, *16*: 592.
408. Peggion, E., Testa, F., and Talamini, G. (1964). *Makromol. Chem.*, *71*: 173.
409. Friis, N., and Hamielec, A. E. (1975). *J. Appl. Polym. Sci.*, *19*: 97.
410. Berens, A. R. (1975). *Angew. Makromol. Chem.*, *47*: 97.
411. Sörvik, E. M. (1977). *J. Appl. Polym. Sci.*, *21*: 2769.
412. Lyngaae-Jorgensen, J. (1971). *J. Polym. Sci. Part C Polym. Symp.*, *33*: 39.
413. Small, K. W., Yearsley, F., and Greaves, J. C. (1971). *J. Polym. Sci. Part C Polym. Symp.*, *33*: 201.
414. Dimov, K., and Slavtcheva, L. (1973). *Polymer*, *14*: 234.
415. Blake, G. J. (1970). Ger. 2,019,833 to Imperial Chemical Industries Ltd.; C.A. (1971), *74*: 42872u.
416. Germar, H., Hellwege, K. H., and Johnsen, U. (1963). *Makromol. Chem.*, *60*: 106.
417. Minato, H. (1969). Jpn. 6,921,344 to Teijin Ltd.; C.A. (1970), *72*: 79645g.
418. Crosato-Arnaldi, A., Talamini, G., and Vidotto, G. (1968). *Makromol. Chem.*, *111*: 123.
419. Wesslen, B., and Wirsen, A. (1975). *J. Polym. Sci. Polym. Chem. Ed*, *13*: 2571.
420. Jisova, V., Kolinsky, M., and Lim, D. (1970). *J. Polym. Sci. Part A-1*, *8*: 1525.
421. Kudrna, M., Jisova, V., and Kolinsky, M. (1977). *J. Polym. Sci. Polym. Chem. Ed.*, *15*: 2367.
422. Guyot, A., and Mordini, J. (1971). *J. Polym. Sci. Part C Polym. Symp.*, *33*: 65.
423. Yamazaki, N., Sasaki, K., and Kambara, S. (1964). *J. Polym. Sci. Part B Polym. Lett.*, *2*: 487.
424. Haszeldine, R. N., Hyde, T. G., and Tait, P. J. T. (1973). *Polymer*, *14*: 215.
425. Ulbricht, J., Giesemann, J., and Gebauer, M. (1968). *Angew. Makromol. Chem.*, *3*: 69.
426. Guzman, I. S., Zavadovskaya, E. N., Chigir, N. N., Sharayev, O. K., Tinyakova, Y. I., and Dolgoplosk, B. A. (1977). *Polym. Sci. USSR*, *19*: 3225.
427. Carenza, M., Tavan, M., and Palma, G. (1979). *J. Polym. Sci. Polym. Chem. Ed.*, *17*: 2087.
428. Russo, S., and Stannett, V. (1971). *Makromol. Chem.*, *143*: 57.

429. Palma, G., Talamini, G., and Tavan, M. (1977). *J. Polym. Sci. Polym. Phys. Ed.*, *15*: 1537.
430. Ellinghorst, G., and Hummel, D. O. (1977). *Angew.* Makromol. *Chem.*, *63*: 157.
431. Ellinghorst, G., and Hummel, D. O. (1977). *Angew. Makromol. Chem.*, *63*: 167.
432. Russo, S., and Stannett, V. (1971). *Makromol. Chem.*, *143*: 47.
433. Thompson, L. F., and Mayhan, K. G. (1972). *J. Appl. Polym. Sci.*, *16*: 2291.
434. Liepins, R., and Sakaoku, K. (1972). *J. Appl. Polym. Sci.*, *16*: 2633.
435. Yasuda, H., and Lamaze, C. E. (1973). *J. Appl. Polym. Sci.*, *17*: 1519.
436. Bier, G. (1965). *Kunststoffe*, *55*: 694.
437. Lambla, M., Valentin, B., Guerrero, S., and Banderet, A. (1977). *J. Macromol. Sci. Chem. Part A*, *11*: 1439.
438. Ghosh, P., Bhattacharyya, A. S., and Maitra, S. (1988). *Angew. Makromol. Chem.*, *162*: 135.
439. Simionescu, C. I., Bulacovschi, V., Macocinschi, D., Stoica, G., and Negulescu, I. I. (1988). *Polym. Bull.*, *19*: 59.
440. Devedjiev, I., Ganev, V., and Borrissov, G. (1988). *Eur. Polym. J.*, *24*: 475.
441. Carroll, W. F., Elcik, R. G., and Goodman, D. (1992). *Plast. Recycl.*, 131.
442. Frey, V. W. (1994). *Kunststoffberater*, *39*: 40.
443. Pfaendner, R., Herbst, H., Hoffmann, K., and Sitek, F. (1995). *Angew. Makromol. Chem.*, *232*: 193.
444. Menges, G. (1996). *Pure and Appl. Chem.*, *68*: 1809.
445. Wuckel, L. (1997). *J. prakt. Chem.*, *339*: 687.
446. Minsker, K. S. (1997). *Polym.-Plast. Technol. Eng.*, *36*: 513.
447. Greenwood, R., Luckham, P. F., and Gregory, T. (1997). *J. Colloid Interface Sci.*, *191*: 11.
448. Sæthre, B., Thorjussen, T., Jacobsen, H., Pedersen, S., and Leth-Olsen, K.-A. (1999). *Plast. Rubber Compos.*, *28*: 170.
449. Hintzer, K. (1987). Houben-Weyl: Methoden der Organischen Chemie, Vol. E20 (Bartl, H., and Falbe, J., eds.), Georg Thieme, Stuttgart, p. 1028.
450. Korinek, P.M. (1990). *Kunststoffe*, *80*: 1137.
451. Swartz, J. (1901). *Bull. Acad Roy. Belg.*, *7*: 383.
452. Starkweather, H. W. (1934). *J. Am. Chem. Soc.*, *56*: 1870.
453. Newkirk, A. E. (1946). *J. Am. Chem. Soc.*, *68*: 2467.
454. Burk, R. E., Coffman, D. D., and Kalb, G. H. (1947). U.S. Pat. 2,425,991 to du Pont; C.A. (1948), *42*: 198f.
455. Coffman, D. D., and Ford, T. A. (1947). U.S. Pat. 2,419,010 to du Pont; C. A. (1947), *41*: 4964c.
456. Kalb, G. H., Coffman, D. D., Ford, T. A., and Johnston, F. L. (1960). *J. Appl. Polym. Sci.*, *4*: 55.
457. Brasure, D., and Ebnesajjad, S. (1989). *Encyclopedia of Polymer Science and Technology*, Vol. 17 (Mark, H. F., Bikales, N. M., Overberger, C.G., Menges, G., Kroschwitz, J.I., eds.), Wiley & Sons, New York, p. 468ff.
458. Sadler, S. R., and Karo, W. (1980). *Progr. Org. Chem. Polym. Syn.*, *29*: 295
459. Sianesi, D., and Caporiccio, G. (1968). *J. Polym. Sci. Part A-1*, *6*: 335.
460. Usmanov, Kh. U., Sirilibaev, T. S., and Yul'chibaev, A. A. (1977). *Russ. Chem. Rev.*, *46*: 462.
461. Reiher, M. (1971). *Kunststoff Handbuch*, Vol. II (Vieweg, R., Reiher, M., and Scheurlen, H., eds.), Carl Hanser, München, p. 411.
462. Coffman, D. D., and Cramer, R. D. (1949). U.S. Pat. 2,461,523 to du Pont; C.A. (1949), *43*: 3437b.
463. Cass, O. W. (1947). U.S. Pat. 2,442,993 to du Pont; C.A. (1948), *42*: 6839f.
464. Söll, J. (1938). U.S. Pat. 2,118,901 to IG Farbenindustrie AG; C.A. (1938), *32*: 5409(6).
465. Usmanov, Kh. U., Yul'Chibaev, A. A., Gafurov, A. Kh., and Kolyodin, V. G. (1973). *Vysokornol. Soedin. Ser. B*, *15*: 124; C.A. (1973), *79*: 19188v.
466. Gafurov, A. Kh. (1971). *Nauchn. Tr. Tashk- Gos. Univ.*, *403*: 98; C.A. (1972), *78*: 72707u.
467. Nishida,T., and Itoi, K. (1970). Jpn. Pat. 70 18,463 to Kurashiki Rayon Co.; C.A. (1970), *73*: 99406d.

468. Neth. Pat. Appl. 6,607,093 (1966) to Pittsburgh Plate Glass Co.; C.A. (1967), *66*: 86116q.
469. Brit. Pat. 607,499 (1948) to British Thomson-Houston Co. Ltd.; C.A. (1949), *43*: 3659i.
470. Coffman, D. D., and Ford, T. A. (1947). U.S. Pat. 2,419,008 to du Pont; C.A. (1947), *41*: 4963i.
471. Coffman, D. D., and Ford, T. A. (1947). U.S. Pat. 2,419,009 to du Pont; C.A. (1947), *41*: 4964b.
472. Johnston, F. L., and Pease, D. C. (1950). U.S. Pat. 2,510,783 to du Pont; C. A. (1950), *46*: 1299d.
473. Fr. Pat. 1,566,920 (1969) to Kureha Chemical Industry Co. Ltd.; C.A. (1969), *71*: 125237e.
474. Fr. Pat. 1,560,029 (1969) to Deutsche Solvay-Werke GmbH; C.A. (1969), *71*: 81896n.
475. Scoggins, L. E. (1971). U.S. Pat. 3,573,242 to Phillips Petroleum Co.; C.A. (1971), *75*: 6639k.
476. Fr. Demande 2,004,758 (1969) to Dynamit Nobel AG; C.A. (1970), *72*: 112058t.
477. Uschold, R. E. (1993). US 5229480 A to Du Pont de Nemours; C.A. (1993), *120*: 31542.
478. Haszeldine, R. N., Hyde, T. G., and Tait, P. J. T. (1973). *Polymer*, *14*: 221.
479. Raucher, D., Demiel, A., Levy, M., and Vofsi, D. (1979). *J. Polym. Sci. Polym. Chem. Ed.*, *17*: 2825.
480. Iikubo, Y., Nishida, T., and Furukawa, Y. (1972). *U.S. Pat.* 3,645,998; C.A. (1972), *76*: 154464k.
481. Haszeldine, R. N., Hyde, T. G., and Tait, P. J. T. (1973). *Polymer*, *14*: 224.
482. Raucher, D., and Levy, M. (1975). *J. Polym. Sci. Chem. Ed.*, *13*: 1339.
483. Korin, A., Levy, M., and Vofsi, D. (1980). *J. Polym. Sci. Polym. Chem. Ed.*, *18*: 109.
484. Usmanov, Kh. U., Yul'chibaev, A. A., and Sirilibaev, T. (1971). *J. Polym. Sci. Part A-1*, *9*: 1779.
485. Usmanov, Kh. U., Yul'chibaev, A. A., Mat'yakubov, A. A., Kuzieva, Kh., and Kazakov, S. (1972). *Dokl. Akad. Nauk Uzb. SSR*, *29*: 37; C.A. (1973). *78*: 44039j.
486. Kobayashi, H., Shen, M., and Bell, A. T. (1974). U.S.N.T.I.S. AD Rep. AD 77 8682/5GA; C.A. (1974), *81*: 152780q.
487. Westwood, A. R. (1971). *Eur. Polym. J.*, *7*: 377.
488. Nakamura, K., Ichimura, M., and Fukushima, Y. (1973). Jpn. Pat. 7,308,755; C.A. (1974), *80*: 109244.
489. Raucher, D., and Levy, M. (1979). *J. Polym. Sci. Polym. Chem. Ed*, *17*: 2675.
490. Rindfleisch, F., DiNoia, T., and Todd, P., and McHugh, M.A. (1996). *J. Phys. Chem.*, *100*: 15581.
491. Wallace, M. L., and Kabayama, M. A. (1966). *J. Polym. Sci. Part A-1*, *4*: 2667.
492. Chapiro, A., Mankowski, Z., and Schmitt, N. (1982). *J. Polym. Sci. Polym. Chem. Ed.*, *20*: 1791.
493. Nguyen, T. (1985). *J. Macromol. Sci. Rev. Macromol. Chem. Part C*, *25*: 227.
494. Chatfield, D. A. (1983). *J. Polym. Sci. Polym. Chem. Ed*, *21*: 1681.
495. Hanes, M., and Lando, J.B. (1993). *J. Appl. Polym. Sci.*, *49*: 1223.
496. Koenig, J. L. (1966). *J. Polym. Sci. Part A-2*, *4*: 401.
497. Koenig, J. L., and Boerio, F. (1969). *Makromol. Chem.*, *125*: 302.
498. Simril, V. L., and Curry, B. A. (1960). *J. Appl. Polym. Sci.*, *4*: 62.
499. Görlitz, M., Minke, R., Trautvetter, W., and Weisgerber, G. (1973). *Angew. Makromol. Chem.*, *29/30*: 137.
500. Golike, R. C. (1960). *J. Polym. Sci.*, *42*: 582.
501. Lando, J. B., Olf, H. G., and Peterlin, A. (1966). *J. Polym. Sci. Part A*, *4*: 941.
502. Bruch, M. D., Bovey, F. A., and Cais, R. C. (1984). *Macromolecules*, *17*: 2547.
503. Willson, C. W., and Santee, E. R., Jr. (1965). *J. Polym. Sci. Part C*, *8*: 97.
504. Ovenall, D.W., and Uschold, R.E. (1991). *Macromolecules*, *24*: 3235.
505. Gafurov, A. Kh., and Yakubov, N. I. (1989). *Plast. Massy*, *7*: 5.
506. Bekiarian, P., and Farnham, W. (2000). PCT Int Appl. WO 2000052060 to Du Pont; C.A. (2000), *133*: 223513.
507. Haruvy, Y. (1987). *Polym. Bull.*, *18*: 137.

508. Fuehrer, J., and Ellinghorst, G. (1981). *Angew. Makromol. Chem.*, *93*: 175.
509. Vierkotten, D., and Ellinghorst, G. (1983). *Angew. Makromol. Chem.*, *113*: 153.
510. Goryacheva, G., Budris, S. V., Tverskoi, V. A., Chikhacheva, I. P., Stavrova, G. K., Karachevtsev, V. G., and Amelina, N. V. (1990). *Plast. Massy*, *3*: 10.
511. Niemöller, A., Scholz, H., Goetz, B., and Ellinghorst, G. (1988) *J. Membr. Sci.*, *36*: 385.
512. Megerle, C. A., and Lewis, K. J. (1982). *Polym. Prepr. Am. Chem. Soc. Div. Polym. Chem.*, *23*: 250. (1983).
513. JP 58091731 to Teijin Ltd. Japan; C.A. (1983), *99*: 213768.
514. Sugaya, Y., Ookubo, H., and Myake, H., (1993). JP 05125205 to Asahi Glass Co Ltd. Japan; C.A. (1993), *K119*: 272848.
515. Ishisaki, T., and Matsushita, K. (1999). JP 11329392 to Nitto Denko Corp., Japan; C.A. (1999), *131*: 353695.
516. Yanai, S., Iga, Katsumi, Matsumoto, Y., and Higo, N. (1996). CA 2178444 to Takeda Chemical Industries Ltd.; C.A. (1996), *126*: 255484.
517. Ford, Th. A., and Hanford, W. E. (1948). U.S. Pat. 2,435,537 to Du Pont; C.A. (1948), *42*: 3215a.
518. Kawai, H. (1969). *Jpn. J. Appl. Phys.*, *8*: 1975.
519. Bergmann, J. G., Jr., McFee, J. H., and Crane, G. R. (1971). *Phys. Lett.*, *18*: 203.
520. Nakamura, K., and Wada, Y. (1971). *J. Polym. Sci. Part A*, *9*: 161.
521. Lovinger, A. J. (1981). *Developments in Crystalline Polymers I* (Bassett, D. C., ed.), Applied Science Publishers, London, p. 195.
522. Jungnickel, B.-J., Russo, S., Pianca, M., Moggi, G., Gregorio, R., Cestari, M., Chaves, N., Nociti, P. S., de Mendonca, J. A., and de Almeida Lucas, A. (1996), in *Polymeric Materials Encyclopedia*, Vol. 9, (Salamone J.C. ed.), CRC Press Boca Raton, p. 7115ff.
523. Humphrey Jr., J. S., Bartoszek, E. J., and Drujon, X. (1996), in *Polymeric Materials Encyclopedia*, Vol. 11 (Salamone J.C. ed.), CRC Press Boca Raton, p. 8585ff.
524. Dohany, J. E. (1994). *Encyclopedia of Chemical Technology* (Kirk-Othmer), 4th ed., Vol. 11 (Kroschwitz, J. I., Howe-Grant, M., eds.), John Wiley & Sons, New York, p. 694ff.
525. Downing, F. B., Benning, A. F., and McHarness, R. C. (1951). U.S. Pat. 2,551,573 to du Pont; C.A. (1951), *45*: 9072e.
526. Kaess, F., Lienhard, K., and Michaud, H. (1969). Ger. Pat. 1,288,085 to Süddeutsche Kalkstickstoff-Werke AG; C. A. (1969), *70*: 77292z.
527. Boalmer, M., and Elsheikh, M. Y. (1991), Eur. Patent Appl. EP 406 748 to *Atochem, Chem. Abstr.* 114, 184780f.
528. Weinhold, S., Litt, M. H., and Lando, J. B. (1979). *J. Polym. Sci. Polym. Lett. Ed.*, *17*: 585.
529. Hauptschein, M. (1965). U.S. Pat. 3,193,539 to Pennsalt Chemicals Corp.; C.A. (1965), *63*: 13443c.
530. Brit. Pat. 1,179,078 to Daikin Kogyo Ltd.; C.A. (1970), *72*: 56095p.
531. Iserson, H. (1966). U.S. Pat. 3,245,971 to Pennsalt Chemicals Corp.; C.A. (1966), *64*: 19822a.
532. McCain, G. H., Semanack, J. R., and Dietrich, J. J. (1968). Fr. Pat. 1,530,119 to Diamond Alkali Co.; C.A. (1969), *71*: 3923e.
533. Fr. Pat. 1,419,741 (1965) to Kureha Chemical Industry Co. Ltd.; C.A. (1965), *65*: 9049b.
534. Liepins, R., Surles, J. R., Morosoff, N., Stannett, V. T., Timmons, M. L., and Wortman, J. J. (1978). *J. Polym. Sci. Polym. Chem. Ed.*, *16*: 3039.
535. Natta, G., Allegra, G., Bassi, I. W., Sianesi, D., Caporiccio, G., and Torti, E. (1965). *J. Polym. Sci. Part A*, *3*: 4263.
536. Gal'perin, E. L., Dubov, S. S., Volkova, E. V., Mlenik, M. P., and Bulygina, L. A. (1966). *Polym. Sci. USSR*, *8*: 2246.
537. Gal'perin, Ye. L., Kosmynin, B. P., Aslanyan, L. A., Mlenik, M. P., and Smirnow, V. K. (1970). *Polym. Sci. USSR*, *12*: 1881.
538. Foster, R. G. (1969). Brit. Pat. 1,149,451 to Imperial Chemical Industries Ltd.; C.A. (1969), *71*: 13541t.

539. Brinati, G., Lazzari, P., and Arcella, V. (1998). EP 816397 A1 to Ausimont S.P.A. Italy. C.A. (1998), *128*: 128408.
540. Brit. Pat. 1,057,088 (1967) to Kali-Chemie AG; C.A. (1967), *66*: 6604 lh.
541. Dohany, J. E., and Humpfrey, J. S. (1989). *Encyclopedia of Polymer Science and Technology*, Vol. 17 (Mark, H. F., Bikales, N. M., Overberger, C.G., Menges, G., Kroschwitz, J.I., eds.), Wiley & Sons, New York, p. 532ff.
542. Kato, M., Naito, S., Ishigaki, H., and Sugama, S. (1992). Jap. Pat. 04 285 602 to Nippon Oil and Fats, *Chem. Abstr.*, *119*: 96401f.
543. McCarthy, T. F., Williams, R., Bitay, J. F., Zero, K., Yang, M. S., and Mares, F. (1998). *J. Appl. Polym. Sci.*, *70*: 2211.
544. Doll, W. W., and Land, J. B. (1970). *J. Appl. Polym. Sci.*, *14*: 1767.
545. Brit. Pat. 1,178,227 (1970) to *Daikin Kogyo Co., Ltd.*; C.A. (1970), *72*: 67510w.
546. Brit. Pat. 1,057,088 (1967) to Kali-Chemie AG; C.A. (1967), *66*: 6604lh.
547. Carlson, D. P. (1969). Ger. Often. 1,806,426 to Du Pont; C.A. (1969), *71*: 13533s.
548. Russo, S., Behari, K., Chengji, S., Pianca, M., Barchiesi, E., and Moggi, G. (1993). *Polymer*, *34*: 4777.
549. Wilson, C. W. (1963). *J. Polym. Sci. Part A*, *1*: 1305.
550. Cais, K. E., and Kometani, J. M. (1984) *Macromolecules*, *17*: 1887.
551. Helfrich, G. F., and Rothermel, E. J., Jr. (1968). U.S. Pat. 3,380,977 to Dow Chemical Co.; C.A. (1968), *68*: 115212s.
552. Belg. Pat. 620-986 (1963) to Deutsche Solvay-Werke GmbH; C.A. (1963), *59*: 2993d.
553. Westwood, A. R. (1971). *Eur. Polym. J.*, *7*: 363.
554. Westwood, A. R. (1971). *Eur. Polym. J.*, *7*: 377.
555. Clark, D. T., and Shuttleworth, D. (1979). *Eur. Polym. J.*, *15*: 265.
556. Ratway, R., and Balik, C.M. (1997). *J. Polym. Sci., Part B: Polym. Phys.*, *35*: 1651.
557. Chasset, R., Legeay, G., Touraine, J.C., and Arzur, B. (1988). *Eur. Polym. J.*, *24*: 1049.
558. Sato, D., Jikei, M., and Kakimoto, M. (1998). *Polym. Prepr. (ACS, Div. Polym. Chem.)*, *39*: 926.
559. Lando, J. B., and Doll, W. W. (1968). *Macromol. Sci. Phys. Part B*, *2*: 205.
560. Moreira, R. L. (1996), in *Polymeric Materials Encyclopedia*, Vol. 11 (Salamone J.C., ed.), CRC Press Boca Raton, p. 8596ff.
561. Pochan, J. M., Hinman, D. F., Froix, M. F., and Davidson, T. (1977). *Macromolecules*, *10*: 113.
562. Wille, R. A., and Burchill, M. T. (1998). WO 9838242 A1 to Elf Atochem. C.A. (1999), *129*: 217043.
563. Watanabe, T., Momose, T., Ishigahi, L., Tabata, T., and Okamoto, J. (1981). *J. Polym. Sci. Polym. Lett., 50. Ed.*, *19*: 599.
564. Gootaert, W. M., Millet, G. H., and Worm, A. T. (1993). *Encyclopedia of Chemical Technology (Kirk-Othmer)*, 4th ed., Vol. 8 (Kroschwitz, J. I., Howe-Grant, M., eds.), Wiley & Sons, New York, p. 990ff.
565. Bonardelli, P., Moggi, G., and Turforro, A. (1986). *Polymer*, *27*: 905.
566. Froix, M. F., Goedde, A. O., and Pochan, J. M. (1977). *Macromolecules*, *10*: 778.
567. Abe, M. (1991). *DE 4031607 A1* to Nippon Mektron Co., Ltd, Japan, C.A. (1991), *115*: 115278.
568. Vierkotten, D., and Ellinghorst, G. (1983). *Angew. Makromol. Chem.*, *113*: 153.
569. de Lacy-Costello, B.P., Evans, P., Guernion, N., Ratcliffe,, N. M., Sivanand, P. S., and Teare, G. C. (2000). *Synth. Met.*, *114*: 181.
570. Eietela, S., Holmber, S., Karjalainer, M., Nasman, J., Paronen, M., Serimaa, R., Sundholm, F., and Vahvaselka, S. (1997). *J. Mater. Chem.*, *7*: 721.
571. Raghavan, S., Jan, B., and Chilkunda, R. (1996). US 5531900 A1 to University of Arizona. C.A. (1996), *125*: 170093.
572. Iwata, H., Oodate, M., Uyama, Y., Amemiya, H., and Ikada, Y. (1991). *J. Membr. Sci.*, *55*: 119.
573. Peterson, E., Betz, N., and Le Moel, A. (1996). *J. Chim. Phys. Phys.-Chim. Biol.*, *93*: 188.

574. Iwata, I., Oodate, M., Uyama, Y., Matsuda, T., Amemiya, H., and Ikada, Y. (1993). *Proc. Int. Conf. Intell. Mater 1st.* 346.

575. Betz, N., Le Moel, A., Durand, J. P., Balanzat, F., and Darnez, C. (1992). *Macromolecules,* 25: 213.

576. Welch, G. J. (1974). *Polymer, 15:* 429.

577. Ali, S., and Raina, A. K. (1978). *Makromol. Chem., 179:* 2925.

578. Chapiro, A., Mankowski, Z., and Schmitt, N. (1982). *J. Polym. Sci. Polym. Chem. Ed., 20:* 1791.

579. Butler, G. B., Olson, K. G., and Tu, C.-L. (1984). *Macromolecules, 17:* 1887.

580. Wilson, C. W., and Santee, E. R., Jr. (1965). *J. Polym. Sci. Part C, 8:* 97h.

581. Bovey, F. A., Schilling, F. C., Kwei, T. K., and Fritsch, H. L. (1977). *Macromolecules, 10:* 559.

582. Lando, J. B., Olf, H. G., and Peterlin, A. (1966). *J. Polym. Sci. Part A, 4:* 941.

583. Tai Chen, L., and Frank, C. W. (1985). *Macromolecules, 18:* 2163.

584. Nguyen, T. (1985). *J. Macromol. Sci. Rev. Macromol. Chem. Part C, 25:* 227.

585. Timmerman, R., and Greyson, W. (1962). *J. Appl. Polym. Sci., 6:* 456.

586. Reiher, M. (1971). *Kunststoff Handbuch,* Vol. 1 1 (Vieweg, R., Reiber, M., and Scheurlen, H., eds.), Carl Hanser, München, p. 403.

587. Lovinger, A. (1991). *ACS Symp. Ser., 475,* 84.

588. Balta Callea, F., Gonzalez Arche, A., Ezquerra, T. A., Santa Cruz, C., Batallan, F., Frick, B., and Lopez Cabarcos, E. (1993). *Adv. Polym. Sci., 108:* 1.

589. Furukawa, T., Date, M., and Fukada, E. (1980). *J. Appl. Phys., 51:* 1135.

590. Furukada, T. (1990). *Ferroelectrics, 104:* 229.

591. Furukada, T. (1989). *Phase Transitions, 18:* 143.

592. Humphrey Jr., J. S., and Drujon, X. (1996), in *Polymeric Materials Encyclopedia* (Salamone J.C., ed.), CRC Press Boca Raton, Vol. 11, p. 8591ff.

593. Furukawa, T., and Johnson, G. E. (1980). *J. Appl. Phys., 52:* 940.

594. Guerra, G., Karaaz, F. E., and MacKnight, W. J. (1986). *Macromolecules, 19:* 1935.

595. Kämpf, G., Freitag, D., Fengler, G., and Sommer, K. (1992). *Adv. Polym. Technol., 3:* 169.

596. Song, J. Y., Wang, Y. Y., and Wan, C. C. (1999). *J. Power Surces, 77:* 183.

597. Boutevin, B., Robin, J. J., and Serdani, A. (1992) *Eur. Polym. J., 28:* 1507.

598. Putnam, R. E. (1989). *Comprehensive Polymer Science,* Vol. 3 (Allen, G., ed.), Pergamon Press, Oxford, p. 321.

599. Eleuterio, H. S. (1960). U.S. Pat. 2,958,685 to Du Pont; C.A. (1961), *55:* 6041c.

600. Miller, W. T. (1951). U.S. Pat. 2,567,956; C.A. (1952), *46:* 1808a.

601. Chandrasekaran, S. (1985). *Encyclopedia of Polymer Science and Engineering,* 2nd ed., Vol. 3 (Mark, H. F., Bikales, N. M., Overberger, C.G., Menges, G., Kroschwitz, J. I., eds.), Wiley, New York, p. 463.

602. Thomas, W. M., and O'Shaughnessy, M. T. (1953). *J. Polym. Sci., 11:* 455.

603. Hamilton, J. M. (1953). *Ind. Eng. Chem., 45:* 1347.

604. Dittman, A. L., Passino, H. J., and Wrighton, J. M. (1954). U.S. Pat. 2,689,241 to M. W. Kellogg Co.; C.A. (1955), *49:* 11680i.

605. Volkova, E. V., Zitnakov, P. V., and Forkin, A. V. (1966). *Dokl. Akad. Nauk. SSSR, 167:* 1057; C.A. (1966). *65:* 811c.

606. Hanford, W. E. (1954). U.S. Pat. 2,820,027 to Minnesota Mining & Manufacturing Co.; C.A. (1958), *52:* 5884f.

607. Jewell, J. W. (1961). U.S. Pat. 3,014,015 to Minnesota Mining & Manufacturing Co.; C.A. (1963), *58:* 6945f.

608. Dittman, A. L., and Wrightson, J. M. (1955). U.S. Pat. 2,705,706 to M. W. Kellogg Co.; C.A. (1955), *49:* 13695d.

609. Young, D. M., and Tompson, B. (1955). U.S. Pat. 2,700,662 to Union Carbide Corp.; C.A. (1955), *49:* 5886i.

610. Mantell, R. M., and Hoyt, J. M. (1962). U.S. Pat. 3,043,823 to Minnesota Mining & Manufacturing Co.; C.A. (1962), *57:* 12719b.

611. Reiher, M. (1971). *Kunststoff Handbuch*, Vol. II (Vieweg, R., Reiher, M., and Scheurlen, H., eds.), Carl Hanser, München, p. 369ff.

612. Lynn, M. M., and Worm, A. T. (1987). *Encyclopedia of Polymer Science and Engineering*, 2nd ed., Vol. 7 (Mark, H. F., Bikales, N. M., Overberger, C. G., Menges, G., Kroschwitz, J. I., eds.), Wiley, New York, p. 257.

613. Brit. Pat. 953,098 (1964) to Du Pont; C.A. (1964), *61*: 16275a.

614. Baird, R. L., and MacLachlan, J. D. (1975). Ger. Offen. 2,457,102 to Du Pont; C.A. (1976), *84*: 18916x.

615. Hintzer, K. (1987). *Houben- Weyl: Methoden der Organischen Chemie*, 4th ed., Vol. E20 (Bartl, H., and Falbe, J., eds.), Georg Thieme, Stuttgart, p. 1028.

616. Elias, H.-G., and Vohwinkel, F., eds. (1983). *Neue Polymere Werkstoffe für die industrielle Anwendung*, Carl Hanser, München, p. 141.

617. (a) Plunkett, R. J. (1941). U.S. 2,230,654 to Kinetic Chemicals, Inc.; C.A. (1941), *35*: 3365,9. (b) Plunkett, R. J. (1986) in: *High Performance Polymers: Their Origin and Development* (Seymour, R. B., and Kirshenbaum, G. S., eds.), Elsevier, Amsterdam.

618. Hanford, W. E., and Joyce , R. M. (1946). *J Am. Chem. Soc.*, *68*: 2082.

619. Berry, K. L., and Peterson, J. H. (1951). *J. Am. Chem. Soc.*, *73*: 5195.

620. Dee, G. T., and Sauer, B. B. (1998). *Adv. Phys. 47*: 161.

621. Avakian, P., Starkweather, H., Fontanella, J. J., and Wintersgill, M. C. (1993). *Polym. Mater. Sci. Eng. 70*: 439.

622. Biswas, S. K., and Vijayan, K. (1992). *Wear, 158*: 193.

623. Thomas, R. R. (1999) in: *Fluoropolymers*, Vol. 2 (Properties), Hougham, G. (ed.), Kluwer Academic/Plenum Press, New York, p. 47.

624. Gangal, S. V., and Grot, W. (1989) *Encycl. Polym. Sci. Eng.*, 2nd ed., Vol. 16 (Mark, H. F., ed.), Wiley, New York, pp. 577–648.

625. Hougham, G. (ed.) (1999) *Fluoropolymers*, Vol. 2 (Properties), Kluwer Academic/Plenum Press, New York.

626. Kerbow, D. L. (1990). *Polym. Prepr. 31*: 362.

627. Bro, M. I., and Sandt, B. W. (1960). U.S. 2,946,763 to DuPont; C.A. (1960), *54*: 26015a.

628. Carlson, J. N. (1970). U.S. 3,528,954 to DuPont; C.A. (1969), *71*: 39639f.

629. Coker, J. N. (1969). U.S. 3,475,391 to DuPont; C.A. (1969), *72*: 22147s.

630. Carlson, D. P. (1970). U.S. 3,624,250 to DuPont; C.A. (1970), *73*: 67223.

631. Kojima, G., and Hisasue, M. (1984). U.S. 4,463,144 to Asahi Glass Co.; C.A. (1984), *101*: 131335q.

632. Nersasian, A. (1969). U.S. 3,467,636 to DuPont; C.A. (1969), *71*: 102940q.

633. Graham, D. P. (1968). U.S. 3,403,191 to DuPont; C.A. (1969), *70*: 3271v.

634. Neth. 6,509,518 (1965) to DuPont; C.A. (1966), *65*: 9292h.

635. Toy, M. S., and Tiner, N. A. (1971). U.S. 3,567,521 to DuPont; C.A. (1971), *74*: 127700m.

636. Tervoort, T., Visjager, J., Graf, B., and Smith, P. (2000). *Macromolecules, 33*: 6460.

637. Smith, P., Visjager, J., Bastiaansen, C., and Tervoort, T. (2000). Int. Appl. WO 00/08071.

638. Dorset, D.L. (1992). *Chemtracts: Macromol. Chem.*, *3*: 198.

639. Yamabe, M. (1994) in: *Organofluorine Chemistry* (Banks, R.E., Smart, B. E., Tatlow, J. C. eds.), Plenum Press, New York, p. 397.

640. Lawson, K. J., and Nicolls, J. R. (1999) in: Hougham, G. (ed.) (1999) Fluoropolymers, Vol. 1 (Synthesis), Kluwer Academic/Plenum Press, New York, p. 313.

641. Biederman, H., and Slavinska, D. (2000). *Surf. Coat. Technol. 125*: 371.

642. Jansen, K. H. M. (1993). *Pitture Vernici Eur.*, *69*: 10.

643. Ruff, O., and Bretschneider, O. (1933). *Z. anorg. Chem.*, *210*: 173.

644. Locke, E. G., Brode, W. R., and Henne, A. L. (1934). *J. Am. Chem. Soc.*, *56*: 1726.

645. Torkington, P., and Thompson, H. W. (1945). *Trans. Faraday Soc.*, *41*: 236.

646. Hals, L. J., Reid, T. S., and Smith, G. H. (1951). *J. Am. Chem. Soc.*, *73*: 4054.

647. Lewis, E. E., and Naylor, M. A. (1947). *J. Am. Chem. Soc.*, *69*: 1968.

648. Atkinson, B., and Trendwith, A.B. (1953) *J. Chem. Soc.*, *75*: 2082.

649. Babb, D.A. in: Hougham, G. (ed.) (1999) *Fluoropolymers*, Vol. 1 (Synthesis), Kluwer Academic/Plenum Press, New York, p. 25.
650. Park, J. D., Benning, A. F., Downing, F. B., Laucius, J. F., and McHamess, R. C. (1947). *Ind. Eng. Chem.*, *39*: 354.
651. Van Bramer, D. J., Shiflett, M. B. and Yokozeki, A. (1994) U.S. 5,345,013.
652. Renfrew, M. M., and Lewis, E. E. (1946). *Ind. Eng. Chem.*, *38*: 870.
653. Putnam, R. E. (1989). *Comprehensive Polymer Science*, Vol. 3, Pergamon Press, Oxford, p. 321.
654. Bankoff, S. G. (1952). U.S. 2,612,484 to DuPont; C.A. (1953), *47*: 3618i.
655. Berry, K. L. (1953). U.S. 2,662,065 to DuPont; C.A. (1954), *48*: 4253b.
656. Lontz, J. F., and Happoldt, W. B. (1952). *Ind. Eng. Chem.*, *44*: 1800; C.A. (1952). *46*: 11756g.
657. Berry, K. L. (1949). U.S. 2,484,483 to DuPont; C.A. (1950), *44*: 1753a.
658. Banard, M. M., and Whipple, G. H. (1957). Ger. 1,016,929 to DuPont; C.A. (1960), *54*: 13735g.
659. Halliwell, R. H. (1963). U.S. 3,110,704 to DuPont; C.A. (1951), *45*: 6950h.
660. Kuhls, J., Steininger, A., and Fitz, H. (1976). Ger. Offen. 2,523,570 to Hoechst AG; C.A. (1977), 86: 44232g.
661. Brit. 781,532 (1957) to DuPont; C.A. (1958), *52*: 1684c.
662. Duddington, J. E., and Sherratt, S. (1959). Brit. 821,353 to ICI Ltd.; C.A. (1961), *55*: 1084a.
663. Hanford, W. E., and Joyce, R. M. (1951). U.S. 2,562,547 to DuPont; C.A. (1952), *46*: 1578i.
664. Brubaker, M. M. (1946). U.S. 2,393,967 to DuPont; C.A. (1946), *40*: 3648, 5.
665. Hartwimmer, R. (1974). Get. 1,720,801 to Hoechst AG; C.A. (1974), *81*: 136771g.
666. Reiher, M., Schleede, D., and Weissermel, K. (1957). Ger. 1,017,308 to Hoechst AG; C.A. (1960), *54*: 11581i.
667. Compto,. J. D., Justice, J. W., and Irwin, C. F. (1950). U.S. 2,510,078 to DuPont; C.A. (1950), *44*: 11176i.
668. Joyce, R. M. (1946). U.S. 2,394,243 to DuPont; C.A. (1946), *40*: 3648, 5.
669. Marks, M., and Thompson, J. B. (1956). U.S. 2,737,533 to DuPont; C.A. (1956), *50*: 7503c.
670. Folda, T., Hoffmann, H., Chanzy, H., and Smith, P. (1988). *Nature*, *333*: 55.
671. Folda, T., Hoffmann, H., and Smith, P. (1987). Ger. Offen. DE 3,718,949 Al to DuPont; C.A. (1988), *108*: 168168h.
672. Malhotra, S. C. (1988). U.S. 4,725,644A to DuPont; C.A. (1988), *108*: 168554f.
673. Beresniewicz, A. (1985). U.S. 4,555,556A to DuPont; C.A. (1986), *104*: 130449e.
674. Cavanaugh, R. J. (1985). U.S. 4,529,781A to DuPont; C.A. (1985), *103*: 124119k.
675. Kupiec, S., Milsowicz, H., and Harodynstri, Z. (1985). Pol. PI, 127,988 BI; C.A. (1987), *106*: 196952m.
676. Jpn. JP60/76516A2 (1985) to Daikin Kogyo Co. Ltd.; C.A. (1985), *103*: 178799a.
677. Suwa, T., Watanabe, T., Seguchi, T., Okamoto, J., and Machi, S. (1979). *J. Polym. Sci. Polym. Chem. Ed.*, *17*: 111.
678. Suwa, T., Watanabe, T., and Okamoto, J. (1979). *J. Polym. Sci. Polym. Chem. Ed.*, *17*: 129.
679. Suwa, T., Takehisa, M., and Machi, S. (1974). *J. Appl. Polym. Sci.*, *18*: 2249.
680. Nakamura, K., Ichimura, M., and Fukushima, K. (1973). Jpn. JP 48/37754 to Asahi Glass Co. Ltd.; C.A. (1974), *81*: 38346c.
681. Nakamura, K., Ichimura, M., and Fukushima, J. (1973). Jpn. JP 48/8755 to Asahi Gel.; C.A. (1974), *80*: 109244k.
682. Suzuki, N. (1974). *J. Polym. Sci. Lett. Ed.*, *12*: 143.
683. Bamford, C. H., Mullik, S. U., and Puddephatt, R. J. (1975). *J. Chem. Soc. Faraday Trans. 1*, *71*: 625.
684. Bamford, C. H., and Mullik, S. U. (1973). *Polymer*, *14*: 39.
685. Hanford, W. E., and Joyce, R. M. (1946). *J. Amer. Chem. Soc.*, *68*: 2082.
686. Banks, R. E., Smart, B. E. and Tatlow, J. C. (1994) *Organofluorine Chemistry: Principles and Commercial Applications*, Plenum Press, New York.
687. Allayarov, S. R., Sumina, 1. V., Barkalov, I. M., and Mikhailov, A. 1. (1987). *Izv. Akad. Nauk. SSSR Ser. Khim.*, *12*: 2698.

688. Berry, K. L. (1955). U.S. 2,662,065 to DuPont; C.A. (1955), *49*: 1814i.

689. Kroll, W. E. (1956). U.S. 2,750,350 to DuPont; C.A. (1956), *50*: 13507.

690. Romack, T. J., Kipp, B. E., and DeSimone, J. M. (1995). *Macromolecules 28*: 8432.

691. DeYoung, J. P., Romack, T. J., and DeSimone, J. M. (1999) in: *Fluoropolymers*, Hougham, G. (ed.), Vol. 1 (Synthesis), Kluwer Academic/Plenum Press, New York, p. 191.

692. Sianesi, D., and Caporiccio, G. (1962). *Makromol. Chem.*, *60*: 213.

693. Bamford, C. H. (1969). *J. Eur. Polym. Suppl.*, *5*: 1.

694. Bamford, C. H., and Mullik, S. U. (1973). *J. Chem. Soc. Faraday Trans. 1*, *69*: 1127.

695. Atkinson, B. (1952). *J. Chem. Soc.*, 2684.

696. Cohen, N., and Heicklen, J. (1965). *J. Chem. Phys.*, *43*: 871.

697. Atkinson, B. (1958). *Experientia*, *14*: 272.

698. Marsh, D. G., and Heicklen, J. (1966). *J. Am. Chem. Soc.*, *88*: 269.

699. Vogh, J. W. (1966). U.S. 3,228,865; C.A. (1966), *64*: 9839e.

700. Haszeldine, R. N. (1949). *J. Chem. Soc.*, 2856.

701. Haszeldine, R. N. (1953). *J Chem. Soc.*, 3761.

702. Sloan, J. P., Tedder, J. M., and Walton, J. C. (1973). *J. Chem. Soc. Faraday Trans. 1*, *69*: 1143.

703. Tedder, J. M., and Walton, J. C. (1966). *Trans. Faraday Soc.*, *62*: 1859.

704. Mungall, W. S., Martin, C. L., and Borgeson, G. C. (1975). *Macromolecules*, *8*: 934.

705. Wright, A. N. (1967). *Nature*, *215*: 953.

706. Wilkus, E. V., and Wright, A. N. (1971). *J. Polym. Sci. Part A1*, *9*: 2097.

707. Davenport, J. E., and Miller, G. M. (1969). *J. Phys. Chem.*, *73*: 809.

708. Gozzo, F., and Camaggi, G. (1966). *Tetrahedron*, *22*: 2181.

709. Bamford, C. H., and Mullik, S. U. (1976). *Polymer*, *17*: 225.

710. Volkova, E. V., Zimakov, P. V., and Fokin, A. V. (1966). *Dokl. Akad. Nauk SSSR*, *167*: 1057.

711. Enslin, S. E., Schnautz, N. G., and Van der Ende, E. (1986). *S. Afr. J. Chem.*, *39*(1): 23.

712. Bruk, M. A., Abkin, A. D., Demidovich, V. V., Yeroshina, L. B., Urman, Y. G., Slonim, I. Y., and Ledeneva, N. V. (1975). *Polym. Sci. USSR*, *17*(1): 1.

713. Kritskaya, D. A., and Ponomarev, A. N. (1982). *Vysokomol. Soedin. Ser.* A, *24*(2): 266.

714. Bruk, M. A., Abkin, A. D., and Khomikovski, P. M. (1963). *Dokl. Akad Nauk SSSR*, *149*: 1322.

715. Tabata, Y., Shibano, H., and Oshima, K. (1967). *J. Polym. Sci. Part C*, *16*: 2403.

716. Bruk, M. A., Chuiko, K. K., Yeroshina, L. V., Aulov, V. A., and Abkin, A. D. (1972). *Polym. Sci. USSR*, *14*: 883.

717. Tabata, Y., Ito, W., and Oshima, K. (1970). *J. Macromol. Sci. Chem. Part A*, *4*(4): 789.

718. Allayarov, S. R., Bol'shakov, A. I., and Barkalov, 1. M. (1987). *VysokomoL Soedin. Ser.* A, *29*(2): 364.

719. Allayarov, S. R., Kirykhin, D. P., Asamov, M. K., and Barkalov, 1. M. (1983). *Polym. Sci. USSR*. *24*(3): 511.

720. Fujioka, S., Shinohara, Y., and Nakamura, T. (1970). *J. Chem. Soc. Jpn. Ind. Chem. Sec.*, *73*: 184.

721. Hisasue, M., Ukihashi, H., and Tabata, Y. (1973). *J. Makromol. Sci. Chem.*, *7*: 795.

722. Tabata, Y., Ito, W., Oshima, K., and Takagi, J. (1970). *J. Macromol. Sci. Chem. Part A*, *4*: 815.

723. Suwa, T., Watanabe, T., Okamoto, J., and Machi, S. (1978). *J. Polym. Sci. Polym. Chem. Ed.*, *16*(3): 2931.

724. Watanabe, T., Suwa, T., Okamoto, J., and Machi, S. (1979). *J. Appl. Polym. Sci.*, *23*: 967.

725. Tabata, Y., and Shimada, T. (1988). Eur. 253,400 A2; C.A. (1988), *108*: 205286t.

726. Tabata, Y., Ishigure, K., Higake, H., and Oshima, K. (1970). *J. Macromol. Sci. Chem. Part A*, *4*(4): 801.

727. Bruk, M. A., Abkin, A. D., Khomikovskii, P. M., and Kotin, B. Y. (1971). *Polym. Sci. USSR*, *15*(1): 549.

728. Watanabe, H., Ito, M., Okamoto, Y., and Machi, S. (1981). *J. Appl. Polym. Sci.*, *26*(3): 3455.

729. Brown, D. W., and Lowry, R. E. (1979). *J. Polym. Sci. Polym. Chem. Ed., 17*: 759.
730. Hollahan, J. R., and Bell, A. T. (1974). *Techniques and Applications of Plasma Chemistry*, Interscience, New York.
731. Mayhan, K. G., Biolsi, M. E., and Havens, M. R. (1976). *J. Vac. Sci. Technol., 3*: 575.
732. Yasuda, H., Vossen, J. L., and Kem, W. (1979). *Thin Film Processes*, Academic Press, New York.
733. Sharma, A. K. (1986). *J. Polym. Sci. Part A Polym. Chem., 24*(11): 3077.
734. Masuoka, T., and Gasuda, H. (1981). *J. Polym. Sci. Polym. Chem. Ed., 19*: 2937.
735. Yasuda, H., Hsu, T. S., Brandt, E. S., and Reilley, C. N. (1978). *J. Polym. Sci. Polym. Chem. Ed., 16*: 415.
736. Lippold, U., Poll, H. U., and Wickleder, K. H. (1973). *Eur. Polym. J., 9*: 1107.
737. Buzzard, P. D., Soong, D. S., and Bell, A. T. (1982). *J. Appl. Polym. Sci., 27*: 3965.
738. Nakajima, K., Bell, A. T., and Shen, M. (1979). *J. Appl. Polym. Sci., 23*: 2627.
739. Yasuda, H., Morosoff, N., and Brandt, E. S. (1979). *J. Appl. Polym. Sci., 23*: 1003.
740. Morosoff, N., Yasuda, H., Brandt, E. S., and Reilley, C. N. (1979). *J. Appl. Polym. Sci., 23*: 3449.
741. Yanagihara, K., and Yasuda, H. (1982). *J. Poly. Sci. Polym. Chem. Ed., 20*: 1833.
742. Legeay, G., Rousseau, J. J., and Brosse, J. C. (1985). *Eur. Polym. J., 21*: 1.
743. Kitade, T., Hozumi, K., and Kitamura, K. (1983). *Yakugaku Zasshi, 103*: 719; C.A. (1983). *99*: 1289263z.
744. Chen, J., Liu, X., and Liu, Y. (1983). *Gaofenzi Tongxun, 1*: 19; C.A. (1983). *99*: 105936k.
745. Hozumi, K., Kitamura, K., and Kitade, T. (1981). *Bull. Chem. Soc. Jpn., 54*: 1392.
746. Ohki, Y., Nakano, T., and Yahagi, K. (1985). *Sym. Proc. Intern. Symp. Plasma Chem., 4*: 1307.
747. Doughty, T. R., Sperati, C. A., and Un, H. (1974). U.S. 3,855,191 to DuPont; C.A. (1975), *82*: 73855k.
748. Khan, A. A., and Morgan, R. A. (1983). Eur. 73,121 to DuPont; C.A. (1983), *99*: 6231s.
749. Berry, K. L. (1949). U.S. 2,478,229 to DuPont; C.A. (1949), *43*: 9528f.
750. Resnick, P. R., and Buck, W. H. (1999) in: *Fluoropolymers, Hougham*, G. (ed.), Vol. 2 (Properties), Kluwer Academic/Plenum Press, New York, p. 25.
751. Smart, B. E., Feiring, A. E., Krespan, C. G., Yang, Z., Hung, M., Resnik, P. R., Dolbier, W. R., and Rong, X. X. (1995) *Macromol. Symp. 98*: 753.
752. Sperati, C. A. (1986) in: *High Performance Polymers: Their Origin and Development* (Seymour, R. B., and Kirshenbaum, G. S., eds.), Elsevier, Amsterdam.
753. Lunkwitz, K., Brink, H. J., Handte, D., and Ferse, A. (1989). *Radiat. Phys. Chem. 33*: 523.
754. Forsythe, J. S., and Hill, D. J. T. (2000). *Prog. Polym. Sci., 25*: 101.
755. Symens, N. K. (1961). *J. Polym. Sci., 51*: 21.
756. Smith, P., and Gardener, K. H. (1985). *Macromolecules, 18*: 1222.
757. Tuminello, W. H. (1999) in: *Fluoropolymers*, Hougham, G. (ed.), Vol. 2 (Properties), Kluwer Academic/Plenum Press, New York, p. 137.
758. Sperati, C. A., and Starkweather, H. W. (1961). *J. Adv. Polym. Sci., 2*: 465.
759. Moynihan, R. E. (1959). *J. Am. Chem. Soc., 81*: 1045.
760. Nishioha, A. (1957). *J. Polym. Sci., 26*: 107.
761. Nishioha, A., and Watanabe, M. (1957). *J. Polym. Sci., 24*: 298.
762. Nishioha, A., Matsumae, K., Watanabe, M., Tayima, M., and Owaki, M. (1959). *J. Appl. Polym. Sci., 2*: 114.
763. Ajroldi, G., Garbuglio, C., and Ragazzini, M. (1970). *J. Appl. Polym. Sci., 14*: 79.
764. Suwa, T., Takehisa, M., and Machi, S. (1973). *J. Appl. Polym. Sci., 17*: 3253.
765. Heise, B. (1966). *Kolloid Z. Polym., 213*: 12.
766. Yeung, C. K., and Jasse, B. (1982). *J. Appl. Polym. Sci., 27*: 587.
767. Lau, S. F., Suzuki, H., and Wunderlich, B. (1984). *J. Polym. Sci. Polym. Phys. Ed., 22*: 379.
768. Starkweather, H. W., Ferguson, R. C., Chase, D. B., and Minov, J. M. (1985). *Macromolecules, 18*: 1684.

769. Starkweather, H. W. (1985). *J. Polym. Sci. Polym. Phys. Ed., 23*: 1177.

770. Starkweather, H. W. (1979). *J. Polym. Sci. Polym. Phys. Ed., 17*: 73.

771. McGeer, P. L., and Duns, H. C. (1952). *J. Chem. Phys., 20*: 1813.

772. Hyndman, D., and Origlio, G. F. (1960). *J. Appl. Phys., 31*: 1849.

773. McCall, D. W., Douglass, D. C., and Falcone, D. R. (1967). *J. Phys. Chem., 71*: 998.

774. Mehring, M., Griffin, R. G., and Waugh, J. S. (1971). *J. Chem. Phys., 55*: 746.

775. Vega, A. J., and English, A. D. (1980). *Macromolecules, 13*: 1635.

776. Hu, W., and Schmidt-Rohr, K. (1999) *Acta Polym. 50*: 271.

777. Brown, R. G. (1964). *J. Chem. Phys., 40*: 2900.

778. Pianca, M., Barchiesi, E., Esposto, G., and Radice, S. (1999) *J. Fluorine Chem., 95*: 71.

779. Bunn, C. W., and Howells, E. R. (1954). *Nature, 174*: 549.

780. Kilian, H. G., and Jenkel, E. (1952). *Z. Elektrochem., 63*: 308.

781. Vancso, G. J. (1996). *Polym. Prepr., 37*: 550.

782. Bunn, C. W. (1955). *J. Polym. Sci., 16*: 323.

783. Davidson, T., Gounder, R. N., Weber, D. K., and Wecker, S. M. (1999) in: *Fluoropolymers* (Hougham, G., ed.), Vol. 2 (Properties), Kluwer Academic/Plenum Press, New York, p. 3.

4

Polymers of Acrylic Acid, Methacrylic Acid, Maleic Acid and their Derivatives

Oskar Nuyken
Technische Universität München, Garching, Germany

I. ACRYLATES AND METHACRYLATES

(This section was prepared by O. Nuyken, G. Lattermann, H. Samarian, U. Schmelmer, C. Strissel, L. Friebe.)

This section is supposed to be a review of the background and possibilities of acrylate and methacrylate polymerization with a main focus on recent developments. Additional information and examples are given in the first edition of this book [1].

A. Introduction

1. Formula and History

The esters of acrylic and methacrylic acid, whose polymerization reactions are described in this chapter, are unsymmetrically substituted ethylenes of the general formula

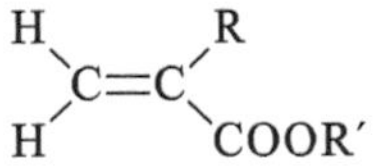

$$
\begin{array}{c}
H \qquad\quad R \\
\diagdown \diagup \\
C{=}C \\
\diagup \diagdown \\
H \qquad\quad COOR'
\end{array}
\tag{1}
$$

with $R = H$ for acrylates and $R = CH_3$ for methacrylates. The substituents R' may be of a great variety: from *n*-alkyl chains to more complicated functional groups. In the following chapters these compounds are generally named *acrylic esters*, although in literature, esters of other α-substituted acrylic acids (e.g., $R = -CN$, $-Cl$, $-C_2H_3$) are sometimes included in this term.

The first report of a polymeric acrylic ester was published in 1877 by Fittig and Paul [2] and in 1880 by Fittig and Engelhorn [3] and by Kahlbaum [4], who observed the polymerization reaction of both methyl acrylates and methacrylates. But it remained to O. Röhm [5] in 1901 to recognize the technical potential of the acrylic polymers. He continued his work and obtained a U.S. patent on the sulfur vulcanization of acrylates in 1914 [6]. In 1924, Barker and Skinner [7] published details of the polymerization of methyl and ethyl methacrylates. In 1927 [8], based on the extensive work of Röhm, the first industrial production of polymeric acrylic esters was started by the Röhm & Haas Company in Darmstadt, Germany (since 1971, Röhm GmbH, Darmstadt). After 1934, the Röhm & Haas Co. in Darmstadt was able to produce an organic glass (Plexiglas) by a cast polymerization process of methyl methacrylate [9]. Soon after, Imperial Chemical Industries (ICI, England), Röhm & Hass Co. (United States), and Du Pont de Nemours followed in the production of such acrylic glasses. Nowadays poly(methyl methacrylate) (PMMA) as homo- or copolymer exceeds by far the combined amount of all other polyacrylic esters produced [10].

2. Monomers

The most common procedure for the technical synthesis of the monomer methyl methacrylate (MMA) is the reaction of acetone cyanhydrine with water and methanol in the presence of concentrated sulfuric acid [11]:

$$\underset{H_3C}{\overset{H_3C}{>}}C\underset{CN}{\overset{OH}{<}} + H_2SO_4 \longrightarrow H_2C=C\underset{CONH_2}{\overset{CH_3}{<}} \cdot H_2SO_4 \qquad (2)$$

$$H_2C=C\underset{CONH_2}{\overset{CH_3}{<}} \cdot H_2SO_4 \xrightarrow{+ CH_3OH} H_2C=C\underset{COOCH_3}{\overset{CH_3}{<}} + NH_4HSO_4 \qquad (3)$$

Many other processes and reactions of the monomer synthesis are described extensively in literature [12–14]. For different acrylic esters, especially on a laboratory scale, the alcoholysis of the corresponding acid chlorides as well as direct esterification reactions of methacrylic acid, but also transesterification reactions of MMA, are often preferred [13–15]. The physical properties of various monomers are well summarized in literature [16,17].

3. Reactions

Acrylic esters have two functional groups, where reactions occur: the ester group and the double bond. Reactions on the ester group are carried out under conditions that prevent polymerization of the double bond (i.e., the use of polymerization inhibitors and low reaction temperatures are necessary). Typical reactions of the ester function are: saponification, transesterification, aminolysis, and Grignard reaction [10,17]. Reactions of the double bond beside polymerization reactions are Diels-Alder reaction; Michael addition; and addition of halogens, dihalocarbenes, hydrogen halogenides, alcohols, ammonia and amines, nitroalkanes, or sulfur compounds such as hydrogen sulfide or mercaptanes [10,17].

Most acrylates are polymerized by both radical and anionic initiations, with the former being the more commonly used. In all cases the heat of polymerization must be carefully controlled to avoid runaway reactions. The values of the heat of polymerization for selected methacrylates are listed in literature [18]. In general, the rate of polymerization and the average molar mass must be controlled by the initiator and monomer concentration and the reaction temperature. In all cases the use of high-purity monomers is important for proper polymerization conditions. Therefore, the removal of inhibitors is necessary. Phenolic inhibitors such as hydroquinone, 4-methoxyphenol, or aromatic amines are usually removed by alkaline or acidic extraction [11,19]. Otherwise, the monomers are distilled from inhibitors of low volatility such as dyes (methylene blue, phenothiazine), aromatic nitro or copper compounds.

To prevent inhibition by dissolved oxygen, acrylic monomers must be carefully degassed before polymerization [19]. After the polymerization step, the isolation of the product is often necessary. Depending on the polymerization technique, this may be achieved by different procedures (e.g., precipitation, spray drying, breakdown of a colloidal system, etc.). Purification of soluble polymers can be achieved by repeated cycles of precipitation, or in the case of water solubility, by dialysis. The removal of solvent may often be very difficult because of strong polymer–solvent interactions. Therefore the polymer is slightly heated above T_g under high vacuum, spray-dried, or freeze-dried. Freeze-drying with benzene, dioxane, or water results in a very dry, highly porous material.

B. Processing

1. Bulk Polymerization

In contrast to acrylic monomers the bulk polymerization of methacrylic esters is very important in manufacturing sheets, rods, tubes, and molding material by cast molding techniques [9–11]. Three important properties are characteristic of the bulk polymerization of acrylates. First, a strong volume contraction, being relatively high compared with other monomers, occurs during the polymerization reaction (see Table 1). It may be overcome either by using 'prepolymers' (i.e., solution of polymers in their monomers, usually prepared by bulk polymerization until a desired viscosity level of the mixture [20]) or by forming rigid polymer networks even at low conversion through cross-linking agents. Second, the polymerization process is accompanied by a considerable reaction heat (see Table 1), which is higher for acrylates than for methacrylates. Therefore, after 20 to 50% conversion, causing an increased viscosity of the system, a drastic autoacceleration process may be possible, known as gel or Trommsdorff effect [11,21,22]. Thus it is necessary to regulate very carefully heat removal during the polymerization in bulk. Third, at high conversion, branching and cross-linking reactions, leading finally to insoluble networks, may occur [23–25]. This is due to chain transfer involving abstraction of hydrogen from the polymer chain, subsequent branching, and combining two branch radicals.

Bulk polymerization is commonly started by radical initiators such as azo compounds and peroxides; however, some examples of thermal self-initiation of bulk

Table 1 Shrinkage and reaction heat of various methacrylates.[a]

Methacrylates	Shrinkage/%	$\Delta H/(kJ/mol)$
Methyl	21.2	54.5
Ethyl	17.8	59.1
Butyl	14.3	56.6
Isobutyl	12.9	

Source: Refs. [26] and [19].
[a]The percent shrinkage can be calculated by using the following equation:
% shrinkage $= 100 \times (D_p - D_m)/D_p$ (D_m = monomer density at 25 °C; D_p = polymer density at 25 °C.

polymerization of MMA [27] and octylacrylate [28] are described. For MMA, which cannot form a Diels-Alder adduct, diradicals are believed to play a role in the thermal initiating mechanism [29–31].

Different descriptions of general procedures for the bulk polymerization of acrylates (sheets, molding material) are given in Refs. [12] and [19]. The bulk polymerization of γ-alkoxy-β-hydroxypropylacrylates is described in Ref. [32]. Bulk atom transfer radical polymerization is reviewed in Ref. [33].

2. Solution Polymerization

Several general disadvantages of bulk polymerization (removal of the reaction heat, shrinkage, nonsolubility of the resulting polymer in the monomer, side reactions in highly viscous systems such as the Trommsdorff effect or chain transfer with polymer) are responsible for the fact that many polymerization processes are carried out in the presence of a solvent. A *homogeneous polymerization* occurs when both monomer and polymer are soluble in the solvent. When the polymer is insoluble in the solvent, the process is defined as *solution precipitation polymerization*. Other *heterogeneous polymerization reactions* in liquid–solid or liquid–liquid systems such as suspension or emulsion polymerizations are described later. Conventional solution polymerization is compared with solution precipitation polymerization for the synthesis of acrylic resins in Ref. [34].

In homogeneous systems including inert solvents, the reaction rate decreases with decreasing monomer concentration. In solution precipitation polymerization, kinetics may deviate from that in homogeneous solution.

In nearly every polymerization system the influence of the solvent on the course of the reaction is important. Thus chain transfer reactions with active chain ends occur in radical polymerization. The solvent can also influence the stereoregularity of the product in anionic polymerizations. The boiling range of the solvents should correspond to that of the monomers and to the decomposition temperature of the initiators. Thus common polymerization temperatures are often between 60 and 120 °C (under reflux of the solvent). A general procedure for the radical homopolymerization of acrylates in solution is given in Ref. [35].

Not only acrylic esters that have intermediate solubility in water due to additional hydroxy or amino groups can be polymerized in water, but also conventional acrylic monomers with a relatively low water solubility (MMA: 15 g/L at room temperature) [36] can be polymerized in water. Acrylate monomers of intermediate solubility in water, such as hydroxyalkyl acrylates and methacrylates or aminoalkyl acrylates or methacrylates, undergo free-radical polymerization with a variety of initiator systems. Both monomer classes have been reviewed in the literature [37]. Highly soluble monomers such as 2-sulfoethyl methacrylates or the corresponding alkali salts are easily polymerized to high molar mass by hydrogen peroxide in aqueous solution [38]. Anionic initiation has been accomplished in a variety of solvents, both polar and nonpolar.

Isolation and purification of the product is performed, for example, by addition of a nonsolvent, leading to polymer precipitation or by removal of the solvent by spray drying or by freeze drying in benzene, dioxane, or water. Polymer precipitation should be quantitative. However, PMMA with a degree of polymerization less than 50 is still soluble even in methanol; thus petroleum ether is necessary to precipitate the low-molar-mass PMMA [39]. Numerous solvents and nonsolvents for polymers are reviewed in Refs. [40] and [41].

In industrial processes it is sometimes advantageous to have a strong solvent–polymer interaction. Thus solution polymerization is often performed for applications in which the solvent remains present (e.g., in protective coatings, adhesives, and viscosity modifiers).

3. Suspension Polymerization

The term *suspension polymerization*, often also called *aqueous suspension polymerization* or *pearl* or *bead polymerization*, means a process where liquid monomer droplets are suspended in an aqueous phase under vigorous stirring. This process can be regarded as a bulk polymerization within the monomer droplets, where the polymerization heat can easily be dissipated by the surrounding water. To prevent the coalescence of the droplets, the presence of suspension stabilizers or suspending agents is necessary.

Two classes of suspension stabilizers are known [42,43]:

1. *Water-soluble polymeric compounds.* These can be natural or modified natural products such as gelatine, starch, or carbohydrate derivatives such as methyl cellulose, hydroxyalkyl cellulose, or salts of carboxymethyl cellulose. Synthetic polymers such as poly(vinyl alcohol), partially hydrolyzed poly(vinyl acetate), sodium salts of poly(acrylic acids), methacrylic acids, and copolymers thereof are widely used in quantities between 0.1 and 1% related to the aqueous phase.
2. *Powdery inorganic compounds.* Earth alkaline carbonates, sulfates, phosphates, aluminum hydroxides, and various silicates (talc, bentonite, Pickering emulgators) are used in quantities between 0.001 and 1%. The initiator systems are the same as for radical bulk or solution polymerization processes (e.g., peroxides or azo compounds). A typical recipe is given in Ref. [44].
3. *Nonaqueous dispersion polymerization* is defined as the polymerization of a monomer, soluble in an organic solvent, to produce an insoluble polymer whose precipitation is controlled by an added stabilizer or dispersant. The resulting stable colloidal dispersion ensures good dissipation of the polymerization heat. Stabilization of the polymeric particles is generally achieved by a lyophilic polymeric additive.

PMMA is mostly homo- or copolymerized in aliphatic hydrocarbon dispersions, using different rubbers, polysiloxanes, long-chain polymethacrylates, or different block and graft copolymers as stabilizers. An interesting variant of the dispersion polymerization of acrylates is carried out in supercritical carbon dioxide [45,46]. Transition-metal-mediated living radical suspension polymerization is discussed in Ref. [47]. Common radical initiators are described in Refs. [48] and [49]. The entire field is reviewed extensively in Ref. [50].

4. Emulsion Polymerization

An *emulsion polymerization* system can comprise three phases: (1) an aqueous phase, containing the water-soluble initiator, the micelle-forming surfactant, and a small amount of the sparingly soluble monomer; (2) monomer droplets; and (3) latex particles, consisting of the polymer and some monomer. The locus of polymerization is predominantly inside the latex particles. Usual free-radical water-soluble initiators are used, such as potassium persulfate for higher reaction temperatures and redox systems [e.g., Fe(III) salts, cumene hydroperoxide] for low-temperature polymerizations.

Three types of surfactants are known: (1) electrostatic (anionic or cationic) low-molecular mass surfactants; (2) steric stabilizers such as poly(vinyl alcohol), or a combination of (1) and (2); and (3) electrosteric stabilizers such as polyelectrolytes. Furthermore, many other additives (protecting agents, cosolvents, chain transfer agents, buffer systems, etc.) are often necessary. The entire field is reviewed in Ref. [51], comprising the special kinetics of particle growth and formation, particle size, and molecular mass distribution.

Various emulsion polymerization procedures for the thermal and redox initiation of acrylic monomers are given in Refs. [52] and [53]. Methyl, ethyl, and n-butyl acrylates and methacrylates are found to form high-molecular-mass compounds quite easily through a plasma-induced emulsion polymerization system [54]. Emulsions are thermodynamically unstable, although they often may have an appreciable kinetic stability. The use of a co-emulsifier (e.g., long-chain alkanes, alkanol or ammonium salts, or block copolymers of ethylene and propylene oxide) can produce *microemulsions*. They are thermodynamically stable systems, exhibiting an average particle size of about 100 nm [55]. Thus transparent microemulsions of MMA can be obtained which have been photo-polymerized together with a photosensitizer [56]. The field of microemulsion is reviewed in Ref. [57].

A emulsifier-free emulsion polymerization of acrylates is possible by the use of 2-hydroxyethyl methacrylate [58]. Acrylate block copolymers (P(MMA-b-MAA)) were used as surfactants in emulsion polymerization of acrylate monomers [59].

5. Irradiation Polymerization

Irradiation-induced bulk polymerization can be divided into two types: solid-state polymerization and polymerization in the liquid state, classified as follows:

1. UV light: the initiation process is thought to occur via a free-radical mechanism.
2. γ-radiation: the induced polymerization process involves free radicals or ionic species, depending on monomer, temperature, dose rate etc. [60].
3. Electron-beam, x-ray, or ion-beam radiation.

Since most of the monomers do not produce initiating species with a sufficiently high yield upon UV exposure, it is necessary to introduce a photosensitive initiator. The photo initiator (PI) will start the polymerization upon illumination. Thus, the PI plays a key role in light-induced polymerization for it absorbs the incident light and generates reactive radicals or ions and it controls the reaction rate and the depth of cure profile within the sample. There are various photoinitiators used in UV-curing applications which can be classified into three categories, depending on the way the initiating species are generated:

1. Radical formation by photocleavage: aromatic carbonyl compounds that undergo homolytic C–C bond cleavage upon UV exposure with formation of two radical fragments like benzoin ether derivatives, hydroxyalkylphenones, α-amino ketones, morpholinoketones (MoK) and bisacylphosphine (BAPO) from Ciba Specialty [61]. Phosphine oxides undergo fast photolysis to generate non-colored products (Scheme 4). Their higher initiation efficiency is caused by a disaggregation that is fast, as the rate of initiation is directly related to the rate of the PI photolysis.

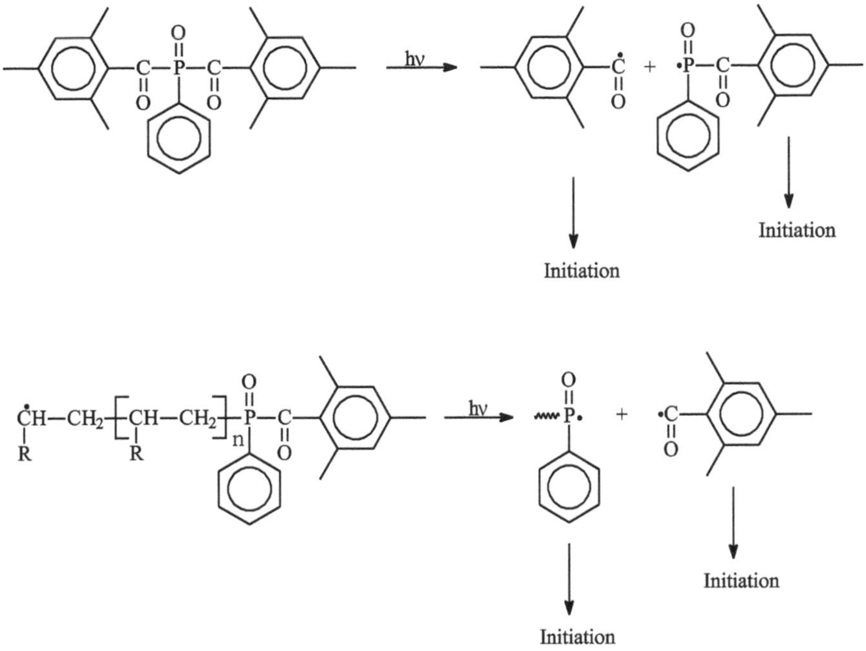

(4)

2. Radical generation by hydrogen abstraction: some photoinitiators tend to abstract a hydrogen atom from a H-donor molecule via an exciplex, to generate a ketyl radical and the donor radical. The H-donor radical initiates the polymerization, the inactive ketyl radical disappears by a radical coupling process (5). This type of photoinitiators includes benzophenone and thioxanthone.

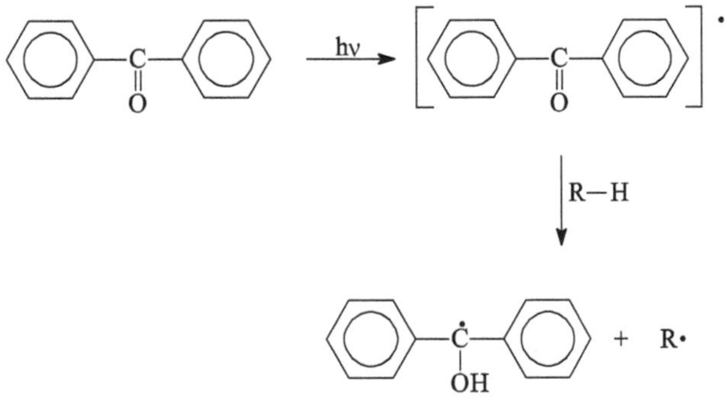

(5)

3. Cationic photoinitiators: like protonic acids.

Oxygen as an initiatior in photo-initiated free-radical polymerization and cross-linking of acrylates is reviewed in Ref. [62].

Methyl methacrylate does not appear to polymerize in the solid state upon simple UV radiation [63,64]. However, under pressure sufficiently high to solidify the monomer at a relatively high temperature or in a 'solid solution' in paraffin wax, polymerization was found to be possible. It is remarkable that the γ-radiation-induced solid-state polymerization is influenced significantly when the polymerization proceeds in tunnel clathrates [1].

Another possibility for irradiation-induced solid state polymerization is that in mono- or multilayers. Thus acrylates or methacrylates with different long-chain ester groups are polymerized by UV light, γ-radiation, or electron-beam radiation [65–67]. The majority of the examples given in the literature for irradiation-induced bulk polymerization deal with monomers in the liquid state as pure compounds. Some examples are given for polymerization in the presence of inclusion compounds or related polymer matrices (see Refs. [60,68–72]). Another possibility has been described as photopolymerization of an oriented liquid crystalline acrylate [73].

Photo- or radiation-initiated bulk polymerization of acrylates is often used for the production of thick coatings or sheets. Demonstration experiments are given in Refs. [12] and [19]. For many purposes (e.g., photocoating, embedding media, etc.) casting resins often contain multifunctional cross-linking compounds [74,75]. A review of the chemistry of photoresists, reacted by UV, eximer laser (deep UV), x-ray, electron-beam, and ionbeam irradiation is given in Ref. [76]. In general, most industrial processes use a large variety of copolymerization reactions.

Besides the above noted polymerization techniques photocuring is a special process that transforms a multifunctional monomer into a crosslinked macromolecule by a chain reaction initiated by reactive species generated by UV irradiation [77]. Three basic components are needed for photocuring:

1. The already mentioned photoinitiator;
2. A functionalized oligomer, which by polymerizing will constitute the backbone of the three dimensional polymer network formed;
3. A mono or multifunctional monomer, which acts as reactive diluent and will thus be incorporated into the network.

UV-curable resins of acrylate and methacrylate monomers gained great commercial success because they offer high reactivity and the possibility of creating a large variety of crosslinked polymers with tailormade properties. On the other hand there are problems like early gelation of the irradiated sample and mobility restrictions of the reactive sites during the preceding reaction and also with increased monomer functionality. Novel acrylate monomers seem to circumvent these problems. Very promising results have been obtained by introducing a carbamate or oxazolidone group into the structural unit of a monoacrylate [77]. As shown by the RTIR profiles, the light-induced polymerization was found to occur faster than with typical monoacrylates or diacrylate monomers.

The UV-cured polymers based on the novel acrylate monomers show some advantages: completely insolubility in organic solvents which makes these very reactive photoresists well suited for imaging applications; high crosslink density; good resistance to moisture, strong acids, weathering and thermal treatment [78].

Photopolymerization in micellar systems is useful for the synthesis of polymers displaying high molecular weights [57]. The model of photopolymerization used to describe a micellar polymerization does not differ from the one in bulk or solution photopolymerization [79].

6. Plasma Polymerization

A general introduction to the field of plasma polymerization is given in Ref. [31]. The plasma used in polymerization processes is the low-temperature plasma or low-pressure plasma, which is usually created by an electric glow discharge caused by, for example, microwave power sources. There are two general methods in use to polymerize pure monomers. First, in *plasma-state polymerization* the plasma reacts directly within the vapor phase of a monomer, resulting in the vacuum deposition of polymers [31,80,81]. Here the course of the initiation reaction depends on the bombardment of the monomer by excited species such as radicals, ions, metastable particles, and on the absorption of UV radiation emitted by the different excited species. Concerning the UV-induced part of plasma polymerization, the propagation will be maintained by a free-radical mechanism. Acrylic monomers are not described as undergoing such processes.

The second way, *plasma-induced polymerization*, is characterized by the formation of initiating species under the influence of a plasma and subsequent polymerization in the condensed phase. One possibility for the initiation process is that it can take place by exposing liquid monomers to a plasma of different gases (helium, argon, nitrogen, NO, CO_2, O_2, CF_4) [82] for several minutes. The presence of radical initiators, photo-initiators, and photosensitizers can influence the course of the polymerization reaction [83–86]. This technique is used to polymerize thin films for coating purposes.

Another possibility in plasma-induced polymerization is to expose the vapor phase over a liquid monomer [31,87], volatile initiator, or monomer solution to the plasma for several seconds only. Chain propagation occurs in the liquid phase during a longer period of postpolymerization in the absence of plasma. The unique feature of this way of plasma-induced polymerization is that the formation of initiating species takes place in the gas phase, presumably creating diradicals with a very long lifetime [31]. In most cases the molar mass increases with reaction time (i.e., conversion). This is not the case in conventional free-radical polymerization, although the tacticity of the resulting acrylic polymers corresponds to that observed in free-radical polymerization. Some similarities of polymer characteristics (gel permeation chromatography, thermogravimetry, differential scanning calorimetry) can be observed between plasma-induced and thermal polymerization, the initiation process of the latter also being caused by diradicals.

C. Mechanism

1. Free Radical Polymerization

The kinetic scheme of this type of polymerization is equivalent to other classical vinyl polymerizations, including initiation, propagation, chain transfer, and termination (Scheme 6).

$$
\begin{array}{llll}
\text{Initiation:} & \text{I} & \xrightarrow{k_d} & 2\text{R}\bullet \\
& \text{R}\bullet + \text{M} & \longrightarrow & \text{P}_1\bullet \\
\text{Propagation:} & \text{P}_1\bullet + n - 1\,\text{M} & \xrightarrow{k_p} & \text{P}_n\bullet \\
\text{Chain transfer:} & \text{P}_n\bullet + \text{M} & \xrightarrow{k_{c,M}} & \text{P}_n + \text{M}\bullet \\
& \text{P}_n\bullet + \text{L} & \xrightarrow{k_{c,L}} & \text{P}_n + \text{L}\bullet \\
\text{Termination:} & \text{P}_n\bullet + \text{P}_m\bullet & \xrightarrow{k_{t,r}} & \text{P}_n - \text{P}_m \quad \text{Recombination} \\
& \text{P}_n\bullet + \text{P}_m\bullet & \xrightarrow{k_{t,d}} & \text{P}_n + \text{P}_m \quad \text{Disproportionation}
\end{array}
\tag{6}
$$

Common solvents include toluene, ethyl acetate, acetone, and 2-propanol. The boiling range of the solvents should correspond to that of the monomers and to the decomposition temperature of the initiators. Thus common polymerization temperatures are often between 60 and 120 °C (under reflux of the solvent).

Most common initiators are compounds decomposing to starting radicals by thermolysis. The main classes for both organic and aqueous media systems are reviewed according to the following main groups:

1. Azo and peroxy like azobisisobutyronitrile (AIBN) and dibenzoyl peroxide (BPO) initiators [88].
2. Redox initiators such as peroxide tertiary amine systems or those based on metals or metal complexes [89].
3. Ylide initiators such as β-picolinium-p-chlorophenacylide or others [90]. This initiating system is especially interesting with respect to alternating copolymers of MMA.
4. Thermal *iniferters* [91], a class of initiators that not only can start a polymeric chain but can also undergo a termination reaction by chain transfer (initiator, transfer agent, chain terminator). The resulting end group is thermally or photochemically labile, being able to undergo reversible homolysis to regenerate a propagating radical. These materials have been applicated in the synthesis of block and graft copolymers.

General conditions for a successful application of radical initiators are [92]:

1. The initiator decomposition rate must be reasonably constant during the polymerization reaction. The 'cage effect' (recombination of initiator radicals before starting a polymer chain) should be small, which is generally more the case with azo compounds than with peroxides.
2. Side reactions of the free radicals (e.g., hydrogen abstraction with dialkyl peroxides and peresters) should be reduced.
3. In addition to initiators, accelerators and chain transfer agents are sometimes used. Thus, with accelerators (often redox activators (e.g., $ZnCl_2$ [93], cobalt salts, tertiary amines [94]), the reaction temperature can be drastically reduced; with chain transfer agents the average molar mass of the resulting polymer can be regulated.

Concerning the growing radicals in polymerization reactions, they can be studied directly by ESR spectroscopy as in the case of triphenylmethyl methacrylate and MMA [95]. In the latter case it was concluded that there are two stable conformations of the propagating radicals. The steric effect of the α-methyl group of MMA is not only responsible for the comparatively low heat of the polymerization reaction, but also for a certain control of the propagation steps. Therefore, in radical solution polymerization the polymethacrylates exhibit in most cases a favored syndiotacticity.

With respect to the termination mechanism in radical acrylate polymerization, some results are reviewed in Ref. [96]. In MMA polymerization the preferred termination mechanism is solvent dependent (e.g., disproportionation is being favored in benzene). For alkyl acrylates termination involves predominantly combination.

As mentioned earlier, a general procedure for the radical homopolymerizaiton of acrylates in solution is given in Ref. [35]. With α-substituted acrylates other than methacrylates, isotacticity is somewhat enhanced [97].

Tacticity of acrylate or methacrylate polymers obtained by radical initiators is an important matter of research, as it influences the physical properties of the acrylate polymers: for example, the higher the syndiotacticity, the higher the glass transition temperature (atactic PMMA: $T_g = 105\,°C$ [68]; highly syndiotactic PMMA: $T_g = 123\,°C$ [97]). The polymerization of MMA by redox initiation within solid particles of stereoregular PMMA affects the configuration of chains [68,94,97–99]. There is a greater configurational disorder in the resulting product than with the PMMA obtained through bulk polymerization without a stereoregular PMMA matrix.

Capek et al. polymerized various alkyl acrylates, methyl (MA), ethyl (EA), butyl (BA), hexyl (HA) and 2-ethylhexyl (EHA) acrylate, and alkyl methacrylates in microemulsion [100]. Microemulsion polymerizations of BA and EHA reached in a short time a conversion close to 100%. In case of PMMA the polydispersity index varied from 2 to 4. This can be taken as evidence that the chain transfer events contribute to the termination mechanism [57].

Cyclic acrylates are known to undergo ring-opening polymerizations according to the following scheme:

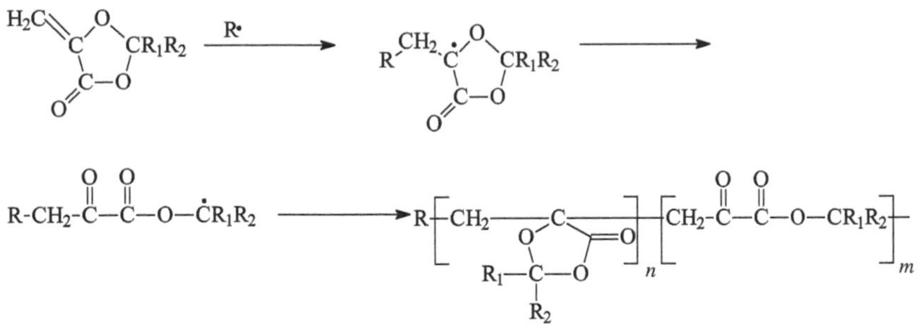

(7)

For several examples with different R_1 and R_2, it was shown that polymerization in bulk gave a copolymer of the structure given above. Quantitative ring opening occurred ($n = 0$) when this reaction was carried out in t-butylbenzene at $140\,°C$ [101].

Mathias et al. explored the chemistry of functional methacrylates and developed a one-step, inexpensive entry via the hydroxymethyl derivatives [102]. The radical polymerization of the esters of alpha-hydroxymethylacrylate (RHMA) and the ether dimers were carried out in solution or bulk. They developed a mild, general synthesis of the ester of alpha-hydroxymethylacrylate. 1,4-Diazabicyclo[2.2.2]octane (DABCO) was the catalyst for the addition of formaldehyde and activated vinyl monomers (Scheme 8).

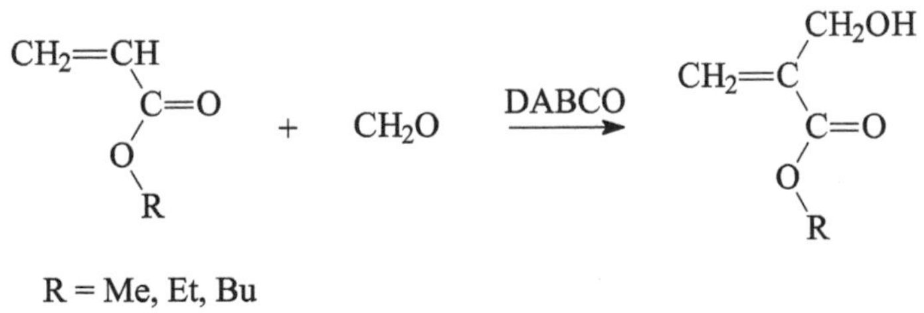

(8)

R = Me, Et, Bu

These alcohol monomers provide a versatile entry to a multitude of multifunctional polymers. Derivatization before and after polymerization allows incorporation of various functional groups such as ester, ether, thioether, amine, and siloxy groups.

Isolated RHMA could be readily converted to the ether in high yield by heating neat with amine base. The ethers were found to be excellent crosslinking agents for organic-soluble monomers such as styrene and commercial acrylates. The hydrolyzed diacid and its salt provided crosslink sites for water-soluble monomers such as acrylic acid. In addition, the ester derivatives of the diacrylate ethers underwent cyclopolymerization (Scheme 9).

(9)

The unexpected dimerization of the alcohol monomers provides new materials capable of cyclopolymerization, crosslinking organic and water-soluble polymers, and Michael polyaddition with dithiols and diamines.

In free-radical copolymerization of two monomers the relationship between the composition of the copolymer and the initial monomer mixture is ruled by the monomer reactivity ratios r_1 and r_2. These ratios are related to an individual system of given comonomers, initiator, and temperature [103]. They are summarized in Ref. [104] for numerous systems.

To estimate the reactivity ratios of new comonomer pairs, their Q and e values, as summarized in Ref. [105], can be compared. The Q and e values are a measure of the reactivities and the polarities in a copolymer system. A special solvent effect has been described in the radical copolymerization of optically active acryloyl-D-phenylglycine methyl ester with MMA or MA in D- or L-ethylmandelate as optically active solvent. The rate of polymerization was higher in the D-ester [106].

The copolymerization behavior of the different acrylates and methacrylates is largely independent of the nature of the ester group if there are no important interactions with the monomer or solvent. Thus copolymerization reactions between different acrylates or between methacrylates yields uniform products in the monomer mixing proportion [107]. The reactivity ratios for the copolymerization of methacrylates (M$_1$) with acrylates (M$_2$) are given in a first approximation as $r_1 = 2.0$ and $r_2 = 0.3$ [107]. If chemical uniform copolymers are desired, the reaction should be stopped at a low conversion value ($\leq 5\%$) [108]. On the other hand, the sequence distribution can be controlled by, for example, changing the addition time of one of the monomers [109]. Despite their structural similarities, different methacrylates or acrylates are often incompatible [107]. A typical recipe for the preparation of a suspension copolymerization of ethyl acrylate and MMA and of an acrylic solution terpolymerization of 2-ethylhexyl acrylate, MMA, and hydroxyethyl methacrylate is described in Ref. [110]. Among the numerous comonomers, styrene and α-methyl styrene are the most important for industrial purposes, as light fastness and chemical resistance of the acrylics can be combined with the higher heat resistance of the styrene compounds. Those copolymers are produced by bulk, solution, or suspension techniques [111].

In principal, all homopolymerization techniques can be applied to random copolymerization. For radical copolymerizations numerous examples have been described before. Some selected typical examples of other polymerization methods are listed in Refs. [112–127]. Methods for the radical and anionic copolymerization of MMA with styrene are given in Ref. [128].

The following examples of alternating copolymerization are given in the literature:

1. MMA with styrene through photopolymerization in the presence of boron trichloride, ethyl boron dichloride, or aluminum tribromide [129]
2. MA or MMA with styrene in the presence of ethylaluminum sesquichloride [130]
3. MMA with styrene, initiated by β-picolinium-*p*-chlorophenacylide [90]
4. MA with isobutylene, initiated by a complex system of Al(ethyl)Cl$_2$ and benzoyl peroxide [131]

Polar side groups are useful to improve the adhesion of copolymers on surfaces, to reduce incompatibility with other polymers and to modify the solubility of polymers, or to synthesize graft copolymers. Common functional monomers for free-radical copolymerization with acrylic monomers are listed in Table 2 [132].

Side groups can be introduced by polymer-modification reactions; for example, a hydroxy group can be converted to halides, *tert*-amino, nitro, sulfane, and disulfane groups and to heterocyclic units [107]. Acrylic monomers, with $C=C$ double-bond containing side groups, can be used for radical and anionic crosslinking.

Acrylates and methacrylates of bi- and polyfunctional alcohols are often used for the direct crosslinking copolymerization. Common diols used to obtain relevant diesters are glycol, 1,4-butanediol, glycerol, 2,2-bis(hydroxymethyl)-1-butanol, oligo(glycol ethers), and oligo(1,2-propane diol ethers). Allyl and vinyl ester are particularly interesting, due to the different reactivity of both polymerizable double bonds. A typical recipe for the radical cross-linking of acrylamide, 2-hydroxyethyl methacrylate, and ethylene dimethacrylate to a copolymer gel is given in Ref. [133].

Radical techniques are also used for the synthesis of graft polymers. The grafting polymerization of MMA or its mixture with other comonomers from diene units containing rubbers, in bulk or suspension [134–140], and from a terpolymer of styrene, MMA, and *t*-butylperoxy acrylate [141]. Furthermore, redox reactions of OH-containing polymers, such as poly(vinyl alcohol) [142–144] or poly(hydroxyethyl methacrylate) [144], but also natural products such as cellulose [145,146] or gelatine [147] with, for example, Ce^{4+} are used to graft MMA side chains. Otherwise, hydroxyl functions in starch have been reacted with methacrylic anhydride. Subsequently, MMA was radically grafted from these sites [148]. Other monomers, such as methacrylonitrile or styrene, have been grafted radically from a copolymer of MMA and an azo side group containing methacrylate [149].

Table 2 Common functional monomers for free-radical copolymerization with acrylic monomers.

Functionality	Monomer
Carboxyl	Acrylic acid, methacrylic acid, itaconic acid
Amino	2-*t*-butylaminoethyl methacrylate, 2-dimethylaminoethyl methacrylate
Hydroxyl	2-Hydroxyethyl methacrylate, 2-hydroxyethyl methacrylate
n-Hydroxymethyl	*n*-Hydroxymethyl acrylamide
Oxirane	Glycidyl methacrylate

True radical 'grafting onto' reactions have not been described for PMMA since radical recombination does not occur separately. On the other hand, 'grafting onto' functional groups with reasonable transfer constants are described for poly(vinyl chloride) or chlorinated rubber [150] or for a poly(diethylamino methacrylate) backbone [151].

2. Anionic Polymerization

'Living' anionic polymerization was first discovered by Szwarc et al. in the fifties [152], and since then, a lot of work has been done in this field as anionic polymerization allows a precise control of the molecular mass and results in a narrow molecular mass distribution. Additionally, the tailoring of block copolymers is possible [153–155]. The living character of anionic polymerization and the higher reaction rates, compared with free radical polymerization, especially in polar solvents, necessitate special experimental techniques. They are well described in Refs. [156–158].

Anionic initiation has been accomplished in a variety of solvents, both polar and nonpolar. Typically, initiation can proceed by electron transfer reactions from alkali or alkaline earth metals, polycyclic aromatic radical anions, or alkali and magnesium ketyls. The other possibility includes the nucleophilic addition of organometallic compounds to the monomers. Related monofunctional initiators comprise alkyl derivatives of alkali metals or organomagnesium compounds such as Grignard reagents. Difunctional species are alkali derivatives of α-methylstyrene tetramer or the dimer of 1,1-diphenylethylene. An overview of the initiation process in carbanionic polymerization is given in Ref. [159].

Ester compounds of the acrylic acid are polymerizable anionically only in certain cases, mostly with only partial conversion. The polymerization of methacrylic esters, however, proceeds with minor problems. The need for strong purification of the monomers, the in general required low reaction temperature, and the tendency for the carbonyl group to participate in the polymerization, particularly during the initiation stage, are serious handicaps for its commercial application. Considering these difficulties and the big interest especially in block copolymers containing methacrylic esters, it is no surprise that permanent efforts were devoted to the development of a 'perfectly' controlled polymerization of these monomers in terms of molecular characteristics like the molecular mass and the molecular mass distribution, regio- and stereoselectivity and the design of block copolymers.

As far as the stereoregularity is concerned, studies of various types of initiation show that methacrylates could be polymerized to give as well as isotactic, syndiotactic atactic polymers. Numerous physical properties are tacticity dependent: for example, the rate of water absorption is higher for syndiotactic than for isotactic polymer [97], the transition temperatures of liquid crystalline methacrylic polymers can be specifically influenced [160–162], and the miscibility of polymer blends is changed [163–165]. In general, the stereoregularity depends on the solvent used, the initiator, and the reaction temperature. Reviews have provided an overview concerning analysis, properties and reactivities of polymers with respect to their tacticity [97,166,167].

Highly isotactic PMMA can be formed in nonpolar solvents with lithium-based initiators or some Grignard reagents [97]. A laboratory recipe for isotactic PMMA (>96%) with narrow molecular mass distribution through polymerization of MMA in toluene with t-butyl-MgBr is given in Refs. [97,156,168]. The polymerization proceeds in a living manner as the molecular mass increases direct proportionally with the conversion and the result is a highly isotactic polymer with narrow molecular mass distribution

Table 3 Isotactic living polymerization of MMA with t-butyl-MgBr in toluene at $-78\,^\circ$C [169].

[M]$_0$/[I]$_0$	Time/h	Yield/%	M_n^b	M_w^c/M_n	Tacticity^d/%		
					mm	*mr*	*rr*
50	24	73	3,660	1.14	96.3	3.6	0.1
50	72	100	4,930	1.10	96.5	3.2	0.3
100	120	100	10,100	1.10	96.8	2.9	0.3
100	145	99	21,200	1.08	96.7	3.0	0.3

^{a}MMA 10.0 mmol, toluene 5.0 mL; ^{b}Determined by VPO; ^{c}Determined by SEC; ^{d}Determined by ^{1}H-NMR; ^{e}MMA 20.0 mmol, toluene 10.0 mL.

(Table 3). In case of polymerization of ethyl (EMA) and n-butyl (n-BuMA) methacrylates under the same conditions, a bimodal molecular mass distribution was observed. The similar isotacticity in both fractions, indicates the existence of two types of active species [169]. The addition of $(CH_3)_3Al$ to the polymerization of EMA recently has been found to have the beneficial effect of allowing the synthesis of highly isotactic PEMA with low polydispersity [167].

Rather high syndiotactic PMMA in general can be achieved in polar solvents [e.g., with bulky alkyllithium compounds in THF at $-78\,^\circ$C (85%)] [97]. In addition, certain types of Grignard reagents [e.g., 3-vinylbenzyl-MgCl in THF at $-110\,^\circ$C (living polymerization)] were used successfully for the preparation of highly syndiotactic (97%) PMMA [170]. Contrary to the above-mentioned rule, highly syndiotactic PMMA (9–8%) with small molecular mass distribution has been described in apolar solvents, too [e.g., with the complex catalyst t-butyl-Li/Al(alkyl)$_3$ in toluene at $-78\,^\circ$C] [171].

More recently, atactic living anionic polymerization has been achieved by using t-C$_4$H$_9$Li and bis(2,6-di-t-butylphenoxy)methylaluminum [MeAl(ODBP)$_2$] (Al/Li = 5) in toluene at low temperature [172,173]. Thereby, the role of MeAl(ODBP)$_2$ is the stabilization of the propagating species and activation of the monomer by coordination. As the stereospecificity of the polymerization strongly depends on the polymerization conditions, e.g., the ratio of the initiator components in the binary initiator system, combinations of t-C$_4$H$_9$Li and MeAl(ODBP)$_2$ can also provide stereoregular statistical copolymers of methacrylates acrylates [174,175] as well as stereoregular block copolymers and block copolymers [176,177] via living polymerization. Replacement of the methyl group in MeAl(ODBP)$_2$ by other alkyl groups (Scheme 10) resulted in an increase of syndiotacticity with the size of the alkyl rest as it is shown in Table 4 polymerization of ethyl methacrylate (EMA) with t-C$_4$H$_9$Li and alkylaluminum bisphenoxide (molar ratio = 1:3) in toluene at $-78\,^\circ$C for 24 h) [178].

$$(10)$$

Kinetic, thermodynamic, and mechanistic aspects of the anionic polymerization process of acrylic esters have been reviewed in several articles [97,158,179]. The control of

Table 4 Polymerization of EMA with t-C$_4$H$_9$Li and alkylaluminum bisphenoxide.[a]

R_1[b]	R_2[b]	Yield/%	M_n[c]	M_w[c]$/M_n$	Tacticity[d]/%		
					mm	*mr*	*rr*
CH$_3$	H	100	7,510	1.13	7.3	87.6	5.1
CH$_3$	CH$_3$	97	6,040	1.12	6.9	67.5	25.6
CH$_3$	t-C$_4$H$_9$	100	8,170	1.10	6.2	84.3	9.5
CH$_3$	Br	100	6,360	1.08	13.8	82.7	4.1
C$_2$H$_5$	H	100	6,490	1.09	0.0	8.1	91.9
i-C$_4$H$_9$	H	30	4,990	1.14	0.3	17.5	82.2

[a]EMA 10 mmol, t-C$_4$H$_9$Li 0.2 mmol, toluene 10 mL, alkylaluminum phenoxide 1.0 mmol; [b]see Scheme (10); [c]Determined by SEC; [d]Determined by ^{13}C-NMR.

this kind of polymerization is often limited by the occurrence of side reactions, including (1) the attack of the initiator at the carbonyl double bond of the monomer or polymer, (2) chain transfer of α-situated protons, (3) 1,4-addition via the enolate oxygen instead of 1,2-addition through the carbanionic centers [see Scheme 11], and (4) coordination of the counterion of the active centers with carbonyl groups. Additionally, the ion pairs tend to aggregate into much less active dimers and higher agglomerates. However, despite those complications, it is possible to obtain polymers of narrow molecular mass distribution and 'ideal' polymerization kinetics under appropriate conditions [179]:

(11)

Therefore, several partially successful strategies have been developed to avoid the mentioned side reactions. One of these is the so called ligated anionic polymerization (LAP). The basic concept of LAP is the use of suitable ligands, which are able to coordinate at the active initiating or propagating ion-pairs. The three major functions of the ligands are (1) to promote a new complexation equilibrium, with ion-pairs and/or aggregates, preferably leading to a single stable active species, (2) to modulate the electron density at the metal enolate ion-pair and thereby influencing stability and reactivity, and (3) to protect the ion-pair by effecting a steric hindrance, and thus avoiding back-biting reactions of the growing anion [180]. Two efficient classes of ligand systems have been investigated quite recently:

- μ-type ligands, such as alkali metal *tert*-alkoxides [181,182], aluminum alkyls [183,184] and some inorganic lithium salts [185]
- μ/σ-type dual ligands, such as lithium 2-methoxyethoxide (MEOLi) [186], lithium 2-(2-methoxyethoxy) ethoxide (MEEOLi) [187,188] and lithium aminoalkoxide [189].

Tert-alkoxides, especially lithium *tert*-butoxide (t-BuOLi), have been used by Vlček et al. in complex initiator systems with alkali metal ester enolates, such as ethyl α-lithioisobutyrate. MMA [181], t-butyl acrylate [190], 2-ethylhexyl acrylate [191] have been prepared with beneficial effects of the additive, but at least a 10-fold excess of the additive with respect to the initiator was necessary to reach low PDIs. An overview is given in Ref. [192]. When aluminum alkyls are used as μ-type ligands for MMA polymerization in toluene a fairly broad molecular mass distribution is observed. Adding Lewis bases as co-solvents, such as methyl pivalate and diisooctyl phthalate resulted in the synthesis of syndiotactic PMMA with low polydispersity, even at 0 °C [183]. Various lithium salts have been investigated as additives in anionic polymerization of MMA. Thereby, LiCl was showed to have a favourable effect on the anionic polymerization, as the initiator efficiency has been kept high and polymers with narrow molecular mass distributions have been obtained. This effect was remarkable only when less sterically hindered initiators like α-methyl styrene have been used [193]. Substitution of LiCl by LiClO$_4$ as μ-type ligand resulted in the synthesis of well defined polymethacrylates due to the better solubility in hydrocarbons [185].

Lithiated alkoxyalkoxides, bidentate ligands of the μ/σ-type (see Scheme 12), have been intensively investigated and they restricted the tendency for back-biting reactions by forming strong complexes with the end of the 'living' chain. Due to this higher stabilizing efficiency, they provide excellent control over polymerization of acrylates as well as methacrylates at low temperatures in THF and toluene. Best results for MMA polymerization were obtained with MEEOLi when the polymerization was performed at very low temperatures in a moderately polar solvent (toluene/THF mixture) [194]. The same observation was made for the polymerization of butyl acrylate [195]. The outstanding role of toluene as solvent for MMA polymerization in the presence of monolithium alkoxyalkoxides has been shown by Müller et al. [196].

Recently, polydentate dilithium alkoxides (dilithium triethylene glycoxides) (Scheme 12) have been shown to be suitable additives for the polymerization of methyl methacrylates, as they provide high initiator efficiencies and narrow molecular weight distributions (1.1–1.3). The addition of dilithium triethylene glycoxide to the anionic polymerization of MMA (THF, (1,1-diphenylhexyl)lithium as initiator) resulted in the synthesis of well controlled polymers even at relatively high temperatures. This beneficial effect could be assigned to a better coordination with the enolate ion pairs, thus slowing down the polymerization rates (Table 5) [197].

(12)

Several reviews of anionic polymerization of methacrylates and acrylates in the presence of stabilizing additives have been published in the last years [198–200]. Additionally, mechanistic studies of the propagating species have been investigated [198,200–203].

Another quite recently developed method for the controlled polymerization of methacrylates via anionic polymerization is the screened anionic polymerization (SAP), investigated by Haddleton et al. The systems are based on lithium aluminum alkyl/phenoxide initiators, which are synthesized in situ following the equation shown

Table 5 Anionic polymerization of MMA in THF at various temperatures using DPHLi[a] as initiator in the presence of DLiTG.[b]

Temp./°C	[DPHLi]/ (mmol/L)	[MMA]/ (mol/L)	[DLiTG]/ [DPHLi]	Yield/ %	$10^{-3} M_{n,calc}$[c]	$10^{-3} M_n$[d]	PDI
−40	1.015	0.228	10	100	22.4	25.0	1.09
−20	0.08	0.224	0	95	28	42.2	1.54
−20	0.952	0.267	4	100	28.0	36.4	1.27
−20	0.335	0.303	10	100	90.5	100.5	1.18
0	1.90	0.312	0	80	16.3	26.6	1.67
0	1.41	0.330	5	90	23.4	26.8	1.34
0	0.47	0.300	10	100	63.8	66.8	1.09

[a]DPHLi = (1,1-diphenylhexyl)lithium (initiator).
[b]DLiTG = dilithium triethylene glycoxide (additive).
[c]$10^{-3} M_{n,calc}$ = (moles of monomer/moles of initiator) × 100.
[d]Determined by SEC.

in Scheme 13. The polymerization was proved to have a 'living' nature by sequential monomer addition experiments [204–206].

$$^tBu_3Al \ + \ HO-\!\!\!\bigcirc\!\!\!-R \ \xrightarrow{\ toluene/0\,°C\ } \ ^tBu_2AlO-\!\!\!\bigcirc\!\!\!-R \qquad (13)$$

R = Me/H

   a. End-functional polymers and copolymers. One advantage of living anionic polymerization is the availabilty of telechelic polymers [207] and macromonomers, which are of specific interest for the preparation of comb-like (if monofunctional) and network (if difunctional) structures [208,209]. In addition, due to its 'living' nature, anionic polymerization provides a versatile synthetic route for the synthesis of a wide range of well defined polymer structures. Thereby, the steadily increasing capability of LAP offers numerous possibilities, e.g., for the preparation of block copolymers.
   Fully methacrylic triblocks, containing a central rubbery poly(alkyl acrylate) block and two peripheral hard poly(alkyl methacrylate) blocks, are potential substitutes for the traditional styrene-diene-based thermoplastic elastomers (TPEs), which have relatively low service temperatures. Fully methacrylic triblock copolymers are able to cover service temperatures due to the varying T_g from −50 °C (poly(isooctyl acrylate)) to 190 °C (poly (isobornyl methacrylate) [210]. Poly(methyl methacrylate)-*b*-poly(*n*-butyl acrylate)-*b*-poly(methyl methacrylate) triblock copolymers, which are precursors for poly(methyl methacrylate)-*b*-poly(alkyl acrylate)-*b*-poly(methyl methacrylate) via selective transalcoholysis, have been synthesized by a three-step sequential polymerization of MMA, *tert*-butyl acrylate (*t*-BuA), and MMA in the presence of LiCl as stabilizing ligand [211,212]. Various diblock copolymers, such as poly(methyl methacrylate)-*b*-poly(*n*-butyl acrylate) and poly(methyl methacrylate)-*b*-poly(*n*-nonyl acrylate), have been synthesized

via LAP with lithium 2-(2-methoxyethoxy) ethoxide (MEEOLi) and diphenylmethyl-lithium and low polydispersities have been observed (1.20–1.35). Sequential anionic polymerization of MMA and n-BuA in the absence of MEEOLi resulted in polymers with molecular masses, significantly differing from the calculated values, and with broader molecular mass distributions (PDI = 2.65) [188]. Additionally, the synthesis of acrylate diblock copolymers was investigated in the presence of *tert*-alkoxides, such as t-BuOLi [192]. Stereoregular block polymers and block copolymers are also described in literature [176,177].

Besides, polystyrene/polyacrylate [193,213] and polydiene/polyacrylate [214] block copolymers have been synthesized via LAP. Thereby, the addition of stabilizing ligands, such as t-BuOLi and LiCl, provided narrow molecular mass distributions of the resulting polymer.

3. Polymerization by Complex Initiators

In this section polymerization reactions in the presence of organometallic systems are summarized. Recent work by Yasuda et al. [215] has revealed the potential of rare earth metal, $[SmH(C_5Me_5)_2]_2$ or $LnMe(C_5Me_5)_2(THF)$ (Ln = Sm, Y and Lu), to initiate polymerization of polar and nonpolar monomers in a living fashion (Table 6). Polymers with high molecular mass and narrow polydispersity can be obtained with high yield. The initiation mechanism was discussed on the basis of x-ray analysis of the 1:2 adduct of $[SmH(C_5Me_5)_2]_2$ with MMA. An eight-membered ring intermediate is formed which stabilizes the enol chain end, also allowing insertion of monomer. Afterwards the chain end coordinates to the metal in an enol form, while the penultimate MMA unit coordinates to the metal at its C=O group (Scheme 14).

(14)

Investigating several different lanthanide metals it was shown that the rate of polymerization increased with an increase of ionic radius of the metals (Sm > Y > Yb > Lu) and decreased with an increase of steric bulk of the auxiliary ligands ($C_5H_5 > C_5Me_5$).

Stereospecific polymerization of ethyl, isopropyl, and *tert*-butyl methacrylates with organolanthanide initiators was also possible. The rate of polymerization and syndiotacticity decreased with increasing bulkiness of the alkyl group in the order Me > Et > iPr $\gg$ tBu. For example high molecular mass isotactic poly(MMA)

Table 6 Results of the organolanthanide initiated polymerization of alkyl methacrylates.[a]

Initiator	Monomer[b]	$10^{-3} M_n$	M_w/M_n	rr/%	Conv./%
[SmH(C$_5$Me$_5$)$_2$]$_2$	MMA	57	1.03	82.4	98
	EtMA	80	1.03	80.9	98
	iPrMA	70	1.03	77.3	90
	tBuMA	63	1.42	77.5	30
LuMe(C$_5$Me$_5$)$_2$(THF)	MMA	61	1.03	83.7	98
	EtMA	55	1.03	81.0	64
	iPrMA	42	1.05	80.0	63
	tBuMA	52	1.53	79.5	20

[a]Polymerization conditions: 0 °C in toluene, initiator concentration: 0.2 mol%.
[b]EtMA: ethyl methacrylate; iPrMA: isopropyl methacrylate; tBuMA: $tert$-butyl methacrylate.

(mm = 97 %, M_n = 500,000, M_w/M_n = 1.12) was for the first time obtained quantitatively by the use of [(Me$_3$Si)$_3$C]$_2$Yb (Scheme 15) [215].

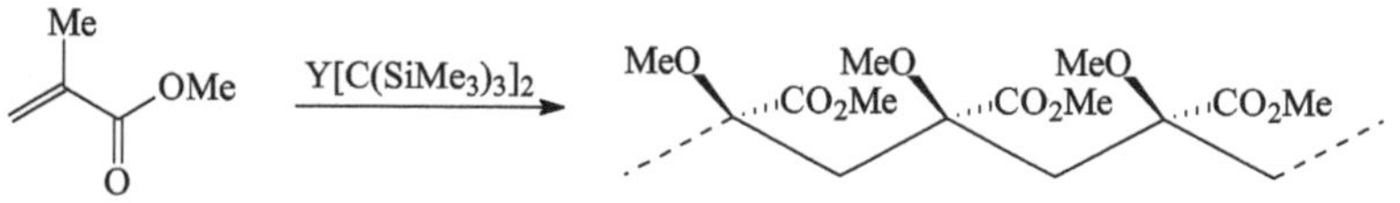

(15)

Polymerization of acrylic esters, i.e. methyl acrylate, ethyl acrylate, butyl acrylate, and $tert$-butyl acrylate, initiated by rare earth metal complexes were non-stereospecific [216].

Various block copolymerizations of hydrophobic and hydrophilic acrylates were also investigated, i.e., ABA type triblock copolymerization of MMA/BuA/MMA, triblock polymerization of MMA/EtA/EtMA, and block copolymerization of MMA/TMSMA.

In recent years metallocene complexes have also been successfully used as polymerization catalysts for methyl methacrylate. Collins et al. and Soga et al. reported that cationic zirconocene complexes catalyse the polymerization of MMA [217–220]. But these metallocene complexes consisted of more components than the only metallocene complex. Höcker et al. investigated some novel single-component zirconocene complexes as catalysts for the stereospecific polymerization of MMA [221]. MMA was polymerized by the cationic bridged zirconocene complex [iPr(Cp)(Ind)Zr(Me)(THF)][BPh$_4$] at temperatures between −20 and 20 °C. The polymerization led to mainly isotactic PMMA due to an enantiomorphic site mechanism and a low polydispersity index (1.12–2.33). Also it has been assumed that the polymerization mechanism is of living character.

Höcker et al. synthesized another zirconocene complex for the polymerization of highly isotactic PMMA, namely Me$_2$CCpIndZrMe(THF)$^+$BPh$_4^-$ (Scheme 16: showing both isomers) [222].

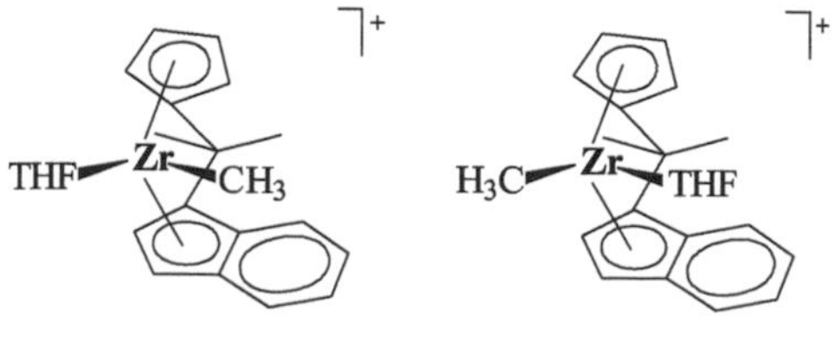

(16)

They also polymerized MMA with $Me_2CCp_2ZrMe(THF)^+BPh_4^-$ at low temperatures yielding syndiotactic PMMA [222]. Investigating the polymerization mechanism it was proposed that a methyl group of a zirconocenium cation is transferred to a coordinated MMA molecule. The resulting cationic ester enolate complex is the active species. It activates the growing chain end as a donor and at the same time an incoming MMA molecule as an acceptor. Thus the catalyst symmetry controls the microstructure of PMMA.

Concerning copolymerization various nickel and palladium-based catalyst systems copolymerize ethylene and acrylates or polar 1-olefins at low pressure [223]. With Brookhart's bisimine palladium complex simultaneous copolymerization and branching was observed. Both polar and non-polar side chains were obtained, the ester side chains can be used as cure sites in branched polyethylene rubbers (Scheme 17).

$$CH_2{=}CH_2 \ + \ CH_2{=}C \!\! \begin{array}{c} COOR \\ \diagdown \\ H \end{array} \quad \xrightarrow{\ cat.\ } \quad \begin{array}{c} (CH_2)_p{-}CH_3 \\ | \\ {+}CH{-}(CH_2)_n{-}CH{-}(CH_2)_m{]}_x \\ | \\ (CH_2)_o \\ | \\ CH_2{-}CH_2{-}COOR \end{array}$$

$$(17)$$

$$\left[\begin{array}{c} R \diagdown \diagup NR \\ \diagup \diagdown \diagup Me \\ Pd \\ \diagup \diagdown \\ R \diagup \diagdown NR \quad OEt_2 \end{array} \right]^+ \quad BAr_4^-$$

4. Metal-free Polymerizations

In 1988 Reetz et al. introduced the concept of metal-free polymerization of acrylates, methacrylates and acrylonitrile [224,225]. Metal-free initiators are salts consisting of a carbanion (A^-) having R_4N^+ as cationic counterions. They are synthesized by the reaction of neutral CH^- or NH-acidic compounds such as malonic acid esters, nitriles, sulfones, nitro-alkanes, cyclopentadiene, fluorene derivates, carbazoles and succinimide. Water is removed azeotropically using toluene.

$$AH + HO^- \, {}^+NR_4 \ \longrightarrow \ A^- \, {}^+NR_4 + H_2O \uparrow \qquad\qquad (18)$$

Scheme (19) shows some examples of the synthesized initiators.

$$(19)$$

Anion and cation are connected to each other via H-bonds. This often leads to dimers in solution and in the solid state. These species are also called 'supramolecular ion pairs'.

These initiator systems are capable of initiating the polymerizations of n-butyl acrylate, methyl methylacrylate and acrylonitrile (PDI 1.1–1.4; molecular mass 1,500–20,000 g/mole). But it must be mentioned that the metal-free polymerization is not a real living process. Backbiting and Hofmann elimination occur to a small but significant extent [226].

Another approach to PMMA is the polymerization of MMA using iodo-malonates in combination with $(n\mathrm{Bu})_4\mathrm{N^+I^-}$ (1:1) as initiators, a new initiator system which is specific for methacrylate, i.e., acrylates are not polymerized (Scheme 20) [227].

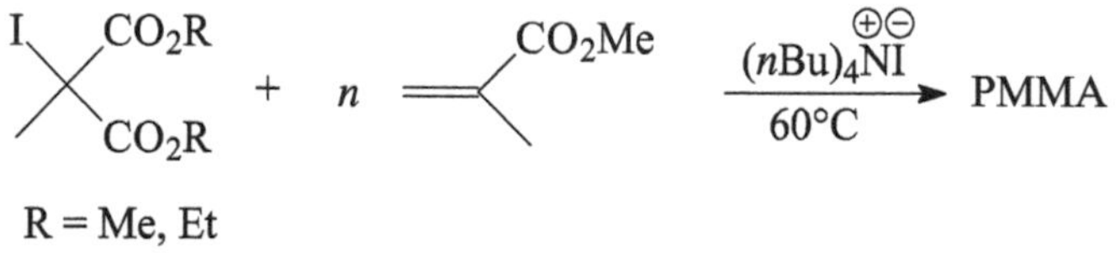

$$R = \mathrm{Me,\ Et}$$

(20)

The molecular mass can be controlled (1,500–20,000 g/mole), polydispersity values in the range of 1.2 to 1.7 could be achieved, however the control of tacticity is not possible.

Zagala et al. investigated the polymerization of methacrylates in the presence of tetraphenylphosphonium (TPP) ion at ambient temperature. The polymerization appears to have living character [228]. In case of MMA number average molecular masses increase linearly with conversion and molecular mass distributions are narrow (< 1.30). Results of ^{1}H, ^{13}C and ^{31}P NMR studies indicated the presence of phosphorylides formed by the addition of the PMMA enolate anion to one of the phenyls of the TPP cation. Müller et al. managed to synthesize another metal-free initiator, namely the salt of the tetrakis[tris(dimethylamino)-phosphoranylideneamino]phosphonium (P_5^+) cation with the 1,1-diphenylhexyl (DPH$^-$) anion, by a metathesis reaction between P_5^+ chloride and 1,1-diphenylhexyllithium (Scheme 21) [229].

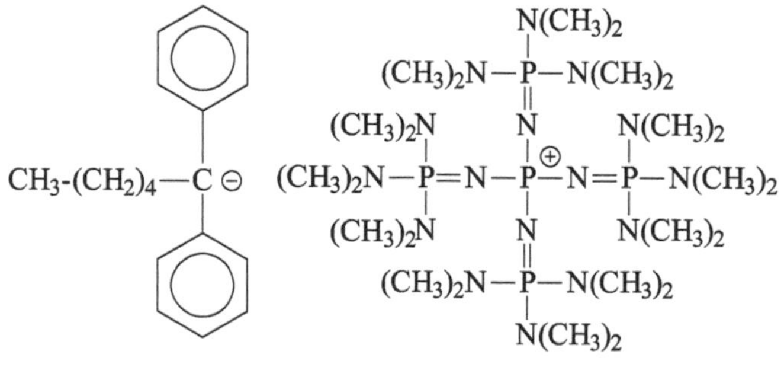

(21)

DPHP$_5$

5. Group Transfer Polymerization

In 1983, Webster et al. reported a new living polymerization method, called group transfer polymerization (GTP) [230]. This process consists of a continuously catalysed Michael addition of a silyl ketene acetal onto α,β-unsaturated ester compounds, mainly acrylic ester monomers. During the polymerization, the silyl group is transferred to the monomer, thus generating a new ketene function:

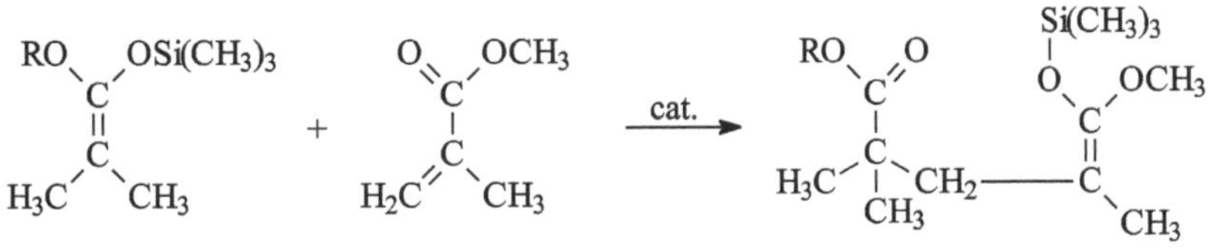

(22)

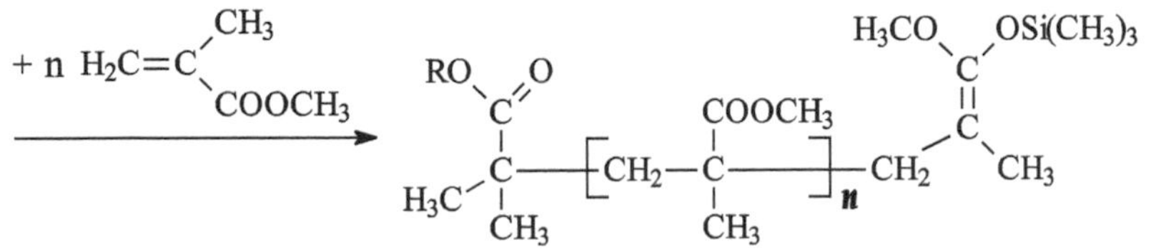

(23)

Beside this transfer mechanism Müller has proposed an associative mechanism, at least for cases involving certain GTP catalyst components [231].

GTP, reviewed briefly in Refs. [156,232–234], is controlled by the stoichiometry of initiator and monomer and shows the characteristics of a living polymerization mechanism. Consequently, polymers with a controlled molecular mass up to 100,000 and a narrow molecular mass distribution are obtained. As an advantage over classical living polymerizations (anionic), GTP proceeds smoothly at room temperature. In general the reaction temperature lies between -100 and $120\,°C$, but 0 to $50\,°C$ is preferred. But GTP does not produce polymers having a high degree of stereoregularity.

Beside the silyl ketene acetal shown above, all silyl derivatives that add to acrylic monomers, subsequently producing ketene acetals, can initiate the GTP (e.g., Me_3SiSMe, Me_3SiSPh, Me_3SiCR_2CN, $R_2P(O)SiMe_3$) [156]. Bifunctional bis(silyl ketene acetals), which are interesting for subsequent block copolymers, have also been used [235]. Stannyl ketene acetals and the corresponding germyl compounds are also known as initiators, although they lead to a somewhat broader molecular mass distribution than do the corresponding silyl derivatives [234,236,237]. Collins has developed an associative group transfer-type polymerization for methyl methacrylate based on zirconocenes [238].

GTP is catalyzed by two different classes of compounds:

1. Anionic catalysts work by coordination to the silicon atom; they are needed in only small amounts ($\approx 0.01\%$ based on the initiator) and are used preferably for methacrylic monomers [232]. The anionic moiety comprises fluoride, azide, and cyanide, but also carboxylates, phenolates, sulfinates, phophinates, nitrite, and cyanates [232,236,237]. These anions are often used in combination with their corresponding acids as biacetate, $H(CH_3COO)_2^-$ or bifluoride, $(HF_2)^-$. The counterions are usually tetraalkyl ammonium or tris(dimethylamino)sulfonium, $[(CH_3)_2N]_3S^+$ (TAS). The most widely used catalysts are $TASHF_2$ and $TASF_2SiMe_3$ [239]. To accelerate the reactivity of potassium bifluoride, KHF_2, a crown ether (18-crown-6)-supported polymerization has been carried out [240].
2. Lewis acids activate the monomer by coordination to the carbonyl group [241]. Lewis acids are used preferably for acrylate monomers [232]. Common catalysts

are zinc halides and organoaluminum compounds (e.g., dialkylaluminum halides and dialkylaluminum oxides) [234]. Mercury compounds such as HgJ_2, $Hg(ClO_4)_2$, or alkyl HgJ also catalyze GTP with good results [242,243].

Detailed descriptions of polymerizations of MMA, ethyl acrylate, and butyl acrylate with either anionic or Lewis acid catalysts are given in Refs. [156] and [234]. Various other monomers, including lauryl, glycidyl, 2-ethylhexyl, 2-trimethylsiloxyethyl, sorbyl, allyl, and 2-(allyloxy)ethyl methacrylates have been employed in GTP [234]. Because of the milder conditions, this polymerization method is generally much more suitable than the classical anionic polymerization for monomers with reactive functional groups.

GTP of MMA with Lewis acid catalysts were reported to give PMMA with a ratio of 2:1 syndiotactic/heterotactic triads, while anionic catalysts such as bifluoride salts lead to a ratio of 1:1 [241]. The influence of the temperature on tacticity is shown in $TASHF_2$/THF systems: With decreasing temperature, syndiotacticity increases from 50% to 80% [234,244,245]. A comparison of the triad distribution for anionic and GT polymerizations of MMA with the same counterions under the same conditions shows that the tacticities of both polymerization types are consistent [97,246]. Some selected examples of the influence of different polymerization parameters on tacticity are given in Ref. [245].

The living character and different characteristic possibilities during GTP allow especially the synthesis of either telechelics or block and graft copolymers. Such characteristic possibilities are:

1. *Functionalized initiators.* Their use leads to terminal functionalized polymers. Thus, with phosphorus-containing ketene silyl acetals, trimethylsilyl methyl sulfide, trimethylsilyl cyanide, dimethylketene-bis(trimethylsilyl)acetal, or dimethylketene-2-(trimethylsiloxy) ethyltrimethyl silyl acetal, terminal phosphoric acid groups, thiomethyl groups, and cyanide, hydroxy, or carboxyl groups are readily introduced [234]. Furthermore, the styrene end group can also be achieved [247].

2. *End-capping reactions.* Reaction of the living end groups with bromine yields an X-bromo ester [248]. With 4-(bromomethyl)styrene a styryl-ended macromonomer is available [249]. Benzaldehyde gives, after hydrolysis, terminal benzhydryl alcohol groups [234]. Terminal monofunctional polymers (e.g., living PMMA with one masked OH end group) can be converted into bifunctional polymers by reacting the living center with bifunctional coupling agents such as 1,4-bis(bromomethyl)benzene [250]. Three- and four-star polymers are obtained when corresponding multifunctional agents were applied [232].

3. *Functionalized monomers.* Since GTP is a much milder process than anionic polymerization, for example, numerous functionalized monomers can be polymerized. Thus trimethylsilyl and 2-(trimethylsiloxy)ethyl, allyl, 2-(allyloxy)ethyl, and 4-vinylbenzene MA give polymers with functional groups along the chain, which were used for further modifications (e.g., for the synthesis of graft copolymers) [232,234,251].

Concerning grafting techniques, in GTP acrylates are much more reactive than methacrylates. Thus 2-methacryloxyethyl acrylate in the presence of $ZnBr_2$ is polymerized exclusively to a polymer with pendant methacrylate groups capable of radical and GTP 'grafting from' polymerizations [73]. Irradiation techniques have often been employed to create active sites on polymer backbones. Thus alkyl acrylates and methacrylates have been grafted from poly(ethylene) [252–254], poly(alkyl methacrylates) [255], or cellophane [256].

6. Catalytic Chain Transfer Polymerization (CCTP)

CCTP has its origins in biochemistry where coenzyme B_{12} is used to conduct many free-radical reactions. Enikolopyan et al. were the first who used analogues of B_{12} for polymerization [257,258]. Methacrylate was polymerized by a catalyzed chain transfer using a cobalt porphyrine. AIBN was used as initiator. Two possible reaction sequences for the 'catalytic' aspect of CCT are described in the following scheme:

$$
\left. \begin{array}{rcl}
R_r + Co\text{-}Por & \longrightarrow & Pr + HCo\text{-}Por \\
HCo\text{-}Por + M & \longrightarrow & R_1 + Co\text{-}Por
\end{array} \right\}
$$

$$
\left. \begin{array}{rcl}
M + Co\text{-}Por & \longrightarrow & (M\text{---}Co\text{-}Por) \\
(M\text{---}Co\text{-}Por) + R_f & \longrightarrow & P_r + Co\text{-}Por + R_i
\end{array} \right\} \quad (24)
$$

In the first sequence the Co complex acts as a chain transfer agent itself; in the second the Co complex catalyses chain transfer to monomer. The disadvantages in using porphyrin reagents are colored reagents, they are expensive, and of limited solubility in polar media. Thus, O'Driscoll et al. replaced the porphyrine with a cobalt(II) dimethyl glyoxime (Co-dmg) [259]. Gridnev used cobaloximes as CCT agents for a number of methacrylates and for styrene [260]. Other CCT agents are pentacyano-cobalt(II) and bimetallic compounds using molybdenum, iron, chromium, or tungsten as the metal [261].

The efficiency of the reagents is influenced by the stabilizing base ligands. A number of bases have been used to enhance the transfer process, ranging from Et_3N, which has the weakest effect, to $(MeO)_3P$, which has the strongest.

7. Living Radical Polymerization

Despite the long-time research in the field of free radical polymerization, this poly-merization technique has been believed to be beyond reach of the precision control that has been achieved in ionic living polymerizations due to the prevention of chain transfer and termination reactions. Nevertheless, many efforts have been made to realize the same control in radical polymerization reactions. The common general principle of the recently developed controlled radical polymerization processes is the temporarily transformation of the radical growing ends into more stable covalent precursors, called dormant species. This dormant species and the active radical are in a dynamic and rapid equilibrium dominated by the covalent species, and thereby suppressing the bimolecular radical termination reactions. As a result, linear increase of the number-average molecular mass M_n of the prepared polymer with respect to conversion as well as narrow molecular mass distributions are observed. Although many systems, such as polymerization in the presence of organocobalt porphyrine complexes [262], were investigated, the two most widely used are stable free radical polymerization (SFRP) and atom transfer radical polymerization (ATRP).

8. Stable Free Radical Polymerization (SFRP)

In the 1990s the groups of Rizzardo and Georges reported a stable free radical polymerization process (SFRP) allowing the preparation of polystyrene with a narrow polydispersity. In the presence of stable free radicals, such as the mainly used 2,2,6,6-tetramethylpiperidine-N-oxyl (TEMPO), macromolecules based on styrene and styrene derivatives with well defined structures were synthesized [263,264].

In contrast, the extension of this promising polymerization process to acrylates proved to be more challenging than expected. Indeed, synthesis of random copolymers of styrene and low amounts of n-butyl acrylate provided high yields and narrow molecular mass distributions, but increasing the level of acrylate resulted in higher polydispersities and a lowering of conversion (Table 7). Additionally to random copolymerization, this method was applied for the synthesis of a poly(styrene-b-(styrene-co-n-butyl methacrylate) block copolymer [265].

The mechanism of SFRP [266] (Scheme 25) involves an equilibrium between nitroxide-capped polymer chains and uncapped polymer chains. Its success relies on the retention of the suitable amount of free nitroxide in the reaction to keep the propagating polymer radical chains at a concentration which allows the polymerization to proceed at a sufficient rate but avoids bimolecular termination by coupling.

$$(25)$$

In nonstyrenic systems a gradual increase in free nitroxide concentration over reaction time seems to inhibit polymerization [267]. Therefore, a successful polymerization of acrylates by the SFRP process requires the reduction of the amount of free nitroxide. Progress was made by removing oxygen from the reaction mixture, as the known radical scavenger molecular oxygen is able to cause chain termination, and as a consequence thereof, free nitroxide is generated. Whereas the polymerization of n-butyl acrylate in the presence of benzoyl peroxide (BPO) and TEMPO is stopped after 1–2 h at a conversion of

Table 7 Effect of increased acrylate level in TEMPO-mediated stable free random copolymerization with styrene.

Mole% n-butyl acrylate	$10^{-3} M_w$	$10^{-3} M_n$	PDI	Conversion/%
24.7	32.5	24.6	1.32	89.3
43.5	38.2	24.8	1.54	83.0
64.1	36.4	24.4	1.49	65.2
81.2	10.6	6.1	1.73	33.4

about 5% and a low molecular mass, generally below 4000, the careful control of the amount of oxygen allows to continue polymerization to higher conversions, rarely exceeding 20%. Similar results were obtained using initiator/nitroxide adducts for the control of the initial amount of excess free nitroxide [268,269] (Scheme 26).

(26)

In contrast, the performance of *n*-butyl acrylate polymerizations in the presence of glucose as radical scavenger and reducing agent and sodium bicarbonate leads to a polymerization process with a 'living' nature to conversions around 60% and a molecular mass of approximately 30,000 in 6.5 h. The living character of the poly(*n*-butyl acrylate) prepared in this manner could be proven by the formation of a block copolymer after the addition of styrene. As ene-diols were believed to be the active species hydroxyacetone was used to substitute the glucose and the polymerization of *n*-butyl acrylate with BPO and hydroxy-TEMPO was performed with yields of 60–70% and molecular masses around 60,000 were obtained after 8.5 h. The living character can be demonstrated by the incremental increase in molecular mass during reaction time [270].

Another class of counter radicals, introduced by Müllen et al. and resulting in a controlled polymerization of acrylates and methacrylates, are triazolinyl radicals. A molecular mass of ca. 60,000 was achieved at a conversion around 35% [271]. Besides, the addition of small amounts of camphorsulfonic acid (CSA) [272] and FMPTS [273,274] was examined. Thereby, a reduction in the concentration of free nitroxide during the polymerization to a level around 5×10^{-6} M resulted in an improvement of polymerization rates and consequently a higher molecular mass, but at a cost in the narrowness of molecular mass distribution.

Additionally, the synthesis of a wide range of block copolymers containing poly(alkyl acrylates) was successfully performed by different groups in 'living' manner by using N-oxyl radicals as radical stabilizing agents [275–281]. In principle, first of all, a nitroxide-terminated macroinitiator is synthesized. Then a second monomer is added and by heating, the relatively weak bond between the macroinitiator chain and the nitroxide end group is broken, which allows stable free radical polymerization of the second monomer to take place. Several examples of synthesized polymers are presented in Table 8. Polystyrene-*block*-poly(alkyl acrylate) copolymers are of particular interest because of their potential application as surface active agents, pigment dispersants, flocculants, and compatibilizers in polymer blends after hydrolysis.

As a solventless route to block copolymers the application of supercritical carbon dioxide in the SFRP process was investigated. This offers additional potential for providing a higher complexity of macromolecular structures in the absence of organic solvents. Due to the increased diffusivity of monomer dissolved in the supercritical CO_2 and the plasticization of the polymer, the rate of polymerization of the second block can be increased, and thereby, a one pot synthesis of block copolymers becomes possible [282,283].

Table 8 Molecular mass and polydispersity of synthesized methacrylate block copolymers.

Macroinitiator	$10^{-4} M_n$	PDI	Comonomer	Time/h	$10^{-4} M_n$	PDI	Ref.
Polystyrene	5.34	1.13	DAMA[a]	2	6.71	1.25	275
Polystyrene	0.15	1.14	n-Butyl methacrylate	3.5	17.35	1.58	279
Polystyrene	0.31	1.15	n-Butyl methacrylate	2	1.44	1.21	280
Polystyrene	0.31	1.15	n-Butyl methacrylate	5	2.69	1.26	280
Polystyrene	0.31	1.15	Ethyl methacrylate	2	3.58	1.20	280
Polystyrene	0.31	1.15	Ethyl methacrylate	5	11.77	1.37	280
Polystyrene	0.31	1.15	Methyl methacrylate	2	1.80	1.21	280
Polystyrene	0.31	1.15	Methyl methacrylate	5	3.34	1.23	280
Polystyrene	0.31	1.15	Octyl methacrylate	8	3.47	1.42	280
Polystyrene	0.31	1.15	DAMA	90	0.97	1.25	280

[a]2-(dimethylamino)ethyl methacrylate.

9. Atom Transfer Radical Polymerization (ATRP)

Another interesting method of controlled radical polymerization has got its roots in organic chemistry's atom transfer radical addition (ATRA) [284,285], so named because it employs atom transfer from an organic halide to a transition-metal complex to generate the reacting radicals. Extending this reaction to the synthesis of polymers, led to the atom transfer radical polymerization (ATRP) [286] which was first developed by Matyjaszewski et al. Compared to other controlled radical systems, ATRP seems to be the most robust system due to its tolerance towards impurities (water, oxygen, inhibitor), and it can be used for a larger number of monomers. Additionally, ATRP is a catalytic system, and therefore, the polymerization rate can be easily controlled by the amount and activity of the catalyst.

The control of the polymerization reaction afforded by ATRP is the result of the formation of 'dormant' alkyl (pseudo)halides. This reduces the instantaneous concentration of the active radicals and thereby suppresses bimolecular termination reactions. The reversible deactivation and activation leads to a slow, but steady growth of the polymer chain with a well defined end group (Scheme 27). Control and properties of the synthesized polymers depend on the stationary concentration of active radicals and the relative rates of propagation and deactivation. When one or less than one monomer unit is incorporated into the polymer chain during one activation step, the polymerization is well controlled. The ATRP equilibrium can be approached from both directions in Scheme 27. Beginning with an alkyl halide and the lower valent metal complex, the process is called direct ATRP. If a conventional thermal initiator like AIBN and the higher valent metal complex are the starting materials, the polymerization process is named reverse ATRP [287].

$$P_n - X \; + \; Cu(I)/2\,bipy \; \underset{k_{deact}}{\overset{k_{act}}{\rightleftharpoons}} \; P_n \cdot \; + \; X - Cu(II)/2\,bipy$$

dormant species

k_p

monomer

(27)

The molecular mass is controlled by the initial monomer-to-initiator ratio and monomer conversion. In case of well controlled polymerization molecular mass increases direct proportionally with conversion.

The multicomponent ATRP system consists of an initiator (alkyl (pseudo)halide, RX), a redox-active transition metal in its lower oxidation state (M_t^n), ligands, a deactivator (XM_t^{n+1} species) and the monomer. ATRP is performed in bulk or in solution at elevated temperatures [288] with the possible use of different additives. One important item to regard is the fact that in ATRP one set of conditions cannot be applied to every monomer class. While neither polyacrylic nor poly(methacrylic) acid can be synthesized with currently available ATRP systems, because the monomers rapidly react with the metal complexes to form metal carboxylates, various acrylate esters can be polymerized by ATRP (Scheme 28) [289]. In analogy to these acrylate esters a wide range of methacrylate esters is expected to undergo ATRP.

$$(28)$$

Alkyls
R=CH$_3$, n-Bu

Acrylic Acid Precursors
R=t-Bu, isobornyl

a. Catalyst System (Transition Metal and Ligands). Several transition metal systems have been reported to control the radical polymerization of acrylic monomers. The metal is supposed to participate in a one-electron transfer redox cycle rather than a two-electron process which would cause side-reactions like oxidative addition followed by reductive elimination. A higher affinity of the metal to group/atom X in comparison with the affinity to hydrogen and alkyl affinity should prevent transfer reactions (e.g., β-H elimination).

While ATRP of methyl acrylate was reported only for the copper catalyst system [290–292] methyl meth(acrylate) was also polymerized with copper [290,293–295], ruthenium/aluminum alkoxide [296,297], iron [298,299] and nickel [300–303] catalyst systems (Table 9). Thereby, it must be noted that in principle, the ruthenium-based system proposed by Sawamoto et al. requires the addition of Lewis acids, e.g., Al(O-i-Pr)$_3$ [297]. Recent investigations showed, that the 'half-metallocene'-type ruthenium(II) chloride Ru(Ind)Cl(PPh$_3$)$_2$ (Ind = indenyl) led to a fast and well controlled polymerization even without the addition of Al(O-i-Pr)$_3$, whereas in case of a polymerization with Ru(Cp)Cl(PPh$_3$)$_2$ (Cp = cyclopentadienyl), the addition of Al(O-i-Pr)$_3$ is necessary. The activity of Ru(II)-catalysts decreases in the order: Ru(Ind)Cl(PPh$_3$)$_2$ > RuCl$_2$(PPh$_3$)$_2$ > Ru(Cp)Cl(PPh$_3$)$_2$ [304].

In the case of the nickel(II)-bromide complexes, such as NiBr$_2$(PPh$_3$)$_2$, additives like Al(O-i-Pr)$_3$ should also be added to improve the control of polymerization [301], whereas for NiBr$_2$(Pn-Bu$_3$)$_2$ such additives are unnecessary [302]. ATRP of MMA and n-BuA catalyzed by NiBr$_2$(PPh$_3$)$_2$ is reported with a reduced control of polymerization [301]. Recently, it has been shown that increasing the monomer concentration is an interesting way to improve the polymerization rate while keeping the actual radical concentration low [304]. The use of bis(ortho-chelated) arylnickel(II) complexes [299] as catalyst was also investigated and the polymerization of MMA without additional Lewis acids was shown to be well controlled, whereas with the recently investigated zerovalent nickel complex, Ni(PPh$_3$)$_4$, the polymerization required the addition of Al(O-i-Pr)$_3$ [303]. Possible catalyst systems are described in several reviews [289,305,306].

In addition to the metal ion, the halide ion has also got an influence on the kinetic of ATRP by affecting the atom transfer equilibrium. The use of copper bromide instead of copper chloride leads to more rapidly decreasing polydispersities (p-toluenesulfonyl chloride/copper chloride (p-TsCl/CuCl) conversion $= 25\%$, $M_n = 8500$, $M_w/M_n = 2$; p-TsCl/CuBr for the same conversion, $M_n = 7800$, $M_w/M_n = 1.18$ [294,295]). This can be assigned to the better efficiency of bromine in the deactivation step [307,308].

The ligand has got an influence on the ATRP by affecting the redox chemistry due to its electronic effects, controlling selectivity by steric and electronic effects and by solubilizing the catalytic system. Thus, the use of bipyridine instead of 4,4′-di-(5-nonyl)-2,2′-bipyridine causes a lower control of the polymerization because of the reduced solubility of the deactivator [309]. In the case of bulk polymerization, well controlled polymer structures are obtained, if substituted bipyridine is used. Nonsubstituted bipyridine as ligands in ATRP, were shown to allow ATRP in ethylene carbonate [310]. Other effective π-accepting ligands like 2-iminopyridines [293] and some aliphatic polyamines have also been described. Replacing the bipyridine ligands with linear and tetraamines resulted in significantly faster and better controlled polymerization [311]. An excess of triphenylphosphine to the polymerization of MMA via ATRP with NiBr$_2$(PPh$_3$)$_2$ as catalyst has got a beneficial effect on the kinetic of this polymerization and results in a linear dependence of M_n on the monomer conversion [304].

b. Alkyl (Pseudo)Halides. In general, any alkyl halide with activating substituents on the α-C-atom, such as aryl, carbonyl, and allyl groups are potential ATRP initiators. The polyhalogenated compounds CCl$_4$ and CHCl$_3$ as well as compounds with weak R–X bonds, such as N–X, S–X, and O–X, can also be used as initiators for ATRP. The wide range and the role of the initiators for polymerization control have been described in several reviews [288,289,306].

The main role of the alkyl halide (RX) is to generate growing chains quantitatively. The structure of the alkyl group R preferably mimics the growing polymer chain. Therefore α-halopropionates are effective initiators for the ATRP of acrylates [288]. The main importance of the choice of the initiator for the polymerization of acrylates and methacrylates is based on the requirement of a fast initiation to obtain molecular mass control. A slow initiation results in higher molecular masses than predicted and a higher polydispersity, which is specified in Table 9 for the polymerization of MMA [299]. Similar results were also observed for the phosphine-based Ni(II) complexes [301,302] and the Ni(0) complex. Here, well controlled radical polymerization was only possible with bromide initiators [303].

As group X bromine and chlorine seem to work good, obviously this group migrates rapidly and selectively between the growing chain and transition metal. Another class of initiating molecules for the polymerization of acrylates and methacrylates are sulfonylchlorides, which were reported by Sawamoto [312] and Percec [313–315].

Table 9 Polymerization of MMA by FeBr$_2$/dNbipy with different initiators.[a]

Initiator	Conversion/%	$10^{-3} M_{n(th)}$	$10^{-3} M_{n(SEC)}$	PDI
Benzyl bromide	59.5	11.9	21.4	1.60
2-Ethyl bromoisobutyrate	72.3	14.5	15.2	1.38
2-Bromopropionitrile	60.6	12.1	12.8	1.25
p-Toluenesulfonyl chloride	53.0	10.6	10.7	1.24

[a] Reaction conditions: 3 h (90 °C/toluene); [monomer]:[initiator]:[FeBr$_2$]:[dNbipy] = 200:1:1:1.

Arenesulfonylchlorides are described as the first universal class of initiators for the functional polymerization of styrene(s), methacrylates and acrylates as they initiate quantitatively and fast regardless of the substituents of these three classes of monomers [313].

 c. *Solvents and Additives.* Typically, ATRPs are performed in bulk, but solvents may be used and are even necessary in case of polymers which are insoluble in their monomers. Solvents used are mostly nonpolar such as benzene, *p*-xylene, *p*-dimethoxybenzene and diphenyl ether. Some polar solvents such as ethylene carbonate, propylene carbonate and water were also used successfully [294,316]. A wide range of additives has been investigated, to study their effects on ATRP. Matyjaszewski et al. showed that moderate concentrations of water, aliphatic alcohols and polar compounds have little or no influence upon copper-mediated ATRP [317], whereas the addition of amine and phosphine ligands leads to an inhibition of ATRP [289]. The addition of various phenols, in contrast, resulted in an acceleration of ATRP of MMA [318]. Sawamoto et al. investigated alcohols such as methanol, 2-butanol, and 2-methyl-2-butanol as solvents for ruthenium-mediated MMA polymerization and they observed that molecular masses grew directly proportional to conversion and that molecular mass distribution was narrow [319].

 Living polymerization in water also led to polymers with a relatively narrow molecular mass distribution (1.1–1.3) and molecular masses, which showed linear increase with conversion, indicating the living character of this polymerization [320]. Recently, Matyjaszewski et al. reported both reverse and direct ATRP of *n*-butyl methacrylate in an aqueous dispersed system via the miniemulsion approach, characterized by a linear increase of the molecular mass with conversion and a narrow distribution of molecular masses [321]. The suspension-type process of living polymerization of MMA in water not only led to well controlled and high molecular masses and low PDIs, but also the polymerization proceeded without the addition of Al(O-*i*-Pr)$_3$ and clearly faster than ATRP in organic solvents [322].

 As an environmentally friendly alternative to organic solvent, the use of supercritical carbon dioxide has recently attracted considerable interest. It offers additional advantages as low solution viscosity and the fact of being effectively chemical inert. Fluorinated methacrylates were successfully polymerized in supercritical carbon dioxide and the 'living' nature was examined by low PDIs and the synthesis of block copolymers [323].

 d. *Temperature and Reaction Time.* As the energy of activation for the propagation in ATRP is higher than that for termination, higher k_p/k_t ratios and therefore better control of polymerization are achieved at higher temperatures. But at elevated temperatures chain transfer and other side reactions are of increased significance. Thus, optimal temperature has to be found for each ATRP system, depending on monomer and catalytic system as well as on the targeted molecular mass and on the desired reaction time. Due to the higher reactivity of acrylate radicals relative to styryl radicals, ATRP of MMA is proceeded at lower temperatures (70–90 °C) than that of styrene ($\approx$110 °C) [289].

 e. *New Materials by ATRP.* A major advantage of ATRP is the fact, that polymers with complex topologies and compositions can be synthesized using a quite simple polymerization technique. The possibilities of obtaining controlled compositions by ATRP have been reviewed by various authors [288,308,324].

 Bifunctional initiators were employed to gain telechelics [304] and trifunctional and tetrafunctional initiators were used for the preparation of star polymers [325,326]. Additionally, halide initiators are helpful for the preparation of end-functionalized PMMAs via ATRP. Thereby, ATRP provides a possibility to attach other functional groups to the chain extremities [304]. Azide displacement reactions are a very successful approach of functionalizing the terminating end of the polymer chains [327].

Successfully homopolymerized methacrylate and acrylate monomers ($H_2C=CRCOOR'$, R = H, CH_3) are: R' = Me, Et, n-Bu, t-Bu, ethylhexyl, 2-hydroxyethyl, glycidyl, fluoroalkyl [288]. Because of relatively similar reactivity of various monomers in radical polymerization a wide range of random copolymers can be synthesized via ATRP [304,328,329]. Due to the 'living' nature of the polymerization, the obtained random copolymers have very similar amounts of comonomers, whereas in conventional free radical copolymerization, the composition of polymer chains within a sample is quite variable from chain to chain [289]. If the reaction medium is slowly alternated from one monomer to another, a compositional gradient along the chain is observed. This method is called gradient copolymerization [330,331]. Physical properties of these gradient copolymers were found to be quite different from those of the corresponding block and random copolymers [332].

Block copolymers have been prepared using ATRP via two ways: by the sequential addition of a second monomer to the polymerization medium after nearly complete consumption of the first monomer [333] or by the synthesis of isolated and purified homopolymers with functional end-groups as macroinitiators [319]. The latter way allows the preparation of ABA block copolymers when bifunctional initiators are used, whereas the first method enables the synthesis of triblock copolymers by addition of the first monomer after consumption of the second monomer [333]. When $RuCl_2(PPh_3)_3$ and Al(O-i-Pr)$_3$ are used as catalyzing system, Sawamoto et al. showed that the second polymerization step proceeds at similar rates as the first polymerization step, and the method resulted in polymers with significantly increased molecular masses and even narrowed molecular mass distributions. This indicates the complete retention of the chlorine end-groups and their suitability for the re-initiation of living polymerization [319].

Additionally, organic–inorganic hybrid polymers were synthesized using ATRP. The potential use of poly(dimethylsiloxane) in block copolymers for applications such as thermoplastic elastomers and pressure sensitive adhesives, resulted in intensive research in this field [334,335].

Several interesting polymer structures were obtained by combining different polymerization methods. A combination of TEMPO-mediated stable free radical polymerization with ATRP was investigated, and thereby, graft copolymers with polystyrene backbones and poly(t-butyl methacrylate) grafts were synthesized [326]. The synthesized polystyrene-precursors, containing suitable initiating groups for ATRP by copolymerization with p-(chloromethyl) styrene are called macroinitiators. Matyjaszewski et al. reported the successful transformation of carbocationic into 'living' radical polymerization, resulting in block copolymers. Thereby, no modification was necessary for the initiation of the second polymerization step [336,337]. This procedure allowed the synthesis of a ABA-block copolymers with a cationically obtained middle block of polyisobutylene (PIB) flanked by methacrylate blocks [337]. The possiblities of the transformation of other living polymerizations to controlled radical polymerization have been reviewed in Ref. [338].

II. ACRYLAMIDE AND METHACRYLAMIDE

(This section was prepared by O. Nuyken, G. Staufer and M. Schäfer.)

A. Introduction

Polyacrylamides (PAAm) and polymethacrylamides (PMAAm) are of great technical and academical importance. The wide range of industrial applications of PAAm and PMAAm is due to their high water solubility. The most important uses for the polymers are as flocculating agents for minerals, coal, industrial waste, and so on; additives in paper manufacturing; thickening agents; agents for water clarifying; and uses in oil recovery [339–346]. In addition, several dozen actual or potential applications have been mentioned [347–355]. Among its many applications, PAAm is most commonly used as a crosslinked hydrogel in electrophoretic separations of biopolymers [356–365].

B. Monomer Synthesis

The first report of polyacrylamide (PAAm) was given in 1894 by C. Moureu [366], who was also the first (one year earlier) to synthesize the acrylamide monomer (AAm), starting from acryloyl chloride and ammonia as reactants. Although acrylamide was known for a long time, commercial production began in 1954 by hydration of acrylonitrile. The starting point is the reaction of acrylonitrile with sulfuric acid and water at 100 °C to form acrylamide sulfate. Several processes have been developed to remove sulfate [367–369]. More detailed information regarding synthesis, properties, and reactions of AAm is given in the literature [347,348,370]. Methacrylamide (MAAm) is prepared in a similar way from methacrylonitrile or directly from acetone cyanohydrine, $(CH_3)_2C(OH)CN$ [371–377]. This reaction is now the most important synthesis of MAAm.

The synthesis of AAm by enzymatic transformation is attracting increasing attention. Microbial nitrile hydratase converts nitriles into AAm. This method has been applied to the industrial kiloton-scale production of AAms [378].

C. General Aspects of Polymerization

Homogenous polymerization of AAm is usually performed in aqueous solution. The radical polymerization leads to a linear polymer of the general structure [379,380] whereby n varies between 20,000 and 300,000. Polymer made by using anionic initiators shows a totally different structure, called nylon-3 or poly(β-alanine) [347] (Figures 1 and 2).

PAAm and PMAAm can also be obtained by polymer analogous step. Polyacrylate esters can react with amines to yield PAAm and PMAAm. Polyacrylonitrile can be saponified to a copolymer of acrylic acid salt and acrylamide.

PAAm is a linear, white, odorless polymer that exhibits very low toxicity. The amorphous polymer shows a glass transition temperature of about 190 °C measured by DTA [381], although higher temperatures obtained by TBA (torsion braid analysis) [382]

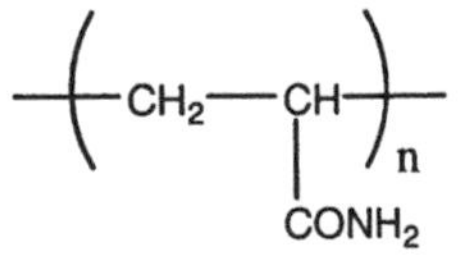

Figure 1 Polyacrylamide.

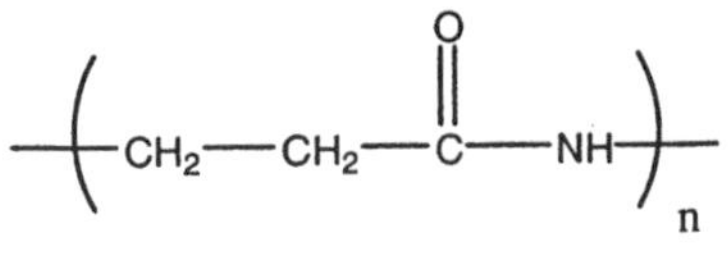

Figure 2 Nylon-3.

and lower temperatures [383,384] are given in the literature. PAA may be crosslinked by imide formation at temperatures >100 °C. The polymer starts to decompose at 220 °C, ammonia is evolved. At 335 °C the second decomposition region begins due to the breakdown of the polymer backbone and the imides to form nitrile units. PAAm is highly soluble in water, whereas it is insoluble in all common organic solvents, such as methanol, chloroform, and tetrahydrofurane.

In contrast to PAAm, polymethacrylamide does not seem to have any great technical importance in applications. Investigations of molecular models show that the presence of the two substituents ($-CH_3$ and $-CONH_2$) leads to an inhibition of rotation about the C–C bonds and to a highly rigid polymer [385]. The solubility of PMAAm is very similar to that of PAAm.

D. Radical Polymerization

Most commercially available polymers are made by radical initiators. Polymerization can be initiated by all types of radical sources, such as peroxides [386,387], persulfates [388,389], azo compounds [390–392], redox systems [393–394], UV light [395–396], x- [397] or γ-radiation [398], electro- [399] or mechanochemically [400]. The radical polymerization shows a strong dependence on temperature, pH, monomer concentration, polymerization medium [392,401], and activators [392]. Water leads to the protonation of the macroradical, which in turn leads to an increase in the reactivity. This is reflected in high values of the chain growth rate constant and therefore the high molecular weight [402] (Figure 3).

A change of solvent (THF, DMSO, DMF) or solvent mixtures (water–methanol, water–DMSO) leads to lower rates of propagation and reduces the molecular mass, bound up with a prolongation of the reaction time to complete conversion. The main reason for that behavior is based on the insolubility of the polymer in the solvents used. The reaction becomes heterogeneous and the polymer precipitates. The propagation rate increases linearly with increasing monomer concentration. Normally, 10 to 30% solutions of monomer are used for polymerization. At higher concentrations deviation from linearity takes place, caused by higher viscosity of the reaction medium. Similar behavior was shown for the increase in initiator concentrations. Up to 5×10^{-3} mol/L the propagation rate increases. At higher initiator concentrations polymerization rate decreases because of a higher termination rate. A survey of characterization methods of PAAm (PMAAm) is given in the literature [403]. The structure is varified by spectroscopic methods. The determination of molecular weight and especially of molecular weight distribution is quite difficult. Gel permeation chromatography is not usable for two main reasons: because of the lack of effective column packings and because water-soluble high-molecular-weight standards are not available. However, several other methods, including light scattering, sedimentation, and viscosimetry, are used successfully to determine the molecular mass of PAAm.

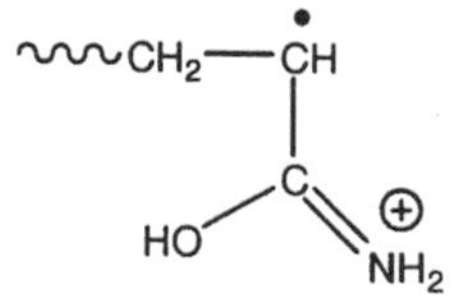

Figure 3 Growing Chain.

E. Anionic Polymerization

Solution polymerization of AAm (MAAm) at high pH, caused by strong bases, yields a polymer with a totally different structure, called poly-β-alanine or nylon-3 [404]. Nylon-3 exhibits interesting properties like high capacity of moisture uptake and high crystallinity. The polymerization is normally carried out in polar solvents with strong bases such as sodium hydroxide as initiators and in the presence of an inhibitor for radical polymerization. Polymerization reactions yield a spectrum of products with fractions soluble in pyridine (A), fractions soluble in water (B), and fractions soluble only in solvents such as formic acid (C). Fraction A consists largely of monomer and dimer. Both other fractions are crystalline polymers with high melting points (325 °C for B and 340 °C for C). Fractions B and C differ only in degree of crystallinity; C is more crystalline, which can be shown by x-ray measurements. Polymers with a molecular weight average of about 80,000 (light scattering in 90% formic acid) show typical polyamide behavior. Wet spinning from formic acid or chloroacetic acid yields fibers, but the polymer has not been produced commercially until now. Another interest in nylon-3 arises from its ability to adopt conformations similar to the characteristic α-helix of polypeptides [405]. For the mechanism of polymerization, two principal routes are discussed [406,407]: The first method of initiation is the reaction of base (B$^-$) with the vinyl double-bond followed by hydrogen transfer.

$$B^{\ominus} + H_2C{=}CHCONH_2 \rightleftharpoons BCH_2{-}\overset{\ominus}{C}HCONH_2 \rightleftharpoons BCH_2{-}CH_2CONH^{\ominus}$$

The active species reacts during propagation with further monomer followed by an intramolecular hydrogen transfer.

$$BCH_2{-}\overset{\ominus}{C}HCONH_2 + H_2C{=}CHCONH_2 \rightleftharpoons {-}CH_2CH_2CONHCH_2\overset{\ominus}{C}HCONH_2$$
$$\rightleftharpoons {-}CH_2CH_2CONHCH_2CH_2CONH^{\ominus}$$

The second method or initiation postulates an acid–base reaction:

$$B^{\ominus} + H_2C{=}CHCONH_2 \rightleftharpoons BH + H_2C{=}CHCONH^{\ominus}$$

The propagation step is described by:

$$H_2C{=}CHCONH^{\ominus} + HC{=}CHCONH_2 \rightleftharpoons H_2C{=}CHCONHCH_2\overset{\ominus}{C}HCONH_2$$

$$\xrightarrow{\;HC{=}CHCONH_2\;} H_2C{=}CHCONHCH_2CH_2CONH_2 + HC{=}CHCONH^{\ominus}$$

The following experimental data establish why the second mechanism presented is preferred by the authors:

1. An unsaturated dimer could be isolated from the reaction mixture.
2. The carbanion of the dimer can abstract a proton from any other species. It is not necessary that the proton transfer in every propagation step is intramolecular.
3. The molecular weight distribution is very broad, which should be expected for polymers formed by the chain transfer.

Electro initiated polymerization of AAm solutions leads to PAAm at the anode and nylon-3 at the cathode. A radical mechanism for anodic polymerization and an anionic mechanism for the polymerization at the cathode has been proposed [408,409].

F. Polymerization Processes

AAm can be polymerized in solution, bulk, inverse emulsion, suspension or as precipitation polymerization [340]. The solution polymerization is the oldest and most common method for production of high molecular weight PAAm and takes place as batch and continuous process. A 10 to 70% solution of deoxygenated monomer in water polymerizes rapidly at low temperatures with all common radical initiators. The polymerization is started by increasing the temperature to 40–80 °C, depending on the initiating system. The monomer concentration is limited by the polymerization enthalpy, the rapid kinetic and the molecular mass of the desired polymer. Therefore transfer agents like isopropanol are often used to reduce molecular weight [347,348]. Many authors have shown that polymerization of AAm (MAAm) is strongly influenced by temperature, solvent, concentration of monomer and initiator, additives (inorganic salts, Lewis acids) and pH value [392,401]. It could be shown that propagation rate increases with rising temperature. A maximum velocity of polymerization is reached at 50 to 60 °C. At higher temperatures the propagation rate decreases because of side reactions (intermolecular imidization) and higher rates of termination.

Bulk polymerization can be divided into two types: polymerization in the solid phase and in the molten phase. Bulk polymerization is interesting for the following reasons: (1) polymerization of crystalline monomer may lead to crystalline and stereoregular polymers, and (2) impurities, such as solvent, catalyst, and initiator, may be avoided. However, only the second reason is realistic since polymer obtained by solid-state polymerization is amorphous and shows no tendency to crystallize. The crystalline matrix is unable to exert any appreciable steric control. Further investigations have shown that propagation takes place at the polymer–monomer interface, controlled by local strains and defects in the crystal. Polymerization in the molten monomer soon becomes heterogeneous because of insolubility of polymer in its own monomer.

AAm can be polymerized by ionizing radiation (x-, γ-, or UV-radiation). Crystals are irradiated continuously during polymerization at temperatures between 0 and 60 °C. Monomer can also be exposed to γ-rays at about −80 °C, then removed from the radiation source and allowed to polymerize at higher temperature with a lower propagation rate. If a limiting conversion is reached at one temperature, chain ends are still reactive. Polymerization can be continued by warming up to higher temperatures. Molecular weight increases with time, a transfer reaction to monomer occurs only to a very limited extend, and reaction with oxygen can be neglected [410].

Inverse emulsion polymerization is used for the preparation of polymers with ultrahigh molecular masses. For this type of polymerization, the expression 'dispersion polymerization' is often used in the literature [410]. A concentrated monomer solution (about 40% monomer in water) is dispersed under intensive stirring in aliphatic or aromatic hydrocarbons in the presence of additives (emulsifiers, protective colloids). Polymerization can be initiated by either water-soluble or oil-soluble initiators [411–418]. The advantage of this process is based on the constant viscosity of the reaction mixture, as the increase of viscosity takes place only in the dispersed phase. By the use of additives (tensides), the dispersion inverts when the emulsion is stirred into water. Precipitation from the aqueous solution yields a polymer with ultrahigh molar mass. The quality of polymer made by inverse emulsion polymerization is influenced by the following factors: (1) species and concentration of initiator, (2) species and concentration of additives (emulsifiers, protective colloids), (3) type of oil phase, and (4) particle size of the dispersed water phase. Because of the easy modification of all these parameters, much attention has been given in recent years to water-in-oil emulsion polymerization of AAm and MAAm.

For suspension polymerization the initial system is obtained by dispersion of an aqueous monomer solution in an organic liquid by mechanical stirring in the presence of stabilizers [402]. The dispersion medium may be represented by aromatic and aliphatic saturated hydrocarbons. The polymerization is initiated by water-soluble initiators, UV or γ-radiation. The process occurs in droplets of an aqueous monomer solution (diameters 0.1–5.0 mm) that act as microreactors [419,420].

Precipitation polymerization takes place in organic solvents or in aqueous organic mixtures, that serve as solvents for the monomer but as precipitates for the polymer. During the process the precipitation of the polymer takes places and polymerization proceeds under heterogeneous conditions. The advantage of precipitation polymerization is that the medium never gets viscous and the polymer is easy to isolate and dry.

6. Chemical Properties

PAAm undergo the general reactions of the aliphatic amide group. The most important reactions are hydrolysis, Hofmann degradation and Mannich reaction. At very extreme pH values hydrolysis occurs. At low pH values (lower than 2.5) inter- and intramolecular imidization occurs [370], which leads to partially insoluble products (Figure 4).

At high pH hydrolysis of PAAm becomes limited by 70% due to the structure of the neighbouring groups (Figure 5).

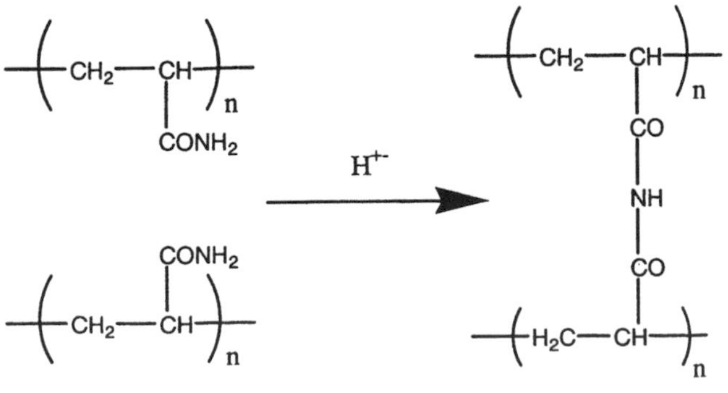

Figure 4 Hydrolysis of PAAm at low pH.

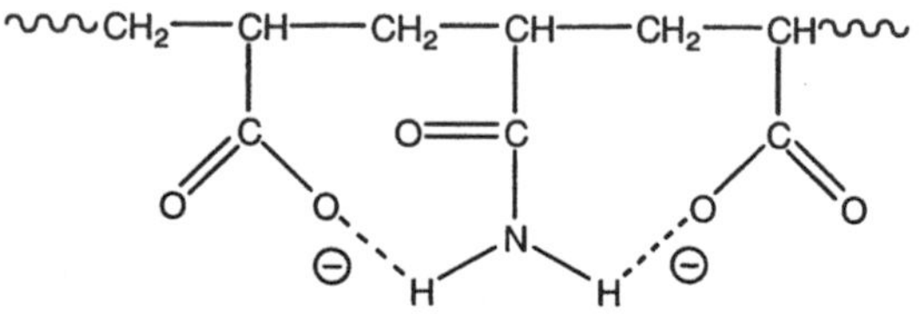

Figure 5 hydrolysis of PAAm at high pH.

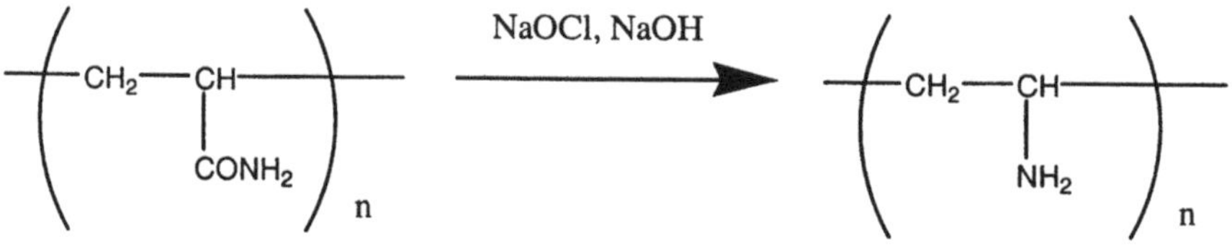

Figure 6 Hofmann degradation.

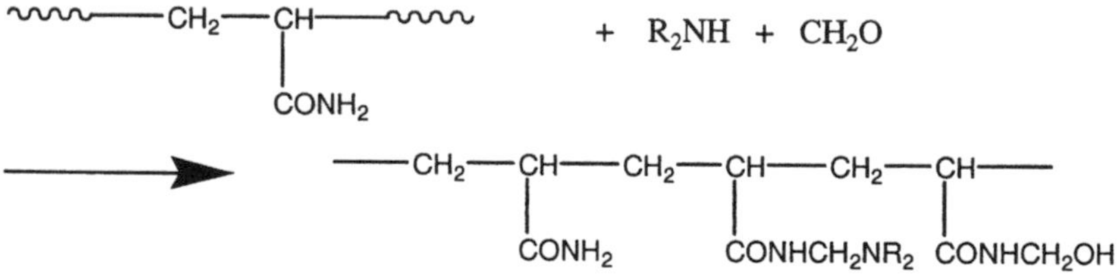

Figure 7 Mannich rection.

The degree of hydrolysis in 10 M NaCl at 100 °C becomes ~95%, but also degradation of the macromolecule occurs.

Hofmann degradation of PAAm leads to polyvinylamine [421–424] (Figure 6).

Polymers containing N-methylol groups can be synthesized by the Mannich reaction of PAAm [425–428] (Figure 7).

In principal it is possible to synthesize AAm derivates first or to change the amide group by polymer analogous reaction. Many P(M)AAm derivates are synthesized and characterized especially in biochemistry [429–433]. This leads to a wide variety of, e.g., antitumor agents [434,435], optical active polymers [436], biorecognizable polymers [437] and gels [438–440].

III. ACRYLIC ACID AND METHACRYLIC ACID

(This section was prepared by O. Nuyken, T. Volkel, and V.-M. Graubner.)

A. Introduction

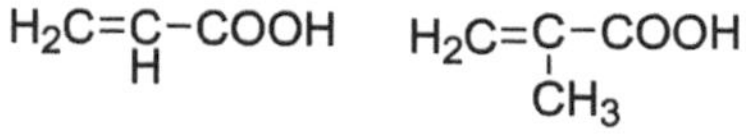

$$H_2C=\underset{H}{C}-COOH \qquad H_2C=\underset{CH_3}{C}-COOH \tag{29}$$

Acrylic and methacrylic acid (propenoic and 2-methylpropenoic acid) are the basic compounds of a large number of derivatives, such as acrylonitrile, acrylamide, methacrylamide, acrylic esters, and methacrylic esters. Homopolymerization of these acids are of minor technical importance; however, they are often used as comonomers to improve special polymer properties. At room temperature glacial acrylic acid and methacrylic acids are clear colorless liquids with sharp penetrating odors that resemble the odor of acetic acid. At lower temperatures they freeze to colorless prismatic crystals [441]. Acrylic acid tends to spontaneous polymerization, which can be explosive. Therefore, an important inhibitor for storage is hydroquinone monomethyl ether and the storage material has to be stainless steel, glass, or ceramic. Rust can start polymerization. To avoid separation of the stabilizer during crystallization, acrylic acid should be storaged above the melting point (13 °C). Above 30 °C, dimerization to 2-carboxyethyl acrylate proceeds (Scheme 30). The stabilization and storage of methacrylic acid are analogous. Some important physical constants of the monomers are listed in Table 10.

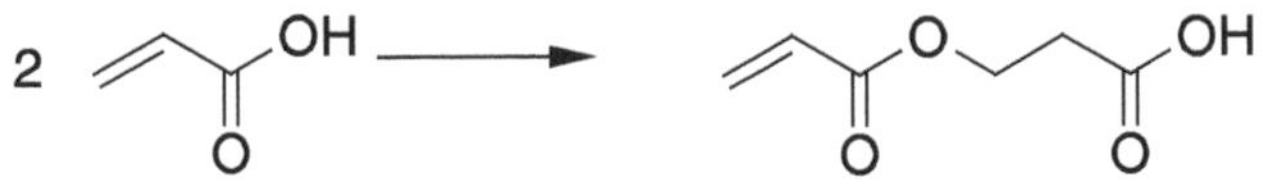

$$\tag{30}$$

Poly(acrylic acid) and poly(methacrylic acid) are hygroscopic, brittle, colorless solids with glass transitions of 106 [443] and 130 °C [444], respectively. Above 200 to 250 °C they lose water and become insoluble cross-linked polymer anhydrides. Poly(methacrylic acid) depolymerizes partially at this temperature. The anhydride is not hydrolyzable by water alone but by aqueous alkaline solutions at room temperature [443]. Decomposition takes place at about 350 °C.

Carefully dried polyacids (e.g., by freeze-drying) dissolve extraordinarily well in water, even with high molar masses. After rigorous drying the solvation rate decreases. Other solvents for these polyacids are dioxane, dimethylformamide, and lower alcohols; nonsolvents are acetone, ether, hydrocarbons, and the monomers. The solubility of poly(acrylic acid) increases with temperature, while the solubility of poly(methacrylic acid) decreases [445]. The solubility of the salts of the polyacids depends in a complex way on the pH value and the counterions. Alkali and ammonium salts are water soluble. Polyvalent cations form in water-swellable gels. The viscosity of aqueous solutions increases with the amount of polymer, to a constant value. Due to this experimental fact, it is not easy to calculate molar masses from the intrinsic viscosities [446].

Table 10 Physical properties of acrylic and methacrylic acids [442].

Property	Acrylic acid	Methacrylic acid
Formula weight (g/mol)	72.06	86.10
Melting point (°C)	13.5	14
Boiling point (°C) at 101 kPa	141	159–163
Vapor pressure (kPa) at 25 °C	0.57	0.13
Density (g/mL) at 25°C	1.045	1.015
Heat of polymerization (kJ/mol)	76.99	56.32
Refractive index, n_D^{25}	1.4185	1.4288
Solubility in water	Miscible	Miscible

Concentrated aqueous solutions of poly(acrylic acids) are thixotropic [447], of poly(methacrylic acid) are rheopectic [448]. Acrylic acid and methacrylic acid easily copolymerize together or with acrylic and methacrylic esters, acrylonitrile, vinylpyrrolidone, styrene, and others. The copolymers are of technical importance. Copolymers with four to six different compounds are quite common.

The simplest and most economical method for preparing polymers is polymerization in aqueous solution. To undergo the problems of handling of the solution and of the removal of the polymerization heat at higher molecular masses and concentrations biphasic systems are used: suspension polymerization, precipitation polymerization, etc. The latter can also be performed in aqueous solution by addition of acids or salts reducing the solubility of the polymers. Also suitable are organic solvents in which the monomers but not the polymers are soluble. Another method is reverse emulsion polymerization [449]. Here one needs emulgators, which form small stable drops of monomers in an inert organic solvent. The size of the resulting polymer particles is variable. Polymers and copolymers of acrylic acid and methacrylic acid are also available by acid or basic hydrolysis of polynitriles, polyesters, and polyamides. Technical significance has the basic hydrolysis of poly(acrylamide).

B. Monomer Synthesis

1. Manufacturing of Acrylic Acid [450]

1. The propylene oxidation process is very attractive because of the availability of highly active and selective catalysts and the relatively low costs of propylene. It proceeds in two steps: the first giving acrolein and the second, acrylic acid [451–454].

$$\diagup\diagdown + O_2 \longrightarrow \diagup\diagdown CHO + H_2O$$

$$\diagup\diagdown CHO + 1/2\, O_2 \longrightarrow \diagup\diagdown COOH \tag{31}$$

2. In 1953 Walter Reppe [455] discovered the reaction of nickel carbonyl with acetylene and water to give acrylic acid. In the commercial process nickel chloride is recovered and recycled to nickel carbonyl.

$$4\, H\!\!-\!\!\equiv\!\!-\!\!H + Ni(CO)_4 + 4\,H_2O + 2\,HCl \longrightarrow 4 \diagup\diagdown COOH + H_2 + NiCl_2 \tag{32}$$

3. The acrylonitrile route is basically a propylene route because acrylonitrile is produced from propylene by ammooxidation [456,457].

$$\diagup\diagdown CN + H_2SO_4 + 2\,H_2O \longrightarrow \diagup\diagdown COOH + NH_4HSO_4 \tag{33}$$

4. The ethylene cyanhydrine route was the first used to manufacture acrylic acid. Ethylene cyanhydrine is formed by addition of hydrogen cyanide to ethylene

oxide [458].

$$\text{HO}\diagdown\diagup\text{CN} + H_2SO_4 + H_2O \longrightarrow \diagup\diagdown\text{COOH} + NH_4HSO_4 \qquad (34)$$

5. A new process is the oxidative carbonylation of ethylene [459,460]. During the reaction the palladium catalyst is reoxidized by a cupric chloride cocatalyst system and oxygen. Selectivity is improved by the addition of a mercury or a tin salt [461].

$$H_2C=CH_2 + CO + 1/2\ O_2 \xrightarrow{\text{Pt-Kat}} \diagup\diagdown\text{COOH} \qquad (35)$$

2. Manufacturing of Methacrylic Acid

1. Commercial production [462] of methacrylates began in 1933 from acetone cyanhydrine, and this is still the basis for essentially all current commercial methacrylate production. The basic materials – acetone, hydrogen cyanide, and sulfuric acid – are available. In the first step, which needs anhydrous materials and conditions, methacrylamide sulfate is formed. The presence of water would form α-hydroxyisobutyramide as the main product. In the second step, methacrylamide sulfate is hydrolyzed by an excess of water to give methacrylic acid.

$$\underset{\text{CN}}{\overset{\text{OH}}{\diagup\diagdown}} + H_2SO_4 \xrightarrow{-H_2O} \diagup\diagdown\overset{\text{O}}{\underset{}{}}\text{NH}_4^+\ \text{HSO}_4^- \qquad (36)$$

$$\diagup\diagdown\overset{\text{O}}{\underset{}{}}\text{NH}_4^+\ \text{HSO}_4^- \xrightarrow{+H_2O} \diagup\diagdown\text{COOH} + NH_4HSO_4$$

2. In a first step the oxidation of isobutene leads to methacrolein and in a second step to methacrylic acid [463–465].

$$\diagup\diagdown + O_2 \longrightarrow \diagup\diagdown\text{CHO} + H_2O \qquad (37)$$

$$\diagup\diagdown\text{CHO} + 3/2\ O_2 \longrightarrow \diagup\diagdown\text{COOH}$$

3. Acid-catalyzed addition of carbon monoxide to propylene gives isobutyric acid, which is dehydrogenated to methacrylic acid [466,467].

$$\diagup\diagdown + CO \xrightarrow{+\ H^+} \diagup\diagdown\text{COOH} \xrightarrow{-H_2} \diagup\diagdown\text{COOH} \qquad (38)$$

4. Condensation of propionic acid and formaldehyde [468].

$$\text{CH}_2=\text{CH}-\text{COOH} + \text{CH}_2\text{O} \longrightarrow \text{CH}_2=\text{C}(\text{CH}_3)-\text{COOH} \tag{39}$$

6. Polymer Synthesis

Acrylic acid and methacrylic acid are described as polymerizing thermally, reacting in an explosive manner. The high heat of polymerization makes it difficult to control polymerization of highly concentrated solutions, and uncontrolled cross-linked polymers may result. For this reason only very small bulk polymerizations should be attempted, with suitable protective measures [469,470]. The resulting products are partially insoluble [471].

1. Polymer Analogous Reactions

The first attempts to synthesize poly(acrylic acid) (PAA) or poly(methacrylic acid) (PMAA) were the hydrolysis of poly(acid derivatives) such as esters, acylchlorides, nitriles, or amides. The hydrolysis has to be quantitative; otherwise, one obtains a copolymer of acid and derivative [472]. On the other hand, hydrolysis in boiling alkaline solution can diminish the molar masses [473]. One possibility is to polymerize methyl methacrylic ester and to hydrolyze the resulting PMMA in acetic acid by the addition of a small amount of p-toluenesulfonic acid as a catalyst. The solution is kept at 120 °C for 18 h and the methylacetate formed is removed by distillation [474]. The degree of hydrolysis depends strongly on the tacticity of the original polymer. Syndiotactic PMMA is hydrolyzed slowly, but isotactic polymer is hydrolyzed very rapidly [475]. The polymers examined had molar masses up to 125,000.

Up to now the only way to get isotactic poly(acrylic acid) or poly(methacrylic acid) has been by hydrolysis of isotactic poly(acrylates) or poly(methacrylates) [476]. Direct routes to get isotactic polymers would be anionic and coordination polymerization. But these polymerizations are not practicable, because the acid function would destroy the initiator. Kargin et al. [477] prepared isotactic poly(acrylic acid) by reaction of isotactic poly(isopropylacrylate) in toluene as solvent with potassium hydroxide in propanol. Propanol and the formed isopropanal were removed after refluxing for 6 h by distillation. Complete hydrolysis was reached after 10 h. Aylward synthesized isotactic and syndiotactic poly(methacrylic acid) by quantitative hydrolysis of poly(trimethylsilyl methacrylate) [478].

Alternatively, studies have been made to hydrolyze poly(acrylamides) at neutral pH and temperatures between 75 and 90 °C. This reaction is autocatalytic, because the generated NH_3 catalyzes the hydrolysis [479].

Head-to-head poly(acrylic acid) is also accessible by hydrolysis. Standing in water at ambient temperature the alternating copolymer of ethylene and maleic anhydride gives the polyacid [480].

2. Radical Polymerization

Acrylic and methacrylic acid are soluble in a large variety of organic solvents [481] – not so their polymers, but PAA is generally more soluble than PMAA. In some instances their solubilities increase at low temperatures. At 26 °C methanol is a theta solvent [444] for PMAA. In aqueous systems the solubility depends on the pH value of the medium [482] and the concentration of dissolved electrolytes. In water, slight decomposition of PMAA was observed [483]. In the presence of small amounts of water acrylic acid tends to popcorn polymerization without the need of initiator. The product is cross-linked and insoluble [484,485].

The most common polymerization of acrylic acid and methacrylic acid [470] is free-radical polymerization. It leads to atactic polymers. Initiators in aqueous solution may be hydrogen peroxide [486–488] (observed M_w 875,000 [489]), persulfate ion [490–492] (observed M_w, 600,000 [493]) or other peroxides [494]. The advantage of hydrogen peroxide over many other initiators is that it does not leave any organic or ionic impurities in the system.

Other solvents and initiators are, for example, butanone [495] or ethyl methyl ketone [470,476] with AIBN [496–499] and dioxane with benzoyl peroxide [474,484,500,503]. In benzene as solvent the resulting polymer forms a slurry [504]. AIBN can be used as an initiator in aqueous solutions by solubilizing the initiator with 4% ethanol [504]. Methacrylic acid polymerizes in nitric acid at 5 to 30 °C to a molar mass of 2×10^6 and the product precipitates in acetone as a white powder [505]. Nitrogen dioxide reacts as an initiator in benzene to synthesize poly(acrylic acid) at 50 °C with molar masses of 48,000 [506]. Sodium bisulfite initiates polymerization of methacrylic acid in an aqueous medium but is ineffective for acrylic acid [507].

Another type of initiation of acrylic acid polymerization is the initiation by redox systems. Some redox systems investigated for the polymerization of acrylic and methacrylic acid are listed in Table 11. W. Kem obtained poly(acrylic acid) on the cathode during electrolysis of an aqueous acrylic acid solution with KCN or $BaCl_2$. Active initiator is the freshly generated hydrogen. The electrode material can be platinum, lead, iron, or mercury [508].

Although the acrylic acid monomers are soluble in water, suspension polymerization is a favorite method. If the aqueous phase contains a high concentration of dissolved electrolytes, the monomer acids are salted out. They are then dispersed by agitation and the suspension polymerization procedure is applied. The resulting product is swellable in water and soluble in aqueous alkaline solutions. Generally, the molar masses are high. This method has the advantage of controlling the particle size by varying the agitation and the electrolyte concentration and of fixing the molecular weight by varying the level of initiator and the polymerization temperature. It is possible to use water-soluble initiators such as potassium persulfate or monomer-soluble initiators such as benzoyl peroxide. Modes of processing and improving this polymerization technique have been described in

Table 11 Redox systems for the polymerization of acrylic and methacrylic acid polymerization.

Polymer	System	Solvent	Molar mass	Refs.
PAA	$Mn(OAc)_3/H_2SO_4$	H_2SO_4/H_2O		[509]
PMAA				
PMAA	Ce^{4+}/glycolic acid	H_2O	195,000	[510]
PMAA	Peroxodiphosphate/sodium thiosulfate	H_2O	170,000	[511]
PAA	Mn^{3+}/isobutyric acid	H_2SO_4		[512]
PMAA				
PAA	$KMnO_4$/oxalic acid	H_2O		[513]
PMAA	$KMnO_4$/oxalic acid	H_2O		[514]
PMAA	Fe^{2+}/H_2O_2	H_2O		[515,516]
PAA				
PAA	Poly(g-mercaptopropylsiloxane)/CCl_4	CCl_4		[517]

numerous patents. For example, it is possible to polymerize methacrylic acid in benzene. The monomer is soluble in this solvent but not the polymer, which forms a slurry [504].

Another method described is that of dispersing an aqueous monomer solution in hydrophobic solvents, such as toluene, hexane, or mixed hydrocarbons. The water may be removed during polymerization as an azeotrope with hydrocarbon, if desired [518,519].

A more recent development is template polymerization [520–522]. When acrylic acid was polymerized in aqueous solution using potassium persulfate as initiator, the polymerization proceeded very slowly. In the presence of poly(vinylpyrrolidone) but under otherwise identical reaction conditions, the rate of polymerization increased dramatically, depending on the amount of PVP. At nearly equimolar concentrations of PVP and monomer, the rate of polymerization reaches a maximum value, because of the strong interaction between poly(vinylpyrrolidone) and acrylic acid in aqueous solution (Scheme 40) [523].

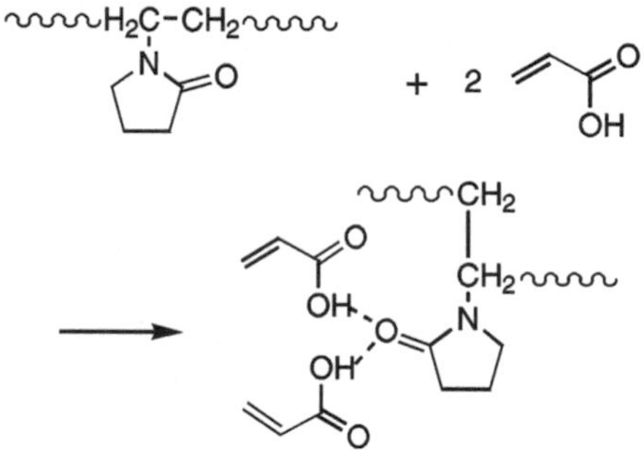

(40)

PVP functions as a template for polymerization of acrylic acid (Scheme 40). The characteristics of a template polymerization are:

1. The structural and conformational features in the template should be reflected in the corresponding polymer.
2. Temperature has an inverse effect on the rate of polymerization.
3. The template and daughter polymers have effectively identical average molecular weights and molecular weight distributions.
4. The addition of a nonpolymerizable acid such as α-hydroxyisobutyric acid decreases the rate of template polymerization.

Comparable results were obtained with the system poly(vinylpyridine) and acrylic acid [524]. Nozaki et al. used dextrine to get optically active poly(methacrylic acid) [525].

3. Radiation-induced Polymerization [525–527]

Ultraviolet [528,529] and visible light [530] and γ-radiation [531,532] initiate the polymerization of acrylic and methacrylic acid. This preparation can be carried out with liquid monomers, in solvents, and with frozen monomers. Ultraviolet light decomposes free-radical initiators such as benzoyl peroxide or AIBN [486]. Another polymerization mechanism was described by Pramanick et al. [533], who irradiated methacrylic acid with 20% of a mixture of tributylamine and CCl_4. Dyes have been used to extend the effective range of wavelengths into the visible region. In the absence of oxygen the quantum efficiency is low. Dyes used are rose bengal (with ascorbic acid and oxygen), fluorescein, eosin, phloxine, and erythrosine [444].

Poly(methacrylic acid) was synthesized by cobalt-60 irradiation [534,535] in various solvents. The stereochemistry of the polymer chain depends on the molecular structure of the solvent. Syndiotacticity increases with decreasing polymerization temperature. Resulting molar masses are in the range 40,000 to 80,000. Using this method, highly disperse poly(methacrylic acid) and poly(acrylic acid) were prepared by Beddows et al. [536]. After 10 h of irradiation with 36 krad/h at 0 °C in the solid state, O'Donnell got polymers with a molar mass of 450,000 [537].

An unexpected result is the fact that irradiation of the liquid monomers at temperatures of 20 to 76 °C leads to the formation of a syndiotactic polymer [538]. In contrast, in the crystalline state, the dry monomer is converted to essentially atactic polymer. It is assumed that like other carboxylic acids, liquid acrylic acid forms association dimers and that in these complexes the two monomer molecules lie in planar symmetry. The polymerization of such dimeric structures could lead to a syndiotactic polymer. The presence of acetic acid, which replaces the monomer in the associations, prevents the formation of a regular alternating polymer.

D. Copolymerization

Acrylic and methacrylic acid are readily copolymerized with many other monomers. Their versatility arises from the combination of their highly reactive double bonds and the miscibility with water- and oil-soluble monomers. Reactivity ratios derived from copolymerization with many comonomers are given in Ref. [539].

Acrylic acid and acrylamide form alternating copolymers in benzene in the presence of AIBN and zinc chloride. When zinc chloride is present, the formation of a charge transfer complex was made responsible for the alternation [540]. In a normal radical polymerization, one obtains random copolymers.

Scherer et al. presented in 1994 a method for gamma radiation-induced graft copolymerization of styrene and acrylic acid monomers into Teflon-FEP (poly(tetra-fluoroethylene-*co*-hexafluoropropylene)) films with a view to develop proton exchange membranes for various applications [541]. This process offers an easy control over the composition of a membrane by careful variation in radiation dose, dose rate, monomer concentration, and temperature of the grafting reaction.

E. Purification and Fractionation [542]

The purification procedures used for PMAA and PAA depend on the stereoregular forms [543]. With conventional polymers a variety of solvent/nonsolvent precipitation systems have been used; for example, methanol/diethyl ether, water/hydrochloric acid, water/butanone, ethanol/diethyl ether. Alternatively, the polymer is isolated by exhaustive dialysis, by freeze-drying [544] and by falling film crystallization [545].

Isotactic polymers require different purification methods because the polymer is insoluble in water. Precipitation is carried out for example in, concentrated sulfuric acid/water or dimethylformamide/water. Dialysis must be made in alkaline aqueous solutions or in dimethylformamide [544]. Direct quantitative estimates of the stereoregularities are obtained by NMR spectra [546,547].

F. Applications [548,549]

In general, applications of poly(acrylic acid) and poly(methacrylic acid) depend on the viscosity and thixotropy that can be generated by low concentrations of these polymers in water, or in their ability to interact with counterions or charged particulate matter. One of the earliest fields of PAA application was as thickening [550–553] as well as binding and coating agents [554–556]. A concentrated solution of the ammonium or sodium salt of poly(acrylic acid), or a dry polymer dissolved and neutralized to any extent desired, are used to form viscous, pourable liquids or gels. Rubber and other lattices can be thickened by these polymers for application to fabrics such as floor coverings (nonslip backing) or waterproof gloves. Toothpaste, cosmetics, hydraulic fluids, and even liquid rocket fuels have been thickened or gelled using acrylic acid polymers [552].

Cross-linked acrylic acid and methacrylic acid polymers provide a useful series of ion-exchange resins [541,557,558]. Acrylic acid or methacrylic acid-rich polymers are used for the dispersion of pigments in paints or for dispersions in cement. Long-chain linear polymers of acrylic acid and methacrylic acid can be used to aggregate suspended particles in the treatment of waste water or potable water [559,560]. The ability to form stable clay aggregates, together with high water adsorption, give the acrylic acid polymers the interesting ability both to improve the tilth of clay soils in agriculture and to modify their water-holding capacity. Other applications are as binders for ceramics, as high-performance dental cements, or as adhesives. Further interesting applications of PAA's are their antiviral activity [561] and as well of PMAAs their use as inhibitors of neoplastic cell growth [562].

Recently, methacrylic acid is developed as a potential functional monomer for non-covalent molecular imprinting. Takeuchi et al. illustrated the use of methacrylic acid with biologically active molecules as model templates [563]. Hereby, the template molecule and functional monomer(s) are covalently or non-covalently bound. After adding the crosslinking agent, the template functional monomer adduct structure is 'frozen' in the polymer network and the template is removed to yield a template fitting cavity as a complementary binding site.

Another interesting topic is the use of weakly cross-linked PAA as an super-absorbent polymer containing ionic functional groups like sulphonate or carboxylate groups [564–566].

IV. ANHYDRIDES AND ACID CHLORIDES OF ACRYLIC AND METHACRYLIC ACID

(This section was prepared by O. Nuyken, P. Strohriegl, and T. Griebel.)

A. Acryloyl Chloride and Methacryloyl Chloride

The first attempt to prepare poly(acryloyl chloride) from poly(acrylic acid) and thionyl chloride or phosphorus pentachloride was that by Staudinger [568]. Unfortunately, the polymer analogous reaction was not complete and he obtained only insoluble, probably highly cross-linked materials. Polymers are more conveniently prepared by free-radical polymerization of acryloyl [569–571] and methacryloyl chloride [572,573]. Linear polymers are obtained from the acyl halides if they are carefully purified and protected from moisture during and after polymerization [574]. Free radical polymerization of acryloyl chloride was conducted in dichloroethane, ethylacetate, THF, dichloromethane, dioxane and cyclohexane. However, poly(acryloyl chloride) with high molecular weights could be obtained only in cyclohexane [575].

Recently, pulsed plasma polymerization of acryloyl chloride allowed formation of films with significant retention of the chloride functionality for subsequent coupling with allylamine vapor to produce amide groups or for introduction of carbon–carbon double bonds [576].

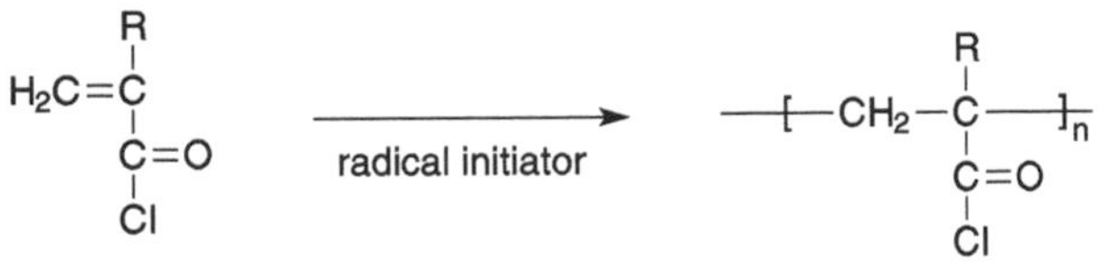

R = - H , - CH₃

For molecular weight determination, poly(acryloyl chloride) was reacted with liquid ammonia. The resulting poly(acrylamide) had a molecular weight of about 30,000. Both acryloyl [571] methacryloyl chloride [577] readily copolymerize with monomers such as acrylate and methacrylate esters or styrene. The corresponding poly(methacryloyl fluoride) has been prepared by free-radical polymerization of methacryloyl fluoride [578] and by the polymer analogous reaction of poly(methacrylic acid) with sulfur tetrafluoride [579]. Because of their reactive acyl chloride groups, poly(acryloyl chloride) and poly(methacryloyl chloride) have been used in polymer-analogous reactions with ammonia [574], primary and secondary amines [580], and alcohols [571]. The Arndt–Eistert reaction of poly(methacryloyl chloride) with diazomethane [572] (Scheme 42) has become one of the classical examples of a neighbor-group effect in polymer chemistry.

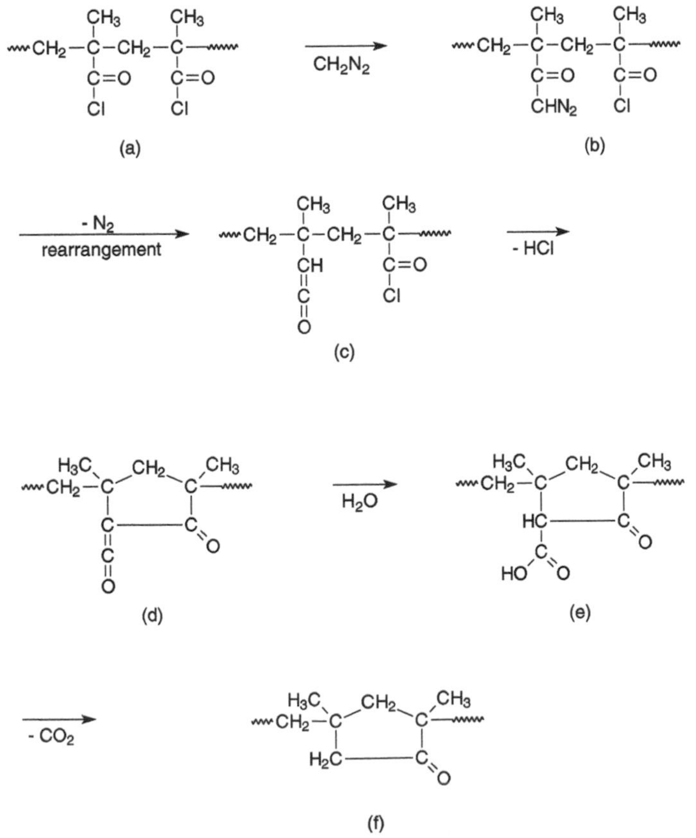

(42)

In the course of the reaction the diazoketone (42b) undergoes a Wolff rearrangement to the ketene (42c). In contrast to the normal Arndt-Eistert reaction, this ketene interacts with a neighboring acyl chloride group with the formation of the β-ketoketene (42d). On hydrolysis, β-ketocarboxylic groups (42e) are formed that lose carbon dioxide, and finally, polymer (42f) with substituted cyclopentanone units is obtained. Although poly(acryloyl chloride) and poly(methacryloyl chloride) have been known for quite a long time, only a few reactions with more complex side groups have been carried out over the years.

Acrylic acid polymers (e.g. acryloyl chloride polymer modified with pyrrolidinone or succinimide) were coupled with insulin, and the insulin reaction products with acrylic acid–acryloylpyrrolidinone or acrylic acid–acryloylsuccinimide polymers were treated with $BaCl_2$, $Cu(OAc)_2$, $Pb(NO_3)_2$ or $Hg(NO_3)_2$ to give metal complexes [581]. Paleos et al. described the synthesis of thermotropic liquid crystalline side-group polymers by the reaction of poly(acryloyl chloride) with mesogenic biphenyl and azobenzene moieties [582,583]. Poly(2-acrylamidobenzoic acid was formed in a polymer-analogue condensation with aminobenzoic acid [584]. Poly(1,4-benzamide) with one terminal amino group has been prepared by polycondensation of 4-sulfinylaminobenzoyl chloride in the presence of aniline. The reaction with poly(acryloyl chloride) and subsequent esterification of the residual acyl chloride groups with various alcohols yielded poly(alkyl acrylate-g-polybenzamide)s [585] with mechanical properties superior to those of the nongrafted homopolymers [586].

The synthesis of a series of polyacrylates and methacrylates with pendant carbazole groups has been reported [587]. The polymers were prepared by the reaction of ω-hydroxyalkylcarbazoles or the corresponding alkoholates with poly(acryloyl chloride) and poly(methacryloyl chloride) (43). GPC, ^{1}H-NMR, and elemental analysis show that high-molecular weight polymers with an almost quantitative degree of substitution are obtained by this polymer analogous reaction.

R = -H, -CH$_3$
X = -H, Na, Li

(43)

B. Acrylic Anhydride and Methacrylic Anhydride

The polymerization of acrylic and methacrylic anhydride is interesting because linear polymers can be obtained in contrast to the cross-linked networks usually formed from difunctional vinyl monomers [588–590]. In a cyclopolymerization reaction, alternating intramolecular and intermolecular propagation steps give rise to anhydride rings along the polymer backbone.

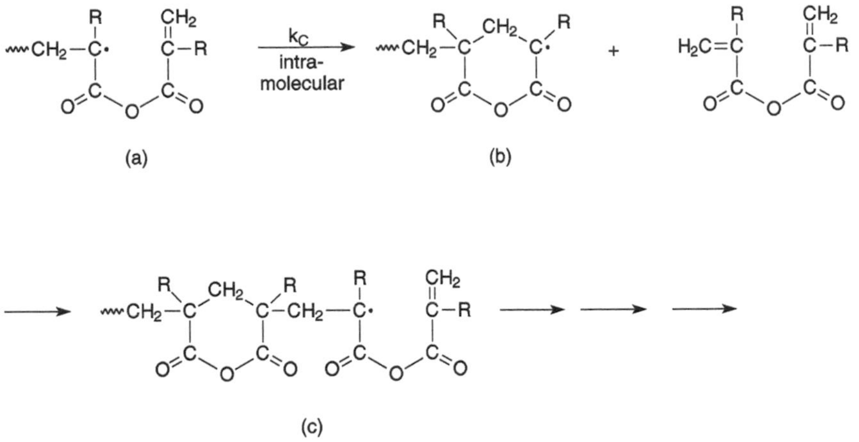

(a) $\xrightarrow[\text{molecular}]{\text{intra-}}$ (b) +

(44)

$\longrightarrow$ (c) $\longrightarrow$ $\longrightarrow$ $\longrightarrow$

R = -H, -CH$_3$

It is also possible for the polymer radical (44a) to add another methacrylic anhydride unit before cyclization takes place. This leads to units with pendant double bonds and in the course of the molecule, the reaction cross-linking may occur. The intermolecular propagation is favored by high monomer concentration. So the polymerization of methacrylic anhydride at concentrations below 20% gives a polymer that is completely soluble in dimethyl sulfoxide, whereas at concentrations above 20% the resulting polymer is partially insoluble [589,591]. A kinetic study [592] carried out on methacrylicanhydride pointed out some important aspects of the cyclopolymerization process. Although activation energy for the intramolecular cyclization step from (44a) to (44b) is higher than for the intermolecular step leading to (45) by approximately 10.9 kJ/mol, the cyclization is considerably faster than the intermolecular propagation. The ratio k_c/k_{11} is 2.4 mol/L and the Arrhenius frequency factor ratio is 256 mol/L in favor of the cyclization step. Methacrylic anhydride has been copolymerized with a variety of common vinyl monomers such as styrene, methyl methacrylate, vinyl acetate, and 2-chloroethyl vinyl ether [590]. By both bulk and solution polymerization with benzoyl peroxide as initiator, soluble copolymers form if the amount of methacrylic anhydride is small compared to the comonomer and if both the concentration of the monomers and the conversion are kept small.

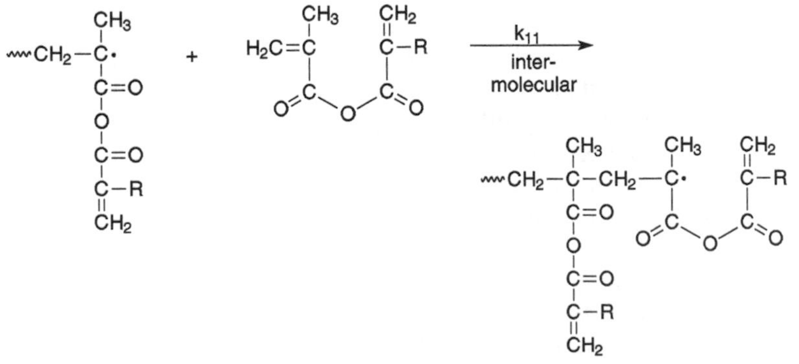

+ $\xrightarrow[\text{molecular}]{\text{inter-}} k_{11}$

(45)

V. ACRYLONITRILE

A. Introduction

Acrylonitrile (AN) was first synthesized in 1893 by Moureu [593], who was also the first (one year later) to report on an acrylonitrile polymer (PAN). The first synthesis of AN was based on the dehydration of ethylene cyanhydrine (1-cyanoethanol) or acrylamide. Early industrial processes for AN production also used ethylene cyanhydrine as starting material, but since 1960 practically the entire AN production has been based on catalytic ammonoxidation of propene. More detailed information on the industrial production processes for AN is given in Ref. [594]. A number of recently published review papers from Japan deals with the ammonoxidation of propane rather than propene to produce acrylonitrile [595–599]. Information on toxicity, mutagenicity, teratogenicity and carcinogenicity of acrylonitrile can be found in Refs. [600–604].

The polymerization of AN differs characteristically from that of other vinyl polymerization reactions: AN itself is soluble in most organic solvents and in water (the azeotrope with water contains 88% of AN) [594]. However, PAN is insoluble in most common organic solvents, in water, and in its monomer. For this reason, the polymerization reaction often becomes heterogeneous even at low conversions and monomer concentrations, and the borders between emulsion and suspension polymerization are not well defined. Heterogeneous AN polymerization shows autoacceleration when an insufficient amount of a surfactant is used. Furthermore, there is obviously no consensus among the authors reporting on AN polymerization as far as use of the terms *solution polymerization*, *dispersion polymerization*, and *precipitation polymerization* is concerned, especially in aqueous systems.

The morphology of PAN is unique [594]. Due to strong repellent dipole–dipole interactions between intramolecular neighboring nitrile groups in parallel position, the polymer backbone is forced into an irregular helical conformation. Strong attractive dipole–dipole interactions between antiparallel nitrile groups of different chains cause parallel orientation of the individual irregular helices.

The usual two-phase model for the structure of polymers (the solid phase consists of crystalline and amorphous regions) can be applied for PAN, but only within limits. Although there is evidence for the presence of crystallinity, it has been shown that there is strong interaction between them. It has been observed that the glass transition occurs at the same temperature as some characteristic change in the x-ray pattern of PAN. In a two-phase structure, the glass transition should not have any effect on the crystalline regions of a polymer. This indicates at least strong interactions between the two phases, and it shows that there is probably less difference between the crystalline and amorphous parts of PAN than in other polymers.

The two-phase concept for the structure of PAN is supported by the observation that the absorption curve (measured as a weight gain of PAN in an aqueous solution of iodine/potassium iodide) shows several steps. This is interpreted as the penetration of the solution into domains of different order. X-ray diffraction patterns of PAN differ from those of other fibers [594,605]. In the patterns of PAN, distinct off-equatorial reflections are absent. This indicates a lack of order along the chain axis. Equatorial reflections are present, and they indicate high order parallel to the fiber axis. These observations support the view of the structure of PAN given above. The mechanical properties of PAN fibers can be improved considerably by spinning from metal ion containing solutions [606].

Nearly two-thirds of the AN production is consumed by the synthesis of PAN fibers; the rest is used for acrylic rubbers, ABS-type polymers, and the production of acrylamide [594]. Industrial processes acrylic rubbers and ABS-type polymers do not have to regard the typical properties of AN polymerization because the AN content is usually low. Therefore, in this chapter only those processes are considered that yield homopolymers of AN or copolymers with up to 15% comonomer content. These polymers are used for fiber manufacture and are referred to as *acrylic fibers* (homopolymer) and *modacrylic fibers* (copolymers with up to 15% comonomer). The only exception to this is the use of acrylonitrile homopolymer as membrane material because of its excellent gas barrier properties: the ratio of the permeabilities of helium and oxygen is 1770, while 58.5 for poly(vinylidene chloride) is already considered as good [594]. Porous PAN membranes are used for ultrafiltration [607–611]. In addition, because of the ability to form complexes with metal ions, PAN membranes are also used as ion conducting membranes in lithium secondary batteries [612,614].

Acrylonitrile is also the monomer for the commercially avaliable ASTRAMOL dendrimers, which are prepared by repetitive Michael-addition of acrylonitrile to amino groups, followed by reduction of the nitrile group to the terminal amino groups of the next generation [615,616].

Usually, one distinguishes among five polymerization methods, which may show different kinetic behavior: bulk, solution, precipitation, suspension, and emulsion polymerization. In general, water is used as the continuous phase in suspension and emulsion polymerization. Due to the solubility of AN and the insolubility of PAN in water, the differences among solution, suspension, and emulsion polymerization of AN are small, and the kinetics differ strongly from those of normal suspension and emulsion polymerization.

B. Bulk Polymerization

Bulk polymerization of AN shows autocatalytic behavior even at low conversions. Thus as temperature control is difficult due to strongly increasing viscosity, an explosion may occur [617–619]. Despite this problem and the other well-known shortcomings of bulk polymerization, several processes for AN bulk polymerization have been developed. A continuous process for industrial application developed by Montedison Fibre utilizes initiating systems that decompose rapidly at the temperature of the reaction medium. Thus a rise in temperature cannot affect the rate of initiation because there is no (or little) initiator left. The rate of initiation can be controlled by feeding the initiator into the reactor at a variable rate [620–623]. A continuous process has also been developed by Mitsubishi Rayon, but it uses AN diluted with water instead of pure AN [624].

For laboratory use, a method for the production of transparent molded pieces exists [617–619]. *p*-Toluenesulfinic acid, *p*-toluenesulfonic acid, AIBN, benzoyl peroxide, and mixtures of these are used as initiators. The reaction media are not stirred. The reaction is carried out at 25 to 50 °C when the size of the experiment does not exceed 3 mL of AN. At bigger sizes, explosions occur easily at temperatures above 4 °C. It was found that for conversion of the mixture from a heterogeneous system to a transparent homogeneous one, an intermediate rate of polymerization is necessary. If the rate is too low, transformation does not take place. If it is too high, thermal runaway occurs.

The kinetic behavior is characterized by a short induction period (which could not be eliminated), followed by an acceleration period [625]. The autoacceleration is supposed to be caused by the increasing number of particles of precipitated polymer. It is assumed that the reaction takes place on the particle surface rather than inside the particles, due to the poor swelling of the particles by monomer. Polymerization in the liquid phase is also considered as unimportant, because polymer radicals precipitate at low molecular masses.

A number of investigations deal with the influence of modifiers (for the control of temperature and degree of polymerization), copolymerization, pH, particle size, pressure, and stirring. Initiators can be redox systems [626–631], peroxides [625,631–635], azo compounds [625,631–634,636,637], γ-radiation [638–642], and plasma [643] (for details see 1st edition of this work).

C. Solution Polymerization

1. Radical-Induced Polymerization

Solution polymerization with radical initiators is one of the two principal methods used for industrial production of PAN. It allows immediate manufacture of spinning dopes, which is very important, as most of the PAN is used for fiber production. Two major drawbacks limit this polymerization method. The first one is that the concentration of the polymer solution is rather low, so wet spinning usually has to be used. The other, much more important problem is that solvents suitable for solution polymerization of AN usually have high transfer constants (Table 12, Refs. [644–647]).

DMF and DMSO are the primary organic solvents for industrial applications. Aqueous solutions of NaSCN, $ZnCl_2$, HNO_3, and others are also widely used as solvents. A list of fiber producers using solution polymerization, together with product name and solvents, is given in Ref. [644]. Recently, low melting salts or mixtures of salts (NaSCN/KSCN or $LiClO_4 \cdot 3H_2O$) have also been used successfully [647].

For industrial use, dyability of the PAN fibers is important. In general, cationic dyes are used, which requires the presence of acidic functions in the polymer chain. This is achieved either by copolymerization with monomers with acidic side groups (sodium

Table 12 Selected solvents and chain transfer constants for polymerization of AN.

Solvent	Transfer constant $\times 10^4$	Temp. (°C)
Benzene	2.46	60
α-Butyrolactone	0.66–0.74	50
Dimethylacetamide (DMAC)	4.95–5.05	50
Dimethylformamide (DMF)	2.7–2.8	50
Dimethyl sulfoxide (DMSO)	0.11–0.8	50
Ethylene carbonate	0.33–0.5	50
Propylene carbonate		
1-Methyl-2-pyrrolidone		
Phosphoric acid tris(dimethylamide)		
Aq. Copper(II)chloride	190,000	35
Aq. Iron(III)chloride	33,300	60
Aq. Lithium bromide		
Aq. Magnesium perchlorate	<0.05	50
Aq. Nitric acid		
Aq. Sodium perchlorate		
Aq. Sodium thiocyanate		
Aq. Sulfuric acid		
Aq. Zinc chloride	0.006	50
Sulfur dioxide (anhydrous)	0	50
NaSCN/KSCN		
$LiClO_4 \cdot 3H_2O$		

styrene sulfonate, etc.), or by using initiators that produce acidic end groups. For this purpose, a persulfate–sulfite system is generally used. In the latter case it is necessary to keep the molecular mass of the polymer at about 40,000 to 50,000 or below. Otherwise, the concentration of dye sites in the fiber is too low. Therefore, the chain transfer activity of some solvents might even be desirable.

The kinetics of AN polymerization in solution follows that of the conventional models for ordinary vinyl polymerization as long as the monomer concentration is not too high. It has been shown that at concentrations above 4 M in DMF and 6 M in ethylene carbonate, the monomer acts as a nonsolvent and the mixture becomes heterogeneous. Polymerization rate and molecular mass proceed through a maximum in ethylene carbonate when the AN concentration is raised [648].

Usually, transfer reactions to the solvent have to be considered, and some specific side reactions may also be important. Branching can also occur during polymerization. Two major mechanisms specific for branching of PAN are given in the literature: Ulbricht suggests that H abstraction from an α-C atom in the polymer chain by a polymer radical produces a location for the start of a side chain [649]. Peebles claims that branching occurs by polymerization through the nitrile group [650]. Side reactions, leading to structural defects, chain scission, and development of color, have been studied [651–659]. The possibility of preventing such reactions by additives has been claimed [660–675].

For applications in the textile industry it is necessary that there be no variation in the color of PAN. Usually, a white-colored fiber (before dying) is desired, and darkening must therefore be prevented. Sometimes, discoloration of the PAN occurs during polymerization. Therefore, several additives have been tested to prevent discoloration. It was found that strong organic or inorganic acids (H_2SO_4 or arsene sulfonic acid) [676], oxalic acid [677], other dicarboxylic acids, their anhydrides, imides, or salts [678], and complexing agents [679] are useful for this purpose.

The precipitation of polymer may take place during an early stage of polymerization, because the monomer acts as a nonsolvent for the polymer until the polymerization has consumed enough monomer to decrease its concentration below 4 mol/L or 6 mol/L, depending on the solvent. It was found that the addition of water can prevent this heterogenization of the reaction mixture in the beginning of the polymerization in organic solvents [680,681].

Detailed procedures for free radical polymerization in various solvents, including kinetic studies and industrial processes, can be found in the following references: DMF [648,682–702], DMAc [703], DMSO [687,691,704–707], ethylene carbonate [648,683, 692,708–712], γ-butyrolactone [687,692], Liquid SO_2 [713], Aq. $ZnCl_2$ solution [714,719], Aq. NaSCN solution [715], Aq. Ca(SCN)$_2$ solution [716], Aq. HNO_3 solution [717,723], Aq. NaClO$_4$ solution [718], Aq. Mg(ClO$_4$)$_2$ solution [720], water [721,724,727, 728], Aq. H_2SO_4 solution [725], NaSCN/KSCN mixture [647], and LiClO$_4 \cdot 3H_2O$ [647]. Peroxides [682,686,688,691,692,700,707,714,721,722,726,727,729,730,735,736], redox systems [683,687,699,703,718,731–734,737–740], azo compounds [648,682,684,685, 689–698,704,705, 709–711,713,715,719], γ-irradiation [701,716], and light [702,706] can be used to initiate the free radical polymerization.

Controlled free radical polymerization of acrylonitrile has been achieved by ATRP (atom transfer radical polymerization) [741–747]. Control of molar mass was shown to be possible up to values of $\overline{M}_n = 30,000$ g/mol with polydispersities as low as 1.06. However, end group control was limited. This was attributed to oxidation of the free radical to an anion by reaction with Cu^I [741–743].

2. Anionic Polymerization

Anionic polymerization of AN is possible, although most of the systems are not living. Not only can the vinyl group of AN be attacked by anions, but its nitrile group as well. Thus anionic polymerization often results in colored and branched products. The nature of the products of side reactions has been studied by NMR [748]. The molecular masses that can be achieved by anionic polymerization are not very high, due to side reactions. Although this method is not used commercially, there are a considerable number of publications available on the anionic polymerization of AN [748–766].

Lithium [749,750,760–762] and sodium [750,760] organic compounds, lithium alcoholates [752,757,760–762], sodiomalonic diesters [755], complex bases from alkali imides and alcohols or alcoholates [756], phosphines [758,759], and others [751,753,754] have been used as initiators. It was found that with THF as solvent and fluorenyllithium or phenyllithium as initiator, molar mass is independent of initiator and monomer concentration. Relatively low masses of 2600 to 4200 were found. With DMF as solvent, the molecular mass increases with the monomer concentration at low (1.5 mmol/L) initiator levels. With cyclopentadienyllithium or cyclopentadienyl sodium at high concentrations (68 mmol/L) and DMF as solvent, the molecular mass increases strongly with the monomer concentration. This is explained on the basis of a polyfunctionality of cyclopentadienyllithium and cyclopentadienyl sodium initiators. This view is supported by ozonolysis of the incorporated initiator, which leads to a decrease in the molar masses only of those polymers that were initiated by cyclopentadienyllithium or cyclopentadienyl sodium [750].

The activity of several complex bases from $NaNH_2$, $LiNH_2$, and KNH_2 with primary alcohols, secondary alcohols, and salts as initiators with or without solvent has been tested. The highest masses (97,000) were achieved with $NaNH_2/Et(OEt)_2ONa$ at $-78\,°C$ [756].

New developments in the anionic polymerization of polar monomers, including acrylonitrile, have been reviewed in 1997 [763]. Remarkable are the achievment of stereoregular poly(acrylonitrile) by anionic polymerization, using diethyl beryllium as initiator by Nakano, Hisatani and Kamide [764], which has not been possible before in any other way, the living polymerization of AN using metal free carbanionic initators [765], and the controlled functionalization by anionic polymerization of AN with an initator system composed of triethylene diamine and an epoxide [766]. Triethylene diamine alone does not initiate the polymerization, but in the presence of an epoxide, the epoxide is introduced at the beginning of the chain. Thus, macromonomers were prepared using glycidyl methacrylate as the epoxide component [766].

Polymerization with phosphines as initiators was proposed to proceed via zwitterions [758], Figure 8.

The zwitterion 1 initiates the polymerization through its negative charge. In Ref. [166], the equilibrium between linear chains ('free ions') and cyclic chains (ion pairs) is discussed. It was found that the amount of phosphorus in the polymer was much lower than one would

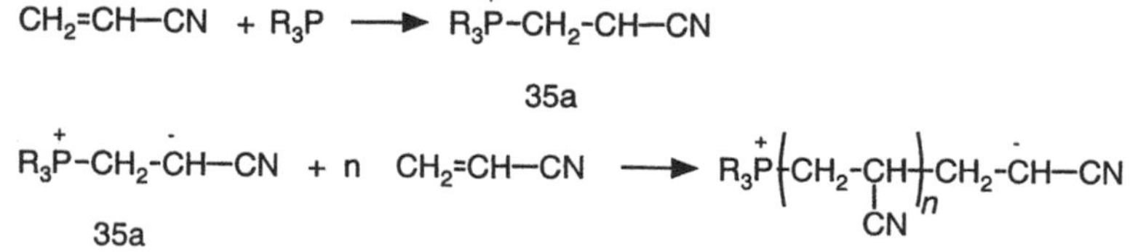

$$CH_2{=}CH{-}CN \;+\; R_3P \;\longrightarrow\; R_3\overset{+}{P}{-}CH_2{-}\overset{-}{C}H{-}CN$$

35a

$$R_3\overset{+}{P}{-}CH_2{-}\overset{-}{C}H{-}CN \;+\; n\;\; CH_2{=}CH{-}CN \;\longrightarrow\; R_3\overset{+}{P}{\Big(}CH_2{-}\underset{CN}{C}H{\Big)}_n CH_2{-}\overset{-}{C}H{-}CN$$

35a

Figure 8 Initiation of acrylonitrile polymerization of phosphines[758].

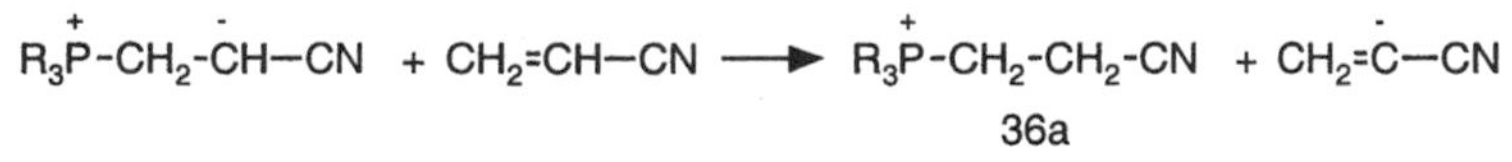

$$R_3\overset{+}{P}\text{-}CH_2\text{-}\overset{-}{C}H\text{—}CN \ + \ CH_2\text{=}CH\text{—}CN \ \longrightarrow \ R_3\overset{+}{P}\text{-}CH_2\text{-}CH_2\text{-}CN \ + \ CH_2\text{=}\overset{-}{C}\text{—}CN$$

36a

Figure 9 Transfer in zwitterionic polymerization of acrylonitrile [759].

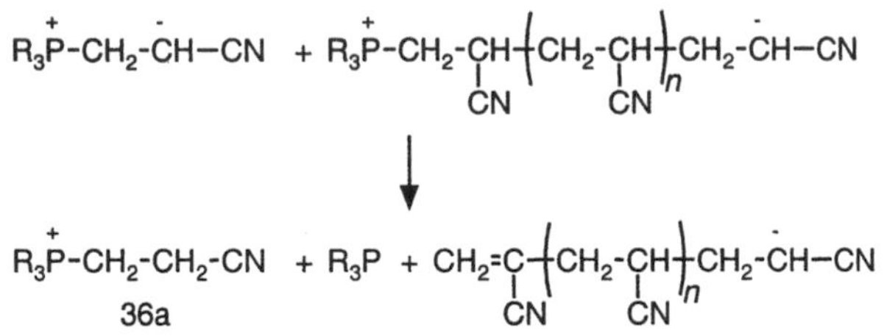

$$R_3\overset{+}{P}\text{-}CH_2\text{-}\overset{-}{C}H\text{—}CN \ + \ R_3\overset{+}{P}\text{-}CH_2\text{-}\underset{CN}{C}H\!\!\left(\!CH_2\text{-}\underset{CN}{C}H\!\right)_{\!n}\!\!CH_2\text{-}\overset{-}{C}H\text{—}CN$$

$$\downarrow$$

$$R_3\overset{+}{P}\text{-}CH_2\text{-}CH_2\text{-}CN \ + \ R_3P \ + \ CH_2\text{=}\underset{CN}{C}\!\!\left(\!CH_2\text{-}\underset{CN}{C}H\!\right)_{\!n}\!\!CH_2\text{-}\overset{-}{C}H\text{—}CN$$

36a

Figure 10 Loss of phosphorous during zwitterionic polymerization of Acrylonitrile.

expect from the mechanism proposed. This observation was explained by a charge transfer of the negative charge from the zwitterion to a monomer molecule [759] (Figure 9).

By GPC it was shown that all phosphorus detected in the raw polymer was either bound to the polymer as a phosphonium ion or existed as cation **2**. Polymer-bound phosphorus was detected only at high temperatures or high concentrations of phosphine (as initiator). This was taken as evidence against the zwitterion mechanism. However, a more detailed look at the data given in Ref. [759] reveals that reactions of the following type could not be excluded (Figure 10).

This reaction can take place at any time during polymerization or polymer recovery. Thus there exists no conclusive proof against the zwitterion mechanism. It is obvious that more work has to be done before this topic can be settled.

3. Transition Metal Catalyzed Polymerization of Acrylonitrile

Heterogeneous Ziegler–Natta-type catalysts based on Gadolinium ($Gd(OCOCCl_3)_3$/(i-Bu)$_3$Al/Et$_2$AlCl) can be used to polymerize polar monomers, including acrylonitrile [767]. In homopolymerizations, the ability to polymerizes decreases from MMA to MA to AN [767]. Sandwich and half-sandwich complexes can also polymerize acrylonitrile. The metallocenes Cp$_2$Co, Cp$_2$Ni, and Cp$_2$Fe polymerize AN with the cobaltocene being the most active catalyst in this series [768]. Half-sandwich complexes based on Sc, Y, and La [769–771] as well as Cr [772] have been used successfully.

4. Stereoregularity of Polyacrylonitrile

Early attempts to prepare stereoregular poly(acrylonitrile) were unsuccessful. Neither free radical [773–776] nor anionic [773] polymerization proceeds stereochemically controlled. However, diethylberyllium as initiator for anionic polymerization of AN allows the preparation of predominantly isotactic PAN [764]. It has also been shown that at low temperatures ($-78\,^\circ$C), when AN is occluded in the channels of urea crystals, irradiation of this complex with γ-rays results in PAN with an elevated content in isotactic bonding of the monomer units [777–786]. NMR spectroscopy allows the identification of sequences up to pentads [778]. The assignment of the signals was confirmed by the investigation of model compounds [777–783]. Cyclodextrin clathrates of acrylonitrile can also be used to prepare isotactic PAN [787].

D. Dispersion, Precipitation, and Emulsion Polymerizations

1. Dispersion Polymerization

Aqueous dispersion polymerization is the most widely used process in industrial production of PAN. Its advantages are that the polymer can easily be recovered by filtration, that the transfer constant of water is nearly zero, and that even at high monomer concentration levels (water/monomer = 1.5 to 2.5) removal of the heat of polymerization causes no problems.

It is generally agreed that at least when inorganic compounds are used for initiation, the initiation and primary steps of polymerization occur in the aqueous phase. Since there is an equilibrium between polymer radicals in the aqueous phase and those absorbed at the surface of the polymer particles, two different loci of polymerization have to be considered. However, with an increasing number of polymer particles, polymerization in the aqueous phase becomes less and less important.

In the absence of particle stabilizing agents, agglomeration of particles becomes more and more probable with increasing particle size. Consequently, after an acceleration period in which the number of particles, and therefore the number of reaction sites, increases, the rate of polymerization reaches a constant value. During this period of constant polymerization rate, the rate of nucleation equals the rate of agglomeration. Thus the number of particles remains constant. Toward the end of the reaction, the rate of polymerization decreases because no new nucleation occurs [644].

Persulfate/bisulfite is the most widely used initiator system in aqueous dispersion polymerization for industrial purposes, because it provides sulfate and sulfonate end groups in the polymer chains. These end groups are used as acidic dye sites. It has indeed been shown that there is a dependence of PAN dye sites on the persulfate/bisulfite ratio of the initiator system [788,789]. For commercial applications, there is generally 8 to 10 times more bisulfite than persulfate in the initiator system used.

The size of the particles is about 5 to 150 pm in general. In precipitation polymerization, the AN concentration is usually about 5 to 10% because of its limited water solubility. Branching is possible of course, but defects, which are possible in organic solution polymerization, such as ketene imine or P-oxonitrile groups are impossible [790]. The kinetics are similar to bulk or emulsion polymerization of AN. It was found that some complexing agents are able to improve the reproducibility of the rate of polymerization and the molecular mass of the product [791–794]. Further information on methods for aqueous dispersion and precipitation polymerization can be found in Refs. [795–864].

2. Precipitation Polymerization of AN in Nonaqueous Systems

Precipitation polymerization of AN is possible in aliphatic, aromatic, and halogenated aromatic solvents. Alcohols yield only PAN with low molecular masses due to their strong transfer abilities. The absence of water allows the same structural defects as bulk polymerization. Depending on monomer concentration, the kinetics of the system are similar to those in bulk or aqueous precipitation polymerization. Some systems for precipitation polymerization of AN in nonaqueous systems are given in Refs. [865–879].

3. Emulsion Polymerization of Acrylonitrile

Compared with other systems, the possibility of preparing an emulsion with a high PAN content is rather limited. This is caused by the comparatively high water solubility of AN, the poor solubility of PAN in its monomer (which is required for the formation of latex particles), the high polarity of PAN, and the strong attractive forces between the nitrile groups on the surface of the polymer particles. Therefore, strong emulsifiers are required in high concentrations [880–883]. Otherwise, only very low contents of solids can be realized [878,884].

Another possibility for producing stable PAN emulsions is to incorporate ionic or hydrophilic groups such as sulfonate or carboxylate groups into the polymer. This can be done by copolymerizing AN with monomers with adequate functions [885,886].

The kinetic behavior of AN emulsion polymerization deviates from the usual emulsion polymerization kinetics. It is generally agreed that these deviations result from the possibility of radical desorption from the latex particles [873,883,887–890]. It has been shown that in certain cases the radical population may be as low as 1 per 1000 particles [891,892].

These disadvantages of the emulsion polymerization of AN are soon overcome when the comonomer content rises. Thus for the production of modacrylics (15 to 50% comonomer), emulsion polymerization is as advantageous as it is for common vinyl polymerization. So for industrial use, it is employed only for copolymerization of AN. Therefore, a number of patents concerning such processes, developed by several companies, exist: Toray Industries [886,902,903], Unitika [895], Du Pont [892,896,897], Bayer [898,899], Mitsubishi Rayon [900,901], Japan Exlan [886,902,903], Kanegafuchi Kagaku [885], Asahi Chem. Corp. [904], Carbide and Carbon Chemicals Corp. [905], Sun Chemical Corp. [906], Soken Chemical Engineering [907], Institutul de Cercetari Chimice ICECHIM (Rom.) [908], and Petrograd Technological Institute [909]. Further information on this topic is given in Refs. [910–914].

VI. MALEIC ACID AND RELATED MONOMERS

(This section was prepared from O. Nuyken, H. Kricheldorf, and M. Burger.)

A. Introduction

The monomers discussed in this section are derived from maleic acid (MA, *cis*-butenedioic acid) or fumaric acid (FA, *trans*-butenedioic acid). Whereas FA is present in the metabolism of numerous plants, MA is not, but MA, in the form of its anhydride (MAH), is technically produced in large quantities in contrast to FA. The technical production of MAH (mp 52 to 53 °C) is based on the oxidation of benzene or butane with air in the presence of vanadium oxide-based catalysts. FA (mp 287 °C) is thermodynamically more stable than MA (mp 138 to 139 °C) by 22.8 kJ/mol [915]. Therefore, FA is easily produced by catalytic isomerization of MA.

The conjugated C=O and C=C bonds of MAH may participate individually in chemical reactions, and because of their high reactivity, MAH is used as a starting material for the technical production of various copolymers. Unsaturated polyesters are produced by polycondensation processes mainly involving the carboxyl groups. Various types of polyimides can be prepared by the polyaddition of difunctional OH, SH, or NH compounds with difunctional maleimides. The addition steps mainly involve the C=C bonds (Michael-type addition). The scope of the present section is limited to polymerization processes involving exclusively the C=C-π bond and yielding polymers with a backbone consisting of C–C-σ bonds.

Whereas homopolymerizations of maleic or fumaric acid derivatives have not yet found technical interest, copolymers derived from MAH and various vinyl monomers are commercialized by several chemical companies. Such commercial copolymers may be based on styrene (Scriptset or SMA), ethene (EMA, Malethamer, VINAC), isobutene (ISOBAM), butadiene (Maldene), vinyl alkyl ethers (Gantrez AN or Viscofres), and vinyl acetate (Lyfron, Benex, Amoco Drilling Aid 420,421, or Baroid X tend).

Owing to the versatility and technical importance of MAH, numerous reviews have appeared, of which a selection is presented in Refs. [915,924].

B. Reactivity and Homopolymerization of Maleic Anhydride

Despite successful oligomerization with benzoyl peroxide [916], MAH was long believed to be reluctant to homopolymerize. A broader report on its homopolymerization appeared in 1961 [925] and demonstrated that MAH can undergo a free-radical-initiated polymerization in concentrated solutions or in the molten state.

Several studies on UV-light-induced polymerization of MAH revealed that this approach only yields oligomers in low yields [917–929]. Solvent, residues, or photo-sensitizers (e.g., cyclic acetals, benzophenone, or 2,4,8,10-tetraoxaspiro[5.5]undecane) are incorporated as end groups. Higher yields and molecular weights are obtainable by peroxide-initiated polymerizations of molten MAH [920,930]. The results reported so far suggest that an activated MAH (excimer) is the active species in free-radical polymerization, whereas the ground state is inactive [922,931]. Reactor conditions and results of individual polymerizations are summarized in Table 13. (For further reports and patents dealing with peroxide-initiated polymerization of MAH, see Refs. [935–949]).

Kinetic and thermodynamic studies of the homopolymerization have been conducted [940,941,943,945]. $\Delta H = 58.5\,kJ/mol$, $\Delta S = 125$ to $145\,kJ/mol$, and a ceiling temperature of $150\,°C$ were reported. Recent studies [950], indicate, that the ceiling temperature might be even higher. The structure of the poly(MAH) was and is a matter of controversy. Some investigators have proposed, that maleic anhydrid polymers are comprised of cyclopentanone derivatives and conjugated ketoolefinic units [951–955]. NMR-studies however did not detect olefinic or CH_2 structures [947,948], indicating the predominance of monomer units (Figure 11).

Several attempts were made to homoplymerize MAH with basic or acidic initiators. Preferentially used were tertiary amines such as triethylamine [956], pyridine [942,957–960], picolines [961], vinyl-pyridine [959,962] and imidazole [963]. Typical for

Table 13 Rection conditions and results of peroxide-initiated polymerizations of MAH.

Initiator	Reaction medium	Temp. (°C)	Yield (%)	Molecular weight	Ref.
Benzoyl peroxide	Melt	55–80		DP = 25–29	[926]
Benzoyl peroxide	Melt	70	34	$[h] = 0.12\,dL/g$	[925]
Benzoyl peroxide[a]	Melt	>80	80	M_W up to 94,000	[934]
Acetyl peroxide	Melt	85–135	>95	M_W 3500–7500	[930]
t-Butyl(2,2-dimethylpropanol) peroxide	Melt	80	63	$[h] = 0.05\,dL/g$	[931]
$H_2O_2 + Ac_2O$	Chlorobenzene	105–107	87		[932]
Diisopropylpercarbonate	Benzene	75	85		[933]

[a]Dropwise addition.

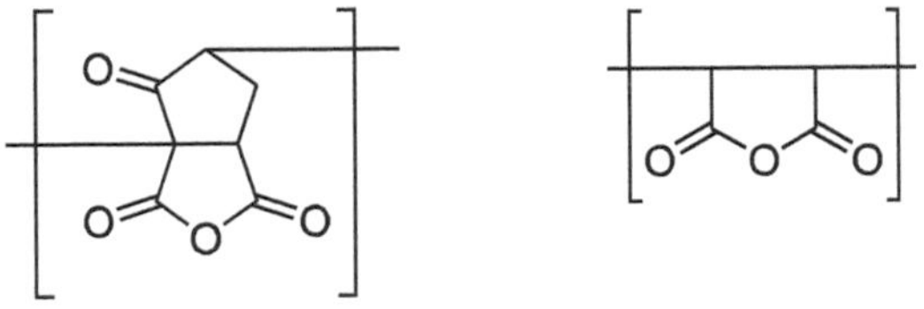

(a) (b)

Figure 11 Proposed structures of polymaleic anhydrid.

these polymerizations is the evolution of CO_2 [951,955,960]. It is proposed, that in this case an ionic mechanism is involved in polymer formation [951,964].

C. Copolymerization with Vinyl Monomers

Although it is diffucult to homopolymerize MAH it can be easily copolymerized with numerous vinyl monomers. Such copolymers have achieved technical importance as coatings, glues, adhesives, thickeners, resins, and engineering plastics. In recent literature these polymers were also investigated as side-chain liquid-crystalline polymers [965], ArF- or 193 nm-photoresists [966–972].

Copolymerizations of MAH have also found great theoretical interest, because they almost every time yield alternating copolymers and numerous 1,2-di-substituted ethens, that are reluctant to homopolymerization, can be copolymerized.

The most widely accepted explanation for these surprising results is the formation of an charge-transfer complex between the electron acceptor MAH and an electron donor comonomer that exclusively or predominantly participates in the chain growth process [973,974].

The charge-transfer concept is backed by the following observations:

- Spontaneous copolymerization of MAH with strong electron donors like 5,6-dihydro-1,4-dioxin [975], 1,1-dimethoxyethene [976], alkyl vinyl sulfides [977], phenyl vinyl sulfide [978] or styrene [979].
- The gradual transition from alternating to random copolymerizations at higher temperatures (e.g., MAH/styrene > 130°C [979] and MAH/α-methylstyrene > 80 °C [980]).
- The existence of charge transfer complexes of MAH in solution. Equilibrium constants in the range 3 to 60×10^2 cm^3/mol were found at 20 to 30 °C in $CHCl_3$ or hexane for comonomers such as styrene, α-methylstyrene, vinyl acetate, vinyl ethers, or vinyl sulfides [21–23].

On the other hand, it is worth noting that the *Qle* scheme of Alfrey and Price [981] predicts more or less alternating copolymers, when the *e*-values of two comonomers show a significant difference as it is true for MAH ($e = 2.5$ to 3) and most electron-rich comonomers ($e \leq 0$). The formation of alternating sequences is then attributed to a favorable electronic interaction between monomer and active chain end. Anyway, the assumption of such a favorable interaction between comonomer and active chain end of MAH (or vice versa) is not a contradiction to the formation and polymerization of charge transfer complexes. Both kinds of electronic interactions might be cooperating in numerous copolymerizations of MAH. Regardless of the mechanism the strong tendency to form 1:1 copolymers with alternating sequence is documented by the low reactivity ratios listed in Table 14. A broader range of reactivity ratios is summarized in Ref. [23].

Table 14 Reactivity ratios of copolymerizations of MAH (r_1) and various vinyl monomers (r_2).

Comonomer	r_1	r_2	Temp. (°C)	Refs.
Acenaphthylene	0.10	0.32	60	[1022]
Acrolein, diethylacetal	0.18	0.07		[1023]
Acrylamide	0.0	0.56		[1024]
Acrylic acid	0.3	6.25	70	[1025]
Acrylic acid, methyl ester	0.0	2.50	60	[1026]
	0.007	2.15	60	[1027]
Acrylonitrile	0.0	6.0	60	[1026]
Allyl acetate	0.018	0.030	60	[1028]
	0.072	0.01		[1029]
Allylidine diacetate	0.5	0.0		[1030]
Butadiene, 2,3-dimethyl	(−0.021)	(−0.057)		[1031,1032]
cis-2-Butene	0.016	0.0	60	[1033]
trans-2-Butene	0.03	0.0	60	[1034]
2-Butene, 2-methyl	0.01	0.035	70	[1035]
Chlormethylstyrene	0.0	0.974	60	[1036]
	0.02	1.08		[1037]
Crotonaldehyde diethylacetal	8.5	0.01		[1038]
cis-Crotonitrile	0.0	0.0	60	[1039]
Cycloheptene	0.068	0.0	60	[1033]
Cyclohexene	0.08	0.0	60	[1023]
Cyclooctene	0.067	0.04		[1023]
1,3-Dithiolane	0.61	0.60	40	[1040]
Ethylene	0.0	0.0	40	[1041]
Ethylene, 1,1,2-trichloro	0.0	3.7		[1042]
1-Hexane	0.15	0.015	70	[1035]
Methacrylic acid, methyl ester	0.01	3.40	60	[1043]
	0.02	0.90	60	[1043]
Methyacrylic acid, trimethylstannyl ester	0.0	0.20		[1044]
Methacrylonitrile	0.017	18.1		[1045]
Naphthalene, 1,2-dihydro	0.026	0.0	60	[1044]
	0.28	0.0	60	[1044]
Norbornene	0.03	0.05	70	[1035]
Norbornene, 2-carbonitrile	0.12	0.04		[1046]
Norbornene, 2-carboxylic acid	0.07	17.3		[1048,1049]
	0.03	12.35		[1050,1051]
	0.36	0.08		[1047]
Norbornene, 2-methoxycarbonyl	0.0	1.1	60	[1042]
	0.26	0.1		[1052,1053]
Pentaerythritol, diallylidene	0.03	0.08		[1054]
Phthalic acid, diallyl ester	0.06	0.015		[1055]
Phenylacetylene	0.065	0.076		[1056]
Propene	0.13	0.008	70	[1057]
Propene, cis-1-chloro	0.41	0.004	60	[1058]
Propene, trans-1-chloro	0.26	0.05	60	[1059]
Propene, 2-chloro	0.06	0.06	60	[1058,1059]
Propene, 3-chloro	0.19	(−0.03)		[1060]
cis-Stilbene	0.08	0.07	60	[1061]
trans-Stilbene	0.03	0.03	60	[1061]

(*continued*)

Table 14 Continued.

Comonomer	r_1	r_2	Temp. (°C)	Refs.
	0.13	0.0	70	[1062]
Styrene	0.0	0.02	60	[1063,1064]
	0.01	0.04	60	[1065,1066]
	0.05	0.13		[1067]
Styrene, α-methyl	0.27	0.005	60	[1068]
Styrene, *m,p*-benzoyl (ratio 60:40)	0.04	0.09	75	[1069]
Styrene, *m,p*-acetyl (ratio 60:40)	0.11	0.10	75	[1069]
Styrene, *m,p*-methyl (ratio 60:40)	0.23	0.10	75	[1069]
Vinyl acetate	0.027	0.01		[1070]
Vinyl butyl ether	0.045	0.0	50	[1071,1072]
Vinyl chloride	0.668	(−0.104)	60	[1073]
Vinylferrocene	0.21	0.02		[1074]
N-Vinylphthalimide	0.01	0.20	65	[999]
	0.003	0.30	90	[1075]
N-Vinylpyrrolione-2	0.074	(−0.027)	30	[1076]
Vinylidene chloride	0	9	60	[1026]
Vinylidene cyanide	0	45		[1077,1078]

A great variety of ternary systems have been polymerized. They were mainly designed to investigate the role of charge-transfer complexes in the mechanism of copolymerizations of MAH. Different system compositions were investigated. First the copolymerization of MAH with two different donor monomers like *trans*-stilbene, styrene [974,982–994], 4-chlorostyrene [987], 2-chloroethylvinyl ether [985,987,995–997], 1,4-dioxene [995–997], vinyl chloride, S-alkylvinyl sulfides [998] or N-vinylphthalimide [999,1000]. These systems can be treated as copolymerizations of two different donor–acceptor complexes. Other systems consist of MAH, a donor monomer and another acceptor monomer, for example: 2-chloroethylvinyl ether/MAH/fumarodinitrile [996], butadiene/MAH/SO₂ [1001], styrene/MAH/N-phenylmaleimide [974], styrene/MAH/dichlorodicyano-benzoquinone [1002], and 2-chloroethylvinyl ether/MAH/7,7,8,8-tetrakis(ethoxycarbonyl)quinodimethane [1003]. This system also can be treated as copolymerizations of two different donor–acceptor complexes. The third systems consist of MAH, a donor monomer and a so called neutral monomer, usually acrylonitrile [995,997,1004–1016]. Results indicate the copolymerization of an charge-transfer complex (MAH/donor) and the neutral monomer.

The tendency of MAH to form copolymers is so strong, that it can even be copolymerized with thiophene, its 2-methyl or 3-methyl derivatives [1017–1019], furan and 2-methylfuran [1020,1021] all stable aromatic heterocycles, that are reluctant to homopolymerize or copolymerize with other monomers. The repeating units consist of structures with 2,5-linkages (furan, thiophen) and 2,3-linkages across the methyl-substituted derivatives (Figure 12).

D. Polymerizations of Monomers Derived from Maleic Acid

Since polymerizations of maleic acid and various alkyl esters were mentioned earlier, the 'derivatives' of maleic acid discussed below are fumaric acid and dialkylfumarates, maleonitrile and fumaronitrile, maleimide and its N-substituted derivatives, and methylene succinic acid and its anhydride (itaconic anhydride).

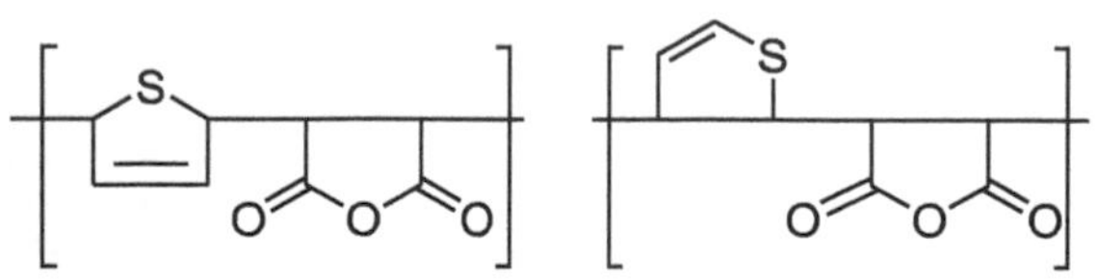

Figure 12 Structures of copolymers of MAH and thiophenes.

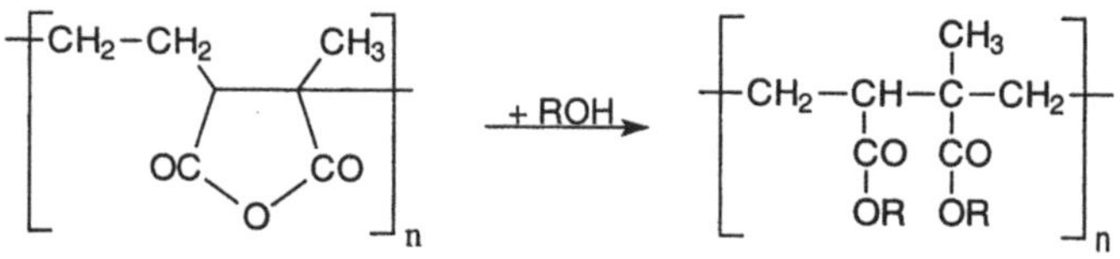

Figure 13 Esterification of copolymers from ethylene and methyl MAH.

Dialkylfumarates, in particular fumaric acid itself, are difficult to homopolymerize. Nevertheless, several radical-initiated polymerizations of dialkyl fumarates have been reported [1079–1082]. Typical reaction temperatures are in the range 60 to 90 °C, typical yields in the range 5 to 35%, and the inherent viscosities vary between 0.1 and 0.4 dL/g (benzene, 60 °C). Synthesis of high-molecular-weight poly(diethylfumarate) was reported [1081].

In contrast to MAH N-substituted maleimides can be homopolymerized with high conversions and up to high molecular weights [1083–1090].

N-substituted maleimides as electron acceptor monomers copolymerize alternatingly with a variety of electron donor monomers like styrene [1091–1098], α-methylstyrene [1099,1100], alkyl (2-chloroethyl) vinyl ethers [1093,1101], cyclohexyl vinyl ketone and its derivatives [1102,1103], isobutylene [1095], 1,3-butadiene [1104] and 2-vinylpyridine [1095]. Maleimides can also be polymerized by means of anionic initiators, such as sodium methoxide, lithium or potassium *tert*-butoxide, and *n*-butyllithium [1083,1105–1107]. Anionic polymerizations proceed at low temperatures (e.g., at −72 °C) and give high yields [1107]. The molecular weights are in general lower than those obtained by radical initiation and increase with the monomer/initiator ratio.

Ethylene can be copolymerized with methyl MAH [1108] (Figure 13). Esterification of the resulting copolymer with alcohols yields head-to-head copolymers of acrylates and methacrylates [1108].

Itaconic acid and its anhydrid can only be homo- and copolymerized by radical initiation [1109–1114]. The polymeric acid is best prepared by hydrolysis of its polyanhydride [1109] because the free-radical polymerization of itaconic acid is accompanied by partial decarboxylation [1110,1111]. M_n values up to 900 g/mol were reported for the polymeric acid [1109].

REFERENCES

1. Nuyken, O., and Lattermann, G. (1991). *Handbook of Polymer Synthesis Part A*, 1st ed, (Kricheldorf, H. R., ed.), Marcel Dekker, New York, p. 223ff.
2. Fittig, R., and Paul, L. (1887). *Justus Liebigs Ann. Chem.*, *188*: 55.
3. Fittig, R., and Engelhorn, F. R. (1880). *Justus Liebigs Ann. Chem.*, *200*: 70.
4. Kahlbaum, G. W. A. (1880). *Ber. Dtsch. Chem. Ges.*, *13*: 2348.

5. Röhm, O. (1901). Ph.D. thesis, Tübingen.
6. Röhm, O. (1914). U.S. 1,121,134; CA (1914), *09*: 395 A.
7. Barker, A. L., and Skinner, G. S. (1924). *J. Am. Chem. Soc.*, *46*: 403.
8. Riddle, E. H. (1954). *Monomeric Acrylic Esters*, Reinhold, New York.
9. Kautter, C. T. (1975). *Kunststoff Handbuch*, Vol. 9 (Vieweg, R., and Esser, F., eds.), Carl Hanser, München, p. 1.
10. Kine, B. B., and Novak, R. W. (1985). *Encyclopedia of Polymer Science and Engineering*, 2nd ed., Vol. 1 (Mark, H. F., Bikales, N. M., Overberger, C. G., and Menges, G., eds.), Wiley, New York, p. 234ff.
11. Kautter, C. T., Kösters, B., Quis, P., and Trommsdorf, E. (1975). *Kunststoff Handbuch*, Vol. 9 (Vieweg, R., and Esser, F., eds.), Carl Hanser, München, p. 8ff.
12. Schröder, G. (1987). *Houben Weyl: Methoden der organischen Chemie*, 4th ed., Vol. E 20, part 2 (Bartl, H., and Falbe, J., eds.), Georg Thieme, Stuttgart, p. 1141 ff.
13. Glavis, F. J., and Specht, E. H. (1963). *Kirk-Othmer Encyclopedia of Chemical Technology*, Vol. 1 (Mark, H. F., McKetta, J. J., Jr., Othmer, D. F., and Standen, A., eds.), Interscience, New York, p. 285ff.
14. Glavis, F. J., and Woodman, J. F. (1967). *Kirk-Othmer Encyclopedia of Chemical Technology*, Vol. 13 (Mark, H. F., McKetta, J. J., Jr., Othmer, D. F., and Standen, A., eds.), Interscience, New York, p. 331ff.
15. Mays, J. W., and Hadjichristidis, N. (1988). *J. Macromol Sci. Rev. Macromol Chem. Phys. Part C*, *28*: 371.
16. Glavis, F. J., and Specht, E. H. (1963). *Kirk-Othmer Encyclopedia of Chemical Technology*, Vol. 1 (Mark, H. F., McKetta, J. J., Jr., Othmer, D. F., and Standen, A., eds.), Interscience, New York, p. 285ff.
17. Glavis, F. J., and Woodman, J. F. (1967). *Kirk-Othmer Encyclopedia of Chemical Technology*, Vol. 13 (Mark, H. F., McKetta, J. J., Jr., Othmer, D. F., and Standen, A., eds.), Interscience, New York, p. 331ff.
18. Glavis, F. J., and Woodman, J. F. (1967). *Kirk-Othmer Encyclopedia of Chemical Technology*, Vol. 13 (Mark, H. F., McKetta, J. J., Jr., Othmer, D. F., and Standen, A., eds.), Interscience, New York, p. 346ff.
19. Sandler, S. R., and Karo, W. (1974). *Polymer Syntheses*, Vol. 1, Academic Press, Orlando, p. 271ff.
20. *The Manufacture of Acrylic Polymers*, Tech. Bull. SP-233, Röhm & Haas Co., Philadelphia, 1962.
21. Trommsdorf, E., Köhle, H., and Lagally, P. (1948). *Makromol. Chem.*, *1*: 169.
22. Detrick, C. A. (1970). *Ind. Eng. Chem. Process Design Develop.*, *9*: 191; C.A. (1970), *72*: 122155m.
23. Kine, B. B., and Novak, R. W. (1985). *Encyclopedia of Polymer Science and Engineering*, 2nd ed., Vol. 1 (Mark, H. F., Bikales, N. M., Overberger, C. G., and Menges, G., eds.), Wiley, New York, p. 269.
24. Matheson, M. S., Auer, E. E., Bevilacqua, E. B., and Hart, E. J. (1951). *J. Am. Chem. Soc.*, *73*: 5395.
25. Mayo, F. R. (1943). *J. Am. Chem. Soc.*, *65*: 2324.
26. Pappas, S. P. (1989). *Comprehensive Polymer Science*, Vol. 4 (Allen, G., and Bevington, J. C., eds.), Pergamon Press, Oxford, p. 337ff.
27. Walling, C., and Briggs, E. R. (1946). *J. Am. Chem. Soc.*, *68*: 1141.
28. Katime, I. A., and Nufio, T. (1988). *Makromol. Chem. Symp.*, *20/21*: 99.
29. Pryor, W. A., Eino, M., and Newkome, G. R. (1977). *J. Am. Chem. Soc.*, *99*: 6003.
30. Moad, G., Rizzardo, E., and Solomon, D. H. (1989). *Comprehensive Polymer Science*, Vol. 3 (Allen, G., and Bevington, J. C., eds.), Pergamon Press, Oxford, p. 141ff.
31. Yasuda, H., and Iriyama, Y. (1989). *Comprehensive Polymer Science*, Vol. 6 (Allen, G., and Bevington, J. C., eds.), Pergamon Press, Oxford, p. 357ff.
32. Chengxun Lu, Nauchun Chen, Zhong Wei Gu, and Xinde Feng (1980). *J. Polym. Sci. Polym. Chem. Ed.*, *18*: 3403.

33. Matyjaszewski, K. (1997). *Polym. Prepr. (Am. Chem. Soc., Div. Polym. Chem.)*, *38*: 383.
34. Diakoumakos, C. D., Xu, Q., Jones, F. N., Baghdachi, J., and Wu, L. (2000). *J. Coat. Technol.*, *72*: 61.
35. Markert, G. (1987). *Houben Weyl: Methoden der organischen Chemie*, Vol. E20, part 2 (Bartl, H., and Falbe, J., eds.), Georg Thieme, Stuttgart, p. 1157.
36. Hamilton, C. J., and Tighe, B. J. (1989). *Comprehensive Polymer Science*, Vol. 3 (Allen, G., and Bevington, J. C., eds.), Pergamon Press, Oxford, p. 261ff.
37. Yocum, R. H., and Nyquist, E. B., eds. (1974). *Functional Monomers*, Vol. 1, Marcel Dekker, New York, p. 2ff.
38. Tan, J. S., and Gasper, S. P. (1975). *J. Polym. Sci. Polym. Phys. Ed.*, *13*: 1705.
39. Stickler, M. (1989). *Comprehensive Polymer Science*, Vol. 3 (Allen, G., and Bevington, J. C., eds.), Pergamon Press, Oxford, p. 59ff.
40. Gnanim, H., and Fuchs, O. (1980). *Lösungsmittel und Weichmachungsmittel*, 8th ed., Vol. 11, Wissenschaftliche Verlagsgesellschaft, Stuttgart.
41. Fuchs, O., and Suhr, H. H. (1975). *Polymer Handbook*, 2nd ed. (Brandrup, J., and Immergut, E. H., eds.), Wiley, New York, p. IV/241ff.
42. Trommsdorff, E. (1975). *Kunststoff Handbuch*, Vol. 9 (Vieweg, R., and Esser, F., eds.), Carl Hanser, München, p. 33ff.
43. Dawkins, J. V. (1989). *Comprehensive Polymer Science*, Vol. 4 (Allen, G., and Bevington, J. C., eds.), Pergamon Press, Oxford, p. 231.
44. Kine, B. B., and Novak, R. W. (1985). *Encyclopedia of Polymer Science and Engineering*, 2nd ed.,Vol. I (Mark, H. F., Bikales, N. M., Overberger, C. G., and Menges, G., eds.), Wiley, New York, p. 282.
45. Fehrenbacher, U., Muth, O., Hirth, T., and Ballauff, M. (2000). *Macromol. Chem. Phys.*, *201*: 1532.
46. Shiho, H., and DeSimone, J.M. (2000). *J. Polym. Sci., Part A: Polym. Chem.*, *38*: 3783.
47. Sawamoto, M., and Kamigaito, M. (2000). *Macromol. Sym.*, *161*: 11.
48. Walbridge, D. J. (1989). *Comprehensive Polymer Science*, Vol. 4 (Allen, G., and Bevington, J. C., eds.), Pergamon Press, Oxford, p. 243.
49. Sandler, S. R., and Karo, W. (1974). *Polymer Synthesis*, Vol. 1, Academic Press, Orlando, Fla., p. 284.
50. Barett, K. E. J., ed. (1975). *Dispersion Polymerization in Organic Media*, Wiley, New York.
51. Napper, D. H., and Gilbert, R. G. (1989). *Comprehensive Polymer Science*, Vol. 4 (Allen, G., and Bevington, J. C., eds.), Pergamon Press, Oxford, p. 171.
52. Sandler, S. R., and Karo, W. (1974). *Polymer Synthesis*, Vol. 1, Academic Press, Orlando, Fla., p. 292.
53. Kine, B. B., and Novak, R. W. (1985). *Encyclopedia of Polymer Science and Engineering*, 2nd ed., Vol. I (Mark, H. F., Bikales, N. M., Overberger, C. G., and Menges, G., eds.), Wiley, New York, p. 281.
54. Osada, Y., and Takase, M. (1983). *J. Polym. Sci. Polym. Lett. Ed.*, *21*: 643.
55. Dunn, A. S. (1989). *Comprehensive Polymer Science*, Vol. 4 (Allen, G., and Bevington, J. C., eds.), Pergamon Press, Oxford, p. 219.
56. Grätzel, C. K., Jirousek, M., and Grätzel, M. (1986). *Langmuir*, *2*: 292; C.A. (1986), 104: 186921v.
57. Capek, I. (1999). *Adv. Colloid Interface Sci.*, *80*: 85.
58. Guo, T., Song, M., Hao, G., Zhao, F., and Zhang, B. (1999). *Chin. J. React. Polym.*, *8*: 38.
59. Urban, D., Gerst, M., Rossmanith, P., and Schuch, H. (1998). *Polym. Mater. Sci. Eng.*, *79*: 440.
60. Balakrishnan, T., Devarajan, R., and Santappa, M. (1984). *J. Polym. Sci. Polym. Chem. Ed.*, *22*: 1909.
61. Decker, C. (1999). *Macromol. Symp.*, *143*: 45.
62. Hageman, H. J. (1989). *Polym. Mater. Sci. Eng.*, *60*: 558; C.A. (1989), III: 24966f.
63. Amagi, Y., and Chapiro, A. (1962). *J. Chim. Phys.*, *59*: 537.
64. Chapiro, A. (1972). *Israel J. Chem.*, *10*: 129.
65. Shibasaki, Y., Nakahara, H., and Fukuda, K. (1979). *J Polym. Sci. Polym. Chem. Ed.*, *17*: 2387.

66. Shibasaki, Y., and Fukuda, K. (1979). *J. Polym. Sci. Polym. Chem. Ed.*, *17*: 2947.
67. Ringsdorf, H., Schlarb, B., and Venzmer, J. (1988). *Angew. Chem. Intern. Ed. Engl.*, *27*: 113.
68. Akkapeddi, M. K. (1979). *Macromolecules*, *12*: 546.
69. Allcock, H. R., and Levin, M. L. (1985). *Macromolecules*, *18*: 1324.
70. Sastre, R., Conde, M., and Mateo, J. L. (1988). *J. Photochem. Photobiol. Part A*, *44,1*: 11.
71. Ikeda, M. (1978). *Netsu Sokutei*, *5*: 100; C.A. (1979), *90*: 104412k.
72. Kopietz, M., Lechner, M. D., Steinmeier, D. G., and Franke, H. (1986). *Makromol. Chem.*, *187*: 2787.
73. Broer, D. J., Finkelmann, H., and Kondo, K. (1988). *Makromol Chem.*, *189*: 185.
74. Klemm, E., Hoerhold, H. H., and Heide, K. (1988). DD 256,139 to Friedrich-Schiller-Universität; C.A. (1989), *110*: 155765s.
75. Bacte, B., and Ballot, D. (1987). Fr. 259,3431AI to ACONM, Université de Caen; C.A. (1988), *108*: 133011z.
76. Turner, S. R., and Daly, R. C. (1989). *Comprehensive Polymer Science*, Vol. 6 (Allen, G., and Bevington, J. C., eds.), Pergamon Press, Oxford, p. 193ff.
77. Hollahan, J. R., and Bell, T. A., eds. (1974). *Techniques and Applications of Plasma Chemistry*, Wiley, New York.
78. Shen, M., ed. (1976). *Plasma Chemistry of Polymers*, Marcel Dekker, New York.
79. Brosse, J. C., Epaillard, F., and Legeay, G. (1983). *Eur. Polym. J.*, *19*: 381.
80. Brosse, J. C., Epaillard, F., and Legeay, G. (1983). *Eur. Polym. J.*, *19*: 743.
81. Brosse, J. C., Epaillard, F., and Legeay, G. (1983). *Eur. Polym. J.*, *19*: 749.
82. Epaillard, F., Brosse, J. C., and Legeay, G. (1987). *Eur. Polym. J.*, *23*: 233.
83. Epaillard, F., Brosse, J. C., and Legeay, G. (1988). *Makromol. Chem.*, *189*: 1035.
84. Osada, Y., Bell, A. T., and Shen, M. (1978). *J. Polym. Sci. Polym. Lett. Ed.*, *16*: 309.
85. Moad, G., and Solomon, D. H. (1989). *Comprehensive Polymer Science*, Vol. 3 (Allen, G., and Bevington, J. C., eds.), Pergamon Press, Oxford, p. 97ff.
86. Bamford, C. H. (1989). *Comprehensive Polymer Science*, Vol. 3 (Allen, G., and Bevington, J. C., eds.), Pergamon Press, Oxford, p. 123ff.
87. Vasishtha, R., Saini, S., Nigam, S. K., and Srivastava, A. K. (1989). *J. Macromol Sci. Rev. Macromol Chem. Phys. Part C*, *29*: 39.
88. Moad, G., Rizzardo, E., and Solomon, D. H. (1989). *Comprehensive Polymer Science*, Vol. 3 (Allen, G., and Bevington, J. C., eds.), Pergamon Press, Oxford, p. 141ff.
89. Walling, C. (1970). *Polym. Prepr. Am. Chem. Soc. Div. Polym. Chem.*, *11*: 721.
90. Malavasic, T., Vizovisek, I., Osredkar, U., and Anzur, I. (1981). *J. Polym. Sci. Polym. Symp.*, *69*: 73.
91. Yau, H., and Stupp, S. J. (1985). *J. Polym. Sci. Polym. Chem. Ed.*, *23*: 813.
92. Bevington, J. C. (1989). *Comprehensive Polymer Science*, Vol. 3 (Allen, G., and Bevington, J. C., eds.), Pergamon Press, Oxford, p. 65ff.
93. Moad, G., and Solomon, D. H. (1989). *Comprehensive Polymer Science*, Vol. 3 (Allen, G., and Bevington, J. C., eds.), Pergamon Press, Oxford, p. 147ff.
94. Hatada, K., Kitayama, T., and Ute, K. (1988). *Progr. Polym. Sci.*, *13*: 189.
95. Tsuruta, T., Makimoto, T., and Kanai, H. (1966). *J. Macromol. Sci.*, *1*: 31.
96. Niezette, J., and Desreux, V. (1971). *Makromol. Chem.*, *149*: 177.
97. Capek, I., Juraničová, V., Bartoň, J., Asua, J. M., and Ito, K. (1997). *Polym. Int.*, *43*: 1.
98. Bailey, W. J., Chou, J. L., Feng, P.-Z., Issari, B., Kuruganti, V., and Zhou, L.-L. (1988). *J. Macromol. Sci. Chem. Part A*, *25*: 781.
99. Mathias, L. J., Kusefoglu, S. H., Kress, A. O., Halley, R. J., and Colletti, R. F. (1989). *Polym. Prepr. (Am. Chem. Soc., Div. Polym. Chem.)*, *30*: 339.
100. Johnson, A. F., Khaligh, B., Ramsay, J., and O'Driscoll, K. (1983). *Polym. Commun.*, *24*: 35.
101. Greenley, Z. R. (1989). *Polymer Handbook*, 3rd ed. (Brandrup, J., and Immergut, E. H., eds.), Wiley, New York, p. 11/153.
102. Greenley, Z. R. (1989). *Polymer Handbook*, 3rd ed. (Brandrup, J., and Immergut, E. H., eds.), Wiley, New York, p. 11/267.

103. Kamachi, M., Kuwae, Y., Nozakura, S., and Hatada, K. (1983). *Polym. Bull*, *10*: 98.
104. Markert, G. (1987). *Houben Weyl: Methoden der organischen Chemie*, Vol. E20, part 2 (Bartl, H., and Falbe, J., eds.), Georg Thieme, Stuttgart, p. 1158ff.
105. Wunderlich, W. (1987). *Houben Weyl: Methoden der organischen Chemie*, Vol. E20, part 2 (Bartl, H., and Falbe, J., eds.), Georg Thieme, Stuttgart, p. 1145.
106. Yamaguchi, S., and Matsumoto, T. (1986). Jpn. 61/151,201 A2 to Nippon Oils and Fats Co., Ltd.; C.A. (1986), *105*: 227567a.
107. Kine, B. B., and Novak, R. W. (1985). *Encyclopedia of Polymer Science and Engineering*, 2nd ed.,Vol. I (Mark, H. F., Bikales, N. M., Overberger, C. G., and Menges, G., eds.), Wiley, New York, p. 279, 282.
108. Kautter, C. T. (1975). *Kunststoff Handbuch*, Vol. 9 (Vieweg, R., and Esser, F., eds.), Carl Hanser, München, p. 36.
109. Okubo, M., TacMka, H., Ando, M., Tange, T., Yamashita, S., and Matsumoto, T. (1982). *Nippon Setchaku Kyokaishi*, *18*: 153; C.A. (1982), *97*: 72878t.
110. Brosse, J. C., Gauthier, J. M., and Lenain, J. C. (1983). *Makromol. Chem.*, *184*: 1379.
111. Ikladious, N. E., Messiha, N. N., and Shaaban, A. F. (1984). *J. Appl. Polym. Sci.*, *29*: 509.
112. Müller, S., Schönhals, A., and Lorkowski, H. J. (1986). *Plaste Kautschuk*, *33*: 167.
113. Kumler, P. L., Kailani, M. H., Schober, B. J., Kolasa, K. A., Dent, S. J., and Boyer, R. F. (1989). *Macromolecules*, *22*: 2994.
114. Bajaj, P., and Padmanaban, M. (1983). *J. Polym. Sci. Polym. Chem. Ed.*, *21*: 2261.
115. Vasishtha, R., Awasthi, S., and Srivastava, A. K. (1990). *Brit. Polym. J.*, *22*: 53.
116. Senga, M., Kondo, S., and Tsuda, K. (1982). *J. Polym. Sci. Polym. Lett. Ed.*, *20*: 657.
117. Tanaka, M., Machida, S., and Uoi, N. (1986). Jpn. 60/250,007 to Idemitsu Kosan Co., Ltd.; C.A. (1986), *105*: 6980c.
118. Capek, I., Mlynarova, M., and Barton, J. (1988). *Makromol Chem.*, *189*: 341.
119. Kress, A. O., Mathias, L. J., and Cei, G. (1989). *Macromolecules*, *22*: 537.
120. Saegusa, T., Niwano, M., and Kobayashi, S. (1980). *Polym. Bull*, *2*: 249.
121. Dharia, J. R., Pathak, C. P., Babu, G. N., and Gupta, S. K. (1988). *J. Polym. Sci. Polym. Chem. Ed.*, *26*: 595.
122. Ghanem, N. A., Messiha, N. N., Ikladious, N. E., and Shaaban, A. F. (1981). *J. Appl. Polym. Sci.*, *26*: 97.
123. Bajaj, P., and Padmanaban, M. (1984). *J. Macromol. Sci. Chem. Part A*, *21*: 519.
124. Ito, K., Kodaira, K., and Onishi, Y. (1981). *J. Appl. Polym. Sci.*, *26*: 423.
125. Braun, D., Cherdron, H., and Kern, W. (1979). *Praktikum der makromolekularen organischen Chemie*, 3rd ed., Hüthig Verlag, Heidelberg, p. 230.
126. Hirai, H., Takeuchi, K., and Komiyama, M. (1985). *J. Polym. Sci. Polym. Chem. Ed.*, *23*: 901.
127. Bamford, C. H., and Hirooka, M. (1984). *Polymer*, *25*: 1791.
128. Wu, G. Y., Qi, Y. C., Lu, G. J., and Wei, Y. K. (1989). *Polym. Bull*, *22*: 393.
129. Kine, B. B., and Novak, R. W. (1985). *Encyclopedia of Polymer Science and Engineering*, 2nd ed., Vol. 1 (Mark, H. F., Bikales, N. M., Overberger, C. G., and Menges, G., eds.), Wiley, New York, p. 268.
130. Sandler, S. R., and Karo, W. (1974). *Polymer Synthesis*, Vol. 1, Academic Press, Orlando, Florida, p. 286.
131. Kobryner, W., and Banderet, A. (1959). *J. Polym. Sci.*, *34*: 381.
132. Swift, P. M. (1958). *J. Appl. Chem. (London)*, *8*: 803; C.A. (1959), *53*: 20878e.
133. Allen, P. W., Ayrey, G., and Moore, C. G. (1959). *J. Polym. Sci.*, *36*: 55.
134. Hibbard, B. B. (1962). U.S. 3,029,223 to Dow Chemical Co.; C.A. (1962), *57*: 2430i.
135. Sexton, B. J., and Curftnan, D. C. (1976). *Encyclopedia of Polymer Science and Technology*, 1st ed., Suppl., Vol. 1 (Mark, H. F., and Bikales, N. M., eds.), Interscience, New York, p. 307ff.
136. Feuer, S. S. (1958). U.S. 2,857,360 to Röhm & Haas Co.; C.A. (1959), *53*: 2686c.
137. Curfman, D. C. (1971). Fr. 2,037,183 to Borg-Warner Corp.; C.A. (1971), *75*: 141620n.
138. Smets, G., and Dysseleer, E. (1966). *Makromol Chem.*, *91*: 160.
139. Mino, G., and Kaizerman, S. (1958). *J. Polym. Sci.*, *31*: 242.

140. Iwakura, Y., and Imai, Y. (1966). *Makromol Chem.*, *98*: 1.
141. Odian, G., and Kho, J. H. (1970). *J. Macromol. Sci.*, *4*: 317.
142. Toda, T. (1962). *J. Polym. Sci.*, *58*: 411.
143. Morsum-Zade, A. A., Goryainova, Y. S., Livshits, R. M., Rogovin, Z. A., and Konkin, A. A. (1964). *Polym. Sci. USSR*, *6*: 1481.
144. Khetarpal, R. C., Gill, K. D., Mehta, I. K., and Misra, B. N. (1982). *J. Macromol. Sci. Chem. Part A*, *18*: 445.
145. Brockway, C. E. (1965). *J. Polym. Sci. Part A*, *3*: i03i.
146. Nuyken, O., Rengel, R., and Kerber, R. (1980). *Makromol Chem.*, *181*: 1565.
147. Rao, S. P., and Santappa, M. (1967). *J. Polym. Sci. Part A-1*, *5*: 2681.
148. Bamford, C. H., Jenkins, A. D., and White, E. F. T. (1959). *J. Polym. Sci.*, *34*: 271.
149. Szwarc, M. (1998). *J. Polym. Sci.: Part A*, *36*: 9.
150. Creutz, S., Teyssié, Ph., and Jérôme, R. (1997), *Macromolecules*, *30*: 5596.
151. Swarc, M. (1983). *Anionic Polymerization: Principles and Practice* (Morton, M., ed.), Academic Press, New York.
152. Lach, R., Grellmann, W., Weidisch, R., Altstädt, V., Kirschnick, T., Ott, H., Stadler, R., and Mehler, C. (2000). *J. Appl. Polym. Sci.*, *78*: 2037.
153. Panke, D. (1987). *Houben Weyl: Methoden der organischen Chemie*, Vol. E20, part 2 (Barti, H., and Falbe, J., eds.), Georg Thieme, Stuttgart, p. 1164ff.
154. Fontanille, M., and Miller, A. H. E. (1989). *Comprehensive Polymer Science*, Vol. 3 (Allen, G., and Bevington, J. C., eds.), Pergamon Press, Oxford, p. 571ff.
155. Van Beylen, M., Bywater, S., Smets, G., Szwarc, M., and Worsfold, D. J. (1988). *Adv. Polym. Sci.*, *86*: 87.
156. Fontanille, M. (1989). *Comprehensive Polymer Science*, Vol. 3 (Allen, G., and Bevington, J. C., eds.), Pergamon Press, Oxford, p. 365ff.
157. Hahn, B., Wendorff, J. H., Portugall, M., and Ringsdorf, H. (1981). *Colloid Polym. Sci.*, *259*: 875.
158. Frosini, V., Levita, G., Lupinacci, D., and Magagnini, P. L. (1981). *Mol. Cryst. Liq. Cryst.*, *66*: 21.
159. Shoji, K., Nakajima, Y., Ueda, E., and Takeda, M. (1985). *Polym. J.*, *17*: 997.
160. Cantow, H.-J., and Schulz, O. (1986). *Polym. Bull.*, *15*: 539.
161. Hatada, K., Ute, K., Kashiyama, T., Nishiura, T., and Miyatake, N. (1994). *Macromol. Symp.*, *85*: 325.
162. Hatada, K. (1999). *J. Polym. Sci.: Part A: Polym. Chem.*, *37*: 245.
163. Bywater, S. (1989). *Comprehensive Polymer Science*, Vol. 3 (Allen, G., and Bevington, J. C., eds.), Pergamon Press, Oxford, p. 441ff.
164. Hatada, K., and Kitayama, T. (2000). *Polym. Int.*, *49*: 11.
165. Hatada, K., Ute, K., Tanaka, T., Kitayama, T., and Okamoto, Y. (1989). *Recent Advances in Anionic Polymerization* (Hogen-Esch, T. E., and Stnid, J., eds.), Elsevier, New York, p. 195.
166. Hatada, K., Ute, K., Tanaka, K., Okamoto, Y., and Kitayama, T. (1986). *Polym. J.*, *18*: 1037.
167. Hatada, K., Nakanishi, H., Ute, K., and Kitayama, T. (1986). *Polym. J.*, *18*: 581.
168. Kitayama, T., Shinozaki, T., Sakamoto, T., Yamamoto, M., and Hatada, K. (1989). *Makromol. Chem. Suppl.*, *15*: 167.
169. Kitayama, T., Zhang, Y., and Hatada, K. (1994). *Polym. Bull.*, *32*: 439.
170. Kitayama, T., Hirano, T., Zhang, Y., and Hatada, K. (1996). *Macromol. Symp.*, *107*: 297.
171. Kitayama, T., Kawauchi T., Tabuchi, M., and Hatada, K. (1999). *Polym. Prepr. Jpn.*, *48*: 190; *Engl. Ed.*: E307.
172. Kitayama, T., Ute, K., and Hatada, K. (1990). *Br. Polym. J.*, *23*: 5.
173. Hatada, K., Kitayama, T., and Ute, K. (1993). *Makromol. Chem. Macromol. Symp.*, *70/71*: 57.
174. Kitayama, T., Fujimoto, N., Yanagida, T., and Hatada, K. (1994). *Polym. Int.*, *33*: 165.
175. Kitayama, T., Hirano, T., and Hatada, K. (1997). *Tetrahedron*, *53*: 15263.
176. Müller, A. H. E. (1989). *Comprehensive Polymer Science*, Vol. 3 (Allen, G., and Bevington, J. C., eds.), Pergamon Press, Oxford, p. 387ff.

177. Wang, J. S., Jérôme, R., and Teyssié, P. (1995). *J. Phys. Organ. Chem.*, *8*: 208.
178. Lochmann, L., Rodová, M., and Trekoval, J. (1974). *J. Polym. Sci.: Part A: Polym. Chem.*, *12*: 2091.
179. Janata, M., Lochmann, L., Vlček, P., Dybal, J., and Müller, A. H. E. (1992). *Makromol. Chem.*, *193*: 101.
180. Schlaad, H., Schmitt, B., and Müller, A. H. E. (1998). *Macromolecules*, *31*: 573.
181. Schmitt, B., Schlaad, H., and Müller, A. H. E. (1998). *Macromolecules*, *31*: 1705.
182. Baskaran, D., and Sivaram, S. (1997). *Macromolecules*, *30*: 1550.
183. Maurer, A., Marcarian, X., and Müller, A. H. E. (1997). *Polym. Prepr. (Am. Chem. Soc., Div. Polym. Chem.)*, *38*: 467.
184. Antoun, S., Teyssié, P., and Jérôme, R. (1997). *J. Polym. Sci.: Part A: Polym. Chem.*, *35*: 3637.
185. Nugay, N., Nugay, T., Jérôme, R., and Teyssié, P. (1997). *J. Polym. Sci.: Part A: Polym. Chem.*, *35*: 1543.
186. Marchal, J., Gnanou, Y., and Fontanille, M. (1997). *Polym. Prepr. (Am. Chem. Soc., Div. Polym. Chem.)*, *38*(1): 473.
187. Janata, M., Lochmann, L., Vlček, and Müller, A. H. E. (1990). *Makromol. Chem.*, *191*: 2253.
188. Vlček, P., Lochmann, L., and Otoupalová, J. (1992). *Makromol. Chem., Rapid Commun.*, *13*: 163.
189. Vlček, P., Dvořánek, L., Otoupalová, J., and Janata, M. (1995). *Macromol. Symp.*, *95*: 27.
190. Varshney, S. K., Hautekeer, J.-P., Jérôme, R., Fayt, R., and Teyssié, P. (1990). *Macromolecules*, *23*: 2618.
191. Bayard, P., Jérôme, R., Teyssié, P., Varshney, S. K., and Wang, J. S. (1994). *Polym. Bull.*, *32*: 381.
192. Nugay, N., Nugay, T., Jérôme, R., and Teyssié, P. (1997). *J. Polym. Sci.: Part A: Polym. Chem.*, *35*: 361.
193. Maurer, A., Marcarian, X., Müller, A. H. E., Navarro, C., and Vuillemin, B. (1997). *Polym. Prepr. (Am. Chem. Soc., Div. Polym. Chem.)*, *38*: 467.
194. Baskaran, D. (2000). *Macromol. Chem. Phys.*, *201*: 890.
195. Vlček, P., and Lochmann, L. (1999). *Progr. Polym. Sci.*, *24*: 793.
196. Jérôme, R., and Tong, J. (1998). *Curr. Opin. Solid State Mater. Sci.*, *3*: 573.
197. Teyssié, P., Bayard, P., Jérôme, R., Varshney, S. K., and Wang, J. S. (1995). *Macromol. Symp.*, *98*: 171.
198. Jérôme, R., Teyssié, P., Vuillemin, B., Zundel, T., and Zune, C. (1999). *J. Polym. Sci.: Part A: Polym. Chem.*, *37*: 1.
199. Zune, C., and Jérôme, R. (1999). *Progr. Polym. Sci.*, *24*: 631.
200. Vlček, P., Otoupalová, Kříž, J., and Schmidt, P. (2000). *Macromol. Symp.*, *161*: 113.
201. Ballard, D. G. H., Bowles, R. J., Haddleton, D. M., Richards, S. N., Sellens, R., and Twose, L. (1992). *Macromolecules*, *25*: 5907.
202. Haddleton, D. M., Hunt, K. H., and Crossman, M. C. (1996). *Macromol. Symp.*, *107*: 177.
203. Davis, T. P., Haddleton, D. M., and Richards, S. N. (1994). *J. Macromol. Sci., Rev. Macromol. Chem. Phys.*, *C34*: 243.
204. Varshney, S. K., Bayard, P., Jacobs, C., Jérôme, R., Fayt, R., and Teyssié, P. (1992). *Macromolecules*, *25*: 5578.
205. Antolin, K., Lamps, J. P., Rempp, P., and Gnanou, Y., (1990). *Polymer*, *31*: 967.
206. Ruckenstein, E., and Zhang, H. M. (1997). *Macromolecules*, *30*: 6852.
207. Cowie, J. M. G., Ferguson, R., Fernandez, M. D., Fernandez, M. J., and McEwen, I. J. (1992). *Macromolecules*, *25*: 3170.
208. Tong, J. D., Leclère, P., Doneux, C., Brédas, J. L., Lazzaroni, R., and Jérôme, R. (2001). *Polymer*, *41*: 2499.
209. Tong, J. D., Moineau, G., Leclère, P., Brédas, J. L., Lazzaroni, R., and Jérôme, R. (2000). *Macromolecules*, *33*: 470.
210. Vlček, P., and Lochmann, L. (1995). *Makromol. Chem., Macromol. Symp.*, *95*: 111.

211. Hautekeer, J.-P., Varshney, S. K., Fayt, R., Jacobs, C., Jérôme, R., and Teyssié, P. (1990). *Macromolecules, 23*: 3893.
212. Yasuda, H. (1999). *Top. Organomet. Chem., 2(Lanthanides)*: 255.
213. Ihara, E., Morimoto, M., and Yasuda, H. (1995). *Macromolecules, 28*: 7886.
214. Collins, S., Ward, D. G., and Suddaby, K. H. (1994). *Macromolecules, 27*: 7222.
215. Soga, K., Deng, H., Yano, T., and Shiono, T. (1994). *Macromolecules, 27*: 7938.
216. Soga, K., Deng, H., and Shiono, T. (1995). *Macromolecules, 28*: 3067.
217. Soga, K., Deng, H., and Shiono, T. (1995). *Macromol. Chem. Phys., 196*: 1971.
218. Stuhldreier, T., Keul, H., and Höcker, H. (2000). *Macromol. Rapid Commun., 21*: 1093.
219. Frauenrath, H., Keul, H., and Höcker, H. (2001). *Macromolecules, 34*: 14.
220. Johnson, L. K., Mecking, S., and Brookhart, M. (1996). *J. Am. Chem. Soc., 118*: 267.
221. Reetz, M. T. (1988). *Angew. Chem., 100*: 1026; *Angew. Chem. Int. Ed. Engl., 27*: 994.
222. Reetz, M. T., Knauf, T., Minet, U., and Bingel, C. (1988). *Angew. Chem., 100*: 1422; *Angew. Chem. Int. Ed. Engl., 27*: 1373.
223. Reetz, M. T., Hütte, S., and Goddard, R. (1995). *J. Phys. Chem., 8*: 231.
224. Reetz, M. T., Herzog, H. M., and Könen, W. (1996). *Macromol. Rapid Commun., 17*: 383.
225. Zagala, A. P., and Hogen-Esch, T. E. (1996). *Macromolecules, 29*: 3038.
226. Baskaran, D., and Müller, A. H. E. (2000). *Macromol. Rapid Commun., 21*: 390.
227. Webster, O. W., Hertler, W. R., Sogah, D. Y., Farnham, W. B., and Rajan Babu, T. V. (1983). *J. Am. Chem. Soc., 105*: 5706.
228. Müller, A. H. E. (1994). *Macromolecules, 27*: 1685.
229. Webster, O. W., and Sogah, D. Y. (1989). *Comprehensive Polymer Science*, Vol. 4 (Allen, G., and Bevington, J. C., eds.), Pergamon Press, Oxford, p. 163.
230. Teyssie, P., Fayt, R., Jacobs, C., Jerome, R., Leemans, L., and Varshney, S. (1988). *Polym. Prepr. Am. Chem. Soc. Div. Polym. Chem., 29*: 52.
231. Sogah, D. Y., Hertler, W. R., Webster, O. W., and Cohen, G. M. (1987). *Macromolecules, 20*: 1473.
232. Bandermann, F., Steinbrecht, K., and Witkowski, R. (1988). *Polym. Prepr. Am. Chem. Soc. Div. Polym. Chem., 29*: 97.
233. Citron, J. D. (1988). U.S. 4,771,116A to E. I. Du Pont de Nemours Co.; C. A. (1989), *110*: 76288k.
234. Hertler, W. R. (1989). Eur. 276,976 A2 to E. I. Du Pont de Nemours Co.; C.A. (1989), *110*: 76281c.
235. Li, Y., Ward, D. G., Reddy, S. S., and Collins, S. (1997). *Macromolecules, 30*: 1875.
236. Sitz, H.-D., Speikamp, H.-D., and Bandermann, F. (1988). *Makromol. Chem., 189*: 429.
237. Chou, S. S. P., and Niu, C. W. (1987). *MRL Bull. Res. Develop., 1*: 33; C.A. (1988), *108*: 113010v.
238. Hertler, W. R., Sogah, D. Y., Webster, O. W., and Trost, B. M. (1984), *Macromolecules, 17*: 1415.
239. Dicker, I. B. (1988). *Polym. Prepr. Am. Chem. Soc. Div. Polym. Chem., 29*: 114.
240. Dicker, I. B. (1988). U.S. 4,732,955A to Columbian Chemicals Co.; C.A. (1988), *109*: 38426a.
241. Wei, Y., and Wnek, G. (1987). *Polym. Prepr. Am. Chem. Soc. Div. Polym. Chem., 28*: 252.
242. Müller, M. A., and Stickler, M. (1986). *Makromol. Chem. Rapid Commun., 7*: 575.
243. Quirk, R. P., and Bidinger, G. P. (1989). *Polym. Bull., 22*: 63.
244. Asami, R., Kondo, Y., and Takaki, M. (1986). *Polym. Prepr. Am. Chem. Soc. Div. Polym. Chem., 27*: 186.
245. Eastmond, G. C., and Grigor, J. (1986). *Makromol. Chem. Rapid Commun., 7*: 375.
246. Asami, R., Takaki, M., and Moriyama, Y. (1986). *Polym. Bull, 16*: 125.
247. Sogah, D. Y., Hertler, W. R., and Webster, O. W. (1984). *Polym. Prepr. Am. Chem. Soc. Div. Polym. Chem., 25*: 3.
248. Pugh, C., and Percec, V. (1985). *Polym. Bull., 14*: 109.
249. Chapiro, A. (1958). *J. Polym. Sci., 29*: 321.

250. Zurakowska-Orszagh, J., Soerjosoeharto, K., Busz, W., and Oldziejewski, J. (1977). *Chem. Zvesti*, *31*: 66; C.A. (1978), *88*: 74575e.
251. Hsiue, G. H., and Huang, W. K. (1985). *J. Appl. Polym. Sci.*, *30*: 1023.
252. Graham, R. K., Gluckman, M. S., and Kampf, M. J. (1959). *J. Polym. Sci.*, *38*: 417.
253. Olenin, A. V., Khainson, A. B., Golubev, V. B., Lachinov, M. B., Zubov, V. P., and Kabanov, V. A. (1980). *Vysokomol. Soedin. Ser. A*, *22*: 2359; C.A. (1981), *94*: 31141s.
254. Smirnov, B. R., Bel'Govskii, I. M., Ponomarev, G. V., Marchenko, A. P., and Enikolopyan, N. S. (1980). *Dokl. Akad. Nauk.*, *254*: 127.
255. Enikolopyan, N. S., Smirnov, B. R., Ponomarev, G. V., and Belgovskii, I. M. (1981). *J. Polym. Sci., Polym. Chem. Ed.*, *19*: 879.
256. Burczyk, A. F., O'Driscoll, K. F., and Rempel, G. L. (1984). *J. Polym. Sci., Polym. Chem. Ed.*, *22*: 3255.
257. Grednev, A. A. (1989). *Polym. Sci. USSR*, *31*: 2369.
258. Janowicz, A. H., U.S. 4,746,713 (1988), CA *107*: 218246.
259. Wayland, B. B., Basickes, L., Mukerjee, S., Wei, M., and Fryd, M. (1997). *Macromolecules*, *30*: 8109.
260. Solomon, D. H., and Rizzardo, E., US 4,581,429, CA *102*: 221335.
261. Georges, M. K., Veregin, R. P. N., Kazmaier, P. M., and Hamer, G. K. (1993). *Macromolecules*, *26*: 2987.
262. Odell, P. G., Listigovers, N. A., Quinlan, M. H., and Georges, M. K. (1998). *ACS Symp. Ser.*, *713*: 80.
263. Georges, M. K., Veregin, R. P. N., Kazmaier, P. M., and Hamer, G. K. (1994). *Trends Polym. Sci.*, *2*: 66.
264. Odell, P. G., Rabien, A., Michalak, L. M., Veregin, R. P. N., Quinlan, M. H., Moffat, K. A., Macleod, P. J., Listigovers, N. A., Honeyman, C. H., and Georges, M. K. (1997). *Polym. Prepr. (Am. Chem. Soc., Div. Polym. Chem.)*, *38*: 414.
265. Hawker, C. J. (1994). *J. Am. Chem. Soc.*, *116*: 11185.
266. Catala, J. M., Bubel, F., and Oulad Hammouch, S. (1995). *Macromolecules*, *28*: 8441.
267. Keoshkerian, B., Georges, M., Quinlan, M., Veregin, R., and Goodbrand, B. (1998). *Macromolecules*, *31*: 7559.
268. Klapper, M., Brand, T., Steenbock, M., and Müllen, K. (2000). *ACS Symp. Ser.*, *768*: 152.
269. Veregin, R. P. N., Odell, P. G., Michalak, L. M., and Georges, M. K. (1996). *Macromolecules*, *26*: 4161.
270. Odell, P. G., Veregin, R. P. N., Michalak, L. M., Brousmiche, D., and Georges, M. K. (1995). *Macromolecules*, *28*: 8453.
271. Odell, P. G., Veregin, R. P. N., Michalak, L. M., and Georges, M. K. (1997). *Macromolecules*, *30*: 2232.
272. Listigovers, N. A., Georges, M. K., Odell, P. G., and Keoshkerian, B. (1996). *Macromolecules*, *29*: 8992.
273. Yousi, Z., Jian, L., Rongchuan, Jianliang, Y., Lizong, D., and Lansun, Z. (2000). *Macromolecules*, *33*: 4745.
274. Burguiere, C., Dourges, M. A., Charleux, B., and Vairon, J. P. (1999). *Macromolecules*, *32*: 3883.
275. Yoshida, E., Ishizone, T., Hirao, A., Nakahama, S., Takata, T., and Endo, T. (1994). *Macromolecules*, *27*: 3119.
276. Lokaj, J., Vlček, P., and Křiž (1997). *Macromolecules*, *30*: 7644.
277. Christie, D., Haremza, S., Brinkmann-Rengel, S., and Raether, R. B., DE 19990819, CAN *134*: 178976.
278. Christie, D., Haremza, S., Brinkmann-Rengel, S., and Raether, R. B., DE 19939328, CAN *134*: 178976.
279. Odell, P. G., and Hamer, G. K. (1996). *Polym. Mater. Sci. Eng.*, *74*: 404.
280. Watkins, J. J., and McCarthy, T. J. (1994). *Macromolecules*, *27*: 4845.
281. Curran, D. P. (1988). *Synthesis*, 489.

282. Bellus, D. (1985). *Pure Appl. Chem.*, *57*, 1827.
283. Wang, J.-S., and Matyjaszewski, K. (1995). *J. Am. Chem. Soc.*, *117*: 5614.
284. Wang, J.-S., and Matyjaszewski, K. (1995). *Macromolecules*, *28*: 7572.
285. Matyjaszewski, K. (1997). *Pure Appl. Chem.*, *A34*(10): 1785.
286. Patten, T. E., and Matyjaszewski, K. (1998). *Adv. Mat.*, *10*: 901.
287. Wang, J.-S., and Matyjaszewski, K. (1995). *Macromolecules*, *28*: 7901.
288. Patten, T. E., Xia, J., Abernathy, T., and Matyjaszewski, K. (1996). *Science*, *272*: 866.
289. Paik, H., and Matyjaszewski, K. (1996). *Polym. Prepr. (Am. Chem. Soc., Div. Polym. Chem.)*, *37*: 274.
290. Haddleton, D., Jasieczek, C. B., Hannon, M. J., and Schooter, A. J. (1997). *Macromolecules*, *30*: 2190.
291. Grimaud, T., and Matyjaszewski, K. (1997). *Macromolecules*, *30*: 2216.
292. Wang, J. L., Grimaud, T., and Matyjaszewski, K. (1997). *Macromolecules*, *30*: 6507.
293. Kato, M., Kamigaito, M., Sawamoto, M., and Higashimura, T. (1995). *Macromolecules*, *28*: 1721.
294. Ando, T., Kato, M., Kamigaito, M., and Sawamoto, M. (1996). *Macromolecules*, *29*: 1070.
295. Ando, T., Kamigaito, M., and Sawamoto, M. (1997). *Macromolecules*, *30*: 4507.
296. Matyjaszewski, K., Wei, M., Xia, J., and McDermott, N. E. (1997). *Macromolecules*, *30*: 8161.
297. Granel, C., Dubois, P., Jérôme, R., and Teyssié (1996). *Macromolecules*, *29*: 8576.
298. Uegaki, H., Kotani, Y., Kamigaito, M., and Sawamoto, M. (1997). *Macromolecules*, *30*: 2249.
299. Uegaki, H., Kotani, Y., Kamigaito, M., and Sawamoto, M. (1998). *Macromolecules*, *31*: 6756.
300. Uegaki, H., Kamigaito, M., and Sawamoto, M. (1999). *J. Polym. Sci.: Part A: Polym. Chem.*, *37*: 3003.
301. Moineau, G., Minet, M., Dubois, P., Teyssié, P., Senninger, T., and Jérôme, R. (1999). *Macromolecules*, *32*: 27.
302. Sawamoto, M., and Kamigaito, M. (2000). *Polymer News*, *25*: 149.
303. Sawamoto, M., and Kamigaito, M. (1998). *Am. Chem. Soc. Symp. Ser.*, *685*: 296.
304. Bengough, W. I., and Fairservice, W. H. (1965). *Trans. Faraday Soc.*, *61*: 1206.
305. Bengough, W. I., and Fairservice, W. H. (1971). *Trans. Faraday Soc.*, *67*: 414.
306. Xia, J., and Matyjaszewski, K. (1996). *Polym. Prepr. (Am. Chem. Soc., Div. Polym. Chem.)*, *37*: 513.
307. Matyjaszewski, K., Nakagawa, Y., and Jasieczek, C. B. (1998). *Macromolecules*, *31*: 1535.
308. Xia, J. H., and Matyjaszewski, K. (1997). *Macromolecules*, *30*: 7697.
309. Matsuyama, M., Kamigaito, M., and Sawamoto, M. (1996). *J. Polym. Sci., Part A: Polym. Chem.*, *28*: 1721.
310. Percec, V., Barboiu, B., and Kim, H.-J. (1998). *J. Am. Chem. Soc.*, *120*: 305.
311. Percec, V., Kim, H.-J., and Barboiu, B. (1997). *Macromolecules*, *30*: 6702.
312. Percec, V., and Barboiu, B. (1997). *Polym. Prepr. (Am. Chem. Soc., Div. Polym. Chem.)*, *38*: 733.
313. Jo, S. M., Gaynor, S. G., and Matyjaszewski, K. (1996). *Polym. Prepr. (Am. Chem. Soc., Div. Polym. Chem.)*, *37*: 272.
314. Matyjaszewski, K., Patten, T. E., and Xia, J. (1997). *J. Am. Chem. Soc.*, *119*: 674.
315. Haddleton, D., Crossman, M. C., Hunt, K. H., Topping, C., Waterson, C., and Suddaby, K. S. (1997). *Macromolecules*, *30*: 3992.
316. Nishikawa, T., Ando, T., Kamigaito, M., and Sawamoto, M. (1997). *Macromolecules*, *30*: 2244.
317. Nishikawa, T., Kamigaito, M., and Sawamoto, M. (1997). *Polym. Prepr. (Am. Chem. Soc., Div. Polym. Chem.)*, *38*: 740.
318. Matyjaszewski, K., Qiu, J., Tsarevsky, N. V., and Charleux, B. (2000). *J. Polym. Sci.: Part A: Polym. Chem.*, *38*: 4724.
319. Nishikawa, T., Kamigaito, M., and Sawamoto, M. (1999). *Macromolecules*, *32*: 2204.
320. Xia, J., Johnson, T., Gaynor, S. G., Matyjaszewski, K., and DeSimone, J. (1999). *Macromolecules*, *32*: 4802.
321. Uegaki, H., Kotani, Y., Kamigaito, M., and Sawamoto, M. (1998). *Macromolecules*, *31*: 6756.

322. Matyjaszewski, K. (1999). *Macromol. Symp.*, *143*: 257.
323. Moschogianni, P., Pispas, S., and Hadjichristidis, N. (2001). *J. Polym. Sci.: Part A: Polym. Chem.*, *39*: 650.
324. Matyjaszewski, K., Nakagawa, Y., and Gaynor, S. G. (1997). *Macromol. Rapid. Commun.*, *18*: 1057.
325. Wang, J. S., Gaynor, S. G., and Matyjaszewski, K. (1995). *Polym. Prepr. (Am. Chem. Soc., Div. Polym. Chem.)*, *36*: 465.
326. Kotani, Y., Kamigaito, M., and Sawamoto, M. (1998). *Macromolecules*, *31*: 5582.
327. Greszta, D., and Matyjaszewski, K. (1996). *Polym. Prepr. (Am. Chem. Soc., Polym. Div.)*, *37*: 569.
328. Arehart, S., Greszta, D., and Matyjaszewski, K. (1997). *Polym. Prepr. (Am. Chem. Soc., Polym. Div.)*, *38*: 705.
329. Matyjaszewski, K., Greszta, D., and Pakula, T. (1997). *Polym. Prepr. (Am. Chem. Soc., Div. Polym. Chem.)*, *38*: 709.
330. Kotani, Y., Kato, M., Kamigaito, M., and Sawamoto, M. (1996). *Macromolecules*, *29*: 6979.
331. Matyjaszewski, K., Miller, P. J., and Nakagawa, Y. (1998). *Polymer*, *39*: 5163.
332. Brown, D. A., and Price, G. J. (2001). *Polymer*, *42*: 4767.
333. Coca, S., and Matyjaszewski, K. (1997). *Polym. Prepr. (Am. Chem. Soc., Polym. Div.)*, *38*: 693.
334. Coca, S., and Matyjaszewski, K. (1997). *Macromolecules*, *30*: 2808.
335. Matyjaszewski, K. (1998). *Macromol. Symp.*, *132*: 85.
336. Sojka, R. E., Lentz, R. D., and Westermann, D. T. (1998). *Soil Sci. Soc. Am. J.*, *62*(2): 1672.
337. Buchholz, F. L. (1992). *The Ullmann's Encyclopedia of Industrial Chemistry*, Vol. A 21, VCH Publishers, Inc.
338. Lee, H. K., and Jong, W. L. (1997). *J. Polym. Res.*, *4*(2): 119.
339. Kapoor, J. N., and Mathur, D. P. (1983). *Fert. News*, *28*(4): 40.
340. Kigel, M. Y., Kofman, M., Vishkina, T. V., and Wekilsky, K. C. (2000). U.S. 6,159,365 to American Envirocare, Inc.; C.A. (2000), *134*: 32740.
341. Moffet, R. H. (2000). WO 2,000,071,471, to E.I. Du Pont De Nemours and Company; C.A. (2000), *133*: 365937.
342. Persson, M., Tokarz, M., Dahlgren, M. I., and Jahansson-vestin, H. (2000). WO 2,000,066,492 to Akzo Nobel N.V., Neth.; Eka chemicals Ab; C.A. (2000), *133*: 337274.
343. Mandell, Kathleen, Darlington, J. W., Jr., and Tomlin, A. S. (2000). WO 2,000,066,187 to Amcol International Corporation, USA; C.A. (2000), *133*: 355299.
344. Thomas, W. M. (1985). *Encyclopedia of Polymer Science and Engineering*, 2nd ed., Vol. 1 (Mark, H. F., Gaylord, N. G., and Bikales, N. M., eds.), Wiley, New York, p. 177.
345. Thomas, W. M., and Wang, D. W. (1985). *Encyclopedia of Polymer Science and Engineering*, 2nd ed., Vol. 1 (Mark, H. F., Gaylord, N. G., and Bikales, N. M., eds.), Wiley, New York, p. 169.
346. Pfefferkorn, E. (1999). *J. Colloid Interface Sci.*, *216*(2): 197.
347. Tumakov, S. A., and Pervushkin, S. V. (1977). *Lab. Delo*, *7*: 408.
348. Hoffman, A. S., and Hayashi, Y. (2000). U.S. 6,165,509 to University of Washington; C.A. (2000), *134*: 61537.
349. Unger, E. C., Fritz, T. A., and Gertz, E. W. (2000). U.S. 6,139,819 to ImarRX Pharmaceutical Corp., USA; C.A. (2000), *133*: 340255.
350. Barnea, E. (2000). WO 2,000,063,231 to Bioincept, Inc., USA: C.A. (2000), *133*: 321002.
351. Schwarz, M., Miller, K., and Kamath, K. (2000). WO 2,000,062,830 to Scimed Life Systems, Inc., USA; C.A. (2000), *133*: 340283.
352. Matz, G. F. (2000). EP 1,046,390 to Calgon Corporation, USA; C.A. (2000), *133*: 325443.
353. Shirota, H., and Castner, E. W. (2000). Abstr. Pap. – Am. Chem. Soc., 220th PHYS-422.
354. Allen, R. C. (1978). *J. Chromatogr.*, *146*(1): 1.
355. Hoffmeister, H., Allen, R. C., and Maurer, H. R. (1974). *Electrophor. Isoelectric Focusing Polyacrylamide Gel, [Proc. Small Conf.]*: 266.

356. Wirth, P. J., and Romano, A. (1995). *Journal of Chromatography, A, 698*(1–2): 123.

357. Allen, R. C. (1978). *Journal of Chromatography, 146(1)*: 1.

358. Kerschmann, R. L., Odom, R., Kuwahara, T., and Reddington, M. (2000). WO 2,000,077,293 to Resolution Sciences Corp., USA; C.A. (2000), *134*: 39062.

359. Nordheim, A., and Cahill, M. (2000). EP 1,059,529; C.A. (2000), *134*: 27252.

360. Mansfield, E. S., Peponnet, C., Bashkin, J. S., and Kautzer, C. R. (2000). U.S. 6,156,178 to Molecular Dynamics, Inc., USA; C.A. (2000), *134*: 27234.

361. Shih, L. B., Vilalta, P., and Williams, M. (2000). WO 2,000,067,009 to Applied Hydrogel Technology Corporation, USA; C.A. (2000), *133*: 331781.

362. Updyke, T. V., and Engelhorn, S. C. (2000). U.S. 6,143,154 to Novex, USA; C.A. (2000), *133*: 331780.

363. Moureu, C. (1894). *Anj. Chim. Phys.*, *2*: 175.

364. Carpenter, E. L., and Davis, H. S. (1957). *J. Appl. Chem.*, *7*: 671.

365. Brit. Pat. (1949). 631,592 to American Cyanamid Co.; C.A. (1950), *44*: 4494b.

366. Jones, G. D. (1950). U.S. Pat. 2,504,074 to General Aniline & Film Corp.; C.A. (1955), *44*: 6677i.

367. Bikales, N. M., and Kolodny, E. R. (1963). *Kirk-Othmer Encyclopedia of Chemical Technology*, Vol. 1 (Mark, H. F., Mc Ketaa, J. J., Jr., Othmer, D. F., and Standen, A., eds.), Wiley, New York, p. 274.

368. Crawford, J. W. C., and Grigor, J. (1936). Brit. Pat. 456,533 to Imperial Chemical Industries; C.A. (1937), *31*: 2230(6).

369. Crawford, J. W. C., and McGrath, J. (1936). Brit. 456,533 to Imperial Chemical Industries; C.A. (1936), *30*: 4180(4).

370. Crawford, J. W. C., and McGrath, J. (1936). Brit. 456,533 to Imperial Chemical Industries; C.A. (1939), *33*: 2536(8).

371. Davis, H. S., Lichtenwalter, M., and Zeischke, W. M. (1947). U.S. 2,431,468 to American Cyanamid; C.A. (1948), *42*: 3429b.

372. Wiley, R. H., and Waddy, W. E. (1949). *Org. Synth. Coll.*, *III*: 560.

373. Lichtenwalter, M., and Wiedeman, O. (1950). U.S. 2,508,279 to American Cyanamid Co.; C.A. (1950), *44*: 8365a.

374. Hegboer, J., and Stavermann, A. J. (1950). *Recl. Trav. Chim. Pays-Bas*, *61*: 787.

375. Kobayashi, M., Nagasawa, T., and Yamada, H. (1992). *Trends Biotechnol.*, *10*: 402.

376. Lancaster, J. E., and O'Connor, M. W. (1982). *J. Polym. Sci. Polym. Lett. Ed.*, *20*: 547.

377. Sawant, S., and Morawetz, H. (1982). *J. Polym. Sci. Polym. Lett. Ed.*, *20*: 385.

378. Klein, J., and Hietzmann, R. (1978). *Makromol. Chem.*, *179*: 1859.

379. Mac Callum, J. R., and MacKerron, D. H. (1982). *Brit. Polym. J.*, *14*: 14.

380. Lewis, O. G. (1968). *Physical Constants of Linear Homopolymers*, Springer-Verlag, New York.

381. Illers, K. J. (1963). *Kolloid Z.*, *190*: 16.

382. Pinner, S. M. (1953). *J. Polym. Sci.*, *10*: 379.

383. Barton, J., and Juranicova, V. (2000). *Polym. Int.*, *49*(11): 1483.

384. Lenka, S., Nayak, P. L., Dash, S. B., and Ray, S. (1983). *Colloid Polym. Sci.*, *261*: 40.

385. Fortenberry, D. I., and Pojman, J. A. (2000). *J. Polym. Sci., Polym. Chem. (A)*, *38*(7): 1129.

386. Riggs, J. P., and Rodriguez, F. (1967). *J. Polym. Sci. Part A-1*, *5*: 3151.

387. Pross, A., Platokwski, K., and Reichert, K.-H. (1998). *Polym. Int.*, *45*(1): 22.

388. Chen, T.-M., Wang, Y.-F., Sakaguchi, T., Li, Y.-J., Kitamura, M., Nakayada, T., and Sakurai, I. (1996). *Eur. Polym. J.*, *32*(11): 1263.

389. Gromov, V. F., Osmanov, T. O., Khomikovskii, P. M., and Abkin, A. D. (1980). *Eur. Polym J.*, *16*: 803.

390. Cakmak, I., Hazer, B., and Yagci, Y. (1991). *Eur. Polym. J.*, *27*(1): 101.

391. Revelskaya, L. G., and Kurlyankina, V. J. (1987). *Vysokomol. Soedin. Ser. A*, *29*: 1205.

392. Saito, R., Ni, X., Ichimura, A., and Ishizu, K. (1998). *J. Appl. Polym. Sci.*, *69*(2): 211.

393. Robert, B., Bolte, M., and Lamaire, J. (1985). *J. Chem. Phys. Phys. Chem. Biol.*, *82*: 361.

394. Restaino, A. J., and Bristowe, W. W. (1974). Ger. Offen. 2,348,400 to ICI Americas Inc.; C.A. (1975), *82*: 33183h.
395. Restaino, A. J. (1970). Ger. Offen. 1,961,099 to Atlas Chemical Industries Inc.; C.A. (1970), *73*: 67331j.
396. Sara, A. S., Yavuz, O., and Sezer E. (1999). *J. Appl. Polym. Sci., 72*(7): 861.
397. Kovarskii, A. L., and Sivergin, Y. M. (1996). *The Polymeric Materials Encyclopedia*, CRC Press, Inc.
398. Kurenkov, V. F., and Myagchenkov, V. A. (1980). *Eur. Polym. J., 16*: 1229.
399. Kurenkov, V. F., and Myagchenkov, V. A. (1996). *The Polymeric Materials Encyclopedia*, CRC Press, Inc.
400. Kulicke, W. M., Kniewske, R., and Klein, J. (1982). *Prog. Polym. Sci., 8*: 373.
401. Thomas, R. A. M. (1983). *Chemistry and Technology of Water Soluble Polymers*, Plenum Press, New York, p. 31.
402. Munoz-Guerra, S. (1991). *Makromol. Chem., Macromol. Symp., 48–49*: 71.
403. Breslow, D. S., Hulse, G. E., and Matlack, A. S. (1957). *J. Am. Chem. Soc., 79*: 3760.
404. Morgenstern, U., and Berger, W. (1992). *Makromol. Chem., 193*(10): 2561.
405. Bhadani, S. N., Prasad, Y. K., and Kundu, S. (1979). *J. Polym. Sci. Polym. Chem., 18*: 1459.
406. Bhadani, S. N., and Prassad, Y. K. (1977). *J. Polym. Sci. Polym. Lett. Ed., 15*: 721.
407. Kishore, K., and Santhamalakshmi, K. N. (1983). *J. Macromol. Sci. Chem. Part A, 20*: 23.
408. Kulicke, W. M. (1987). Houben-Weyl: *Makromolekulare Stoffe*, Vol. E20, Georg Thieme, Stuttgart, p. 1176.
409. Friedrich, R. E., McConnell, R., and Garrets, W. L. (1960). Brit. 843,374 to Dow Chemicals: C.A. (1961), *55*: 6031h.
410. Vanderhoff, I. W., and McConnell, R. (1960). Brit. 841,127 to Dow Chemicals: C.A. (1961), *54*: 26006e.
411. Nachtigall, G. W., and Scanley, C. S. (1976). DOS 2,431,794 to American Cyanamid Co.; C.A. (1976), *84*: 122913e.
412. Schenck, H. U., Krapf, H., and Oppenlaender, K. (1977). DOS 2,554,082 to BASF AG; C.A. (1977), *87*: 69227t.
413. Schenck, H. U., Krapf, H., and Oppenlaender, K. (1977). DOS 2,557,324 to BASF AG; C.A. (1977), *87*: 69137p.
414. Anderson, D. B., Forder, J., and Johnson, I. M. (1977). Brit. Pat. 1482,515 to Allied Colloids Ltd.; C.A. (1978), *88*: 23811W.
415. Fink, H., Pennewiss, H., Plainer, H., Trabitzsch, H., Frieser, J., and Masanek, J. (1974). DOS 2,322,883 to Roehm GmbH: C.A. (1975), *82*: 126245v.
416. Ni, X., Bennet, D. C., Symes, K. C., and Grey, B. D. (2000). *J. Appl. Polym. Sci., 76*(11): 1669.
417. Kiatkamjornwong, S., and Phunchareon, P. (1999). *J. Appl. Polym. Sci., 72*(10), 1349.
418. Yamamoto, Y., and Sefton, M. V. (1998). *Polym. Tissue Eng.*, 189.
419. Yuan, Z., Shen, L. H., Yang, D. F., and He, B. (1997). *Gaodeng Xuexiao Huaxue Xuebao, 18*(1): 154.
420. Bicak, N., Sarac, A., Koza, G., Atay, T., and Senkal, F. (1993). *React. Polym., 21*(1–2): 135.
421. Hansen, L. D., and Eatough, D. J. (1987). *Thermochim. Acta, 111*: 57.
422. Öz, N., and Akar, A. (2000). *J. Appl. Polym. Sci., 78*(4): 870.
423. Shao-Jie, L., and Xi-Ming, S. (2000). *Trans. Tianjin Univ., 6*(1): 90.
424. Lun, N., Wang, X., and Li, Y. (1999). *Shandong Jiancai Xueyuan Xuebal, 13*(2): 114.
425. Ryan, M., and Pawlowska, L. (1998), U.S. 5,789,472 to Cytec Technology Corp., USA; C.A. (1998), *129*: 162251.
426. Jaeger, W., and Hahn, M. (1996). *Macromol. Symp., 111*: 95.
427. Kim, J., Deike, I., Dingenouts, N., Norhausen, Ch., and Ballauff, M. (1999). *Macromol. Symp., 142*: 217.
428. Schild, G. (1992). *Prog. Polym. Sci., 17*: 163.

429. Gehrke, S. H. (1993). *Advances in Polymer Science*, Vol. 110, Springer-Verlag, Berlin, Heidelberg.
430. Kobayashi, M., Nagasawa, T., and Yamada, H., (1992). *Trends Biotechnol.*, *10(11)*: 402.
431. Putnam, D., and Kopecek, J. (1995). *Advances in Polymer Science*, Vol. 122, Springer-Verlag, Berlin, Heidelberg.
432. Duncan, R., and Ulbrich, K. (1993). *Makromol. Chem., Macromol. Symp.*, *70/71*: 157.
433. Okamoto, Y., Nakano, T., Habaue, S., Shiohara, K., and Maeda, K. (1997). *Journal Macromol. Sci. Pure Appl. Chem.*, *A34(10)*: 1771.
434. Kopecek, J., Kopeckova, P., and Konak, C. (1997). *Journal Macromol. Sci. Pure Appl. Chem.*, *A34(10)*: 2103.
435. Simo-Alfonso, E., Gelfi, C., Sebatiano, R., and Citterio, Righetti, P. G. (1996). *Electrophoresis*, *17*: 723.
436. Simo-Alfonso, E., Gelfi, C., Sebatiano, R., Citterio, and Righetti, P. G. (1996). *Electrophoresis*, *17*: 732.
437. Chiari, M., Micheletti, C., Nesi, M., Fazio, M., and Righetti, P. G. (1994). *Electrophoresis*, *15*: 177.
438. Glavis, F. J. (1963). *Kirk-Othmer Encyclopedia of Chemical Technology*, 2nd ed., Vol. I (Grayson, M. S., ed.), Wiley-Interscience, New York, p. 285ff.
439. Nemec, J. W., and Bauer, W., Jr. (1985). *Encyclopedia of Polymer Science and Engineering*, 2nd ed., Vol. 1 (Mark, H. F., Bikales, N. M., Overberger, C. G., and Menges, G., eds.), Wiley, New York, p. 212.
440. Hughes, L. T. J., and Fordyce, D. B. (1956). *J. Polym. Sci.*, *22*: 509.
441. Sandler, S. R., and Karo, W. (1977). *Organic Chemistry*, Vol. 29, Academic Press, Orlando, Fla., p. 264.
442. Silberberg, A., Eliassaf, J., and Katschalsky, A. (1957). *J. Polym. Sci.*, *23*: 259.
443. Wiederhom, N. M., and Brown, A. R. (1952). *J. Polym. Sci.*, *8*: 651.
444. Sakamoto, R., and Yoshioka, K. (1962). *Nippon Kagaku Zasshi*, *83*: 517.
445. Eliassaf, J., Silberberg, A., and Katschalsky, A. (1955). *Nature (London)*, *176*: 1119.
446. Arai, K., Maseki, Y., and Ogiwara, Y. (1986). *Makromol. Chem. Rapid Commun.*, *7*: 655.
447. Miller, M. L. (1964). *Encyclopedia of Polymer Science and Technology*, Vol. 1 (Mark, H. F., Gaylord, N. G., and Bikales, N. M., eds.), Wiley, New York, p. 197ff.
448. Adams, C. R. (1970). *Chem. Ind. (NY)*, *26*: 1644.
449. Nemec, J. W., and Schlaefer, F. W. (1970). Ger. 1,961,894 to Röhm & Haas Co.; C.A. (1970), *73*: 88391u.
450. Krabetz, R., and Engelbach, H. (1971). Ger. 1,908,965 to Badische Anilin-und Sodafabrik AG; C.A. (1971), *75*: 36958f.
451. Naito, H., Ootani, M., Ookita, M. and Shiotani, T. (1993). Jpn. 05317713 to Mitsubishi Rayon Co.
452. Reppe, W. (1953). *Justus Liebigs Ann. Chem.*, *582*: 1.
453. Matsuda, F., and Kato, T. (1976). Jpn. to Mitsui Toatsu Chemicals, Inc.; C.A. (1977), *86*: 89183h.
454. Dunn, K. A. (1962). U.S. 3,069,433 to Celanese Corp. of America; C.A. (1963), *58*: 10087f.
455. Luke, O. V., Robeson, M. O., and Taylor, W. E. (1958). U.S. 2,820,058 to Celanese Corp. of America; C.A. (1958), *52*: 4681b.
456. Fenton, D. M., and Olivier, K. L. (1972). *Chem. Technol*, Apr., p. 220.
457. Fenton, D. M. (1968). U.S. 3,397,225 to Union Oil of California; C.A. (1968), *69*: 76660k.
458. Slejko, F. L., and Clovis, J. S. (1974). Ger. 2,546,741 to Röhm und Haas Co.; C.A. (1976), *85*: 46000i.
459. Nemec, J. W., and Kirch, L. S. (1981). *Kirk-Othmer Encyclopedia of Chemical Technology*, 3rd ed., Vol. 15 (Grayson, M. S., ed.), Wiley-Interscience, New York, p. 346ff.
460. Sikahura, Y., Sakai, F., and Shimizu, H. (1976). Jpn. 76/63,112 to Nippon Kagaku Co. Ltd.; C.A. (1976), *85*: 1233352d.

461. Kirch, L. S., and Kennelly, W. J. (1981). Eur. 27,351 to Roehm und Haas Co.; C.A. (1981), *95*: 97060s.
462. Misono, M., and Nojiri, N. (1990). *Appl. Catal.*, *64*: 1.
463. Daniel, C., and Brusky, P. L. (1981). U.S. 4,299,980 to Ashland Oil, Inc.; C.A. (1982), *96*: 69583s.
464. Akimoto, M., Tsuchida, Y., Sato, K., and Echigoya, E. (1981). *J. Catal.*, *72*: 83.
465. Gaenzler, W., Kals, K., and Schroeder, G. (1977). Ger. 2,702,187 to Roehm GmbH; C.A. (1978), *89*: 164170u.
466. Breitenbach, J. W., Kauffmann, H. F., and Zwilling, G. (1976). *Makromol. Chem.*, *177*: 2787.
467. Loebl, E. M., and O'Neill, J. J. (1963). *J. Polym. Sci. Part B*, *1*: 27.
468. Monjol, P., and Champetier, G. (1972). *Bull. Soc. Chim. Fr.*, *47*: 1302.
469. Pinner, S. H. (1953). *J. Polym. Sci.*, *10*: 379.
470. Glavis, F. J. (1954). *J. Polym. Sci.*, *36*: 547.
471. Katchalsky, A., and Eisenberg, H. (1951). *J. Polym. Sci.*, *6*: 145.
472. Kargin, V. A., Kabanov, V. A., Mirlina, S. Y., Mikheleva, G. A., and Viasov, A. V. (1960). *Dokl. Akad. Nauk. SSSR*, *135*: 893; C.A. (1962), *56*: 6159f.
473. Loebl, E. M., and O'Neill, J. J. (1960). *J. Polym. Sci.*, *45*: 538.
474. Kabanov, V. A., Mirlina, S. Y., and Vlasov, A. V. (1962). *Polym. Sci. (USSR)*, *3*: 28.
475. Aylward, N. N. (1970). *J. Polym. Sci., Part 4–1*, *8*: 319.
476. Kheradmand, H., Francois, J., and Placanet, V. (1988). *Polymer*, *29*: 860.
477. Otsu, T., and Quach, L. (1981). *J. Polym. Sci. Polym. Chem. Ed.*, *19*: 2377.
478. Hadley, D. J., and Evans, E. M. (1973). *Propylene and Its Industrial Derivatives* (Hancock, E. G., ed.), Wiley, New York, p. 367.
479. Katchalsky, A., and Blauer, G. (1951). *Trans. Faraday Soc.*, *47*: 1360.
480. Priel, Z., and Silberberg, A. (1970). *J. Polym. Sci. Part A-2.*, *8*: 689.
481. Breitenbach, J. W., and Kauffmann, H. F. (1974). *Makromol. Chem.*, *175*: 2597.
482. Breitenbach, J. W., Kauffmann, H. F., and Zwilling, G. (1976). *Makromol. Chem.*, *177*: 2787.
483. Priel, Z., and Silberberg, A. (1970). *J. Polym. Sci. Part A-2*, *8*: 689.
484. Alexandrowicz, Z. (1959). *J. Polym. Sci.*, *40*: 91.
485. Leyte, J. C., and Mandel, M. (1964). *J. Polym. Sci. Part A*, *2*: 1879.
486. Fakirov, S., Simov, D., Baldijeva, R., and Michailov, M. (1970). *Makromol. Chem.*, *138*: 27.
487. Pinner, S. H. (1952). *J. Polym. Sci.*, *9*: 282.
488. Van der Trift, W. P. J. T., and Overbeck, J. T. G. (1979). *Recl. Trav. Chim. Pays-Bas*, *98*: 81.
489. Creszenzi, V., Delben, F., and Quadrifoglio, F. (1972). *J. Polym. Sci. Part A-2*, *10*: 357.
490. Tsuchida, E., Osada, Y., and Abe, K. (1974). *Makromol. Chem.*, *175*: 583.
491. Sarasvathy, S., and Venkatarao, K. (1981). *Makromol. Chem. Rapid Commun.*, *2*: 219.
492. Kay, P. J., and Trelor, F. E. (1974). *Makromol. Chem.*, *175*: 3207.
493. Greger, G., and Egle, G. (1960). *Makromol. Chem.*, *40*: 1.
494. Monjol, P., and Champetier, G. (1972). *Bull. Soc. Chim. Fr.*, *4*: 1302.
495. Courtland, L. A. (1945). U.S. 2,367,483 to E. I. du Pont de Nemours Co.; C.A. (1945), *39*: 3703(5).
496. Riggs, J. P., and Rodriguez, F. (1967). *J. Polym. Sci. Polym. Chem. Ed.*, *5*: 3151.
497. Roy-Chowdhury, P. (1968). *J. Appl. Polym. Sci.*, *12*: 751.
498. Newman, S., Krigbaum, W. R., Langier, C., and Flory, P. J. (1954). *J. Polym. Sci.*, *14*: 451.
499. Katchalsky, A., and Spitnik, P. (1947). *J. Polym. Sci.*, *2*: 432.
500. Graves, G. D. (1940). U.S. 2,205,882 to E. I. du Pont de Nemours Co.; C.A. (1940), *34*: 7656(9).
501. Barrett, G. R. (1956). Ger. 943,676 to Monsanto Chem. Co.; C.A. (1958), *52*: 12456c.
502. Bugni, E. A., Lachtertnacher, M. G., Montiero, E. E. C., Mano, E. B., and Overberger, C. G. (1986). *J. Polym. Sci. Polym. Chem. Ed.*, *24*: 1463.
503. Mishra, M. K., and Bhadani, S. N. (1983). *Makromol. Chem.*, *184*: 955.

504. Mukherjee, A. R., Ghosch, P., Chadha, S. C., and Palit, S. R. (1964). *Makromol. Chem.*, *80*: 208.
505. Kern, W., and Quast, H. (1953). *Makromol. Chem.*, *10*: 202.
506. Kaliyamurthy, K., Elayaperumal, P., Balakrishnan, T., and Santappa, M. (1982). *J. Macromol. Sci. Part A*, *18*: 219.
507. Misra, G. S., and Arya, B. D. (1984). *J. Polym. Sci. Polym. Chem. Ed.*, *22*: 3563.
508. Lenka, S., Nayak, P. L., and Ray, S. (1984). *J. Polym. Sci. Polym. Chem. Ed.*, *22*: 959.
509. Elayaperumal, P., Balakrishnan, T., Santappa, M., and Lenz, R. W. (1980). *J. Polym. Sci. Polym. Chem. Ed.*, *18*: 2471.
510. Misra, G. S., and Narain, H. (1968). *Makromol. Chem.*, *113*: 85.
511. Misra, G. S., and Narain, H. (1968). *Makromol. Chem.*, *114*: 234.
512. Arnold, R., and Caplan, S. R. (1955). *Trans. Faraday Soc.*, *51*: 857.
513. Oth, A., and Doty, P. (1952). *J. Phys. Chem.*, *56*: 43.
514. Huang, M.-Y., Wu, R., and Jiang, L.-R., *Polym. Bull.*, *9*: 5.
515. Trapasso, L. E. (1976). U.S. 3,975,341 to Celanese Corp.; C.A. (1976), *85*: 160846p.
516. Kojima, K. (1977). Jpn. 77 66,59 1; C.A. (1977), *87*: 85782a.
517. Srivastava, A. K., Nigam, S. K., Shuda, A. K., Saini, S., Kumar, P., and Trevari, N. (1987). *J. Macromol. Sci. Rev. Macromol. Chem. Phys. Part C*, *27*: 171.
518. Fujimori, K. (1979). *Makromol. Chem.*, *180*: 1743.
519. Eliassaf, J., Eriksson, F., and Eirich, F. R. (1960). *J. Polym. Sci.*, *47*: 193.
520. Ferguson, J., Al-Alawi, S., and Granmayeh, R. (1983). *Eur. Polym. J.*, *19*: 475.
521. Nogaki, K., Matsubara, Y., Yoshihara, M., and Maeshima, T. (1984). *Makromol. Chem. Rapid Commun.*, *5*: 723.
522. Alexander, P., and Hitch, S. F. (1952). *Biochem. Biophys. Acta*, *9*: 219.
523. Dainton, F. S. (1957). *Trans. Faraday Soc.*, *53*: 499, 666.
524. Oster, G. K. (1957). *J. Am. Chem. Soc.*, *79*: 595.
525. Bamford, C. H., Eastmond, G. C., and Ward, G. C. (1963). *Proc. Roy. (London) Ser. A*, *271*: 357.
526. Monjol, P. (1967). *C. R? – Acad. Sci. Ser. C*, *265*: 1426.
527. Evans, M. G., Santappa, M., and Uri, N. (1951). *J. Polym. Sci.*, *7*: 243.
528. Conio, G., Patrone, E., Russo, S., and Trefiletti, V. (1976). *Makromol. Chem.*, *177*: 49.
529. Morawetz, H., and Kandanian, A. Y. (1966). *J. Phys. Chem.*, *70*: 2995.
530. Pramanick, D. (1979). *Colloid. Polym. Sci.*, *41*: 257.
531. Lando, J. B., and Semen, J. (1972). *J. Polym. Sci. Polym. Chem. Ed.*, *10*: 3003.
532. Collinson, E., Dainton, F. S., and McNaughton, G. S. (1957). *Trans. Faraday Soc.*, *53*: 476, 489.
533. Beddows, C. G., Barkee, P. S., and Guthrie, J. T. (1981). *Polym. Bull.*, *4*: 149.
534. O'Donnell, J. H., and Sothman, R. D. (1980). *J. Macromol. Sci. Chem. Part A*, *14*: 879.
535. Laborie, F. (1977). *J. Polym. Sci. Polym. Chem. Ed.*, *15*: 1255.
536. Brandrup, J., and Immergut, E. H. (1989). *Polymer Handbook*, 3rd ed., Wiley-Interscience, New York.
537. Ferguson, J., and Shah, S. A. (1968). *Eur. Polym. J.*, *4*: 343.
538. Gupta, B., and Scherer, G. G. (1994). *Chimia*, *48*: 127.
539. Thomas, R. A. M. (1983). *Chemistry and Technology of Water Soluble Polymers* (Finch, N. C., ed.), Plenum Press, New York, p. 75ff.
540. Miller, M. L., O'Donnell, K., and Skogman, J. (1962). *J. Colloid Sci.*, *17*: 649.
541. O'Neill, J. J., Loebl, E. M., Kandanian, A. Y., and Morawetz, H. (1965). *J. Polym. Sci. Part A*, *3*: 4201.
542. Fischer, O., and Kuszlik, A. (1996). *Chem. Ind. (Duesseldorf)*, *119(10)*: 36.
543. Bovey, F. A. (1971). *Progress in Polymer Science*, Vol. 3, Pergamon Press, Oxford, Chap. 1.
544. Klesper, E., Strasilla, D., and Regel, W. (1974). *Makromol. Chem.*, *175*: 523.
545. Glavis, F. J. (1968). *Water Soluble Resins*, 2nd ed. (Davidson, R. L., and Sittig, M., eds.), Van Nostrand Reinhold, New York, Chap. 8.

546. Thomas, W. M., and Wang, D. W. (1985). *Encyclopedia of Polymer Science*, Vol. 1, Wiley, New York, p. 211.

547. Carlson, E. J. (1953). U.S. Pat. 2,657,197 to B.F. Goodrich Co.; C.A. (1954), *48*: 3722i.

548. Schroeder, W. D., and Brown, G. L. (1951). *Rubber Age*, *69*: 433.

549. Greenwald, H. L., and Luskin, L. S. (1980). *Handbook of Watersoluble Gums and Resins*, (Davidson, R. L., ed.), McGraw-Hill, New York.

550. Bere, J. (1982). *Przem. Chem.*, *61*: 300.

551. Swift, G. (1994). *Polym. Degrad. Stab.*, *45*: 215.

552. Palenik, K., and Raczynska, Z. (1992). *Przegl. Papier.*, *48(5)*: 152.

553. Palenik, K. (1989). *Przegl. Papier.*, *45(12)*: 463.

554. Howe, E. E., and Putter, I. (1951). U.S. 2,541,320 to Merck & Co. and Röhm & Haas Co.; C.A. (1951), *45*: 4893g.

555. Zhi-Li, X. (1987). *Desalination*, *62*: 259.

556. Hopkins, R. P. (1955). *Ind. Eng. Chem.*, *47*: 2258.

557. Carr, C. E., and Greenland, D. J. (1974). *Rep. Prog. Appl. Chem.*, *59*: 269.

558. Mueck, K. F., Rolly, H., and Burg, K. (1977). *Makromol. Chem.*, *178*: 2773.

559. Bolewski, K. (1970). *Farm. Pol.*, *26(6)*: 443.

560. Takeuchi, T., and Matsui, J. (1998). *ACS Symp. Ser.*, *703*: 119.

561. Buchholz, F. L. (1993). *Polym. Mater. Sci. Eng.*, *69*: 489.

562. Buchholz, F. L., and Burgert, J. H. (1996). *Spec. Publ.-R. Soc. Chem.*, *186*: 92.

563. Grasselli, R. K., Centi, G., and Trifido, F. (1989). *Roc. Int. Symp. Uses Selenium Tellurium*, 4th, 609.

564. Misono, M., and Nojiri, N. (1990). *Appl. Catal.*, *64*: 1.

565. Staudinger, H., and Urech, E. (1929). *Helv. Chim. Acta*, *12*: 1107.

566. Marvel, C. S., and Levesque, C. L. (1939). *J. Am. Chem. Soc.*, *61*: 3244.

567. Vrancken, M., and Smets, G. (1954). *J. Polym. Sci.*, *14*: 521.

568. Boyer, S., and Rondeau, A. (1958). *Bull. Soc. Chim. Fr.*, *25*: 240.

569. Rondou, S., Smets, G., and DeWilde-Delvaux, M. C. (1957). *J. Polym. Sci.*, *24*: 261.

570. Blatz, P. E. (1962). *J. Polym. Sci.*, *58*: 755.

571. Schulz, R. C., Elzer, P., and Kern, W. (1961). *Makromol. Chem.*, *42*: 189.

572. Yang, Y. S., Qi, G. R., Qian, J. W., and Yang, S. L. (1998). *J. Appl. Polym. Sci.*, *68*: 665.

573. Timmons, R. B. (1999). *Surf. Coat. Aust.*, *36*: 10.

574. Hall, L. A. R., Belanger, W. J., Kirk, W., Jr., and Sundstrom, Y. V. (1969). *J. Appl. Polym. Sci.*, *2*: 246.

575. Howk, B. W., and Jacobsen, R. A. (1944). U.S. 2,440,090 to E.I. du Pont de Nemours; C.A. (1948), *42*: 4794h.

576. Khardin, A. P., Protopopov, P. A., and Solomina, T. I. (1985). *Macromolecular Synthesis*, Vol. 9 (Moore, J. A., ed.), Wiley, New York, p. 65.

577. Schulz, R. C., Elzer, P., and Kern, W. (1961). *Makromol. Chem.*, *42*: 197.

578. Biedermann, H. G., Eiband, M., Gunter, H., and Kolb, H. (1977). *Z. Naturforsch.*, B: 32B: 1455.

579. Paleos, C. M., Filippakis, S. E., and Margomenou-Leonidopoulou, G. (1981). *J. Polym. Sci. Chem. Ed.*, *19*: 1427.

580. Paleos, C. M., Margomenou-Leonidopoulou, G., Filippakis, S. E., Malliaris, A., and Dais, P. (1982). *J. Polym. Sci. Chem. Ed.*, *20*: 2267.

581. Diab, M. A., El-Sonbati, A. Z., El-Sanabari, A. A., and Taha, F. I. (1990). *Acta Polym.*, *41*: 45.

582. Yahagi, I., Watanabe, M., Sanui, K., and Ogata, N. (1987). *J. Polym. Sci. Chem. Ed.*, *25*: 727.

583. Ogata, N., Sanui, K., Watanabe, M., and Yahagi, I. (1985). *J. Polym. Sci. Chem. Ed.*, *23*: 349.

584. Stohriegl, P. (1990). *Mol. Cryst. Liq. Cryst.*, *183*: 261.

585. Jones, J. F. (1985). *J. Polym. Sci.*, *33*: 15.

586. Gibbs, W. E., and Murray, J. T. (1962). *J. Polym. Sci.*, *58*: 1211.

587. Hwa, J. C. H., and Miller, L. (1961). *J. Polym. Sci.*, *55*: 197.

588. Aso, G. (1960). *Resins Rubbers Plast.*, *14*: 867.
589. Gray, T. F., Jr., and Butler, G. B. (1975). *J. Macromol. Sci. Chem.*, *9*: 45.
590. Moureu, C. (1893). *Ann. Chem. Phys.*, *2*: 186.
591. Peng, F. M. (1985). *Encyclopedia of Polymer Science and Engineering*, 2nd ed., Vol. 1 (Mark, H. F., Bikales, N. M., Overberger, C. G., Menges, G., and Kroschwitz, J. J., eds.), Wiley, New York, p. 426.
592. Centi, G., Perathoner, S., and Trifiro, F. (1997). *Appl. Catal.*, *A*, *157*: 143.
593. Ueda, W., Morooka, Y., and Sasaki, Y. (1992). *Petrotech (Tokyo)*, *15*: 346.
594. Quingling, C., Xin, C., Liansheng, M., and Wencai, C. (1999). *Catal. Today*, *51*: 141.
595. Matsuura, I. (1999). *Kemikaru Enjiniyaringu*, *44*: 271; C.A. (1999), *131*: 158000.
596. Fuzikawa, N. (1998). *Idemitsu Giho*, *41*: 141; C.A. (1998), *128*: 270887.
597. Leonard, A., Gerber, G. B., Stecca, C., Rueff, J., Borba, H., Farmer, P. B., Sram, R. J., Czeizel, A. E., and Kalina, I. (1999). *Mutat. Res.*, *436*: 263.
598. Whysner, J., Ross, P. M., Conaway, C. C., Verna, L. K., and Williams, G. M. (1998). *Regul. Toxicol. Pharmacol.*, *27*: 217.
599. Ojajarvi, I. A., Partanen, T. J., Ahlbom, A., Boffetta, P., Hakulinen, T., Jourenkova, N., Kauppinen, T. P., Kogevinas, M., Porta, M., Vainio, H. U., Weiderpass, E., and Wesseling, C. H. (2000). *Qccup. Environ. Med.*, *57*: 316.
600. Saillenfait, A. M., and Sabate, J. P. (2000). *Toxicol. Appl. Pharmacol.*, *163*: 149.
601. Fahmy, M. A. (1999). *Cytologia*, *64*: 1.
602. Allen, R. A., and Ward, I. M. (1994). *Polymer*, *35*: 2063.
603. Cho, S. H., Park, J. S., Jo, S. M., and Chung, I. J. (1994). *Polym. Int.*, *34*: 333.
604. Henmi, M., and Yoshioka, T. (1993). *J. Membr. Sci.*, *85*: 129.
605. Fritzsche, A. K., Arevalo, A. R., Moore, M. D., and O'Hara, C. (1993). *J. Membr. Sci.*, *81*: 109.
606. Paul, D., Kamusewitz, H., Hicke, H. G., and Buschatz, H. (1992). *Acta Polym.*, *43*: 353.
607. Ulbricht, M., Riedel, M., and Schmidt, C. (1996). *GIT Fachz. Lab.*, *40*: 776.
608. Orchard, A. C., and Bates, L. J. (1993). *BHR Group Conf. Ser. Publ.*, *3*: 59.
609. Campet, G., Treuil, N., Poquet, A., Hwang, S. J., Labrugere, C., Deshayes, A., Frison, J. C., Portier, J., Reau, J. M., and Choy, J. H. (1999). *Bull. Korean Chem. Soc.*, *20*: 885.
610. Scrosati, B. (1994). *Solid State Ionic Mater.*, *Proc. Asian Conf. Solid State Ionics*, *4*: 111.
611. Ferry, A., Edman, L., Forsyth, M., MacFarlane, D. R., and Sun, J. (1999). *J. Appl. Phys.*, *86*: 2346.
612. De Brabander, E. M. M., Brackman, J., Mure-Mak, M., de Man, H., Hogeweg, M., Keulen, J., Scherrenberg, R., Coussens, B., Mengerink, Y., and van der Wal, S. (1996). *Macromol. Symp.*, *102*: 9.
613. Froehling, P., and Brackman, J. (2000). *Macromol. Symp.*, *151*: 581.
614. Shavit, N., and Konigsbuch, M. (1967). *J. Polym. Sci. Part C*, *16*: 43.
615. Amdur, S., and Shavit, N. (1967). *J. Polym. Sci. Part C*, *16*: 1231.
616. Shavit, N., Konigsbuch, M., and Oplatka, A. (1967). *J. Polym. Sci. Part C*, *16*: 1247.
617. Melacini, P., Patron, L., Moretti, A., and Tedesco, R. (1972). Ital. 903,309 to Chatillon, SA); C.A. (1977), *86*: 107252.
618. Melacini, P., Patron, L., Moretti, A., and Tedesco, R. (1971). Ger. Offen. 2,120,337 to Chatillon, SA; C.A. (1972), *76*: 86433x.
619. Melacini, P., Patron, L., Moretti, A., and Tedesco, R. (1973). Ger. Offen. 2,326,063 to Chatillon, SA; C.A. (1974), *81*: 38088v.
620. Patron, L., Moretti, A., and Tedesco, R. (1974). Ger. Offen. 2,400,043 to Montedison SPA; C.A. (1975), *82*: 4754q.
621. Ito, S. (1986). *Kobunshi Ronbunshu*, *43*: 1; C.A. (1986), *104*: 130368c.
622. Thomas, W. M., and Pellon, J. J. (1954). *J. Polym. Sci.*, *13*: 329.
623. Melacini, P., Patron, L., Moretti, A., and Tedesco, R. (1972). Ital. 903,309 to Chatillon, SA); C.A. (1977), *86*: 107252.

624. Melacini, P., Patron, L., Moretti, A., and Tedesco, R. (1971). Ger. Offen. 2,120,337 to Chatillon, SA; C.A. (1972), *76*: 86433x.

625. Melacini, P., Patron, L., Moretti, A., and Tedesco, R. (1973). Ger. Offen. 2,326,063 to Chatillon, SA; C.A. (1974), *81*: 38088v.

626. Patron, L., Moretti, A., and Tedesco, R. (1974). Ger. Offen. 2,400,043 to Montedison SPA; C.A. (1975), *82*: 4754q.

627. Ito, S. (1986). *Kobunshi Ronbunshu, 43*: 1; C.A. (1986), *104*: 130368c.

628. Helm, E., Berger, W., and Pippel, W. (1982). *Acta Polym., 33*: 300.

629. Shavit, N., and Konigsbuch, M. (1967). *J. Polym. Sci. Part C, 16*: 43.

630. Amdur, S., and Shavit, N. (1967). *J. Polym. Sci. Part C, 16*: 1231.

631. Shavit, N., Konigsbuch, M., and Oplatka, A. (1967). *J. Polym. Sci. Part C, 16*: 1247.

632. Melby, L. R., Janowicz, A. H., and Ittel, S. D. (1986). Eur. Appl. EP 199,436 Al; C.A. (1987), *106*: 157018s.

633. Amdur, S. (1971). *J. Polym. Sci. Polym. Chem. Ed., 9*: 175.

634. Okamoto, M., Suenaga, K., and Ishizuka, O. (1986). *Bull. Chem. Soc. Jpn., 59*: 1545.

635. Wada, T., Watanabe, T., and Takehisa, M. (1973). *Polym. J., 4*: 136.

636. Tabata, Y., and Oizumi, C. (1970). Jpn. 70/38,629 to Japan Atomic Energy Research Institute; C.A. (1971), *74*: 88372j.

637. Bensasson, R., Dworkin, A., and Marx, R. (1963). *J. Polym. Sci. Part C, 4*: 881.

638. Chapiro, A., and Mankowski, Z. (1978). *Eur. Polym. J., 14*: 15.

639. Shu, S., Tabata, Y., and Oshima, K. (1968). *Kobunshi Kagaku, 25*: 425; C.A. (1968), *69*: 9723 Im.

640. Siffiionescu, C. I., and Simionescu, B. C. (1982). *Rev. Roum. Chim., 27*: 141; C.A. (1982), *97*: 6882p.

641. Frushour, B. G., and Knorr, R. S. (1985). *Handbook of Fiberscience and Technology*, Vol. 4 (Lewin, M., and Pearce, E. M., eds.), Marcel Dekker, New York, p. 171.

642. Korte, S. (1987). *Houben-Weyl: Methoden der organischen Chemie*, Vol. E20 (Bartl, H., and Falbe, J., eds.), Georg Thieme, Stuttgart, p. 1195.

643. Fuchs, O., and Suhr, H.-H. (1975). *Polymer Handbook*, 2nd ed. (Brandrup, J., and Immergut, E. H., eds.), Wiley, New York, p. IV-241.

644. Hettrich, K., Fischer, S., Brendler, E., and Voigt, W. (2000). *J. Appl. Polym. Sci., 77*: 2113.

645. Vidotto, G., Grosato-Amaldi, A., and Talamini, G. (1969). *Makromol. Chem., 122*: 91.

646. Ulbricht, J. (1962). *Z. Phys. Chem., 5*: 346.

647. Peebles, L. H., Jr. (1958). *J. Am. Chem. Soc., 80*: 5603.

648. Kirby, J. R., Brandrup, J., and Peebles, L. H., Jr. (1968). *Macromolecules, 1*: 53.

649. Brandrup, J., Kirby, J. R., and Peebles, L. H., Jr. (1968). *Macromolecules, 1*: 59.

650. Chen, C., Colthup, M., Deichert, W., and Webb, R. L. (1969). *J. Polym. Sci., 45*: 247.

651. Levin, C. A., and Harris, G. H. (1962). *J. Polym. Sci., 62*: 100.

652. Patron, L., and Bastianelli, U. (1974). *Appl. Polym. Symp., 25*: 105.

653. Patron, L., Mazzolini, C., and Moretti, A. (1973). *J. Polym. Sci. Polym. Symp., 41*: 407.

654. Guyot, A., Dumont, M., Graillat, C., Guillot, J., and Pichot, C. (1975). *J. Macromol. Sci. Chem. Part A, 9*: 483.

655. Peebles, L. H., Jr., and Brandrup, J. (1966). *Makromol. Chem., 98*: 189.

656. Ayrey, G., Chadda, S., and Poller, R. C. (1982). *J. Polym. Sci. Polym. Chem. Ed., 20*: 2249.

657. Brit. 1,341,868 (1973) to Mitsubishi Rayon; C.A. (1974), *80*: 146892r.

658. Yokouchi, N., Kawamura, T., and Tokitada, T. (1971). Jpn. 46/38,561 to Mitsubishi Rayon; C.A. (1972), *76*: 100408b.

659. Kirby, J. R. (1974). U.S. 3,784,511 to Monsanto; C.A. (1975), *82*: 44904v.

660. Krebs, K. W., Engelhardt, H., and Nischk, G. (1970). Ger. Offen. 2,059,948 to Bayer AG; C.A. (1972), *77*: 103163z.

661. Fr. 2,029,457 (1970) to American Cyanamid; C.A. (1971), *74*: 143207m.

662. Okada, H., Kawamura, T., Kawata, H., Yoneyama, H., and Mimura, K. (1972). Ger. Offen. 2,138,839 to Mitsubishi Rayon; C.A. (1972), *77*: 127928d.

663. Matsumura, S., and Kaneniitsu, C. (1971). Jpn. 46/26,333 to Teijin; C.A. (1972), *76*: 25792s.

664. Nakao, S., Numata, N., and Yamamoto, T. (1973). Jpn. 48/29,874 to Kanebo; C.A. (1974), *81*: 323f.

665. Sakai, H., Izumi, Z., Kitagawa, H., Hamada, S., and Hoshira, M. (1972). Brit. 1,295,529 to Kanebo; C.A. (1973), *78*: 98268b.

666. Sakai, H., Hoshina, M., Hamada, S., Izumi, Z., and Kitagawa, H. (1970). Jpn. 45/37,553 to Toray Industries; C.A. (1971), *74*: 77075p.

667. Nakao, S., Numata, N., and Yamamoto, T. (1973). Jpn. 48/29,872 to Kanebo; C.A. (1974), *80*: 109266u.

668. Ito, I., Izuffii, Z., and Kitagawa, H. (1968). Jpn. 43/28,114 to Toyo Rayon; C.A. (1969), *70*: 97394n.

669. Kitagawa, H., and Aragane, T. (1968). Jpn. 43/28,470 to Toyo Rayon; C.A. (1969). *70*: 88436a.

670. Sakai, H., Izumi, Z., and Kitagawa, H. (1968). Jpn. 43/28,466 to Toyo Rayon; C.A. (1969), *70*: 78548z.

671. Chen, S. S., Herms, J., Peebles, L. H., Jr., and Uhlmann, D. R. (1981). *J. Mater. Sci.*, *16*: 1490.

672. Peters, R. H., and Still, R. H. (1979). *Applied Fiber Science*, Vol. 2 (Happy, F., ed.), Academic Press, New York, p. 321.

673. Szita, J., Unger, O., Marzolph, H., and Nischk, G. (1970). U.S. 3,511,800 to Bayer AG; C.A. (1972), *77*: 21448g.

674. Thompson, R. B., Jr., and Wilson, W. K. (1964). U.S. 3,153,024 to Monsanto; C.A. (1965), *62*: 1761e.

675. Roth, E., and Reifigerste, E. (1963). Belg. 624,366 to Agfa Wolfen VEB Filmfabrik; C.A. (1963), *59*: 5307b.

676. Yonemura, A., Shishido, Y., and Tamura, J. (1968). Jpn. 72/1229 to Teijin; C.A. (1973), *77*: 7144u.

677. Goodman, A., Grandine, L. D., and Vosburgh, W. G. (1962). U.S. 3,060,157 to E.-I. du Pont de Nemours Co.; C.A. (1964), *60*: 12166b.

678. Kitagawa, H., Mukoyama, F., Suzuki, Z., Kato, T., Hosaka, S., Sakai, H., Hamada, S., Yamanaka, Y., Ito, I., and Izumi, Z. (1972). Ger. Offen. 2,140,463 to Toray Industries; C.A. (1972), *76*: 155468b.

679. Fr. 1,517,989 (1968) to SNIA Viscosa SPA; C.A. (1969), *70*: 88752a.

680. Krebs, K. W., Engelhard, H., Miller, E. H., and Nischk, G. (1972). Ger. Offen. 2,032,548 to Bayer AG; C.A. (1972), *76*: 127799s.

681. Yoshida, M., Kaneko, H., and Edo, T. (1972). Jpn. 47/21,577 to Toho Beslon; C.A. (1972), *77*: 127259m.

682. Balitrand, G. C., Mison, A., Roget, J., and Tarbouriech, P. (1970). Ger. Offen. 1,938,820 to Rhone-Poulenc; C.A. (1970), *72*: 91409r.

683. Aurich, J., and Peter, E. (1967). Ger. Offen. 1,669,565 to VEB Chemiefaserkombinat Friedrich Engels; C.A. (1969), *70*: 38807b.

684. Caldwell, J. R., and Dannelly, C. C. (1965). Fr. 1,394,374 to Eastman Kodak; C.A. (1966), *64*: 2235f.

685. Beniska, J., and Staudner, E. (1973). *J. Polym. Sci. Polym. Symp.*, *42*: 429.

686. Czajlik, I., Földes-Berezsnich, T., Tüdös, F., and Szakacs, S. (1978). *Eur. Polym. J.*, *14*: 1059.

687. Czajlik, I., Földes-Berezsnich, T., Tüdös, F., and Vertes, E. (1980). *J. Macromol. Sci. Chem. A.*, *14*: 1243.

688. White, E. F. T., and Zissel, M. J. (1963). *J. Polym. Sci. Part A*, *1*: 2189.

689. Ulbricht, J. (1959). *Faserforsch. Textiltech.*, *10*: 166.

690. Bamford, C. H., Jenkins, A. D., and Johnston, R. (1957). *Proc. Roy. Soc. (London) Ser. A*, *239*: 214.

691. Bamford, C. H., and White, E. F. T. (1956). *Trans. Faraday Soc.*, *52*: 716.

692. Huff, T., and Perry, E. (1963). *J. Polym. Sci. Part A, 1*: 1553.
693. Bamford, C. H., Jenkins, A. D., and Johnston, R. (1959). *Trans. Faraday Soc., 55*: 1451.
694. Thomas, W. M., Gleason, E. H., and Pellon, J. J. (1955). *J. Polym. Sci., 17*: 275.
695. Chen, C. Y., Kuo, J. F., Shieh, S. Y., and Tomt, J. G. (1986). *J. Chinese Inst. Eng., 9*: 259; C.A. (1986), *105*: 115454e.
696. Shumnyi, L. V., Kuznetsova, T. A., Konovalenko, V. V., and Ivanchev, S. S. (1983). *Vysokomol. Soedin. Ser. B., 25*: 759.
697. Mohanty, B., Palit, S. K., and Biswas, M. (1987). *J. Polym. Sci. Polym. Lett. Ed., 25*: 187.
698. Batty, N. S., and Guthrie, J. T. (1980). *J. Appl. Polym. Sci., 25*: 2539.
699. Kubota, H., and Ogiwara, I. (1982). *J. Appl. Polym. Sci., 27*: 2683.
700. Vrancken, A., and Van Eygen, C. (1970). Ger. Offen. 1,950,195 to Algemene Kunstzijde Unie NV; C.A. (1970), *73*: 4815z.
701. Fr. 2,097,570 (1970) to Toray Industries; C.A. (1972), *77*: 127240y.
702. Tamaoki, H., Ueda, F., and Tanaka, H. (1986). Jpn. 61/4706 A2 to Toray Industries; C.A. (1986), *105*: 24806e.
703. Liu, D., Ye, D., Li, G., Liu, D., and Qin, J. (1983). *Wuhan Daxue Xuebao Ziran Kexueban, 4*: 121; C.A. (1984), *101*: 38892j.
704. Kitagawa, H. (1963). *Kobunshi Kagaku, 20*: 5; C.A. (1964), *61*: 1942e.
705. Vidotto, G., Brugnaro, S., and Talamini, G. (1970). *Makromol. Chem., 140*: 263.
706. Peebles, L. H., Jr. (1965). *J. Polym. Sci. Part A, 3*: 341.
707. Peebles, L. H., Jr. (1965). *J. Polym. Sci. Part A, 3*: 353.
708. Peebles, L. H., Jr. (1965). *J. Polym. Sci. Part A, 3*: 361.
709. Otani, T., Kobayashi, T., and Okada, H. (1978). Jpn. 53/30,683 to Mitsubishi Rayon; C.A. (1978), *89*: 148068m.
710. Tokura, N., Matsuda, M., and Yazaki, F. (1960). *Makromol. Chem., 42*: 108.
711. Taniyama, M. K., and Yoshida, M. (1961). Jpn. 11,248 to Toyo Rayon; C.A. (1962), *56*: 3683a.
712. Schmidt, W. G. (1960). Brit. 831,049 to Courtaulds, Ltd.; C.A. (1960), *54*: 15951h.
713. Haneda, T., and Ishihara, M. (1959). Jpn. 9,597 to Toyo Spinning; C.A. (1960), *54*: 8102j.
714. Aitaku, S., and Kobayashi, H. (1970). Jpn. 45/41233 to Asahi; C.A. (1971), *74*: 126720f.
715. Miller, M. L. (1960). U.S. 2,963,457 to American Cyanan-dd; C.A. (1961), *55*: 9953c.
716. Yoshida, M., and Tanouchi, K. (1963). *Kobunshi Kagaku, 20*: 545; C.A. (1964), *60*: 14609f.
717. Ulbricht, J. (1962). *Z. Phys. Chem., 221*: 346.
718. Watanabe, M., and Kiuchi, H. (1962). *J. Polym. Sci., 58*: 103.
719. Koval'chuk, E. P., and Aksiment'eva, E. I. (1981). *Zh. Prikl. Khim. (Leningrad), 54*: 348; C.A. (1981), *94*: 1922780.
720. Wu, J., Li, W., and Zhou, B. (1988). *Huagong, Jixie, 15*: 117; C.A. (1989), *109*: 13596p.
721. Elbing, E., McCarthy, S. J., Coller, B. A. W., and Wilson, J. (1987). *J. Chem. Soc. Faraday Trans. 1, 83*: 657.
722. Aksiment'eva, E. I., Mirkind, L. A., Koval'chuk, E. P., and Bogoslovskii, K. G. (1986). *Electrokhimiya, 22*: 684; C.A. (1986), *105*: 14106n.
723. Misra, S., and Sahu, G. (1983). *J. Macromol. Sci. Chem. Part A, 19*: 129.
724. Qiu, K., and Ma, J. (1985). *Hauaxue Tongbao, 11*: 15; C.A. (1986), *104*: 34402p.
725. Ouchi, T., Nomoto, K., Hosaka, Y., Imoto, M., Nakaya, T., and Iwamoto, T. (1984). *J. Macromol. Sci. Chem. Part A, 21*: 859.
726. Graczyk, T., and Hornof, V. (1985). *J. Macromol. Sci. Chem. Part A, 22*: 1209.
727. Ham, G. E. (1956). *J. Polym. Sci., 21*: 337.
728. Mazzolini, C., Patron, L., Moretti, A., and Campanelli, M. (1970). *Ind. Eng. Chem. Prod. Res. Develop., 9*: 504; C.A. (1971), *74*: 42988m.
729. Antonov, V. N., Belonovskaya, G. P., Darvin, V. V., Glazornitskii, K. L., Gol'tsin, B. E., Koton, M. M., Kulev, E. A., Minkova, R. M., Roskin, E. S., and Rostovskii, E. (1969). U.S.S.R. 256,935; C.A. (1970), *73*: 16152u.

730. Konig, J., and Schiefer, E. (1973). U.S. 3,720,745 to Davy-Ashmore AG; C.A. (1973), *78*: 148875v.

731. Nakanome, I., Takeya, K., and Suzuki, H. (1972). U.S. 3,632,543 to American Cyanamid; C.A. (1972), *76*: 128669m.

732. Hunyar, A., Kubanczyk, M., and Ulbricht, J. (1965). Ger. Offen. 1, 195,051 to Deutsche Akademie der Wissenschaften, Berlin; C.A. (1965), *63*: 5780d.

733. Kitagawa, H., and Izumi, Z. (1968). Ger. Offen. 1,272,546 to Toyo Rayon; C.A. (1968), *69*: 78379z.

734. Krebs, K. W., Engelhardt, H., Mueller, E. H., and Nischk, G. (1970). Ger. Offen 2,032,548 to Bayer AG; C.A. (1972), *76*: 127799s.

735. Szita, J., Unger, O., Marzolph, H., and Nischk, G. (1967). Brit. 1,097,157 to Bayer AG; C.A. (1968), *68*: 40904h.

736. Ulbricht, J. (1961). *Faserforsch. Textiltech.*, *12*: 108.

737. Siclari, F., Calgari, S., Jacuone, R., Cesano, M., and Rossi, P. P. (1969). DOS 1,940,074 to SNIA Viscosa SPA; C.A. (1970), *73*: 26508d.

738. Matyjaszewski, K., Jo, S. M., Paik, H., and Gaynor, S. G. (1997). *Macromolecules, 30*: 6398.

739. Matyjaszewski, K., Jo, S. M., Paik, H., and Shipp, D. A. (1999). *Macromolecules, 32*: 6431.

740. Patten, T. E., and Matyjaszewski, K. (1998). *Adv. Mater., 10*: 901.

741. Matyjaszewski, K., Coca, S., Gaynor, S. G., Nakagawa, Y., and Seong, M. (1998). WO 9801480 (to Carnegie Mellon University). C.A. (1998), *128*: 115391.

742. Hawker, C. J., Benoit, D., Rivera, F. Jr., Chaplinski, V., Nilsen, A., and Braslau, R. (1999). *Polym. Mat. Sci. Eng. 80*: 90.

743. Jo, S. M., Gaynor, S. G., and Matyjaszewski, K. (1996). *Polym. Prepr. (Am. Chem. Soc., Div. Polym. Chem.)*, *37*(2): 272.

744. Jo, S. M., Paik, H., and Matyjaszewski, K. (1997). *Polym. Prepr. (Am. Chem. Soc., Div. Polym. Chem.)*, *38*(1): 697.

745. Ono, H., Hisatani, K., and Kamide, K. (1993). *Polym. J. (Tokyo)*, *25*: 245.

746. Erussalimsky, B. L., and Novoselova, A. V. (1975). *Faserforsch. Textiltech.*, *26*: 293.

747. Ottolenghi, A., Barzakay, S., and Zilkha, A. (1963). *J. Polym. Sci. Part A*, *1*: 3643.

748. Zilkha, A., Neta, P., and Frankel, M. (1960). *J. Chem. Soc.*, 3357.

749. Novoselova, A. V., Orlova, G., Erussalimsky, B. L., Adler, H. J., and Berger, W. (1985). *Acta Polym.*, *36*: 599.

750. Ulbricht, J., and Sourisseau, R. (1961). *Faserforsch. Textiltech.*, *12*: 547.

751. Suling, C., and Logemann, H. (1968). Brit. 1,110,998 to Bayer AG; C.A. (1968), *69*: 3328y.

752. Cundall, R. B., Eley, D. D., and Worrall, J. (1962). *J. Polym. Sci.*, *58*: 869.

753. Raynal, S. (1986). *Eur. Polym. J.*, *22*: 559.

754. Riedel, S., Dreyer, R., Lueck, S., Lehmann, D., and Wunderlich, G. (1982). *Isotopenpraxis*, *18*: 262.

755. Markevich, M. A., Kochetov, E. V., Rangogajec, F., and Enikolopyan, N. S. (1974). *J. Macromol. Sci. Chem. Part A*, *8*: 265.

756. Eisenbach, C. D., Jaacks, V., Schnecko, H., and Kern, W. (1974). *Makromol. Chem.*, *175*: 1329.

757. Malch, G., Dautzenberg, H., Krippner, W., Scheller, D., Fritsche, P., and Berger, W. (1981). *Acta Polym.*, *32*: 758.

758. Berger, W., and Adler, H. J. (1986). *Makromol. Chem. Makromol Symp.*, *3*: 301.

759. Polotskaya, G. A., Kuznetsov, Y. P., Ulyanova, N. N., Beponovskaya, G. P., and Baranovskaya, G. P. (1986). *Acta Polym.*, *37*: 83.

760. Hirao, A. (1997). *Desk Ref. Funct. Polym.*, 19.

761. Nakano, Y., Hisatani, K., and Kamide, K. (1994). *Polym. Int.*, *35*: 207.

762. Sivaram, S., Dhal, P. K., Kashikar, S. P., Khisti, R. S., Shinde, B. M., and Baskaran, D. (1991). *Macromolecules*, *24*: 1697.

763. Ikeda, I., Mitsumoto, S., Yamada, Y., and Suzuki, K. (1991). *Polym. Int.*, *26*: 115.

764. Kaita, S., Otaki, T., Kobayashi, E., Aoshima, S., and Furukawa, J. (1997). *J. Polym. Sci., Part A: Polym. Chem.*, *35*: 2591.
765. Woo, H.-G., Hong, L.-Y., Yang, S.-Y., Kim, B.-H., Kang, H.-G., Chae, H.-N., Choi, J.-Y., Park, J.-H., and Ham, H.-S. (1998). *Bull. Korean Chem. Soc.*, *19*: 580.
766. Okuda, J., Hultzsch, K., and Gepraeges, M. (2000). DE 19,836,819 to BASF AG; C.A. (2000), *132*: 166730.
767. Okuda, J., Hultzsch, K., and Gepraeges, M. (2000). DE 19,836,926 to BASF AG; C.A. (2000), *132*: 152321.
768. Okuda, J., Amor, F., Eberle, T., Hultzsch, K. C., and Spaniol, T. P. (1999). *Polym. Prepr. (Am. Chem. Soc., Div. Polym. Chem.)*, *40*(1): 371.
769. Messere, R., Noels, A. F., Dournel, P., Zandona, N., Breulet, J., and Solvay, S. A. (1996). *Metallocenes '96, Proc. Int. Congr. Metallocene Polym.*, 2nd.: 309.
770. Bargon, J., Hellwge, K. H., and Johnson, U. (1966). *Kolloid Z. Z. Polym.*, *213*: 51.
771. Yamadera, R., and Murano, M. (1967). *J. Polym. Sci. Part A-1*, *5*: 1059.
772. Elias, H. G., Goeldi, P., and Johnson, B. L. (1973). *Adv. Chem. Ser.*, *128*: 21.
773. Murano, M., and Yamadera, R. (1967). *J. Polym. Sci. Polym. Lett. Ed.*, *5*: 333.
774. Matsuzaki, K., Uryu, T., Okada, M., and Sliitoki, H. (1968). *J. Polym. Sci. Polym. Chem. Ed.*, *6*: 1475.
775. Kamide, K., Yamazaki, K., Okajima, K., and Hikichi, K. (1985). *Polym. J.*, *17*: 1291.
776. Matsuzaki, K., Okada, M., and Uryu, T. (1971). *J. Polym. Sci. Polym. Chem. Ed.*, *9*: 1701.
777. Kamide, K., Yamazaki, H., Okajima, K., and Hikichi, K. (1985). *Polym. J.*, *17*: 1233.
778. Balard, H., Fritz, H., and Meybeck, J. (1977). *Makromol. Chem.*, *178*: 2393.
779. Clark, H. G. (1963). *Makromol. Chem.*, *63*: 69.
780. Murano, M., and Yarnadera, N. (1968). *J. Polym. Sci. Part A-1*, *6*: 843.
781. Minagawa, M., Okada, Y., Nouchi, K., Sato, Y., and Yoshii, F. (2000). *Colloid Polym. Sci.*, *278*: 757.
782. Minagawa, M., Hashimoto, K., Shirai, H., Morita, T., and Yoshii, F. (2000). *Colloid Polym. Sci.*, *278*: 352.
783. Minagawa, M., Taira, T., Kondo, K., Yamamoto, S., Sato, E., and Yoshii, F. (2000). *Macromolecules*, *33*: 4526.
784. Ravi, P., and Divakar, S. (1999). *J. Macromol. Sci. – Pure Appl. Chem.*, *A36*: 1935.
785. Peebles, L. H., Jr., Thompson, R. B., Jr., Kirby, J. R., and Gibson, M. E. (1972). *J. Appl. Polym. Sci.*, 16: 3341.
786. Tsuda, J. (1961). *J. Appl. Polym. Sci.*, *5*: 104.
787. Minagawa, M. (1980). *J. Polym. Sci. Polym. Chem. Ed.*, *18*: 2307.
788. LoMonaeo, S., and Patron, L. (1970). Brit. 1,215,351 to Chatillon SA; C.A. (1971), *74*: 43454w.
789. König, J., and Süling, C. (1978). *Angew. Makromol. Chem.*, *71*: 91.
790. König, J., and Süling, C. (1973). Ger. Offen. 2,318,609 to Bayer AG; C.A. (1975), *83*: 81055k.
791. Dennstedt, I. (1963). Ger. Offen. 1,152,262 to Bayer AG; C.A. (1963), *59*: 15403b.
792. Bero, M., and Rosner, T. (1970). *Makromol. Chem.*, *136*: 1.
793. Anthes, H. J. (1958). U.S. 3,255,158 to E. I. du Pont de Nemours Co.; C.A. (1966), *65*: 10687a.
794. Lunney, T. W. (1956). U.S. 3,123,588 to E. I. du Pont de Nemours Co.; C.A. (1964), *60*: 14685h.
795. Sugimori, T., and Nishikawa, S. (1972). *Kobunshi Kagaku*, *29*: 817; C.A. (1973), *78*: 72717x.
796. Tamura, H. (1974, 1975). U.S. 3,915,942 to Mitsubishi Rayon; C.A. (1975), *82*: 141017r.
797. Weber, H., and Sommer, K. (1978). Belg. 860,352 to Benckiser-Knapsack; C.A. (1978), *88*: 191787m.
798. Ohfuka, T., Shinoide, K., and Ichikawa, I. (1965). U.S. 3,432,482 to Asahi Chemical Industries; C.A. (1969), *71*: 50640v.
799. Nakayama, C., Shirode, K., Teiichi, K., Kazuo, J., and Yoshihisa, F. (1960). Jpn. 3136 to Asahi Kasei K, K; C.A. (1963), *58*: 12701c.
800. Patron, L., and LoMonaco, S. (1967). Fr. 1,534,521 to Monsanto; C.A. (1969), *71*: 13539y.

801. Rinkler, H., and Hurm, K. (1972). Ger. Offen. 2,046,092 to Bayer AG; C.A. (1972), *77*: 35249e.
802. Kobashi, T., and Takayi, S. (1986). Ger. Offen. 3,427,272 A1 to Japan Exlan; C.A. (1986), *104*: 169087k.
803. Jpn. 59/191,704 (1984) to Japan Exlan; C.A. (1985), *102*: 63516t.
804. Guess, A. P., and McCaskill, W. B. (1954). U.S. 2,693,462 to E. I. du Pont de Nemours Co.; C.A. (1955), *49*: 2779g.
805. Mallison, W. C. (1955). Brit. 722,4512 to American Cyanamid; C.A. (1955), *49*: 10665g.
806. Mallison, W. C. (1959). U.S. 2,847,405 to American Cyanamid; C.A. (1958), *53*: 2692f.
807. Chaney, D. W. (1951). U.S. 2,537,030 to American Viscose; C.A. (1951), *45*: 3652b.
808. Chaney, D. W. (1951). U.S. 2,537,031 to American Viscose; C.A. (1951), *45*: 4093f.
809. Hunyar, and Reichert, H. (1954). *Faserforsch. Textiltech.*, *5*: 4.
810. Fritsche, P., and Ulbricht, J. (1963). *Faserforsch. Textiltech.*, *14*: 517.
811. Fritsche, P., and Ulbricht, J. (1964). *Faserforsch. Textiltech.*, *15*: 93.
812. Peebles, L. H., Jr. (1973). *J. Appl. Polym. Sci.*, *17*: 113.
813. Gerrens, H., and Stein, D. (1965). *Makromol. Chem.*, *87*: 228.
814. Ito, S., and Plant, O. (1986). *J. Appl. Polym. Sci.*, *31*: 849.
815. Ito, S., and Yoshida, K. (1983). *Kobunshi Ronbunshu*, *40*: 307; C.A. (1983), *99*: 6127n.
816. Thomas, W. M., Gleason, E. H., and Mino, G. (1957). *J. Polym. Sci.*, *24*: 43.
817. Dainton, F. S., and Seaman, P. H. (1959). *J. Polym. Sci.*, *39*: 279.
818. Dainton, F. S., et al. (1959). *J. Polym. Sci.*, *34*: 209.
819. Dainton, F. S., and James, D. G. L. (1959). *J Polym. Sci.*, *39*: 299.
820. Dainton, F. S., and Eaton, R. S. (1959). *J. Polym. Sci.*, *39*: 313.
821. Lenka, S., and Dhal, A. K. (1981). *J. Polym. Sci. Chem. Ed.*, *19*: 2115.
822. Bacon, R. G. R. (1946). *Trans. Faraday Soc.*, *42*: 140.
823. Dainton, F. S. (1948). *J. Phys. Colloid Chem.*, *52*: 490.
824. Evans, M. G., Santappa, M., and Uri, N. (1951). *J. Polym. Sci.*, *7*: 243.
825. Dainton, F. S., and James, D. G. L. (1958). *Trans. Faraday Soc.*, *54*: 649.
826. Saldick, J. (1956). *J. Polym. Sci.*, *19*: 73.
827. Mahadevan, V., and Santappa, M. (1961). *J. Polym. Sci.*, *50*: 361.
828. Kimura, S., and Imoto, M. (1960). *Makromol. Chem.*, *42*: 140.
829. Parts, A. G. (1959). *J. Polym. Sci.*, *37*: 131.
830. Moore, D. E., and Parts, A. G. (1960). *Makromol. Chem.*, *37:* 108.
831. Kern, W., and Quast, H. (1953). *Makromol. Chem.*, *10*: 202.
832. Kolthoff, J. M., and Ferstanding, L. L. (1951). *J. Polym. Sci.*, *6*: 563.
833. Bouizem, Y., Chao, F., Costa, M., Tadjeddine, A., and Lecayon, G. (1984). *J. Electroanal. Chem. Interfacial Electrochem.*, *172*: 101; C.A. (1984), *101*: 152411b.
834. Bouizem, Y., Chao, F., Costa, M., and Tadjeddine, A. (1983). *J. Phys. Colloq. Part C*, *10*: 509; C.A. (1984), *100*: 182101x.
835. Dainton, F. S. (1947). *Nature*, *160*: 268.
836. Benasson, R., and Prevot-Bemas, A. (1956). *J. Chim. Phys.*, *53*: 93.
837. Arthur, J. C., Jr., Denimt, R. J., and Pittman, R. A. (1959). *J. Phys. Chem.*, *63*: 1366.
838. Ishihara, M., and Haneda, T. (1957). *Genshiryoku Shimp. Hobunshu*, *4*: 146; C.A. (1960), *54*: 25954c.
839. Collinson, E., and Dainton, F. S. (1952). *Discussions Faraday Soc.*, *12*: 212.
840. Edgecombe, E. H. C., and Norrish, R. G. W. (1963). *Nature*, *197*: 282.
841. Copestake, T. B., and Uri, N. (1955). *Proc. Royal Soc. (London) Ser. A.*, *252*: 228.
842. Oster, G. (1954). *Nature*, *173*: 300.
843. Watanabe, A., and Koizumi, M. (1961). *Bull. Chem. Soc. Jpn.*, *34*: 1086.
844. Morgan, L. B. (1946). *Trans. Faraday Soc.*, *42*: 169.
845. Smeltz, K. C., and Dryer, E. (1952). *J. Am. Chem. Soc.*, *74*: 623.
846. Nonhebel, D. C., and Waters, W. A. (1957). *Proc. Royal Soc. (London) Ser. A.*, *242*: 16.
847. Parravano, G. (1950). *J. Am. Chem. Soc.*, *72*: 3856.

848. Parravano, G. (1950). *J. Am. Chem. Soc.*, *72*: 5546.
849. Gritsenko, T. M., and Medvedev, S. S. (1956). *Zhur. Fiz, Khim.*, *30*: 1238; C.A. (1957), *51*: 7118e.
850. Baxendale, J. H., Evans, M. G., and Park, G. S. (1946). *Trans. Faraday Soc.*, *42*: 155.
851. Barb, W. G., Baxendale, J. H., George, P., and Hargrave, K. R. (1951). *Trans. Faraday Soc.*, *47*: 462.
852. Roskin, E. S. (1957). *Zh. Prikl. Chem.*, *30*: 1030; C.A. (1957), *51*: 18692h.
853. Palit, S. R., and Biswas, M. (1961). *J. Sci. Ind. Res.*, *20B*: 160; C.A. (1961), *55*: 24090b.
854. Yuguchi, S., and Hoshina, M. (1962). *Kobunshi Kagaku*, *18*: 381; C.A. (1961), *55*: 27945h.
855. Drummond, A. Y., and Waters, W. A. (1953). *J. Chem. Soc.*, 2836.
856. Davis, P., Evans, M. G., and Higginson, W. E. (1951). *J. Chem. Soc.*, 2563.
857. Marvel, C. S., Friedlander, H. Z., Swann, S., Jr., and Inskip, H. K. (1953). *J. Am. Chem. Soc.*, *75*: 3846.
858. Homer, L., and Podschus, G. (1951). *Angew. Chem.*, *63*: 531.
859. Yuguchi, S., and Watanabe, M. (1961). *Kobunshi Kagaku*, *18*: 273; C.A. (1961), *55*: 27945f.
860. Roskin, E. S., et al. (1958). U.S.S.R. 114,584; C.A. (1959), *53*: 11890g.
861. Uri, N. (1952). *Chem. Rev.*, *50*: 375.
862. Korte, S., Neukam, T., and Sueling, C. (1981). *Angew. Makromol. Chem.*, *98*: 113.
863. Sueling, C., Korte, S., and Neukam, T. (1978). Ger. Offen. 2,833,143 to Bayer AG; C.A. (1980), *92*: 182070c.
864. Sueling, C., Neukam, T., and Korte, S. (1978). Ger. Offen. 2,843,157 to Bayer AG; C.A. (1980), *93*: 72592n.
865. König, J., and Korte, S. (1958). DOS 3,342,694 to Bayer AG; C.A. (1985), *103*: 105449b.
866. Thompson, M. W. (1968). Fr. 1,571,212 to ICI; C.A. (1970), *73*: 110523n.
867. Kobashi, T., and Naka, H. (1984). U.S. 4,546,146 to Japan Exlan; C.A. (1985), *103*: 142850w.
868. Kobashi, T., and Naka, H. (1985). Jpn. 60/133,012 A2 to Japan Exlan, C.A. (1986), *104*: 169076f.
869. Yamaskita, M., and Kunii, N. (1986). Jpn. 61/275,317 A2 to Asahi Glass; C.A. (1987), *107*: 40569b.
870. Taylor, G. A., and Hoy, K. L. (1985). Eur. Appl. EP 131,209 A2 to Union Carbide; C.A. (1987), *106*: 85678g.
871. Hoffman, D. K., and Arends, C. B. (1985). PCT Int. Appl. WO 8500/610 to Dow Chemical; C.A. (1985), *103*: 7285a.
872. Imoto, M., and Takatsugi, H. (1957). *Makromol. Chem.*, *23*: 119.
873. Imoto, M., and Takatsugi, H. (1958). *J. Polym. Sci.*, *31*: 195.
874. Barrett, K. E. J. (1975). *Dispersion Polymerization in Organic Media*, Interscience, New York, pp. 108, 231.
875. Elbing, E., McCarthy, S., Snowdon, S., Coller, B. A. W., and Wilson, J. R. (1986). *J. Chem. Soc. Faraday Trans.*, *82*: 943.
876. Everett, D. H., and Rojas, F. (1988). *J. Chem. Soc. Faraday Trans.*, *84*: 1455.
877. Sandler, S. R., and Karo, W. (1974). *Polymer Synthesis*, Vol. 1, Academic Press, New York, p. 302.
878. Ono, H., EM, J., and Fuji, A. (1975). *Z Phys. Chem.*, *79*: 2020.
879. Izumi, Z., Kiuchi, H., and Watanabe, M. (1967). *J. Polym. Sci. Part A-1*, *5*: 455.
880. Izumi, Z. (1967). *J. Polym. Sci. Part A-1*, *5*: 469.
881. Morris, C. E. M., and Parts, A. G. (1976). *Makromol. Chem.*, *177*: 1433.
882. Kozuka, K. T., Kobayashi, S., and Watanabe, A. (1973). Ger. Offen. 2,347,438 to Kanegafuchi Kagaku; C.A. (1977), *86*: 30277.
883. Kobashi, T., Shiota, H., and Umetani, H. (1977). Ger. Offen 2,709,503 to Japan Exlan; C.A. (1977), *87*: 168791.
884. Gershberg, D. (1965). *New Chem. Eng. Prob. Util Water Proc. Symp. Issue*, *3*: 4; C.A. (1968), *68*: 87614.

885. O'Neill, T., and Stannett, V. (1974). *J. Macromol. Sci. Chem. Part A*, 8: 949.

886. Nomura, N. (1982). *Emulsion Polymerization* (Piirma, I., ed.), Academic Press, New York, p. 191.

887. Thompson, R. W., and Stevens, J. D. (1975). *Chem. Eng. Sci.*, 30: 663.

888. Ugelstad, J. (1981). *Pure Appl. Chem.*, 53: 323.

889. Turner, J. J. (1975). U.S. Pat. 3,873,508 to E. I. du Pont de Nemours Co.; C.A. (1975), 83: 29055w.

890. Saka, H., Hamada, S., Inoue, H., and Hosaka, S. (1973). Jpn. 90,380 to Toray Industries; C.A. (1974), 81: 38755k.

891. Saka, H., Hamada, S., Kitagawa, H., and Mukoyama, E. (1974). Jpn. 49/14,546 to Toray Industries; C.A. (1975), 82: 5279a.

892. Mata, I., Sakai, I., and Tomita, M. (1973). Jpn. 48/19,821 to Unitika; C.A. (1973), 79: 32669x.

893. Bechtold, M. F. (1961). U.S. 2,972,511 to E. I. du Pont de Nemours Co.; C.A. (1961), 55: 12876h.

894. Saxton, R. L. (1976). U.S. 3,963,687 to E. I. du Pont de Nemours Co.; C.A. (1976), 85: 63664v.

895. König, J., Süling, C., and Böhmke, G. (1977). Ger. Offen. 2,604,630 to Bayer AG; C.A. (1977), 87: 119170s.

896. König, J., and Süling, C. (1974). Ger. Offen. 2,300,713 to Bayer AG; C.A. (1975), 82: 48491v.

897. Iwata, H., Otani, T., and Kobayashi, T. (1978). Jpn. Pat. 104,689 to Mitsubishi Rayon; C.A. (1979), 90: 73213q.

898. Jpn. Pat. 57/42,703 (1982) to Mitsubishi Rayon; C.A. (1982), 97: 39525c.

899. Kohashi, N., Shioda, H., and Umetani, H. (1977). Jpn. 107,045 to Japan Exian; C.A. (1978), 88: 51891r.

900. Kobashi, T., and Masuhara, K. (1976). Ger. Offen. 2,550,871 to Japan Exlan; C.A. (1976), 85: 95671m.

901. Shichijo, Y., Sato, H., Iwasa, T., and Uchida, Y. (1970). Ger. Offen. 2,154,676 to Asahi Chem.; C.A. (1973), 77: 63216g.

902. Shriver, L. C., and Fremon, G. H. (1947). U.S. 2,420,330 to Carbide and Carbon Chemicals Corp.; C.A. (1947), 41: 5343g.

903. Lindemann, M. K., and Deacon, K. (1986). U.S. 4,616,057Ato Sun Chemical Corp.; C.A. (1986), 105: 227574a.

904. Kawase, S. (1986). Jpn. 61/14,201 to Soken Chemical Engineering Co.; C.A. (1982), 105: 43555p.

905. Lupu, A., Butaciu, F., and Zenide, F. (1973). Rum. 55,795 to Institutul de Cercetari Chimice ICECHIM; C.A. (1974), 81: 79226g.

906. Kukushkina, N. P., Belogorodskaya, K. V., Nikolaev, A. F., Lavrov, N. A., and Soloveva, V. P. (1982). U.S.S.R. SU 897,775 Al to Leningrad Technological Institute; C.A. (1982), 97: 39525c.

907. Elbing, E., Tan, W. K., Lyons, C. J., Coller, B. A. W., and Wilson, I. R. (1987). *J. Chem. Soc. Faraday Trans. 1*, 83: 645.

908. Omi, S., Negishi, W. M., Fujitake, M., and Iso, M. (1986). *Zairyo Gijutsu*, 4: 130; C.A. (1987), 106: 885149k.

909. Patsiga, R. A., Lerdthusnee, W., and Marawi, I. (1984). *Ind. Eng. Chem. Prod. Res. Div.*, 23: 238; C.A. (1984), 100: 192347w.

910. Patsiga, R. A., Lerdthusnee, W., and Marawi, I. (1983). *Org. Coat. Appl. Polym. Sci. Proc.*, 48: 790; C.A. (1984), 100: 157030r.

911. Polikarpov, V. V., and Lukhovitskii, V. I. (1982). *Proc. Tihany Symp. Radiat. Chem.*, 5: 751; C.A. (1983), 99: 140495j.

912. *Ind. Chem. News* (1986), 7: 10.

913. *Chem. Eng. News* (1985), 27 (June 10).

914. Norton, J. A. (1942). *Chem. Rev.*, 31: 319.

915. Flett, L. H., and Gardner, W. H. (1952). *Maleic Anhydride Derivatives*, Wiley, New York.

916. Flett, L. H., and Gardner, W. H. (1953). *Encyclopedia of Chemical Technology*, Vol. 8 (Standem, A., and Scott, J., eds.), Interscience, New York, pp. 680–696.
917. Brownell, G. L. (1964). *Encyclopedia of Polymer Science and Technology*, Vol. I (Mark, H. F., and Gaylord, N. G., eds.), Wiley, New York, pp. 67–95.
918. Trivedi, B. C., and Culbertson, B. M. (1982). *Maleic Anhydride*, Plenum Press, New York, p. 871.
919. Sackmann, G. (1987). *Houben-Weyl: Methoden der organischen Chemie*, Vol. E20/2 (Bartl, H., and Falbe, I., eds.), Georg Thieme, Stuttgart, pp. 1234–1255.
920. Culbertson, B. M. (1987). *Encyclopedia of Science and Engineering*, Vol. 9 (Mark, H. F., and Bikales, N. M., eds.), Wiley, New York, pp. 225–294.
921. Cowie, J. M. G. (1989), *Comprehensive Polymer Science*, Vol. 4 (Eastmont, G. G., Ledwith, A., Rasso, S., and Signalt, P., eds.), Pergamon Press, Oxford, pp. 385–394, 409–419.
922. Lang, J. L., Pavelich, W. A., and Clarey, H. D. (1961). *J. Polym. Sci.*, *55*: 31.
923. Bartlett, P. D., and Nozaki, K. (1946). *J. Am. Chem. Soc.*, *68*: 1495.
924. Nagahiro, I., Nishihara, K., and Sakota, N. (1974). *J. Polym. Sci.*, *12*: 785.
925. Ouchi, T., and Azuma, T. (1982). *Eur. Polym. J.*, *18*: 809.
926. Bryce-Smith, D., Gilbert, A., and Vickery, B. (1962). *Chem. Ind.*: 2060.
927. Pat. 1,073,323 (1967) to Procter and Gamble Co.; C.A. (1967), *67*: 64903n.
928. Gaylord, N. G., and Maiti, S. (1973). *J. Polym. Sci. Polym. Chem. Ed.*, *11*: 253.
929. Lancelot, C. J., Blumbergs, H. H., and Machellar, D. G. (1971). Ger. Offen. 2,154,510 to FMC Corp.; C.A. (1973), *79*: 54046.
930. Heuck, C., and Lederer, M. (1969). U.S. 3,457,240 to Hoechst AG; C.A. (1969), *71*: 71220g.
931. Merijan, A. (1968). U.S. 3,385,834 to GAF Co.; C.A. (1968), *69*: 19805n.
932. Fr. Pat. 1,544,728 (1968) to Kao Soap Co.; C.A. (1969), *71*: 50690m.
933. Pemer, J. (1981). U.S. Pat. 4,260,724 to BASF AG; C.A. (1980), *93*: 99246.
934. Richardson, N. J., Thomas, I., and Miles, P. (1974). Ger. Offen. 2,405,284 to CIBA Geigy AG; C.A. (1975), *82*: 44022f.
935. Birell, R. N., and Royle, N. E. (1978). Ger. 2,732,628 to CIBA Geigy AG; C.A. (1978), *88*: 106219k.
936. Joshi, R. M. (1962). *Makromol. Chem.*, *53*: 33; *55*: 35.
937. Kellou, M. S., and Jenner, G. (1976). *Eur. Polym. J.*, *12*: 883.
938. Deb, P. (1975). *Eur. Polym. J.*, *11*: 31.
939. Braun, D., EI Sayed, I. A., and Pomakis, J. (1969). *Makromol. Chem.*, *124*: 249.
940. Shantarovich, P. S., and Sosnovskoya, L. N. (1970). *Jzv. Akad Nauk. SSSR Ser. Khim.*, *2*: 358; C.A. (1970), *73*: 15345x.
941. Kuliev, V. B., Shakhmaliev, A. M., Novichkova, L. M., and Rostovskii, E. N. (1971). *Uch. Zap. Azerb. Inst. Nefti Khim.*, *9*: 70; C.A. (1974), *81*: 106049k.
942. Sharabash, M. M., and Guile, R. L. (1976). *J. Macromol. Sci. Part A*, *10*: 1017.
943. Bevington, J. C., and Johnson, M. (1966). *Eur. Polym. J.*, *2*: 185.
944. Bacski, R. (1976). *C*, *14*: 1797.
945. Regel, W., and Schneider, C. (1981). *Makromol. Chem.*, *182*: 237.
946. Nakagawa, K., and Nakada, T. (1968). Jpn. 21,747 to Shionogi Co.; C.A. (1969). *70*: 68903x.
947. De Roover, B., Devaux, J., and Legras, R. (1996). *J. Polym. Sci. Part A*, *34*: 1195.
948. Hess, A. N., and Chin, Y.-P., (1996). *Colloids Surfaces A: Physicochem. Eng. Aspects*, *107*: 141.
949. Braun, D., Aziz El Sayed, I. A., and Pomakis, J. (1969). *Makromol. Chem.*, *124*: 249.
950. Joshi, R. M. (1962). *Makromol. Chem.*, *53*: 33.
951. Bhadani, S. N. and Prasad, J. (1977). *Makromol. Chem.*, *178*: 187.
952. Zweifel, H., and Voelker, T. (1973). *Makromol. Chem.*, *170*: 141.
953. Rittenberg, D., and Ponticorvo, L. (1960). *Proc. Natl. Acad. Sci. USA*, *46*: 822.
954. Braun, D., and Pomakis, J. (1974). *Makromol. Chem.*, *175*: 1411.
955. Schopov, V. I. (1970). *Makromol. Chem.*, *137:* 293.
956. Hallensleben, M. L. (1971). *Makromol. Chem.*, *142*: 303.
957. Zweifel, H., Löliger, J., and Völker, T. (1972). *Makromol. Chem.*, *153*: 125.

958. Bhadani, S. N., and Prasad, J. (1977). *Makromol. Chem.*, *178*: 1651.
959. Papisov, I. M., Garina, E. S., Kabanov, V. A., and Kargin, V. A. (1969). *Vysokomol. Soedin. Ser. B, 11*: 614.
960. Araki, M., Kato, K., Kayanagi, T., and Machida, S. (1977). *J. Macromol. Sci. Chem. Part A, 11*: 1039.
961. Wurm, H., Hallensleben, M. L., and Regel, W. (1979). *Makromol. Chem.*, *180*: 1589.
962. Nieuwhof, R. P., Marcelis, A. T. M., and Sudhölter, E. J. R. (1999). *Macromolecules, 32*: 1398.
963. Lee, K. K., Jung, J. C., and Jhon, M. S. (1998). *Polymer, 39*: 4457.
964. Kim, J.-B., Lee, B.-W., Kang, J.-S., Seo, D.-C., and Roh, C.-H. (1999). *Polymer, 40*: 7423.
965. Kim, J.-B., Lee, B.-W., Yun, H.-J., and Kwon, Y.-G. (2000). *Chem. Lett.*, *4*: 414.
966. Rushkin, I. L., and Houlihan, F. M. (1999). *Proc. SPIE-Int. Soc. Opt. Eng.*, *3678*: 44.
967. Douki, K., Kajita, T., and Shimokawa, T. (2000). *Proc. SPIE-Int. Soc. Opt. Eng.*, *3999*: 1128.
968. Kim, H.-W., Lee, S.-H., Kwon, K.-Y., Jung, D.-W., Lee, S., Yoon, K.-S., Choi, S.-J., Woo, S.-G., and Moon, J.-T. (2000). *Proc. SPIE-Int. Soc. Opt. Eng.*, *3999*: 1100.
969. Choi, S.-J., Kim, H.-W., Woo, S.-G., and Moon, J.-T. (2000). *Proc. SPIE-Int. Soc. Opt. Eng.*, *3999*: 54.
970. Cowie, J. M. G. (1985). *Alternating Copolymers*, Plenum Press, New York, London.
971. Rzaev, Z. M. O. (2000). *Prog. Polym. Sci.*, *25*: 163.
972. Kokubo, T., Iwatsuki, S., and Yamashita, Y. (1969). *Makromol. Chem.*, *123*: 256.
973. Gaylord, N. G., and Patnaik, B. (1970). *J. Polym. Sci. Part B, 8*: 549.
974. Otsu, T., and Inoue, H. (1969). *Makromol. Chem.*, *128*: 31.
975. Otsu, T., and Inoue, H. (1969). *Makromol. Chem.*, *128*: 31.
976. Matsuda, M., and Abe, K. (1968). *J. Polym. Sci. Polym. Chem. Ed.*, *6*: 1441.
977. Seymour, R. B., and Gamer, D. P. (1976). *J. Coat. Technol*, *48*: 41.
978. Alfrey, T. J., Bohrer, J. J., and Mark, H. (1952). *Copolymerization*, Wiley-Interscience, New York.
979. Kawai, W. (1968). *J. Polym. Sci. Part A-1, 6*: 1945.
980. Coleman, L. E., Bork, J. F., Wyman, D. P., and Hoke, D. I. (1965). *J. Polym. Sci.*, *3*: 1601.
981. Jaoni, I., Fles, D., and Vukovic, R. (1982). *J. Polym. Sci. Polym. Chem. Ed.*, *20*: 977.
982. Tuschida, E., and Tomono, T. (1971). *Makromol. Chem.*, *141*: 265.
983. Kawai, W. (1967). *J. Polym. Sci. Part B, 5*: 1103.
984. Iwatsuki, S., Itoh, T., Shimizu, M., and Ishikawa, S. (1983). *Macromolecules, 16*: 1407.
985. Rzaev, Z. M., Bryksina, L. V., Kyazimov, S. K., and Sadykh-Zade, S. I. (1972). *Vysokomol. Soedin. Ser. A, 14*: 259.
986. Rzaev, Z. M., Sadykh-Zade, S. I., and Bryksina, L. V. (1974). *Vysokomol. Soedin. Ser. B., 16*: 8.
987. Rzaev, Z. M., Kyazimov, S. K., and Sadykh-Zade, S. I. (1975). *Epoksidnye Monomery Epoksidnye Snioly.*, 175; C.A. (1972). *76*: 154189z.
988. Rzaev, Z. M., Bryksina, L. V., and Dhzafarov, R. V. (1975). *Vysokomol. Soedin. Ser. A, 17*: 2371.
989. Shantarovich, P. S., Sosnovskaya, L. N., and Potapova, T. P. (1970). *Dokl. Akad Nauk. SR, 191*: 100; C.A. (1970). *73*: 15333s.
990. Dzhafarov, R. V., and Rzaev, Z. M. (1976). *Azerb. Khim.*, *1*: 70; C.A. (1978). *79*: 79289p.
991. Kojima, K., Iwabuchi, S., Nakahira, T., Uchiyama, T., and Koshiyama, Y. (1976). *J. Polym. Sci. Polym. Lett. d., 14*: 143.
992. Iwatsuki, S., and Itoh, T. (1979). *Makromol. Chem.*, *180*: 663.
993. Iwatsuki, S., and Yamashita, Y. (1967). *Makromol. Chem.*, *104*: 263.
994. Kokubo, T., Iwatsuki, S., and Yamashita, Y. (1968). *Macromolecules, 1*: 482.
995. Inoue, H., and Otsu, T. (1972). *Makromol. Chem.*, *153*: 37.
996. Nikolaev, A. F., and Andreeva, M. A. (1969). *Vysokomol. Soedin. Ser. B, 11*: 300.
997. Nikolaev, A. F., and Andreeva, M. A. (1967). *Vysokomol. Soedin. Ser. A, 9*: 1720.

998. Iwatsuki, S., Amano, S., and Yamashita, Y. (1967). *Kogyo Kagaku Zasshi*, *70*: 2027; C.A. (1968). *68*: 86623u.

999. Iwatsuki, S., Itoh, T., and Hiraiva, A. (1981). *Makromol. Chem.*, *182*: 2161.

1000. Iwatsuki, S., Itoh, T., and Sato, T. (1986). *Macromolecules*, *19*: 1800.

1001. Iwatsuki, S., and Yamashita, Y. (1965). *Kogyo Kagaku Zasshi*, *68*: 1138; C.A. (1966). *64*: 6764c.

1002. Kokubo, T., Iwatsuki, S., and Yamashita, Y. (1969). *Makromol. Chem.*, *123*: 256.

1003. Ko, T., Iwatsuki, S., and Yamashita, Y. (1970). *Macromolecules*, *3*: 518.

1004. Iwatsuki, S., and Yamashita, Y. (1967). *J. Polym. Sci. Part A-1*, *5*: 1753.

1005. Fujimori, K., and Wickramasinghe, N. A. (1980). *Aust. J. Chem.*, *33*: 1289.

1006. Raetzsch, M., Borman, G., and Ruetzel, S. (1976). *Plaste Kautschuk*, *23*: 239; C.A. (1976). *85*: 6148f.

1007. Sander, B., Klug, P., and Groebe, V. (1981). *Acta. Polym.*, *32*: 266.

1008. Sitninoescu, C., Asondei, N., and Negulescu, I. (1973). *Acta. Chim. Budapest*, *76*: 319.

1009. Yamaguchi, T., Nagai, K., and Itaba-Shi, O. (1973). *Kobusshi Hagaku*, *30*: 464; C.A. (1974). *80*: 71156t.

1010. Yamaguchi, T., and Nagai, K. (1971). *Kobusshi Kagaku*, *28*: 725; C.A. (1971). *75*: 49803b.

1011. Tomescu, M., and Pusztai, K. (1974). *Mater. Plast. Bucharest*, *11*: 178; C.A. (1975). *82*: 17420c.

1012. Furukawa, J., Kobayashi, E., and Nakamura, M. (1974). *J. Polym. Sci. Part A-1*, *12*: 2789.

1013. Konsulov, V., Grozeva, Z., Piruleva, Y., and Sefanova, R. (1985). *Khim. Ind. Sofia*, *57*: 245; C.A. (1986). *104*: 69489z.

1014. Cardon, A., Goethals, A. (1971). *J. Macromol. Sci. Chem.*, *A5*: 1021.

1015. Ishigaki, I., Vatanabe, Y., Ito, A. and Hayashi, H. J. (1978). *J. Macromol. Sci. Chem.*, *A12*: 837.

1016. Ragab, Y. A., (1990). *J. Polym. Sci. Polym. Lett.*, *C28*: 289.

1017. Ragab, Y. A. and Butler G. B. (1981). *J. Polym. Sci. Polym. Chem. Ed.*, *19*: 1175.

1018. Gaylord, N. G., Maiti, S., Patnaik, B. and Takahashi, A. (1972). *J. Macromol. Sci. Chem.*, *A6*: 1459.

1019. Romani, M. N., and Weali, K. E. (1973). *Eur. Polym. J.*, *5*: 389.

1020. Sastre, R., Mateo, J. L., and Acosta, J. (1978). *Angew. Makromol Chem.*, *73*: 25.

1021. Tomescu, M., and Cretu, S. F. (1982). *BuL Inst. Politeh. Bucuresti Ser. Chim. Metal*, *44*: 67.

1022. Yakolewa, M. K., Sheinker, A. P., and Abkiu, A. D. (1969). *Vysokomol. Soedin. Ser. A*, *11*: 282.

1023. Mayo, F. R. Lewis, F. M., and Walling, C. (1948). *J. Am. Chem. Soc.*, *70*: 1529.

1024. Abramyan, R. K., Anisimova, N. N., and Fillipychev, G. F. (1971). *Uch. Zap. Yaroslavok-Teknol. Inst.*, *27*: 135; C.A. (1973), *78*: 72704r.

1025. Niwa, M., Genno, H., and Noma, K. (1968). *Doshisha Daigaku Rikogaku Kenkya Kokoku*, *9*: 199; C.A. (1969), *70*: 115591f.

1026. Imoto, E., and Horiuchi, H. (1951). *Chem. High Polym. Jpn.*, *8*: 463.

1027. Oota, T., Otsu, T., and Imoto, M. (1968). *Kogyo Kagaku Zasshi*, *71*: 736; C.A. (1968), *69*: 67781g.

1028. Kato, M., and Nakano, Y. (1972). *J. Polym. Sci. Part B*, *10*: 157.

1029. Stoyachenko, I., Gergiev, G. S., Golubev, V. B., Zulov, W. P., and Kabanov, V. A. (1973). *Vysokomol. Soedin. Ser. B*, *15*: 1899.

1030. Murahashi, S., Nozakura, S., and Yasufuku, K. (1965). *Bull. Chem. Soc. Jpn.*, *38*: 2082.

1031. Murahashi, S., Nozakura, S., and Yasufuku, K. (1966). *Bull. Chem. Soc. Jpn.*, *39*: 1338.

1032. Kellou, M., and Jenner, G. (1979). *Makromol Chem.*, *180*: 1687.

1033. Bauduin, G., Boutevin, B., and Malek, F. (1992). *Eur. Polym. J.*, *28*: 1237.

1034. Umarova, M. A., and Tashk, T. (1974). *Politekh. Inst.*, *95*: 132.

1035. Ohta, T. (1965). *Nippon Kagaku Zashi*, *86*: 850.

1036. Nishikubo, T., Ichijo, T., Imaura, M., Tado, T., and Takaoka, T. (1974). *Nippon Kagaku Kaishi*, *1*: 45; C.A. (1969), *71*: 90715e.

1037. Ouchi, T., and Aso, M. (1979). *J. Polym. Sci. Polym. Chem. Ed.*, *17*: 2639.
1038. Machi, S., Sakai, T., Gotoda, M., and Kagiya, T. (1966). *J. Polym. Sci. Part A-1*, *4*: 821.
1039. Siddiqui, R. A., and Quddus, M. A. (1971). *Pakistan J. Sci. Ind. Res.*, *14*: 197; C.A. (1972), *76*: 100134j.
1040. Isuchida, E., Shimomura, T., Fujimori, K., Ohtani, Y., and Shiwohara, T. (1967). *Kogyo Kagaku Zasshi*, *70*: 566, 666.
1041. Murahashi, S., Nozakura, S., Emura, K., and Yasufuku, K. (1966). *Kobunshi Kagaku*, *23*: 361; C.A. (1965), *63*: 16477h.
1042. Kokubo, T., Iwatsuki, S., and Yamashita, Y. (1970). *Macromolecules*, *3*: 518.
1043. Nyitrai, K., and Hardy, G. (1967). *Acta Chim. Acad. Sci. Hung.*, *52*: 99; C.A. (1967), *67*: 54473.
1044. Nyitrai, K., and Hardy, G. (1960). *Magy. Kem. Folyoirat*, *72*: 491; C.A. (1961), *55*: 26507a.
1045. Braun, D., and Pomakis, J. (1974). *Eur. Polym. J.*, *10*: 357.
1046. Gyula, H., Kavolay, N., and Czaba, C. (1966). *Magy. Kem. Folyoirat*, *72*: 517.
1047. Hardy, G., Nyitrai, K., and Sarlo, C. (1968). *Periodica Polytech. Chem. Eng.*, *12*: 13.
1048. Hardy, G., and Nitrai, K. (1966). *Proceedings of the 2nd Tihany Symposium on Radiation Chemistry*, Tihany, Hungary, p. 613.
1049. Ismailov, A. G., Rzaev, Z. M., Movsum-Zade, A. A., Rasulov, N. S., and Bryksina, L. V. (1973). *Azerb. Khim. Zh.*, *3*: 114; C.A. (1974), *81*: 10103d.
1050. Abayasekara, D. R., and Ottenbrite, R. M. (1985). *Polym. Prepr. Am. Chem. Soc. Div. Polym. Chem.*, *26*: 285.
1051. Ishigaki, I., Watanaba, Y., Itoh, A., and Mayashi, K. (1978). *J. Macromol. Sci. Chem. Part A*, *12*: 837.
1052. Chang, H. C., Yin, T., Tsao, W. H., and Feng, H. T. (1965). *Koo Fen Tzu Tsung Hsun*, *7*: 415; C.A. (1966), *64*: 14352b.
1053. Novikov, S. N., Pebalk, D. V., Vasyanina, K. L., and Pravednikov, A. N. (1976). *Vysokomol Soedin. Ser. A.*, *18*: 2333.
1054. Frank, H. P. (1968). *Makromol. Chem.*, *114*: 113.
1055. Otsu, T., Shimizu, A., and Imoto, M. (1965). *J. Polym. Sci. Part A*, *3*: 615.
1056. Akihiko, S., and Takayuki, O. (1964). *Kogyo Kagaku Zasshi*, *67*: 966.
1057. Joshi, R. M. (1962). *Makromol. Chem.*, *53*: 33.
1058. Lewis, F. M., and Mayo, F. R. (1948). *J. Am. Chem. Soc.*, *70*: 1534.
1059. Kellou, M., and Jenner, G. (1977). *Eur. Polym. J.*, *13*: 9.
1060. Bammford, C. H., and Bark, W. G. (1953). *Discuss. Faraday Soc.*, *14*: 208.
1061. Chapman, C. B., and Valentine, L. (1959). *J. Polym. Sci.*, *34*: 319.
1062. Sheremeteva, T. V., and Larina, G. N. (1965). *Dokl. Akad. Nauk. SSSR*, *162*: 1323; C.A. (1965), *63*: 11713c.
1063. Noma, K., Niwa, M., and Iwasaki, K. (1963). *Kobunshi Kagaku*, *20*: 646; C.A. (1964), *60*: 14618c.
1064. Baramboim, N. K., and Antonova, L. A. (1976). *VysokomoL Soedin. Ser. A.*, *18*: 675.
1065. Price, C. C., and Walsh, J. G. (1951). *J. Polym. Sci.*, *6*: 239.
1066. Zerroukhi, A., Cincu, C., and Montheard, J. P. (1999). *J. Appl. Polym. Sci.*, *71*: 1447.
1067. Imoto, E., and Horiuchi, H,. (1951). *Chem. High Polym. Jpn.*, *8*: 463.
1068. Noma, K., Utsumi, H., and Niwa, M. (1969). *Kobunshi Kagaku*, *26*: 889.
1069. Smirnov, A. I., Deryabina, G. I., Kalabina, A. V., Ratovskii, G. V., and Beloborodov, V. L. (1975). *Vysokomol. Soedin. Ser. B*, *17*: 828.
1070. Tsuchida, E., and Kawagoe, T. (1968). *Enka Biniru To Porima*, *8*: 21; C.A. (1968), *69*: 36467w.
1071. Pittman, C. V., Voges, R. L., and Elder, J. (1971). *J. Polym. Sci. Part B*, *9*: 191.
1072. Hopff, H., and Becker, G. (1970). *Makromol. Chem.*, *133*: 1.
1073. Semenova, A. S., Leitman, M. I., Stefanovich, L. G., and Shalaeva, L. F. (1972). *Vysokomol. Soedin. Ser. A.*, *14*: 2102.
1074. Kelen, T., and Tüdös, F. (1975). *J. Macromol. Sci. Chem. Part A*, *9*: 1.

1075. Gilbert, H., Miller, F. F., Averill, S. J., Carlson, E. J., Folt, V. L., Heller, H. J. Stewart, F. D., Schmidt, R. F., and Trumbull, H. L. (1956). *J. Am. Chem. Soc., 78*: 1669.
1076. Otsu, T., Minai, H., and Toyoda, N. (1985). *Makromol Chem. Suppl., 12*: 133.
1077. Otsu, T., Ito, O., and Toyoda, N. (1983). *J. Macromol. Sci. Chem. Part A, 19*: 27.
1078. Bengough, W. I., Park, G. B., and Young, R. A. (1975). *Eur. Polym. J., 11*: 305.
1079. Murata, Y., and Hirano, J. (1985). *Chem. Econ. Eng. Rev., 17*: 18; C.A. (1986), *104*: 149806r.
1080. Cubbon, R. C. P. (1965). *Polymer, 6*: 419.
1081. Yamada, M., Takase, I., and Kobayashi, M. (1972). *Kobunshi Kagaku, 29*: 144; C.A. (1972), 7720159b.
1082. Hagiwara, T., Mizota, J., Hamana, H., and Narita, T. (1985). *Makromol. Chem. Rapid Commun., 6*: 169.
1083. Shima, K., and Yamamoto, R. (1969). *Nippon Kagaku Zasshi, 90*: 1168; C.A. (1969), *72*: 67616k.
1084. Florianczyk, T., Sullivan, C., Janovic, Z., and Vogl, O. (1981). *Polym. Bull., 5*: 521.
1085. Matsumoto, A., Kubota, T., and Otsu, T. (1990). *Macromolecules, 23*: 4508.
1086. Oishi, T., Onimura, K., Isobe, Y., Yanagihara, H., and Tsutsumi, H. (2000). *J. Polym. Sci. Part A, 38*: 310.
1087. Ameduri, B., Boutevin, B., and Malek, F. (1994). *J. Polym. Sci. Part A, 32*: 3161.
1088. Barrales-Rienda, J. M., Gonzales de la Campa, J. I., and Ramos, J. G. (1977). *J. Macromol. Sci. Chem., A11*: 267.
1089. Mohamed, A. A., Jebrael, F. H., and Elsabeé, M. Z. (1986). *Macromolecules, 19*: 32.
1090. Prementine G. S., Jones, S. A., and Tirrell, D. A. (1989). *Macromolecules, 22*: 770.
1091. Matsumoto, A., Kubota, T., and Otsu, T. (1990). *Macromolecules, 23*: 4508.
1092. Otsu, T., Matsumoto, A., and Kubota, T. (1991). *Polym. Int., 25*: 179.
1093. Rzaev, Z. M., and Dzhafarov, R. V. (1983). *Azerb. Khim. Zh., 6*: 89.
1094. Rzaev, Z. M., and Dzhafarov, R. V. (1984). *Chem. Abstr., 101*: 231079c.
1095. Mamedova, S. G., Rzaev, Z. M., Medyakova, L. V., Rustamova, F. B., and Askerova, N. A. (1987). *Polym. Sci. USSR, A29*: 2111.
1096. Fles, D. D., Vukovic, R., and Ranogajec, F. (1989). *J. Polym. Sci. Polym. Chem., A27*: 3227.
1097. Fles, D. D., Vukovic R., and Kuresevic, V. J. (1991). *J. Macromol. Sci. Chem., A28*: 977.
1098. Olson, K. G., and Butler, G. B. (1984). *Macromolecules, 17*: 2486.
1099. Rasulov, N. S., Medyakova, L. V., Kuliyeva, E. Y., Rzaev, Z. M., and Zubov, V. P. (1986). *Polym. Sci. USSR, A28*: 2887.
1100. Rzaev, Z. M., Rasulov, N. S., Medyakova, L. V., Lezgiyev, N. Y., Kulieva, E. Y., and Zubov, V. P. (1987). *Polym. Sci. USSR, 29*: 540.
1101. Hynkova, V., and Frank, F. (1976). *J. Polym. Sci. Polym. Chem. Ed., 14*: 2587.
1102. Cubbon, R. C. P. (1965). *Polymer, 6*: 419.
1103. Yamada, M., Takahase, I., and Kobayashi, M. (1972). *Kobunshi Kagaku, 29*: 144; C.A. (1972), 7720159b.
1104. Hagiwara, T., Mizota, J., Hamana, H., and Narita, T. (1985). *Makromol. Chem. Rapid Commun., 6*: 169.
1105. Quach, L., and Otsu, T. (1981). *J. Polym. Sci. Polym. Chem. Ed., 19*: 2405.
1106. Yokota, K., Hirabayashi, T., and Takashima, T. (1975). *Makromol. Chem., 176*: 1197.
1107. Braun, D., and Azis El Sayed, J. A. (1966). *Makromol. Chem., 96*: 100.
1108. Tate, B. E. (1967). *Makromol. Chem., 109*: 176.
1109. Nakamoto, H., Ogo, Y., and Imoto, T. (1968). *Makromol. Chem., 111*: 104.
1110. Higuchi, T., Tsutsui, K., Shimada, A., Bando, Y., and Minoura, Y. (1978). *Polym. J., 10*: 111.
1111. Kaetsu, I., Tsuji, K., Hayashi, K., and Okamura, S. (1967). *J. Polym. Sci. Part A-1, 8*: 1899.

5
Polymeric Dienes

Walter Kaminsky and B. Hinrichs
University of Hamburg, Hamburg, Germany

I. INTRODUCTION

Homopolymers of conjugated dienes such as 1,3-butadiene, isoprene, chloroprene, and other alkylsubstituted 1,3-butadienes, as well as copolymerzs with styrene and acrylonitrile, are of great economical importance [1–3]. The conjugated dienes can polymerize via 1,4 or 1,2 linkage of monomeric units. In addition to this, 3,4 linkage occurs with butadienes bearing substituents in the 2-position. In the case of 1,4 linkage the polymer chain can exist as *cis* or *trans* type:

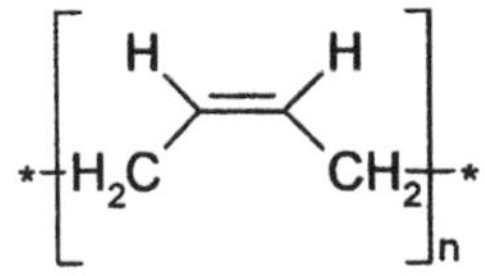

cis-type 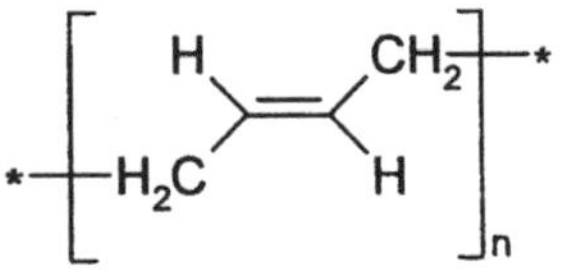   *trans*-type

1,4-linkage

$$(1-3)$$

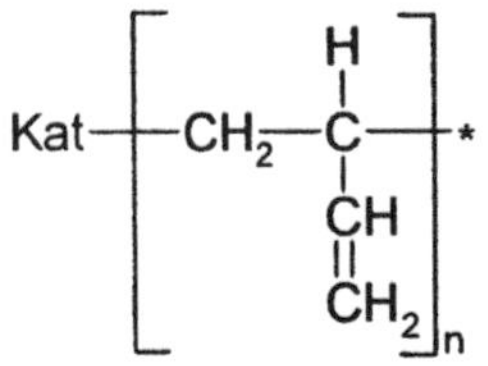

1,2-linkage

1,2 Linkage yields a tertiary carbon atom, thereby making it possible to form isotactic, syndiotactic, and atactic polybutadiene (3), in analogy to polypropene. The rare 3,4 linkage also gives isotactic, syndiotactic, or atactic configuration. This applies only to high stereoselectivities. Further isomeric structures are formed when next to head-to-tail linkages; head-to-head and tail-to-tail linkages also occur. The polymerization of dienes can be initiated ionically by coordination catalysts or by radicals [4–10].

II. POLYBUTADIENE

Polybutadiene belongs to the most important rubbers for technical purposes. In 1999 more that 2 million tons were produced worldwide, that is about 20% of all synthetic rubbers [11,12]. The *cis* type made by 1,4-addition is economically the most important polybutadiene [13,14]. *Trans*- as well as isotactic, syndiotactic, or atactic 1,2-polybutadiene can also be synthesized in good purity with suitable catalysts. For anionic polymerization with butyllithium or the coordinative process with Ziegler catalysts, 1,3-butadiene must be carefully purified from reactive contaminants such as acetylene, aldehydes, or hydrogen sulfide.

A. Anionic Polymerization

Metal alkyls, preferably of alkali metals, are used as initiators. The polarization of the catalyst exerts a strong influence on the stereospecifity (Table 1) [15,16]. Lithium alkyls give a polymer with the greatest *trans*-1,4-portion. The stereospecifity is also influenced by catalyst concentrations, temperatures, and associative behavior [17–34]. In more concentrated solutions, alkyllithium, especially butyllithium, which is the preferred initiator, forms hexameric associates that are dissociated in several steps to finally give monomers [35–53]. Only monomeric butyllithium is suited for the insertion. Isobutyl-lithium shows an association grade of 4 in cyclohexane [36]. Branched alkyl groups gave higher activities than those with *n*-alkyl groups. As postulated by the kinetic model for very weak initiator concentration, the reaction order is 1 and less than 1 for higher concentrations [54–62]. This results in a series of reactions:

$$\text{Dissociation:} \quad (CH_3-(CH_2)_3-Li)_6 \rightarrow 6CH_3-(CH_2)_3-Li \tag{4}$$

$$\text{Start:} \quad \begin{aligned} &CH_3-(CH_2)_3-Li + CH_2=CH-CH=CH_2 \\ &\rightarrow CH_3-(CH_2)_3-CH_2-CH=CH-CH_2-Li \end{aligned} \tag{5}$$

$$\text{Propagation:} \quad \begin{aligned} &CH_3-(CH_2)_4-CH=CH-CH_2-Li + CH_2=CH-CH=CH_2 \\ &\rightarrow CH_3-(CH_2)_3-(CH_2-CH=CH-CH_2)_2-Li \end{aligned} \tag{6}$$

Table 1 Microstructure of poly(1,3-butadiene) in relation to the initiator.

Initiator	Solvent	Microstructure (%)		
		cis	*trans*	1,2
C_2H_5Li	Hexane	43	50	7
C_2H_5Li	THF	0	6	91
C_4H_9Li	Hexane	35	55	10
$C_{10}H_8Li$	THF	0	3.6	96.4
$C_{10}H_8Na$	THF	0	9.2	90.8
$C_{10}H_8K$	THF	0	17.5	82.5
$C_{10}H_8Rb$	THF	0	24.7	75.3
$C_{10}H_8Cs$	THF	0	25.5	74.5

Source: Refs. [15] and [17].

With hydrocarbons as solvents, the rate of the starting reaction is up to a factor of 100 smaller than that of the propagation step. This difference is caused by the absence of a double bond in conjugation to lithium in butyllithium. In contrast to this, the use of ether accelerates the starting reaction such that propagation becomes the rate-determining step [63–67].

In the absence of chain transfer reagents, the molecular weight increases steadily with increasing conversion of monomer. In this way one gets living polymers with very narrow molecular weight distribution when the starting reaction is fast or lithium octenyl is used as a starter (Poisson distribution). The average degree of polymerization is equal to the ratio of converted moles of monomer (starting concentration $[M]_0$) over the number of moles of initiator $[I]$ reacted:

$$\overline{P}_n = \frac{[M]_0}{[I]_0 - [I]} > \frac{[M]_0}{[I]_0} \tag{7}$$

At the end of the polymerization when no more unreacted initiator is present ($[I] = 0$), the number average of the molecular weight can be calculated as follows:

$$\overline{M}_n = \frac{[M]_0}{[I]_0} \times 54 \tag{8}$$

Equation (9) is valid as long as there is still some monomer in the reaction mixture:

$$[M] - [M]_0 = ([I]_0)\left(1 - \frac{kw}{kg}\right) + \frac{kw}{kg}[I]_0 \ln \frac{[I]}{[I]_0} \tag{9}$$

To improve the processibility of linear polybutadiene with its narrow molecular weight distribution, one can continuously add initiator in the course of the polymerization, vary the reaction temperature, or force long-chain branching by addition of divinyl compounds [68–74]. Addition of small amounts of ethers or tertiary amines alters the vinyl content from some 12% to more than 70% (Table 2). Bis(2-methoxy)ethyl ether and 1,2-bis(dimethylamino)ethane as well as crown ethers [75,76] are particularly effective. The microstructures of the products are determined by IR [77–87], NMR [88–99], x-ray diffraction, and other methods [100,101].

The anionic poymerization of 1,3-butadiene is normally carried out in solvents [102–109]. Aliphatic, cycloaliphatic, aromatic hydrocarbons, or ethers as solvents could be used. Working in ethers requires low temperatures because of the high reactivity and low stability of the lithium alkyl in this solvent. Using n-hexane as solvent, a butadiene concentration of 25 wt% and a polymerization temperature of 100 to 200 °C is preferred.

Low-molecular-weight polybutadiene oils result when the polymerization is catalyzed by a mixed system of butyllithium, 1,2-bis(dimethylamino)ethane, and potassium t-butanolate [110–112]. With 1,4-dilithium-1,1-4,4-tetraphenylbutane it is possible to get bifunctional living polymers (seeding technique) [113–118].

B. Coordination Catalysts

A large number of complex metal catalysts have been employed in the polymerization of conjugated dienes [119–139]. Table 3 shows a selection of catalyst systems that have

Table 2 Influence of polar compounds on the microstructure (1,2 content)[a].

Polar compound	Molar ratio	1,2 Structure (wt%) for polymerization temperature		
		30 °C	50 °C	70 °C
$(H_3C–O–CH_2–CH_2)_2O$	0.10:1	51	24	14
	0.45:1	77	56	28
	0.80:1	77	64	40
$(H_3C)_2N–(CH_2)_2–N(CH_3)_2$	0.06:1	26	14	13
	0.60:1	57	47	31
	1.14:1	76	61	46

Source: Ref. [73].
[a]Catalyst: C_4H_9Li.

Table 3 Catalysts for the polymerization of 1,3-butadiene.

Catalyst	Microstructure (%)			
	cis	*trans*	1,2	Refs.
$TiCl_4/R_3Al$	65	35		[123]
$TiJ_4(R_3Al$	95	2	3	[124]
$Co(O\text{-}CO\text{-}R)_2/(H_5C_2)_2Al\text{-}Cl/H_2O$	96			[125]
$Ni(O\text{-}CO\text{-}R)_2/F_3B\text{-}O(C_2H_5)_2/R_3Al$	97			[126]
$Ce(O\text{-}CO\text{-}R)_3/(H_5C_2)_3Al_2CL_3/R_3AL$	97			[127]
$U(OCH_3)_4/AlBr_3/R_3AL$	98.5	1	0.5	[128]
$U(O\text{-}CO\text{-}C_7H_{15})_4/AlBr_3/R_3Al$	98.2	1.1	0.7	[129]
$Nd(O\text{-}CO\text{-}R)/R_nAlCl_{3\text{-}n}/R_3Al$	98	1.5	0.5	[130,132]
$VCl_3(VOCl_3)/R_3Al$		99	1	[133,134]
$Cr(C_5H_7O_2)_3/R_3Al$	8	2	90	[135]
$Rh(C_5H_7O_2)_3{}^a/R_3Al$		98		[136]
$Cr(allyl)_3$		10	90	[120]
$Nb(allyl)_3$	1	2	97	[121]
$Cr(allyl)_2Cl$	90	5	5	[144]

[a]2,4-Pentandionato.

been used for the polymerization of butadiene. Some systems yield polymers with a high percentage of *cis*-1,4 linkage, while others favor the formulation of *trans*-1,4 or *trans*-1,2 linkages. As in the case of Ziegler–Natta catalysis of propene, the active centers are transition metal-carbon bonds. They normally form a η^3-alloyl bond [140]:

$$(10)$$

[M] = Transition metal

The propagation reaction proceeds via insertion into these carbon–transition metal bonds after the diene has been coordinated as a π-complex:

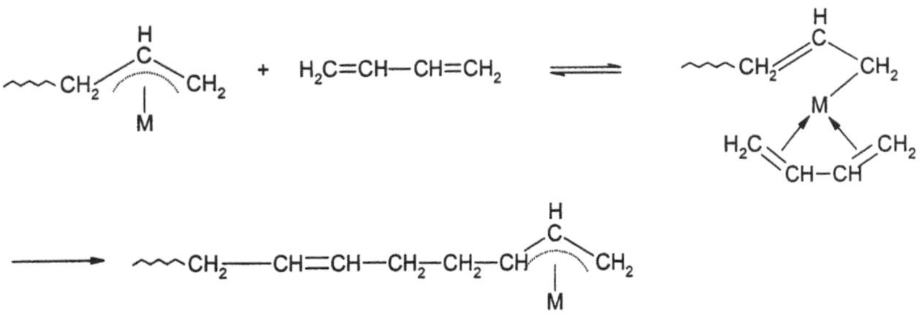

$$(11, 12)$$

In the transition state a short-lived σ-allyl bond is formed, which in the case of *cis* migration, restores an alkyl-transition metal bond [141–143].

Various mechanisms for the control of the *cis* linkage in the propagation step are discussed [144,145]. Allyl compounds can occur in *syn* or *anti* form [Structures (13–16)], from which double bonds with *trans* or *cis* configuration are formed [146,147], respectively. Solvents or cocatalysts as ligands are of great importance for the equilibration.

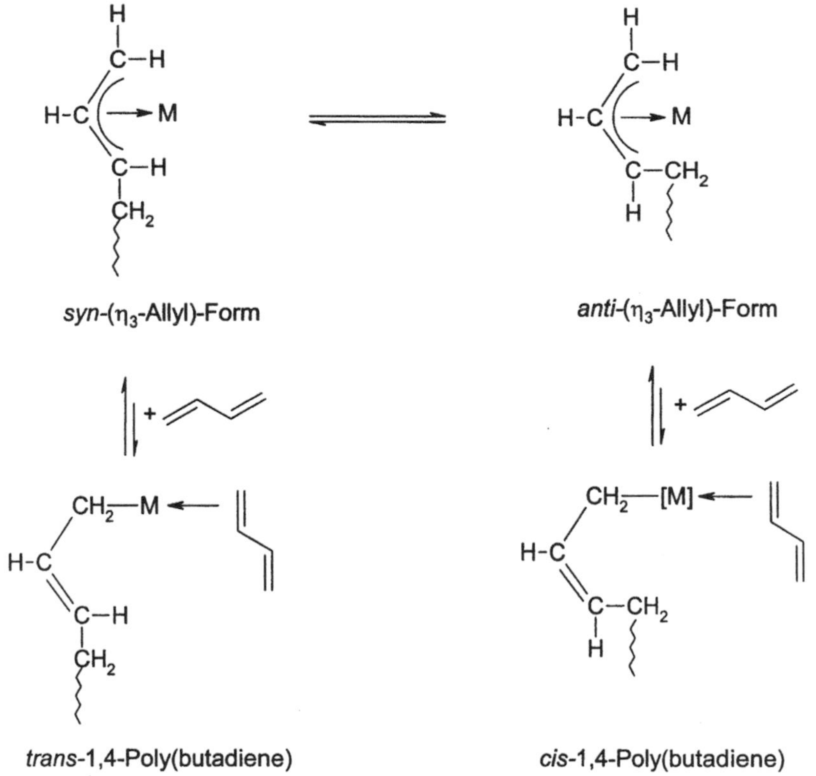

syn-(η₃-Allyl)-Form

anti-(η₃-Allyl)-Form

$$(13–16)$$

trans-1,4-Poly(butadiene)

cis-1,4-Poly(butadiene)

C. *cis*-1,4-Polybutadiene

cis-1,4-Polybutadiene is preferrentially produced with mixed catalysts. Systems on the basis of titanium (IV) iodine/trialkylaluminum are employed [148–150]. For better dosing a mixture of $TiCl_4/I_2/R_3Al$, $TiCl_4/R_2AlI$, or $Ti(OR)I_3/TiCl_4/(C_2H_5)_3Al$ in which all compounds are soluble in hydrocarbons, is used. It is essential for a high *cis* content of the products that the catalyst contains iodine. Those of $TiCl_4$ and R_3Al only lead predominantly to the formation of *trans*-1,4-polybutadiene. Aromatic hydrocarbons (benzene, toluene) are used as solvents. The polymerization is a first-order reaction with respect to the 1,3-butadiene concentration [150,151]. As $TiCl_4$ gives living polymers, the molecular weight increases almost linearly with the conversion of monomer [152]. At higher degrees of conversion, the molecular weight can be controlled by varying the catalyst concentration or composition. The molecular weight distribution M_w/M_n ranges from 2 to 4 with a *cis* content between 90 and 94%. Regulation of molecular weights can be achieved by the addition of 1,5-cyclo-octadiene [153].

Supported Ziegler catalysts are also used [154–156]. High *cis* contents up to 98% can be obtained with cobalt salts [cobalt octanoate, cobalt naphthenate, tris(2,4-penta-dionato) cobalt] in combination with alumoxanes which are synthesized in situ by hydrolysis of chlorodiethylaluminum or ethylaluminum sesquichloride. Only 0.005 to 0.02 mmol of cobalt salt is needed for the polymerization of 1 mol of 1,3-butadiene [157–159]. At 5 °C the molecular weight varies from 350 000 to 750 000 depending on the alkylaluminum chloride, while at 75 °C the variation is between 20 000 and 200,000. The polymerization rates are fast over a considerable range of chloride content. The *cis*-1,4-structure increases with chloride content. The molecular weight increases with the chloride level [160].

Nickel compounds can also be employed as catalysts [161–170]. A three-component system consisting of nickel naphthenate, triethyl-aluminum, and boron trifluoride diethyletherate is used technically. The activities are similar to those of cobalt systems. The molar Al/B ratio is on the order of 0.7 to 1.4. Polymerization temperatures range from −5 to 40 °C. On a laboratory scale the synthesis of *cis*-1,4-polybutadiene with allylchloronickel giving 89% *cis*, 7.7% *trans*, and 3.4% 1,2-structures is particularly simple [8]. In nickel compounds with Lewis acids as cocatalysts, complexes with 2,6,10-dodecatriene ligands are more active than those with 1,5-cyclooctadiene (Table 4) [171].

The influence of the ligand on *cis* or *trans* insertion is particularly obvious for η^3-allyl nickel systems.

$$(17\text{--}20)$$

Table 4 Polymerization of 1,3-butadiene.[a]

Cocatalyst	Molar ratio, HX/Ni	Reaction time (h)	Yield (%)	cis-1,4 (%)	trans-1,4 (%)	1,2 (%)
HCl	1	3	13	84	13	3
HBr	1	3	4	72	25	3
HJ	1	6	30	0	100	0

Source: Ref. [161].
[a]3,4 mol butadiene, 0.014 mol of 2,6,10-dodecatrienylchloronickel at 55 °C in heptane.

Alkanolates or carboxylates of lanthanides and actinides, especially uranium, are particularly well suited for the production of cis-1,4-polybutadiene [172–187]. Of the lanthanides, compounds of cerium, praseodymium, and neodymium are combined with trialkylaluminum and a halogen containing Lewis acid [188,189]. The polymerization can also be carried out in aliphatic solvents at 20–90 °C [190].

The microstructures are influenced primarily by the nature of the alkylaluminum compound. With triethylaluminum the portion of trans-1,4 double bonds reaches a relatively high level of 10%, while tris(2-methylpropyl)aluminum and bis(2-methylpropyl) aluminum hydride yield cis-1,4 contents as high as 99% [190]. Similarly, high cis-1,4 portions are obtained in the polymerization of 1,3-butadiene with η^3-allyluranium complexes. The osmometric measured mole mass ranges from 50 to 150 000, the molecular mass distribution between 3 and 7. The extremely high temperature-induced crystallization rate of uranium polybutadiene in comparison with titanium or cobalt polybutadiene corresponds to a greater tendency toward expansion-induced crystallization. A technical application, however, is in conflict with the costly removal of weakly radioactive catalyst residues from the products [132].

1. Metallocene-catalysts

Different methyl substituted cyclopentadienyl titanium compounds can be employed as catalysts (Table 5) [191].

At a polymerization temperature of 30 °C the chlorinated and the fluorinated complexes show nearly the same activity. Only the highly substituted fluorinated compounds (tetra- and pentamethylcyclopentadienyl titanium trichloride Me_4CpTiF_3, Me_5CpTiF_3) are significantly more active than the corresponding chlorinated ones. At higher polymerization temperatures a corresponding behavior can be observed, however with increasing polymerization temperature also the activity of the complexes increase. The activities of the 1,3-dimethylcyclopentadienyl titanium trihalides are the highest and reach about 700 kg Br/mol Ti*h. It makes no difference if one of the fluorides is substituted by another ligand like perfluoroacetic or perfluorobenzoic acid (Me_5CpTi-$F_2(OCOCF_3)$, $Me_5CpTiF_2(OCOC_6F_5)$).

The activity reaches a maximum value for all catalysts after a short induction period of 5 to 10 min. After this, the activity decreases to a value being constant for a longer period of time of up to about 1 h.

The substitution pattern influences the induction period. The most active compounds show the shortest induction period, whereas the less active ones need a clearly longer period.

Table 5 Activities of titanium complexes for the polymerization of 1,3-butadiene in 100 ml toluene, 10 g 1,3-butadiene, 0.29 g MAO, $[Ti] = 5 \times 10^{-5}$ mol/l, $Al/Ti = 1000$, $T = 30\,°C$, polymerization time $= 20$ min.

Catalyst	Activity[a]	Catalyst	Activity[a]
CpTiCl$_3$	260	CpTiF$_3$	260
MeCpTiCl$_3$	300	MeCpTiF$_3$	310
Me$_2$CpTiCl$_3$	750	Me$_2$CpTiF$_3$	605
Me$_3$CpTiCl$_3$	340	Me$_3$CpTiF$_3$	350
Me$_4$CpTiCl$_3$	165	Me$_4$CpTiF$_3$	350
Me$_5$CpTiCl$_3$	60	Me$_5$CpTiF$_3$	350
IndTiCl$_3$	310	Cp*TiF$_2$(OCOCF$_3$)	330
PhCpTiCl$_3$	325	Cp*TiF$_2$(OCOC$_6$F$_5$)	340

[a]Activity: kg BR/mol Ti*h.

The activity increases linear with increasing butadiene concentrations in the starting phase of the polymerization. The kinetic order of the butadiene concentration is 1. At constant Al:Ti ratio the polymerization rate is given by

$$r_p = k_p \cdot c_{cat} \cdot c_\beta \tag{21}$$

where c_β is the concentration of butadiene. The activity increases with an increasing Al:Ti ratio, reaches a maximum at an Al:Ti ratio of about 700 and decreases slowly with increasing Al:Ti ratios.

High molecular weights are obtained for the polybutadienes produced with these catalysts.

The di- and trimethylcyclopentadienyl titanium trichlorides give the highest molecular weights while the fluorinated compounds have significantly lower molecular weights, even if their activity is higher, as shown for the Me$_4$CpTiF$_3$ and Me$_5$CpTiF$_3$ complexes (Table 6).

The glass transition temperatures range of -90.1 and $-96.9\,°C$. The polybutadienes produced with the most active catalysts have the highest content of *cis*-1,4 units and the lowest glass transition temperature.

For all catalysts, the *cis*-1,4 structure units of the polybutadiene range between a content of 74 and 85.8%, the *trans*-1,4 between 0.5 and 4.2%, and the 1,2-units between 13.7 and 22.6% (Table 7). The most active systems generate the polymer with the highest content of *cis*-1,4 and the lowest content of *trans*-1,4 and 1,2-units. The fluorinated compounds show a similar behavior. A mechanism for the formation of these microstructures is published by Porri [192].

There is no dependence of the microstructure on the polymerization time (between 10 and 120 min the *cis* content is $81.8 \pm 0.3\%$ for MeCpCl$_3$) and on the Al:Ti ratio (between Al:Ti $= 500$ and Al:Ti $= 10\,000$ the *cis* content is about 80.7 ± 1.2 for MeCpTiF$_3$).

D. *trans*-1,4-Polybutadiene

Butadiene can be polymerized with Ti/Al catalyst systems. A sharp change in structure of polybutadiene can be seen by varying the mole ratio of TiCl$_4$ to R$_3$Al. At Ti/Al ratios of

Table 6 Molecular weights of the polybutadienes produced with fluorinated and chlorinated catalysts.

Catalyst	X = Cl Molar mass M_n [g/mol $\times 10^6$]	X = F Molar mass M_n [g/mol $\times 10^6$]
CpTiX$_3$	1.2	0.97
MeCpTiX$_3$	1.6	1.22
Me$_2$CpTiX$_3$	3.1	1.28
Me$_3$CpTiX$_3$	3.6	1.25
Me$_4$CpTiX$_3$	3.3	1.5
Me$_5$CpTiX$_3$	2.6	1.4
IndTiCl$_3$	1.25	–
PhCpTiCl$_3$	0.86	–
CyCpTiCl$_3$	1	–
(Me$_3$Si,MeCp)TiCl$_3$	1.5	–

Table 7 Microstructure and glass transition temperatures of polybutadienes produced with chlorinated and fluorinated catalyst precursors.

Catalyst	cis-1,4 [%]	trans-1,4 [%]	1,2 [%]	T_g [°C]
CpTiCl$_3$	81.7	1.1	17.2	− 95.1
MeCpTiCl$_3$	81.9	1.1	17	− 95.3
Me$_2$CpTiCl$_3$	85.8	0.5	13.7	− 96.9
Me$_3$CpTiCl$_3$	83.8	1.1	15.2	− 95.6
Me$_4$CpTiCl$_3$	80	1.7	18.3	− 91.5
Me$_5$CpTiCl$_3$	74.8	2.6	22.6	− 91
IndTiCl$_3$	74.3	4.2	21.5	− 90.1
PhCpTiCl$_3$	80.9	2.1	16.9	− 95.8
CyCpTiCl$_3$	82.6	0.8	16.7	− 95
CpTiF$_3$	81.8	1.4	16.8	− 95
MeCpTiF$_3$	81.9	1.2	16.9	− 92.7
Me$_2$CpTiF$_3$	82	2	16	− 95
Me$_3$CpTiF$_3$	84	1.1	14.9	− 94.1
Me$_4$CpTiF$_3$	80.4	1.9	17.7	− 89.9
Me$_5$CpTiF$_3$	74.6	2.8	22.5	− 87.9

0.5 to 1–5, the *cis* content of the 1,4-polybutadiene increases to about 70% at a ratio of 1, and then falls off so that *trans*-1,4-polybutadiene is obtained at Ti/Al ratios of 1.5 to 3. Under these conditions it is a good catalyst for preparing *trans*-1,4-polybutadiene. Also heterogeneous catalysts consisting of TiCl$_4$ immobilized on MgCl$_2$ have been reported [193].

Other catalysts contain the transition metals vanadium, chromium, cobalt, and nickel as their main components [194–202]. The polymerization activity is usually far lower than in the synthesis of *cis* polymers (see Table 2). Addition of a donor such as tetrahydrofuran, which directs the bonds into a *trans*-position to the catalyst of titanium tetraiodide and triethylaluminum, results in the formation of a polybutadiene with 80% *trans*-1,4-double bonds [197].

Another possibility is anionic polymerization with alkyllithium in combination with barium compounds such as barium 2,4-pentanedionate [192–194, 203–205]. Also, cobalt(II) chloride in combination with diethylaluminum chloride and triethylamine is used, yielding a polymer with 91% *trans*-1,4 and 9% 1,2 structures.

E. 1,2-Polybutadiene

The synthesis of crystalline, syndiotactic 1,2-polybutadiene is also successful with compounds of titanium, cobalt, vanadium, and chromium [194,206–210]. Alcoholates [e.g., cobalt(II) 2-ethylhexanoate or titanium(III) butanolate] with triethylamine as cocatalyst, are especially well suited for this purpose. They are capable of producing polymers with up to 98% 1,2 structure. Amorphous 1,2-polybutadiene is produced with molybdenum(V) chloride and diethylmethoxyaluminum [211]. Addition of esters of carboxylic acids raises the vinyl content of the products [212]. The influence of the coordination at the center atom is remarkable. Trisallylchromium polymerizes 1,3-butadiene to 1,2-polybutadiene, while bisallylchromchloride gives 1,4-polybutadiene.

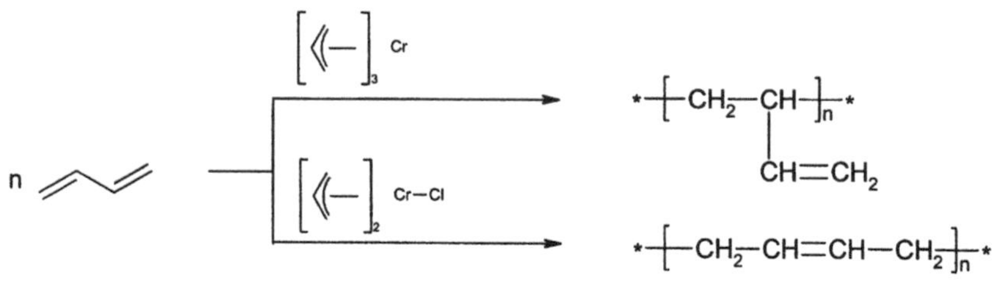

(22–23)

1. Polymerization Processes

Polybutadiene can be produced in nonaqueous media or by a radical mechanism in an aqueous emulsion. The field of homopolymerizations is dominated by the processes in nonaqueous media, as described. Emulsion polymerization is characterized by good dissipation of the reaction heat. The monomer concentration is on the order of 50 wt%. The reaction is initiated by free radicals, which are preferably formed from organic hydroperoxides such as *p*-menthane hydroperoxide [213,214]. Sodium formaldehyde sulfoxylate and iron(II) complexes are employed as reducing agents. At reaction temperatures below 5°C the polymerization is discontinued at a degree of conversion between 50 and 60%, to avoid cross-linking. The product features low stereospecifity (14% *cis*-1,4, 69% *trans*-1,4, and 17% 1,2 structures). At higher temperatures degradation of the polybutadiene lowers the molecular weight [215,216].

III. POLYISOPRENE

The homopolymerization of isoprene

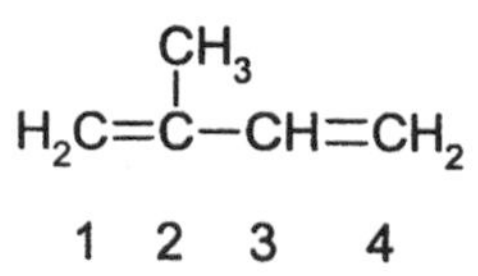

(24)

can take place with a *cis*-1,4, *trans*-1,4, 1,2, or 3,4 connection.

1,2 structure	3,4 structure	*cis*-1,4 structure	*trans*-1,4 structure
(25)	(26)	(27)	(28)

In addition, the 3,4- and 1,2-polyisoprenes can both exist in three forms: isotactic, syndiotactic, and atactic. Thus there are eight possible structures if we disregard head-to-head possibilities. The part of the structure elements in the polymer depends on the catalysts. In general, the polymerization activity is lower compared to polybutadiene. Of the various structures of polyisoprene, only *cis*- and *trans*-1,4-polyisoprene and atactic 3,4-polyisoprene are important (Table 8) [217–219].

A. *cis*-1,4-Polyisoprene

Natural rubber (hevea) is 98% *cis*-1,4-polyisoprene with 2% 3,4-structure. It can be synthesized by anionic polymerization with alkyllithium compounds or with Ziegler–Natta catalysts [220–225]. The polymerization is carried out in solvents. Impurities such as acetylenes, carbonyl compounds, hydrogen sulfide, and water have to be removed [217,226–228].

1. Anionic Polymerization

cis-Polyisoprene can be obtained with butyllithium under certain condition. The *cis* content depends on the initiator and monomer concentrations as well as on the temperature [23,49]. In aliphatic solvents up to 97% 1,4-*cis* polymer could be obtained (Table 9). The strong influence of the initiator concentration is explained by a two step mechanism [229].

Table 8 Homopolymerization of isoprene; microstructure of polyisoprenes.

Catalyst	*cis*-1,4 (%)	*trans*-1,4 (%)	1,2 (%)	3,4 (%)
LiC_4H_9 in heptane	93	0	0	7
LiC_4H_9 in THF	0	30	16	54
$(i\text{-}Bu)_3Al/TiCl_4$	97	0	0	3
$Nd(2\text{-ethylhexanoate})/Et_2AlCl/THF$	96.9	0	0	3.1
Et_3Al/VCl_3	0	98	0	2
Na dispersion	29	29	0	42
$K_2S_2O_8$	22	65	6	7

Source: Refs: [217] and [218].

Table 9 Dependence of polyisoprene microstructure on butyllithium concentration.

Butyllithium (mmmol/l)	Microstructure (%)		
	cis-1,4	*trans*-1,4	3,4
61.2	74	18	8
1	78	17	5
0.1	84	11	5
0.008	97	0	3

Source: Ref. [23].

(29–33)

First, dissociation of the lithium alkyl association [Structure (29)] takes place, followed by activation by complexing of the monomer lithium alkyl with the *cis*-isoprene [Structure (31)]. For insertion in a second step, a dimer alkyllithium is necessary [Structures (32) and (33)].

The living polymerization shows no breaking-off or transfer reactions and therefore gives polymers with a narrow molecular weight distribution [230]. The molecular weight can be calculated as follows:

$$M_{cal} = \frac{[\text{isoprene}]}{[\text{RLi}]} \times 68 \tag{34}$$

The polymer is highly linear without branching. For the synthesis of polyisoprenes with an extremely narrow molecular weight distribution ($M_w/M_n = 1.05$) a vacuum or seeding technique could be used [231, 232]. In the second case the polymerization is started with separately prepared polyisoprene of low molecular weight. Polar solvents such as ethers and amines have an influence on the microstructure [233–236]. The initiation step increases in relation to the propagation step [237].

The anionic polymerization leads to polymers with an active lithium end group. This can be used for further reactions. By treatment with chlorsilanes such as 1,2-bis(dichloromethylsilyl)ethane, a four-star polymer results; with 1,2-bis(trichlorosilyl) ethane, a six-star polymer. Aromatic divinyl compounds used for the same purpose have been described [238–240].

2. Coordinative Catalysts

Titanium tetrachloride in combination with aluminum trialkyl (ratio 1:1) gives optimum activity in isoprene polymerization. The Ziegler system TiX_4/R_3Al (X = halides) yields either cis-1,4, trans-1,4 or 3,4-polyisoprene, while the unmodified lithium systems produce predominantly cis-1,4-polyisoprene (Table 10). Using $TiCl_4$ and R_3Al cis-1,4-polyisoprene is obtained at Ti/Al ratios of 0.5 to 1.5 [160]. At lower Ti/Al ratios, oligomers are formed. At ratios of 1.3 to 1.6, mixed cis/trans polymers are obtained; at 1.6 to 2, trans-1,4-polyisoprenes. Ratios above 2 give resinous materials that are cyclized trans-polymers. The other titanium halides were found to be equivalent to $TiCl_4$ in these reactions. Catalyst efficiency is increased by complexing the R_3Al with ethers and tertiary amines.

It is important to mix the catalyst components and alter the heterogeneous system before adding the monomer [241, 242]. Titanium (II) seems to be inactive. Therefore the catalyst could be stabilized by addinng electron donors such as ethers and esters [243–246]. Instead of alkylaluminum, alane etherates such as $HAlCl_2 \cdot O(C_2H_5)$ are used [247–251].

The best results in obtaining high yields of cis-1,4-polyisoprene are given by rare earth catalysts [252–257]. Similar to the polymerization of butadiene, three component catalysts (transition metal compound, Lewis acids, and alkyl aluminum) are used. It is necessary to have an excess of 4 to 10 times of the aluminum component. Most attractive

Table 10 Ziegler catalysts for isoprene polymerization: influence of the Ti/Al ratio on the microstructure.

Catalyst	Molar ratio Ti/Al	Microstructure (%)			
		1,2	3,4	cis-1,4	trans-1,4
$TiCl_4/R_3Al$	0.5–1.5	<1	3	97	<1
$TiCl_4/R_3Al$	1.3–1.6	<1	3	56	41
$TiCl_4/R_3Al$	1.6–2.0	<1	3	8	97
$TiCl_4/R_2AlCl/R_3N$	0.1–1.5	<1	3	97	<1
$Ti(OR)_4/RAlCl_2$	1	<1	99	<1	<1

Source: Ref. [160].

is neodymium salt. With increasing temperature and the Al/Nd ratio, the molecular weight of the polymer decreases. The *cis*-1,4 content is higher than 95% and the 3,4 part is less than 5% (see Table 8).

B. trans-1,4-Polyisoprene

The natural products gutta-percha and balata consist of *trans*-1,4-polyisoprene. With the aid of vanadium trichloride and triethylalumium, *trans*-1,4-polyisoprene can be produced with 98% *trans*-1,4 enchainments [133,258]. The optimal Al/V ratio is the range of 5 to 7. The activity can be increased by the addition of small amounts of ether, heterogenerization on supports (kaolin, TiO_2), or blending with titanium(III) chloride or titanium alcoholates [259–261]. Further catalysts featuring lower activity, however, are allylnickel iodide, trisallylchormium on silica, or complexes of neodymium [262–265].

$$Li[Nd(C_3H_5)_4] \cdot 1,4\text{-dioxane} \tag{35}$$

$$(H_5C_6-CH_2)_3Nd \tag{36}$$

Pure *trans*-1,4-polyisoprenes as well as *trans*-1,4-polybutadienes can be synthesized by polymerization in inclusion compounds [266–269]. As typical hosts for this dienes, the inclusion compounds or clathrates of urea, thiourea, or perhydrotriphenylene [PHTP; Eq. (36)] are used [270,271]. The host forms the frame of the crystal and the guest is placed in the cavities existing in the lattice. Polymerization is generally started by subjecting the inclusion compound to irradiation with α-, γ-, or x-rays and proceeds by a radical mechanism [272,273]. Also, free radical initiators such as di-*tert*-butylperoxide could be used [274]. Inclusion in urea yields crystalline *trans*-1,4 polymers, whereas *trans*-1,4-polyisoprene obtained in PHTP is amorphous. There is no trace of 1,4-*cis* units or of 1,2, 3,4, and cyclic units. The reason for the amorphous product is the presence of a substantial number of head-to-head and tail-to-tail junctions in addition to head-to-jail junctions [275, 276].

C. 1,2- and 3,4-Polyisoprene

Pure 1,2-polyisoprene cannot be synthesized. With alkyllithium in polar solvents (THF) or upon addition of bases such as 1,2-bis(dimethylamino)ethane (TMEDA), a large amount of 3,4 structures is formed, which can exceed 70% [33,277–279]. Polymers with a dominating portion of 1,2 structures have not been described so far. Catalysts on the basis of tris(2,4-pentanedionato)chromium and triethylaluminum are used but feature low activities [280].

Addition of TMEDA to butyllithium results in the formation of polyisoprenes with *cis*-1,4 and 3,4 structures (see Table 8). These equibinary polyisoprenes can also be synthesized with cobalt halogenides in connection with phenylmagnesium bromide. The addition of alcohols (e.g., $CoI_2/H_5C_6MgBr/octanole$ at a ratio of 1:2.2:2) leads to an increase in activity [281–283]. Polyisoprene with over 99% 3,4 structures can be generated with $(C_4H_9O)_4Ti/AlR_3$ catalysts. The product is isotactic.

D. Metallocene-catalysts

Also, half sandwich titanium compounds can polymerize isoprene (Table 11).

Table 11 Homopolymerization of isoprene. Polymerization conditions: 50 ml toluene, 50 ml isoprene. [Ti] = 5×10^{-5} mol/l, Al/Ti = 200, Tp = 30 °C, t_p = 5–24 h.

Catalyst	Activity [g IR/mol Ti-h]	T_g [°C]
CpTiCl$_3$	'28	−52
Me$_5$CpTiCl$_3$	8	n.b.
CpTiF$_3$	840	−50.3
MeCpTiF$_3$	250	n.b.
Me$_5$CpTiF$_3$	29	n.b.

As of steric effects the unsubstituted cyclopentadienyl compound is more active than the substituted ones. The fluorinated compounds are much more active (up to a factor of 30) than the chlorinated ones. The glass transition temperature of the polyisoprenes is about −52 °C.

IV. CHLOROPRENE

The polymerization of 2-chloro-1,3-butadiene(chloroprene), which is made from acetylene or 1,3-butadiene [284–287] is strongly exothermic (75 kJ/mol). It can be initiated radically, anionically, cationically, or with Ziegler catalysts [288]. Only the free-radical process, which is usually run as an emulsion polymerization, is of technical importance [289–294]. Compared with polybutadiene and polyisoprene, polychloroprene features improved gasoline and aging resistance, low-temperature flexibility, and is less combustable [295–297].

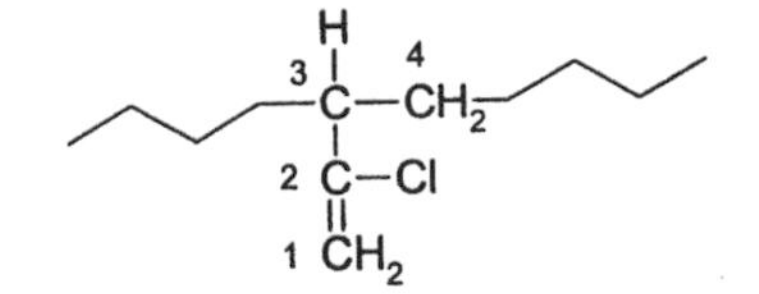

trans-1,4

(37)

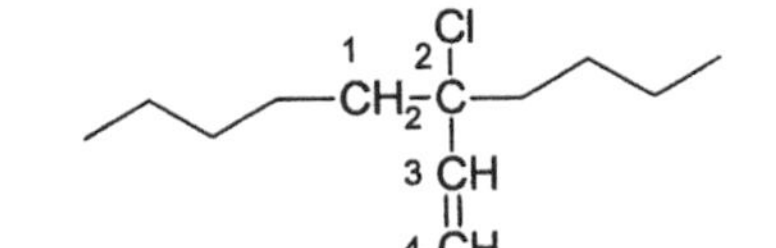

cis-1,4

(38)

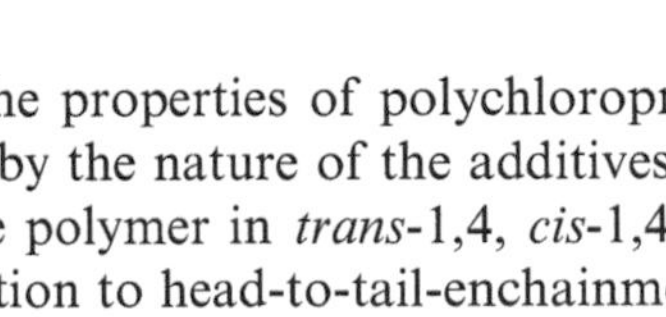

3,4

(39)

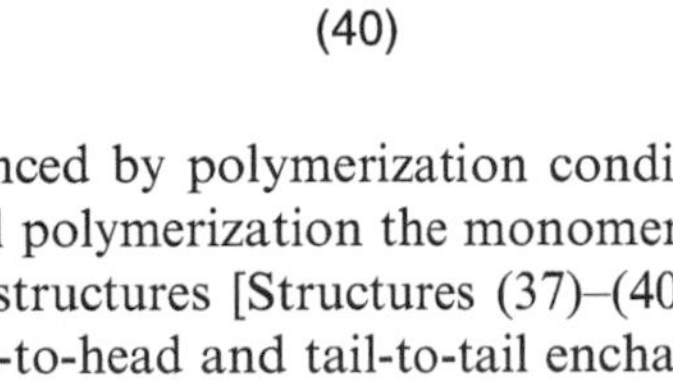

1,2

(40)

The properties of polychloroprene are influenced by polymerization conditions as well as by the nature of the additives. In the radical polymerization the monomer is built into the polymer in *trans*-1,4, *cis*-1,4, 1,2, and 3,4 structures [Structures (37)–(40)] [298]. In addition to head-to-tail-enchainments, also head-to-head and tail-to-tail enchainments occur, with a probability of 10 to 15% (Table 12) [297, 299]. Polymers with a high *trans* (>98%) or *cis* content (>95%) are both possible [300, 301]. The glass transition temperature for the *trans*-polychloroprene is − 45 °C; that for the *cis* polymer is − 20 °C at a degree of crystallinity of about 12%.

Table 12 Microstructure of polychloroprene as related to polymerization temperature.

Temp.(°C)	trans-1,4		cis-1,4 (%)	1,2 (%)	3,4 (%)
	Head/tail (%)	Head/head, tail/tail (%)			
12	83.0	11.5	3.8	1.0	0.8
30	81.5	12.0	4.5	1.2	1.0
42	80.5	12.0	5.2	1.2	1.1
57	80.5	11.0	5.8	1.4	1.3
70	75.0	13.5	8.4	1.5	1.4

Source: Refs. [297] and [299].

Products with a low degree of crystallinity that can be decreased by comonomers such as 2,3-dichloro-1,3-butadiene are suitable for applications as rubbers, whereas more crystalline polymers which are produced at lower polymerization temperatures are used in applications as components of adhesives. Due to the increased reactivity of the chlorine atoms in 1,2 structures, they tend to trigger aging reactions [302].

Chloroprene is mainly produced by emulsion polymerization [303–308].

A. Sulfur Modified Chloroprene

Chloroprene is very reactive and can be polymerized with elemental sulfur. The resulting block copolymer consists of chloroprene and sulfur segments of various lengths. The sulfur can either be dissolved in the liquid monomer or added as a dispersion. The sulfur bridges are easily cleaved by iodoform or other additives, thus permitting a variation of molecular weights over a wide range [297].

V. SUBSTITUTED POLYBUTADIENES

Next to isoprene, pentadienes and 2,3-dimenthyl-1,3-butadiene are produced as alkylbutadienes on a large scale. Poly-2,3-dimethyl-1,3-butadiene was one of the first synthetical rubbers [309, 310]. Terminally substituted 1,3-butadienes give 1,4 monomeric units each of which contains one or two asymmetric carbon atoms [R = H or alklyl group; R' = alkyl group; (41) and (42)]. Therefore, monomers of this type can lead to different stereoregular 1,4-polymers: *cis*-1,4 iso- or syndiotactic, *trans*-1,4 iso- or syndiotactic [311,312].

$$CHR{=}CH{-}CH{=}CHR' \longrightarrow {-}CH{-}CH{=}CH{-}CH{-}$$

$$\begin{array}{cc} & | \qquad\qquad\quad | \\ & R \qquad\qquad\quad R' \end{array}$$

(41)   (42)

A. Poly(2,3-dimethyl-1,3-butadiene)

2,3-Dimethyl-1,3-butadiene is produced by dehydration of pinacol, which in turn is made by reductive coupling of acetone, followed by purification via sulfur dioxide adducts [313]. It can be polymerized radically (emulsion polymerization), anionically, cationically, or by coordinative catalysts [314–319]. Due to the sterical hindrance of two methyl groups,

1,2 enchainment is hindered in comparison to 1,3-butadiene [320] [*cis*-1,4-poly(2,3-dimethyl-1,3-butadiene)].

$$
\begin{array}{c}
\text{H}_3\text{C}\quad\text{CH}_3 \\
\text{H}_3\text{C}-\overset{|}{\text{C}}-\overset{|}{\text{C}}-\text{CH}_3 \\
\overset{|}{\text{HO}}\quad\overset{|}{\text{OH}}
\end{array}
\quad\xrightarrow[-\,2\,\text{H}_2\text{O}]{\text{HBr}}\quad
\begin{array}{c}
\text{H}_3\text{C}\diagdown\qquad\diagup\text{CH}_2 \\
\qquad\text{C}-\text{C} \\
\text{H}_2\text{C}\diagup\qquad\diagdown\text{CH}_3
\end{array}
\qquad (43)
$$

By analogy with the polymerization of isoprene with Ziegler catalysts, the microstructure of the polymer is determined by the aluminum/titanium ratio [321–323]. At Al/Ti ratios smaller than 1 the portion of *trans*-1,4 structures goes up to 75%, while the formation of a *cis*-1,4 polymer requires a Al/Ti ratio of at least 1. In either case some 10% of 1,2 structures are formed in the reaction [324]. The polymerization is carried out in benzene, toluene, or hexane as solvents. The *trans*-1,4 polymer has a higher melting point of 260 °C, compared to 190 °C for the *cis*-1,4 polymer. Since this is connected with a high degree of crystallinity, these polymers do not possess any rubber elasticity.

Catalysts on the basis of complexes of cobalt or iron salts [e.g., cobalt(II) chloride/pyridine, cobalt(II) acetate/AlR$_2$Cl] yield mixed structures with more than 20% 1,2 double bonds and rubber elastic-like polymers [325,326]. Rare earth catalysts have also been described [327]. A crystalline *cis*-1,4 polymer with a melting point of 198 °C and a molecular weight of 100 000 is obtained with aluminum alkyls/neodymium compounds at a molar ratio of 31:1. The yield is in the range of 30%. Cobalt(II) acetate in combination with diethylaluminum chloride or rhodium salts also yields a *cis*-1,4 polymer [328,329].

Large amounts of *trans*-1,4-poly(2,3-dimethyl-1,3-butadiene) can be prepared by inclusion polymerization [330–332]. Urea or thiourea are used as templates. *Trans*-1,4 polymer (99 °C) is also obtained with π-allylnickel chlorides in combination with tetrachlor-1,4-benzoquinone. Anionic polymerization by butyllithium allows good control of the products microstructure over a wide range [97].

B. Poly(alkyl-1,3-butadienes)

Polymers of some of the higher 2-alkyl-1,3-butadienes give vulcanizates with tensile strength and elasticity comparable to that of natural rubber. Poly(2-ethylbutadiene) and poly(2-phenylbutadiene) are most important. 2-Ethyl-1,3-butadiene can be polymerized in the same way as isoprene [333,334]. The polymer has a glass transition temperature of −76 °C [335]. A polymer rich in *trans*-1,4 structures is obtained by catalysis with vanadium(III) chloride/triisobuthylaluminum. In contrast to *trans*-1,4-polyisoprene, the product can be used as rubber, due to its reduced tendency to crystallization [336, 337].

Additional alkyl-substituted polybutadienes are listed in Table 13. Parallel to an increase in the alkyl substituents' volume and electron donor properties, there is a decrease in selectivity and activity for *cis*-1,4 insertions, although the vulcanizing properties of the products are improved [338–345].

Cationic polymerization of substituted alkyl-1,3-butadienes is accompanied by a considerable loss of double bonds (up to 80%) due to the formation of cyclic products [346]. Tin(IV) chloride in trichloracetic acid, tungsten(VI) chloride, and boron trifluoride etherate have been tested as cationic catalysts [347, 348]. In addition to polymerization, isomerizations are observed with these catalysts.

Table 13 Polymerization of 2-alkyl-1,3-butadienes with various: influence on the microstructure.

Monomer	Catalyst	Temp. (°C)	Time (h)	Yield (%) (loss of double bond)	Polymer microstructure (%)			
					cis-1,4	trans-1,4	3,4	1,2
2,2-Dimethyl-1,3-butadiene	H₉C₄Li	0	24	98	52		0	48
2-Ethyl-1,3-butadiene	AlR₃/TiCl₄/BF₃	10	5	100	98	0	2	0
	H₉C₄Li	40	18	97	78	14	8	0
	AlR₃/VOCl₃	25	18	16	39	53	8	0
	Allyl-Ni-J	25	40	47(3)	0	85	14	1
2-Propyl-1,3-butadiene	AlR₃/TiCl₄/BF₃	10	18	49	95	0	5	0
	H₉C₄Li	40	18	91	91	4	5	0
2-Isopropyl-1,3-butadiene	AlR₃/TiCl₄/BF₃	10	18	50	91	9	0	0
	Allyl-Ni-J	25	40	29	0	76	22	2
2-Butyl-1,3-butadiene	AlR₃/TiCl₄/BF₃	10	18	27	60	33	7	0
	H₉C₄Li	40	18	88	62	35	3	0
1,3-Pentadiene	WCl₆	0	1	(54)	30		0	16
5-Methyl-1,3,hexadiene	SnCl₄/CCl₃COOH	0	1	(30)	26		44	0
5,5-Dimethyl-1,3-hexadiene	SnCl₄/CCl₃COOH	0	1	(27)	21		52	0

Source: Ref. [342].

C. Phenyl-1,3-butadienes

Poly(2-phenylbutadienes) with a high *cis* content are also produced with the triisobuthyl-aluminum/titanium tetrachloride catalysts [349]. Phenyl-1,3-butadienes can also be considered as vinyl-substituted styrenes, which explains the effects on activities and microstructures. Poly(2-phenyl butadienes) occur in *trans*-1,4, *cis*-1,4, 3,4, and 1,2 structures. Maximum conversions are achieved with a molar Al/Ti ratio of 1, which leads to the formation of 73% *cis*-1,4 and 27% 1,2 structures. At higher Al/Ti ratios the *cis*-1,4 content goes up to 96%. The molecular weights are low, ranging from 2000 to 18,000.

In contrast to this, the polymerization of 1-phenyl-1,3-butadiene was found to produce polymers with high contents of 3,4- but no 1,2-structures.

Generally, the molecular weights are low, with a ceiling of 10 000. The more crowded the positions of phenyl residue and methyl group, the higher is the 3,4 content. At the same time, there is an increasing tendency towards partial cyclization of the polymers via 3,4 structures (Table 14).

D. Polybutadienes with Heteroatoms

Next to chloroprene, numerous other polybutadienes with different substitution patterns and substituents have been synthesized, although they are of no commercial importance (Table 15) [350–358].

VI. POLY(1,3-PENTADIENE)S

A. Poly-1,3-Pentadiene

1,3-Pentadiene is the most studied terminally substituted butadiene. It exists in two geometric isomers, which have different conformers:

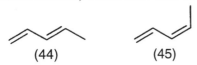

(44)   (45)

The polymerization leads to four ditactic isomers [359,360]: isotactic *trans*-1,4 (46), isotactic-*cis*-1,4 (47), syndiotactic *trans*-1,4 (48) and syndiotactic *cis*-1,4 (49). In addition, there are also the *cis*- and *trans*-1,2 and 3,4 polymers, which can have an iso- or syndiotactic structure [(50) and (51)].

Table 14 Microstructure of poly(phenyl-1,3-butadienes).[a]

Polymer	Microstructure (%)		
	1,4	3,4	Cyclic
Poly(4-phenyl-1,3-pentadiene)	7	80	13
Poly(2-methyl-1-phenyl-1,3-butadiene)	29	62	9
Poly(3-methyl-1-phenyl-1,3-butadiene)	82	16	2
Poly(1-phenyl-1,3-pentadiene)	100	0	0
Poly(2-phenyl-1,3-pentadiene)	100	0	0
Poly(3-phenyl-1,3-pentadiene)	100	0	0

Source: Ref. [347].
[a]Catalysts $ZnCl_4$ and CCl_3COOH in dichloromethane at $-78\,°C$.

Table 15 Polymerization of different monomers leading to substituted polybutadienes.

Polymer	Monomer (structure)
Poly-(1-chlor-1,3-butadiene)	Cl—CH=CH—CH=CH₂
Poly(3-chlor-2-methyl-1,3-butadiene)	H₂C=C(Cl)—C(CH₃)=CH₂
Poly(1,2-dichlor-1,3-butadiene)	Cl—CH=C(Cl)—CH=CH₂
Poly(2-chlor-1-phenyl-1,3-butadiene)	H₅C₆—CH=C(Cl)—CH=CH₂
Poly-(2,3-dichlor-1,3-butadiene)	H₂C=C(Cl)—C(Cl)=CH₂
(Poly(2-fluor-1,3-butadiene)	H₂C=C(F)—CH=CH₂
Poly(1-cyan-1,3-butadiene)	H₂C=CH—CH=CH—CN
Poly(2-cyan-1,3-butadiene)	H₂C=C(CN)—CH=CH₂
Poly(1-methoxy-1,3-butadiene)	H₂C=CH—CH=CH—O—CH₃
Poly(2-methoxy-1,3-butadiene)	H₂C=C(O—CH₃)—CH=CH₂
Poly(1-dimethoxyphosphoryl-2-methyl-1,3-butadiene)	H₂C=CH—C(CH₃)=CH—P(O)(OCH₃)₂
Poly[2,3-bis(diethylphosphano)-1,3-butadiene]	H₂C=C(P(C₂H₅)₂)—C(P(C₂H₅)₂)=CH₂

Trans-1,4 isotactic, *cis*-1,4 isotactic and *cis*-1,4 syndiotactic polypentadienes have been prepared. The *cis*-1,4-polypentadienes are of technical interest [361,362].

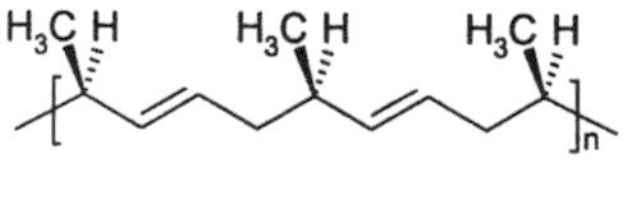

(46)

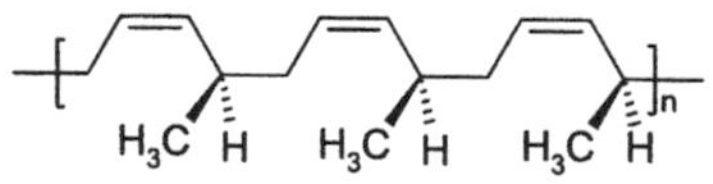

(47)

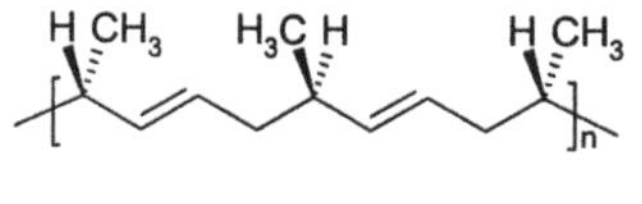

(48)

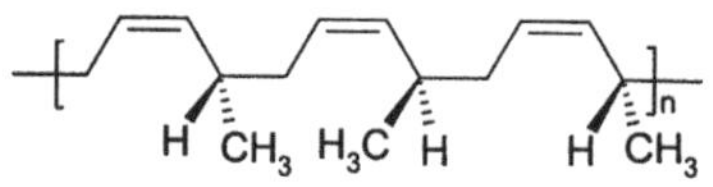

(49)

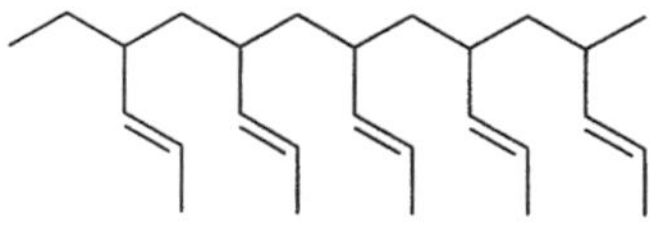

1,2-sequence
(50)

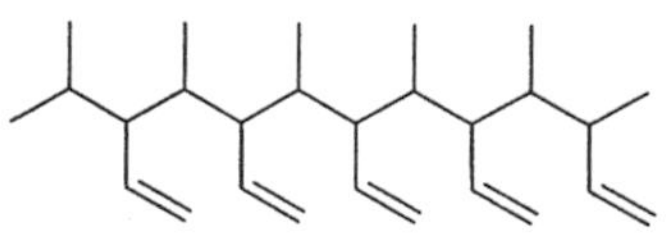

3,4-sequence
(51)

B. 1,4-Poly(1,3-Pentadiene)

The polymerization of 1,3-pentadiene with cobalt acetylacetonate and chloralumoxane (52) or diethylaluminumchloride

$$\underset{H_5C_2}{\overset{Cl}{\diagdown}}Al—O—Al\underset{Cl}{\overset{C_2H_5}{\diagup}} \tag{52}$$

leads to syndiotactic cis-1,4-poly(1,3-pentadiene) (Table 16) [312]. The catalyst is only able to polymerize the *trans*-isomer of 1,3-pentadiene (61). The addition of thiophene or pyridine decreases the amorphic part, which contains a high number of 1,2 structures. In the reaction between AlEt$_2$Cl and Co(acac)$_2$, all the acetylacetonato groups are displaced from the cobalt atom with the formation of a Co(I) species. The polymerization-active cobalt system (and the nickel system as well) is a cationic system in benzene [363–367]. In systems of this type the mode of presentation of the monomers is probably determined by the steric interaction between the butenyl group and the incoming monomer and forms by minimizing the steric interaction the syndiotactic polymer [368].

Neodymium catalysts show a different behavior. In this catalytic complex the neodymium is probably in the trivalent state and at least one Nd–Cl bond is present [369]. The AlEt$_3$-Ti(OR)$_4$ system is also a catalyst in which some alkoxy groups remain bonded to the titanium. Both catalysts give cis-1,4 isotactic polypentadiene. The anionic ligands bonded to the neodymium or the titanium atom of the catalytic species force the new monomer to react to the isotactic structure.

With these catalysts a mixture of cis- and trans-1,3-pentadienes in a wide range could be polymerized. But the polymer obtained from the *trans* isomer is more clean and crystalline. In the titanium catalyst the Al/Ti ratio plays an important role for the molecular weight. With an increasing Al/Ti ratio, the molecular weight of the polymer decreases [370,371]. An optimal values is Al/Ti = 7.

With the optically active aluminum triethyltitanium tetramenthoxide system, an optically active cis-1,4 isotactic polypentadiene was obtained, a fact that can be accounted for by assuming that at least one menthoxy group is bonded to the titanium atom of the catalytic complex [370]. The melting point of cis-1,4-poly(1,3-pentadiene) is in the range of 40 to 53 °C depending on the cristalline part, and the molecular weight in the range of 20 000 to 400 000.

Isotactic trans-1,4-poly(1,3-pentadiene) shows a melting point of 95 °C [373]. It can be synthesized with the heterogeneous catalyst triethylaluminum/vanadium trichloride [311]. The trans-1,4 units reach nearly 100%. Small amounts (10 to 30%) of an amorphic polymer can be extracted with ether. AlR$_3$/α-TiCl$_3$ could also be used. This system gives polymers consisting of trans-1,4-units; hence the butenyl group has a *syn* configuration [368]. A *syn*-butenyl group can derive either from coordination of the monomer with only one double bond or from coordination with the two double bonds in the cisoid conformation, forming an *anti*-butenyl group, followed by an *anti–syn* isomerization. The heterogeneous systems AlEt$_3$/α-TiCl$_3$ and AlEt$_3$/VCl$_3$ are capable of polymerizing both the *trans*- and the *cis*-isomers of 1,3-pentadiene. For the *cis*-isomer the cisoid conformation is unfavored for steric reasons. It is therefore likely that for these catalysts the coordination of the monomer occurs with only one double bond for both isomers of pentadiene. In the presence of CrO$_3$ only the *trans*-isomer of pentadiene is polymerized to a crystalline polymer with 80% trans-1,4 and 20% 1,2 structures.

Table 16 Polymerization of 1,3-pentadiene with various catalysts in benzene.

Catalyst	Al/Metal (mol/mol)	Time (h)	Yield (%)	Microstructure (%)		
				cis-1,4(1,2)	trans-1,2	3,4
Bis(pentandithionato)-cobalt/Al(C₂H₅)₂Cl	600	20	78	45	55	0
	300	15	50	65	35	0
Al(C₂H₅)₂Cl₂ (heptane)	600	20	57	(5)	95	0
Al₂O(C₂H₅)₂Cl₂	600	20	85	82	18	0
Al₂O(C₂H₅)₂Cl₂ + thiophene	600	20	82	93	7	0
Ti(OC₄H₉)₄/Al(C₂H₅)₃	3	26	2	61	29	10
	7	26	31	68	20	12

Source: Ref. [312].

Cationic polymerization provides, independent of the isomer of the 1,3-pentadiene, high *trans*-1,4 and *trans*-1,2 microstructures (Table 17). Studies on the insertion mechanism and the various side reactions have been carried out [373,374].

In principle, 1,4-disubstituted butadienes can give different types of 1,4-stereoregular polymers: erythro (or threo) *trans*-1,4 iso- or syndiotactic.

With the mode of presentation indicated, the new monomer gives, after insertion, a butenyl group superimposable on the preceding one; hence a diisotactic polymer will be formed.

C. Poly(methyl-1,3-pentadiene)

2-Methyl-1,3-pentadiene and 4-methyl-1,3-pentadiene are easily polymerized via a cationic route [376,377]. Even weak acids at low temperatures give high-molecular-weight polymers that are not cyclized and contain a large *trans*-1,4 portion. However, 4-methyl-1,3-pentadiene can also give mostly 1,2 structures at low yields. The same compounds that are used for the polymerization of 1,3-pentadiene are employed as acids (Table 18).

In comparison, the cationic polymerization of 3-methyl-1,3-pentadiene yields up to 70% cyclized poly(3-methyl-1,3-pentdiene) with 1,2- and 1,4-microstructure of the remaining double-bond portion [378].

Also, Ziegler–Natta catalysts can be used with reduced activities. Most of the polymerizates feature low degrees of crystallinity. *Trans*-2-methyl-1,3-pentadiene polymerizes with the homogeneous $Ti(OR)_4/VCl_3/AlR_3$ catalyst to give amorphous *trans*-1,4-poly(2-methyl-1,3-pentadiene), whereas the heterogeneous system consisting of $TiCl_4/AlR_3$ produces a partly crystalline *cis*-1,4-polymer. In this process the 4-methyl-1,3-pentadiene is converted almost exclusively to isotactic 1,2-sequences. The *trans*-3-methyl-1,3-pentadiene polymerizes with the titanium catalyst to *cis*-1,4-poly(3-methyl-1,3-pentadiene) with high molecular weights and melting points between 79 and 94 °C. In the anionic polymerization of 2- and 4-methyl-1,3-pentadiene with butyllithium only the *trans*-isomers give polymers with 60% *cis*-1,4 and 40% *trans*-1,4 structures [376]. Optically active poly(2-methyl-1,3-pentadiene) with up to 100% *trans*-1,4-double bonds is obtained by inclusion polymerization in desoxy- and apocholic acid [273].

Using the metallocene-catalysts $CpTiCl_3/MAO$ it is possible to polymerize 4-methyl-1,3-pentadiene to a mainly syndiotactic polymer [379].

VII. MISCELLANEOUS DIENES

A. Poly(2,4-Hexadiene)

Hexadiene occurs also in several isomers, of which the *trans–trans* isomer is most reactive with Ziegler catalysts. Results of polymerization reactions with a number of different catalysts are compiled in Table 19 [380]. Although Ziegler catalysts are normally not capable of polymerizing olefins with internal double bonds, this is successful in the case of 2,4-hexadiene, leading to crystalline polymers with high molecular weights [381]. Also, cationic polymerization gives high-molecular-weight polymers [382]. Anionic polymerization yields only oligomers [383,384]. Exclusively *trans*-1,4-three diisotactic microstructure is found in crystalline poly(2,4-hexadiene) [385,386]. The melting point is near 87 °C.

Table 17 Polymerization of 1,3-pentadiene with cationic catalysts in benzene at 20 °C.

Catalyst	mmol	Monomer isomer	H_2O/catalyst (mol/mol)	Gel (%)	Double bond (%)	Microstructure (%)		
						1,4	1,2	3,4
$SnCl_4$	0.38	*trans*	1.00	12	79	47.1	18.6	0
$SnCl_4$	0.38	*cis*	1.00	15	85	52.5	19.5	0
$TiCl_4$	0.53	*trans*	0.65	0	89	46.5	21.0	0
$TiCl_4$	0.53	*cis*	0.70	16	78.5	23.8	21.2	0
$AlCl_3$	16	*trans*	0.02	0	76	40	16	0
$AlEtCl_2$	0.79	*trans*	0.01	0	91.6	60.4	25.6	0
$AlEtCl_2$	0.79	*trans*	0.10	15	60.5	41.5	12.5	3

Source: Ref. [375].

Table 18 Polymerization of methyl-1,3-pentadiene with various catalysts.

Monomer	Solvent	Catalyst	(mmol)	Microstructure (%)		
				cis-1,4	trans-1,4	1,2
2-Methyl-	Benzene	AlEt$_2$Cl/H$_2$O	0.12		90	>5
	Heptane	TiCl$_4$	0.43	<5	90	<5
	Heptane	VOCl$_3$/AlEt$_3$	0.11		90	10
	Benzene	TiCl$_4$/Al(i-Bu)$_3$	0.5	90	10	
	Benzene	TiCl$_4$/AlEt$_3$	0.5	50	45	<5
	Heptane	H$_9$C$_4$Li	3.4	52	48	
4-Methyl-	Heptane	AlEt$_2$Cl/H$_2$O	0.1		93	<7
	Heptane	TiCl$_3$/AlEt$_3$	3.2	Crystalline, mp = 166 °C		
	Benzene	H$_9$C$_4$Li	1.7		83	17

Source: Refs. [7,376,377].

B. Polyterpenes

Myrcenes and ocimenes are isoprenoids that are occur in plants. They can be considered as multiply substituted 1,3-butadienes [structures (53)–(57)].

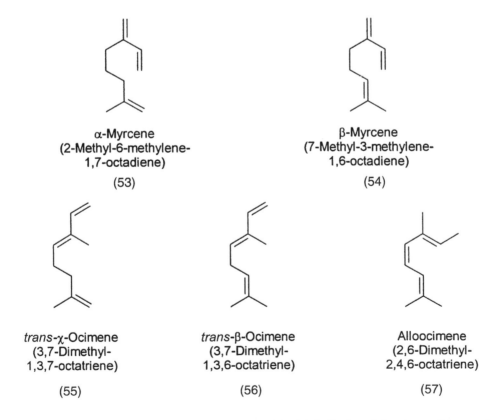

α-Myrcene
(2-Methyl-6-methylene-
1,7-octadiene)

(53)

β-Myrcene
(7-Methyl-3-methylene-
1,6-octadiene)

(54)

trans-χ-Ocimene
(3,7-Dimethyl-
1,3,7-octatriene)

(55)

trans-β-Ocimene
(3,7-Dimethyl-
1,3,6-octatriene)

(56)

Alloocimene
(2,6-Dimethyl-
2,4,6-octatriene)

(57)

With Ziegler catalysts such as titanium(IV) chloride/triisobutylaluminum they can be polymerized to rubbery products with high molecular weights [387,388]. Also, Lewis acids

Table 19 Polymerization of 2,4-hexadiene with various catalysts.

Catalyst	Solvent	Time (h)	Yield (%)	Melting point (°C)
H_9C_4-Li	THF	20	11	Liquid oligomer
$TiCl_4$	CH_2Cl_2	19	74	Amorphic
$TiCl_4/(i\text{-}Bu)_3Al$	Hexane	19	47	78
Cp_2TiCl_2/Et_2AlCl	Benzene	3	52	Amorphic
Tris/(2,4-pentandionato)-cobalt/Et_2Al	Benzene	14	32	78

Source: Ref. [380].

Table 20 Polymerization of trienes and tetraenes.

Polymer	Monomer (structure)	Refs.
Poly(1,3,5-hexatriene)	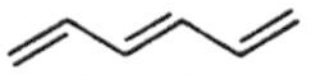	[396,398]
Poly(2,3,4,5-tetrachloro-1,3,5-hexatriene)	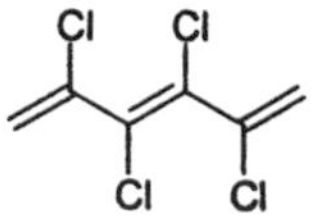	[397]
Poly(2,5-dimethyl-1,3,5-hexatriene)	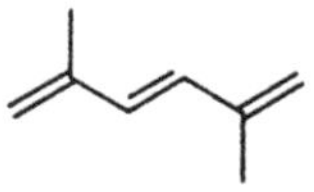	[396,398]
Poly(2,4,6-octatriene)	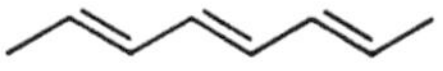	[396,398]
Poly(1,3,5,7-octatetraene)	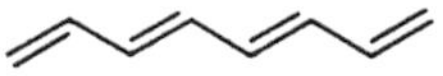	[399]
Poly(3-methyl-1,4,6-heptatriene)	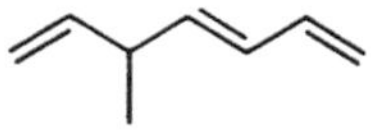	[400]
Poly(1,3,6-octatriene)	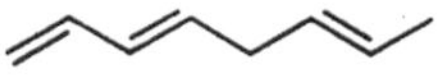	[401]

are described as catalysts [389–391]. The content of 1,4-structures is high [392,393]. Myrcene yields only low-molecular-weight products with mostly (93%) 1,4 and few 3,4 structures when it is polymerized with butyllithium [Structures (58) and (59)] [393–395]. Polymers of other polyenes are compiled in Table 20 [396–401].

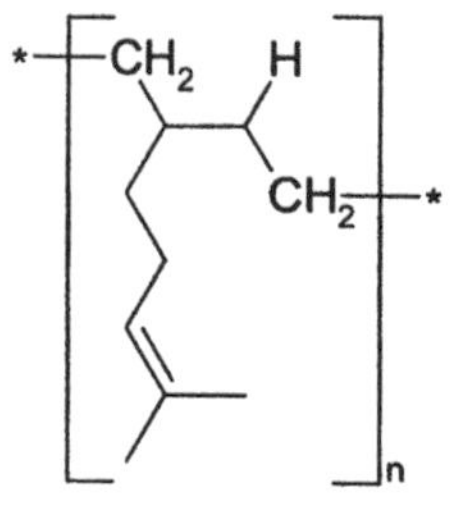

1,4-microstructure

(58)

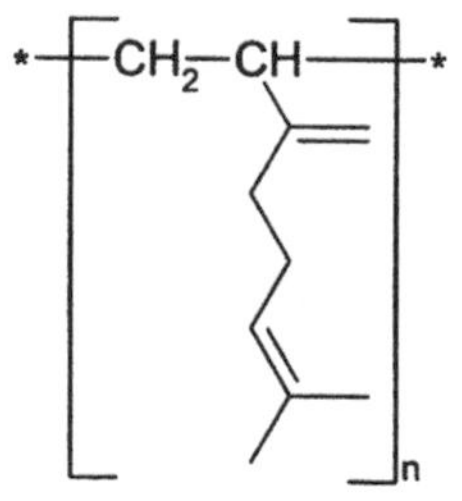

3,4-microstructure

(59)

C. Polycyclodienes

The cationic polymerization of cyclopentadiene is catalyzed by tin-tetrachloride/ trichloroacetic acid/boron trifluoride [402–407]. The polymers that are partially insoluble in hydrocarbons feature 40 to 60% 1,2-structures next to *trans*-1,4 portions [Structures (60) and (61)]. With $Ti(OC_4H_9)Cl_3$, soluble high-molecular-weight polycyclopentadiene is formed. Due to H-atoms in allyl and tertiary positions of the chain, the polymer is extremely sensitive to oxidation. This can be overcome by chlorination of the double bonds [408,409].

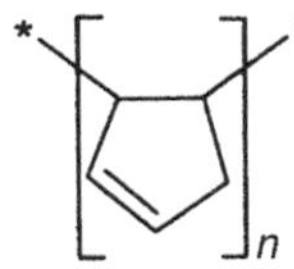

(60) 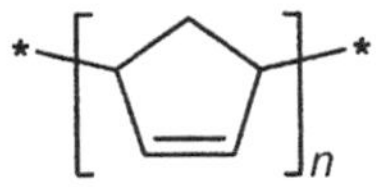   (61)

The cationic polymerization of 1,3-dimethylcyclopentadiene can be initiated by boron trifluoride/diethyl ether/tin(IV) chloride, titanium(IV) chloride, and triethylaluminum/titanium(IV) chloride [410]. The poly (1,3-dimethylcyclopentadienes) that are obtained at $-78\,^\circ$C are easily soluble with identical portions of 1,4 and 1,2 structures [411]. Under the same conditions, 1-methyl- and 2-methylcyclopentadienes are polymerized to powdery polymers [412]. The same applies to allylcyclopentadiene and allylmethylcyclopentadiene. The polymerization proceeds exclusively via the C–C double bond in the cyclopentadiene ring with increasing portions of 1,4-structures in the polymers for monomers with growing steric hindrance.

The polymerization of 1,3-cyclohexadiene can be initiated by free radicals, Ziegler–Natta catalysts, and transition metal catalysts, both cationically and anionically [413–416]. The synthesis of high-molecular-weight poly(1,3-cyclohexadiene) containing 1,4-structures (62), especially has been of great interest, as unbranched poly(*p*-phenylenes) are accessible from it after dehydrogenation [417–419].

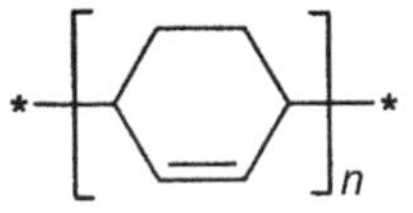

(62)

The cationically produced amorphous poly(1,3-cyclohexadiene) contains 1,4 and 1,2 structures and has a low molecular weight of 3000 and a softening range of 114 to 130 °C [420]. The anionic polymerization of 1,3-cyclohexadiene was carried out with alkyllithium and naphthalene alkali metals (Table 21) [421]. The effect of the solvent on the microstructure is analogous to that in the polymerization of other dienes [422–424]. Even in polar solvents the content of 1,2 sequences cannot be increased beyond 50%. 1,4 Structures are present in any case. Steric hindrance prevents the formation of longer 1,2 sequences. As a rule, the polymers formed features a narrow molecular weight distribution. The tendency toward chain transfer reactions depends on the counterion and increases in going from lithium to potassium. With butyllithium high-molecular-weight polymers with degrees of polymerization $P_n > 700$ are formed. In contrast to this,

Table 21 Polymerization of 1,3-cyclohexadiene with napthalene alkali metals.

Catalyst metal	Catalyst (mol/l)	Solvent	Temp (°C)	Time (h)	Yield (%)	M_n	M_w/M_n
Li	4.0	THF	25	2	48	13 000	1.48
Li	4.3	THF	−20	1.2	96	12 600	1.19
Na	7.95	THF	25	1.6	62	4 300	1.57
K	7.52	THF	25	0.2	61	2 400	1.48
Li	1.7	DME	−20	13.5	95	19 000	
Na	2.34	DME	29	19	48	570	1.44
Na	3.28	DME	−73	2.5	81	38 700	1.45

Source: Ref. [421].

Ziegler–Natta catalysts produce lower-molecular-weight polymers consisting mainly of 1,4 structures [418].

Allyl nickel complexes yield crystalline poly(1,3-cyclohexadienes) with 90% 1,4-structures and melting points between 180 and 270 °C [425]. Thiourea inclusion templates give polymers with purely 1,4 structures and unusually high melting points, between 370 and 380 °C. Additional polymers of cyclic dienes are listed in Table 22 [331,426–429].

D. Nonconjugated Polydienes

Nonconjugated dienes could have double bonds in which one is more reactive than the other [450–453] (e.g., 1,4-hexadiene and 5,7-dimethyl-1,6-octadiene). In this case only the reactive double bond is polymerized to give linear polymer chains. There are nearly no cyclic rings in the polymer. The situation is different if both double bonds have similar reactivities (e.g., 1,5-hexadiene and 1,6-heptadiene) [454–461]. In this case polymers with a high proportion of cyclic units are found:

$$(63)$$

Cyclopolymerization of 1,5-hexadiene was reported by Marvel and Stille [462] and Makowski [463]. They use catalysts derived from $TiCl_4/Al(i\text{-}Bu)_3$ and $TiCl_4/AlEt_3$ which show low activities and incomplete cyclization of the diolefin. With $TiCl_3/AlEt_2Cl$ complete cyclization and a 1:1 ratio of *trans*- and *cis*-cyclopentane rings in the polymer were found [454].

The highest activities give the homogeneous catalysts Cp_2ZrCl_2 or $Cp_2Zr(CH_3)_2$ with methylalumoxane [464]. Polymerization in bulk monomer proceeded with 100% conversion after 1 h, greater than 99% cyclization has taken place and 59% of the polymer ($M_w \sim 27\,000$) was insoluble in benzene, suggesting that some cross-linking may have occurred. At a polymerization temperature of 80 °C, ^{13}C-NMR analysis indicates that

Table 22 Polymerization of cyclic dienes.

Monomer	Polymer		Catalyst	Refs.
Methylen-cyclobutene	1,5-linkage		Anionic Cationic	[430,431]
1-Methyl-3-methylen-cyclobutene	1,2-linkage		Anionic Cationic	[432]
1,2,Bis(methylen)-cyclobutane	1,4-	1,2-linkages	Cationic	[431]
Spiro(2,4)hepta-4,6-diene	1,4-	1,2-linkages	Cationic	[433,434]
cis-1,2-Bis-(vinyl)-cyclohexane			Ziegler–Natta	[435]
3-Methylen-cyclohexene	3,1-linkage		Cationic Ziegler–Natta	
1,3-Cycloheptadiene	3,4-	1,2-linkages	Ionic	
1,3-Cyclooctadiene	1,2-	1,4-linkages	Cationic	

there is a 1:1 ratio between *trans* and *cis* five-membered rings:

$$trans/cis \qquad (64)$$

At a polymerization temperature of 22 °C, approximately 80% of the cyclopentane rings in the polymer are *trans*. Table 23 shows some other examples of the polymerization of nonconjugated dienes to cyclic units [320].

VIII. COPOLYMERIZATION OF DIENES

A. 1,3-Butadiene-Styrene-Copolymers

Copolymers of 1,3-butadiene and styrene (SBR) are elastomers of great technical importance that are used for automobile tires [465–474]. In addition to a free-radical process, they can be made by anionic initiation with alkyllithium compounds. In polar solvents the reaction rate of styrene anions with 1,3-butadiene is greater than with styrene, whereas in polar solvents this is just the other way around. The copolymerization parameter r_7 for styrene-butadiene is 0.03 in hexane and 8 in THF; r_2 is calculated as 12.5 in hexane and 0.2 in THF [465]. Therefore, a strong dependence of the styrene content of the polymers on the degree of conversion is observed in discontinuous polymerizations.

At first only little styrene is converted, and it is not until the monomer mixture is largely depeleted of 1,3-butadiene that the principal amount of styrene polymerizes forming a block consisting mainly of styrene. In THF, also, the content of vinyl double bonds increases and thus the glass transition temperature is raised [475]. A uniform copolymer composition can be achieved by a continuous process or by programmed addition of butadiene to the reaction mixture [476,477]. Copolymers with a high *trans*-1,4 content are obtained with barium compounds [478]. In contrast, the coordinative copolymerization yields products with a high *cis*-1,4 content.

The copolymerization of 1,3-butadiene and styrene in an emulsion process is initiated by free radicals [121]. For this purpose radical-forming initiators such as azoperoxy compounds are added to the emulsion. When the entire monomer mixture is in the reactor at the start of the polymerization, the monomeric units are converted more or less statistically, depending on diffusion rate and reactivity.

Addition of one monomer at a later point of the reaction results in the formation of latex particles of the core-shell type [479]. For special lattices, acrylonitrile and methacrylic acid are added [480,481]. Heat-sensitive lattices contain such additives as quaternary ammonium compounds, sulfonamides, polyether urethanes, and organopolysiloxanes, which affect the coagulation of the latex above a certain temperature [482].

Potassium laurate and sodium abietate are often used as emulsifiers that are attached to the surfae of the latex particles. There are also emulsifiers such as maleic acid monoesers, methacrylic acid, and vinyl benzene sulfonic acid that are built into the polymer, thus improving the tensile strength of the final rubber coating [483]. The nature and concentration of the emulsifier is also a means to controlling the particle size distribution of the products. In monodisperse lattices the particle size varies by less than 1% which can be achieved by short polymerization times [484,485]

Table 23 Polymerization of cyclic dienes.

Monomer	Polymer	Catalyst	Refs.
1,5-Hexadiene	3,1-linkage	Ziegler–Natta	[436,437]
2,5-Dimethyl-1,5-hexadiene	3,1-linkage	Ziegler–Natta	[436,437]
2-Methyl-1,5-hexadiene	3,1-linkage	Ziegler–Natta	[438]
1,5-Heptadiene		AlEt$_3$/TiCl$_4$	[439]
1,6-Heptadiene	3,1-linkage	Al(i-Bu)$_3$/TiCl$_4$	[440]
1,(ω – 1)Alkadiene (CH$_2$)$_n$ n=4–18		Ziegler–Natta	[441]

(continued)

Table 23 Coninued.

Monomer	Polymer	Catalyst	Refs.
2,(ω−1)Diphenyl-1,(ω−1)-alkadiene		Radical, anionic, Cationic	[442,443]
1,4-Bis-(methylen)-cyclooctadiene		BF$_3$	[444]
1,5-Cyclooctadiene	2,6-linkage	Ziegler–Natta	[445]
cis,cis-1-Methyl-1,5-cyclooctadiene	2,6-linkage	Ziegler–Natta	[446]
Norbomadiene	3,5-	Radical Cationic	[447,448]
5-Methylen-bicyclo-(2.2.1)heptene	5,2-linkage	AlBr$_3$	[449]

The styrene–butadiene elastomers must also be blended stabilizers and antioxiants [486,487]. Monomeric stabilizers [Structures (65–67)] can also be added to the reaction mixture. In this way they are incorporated into the products, which prevents them from being washed out [488].

$$(65-67)$$

B. Block Copolymers

Due to the absence of termination reaction in the anionic polymerization with alkyllithium, it is possible to produce styrene/1,30butadiene/styrene triblock polymers [489–495]. To do this, one starts by polymerizing styrene in a polar solvent, then adding butadiene, and finally after addition of a polar solvent, adding styrene again. To evade a twofold addition of monomer processes have been developed in which bifunctional starters make it possible first to polymerize butadiee and then to build the styrene blocks on both ends of the polybutadiene chain by adding styrene [496,497]. Another method used is based on merging two living stock copolymers by means of bifunctional coupling reagents [(e.g., bis(bromoethyl) ether and 1,2-dibromoethane)] [489].

C. Copolymers with Dienes and Olefins

By using rare earth metals or radicals it is possible to copolymerize 1,3-butadiene and other dienes with *cis*-1,4 linkage [3,498]. Polymers of 1,3-butadiene and isoprene at any ratio can be obtained. Copolymes of 1,3-butadiene and 1,3-pentadiene can be produced with catalysts on the basis of vanadium chelates. 1,3-Butadiene is almost completely converted to *trans*-1,4 units, whereas 1,3-pentadiene yields 50 to 60% 1,4-addition and 40 to 50% 1,2-addition products. At a 1,3-pentadiene content of 26 to 45 wt%, the copolymers are amorphous, featuring high rigidity [499–501]. Diethylaluminum chloride, nickel naphthenate, and water catalyze the copolymerization of 1,3-butadiene and acetylene. The low-molecular-weight copolymers contain mostly *cis*-C–C double bonds [502].

Alternating copolymers of 1,3-butadiene with ethene, propene, acrylonitrile, isobutene, and 2-methylacrylic acid methyl ester are known [503–513]. Suitable catalysts are titanium and vanadium compounds. The stereoregularity of the butadiene units is relatively low, with 85% *trans*-1,4 linkage. In contrast to this, the bis(2,2-dimethyl-propyloxy)vanadium oxide chloride/triisobutylaluminum system forms strictly alternating copolymers with more than 98% *trans*-1,4 structures [507,514]

Statistical copolymers of 1,3-butadiene and ethene are obtained with bis(cyclopen-tadienyl)zirconium dichloride and aluminoxane. The copolymers contain up to 13% *trans*-1,4 bonded butadiene [515]. 1,3-Butadiene forms alternating copolymers with acrylonitrile and esters of acrylic acid when the latter is complexed by strong Lewis acids (e.g., alkylaluminum halogenides and zinc chloride) [516,517]. Selective hydrogenation of the double bond yields nitrile rubbers with a low swelling capacity.

D. Isoprene Copolymers

With the aid of Ziegler catalysts it is possible to copolymerize isoprene with ethene and other α-olefins. Just like the analogous butadiene copolymers, the products are of alternating structure [518–521].

Copolymerization of isoprene with acrylic monomers is also possible [516,522,523]. The copolymers can be synthesized with statistical as well as strictly alternating structure [524]. Ethylaluminum dichloride is used as a catalyst. The dine units are incorporated predominantly with *trans*-1,4 enchainment. Addition of transition metal compounds or radical-forming compounds increases the polymerization rate [525,526]. Cationic palladium complexes are used as highly active catalysts in the polymerization of isoprene with diethylamine [527].

It is possible to synthesize copolymers of 1,3-butadiene and isoprene by using a MeCpTiF$_3$/MAO catalyst. The copolymer has a glass transition temperature of $-60\,°C$.

E. Copolymers of Other Dienes

1-Chloro-1,3-butadiene can be polymerized with styrene [528]. The anionic block copolymerization of 1- or 2-phenyl-1,3-butadiene with styrene leads to block polymers of low molecular weight [529]. Similar copolymers are described of 1,3-pentadiene with styrene. With alkyllithium there is no reaction of 1,4-diphenyl-1,3-butadiene with styrene [530].

Quirk [531] reports on the copolymerization of myrcene and styrene. Block copolymers with molecular weight of more than 100 000 are obtained. Many combinations of substituted 1,3-butadienes and cyclodienes with other dienes, olefins, and styrene have been described [532–542]. Cyclopentadiene or 1,3-cyclohexadiene can be copolymerized with α-methylstyrene [543–545], isobutene [546,547], 1,3-butadiene [548,549], acrylonitrile [550,551], SCl$_2$ [552], SO$_2$ [553], and compounds of maleic acid [554–557].

Nonconjugated dienes as ethylidene norbornene, dicyclopentadiene, and 1,4-hexadienes are used as diene components in ethene-propene-diene-monomer (EPDM) elastomers [558–564]. The copolymers are synthesized with Ziegler–Natta catalysts. Vanadium compounds also give living copolymers of propene and 1,5-hexadiene [565]. It was shown that soluble Ziegler catalysts containing ethylene bis(indenyl)zirconium dichloride and methylalumoxane can be used to polymerize 1,5-hexadiene as a more simple diene for terpolymerization with ethene and propene [566,567]. It is possible to incorporate up to 6 mol% of the diene. Higher 1,5-hexadiene concentrations lead to cross-linked polymers. At a polymerization temperature of 25 °C, the molecular weight is 260 000; the molecular weight distribution $M_w/M_n = 6.3$.

REFERENCES

1. Elias, H.-G. (2001). *Makromoleküle*, Wiley-VCH, Weinheim.
2. Henderson, J. N., and Throckmorton, M. C. (1987). *Encycl. Polym. Sci. Eng.*, *10*: 811.
3. Schlüter, A.-D. (ed.) (1999). *Synthesis of Polymers*, Wiley-VCH, Weinheim.
4. Harries, C. H. (1911). *Justus Liebigs Ann. Chem.*, *383*: 184.
5. Porri, L., Giarrusso, A., and Ricci, G. (1994). *Polym. Sci., Ser. A.*, *36*: 1421.
6. Bywater, S., and Worsfold, D. J. (1967). *J. Organomet. Chem.*, *10*: 1.
7. Porri, L., Natta, G., and Gallazzi, M. C. (1967). *J. Polym. Sci., Part C*, *16*: 2525.
8. Marconi, W., Santostasi, M. L. and De Malde, M. (1962). *Chim. Ind. (Milan)*, *44*: 235.
9. Ghosh, P. (1971). *Makromol. Rev.*, *5*: 204.
10. Barlett, P. D., and Kwart, W. (1950). *J. Am. Chem. Soc.*, *20*: 457.
11. Reports (1998). *KGK Kautschuk-Gummi Kunststoffe*, *51*: 140.
12. Duckemin, F., Bennerault-Celton, V., Cheradame, H., Merienne, C., and Macedo, A. (1998). *Macromolekules*, *31*: 7627.
13. Racanelli, P., and Porri, L. (1970). *Eur. Polym. J.*, *6*: 751.
14. Szwarc, M. (1982). *Adv. Polym. Sci.*, *47*: 1.
15. Adams, H. E., Bebb, R. L., Foreman, L. E., and Wakefield, L. B. (1972). *Rubber Chem. Technol.*, *45*: 1252.
16. Morton, M. (1983). *Anionic Polymerization; Principles and Practice*, Academic Press, New York.
17. Tobolsky, A. V., and Rogers, C. E. (1959). *J. Polym. Sci.*, *40*: 73.
18. Foremann, L. E. (1969). *Polymer Chemistry of Synthetic Elastomers* (Kennedy, J. P., and Tornqvist, E., eds.), Part II, Wiley, New York, p. 567.
19. Morton, M., and Fetters, L. J. (1975). *Rubber Chem. Technol.*, *48*: 359.
20. Fetters, C. J., Kiss, A. D., Pearson, D. S., Quack, G. F., and Vitus, F. J. (1993). *Macromolecules*, *26*: 647.
21. Ziegler, K. (1936). *Angew. Chem.*, *49*: 499.
22. Pata, F. (1962). *Pure and Appl. Chem.*, *4*: 333.
23. Gebert, W., Hinz, J., and Sinn, H. (1971). *Makromol. Chem.*, *144*: 97.
24. Bandermann, F. (1971). *Angew. Makromol. Chem.*, *18*: 137.
25. Cowie, J. M. C., Elexpuru, E. M., Harris, J. H., and McEwen, I. J. (1989). *Makromol. Chem. Rapid Comm.*, *10*: 687.
26. Nel, J. G., Wagener, K. B., and Boncella, J. M. (1989). *Polym. Prepr.*, *30*: 130.
27. Jin, G., Fan, L., and Yao, W. (1987). *J. Polym. Mater.*, *4*: 215.
28. Jin, G., Yao, W., and Fan, L. (1988). *J. Polym. Mater*, *5*: 151.
29. Weerts, P. A., van D. Loss, J. L. M., and German, A. L. (1988). *Polym. Commun.*, *29*: 278.
30. Jin, G., and Lu, T. (1987). *J. Polym. Mater*, *4*: 41.
31. Jin, G., and Yang, W. (1989). *J. Polym. Mater*, *6*: 31.
32. Bywater, S., Warsfold, D. G., and Black, P. (1989). *Makromol. Chem. Suppl.*, *15*: 31.
33. Brydson, J. A. (1988). *Rubbery Materials and their Compounds*, Elsevier Applied Science, London.
34. Elvers, B., Hawkins, S., Russey, W., and Schulz, G. (Eds). *Ullmanns' Encyclopedia of Industrial Chemistry*, vol. 13, 4th edn., VCH, Weiheim, p. 595.
35. Hsieh, H. (1965). *J. Polym. Sci., Part A.*, *3*: 163.
36. Lewis, H. L., and Brown, T. L. (1970). *J. Amer. Chem. Soc.*, *92*: 4664.
37. Wittig, G., Meyer, F. J., and Lange, G. (1951). *Liebigs Ann. Chem.*, *571*: 167.
38. Brown, T. L., and Rogers, M. T. (1957). *J. Amer. Chem. Soc.*, *79*: 1859.
39. Brown, T. L., Ladd, J. A., and Newman, G. N. (1965). *J. Organometal. Chem.*, *3*: 1.
40. Hein, F., and Schramm, H. (1930). *Z. Phys. Chem.*, *151A*: 234.
41. Morton, M., Bostick, E. E., Livigni, R. A., and Fetters, L. J. (1963). *J. Polym. Sci.*, *A1*: 1735.
42. Morton, M., and Fetters, L. J. (1964). *J. Polymer Sci.*, *A2*: 3311.

43. Worsfold, D. J., and Bywater, S. (1972). *Macromolecules, 5*: 393.
44. Guyot, A., and Vialle, J. (1970). *J. Macromol. Sci. Chem., A4*: 79.
45. Margerison, D., Bishop, D. M., East, G. C., and McBride, P. (1968). *Trans Faraday Soc., 64*: 1872.
46. Worsfold, D. J., and Bywater, S. (1964). *Can. J. Chem., 42*: 2884.
47. Worsfold, D. J. (1967). *J. Polym. Sci. A-1, 5*: 2783.
48. Johnson, A. F., and Worsfold, D. J. (1965). *J. Polymer Sci. A, 3*: 449.
49. Sinn, H., Lundborg, C., and Onsager, O. T. (1964). *Makromol. Chem., 70*: 222.
50. Rodionov, A. N., Talalaeva, T. V., Shigorin, D. N., Timofeyuk, G. N., and Kocheshkov, K. A. (1963). *Dokl. Akad. Nauk. SSSR, 151*: 1131.
51. Makowski, H. S., and Lynn, M. (1966). *J. Macromol. Chem., 1*: 443.
52. Makowski, H. S., Lynn, M., and Bogard, A. N. (1968). *J. Macromol. Sci. Chem., A-2*: 665.
53. Bywater, S. (1965). *Adv. Polym. Sci., 4*: 66.
54. Hernandez, H., Semel, J., Brücker, H. Ch., Zachmann, H.-G., and Sinn, H. (1980). *Makromol. Chem. Rapid Comm., 1*: 75.
55. Roovers, J. E. L., and Bywater, S. (1968). *Macromolecules, 1*: 328.
56. Morton, M., Rembaum, A. A., and Hall, J. L. (1963). *J. Polym. Sci., Part A, 1*: 461.
57. Bywater, S., Mackeron, D. H., ad Worsfold, D. J. (1985). *J. Polym. Sci. Polym. Chem. Ed., 23*: 1997.
58. Lundborg, C., and Sinn, H. (1960). *Makromol. Chem., 41*: 242.
59. Sinn, H., and Onsager, O. T. (1962). *Makromol. Chem., 55*: 167.
60. Morton, M., Fetters, L. J., Pett, R. A., and Meier, J. F. (1970). *Macromolecules, 3*: 327.
61. Glaze, W. H., Hanicak, J. E., Moore, M. L., and Chadheri,, J. (1972). *J. Organomet. Chem., 44*: 39.
62. Morton, M., and Ruppert, J. (1983). *Initiation of Polymerization* (Bailey, F. E., ed.), ACS Symposium Series Nr. 212, Am. Chem. Soc., Washington.
63. Waack, R., and West, P. (1964). *J. Amer. Chem. Soc., 86*: 4484.
64. Wittig, G., Meyer, F. J., and Lange, G. (1951). *Liebigs Ann. Chem., 571*: 167.
65. Eastham, J. F., and Gibson, G. W. (1963). *J. Am. Chem. Soc., 85*: 2171.
66. Cheema, Z. K., Gibson, G. W., and Eastham, J. F. (1963). *J. Amer. Chem. Soc., 85*: 3517.
67. Guyot, A., and Vialle, J. (1968). *J. Polym. Sci. B, 6*: 403.
68. Hshieh, H. L., Farrar, R. C., and Udipi, K. (1981). *Chemtech., 11*: 626.
69. Antkowiak, T. A., Obester, A. E., Halasa, A. F., and Tate, B. P. (1972). *J. Polym. Sci. Part A-1, 10*: 1319.
70. Minekawa, S., Yonekawa, S., Tabata, H., Ishida, T., Tsuchida, S., and Yamada, K. (1966). DAS 1720249 to Asahi Kasei Kogiyo. *C.A.* (1969) 70: 88696.
71. Witte, J., Pampus, G., and Becker, W. (1962). DAS 1128666 to Bayer.
72. Zelinski, R. P., and Wofford, C. F. (1965). *J. Polym. Sci., Part A3*: 95.
73. Oberster, A. E., Bouton, T. C., and Valaitis, J. K. (1973). *Angew. Makromol. Chem., 29/30*: 291.
74. Nordsiek, K. H. (1972). *Kautsch. Gummi Kunststoffe, 25*: 87.
75. Pederson, C. J., and Freusdorff, K. H. (1978). *Angew. Chem. Int. Ed. Eng., 11*: 16.
76. Lehn, J. M. (1978). *Acc. Chem. Res., 11*: 49.
77. Binder, J. L. (1954). *Anal. Chem., 26*: 1877.
78. Binder, J. L., and Ransaw, H. (1957). *Anal. Chem., 29*: 503.
79. Binder, J. L. (1963). *J. Polymer Sci., A, 1*: 37.
80. Golub, M. A. (1970). *Spectrochim. Acta, 26A*: 1883.
81. Hampton, R. R. (1949). *Anal. Chem., 21*: 923.
82. Kimmer, W., Sehan, R., and Schellenberger, A. (1965). *Plaste Kautschuk, 12*: 104.
83. Morero, D., Santamborgio, Porri, L., and Ciampelli, F. (1959). *Chim. Ind.* (*Milan*), *41*: 759.
84. Richardson, W. S. (1954). *J. Polym. Sci., 13*: 229.
85. Silas, R. S., Yates, J., and Thornton, V. (1959). *Anal. Chem., 1*: 529.
86. Sinak, R., and Fahrbach, G. (1970). *Angew. Makromol. Chem., 12*: 73.

87. Tanaka, Y., Takeuchi, Y., Kobayashi, M., and Tadokoro, H. (1971). J. Polymer Sci. A-2, 9: 43.
88. Mochel, V. D. (1972). *J. Polymer Sci. A-1*, *10*: 1009.
89. Mochel, V. D. (1972). *Rubber Chem. Technol.*, *45*: 1283.
90. Chen, H. Y. (1962). *Anal. Chem.*, *34*: 1134.
91. Chen, H. Y. (1966). *J. Polymer Sci.*, *Part B*, *4*: 891.
92. Clague, A. D. H., van Broekhoven, J. A. M., and De Haan, J. W. (1973). *J. Polymer Sci. Part B*, *11*: 299.
93. Duch, M. W., and Grant, D. M. (1970). *Macromolecules*, *3*: 165.
94. Elgert, K. F., Stützel, B., Frenzel, P., Cantow, H.-J., and Streck, R. (1973). *Makromol. Chem.*, *170*: 257.
95. Morton, M., Sanderson, R. D., and Sakata, R. (1973). *Macromolecules*, *6*: 181.
96. Thomassin, M., Walchiers, E., Warin, R., and Teyssie, P. (1973). *J. Polym. Sci. Part B*, *11*: 229.
97. Youn, R. N., Quirk, R. P., and Fetters, I. J. (1984). *Adv. Polym. Sci.*, *56*: 43.
98. Morton, M., Sanderson, R. D., and Sakata, R. (1971). *J. Polym. Sci. Part B*, *9*: 61.
99. Shen, Z., Long, Z., and Zhong, C. (1980). *Sci. Sin*, *23*: 734.
100. Glaze, W. H., and Adams, J. M. (1966). *J. Am. Chem. Soc.*, *88*: 4653.
101. Kraft, M. (1973). *Struktur und Absorptionsspektroskopie der Kunststoffe*, Verlag Chemie Weinheim.
102. Morton, M., Pett, R. A., and Fethers, L. J. (1970). *Macromolecules*, *3*: 333.
103. Szwarc, M. (1968). *Carboanious, Living Polymers and Electron Transfer Processes*, Wiley Interscience, New York.
104. Hostalka, H., Figini, R. V., and Schulz, G. V. (1964). *Makromol. Chem.*, *71*: 198.
105. Szwarc, M. (1983). *Adv. Polym. Sci.*, *49*: 1.
106. Santee, E. R., Malotky, L. O., and Morton, M. (1971). *Rubber Chem. Technol.*, *46*: 1156.
107. Bywater, S. (1985). *Encyclopedia of Polymer Science and Engineering*. Vol. 2, John Wiley and Sons, New York, p. 1.
108. Halasa, A. F., Lohr, D. F., and Hall, J. E. (1981). *J. Polym. Sci. Polym. Chem. Ed.*, *19*: 1357.
109. Nordsiek, K. H. (1972). *Kautsch. Gummi Kunstst.*, *25*: 87.
110. Kume, S. (1966). *Markomol. Chem.*, *98*: 120.
111. Lodemann, L., Pospigil, J., and Lim, D. (1966). *Tetrahedron Lett.*, *7*: 257.
112. Luxton, A. R. (1981). *Rubber Chem. Tech.*, *54*: 596.
113. Journ, J., and Widmaier, J. M. (1977). *Eur. Polym. J.*, *13*: 379.
114. Fetters, L. J., and Morton, M. (1969). *Macromolecules*, *2*: 453.
115. Lutz, P., Franta, E., and Rempp, P. (1976). *C. R. Acad. Sci.*, *283*: 123.
116. Lutz, P., Frantza, E., and Rempp, P. (1982). *Polymer*, *23*: 1953.
117. Höcker, H., and Schulz, G. (1979). DEOS 2938658 to BASF. *C.A.* (1982) *95*: 133 137.
118. Bandermann, F., Spekamp, H. D., and Weigel, L. (1985). *Makromol. Chem.*, *186*: 2017.
119. Witte, J. (1981). *Angew Makromol. Chem.*, *94*: 119.
120. Dalgoplosk, B. A., and Tinyakova, E. I. (1984). *Russ. Chem. Rev.*, *53*: 2.
121. Witte, J. (1986). *Methoden der organischen Chemie, Makromolekulare Stoffe* (Houben Weyl), Vol. E 20/1, G. Thieme Press, Stuttgart, New York, p. 94.
122. Mazzei, A. (1981). *Makromol. Chem. Suppl.*, *4*: 61.
123. Natta, G., Porri, L., and Mazzei, A., (1959). *Chim. Ind. (Milan)*, *41*: 398.
124. Kraus, G., Short, J., and Thornton, V. (1957). *Rubber Plast., Age*, *38*: 880.
125. Tucker, H. (1958). DEP 1128143 to Goodrich Gulf. *C. A.* (1963) *58*: 1614.
126. Maeda, K. Onishi, A., and Ueda, K. (1959). GB. P. 906334 to Bridgestone Tire, *C.A.* (1962) *57*: 1011.
127. Throckmorton, M. C. (1969). *Kautsch. Gummi Kunstst.*, *22*: 293.
128. Keim, W. (1990). *Angew. Chem.*, 102d: 251.
129. Sylvester, G., Witte, J., and Marwede, G. (1976). DEOS 2625390 to Bayer AG, *C.A.* (1978) *88*: 75165.
130. Shen, Z., Long, Z., and Zhong, C. (1964). *Sci. Sin*, *13*: 1339.

131. Kaminsky, W. (1986). *Methoden der Organischen Chemie* (*Houbern-Weyl*), Vol. E. 1812, G. Thieme Press, Stuttgart, New York, p. 978.
132. Sylvester, G., Witte, J., and Marwede, G. (1980). DEOS 2848964 to Bayer AG, *C.A.*, (1980) 93, 96555.
133. Natta, G., Porri, L., Corradini, P., and Morero, D. (158). *Chim. Ind.* (*Milan*), *40*: 362.
134. Natta, G., Porri, L., Fiore, L., and Zanini, G. (1958). *Chim. Ind.* (*Milan*), *40*: 116.
135. Natta, G., Porri, L., Carbonaro, A., and Corradini, P. (1962). *Makromol. Chem.*, *51*: 229.
136. Zachoval, J., and Vernovic, B. (1966). *J. Polym. Sci, Part B*, *4*: 965.
137. Bawn, C. E. H., North, A. M., and Walker, J. S. (1964). *Polymer*, *5*: 419.
138. Hsieh, H. L., and Yeh, H. C. (1985). *Rubber Chem. Technol.*, *58*: 117.
139. Pragliola, S., Forlenza, E., and Longo, P. (2001). *Macromol. Rapid Commun.* *22*: 783–786.
140. Hughes, R. P., and Power, J. (1972). *J. Am. Chem. Soc.*, *94*: 7723.
141. Lobach, M. I., Kormer, V. A., Tsereteli, I. K., Kondratenkov, G. P., Babitskii, B. D., and Klepikova, V. I. (1971). *J. Polym. Sci. Part B*, *9*: 71.
142. Klepikova, V. I., Kondratenkov, G. P., Kormer, V. A., Lobach, M. I., and Churlyaera, L. A. (1973). *J. Polym. Sci. Part B.*, *11*: 193.
143. Wilke, G., Bogdanovic, B., Hardt, P., Heimbach, P., Keim, W., Kröner, M., Oberkirch, W., Tanaka, K., Steinbrücke, E., Walter, D., and Zimmermann, H. (1966). *Angew. Chem.* 78: 157. Engl. Ed. *5*: 151.
144. Dawans, F., and Teyssie, Ph. (1971). *Ind. Eng. Chem. Prod. Res. Dev.*, *10*: 261.
145. Ricci, G., Botta, G., and Porri, L. (1986). *Makromol. Chem. Rapid Commun.*, *7*: 355.
146. Otsuka, S., and Kawakami, M. (1965). *Kogyo Kagaku Zasshi*, *68*: 874.
147. Tolman, C. A. (1970). *J. Am. Chem. Soc.*, *92*: 6777.
148. Zelinski, R., and Smith, D. (1956). GB.P. 848065 to Phillips Petroleum Comp., *C.A.* (1961) 55: 15982.
149. Stewart, R., Darey, J., and McLeod, L. (1959). US.P. 3409604 to Polym. Corp., *C.A.* (1963), *58*: 1554.
150. Schön, N., Pampus, G., and Witte, J. (1962) to Bayer AG.
151. Mazzei, A., Avaldi, M., Marconi, W., and de Malde, M. (1965); *J. Polym. Sci. Part A*, *3*: 753.
152. Harwarth, M., Gehrke, K., and Ringel, M. (1975). *Plaste Kautschuk*, *22*: 1233.
153. Giachetti, E., and Bortolini, W. (1963) DAS 1193249 to Montecatini, *C.A.* (1965) *63*: 4495.
154. Chaplin, R. P., Burford, R. P., Tory, G. J., and Kirby, S. (1987).*Polymer*, *28*: 1418.
155. Honig, J. A., Gloor, P. E., McGregor, J. F., and Haniefel, A. E. (1987). *J. Appl. Polym. Sci.*, *34*: 829.
156. Sun, L., Lu, Z., Lu, Y., and Lin, S. (1988). *J. Polym. Sci., Part B.*, *26*: 2113.
157. Brockway, C. E., and Ecker, A. F. (1955) US. P. 2977349 to Goodrich Gulf, *C. A.* (1961) 55: 16012.
158. Saltman, W. M., and Kuzma, L. J. (1973). *Rubber Chem. Technol.*, *46*: 1055.
159. Lasis, E. (1978) US.P. 4242478 to Polysar, *C.A.* (1980) *92*: 60147.
160. Horne, S. E. (1983). *Transsition Metal Catalyzed Polymerizations* (Quirk, R. P., ed.), MMI Press, Harwood Academic Publ., New York, p. 527.
161. Ichigawa, T., Ikeda, M., Yamamoto, H., Ozawa, N., Yasunaga, H., Ueda, K., and Yokkaichi, M. (1965). DAS 1620927 to Japan Synthetic Rubber Co., *C.A.* (1968) *68*: 13886.
162. Yoshimoto, T., Komatsu, K., Sakata, R., Yamamoto, K., Takenchi, Y., Onishii, A., and Ueda, K. (1970). *Makromol. Chem.*, *139*: 61.
163. Sakata, R., Honoso, J., Onishi, A., and Ueda, K. (1970). *Makromol. Chem.*, *139*: 73.
164. Tkac, A., and Stasko, A. (1973). *Collect Czech. Chem. Comm.*, *38*: 1346.
165. Novikova, E. S., Parenago, O. P., Frolov, V. M., and Dolgoplosk, B. A. (1976). *Kinet. Katal.*, *17*: 928; *C.A.* (1976) *85*: 160654.
166. Wilke, G. (1963). *Angew. Chem.*, *75*: 10. Engl. Ed. *2*: 105.
167. Byrikhin, V. S., and Kadantseva, A. I. (1974). *Vysokomol. Soedin. Ser. B.*, *16*: 899, *C.A.* (1975) *82*: 98517.
168. Gebauer, U., Ludwig, J., and Gehrke, K. (1988).*Acta Polym.*, *39*: 368.

169. Gebauer, U., and Gehrke, K. (1988). *Plaste Kautsch.*, *36*: 109.
170. Gebauer, U., Reggentin, M., and Gehrke, K. (1988). *Plaste Kautsch.*, *35*: 78.
171. Durand, J. P., Dawans, F., and Teyssie, P. (1970). *J. Polym. Sci., Part A*, *1*: 979.
172. Von Dohlen, W. C., Wilson, T. P., and Caflish, E. G. (1963), US.P. 3297667 to Union Carbide Corp., *C.A.* (1965) *63*: 58746.
173. Tse-chuan, S., Chung-yuan, G., Chung-chin, C., and Chung, O. (1964). *Sci. Sin.*, *8*: 1339.
174. Witte, G. (1982). *Angew. Makromol. Chem.*, *94*: 119.
175. Pedretti, U., Lugli, G., Poggio, S., and Mazzei, A. (1977), DE OS 2833721 to Anic, *C.A.* (1979) *90*: 169316.
176. Chen, W., Jin, Z., Xing, Y., Fan, Y., and Yang, G. (1987). *Inorg. Chim. Acta*, *130*: 125.
177. Murinov, Y. I., and Monakov, Y. B. (1987). *Inorg. Chim. Acta*, *140*: 125.
178. Sabirov, Z. M., Minchenkova, N. K., and Monakov, Y. B. (1989). *Inorg. Chim. Acta*, *160*: 99.
179. Kozlov, U. G., Marina, N. G., and Saveleva, J. G. (1988). *Inorg. Chim. Acta*, *154*: 239.
180. Shan, C., Liu, Y., Wang, M., and Shi, E. (1987). *Kexue Tongbao (Foreign Lang. Ed.)*, *32*: 964.
181. Zhang, X., Jin, Y., and Pei, F. (1987). *Kexue Tongbao (Foreign Lang. Ed.)*, *32*: 821.
182. Li, Y., and Ouyang, J. (1987). *J. Macromol. Sci. Chem., Part A*, *24*: 227.
183. Lee Ho D., Lee Hee D., and Ahn Tae O. (1988). *Polymer*, *29*: 713.
184. Li, Y., Liu, G., and Yu, G. (1989). *J. Macromol. Sci. Chem., Part A*, *26*: 405.
185. Horne, S., in Quirk, R. P. (ed.) (1983). *Transition Metal Catalyzed Polymerizations*, Vol. 4, part B, Harwood Academic Publishers, New York, p. 527.
186. Burford, R. P. (1982). *J. Macromol. Sci. Chem.*, A*17*: 123.
187. Bhowmick, A. K., and Stevens, H. L. (1988). *Handbook of Elastomers, New Developments and Technology*, Dekker, New York.
188. Takenchi, Y., Sakakibara, M., ad Shibata, T. (1981). G.B.P. 2101616 to Japan Synthetic Rubber Co., *C.A.* (1983) *99*: 54873.
189. Sylvester, G., and Wieder, W. (1982). *ACS Symp. Ser.*, *193*: 57.
190. Lugli, G., Mazzei, A., and Brandi, G. (1972). DE.P. 2359581 to Snam-Progetti, *C.A.* (1975) *82*: 18346.
191. Kaminsky, W., and Scholz, V. (2001). *Organometallic Catalysis and Olefin Polymerization: Catalyts for a New Millenium* (Blorn, Follestad, Rytter, Tilset, Ystenes eds.), Springer, Berlin–Heidelberg, p. 346.
192. Porri, L., Giarrusso, G., and Ricci, G. (2000). *Metallocene-based Polyolefins* (Scheirs, J., and Kaminsky, W., eds.), Wiley Series, Vol. 2, Chichester, p. 115.
193. Antipov, E., and Podolsky, Y. (1998). *Journal Macromol. Sci. Phys.*, *B37*(04): 431–450.
194. Natta, G., Porri, L., and Mazanti, G. (1955). DOS 1420553 to Montecatini, *C.A.* (1959) *53*: 3756.
195. Gaylord, N., Kwei, T., and Mark, H. (1960). *J. Polym. Sci.*, *17*: 417.
196. Van Amerongen, G. (1966). *Adv. Chem. Ser.*, *52*: 136.
197. Tornqvist, E., and Cozewith, C. (1973). DOS 2441015 to Exxon *C.A.* (1975) *82*: 157574.
198. Dolgoplosk, B., Tinyakowa, E., Stefanovskaya, N., Oreshkin, I., and Shnonina, V. (1974). *Eur. Polym. J.*, *10*: 605.
199. Degler, G., and Hank, R. (1963). DAS 1262019 to Dunlop, *C.A.* (1968) 79392.
200. Cooper, W., Eaves, D., and Vaughan, G. (1966). *Adv. Chem. Ser.*, *52*: 46.
201. Halasa, A., and Hall, J. (1978). DOS 2900800 to Firestone, *C.A.* (1979) *91*: 124234.
202. Teyssie, P., Julemont, M., Thomassin, J., Walckiers, E., and Warin, R. (1975). *Coordination Polymerization* (Chien, J., ed.), Academic Press, New York, p. 327.
203. Hargis, I., and Livigni, R. (1974). US.P. 3903019 to General Tire, *C.A.* (1975) *83*: 207374.
204. DOS 2524849 (1974) to Michelin, *C.A.* (1977) *86*: 91448.
205. Hattori, Y., Ikematu, T., Ibaragi, T., and Honda, M. (1978). DOS 2936653 to Asahi, *C.A.* (1980) *92*: 199065.
206. Natta, G., Porri, L., and Carbonaro, A. (1964). *Makromol. Chem.*, *77*: 126.
207. Ichikawa, M., Kurita, H., and Kogure, A. (1966). US.P. 3498963 to Japan Synthetic Rubber, *C.A.* (1969) *71*: 82394.

208. Kampf, W., and Nordsiek, K. (1976). DOS 2447203 to Chemische Werke Huels, *C.A.* (1976) *85*: 47938.
209. Sugiura, S., Ueno, H., and Hamada, H. (1970). US.P. 3778424 to Ube Ind., *C.A.* (1971) *75*: 65081.
210. Ashitaka, H., Ishikawa, H., Ueno, H., and Nagasaka, A. (1983). *J. Polym. Sci., Polym. Chem. Ed., 21*: 1853.
211. Dawans, F., and Teyssie, P. (1965). US.P. 3451987 to Institut Francais du Petrole, *C.A.* (1968) *68*: 3663.
212. Dawans, F., and Teyssie, P. (1970). DOS 2157004 to Insitut Francais du Petrole, *C.A.* (1972) *77*: 89090.
213. Dolgoplosk, B. A., and Tingakova, E. (1970). *Izv. Akad. Nauk SSSR, Ser. Khim*: 344; C.A. (1970) *73*: 25939.
214. Weerts, P. A., van der Loos, J. L. M., and German, A. L. (1989). *Makromol. Chem., 190*: 777.
215. Lueddeckens, G., and Wendler, K. (1987). *Plaste Kautsch., 34*: 287.
216. Chiantore, O., die Cortemiglia, M. P. C., Guaito, M., and Rendina, G. (1989). *Makromol. Chem., 190*: 3143.
217. Witte, J. (1987). *Houben Weyl, Methoden der Organischen Chemie, Makromol. Stoffe*, Vol. E 20/2, G. Thieme Press, Stuttgart, New York, p. 822.
218. Brock, M. J., and Hackathorn, M. J. (1972). *Rubber Chem. Technol., 45*: 1303.
219. Bean, A. R., Himes, G. R., Holden, G., Houston, R. R., Langton, J. A., and Mann, R. H. (1967). *Encyclopedia of Polymer Science and technology*, Vol. 7, J. Wiley and Sons, New York, p. 786.
220. Cooper, W. (1977). *The Stereo Rubbers* (Saltman, W. M., ed.), Wiley Interscience, New york, Cap. 2, p. 48.
221. Mayr, H., Schneider, R., and Schade, C. (1988). *Makromol. Chem., Macromol. Symp., 13/14*: 43.
222. Tappe, R., and Bandermann, F. (1988). *Angew. Makromol. Chem., 160*: 117.
223. Sinn, H., and Lundborg, C. (1961). *Makromol. Chem., 47*: 86.
224. Bandermann, F., and Sinn, H., (1966). *Makromol. Chem., 96*: 150.
225. S. N. Chakravarty, S. K. Mustafi, and A. K. Mukherjee (1989). *Rubber News* 28, July, 33.
226. Graulic, W., Swodenk, W., and Theisen, P. (1972). *Hydrocarbon Process, 51*: 12.
227. Joly, D. (1968). *Z. Anal. Chem., 236*: 259.
228. Fowler, R., and Barker, D. (1971). *Chem. Eng. (London)*: 322.
229. Patat, F., and Sinn, H. (1958). *Angew. Chem., 70*: 496.
230. Beattie, W. H., and Booth, C. (1963). *J. Appl. Polym. Sci., 7*: 507.
231. Fetters, L. J., and Morton, M. (1975). *Rubber Chem. Technol., 48*: 359.
232. Morton, M., Bostick, E. E., and Clarke, R. G. (1963). *J. Polym. Sci., Part A, 1*: 475.
233. Roovers, J. E. L., and Bywater, S. (1975). *Macromolecules, 8*: 251.
234. Burwell, R. L. (1954) *Chem. Rev., 54*: 615.
235. Gilman, H., Haubein, A. H., and Hartzfeld, H. (1954). *J. Org. Chem., 19*: 1034.
236. Worsfold, D. J., and Bywater, S. (1963). *Makromol. Chem., 65*: 245.
237. Bywater, S., and Worsfold, D. J. (1966). *Adv. Chem. Serv., 52*: 36.
238. Fetters, L. J., Morton, M. (1974). *Macromolecules, 7*: 552.
239. Hadjichristidis, N., and Roovers, J. E. L. (1974). *J. Polym. Sci., Polym. Phys. Ed., 12*: 2521.
240. Zilliox, J., Rempp, P., and Parrod, J. (1968). *J. Polym. Sci., Part C., 22*: 145.
241. Yamazaki, N. T., Luminoe, T., and Kambara, S. (1963). *Macromol. Chem., 65*: 157.
242. Schönberg, E., Chalfant, D. L., and Hanlon, T. L. (1996). *Adv. Chem. Ser., 52*: 6.
243. Tolstopyatov, G. M., Raitses, B. F., Kornia, R. A., and Bresler, R. S. (1978). *Kinet Katal., 18b*: 1136.
244. Haas, F., Kuntz, E., Pampus, G., Schön, N., and Witte, J. (1968). DIS 1720722 to Bayer AG, *C.A.* (1970) *72*: 44805.
245. Gippin, M. (1970). *J. Appl. Polym. Sci., 44*: 1807.
246. Schoenberg, E. (1963). GB. P. 992189 to Goodyear, *C.A.* (1964) *61*: 14873.
247. Marconi, W., Mazzei, A., Cesca, S., and de Malde, M. (1969).*Chim. Ind. (Milan), 51*: 1084.

248. Marconi, W., Mazzei, A., Cucinella, S., and de Malde, M. (1964). *Makromol. Chem.*, *71*: 118.
249. Mazzei, A., Cucinella, S., and Marconi, W. (1969). *Makromol. Chem.*, *122*: 168.
250. Balducci, A., Bruzzone, M., Cucinella, S., and Mazzei, A (1975). *Rubber Chem. Technol.*, *48*: 736.
251. Cucinella, S. (1977). *Chim. Ind. (Milan)*, *59*: 696.
252. Bruzzone, M., Mazzei, A., and Giuliani, G. (1974). *Rubber Chem. Technol.*, *47*: 1175.
253. Takeuchi, y., Sakakibara, M. and Shibata, T. (1982). Birt Pat 2, 101, 616 to Japan Synthetic Rubber, *C.A.* (1983), *99*: 54873.
254. Baogong, Q., Fuseng, Y., Rongshi, C., Wanjum, R., Shentian, l., Yashon, Z., and Yuhua, Y. (1985). *J. Appl. Polym. Sci.*, 30–375.
255. Honeychuck, R. V., Bonnesen, P. V., Farahi, J., and Hersh, W. H. (1987). *J. Org. Chem.*, *52*: 5293.
256. Lee, D. H., Jang, J. K., and Ahn, T. O. (1987). *J. Polym. Sci. Part A. Polym. Chem.*, *25*: 1457.
257. Gebauer, U., Engelmann, S., and Gehrke, K. (1989). *Acta Polym.*, *40*: 341.
258. Lovering, E. G., and Wright, W. B. (1968). *J. Polym. Sci. Part A-I*, *6*: 2221.
259. Cooper, W., and Vaughan, G. (1962). Brit. Pat. 1,024,179 to Dunlop Rubber Co., *C.A.* (1964) *61*: 7209.
260. Cooper, W., Gaves, D. E., Owen, G. D. T., and Vaughan, G. (1964). *J. Polym. Sci. Part C*, *4*: 211.
261. Lasky, J. S., Garner, H. K., and Ewart, R. H. (1962). *Ind. Eng. Chem. Prod. Res. Develop.*, *1*: 82.
262. Soboleva, T. V., Jakovler, V. A., Tinyakova, E. T., and Dolgoplosk, B. A. (1973). *Dokl. Akad Nauk SSR*, *212*: 893; *C.A.* (1974), *80*: 84422.
263. Shominina, V. L., Stetanovskaya, N. N., Tinyakova, E. T., and Dolgoplosk, B. A. (1974). *Eur. Polym. J.*, *10*: 605.
264. Chigir, N. N., Guzman, I. S., Sharnev, O. K., Tinyakova, E. T., and Dolgoplosk, B. A. (1982). *Dokl. Akad Nauk SSR*, *263*: 375, *C.A.* (1982) *97*: 56281.
265. Jenkins, D. K. (1985). *Polymer*, *26*: 147.
266. Farina, M. (1987). *Recent Advances in Mechanistic and Synthetic Aspects of Polymerization* (Fantanille, M., and Guyot, A., eds.), Nato ASI Series, Reidel, Dordrecht, The Netherlands, p. 261.
267. Chatani, Y. (1974). *Progr. Polym. Sci. Jpn.*, 7, 149.
268. Takemoto, K., and Miyataa, M. (1980). *J. Macromol. Sci. Rev. Macromol. Chem. Part C*, *18*: 83.
269. Farina, M. (1981). *Makromol. Chem. Suppl.*, *4*: 21.
270. Farina, M., Silvestro, G. D., and Sozzani, P. (1983). *Mol. Tryst. Liq. Cryst.*, *93*: 169.
271. Farina, M., Allegra, G., and Natta, G. (1964). *J. Am. Chem. Soc.*, *86*: 516.
272. Chantani, Y., Nakatani, S., and Tadokoro, H. (1970). *Macromolecules*, *3*: 481.
273. Miyata, M., Kitahara, Y., and Takemoto, K. (1981). *Polym. J.*, *13*: 111.
274. Miyata, M., Noma, F., Osaki, Y., Takemoto, K., and Kamachi, M. (1986). *J. Polym. Sci. Polym. Lett.*, *24*: 457.
275. Farina, M., Andisio, G., and Gramegna, M. T. (1972). *Macromolecules*, *5*: 617.
276. DiSilvestro, G., Sozzani, P., and Farina, M. (1987). *Macromolecules*, *20*: 999.
277. Davidjan, A., Nikolaew, N., Sgomnik, V., Belenkii, B., Nesterow, V., and Erussalinsky, B. (1976). *Makromol. Chem.*, *177*: 2469.
278. Moustafa, A. B. (1970). *Ind. J. Techn.*, *8*: 290.
279. M. Morton (1983). *Anionic Polymerization: Principles and Practice*, Academic Press, New York.
280. Deshpande, A. B., Subbramanian, R. V., and Kapar, S. L. (1967). *J. Polym. Sci. Part A-1*, *5*: 761.
281. Dawans, F., and Theyssie, P. (1967). *Makromol. Chem.*, *109*: 68.
282. Delheye, G., and Dawans, F. (1966). *Makromol. Chem.*, *98*: 164.
283. Dawans, F., and Theyssie, P. (1969). *Eur. Polym. J.*, *5*: 541.

284. Stewart, C. A. (1971). *J. Am. Chem. Soc.*, *93*: 4815.
285. Sufeak, M. (1970). *J. Appl. Polym. Sci.*, *14*: 1103.
286. Sufeak, M. (1968). *J. Appl. Polym. Sci.*, *12*: 2193.
287. Sufeak, M. (1971). *J. Appl. Polym. Sci.*, *15*: 2555.
288. Johnson, P. R. (1976). *Rubber Chem. Technol.*, *49*: 650.
289. Miyata, M., Noma, F., and Okanishi, K. (1987). *J. Inclusion Phenom.*, *5*: 249.
290. Eastmono, G. C., Parr, K. J., and Woo, J. (1988). *Polymer*, *29*: 950.
291. Miyata, M., Akizuki, S., Tsatsumi, H., and Takemoto, K. (1988). *J. Polym. Sci. Part C Polym. Lett.*, *26*: 229.
292. Miyata, M., Tsuzuki, T., Noma, F., Takemoto, K., and Kamachi, M. (1988). *Makromol. Chem. Rapid Commun.*, *9*: 45.
293. Stewart, C. A., Takeshita, T., and Coleman, M. L. (1985). *Encycl. Polym. Sci. Eng. 3*: 441–462.
294. International Institute of Synthetic Rubber Producers, Inc. (IISPR) (1990). *Worldwide Rubber Statistics 1991, Rubber Handbook*, 8th ed., Swedish Institution of Rubber Technology (SFG).
295. Göbel, W., Rohde, E., and Schwinum, E. (1982). *Kautschuk Gummi*, *35*: 942.
296. Dollhausen, M. (1982). *Adhaesion, 10*: 23.
297. Obrecht, W. (1987). *Houben Weyl: Methoden der organischen Chemie: Makromolekulare Stoffe*, Vol. E20/2 (Bartl, H., and Falbe, J., eds.), Georg Thieme, Stuttgart, p. 842.
298. Bauchwitz, P. S., Finlay, J. B., and Stewart, C. A. (1971). *Vinyl and Diene Monomers*, Part II (Leonard, E. C., ed.), Wiley, New York, p. 1149.
299. Petiand, R., and Tho Pham, Q. (1985). *J. Polym. Sci. Polym. Chem. Ed., 23*: 1333.
300. Garrett, R. R., Hargreaves, C. A., and Robinson, D. N. (1970). *J. Macromol. Sci. Chem., 4*: 1679.
301. Aufdermarsch, C. A., and Pariser, R. (1964). *J. Polym. Sci. Part A, 2*: 4727.
302. Pariser, R. (1960). *Kunststoffe, 50*: 623.
303. Turner, N. L. (1972). DEOS 22 10, 957 to Petro Tex. Chem. Corp., *C.A.* (1973) *78*: 59545.
304. Aho, C. E. (1969). U.S. Pat. 3,422,045 to Du Pont, *C.A.* (1969) *70*: 58750.
305. Dohi, M., Sumida, T., and Yokobori, K. (1974). DEAS 251,975 to Denki Kagaku Kogo; *C.A.* (1976) *84*: 60867.
306. Branlard, P., and Modiano, J. (1974). DEOS 2,255,232 to Distugil, *C.A.* (1974) *80*: 146816.
307. Nolte, W., and Esser, H. (1982). DEOS 3,111,138 to Bayer AG, *C.A.* (1983) *98*: 35846.
308. Musch, R., Pampus, G., Müller, P., Eisele, U., Konter, W., and Göbel, W. (1982). Eur. Pat. 657,718 to Bayer AG, *C.A.* (1983) *98*: 180875.
309. Törnquist, E. G. M. (1968). *Polymer Chemistry of Synthetic Elastomers* (Kennedy, J. P., and Törnquist, E. G. M., eds.), Wiley, New York, p. 66.
310. Dunbrook, R. F. (1954). *Synthetic Rubber* (Whitby, G. S., Davis, C. C., and Dunbrook, R. F., eds.), Wiley, New York, p. 32.
311. Natta, G., Porri, L., Corradini, P., Zanini, G., and Ciampelli, F. (1961). *J. Polym. Sci., 51*: 463.
312. Porri, L., di Corato, A., and Natta, G. (1969). *Eur. Polym. J., 5*: 1.
313. Allen, C. F., and Bell, A. (1955). *Organic Synthesis*, Vol. 11, Wiley, New York, p. 312.
314. Toman, L., and Marek, M. (1981). *J. Macromol. Sci. Chem., 15*: 1533.
315. Spevacek, J., Toman, C., and Marek, M. C. (1981). *J. Macromol. Sci. Chem., 16*: 645.
316. Konigsberger, C., and Salomon, G. (1946). *J. Polym. Sci., 1*: 353.
317. Marvel, C. S., and Williams, J. R. (1949). *J. Polym. Sci., 4*: 264.
318. Gilbert, R. D., and Williams, H. L. (1952). *J. Am. Chem. Soc., 74*: 4114.
319. Morton, M., and Gibbs, W. E. (163). *J. Polym. Sci. Part A, 1*: 2679.
320. Frauendorf, B. (1986). *Houben Weyl: Methoden der organischen Chemie: Makromolekulare Stoffe*, Vol. E20/2 (Bartl, H., and Falbe, J., eds.), Georg Thieme, Stuttgart, p. 858.
321. Yen, T. F. (1959). *J. Polym. Sci., 35*: 533.
322. Gordon, B., and Blumenthal, M. (1985). *Polym. Bull., 14*: 69.

323. Gaylord, N. G., Stolka, M., Stepan, V., and Kössler, I. (1968). *J. Polym. Sci. Part C.*, *23*: 317.
324. Gaylord, N. G., and Kössler, I. (1968). *J. Polym. Sci. Part C*, *16*: 3097.
325. Throckmorton, C. M. (1976). DEOS 2,549,166 to Goodyear, *C.A.* (1977) *87*: 54314.
326. Zimmermann, M., and Marwede, G. (1977). DEOS 2,541,004 to Bayer AG, *C.A.* (1977) *86*: 156808.
327. Yeh, H. C., and Hsieh, M. L. (1985). *Polym. Sci. Technol.*, *31*: 483.
328. Cabassi, F., Italia, S., Giarusso, A., and Porri, L. (186). *Makromol. Chem.*, *187*: 913.
329. Entezami, A., Gerandelle, A., Kaufmann, F., Schue, F., Deluzarche, A., and Maillard, A. (1971). *Eur. Polym. J.*, 7.889.
330. Stepan, V., Vedehnal, J., Kössler, I., and Gaylord, N. G. (1967). *J. Polym. Sci. Part A-1*, *5*: 503.
331. Brown, J. F., and White, D. M. (1960). *J. Am. Chem. Soc.*, *82*: 5671.
332. Chatani, Y., and Nakatani, S. (1972). *Macromolecules*, *5*: 597.
333. Ohno, R., Tanaka, Y., and Kawakami, M. (1973). *Polym. J.*, *4*: 56.
334. Livishits, I. A., and Korobova, L. M. (1958). *Dokl. Akad. Nauk SSSR*, *121*: 474.
335. Livishits, I. A., Kovobova, L. M., and Siborovich, E. A. (1967). *Meth. Polim. 4*: 596.
336. Henderson, J. N. (1975). MS Pat. 3,928,301 to Goodyear, *C.A.* (1976) *84*: 91373.
337. Henderson, J. N. (1971). DEOS 1,945,795 to Goodyear, *C.A.* (1971) *75*: 7114.
338. Kormer, V. A., Babitskii, B. D., and Lobach, M. I. (1969). *Adv. Chem. Ser.*, *91*: 306.
339. Lobach, M. I., Kormer, V. A., Yu Tseretely, I., Kondratenkov, G. P., Babitskii, B. D., and Klepikova, V. I. (1971). *Dokl. Akad. Nauk-SSSR*, *196*: 114, *C.A.* (1971) *75*: 7114.
340. Lobach, M. I., Kormer, V. A, Yu Tseretely, I., Kondratenkov, G. P., Babitskii, B. D., and Klepikova, V. I. (1971). *J. Polym. Sci. Part B.*, *9*: 71.
341. Klepikova, V. I., Kondratenko, G. P., Kormer, V. A., Lobach, M. I., and Churillyaeva, L. A. (1973). *J. Polym. Sci. Part B*, *11*: 193.
342. Visiliev, V. A., Kalinicheva, N. A., Kormer, V. A., Lobach, M. I., and Kepikova, V. 1. (1973). *J. Polym. Sci. Polym. Chem. Ed.*, *11*: 2489.
343. Druz, N. N., Zak, A. V., Lobach, M. I., Shpakor, P. P., and Kormer, V. A. (1977). *Eur. Polym. J.*, *13*: 875.
344. Druz, N. N., Zak, A. V., Lobach, M. I., Visiliev, V. A., and Kormer, V. A (1978). *Eur. Polym. J.*, *14*: 21.
345. Druz, N. N., Lobach, M. I., Khatchaturov, A. S., Klepikova, V. I., and Kormer, V. A. (1978). *Eur. Polym. J.*, *14*: 743.
346. Churlyaeva, L. A., Valrue, V. I., Bodrova, V. S., Duitrieva, T. S., Lobach, M. I., and Kormer, V. A. (1979). *Macromolecules*, *12*: 38.
347. Hasegawa, K., Asami, R., and Higashimura, T. (1977). *Macromolecules*, *10*: 592.
348. Higashimura, T., and Hasegawa, H. (1979). *J. Polym. Sci. Polym. Chem. Ed*, *17*: 2439.
349. Stille, J. K., and Vessel, E. D. (1961). *J. Polym. Sci.*, *49*: 419.
350. Ianni, J. D. (1954). *Synthetic Rubber* (Whitby, G. S., Davis, C. C., and Dunbrook, R. F., eds.), Wiley, New York, p. 612.
351. Brown, J. F., and White, D. M. (1960). *J. Am. Chem. Soc.*, *82*: 5671.
352. Miyata, M., and Takemoto, K. (1976). *J. Polym. Sci. Polym. Symp.*, *55*: 279.
353. Taft, W. K., and Tiger, G. Y. (1954). *Synthetic Rubber* (Whitby, G. S., Davis, C. C., and Dunbrook, R. F., eds.), Wiley, New York, p. 682.
354. Montermozo, J. C. (1961). *Rubber Chem. Technol.*, *34*: 1521.
355. Starkweather, J. W., Bare, P. O., Cartere, A. S., and Hill, F. B. (1947). *Ind. Eng. Chem.*, *39*: 210.
356. Heck, R. F., and Breslow, D. S. (1959). *J. Polym. Sci.*, *41*: 521.
357. Kuzina, N. g., Mashlyakovskii, L. N., Podol'skii, A. F., and Okhirmenko, I. S. (1980). *Polym. Sci. USSR*, *22*: 1394.
358. Van Landuyt, D. C. (1972). *J. Polym. Sci. Part B*, *10*: 125.
359. Natta, G., and Porri, L. (1966). *Adv. Chem. Ser.*, *52*: 24.
360. Napolitano, R. (1988). *Macromolecules*, *21*: 622.

361. Natta, G., Porri, L., Carbonaro, A., and Stoppa, G. (1961). BE Pat. 617, 545 to Montecatini, *C.A.* (1963) *58*: 6943.
362. Brit. Pat. (165) 993, 485 to Phillips Petroleum, *C.A.* (1965) *63*: 4479.
363. Porri, L., and Carbonaro, A. (1963). *Makromol. Chem.*, *177*: 1465.
364. Fischer, E. O., and Lindner, H. H. (1964). *J. Organomet. Chem.*, *1*: 307.
365. Scott, H., Frost, R. E., Bet, R. F., and Reilly, D. E. O. (1964). *J. Polym. Sci. Part A*, *2*: 3233.
366. Wilke, G., Bogdanovic, B., Hardt, P., Heimbach, P., and Keim, W. (1966). *Angew. Chem. Intern. Ed. Engl.*, *5*: 151.
367. Bogadanovic, B., Heimbach, P., Kröner, M., and Wilke, G. (1969). *Justus Liebigs Ann. Chem.*, *727*: 143.
368. Porri, L., and Gallazzi, MN. C. (1983). *Transition Metal Catalyzed Polymerizations: Alkenes and Dienes* (Quirk, R. P., ed.), MMI Press, Harwood Academic, New York, p. 555.
369. Zhiquan, S. (1980). *J. Polym. Sci. Polym. Chem. Ed.*, *18*: 3345.
370. Natta, G., Porri, L., and Valenti, S. (1963). *Makormol. Chem.*, *67*: 225.
371. Bujadoux, K., Clement, R., Jozefouvicz, J., and Neel, J. (1973). *Eur. Polym. J.*, *9*: 189.
372. Bruckner, S., Sozzani, P., Boeffel, C., Destri, S., and Disilvestro, G. (1989). *Macromolecules*, *22*: 607.
373. Duchemin, F., and Bennevault-Celton, V. (1998). *Macromol. Chem. Phys.*, *199*(11): 2533–2539.
374. Duchemin, F., and Bennevault-Celton, V. (1998). *Macromolecules*, *31*(22): 7627–7635.
375. Denisova, T. T., Livshits, I. A., and Gershtein, Y. R (1974). *Vysokomol. Soedin. Ser. A*, *16*: 880.
376. Cuzin, D., Chauvin, Y., and Lefebvre, G. (1967). *Eur. Polym. J.*, *3*: 367.
377. Cuzin, D., Chauvin, Y., and Lefebvre, G. (1969). *Eur. Polym. J.*, *5*: 283.
378. Hawegawa, K. I., and Asami, R. (1978). *J. Polym. Sci. Polym. Chem. Ed.*, *16*: 1449.
379. Zambelli, A. (1996). *Macromolecules*, *29*(16): 5500–5501.
380. Kamachi, M., Wakabayashi, N., and Murahashi, S. (1974). *Macromolecules*, *7*: 744.
381. Murahashi, S., Kamachi, M., and Wakabayashi, N. (1969). *J. Polym. Sci. Polym. Lett. Ed.*, *7*: 135.
382. Warin, R., Teyssie, Ph., Baurdaudury, P., and Dawans, F. (1973). *J. Polym. Sci. Part B*, *11*: 177.
383. Al-Harrah, M. M. F., and Young, R. N. (1980). *Polymer*, *21*: 119.
384. Morton, M., Falvo, L. A., and Fetters, L. J. (1972). *J. Polym. Sci. Polym. Lett. Ed.*, *10*: 561.
385. Wang, F., Bolognesi, A., Immirzi, A., and Porri, L. (1981). *Makromol. Chem. Rapid Commun.*, *1*: 293.
386. Destri, S., Gallazzi, M. C., Giarrusso, A., Porri, L. (1980). *Makromol. Chem. Rapid Commun.*, *1*: 293.
387. Jones, A. M., Planchard, J. A., Speed, R. A., and Claybough, B. E. (1968). U.S. Pat. 3,414,637 to ESSO Co., *C.A.* (1969) *70*: 29897.
388. Rummelsburg, A. L. (1945). U.S. Pat. 2,37773, 419 to Hercules Corp., *C.A.* (1945) *39*: 4528.
389. Carmody, M., and Carmody, W. (1937). *J. Am. Chem. Soc.*, *59*: 1312.
390. Smachinskii, V. F. (1937). *Zh. Prikl. Khim.*, *10*: 2028.
391. Jones, J. F. (1958). *J. Polym. Sci.*, *33*: 513.
392. Marvel, C. S., and Kiener, P. E. (1962). *J. Polym. Sci.*, *61*: 311.
393. Marvel, C. S., Kiener, P. E., and Vessel, E. D. (1959). *J. Am. Chem. Soc.*, *81*: 4694.
394. Sivola, A. (1977). *Acta Polytech Scand.*, *134*: 1.
395. Quirk, R. P., and Huang, T. L. (1984). *Polym. Sci. Technol.*, *25*: 329.
396. Bell, V. L. (1964). *J. Polym. Sci. Part A*, *2*: 5291.
397. Akopyan, A. N., Krbekyan, g. Y., and Sinanyan, E. G. (1963). *Vyskomol. Soedin.*, *5*: 681.
398. Priola, A., Corno, C., Bruzzone, M., and Cesca, S. (1981). *Polym. Bull.*, *4*: 735.
399. Rhodes, R. P., and Guthrie, D. A. (1967). U.S. Pat. 3,365,554 to ESSO Co., *C.A.* (1968) *68*: 30890.

400. Verobeva, A. I., Gurevich, M. A., Leplyanin, G. V., Ratikov, S. R., Timirova, R. G., and Odinokov, V. N. (1974). *Vysokomol. Soedin. Se. A*, *16*: 1826.
401. Monakov, Y. B., Glukhov, Y. A., Farkhiyeva, I. T., Kudashev, R. K., Minsker, K. S., Dzhemilev, U. M., Tolstikov, G. A.,and Rafikov, S. R. (1980). *Vysokomol Soedin. Ser. A*, *22*: 385.
402. Staudinger, H., and Bruson, H. A. (1926). *Justus Liebigs Ann. Chem.*, 447.97.
403. Aso, C., Kunitake, T., and Ishimoto, Y. (1968). *J. Polym. Sci. Part A-1*, *6*: 1163.
404. Imanshi, y., Kohjiya, S., and Okamura, S. (1968). *J. Makromol. Sci. Chem. Part A*, *2*: 471.
405. Sigwalt, D., and Vairon, J. P. (1964). *Bull. Soc. Chim. Fr.*, *1964*: 482.
406. French, P. V., Roubinek, L., and Wassermann, A. (1960). *Proc. Chem. Soc. London*, 248.
407. Heublein, G., and Barth, O. (1974). *J. Prakt. Chem.*, *316*: 649.
408. Heublein, G., and Freitag, W. (1977). *J. Prakt. Chem.*, *319*: 968.
409. Meyeresen, K., and Wang, J. Y. C. (1967). *J. Polym. Sci. Part A-1*, *5*: 725.
410. Aso, H., and Ohara, O. (1969). *Makromol. Chem.*, 127.78.
411. Kohjiya, S., Imanishi, Y., and Okamura, S. (1968). *J. Polym. Sci. Part A-1*, *6*: 809.
412. Mitchell, R. S., McLean, S., and Guillet, J. E. (1968). *Macromolecules*, *1*: 417.
413. Nakata, T., and Choumei, N. (1967). *J. Makromol. Sci. Chem. Part A*, *1*: 1433.
414. Hong, K., Wan, Y., and Mays, J. W. (2001). *Macromolecules*, *34*(8): 2482–2487.
415. Hong, K., and Mays, J. W. (2001). *Macromolecules*, 34(4): 782–786.
416. Williamson, D. T., Elman, J. F., Madison, P. H., Pasquale, A. J., and Long, T. E. (2001). *Macromolecules*, 34(7): 2108–2114.
417. Marvel, C. S., and Hartzell, G. E. (1959). *J. Am. Chem. Soc.*, *81*: 448.
418. Freg, D. A., Hasegawa, G. E., and Marvel, C. S. (1963). *J. Polym. Sci. Part A*, *1*: 2057.
419. Lefebvre, G., and Dawans, F. (1964). *J. Polym. Sci. Part A*, *2*: 3277.
420. Imanishi, Y., Yamane, T., Kohjiya, S., and Okamura, S. (1969). *J. Makromol. Sci. Chem.*, *3*: 223.
421. Sharaby, Z., Jagur-Grodzinski, J., Martan, M., and Vofsi, D. (1982). *J. Polym. Sci. Polym. Chem. Ed.*, *20*: 901.
422. Mango, L. A., and Lenz, R. W. (1973). *Makromol. Chem.*, *163*: 13.
423. Naumova, S. F., Yurina, O. D., and Erofeev, B. V. (1975). *Dokl. Akad Nauk Belorussk. SSR*, *19*: 718.
424. Sharaby, Z., Martan, M., and Jagur-Grodzinski, J. (1982). *Macromolecules*, *15*: 1167.
425. Dolgoplosk, B. A., Beilin, S. I., Korshak, Y. V., CHeinenko, G. M., Vardanyan, L. M., and Teterina, M. P. (1973). *Eur. Polym. J.*, *9*: 895.
426. Sozzani, P., and Oliva, C. (1985). *J. Magn. Reson.*, *63*: 115.
427. Farina, M., Silvestro, G. D., Sozzani, P., and Savare, B. (1985). *Macromolecules*, *18*: 923.
428. Silvestro, G. D., Sozzani, P., Savare, B., and Farina, M. (1985). *18*: 928.
429. Farina, M. (1987). *Top Stereochem.*, *17*: 1.
430. Wu, C. H., and Lenz, R. W. (1972). *J. Polym. Sci. Polym. Chem. Ed.*, *10*: 3529.
431. Tsao, J. H., and Lenz, R. W. (1979). *J. Polym. Sci. Polym. Chem. Ed.*, *17*: 331.
432. Wu, C. H., and Lenz, R. W. (1972). *J. Polym. Sci. Polym. Chem. Ed.*, *10*: 3529.
433. Ohara, O., Aso, C., and Kunitake, T. (1973). *J. Polym. Sci. Polym. Chem. Ed.*, *11*: 1917.
434. Kunitake, T., Ochiai, T., and Ohara, O. (1975). *J. Polym. Sci. Polym. Chem. Ed.*, *13*: 2581.
435. Aso, C., and Uchio, H. (1970). DOS 1,955,519 to Asahi Kasei, *C.A.* (1971) 74: 54411.
436. Makowski, H. S., Shim, B. K. C., and Wilshinsky, Z. W. (1964). *J. Polym. Sci. Part A.*, *2*: 1549.
437. Marek, M., Roosova, M., and Doskocilova, D. (1967). *J. Polym. Sci. Part C*, *16*: 971.
438. Anderson, W. S. (1963). *Proceedings of the Battelle Symposium on Thermal Stability of Polymers*, Columbus, Ohio, *C.A.* (1964) *60*: 9374.
439. Romanov, L. M., Verkhoturova, A. P., Kissin, Y. V., and Rakara, G. V. (1963). *Vysokomol Soedin.*, *5*: 719.
440. Marvel, C. S., and Stille, J. K. (1958). *J. Am. Chem. Soc.*, *80*: 1740.
441. Marvel, C. S., and Garrison, W., (1959). *J. Am. Chem. Soc.*, *81*: 4737.

442. Field, N. (1960). *J. Org. Chem.*, *25*: 1006.
443. Marvel, C. S., and Gall, E. J. (1960). *J. Org. Chem.*, *25*: 1784.
444. Ball, L. E., and Harwood, H. J. (1961). *Khim, Tekhnol. Polim.*, *11*: 24.
445. Reichel, B., Marvel, C. S., and Greenley, R. Z. (1963). *J. Polym. Sci. Part A*, *1*: 2935.
446. Valvassi, A., Sartori, G., Turba, V., and Lachi, M. P. (1967). *J. Polym. Sci. Part C*, *16*: 23.
447. Zutty, N. L. (1963). *J. Polym. Sci. Part A*, *1*: 2231.
448. Roller, M. B., Gillham, J., and Kennedy, J. P. (1973).*J. Appl. Polym. Sci.*, *17*: 2223.
449. Sartori, G., Valvassori, A., Turba, V., and Lachi, M. P. (1963). *Chim. Ind. (Milan)*, *45*: 1529.
450. Lal, J., Chen, H. Y., and Sandstrom, P. H. (1979). *J. Polym. Sci. Polym. Lett. Ed.*, *17*: 95.
451. Lal, J., Hanlon, T. L., and Chen, H. Y. (1980). *J. Polym. Sci. Polym. Chem. Ed.*, *18*: 2921.
452. Lal, J., Hanlon, T. L., Chen, H. Y., and Thudium, R. N. (1982). *ACS Symp. Ser.*, *193*: 171.
453. Wilbur, J. M., and Marvel, C. S. (1964). *J. Polym. Sci. Part A*, *2*: 4415.
454. Cheng, H. N., and Khasat, N. P. (1988). *J. Appl. Polym. Sci.*, *35*: 825.
455. Koton, M. M. (1960). *Tekhnol. Polim.*, *7/8*: 54.
456. Simpson, W., Holt, T., and Zetic, R. J. (153). *J. Polym. Sci.*, *10*: 489.
457. Howard, R. N. (1954). *J. Polym. Sci.*, *14*: 535.
458. Simpson, W., and Holt, T. (1955). *J. Polym. Sci.*, *18*: 335.
459. Gordon, M., and Row, R. (1956). *J. Polym. Sci.*, *21*: 27.
460. Butler, G. B., and Angelo, R. J. (1957). *J. Am. Chem. Soc.*, *79*: 3128.
461. Butler, G. B., Crawshow, A., and Miller, M. L. (1958). *J. Am. Chem. Soc.*, *80*: 3615.
462. Marvel, C. S., and Stille, J. K. (1958). *J. Am. Chem. Soc.*, *80*: 1740.
463. Makowski, H. S., Shim, B. K. C., and Wilshinsky, Z. W. (1964). *J. Polym. Sci. Part A*, *2*: 1549.
464. Resconi, L., and Waymouth, R. M. (1990). *ACS Prep. Polym. Develop.*
465. Bywater, S. (1985). *Encyclopedia of Polymer Science and Engineering*, 2nd ed., Vol. 2 (Mark, H. F., ed.), Wiley, New York, p. 1.
466. Gehrke, K., Pohlmann, H. W., and Harwart, M. (1989). *Acta Polym.*, *40*: 60.
467. Wendler, K., Seifert, C., and Herzog, C. (1988). *Plaste Kautschuk*, *35*: 337.
468. Bartenev, G. M., Shut, N. I., and Lazorenko, M. V. (1985). *Acta Polym.*, *36*: 278.
469. Kanetakis, J., Wong, F. Y. C., Hamielec, A. E., and McGregor, J. F. (1985). *Chem. Eng. Commun.*, *35*: 123.
470. Braun, H. G., and Rehage, G. (1985). *Angew. Makromol. Chem.*, *131*: 107.
471. Goldberg, A., Lesner, D. R., Stone, J. C., and Patt, J. (1985). *Energy Res.Abstr.*, *10*: 13667.
472. Kaushik, V. K., and Sharma, J. N. (1985), *Polym.Bull.*, *13*: 373.
473. Mirely, C. L., Sung, N. H., and Schneider, N. S.(1985). *Polym. Master. Sci. Eng.*, *53*: 325.
474. Marwede, G. W., Stolifuß, B., and Sumner, A. J. M. (1993). *Kautsch. Gummi Kunstst.*, *46*: 380–388.
475. Hsieh, H., and Glaze, W. (1979). *Rubber Chem. Technol.*, *43*: 22.
476. Keckler, N. (1963). DAS 1,300,239 to Firestone, *C.A.* (1964) *61*: 808.
477. Bouton, T., and Futamura, S. (1974). *Rubber Age (NY)*, *106*: 33.
478. Hargis, I., Livigni, R., and Aggarwal, S. (1982). *ACS Symp. Set.*, *193*: 73.
479. Lee, D. I., and Ishikawa, T. (1983). *J. Polym. Sci.* Polym. Chem. Ed., *21*: 147.
480. Mark, H. F., Gaylord, N. G., and Bikales, N. M., eds. (1964). *Encyclopedia of Polymer Science and Technology*, Vol. 1, Interscience, New York, p. 436.
481. Ishigure, K., Neill, T. O., Stahel, E. P., and Stannett, V. (1974). *J. Macromol. Sci. Chem.*, *8*: 353.
482. Heins, F., Matner, M., and Bross, H. J. (1981). DE Pat. 3,119,612 to Bayer AG, *C.A.* (1983) *98*: 127500.
483. Heins, F. (1978). DE Pat. 2,830,393 to Bayer AG, *C.A.* (1980) *92*: 130366.
484. Poehlein, G. W., Ottewill, R. H., and Goodwin, J. W. (1983). *Science and Technology of Polymer Colloids*, Vol. 1, Martinus Nijhoff, Hingham, Mass., p. 88.
485. Vanderhoff, J. W. (1985). *J.Polym. Sci. Polym. Symp.*, *72*: 161.
486. Meyer, G. E., Karchok, R. W., and Naples, F. J. (1973). *Rubber Chem. Technol.*, *46*: 106.

487. Nowakowska, M. (1985). *Polym. Photochem.*, 6: 303.
488. Hofmann, W. (1984). *Kautschuk Gummi Kunstst.*, 37: 753.
489. Fetters, L. (1969). *J. Polym. Sci. Part C*, 26: 1.
490. Efremkin, A. F., Ivanov, V. B., and Shlyapimtokh, V. J. (1985). *Eur. Polym. J.*, 21: 769.
491. David, C., and Zabeau, F. (1985). *Eur. Polym. J.*, 21: 343.
492. Lattimer, R. P., Scheer, K. M., Windig, W., and Menzelaar, H. L. C. (1985). *J. Anal. Appl. Pyrolysis*, 8: 95.
493. Fukumori, K., and Kuranchi, T. (1985). *J. Mater. Sci.*, 20: 1725.
494. Mirely, C. L., Sung, N. H., and Schneider, N. S. (1985). *Polym. Master. Sci. Eng.*, 53: 325.
495. Xie, H., and Ma. L. (1985). *J. Macromol. Sci. Chem. Part A*, 22: 1333.
496. Fetters, L. (1979). *Macromolecules*, 12: 344.
497. Guyot, P., Favier, J., Uyterhoeven, H., Fontanielle, M., and Sigwalt, P. (1981). *Polymer*, 22: 1724.
498. Weerts, P. A., Van Loos, J. L. M., and German, A. L. (1988). *Polym. Commun.*, 29: 278.
499. Carbonaro, A., Zamboni, V., Novara, G., and Dall'Asta, G. (1973). *Rubber Chem. Technol.*, 46: 1274.
500. Bruzzone, M., Carbonaro, A., and Gargani, L. (1978). *Rubber Chem. Technol.*, 51: 907.
501. Henderson, J. N., and Throckmorton, M.C. (1987). *Encycl. Polym. Sci. Eng.*, 10: 811.
502. Furukawa, J., Furuno, N., Matsumura, A., and Kuwajima, T. (1980). DOS 3,122,079 to Nippon Paint Co., *C.A.* (1982) 96: 123516.
503. Furukawa, J. (1973). *Gummi Asbest Kunstst.*, 10: 837.
504. Furukawa, J., Hirai, R., and Nakaniwa, M. (1969). *Polym. Lett.*, 7: 671.
505. Furukawa, J., Amano, H., and Hirai, R. (1972). *J. Polym. Sci. Part A-1*, 10: 681.
506. Furukawa, J. (1972). *Angew. Makromol. Chem.*, 23: 189.
507. Wieder, W., Krömer, H., and Witte, J. (1982). *J. Appl. Polym. Sci.*, 27: 3639.
508. Furukawa, J., Iseda, Y., Haga, K., and Kataoka, N. (1970). *J. Polym. Sci. Part A-1*, 8: 1147.
509. Furukawa, J., Kobayashi, E., Iseda, Y., and Arai, Y. (1970). *Polym. J.*, 1: 442.
510. Priola, A., Corno, C., Ferraris, G., and Cesca, S. (1979). *Makromol. Chem.*, 180: 2859.
511. Jiao, S., and Yu, D. (1985). *Polym. J.*, 17: 899.
512. Wang, S., Li, Z., and Wang, F. (1984). *Jaofenzi Tongxun*, 6: 420.
513. Mülhaupti, R., Ovenall, D. W., and Ittel, S. D. (1988). *J. Polym. Sci. Polym. Chem. Part A*, 26: 2487.
514. Jiao, S., Su, D., Yu, D., and Hu, L. (1988). *Chinese J. Polym. Sci.*, 6: 135.
515. Kaminsky, W., and Schlobohm, M. (1986). *Makromol. Chem. Macromol. Symp.*, 4: 103.
516. Furukawa, J., and Iseda, Y. (1969). *Polym. Lett.*, 7: 47.
517. Thörmer, J., Marwede, G., and Buding, H. (1983). *Kaustchuk Gummi Kunstst.*, 36: 269.
518. Kawasaki, A., Maruyama, I., Taniguchi, M., Hirai, R., and Furukawa, J. (1969). *J. Polym. Sci. Part B*, 7: 613.
519. Furukawa, J., and Hirai, R. (1972), *J. Polym. Sci. Polym. Chem. Ed.*, 10: 2139.
520. Khurshio, A., Toppare, L., and Akbulut, U. (1987). *Polym. Commun.*, 28: 269.
521. Kuntz, J., Powers, K. W., Hsu, C. S., and Rose, K. D. (1988). *Makromol. Chem. Macromol. Symp.*, 13/14: 337.
522. Bertram, H. H. (1981). *Developments in Rubber Technology, Part 2* (Whelan, A., and Lee, K. S., eds.), Applied Science, London, p. 63.
523. Hirai, H. (1976). *J. Polym. Sci. Macromol. Rev.*, 11: 49.
524. Takahashi, A., and Gaylord, N. G. (1970). *J. Macromo. Sci. Che.*, 4: 127.
525. Gaylord, N. G., and Matiska, B. (1970). *J. Macromol. Sci. Chem.*, 4: 1519.
526. Bamford, C. H., Han, X., and Malley, P. J. (1982). *Plaster Kautschui*, 29: 137.
527. Roeper, M., and Schieren, H.R. (1985), *J. Mol. Catal.*, 31: 335.
528. Kohjiya, S., (1985). *Polym. J. (Tokyo)*, 17: 661.
529. Tsuji, Y., Suzuki, T., Watanabe, Y., and Takegumi, Y. (1981). *Macromolecules*, 14: 1194.
530. Hsieh, H. L., and Kraus, G. (1973). *J. Polym. Sci. Polym. Chem. Ed.*, 11: 453.
531. Qurik, R. P., and Chen, W. C. (1982). *Makromol. Chem.*, 183: 2071.

532. Furukawa, J., Kobayashi, E., and Nagata, S. (1977). *Polym. J.*, *9*: 633.
533. Sigwalt, P. (1961). *J. Polym. Sci.*, *52*: 15.
534. Nagai, K., Machida, S., and Nonaka, T. (1981). *J. Polym. Sci. Polym. Chem. Ed.*, *19*: 773.
535. Kobuke, Y., Fueno, T., and Furukawa, J. (1970). *J. Am. Chem. Soc.*, *92*: 6548.
536. Cantello, B. C. C., and Mellor, J. M. (1968). *Tetrahedron Lett.*, 5179.
537. Moore, J. A., and Partain, E. M. (1985). *J. Polym. Sci. Polym. Chem. Ed.*, *23*: 591.
538. Nagai, K., and Yonezawa, H. (1983). *J. Polym. Sci. Polym. Lett. Ed.*, *21*: 115.
539. Huvard, G. S., Nicholas, P. P., and Horre, S. E. (1985). *J. Polym. Sci. Polym. Chem. Ed.*, *23*: 2005.
540. Hoover, J. M. Ward, T. C., and McGrath, J. E. (1985). *Polym. Prepr. ACS Div. Polym. Chem.*, *26*: 253.
541. Zurkova, E., Bouchal, K., Kalal, J., Chvatal, Z., and Dedek, V. (1987). *Angew. Makromol. Chem.*, *155*: 101.
542. Murinov, Y. I., and Monakov, Y. B. (1987). *Inorg. Chim. Acta*, *140*: 25.
543. Heubelin, G., and Albrecht, G. (1982). *Acta Polym.*, *33*: 505.
544. Imahishi, Y., Hara, K., Kohjiya, S., and Okamura, S. (1968). *J. Macromol. Sci. Chem.*, *2*: 1423.
545. Heublein, G., Albrecht, G., Kümpfel, W., and Böse, M. (1984). *Acta Polym.*, *35*: 642.
546. Thaler, W. A., and Buckley, D. J. (1976). *Rubber Chem. Techn.*, *49*: 960.
547. Bötke, M., Knöppel, G., Hallpap, P., Stadermann, D., and Heublein, G. (1989). *Makromol. Chem. Rapid Commun.*, *10*: 157.
548. Heublein, G., and Freitag, W. (1979). *J. Prakt. Chem.*, *321*: 544.
549. Heublein, G., Freitag, W., and Mock, W. (1980). *Makromol. Chem.*, *181*: 267.
550. Furukawa, J., Kobayashi, E., and Wakui, T. (1983). *Polym. J.*, *15*: 435.
551. Ferdinandi, E. S., Garby, W. P., and James, D. G. L. (1964). *Can. J. Chem.*, *42*: 2568.
552. Meyers, R. A., and Wilson, E. R. (1968). *J. Polym. Sci. Part B*, *6*: 581.
553. Yamaguchi, T., Nagai, K., and Ono, T. (1969). *Kobunshi Kagaku*, *26*: 463, *C.A.* (1969) *71*: 91944.
554. Nogai, K., and Machida, S. (1977). *Polym. Prepr. Jpn.*, *26*: 768.
555. Gaylord, N. G., Deshpande, A. B., and Martan, M. (1976). *J. Polym. Sci. Part B*, *14*: 679.
556. Gaylord, N. G., and Deshpande, A. B. (1977). *J. Macromol. Sci. Chem.*, *11*: 1795.
557. Gaylord, N. G., Solomon, O., Stolka, M., and Patnaik, B. K. (1974). *J. Makromol. Sci. Chem.*, *8*: 981.
558. Natta, G., Crespi, G., Valvasori, A., and Sartori, G. (1963). *Rubber Chem. Technol.*, *36*: 1583.
559. Bushnik, R. D. (1969), *J. Polym. Sci. Part A*, *3*: 2047.
560. Kontos, E. G., Easterbrook, E. K., and Gilbert, R. D. (1962). *J. Polym. Sci.*, *61*: 69.
561. Kaminsky, W., and Miri, M. (1985). *J. Polym. Sci. Polym. Chem. Ed.*, *23*: 2151.
562. Lehr, M. H., and Corman, C. J. (1969). *Macromolecules*, *2*: 217.
563. Yu, Z., Marques, M., Rausch, M. D., and Chien, J. C. W. (1995). *J. Polym. Sci. Part A: Polym. Chem.*, *33*: 2795.
564. Nentwig, W., 5th ed., in: Elvers, B. Hawkins, S., Russey, W., and Schulz (Eds.) (1993). *Ullmanns'Encyclopedia of Industrial Chemistry, Vol. A23*, VCH, Weinheim, pp. 276–288.
565. Doi, Y., Tokuhiro, N., and Soga, K. (1989). *Makromol. Chem.*, *190*: 643.
566. Kaminsky, W., and Drögemüller, H. (1989). *Polymer Reaction Engineering* (Reichert, G. H., and Geiseler, W., eds), VCH Publishers, Weinheim, p. 372.
567. Kaminsky, W., and Drögemüller, H. (1990). *Makromol. Chem. Rapid Commun.*, *11*: 89.

6

Metathesis Polymerization of Cycloolefins

Ulrich Frenzel, Bettina K. M. Müller and Oskar Nuyken
Technische Universität München, Garching, Germany

I. INTRODUCTION

The term 'olefin metathesis' refers to an interchange reaction of alkylidene groups between alkenes. The total number of double bonds remains unchanged [1].

The history of olefin metathesis started in the mid 1950s, when Anderson and Merckling (Du Pont) – during their work on the Ziegler–Natta polymerization of norbornene (NBE) – received by means of $TiCl_4/EtMgBr$ catalysts a novel polymer [2,3]. In 1957 Eleuterio (Du Pont) filed a patent, which describes the polymerization of several cyclic olefins employing, among others, $LiAlH_4$-activated MoO_3/Al_2O_3 catalysts [4]. Ozonolysis of a norbornene polymer yielded *cis*-cyclopentane-1,3-dicarboxylic acid, thus demonstrating the novel and unexpected nature of this polymerization reaction [4,5].

(1)

In the same year Peters and Evering patented a 'disproportionation' reaction of propene yielding ethene and 2-butene with $Al(i\text{-}Bu)_3 + MoO_3/Al_2O_3$-catalysts as the first metathetical conversion of acyclic alkenes [6]. The first report on the metathesis of acyclic olefins in the open literature appeared in 1964. It describes the 'disproportionation' of olefins into homologs of higher and lower molecular weight using $Mo(CO)_6/Al_2O_3$ catalysts [7].

At this time ring-opening metathesis polymerization and metathesis of acyclic olefins – originally considered as 'olefin disproportionation' [7] – were regarded as two different reactions. Calderon recognized in 1972 that these both are two sides of the same coin and introduced the term 'olefin metathesis' for this reaction type [8–11].

From these very beginnings the olefin metathesis reaction is a central topic of industrial as well as academic research due to its great synthetic applicability. Many reviews and monographs about this topic were published since then [1,12–31]. The most important metathesis reaction pathways including cyclic olefins, ring-closing metathesis (RCM, [14–19]), ring-opening metathesis (ROM, [17,18]) and ring-opening metathesis polymerization (ROMP, [4,20–29]) are schematically shown in structure (2).

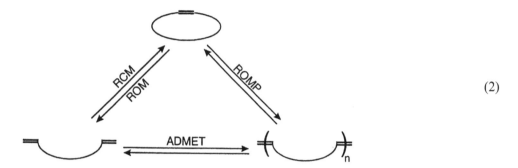

(2)

The present contribution deals primarily with the polymer synthesis via ring-opening metathesis polymerization. Acyclic diene metathesis (ADMET) [20,32–38], the other metathetic route to polymers is omitted. This article intends to give a brief overview, for more details and further applications see, e.g., Refs. [1,4,20–29].

II. GENERAL MECHANISTIC ASPECTS

As mentioned above, Calderon recognized in 1972 that metathesis polymerization and metathesis of acyclic olefins are two aspects of the same reaction [10]. As early as 1968 he had identified the double bonds as the reactive centers in the metathesis of acyclic olefins. Apart from the educts the metathesis reaction of d_8-2-butene with 2-butene yielded only d_4-2-butene, so he could exclude the cleavage of any single bond [39,40]. Dall' Asta and Motroni drew an analogous conclusion for ROMP by copolymerization of 1-^{14}C-cyclopentene and cyclooctene (3). After ozonolytic degradation of the polymers the complete radioactivity was found in the C_5-fraction, showing the exclusive cleavage of the double bonds (pathway (3b)) [41,42].

(3)

In early mechanistic theories several pairwise mechanisms were proposed with, e.g., various quasi-cyclobutane (4b) [43–45], metal tetracarbene (4a) [46] or metallacyclopentane [47,48] intermediates respectively transition states [49].

$$
\begin{array}{c}
\text{CHR}^1 \\
\| \\
\text{R}^1\text{HC}=\text{[M]}=\text{CHR}^2 \\
\| \\
\text{CHR}^2 \\
\text{(a)}
\end{array}
$$

$$
\begin{array}{c}
\text{R}^1\text{HC} \quad\quad \text{CHR}^1 \\
\| \rightarrow \text{[M]} \leftarrow \| \\
\text{R}^2\text{HC} \quad\quad \text{CHR}^2
\end{array}
\qquad
\begin{array}{c}
\text{R}^1\text{HC}=\!=\text{CHR}^1 \\
\text{[M]} \\
\text{R}^2\text{HC}=\!=\text{CHR}^2
\end{array}
\qquad (4)
$$

$$
\begin{array}{c}
\text{R}^1\text{HC}-\text{-CHR}^1 \\
\vdots \ \text{[M]} \ \vdots \\
\text{R}^2\text{HC}-\text{-CHR}^2 \\
\text{(b)}
\end{array}
$$

Chauvin and Hérisson found in 1970, that the initial product distribution in the cross metathesis of cyclopentene and 2-pentene is not in accordance with such a simple pairwise mechanism [30,50]. Therefore, they proposed a novel non-pairwise mechanism with metal carbene complexes as intermediates (5) [50].

This so-called metallacyclobutane mechanism is further supported by the fact that ROM polymerizations yield high molecular weight polymers already at low yields [51]. In the case of a step growth polymerization, as suggested by a simple pairwise mechanism, polymers with high molecular weight should yield only at high conversions. However, both findings may also be explained by a modified pairwise mechanism [30,49], but Katz et al. and Grubbs et al. demonstrated by highly sophisticated experiments using isotope labeled olefins that a pairwise mechanism is improbable [52–55].

$$
\begin{array}{c}
\text{H}_2\text{C}=\text{CH}_2 \\
+ \\
\text{[M]}=\text{CHR}
\end{array}
\ \rightleftharpoons\
\begin{array}{c}
\text{H}_2\text{C}-\text{CH}_2 \\
| \quad\quad | \\
\text{[M]}-\text{CHR}
\end{array}
\ \rightleftharpoons\
\begin{array}{c}
\text{CH}_2 \\
\| \\
\text{[M]}
\end{array}
+
\begin{array}{c}
\text{CH}_2 \\
\| \\
\text{CHR}
\end{array}
\qquad (5)
$$

Dolgoplosk's finding that metal carbene-generating diazoalkanes [56] may act as highly efficient cocatalysts supported Chauvin's mechanism [51]. The first metathesis polymerization using a well-defined metal carbene complex as initiator was performed in 1976 by Katz [57] with $(CO)_5W=CPh_2$ [58]. Since these initial investigations a broad variety of isolable metal carbene complexes has been synthesized and employed with great success as metathesis initiators. Furthermore, the metallacyclobutane mechanism was supported by many other investigations, e.g., the characterization of intermediate metall-acyclobutanes [59–61] or olefin-π-metal carbene complexes [62,63] and it is now generally accepted [1,30]. Four basic steps are proposed: coordination of the olefin to the metal center of a carbene complex, [2 + 2]-cycloaddition forming the metallacyclobutane intermediate, cycloreversion and finally de-coordination of the olefin. All these reactions are reversible as shown in Scheme (5).

In contrast to these success many details of the mechanism still remain unclear until now. For example Rooney et al. recently reported the presence of persistent metal anion radicals in metathesis reactions using the Grubbs catalyst $(PCy_3)_2Cl_2Ru=CHPh$ and proposed for this initiator a novel mechanism involving radicals (Scheme 6) [64,65].

(6)

III. CATALYSTS

A. General Aspects

Most metathesis catalysts are based on compounds of Ti, Ta, Mo, W, Re, Ru, Os and Ir. Only a few reports on the use of Nb, Zr, V, Cr, Tc, Co and Rh systems appeared in literature [1]. But even $MgCl_2$ has been reported recently to be an active catalyst for polymerization of strained olefins, i.e. norbornenes [66].

Metathesis catalysts may be divided formally into three groups: homogeneous, heterogeneous and immobilized homogeneous catalysts. In general the former are utilized for metathetic polymerizations and only few reports on ROMP with heterogeneous or immobilized catalysts were published until now (see, e.g., [6,7,67–72]).

Early homogenous metathesis catalysts – often called 'classical catalysts' – are formed in situ from a transition metal halide and a main group metal alkyl co-catalyst. Typical examples of such multicomponent catalysts are carbonyl, nitrosyl, chlordie or oxychloride complexes of molybdenum, tungsten or rhenium in combination with lithium, aluminium or tin organyl compounds. Often also promoters, mostly containing oxygen, are added [1]. It has been reported that oxo ligands formed from traces of moisture or oxygen are of crucial importance for the activity of some classical catalysts, e.g., WCl_6/BuLi [73,74].

Such binary and ternary catalyst systems including for example $MoCl_5$/$SnPh_4$ [75], $MoCl_2(NO)_2(C_2H_5N)_2$/$EtAlCl_2$ [76], the so-called Calderon catalyst WCl_6/$EtAlCl_2$/EtOH [11], $ReCl_5$/Bu_4Sn [77] or $ReCl(CO)_5$/$EtAlCl_2$ [78], can catalyze metathesis reactions of cyclic and acyclic olefins with great success. However, also the monomer itself may act as co-catalyst. Various mechanisms involving monomer molecules were proposed for the generation of the propagating species in these systems [1,79,80].

The catalyst systems mentioned above are widely used in the commercial applications of metathesis polymerization due to their low costs and simplicity of preparation (see Section VII). However, the harsh reaction conditions and strong Lewis acids often required limit the utility of such catalysts [81]. These may cause side-reactions and make them incompatible with most functional groups [82]. The propagating species are poorly defined and often neither quantitatively formed nor uniform. Hence, there is often a lack of reaction control using these systems. Moreover, for the polymerization of functionalized monomers it is often necessary to use tin organyls instead of aluminium alkyls. These are more expensive and may cause severe injuries of health [25,83].

As a consequence of a better understanding the mechanistical aspects of olefin metathesis and the synthesis of the first metal carbene complexes by Fischer [84] and Schrock [85] the situation has dramatically changed. These findings triggered the development of highly active unicomponent homogeneous catalysts [81]. Early examples are $(CO)_5W=CPh_2$ reported by Katz and Casey [57,58,86], the Tebbe reagent (7a) [87–89] and the titanacyclobutanes developed by Grubbs (7b–e) [60,90]. This trend towards well-defined, isolable single component initiators continues in the field of olefin metathesis. The cocatalyst-free alkylidenes combine a fast initiation with high catalytic activitiy. Their high degree of reaction control allows to perform living polymerizations, i.e., precise adjusting of molecular weight by the monomer/initiator-ratio and a low polydispersity [1].

An alternative approach using diazo compounds for the activation of suitable transition metal complexes is worth to be mentioned, too [51,91–98]. The formation of alkylidenes in situ avoids the expensive multi-step synthesis and isolation of well-defined initiators. However, these systems are also ill-defined and Noels reported for $[RuCl_2(p$-cymene)]_2/PCy_3$ (2eq.)/trimethylsilyldiazomethane that only 15–20 mol% of the employed Ru become catalytically active [97]. Nevertheless, such systems can exhibit exceptionally high catalytic activities [94,95].

The development of highly active and robust catalysts, which tolerate additionally functional groups is an important goal in transition metal catalyzed polymerizations. In metathesis that may be gained by the use of catalysts which are based on late transition metals. As shown in the table, ruthenium has unique properties in this respect [1,81,99]. Due to their remarkable stability and activity ruthenium based catalysts were focused during the last decade. These catalysts are remarkable tolerant towards oxygen and moisture than early transition metals. Moreover, the polymerization of a broad variety of monomers bearing polar protic functionalities became possible. Some ruthenium systems even enable polymerizations in polar protic solvents, e.g., alcohols or water [94,100–109]. Also emulsion polymerizations by means of ROMP became possible with suitable Ru compounds, e.g., floc free lattices in high yields were obtained via emulsion ROM polymerization of norbornene using a water soluble Grubbs-type catalyst [98]. General trends in tolerance towards functional groups for transition metal based metathesis catalysts are listed in the following table [81,99].

Titanium	Tungsten	Molybdenum	Ruthenium	
Acids	Acids	Acids	**Olefins**	
Alcohols, water	Alcohols, water	Alcohols, water	Acids	
Aldehydes	Aldehydes	Aldehydes	Alcohols, water	↑ Increasing reactivity
Ketones	Ketones	**Olefins**	Aldehydes	
Ester, amides	**Olefins**	Ketones	Ketones	
Olefins	Esters, amides	Easters, amides	Easter, amides	

The catalytic activity of the originally employed $RuCl_3$ hydrate-based systems [110] was in comparison low, but the highly sophisticated modern Ru-alkylidene initiators can exhibit as high activities as Mo based systems [111].

B. Titanium-Based Initiators

Grubbs et al. synthesized and characterized a series of titanacyclobutanes (e.g., 7b–e), which enabled for the first time living ROM polymerizations of cyclic olefins. For instance, the molecular weights were adjustable, the PDIs low and moreover, the chain carrying intermediates were well characterized [29,60,90,112,113]. These complexes were obtained by reaction of the Tebbe reagent (7a) [87–89] with suitable olefins in the presence of a Lewis base, e.g., pyridine or N,N-dimethylaminopyridine [60,90,112].

$$(7)$$

(a)   (b)   (c)   (d)   (e)

Living polymerization using titanacyclobutane initiators enabled also the preparation of block copolmers by sequential addition of different monomers [114–116] and synthesis of highly conjugated polymers and block copolymers of 3,4-diisopropylidene-cyclobutene [116].

C. Tantalum-Based Initiators

Schrock and co-workers reported a series of tantalum alkylidenes with the general formula $Ta(=CH-R)X_3(solv)$ ($R = t$-Bu, etc., $X = O$-2,6-i-Pr_2-C_6H_3, O-2,6-Me_2-C_6H_3, S-2,4,6-i-Pr_3-C_6H_2; solv = py, THF). The complexes were used for ROM polymerizations of norbornene and additionally tantalacyclobutane intermediates were isolated and characterized [117]. Several other Ta based initiators were synthesized and characterized [118,119]. However, the propagating species of Ta-based catalysts are often short living [117,120] and may react with functional groups containing heteroatoms [29]. Therefore, tantalum systems never gained the importance of well-defined Ru, Mo and W initiators.

D. Molybdenum-Based Initiators

The synthesis of high-oxidation-state molybdenum alkylidenes was reported by Schrock in 1987 [121]. Due to their improved tolerance towards functional groups (table) their better reaction profile and their lower costs well-defined molybdenum based initiators are now preferred over the related systems containing tungsten [122].

A broad variety of complexes with the general formula $Mo(NAr)(CHR^1)(OR^2)_2$ (9) have been synthesized successfully, e.g., starting from $Mo(NAr)(CHR^1)(OTf)_2(dme)$ ($Tf = SO_2CF_3$; dme = 1,2-dimethoxyethane) (8) [20,121–124]. The 'universal precursors' of type (8) are readily accessible even in large-scale syntheses and storable under inert atmosphere at room temperature [122–124].

$$(8)$$

It is crucial to prevent a bimolecular decomposition of the 14e⁻-species' $Mo(NAr)(CHR^1)(OR^2)_2$ by sterical shielding of the metal center. Consequently, it is necessary to use bulky NAr, OR^2 and $=CHR^1$ ligands. Hence, neopentylidene and neophylidene

Table 1 Examples of well-defined molybdenum-based metathesis initiators.

Complex	Remarks	Refs.	#
	Ar = Ph; 2,6-Me$_2$-C$_6$H$_3$; 2,6-i-Pr$_2$-C$_6$H$_3$, etc. R^1 = CMe$_2$Ph; t-Bu; SiMe$_3$, etc. R^2 = t-Bu, CMe$_2$CF$_3$, CMe(CF$_3$)$_2$, C(CF$_3$)$_3$, aryl, etc. Numerous combinations were realized	[121–124] [121,122,123,124]	(9)
		[146]	(10)

ligands are commonly employed, since these substituents generally yield stable, isolable species, as long as the NAr and OR groups themselves are relatively bulky (cf. Table 1, (9)) [122]. These Mo(NAr)(CHR^1)(OR^2)$_2$ complexes (9) and the analogous tungsten systems are now commonly called 'Schrock catalysts' [20].

Schrock's highly active Mo catalysts enable the polymerization of a broad variety of monomers often in a living manner. Variations of the electronic and steric properties of the particularly used ligands enable to tailor the microstructure of the resulting polymers [20,125,126,128–130]. Many reports dealing with this topic and with the influence of the imido or alkoxy ligands in particular appeared [20]. Even highly tactic all-*cis* ROM polymers can be accessible with initiators bearing suitable chiral ligands, e.g, BINO derivatives. Chiral Schrock systems were used not only in ROMP to yield highly tactic polymers [128,129,131,132], but also for asymmetric RCM (ARCM) and related metathesis reactions [133–141].

The major drawback of Schrock's systems in their high sensitivity towards oxygen and moisture. On the other hand they possess a remarkable tolerance towards numerous functionalities and successful polymerizations of monomers with, e.g., cyano [59,142], ester [59,142], carboxylic acid anhydride [143], amide [144] and ether [59] functionalities were reported [20]. The addition of aldehydes allows for quenching the metathesis reaction and cleaves the polymer chain from the metal via a Wittig-like reaction (11a) [20]. Related coupling reactions involving molecular oxygen can cause a fraction of polymers having the double molecular weight as expected (11b) [1,20,145].

A related complex, Mo(N-*t*-Bu)(CH-*t*-Bu)(OCMe(CF$_3$)$_2$)$_2$ (10), was synthesized by Osborn et al. and investigated for the ROMP of norbornene and acyclic internal olefins [146]. Boncella performed metathesis reactions using tris(pyrazolyl)borate stabilized molybdenum complexes in combination with AlCl$_3$ [147].

Two further reports on Schrock-type catalysts are worth mentioning: Feher et al. used sesquisiloxanes as ligands [148] and Stelzer et al. heterogeneized them on a γ-Al$_2$O$_3$ support using hexafluorobisphenol-A linkers [70].

E. Tungsten-Based Initiators

The first report on the use of an isolated alkylidene as initiator has been published by Katz in 1976 [57,86] using Casey's (CO)$_5$ WCPh$_2$ [58] (Table 2).

The first well-defined tungsten(VI) alkylidenes which serve as highly active metathesis initiors were reported by Osborn and co-workers in 1982 [149]. W(=CH-*t*-Bu)(OCH$_2$-*t*-Bu)$_2$X$_2$ (13) in combination with AlBr$_3$ or GaBr$_3$ polymerizes a variety of cycloolefins [61,62,149–151]. Later it has been reported that the related cyclopentylidene complex W(=C(CH$_2$)$_4$)(OCH$_2$-*t*-Bu)$_2$Br$_2$ polymerizes numerous cycloolefins, e.g., various methoxycarbonyl derivatives of norbornene, even without addition of a Lewis acidic cocatalyst [152–154].

Early examples of well-defined Lewis acid-free initiators were reported by Basset in 1985: tungsten(VI) alkylidenes of the type W(=CH-*t*-Bu)(OAr)$_2$Cl(CH$_2$-*t*-Bu)*(OR$_2$) (14) displayed high activity in metathesis of cyclic and acyclic olefins whilst avoiding the disadvantages of Lewis acid addition [155].

Basset et al. synthesized the highly active aryloxy-alkyloxy tungsten initiator (15) [156] and performed ROM polymerizations [156,157] and RCM reactions [158] with it.

Oxoalkylidene complexes of the type (W=CH-CH=CPh$_2$)(O)[OCMe(CF$_3$)$_2$]$_2$*L (16) were obtained by Grubbs and utilized for ROMP of norbornene [159]. His synthetic strategy employed 3,3-diphenylcyclopropene for the synthesis of the alkylidene moiety [159,160]. This method was later adapted for the synthesis of the first well-defined ruthenium metathesis initiators [161]. Furthermore, (OAr)$_2$W(CH-*t*-Bu)(O)(PMe$_3$) (17) was synthesized by Schrock et al. and reported to polymerize 2,3-dicarbomethoxynorbornadiene in a living manner [162].

A broad variety of alkoxy-imido tungsten alkylidenes (NAr)W(CHR^1)(OR^2)$_2$ (18) were developed by Schrock and coworkers since 1986 [163–165]. These complexes serve as highly active initiators and were utilized, e.g., for the living polymerization of *endo,endo*-5,6-dicarbomethoxynorbornene [164] and other monomers [166]. But for most applications these highly active metathesis initiators were replaced by the related molybdenum systems [122]. Similar tungsten-based systems having an ether functionalized chelating benzylidene ligand were elaborated by Grubbs et al. (19) [167].

The diamido tungsten(VI) complex (20) was reported by Boncella et al. But this initiator exhibited only low metathesis activity, probably due to the high stability of the W–L bond [169].

A report by van der Schaaf et al. is further worth to be mentioned: [Me(CF$_3$)$_2$CO]$_2$ (NPh)W(CH$_2$SiMe$_3$)$_2$ and Cl(NPh)W(CH$_2$SiMe$_3$)$_3$ are transformed into Schrock-type initiators by irridation and were used for the photoinduced ROMP (PROMP) of norbornene and dicyclopentadiene [168]. The main advantage of these thermally very stable PROMP systems is their latency in pure monomers in the absence of light and the easier synthesis [26,27].

Table 2 Examples of well-defined tungsten-based metathesis initiators.

Complex	Remarks	Refs.	#
$(CO)_5W=CPh_2$		[57,58]	(12)
t-BuCH$_2$O, *t*-BuCH$_2$O—W(X)$_2$=CHR (structure)	Cocatalyst: AlBr$_3$ or GaBr$_3$ X = Cl, Br R = *t*-Bu and others	[149,151]	(13)
$(OAr)_2W(=CH\text{-}t\text{-}Bu)Cl(CH_2CMe_3)(OR_2)$	R = 2,6-Et, *i*-Pr Ar = 2,6-Ph$_2$-C$_6$H$_3$	[155]	(14)
(structure with CH-*t*-Bu, O—W—OEt$_2$, RO, Cl)	R = 2,6-Ph$_2$-C$_6$H$_3$	[156]	(15)
(structure: (CF$_3$)$_2$MeCO, (CF$_3$)$_2$MeCO—W(=O)(L)=CH—CH=CPh$_2$)	L = P(OMe)$_3$, THF	[159]	(16)
$W(CH\text{-}t\text{-}Bu)(O)(OAr)_2(PMe_3)$	Ar = 2,6-Ph$_2$-C$_6$H$_3$	[162]	(17)
(structure: Ar—N=W(OR^2)$_2$=CH(R^1))	Ar = 2,6-*i*-Pr$_2$-C$_6$H$_3$, etc. R^1 = CMe$_2$Ph; *t*-Bu, etc. R^2 = *t*-Bu, etc.	[163,164,165]	(18)
(structure: Me(CF$_3$)$_2$CO, Me(CF$_3$)$_2$CO—W(=N-Ar)(THF)=CH...O)	Ar = 2,6-Ph$_2$-C$_6$H$_3$	[167]	(19)
(structure with N—TMS, NPh, N—W(L)=CH-*t*-Bu, TMS)	L = PMe$_3$; PEt$_3$; TMS = SiMe$_3$	[169]	(20)
$W(NPh)[OCMe(CF_3)_2]_2(CH_2\text{-}SiMe_3)_2$	Used for photoinduced ROMP	[168]	(21)
$W(NPh)Cl(CH_2SiMe_3)_3$	Used for photoinduced ROMP	[168]	(22)

F. Ruthenium-Based Initiators

$RuCl_3$ hydrate is known to be a metathesis initiator since many years [170–172] and it is used in combination with HCl in butanol for the polymerization of norbornene in industrial scale (see Section VII) [81]. Its advantage is the tolerance towards functional groups, however, the induction periods are long and only a small amount of the employed Ru becomes catalytically active [100,171].

An important milestone in the way toward modern Ru initiators was the synthesis of $Ru(tos)_2(H_2O)_6$. This Ru(II)-based initiator exhibited higher activity and much better initiation [80,100,102,103,173,174]. Even ROM polymerizations in water [103] and CO_2 [175,176] were performed with this initiator. However, despite the characterization of some olefin-ruthenium(II) complexes, the actual propagating species in such systems is still ill-defined [20].

Later Noels reported the activation of, e.g., $[Ru(p\text{-cymene})Cl_2]_2$ and $RuCl_2$ (p-cymene)(PCy_3) with diazo compounds in situ. But as mentioned above only a part of the employed ruthenium becomes catalytically active [97].

The major breakthrough was the synthesis of $(PPh_3)_2Cl_2Ru=CH-CH=CPh_2$ as the first well-defined, unicomponent Ru based metathesis initiator by Grubbs et al. in 1992 [161]. Further investigations showed that benzylidene complexes of type $(PR_3)_2Ru(CHPh)Cl_2$ initiate significantly faster than $(PR_3)_2Cl_2Ru=CH-CH=CPh_2$ [177,178]. Furthermore, the activity of these initiators could be strongly improved by using PCy_3-ligands instead of PPh_3 [179]. Many other phosphines were tested, but PCy_3 substituted initiators were the optimal ones [81,180]. The reason for this behavior is probably the high e^--donation ability and optimal steric demand of PCy_3. With this ligand stable, isolable complexes are formed and nevertheless the dissociation of one ligand during the formation of the propagating species is possible. Two pathways were considered for metathesis reactions with these catalysts: an associative one with both posphines bond to the metal center and a dissociative one with only one PR_3-ligand [81,180]. There is now much evidence from NMR [181,182], MS [183–185] and theoretical [186,187] investigations that these catalysts propagate mainly or even exclusively via a dissociative mechanism. This is additionally supported by the facts that the addition of Cu(I)-salts which may act as phosphine scavengers improves catalytic activity and the presence of additional PCy_3 diminishes the metathesis activity [180,188]. The influence of the alkylidene moiety on metathesis activity has been studied in detail [177].

Grubbs' catalysts were utilized for the ROM polymerizations of a broad variety of olefins. These reactions are not as controlled as with Schrock's Mo based systems but a high degree of reaction control is generally given and PDIs are often low. Moreover, Grubbs' versatile systems tolerate many functional groups and were successfully used even for ROM polymerizations in polar protic solvents. Especially complexes with ionic ligands (24) [106–109] are very useful in this respect and were employed for polymerizations in alcohols, water or emulsion [98,106–109].

A chain termination with aldehydes as with the Schrock systems is not possible if using Grubbs' catalysts. Vinylethers are used to cleave the polymer chain from the metal via formation of Fischer-type carbene complexes, which are metathesis inactive as reported by Grubbs, and an olefinic end group [189,190].

Gibson reported that the analogous ruthenium complex bearing $PCy_2(CH_2SiMe_3)$ ligands exhibits a better initiation/propagation ratio but much slower propagation than the parent Grubbs complex [191].

Just recently a broad variety of mononuclear *N*-heterocyclic carbene (NHC) substituted Ru alkylidenes have been reported (25–27) [192–199,202–209]. NHCs were utilized as ligands for metathesis catalysts for the first time by Herrmann's group in 1998 [192] and gained much attention since then. As pointed out by Grubbs mixed NHC/ phosphine complexes (26–27) may reach the activity of Mo-based catalyst systems [111] while conserving the unique reaction profile of Ru based systems [111,221]. ROM polymerizations with these initiators are well known [111,221,222]. Additionally these initiators permitted the synthesis of numerous cyclic compounds having a tri- or even tetrasubstituted double bond [219,195,196,200,204,206,207]. The higher activity of these systems in comparison to the parent Grubbs catalyst has been explained by their higher selectivity for binding π-acidic substrates instead of σ-donating phosphines [210]. The NHC substituents enable also an efficient heterogenization of these catalysts [223]. Selected examples of mononuclear ruthenium alkylidenes employed as metathesis initiators are summarized in Table 3.

Table 3 Examples of ruthenium-based metathesis initiators.

Complex	Remarks	Refs.	#
	$R^1 = CH = CPh_2$; $R^2 = Ph$, Cy, etc. $R^1 = Ph$; $R^2 = Ph$, Cy, etc. $R^1 = Ph$; $PR^2_3 = PCy_2(CH_2SiMe_3)$	[161] [177,178,180] [191]	(23)
	$L = $	[106–109]	(24)
	R = *i*-Pr, Cy, CHMePh, CHMeCy, CHMe(Naphthyl) etc.	[192,195]	(25)
	R = Cy, CHMePh, CHMe(Naphthyl), etc. 2,4,6-Me$_3$-C$_6$H$_2$, 4-Me-C$_6$H$_4$, 4-Cl-C$_6$H$_4$ 2,4,6-Me$_3$-C$_6$H$_2$ 2,6-*i*-Pr$_2$-C$_6$H$_3$	[193–195] [197] [196,201,208] [202]	(26)
	R = 2,4,6-Me$_3$-C$_6$H$_2$	[198,199,203,205]	(27)

Several complexes with bidentate phosphines were utilized for ROMP, but the activity of (28a) was not as high as with other Ru based initiators [211], whereas ionic complexes of type (28b) serve as highly active initiators [212,213]. Amoroso and Fogg [214] and Leitner et al. [215] reported ROMP reactions with further initiators bearing bidentate phosphine ligands.

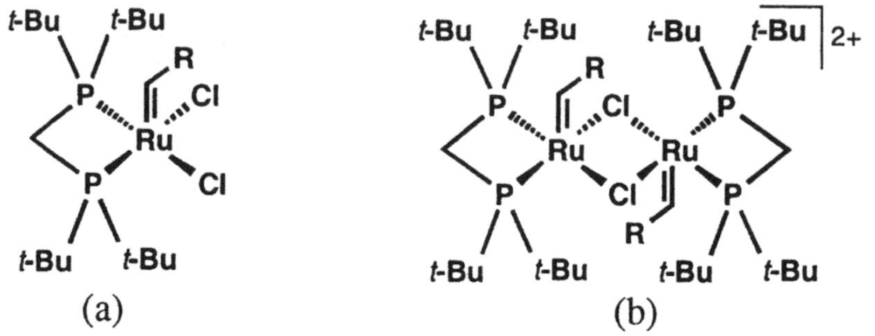

(28)

(a)   (b)

Furthermore, a variety of complexes with a labile non-phosphine ligand, e.g., a Schiff-base (29a) [216] or a chelated ether-functionalized alkylidene moiety (29b–c) [217–219], were synthesized. Structure (29a) was reported to be active in RCM reactions, but it is at room temperature less reactive than the parent Grubbs catalyst [216]. On the other hand (29b) is more active, but initiation with this complex is about 30 times slower [217]. Structure (29c) is highly active in RCM reactions [218,219].

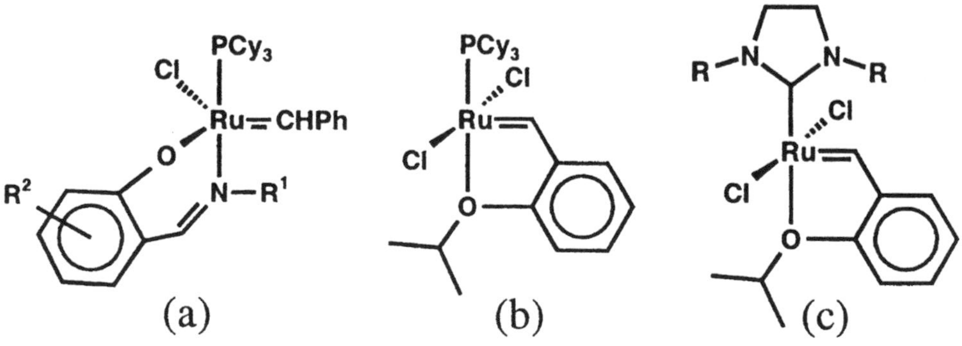

(a)   (b)   (c)

(29)

A further approach for making available a vacant coordination site for olefins at the metal was the synthesis of various bimetallic complexes with a weakly bond metal fragment (30a–c) by Grubbs [188]. These complexes displayed high activity in ROM polymerizations [188]. Herrmann et al. reported a series of analogous NHC complexes (30d–f) [193–195] with improved activity [194,220] and stability [220]. A comparative report on the stability of various metathesis initiators in RCM is given in Ref. [220].

(30)

G. Concluding Remarks

Actually in the field of olefin metathesis the development of new initiators is very rapid particularly for Ru-based systems. Therefore, in this article we could only present a brief overview. For further Ru based initiators see, e.g., Refs. [224–229]. Moreover, a variety of difunctional initiators were synthesized and employed for the synthesis of triblock copolymers [230,231].

The Schrock catalysts, especially the Mo based systems, and the Grubbs catalyst are now well established as well-defined standard initiators enabling the ROM polymerization of many cycloolefins. Both systems guarantee mostly a high degree of reaction control and are commercially available.

The control over the molecular weight is possible in particular either by addition of acyclic olefins acting as chain transfer agents [111,232,233] or by adjusting the monomer/initiator ratio [145,181]. The former strategy enables also a simple and efficient synthesis of telechelic ROM polymers [234–238]. Moreover, telechelics were synthesized by degradation of suitable ROM copolymers [239].

It is worth mentioning that such well-defined initiators reported above are too expensive for most industrial applications [83]. Therefore numerous papers dealing with alternative catalyst systems appeared: for example Nubel et al. used $RuCl_3$ or $RuBr_3$ in combination with various phosphines and alkynes preferably under H_2-atmosphere [240], Mühlebach et al., e.g., $(p\text{-cymene})PCy_3RuCl_2$ [241,242] and Grubbs et al. a NHC bearing Ru system [243]. For the same reason alternative syntheses of ruthenium alkylidenes and related species avoiding the use of diazo compounds were thoroughly investigated [244–246].

IV. THERMODYNAMIC ASPECTS

ROM polymerizations, as well as other reactions, occur only if ΔG under the particular reaction conditions is negative. In general, basically due to the loss of translational entropy, reaction entropy ΔS in the ring-opening polymerization of olefins is negative. According to $\Delta G = \Delta H - T\Delta S$, the consequently positive entropy term $(-T\Delta S)$ has to be counterbalanced by an adequate negative ΔH to enable the reaction [247–249].

During ROM polymerizations the nature and number of the bonds remain unchanged. Therefore ΔH and thus the metathetic polymerizability of any given cycloalkene are crucially affected by its ring strain and consequently ring size. Indeed, very large rings are virtually unstrained, so ΔH becomes approximately zero. But ΔS is positive now and accordingly ΔG still negative, as pointed out by Ivin and Dainton [247,250].

Other important chemical factors influencing ΔG are for example substituents or the microstructure of the formed polymer (e.g., cis/trans-isomerism, tacticity) [1,248]. As shown in Table 4 formation of trans-polymers is thermodynamically preferred to cis-polymers.

Under standard conditions all unsubstituted cycloolefins apart from cyclohexene are metathetically polymerizable (Table 4). In fact, polymers of (Z,E)-cyclodeca-1,5-diene [251] or 1,3-cyclooctadiene [252] containing 1,7-octadiene units pinch off cyclohexene in the presence of suitable metathesis catalysts. But even cyclohexane has recently been oligomerized at lower temperature [253].

Due to the low strain of 5-membered rings, the polymerizability of cyclopentene derivatives is strongly affected by substituents. For example, 1-methylcyclopentene or 3-isopropylcyclopentene do not polymerize under all reaction conditions tested until now, probably because ΔG is positive [1]. On the other hand 3-methylcyclopentene and 1-(dimethyl)silacyclopent-3-ene were polymerized successfully [1]. Substituents in general make ΔG less negative or even positive [1,248,254]. However, bridges can strongly increase the ring strain and thus improve the polymerizability. Norbornene for example is readily polymerizable despite of its 6-membered ring.

Apart from the above-mentioned chemical factors of course physical factors such as temperature or solvent are very important, too. Generally, as in most other ring-opening polymerizations a ceiling temperature at a given concentration and an equilibrium monomer concentration at a given temperature exists.

In particular, if polymerizing the low-strained cyclopentene or its derivatives the equilibrium concentration $[M_e]$ of monomer is not negligible, as theoretically predicted: $\ln[M_e] = \Delta H^\circ/RT - \Delta S^\circ/R$. $[M_e]$ has been estimated at 3.2 mol/l for the

Table 4 Thermodynamic parameters for ROMP of cycloalkenes under standard conditions.

Monomer	Polymer	$-\Delta H^\circ$ kJ/mol	$-\Delta S^\circ$ J/(K·mol)	ΔG° kJ/mol	Refs.
Cyclobutene	cis	121	52	−105	[255]
Cyclopentene	cis	15.4	51.8	−0.3	[255]
	trans	18	52	−2.6	[255]
Cyclohexene[a]	cis	3 to −1	31	6.2	[256]
	trans	1 to 3	28	7.3	[256]
Cycloheptene	70% trans	18	37	−7	[255]
Cyclooctene	48% trans	13	9	−13	[255]
1,5-Cyclooctadiene[a]	cis	25	5	−19	[256]
	trans	33	5	−24	[256]

[a]Semi-empirical data for ROMP of liquid cycloalkene to solid amorphous polymer.

formation of highly *cis* polypentenamer at $-10\,^\circ$C [257] and 0.51 mol/l for the formation of *trans* polypentenamer at $0\,^\circ$C [258]. As expected, $\ln[M_e]$ decreases linearly with reciprocal temperature [258].

V. FORMATION OF CYCLIC BY-PRODUCTS

It is a well known phenomenon in ring-opening polymerizations that the polymerization process is accompanied by formation of cyclic oligomers [248,259,260].

In ROMP these by-products arise from back-biting processes, because already formed polymer chains as well as monomers may approach the propagating carbene species with their double bonds. If these secondary metathesis reactions proceed intramolecularly cyclic products are formed as shown in structure (31). The cyclic structure of such oligomers has been proven by mass spectroscopy [261,262].

$$\text{(31)}$$

This formation of cyclic oligomers has been studied in detail for various monomers (see [1]). Some selected references dealing with important monomers are summarized in Table 5.

Only cyclic oligomers instead of polymer are formed, if the initial concentration of monomer lies below a certain minimum concentration, the so-called critical concentration $[M_c]$ [261].

Of course it is necessary to distinguish carefully between the kinetical and the thermodynamical spectra of reaction products. The rate of back-biting and therefore the kinetical fraction of cyclic by-products depends, among others, on the particular reaction conditions, the used catalyst and the accessibility of the double bonds of the polymer backbone. For example, the tendency of back biting is reported to be small for the

Table 5 Examples of investigations on the formation of cyclic oligomers for various monomers.

Monomer	Catalyst/Solvent	Ref.
Cyclopentene	Various W-based classical catalysts in toluene	[263]
	No information given	[258]
Cyclooctene	$WCl_6/Sn(CH_3)_4$ in PhCl	[264]
	Calderon catalyst/cholorobenzene	[261]
	$WCl_6/EtAlCl_2EtOH$ in benzene, heptane, cyclohexane	[265]
1,5-Cyclooctadiene[a]	Various W-based/chlorobenzene	[266]
Cyclobutene[b]	No information given	[267]
	Schrock catalyst in toluene	[270–272]
Norbornene	$WCl_6/Sn(CH_3)_4$ in PhCl	[264,273]

[a]Oligomers possess formula $(C_4H_6)_n$.
[b]Same equilibrium composition as starting from 1,5-cyclooctadiene.

polymerization of norbornene [274,275], probably due to steric shielding by the cyclopentylene rings [275].

In equilibrium the product composition is unaffected by the starting material (monomer, oligomer, polymer) and typical for each monomer respectively polymer under the particular reaction conditions [277]. Therefore the metathetic degradation of polymers may yield oligomers, too [268,269].

The equilibrium between linear chains M_y and cyclic species with x repeating units represented by $c - M_x$ is:

$$M_y \leftrightarrow M_{y-x} + c - M_x$$

The equilibrium constant is $K_x = ([c - M_x][M_{y-x}])/[M_y]$ and for high degrees of polymerization approximately: $K_x = [c - M_x]$.

Assuming that all rings are free of strain, the Jacobson–Stockmayer theory predicts that K_x is proportional $x^{-2.5}$ and independent of temperature [248,278,279].

As proposed by this theory, in ROM polymers the fraction of cyclic oligomers decreases with increasing ring size [278]. Larger rings are virtually unstrained and plotting $\ln K_x$ vs. $\ln x$ fits very well the predicted straight line with slope -2.5 [261,264,265,277,280]. For smaller rings there are distinct discrepancies and generally the predicted absolute values are too high for all ring sizes. Therefore Suter and Höcker developed an improved description using a rotational isomeric state (RIS) model [281]. Kornfield and Grubbs advanced a theory, in which as well entropic as enthalpic terms are considered and which enables better predictions [267].

These secondary metathesis reactions of course may also affect the microstructure of the polymers, e.g., the fraction of *cis* double bonds [275,283]. The extend of these reactions depends in particular on steric factors and the activity of the applied catalyst. The sterical shielding of polynorbornenes' backbone by the 5-membered rings is probably responsible for the low reactivity in degradation reactions via cross methathesis with acyclic olefins, too [276].

VI. POLYMER MICROSTRUCTURE AND MECHANISTICAL CONSIDERATIONS

It is well known that the physical properties of many polymers are strongly influenced by their microstructures. The determination and beyond that the specific control of microstructure is therefore an important topic. Furthermore, these investigations may enable deeper insights into the polymerization mechanism and the structure of the propagating species.

Numerous papers dealing with the microstructure of ROM polymers and its origin appeared in the recent years. In this chapter some of the most important aspects are briefly discussed.

The first point to consider is the stereochemistry of the double bonds in the polymer main chain. In ROM polymers made from unsubstituted monocyclic monomers this is the only microstructural feature. Such polymers were investigated thoroughly using NMR spectroscopy [1,282,283,301].

Most of the microstructure investigations beyond these simple cycloalkenes were carried out on polymers of norbornene, norbornadiene and their derivatives (Table 6) [284].

Table 6 Selected investigations on tacticity of ROM polymers using NMR techniques.

Monomer	Ref.
Norbornene[a]	[285]
1-Methylnorbornene	[286]
7-Methylnorbornene	[287]
5,5-Dimethylnorbornene	[289,290]
(±)-5,6-Dimethylnorbornene (*endo,endo/endo,exo/exo,exo*)	[291,130]
(±)-*endo*-Norbornene-2-acetate	[294]
7-Methylnorbornadiene	[292]
7-Phenylnorbornadiene	[288]
7-*tert*-Butoxynorbornadiene	[287,292,293]
Benzonorbornadiene	[288]
2,3-Bis(menthyloxy)carbonyl)norbornadiene	[132]
2,3-Bis(pantalactonyloxy)carbonyl)norbornadiene	[132]

[a]Tacticity determined from hydrogenated polymer.

Tacticity arises in these polymers from the relative orientation of the five membered rings in the polymer backbone as shown in structure (32). Thus, depending on the configuration of the adjacent double bond *cis*-syndiotactic (*cis* racemic, c_r), *cis* isotactic (*cis* meso, c_m), *trans* syndiotactic (*trans* racemic, t_r) and *trans*-isotactic (*trans* meso, t_m) dyads result.

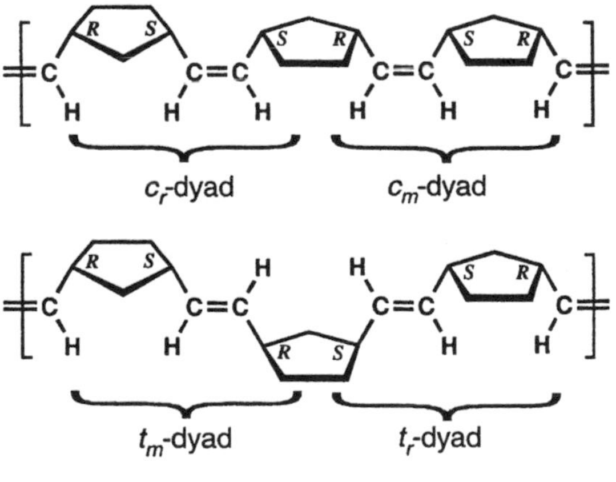

c_r-dyad c_m-dyad

(32)

t_m-dyad t_r-dyad

Dyads associated with *cis* or *trans* double bonds may independently tend to be isotactic, syndiotactic or atactic in one given polymer (Table 7) [1,284].

For substituted derivatives it may be necessary to take additionally head/tail [286,289,290] and/or *syn/anti* [287,292,293] isomerisms into account.

^{13}C-NMR techniques were used with great success for the determination of all these structures [1,284]. However, the microstructure of the particularly formed polymers is not characteristic for one given catalyst, but may vary with monomer, temperature, dilution and solvent [1].

It is assumed that monomers having norbornene structure approach the metal-carbene exclusively with the *exo* face of their double bond (structure 33a) [1,284]. This is probably caused by steric hindrance [299] and/or electronic effects [300].

Table 7 Microstructure of poly(7-methylnorbornene) and poly(7-methylnorbornadiene) samples prepared using various catalysts.

Monomer	Catalyst	σ_{cis}[a]	cis-Dyads[b]	trans-Dyads[b]	Regio-selectivity	Ref.
7-Methyl-norbornene	(25,26)	0.34–0.56[c]	Isotactic	Syndiotactic	All anti	[295]
	(PCy$_3$)$_2$Cl$_2$Ru=CHPh	0.2	Isotactic	Isotactic	All anti	[295,296]
	(t-BuO)$_2$Mo(NAr)=CHR	0.1–0.21[f]	Not det.	Isotactic	[e]	[287,297]
7-Methyl-norbornadiene	(25,26)	0.48–0.53[c]	Syndiotactic	Atactic	All anti	[295]
	(PCy$_3$)$_2$Cl$_2$Ru=CHPh	0.29	Syndiotactic	Syndiotactic	All anti	[295,296]
	(t-BuO)$_2$Mo(NAr)=CHR	0.68	Syndiotactic	Isotactic	6.5–7.5%	[298]
	OsCl$_3$	0.97	Syndiotactic	Isotactic	syn units	[292]

For reaction conditions, see the particular references. [a]Fraction of cis double bonds in the polymer main chain; [b]Tacticity bias of the cis- or trans-centred dyads; [c]Depends on the particular N-substituents of the N-heterocyclic carbene ligands(s), see Ref. [295]; [d]Data given for the initially yielded poly(anti-7-methylbornene); [e]Syn-iosmer is polymerized slowly after complete consumption of anti isomer, thus forming a block copolymer; [f]Depends on monomer-initiator ratio.

As shown in (33a) by way of example for the polymerization of norbornene (NBE) the configuration of the resulting double bond is locked already in the metallacyclobutane intermediate. If forming this metallacycle the relative orientation of the approaching norbornene monomer and the last incorporated NBE unit determines the stereochemistry of the resulting double bond as well as the tacticity of the dyad (33b). In the polymer the kinetical fraction of double bonds having cis configuration (σ_{cis}) depends on the relative ease of forming the respective metallacyclobutane [1]. Steric factors and the reactivity of monomer and catalyst were identified as important determinants influencing σ_{cis} [1]. Additionally, the configuration of a double bond is influenced by the stereochemistry of the last formed double bond, manifested in the blocky distribution of cis and trans double bonds which is often recognized in ROM polymers [1,308].

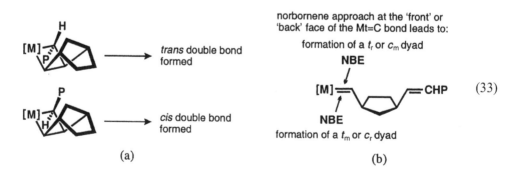

(a)   (b)   (33)

The formation of tactic polymers is a well known phenomenon in ROMP (cf. Table 7). This is due to the inequality of the two faces of the propagating alkylidenes' Mt=C bonds which has been explained assuming either chain end control or enantiomorphic site control mechanisms.

In mechanisms assuming enantiomorphic site control a stereo selection at a chiral metal center is proposed. The basic theory reported by Ivin is based on the assumption of two enantiomeric propagating species P_l and P_r as shown in structure (34) [301].

Both possess octahedral geometry as indicated by lines and one vacant coordination site for monomer coordination represented by the symbol '□'. The influence of the other ligands is neglected. Moreover, other geometries of the propagating complexes, than the proposed octahedral, are possible too. When forming a *cis* double bond P_l is converted into P_r and vice versa, whilst when forming a *trans* double bond the chirality remains unchanged (structure 34). Thus preferably *cis*-syndiotactic and *trans*-isotactic dyads will result if propagation is faster than epimerization of P_l and P_r or ligand exchange. Isomerization occurs by rotation about Mt=C [dotted lines in (34)]. This leads to the subsequent formation of a c_m or t_r dyad, consequently the monomer now approaches at the other face of the Mt=C bond [cf. (33b)].

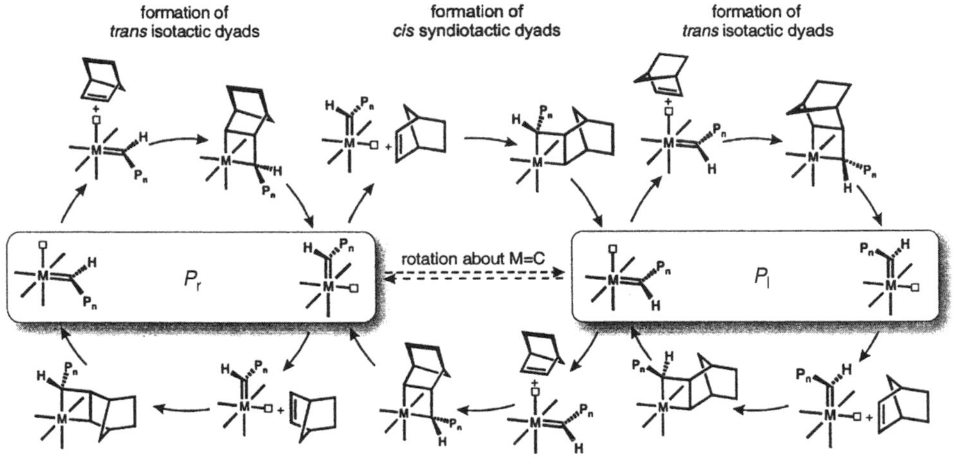

formation of
trans isotactic dyads

formation of
cis syndiotactic dyads

formation of
trans isotactic dyads

P_r

rotation about M=C

P_l

(34)

This mechanism has been modified later, proposing two main kinds of propagating species differing in whether the last formed double bond is still coordinated to the metal center or not. Furthermore, the stereochemistry of this coordinated bond must be considered. Hence species like the chiral P_c respectively P_t, P and the more relaxed achiral P' are predicted (structure 35). Depending on the microstructure of the particularly formed polymer various of these species must be taken into account as chain carriers and different reaction pathways were proposed [284,290,291,302–304]. This proposal of such kinetically distinct propagating species provides also a rationale for the often recognized blocky distribution of *cis* and *trans* double bonds in ROM polymers.

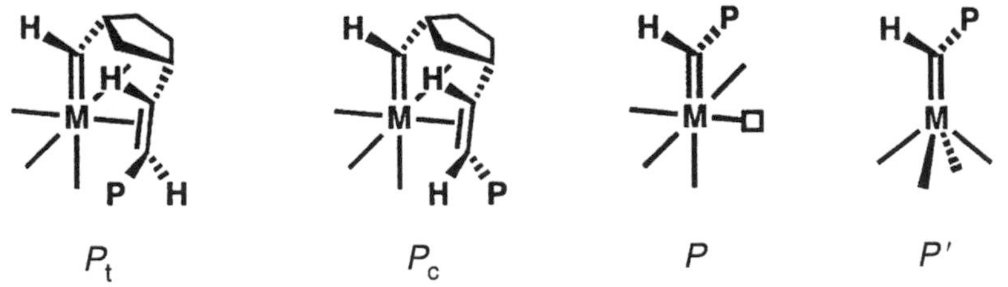

P_t

P_c

P

P'

(35)

Rotational isomers about the Mt=C bond are of crucial importance in these models. Such rotamers were studied using variable-temperature NMR techniques and rotation barriers were estimated [151,305,306].

Most of the deeper investigations mentioned above were carried out on polymers made with classical catalysts. Only few papers on the microstructure of polymers prepared with modern unicomponent well-defined initiators appeared until now [113,157,173, 295,296].

An exception to that rule is the Schrock catalyst, which was thoroughly investigated [127,128,130–132]. These complexes exist in the form of two rotational isomers. The so-called *syn* and *anti* rotamers differ in the relative orientation of the *t*-butyl group and the imido ligand and may interconvert (structure 36) [127,128].

$$K_{eq} = k_{a/s}/k_{s/a}$$

(36)

Schrock assumes that *cis* double bonds are formed if propagation proceeds via the *syn* and *trans* double bonds via the *anti* rotamer [126,128,129]. Reaction rates and equilibrium constants for the interconversion of *syn* and *anti* rotamers were estimated for a variety of complexes and it appeared that they depend strongly on the particular alkoxy ligands. The stereochemistry of the formed double bonds depends therefore crucially on the OR-ligands and its seems that O*t*-Bu substituted complexes form high-*trans* polymers, while OMe(CF$_3$)$_2$ substituted complexes form high-*cis* polymers [129,131]. Block copolymers of *cis* and *trans* poly(2,3-bistrifluoronorbornadiene) were obtained by exchange of the alkoxy ligands of the living chain end [307].

The two CNO-faces (the faces of the Mt=C bond) of each rotamer are equivalent in the initiator, but not in the propagating species by virtue of the chiral C$_\beta$ of the growing polymer chain. Thus, Schrock proposes for these well defined catalysts a chain end control mechanism.

Schrock catalysts bearing chiral ligands may enable the formation of highly tactic high-*cis* polymers (Table 8) [128,129,131,132]. Some suitable ligands are shown in structure (37). Enantiomerically pure initiators are not necessarily required for the synthesis of these highly tactic polymers, racemic complexes may also be used. If polymerizing an enantiomerically pure monomer with a racemic initiator a bimodal distribution of molecular mass may arise from different reaction rates at the two enantiomeric sites of the catalyst [128,132].

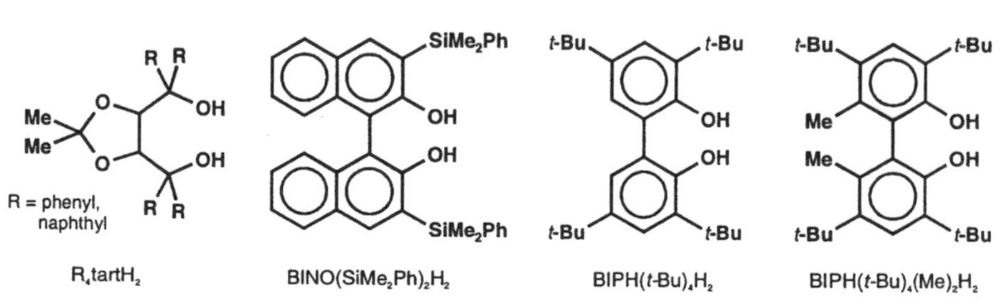

(37)

Table 8 Some examples of the formation of tactic polymers using Schrock catalysts.

Monomer	Catalyst	σ_{cis}	Tacticity	Ref.
	$(t\text{-BuO})_2\text{Mo(NAr)(=CHR)}$	0.06	Syndiotactic	[132]
2,3-Bis[(1R,2S,5R-(−)-	$[(\text{CF}_3)_3\text{CO}]_2\text{Mo(NAr)(=CHR)}$	0.99	Isotactic	
menthyloxy)-	$[\text{BIPH}(t\text{-Bu})_4]\text{Mo(NAr}')(=\text{CHR})$	0.99	Isotactic	
carbonyl]norbornadiene	$[(\pm)\text{-BINO(SiMe}_2\text{Ph})_2]\text{Mo(NAr}')(=\text{CHR})$	0.99	Isotactic	
	$[\text{Me(CF}_3)_2\text{CO}]_2\text{Mo(NAr)(=CHR)}$	0.97	$(\sigma_\text{T})_c = 0.74$	
	$((+)\text{-Ph}_4\text{-tart}]\text{Mo(NAr)(=CHR)}$	0.98	$(\sigma_\text{T})_c = 0.88$	
2,3-Bis(trifluoromethyl)	$[(+)\text{-Nap}_4\text{tart}]\text{Mo(NAr)(=CHR)}$	0.97	$(\sigma_\text{T})_c = 0.97$	[131]
nor-bornadiene	$[(\pm)\text{-BINO(SiMe}_2\text{Ph})_2]\text{Mo(NAr)(=CHR}')$	0.71	$(\sigma_\text{T})_c = 0.86$	
	$[(\pm)\text{-BINO(SiMe}_2\text{Ph})_2]\text{Mo(NAr}')(=\text{CHR})$	>0.99	$(\sigma_\text{T})_c > 0.99$	
	$[\text{BIPH}(t\text{-Bu})_4(\text{Me})_2]\text{Mo(NAr}')(=\text{CHR})$	0.96	$(\sigma_\text{T})_c > 0.99$	[128]

$R = \text{CMe}_2\text{Ph}$; $R' = \text{CH-}t\text{-Bu}$; $\text{Ar}=2,6\text{-}i\text{-Pr}_2\text{-C}_6\text{H}_3$; $\text{Ar}' = 2,6\text{-Me}_2\text{-C}_6\text{H}_3$; $(\sigma_\text{T})_c = $ tacticity of all-*cis* triads (it was not possible to decide safely whether the *cis* centered dyads of these poly(2,3-bis(trifluoromethyl)norbornadiene) samples were isotactic or syndiotactic biased [131]. It is suggested that they are probably syndiotactic biased [128,129,131]).

On the other hand if using NHC substituted complexes the presence of a chiral environment at the metal center has no significant effect on the tacticity of the formed polymers under the conditions tested until now [295].

VII. INDUSTRIAL APPLICATIONS

A. Norsorex

Polynorbornene, the first metathesis polymer produced in industrial scale, was marketed in 1976 by CdF Chimie under the trade name Norsorex®. The monomer is produced by Diels–Alder reaction of cyclopentadiene and ethylene and polymerized in *n*-butanol using an RuCl_3/HCl catalyst. Norsorex is a very high molecular weight ($M_n > 2 \times 10^6\,\text{g/mol}$), thermoplast ($T_g = 35\,°\text{C}$) with approximately 90% *trans*-double bonds. The polymer is compatible with high loads of extending oils and plasticizers (up to 700%) and easily vulcanizable. By addition of suitable amounts of plasticizers the polymer is converted into an elastomer ($T_g = -60\,°\text{C}$).

Vulcanized Norsorex is sued specially for noise or vibration damping and for low hardness gaskets, rollers and bump stops. At the present Elf Atochem produces up to 5000 t/a polynorbornene at a plant in Carling/Lothringen [1,4,25,309–311].

B. Vestenamer

In 1980 the Chemische Werke Hüls started the production of polyoctenamers, which are put on the market under the trade name Vestenamer®. Two types are available now: Vestenamer 8012 (80% *trans*; $M_w = 75,000\,\text{g/mol}$; $T_g = -65\,°\text{C}$) and Vestenamer 6213 (60% *trans*; $M_w = 95,000\,\text{g/mol}$; $T_g = -75\,°\text{C}$). The polymerization process probably uses a homogeneous W-based catalyst and has a capacity of approx. 12,000 tons a year. Special properties of these short chain, partly crystalline (Vestenamer 8012: 30% at 23 °C) polymers are caused by the simultaneous formation of unbranched chains as well as macrocyclic products and the high fraction of *trans*-double bonds.

Vestenamers are especially used as processing aids in blends with other rubbers (e.g., CR, SBR, NBR, EPDM, IR). This compounding leads to a lower Mooney viscosity during processing, a better incorporation and dispersion of fillers and a higher green strength. The Vestenamer is incorporated into the network during vulcanization. Properties of the vulcanizates are usually less effected, in some cases, e.g., tear resistance is improved or swelling reduced. By addition of polyoctenamers the compatibility of polar and non-polar rubbers is often improved [312–314].

C. Poly(dicyclopentadiene)

Dicyclopentadiene (DCPD) is obtained as a by-product from the C_5-cut of naphta cracking. Its cheapness and high reactivity make it a very attractive monomer for ROMP. Liquid resins for the production of cross-linked poly(DCPD) are marked as Metton® and Telene®. Reaction injection molding (RIM) technique is used to produce objects. The Metton resin system consists of two compounds as shown in Table 9, one containing the tungsten-based catalyst and one with an Al-alkyl as cocatalyst. The viscosity of both is adjusted by addition of elastomers. After rapid mixing of these compounds the liquid resin is injected directly in the mold to polymerize. The product is a crosslinked polymer with high E-modulus, excellent impact strength even at low temperatures and low water absorption [1,4,74,241,315]. A hard oxide layer on the surface of the material limits oxygen diffusion into the polymer bulk and protects the underlaying material [315,316].

Ciba Spezialitätenchemie researched thoroughly the use of ruthenium-based catalyst for the polymerization of dicyclopentadiene. These catalysts are more tolerant towards moisture, oxygen and fillers. Possible applications of the filled or non-filled duromers are in the sector of electro casting and insulation [241].

Despite their industrial relevance the structure of these polymers is not completely cleared up until now. Metathetic conversion of the cyclopentene rings and/or olefin addition are proposed as mechanisms for cross-linking [4,74,241,317–319].

D. Zeonex

Zeonex®, obtained by ROM-polymerization of norbornene-type monomers with subsequent complete hydrogenation of double bonds, is produced by Nippon Zeon since 1991. The high glass transition temperature ($T_g = 140$–$160\,°C$ [1]), low water absorption and excellent optical properties make it especially suitable for optical discs, lenses and pharmaceutical packages [320].

E. Others

In addition to the above mentioned processes realized in industrial scale, many other applications, (e.g., the (co-)polymerization of cyclopentene [1] or cyanonorbornene

Table 9 Ingredients of the two components of the 'Metton' liquid resin for RIM polymerization [4].

Component A		Component B	
DCPD		DCPD	
Elastomer		Elastomer	
Catalyst:	Al Alkyl	*Catalyst*:	$WCl_6/WOCl_4$
	Lewis base		nonylphenol
Additives (antioxidants, others)			acetylacetone
Fillers		Additives, fillers	

[5,321–323] and the cross-metathesis of polymers [324,325] were investigated due to their potential industrial importance.

VIII. FURTHER APPLICATIONS

Numerous other possible applications for ROM polymers were reported during the last years. Many of them became possible since the development of Grubbs' robust versatile ruthenium based systems. Some examples are briefly listed below:

- liquid crystalline polymers [20,326–441];
- polymers with bioactive oligopeptide side chains [344,345];
- sugar-substituted poly(norbornenes) [346];
- polymers of porphyrazine benzonorbornadiene derivatives [347];
- conjugated polymers [20,252–357];
- stationary phases [143,358];
- hydrogels and polyelectrolytes [359–361];
- photoresists [362–365];
- high-temperature polymers [366];
- polymerization from surfaces [367,368].

IX. SELECTED EXAMPLES FOR THE POLYMERIZATION OF CYCLIC OLEFINS

This chapter intends to give a brief overview over the ROMP of some selected cycloolefins to demonstrate the great utility of ring-opening metathesis polymerization.

A. Monocyclic Olefins

1. Cyclobutene and Derivatives

Dell'Asta et al. performed ROM polymerization of cyclobutene in 1962 for the first time [369]. Employing a titanium-based system they received a high-*cis* poly(butadiene). Since then this monomer has been polymerized using a broad variety of catalysts. The spectrum of utilized catalysts ranges from classical systems [370] to modern well-defined initiators [371,372]. Of course the microstructure of the resulting polymers depends on the particularly utilized catalyst. For example, $TiCl_4/AlEt_3$ (1:3) has been reported to yield a high-*cis* polymer [369] and $RuCl_3$ in EtOH a high-*trans* polymer [22,42d].

Furthermore, various 3-substituted cyclobutenes were polymerized metathetically [181]. 1-Methylcyclobutene has been polymerized with a WCl_6 [370,373], $RuCl_3$ [42d] or $Mo(N-2,6-i-Pr_2-C_6H_3)(=CHCMe_2R)(OCMe_n(CF_3)_{3-n})_2$ (R = Me, Ph) [374,375] initiator. Many other functionalized derivatives were successfully ROM polymerized, too [22,145, 181,376,377].

2. Cyclopentene and Derivatives

Metathesis catalysts, e.g., tungsten chlorides in combination with $Si(CH_2CH=CH_2)_4$ [378] or $AlEt_3$/benzoyl peroxide [379], lead to high-*cis* polymers in the ROMP of cyclopentene. A high *cis*-polypentenamer is also obtained under use of $MoCl_5$ and $AlEt_3$ [380].

On the other hand, a high-*trans* polymer is formed with $TiCl_4/AlEt_3$ [370,380] or WCl_6 in presence of suitable cocatalysts (e.g., $AlEt_3$) [380,381]. Not only the catalyst and cocatalyst choosen are crucially for the microstructure of the resulting polypentenamers, but also the particular reaction conditions, especially the temperature or concentration [379]. Selected examples for the ring-opening metathesis polymerization of cyclopentene are listed in Table 10.

3. Cyclooctene and Derivatives

The first ROM polymerization of cyclooctene was performed by Natta et al. [370]. This initial report was followed by numerous others, only selected polymerizations of cyclo-octene are summarized in Table 11.

Functionalized cyclooctenes have been polymerized by Grubbs et al. employing a Ru-based catalyst [111,394]. Furthermore, other authors reported on the ROMP of various substituted cyclooctenes with several metathesis catalysts and on their influence of the configuration [95,395–397].

4. Cyclooctadiene and Derivatives

Not only ROMP of 1,5-cyclooctadiene and its derivatives is reported, but also the polymerization of substituted and unsubstituted cyclooctapolyenes [1,20]. Classical catalysts, e.g., WCl_6 in combination with various cocatalysts can polymerize these

Table 10 Some examples of ROMP of cyclopentene.

Catalyst	σ_{cis}	Remarks	Ref.
WCl_6/PhCCH (1:1)	0.8	other cocatalysts are possible	[382,383]
[(Cyclooctene)$_2$Ir(OCOCF$_3$)]$_2$		Predominantly *trans* configuration[a]	[384,402]
W(CH-t-Bu)(N-2,6-i-Pr$_2$-C$_6$H$_3$)(O-t-Bu)$_2$		$T = 25\,^\circ$C; $M_w/M_n = 1.48$	[385]
RuCl$_2$(p-cymene)(PCy$_3$)[b]	0.18	$M_w/M_n = 1.66$	[386][c]

[a]Reaction temperature $\sim 55\,^\circ$C.
[b]TMSD: 2×10^{-4} mol.
[c]Ru catalysed ROMP of cyclopentene see [111,179].

Table 11 Few examples of ROMP of cyclooctene.

Catalyst	Config.	σ_{cis}	Remarks	Ref.
WCl_6/PhCCH	*cis*	~ 0.8	Small yields	[387][a]
Ru(=CH-t-Bu)Cl$_2$(P-i-Pr$_3$)$_2$	*cis*			[439][b]
MoCl$_2$(PPh$_3$)$_2$(NO)$_2$/AlCl$_2$Et	*trans*	0.43	Predominantly *trans* polyoctenamer	
	cis	0.67	Predominantly *cis* cyclooctenamer instead of AsCy$_3$ other ligands have	[392][c]
			Been used (PCy$_3$, SbCy$_3$ or N-heterocyclic	
RuCl$_2$(p-cymene)(AsCy$_3$)	*cis*	0.76	carbene ligands)	[386,393]

[a]For other W-based catalysts see [62,264,376,381,388–391].
[b]For the ROMP of cyclooctene with Ru-based systems see also Refs. [92–95,104,111,177,179,192].
[c]For other Mo catalysts see Refs. [70,147].

compounds metathetically [1,22,398]. Additionally, cyclooctapolyenes and derivatives are also polymerizable with well-defined catalyst [20,70,111,243].

ROMP of cyclooctatetraene yields polyacetylene. This was reported by Korshak et al. for the first time [399]. Highly conjugated polymers are available via ROMP of cyclooctatetraene derivatives using a W- or Mo-based Schrock catalyst, too [20]. The physical properties of the polymers e.g., the conductivity after a doping process, may enable interesting applications of these materials [1,348].

B. Bi- and Tricyclic Olefins

1. Norbornene

The number of publications on the ROM polymerization of norbornene is tremendous. Due to its high reactivity and low prize this molecule became one of the standard monomers if testing the activity of a potential metathesis catalyst [1]. Polymerizations were reported using a wide range of initiators from simple metal halides, e.g., $TiCl_4$ [7], $RuCl_3$ [161,400], $MoCl_5$ [401] or $MgCl_2$ [66] to well-defined carbene initiators [166,177,439]. The first living ROMP was reported in 1986 using a titanium metallacycle as catalyst [112,113]. Table 12 summarizes selected reports on the polymerization of norbornene with various metathesis catalysts.

2. Norbornene Derivatives

a. Norbornenes with an Alkyl Substituent. The norbornene derivatives bearing methyl substituents shown in scheme (38) were all successfully polymerized utilizing various metathesis catalysts [1,22].

Table 12 Examples of ROMP of norbornene.

Catalyst	σ_{cis}	Remarks	Ref.
$ReCl_5$	0.75	σ_{cis} depends on the monomer concentration	[301][a]
$[IrCl(C_8H_{14})_2]_2$			[402]
$OsCl_3/PhCCH$	~0.9	Syndiotactic polymer	[422]
$WCl_6/SnMe_4$		Atactic polymer	[422]
$Re(CO)_5Cl/EtAlCl_2$		Various temperatures; $M_w/M_n = 1.45$	[407]
$ReO_3Me(MTO)/R_nAlCl_{3-n}$ (R = Me, Et; $n = 1,2$)	0.84	The *cis* content variate with the cocatalyst	[403]
Cp_2TiMe_2	0.37		[112,404]
$(Me_3SiCH_2)_4Ta_2(\mu\text{-}CSiMe_3)_2$		Higher *trans* content; E:Z ratio 1:2.1	[405]
$W(CH\text{-}t\text{-}Bu)(N\text{-}2,6\text{-}i\text{-}Pr_2\text{-}C_6H_3)(OCMe(CF_3)_2)_2$	0.95	$M_w/M_n = 1.03$	[166,406][c]
$Mo(CHCMe_3)(N\text{-}2,6\text{-}i\text{-}Pr_2\text{-}C_6H_3)(OCMe_3)_2$	~0.35	$M_w/M_n = 1.10\text{-}1.15$	[142,406]
$Ru(=CH\text{-}p\text{-}C_6H_4X)(PPh_3)_2Cl_2$ X = H, NMe_2, OMe, Cl	0.10		[177,231][d]

[a]For more details see also [407].
[b][408].
[c]ROMP of NBE with tungsten-based catalysts see [409].
[d]For the Ru-based ROMP of NBE see [1,92–95,98,109,192,386,439].

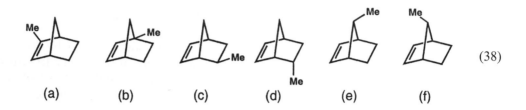

$$(38)$$

(a)   (b)   (c)   (d)   (e)   (f)

ROMP of racemic (38a) yields a polymer with an all-HT-structure [57d]. Strongly HT-biased polymers are reported for poly(38b) too [286,296]. ROMP of (38c) or (38d) leads to polymers with randomly oriented methyl-substituents [414–416]. W(CO)$_3$ (mesitylene) in combination with AlEtCl$_2$ polymerizes (38e) as well as (38f) [417]. These two prochiral monomers (38e/f) have different reactivities. The polymerization of these monomers results in polymers of varying tacticity, depending on the catalyst [152,287,418–421]. With the tungsten-based catalyst (W(C(CH$_2$)$_3$CH$_2$)(OCH$_2$CMe$_3$)$_2$Br$_2$) a mixture of *syn*- and *anti*-7-methylbicyclo[2.2.1]hept-2-ene (38f/e) undergo a ROMP reaction. In this case only the *syn*-isomer polymerizes, i.e., the polymer is completely free of the *anti*-isomer [152].

Structure (38b) and a mixture of (38e) and (38f) were polymerized utilizing as well classical catalysts, e.g.,WCl$_6$ in combination with SnBu$_4$ or SnPh$_4$ [286,421], as the modern unicomponent initiator Ru(=CHPh)(PCy$_3$)$_2$Cl$_2$ [296].

Examples of ROMP of the methyl substituted norbornenes (38a–f) are listed in Table 13.

Substituted norbornenes with R = CHMe$_2$ [425], aryl R = C$_n$H$_{2n+1}$ (n = 8–12) [427]/ R = C$_6$H$_5$ [428] or alkenyl R = CHMe$_2$ [425]/R = CH$_2$CH$_2$CH=CH$_2$ [426] can also undergo ROMP.

Additionally, Table 14 shows selected examples of dimethylnorbornenes, which were polymerized metathetically.

b. Norbornenes with a Silicon-containing Substituent. Silicon-containing norbornenes (*exo/endo* mixtures) undergo a ROMP reaction with various catalysts as shown in Table 15.

Norbornenes bearing a Si(OEt)$_3$ group can be heterogenized on SiO$_2$ particles or other surfaces with suitable binding units, e.g., OH-groups [412,413,433]. These methods are useful for the preparation of specific polymer architectures [433,434].

c. Norbornenes with a COOR or a COR Substituent. Some examples of the ROMP of mono COOR or COR substituted norbornenes are listed in Tables 16 and 17.

In addition to the polymerization of mono-substituted norbornene esters numerous ROMP reactions of bi-substituted norbornene esters are reported. These derivates have been polymerized not only with classical systems [435,461] but also with alkylidene complexes [23,123,153,423,132,438,439].

ROMP or norbornene esters bearing mesogenic side groups yields side-chain liquid-crystalline polymers [440,441]. Norbornenes which are functionalized in the 5- or in the 5- and 6-position with ester groups have been polymerized with both molybdenum and ruthenium-based initiators.

Examples for ROMP of liquid-crystalline monomers can be found in literature [326–343,449].

Norbornenes bearing a chain substituent in 5-position and a methacrylate end-group have also been polymerized to give polymers with electro-optical applications [442].

Tables 17 and 18 show some ring-opening metathesis polymerization reactions of norbornenes which were functionalized in the 2-position.

Table 13 ROMP of norbornenes bearing methyl substituents.

Catalyst	Monomer	σ_{cis}	Remarks	Ref.
$(CO)_5W{=}CPh_2$	(38a)		all-HT polymer	[57d]
[Ti/Al CH_3 structure]	(38b)		High *trans* conformation	[179][a]
$Ru({=}CHPh)(PCy_3)_2Cl_2$	(38b)	0.22	all HT	[296]
$ReCl_5$	(38b)	1.00		[286]
$RuCl_3{\cdot}3H_2O$	(38b)	0		[286][b]
[W structure: CH-*t*-Bu, OEt$_2$, RO, Cl]	(38b)	1.00	RO=OAc	[156,284]
$ReCl_5$	(38c)	1.00	All *cis* polymer end with a fully syndiotactic ring sequence	[414,410]
$WCl_6/EtAlCl_2$ (1:4)	(38c)	0.52	Atactic polymer	[411]
$ReCl_5/PhCl$	(38d)	1.00		[415]
$WCl_6/n\text{-}Bu_4Sn$ (1:2)	(38d)	0.63		[415]
[W structure: Br, Me$_2$H$_2$CO, Me$_2$H$_2$CO, Br]	(38f)	0.15	Cocatalyst: $GaBr_3$	[152,423]
WCl_6/Me_4Sn	(38e) and (38f)		Only the *anti* isomer (e) reacts; (f) can be isolated as utilized	[421,422][c]

(continued)

Table 13 Continued

Catalyst	Monomer	σ_{cis}	Remarks	Ref.
W(=CHCMe$_3$)Br$_2$(OCH$_2$CMe$_3$)$_2$/ GaBr$_3$	(38e) and (38f) also (38b) and (38d)		The less reactive monomer (f) will tend to react last The anti-isomer (e) reacts preferentially at the beginning, later the syn-isomer reacts to give its own distinctive	[424]
W(=CHCMe$_3$)(N-2,6-i-Pr$_2$-C$_6$H$_3$)(OCMe$_3$)$_2$	(38e) and (38f)		propagating species	[287,297]
Mo(=CHR^1)(N-2,6-i-Pr$_2$-C$_6$H$_3$)(OR^2)$_2$ R^1 = t-Bu, CMe$_2$Ph R^2 = CMe$_3$, CMe(CF$_3$)$_2$	(38e) and (38f) (38e) and (38f)		High trans conformation anti 7-methylnorbornene (e) is much more reactive than its syn isomer	[457]
Ru(=CHPh)(PCy$_3$)$_2$Cl$_2$ NHC with Cy^d (or other rests, e.g., R-NaphEt, R-CyEt) Ru-bu-bimetallic/NHC	(38e) and (38f) (38e) (38e)	0.2 0.34 0.44	Syndiotactic bias	[293,295,296] [295] [295]

[a] For more information see [286,287,420,113].
[b] A large variety of catalysts can also polymerize (b) under an ROM reaction [296,422].
[c] For more details see [152,287,297,417].
[d] In literature [295] more catalysts are shown which make an metathesis polymerization.

Table 14 ROMP of various dimethylnorbornenes (DMNBE).

Catalyst	Monomer/Rest	σ_{cis}	Ref.
OsCl$_3$/PhCCH	*endo,exo*-5,6-DMNBE	0.61	[429]
Mo(=CHCMe$_3$)(N-2,6-*i*-Pr$_2$-C$_6$H$_3$)(OCMe$_3$)$_2$	*endo,exo*-5,6-DMNBE	0.05	[130][a]
W(=CHCMe$_3$)Br$_2$(OCH$_2$CMe$_3$)$_2$/GaBr$_3$ (2:1)	5,5-DMNBE		[424]
RuCl$_3$	5,5-DMNBE	0.05	[290]
ReCl$_5$	5,5-DMNBE	1.00	[290]
Ru=CHPh(PCy$_3$)$_2$Cl$_2$	5,5-DMNBE	0.19	[296]

[a]OCMe$_2$(CF$_3$) or OCMe(CF$_3$)$_2$-substituted Schrock catalysts were utilized, too [132].

Table 15 Some examples of the polymerization of silicon-containing norbornene derivatives.

Catalyst system	Monomer with R:	#	Remarks	Ref.
Re$_2$O$_7$/Al$_2$O$_3$, promoted with *n*-Bu$_4$Sn, Et$_4$Pb or a few other co-catalysts	SiMe$_3$	(a)		[72]
			Various co-catalysts, e.g., PhCCH (1:1),	
WCl$_6$	SiMe$_3$		Ph$_4$Sn (1:2)	[72,430]
WCl$_6$/PhCCH (1:1)[a]	SiME$_2$CH$_2$SiMe$_3$	(b)		[430]
RuCl$_2$(PPh$_3$)$_3$	SiMe$_2$((CH$_2$)$_3$-9-carbazolyl)	(c)		[431]
RuCl$_3$*3H$_2$O[b]	CH$_2$SiMe(CH$_2$CH$_2$CH$_2$)	(d)		[430]
WCl$_6$/AlEt$_3$	SiMeCl$_2$	(e)[c]		[432]
WCl$_6$/*i*-Bu$_2$AlCl (1:5)	SiCl$_3$	(f)[c]		[430]
RuCl$_3$*3H$_2$O[a,b]	Si(OMe)$_3$[d]	(g)[c]		[430]

[a]Re$_2$O$_7$/Al$_2$O$_3$, *n*-Bu$_4$Sn (3 wt%) can also be used as catalyst [324].
[b]Another possible catalyst is WCl$_6$/*i*-Bu$_2$AlCl (1:5) [430].
[c]See additionally [324].
[d]The monomer NBE-Si(OEt)$_3$ (h) can be polymerized with the same Ru-based catalysts.

Classical catalysts like Re- and Ir-based systems can also catalyze ROMP of 5-norbornene-2-carboxyaldehyde [22].

d. Norbornenes with a OCOR, Hydroxy or Alkoxy Substituent. The ROMP of NBE which is substituted in the 2-position with an OCOR group can be achieved with various catalysts.

In addition, classical catalysts, e.g., Mo- and W-based systems can also polymerize the OCOR functionalized norbornenes [22,446]. Monomers containing 5- and 6-substituents and bearing a OCOR-group have also been prepared and polymerized [23,123,438].

A special SH-bearing linkers was fixed at the hydroxy functionalized NBE, which allowed the anchoring of this NBE unit to a gold surface. This NBE derivative has been polymerized with an ruthenium-based initiator for coating purposes [367].

Other groups have been bound in 2-position on the CH$_2$OH-functionalized NBE. The H-atom was replaced by a variety of functional substituents, e.g., methyl [450] and

Table 16 ROMP of norbornenes with a COOR substituent in 2-position.

Catalyst	COOR/configuration	σ_{cis}	Remarks	Ref.
NHC with isopropyl	H/mixture (exo/endo)			[221]
RuCl$_2$(PPh$_3$)$_3$	endo-Me	0.25		[435]
Mo(=CHCMe$_2$Ph)(OCMe$_3$)(N-2,6-i-Pr$_2$-C$_6$H$_3$)	endo-Me	0.3		[436]
Mo(=CHCMe$_2$Ph)(OCMe(CF$_3$)$_2$(N-2,6-i-Pr$_2$-C$_6$H$_3$)	endo-Me	0.87	cis-centered dyads are biased towards isotactic	[436]
WCl$_6$/SnBu$_4$ (1:1)	endo-Me	0.85		[435,437]
WCl$_6$/SnBu$_4$ (1:2)	exo-Me	0.5		[435]
Me$_2$H$_2$CO, Me$_2$H$_2$CO — W(Br)(Br)=cyclopentylidene	exo-Me[a]	0.4	Similar polymerization behavior as the endo isomer	[153][b]
RuCl$_3$	endo-Et	0.15		[435]
OsCl$_3$	exo-Et	0.3		[435]

[a]The endo-COOMe functionalized norbornene can also be polymerized with the same catalyst [153].
[b]A cocatalyst is needed (GaBr$_3$).

Table 17 ROMP of norbornenes with COR substituent.

Catalyst	exo/endo mixture of COR/configuration	Remarks	Ref.
RuCl$_3$·xH$_2$O/alcohol	H		[444]
NHC with isopropyl	H	$M_w/M_n = 2.3$	[221]
NHC with isopropyl	Me	$M_w/M_n = 1.4$	[221]

Table 18 Polymerization of norbornenes with OCOR substituent in 2-position.

Catalyst	exo/endo mixture of OCOR/configuration	σ_{cis}	Remarks	Ref.
WCl$_6$/SnMe$_4$ (1:1)	Me	0.71		[294]
Mo(=CHMe$_2$Ph)(OCMe(CF$_3$)$_2$)$_2$(N-2,6-i-Pr$_2$-C$_6$H$_3$)	endo-Me	~0.9		[436]
W(=CHCMe$_3$)Cl(CH$_2$CMe$_3$)(OAr)$_2$(O(CHMe$_2$)$_2$)	Me			[155]
Mo(CH-t-Bu)(NAr)(O-t-Bu)$_2$	endo-Bu[a]		Optically active	[445]
Mo(CH-t-Bu)(NAr)(O-t-Bu)$_2$	endo-Bz[b]		Optically active	[445]
NHC with isopropyl	Me		$M_w/M_n = 2.0$	[221]

[a]Bu = Butyrate.
[b]Bz = Benzoate.

Table 19   ROMP of alcohol functionalized norbornenes.

Catalyst	exo/endo mixture of R/configuration	σ_{cis}	Remarks	Ref.
MoCl$_5$/AlEt$_3$ (1:1)	OMe	0.48	Random orientation of substituents	[451]
ReCl$_5$	CH$_2$OMe	~1.0	Syndiotactic bias	[450]
NHC with isopropyl	OH		Insoluble	[221]
NHC with isopropyl	CH$_2$OH		Insoluble	[221]

other sterically demanding groups [447,448]. ROMP of monomers bearing a mesogenic group results in liquid crystalline polymers [449].

Hydroxy functionalized norbornenes have been polymerized by ring-opening metathesis polymerization using various catalysts. Some selected compounds which undergo ROMP are listed in Table 19.

  e. Norbornenes Bearing a Cyano Substituent. Classical catalysts as well as well-defined W- [155,423,453,454] and Mo-based initiators were utilized for the polymerization of 2-cyanonorborn-5-ene [142,455]. This monomer could not successfully be polymerized employing Ru-based systems [81,221].

Depending on the particularly used catalyst the *exo* and *endo* isomers may have different reactivities. For example the *syn/anti* ratio of polymers made with WCl$_6$/AlEt$_2$OEt correspond to the *exo/endo* ratio of the monomer [1]. On the other hand the *exo* isomer is much faster consumed if WCl$_6$/AlEt$_2$Cl is applied [321].

Poly(2-cyanonorborn-5-ene) possesses a thermoplastic behaviour [454] and is of technical interest [5].

  f. Halogen-substituted Norbornenes. Fluorinated norbornenes have been polymerized with classical systems, e.g., WCl$_6$/SnMe$_4$ or MoCl$_5$/SnMe$_4$ [456] but also with alkylidene compounds, e.g., Schrock catalysts [457,458]. Fluorinated norbornadienes show the same behaviour [458].

Chlorinated and brominated norbornene derivatives are also polymerizable in a ring-opening reaction in the presence of WCl$_6$/AlEt$_2$Cl [23,430,459,22]. These polymerizations had been first been reported in several patents [1,437,460].

  g. Norbornenes with an Amine or Amide Substituent. The metathetical polymerizability of amine or amide functionalized substrates depends strongly on the particular monomer and reaction conditions. Not all metathesis catalysts are compatible with amine substituted NBE. Recently, Grubbs et al. reported, that ruthenium-based catalysts are limited by the incompatibility with amine groups [81]. However, classical catalysts like W(=CPh$_2$)(CO)$_5$/EtAlCl$_2$/O$_2$ polymerize functionalized NBEs which contain a CH$_2$NR (R = H, Me or CHMe$_2$) group [1]. Ru(=CHPh)Cl$_2$(PCy$_3$)$_2$ could polymerize a carbazole functionalized norbornene (5-(N-carbazol-methylene)-2-norbornene) [464]. ROMP can be extended to oligopeptide functionalized NBE [345]. Pure amide substituted and CONMe$_2$ groups functionalized norbornenes could also undergo a ring-opening metathesis reaction in the presence of WCl$_6$ in combination with an aluminium alkyl (AlEt$_3$ or AlEt$_2$Cl) [22].

  h. 7-Oxanorbornene and its Derivatives. Another promising field in the ROMP of cycloolefines are bicyclo[2.2.1] compounds containing a heteroatom, here oxygene, in the 7-position. Scheme (39) shows a couple of derivatives of 7-oxanorbornene, which are listed in Table 20.

Table 20 Selected examples for the polymerization of 7-oxanorbornene and its derivatives.

Catalyst system	Monomer	R^1	R^2	σ_{cis}	Remarks	Ref.
$Ru(H_2O)_6 \cdot (tos)_3$	(a)	H	H			[173]
$((CH_3)_3CCH_2O)_2W(CH\text{-}t\text{-}Bu)Br_2$	(a)	$Me^a endo/exo = 3/1$	H		Various catalysts can be used	[170]
$RuCl_3$	(b)	OMe	OMe	0.07^b		$[170]^c$
$RuCl_3$	(c)	OMe^d	OMe	0.30^b		[466]
$OsCl_3 \cdot 3H_2O$	(d)	H	H	0.65	ROMP in the presence of water; Ethanol/water;	[467]
$K_2[RuCl_5 \cdot H_2O]^e$	(d)	Me	Me	^{b}	PDI = 1.89	[468]
$Ru(=CHPh)(PCy_3)_2Cl_2$	(d)	Me	Me	0.5		$[296]^f$
$Mo(=CHCMe_2Ph)(N\text{-}2,6\text{-}i\text{-}Pr_2\text{-}C_6H_3)(OCMe_2(CF_3))_2$	(e)	–	–			[59]
$RuCl_3$	(f)	–	–	0.25		[466]
$Ru(=CHCHCPh_2)(PCy_3)_2Cl_2$	$(f)^g$	–	–		Insoluble products	[189]
$Ru(=CHCHCPh_2)(PCy_3)_2Cl_2$	$(g)^g$	Me	–			$[189]^h$

[a]Other rests are possible for the polymerization ($R^1 = R^2 = CH_2OH$) [170,469].
[b]The *cis*-content and the configuration of the resultant polymer is dependent on the solvent [170,470].
[c][173,466].
[d]Other substituents like OCOMe can also be used [170].
[e]Another possible catalyst is $Ru(H_2O)_6 \cdot (tos)_3$ (tos = *p*-toluenesulfonate).
[f]For different Ru-based catalysts see [93,439].
[g]In general, an *endo* isomer is much less reactive than the corresponding *exo* isomer [1].
[h]See also [105,109].

(39)

Table 21 ROMP of N-substituted norbornene dicarboxyimides.

Catalyst	exo/endo mixture of dicarboxyimide/R	σ_{cis}	Ref.
WCl$_6$/SnMe$_4$	exo/(CH$_2$)$_4$CH$_3$	~0.37	[471]
MoCl$_5$/SnMe$_4$	exo/(CH$_2$)$_9$CH$_3$	~0.15	[471]
Mo(CHCMe$_3$)(N-2,6-i-Pr$_2$-C$_6$H$_3$)(OCMe(CF$_3$)$_2$)$_2$	Mixture[a]/(CH$_2$)$_9$CH$_3$	~0.70	[471,472]
Mo(CHCMe$_3$)(N-2,6-i-Pr$_2$-C$_6$H$_3$)(OCMe$_3$)$_2$	exo/C$_6$H$_5$		[142,472]
Ru(=CHPh)(PCy$_3$)$_2$Cl$_2$	exo/(CH$_2$)$_5$CH$_3$	0.18	[471][b]
Ru(=CHPh)(PCy$_3$)$_2$Cl$_2$	endo/(CH$_2$)$_5$CH$_3$	0.0	[471][b]
Ru(=CHPh)(PCy$_2$CH$_2$SiMe$_3$)$_2$Cl$_2$	exo/CH$_2$CO$_2$Me	0.15	[191]

[a] The endo-adduct also undergo ROMP using Mo-based catalyst [472].
[b] See also Ref. [105].

    i. N-substituted Norbornene Dicarboxyimides. Classical catalysts as well as alkylidene complexes are utilized for the metathetical polymerization of N-substituted norbornene dicarboxyimides [see Scheme (40)] [471]. Some examples are listed in Table 21.

(40)

REFERENCES

1. Ivin, K. J., and Mol, J. C. (1997). *Olefin Metathesis and Metathesis Polymerization*, Academic Press, San Diego.
2. Anderson, A. W., and Merckling, N. G. (1956) US 2721189 1955. Du Pont de Nemours & Co., *C.A. 50*: 3008.
3. For the story of the discover of olefin metathesis see: (a) Eleuterio, H. S. (1991). *J. Mol. Catal. 65*: 55; (b) Banks, R. L. (1986). *CHEMTECH, 16*: 112 and [1,4,5].
4. Eleuterio, H. S. (1960). US 3,074,918 1957, Ger. 1,072,811, Du Pont de Nemours & Co., *C.A. 55*: 16005.
5. Truett, W. L., Johnson, D. R., Robinson, I. M., and Montague, B. A. (1960). *J. Am. Chem. Soc., 82*: 2337.
6. Peters, E. F., and Evering, B. L. (1961). US 2, 963, 447 1960, Standard Oil of Indiana, *C. A. 55*: 5810.
7. Banks, R. L., and Bailey, G. C. (1964). *Ind. Eng. Chem., Prod. Res. Dev., 3*: 170.
8. Calderon, N. (1972). *Acc. Chem. Res., 5*: 127.
9. Calderon, N., Chen, H. Y., and Scott, K. W. (1967). *Tetrahedron Lett., 34*: 3327.
10. Calderon, N. (1967). *Chem. Eng. News, 45*: 51.
11. The term "metathesis" is derived from the greek μετατιθεναι (to interchange).
12. Breslow, D. S. (1993). *Prog. Polym. Sci., 18*: 1141.
13. Streck, R., (1989). *CHEMTECH, 19*: 498.
14. Grubbs, R. H., and Chang, S. (1998). *Tetrahedron, 54*: 4413.
15. Ivin, K. J. (1998). *J. Mol. Catal., 133*: 1.
16. Fürstner, A. (2000). *Angew. Chem., 112*: 3140. *Angew. Chem. Int. Ed. Engl., 39*: 3012.

17. Schuster, M., and Blechert, S., (1997). *Angew. Chem. 109*: 2124; (1997). *Angew. Chem. Int. Ed. Engl., 36*: 2121.
18. Schuster, M., and Blechert, S. (1998). "2.17 Application of Olefin Metathesis". In *Transition Metals for Organic Synthesis*, Vol. 1 (Beller, M., and Bolm. C., eds.), Wiley-VCH, Weinheim, 275 pp.
19. Grubbs, R. H., Miller, S. J., and Fu, G. C. (1995). *Acc. Chem. Res., 28*: 446.
20. Buchmeiser, M. R. (2000). *Chem. Rev., 100*: 1565.
21. Feast, W. J. (1989) "Metathesis Polymerization: Applications". In *Comprehensive Polymer Science*, Vol. 4 (Eastmond, G. C., Ledwith, A., Russo, S., Sigwalt, P., Allen, G., and Berington, J.C., eds.), Pergamon Press, Oxford, 135 pp.
22. Ivin, K. J. (1987). "Metathesis Polymerization". In *Encyclopedia of Polymer Science and Engineering*, 2nd ed., Vol. 9 (Mark, H. F., Bikales, N. M., Overberger, C. G., Menges, G., and Kroschwitz, J. I., eds.), John Wiley & Sons Inc., New York, 634 pp.
23. Moore, J. S. (1995). "Transition Metals in Polymer Synthesis. Ring-opening Metathesis Polymerization and Other Transition Metal Polymerization Techniques". In *Comprehensive Organometallic Chemistry II* (Abel, E. W., Stone, F. G. A., Wilkinson, G., and Hegedus, C. S. eds.), Elsevier Science Ltd., Oxford, 1209 pp.
24. Amass, A. J. (1989). "Metathesis Polymerization. Chemistry". In *Comprehensive Polymer Science*, Vol. 4 (Eastmond, G. C., Ledwith, A., Russo, S., Sigwalt, P., Allen, G., and Berington, J. C., eds.), Pergamon Press, Oxford, 109 pp.
25. Ivn, K. J. (1984). "Cycloalkenes and Bicycloalkenes". In *Ring-Opening Polymerization* (Ivin, K. J., and Saegusa, T., eds.), Elsevier Applied Science Publishers, London/New York, 121 pp.
26. Hafner, A., van der Schaaf, P. A., Mühlebach, A., Bernhard, P., Schaedeli, U., Karlen, T., and Ludi, A. (1997). *Prog. Org. Coat., 32*: 89.
27. Hafner, A., van der Schaaf, P. A., and Mühlebach, A. (1996). *Chimia, 50*: 131.
28. Gibson, V. C. (1994). *Adv. Mat., 6*: 37.
29. Novak, B. M., Risse, W., and Grubbs, R. H. (1992). *Adv. Polym. Sci., 102*: 47.
30. Grubbs, R. H. (1982). "Alkene and Alkyne Metathesis Reactions". In *Comprehensive Organometallic Chemistry*, Vol. 8 (Wilkinson, G., ed.), Pergamon Press Ltd., Oxford, 499 pp.
31. Mol. J. C. (1996). "Metathesis". In *Applied Homogenous Catalysis with Organometallic Compounds* (Cornils, B., and Herrmann, W. A., eds.), VCH, Weiheim, 318 pp.
32. Wagener, K. B., Boncella, J. M., and Nel, J.G. (1991). *Macromolecules, 24*: 2649.
33. Smith, J. A., Brzezinska, K. R., Valenti, D. J., and Wagener, K. B. (2000). *Macromolecules, 33*: 3781.
34. Wagener, K. B., Brzezinska, K., Anderson, J. D., Younkin, T. R., Steppe, K., and DeBoer, W. (1997). *Macromolecules, 30*: 7363.
35. Gómez, F. J., and Wagener, K. B. (1998). *Macromol. Chem. Phys., 199*: 1581.
36. Wagener, K. B., Brzezinska, K., and Bauch, C. B. (1992). *Makromol. Chem., Rapid. Commun., 13*: 75.
37. Wolfe, P. S., and Wagener, K. B. (1998). *Macromol. Rapid. Commun., 19*: 305.
38. Wolfe, P. B., Gomez, F. J., and Wagener, K. B. (1997). *Macromolecules, 30*: 714.
39. Calderon, N., Ofstead, E. A., Ward, J. P., Judy, W. A., and Scott, K. W. (1968). *J. Am. Chem. Soc., 90*: 4133.
40. Same results were independently found by metathesis of ^{14}C-labeled propene, see: Mol, J. C., Moulijn, J. A., and Boelhouwer, C. J. (1968). *Chem. Soc., Chem. Commun.*, 633.
41. Dall'Asta, G., and Motroni, G. (1971). *Eur. Polym. J., 7*: 707.
42. The cleavage of any other bond than the double bonds or α-Single bonds could be excluded before, see: (a) Natta, G., Dall'Asta, G., Mazzanti, G., and Motroni, G. (1963). *Makromol. Chem., 69: 163*; (b) Natta, G., Dall'Asta, G., and Porri, L. (1965). *Makromol. Chem., 81*: 253; (c) Dall'Asta, G., and Motroni, G. J. (1968). *Polym. Sci. A1, 6*: 2405; (d) Truett, W. L., Johnson, D. R., Robinson, I. M., and Montague, B. A. (1960). *J. Am. Chem. Soc., 82*: 2337.
43. Bradshaw, C. P. C., Howman, E. J., and Turner, L. J. (1967). *Catal., 7*: 269.
44. Mango, F. D., and Schachtschneider, J. H. (1967). *J. Am. Chem. Soc., 89*: 2484.

45. Mango, F. D., and Schachtschneider, J. H. (1971). *J. Am. Chem. Soc.*, *93*: 1123.
46. Lewandos, G. S., and Pettit, R. (1971). *J. Am. Chem. Soc.*, *93*: 7087.
47. Grubbs, R. H., and Brunck, T. K. (1972). *J. Am. Chem. Soc.*, *94*: 2538.
48. Biefield, C. G., Eick, H. A., and Grubbs, R. H. (1973). *Inorg. Chem.*, *12*: 2166.
49. Calderon, N., Ofstead, E. A., and Judy, W. A. (1976). *Angew. Chem.*, *88*: 433.
50. Hérisson, J.-L., and Chauvin, Y. (1970). *Makromol Chem.*, *141*: 161.
51. Dolgoplosk, B. A., Makovetsky, K. L., Golenko, T. G., Korshak, Y. V., and Tinyakova, E. I. (1974). *Eur. Polym. J.*, *10*: 901.
52. Katz, T. J., and Rothchild, R. (1976). *J. Am. Chem. Soc.*, *98*: 1519.
53. Katz, T. J., and McGinnis, J. (1975). *J. Am. Chem. Soc.*, *97*: 1592.
54. Grubbs, R. H., Carr, D. D., Hoppin, C., and Burk, P. L. (1976). *J. Am. Chem. Soc.*, *98*: 3478.
55. Grubbs, R. H., Burk, P. L., and Carr, D. D. (1975). *J. Am. Chem. Soc.*, *97*: 3264.
56. (a) Herrmann, W. A. (1978). *Angew. Chem.*, *90*: 855; (1978). *Angew. Chem. Int. Ed. Engl.*, *17*: 800; (b) Herrmann, W. A., Schweizer, I., Creswick, M., and Bernal, I. J. (1979). *Organomet. Chem.*, *165*: C17; (c) Braun, T., Gevert, O., and Werner, H. (1995). *J. Am. Chem. Soc.*, *117*: 7291.
57. (a) Katz, T. J., Lee, S. J., and Acton, N. (1976). *Tetrahedron Lett.*, *47*: 4247; b) Katz, T. J., and Acton, N. (1976). *Tetrahedron Lett.*, *47*: 4251; (c) Katz, T. J., Mc Ginnis, J., and Altus, C. (1976). *J. Am. Chem. Soc.*, *98*: 608; (d) Katz, T. J., Lee, S. J., and Shippey, M. A. (1980). *J. Mol. Catal.*, *8*: 219.
58. Casey, C. P., and Burkhardt, T. J. (1973). *J Am. Chem. Soc.*, *95*: 5833.
59. Bazan, G. C., Oskam, J. H., Cho, H.-N., Park, L. Y., Schrock, R. R. (1991). *J. Am. Chem. Soc.*, *113*: 6899.
60. Howard, T. R., Lee, J. B., and Grubbs, R. H. (1980). *J. Am. Chem. Soc.*, *102*: 6876.
61. Kress, J., Osborn, J. A., Greene, R. M. E., Ivin, K. J., and Ronney, J. J. (1987). *J. Am. Chem. Soc.*, *109*: 899.
62. Kress, J., and Osborn, J. A. (1992). *Angew. Chem.*, *104*: 1660; (1992). *Angew. Chem. Int. Ed. Engl.*, *31*: 1585.
63. Tallarico, J. A., Bonitatebus, P. J., Jr., and Snapper, M. L. (1997). *J. Am. Chem. Soc.*, *119*: 7157.
64. Amir-Ebrahimi, V., Hamilton, J. G., Nelson, J., Rooney, J. J., Thompson, J. M., Beaumont, A. J., Rooney, A. D., and Harding, C. J. (1999). *Chem. Commun.*, 1621.
65. Amir-Ebrahimi, V., Hamilton, J. G., Nelson, J., Rooney, J. J., Rooney, A. D., and Harding, C. J. (2000). *J. Organomt., Chem.*, *606*: 84.
66. Buchacher, P., Fischer, W., Aichholzer, K. D., and Stelzer, F. (1997). *J. Mol. Catal.*, *115*: 163.
67. McCann, M., Coda, E. M. G., and Maddock, K. (1994). *J. Chem. Soc., Dalton Trans.*, 1489.
68. Sato, H., Tanaka, Y., and Taketomi, T. (1977). *Makromol. Chem.*, *178*: 1993.
69. Saito, K., Yamaguchi, T., Tanabe, K., Ogura, T., and Yagi, M. (1979). *Bull. Chem. Soc. Jpn.*, *52*: 3192.
70. Preishuber-Pflügl, P., Buchacher, P., Eder, E., Schitter, R. M., and Stelzer, F. (1998). *J. Mol. Catal.*, *133*: 151.
71. Herrmann, W. A., Stumpf, A. W., Priermeier, T., and Bogdanovic, S. (1996). *Angew. Chem.*, *108*: 2978; (1996). *Angew. Chem. Int. Ed. Engl.*, *35*: 2803.
72. Finkel'shtein, E. S., Makovetskii, K. L., Yampol'skii, Y. P., Portnykh, E. B., Ostrovskaya, I. Y., Kaliuzhnyi, N. E., Pritula, N. A., Golberg, A. I., Yatsenko, M. S., and Platé, N. A. (1991). *Makromol. Chem.*, *192*: 1.
73. (a) Mocella, M. T., Rovner, R., and Muetterties, E. L. (1976). *J. Am. Chem. Soc.*, *98*: 4689; (b) Mocella, M. T., Busch, M. A., and Muetterties, E. L. (1976). *J. Am. Chem. Soc.*, *98*: 1283; (c) Basset, J., Taarit, Y. B., Coudurier, G. G., and Pralaiud, H. (1974). *J. Organomet. Chem.*, *74*: 167.
74. Breslow, D. S. (1990). *CHEMTECH*, *20*: 541.
75. van Dam, P. B., and Boelhouwer, C. (1974). *React. Kinet. Catal. Lett.*, *1*: 165.
76. Hughes, W. B. (1970). *J. Am. Chem. Soc.*, *92*: 532.
77. Moulijn, J. A., and Boelhouwer, C. J. (1971). *Chem. Soc. Chem. Commun.*, 1170.

78. Farona, M. F., and Greenlee, W. F. J. (1975). *Chem. Soc. Chem. Commun.*, 759.
79. Laverty, D. T., and Rooney, J. J. (1983). *J. Chem. Soc., Faraday Trans.*, *79*: 869.
80. Mühlebach, A., Bernhard, P., Bühler, N., Karlen, T., and Ludi, A. (1994). *J. Mol. Catal.*, *90*: 143.
81. Trnka, T. M., and Grubbs, R. H. (2001). *Acc. Chem. Res.*, *34*: 18.
82. Lindemark-Hamberg, M., and Wagener, K. B. (1987). *Macromolecules*, *20*: 2949.
83. Streck, R. (1992). *J. Mol. Catal.*, *76*: 359.
84. Fischer, E. O., Maasböl, A. (1964). *Angew, Chem.*, *76*: 645.
85. Schrock, R. R. (1974). *J. Am. Chem. Soc.*, *96*: 6796.
86. Katz, T. J., and Sivavec, T. M. (1985). *J. Am. Chem. Soc.*, *107*: 737.
87. Tebbe, F. N., Parshall, G. W., and Reddy, G. S. (1978). *J. Am. Chem. Soc.*, *100*: 3611.
88. Tebbe, F. N., Parshall, G. W., and Ovenall, D. W. (1979). *J. Am. Chem. Soc.*, *101*: 5074.
89. Klabunde, U., Tebbe, F. N., Parshall, G. W., and Harlow, R. L. (1980). *J. Mol. Catal.*, *8*: 37.
90. Grubbs, R. H., and Tumas, W. (1989). *Science*, *243*: 907.
91. Noels, A. F., Demonceau, A., Carlier, E., Hubert, A. J., Márquez-Silva, R.-L., and Sánchez-Delgado, R. A. (1988). *J. Chem. Soc. Chem. Commun.*, 783.
92. Demonceau, A., Noels, A. F., Saive, E., and Hubert, A. J. (1992). *J. Mol. Catal.*, *76*: 123.
93. Stumpf, A. W., Saive, E., Demonceau, A., and Noels, A. F. (1995). *J. Chem. Soc., Chem. Commun.*, 1127.
94. Hermann, W. A., Schattenmann, W. C., Nuyken, O., and Glander, S. C. (1996). *Angew. Chem.*, *108*: 1169; (1996). *Angew. Chem. Int. Ed. Engl.*, *35*: 1087.
95. Demonceau, A., Stumpf, A. W., Saive, E., and Noels, A. F. (1997). *Macromolecules*, *30*: 3127.
96. Delaude, L., Demonceau, A., and Noels, A. F. (1999). *Macromolecules*, *32*: 2091.
97. Demonceau, A., Stumpf, A. W., Saive, E., and Noels, A. F. (1995). *J. Chem. Soc. Chem. Commun.*, 1127.
98. Claverie, J. P., Viala, S., Maurel, V., and Novat, C. (2001). *Macromolecules*, *34*: 382.
99. Grubbs, R. H. (1994). *J. Macromol. Sci. – Pure Appl. Chem.*, *A31*: 1829.
100. Novak, B. M., and Grubbs, R. H. (1998). *J. Am. Chem. Soc.*, *110*: 7542.
101. Feast, W. J., and Harrison, D. B. (1991). *J. Mol. Catal.*, *65*: 63.
102. Grubbs, R. H., Novak, B. M., McGrath, D. M., Benedicto, A., France, M., Nguyen, S. T. (1992). *Polym. Prepr. (Am. Chem. Soc., Div. Polym. Chem.)*, *33*: 1225.
103. Hillmyer, M. A., Lepetit, C., McGrath, D. V., Novak, B. M., and Grubbs, R. H. (1992). *Macromolecules*, *25*: 3345.
104. France, M. B., Paciello, R. A., and Grubbs, R. H. (1993). *Macromolecules*, *26*: 4739.
105. Lynn, D. M., Kanaoka, S., and Grubbs, R. H. (1996). *J. Am. Chem. Soc.*, *118*: 784.
106. Mohr, B., Lynn, D. M., and Grubbs, R. H. (1996). *Organometallics*, *15*: 4317.
107. Kirkland, T. A., Lynn, D. M., and Grubbs, R. H. (1998). *J. Org. Chem.*, *63*: 9904.
108. Lynn, D. M., Mohr, B., and Grubbs, R. H. (1998). *J. Am. Chem. Soc.*, *120*: 1627.
109. Lynn, D. M., Mohr, B., Grubbs, R. H., Henling, L. M., and Day, M. W. (2000). *J. Am.Chem. Soc.*, *122*: 6601.
110. Michelotti, F. W., and Keaveney, W. P. (1965). *J. Polym. Sci. A*, *3*: 895.
111. Bielawski, C. W., and Grubbs, R. H. (2000). *Angew. Chem.*, *112*: 3025; (2000). *Angew. Chem. Int. Ed.*, *39*: 2903.
112. Gilliom, L. R., and Grubbs, R. H. (1986). *J. Am. Chem. Soc.*, *108*: 733.
113. Gailliom, L. R., and Grubbs, R. H. (1988). *J. Mol. Catal.*, *46*: 255.
114. Cannizzo, L. F., and Grubbs, R. H. (1988). *Macromolecules*, *21*: 1961.
115. Risse, W., and Grubbs, R. H. (1991). *J. Mol. Catal.*, *65*: 221.
116. Swager, T. M., and Grubbs, R. H. (1987). *J. Am. Chem. Soc.*, *109*: 894.
117. Wallace, K. C., Liu, A. H., Dewan, J. C., and Schrock, R. R. (1988). *J. Am. Chem. Soc.*, *110*: 4964.
118. Mashima, K., Kaidzu, M., Tanaka, Y., Nakayama, Y., Nakamura, A., Hamilton, J. G., and Rooney, J. J. (1998). *Organometallics*, *17*: 4183.

119. Schrock, R. R., Rocklage, S., Wengrovius, J., Rupprecht, G., and Fellmann, J. (1980). *J. Mol. Catal.*, *8*: 73.
120. Rocklage, S. M., Fellmann, J. D., Rupprecht, G. A., Messerle, L. W., Schrock, R. R. (1981). *J. Am. Chem. Soc.*, *103*: 1440.
121. Murdzek, J. S., and Schrock, R. R. (1987). *Organometallics*, *6*: 1373.
122. Schrock, R. R. (1999). *Tetrahedron*, *55*: 8141.
123. Schrock, R. R. (1990). *Acc. Chem. Res.*, *23*: 158.
124. Schrock, R. R., Murdzek, J. S., Bazan, G. C., Robbins, J., DiMare, M., and O'Regan, M. (1990). *J. Am. Chem. Soc.*, *112*: 3875.
125. Schrock, R. R. (1994). *Pure Appl. Chem.*, *66*: 1447.
126. Oskam, J. H., and Schrock, R. R. (1993). *J. Am. Chem. Soc.*, *115*: 11831.
127. Oskam, J. H., and Schrock, R. R. (1992). *J. Am. Chem. Soc.*, *114*: 7588.
128. Totland, K. M., Boyd, T. J., Lavoie, G. G., Davies, W. M., and Schrock, R. R. (1996). *Macromolecules*, *29*: 6114.
129. Schrock, R. R., Lee, J.-K., O'Dell, R., and Oskam, J. H. (1995). *Macromolecules*, *28*: 5933.
130. Sunaga, T., Ivin, K. J., Hofmeister, G. E., Oskam, J. H., and Schrock, R. R. (1994). *Macromolecules*, *27*: 4043.
131. McConville, D. H., Wolf, J. R., and Schrock, R. R. (1993). *J. Am. Chem. Soc.*, *115*: 4413.
132. O'Dell, R., McConville, D. H., Hofmeister, G. E., and Schrock, R. R. (1994). *J. Am. Chem. Soc.*, *116*: 3414.
133. Zhu, S. S., Cefalo, D. R., La, D. S., Jamieson, J. Y., Davis, W. M., Hoveyda, A. H., and Schrock, R. R. (1999). *J. Am. Chem. Soc.*, *121*: 8251.
134. Alexander, J. B., Schrock, R. R., Davis, W. M., Hultzsch, K. C., Hoveyda, A. H., and Houser, J. H. (2000). *Organometallics*, *19*: 3700.
135. Fujimura, O., and Grubbs, R. H. (1998). *J. Org. Chem.*, *63*: 824.
136. Fujimura, O., and Grubbs, R. H. (1996). *J. Am. Chem. Soc.*, *118*: 2499.
137. Fujimura, O., de la Mata, F. J., and Grubbs, R. H. (1996). *Organometallics*, *15*: 1865.
138. Weatherhead, G. S., Ford, J. G., Alexanian, E. J., Schrock, R. R., and Hoveyda, A. H. (2000). *J. Am. Chem. Soc.*, *122*: 1828.
139. Alexander, J. B., La, D. S., Cefalo, D. R., Hoveyda, A. H. and Schrock, R. R. (1998). *J. Am. Chem. Soc.*, *120*: 4041.
140. Cefalo, D. R., Kiely, A. F., Wuchrer, M., Jamieson, J. Y., Schrock, R. R., and Hoveyda, A. H. (2001). *J. Am. Chem. Soc.*, *123*: 3139.
141. Hoveyda, A. H., and Schrock, R. R. (2001). *Chem. Eur. J.*, *7*: 945.
142. Bazan, G. C., Schrock, R. R., Cho, H.-N., and Gibson, V. C. (1991). *Macromolecules*, *24*: 4495.
143. Buchmeiser, M. R., Atzl, N., and Bonn, G. K., (1997). *J. Am. Chem. Soc.*, *119*: 9166.
144. Sinner, F., Buchmeiser, M. R., Tessadri, R., Mupa, M., Wurst, K., and Bonn, G. K. (1998). *J. Am. Chem. Soc.*, *120*: 2790.
145. Perrott, M. G., and Novak, B. M. (1995). *Macromolecules*, *28*: 3492.
146. Schoettel, G., Kress, J., and Osborn, J. A. (1989). *J. Chem. Soc. Chem. Commun.*, 1062.
147. Vaughan, W. M., Abboud, K. A., and Boncella, J. M. (1995). *Organometallics*, *14*: 1567.
148. Feher, F. J., and Tajima, T. L. (1994). *J. Am. Chem. Soc.*, *116*: 2145.
149. Kress, J., Wesolek, M., and Osborn, J. A. (1982). *J. Chem. Soc. Chem. Commun.*, 514.
150. Kress, J., and Osborn, J. A. (1983). *J. Am. Chem. Soc.*, *105*: 6346.
151. Kress, J., Osborn, J. A. (1987). *J. Am. Chem. Soc.*, *109*: 3953.
152. Kress, J., Ivin, K. J., Amir-Ebrahimi,V., and Weber, P. (1990). *Makromol. Chem.*, *191*: 2237.
153. Ivin, K. J., Kress, J., and Osborn, J. A. (1992). *Makromol. Chem.*, *193*: 1695.
154. Ivin, K. J., Kress, J., and Osborn, J. A. (1988). *J. Mol. Catal.*, *46*: 351.
155. Quignard, F., Leconte, M., and Basset, J.-M. (1985). *J. Chem. Soc. Chem. Commun.*, 1816.
156. Couturier, J.-L., Paillet, C., Leconte, M., Basset, J.-M., and Weiss, K. (1992). *Angew. Chem.*, *104*: 622; (1992). *Angew. Chem. Int. Ed. Engl.*, *31*: 628.

157. Basset, J.-M., Leconte, M., Lefebvre, F., Hamilton, J. G., and Rooney, J. J. (1997). *Macromol. Chem. Phys.*, *198*: 3499.
158. Pagano, S., Mutch, A., Lefebvre, F., and Basset, J.-M. (1998). *J. Mol. Catal.*, *133*: 61.
159. de la Mata, F. J., and Grubbs, R. H. (1996). *Organometallics*, *15*: 577.
160. Johnson, L. K., Grubbs, R. H., and Ziller, J. W. (1993). *J. Am. Chem. Soc.*, *115*: 8130.
161. Nguyen, S. T., Johnson, L. K., and Grubbs, R. H. (1992). *J. Am. Chem. Soc.*, *114*: 3974.
162. O'Donoghue, M. B., Schrock, R. R., LaPointe, A. M., and Davis, W. M. (1996). *Organometallics*, *15*: 1334.
163. Schaverien, C. J., Devan, J. C., and Schrock, R. R. (1986). *J. Am. Chem. Soc.*, *108*: 2271.
164. Schrock, R. R., Krouse, S. A., Knoll, K., Feldman, J., Murdzek, J. S., and Yang, D. C. (1988). *J. Mol. Catal.*, *46*: 243.
165. Schrock, R. R., DePue, R. T., Feldman, J., Yap. K. B., Yang, D. C., Davis, W. M., Park, L. Y., DiMare, M., Schofield, M., Anhaus, J., Walborsky, E., Evitt, E., Krüger, C., and Betz, P. (1990). *Organometallics*, *9*: 2262.
166. Schrock, R. R., Feldman, J., Cannizzo, L., and Grubbs, R. H. (1987). *Macromolecules*, *20*: 1169.
167. Johnson, L. K., Virgil, S. C., Grubbs, R. H., and Ziller, J. W. (1990). *J. Am. Chem. Soc.*, *112*: 5384.
168. van der Schaaf, P. A., Hafner, A., and Mühlebach, A. (1996). *Angew. Chem.*, *108*: 1974.
169. VanderLende, D. D., Abboud, K. A., and Boncella, J. M. (1994). *Polym. Prepr. (Am. Chem. Soc., Div. Polym. Chem.)*, *35*: 691.
170. Novak, B. M., and Grubbs, R. H. (1988). *J. Am. Chem. Soc.*, *110*: 960.
171. Lu, S.-Y., Quayle, P., Heatley, F., Booth, C., Yeates, S. G., and Padget, J. C. (1992). *Macromolecules*, *25*: 2692.
172. Feast, W. J., and Herrison, D. B. (1991). *Polym. Bull.*, *25*: 343.
173. Benedicto, A. D., Novak, B. M., and Grubbs, R. H. (1992). *Macromolecules*, *25*: 5893.
174. France, M. B., Grubbs, R. H., McGrath, D. V., and Paciello, R. A. (1993). *Macromolecules*, *26*: 4742.
175. Hamilton, J. G., Rooney, J. J., DeSimone, J. M., and Mistele, C. (1998). *Macromolecules*, *31*: 4387.
176. Mistele, C. D., Thorp, H. H., and DeSimone, J. M. (1996). *J. Macromol. Sci., Pure Appl. Chem. A*, *33*: 953.
177. Schwab, P., Grubbs, R. H., and Ziller, J. W. (1996). *J. Am. Chem. Soc.*, *118*: 100.
178. Schwab, P., France, M. B., Ziller, J. W., and Grubbs, R. H. (1995). *Angew. Chem.*, *34*: 2039; (1995). *Angew. Chem. Int. Ed. Engl.*, *34*: 2039.
179. Nguyen, S. T., Grubbs, R. H., and Ziller, J. W. (1993). *J. Am. Chem. Soc.*, *115*: 9858.
180. Dias, E. L., Nguyen, S. T., and Grubbs, R. H. (1997). *J. Am. Chem. Soc.*, *119*: 3887.
181. Maughon, B. R., and Grubbs, R. H. (1997). *Macromolecules*, *30*: 3459.
182. Ivin, K. J., Kenwright, A. M., and Khosravi, E. (1999). *Chem. Commun.*, 1209.
183. Adlhart, C., Hinderling, C., Bauman, H., and Chen, P. (2000). *J. Am. Chem. Soc.*, *122*: 8204.
184. Adlhart, C., and Chen, P. (2000). *Helv. Chim. Acta.*, *83*: 2192.
185. Hinderling, C., Adlhart, C., and Chen, P. (1998). *Angew. Chem.*, *110*: 2831; (1998). *Angew. Chem. Int. Ed. Engl.*, *37*: 2685.
186. Aagaard, O. M., Meier, R. J., and Buda, F. (1998). *J. Am. Chem. Soc.*, *120*: 7174.
187. Meier, R. J., Aagaard, O. M., and Buda, F. (2000). *J. Mol. Catal.*, *160*: 189.
188. Dias, E. L., and Grubbs, R. H. (1998). *Organometallics*, *17*: 2758.
189. Kanaoka, S., and Grubbs, R. H. (1995). *Macromolecules*, *28*: 4707.
190. Wu, Z., Nguyen, S. T., Grubbs, R. H., and Ziller, J. (1995). *J. Am. Chem. Soc.*, *117*: 5505.
191. Svensson, M., Alexandridis, P., and Linse, P. (1999). *Macromolecules*, *32*: 637.
192. Weskamp, T., Schattenmann, W. C., Spiegler, M., and Herrmann, W. A. (1998). *Angew. Chem.*, *110*: 2631; (1998). *Angew. Chem. Int. Ed. Engl.*, *37*: 2490; (1999). *Corrigendum*: *Angew. Chem.*, *111*: 277; (1999). *Angew. Chem. Int. Ed. Engl.*, *38*: 262.
193. Weskamp, T., Kohl, F. J., and Herrmann, W. A. (1999). *J. Organomet. Chem.*, *582*: 362.

194. Weskamp, T., Kohl, F. J., Hieringer, W., Gleich, D., and Herrmann, W. A. (1999). *Angew. Chem.*, *111*: 2573; (1999). *Angew. Chem. Int. Ed. Engl.*, *38*: 2416.
195. Ackermann, L., Fürstner, A., Weskamp, T., Kohl, F. J., and Herrmann, W. A. (1999). *Tetrahedron Lett.*, *40*: 4787.
196. Scholl, M., Trnka, T. M., Morgen, J. P., and Grubbs, R. H. (1999). *Tetrahedron Lett.*, *40*: 2247.
197. Huang, J., Schanz, H.-J., Stevens, E. D., and Nolan, S. P. (1999). *Organometallics*, *18*: 5375.
198. Scholl, M., Ding, S., Lee, C. W., and Grubbs, R. H. (1999). *Org. Lett.*, *1*: 953.
199. Lee, C. W., and Grubbs, R. H. (2000). *Org. Lett.*, *2*: 2145.
200. Briot, A., Bujard, M., Gouverneur, V., Nolan, S. P. and Mioskowski, L. (2000). *Org. Lett.*, *2*: 1517.
201. Huang, J., Stevens, E. D., Nolan, S. P., and Petersen, J. L. (1999). *J. Am. Chem. Soc.*, *121*: 2674.
202. Jafarpour, L., Stevens, E. D., and Nolan, S. P. (2000). *J. Organomet. Chem.*, *606*: 49.
203. Morgan, J. P., and Grubbs, R. H. (2000). *Org. Lett.*, *2*: 3153.
204. Chatterjee, A. K., Morgan, J. P., Scholl, M., and Grubbs, R. H. (2000). *J. Am. Chem. Soc.*, *122*: 3783.
205. Chatterjee, A. K., and Grubbs, R. H. (1999). *Org. Lett.*, *1*: 1751.
206. Fürstner, A., Thiel, O. R., Ackermann, L., Schanz, H.-J., and Nolan, S. P. (2000). *J. Org. Chem.*, *65*: 2204.
207. Jafarpour, L., Schanz, H.-J., Stevens, E. D., and Nolan, S. P. (1999). *Organometallics*, *18*: 5416.
208. Jafarpour, L., and Nolan, S. P. (2000). *Organometallics*, *19*: 2055.
209. Jafarpour, L., and Nolan, S. P. (2001). *J. Organomet. Chem.*, *17*: 617.
210. Sanford, M. S., Ulman, M., and Grubbs, R. H. (2001). *J. Am. Chem. Soc.*, *123*: 749.
211. Hansen, S. M., Rominger, F., Metz, M., and Hofmann, P. (1999). *Chem. Eur. J.*, *5*: 557.
212. Hansen, S. M., Volland, M. A. O., Rominger, F., Eisenträger, F., and Hofmann, P. (1999). *Angew. Chem.*, *111*: 1360; (1999). *Angew. Chem. Int. Ed. Engl.*, *38*: 1273.
213. Adlhart, C., Volland, M. A. O., Hofmann, P., and Chen. P. (2000). *Helv. Chim. Acta*, *83*: 3306.
214. Amoroso, D., and Fogg, D. E. (2000). *Macromolecules*, *33*: 2815.
215. Six, C., Beck, K., Wegner, A., and Leitner, W. (2000). *Organometallics*, *19*: 4639.
216. Chang, S., Jones, L. II., Wang, C., Henling, L. M., and Grubbs, R. H. (1998). *Organometallics*, *17*: 3460.
217. Kingsbury, J. S., Harrity, J. P. A., Bonitatebus, P. J., Jr., and Hoveyda, A. H. (1999). *J. Am. Chem. Soc.*, *121*: 791.
218. Gessler, S., Randl, S., and Blechert, S. (2000). *Tetrahedron Lett.*, *41*: 9973.
219. Garber, S. B., Kingsbury, J. S., Gray, B. L., and Hoveyda, A. H. (2000). *J. Am. Chem. Soc.*, *122*: 8168.
220. Ulman, M., and Grubbs, R. H. (1999). *J. Org. Chem.*, *64*: 7202.
221. Frenzel, U., Weskamp, T., Kohl, F. J., Schattenmann, W. C., Nuyken, O., and Herrmann, W. A. (1999). *J. Organomet. Chem.*, *586*: 263.
222. Bielawski, C. W., Scherman, O. A., and Grubbs, R. H. (2001). *Polymer*, *42*: 4939.
223. Schürer, S. C., Gessler, S., Buschmann, N., and Blechert, S. (2000). *Angew. Chem.*, *39*: 3898; (2000). *Angew. Chem. Int. Ed. Engl.*, *39*: 3898.
224. Fürstner, A., Liebel, M., Lehmann, C. W., Piquet, M., Kunz, R., Bruneau, C., Touchard, D., and Dixneuf, P. (2000). *Chem. Eur. J.*, *6*: 1847.
225. Stüer, W., Wolf, J., Werner, H., Schwab, P., and Schulz, M. (1998). *Angew. Chem.*, *110*: 3603; (1998). *Angew. Chem. Int. Ed. Engl.*, *37*: 3421.
226. (a) Fürstner, A., Picquet, M., Bruneau, C., and Dixneuf. P. H. (1998). *J. Chem. Soc., Chem. Commun.*, 1315; (b) Piquet, M., Bruneau, C., and Dixneuf, P. H. (1998). *J. Chem. Soc., Chem. Commun.*, 2249; (c) Fürstner, A., Hill, A. F., Liebl, M., Wilton-Ely, J. D. E. T. (1999). *J. Chem. Soc., Chem. Comm.*, 601.

227. (a) Katayama, H., Yoshida, T., and Ozawa, F. (1998). *J. Organomet. Chem.*, *562*: 203; (b) Katayama, H., and Ozawa, F. (1998). *Chem. Lett.*, 67; (c) Katayama, H., Urushima, H., and Ozawa, F. (1999). *Chem. Lett.*, 369; (d) Katayama, H., and Ozawa, F. (1998). *Organometallics*, *17*: 5190.

228. Schanz, H.-J., Jafarpour, L., Stevens, E. D., and Nolan, S. P. (1999). *Organometallics*, *18*: 5187.

229. Ulman, M., Belderrain, T. R., and Grubbs, R. H. (2000). *Tetrahedron Lett.*, *41*: 4689.

230. Tebbe bifunktionell.

231. Weck, M., Schwab, P., and Grubbs, R. H. (1996). *Macromolecules*, *29*: 1789.

232. Kelly, W. J., and Calderon, N. (1975). *J. Macromol. Sci.-Chem. A.*, *9*: 911.

233. Ofstead, E. A., Lawrence, J. P., Senyek, M. L., and Calderon, N. (1980). *J. Mol. Catal.*, *8*: 227.

234. Hillmyer, M. A., Nguyen, S. T., and Grubbs, R. H. (1997). *Macromolecules*, *30*: 718.

235. Hillmyer, M. A., and Grubbs, R. H. (1995). *Macromolecules*, *28*: 8662.

236. Maughon, B. R., Morita, T., Bielawski, C. W., and Grubbs, R. H. (2000). *Macromolecules*, *33*: 1929.

237. Morita, T., Maughon, B. R., Bielawski, C. W., and Grubbs, R. H. (2000). *Macromolecules*, *33*: 6621.

238. Bielawski, C. W., Scherman, O. A., and Grubbs, R. H. (2001). *Polymer*, *42*: 4939.

239. Fraser, C., Hillmyer, M. A., Gutierrez, E., and Grubbs, R. H. (1995). *Macromolecules*, *28*: 7256.

240. Nubel, P. O., and Hunt, C. L. (1999). *J. Mol. Catal.*, *145*: 323.

241. Hafner, A., Mühlebach, A., and van der Schaaf, P. A. (1997). *Angew. Chem.*, *109*: 2213; (1997). *Angew. Chem. Int. Ed. Engl.*, *36*: 2121.

242. A related NHC substituted complex has also been reported, see: Jafarpour, L., Huang, J., Stevens, E. D., and Nolan, S. P. (1999). *Organometallics*, *18*: 3760.

243. Louie, J., and Grubbs, R. H. (2001). *Angew. Chem.*, *113*: 253; (2001). *Angew. Chem. Int. Ed. Engl.*, *40*: 247.

244. Wilhelm, T. E., Belderrain, T. R., Brown, S. N., and Grubbs, R. H. (1997). *Organometallics*, *16*: 3867.

245. Belderrain, T. R., and Grubbs, R. H. (1997). *Organometallics*, *16*: 4001.

246. Wolf, J., Stüer, W., Grünwald, C., Werner, H., Schwab, P., and Schulz, M. (1998). *Angew. Chem.*, *110*: 1165; (1998). *Angew. Chem. Int. Ed. Engl.*, *37*: 1124.

247. Ivin, K. J. (1991). *Makromol. Chem. Maromol. Symp.*, *42/43*: 1.

248. Ivin, K. J., and Saegusa, T., (1984). "General Thermodynamic and Mechanistic Aspects of Ring-opening Polymerization". In *Ring-Opening Polymerization* (Ivin, K. J., and Saegusa, T., eds.), Elsevier Applied Science Publishers, London/New York, pp. 1.

249. Ivin, K. J., and Busfield, W. K. (1987). "Polymerization Thermodynamics". In *Encyclopedia of Polymer Science and Engineering*, 2nd ed., Vol. 12 (Mark, H. F., Bikales, N. M., Overberger, C. G., Menges, G., and Kroschwitz, J. I., eds.), John Wiley & Sons Inc., New York, pp. 555.

250. Dainton, F. S., and Ivin, K. J. (1958). *Quart. Rev. (London)*, *12*: 61.

251. Hocks, L., Berck, D., Hubert, A. J., and Teyssie, P. (1975). *J. Polym. Sci. Polym. Lett.*, *13*: 391.

252. Klavetter, F. L., and Grubbs, R. H. (1988). *J. Am. Chem. Soc.*, *110*: 7807.

253. Patton, P. A., Lillya, C. P., and McCarthy, T. J. (1986). *Macromolecules*, *19*: 1266.

254. Dainton, F. S., Devlin, T. R. E., and Small. P. A. (1955). *Trans. Faraday Soc.*, *51*: 1710.

255. Lebedev, B., and Smirnova, N. (1994). *Macromol. Chem. Phys.*, *195*: 35.

256. Cherednichenko, V. M. (1979). *Polymer Sci. USSR.*, *20*: 1225.

257. Oreshkin, I. A., Redkina, L. I., Kershenbaum, I. L., Chernenko, G.M., Makovetsky, K. L., Tinyakova, E. I., and Dolgoplosk, B. A. (1977). *Eur. Polym. J.*, *13*: 447.

258. Ofstead, E. A., and Calderon, N. (1972). *Makromol. Chem.*, *154*: 21.

259. Semlyen, J. A. (1976). *Adv. Polym. Sci.*, *21*: 41.

260. Andrews, J. M., Jones, F. R., and Semlyen, J. A. (1974). *Polymer*, *15*: 420 and references therein.

261. Höcker, H., Reimann, W., Reif, L., and Riebel, K. (1980). *J. Mol. Catal.*, *8*: 191.
262. Wasserman, E., Ben-Efraim, D. A., and Wolovsky, R. (1968). *J. Am. Chem. Soc*, *90*: 3286.
263. Witte, J., and Hoffmann, M. (1978). *Makromol. Chem.*, *179*: 641.
264. Reif, L., and Höcker, H. (1984). *Macromolecules*, *17*: 952.
265. Höcker, H., Reimann, W., Riebel, K., and Szentivanyi, Z. (1976). *Makromol. Chem.*, *177*: 1707.
266. Chauvin, Y., Commereuc, D., and Zaborowski, G. (1978). *Makromol. Chem.*, *179*: 1285.
267. Chen, Z.-R., Claverie, J. P., Grubbs, R. H., and Kornfield, J. A. (1995). *Macromolecules*, *28*: 2147.
268. Thorn-Csányi, E., Hammer, J., Pflug, K. P., and Zilles, J. U. (1995). *Macromol. Chem. Phys.*, *196*: 1043.
269. Thorn-Csányi, E., and Hammer, J., Zilles, J. U. (1994). *Macromol. Rapid Commun.*, *15*: 797.
270. Thorn-Csányi, E., and Ruhland, K. (1999). *Macromol. Chem. Phys.*, *200*: 2606.
271. Thorn-Csányi, E., and Ruhland, K. (1999). *Macromol. Chem. Phys.*, *200*: 2245.
272. Thorn-Csányi, E., and Ruhland, K. (1999). *Macromol. Chem. Phys.*, *200*: 1662.
273. Reif, L., and Höcker, H. (1981). *Makromol. Chem., Rapid. Comm.*, *2*: 183.
274. Taghizadeh, N., Quignard, F., Leconte, M., Basset, J. M., Larroche, G., Laval, J. P., and Lattes, A. (1982). *J. Mol. Catal.*, *15*: 219.
275. Ivin, K. J., O'Donnell, J. H., Rooney, J. J., and Stewart, C. D. (1979). *Markomol. Chem.*, *180*: 1975.
276. Wagener, K. B., Puts, R. D., and Smith, D. W., Jr. (1991). *Makromol. Chem. Rapid. Commun.*, *12*: 419.
277. Höcker, H. (1981). *Angew. Markomol. Chem.*, *100*: 87.
278. Jacobson, H., and Stockmayer, W. H. (1950). *J. Chem. Phys.*, *18*: 1600.
279. For the complete theory and the other preconditions, see e.g.: [278], [248].
280. Thorn-Csányi, E., and Ruhland, K. (2000). *Macromol. Symp.*, *153*: 145.
281. Suter, U. W., and Höcker, H. (1988). *Makromol. Chem.*, *189*: 1603.
282. Dounis, P., Feast, W. J., and Kenwright, A. M. (1995). *Polymer*, *36*: 2787.
283. Ivin, K. J., Rooney, J. J., Bencze, L., Hamilton, J. G., Lam, L.-M., Lapienis, G., Reddy, B. S. R., and Ho, H. T. (1982). *Pure & Appl. Chem.*, *54*: 447.
284. Hamilton, J. G. (1998). *Polymer*, *39*: 1669.
285. Al-Samak, B., Amir-Ebrahimi, V., Carvill, A. G., Hamilton, J. G., and Rooney, J. J. (1996). *Polymer, Int.*, *41*: 85.
286. Hamilton, J. G., Ivin, K.J., McCann, G. M., and Rooney, J. J. (1985). *Makromol. Chem.*, *186*: 1477.
287. Hamilton, J. G., Ivin, K. J., and Rooney, J. J. (1985). *J. Mol. Catal.*, *28*: 255.
288. Hamilton, J. G., Rooney, J. J., and Snowden, D. G. (1995). *Macromol. Chem. Phys.*, *196*: 1031.
289. Devine, G. I., Ho, H. T., Ivin, K. J., Mohamed, M. A., and Rooney, J. J. (1982). *J. Chem. Soc. Chem. Commun.*, 1229.
290. Ho, H. T., Ivin, K. J., and Rooney, J. J. (1982). *J. Mol. Catal.*, *15*: 245.
291. Greene, R. M. E., Ivin, K. J., McCann, G. M., and Rooney, J. J. (1986). *Makromol. Chem.*, *187*: 619.
292. Hamilton, J. G., Rooney, J. J., and Snowden, D. G. (1993). *Makromol. Chem.*, *194*: 2907.
293. Hamilton, J. G., and Rooney, J. J. (1992). *J. Chem. Soc., Chem. Commun.*, 370.
294. Ivin, K. J., Lam, L.-M., and Rooney, J. J. (1993). *Makromol. Chem.*, *194*: 3203.
295. Hamilton, J. G., Frenzel, U., Kohl, F. J., Weskamp, T., Rooney, J. J., Herrmann, W. A., and Nuyken, O. (2000). *J. Organomet. Chem.*, *606*: 8.
296. Amir-Ebrahimi, V., Corry, D. A., Hamilton, J. G., Thompson, J. M., and Rooney, J. J. (2000). *Macromolecules*, *33*: 717.
297. Feast, W. J., Gibson, V. C., Ivin, K. J., Khosravi, E., Kenwright, A. M., Marshall, E. L., and Mitchell, J. P. (1992). *Makromol. Chem.*, *193*: 2103.
298. Ivin, K. J., Kenwright, A. M., Khosravi, E., and Hamilton, J. G. (2000). *J. Organomet. Chem.*, *606*: 37.

299. Corey, E. J., Hartmann, R., and Vatakencherry, P. A. (1962). *J. Am. Chem. Soc.*, *84*: 2611.

300. Irngartinger, H., Oeser, T., Jahn, R., and Kallfaß, D. (1992). *Chem. Ber.*, *125*: 2067.

301. Ivin, K. J., Laverty, D. T., and Rooney, J. J. (1977). *Makromol. Chem.*, *178*: 1545.

302. Al Samak, B., Carvill, A. G., Hamilton, J. G., Rooney, J. J., and Thompson, J. M. (1997). *J. Chem. Soc., Chem. Comm.*, 2057.

303. Amir-Ebrahimi, V., Carvill, A. G., Hamilton, J. G., Rooney, J. J., and Tuffy, C. (1997). *J. Mol. Catal.*, *115*: 85.

304. Hamilton, J. G., Ivin, K. J., McCann, G. M., and Rooney, J. J. (1984). *J. Chem. Soc., Chem. Commun.*, 1379.

305. Schrock, R. R., Messerle, L. W., Wood, C. D., and Guggenberger, L. J. (1978). *J. Am. Chem. Soc.*, *100*: 3793.

306. Guggenberger, L. J., and Schrock, R. R. (1975). *J. Am. Chem. Soc.*, *97*: 6578.

307. Broeders, J., Feast, W. J., Gibson, V. C., and Khosravi, E. (1996). *Chem. Commun.*, 343.

308. Ivin, K. J., Laverty, D. T., O'Donnel, J. H., Rooney, J. J., and Stewart, C. D. (1979). *Makromol. Chem.*, *180*: 1989.

309. Ohm, R. F. (1980). *CHEMTECH*, *10*: 183.

310. Marbach, A. (1993). "Rubber, 4.3 Polynorbornene". In *Ullmann's Encyclopedia of Industrial Chemistry*, 5th ed., Vol. A23 (Elvers, E., Hawkins, S., Russey, W., and Schulz, G., eds.), VCH Verlagsgesellschaft mbH, Weinheim, 299 pp.

311. Manager, H.-D. (1979). *Kautschuk Gummi Kunststoffe*, *32*: 572.

312. Dräxler, A. (1981). *Kautschuk Gummi Kunststoffe*, *34*: 185.

313. Dietrich, K. M. (1988). "Rubber, 4.4 Polyoctenamers". In *Ullmann's Encyclopedia of Industrial Chemistry*, 5th ed., Vol. A23 (Elvers, E., Hawkins, S., Russey, W., and Schulz, G., eds.), VCH Verlagsgesellschaft mbH, Weinheim, 302 pp.

314. Schneider, W. A., Müller, M. F. (1988). *J. Mol. Catal.*, *46*: 395.

315. Mühlebach, A., van der Schaaf, P. A., Hafner, A., and Setiabuti, F. J. (1998). *Mol. Catal.*, *132*: 181.

316. Yang, Y. S., Lafontaine, E., and Mortaigne, B. (1996). *J. Appl. Polym. Sci.*, *60*: 2419.

317. Davidson, T. A., Wagener, K. B., and Priddy, D. B. (1996). *Macromolecules*, *29*: 786.

318. Davidson, T. A., and Wagener, K. B. (1998). *J. Mol. Catal.*, *133*: 67.

319. Fisher, R. A., and Grubbs, R. H. (1992). *Makromol. Chem. Macromol. Symp.*, *63*: 271.

320. Kohara, T. (1996). *Macromol. Chem., Macromol. Symp.*, *101*: 571.

321. Asrar, J., and Curran, S. A., (1992). *Makromol. Chem.*, *193*: 2961.

322. Asrar, J., and Curran, S. A. (1991). *J. Mol. Catal.*, *65*: 1.

323. a) Chen, Y., Dujardin, R., Bruder, F.-K., and Rechner, J. (1998). EP 0,850,965, Bayer AG, C. A. *129*: 109461j; b) Chen, Y., Dujardin, R., Pielartzik, H., and Franz, U. W. EP 0,850,966, Bayer AG.

324. Streck, R. J. (1982). *Mol. Catal.*, *15*: 3.

325. a) Streck, R., Nordsieck, K.-H., Weber, H., and Meyer, K. (1973). DE 2,131,355 1971, Chemische Werke Hüls AG, C. A, *78*: 112522p; b) Meyer, K., Streck, R., and Weber, H. (1974). DE 2,242, 794 1972, Chemische Werke Hüls AG, C. A., *81*: 122366s.

326. Ungerank, U., Winkler, B., Eder, E., and Stelzer, F. (1997). *Makromol. Chem. Phys.*, *198*: 1391.

327. Maughon, B. R., Weck, M., Mohr, B., and Grubbs, R. H. (1997). *Macromolecules*, *30*: 257.

328. Weck, M., Mohr, B., Maughon, B. R., and Grubbs, R. H. (1997). *Macromolecules*, *30*: 6430.

329. Ungerank, M., Winkler, B., Eder, E., and Stelzer, F. (1995). *Macromol. Chem. Phys.*, *196*: 3623.

330. Ungerank, M., Winkler, B., Eder, E., and Stelzer, F. (1997). *Macromol. Chem. Phys.*, *198*: 1391.

331. Winkler, B., Ungerank, M., and Stelzer, F. (1996). *Macromol. Chem. Phys.*, *167*: 2343.

332. Winkler, B., Rehab, A., Ungerank, M., and Stelzer, F. (1997). *Macromol. Chem. Phys.*, *198*: 1417.

333. Laschewsky, A., Schulz-Hanke, W., (1993). *Makromol. Chem. Rapid Commun.*, *14*: 683.

334. Komiya, Z., and Schrock, R. R. (1993). *Macromolecules, 26*: 1393.
335. Komiya, Z., and Schrock, R. R. (1993). *Macromolecules, 26*: 1387.
336. Komiya, Z., Pugh, C., and Schrock, R. R. (1992). *Macromolecules, 25*: 3609.
337. Komiya, Z., Pugh, C., and Schrock, R. R. (1992). *Macromolecules, 25*: 6586.
338. Pugh, C., and Schrock, R. R. (1992). *Macromolecules, 25*: 6593.
339. Arehart, S. V., Pugh, C. (1997). *J. Am. Chem. Soc., 119*: 3027.
340. Pugh, C., Dharia, J., and Arehart, S. V. (1997). *Macromolecules, 30*: 4520.
341. Pugh, C. (1994). *Macromol. Symp., 77*: 325.
342. Pugh, C., and Kiste, A. L. (1997). *Prog. Polym. Sci., 22*: 601.
343. Percec, V., and Schlueter, D. (1997). *Macromolecules, 30*: 5783.
344. Maynard, H. D., Okada, S. Y., and Grubbs, R. H. (2000). *Macromolecules, 33*: 6239.
345. Maynard, H. D., Okada, S. Y., and Grubbs, R. H. (2001). *J. Am. Chem. Soc., 123*: 1275.
346. Fraser, C., and Grubbs, R. H. (1995). *Macromolecules, 28*: 7248.
347. Montalban, A. G., Steinke, J. H. G., Anderson, M. E., Barrett, A. G. M., and Hoffman, B. M. (1999). *Tetrahedron Lett., 40*: 8151.
348. Sailor, M. J., Ginsburg, E. J., Gorman, C. B., Kumar, A., Grubbs, R. H., and Lewis, N. S. (1990). *Science, 249*: 1146.
349. Ginsburg, E. J., Gorman, C. B., Marder, S. R., and Grubbs, R. H. (1989). *J. Am. Chem. Soc., 111*: 7621.
350. Conticello, V. P., Gin, D. L., and Grubbs, R. H. (1992). *J. Am. Chem. Soc., 114*: 9708.
351. Swager, T. M., and Grubbs, R. H. (1989). *J. Am. Chem. Soc., 111*: 4413.
352. Wagaman, M. W., and Grubbs, R. H. (1997). *Macromolecules, 30*: 3978.
353. Hamilton, J. G., Marquess, D. G., O'Neill, T. J., and Rooney, J. J. (1990). *J. Chem. Soc., Chem. Commun.*, 119–21.
354. Feast, W. J., and Winter, J. N. (1985). *J. Chem. Soc., Chem. Commun.*, 202.
355. Knoll, K., Krouse, S. A., and Schrock, R. R. (1988). *J. Am. Chem. Soc., 110*: 4424.
356. Dounis, P., Feast, W. J., and Widawski, G. (1997). *J. Mol. Catal., 115*: 51.
357. Langsdorf, B. L., Zhou, X., and Lonergan, M. C. (2001). *Macromolecules, 34*: 2450.
358. Mayr, B., Sinner, F., and Buchmeiser, M. R. (2001). *J. Chromatogr. A, 907*: 47.
359. Bell, B., Hamilton, J. G., Law, E. E., and Rooney, J. J. (1994). *Macromol. Rapid Commun., 15*: 543.
360. Hamilton, J. G., Kay, J., and Rooney, J. J. (1998). *J. Mol. Catal., 133*: 83.
361. Schitter, R. M. E., Jocham, D., Stelzer, F., Moszner, N., and Völkel, T. (2000). *J. Appl. Polym. Sci., 78*: 47.
362. Meyer, U., Kern, W., Hummel, K., and Stelzer, F. (1998). *Eur. Polym. J., 35*: 69.
363. Okoroanyanwu, U., Byers, J., Shimokawa, T., and Willson, C. G. (1998). *Chem. Mater., 10*: 3328.
364. Okoroanyanwu, U., Shimokawa, T., Byers, J. D., and Willson, C. G. (1998). *J. Mol. Catal., 133*: 93.
365. Mühlebach, A., Schaedeli, U., and Bernhard, P. (1994). *Polym. Prepr. (Am. Chem. Soc., Div. Polym. Chem.), 35*: 963.
366. Asrar, J. (1994). *Macromolecules, 27*: 4036.
367. Weck, M., Jackiw, J. J., Rossi, R. R., Weiss, P. S., and Grubbs, R. H. (1999). *J. Am. Chem. Soc., 121*: 4088.
368. Juang, A., Scherman, O. A., Grubbs, R. H., and Lewis, N. S. (2001). *Langmuir, 17*: 1321.
369. Dall'Asta, G., Mazzanti, G., Natta, G., and Porri, L. (1962). *Macromol. Chem., 56*: 224.
370. Natta, G., Dall'Asta, G., Bassi, I. W., and Carella, G. (1966). *Macromol. Chem., 91*: 87.
371. Wu, Z., Wheller, D. R., and Grubbs, R. H. (1992). *J. Am. Chem. Soc., 114*: 146.
372. Wu, Z., and Grubbs, R. H. (1994). *Macromolecules, 27*: 6700.
373. Kormer, V. A., Dolinskaya, E. R., and Khatchaturov, A. S. (1980). *Makromol. Chem. Rapid Commun., 1*: 531.
374. Wu, Z., and Grubbs, R. H. (1994). *J. Mol. Catal., 90*: 39.
375. Wu, Z., and Grubbs, R. H. (1995). *Macromolecules, 28*: 3502.

376. Katz, T. J. (1978). *Adv. Organomet. Chem.*, *16*: 283.
377. Perrott, M. G., and Novak, B M. (1996). *Macromolecules*, *29*: 1817.
378. Pakuro, N. I., Makovetsky, K. L., Gantmakher, A. R., and Dolgoplosk, B. A. (1982). *Bull. Acad. Sci. USSR, Chem. Sci.*, *31*: 456; (1982). *Izv. Akad. Nauk SSSR, Ser. Khim*, *31*: 509.
379. Minchak, R. J., and Tucker, H. (1972). *Am. Chem. Soc. Polymer Prepr.*, *13*: 885.
380. Natta, G., Dall'Asta, G., and Mazzanti, G. (1964). *Angew. Chem. Int. Ed. Engl.*, *3*: 723; (1964). *Angew. Chem.*, *18*: 765.
381. Fischer, H., Hellwege. K.-H., Johnsen, U., and Langbein, W. (1966). *Makromol. Chem.*, *91*: 107.
382. Makovetsky, K. L., Gorbacheva, L. I., Ostrovskaya, I. Y., Golberg, A. I., Mikaya, A. I., Zakharian, A. A., and Filatova, M. P. (1992). *J. Mol. Catal.*, *76*: 65.
383. Ceausescu, E., Cornilescu, A., Nicolescu, E., Popescu, M., Coca, S., Cuzmici, M., Hubca, G. H., Teodorescu, M., Dragutan, V., and Chipara, M. (1988). *J. Mol. Catal.*, *46*: 423.
384. Porri, L., Rossi, R., Diversi, P., and Lucherini, A. (1972). *Am. Chem. Soc. Polymer Prepr.*, *13*: 897.
385. Schrock, R. R., Yap, K. B., Yang, D. C., and Sitzmann, H. (1989). *Macromolecules*, *22*: 3191.
386. Jan, D., Delaude, L., Simal, F., Demonceau, A., and Noels, A. F. (2000). *J. Organomet. Chem.*, *606*: 55.
387. Katz, T. J., and Han, C.-C. (1982). *Organometallics*, *1*: 1093.
388. Calderon, N., Ofstead, E. A., and Judy, W. E. (1967). *J. Polymer Sci.*, *5*: 2209.
389. Blosch, L. L., Abboud, K., and Boncella, J. M. (1991). *J. Am. Chem. Soc.*, *113*: 7066.
390. Kress, J. J. (1995). *Mol. Catal.*, *102*: 7.
391. Gamble, A. S., and Boncella, J. M. (1993). *Organometallics*, *12*: 2814.
392. Larroche, C., Laval, J. P., Lattes, A., Leconte, M., Quignard, F., and Basset, J. M. (1983). *J. Chem. Soc., Chem. Commun.*, 220.
393. Delaude, L., Demonceau, A., and Noels, A. F. (2001). *Chem. Commun.*, 986.
394. Hillmyer, M. A., Laredo, W. R., and Grubbs, R. H. (1995). *Macromolecules*, *28*: 6311.
395. Chung, T. C. (1991). *Polymer*, *32*: 1336.
396. Chung, T. C. (1992). *J. Mol. Catal.*, *76*: 15.
397. Cho, I., and Hwang, K. M. (1993). *J. Polym. Sci. Polym. Chem.*, *31*: 1079.
398. Makovetsky, K. L., Gorbacheva, L. I., Ostrovskaya, I. Y., Golberg, A. I., Mikaya, A. I., Zakharian, A. A., and Filatova, M. P. (1992). *J. Mol. Catal.*, *76*: 65.
399. Korshak, Y. V., Korshak, V. V., Kanischka, G., and Höcker, H. (1985). *Makromol. Chem., Rapid Commun.*, *6*: 685.
400. Gillan, E. M. D. Hamilton, J. G., Nicola, O., Mackey, D., and Rooney, J. J. (1988). *J. Mol. Catal.*, *46*: 359.
401. Ivin, K. J., Laverty, D. T., O'Donnell, J. H., and Rooney, J. J. (1979). *Makromol. Chem.*, *180*: 1989.
402. Porri, L., Rossi, R., Diversi, P., and Lucherini, A., (1974). *Makromol. Chem.*, *175*: 3097.
403. Herrmann, W. A., Wagner, W., Flessner, U. N., Volkhardt, U., and Komber, H. (1991). *Angew. Chem.*, *103*: 1704.
404. Petasis, N. A., and Fu, D.-K. (1993). *J. Am. Chem. Soc.*, *115*: 7208.
405. Eilerts, N. W., Heppert, J. A., Morton, M. D. (1992). *J. Mol. Catal.*, *76*: 157.
406. Schrock, R. R., Krouse, S. A., Knoll, K., Feldman, J., Murdzek, J. S., and Yang, D. C. (1988). *J. Mol. Catal.*, *46*: 243.
407. Johnston, J. A., Tokles, M., Hatvany, G. S., Rinaldi, P. L., and Farona, M. F. (1991). *Macromolecules*, *24*: 5532.
408. Heroguez, V., and Fontanille, M. (1994). *J. Polym. Sci. Polym. Chem.*, *32*: 1755.
409. van der Schaaf, P. A., Wilberth, J. J., Spek, A. L., and van Koten, G. J. (1992). *Chem. Soc. Chem. Commun.*, 717.
410. Hamilton, J. G., Rooney, J. J., and Waring, L. C. (1983). *J. Chem. Soc., Chem. Commun.*, 159.
411. Ivin, K. J., Lapienis, G., Rooney, J. J., and Stewart, C. D. (1980). *J. Mol. Catal.*, *8*: 203.

412. Albagli, D., Bazan, G. C., Schrock, R. R., and Wrighton, M. S. (1993). *J. Am. Chem. Soc.*, *115*: 7328.

413. Kim, N. Y., Jeon, N. L., Choi, I. S., Takami, S., Harada, Y., Finnie, K. R., Girolami, G. S., Nuzzo, R. G., Whitesides, G. M., and Laibinis, P. E. (2000). *Macromolecules*, *33*: 2793.

414. Ivin, K. J., Lapienis, G., and Rooney, J. J. (1980). *Polymer*, *21*: 436.

415. Ivin, K. J., Lam, L.-M., and Rooney, J. J. (1981). *Makromol. Chem.*, *182*: 1847.

416. Ho, H. T., Ivin, K. J., Reddy, B. S. R., and Rooney, J. J. (1989). *Eur. Polymer J.*, *25*: 805.

417. Hamilton, J. G., Ivin, K. J., and Rooney, J. J. (1980). *Brit. Polym., J.*, *21*: 237.

418. Greene, R. M. E., Ivin, K. J., McCann, G. M., and Rooney, J. J. (1987). *Brit. Polym. J.*, *19*: 339.

419. Mashima, K., Tanaka, Y., Kaidzu, M., and Nakamura, A. (1996). *Organometallics*, *15*: 2431.

420. Greene, R. M. E., Ivin, K. J., Kress, J., Osborn, J. A., and Rooney, J. J. (1989). *Brit. Polym., J. 21*: 237.

421. Hamilton, J. G., Rooney, J. J., and Snowden, D. G. (1994). *J. Polym. Sci. Polym. Chem.*, *32*: 993.

422. Samak, B. A., Amir-Ebrahimi, V., Corry, D. G., Hamilton, J. G., Rigby, S., Rooney, J. J., and Thompson, J. M. (2000). *J. Mol. Catal.*, *160*: 13.

423. Greene, R. M. E., Ivin, K. J., and Rooney, J. J. (1988). *Makromol. Chem.*, *189*: 2797.

424. Kress, J., Osborn, J. A., Greene, R. M. E., Ivin, K. J., and Rooney, J. J. (1985). *J. Chem. Soc., Chem. Commun.*, 874.

425. Tenney, L. P., and Lane, P. C. (1979). US Pat. 4,136,247; *C.A.* 90, 187884.

426. Dekking, H. C. G. (1961). *J. Polym. Sci.*, *55*: 525.

427. Tenney, L. P. (1979). US Pat. 4,136,248; C. A. 90, 169556.

428. (a) Tanaka, Y., and Ueshima, T., and Kobayashi, S. (1976). *C. A.*, *84*: 165412; (b) Komatsu, K., Matsumoto, S., and Aotani, S. (1977). *C. A.*, *86*: 56201.

429. Ivin, K. J., Kenwright, A. M., Hofmeister, G. E., McConville, D. H., Schrock, R. R., Amir-Ebrahimi, V., Carvill, A.G., Hamilton, J.G., and Rooney, J. J. (1998). *Macromol. Chem. Phys.*, *199*: 547.

430. Makovetsky, K. L., Finkel'shtein, E. S., Ostrovskaya, I. Y., Portnykh, E. B., Gorbacheva, L. I., Golberg, A. I., Ushakov, N. V., and Yampolsksy, Y. P. (1992). *J. Mol. Catal.*, *76*: 107.

431. Finkel'shtein, E. S., Portnykh, E. B., Ushakov, N. V., Greengolts, M. L., Fedorova, G. K., Platé, N. A. (1994). *Macromol. Rapid Commun.*, *15*: 155.

432. Zimmermann, M., Pampus, G., Maertens, D. Chem. Abstr. 85, 95539 (1976). Ger. Offen, 2,460,911.

433. Mayr, B., and Buchmeiser, M. R. (2001). *J. Chromatography A.*, *907*: 73.

434. Seeber, G., Brunner, P., Buchmeiser, M. R., and Bonn, G. K. (1999). *J. Chromatography A.*, *848*: 193.

435. Ho, H. T., Ivin, K. J., Reddy, B. S. R., and Rooney, J. J. (1989). *Eur. Polymer J.*, *25*: 805.

436. Steinhäusler, T., and Stelzer, F. (1994). *J. Mol. Catal.*, *90*: 53.

437. P. Hepworth, Chem. Abstr. 78, 98287 (1973): Ger. Offen. 2,231,995.

438. Piotti, M. E. (1999). *Current Opinion in Solid State and Mat. Sci.*, *4*: 539.

439. Katayama, H., Ursushima, H., and Ozawa, F. (2000). *J. Organomet. Chem.*, *606*: 16.

440. Ungerank, M., Winkler, B., Eder, E., and Stelzer, F. (1997). *Macromol. Chem. Phys.*, *198*: 1391.

441. Weck, M., Mohr, B., Maughon, B R., and Grubbs, R. H. (1997). *Macromolecules*, *30*: 6430.

442. Betterton, K., Ebert, M., Haeussling, L., Lux, M. G., Twieg, R. J., Willson, C. G., Yoon, D., Burns, E. G., and Grubbs, R. H. (1992). *Polymer Mater. Sci. Eng.*, *66*: 312.

443. Eder, K., Sinner, F., Mupa, M., Huber, C. G., and Buchmeiser, M. R. (2001). *Electrophoresis*, *22*: 109.

444. Mutch, A., Leconte, M., Lefebvre, F., and Basset, J.-M. (1998). *J. Mol. Catal.*, *133*: 191.

445. Steinhäusler, T., Stelzer, F., and Zenkl., E. (1994). *Polymer*, *35*: 616.

446. Thorn-Csányi, E., and Ruhland, K. (1992). *Macromol. Chem. Phys.*, *76*: 93.

447. Cummins, C. C., Schrock, R. R., and Cohen, R. E. (1992). *Chem. Mater.*, *4*: 27.

448. Tlenkopatchev, M. A., Miranda, E., Canseco, M. A., Gavino, R., and Ogawa, T. (1995). *Polymer Bull.*, *34*: 385.
449. Kim, S.-H., Lee, J.-H., Jin. S.-H., Cho, H.-N., and Choi, S.-K. (1993). *Macromolecules*, *26*: 846.
450. Ivin, K. J., Lam, L.-M., and Rooney, J. J. (1994). *Makromol. Chem. Phys.*, *195*: 1189.
451. Ivin, K. J., Lam, L.-M., and Rooney, J. J. (1993). *Makromol. Chem.*, *194*: 3493.
452. Michelotti, F. W., and Carter, J. H. (1965). *Am. Chem. Soc., Polymer Prep.*, *6*: 224.
453. Matsumoto, S., Nakamura, R., Suzuki, K., Komatsu, K. (1976). *Chem. Abstr.*, *85*: 33724.
454. Thorn-Csányi, E., and Harder, C. (1991). *Angew. Makromol. Chem.*, *185/186*: 283.
455. Bazan, G. C., and Schrock, R. R. (1991). *Macromolecules*, *24*: 817.
456. Feast, W. J., Millichamp, I. S., and Harper, K. J. (1985). *Mol. Catal.*, *28*: 331.
457. Feast, W. J., Gibson, V. C., Khosravi, E., and Marshall, E. L. (1994). *J. Chem. Soc., Chem. Commun.*, 1399.
458. Feast, W. J., and Khosravi, E. J. (1999). *Fluorine Chem.*, *100*: 117.
459. Buchacher, P., Fischer, W., Aichholzer, K. D., and Stelzer, F. (1997). *J. Mol. Catal.*, *115*: 163.
460. Kobayashi, M. (1978). *Chem. Abstr.*, *88*: 38328.
461. Amir-Ebrahimi, V., Corry, D. A. K., Hamilton, J. G., and Rooney, J. J. (1998). *J. Mol. Catal.*, *133*: 115.
462. Bazan, G. C., Oskam, J. H., Cho, H.-N., Park, L. Y., and Schrock, R. R. (1991). *J. Am. Chem. Soc.*, *113*: 6899.
463. Bazan, G. C., Khosravi, E., Schrock, R. R., Feast, W. J., Gibson, V. C., O'Regan, M. B., Thomas, J. K., and Davis, W. M. (1990). *J. Am. Chem. Soc.*, *112*: 8378.
464. Liaw, D.-J., and Tsai, C.-H., (2000). *Polymer*, *41*: 2773.
465. Hamilton, J. G., and Rooney, J. J. (1992). *J. Chem. Soc. Chem. Commun.*, 370.
466. Lu, S.-Y., Amass, J. M., Majid, N., Glennon, D., Byerley, A., Heatley, F., Quayle, P., Booth, C., Yeates, S. G., and Padget, J. C. (1994). *Macromol. Chem. Phys.*, *195*: 1273.
467. Feast, W. J., and Harrison, D. B. (1991). *Polymer*, *32*: 558.
468. Zenkl, E., and Stelzer, F. (1992). *J. Mol. Catal.*, *76*: 1.
469. Ellsworth, M. W., and Novak, B. M. (1991). *J. Am. Chem. Soc.*, *113*: 2756.
470. Gilbert, M., and Herbert, I. R. (1993). *Polymer Bull.*, *30*: 83.
471. Khosravi, E., Feast W. J., Al-Hajaji, A. A., Leejarkpai, T. J. (2000). *Mol. Catal.*, *160*: 1.
472. Khosravi, E., and Al-Hajaji, A. A. (1998). *Polymer 39*: 5619.

7
Aromatic Polyethers

Hans R. Kricheldorf
Institute of Technical and Molecular Chemistry, University of Hamburg, Hamburg, Germany

I. INTRODUCTION

Aromatic polyethers are a group of high-performance engineering plastics which were developed by several chemical companies over the past forty years. Poly(phenylene ether)s mostly prepared by oxydative polycondensation of phenols, were the first class of aromatic polyethers which was technically produced and commercialized [1–3]. Perfectly linear poly(arylene ether)s free of side chains can only be obtained by oxidative coupling of 2,6-disubstituted phenols. Therefore, the poly(2,6-dimethylphenylene-oxide), usually called PPO, is the most widely used poly(arylene ether). Its outstanding property is its compatibility with polystyrene (and a few other polymers) so that it is mainly used as component of blends [4]. The oxidative coupling of phenols has intensively been studied between 1950 and 1980 (as discussed in the 1st edition of this handbook) but only few research activities were observed during the past 10 years discussed in the subchapter 'Various Aromatic Polyethers'. Therefore the literature evaluated and discussed below mainly concerns poly(ether sulfone)s and poly(ether ketone)s. Furthermore, semiaromatic polyethers and aromatic polysulfides are included in this chapter which mainly covers the literature of the years 1990 through spring 2000 (complementary to the first edition).

The characteristic properties and advantages of aromatic polyethers when compared to aliphatic engineering plastics based on an aliphatic main chain are as follows. Aromatic polyethers are less sensitive to oxydation at all temperatures, and thus, also less inflammable (poly(vinyl chloride) and poly(tetrafluoroethylene) are, of course, exceptions in the case of aliphatic polymers). The thermostability of aromatic polyethers is higher. For all these reasons the maximum service temperature of aromatic polyethers (200–260 °C) is twice as high as that of aliphatic polymers. Furthermore, aromatic polyethers possess higher glass-transition temperatures (T_gs) which may be as high as 230 °C for commercial poly(ether sulfone)s. The T_gs of commercial poly(ether-ketone)s are lower (typically around 140–150 °C) but all commercial poly(ether-ketone)s are semicrystalline materials having melting temperatures in the range of 280–420 °C. Either due to high T_g or due to a high T_m the heat distortion temperature of aromatic polyethers significantly higher than that of aliphatic polymers. A few characteristic disadvantages should also be mentioned. Most poly(ether-sulfone)s reported so far and all commercial examples are amorphous with the advantage of a high transparency and the shortcoming of a high sensitivity to the attack of organic solvents. The crystalline poly(ether ketone)s are rather insensitive to organic solvents, but they are sensitive to a cleavage by UV-irradiation quite analogous to low molar mass benzophenones. Finally, it should be mentioned that a typical application of poly(ether sulfone)s and poly(ether-ketone)s is that as matric material in composites. Glass fiber or carbon fiber are used as reinforcing components.

II. POLY(ETHER SULFONE)S

Poly(ether-sulfone)s, PESs, may in principle be prepared via four different strategies:

1. Polycondensation of suitable monomers involving an electrophilic substitution of an aromatic ring.
2. Polycondensation of suitable monomers involving a nucleophilic substitution of a chloro, fluoro or nitroaromat activated by a sulfonyl group in para-position.
3. Chemical modification of suitable precursor polymers.
4. Ring-opening polymerization of cyclic oligo(ether-sulfone)s

The discussion of synthetic methods and structures presented below will follow this order.

A. Syntheses via Electrophilic Substitution

The oldest approach known for the preparation of PESs is a polycondensation process involving the electrophilic substitution of a phenyl ether group by an aromatic sulfonyl chloride group [3,5,6]. Such polycondensations may be based, either on monomers containing both functional groups in one molecule (Eq. (1)) or by a combination of a nucleophilic and an electrophilic monomer (Eq. (2)). These polycondensations need to be catalyzed by strong Lewis acids such as $FeCl_3$, $AlCl_3$, or BF_3. Characteristic disadvantages of this approach are the need of an expensive inert reaction medium and side reactions such as substitution (including branching) in ortho position of the nucleophilic monomer. Furthermore, this approach is not versatile and limited to a few monomers. Previous research activities in this field were reported in the 1st edition of the *Handbook* [3], but more recently new activities were not observed.

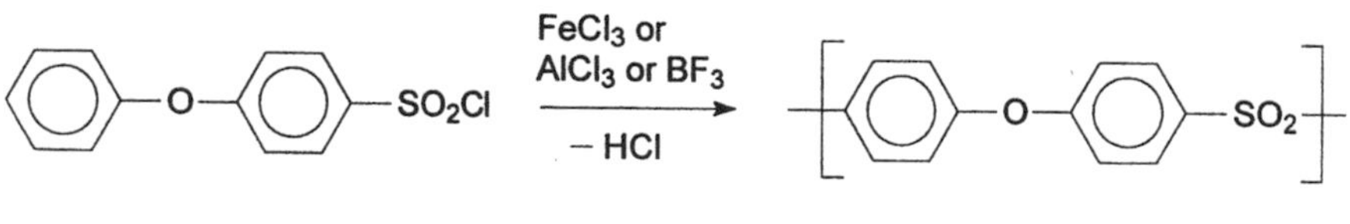

(1)

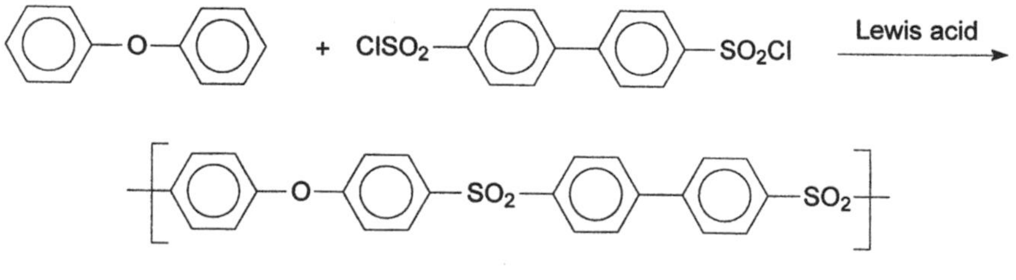

(2)

B. Syntheses Via Nucleophilic Substitution

The most widely used approach to the preparation of PESs in both academic research and technical production is a polycondensation process involving a nucleophilic substitution of an aromatic chloro- or fluorosulfone by a phenoxide ion (Eq. (3)). Prior to the review of new PESs prepared by nucleophilic substitution publications should be mentioned which were concerned with the evaluation and comparison of the electrophilic reactivity of various mono- and difunctional fluoro-aromats [7–10]. The nucleophilic substitution of aromatic compounds may in general proceed via four different mechanism. Firstly, the S_{N1} mechanism which is, for instance, characteristic for most diazonium salts. Secondly, the elimination-addition mechanism involving arines as intermediates which is typical for the treatment of haloaromats with strong bases at high temperature. Thirdly, the addition–elimination mechanism which is typical for fluorosulfones as illustrated in equations (3) and (4). Fourthly, the S_{NAR} mechanism which may occur when poorly electrophilic chloroaromats are used as reaction partners will be discussed below in connection with polycondensations of chlorobenzophenones.

In the case of the addition–elimination mechanism the addition step with the formation of a short lived Meisenheimer complex (Eq. (3)) is the rate determining step. Hence, the electron density of the carbon directly bound to the fluorine (ipso position) is decisive for the reactivity and thus, for the rate of the reaction. In two publications [7,8] the ^{13}C NMR chemical shifts of various fluoroaromats were determined, compared and shown to be useful indicators of the reactivity of the ipso-carbon. This conclusion was confirmed by calculation of the electron density via the quantum semiempirical PM 3 method in the MOPAC software. In fact, a linear correlation between the calculated electron density and the ^{13}C NMR δ values was obtained. Furthermore, the ^{19}F NMR chemical shifts were determined for numerous electrophilic fluoroaromats and again a linear correlation with the calculated electron densities, on the one hand, and with the ^{13}C NMR chemical shifts, on the other, was found [7,8]. These studies proved that the SO_2 group is the strongest activating divalent group. Only the monovalent nitrogroup has a stronger electron-withdrawing effect. The strong electron-withdrawing effect of the SO_2 group has also the consequence that the C-atom directly attached to it is sensitive to a nucleophilic attack. With KF as reagent the cleavage of the PES backbone (back-reaction of Eqs. (3) and (4) was observed at $280\,°C$ [9], but it is not clear if the cleavage will be more favored by other cations such as $Cs^{\oplus}$. Finally, a publication should be mentioned [10] reporting on a partial desulfonylation during the polycondensation of a special ketone-sulfone type monomer.

The standard procedure used by most authors for syntheses of new poly(ethersul-fone)s is based on the reaction of equimolar amounts of a difluoro (or dichloro)- sulfone and a bisphenol with dry K_2CO_3 (equimolar or slight excess) in polar aprotic solvents such as N-methylpyrrolidone (NMP), dimethylacetamide (DMAc) dimethylsulfoxide (DMSO) or sulfolane. In the paper [11] stoichiometric amounts of CsF were applied instead of K_2CO_3. However, CsF has no advantage, but it is significantly more expensive. Following the standard procedure with K_2CO_3 two research groups used commercial 4,4′-dichlorodiphenylsulfone (DCDPS) for the preparation of the PESs (5a) [12] and (5b) [13]. The DCDPS was also taken as electrophilic reagent for the preparation of the functionalized oligo(ether-sulfone)s which were modified at the chloro endgroups (6) [14]. Another class of functional PES (7) was synthesized from commercial 4,4′-Difluorodiphenylsulfone (DFDPS) and 1,1-bis(4-hydroxydiphenyl)ethene [15]. The pendant methylene group allows for thermal crosslinking of these PESs. DFDPS in combination with various diphenols and 4-fluoro-4′-hydroxydiphenylsulfone served as comonomers for the preparation of copoly(ether-sulfone)s having the structure (8). Their properties were evaluated and correlated with their composition and sequence [16].

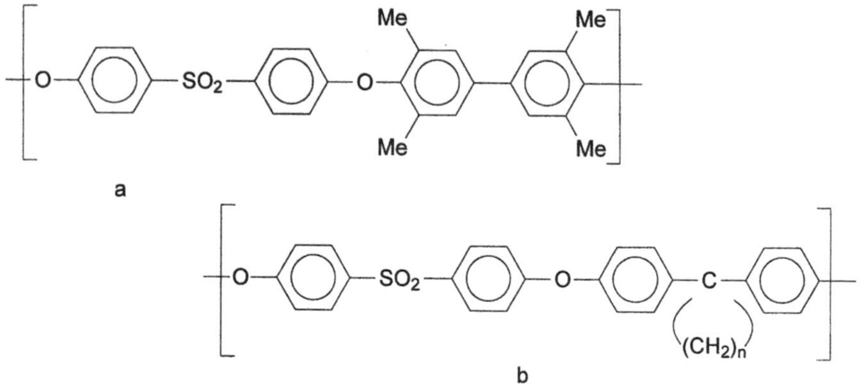

(5)

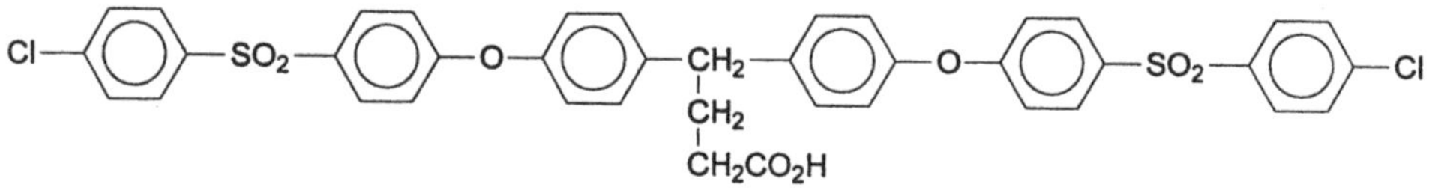

(6)

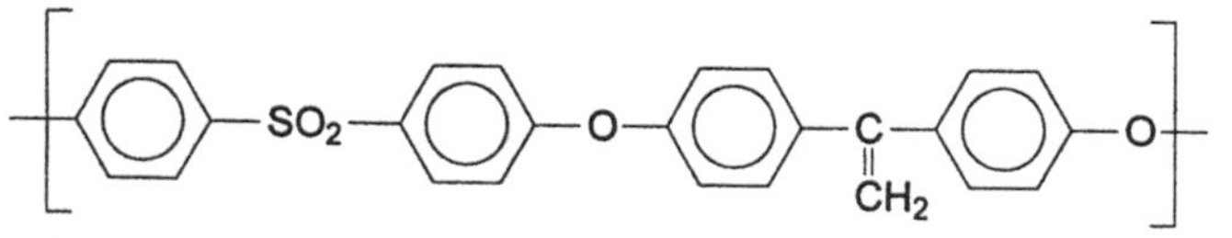

(7)

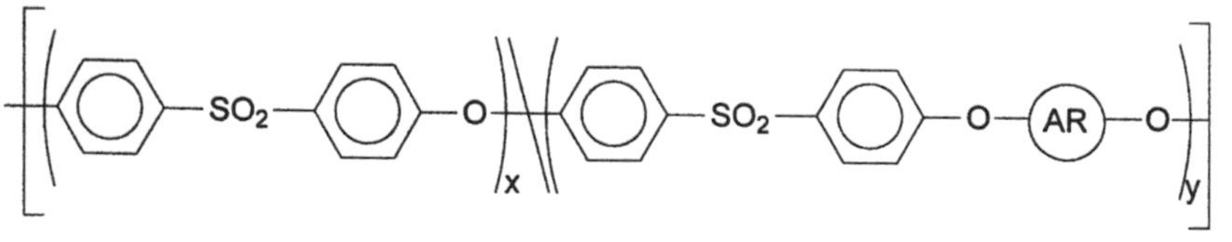

(8)

AR = various aromatic units

Most new PESs reported during the past ten years were prepared from new sulfone type monomers or at least from noncommercial monomers. For instance, the difluoroketone-sulfone (9a) was polycondensed with the dihydrodiphenylketone-sulfone (9b) [17]. The same monomers were later used by another research group together with a variety of new fluoro ketone type monomers [19]. Other authors synthesized the naphthalene containing difluordiphenylsulfones (10a) and (10b) as reaction partners of methyl substituted 4,4'-dihydroxybiphenyls [19]. Dichloro- or difluoro-diphenylsulfones having a central biphenyl unit (11a) were polycondensed with various commercial diphenols [20,21]. In one of these papers [20] PES derived from the bisphenol (11b) were studied in detail.

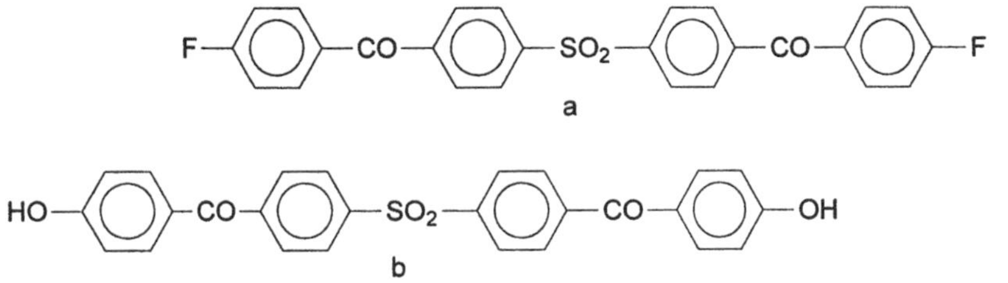

(9)

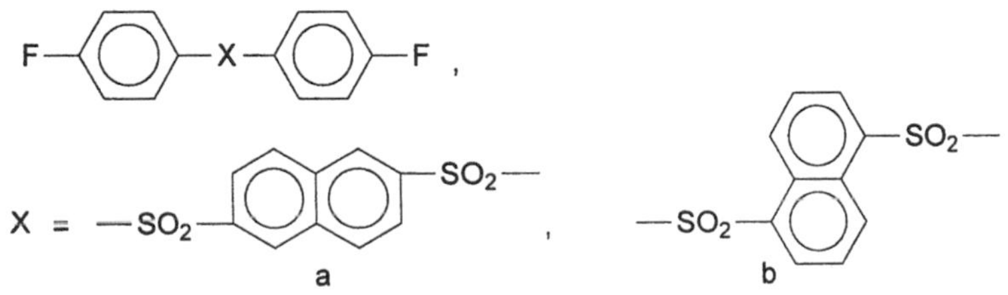

(10)

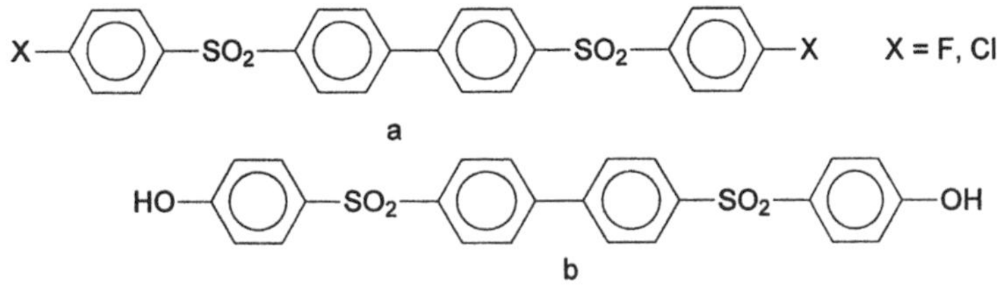

(11)

Thiophene based poly(arylene-ether-sulfone)s were synthesized from monomer (12) and bisphenol-A [22]. Another class of unconventional monomers is outlined in the formulas (13a) and (13b) [23,24]. These monomers have the advantage that its structure can easily be varied at the imide ring, and reactions at the imide ring allow also a modification of the PES itself. In any case these terphenyl monomers considerably raise the glass transition temperature (Tg). Another kinked structure is that of the indan derivatives (14) [25]. They were polycondensed with bisphenol-A and other common diphenols. Two research groups reported on syntheses of more or less fluorinated PESs. Two different synthetic strategies were elaborated. In the first case a difluoro diphenyl disulfone having a fluorinated aliphatic chain segment (15b) was synthesized by oxidation of the corresponding disulfide (15a) [26]. The disulfone (15b) was then polycondensed with a variety of diphenols by means of sodium carbonate in DMAc. The second strategy is characterized by the preparation of PESs having pendant trifluoromethyl groups by polycondensation of the monomers (16a), (16b) and (16c) [27,28].

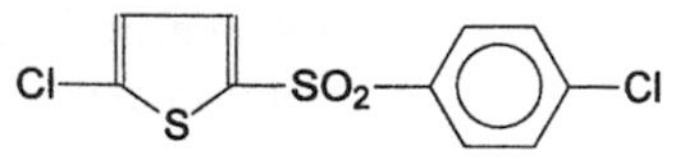

(12)

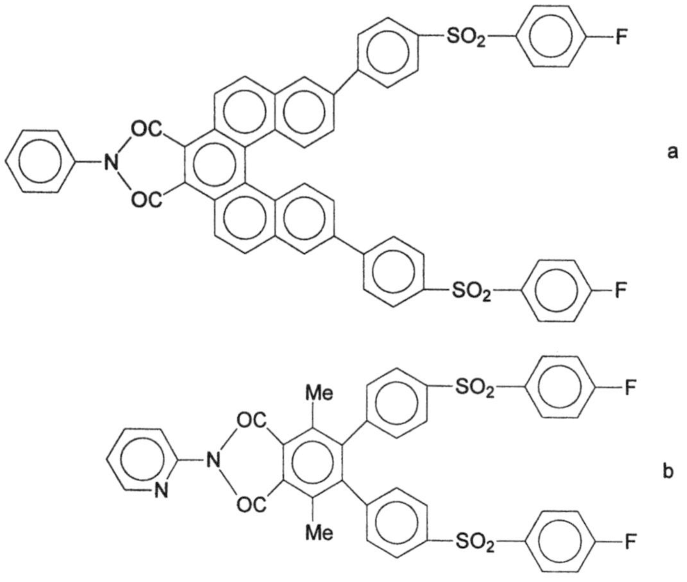

(13)

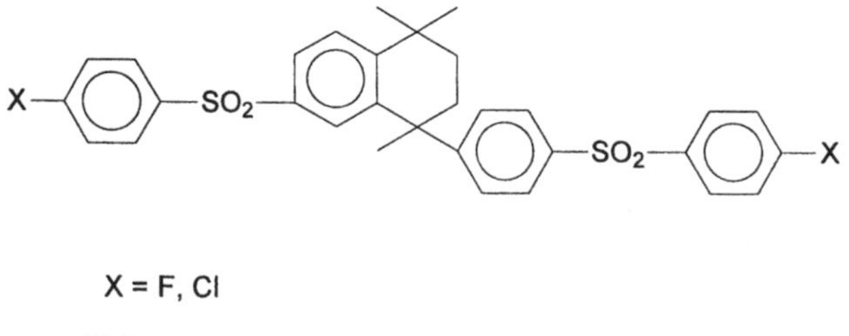

(14)

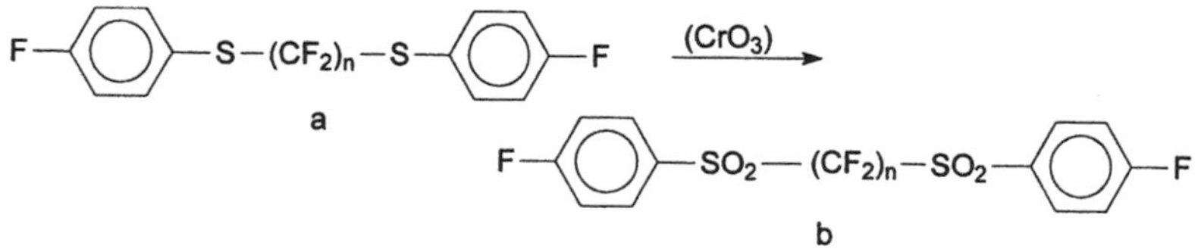

X = F, Cl

(15)

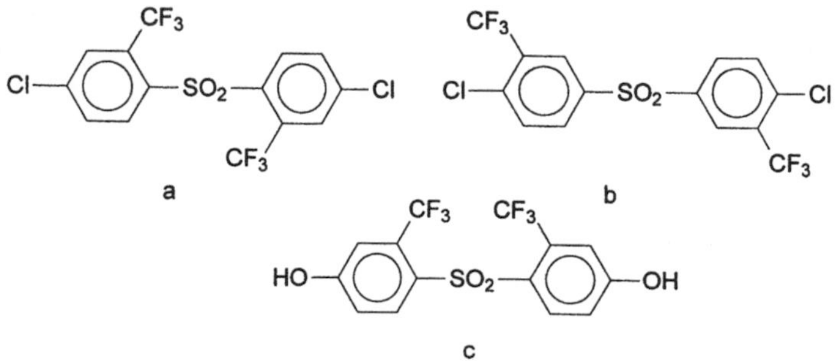

(16)

C. Chemical Modification

The normal route of nucleophilic substitution was also applied to syntheses of PESs having a broad variety of pendant functional groups. In most publications dealing with functional PES the reactive substituents were subjected to further modifications. Two research groups were interested in sulfonated PES which may find potential application as proton transporting membranes in fuel cells. Two different synthetic approaches were explored. The first one is based on polycondensations of a sulfonated DCDPS (17a) with preformed potassium salts of diphenols in DMAc at 170 °C [29]. The second approach consists of the sulfonation of preformed PESs [30,31]. When a PES derived from hydroquinone was sulfonated exclusive monosulfonation of the hydroquinone unit was observed (18a). Increasing reactivity of the sulfonating agent did not influence the degree of substitution but the stability of the PES chain. With 91% sulfuric acid no degradation was observed at room temperature, whereas chlorosulfonic acid and oleum caused severe degradation. Furthermore, PES derived from methyl hydroquinone, dimethylhydro-quinone and trimethylhydroquinone were sulfonated with concentrated sulfuric acid. Complete monosubstitution was found for mono- and dimethyl hydroquinone (18b) and (19a), whereas the sulfonation of the trimethyl hydroquinone units (19b) remained incomplete [31].

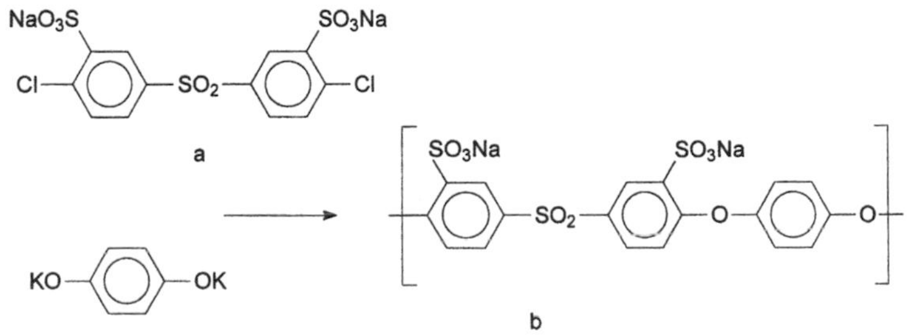

(17)

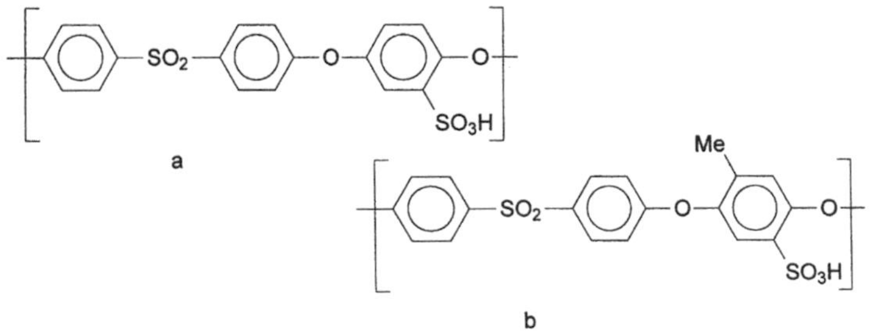

(18)

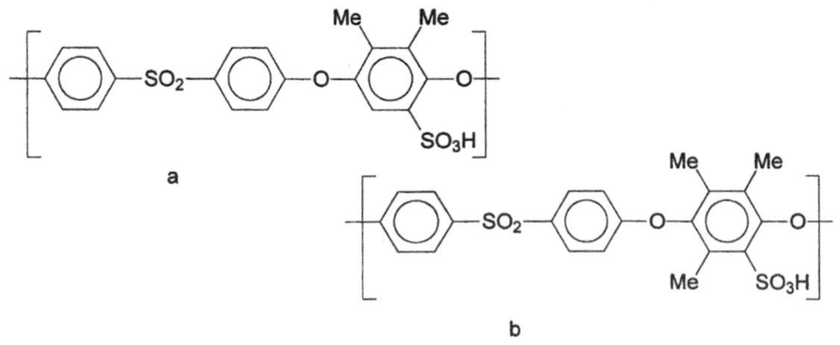

(19)

Several studies dealt with syntheses of PESs having pendant amino groups. Quite analogous to the syntheses of sulfonated PESs two strategies were explored, (a) polycondensation of aminated monomers, and (b) modification of preformed PES. The first strategy was realized with synthesis and polycondensation of the dichlorodiamino-sulfone (20a) which was synthesized by hydrogenation of 2,2'-dinitro-4,4'-dichlorodiphe-nylsulfone [32]. Another research group used the corresponding difluorodiaminosulfone [33]. A further difluoromonomer having a pendant amino group is the phosphine oxide (20b) which was used as comonomer together with DFDPS and various diphenols [34]. The second synthetic strategy was realized in such a way that performed PESs were nitrated at the hydroquinone unit and the nitrogroup was reduced by means of sodium dithionite (21) [35]. Another approach is based on the synthesis of PES, having pendant imide groups (22) [36–38]. Variation of the amine (for instance via transimidization) allows a broad variation of the pendant functional groups including the introduction of an amino group.

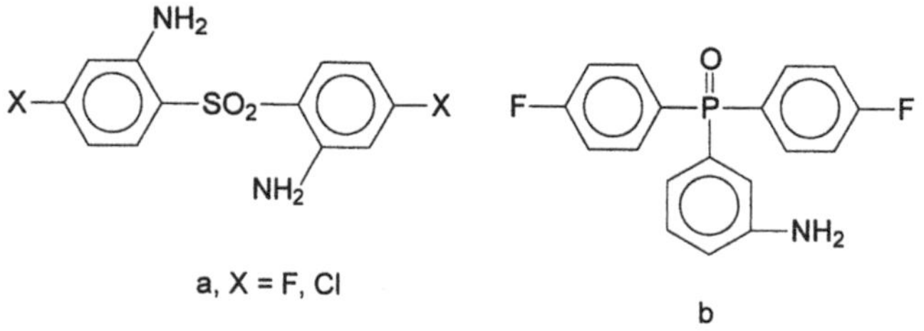

(20)

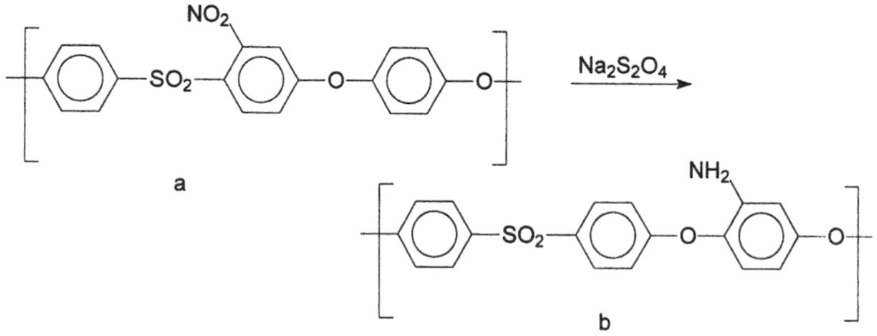

a, X = F, Cl

b

(21)

a

b

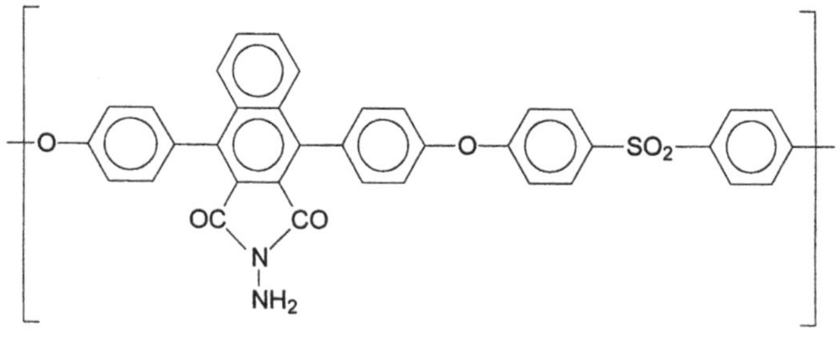

(22)

Five more papers reported on various modifications of PES involving introduction and substitution of chloro or bromoatoms. For instance, bromination of the bisphenol-A unit in a commercial PES yielded the dibromoproduct (23), which was treated with butyllithium. The lithiated PES was then reacted with methyliodide [39] or with tosylazide [40]. The resulting azide groups were finally reduced to amino groups (24). Another modification of brominated PESs utilized palladium complexes as catalysts for the introduction of alkin-type substituents (25). These substituents have the purpose to enable a thermal cure via cyclization or polymerization of the alkin groups [41].

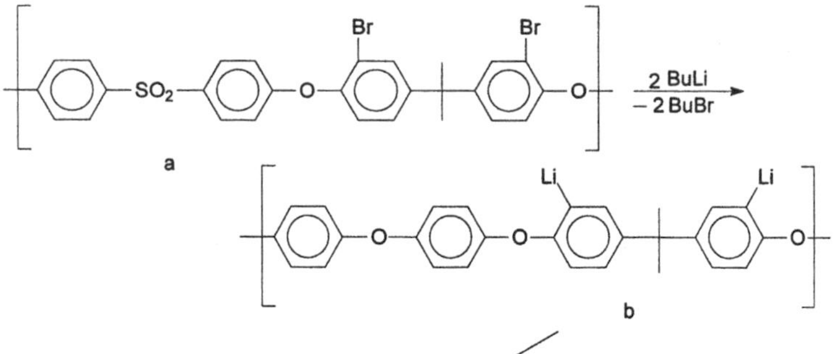

(23)

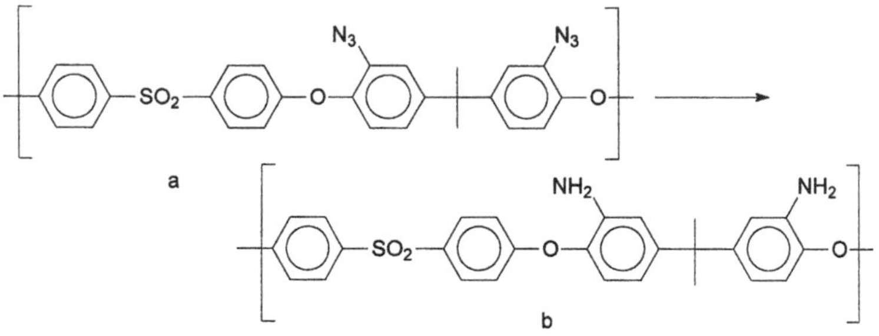

(24)

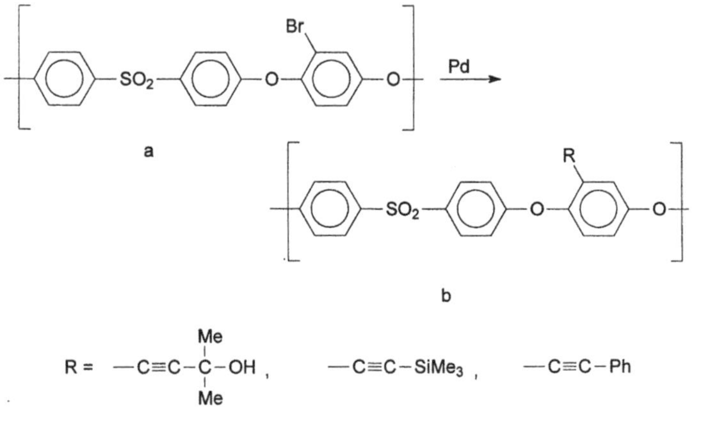

(25)

$$R = \quad -C\equiv C - \overset{\displaystyle Me}{\underset{\displaystyle Me}{\overset{|}{\underset{|}{C}}}} - OH \ , \qquad -C\equiv C - SiMe_3 \ , \qquad -C\equiv C - Ph$$

Two papers reported on the chloromethylation of PESs and the further modification of the chloromethyl groups. In the first paper [42] the combination of octylchloromethyl ether and $SnCl_4$ was used to introduce the CH_2Cl groups, the combination of octylbromomethyl ether and $SnBr_4$ yielded CH_2Br groups (26a) and combinations of chloromethylether + $SnBr_4$ or bromomethyl ether and $SnCl_4$ producing a statistical array of chloro and bromomethyl substituents. Reactions with potassium tert.-butoxide yielded pendant tert.-butyl ether groups (26b) with sodium acetate pendant acetate groups were obtained and after alkaline saponification CH_2OH groups (26b). Furthermore, pendant tosylate groups (27a) and diethylphosphonates (27b) were prepared. With sodium cyanide pendant nitrile groups were formed (28a) which were saponified to yield CH_2CO_2H

groups. Finally the oxidation of chloromethyl groups with dimethylsulfoxid/NaHCO$_3$ or with Cr$_2$O$_7^{2-}$ was studied (yielding aldehyde groups (28b) [42]. In the second paper triflicacid was used as solvent and catalyst in combination with butyl or octyl chloromethyl ether. This system is of course too expensive for any large scale experiments or technical production of functionalized PES. For homo- or copolyether containing hydroquinone an exclusive monosubstitution of the hydroquinone unit was found (29a), and finally the transformation of the chloromethyl groups into triethylammonium groups (29b) was studied [43]. PESs having pendant aldehyde groups were prepared from (co-)polycondensations of the diphenol (30a). The aldehyde groups were almost quantitatively transformed into azomethine groups (30b) [44]. In another publication [45] unsaturated PESs were prepared from 4,4'-dihydroxy-trans-stilbene and DFDPS and treated with H$_2$O$_2$ in the presence of a tungsten catalyst whereby epoxide groups suitable for chemical or thermal crosslinking were obtained (31). Finally, a publication dealing with the influence of energy rich irradiation (x-ray, electrons, AR$^{\oplus}$ and N$_2^{\oplus}$) on PES should be mentioned [46].

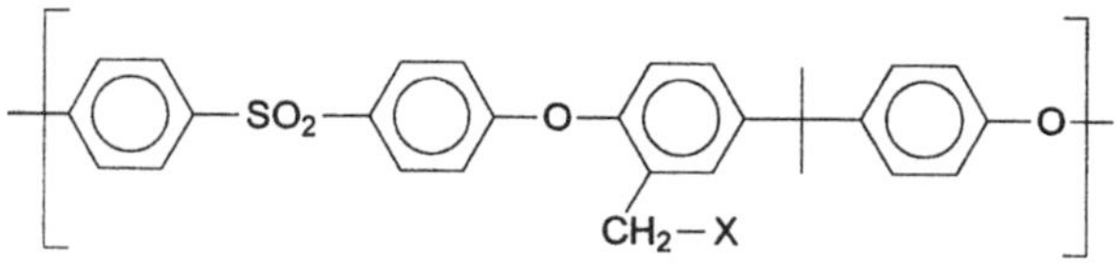

a: X = Cl, Br

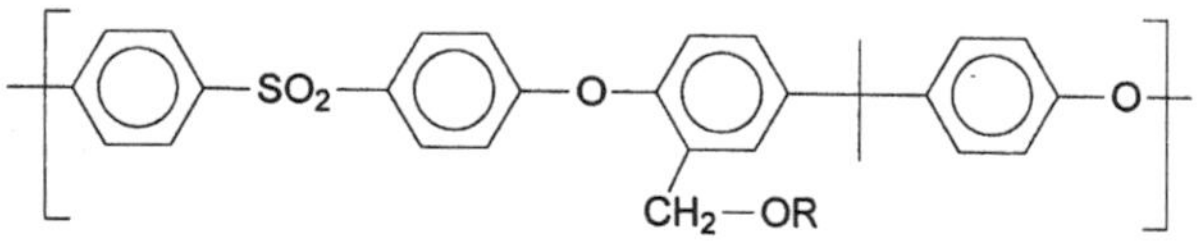

(26)

b: R = H, COCH$_3$, CMe$_3$

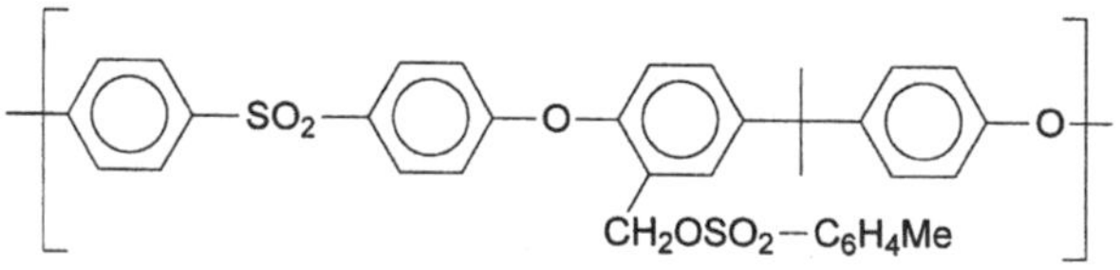

a

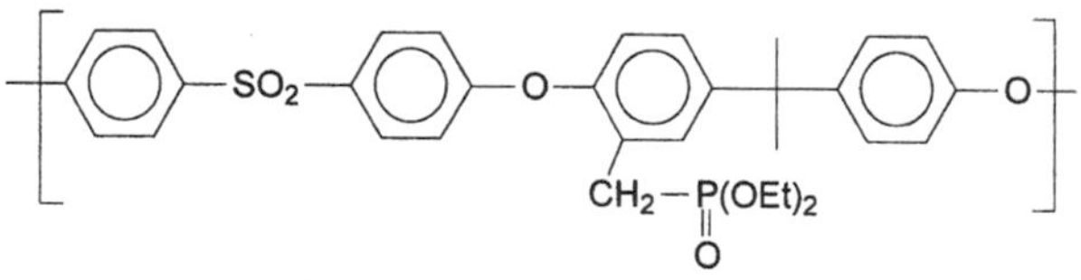

b

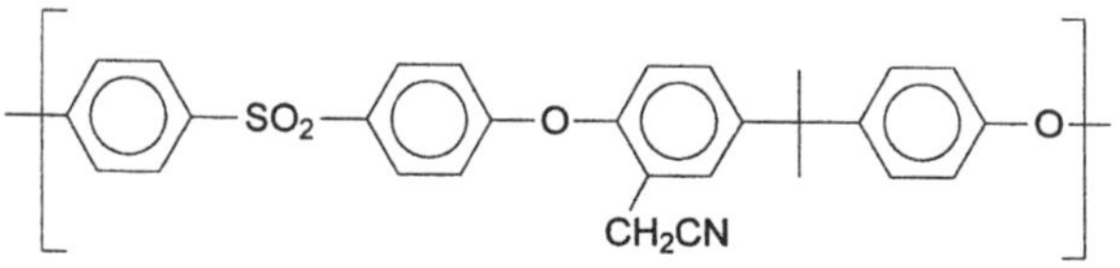

(27)

a

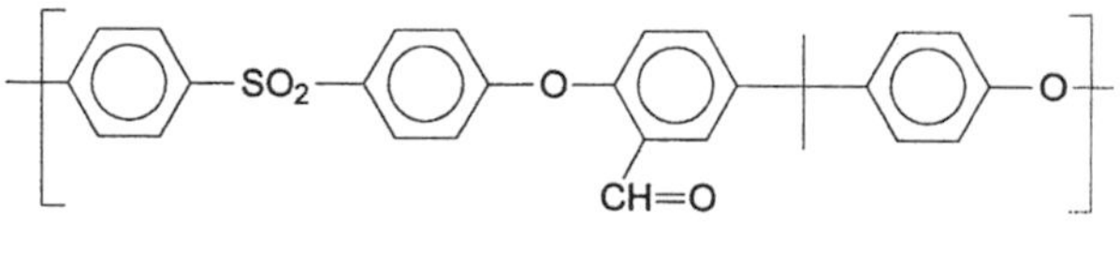

b

(28)

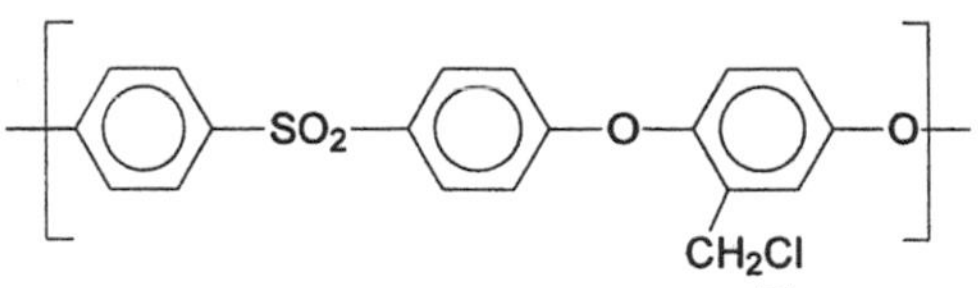

(29)

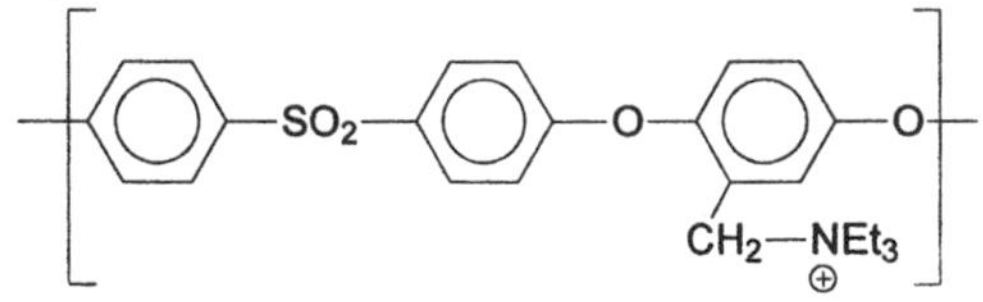

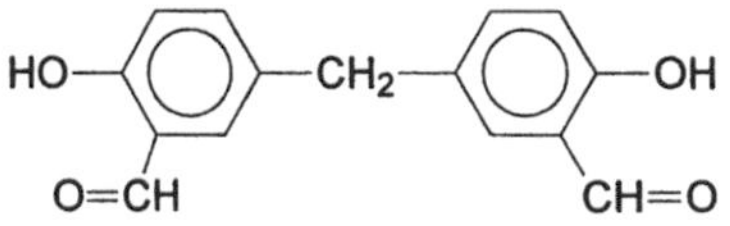

(30)

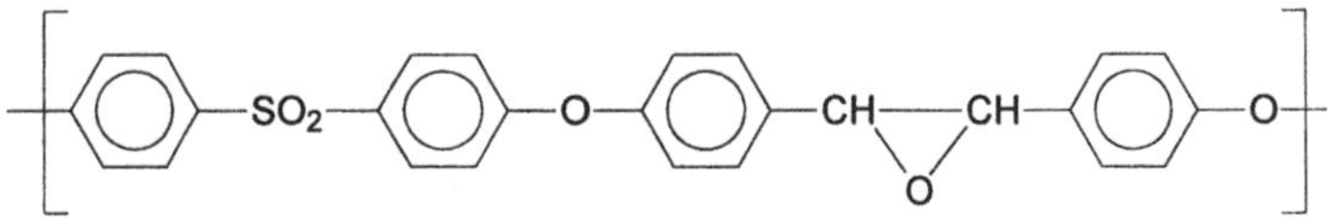

(31)

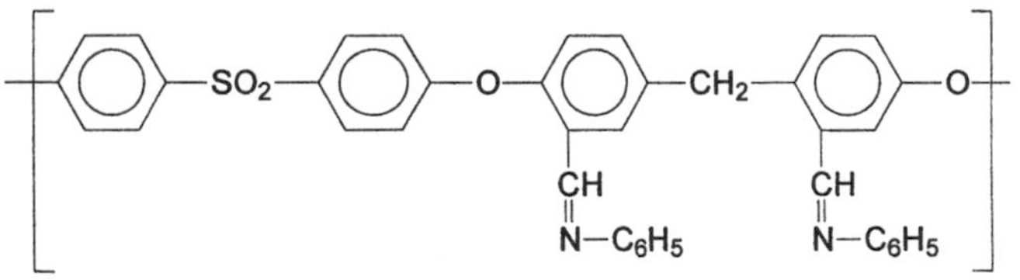

D. Various Synthetic Methods

Numerous publications describe syntheses and characterization of telechelic oligo(ether sulfone)s which served as building blocks of triblock copolymers, multiblock copolymers or networks. OH-terminated oligomers (32) were prepared by polycondensations of DCDPS with an excess of bisphenol-A [47–49]. These oligo(ether sulfone)s were reacted with commercial bisepoxides to yield epoxy networks [47]. They also proved to be useful for syntheses of multiblock poly(ether-esters) [48,49]. The polyester blocks either consisted of poly(ethylene terephthalate) or of LC-poly(ester-imide)s (33). The LC-blocks played the role of a reinforcing component in the PES matrix and showed interesting mechanical properties. Telechelic poly(ether-sulfone)s having C–F endgroups were also prepared by copolycondensation of 4-fluoro-4'-trimethylsiloxy-diphenylsulfone with small amounts of DFDPS [50]. The molecular weight was controlled by the feed ratio of DFDPS. When small amounts of 4,4'-bis(2,4-difluorobenzoyl)diphenyl ether were used as comonomers, four armed stars with C–F endgroups were obtained. Small amounts of silylated 1,3,5-trihydroxybenzene as comonomer yielded three-armed stars having OSiMe$_3$ endgroups [50].

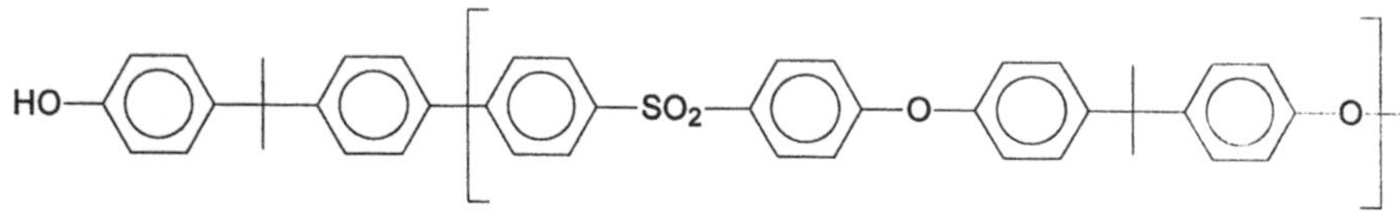

(32)

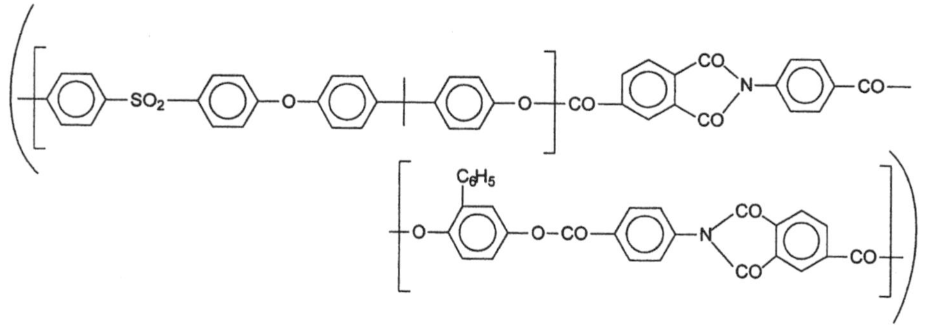

(33)

Several research groups prepared oligo(ether-sulfone)s having primary amino endgroups [51–55]. *m*-Aminophenol, *p*-aminophenol or a higher aminophenol (34) served as endcapping agents. These oligomeric diamines were polycondensed with various aromatic dicarboxylic acid dichlorides to yield polyamides [53] or they were polycondensed with bisanhydrides yielding polyamides such as (35) [54]. Poly(ether-sulfone-amide)s were also prepared by the inverse approach [56]. In this case two 'sulfone dicarboxylic acids' (36) were synthesized and polycondensed with various aromatic diamines via the triphenylphosphite pyridine method.

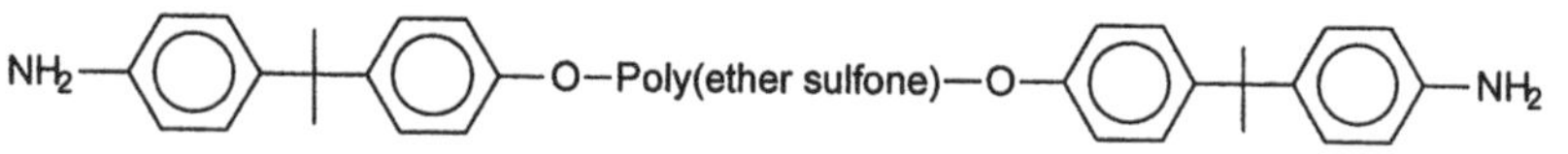

(34)

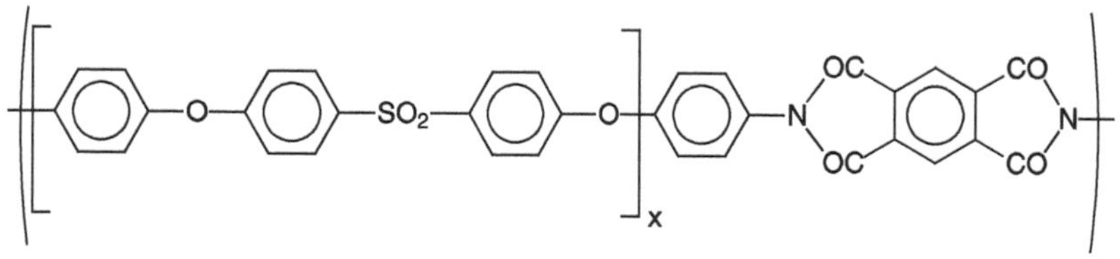

(35)

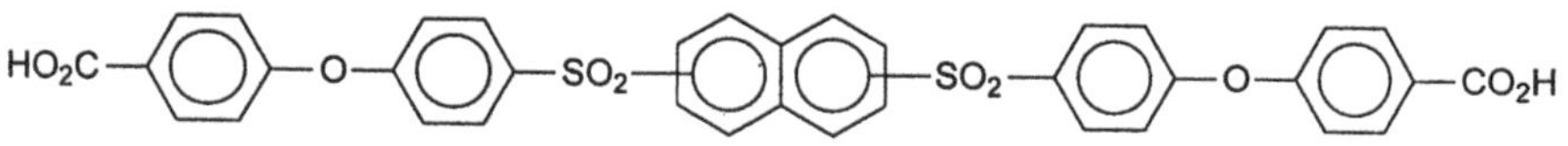

(36)

(1,5 or 2,6)

Triblockcopolymers derived from a central block of PES and two wings of poly(phenylene oxide) were also reported [57]. For this purpose oligo(ether sulfone)s were synthesized with an excess of DFDPS, so that two C–F endgroups were obtained (37). These oligomers were then polycondensed with poly(phenylene oxide)s having one trimethylsiloxy endgroup whereby CsF served as catalyst (38). Multiblock copolymers consisting of PES and polydimethylsiloxane blocks were prepared from OH terminated oligo(ether sulfone)s and diethylamine terminated siloxane blocks (39,40) [58]. The same approach was reported to yield the poly(ether sulfone disilane)s (41) starting from a bis(diethylamino)disilane [59]. These polysilanes deserve interest because of their photosensitivity. Grafting of polysiloxane blocks onto PES (based on bisphenol-A) was achieved in such a way that the PES was lithiated with nBuLi and reacted with chlorodimethylvinylsilane [60]. The Si-H endgroup of a polysiloxane was then added onto the pendant vinyl group (42,43). In this connection a work describing the radical grafting of styrene onto maleimide or nadimide endcapped oligo(ether sulfone)s should be mentioned [61]. In this way PES reinforced polystyrene foams with open pores were obtained. In two papers oligo(ether sulfone)s and oligo(etherketone)s having two acetylenic endgroups were described [62,63]. These oligomers (45) designed to yield thermostable networks upon thermal cure were prepared by means of the new endcapping agents (44).

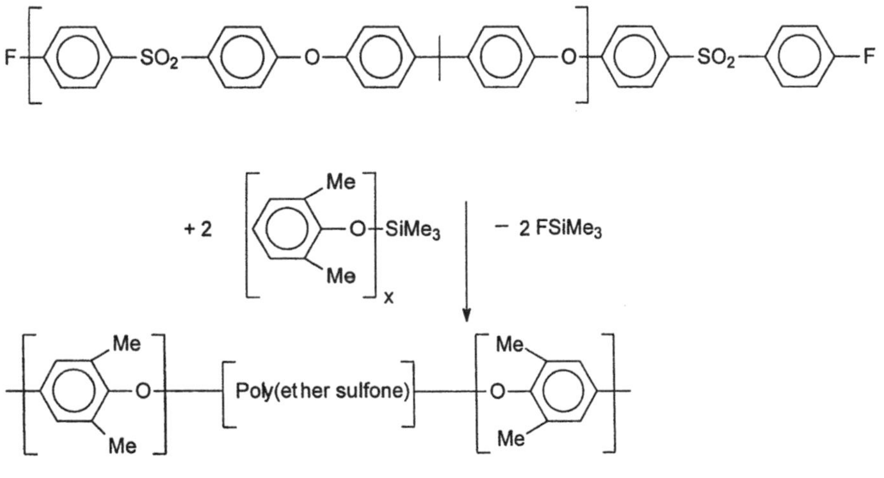

(37)

(38)

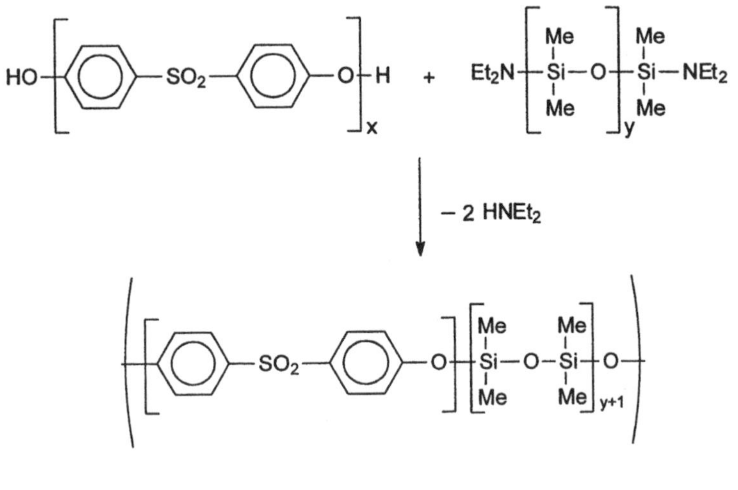

(39)

(40)

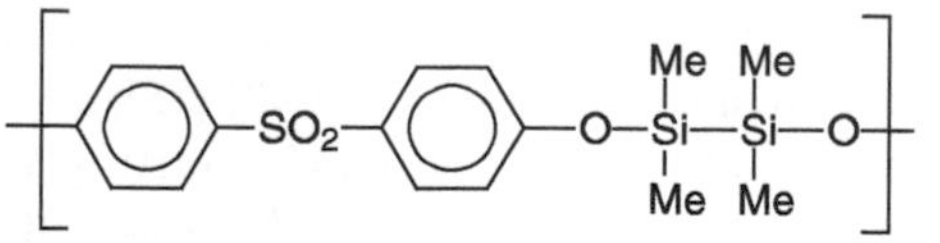

(41)

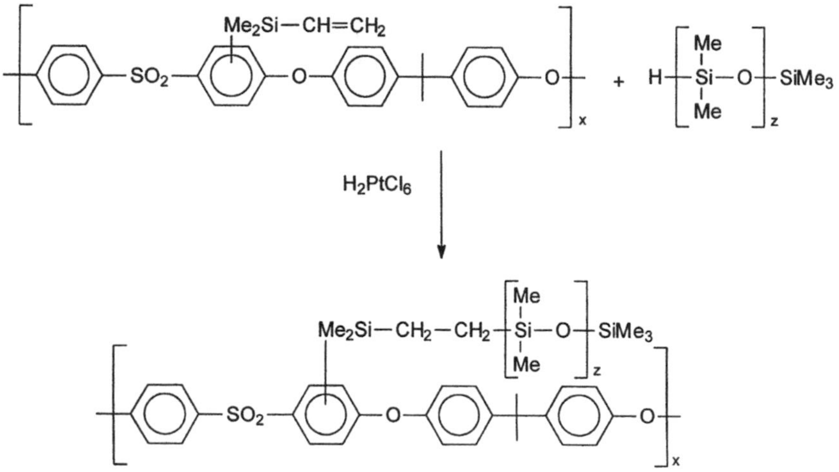

(42)

H_2PtCl_6

(43)

X = Cl, F, NO$_2$

Z = SO$_2$, CO

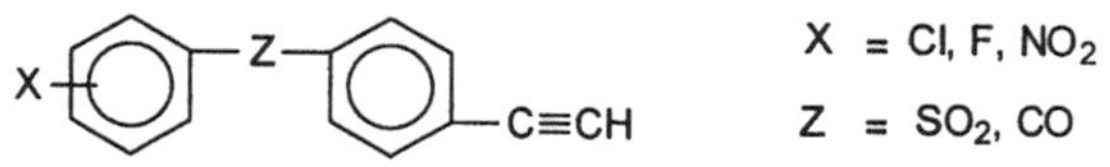

(44)

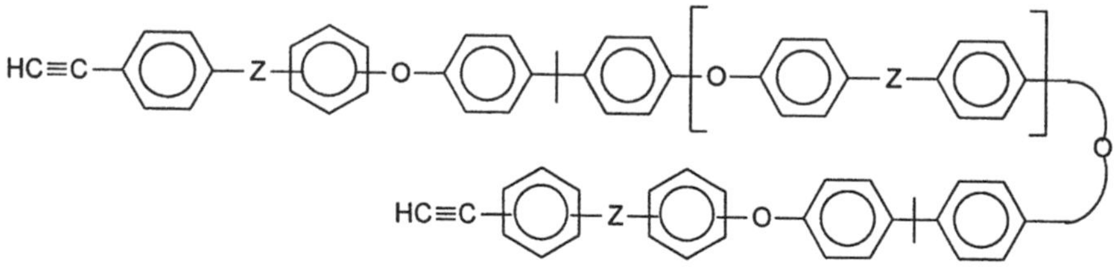

(45)

An entirely new synthetic approach is based on the oxidative coupling of 4,4′-diphenoxydiphenylsulfones (46a) or 4,4′-diphenylsulfidodiphenylsulfone (46a), (R_1–R_4 = H) [64]. These studies were extended to bisnaphthyloxysulfones of structure (47) [65]. This approach involves a radical-cationic mechanism resulting from the single electron transfer reaction (SET) of a π-electron from the aromatic monomers to FeCl$_3$ which plays the role of oxidation and coupling catalyst (so-called Sholl reaction). Nitrobenzene served as the standard solvent and the temperature was varied from 20 to 100 °C, but the molecular weights remained low (M_n < 3000 Da) for all monomers of structure (46a), whereas high molecular weights (M_n ∼ 38,000 Da) were obtained from the polycondensations of monomers (47). Finally, a new polycondensation method should be mentioned yielding poly(ketone sulfone)s free of ether groups (48) and (49) [66].

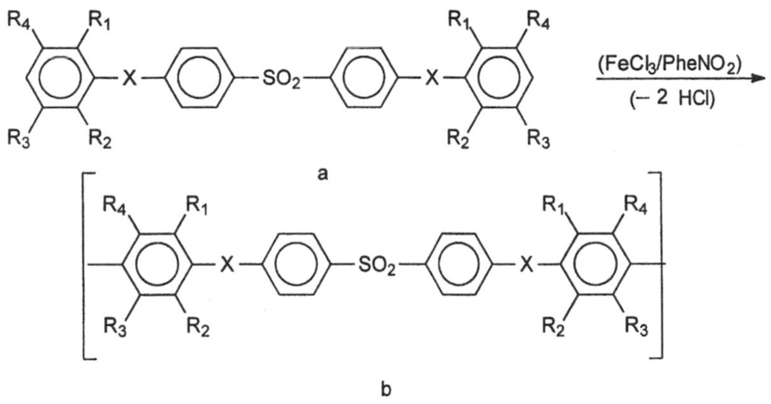

$$\xrightarrow{\text{(FeCl}_3/\text{PheNO}_2)} \atop (-\ 2\ \text{HCl})$$

a

(46)

b

$$X = O, S$$

$R_1, R_2 = H$	$R_3, R_4 = H$
$R_1, R_2 = H$	$R_3, R_4 = Me$
$R_1, R_2 = H$	$R_3, R_4 = CMe_3$
$R_1, R_2 = OMe$	$R_3, R_4 = H$
$R_1, R_2 = C_6H_5$	$R_3, R_4 = H$

(47)

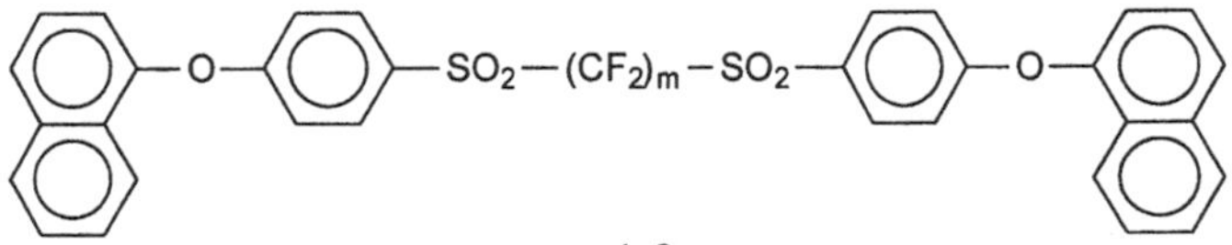

$$SO_2 - (CF_2)_m - SO_2$$

$$m = 4, 8$$

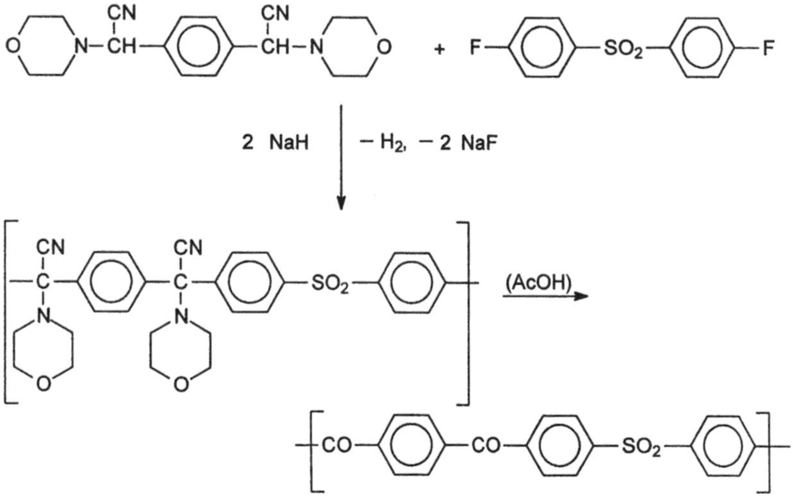

(48)

$$2\ \text{NaH} \quad | \quad - H_2, - 2\ \text{NaF}$$

(AcOH)

(49)

E. Ring-Opening Polymerization (ROP)

All synthetic strategies discussed above have in common to be step growth polymerizations. Over the past ten years a new strategy was elaborated and explored based on the ring-opening polymerization of cyclic oligo(ether sulfone)s, OESs. This chain growth polymerization has the following advantages and disadvantages when compared to step-growth polymerizations. The main problem is the synthesis of the cyclic monomers, above all, when large quantities are needed. The advantages are, firstly, a polymerization process which does neither need solvents, nor produce byproducts. Therefore, the ROP approach is well suited for the reaction-injection molding (RIM) technology. Secondly, the ROP of strained cyclic OESs offers the chance to prepare PESs with very high molecular weight ($M_n > 10^5$ Da). Thirdly, sequential copolymerizations with other cyclic monomers may yield a variety of block copolymers.

Cyclic OESs were prepared in four different ways. Firstly, an electrophilic acylation of 1,4-bisphenoxybenzene was performed under high dilution (50) [67]. Secondly, 1-chloro-4′-hydroxy diphenylsulfone was dimerized and cyclized in the presence of K$_2$CO$_3$ (51) [68]. Thirdly, several cyclic OESs were prepared by condensation of diphenols and dihalosulfones via nucleophilic substitution under high dilution (52) [69–75]. Either mixtures of cyclic OES were isolated and used for ring-opening polymerizations [69,70] or monodisperse cycles were isolated and characterized [71–75]. Fourthly, preformed PES was subjected to back-biting degradation catalyzed by CsF in DMF at 155 °C. At low concentrations, large fractions of cyclic OES were obtained and monodisperse cycles (from the dimer to the hexamer) were isolated by column chromatography [76,77]. Two papers [73,75] describe detailed studies of ring-opening polymerizations conducted in bulk at high temperatures. Unfortunately, cyclic OESs possess high melting temperatures (up to 500 °C), and only in cases of nonsymmetrical cycles the reaction temperatures could be lowered to 290 °C. Such high temperatures have two disadvantages. Firstly, high fractions of cycles remain unreacted for thermodynamic reasons, and gel particles are formed due to partial crosslinking. Anionic polymerizations in concentrated solutions below 250 °C have not been studied yet.

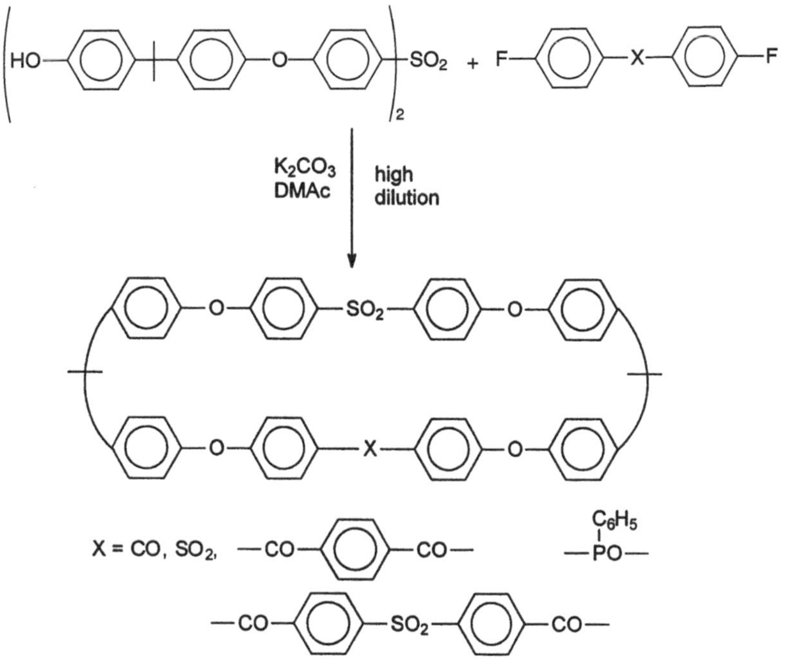

(52)

Finally, a synthetic approach should be mentioned yielding polyamides containing cyclic OESs as part of the repeating unit [78]. The cyclic dicarboxylic acid (53c) was prepared from (53a) and (53b) by a conventional procedure and polycondensed with 4,4'-diamino-diphenylmethane using the triphenylphosphite-pyridine method.

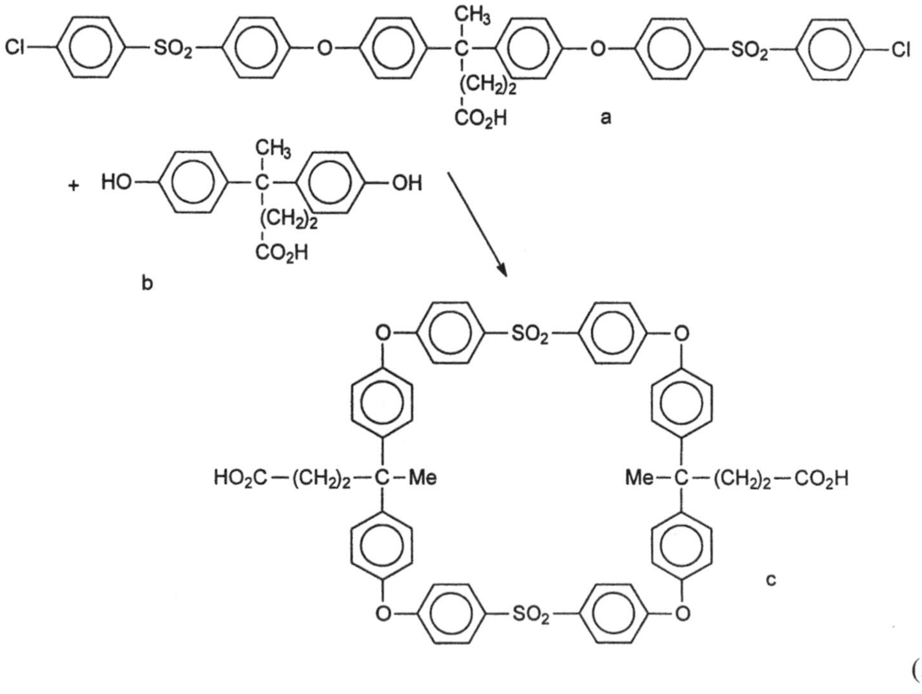

(53)

III. POLYETHERKETONES

Most research activities in the field of aromatic polyethers published over the past ten years concern poly(etherketone)s (in this review the abbreviation PEK is used for all poly(etherketone)s, not only for those having one ether and one keto group in the repeating unit). Quite analogous to PESs the synthetic methods reported for PEKs may be subdivided into three groups:

1. polycondensations involving an electrophilic substitution (i.e., acylation of a phenoxy group)
2. Polycondensations involving a nucleophilic substitution of a chloro-, fluoro- or nitro-aromat activated by a keto group in para position)
3. modification of suitable precursor polymers.

A. Syntheses Via Electrophilic Substitution

Various PEKs were prepared via electrophilic substitution processes such as that exemplarily outlined in equation (54) [79]. The problems of this approach are in principle the same as in the case of PESs. An inert expensive solvent is needed, it is difficult to reach high conversions without side reactions and the number of useful monomers is lower than in the case of syntheses based on nucleophilic substitution reactions. The electrophilic polycondensations may be subdivided into two different methods. Firstly, acid chlorides are used as electrophilic monomers in combination with a Lewis acid. Secondly, free carboxylic acid served as monomers in combination with an acidic dehydrating agent. None of the polycondensation methods described in this section is new, and origin and early exploration of these methods has been reviewed in the 1st edition of this handbook (Chapter 9).

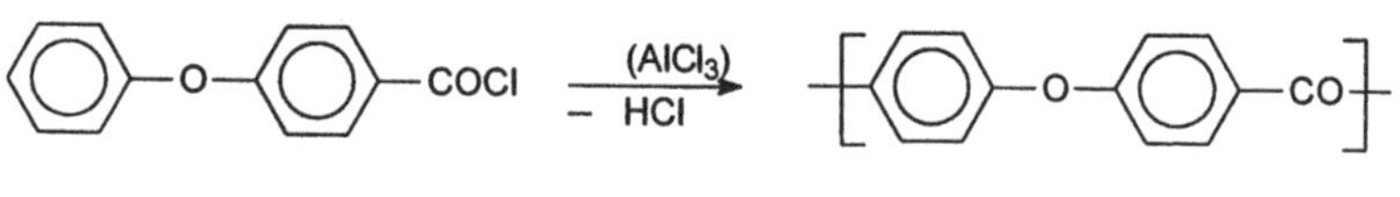

$$\text{(54)}$$

In a publication of 1988 [79] (not reviewed before) polycondensations of phenoxy-benzoyl chloride (Eq. (54)) or polycondensations of diphenylether with isophthaloylchloride, terephthaloyl chloride and a phenolphthalein dicarboxylic acid dichloride (55) were studied. AlCl$_3$ served as catalyst and the solvent was varied. It was found that the homogeneous polycondensation in nitrobenzene gave lower molecular weights than the heterogeneous reaction in CH$_2$Cl$_2$ or (ClCH$_2$)$_2$, (DCE). The chain growth continues in the precipitated AlCl$_3$-oligomer or -polymer complexes. Similar results were found by another group [80] which did not know the first paper [79]. Other authors [81], compared two reaction media: AlCl$_3$/CH$_2$Cl$_2$ and H$_2$F$_2$/BF$_3$. The faster reaction (but similar mol-weights) were found in the H$_2$F$_2$ system. The influence of HCl or Lewis bases on AlCl$_3$ catalyzed polycondensatins of diphenylether and terephthaloylchloride was also studied [81]. New structures were obtained by AlCl$_3$ catalyzed polycondensations of the oligo-ethers (56a,b) with isophthaloylchloride or adipoylchloride [82]. Two other research groups studied AlCl$_3$ catalyzed polycondensations of isophthaloyl chloride, terephthaloyl chloride or naphthalene-1,6-dicarbonyl chloride with diphenyl ether or with the oligoether (57) in much detail [83–88]. Potential defects in the chemical structure (as revealed by ^{1}H and

^{13}C NMR spectroscopy) and the morphology of the particles which crystallized directly from the reaction mixture were intensively investigated [84–88].

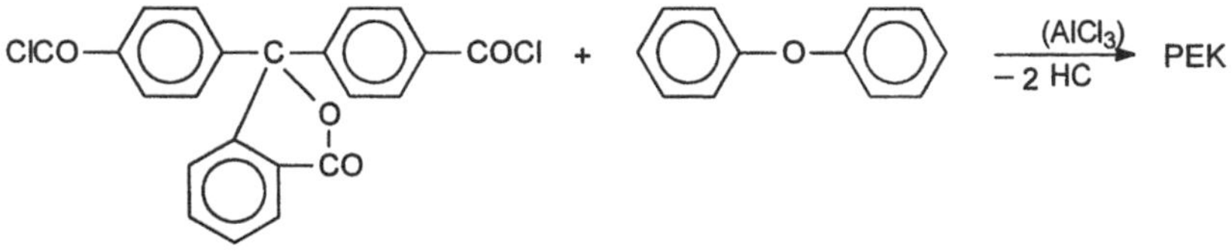

(55)

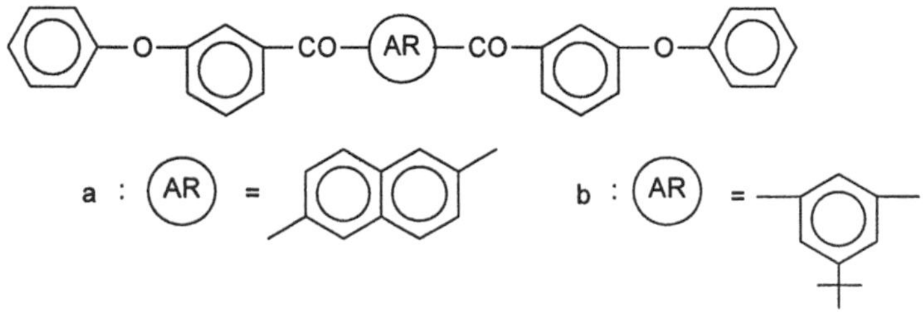

(56)

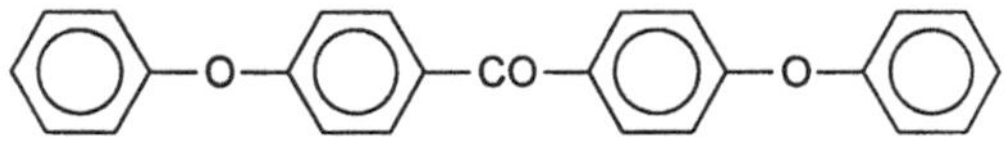

(57)

Polycondensations of dicarboxylic acids with dehydrating agents may be subdivided into two methods, either a solution of P_4O_{10} in methanes sulfonic acid (Eatons reagent) [89–93] or neat triflic acid were used [94–97]. By means of Eatons reagent 3-phenoxyenzoic acid was polymerized (58) [89], and the dicarboxylic acids (59a–d) were polycondensed with the oligoethers (60a,b) [90]. Furthermore a small series of polyethers were reported having alternating sequences of keto, ether and sulfone groups between para functionalized benzene rings [10,91]. PEK's containing CF_3 groups were prepared from the dicarboxylic acid (61a) and diphenyl [92], and PEKs containing phosphazene rings were synthesized analogously from (61b) [93]. All these polycondensations were conducted at 80–120 °C and gave only low ($M_n < 5000$ Da) to moderate ($M_n < 10,000$ Da) molecular weights. Using triflic acid as catalyst and reaction medium the polycondensation of terephthaloyl chloride with diphenyl ether was studied [94]. The most interesting result of this study is the finding that addition of triflic anhydride or P_4O_{10} enhances the molecular weight by a factor of 4 or 5. Another group conducted systematic studies of structure reactivity relationships of various monomers or model compounds in triflic acid catalyzed (poly)condensations [95]. Selected monomers used in this study are presented in the formulas (62a,b) and (63a,b). This work was extended to cocondensations of the carborane monomers (64a,b) with monomers of the structure (62a,b) [96]. The physical and thermal properties of the carborane containing PEKs were also studied. Finally ferrocene containing PEKs (65) should be mentioned which were prepared from ferrocene dicarboxylic acid by means of Eatons reagent or triflic acid [97].

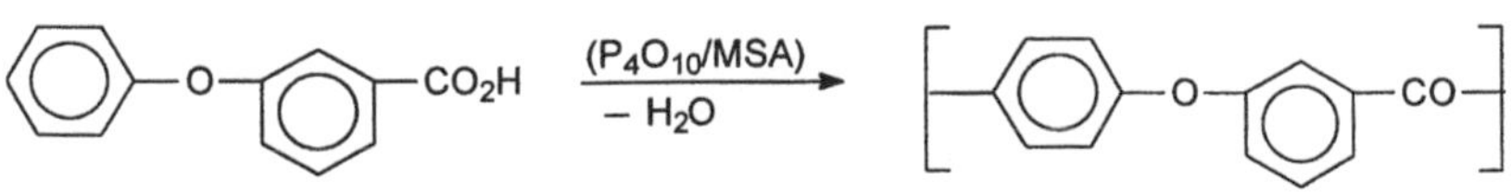

(58)

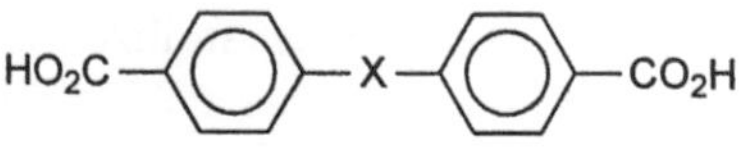

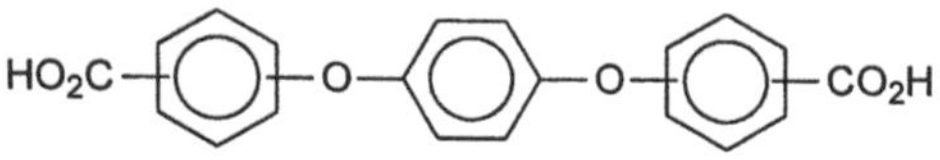

a : X = O, b : X = S　　　　　c : meta,　　d : para

(59)

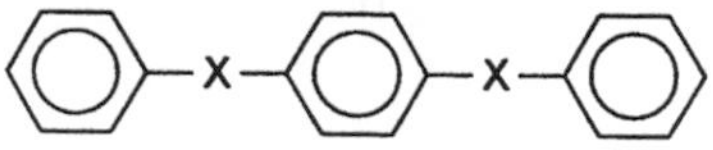

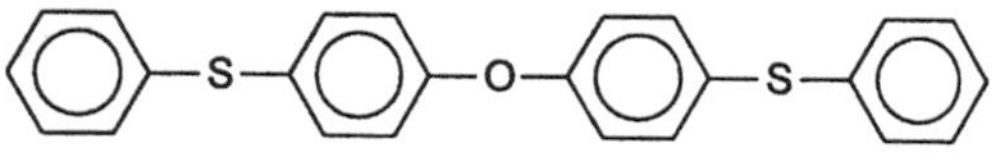

a : X = O,　b : X = S　　　　　　c

(60)

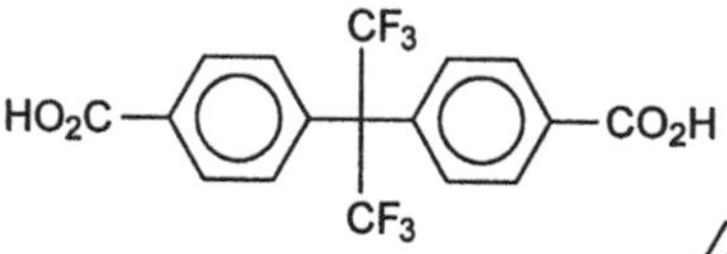

a

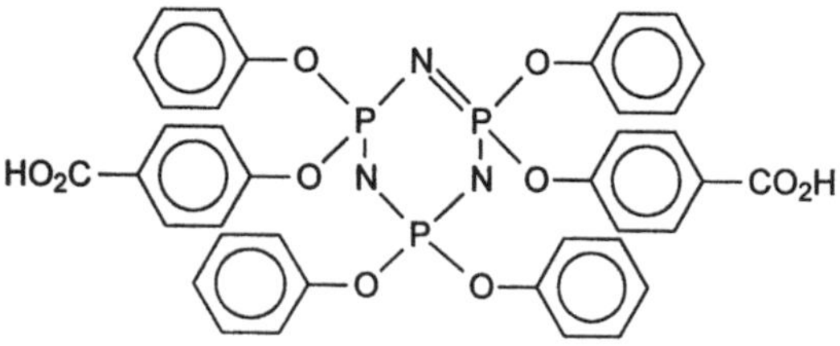

b

(61)

a

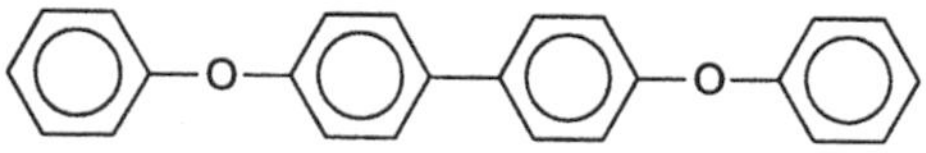

b

(62)

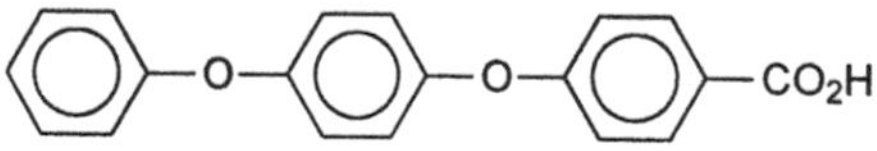

c

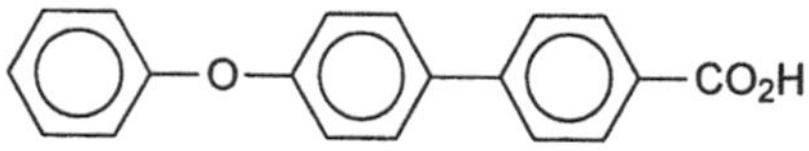

b

(63)

a

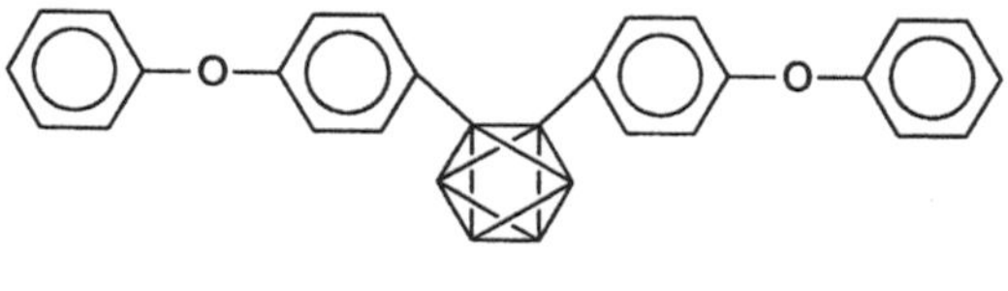

b

(64)

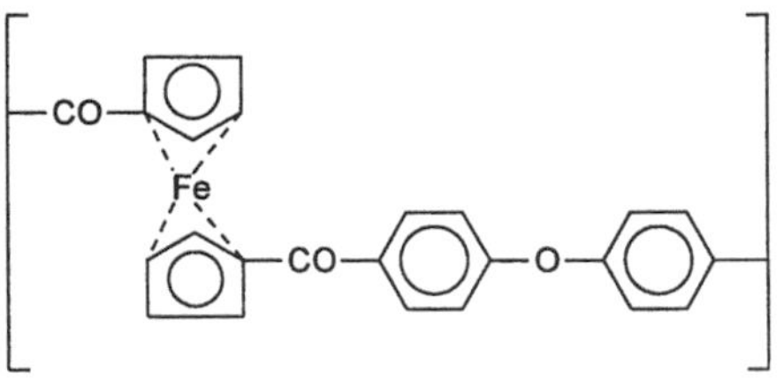

(65)

B. Syntheses Via Nucleophilic Substitution

1. Mechanistic Studies

Numerous PEKs were synthesized by the nucleophilic substitution of aromatic fluoro- or chloroketones, and in this connection several research groups conducted detailed mechanistic studies [7,8,98–110]. Using ^{13}C and ^{19}F NMR spectroscopy combined with model reactions and computer calculations of electron densities the activating power of CO-groups for F in para position was compared to that of other electron-withdrawing groups [7,8] and a significantly weaker activation effect was found. When the reactivities of the dihalobenzonaphthones (66a–d) were compared in polycondensation process with diphenols the reactivity decreased in the given order (a → d). The failure of the dichloro compound (66d) to yield PEKs was attributed to the low reactivity (i.e. electrophilicity) of the Cl–C bond in aromatic nucleophilic substitution (S_NAR). However, several authors found in detailed mechanistic studies [99–103] that the chlorobenzophenones easily undergo a radical side reaction (67–69) yielding saturated chain ends. The extent of this side reaction depends very much on the solvent and to a lesser extent on the redox potential of the phenoxide ions. The following order of decreasing usefulness (i.e. decreasing molecular weights of the isolated PEKs) of polar solvents was found:

$$DPSU > DMAc > NMP > TMU > DMPU$$

(diphenylsulfone, dimethylacetamide, N-methylpyrrolidone, tetramethyl urea, 1,3-dimethyl-perhydropyrimidinone 2).

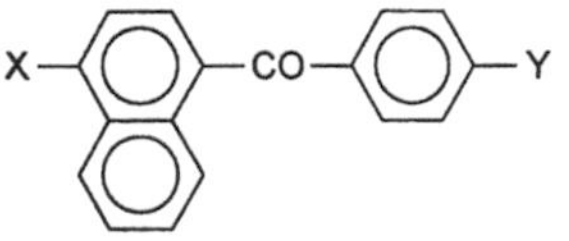

a : X = F, Y = F
b : X = Cl, Y = F
c : X = F, Y = Cl
d : X = Cl, Y = Cl

(66)

(67)

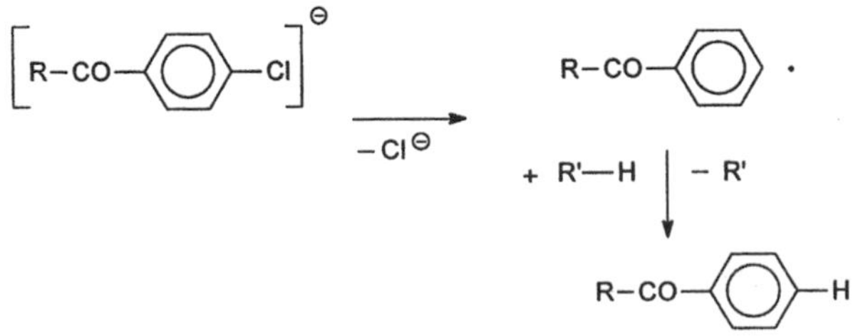

(68)

(69)

In DPSU the radical side reactions are almost completely avoidable [10]. Furthermore, addition of a radical scavenger may be helpful to raise the molecular weights [100]. Another approach consists of the use of special phase transfer catalysts (70a,b) which promote the polycondensation of chlorobenzophenones and diphenols in the presence of K_2CO_3 [104]. These pyridinium salts were selected because they are stable up to 300 °C even under alkaline conditions. Another version of this approach is the combination of these pyridinium salts with an amount of KF. Activation of the phenolic OH-groups and under certain reaction conditions a halogen exchange takes place so that the far more reactive fluoroketones are formed as reaction intermediates [105]. However the activation of KF by means of the phase-transfer catalysts (68a–c) may have the additional effect, that the fluoride ions begins to cleave the PEK backbone at temperatures as low as 160 °C. From other studies [10,106,107] it was known that KF alone attacks the PEK chains only at temperatures $\geq$300 °C. Transetherification, catalyzed by phenoxide ions was also studied by several authors [105–108]. Another important aspect investigated in two papers [109,110] is the influence of the reaction medium on the molecular weight in polycondensations exclusively involving fluoroketones and the S_NAR mechanism. In the first paper [109] difluorobenzil (71a) was polycondensed with free diphenols and K_2CO_3 in four different solvents DMSO and sulfolane gave the best results, whereas cleavage of the PEK backbone was found in NMP and DMPU. However, excellent molecular weights were obtained in NMP when silylated diphenols (71b) and a catalytic amount of CsF were used as reaction partners of (71a). In the second paper it was reported that DMPU is advantageous over NMP when less reactive electrophiles than fluoroketones or fluorosulfones are used (see Section III.F).

$$ (70) $$

a : R = n−C$_4$H$_9$
b : R = n−C$_6$H$_{13}$

$$ (71) $$

C. Various Structures

Most papers reporting on syntheses of PEKs deal with a systematic variation of their structure with the purpose to elucidate structure property relationships. In the present review the discussion of these papers has been subdivided into the following groups:

1. PEKs prepared from 4,4′-difluorobenzophenone (DFBP) and various diphenols [111–122]

2. PEKs prepared from new fluoroketone monomers and commercial diphenols [123–148]
3. Fluorinated PEKs [149–154]
4. Liquid-crystalline PEKs [155–157]
5. Telechelic oligomers, block-copolymers and networks
6. Hyperbranched PEKs.

All these PEKs were synthesized via the standard procedure, K_2CO_3 (rarely in combination with Na_2CO_3) was used as catalyst and HX acceptor in polar solvents such as, DMSO, DMAc or NMP combined with toluene for the azeotropic removal of water. However in Section III.c.2 additional chain extension methods will be discussed.

In two publications [111,112] polycondensation of DFBP with hydroquinone, resorcinol or 4,4′-dihydroxybenzophenone were described. In addition to the homopolymers a series of copolymers with systematic variation of the hydroquinone/resorcinol ratio was studied. Another publication [113] reported on an analogous series of copolymers prepared from DFBP and mixtures of hydroquinone and 1,5-dihydroxynaphthalene. The role of various side reactions was also discussed. In several publications substituted diphenols were used, mainly with the purpose to improve the solubility and to reduce the melting temperature. Typical examples are PEKs derived from resorcinol having a pendant adamantly group (72a) [114] or PEKs prepared from the substituted hydroquinones (72b–e) [115]. Polyelectrolytes of structure (73) were prepared by copolycondensation of a sulfonated hydroquinone and unsubstituted hydroquinone [116]. The free sulfonic acid served as binding site for the fixation of basic NLO chromophors, such as (74a,b) and (75a,b) in the form of their pyridinium salts. Several PEKs showing improved solubilities compared to their unsubstituted analogs resulted from polycondensations of DFBP and methylated dihydroxybiphenyls (76a,b) [117]. Phenyl substituted biphenyl diols (77a–c) were also used as comonomers of DFBP, and these monomers imparted high T_gs combined with good solubilities and high thermostabilities into the PEKs [118]. In another paper several PEKs and PES were prepared from DFBP (or DFDPS) and 4,4′-dihydroxy m-terphenyl [119] to obtain amorphous thermostable polyethers. Amorphous, but also fluorescent PEKs were the result, when phenolphthaleine and the substituted phenolphthaleins (78a,b) were used as comonomers of DFBP [120]. Particularly bulky cardomonomers, such as the fluorene derivatives (79a,b) have, of course, again the consequence that the pertinent PEKs are amorphous, soluble in numerous solvents and highly thermostable [121]. Finally, PEKs derived from the hydroxyphenylphthalazin (79c) need to be mentioned [122]. In this case the PEK backbone includes C–N bonds in addition to ether groups.

$$b: R^1 = H, \quad R^2 = H$$
$$: R^1 = H, \quad R^2 = Me$$
$$c: R^1 = H, \quad R^2 = Ph$$
$$: R^1 = Ph, \quad R^2 = Ph$$

(72)

(73)

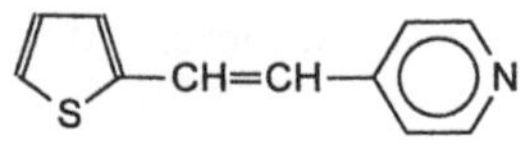

(74)

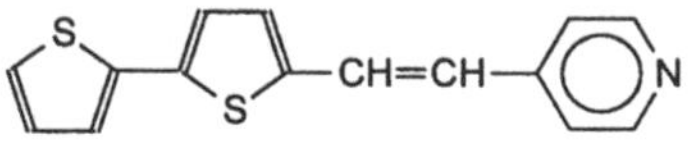

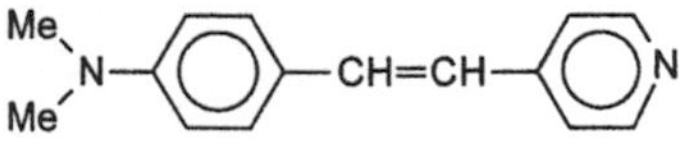

(75)

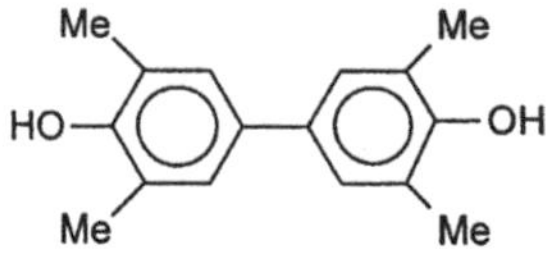

(76)

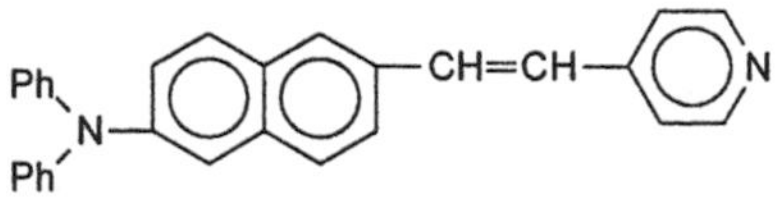

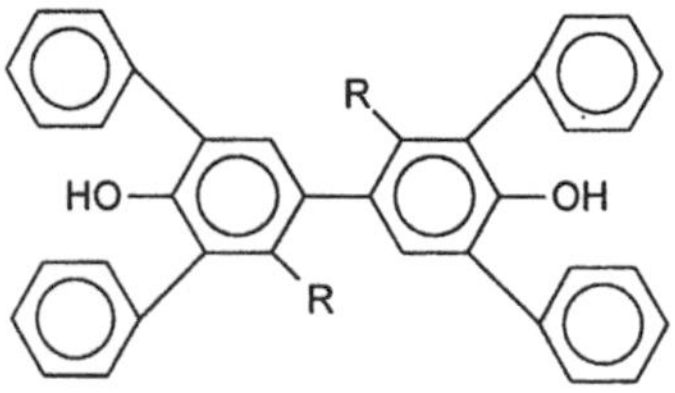

a : R' = H
b : R = Me
c : R = Ph

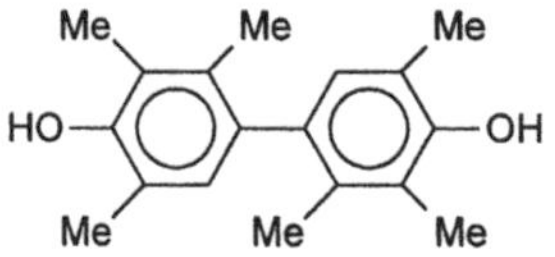

(77)

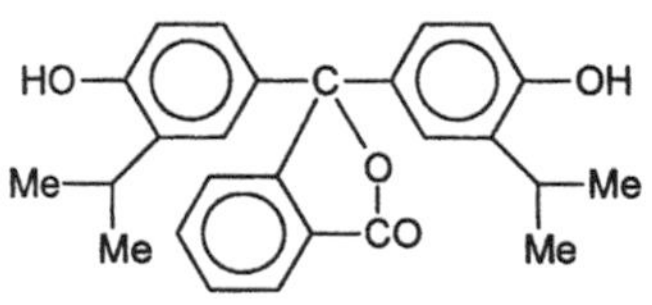

(78)

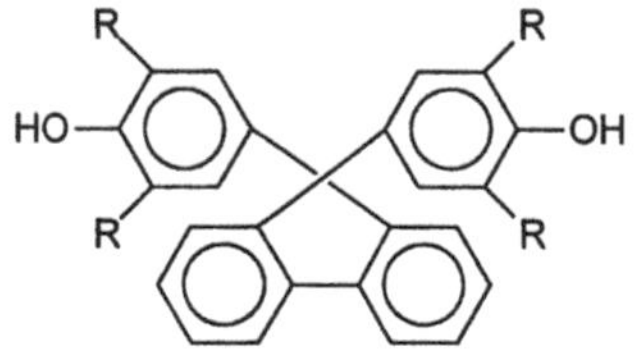

a : R = H,   b : R = Ph 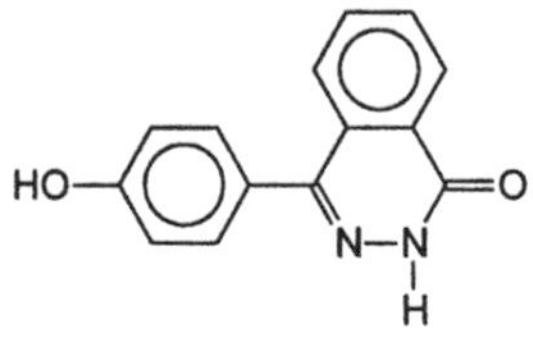   c

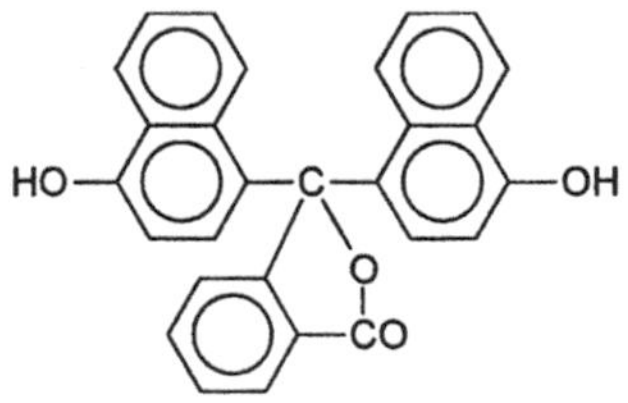

(79)

Most syntheses of PEKs showing new structural elements were based on new or noncommercial 'fluoromonomers'. A difluorodiketone (80a) with a kinked structure designed to reduce the melting temperatures of the PEKs derived from it was prepared from isophthaloyl chloride and fluorobenzone [123]. The extremely kinked group of monomers having structure (80b–d) required a more cumbersome synthesis. The resulting PEKs were amorphous and possessed high glass transition temperatures (T_gs) when derived from (80d) [124,125]. Two research groups [126–128] reported on alkyl substituted PEKs prepared from the lengthy 'fluoromonomers' (81a–c). This substitution pattern considerably improves the solubility and eliminates the crystallinity, but the T_gs remain rather low (around 150 °C). Further studies concerned 'fluoromonomers' derived from the biphenyl moieties (82a,b) [129], (83a) [130], (83b) [131]. Several publications reported on syntheses and polycondensations of new fluoromonomers derived from naphthalene. For instance, the 2,6-substituted naphthalene monomers (84a), were described in Refs. [132] and [133]. Syntheses and polycondensations of the 1,5-substituted naphthalenes (84b,c) were reported in Refs. [134–136]. The chemistry of the '1,8-naphthalene monomer' (85a) was described in Ref. [137]. From the tetrasubstituted monomer (85b) a kind of comb-like PEK was prepared [138]. PEKs derived from a monomer having pendant naphthyl groups were prepared from (86) and had high T_gs [139]. New 'fluoromonomers' derived from indane (87a,b) were synthesized from 4-methyl-α-methylstyrene [140,141]. The PEKs derived from them were as expected amorphous. Two research groups were interested in polyethers having alternating sequences of ketone, ether and sulfone groups [142,143]. For their syntheses mainly the monomers (88a or b) were used. Another research group [144] reported on syntheses of phosphorous containing PEKs from monomer (89a). In this work and in publications discussed below the fluorinated bisphenol-A (89b) was used as one of the comonomers. Several authors had interest in PEKs containing heterocycles in the backbone. Thiophene containing PEKs were obtained from monomers (90a or b) [145], [146], and benzofurane based PEKs or PESs were prepared from monomers (91a or b) [147]. PEKs having pyridine or isoquinoline rings in their backbones were obtained by polycondensations of the monomers (92a) [133] or (92b) [132] and (92c) [148].

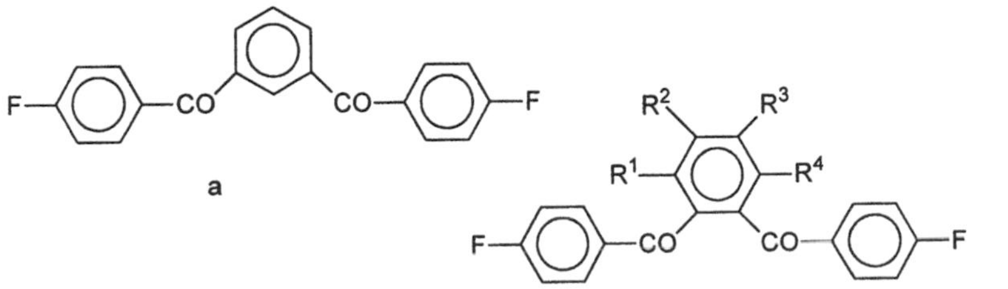

(80)

b : R^1, R^2, R^2, R^4 = H
c : R^1, R^4 = H, R^2, R^3 = Ph
d : R^1, R^2, R^3, R^4 = Ph

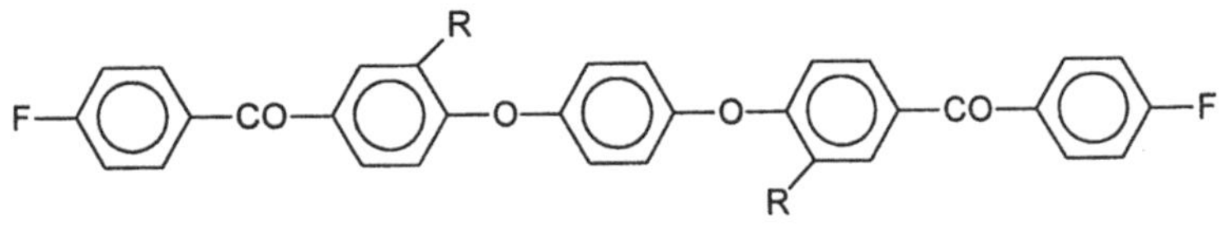

(81)

a : R = Me,   b : R = Et,   c : R = CHMe$_2$

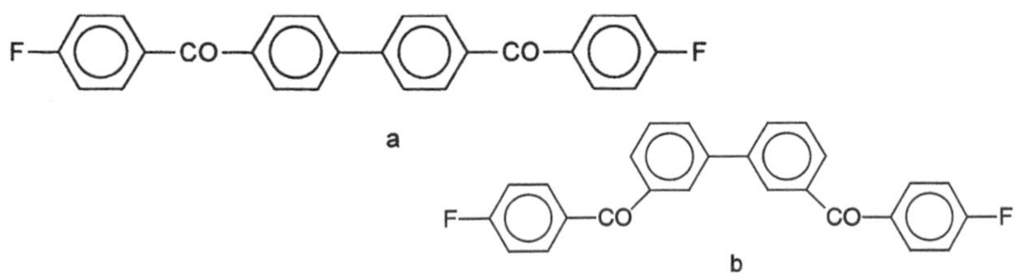

(82)

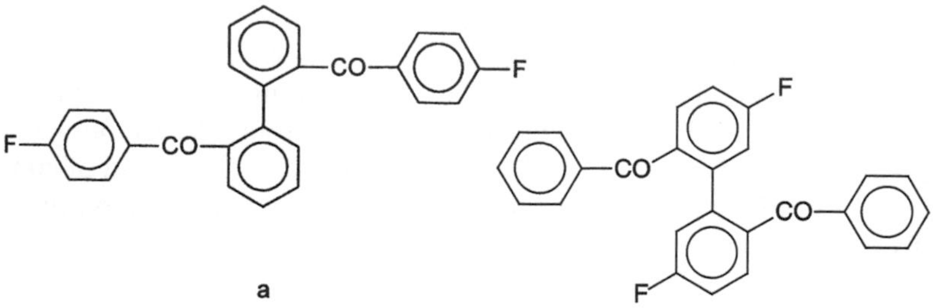

(83)

a   b

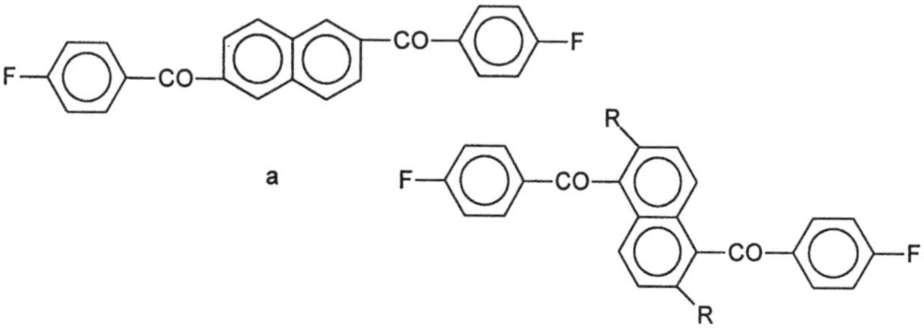

(84)

a   b : H   c : CH$_3$

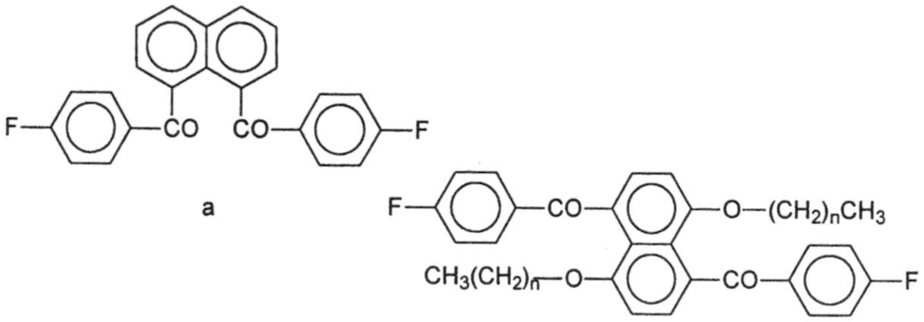

(85)

a

b

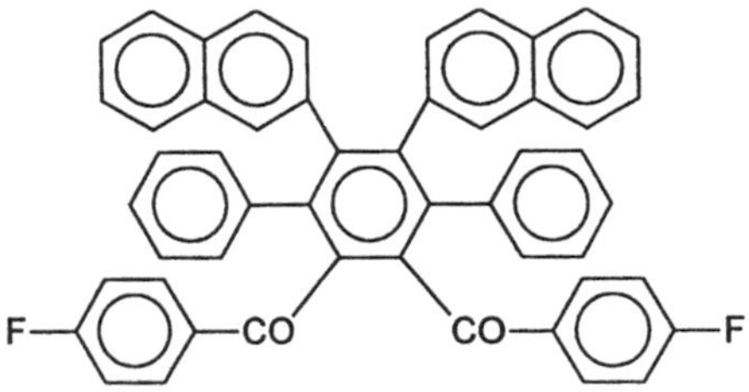

(86)

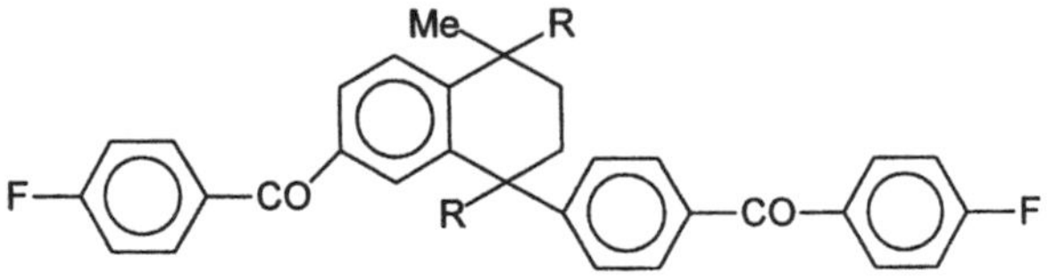

(87)

a : R = Me,   b : R =

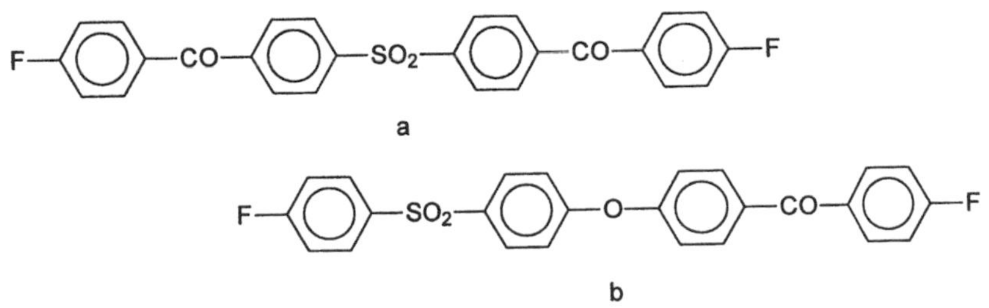

F—⬡—CO—⬡—SO₂—⬡—CO—⬡—F

a

(88)

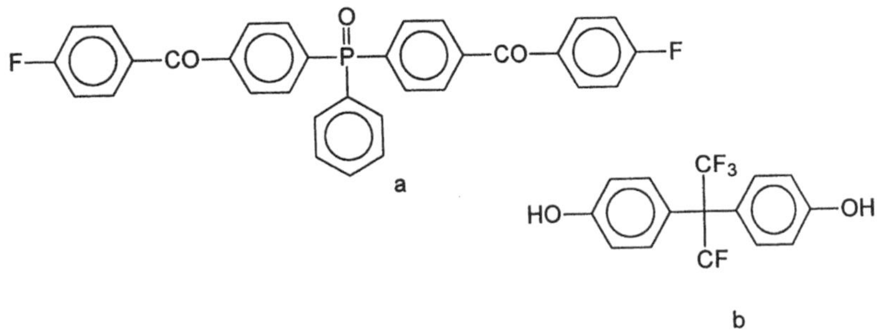

F—⬡—SO₂—⬡—O—⬡—CO—⬡—F

b

F—⬡—CO—⬡—P(=O)—⬡—CO—⬡—F

a

(89)

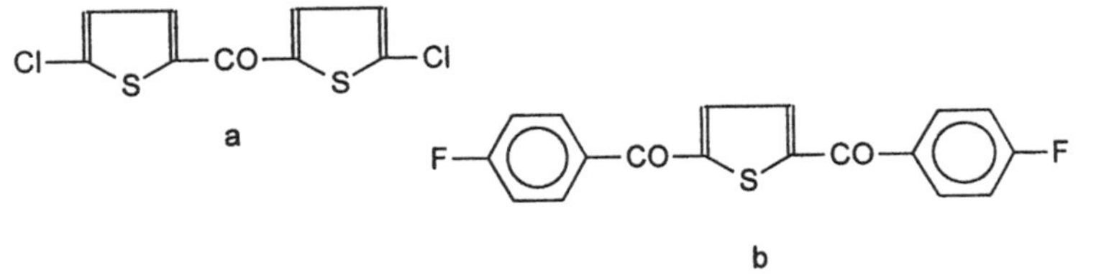

HO—⬡—C(CF₃)(CF)—⬡—OH

b

Cl—[S]—CO—[S]—Cl

a

(90)

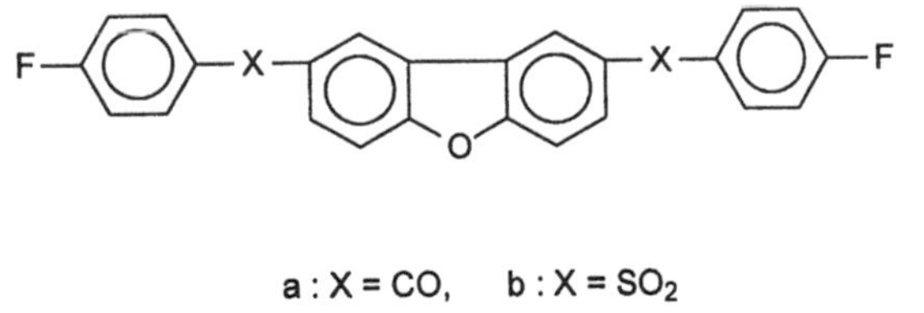

F—⬡—CO—[S]—CO—⬡—F

b

F—⬡—X—[dibenzofuran]—X—⬡—F

(91)

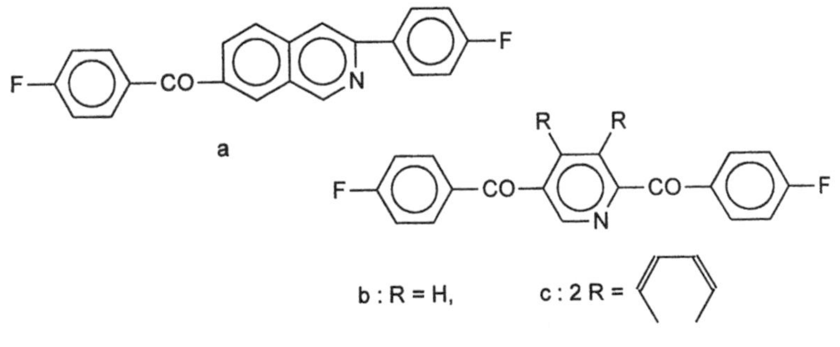

a : X = CO,   b : X = SO₂

F—⬡—CO—[isoquinoline]—⬡—F

a

(92)

F—⬡—CO—[pyridine with R,R]—CO—⬡—F

b : R = H,   c : 2 R = ⬡

Selectively fluorinated PEKs were prepared by polycondensations of decafluoro-benzophenone (93a) with the sodium salts of several diphenols including (89b) [149]. No side reactions (e.g. branching) were found by ^{13}C or ^{19}F NMR spectroscopy. PEKs containing pendant CF_3 groups resulted from polycondensations of the monomers (93b) [150], (94) [151], (95a) [152] and (95b) [153]. In the latter case the less common nitro-displacement reaction was successfully applied using K_2CO_3 in DMSO/toluene according to the 'standard procedure'. Amorphous PEKs with high T_g were, as expected, isolated from polycondensations of the monomers (95c,d) [154].

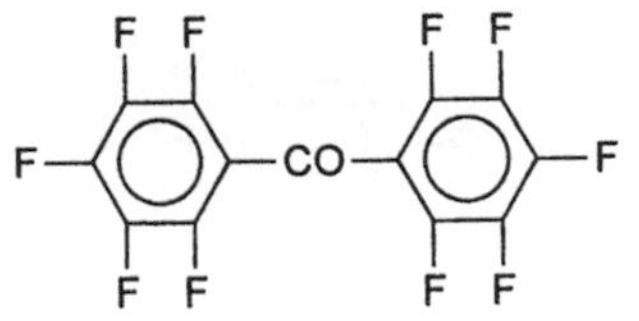

a

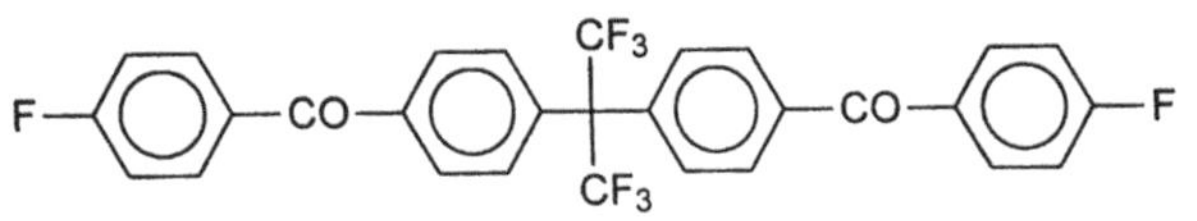

(93)

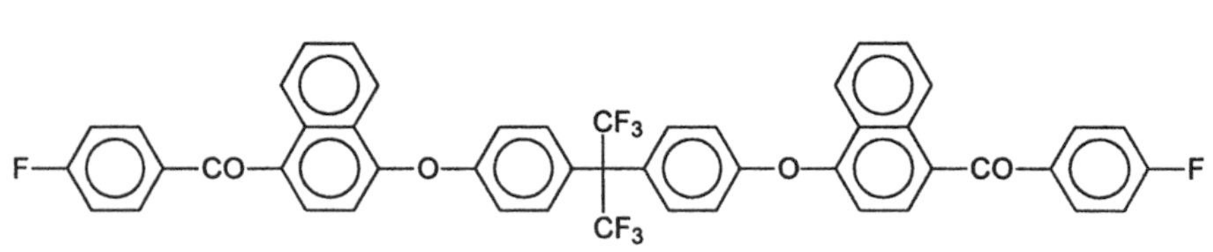

b

(94)

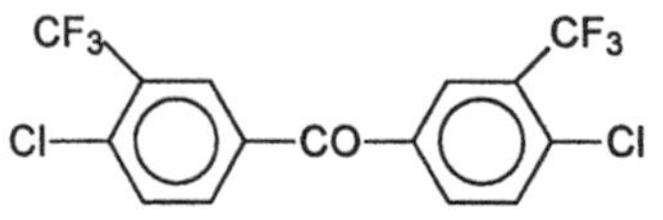

a

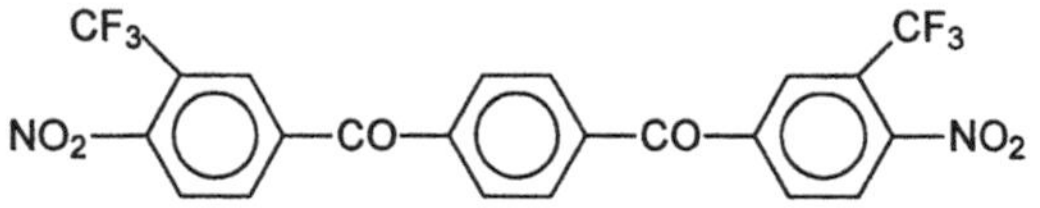

b

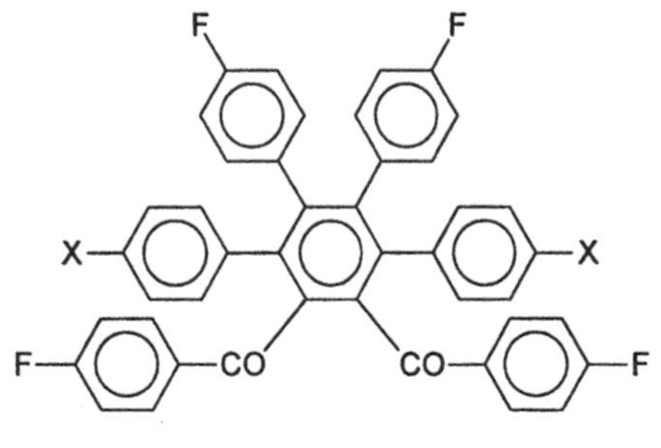

(95)

c : X = H,   d : X = F

1.   Liquid-crystalline PEKs?

One research group reported in three papers [155–157] on synthesis and characterization of seemingly liquid-crystalline PEK. All these PEKs are random copolymers of 4,4′-biphenyl diol and substituted hydroquinones (96) polycondensed with DFBP or 1,4′-bis(4-fluorobenzoyl)benzene. However, any high-temperature micrograph of a typical LC-phase is lacking, any Figure of DSC-measurements and any high-temperature x-ray measurements are missing. Furthermore, the authors claim a smectic-A phase for a random copolymer which is a contradiction. In summary, the liquid-crystalline character of these PEKs is doubtful.

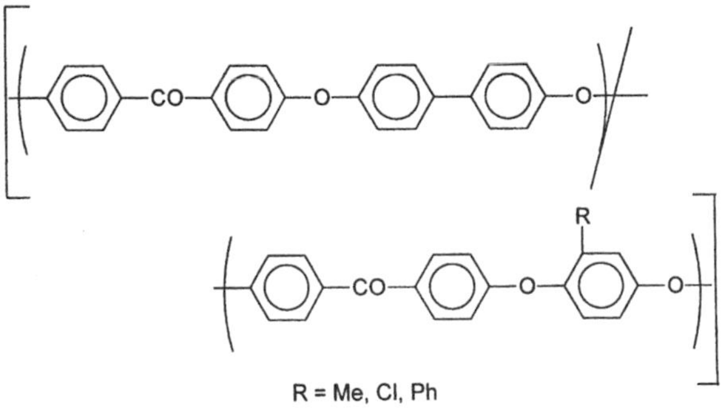

(96)

R = Me, Cl, Ph

2. Telechelic oligomers (tOEKs), block-copolymers and networks

A broad variety of telechelic oligo(ether ketone)s, tOECs, was prepared over the past ten years and used for chain extension via amide, imide or ether groups and for syntheses of block copolymers or thermostable networks [158–179]. Most research groups have concentrated their interest on tOEKs having amino-endogroups [158–166]. One group [158–159] synthesized a monodisperse tOEK (97a) which was polycondensed with the dicarboxylic acid dichloride (97b). The thermal and photochemical *cis–trans* isomerization of the azogroups in the resulting polyamides was investigated. The influence of randomly incorporated 2,2'-bisnaphthyl 'kinking units' was also studied. Most amine-terminated tOEKs were prepared in a 'one-pot procedure' from mixtures so DFBP a diphenol and meta- or para-aminophenol. The tOEK prepared from monomer mixture (98) were chain extended with benzophenone-tetracarboxylic anhydride, and the thermal crosslinking was studied and ascribed to the formation of ketomine groups (99) [160]. Two other research groups [161–164] found that the 'one-pot procedure' does not give satisfactory results due to side reactions of the amino groups and developed alternative strategies. Either, a N-protected 3-aminophenol was used (100) [161,162] or a fluoroterminated tOEK was isolated after the polycondensation (101) treated with the K-salt of 3-aminophenol in a separate step [163,164]. Detailed NMR spectroscopic analyses were reported. The tOEKs isolated from reaction (100) were chain extended with pyromellitic anhydride, whereas the amine terminated tOEKs derived from (101) were used as reinforcing component in epoxy resins [162]. In two publications [165,166], tOEKs endcapped by maleimido groups were described. Two synthetic strategies were applied. Firstly, amine-terminated tOEKs were prepared from monomer mixtures such as (98) or from analogous monomer mixtures containing (102a) instead of DFBP [165]. The amino endgroups of the isolated tOEKs were then reacted with maleic anhydride. Secondly, the polycondensation process was performed in the presence of the maleimido phenol (102b) [165]. Finally, the thermal cure and the physical properties of the resulting networks were studied.

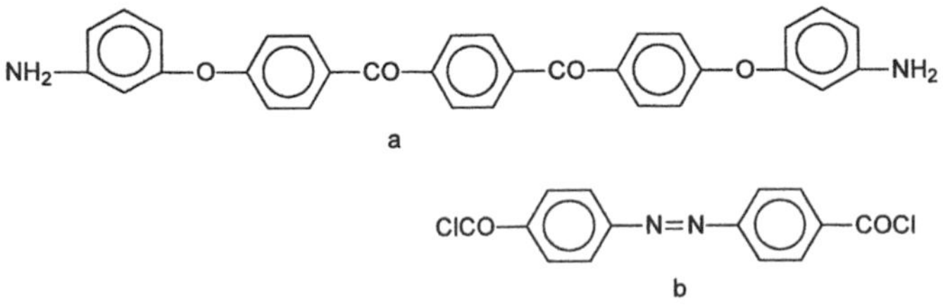

a

(97)

b

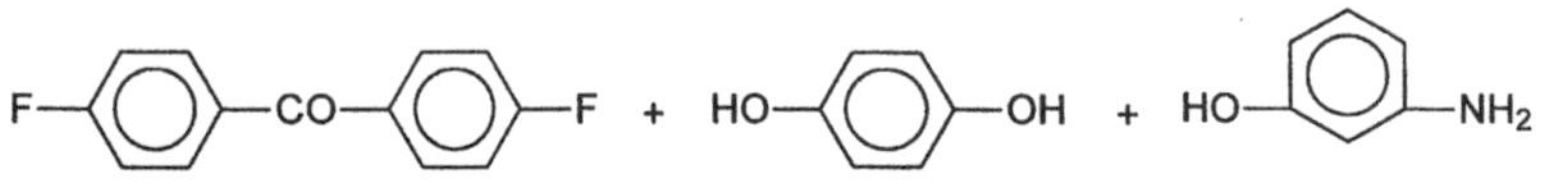

(98)

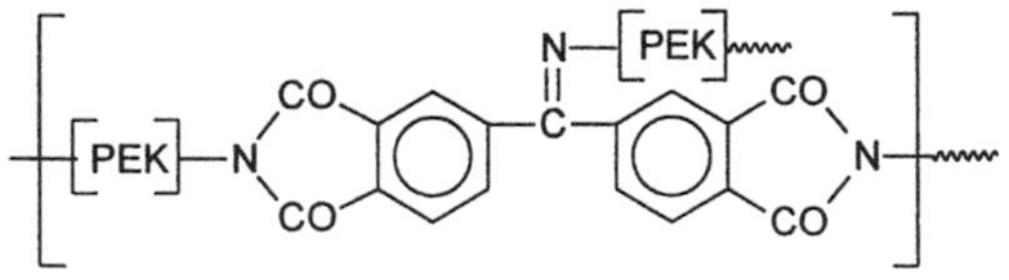

(99)

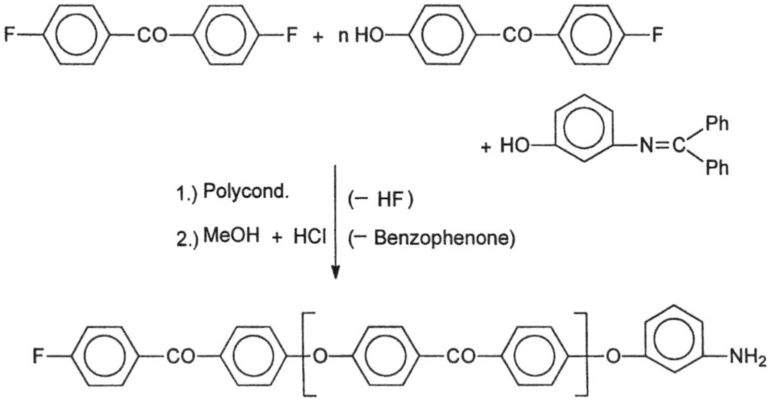

(100)

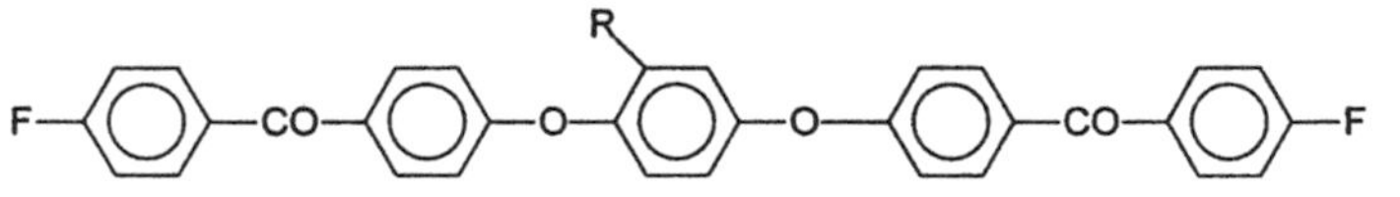

(101)

R = H, —Me, —CMe₃

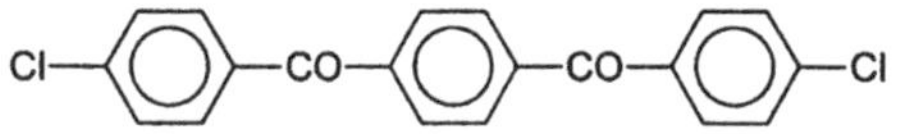

a

(102)

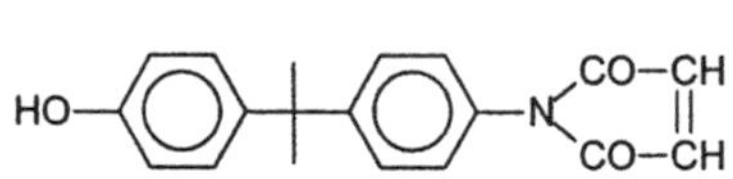

b

Thermostable networks were again the research interest of several groups which prepared tOEKs having acetylenic endgroups [166–171]. These endgroups were either introduced in the form of nucleophilic reagents (103a–c), [166–168] or they were incorporated as electrophilic reagent (104a,b) [170,171]. The acid chloride (104b) was used to esterify the OH-terminated tOEKs. In this connection tOEKs containing styrene endgroups (105) should be mentioned which are suited for radical crosslinking [172].

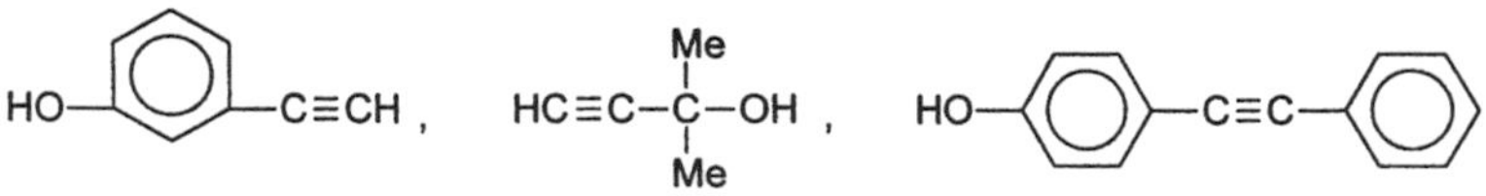

(103)

a   b   c

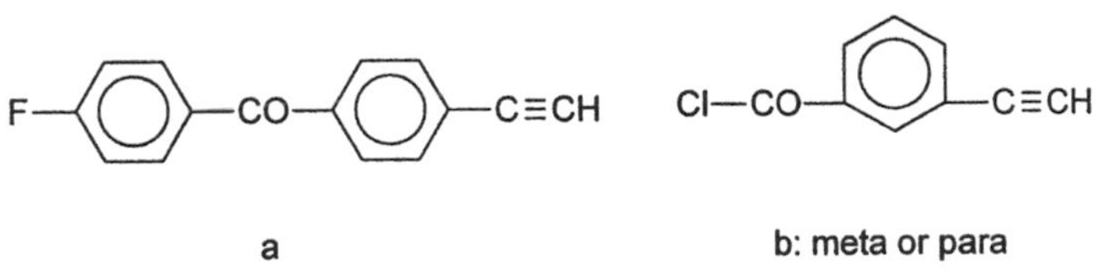
F—CO—C≡CH
Cl—CO—C≡CH
a
b: meta or para

(104)

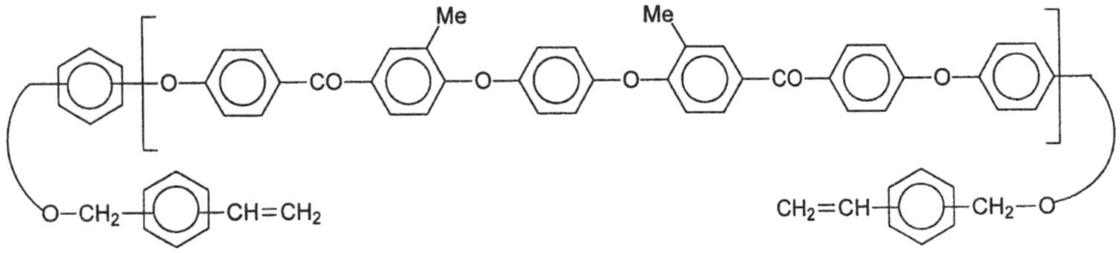
Me
Me
O
CO
O
O
CO
O
O—CH2—CH=CH2
CH2=CH—CH2—O

(105)

Several research groups elaborated strategies for syntheses of tOEKs having two reactive fluoro endgroups. The 'electrophilic substitution strategy' was realized either with monomer mixtures containing 4-fluorobenzoyl chloride as endcapping agent (106) or with fluorobenzene (107) [173–175]. The resulting tOEKs were chain extended with various short or long (oligomeric) diphenols. Furthermore, blockcopolymers containing PES segments (108) were prepared by copolycondensation with DFDPS and 4,4′-dihydroxydiphenylsulfone. tOEKs having C–F endgroups were prepared via nucleophilic substitution by copolycondensation of 4-fluoro-4′-trimethylsiloxy benzophenone and small amounts of DFDPS [176]. The molecular weights were controlled by the feed ratio of DFDPS. F-terminated tOEKs were also synthesized via nucleophilic substitution from hydroquinone and an excess of DFBP (109). They were used in turn for syntheses of PES block copolymers (110) [177,178]. Blockcopolymers containing dimethyl siloxane segments were realized by heating OH-terminated tOEKs (111a) with amine terminated oligosiloxanes (111b) in a two phasic solvent system [179]. Finally the multistep synthesis of A-B-A-triblock copolymers having a central OEK block should be mentioned. Copolycondensation of the monomer mixture (112) yielded tOEK having two Me_3SiO endgroups which were acetylated with a large excess of acetylchloride. The resulting tOEK bisacetates (113a) were polycondensed with silylated 3,5-bisacetoxy benzoic acid (113b) to yield hyperbranched polyester A-blocks (114) [180].

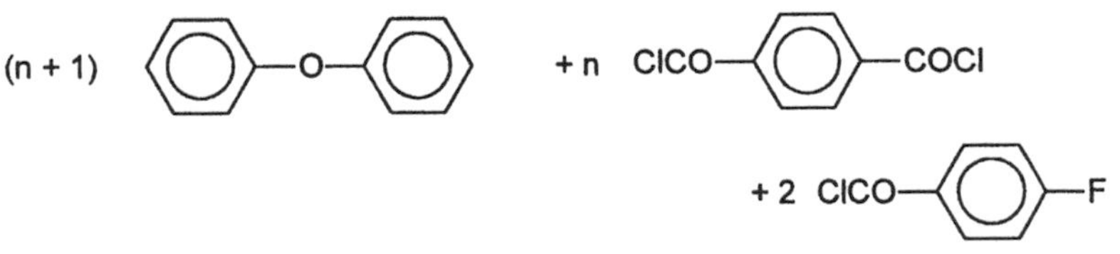
(n + 1)
O
+ n ClCO—COCl
+ 2 ClCO—F

(106)

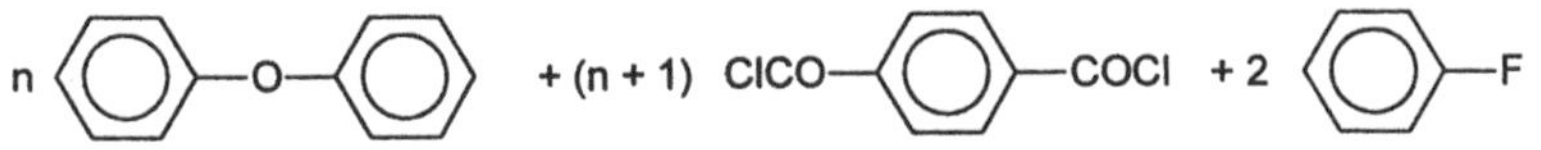
n
O
+ (n + 1) ClCO—COCl + 2
F

(107)

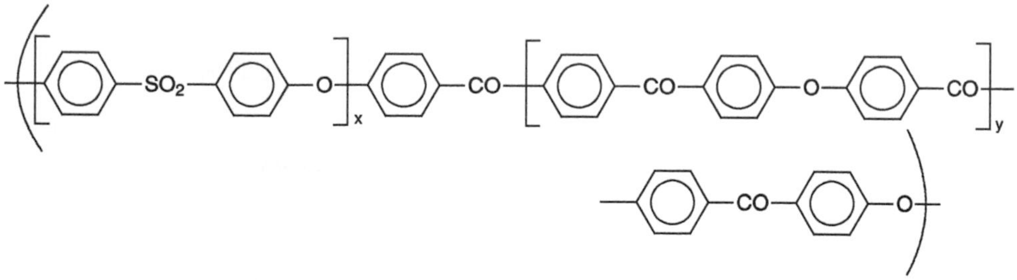

(108)

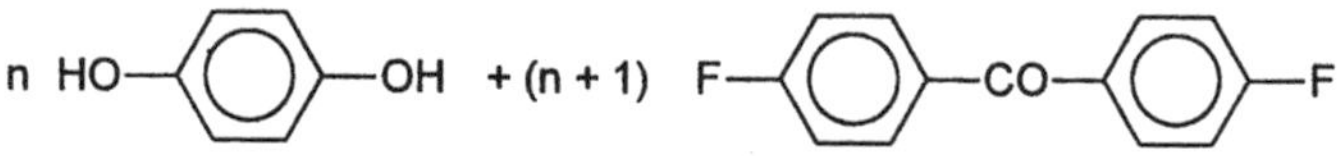

(109)

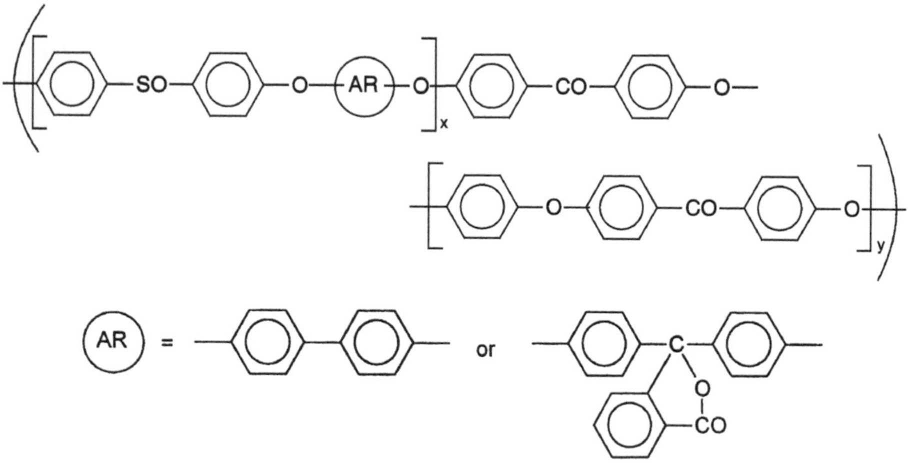

(110)

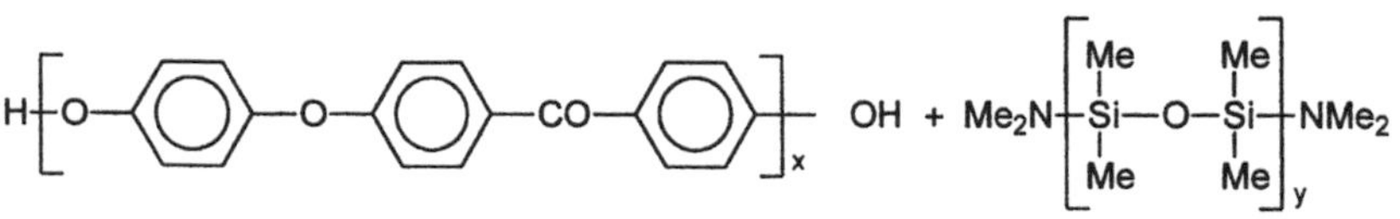

(111)

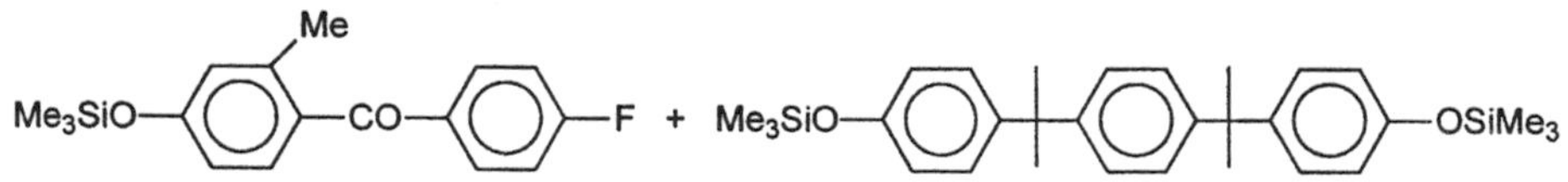

(112)

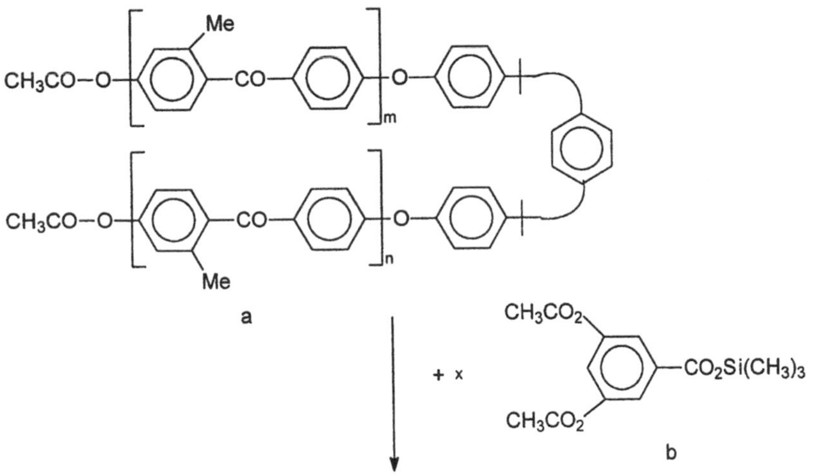

(113)

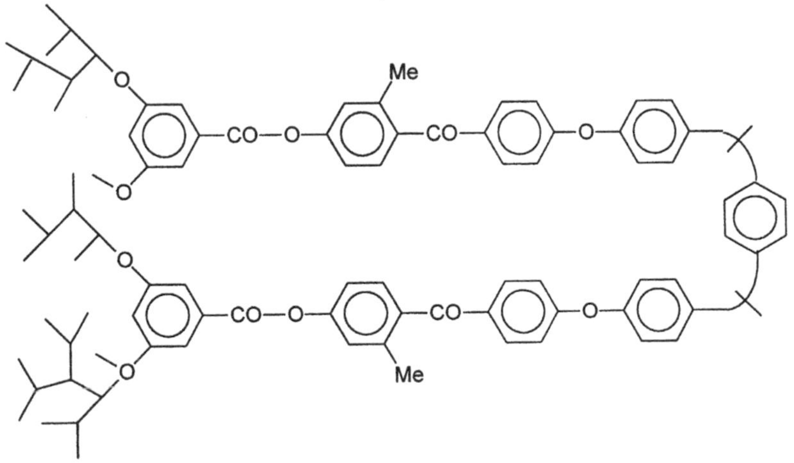

(114)

3. Hyperbranched PEKs

For the sake of completeness six publications dealing with synthesis and characterization of hyperbranched PEKs should be mentioned [181–186]. However a detailed discussion of hyperbranched polymers will be presented in a separate chapter of this handbook.

D. Unusual Synthetic Methods

The discussion of unusual methods reported for syntheses of PEKs is subdivided into two strategies: (1) polycondensation methods, (2) ring-opening polymerization (ROP) of cyclic oligo(ether ketone)s, cOEKs.

1. Polycondensation Methods

The synthetic methods discussed in this section have in common that they deviate significantly from both standard methods. A new polycondensation process based on nucleophilic substitution steps is schematically outlined in (115). In contrast to the standard procedure no diphenols are required only difluoro- or the less expensive dichloroketones are needed (116a–c). Sodium or potassium carbonate serve as source of oxide ions yielding the ether bonds under the catalytic influence of silica which is doped with traces of CuCl, CuCl$_2$ or CuO [187,188]. High molecular weights require high temperatures (up to 320 °C) in DPS as reaction medium. The molecular weights decrease with increasing amounts of silica. Silylated phenolate groups formed on the surface of the catalyst were claimed as reactive intermediates. Another new method involving the formation of biphenyl units by C–C coupling is also based on dichloroketones as starting materials (117) [189,190]. Nickel(II) chloride complexed by triphenylphosphine and bipyridine catalyzes the coupling step at moderate temperatures (80 °C). Low to moderate molecular weights were obtained. A problem is the low solubility of the resulting PEKs at these low reaction temperatures.

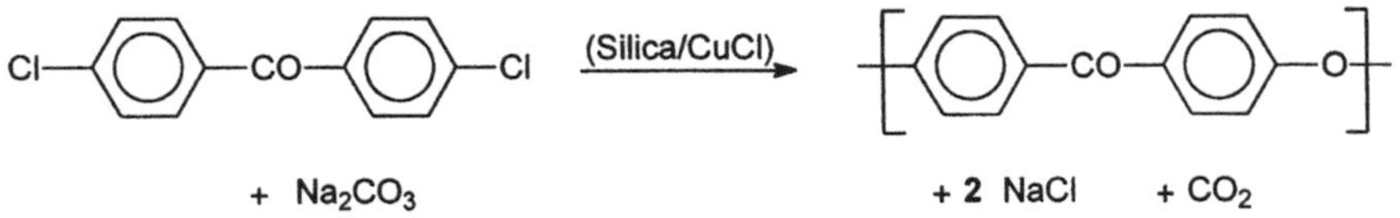

(115)

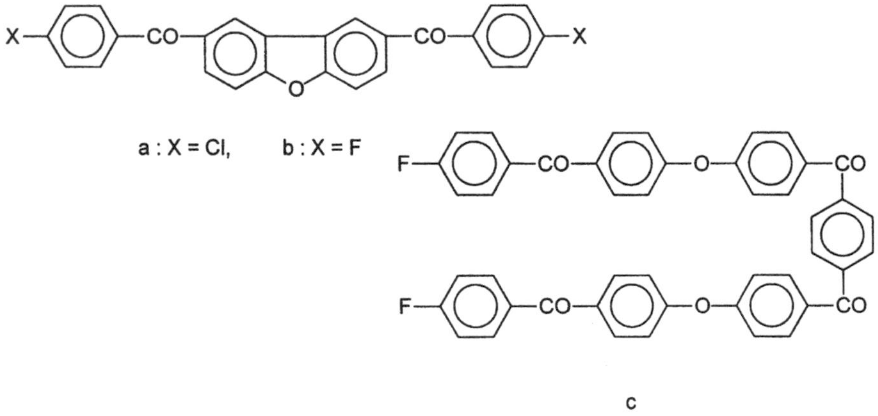

a : X = Cl,   b : X = F

(116)

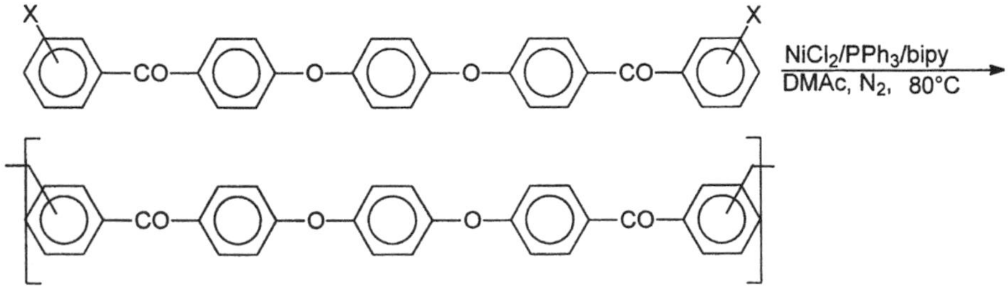

(117)

Finally two new polycondensation methods should be reported, which yield aromatic polyketones free of ether groups. In the first case the formation of the C–C bonds proceeds via a nucleophilic substitution involving the bisanions of bis-(α-aminonitrile)s (118,119) [191–194]. This approach allows a broad variation of the electrophilic monomer, but apparently the expensive fluoroaromats are needed in contrast to the first two methods. The second approach is based on the acylation of stannylated benzene (120). This coupling method requires palladium catalysts and may give M_ns above 20,000 Da, when the polymers are soluble in the reaction medium [195] as it is true for the tert.butylsubstituted PEK of (120).

$$(118)$$

$$(119)$$

$$(120)$$

2. Ring-opening Polymerization

Numerous publications reported on syntheses of cyclic oligo(ether ketone)s, cOEKs, designed to serve as monomers for ring-opening polymerizations. One class of cOEKs (121b) was prepared by reductive condensation of a bischloroketone (121a) with Zn using a Ni⁰-complex as catalyst [196]. However, most syntheses of cOEKs are based on (poly) condensations involving the standard nucleophilic substitution process discussed above with the difference that these ring syntheses were conducted under high dilution (pseudo high dilution method). In one paper [197] a new version of this synthetic approach was studied, using a difluoroketimine as electrophilic monomer (122). The iminogroup is easily hydrolyzed by acid catalysis (123). As discussed in the section 'Modifications' below fluoroketimines are useful monomers for the preparation of crystalline, insoluble PEKs, via an amorphous, soluble precursor polymer, but no advantages are in sight for synthesis of cOEKs.

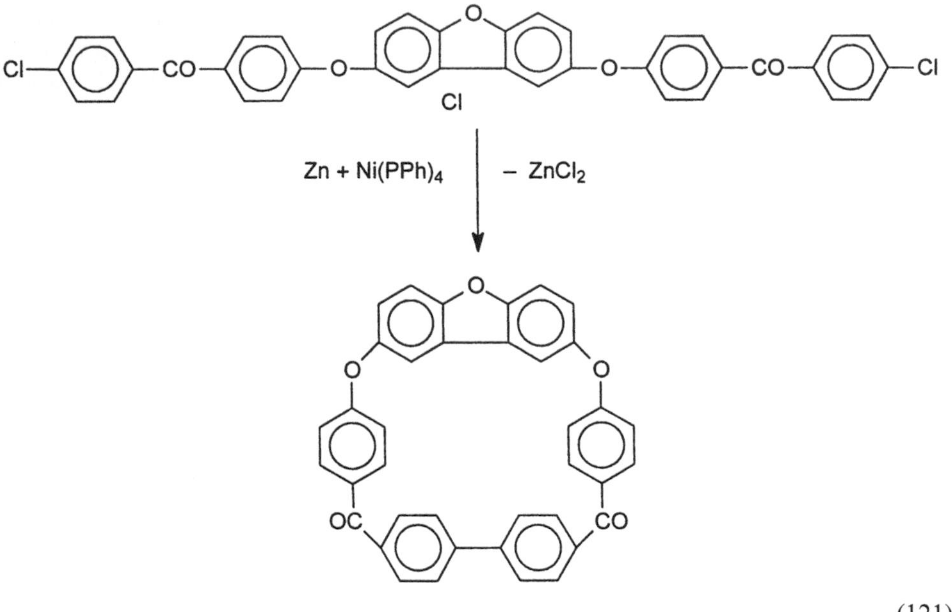

Zn + Ni(PPh)$_4$ − ZnCl$_2$

(121)

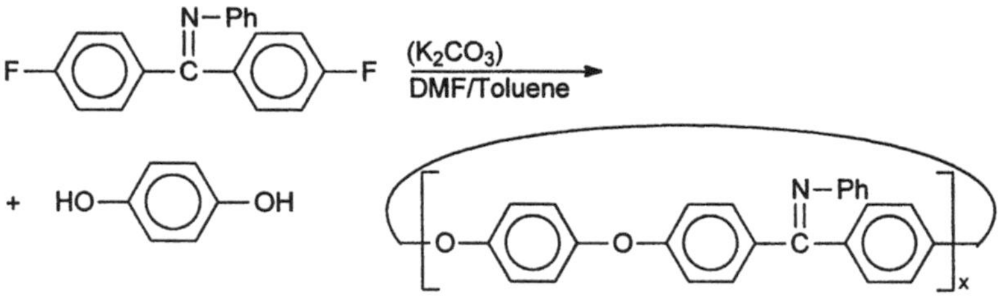

(K$_2$CO$_3$)
DMF/Toluene

(122)

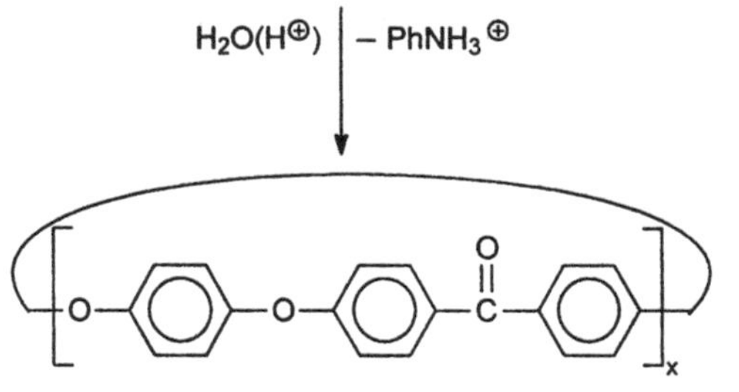

H$_2$O(H$^\oplus$)   − PhNH$_3$ $^\oplus$

(123)

The majority of papers report on syntheses and properties of mixtures of cOEKs (typically DP = 2–6) [197–207], whereas five research groups [208–213] describe isolation of individual macrocycles. From the numerous mixtures of cOEKs the following examples deserve a short comment. When diphenols containing cyclopropane rings (124a,b) were cyclized with DFBP and other fluoroketones, cOEKs (and the corresponding PEKs) were obtained allowing for thermal crosslinking [199]. cOEKs (and PEKs) suitable for thermal cure were again obtained, where a diphenol with a central acetylene group (124a) was used as nucleophilic monomer [203]. cOEKs synthesized from 1,2-bis(4-fluorobenzoyl)benzene (125) were transformed into the corresponding cyclic phthalazines (126a) [200]. However studies of ROP have not been reported yet.

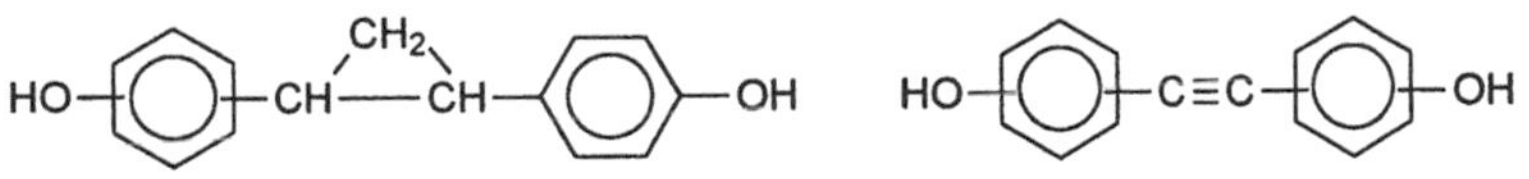

(124)

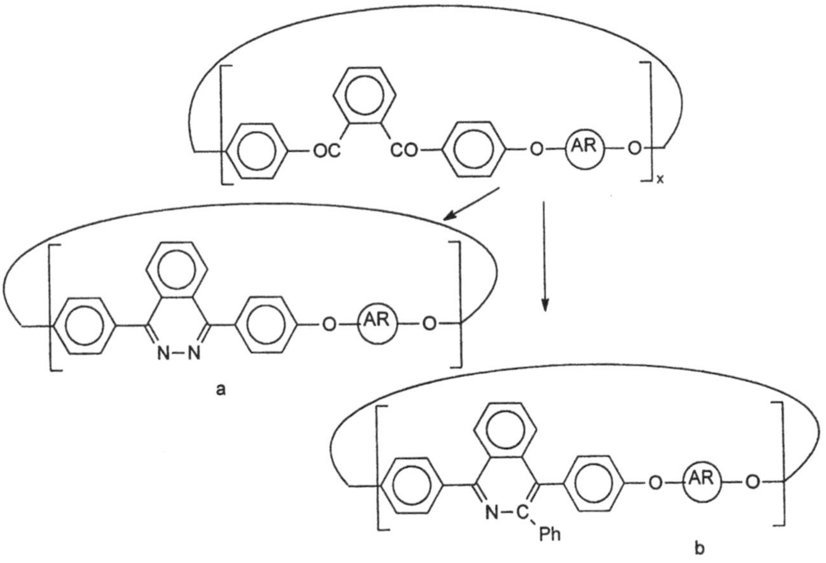

(125)

(126)

The few monodisperse cOEKs reported so far have the structures (127a) [208], (127b) [209], (128a,b) [210], (128c) [211], (129) [212] and (130) [213]. In most publications only simple Ro-polymerizations were described based on the heating of a cOEK powder with CsF (≤ 1 wt%) to temperatures between 260 and 395 °C [197,201,207,210,211]. IR-spectra and DSC measurements were the only characterization of the resulting PEKs, inasmuch, as most PEKs were insoluble due to partial crosslinking. Detailed studies of the RO polymerizations were also reported [206,208,210]. A broad variety of catalysts were tested. For instance, a mixture of cOEKs having structure (125), which was polymerized with Na-, K-, and Cs-phenoxide at 340 °C and similar weights were found. Furthermore, the K salts of various phenols were compared or their concentration was varied. In most polymerizations M_ns in the range of 26,000–34,000 Da (based on PS calibrated GPC) were achieved. Partial crosslinking and a rather high percentage of unreacted cycles due to the thermodynamic equilibrium are characteristic features of ROP at temperatures around or above 340 °C. Using the K and Cs salts of 4,4′-biphenyldiol the cOEKs of structure (125) were also polymerized in refluxing DMF. In this way crosslinking was avoided and lower M_ns (15,000 Da) with high polydispersities were found [206]. When the cyclic dimer (127a) was polymerized in the melt at 260–275 °C crosslinking was again avoided and M_ns in the range of 14,000–50,000 Da were obtained with catalyst such as CsF, K_2CO_3 or Cs_2CO_3 [208]. Lower M_ns resulted from initiation with K- or Cs-phenoxides.

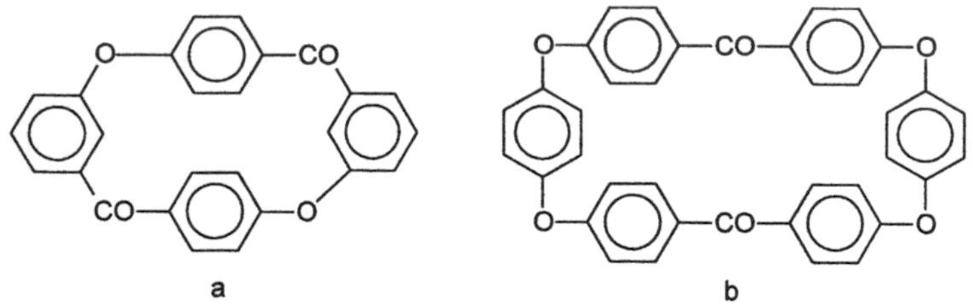

(127)

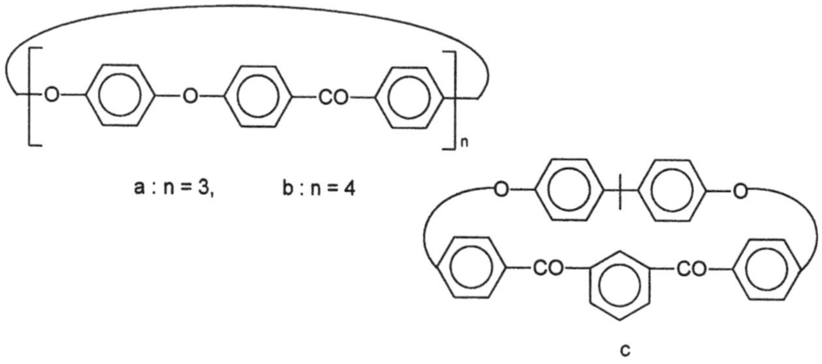

a : n = 3,   b : n = 4

(128)

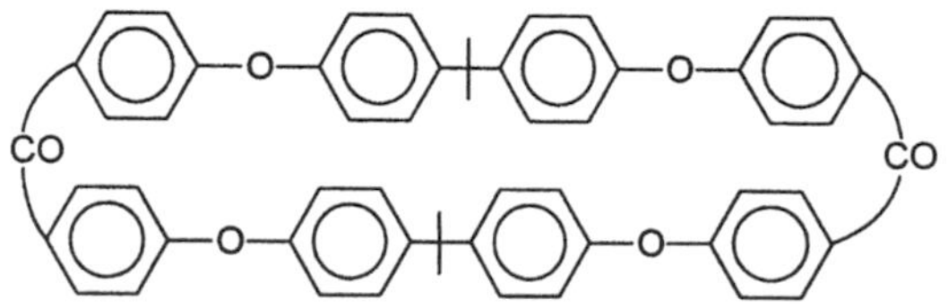

c

(129)

X = σ, O, C(CF$_3$)$_2$

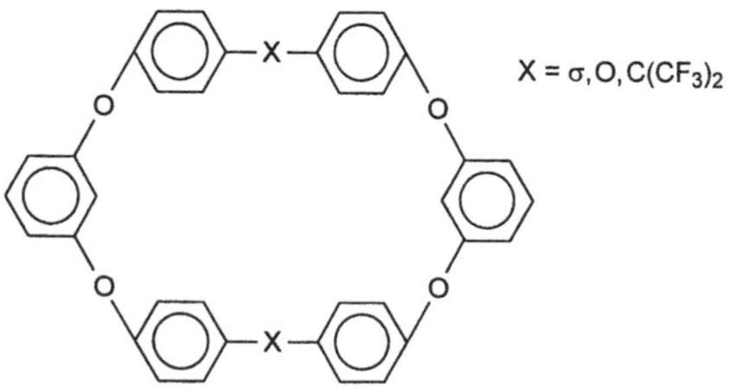

(130)

A special case represents the cOEK mixtures of structure (131). In this case the C–S bond proved to be sensitive to radical cleavage, so that these cOEKs were polymerized by heating with elemental sulfur [205] via a radical polymerization mechanism. In summary, the ROP of cOEKs is certainly a new and highly interesting approach, but it has not demonstrated yet to be useful for a technical production of PEKs.

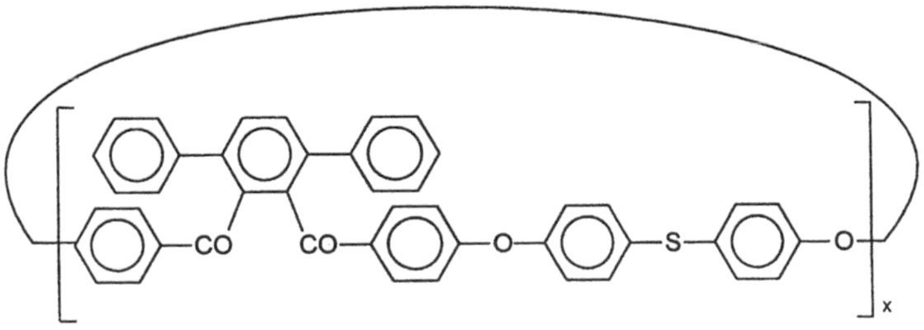

(131)

Finally, it should be mentioned that the cOEKs (130) were also used as nucleophilic monomers in a Friedel–Crafts type polycondensation (132) [213].

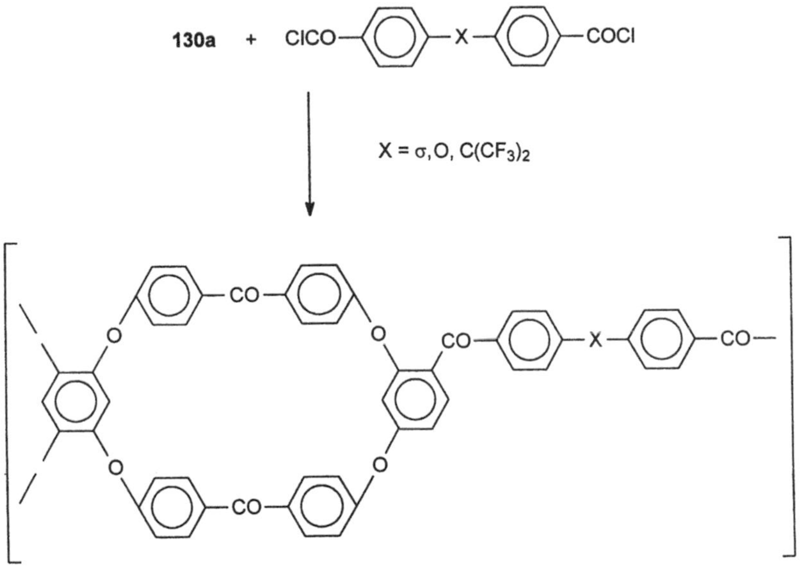

$X = \sigma, O, C(CF_3)_2$

(132)

E. Functionalized poly(ether ketone)s

In this section syntheses and modifications of PEKs having functional substituents should be summarized. At first, a synthetic strategy called 'precursor method' should be mentioned. The insolubility and high melting temperatures of crystalline PEKs, may have the disadvantage of an early precipitation from the reaction mixture and of a difficult processing. Soluble low melting precursor polymers which can easily be transformed into the final crystalline PEKs are an interesting group of polymers. Two precursor routes were recently explored. Firstly, synthesis of PEKs having pendant *t*-butyl groups [194,214,215] which can be eliminated as isobutylene, and secondly, the 'ketimine approach' [216–219]. Tert-butyl groups eliminate isobutene upon heating above 350 °C or more efficiently in strong acids (133) [214]. In the case of PEKs derived from tert-butyl isophthalic acid [195,215] the elimination of isobutylene has not been studied yet.

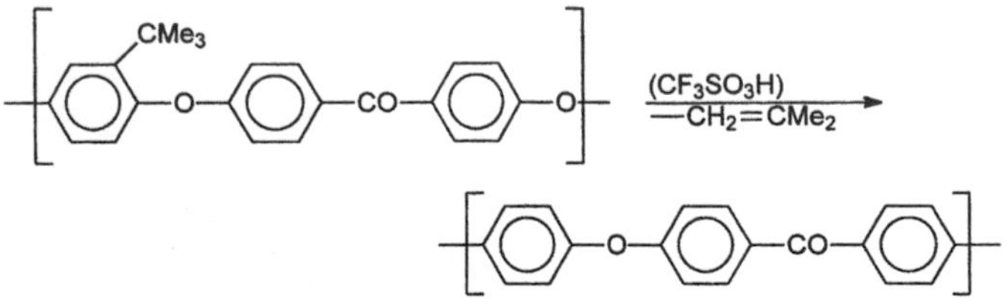

(133)

The 'ketimine approach' follows the scheme outlined for syntheses of cOEKs in (122) and (123) [216–219]. The imine derivatives of PEKs are usually amorphous and soluble in numerous organic solvents. The imino group is easily hydrolyzed in acidic water or methanol. However, it is obvious that both precursor methods are too expensive for a technical production.

PEKs containing pendant sulfonate groups are of interest as membrane materials like the sulfonated PESs [29–31]. The sulfonated PEKs described so far were prepared by the standard method of nucleophilic substitution. The sulfonated DFBP (134) was used as monomer mainly in combination with unsubstituted DFBP [220,221]. PEKs having pendant amino groups were prepared in two ways. One research group used an amine functionalized difluoro monomer (135a) for syntheses of PEKs and the amino groups were later acetylated (135b) [222]. Two other research groups have concentrated their efforts on the surface modification of PEK films including model reactions of low molar mass compounds [223,224]. The first approach [225] is based on the reduction of the keto group with NaBH$_4$ in DMSO yielding benzhydrol units in the polymer backbone (135a). The benzhydrol groups were then transformed into free amino groups (135b), amino groups of γ-aminopropan sulfonic acid (136c) or glutamine groups (136d). In the second work benzhydrol units were again produced and modified, but in addition to this approach a variety of direct modification reactions was studied, such as the formation of oximes (137a), hydrazones (137b), methylene groups (138a) or siloxy nitriles (138b) [226].

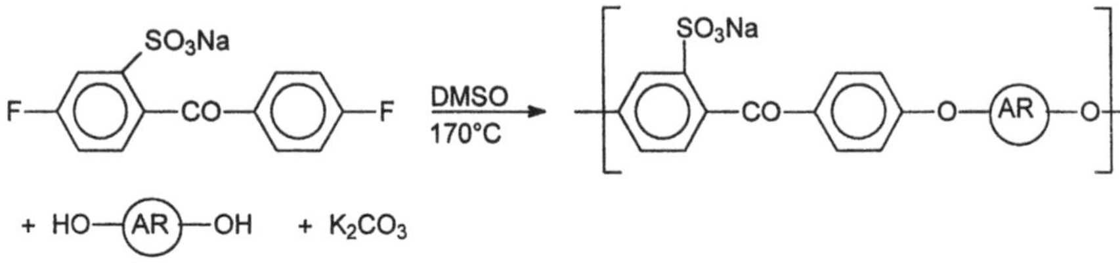

(134)

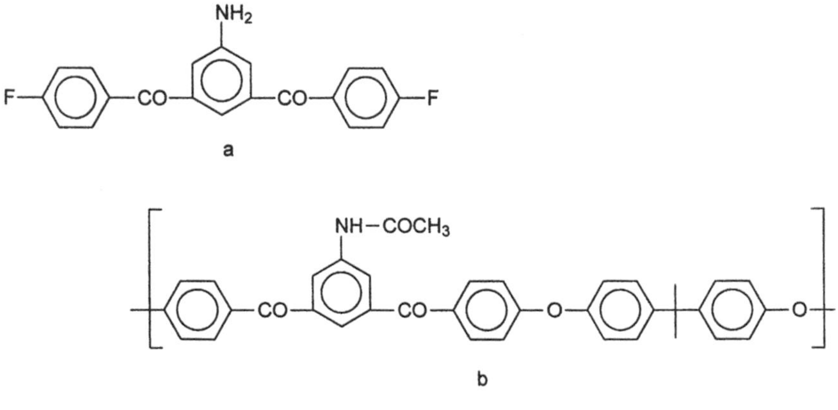

(135)

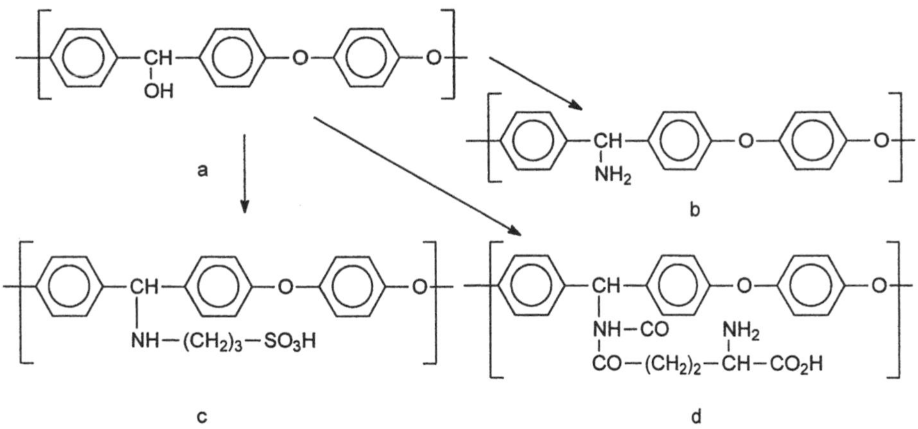

(136)

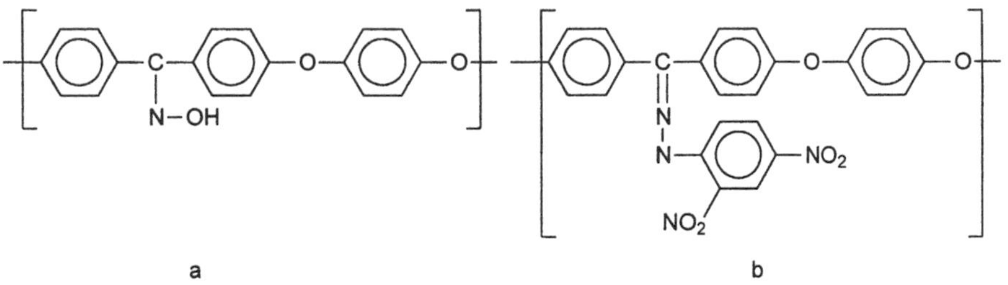

(137)

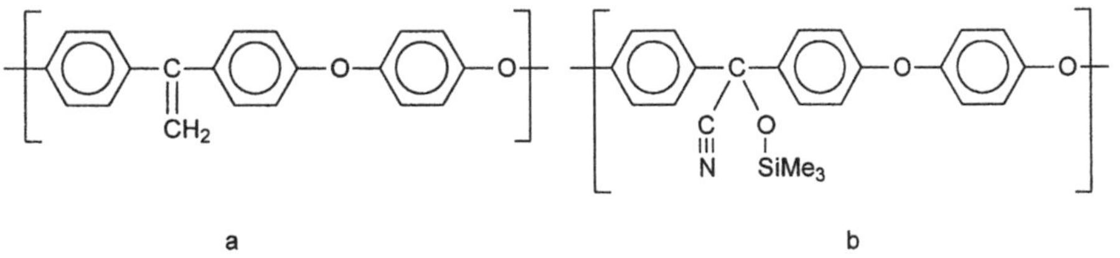

(138)

Three publications [225–227] deal with the radical bromination (using Br_2) of methyl substituted PEKs. In the methyl hydroquinone or 3,3′-dimethyl biphenyl diol both mono- and dibromination of the CH_3 groups was feasible (139a,b). However, due to steric hindrance only a monobromination of the neighboring methyl groups was achieved when 2,3-dimethyl or trimethylhydroquinone units were present (140a,b). The PEKs derived from brominated methylhydroquinone were subjected to various modifications. For instance alcohol, ether and ester groups of structure (141a) or amines of structure (141b) were prepared from the monobrominated precursor. The dibrominated PEK was transformed into aldehydes (142a) or carboxylic acid esters (142b). Another research group produced PEKs with pendant amide groups by nitrodisplacement polycondensation of monomer (143a) [228]. Polyethers having both pendant carboxylic acid and amide groups were obtained by polycondensation of the bisphenol (143b) [229].

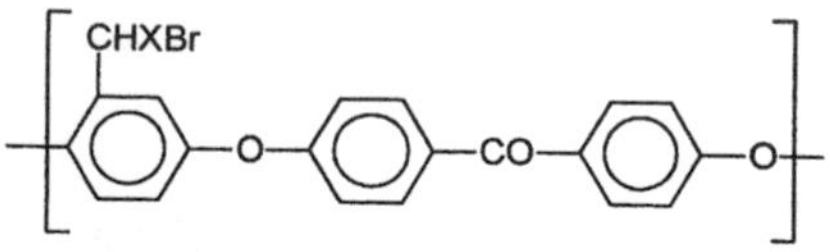

a : X = H or Br

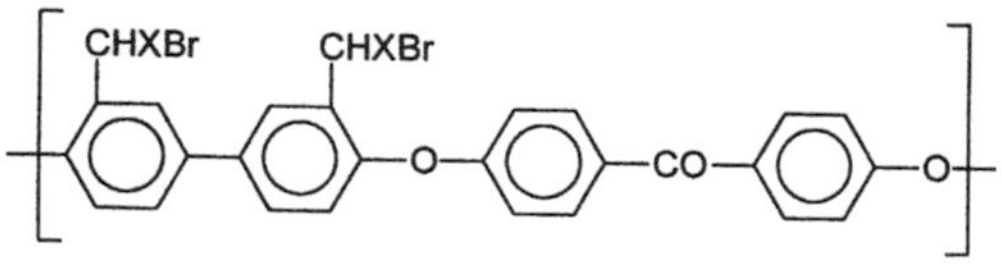

(139)

b : X = H, Br

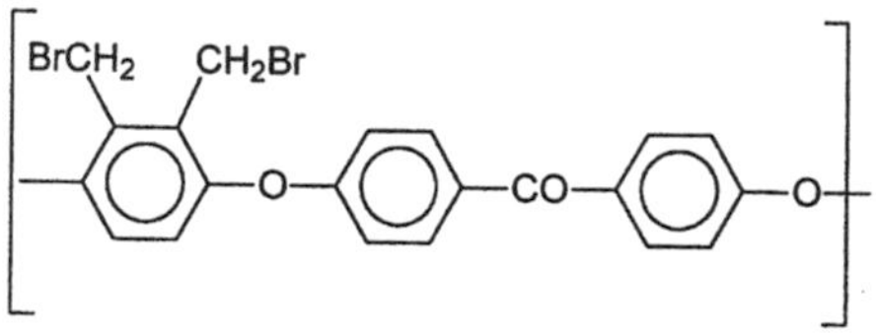

a

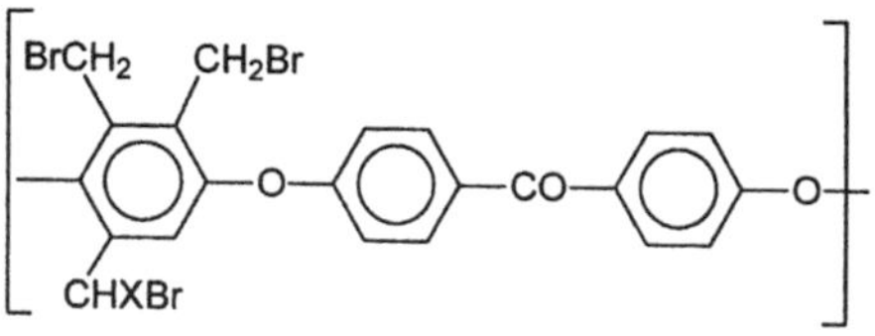

(140)

b, X = H, Br

b

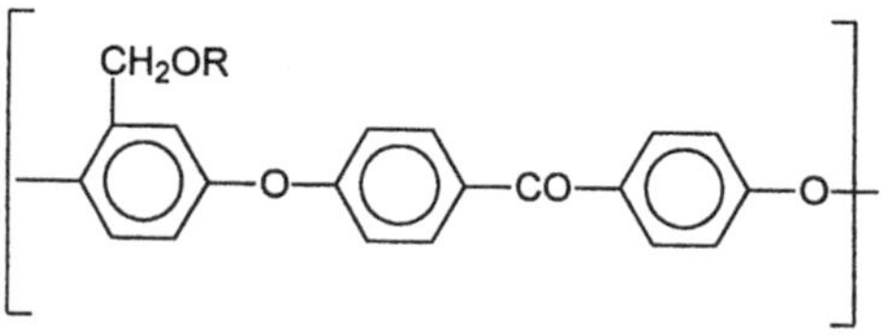

a: R = H, CH$_3$, COCH$_3$

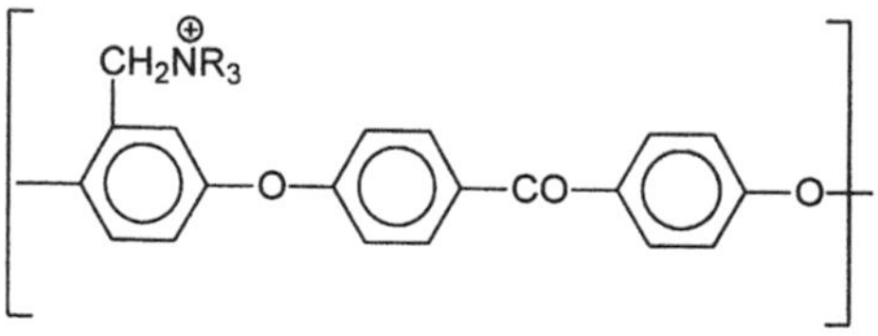

b : R = H, CH$_3$

(141)

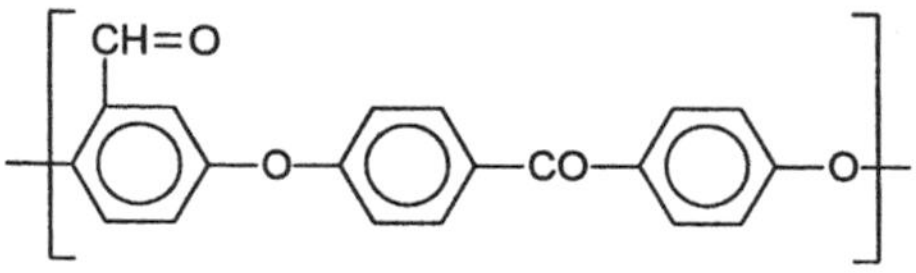

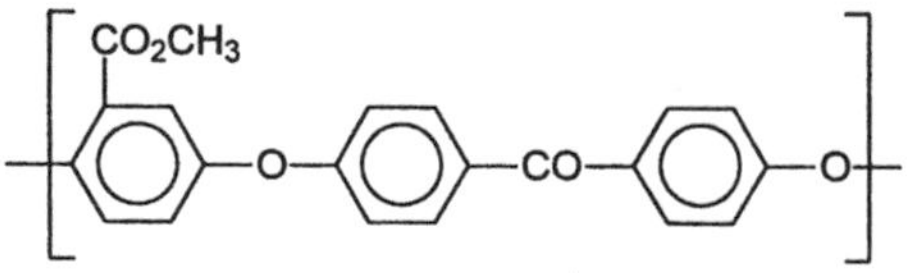

a

b

(142)

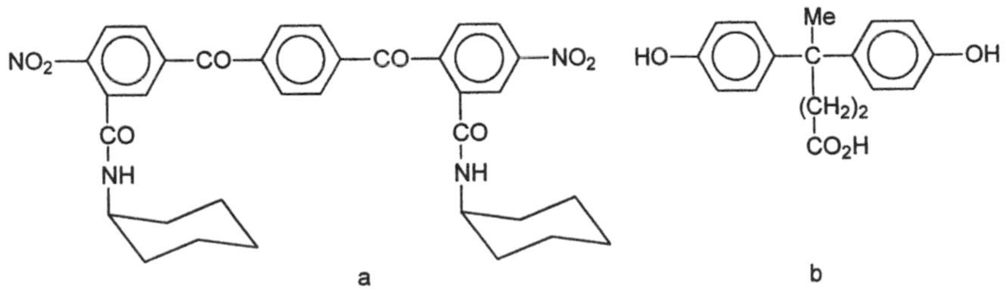

(143)

PEKs with pendant thermolabile substituents allowing for thermal cure were studied by two research groups. Polycondensations involving nucleophilic substitution steps were used in all cases. However, in the first case a thermolabile electrophilic monomer (144a) was used [230], whereas alkine substituted diphenols (144b) served as thermolabile monomers in the second study [231]. Finally, a paper dealing with the grafting of anionically polymerized styrene (145) on a bisphenol-A PEK (146) should be mentioned [232].

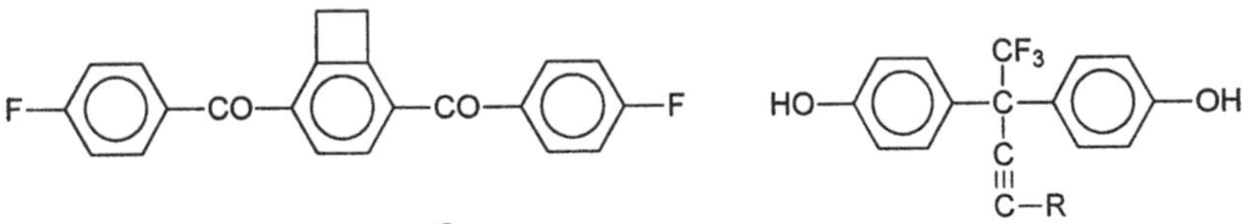

(144)

b : R = H, C$_4$H$_9$, Ph

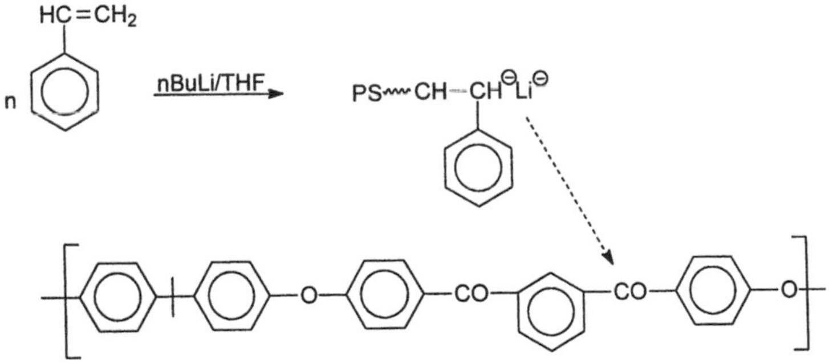

(145)

(146)

F. Various Aromatic Polyethers

1. Polyphenyleneoxides

The research on the reaction mechanism of the oxidative polycondensation of 2,6-dimethylphenol (DMP) was also continued in the years around 1990. In several publications a Dutch research group reported on kinetic studies dedicated to the O$_2$-promoted polycondensation of DMP catalyzed by copper tetramethyl 1,2-diamino-ethane complexes [233–236] or by copper complexes of imidazole [237,238]. The speculative mechanistic scheme was based on dimeric copper complexes such as (147a) which were assumed to incorporate a DMP anion (147b) which was oxidized to yield a phenoxy cation and to coordinate another DMP molecule (148a). The growing step was then assumed to consist of a nucleophilic substitution at the phenoxy cation (149) with liberation of a reduced dimeric copper complex (148b). This complex was believed to be oxidized by oxygen via the bridged complex (150a). Surprisingly the Dutch authors totally ignored the radical mechanism previously elaborated by several research groups [239], and it is strange to see a phenoxy cation attached to a copper cation (148a).

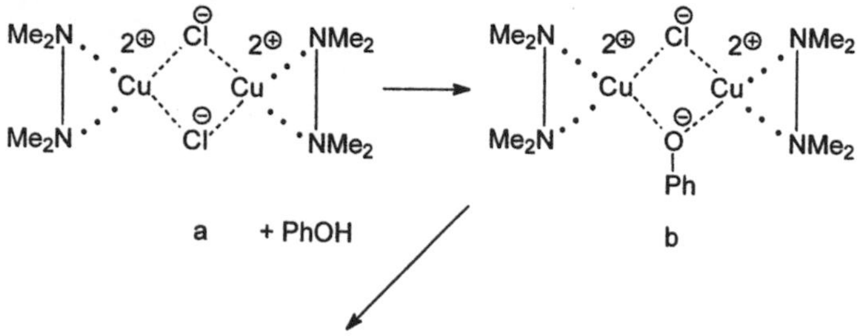

(147)

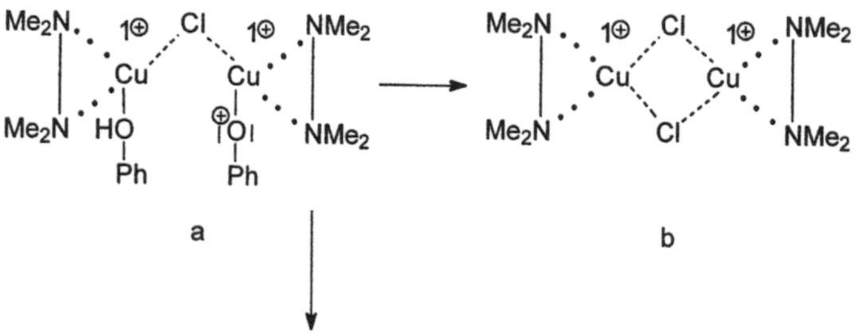

a   + PhOH

b

(148)

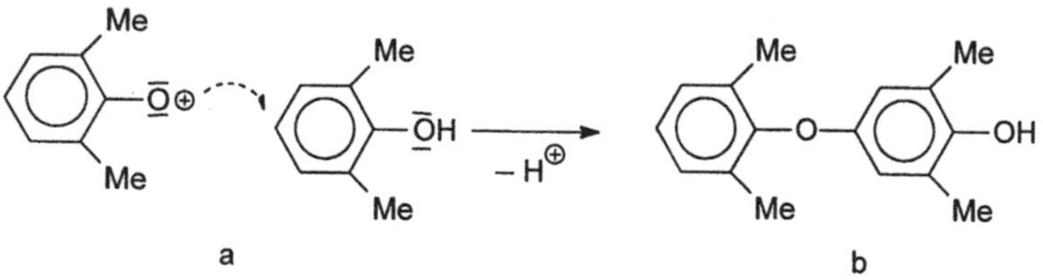

a

b

(149)

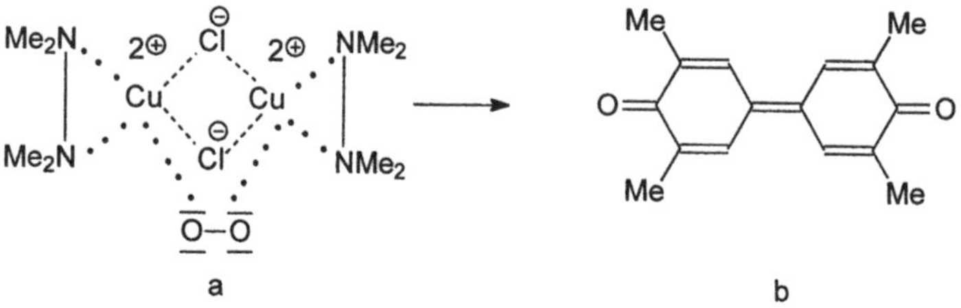

a

b

(150)

a

b

In 1991 a paper reported on the phase-transfer catalyzed polymerization of 4-bromo-2,6-dimethylphenol in the presence of O_2 as oxidizing agent, but in the absence of any redox catalyst [240]. 2,4,6-Trimethylphenol or 4-tert.butyl-2,6-dimethylphenol were added as initiators (or chain terminations) to control the molecular weights (151). In another publication polyphenylene oxides having aromatic substituents in 2,6-position (152) were polycondensed by means of the classical 'CuCl/amine + O_2' system [241]. Copolyethers and blends were also prepared. A further research group [242] prepared and studied thin films of substituted PPOs designed for non-linear optical properties (153). These PPOs were obtained by radical bromination of the CH_3 groups followed by a nucleophilic substitution with a suitable chromophor. Furthermore, commercial PPO was reacted with phenylacrylates, or diphenyl fumarate in an extruder and the reaction products were used as coinitiators of anionic polymerizations of ε-caprolactam. In this way Nylon-grafted PPOs were obtained and characterized [243]. Moreover, four publications dealing with polymerizations (via dehydrohalogenation) of 2,4,6-trihalophenols should be mentioned [244–247]. Well defined complexes such as (154a,b) were synthesized and the course of the thermal or the electrochemical polymerization was investigated. Low molecular weights and substitution in ortho position were observed in most cases.

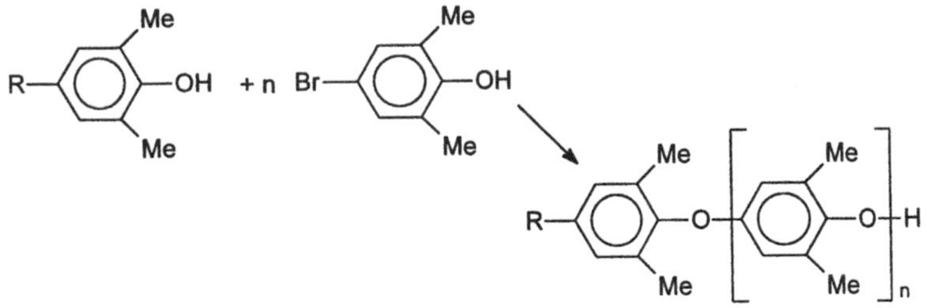

(151)

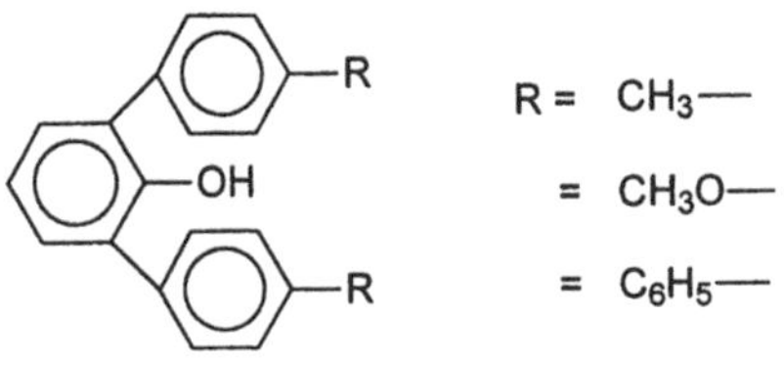

(152)

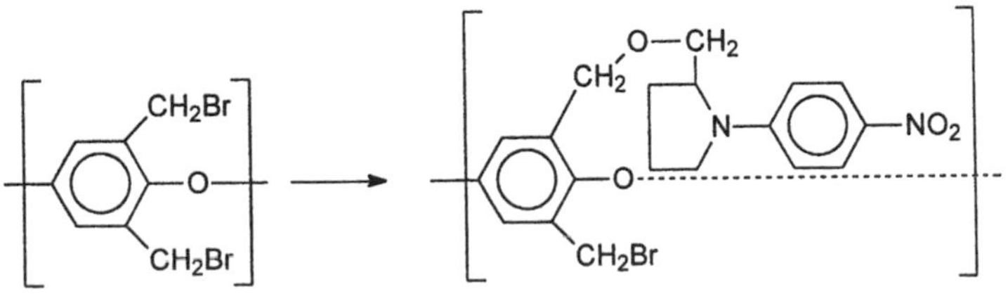

(153)

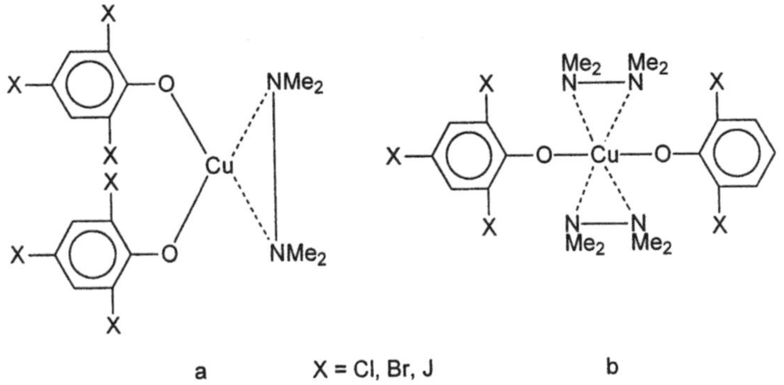

(154)

2. Polyethers with heterocycles in the main chain

Poly(arylene ether)s containing heterocycles in their backbones were prepared in three ways:

1. from difluoromonomers containing a heterocycle
2. from diphenols or diamines containing a heterocycle
3. by chemical modification of suitable precursor polymers.

Poly(arylene ether)s containing the oxazole (or thiazole) ring and in most cases pendant trifluoromethyl groups were prepared from the fluoromonomers (155a,b), (156a,b) and (157a,b). The normal nucleophilic substitution procedure was applied in combination with K_2CO_3 and bisphenol-A or other commercial diphenols. NMP and 1,3-dimethylpyrimidinone-2 were found to be the best reaction media and M_ns in the range of $10^4-4 \times 10^4$ Da were obtained [248–250]. Several publications deal with syntheses of poly(arylene-ether)s based on 1,3,4-oxadiazole rings [251–253] or based on 1,3,4-triazoles [253,254].

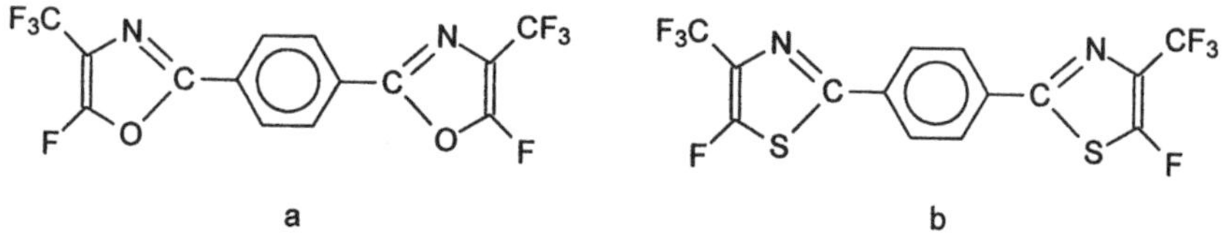

(155)

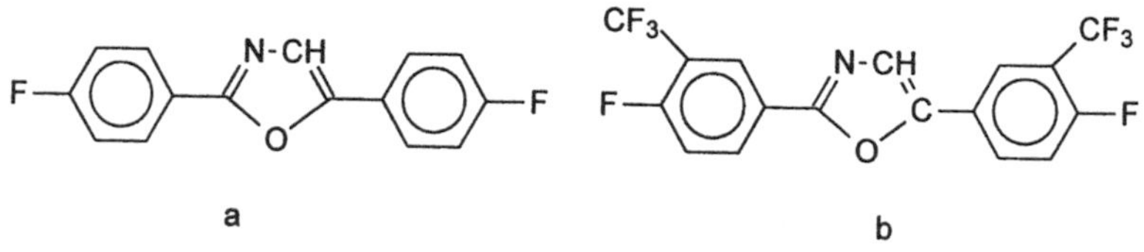

(156)

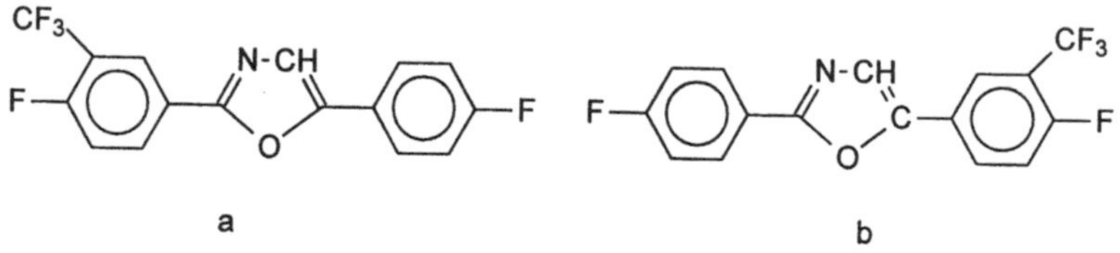

(157)

The electrophilic oxadiazole monomers (158a) were all synthesized by cyclization of the symmetrical bishydrazides; and they proved to be reactive enough to give satisfactory molecular weights, when polycondensed with various bisphenols under standard conditions in NMP/N-cyclohexylpyrrolidone mixtures at 180 °C [251]. The monomer (158, X = F) was also used to synthesize the diamines (159a,b) which were mixed with 4,4'-oxydianilines and chain extended with pyromellitic dianhydride. In this way thermostable imide-aryl ether copolymers were obtained [252]. Numerous poly(ether oxadiazole)s were also prepared by polycondensation of the diphenol (158b) with a broad variety of fluoromonomers including DFDPS and DFBP [253]. The same research group also prepared an analogous series of polyethers from the triazole diphenol (160b). The alternative approach towards a synthesis of polyethers containing 1,3,4-triazole rings was explored by another group of authors [254]. These authors started with the synthesis of eight fluoro monomers of structure (160a) which were then polycondensed with commercial diphenols in DMPU. Polyethers having T_gs above 200 °C were obtained. Another class of electrophilic monomers are the pyrimidines (161a) which were polycondensed with various diphenols in DMAc containing K_2CO_3. Polyethers showing T_gs up to 303 °C were isolated. However, the low inherent viscosities disagree with the seemingly high M_gs (up to 48,000 Da) determined by GPC [255]. Numerous poly(ether ketone)s, poly(ether sulfone)s and poly(ether phosphinoxide)s containing an imidazole ring were prepared from the diphenol (161b) [256]. A difluoromonomer containing an imidazole ring has not been described yet. In this connection cyclic polyethers should be mentioned [257] which were obtained by polycondensations of the 'difluoro phosphinoxide' (162a) with the nucleophilic heterocycle (162b) or with various other diphenols at low concentration.

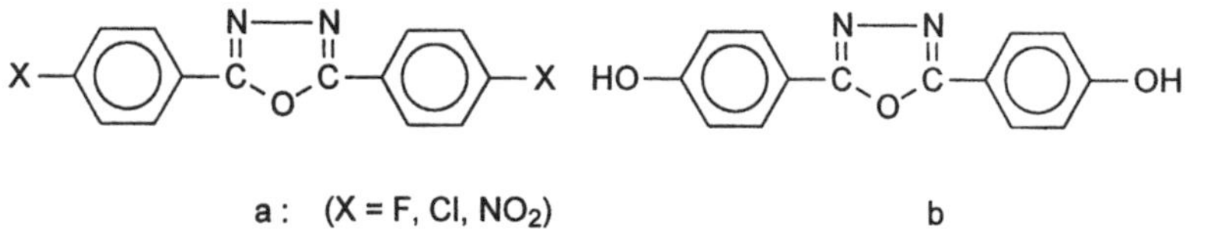

$$a: \quad (X = F, Cl, NO_2) \qquad b \tag{158}$$

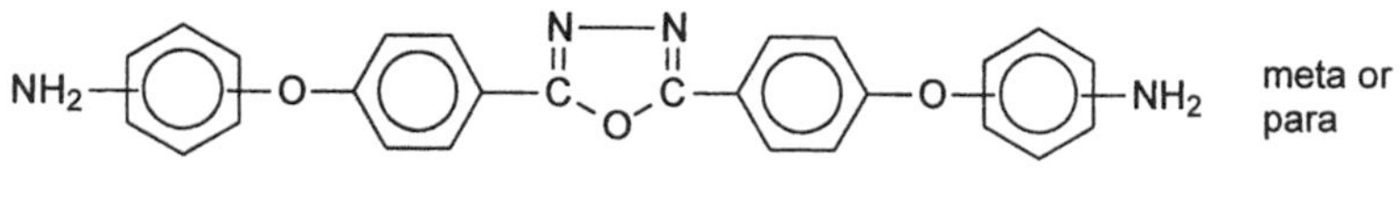

meta or para (159)

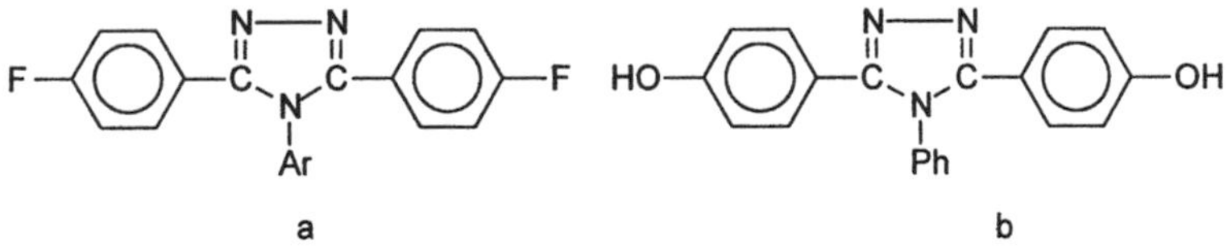

a   b (160)

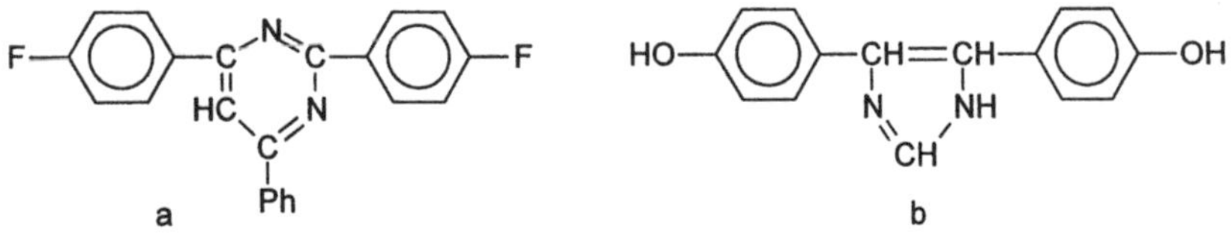

a   b (161)

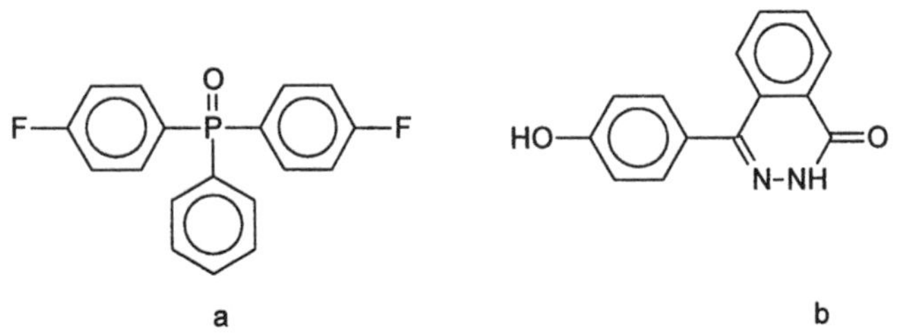

a   b (162)

    Several publications reported on polyethers containing benzoxazole or benzothiazole rings [258–262]. In most studies electrophilic fluoromonomers such as (163a–c) were used to implant the heterocycles into the polyether chains. However, in one paper [259] polycondensations of heterocyclic diphenols (164a,b) were described. On the basis of the bisfluorobenzobisthiazole (163a) the diamines (165) were synthesized and polycondensed with pyromellitic anhydride to yield the corresponding polyimides [262].

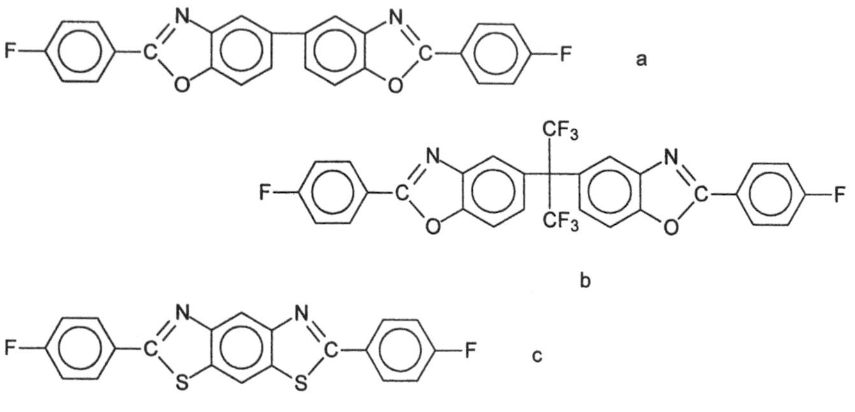

(163)

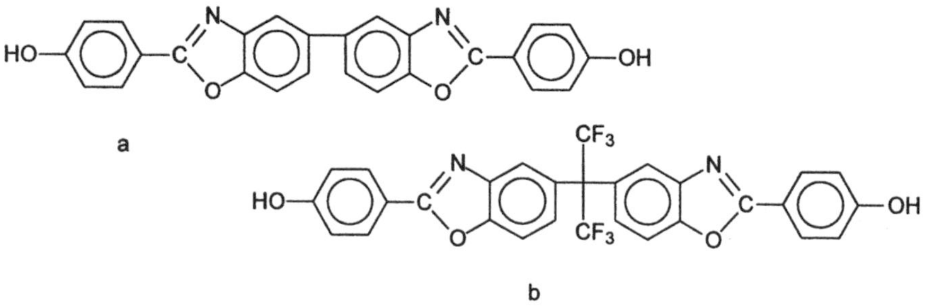

(164)

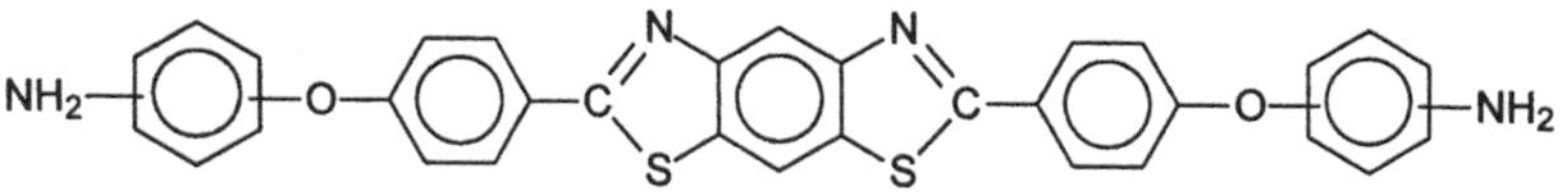

meta or para

(165)

Seven publications were devoted to polyethers containing phenylquinoxaline groups in the backbone [263–269]. For most syntheses difluoromonomers such as (166a,b) [263–266] or (167a,b) [265] used were in combination with commercial diphenols. The difluoromonomers of structure (166a,b) a mixture of isomers analogous to the isomerism of the hydroxyfluoromonomers (168a,b) [266]. Furthermore, several polyethers were prepared from mixtures of the isomeric diphenols (169a,b) and electrophilic monomers such as DFDPS or DFBP [267]. Based on the fluoromonomers (166a,b) monodisperse diamines such as (170) [268] or polydisperse oligomeric diamines [269] were synthesized. These diamines were polycondensed with biphenyl tetracarboxylic anhydride [268] or pyromellitic dianhydride [269] to yield poly(ether-imide)s. All these poly(phenylquinoxalin ether-imide)s are amorphous and show high T_gs in combination with high thermostabilities. In this connection a publication is worth mentioning dealing with the synthesis of cyclic oligo(ether-imide)s from silylated diphenols and arylene bis(fluorophthal-imide)s [270]. These macrocyclic ethers were designed to serve as monomers for the preparation of poly(ether-imide)s via ring-opening polymerization.

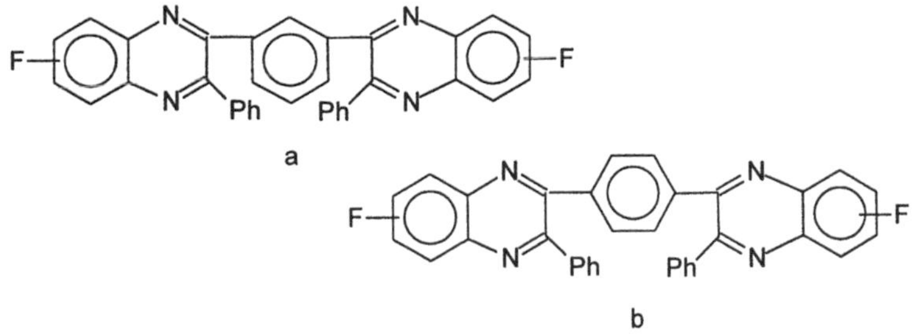

(166)

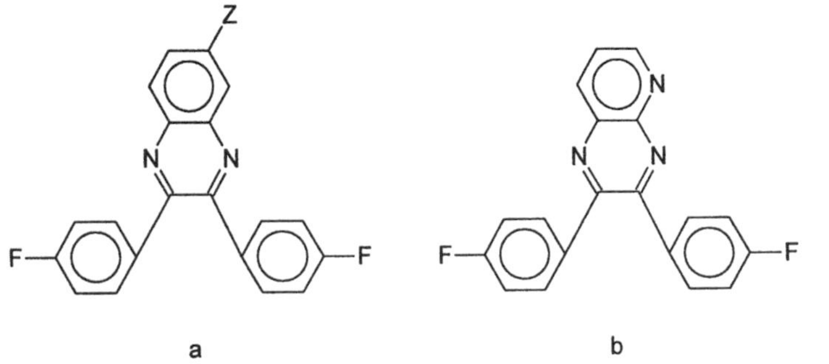

(167)

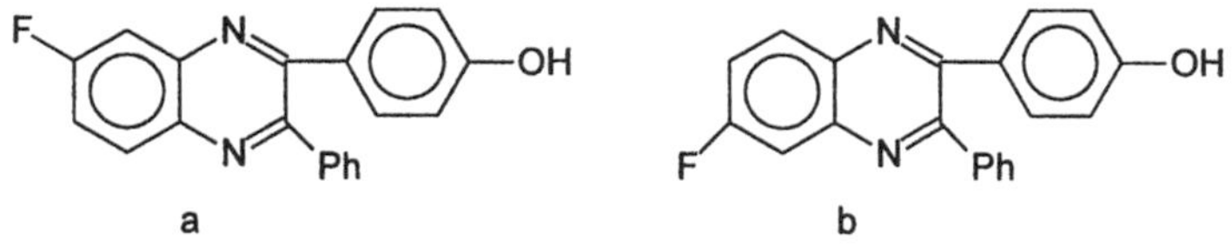

(168)

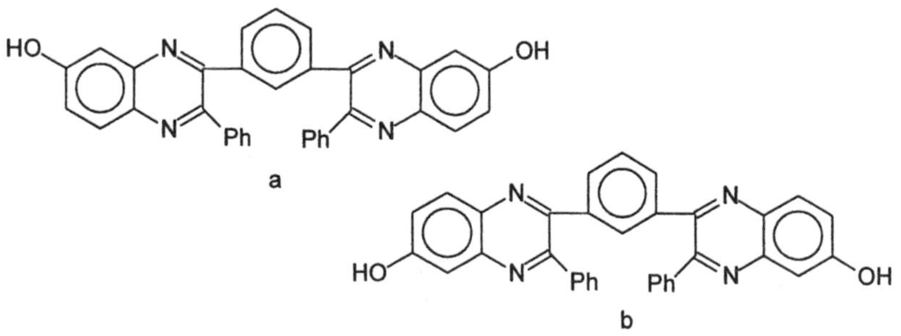

(169)

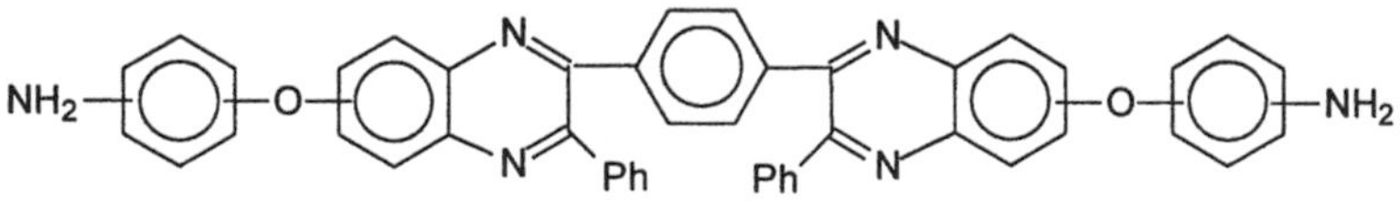

(170)

Finally, syntheses of poly(arylene ether)s by modification of precursors should be mentioned. All the syntheses described in [271–273] were based on poly(ether ketone)s derived from phthaloyl bisfluorobenzene (172). The tertiary amine catalyzed condensation with benzylamine yielded the poly(isoquinoline ether)s (171), whereas condensation with hydrazine produced the polyphthalazines (173). High glass transitions and high thermostabilities are characteristic for all these poly(arylene ether)s.

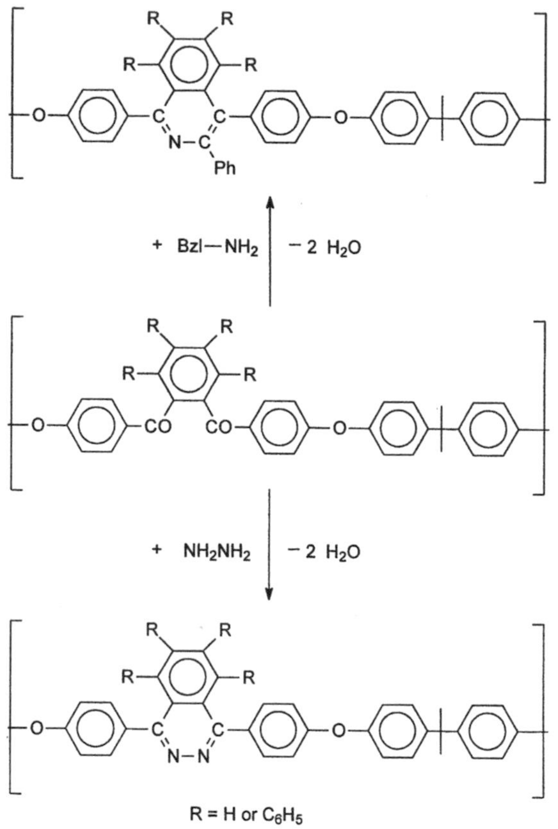

(171)

+ Bzl—NH₂ − 2 H₂O

(172)

+ NH₂NH₂ − 2 H₂O

(173)

R = H or C₆H₅

G. Various Polyethers

A group of polyethers resembling the PEKs in that the electrophilicity of fluoroaromatic groups is activated by keto groups are polyethers derived from 2,6-difluorobenzophenones or from the difluorodiketones (174a). Whereas polyethers prepared from 2,6-difluoro-benzophenones were described in the 1st edition of this handbook (Chapter 9, p. 542), polycondensations of the monomers (174a) with hydroquinone or methylhydroquinone were reported quite recently [274–276]. Diphenylsulfone served as reaction medium (with temperatures up to 320 °C) when hydroquinone was the comonomer to avoid crystallization and precipitation of oligomers.

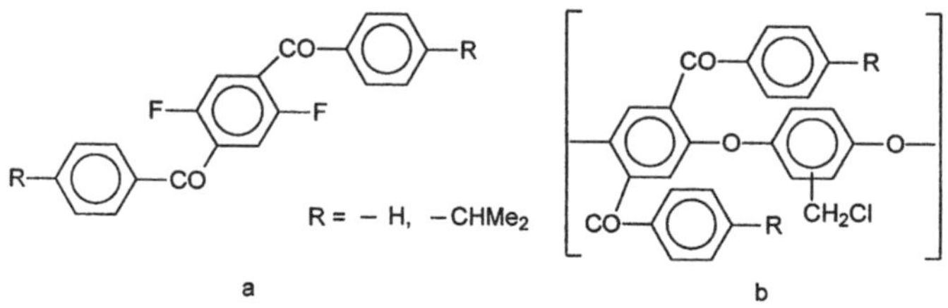

(174)

R = − H, −CHMe₂

a   b

Furthermore, the partial chloromethylation with SnCl₄/bis-1,4-(chloromethoxy)-butane (yielding (174b)) was studied [276]. Another class of monomers containing pendant carbonyl groups for activation of C–Cl or C–Br bonds are the bisimides (175a). They were polycondensed with the sodium salt of bisphenol-A in DMAc containing 18 crown-6 [277].

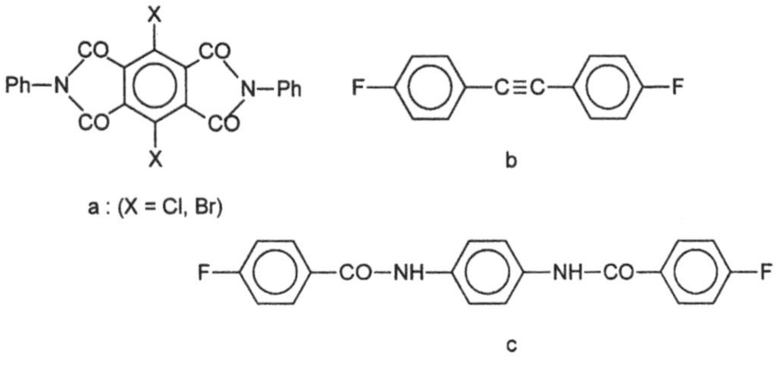

(175)

Three new classes of difluoromonomers having rather weakly activating groups in para-position to the F–C bonds are presented in formulas (175a) [278], (175b) [279] and (175c) [280]. The incorporation of acetylenic groups into the polyether chain had, of course, the purpose to obtain thermally curable materials. For both polyethers described in [278] the maximum of the exotherm resulting from the crosslinking process was found at 380–390 °C. Polycondensations of the difluoromonomer (175c) yielded poly(ether amide)s with high T_gs (up to 254 °C) illustrating the influence of the hydrogen bonds [279]. Another approach designed to yield poly(ether amide)s in a 'one-pot procedure' is based on the reaction of silylated 4-aminophenols with isophthaloyl chloride or terephthaloyl chloride [280]. Due to the much higher reactivity of the silylated amino groups the silylated diphenols (176) and (177) were exclusively acylated at the amino groups, so that new silylated diphenols were obtained. Their polycondensation with reactive fluoromonomers in situ yielded the desired poly(ether amide)s (178). For polycondensations of the difluoroazomethines (179) again silylated (commercial) diphenols were preferred in combination with CsF as catalyst [281]. Regardless, whether CsF or K_2CO_3 are used as catalysts silylated diphenols have the general advantage that no water is liberated [282] which may hydrolyze azomethine groups or C–F bonds.

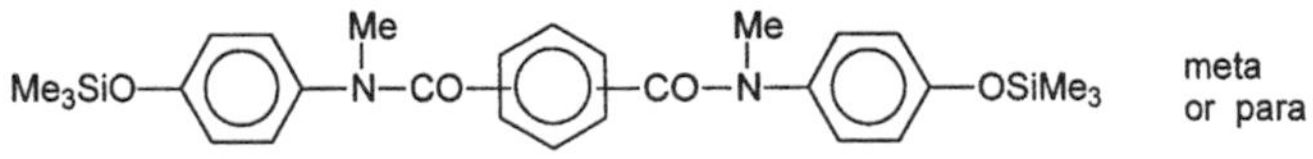

(176)

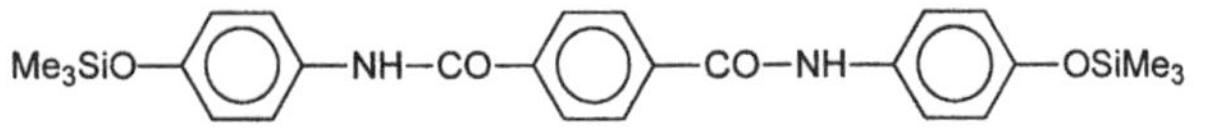

(177)

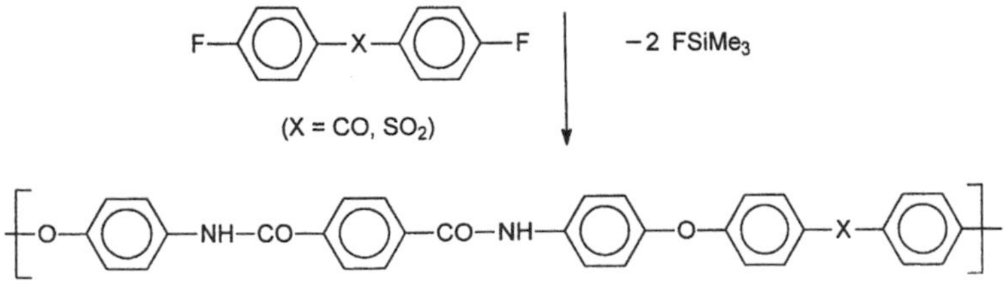

(178)

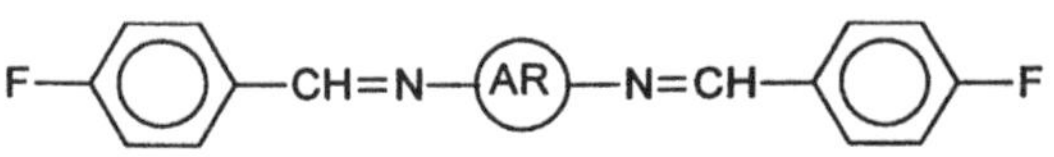

(179)

Several publications deal with polyethers having pendant nitrile (cyano) groups. With 2,6-dichlorobenzonitrile and silylated diphenols high molecular weights were obtained with an excess of K$_2$CO$_3$ in NMP at 180 °C [283]. Even higher molecular weights were found when the more reactive 2,6-difluorobenzonitrile was polycondensed with free diphenols [284]. In two later papers 2-fluoro-6-chlorobenzonitrile was used for syntheses of monodisperse telechelic oligomers such as (180), [285], (181) and (182), [286]. These oligomers were polycondensed with various commercial diphenols under standard conditions. Furthermore, several synthetic strategies were explored designed to yield poly(ester ether)s with pendant nitrile groups when 2,6-difluorobenzonitrile serves as electrophilic monomer [287,288]. The same synthetic strategies also yield poly(ester-ether ketone)s or poly(ester-ether sulfone)s, when DFBP or DFDPS are used as electrophilic monomers. The first strategy consists of the copolycondensation of silylated diphenols and silylated hydrox acids with a fluoromonomer (183) [287]. It should be emphasized that silated dicarboxylic acids are not nucleophilic enough to serve as ester forming comonomers in this approach. The second strategy is based on homo- or copolymerisations of a preformed silylated diphenol containing an ester group such as (184) [288]. Characteristic for the third strategy is the formation of silylated polydisperse oligoesters (185) followed by an in situ polycondensation with a fluoro-monomer such as 2,6-difluorobenzonitrile [287].

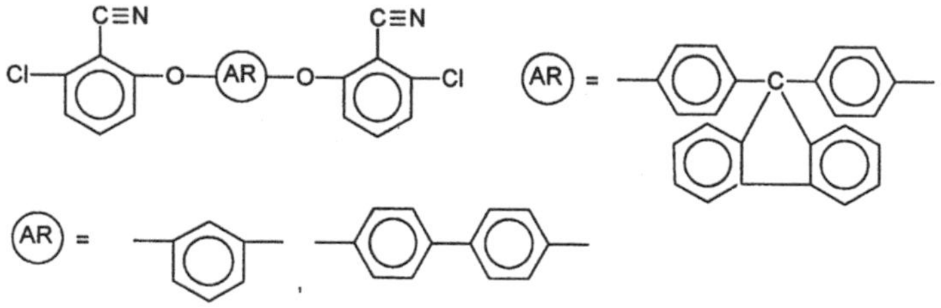

(180)

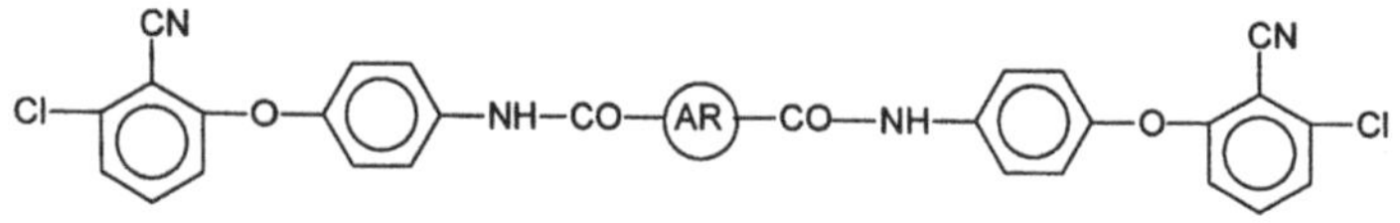

(181)

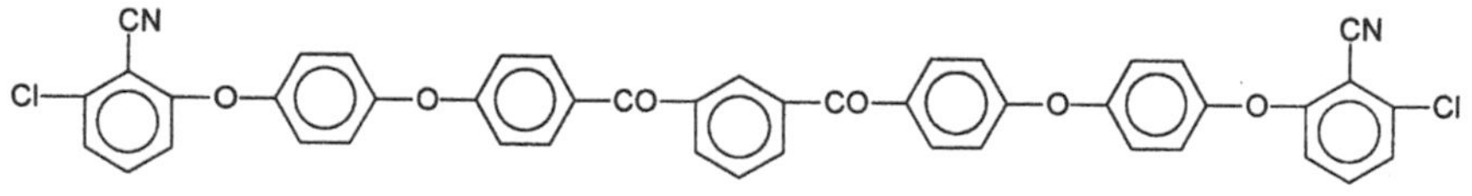

(182)

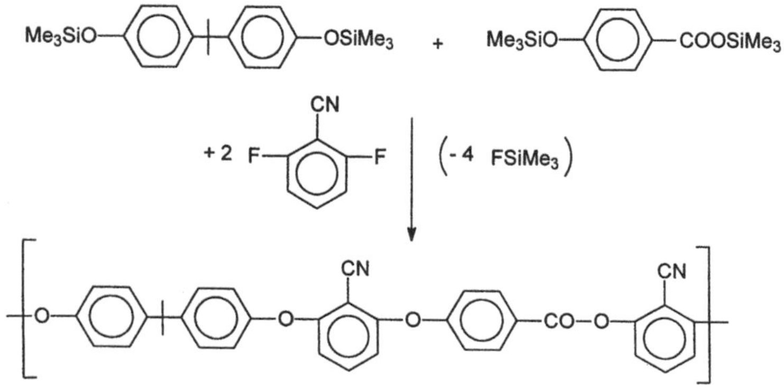

(183)

$$Me_3SiO-\langle\bigcirc\rangle-O-CO-\langle\bigcirc\rangle-OSiMe_3 \qquad (184)$$

$$(X+1)\ \ Me_3SiO-\langle\bigcirc\rangle\!|\!\langle\bigcirc\rangle-OSiMe_3\ +\ x\ ClCO-\langle\bigcirc\rangle-COCl$$

$$(Cl^\ominus)\Big\downarrow\ 2\ x\ ClSiMe_3$$

$$Me_3Si\!\left[\!O-\langle\bigcirc\rangle\!|\!\langle\bigcirc\rangle-O-CO-\langle\bigcirc\rangle-CO\!\right]_x\!\!O-\langle\bigcirc\rangle\!|\!\langle\bigcirc\rangle-OSiMe_3$$

$$(185)$$

Polyethers containing pyridine rings were prepared by polycondensations of 2,6-dichloropyridine with free diphenols [289] or silylated diphenols [283]. In both cases only low to moderate molecular weights were obtained. Higher molecular weights resulted from CsF catalyzed polycondensations of 2,6-difluoropyridine and silylated diphenols in bulk [290]. On the basis of this latter approach also poly(pyridine ester ether)s (186) [288] and poly(pyridineether sulfone)s (187) [291] were obtained. The alkylation of the copolyethers was studied in detail, and the alkylated copolyethers (188) proved to be useful as interesting gas-separating membranes [291]. Furthermore, poly(pyrazine ether)s were prepared from 2,6-dichloropyrazine and alkylated with methyl triflate (189a,b) [292]. These polyethers and those derived from 3,6-dichloropyridazine [283] only had low to moderate molecular weights.

$$\left[\!O-\langle\bigcirc\rangle\!\langle\bigcirc\rangle-O-\langle\!N\!\rangle\!/\!O-\langle\bigcirc\bigcirc\rangle-CO-O-\langle\!N\!\rangle\!\right] \qquad (186)$$

$$\left[\!O-\langle\bigcirc\rangle\!|\!\langle\bigcirc\rangle-O-\langle\!N\!\rangle\!/\!O-\langle\bigcirc\rangle\!|\!\langle\bigcirc\rangle-O-\langle\bigcirc\rangle-SO_2-\langle\bigcirc\rangle\!\right] \qquad (187)$$

$$\left[\!O-\langle\bigcirc\rangle\!|\!\langle\bigcirc\rangle-O-\overset{\overset{CH_3}{|}}{\underset{\oplus}{N}}\!/\!O-\langle\bigcirc\rangle\!|\!\langle\bigcirc\rangle-O-\langle\bigcirc\rangle-SO_2-\langle\bigcirc\rangle\!\right] \qquad (188)$$

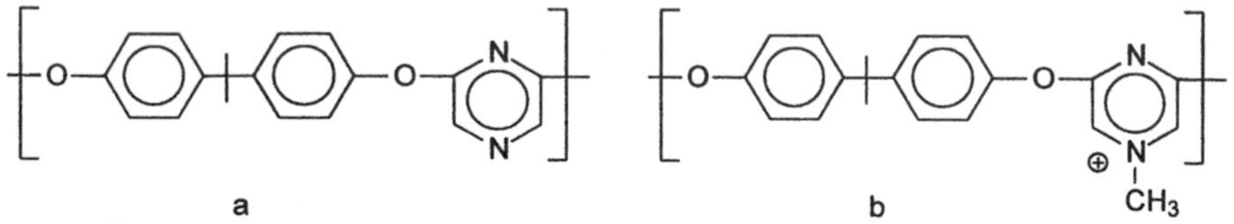

(189)

A broad variety of highly fluorinated polyethers (e.g. (190a)) were described by three research groups. Either, fully aromatic fluoro monomers such as 1,2,4,5-tetrafluoroben-zene, hexafluorobenzene and decafluorobiphenyl were used as electrophilic reactants [293, 294] or the flexible monomer (190b) [295]. Surprisingly, in all the cases, rather clean polycondensations without significant formation of gel particles were achieved.

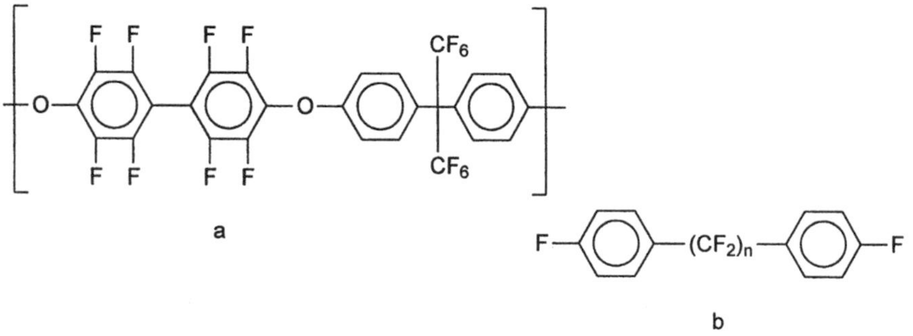

(190)

Numerous publications deal with unusual condensation and chain extension methods. For instance various 4,4'-metal derivatives of diphenyl ether (191) were polycondensed with dibromoaromats. Palladium complexes served as catalysts, but the molecular weights were below 5000 Da in almost all cases [296]. Low molecular weight polyethers were also obtained by an 'Ullmann type condensation' of the 'dibromo monomer' (192) with bisphenol-A and bisphenol-AF [297]. Another research group presented an intensive and detailed study of polyether syntheses via the 'Scholl Reaction' [298–301]. This reaction, schematically outlined in (197) consists of a $FeCl_3$ promoted 'polyoxydation' of nucleophilic aromatic ethers. This 'Scholl Reaction' involves single electron transfer steps with the intermediate formation of radical cations. In most cases M_ns < 6000 Da were obtained, but a few polyethers had M_ns in the range of 10,000–20,000 Da.

A special class of soluble poly(ether ketone)s (194) served as precursor polymers for the preparation of polyethers containing phenanthrene moieties (195) [302–304]. In addition to hexacyclohexyl distannathiane Lawesson's reagent was used for the reductive cyclization-liquid crystalline poly(ester ether)s were prepared from the hydroxy acids (196a,b) [305]. These new monomers were synthesized in an unusual way, namely by etherification of 1,3-dichlorobenzene activated in the form of an iron cyclopentadienyl complex. Silicon-containing polyethers were easily obtained with low to moderate molecular weights by polycondensation of various diphenols with dianilinodiphenylsilane (197) [306]. Several publications were devoted to synthesis and characterization of poly(arylene ether) networks [307–312]. Telechelic oligoethers such as (198) and (199) having two cyanate endgroups served as monomers. Finally a study of hybride ceramer

materials prepared from titanium tetraisopropoxide and the oligoether (200) should be
mentioned [312].

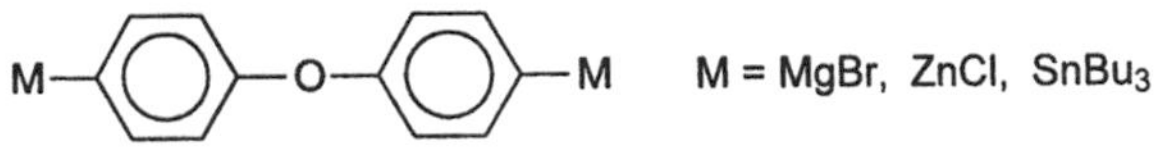

(191)

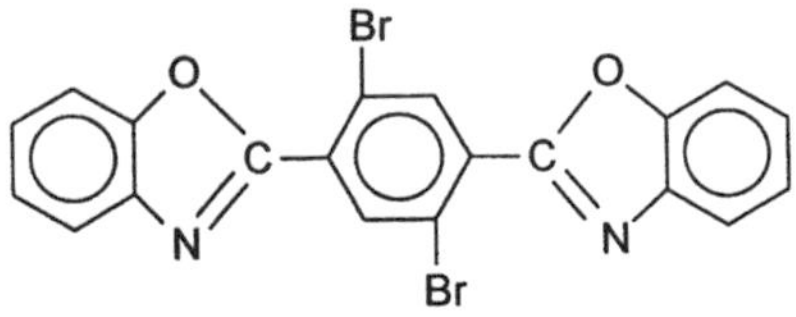

(192)

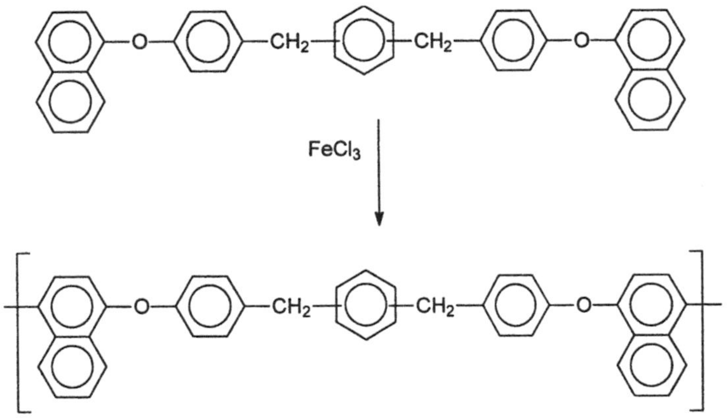

(193)

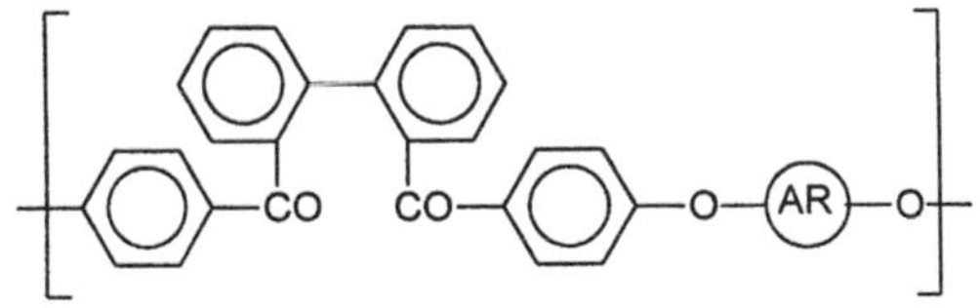

(194)

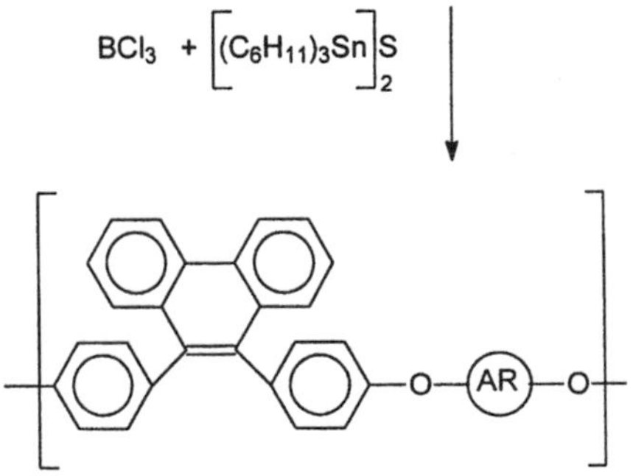

(195)

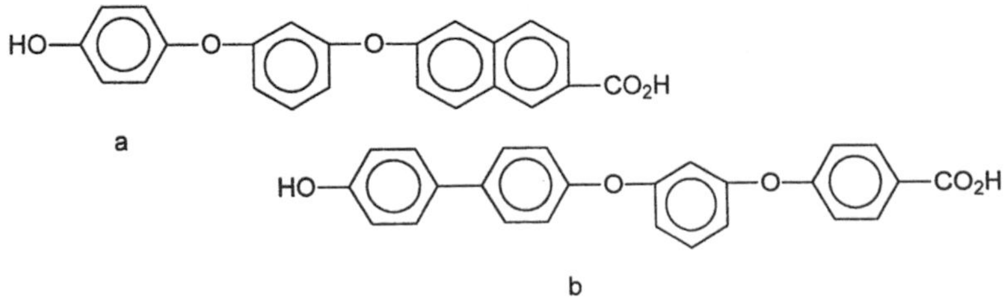

(196)

Ph—NH—Si(Ph)(Ph)—NH—Ph $\xrightarrow{\text{HO—(AR)—OH}}$ [O—Si(Ph)(Ph)—O—(AR)]   (197)

NCO—⬡—⬡—[O—⬡—X—⬡—O—⬡—⬡]—OCN   (198)

NCO—⬡—O—⬡—P(=O)(Ph)—⬡—O—⬡—OCN   (199)

NH₂—⬡—O—[⬡—P(=O)(Ph)—⬡—O—⬡—⬡]ₓ   (200)

—O—⬡—P(=O)(Ph)—⬡—O—⬡—NH₂

H. Aliphatic-Aromatic Polyethers

This subchapter summarizes publications dealing with polyethers having ether oxygens attached to aromatic and aliphatic moieties. Since most of the studies reported in this field concern liquid-crystalline polyethers this section is subdivided into two subsections, the first one entitled 'Various Structures and Synthetic Methods' and the second one entitled 'Liquid-Crystalline Polyethers'.

1. Various Structures and Synthetic Methods

As reported in the 1st edition of this handbook the most widely used approach to syntheses of mixed aliphatic aromatic polyethers involve the nucleophilic substitutions of chloromethyl or bromomethyl aromats. The halomethyl groups attached to aromatic rings are far more electrophilic than n-alkyl halides and react quickly with phenoxide ions. Based on this approach various poly(ether sulfone)s of structure (201) were synthesized from the potassium salt of 4,4'-dihydroxydiphenylsulfone [313–314]. Two series of polyethers were prepared (and characterized) from various commericial diphenols by polycondensation with the α,α'-nitroxylene halides (202a,b) [315]. Phosphorus containing flame retardant polyethers were obtained by polycondensation of the phosphine oxide (203a) with the sodium salts of diphenols in hexamethylphosphorus triamide. In addition to various physical properties the oxygen indices were determined [316]. The monomer (203b) was polycondensed by means of phase-transfer catalysts through nitrodisplacement. Systematic optimization of the reaction conditions yielded polyethers having M_ns up to 32,000 Da [317].

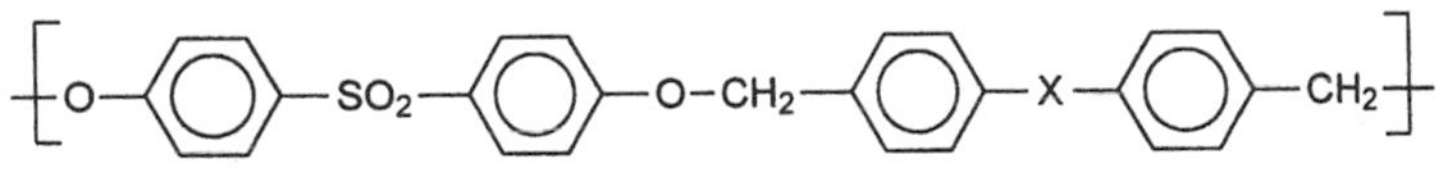

(201)

X = OS, CH$_2$, CO, SO$_2$, σ-bond

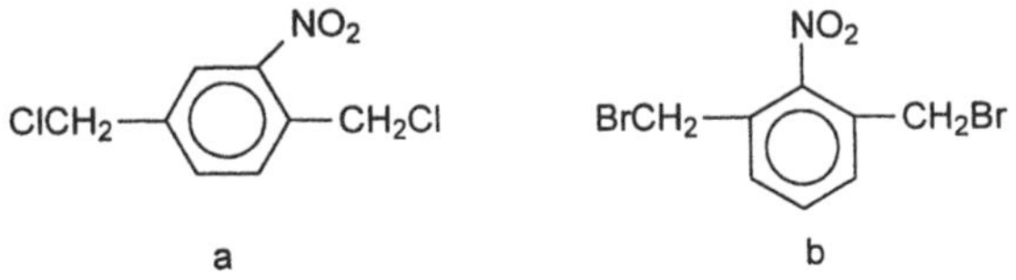

(202)

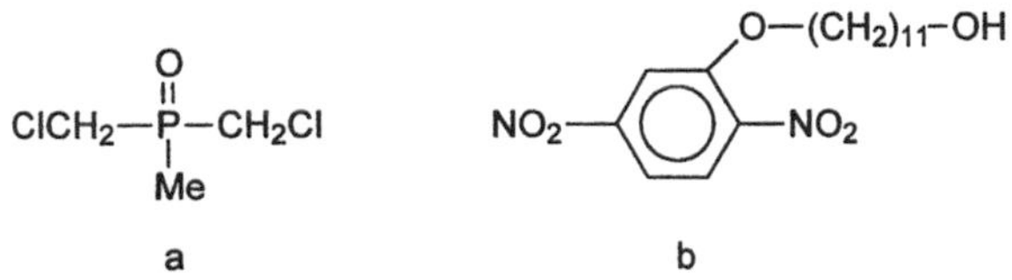

a

b

(203)

a

b

A new synthetic approach was elaborated by alkylation of silylated diphenols with alkylene disulfonates such as (204) [318]. This polycondensation is best promoted by K$_2$CO$_3$ in NMP and has the advantage that no water is liberated. Therefore a variety of diphenols having functional groups could be used as nucleophilic monomers (205a–c) and (206a–c). On the other hand, it was found that silylated aliphatic diols (e.g. (207a,b)) react with DFDPS and K$_2$CO$_3$ yielding polyethers which may be chiral when (207b) is used as monomer [319]. However, this method does not work well, when less reactive difluoro-aromats are used.

X—CH$_2$CH$_2$—O—⟨ ⟩—O—CH$_2$CH$_2$—X

X = CH$_3$SO$_3$—

X = CH$_3$—C$_6$H$_4$—SO$_3$—

X = NO$_2$—C$_6$H$_4$—SO$_3$—

(204)

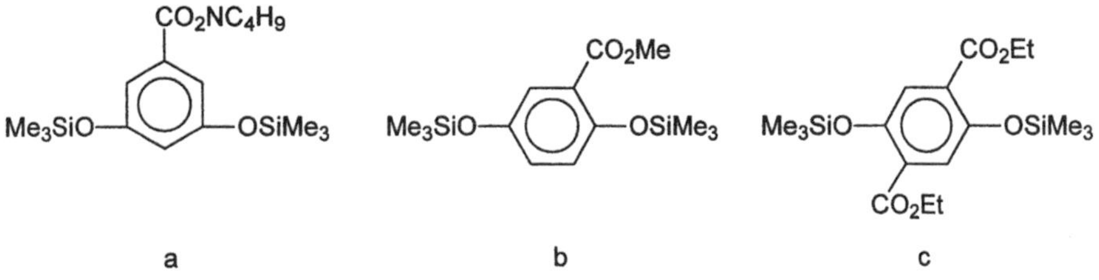

a

b

c

(205)

a

b

c

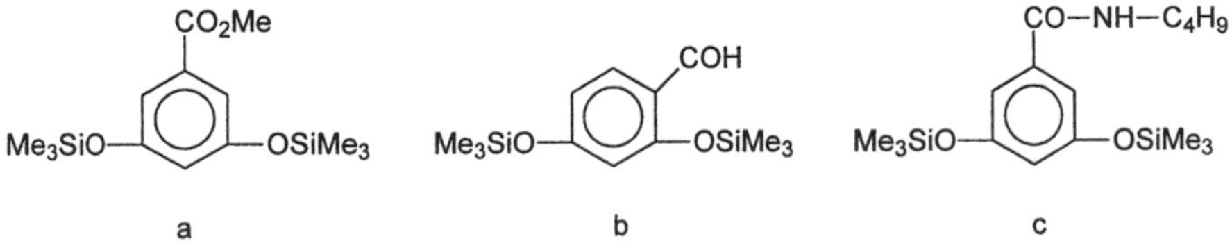

(206)

$$Me_3SiOCH_2CH_2O-\langle\bigcirc\rangle-OCH_2CH_2OSiMe_3 \qquad (207)$$

Several publications deal with syntheses of oligo- and polyethers capable of forming networks or hyperbranched structures. For instance, monodisperse vinyl-terminated oligoethers of structure (208) were synthesized from bisphenol-A and 4-chloromethyl styrene as tetravalent comonomers for radical crosslinking with various vinyl monomers [320]. The same type of Williamson's ether formation was used for the synthesis of a styrene terminated hyperbranched macromer from 3,5-dihydroxybenzylbromide, phloroglucinol and chloromethyl styrene (209) [321]. Using the same trifunctional monomer 3,5-dihydroxybenzylbromide hyperbranched polyethers were also prepared in a 'one-pot procedure' [322].

$$CH_2=CH-\langle\bigcirc\rangle-CH_2-O-\langle\bigcirc\rangle-\langle\bigcirc\rangle-O-CH_2-\langle\bigcirc\rangle-CH=CH_2 \qquad (208)$$

(209)

2. Liquid-Crystalline (LC) Polyethers

Almost all LC-polyethers mentioned in this section were prepared via the Williamson ether synthesis from a mesogenic diphenol and (di)bromoalkanes. For instance a linear LC-polyether was obtained from the cesium salt of a 'diazine-diphenol' and 1,10-dibromodecane in NMP (210) [323]. In two cases, (211) [324] and (209) [325] interfacial polycondensation were performed using sodium hydroxide in combination with a phase-transfer catalyst and an organic solvent. Almost all LC-polyethers prepared in this way [323–325] had M_ns $\leq$ 10,000 Da. A somewhat different synthetic route was used for the preparation of LC-poly(ether-imide)s (213) [326]. Here NH$_2$-terminated spaces were synthesized from 4-nitrophenol and α,ω-dibromoalkanes followed by reduction of the nitrogroups. The polycondensation proceeded via the formation of imide groups.

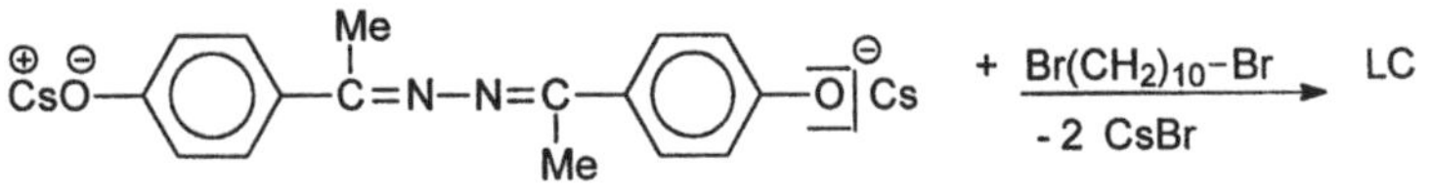

(210)

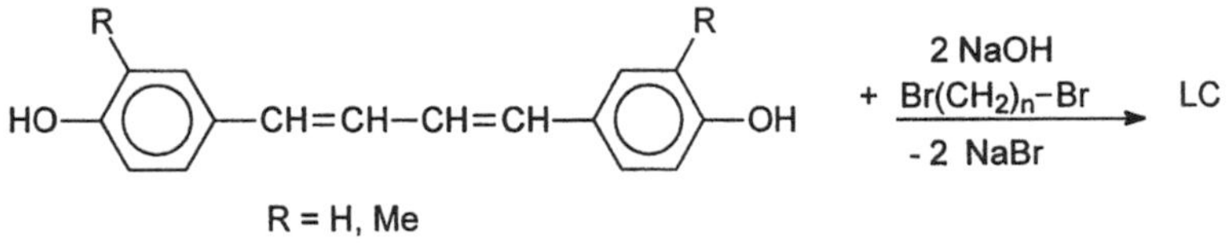

(211)

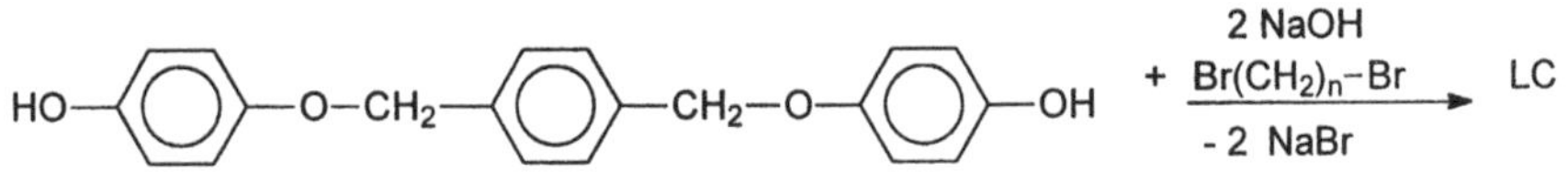

R = H, Me

(212)

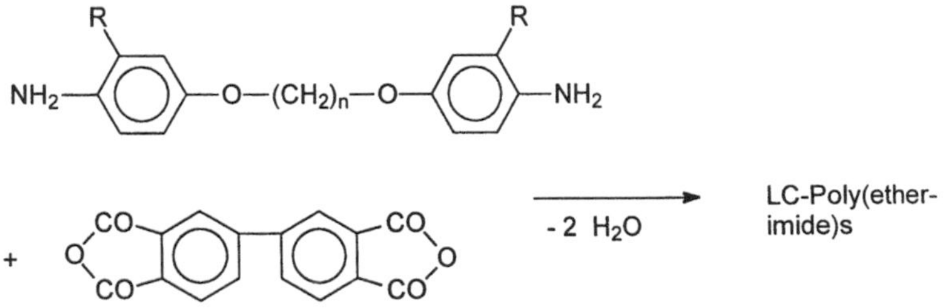

(213)

LC-Poly(ether-imide)s

Another research group published numerous studies of syntheses and physical properties of LC-polyethers having linear chains, hyperbranched and cyclic structures or containing discotic mesogens [327–341]. All these different LC-polyethers were prepared via the normal Williamson ether synthesis involving the nucleophilic attack of a phenoxide ion onto a bromoalkane. For instance, numerous linear LC-polyesters were prepared from the mesogenic diphenols outlined in (214) and formulas (215a,b) [327,337].

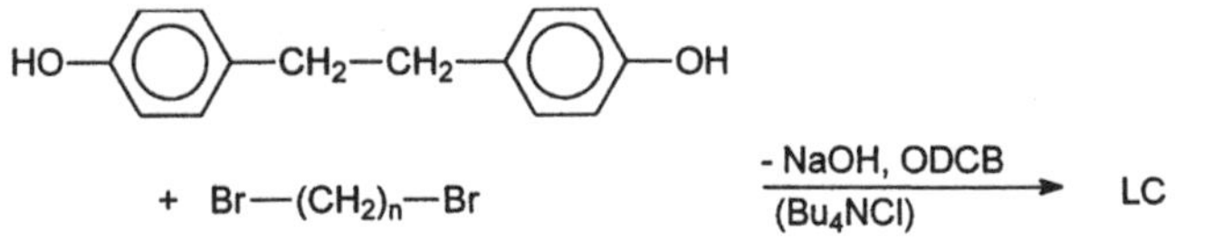

(214)

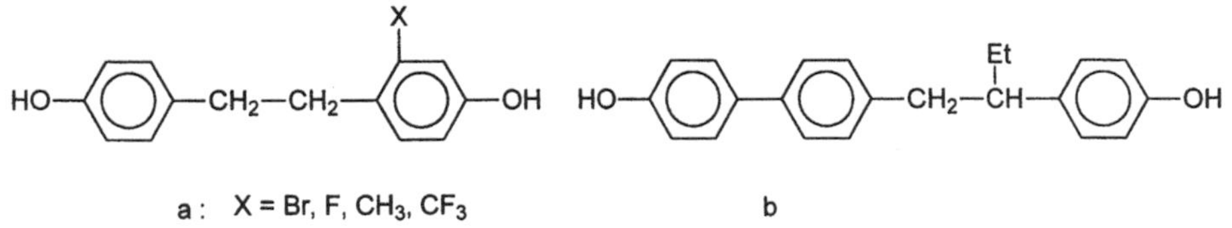

(215)

a: X = Br, F, CH$_3$, CF$_3$

b

The mesogenic diphenol (215b) was also used as component of the macrocyclic LC-ethers (216). The ring size of these macrocycles was varied by variation of the concentration of the reactants [338]. LC-Polyethers containing crown ethers as members of the polymer backbone were prepared by polycondensation of the monomers (217a,b) [339]. LC-Polyethers with a hyperbranched structure were obtained by polycondensation of the trifunctional monomer (218) [337]. Finally, linear oligoethers (219) or branched polyethers containing cyclotetraveritrylene moieties as disk-like mesogens should be mentioned [341].

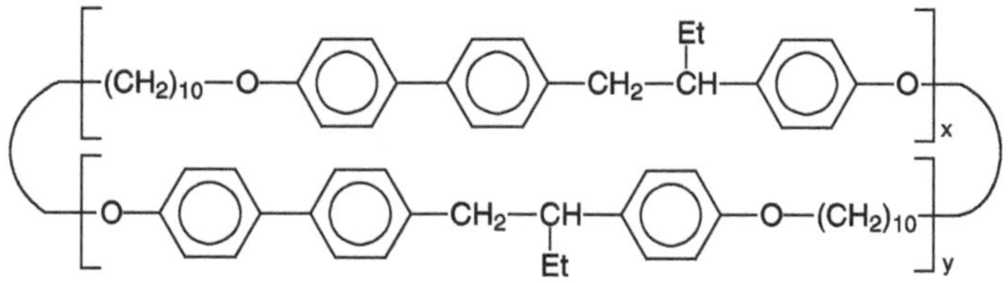

(216)

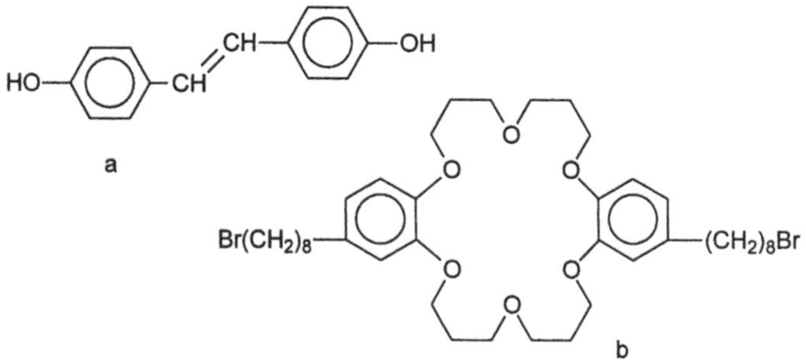

a

(217)

b

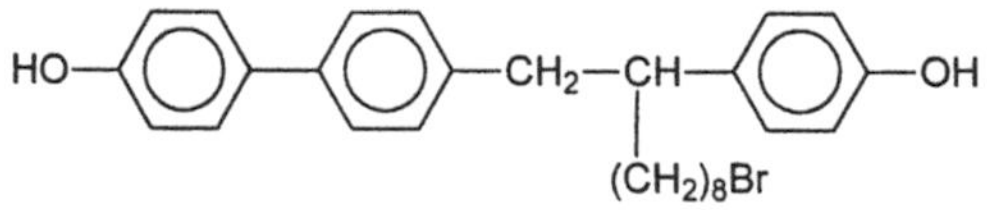

(218)

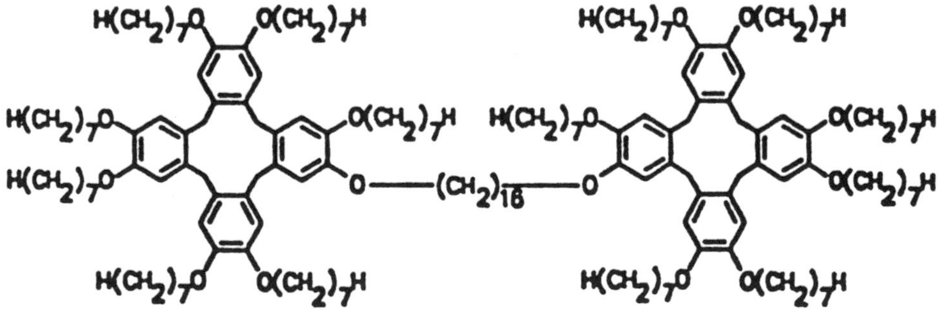

(219)

I. Aromatic Polysulfides

Poly(phenylene-sulfide) has proven over several decades to be an important and versatile high performance engineering plastic and particularly useful as matrix of composites containing glass fiber, carbon fiber, electro conductive particles, etc. The classical syntheses are based on the polycondensation 1,4-dichlorobenzene with Na$_2$S or on the polycondensation of 4-chlorothiophenol salts (220). In this connection intensive studies of the reaction mechanism were performed by several research groups and three polymerization mechanisms were proposed:

1. the normal nucleophilic substitution (S_NAR) involving a Meisenheimer complex as transitional state (221)
2. a $S_{RN}1$ mechanism involving aromatic radical anions
3. a radicalcation mechanism.

$$(220)$$

$$(221)$$

A detailed study (including model reactions) of a research team of the Phillips Petroleum Company reached the clearcut decision that the chain growth steps are exclusively based on the S_NAR mechanism (221) [342].

The nucleophilic substitution approach was also used by several research groups for syntheses of polysulfides with broad variation of the chemical structure and of the reaction conditions [343–352]. For instance, a poly(thioether-ketone) was prepared from DFBP and dry Na_2S with variation of the reaction medium [337]. N-Cyclohexylpyrrolidone was found to yield the highest molecular weights. In another publication random copoly(ketone sulfone sulfide)s were prepared by copolycondensation of 4,4'-dichlorobenzophenone and 4,4'-dichlorodiphenylsulfone with NaSH (222) [344]. The crystallinity was found to depend on the molar fraction of benzophenone moieties.

$$(222)$$

Furthermore, silylated bisthiophenols, such as (223a,b) were polycondensed with numerous activated difluoro- or dichloro monomers [245,247] and polysulfides of low to moderate molecular weights were obtained. These polycondensations were conducted in bulk with C_SF as catalyst. Polymers with an alternating sequence of ether and sulfide bonds were prepared by polycondensation of the diphenol (224a,b) which were used in the bistrimethylsilyl derivatives. The diphenols (224a,b) were synthesized from 4-mercaptophenol and 2,6-dichloropyridine or 1,4-dichloropyridazine [347].

$$(223)$$

a (meta or para)   b (X = O or σ-bond)

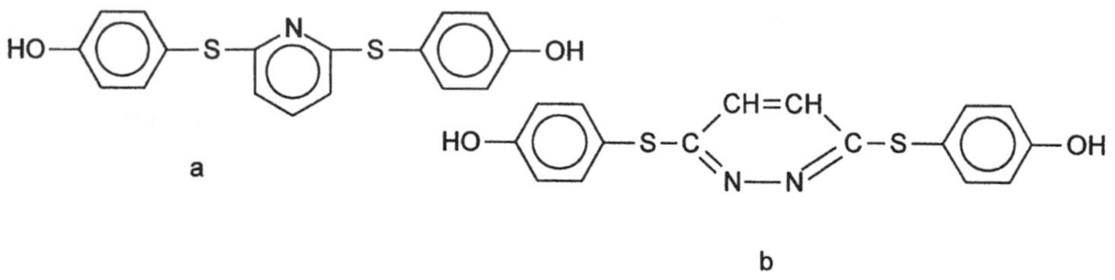

a

b

(224)

Another class of disulfide precursor monomers, namely biscarbamates of dithio-phenols, were described in Refs. [348–350]. When the bisthiophenol is commercial the biscarbamate is easily obtained by addition of isocyanates, which protect the SH-groups against oxidation (225). The isocyanates are liberated in the course of the polycondensation with DPDPS and K_2CO_3 in NMP [348]. More useful and versatile is the acylation of diphenols with Me_2N-CS-Cl followed by a thermal rearrangement which yields the carbamoyl protected bisthiophenols (226), (227). These biscarbamoyl monomers were polycondensed with DFBP and $Cs_2CO_3/CaCO_3$ mixtures in benzophenone at 230–300 °C [348,350].

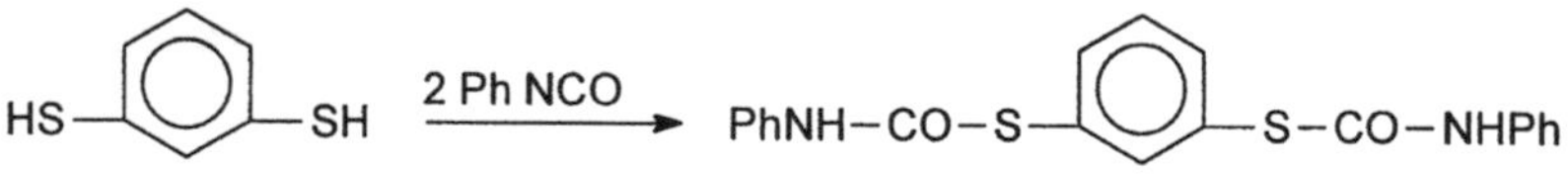

(225)

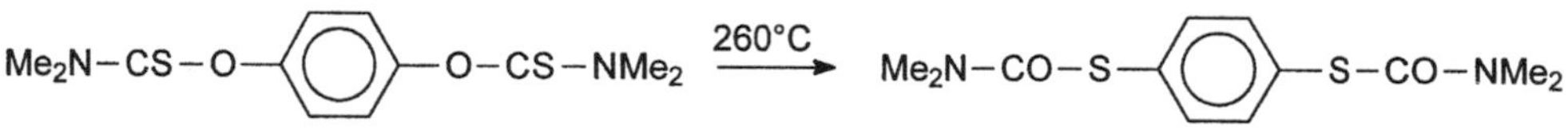

(226)

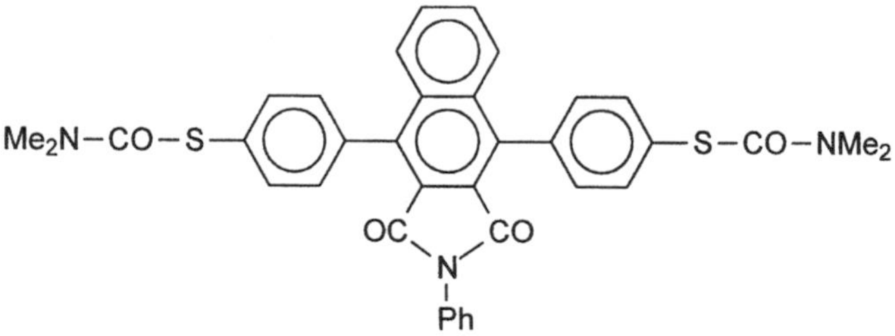

(227)

Several studies deal with polycondensations of diaryl disulfides [351–356]. In most cases the diphenyldisulfide was activated in the form of a sulfenium ion by an oxidizing agent such as $SbCl_5$ (228). Instead of a free sulfenium ion a stabilized form (229a) was proposed as reactive intermediate. In addition of $SbCl_5$, 1,4-benzoquinone, 2,3-dichloro-4,5-dicyanobenzoquinone [353] and O_2/V_2O_5 [354] were used as oxidation agents and catalysts. Furthermore, a broad variety of substituted diphenyldisulfides, (230a,b) and (231a,b), were used as monomers. Regardless of monomer structure and catalyst, the M_ns of the soluble polysulfides (usually containing an insoluble fraction) were below 5000 Da [355]. In this connection a thermal polymerization of 4,4′-diiododiphenyl disulfide should be mentioned (232) [356]. This unusual redox process yields poly(phenylene sulfide) with high molecular weight along with elemental iodine. 4,4′-dichloro- or

4,4′-dibromodiphenyldisulfide may also serve as monomers, when iodide ions are present (233).

$$\text{(228)}$$

$$\text{(229)}$$

$$\text{(230)}$$

a (o, m, p,)   b

$$\text{(231)}$$

a   b

$$\text{(232)}$$

$$\text{(233)}$$

Poly(ether sulfide)s were also prepared by a precursor route, 4,4′-Difluorodiphe-nylsulfoxide (234) was used as electrophilic monomer in combination with hydroquinone or 4,4′-dihydroxybiphenyl. The resulting poly(ether sulfoxide)s were then reduced in tetrachloroethane by means of oxalylchloride and tetrabutylammonium fluoride (235) [357]. Furthermore, the preparation of sulfonated poly(phenylene-sulfide) should be mentioned. SO_3 served as sulfonating agent and the sulfonated polysulfide was treated with $SOCl_2$, whereby a polysulfide with SO_2Cl and Cl substituents was obtained [358].

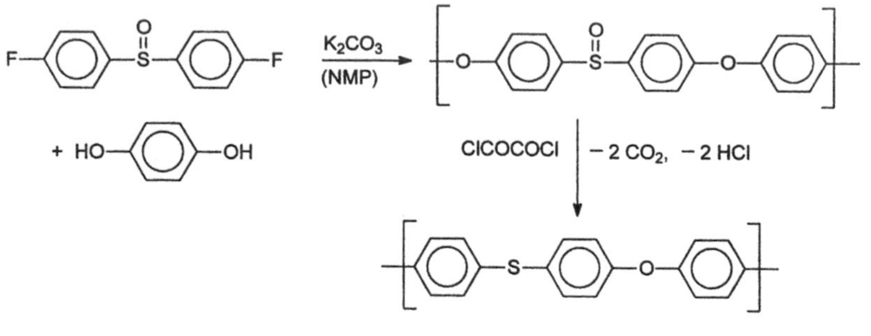

(234)

(235)

A new approach to the synthesis of aromatic polysulfides and polydisulfides consists of the preparation of polydisperse cyclic oligosulfides (or disulfides) followed by ring-opening polymerization. In several publications [358–365] a Canadian research group has elaborated and described this approach in much detail. This approach may be subdivided into two classes of cycles and ring opening mechanisms:

1. macrocylic sulfides (containing C–S–C bond)s
2. macrocyclic disulfides (containing C–S–S–C bonds).

Three different synthetic methods were explored for the preparation of macrocyclic sulfides [359–361]. The first method is based on the normal synthesis of cyclic oligoether via nucleophilic substitution steps involving 4,4'-difluorodiphenylsulfoxide as electrophilic monomer (236). The isolated sulfoxide macrocycles were reduced to the sulfide cycles by means of oxalylchloride (237). Another variant of this method consists of the use of monomers (238) as nucleophilic reaction partners instead of free diphenols. In this way macrocycles containing three sulfide bonds in the repeating unit were obtained [359]. The second method is again based on the nucleophilic substitution yielding cyclic oligoethers which, when derived from 4,4'-dihydroxy diphenyl sulfide (239), contain an aromatic sulfide group. The third method results from a thermal redox rearrangement of copper 4-bromothiophenolate (240). All these reactions were conducted under high dilution to favor the formation of small macrocycles. The ring opening polymerization was performed in bulk or in *m*-terphenyl at temperatures $\geq 300\,^{\circ}\mathrm{C}$. Elemental sulfur or diphenylsulfide were used as initiators and a polymerization mechanism involving free radicals was assumed.

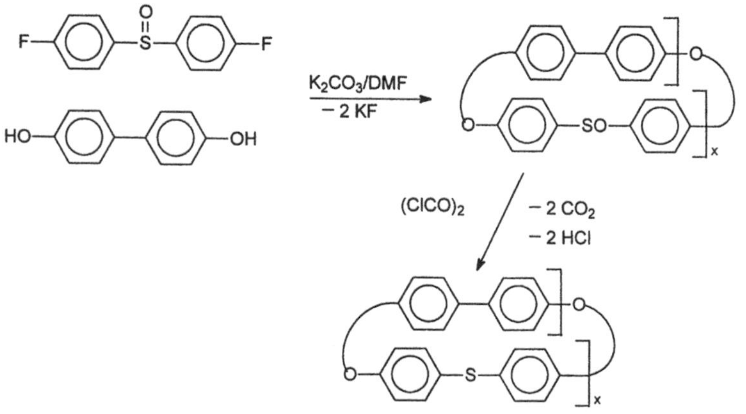

(236)

(237)

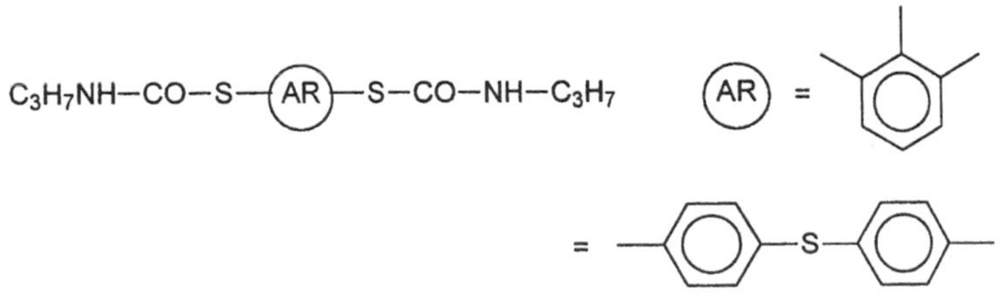

C3H7NH—CO—S—(AR)—S—CO—NH—C3H7   (AR) =

(238)

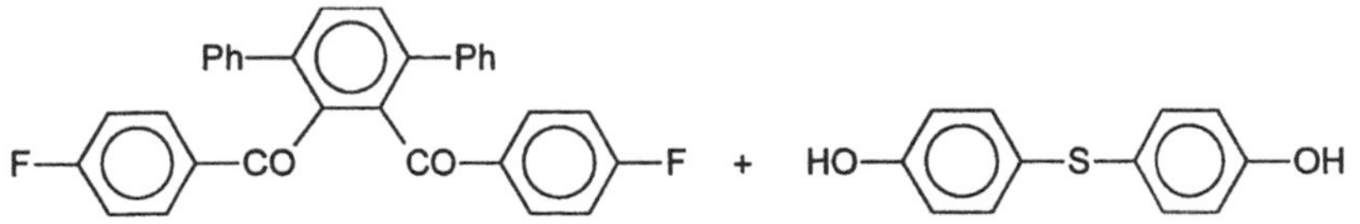

=

(239)

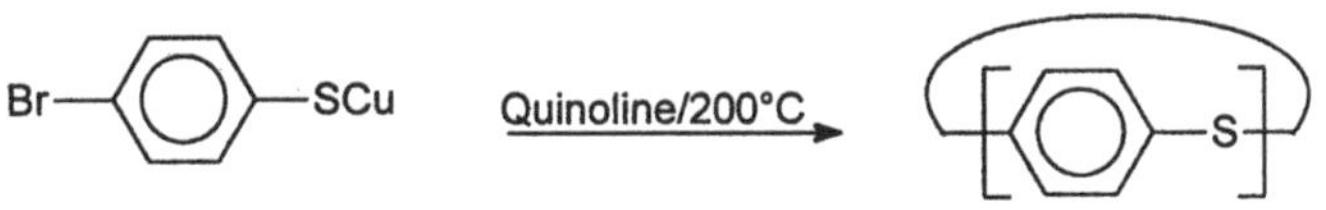

Ph— —Ph

F— —CO   CO— —F   +   HO— —S— —OH

Br— —SCu   Quinoline/200°C⟶

(240)

The macrocyclic disulfides were all prepared by Cu-catalyzed oxidation of aromatic dithiols under high dilution (241). However, three different methods were used for the preparation of numerous dithiophenols [362]. The polymerizations were conducted in bulk or in diphenyl ether between 150 and 250 °C, because at higher temperatures considerable side reactions were observed. Again a free radical polymerization mechanism was assumed starting with the thermal dissociation of the disulfide bond (242). A particularly interesting kind of copolymerization was found, when macrocyclic disulfides were heated together with dibromo- or diiodoaromats in the presence of KI. Due to a radical redox process between S–S and C–I bonds polysulfides were formed along with free iodine (243,244). In this way polysulfides with a regular sequence of both aromatic moieties were obtained [361]. Both polymerization methods the thermal homopolymerization of cyclic disulfides and the copolymerization with dihaloaromates yielded high molecular weight soluble polymers. Finally, the copolymerization of macrocyclic disulfides with elemental sulfur should be mentioned [365]. In summary syntheses and ROP of macrocyclic sulfides and disulfides proved to be a versatile and successful approach to the preparation of high molecular weight aromatic polymers.

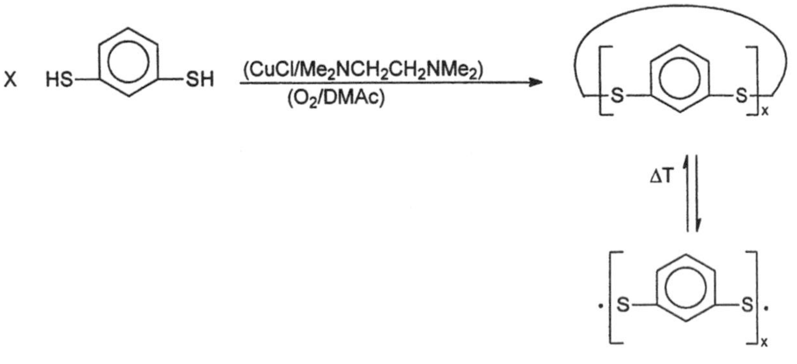

X   HS— —SH $\frac{(CuCl/Me_2NCH_2CH_2NMe_2)}{(O_2/DMAc)}$ ⟶

(241)

ΔT

(242)

$$(243)$$

$$(244)$$

REFERENCES

1. Hay, A. S. (1967). *Adv. Polym. Sci.*, 4: 496.
2. Elias, H.-G. (1975). *Neue Polymere Werkstoffe*, Carl Hauser Publ., München, pp. 106–111.
3. Kricheldorf, H. R. (1992). In *Handbook of Polymer Synthesis* (Kricheldorf, H. R. ed.), Marcel Dekker, New York, Chapter 9.
4. Vogtlänger, V. (1980). *Kunststoffe*, 70: 645.
5. Jenninger, B. E., and Jones, M. E. B. (1967). *J. Polym. Sci., Part C*, 16: 715.
6. Rose, J. B. (1974). *Chimia*, 38: 561.
7. Lonzano, A. E., Jimeno, M. L., de Abajo, J., and de la Campa, J. G. (1994). 27: 7164.
8. Carter, K. R. (1995). *Macromolecules*, 28: 6462.
9. Carlier, U., Devaux, J., Legras, R., and MaGrail, P. T. (1993). *Polymer* 34: 167.
10. Mani, R. S., and Mohanty, D. K. (1995). *Macromolecular Reports A 32 (Suppl. 8)*: 1189.
11. Imai, Y., Ishikawa, H., Park, K.-H., and Kakimoto, M. (1997). *J. Polym. Sci., Part A, Polym. Chem.*, 35: 2055.
12. Kutoku, F., Kakimoto, M., Imai, Y. (1994). *J. Polym. Sci., Part A, Polym. Chem.*, 32: 317.
13. Butuc, E., Rusa, M., Cozan, U., Stoleriu, A., and Simonescu, B. C. (1998). *J.M.S.–Pure Appl. Chem.*, A35: 175.
14. Ritter, H., and Bodewald, B. (1994). *Macromolecular Reports (Suppl. 2) A*, 33: 103.
15. Gao, Ch., and Hay, A. S. (1995). *J. Polym. Sci., Part A, Polym. Chem.*, 33: 2347.
16. Attwood, T. E., Cinderey, M. B., and Rose, J. B. (1993). *Polymer*, 34: 1322.
17. Staniland, P. A., Wilde, C. J., Bottino, F. A., Di Pasquale, G., Pollicino, A., and Recca, A. (1992). *Polymer*, 33: 1976.
18. Yang, J., and Gibson, H. (1999). *Macromolecule*, 32: 8740.
19. Bottino, F. A., Di Pasquale, G., Leonardi, N., and Pollicino, A. (1995). *J.M.S–Pure Appl. Chem.*, A32: 1947.
20. Cummings, D. R., Mani, R. S., Bolanida, P. B., Howell, B. A., and Mohauty, D. K. (1991). *J.M.S.–Pure Appl. Chem.*, A28: 793.
21. Andrews, S. M. (1992). *J. Polym. Sci. Part. A. Polym. Chem.*, 30: 221.
22. Archibald, R. S., Sheares, V. V., Samulski, E. T., and de Simone, P. M. (1993). *Macromolecules*, 26: 7083.
23. McKinnon, S. M., Bander, T. P., and Wang, Z. Y. (2000). *J. Polym. Sci., Part A Polym. Chem.*, 38: 9.
24. McKinnon, S. M., and Wang, Z. Y. (1998). *Macromolecules*, 31: 7970.

25. Maier, G., Yang, D., and Nuyken, O. (1994). *Macromol. Chem. Phys. 195*: 3721.
26. Feiring, A. E., Wonderba, E. R., and Arthur, S. D. (1990). *J. Polym. Sci., Part A, 28*: 2809.
27. Clark, J. H., and Deness, J. E. (1994). *Polymer, 35*: 2432.
28. Clark, J. H., Deness, J. E., Wynd, A. J., and McGrail, T. M. (1994). *J. Polym. Sci., Part A, Polym. Chem., 32*: 1185.
29. Ueda, M., Toyota, H., Quchi, T., Sugiyama, J.-T., Yonetake, K., Masuko, T., and Teramoto, T. (1993). *J. Polym. Sci., Part A, Polym. Chem., 31*: 853.
30. Bunn, A., and Rose, J. B. (1993). *Polymer, 34*: 1114.
31. Al-Omran, A., and Rose, J. B. (1996). *Polymer, 37*: 1735.
32. Cunningham, P., Roach, R. J., Rose, J. B., McGrail, P. T. (1992) *Polymer, 33*: 3951.
33. Bottino, F. A., Mamo, A., Recca, A., Brady, J., Street, A. C., and McGrail, P. T. (1993). *Polymer, 34*: 2901.
34. Pak, S. J., Lyle, G. D., Mercier, R., and McGrath, J. E. (1992). *Polym. Bull., 29*: 477.
35. Naik, H. A., Parsons, I. W., McGrail, P. T., and McKenzie, P. D. (1991). *Polymer, 32*: 240.
36. Strukelj, M., and Hay, A. S. (1991). *Macromolecules, 24*: 6870.
37. Strukelj, M., and Hay, A. S. (1992). *Macromolecules, 25*: 4721.
38. Herbert, C. G., Ghassemi, H., and Hay, A. S. (1997). *J. Polym. Sci., Part A, Polym. Chem., 35*: 1095.
39. Guiver, M. D., Kutowy, O., and ApSimon, J. W. (1989). *Polymer, 30*: 1137.
40. Guiver, M. D., Robertson, G. P., and Foley, S. (1995). *Macromolecules, 28*: 7612.
41. Esser, J. C., Naik, H. N., Parsons, T. W. (1993). *Euro. Polym. J., 29*: 193.
42. Warshawsky, A., Kahana, N., Deshe, A., Gottlieb, H. E., Arad-Yellin, R. (1990). *J. Polym. Sci., Part A, Polym. Chem., 28*: 2885.
43. Naik, H. A., McGrail, P. T., McKenzie, P. D., and Parsons, T. W. (1992). *Polymer, 33*: 166.
44. Cozan, V., Butuc, E., Stoleriu, A., and Cascaval, A. (1995). *J.M.S.–Pure Appl. Chem., A32* 1067.
45. Gao, Ch., and Hay A. S. (1995). *J. Polym. Sci., Part A, Polym. Chem., 33*: 2731.
46. Marletta, G., Pignataro, S., Tóth, A., Bertóti, I., Székely, T., and Keszler, B. (1991). *Macromolecules, 24*: 99.
47. Hedrick, J. L., Yilgor, I., Jurek, M., Hedrick, J. C., Wilkens, G. L., and McGrath, J. E. (1991). *Polymer, 32*: 2020.
48. Pospiech, D., Häußler, L., Meyer, E., Jehnichen, D., Jahnke, A., Böhme, S., and Kricheldorf, H. R. (1988). *Designed Mon. Polym., 1*: 103.
49. Pospiech, D., Häußler, L., Eckstein, K., Komber, H., Voigt, D., Jehnichen, D., Meyer, E., Jahnke, A., and Kricheldorf, H. R. (1998). *Designed Mon. Polym., 1*:187
50. Maes, C., Decaux, J., Legras, R., Parsons, I. W., and McGrail, P. T. (1994). *J. Polym. Sci., Part A, Polym. Chem., 32*: 3171.
51. Kricheldorf, H. R., and Adebahr, T. (1993). *Makromol. Chem., 194*: 2103.
52. Jurek, M. J., and McGrath, J. E. (1989). *Polymer, 30*: 1552.
53. Oishi, Y., Nakata, S., Karkimoto, M., and Imai, Y. (1993). *J. Polym. Sci., Part A, Polym. Chem., 31*: 111.
54. Oishi, Y., Nakata, S., Kakimoto, M., and Imai, Y. (1993). *J. Polym. Sci., Polym. Chem., 31*: 933.
55. Ibieta, J. B., Kalika, D. S., and Penn, L. S. (1998). *J. Polym. Sci., Part A, Polym. Chem., 36*: 1309.
56. Bottino, F. A., DiPasquale, G., Leonardi, N., and Pollicino, A. (1996). *J. Polym. Sci., Part A, Polym. Chem., 34*: 1305.
57. Hedrick, J. L., Brown, H. R., Hofer, D. C., and Johason, R. D. (1989). *Macromolecules, 22*: 2048.
58. Collyer, A. A., Clegg, D. W., Morris, M., Parker, D. G., Wheatley, G. W., and Corfield, G. C. (1991). *J. Polym. Sci., Part A, Polym. Chem., 29*: 193.
59. Padmanaban, M., Kakimoto, M., and Imai, Y. (1990). *J. Polym. Sci., Part A, Polym. Chem., 28*: 2997.

60. Nagasi, Y., Naruse, A., and Matsui, K. (1989). *Polymer*, *30*: 1931.
61. Cameron, N. R., and Sherrington, D. C. (1997). *Macromolecules*, *30*: 5860.
62. Delfort, B., Lucotte, G., and Cormier, L. (1990). *J. Polym. Sci., Part A, Polym. Chem.*, *28*: 2451.
63. Lucotte, G., Cormier, L., and Delfort, B. (1991). *J. Polym. Sci., Part A, Polym. Chem.*, *29*: 897.
64. Percec, V., Wang, J. H., and Oishi, Y. (1991). *J. Polym. Sci., Part A, Polym. Chem.*, *29*: 949.
65. Percec, V., Wang, J. H., Oishi, Y., and Feiring, A. E. (1991). *J. Polym. Sci., Part A*, *29*: 965.
66. Pandya, A., Yang, J., and Gibson, H. W. (1994). *Macromolecules*, *27*: 1387.
67. Chen, M., Guzei, J., Rheingold, A. L., and Gibson, H. W. (1997). *Macromolecules*, *30*: 2516.
68. Attwood, T. E., Dawson, P. C., Freeman, J. L., Hoy, C. R. J., Rose, J. B., and Staniland, P. A. (1981). *Polymer*, *22*: 1096.
69. Mullins, M. J., Woo, E. P., Chen, C. C., Murray D. J., Bishop, M. T., and Balon, K. E. (1991). *Polym. Prepr. (Am. Chem. Soc. Div. Polym. Chem.)*, *32*: 174.
70. Mullins, M. J., Woo, E. P., Murray, D. J., and Bishop, M. T. (1993). *CHEMTECH*, 25.
71. Ganauly, S., and Gibson, H. W. (1993). *Macromolecules*, *26*: 2408.
72. Xie, D., and Gibson, H. W. (1996). *Macromol. Chem. Phys.*, *197*: 137.
73. Xie, D., Ji, Q., and Gibson, H. W. (1997). *Macromolecules*, *30*: 4814.
74. Jiang, H., Chen, T., Bo, S., and Xu, J. (1998). *Polymer*, *39*: 6079.
75. Qui, Y., Song, N., Chen, T., Bo, S., and Xu, J. (2000). *Macromol. Chem. Phys.*, *201*: 840.
76. Ben-Haida, A., Colquhoun, H. M., Hodge, P., Kohnke, F. H., and Williams, D. J. (1997). *Chem. Commun.* 1533.
77. Baxter, I., Ben-Haida, A., Colquhoun, H. M., Hodge, P., Kohnke, F. H., and Williams, D. J. (1998). *Chem. Commun.* 2213.
78. Bodewald, B., and Ritter, H. (1997). *Macromol. Rapid Commun.*, *18*: 817.
79. Gileva, N. G., Zolotukhin, M. G., Salaskin, S. N., Sultanova, V. S., Hörhold, H. H., and Raabe, D. (1988). *Acta Polym.*, *39*: 452.
80. Sakaguchi, Y., Tokai, M., and Kato, Y. (1993). *Polymer*, *34*: 1512.
81. Janson, V., and Dahl, c. (1991). *Makromol. Chem. Macromol. Symp.*, *51*: 87.
82. Özarslan, Ö., Yildiz, E., Yilmaz, T., Güngör, A., and Kuyulu, A. (1998). *Macromol. Chem. Phys.*, *199*: 1787.
83. Tokai, M., Sakaguchi, Y., and Kato, Y. (1995). *High Perform. Polym.*, *7*: 267.
84. Zolothukin, M. G., Dosiere, M., Fougnies, C., Villers, D., Gilveva, N. G., and Fatykhov, A. A. (1995). *Polymer*, *36*: 3575.
85. Zolotukhin, M. G., Ruida, D. R., Baltá-Calleja, F. J., Bruix, M., Cagiao, M. E., Bulai, A., and Gileva, N. G. (1997). *Macromal. Chem. Phys.*, *198*: 1131.
86. Zolotukhin, M. G., Rueda, D. R., Baltá-Calleja, F. J., Cagiao, M. E., Bruix, M., Sedova, E. A., and Gileva, N. G. (1977). *Polymer*, *38*: 1471.
87. Zolotukhin, M. G., Rueda, D. R., Bruix, M., Cagiao, M. E., Baltá-Calleja, Bulai, A., Gileva, N. G., and Van der Elst, L. (1997). *Polymer*, *38*: 3441.
88. Zolotukhin, M. G., Baltá-Calleja, F. J., Rueda, D. R., and Palacios, J. M. (1997). *Acta Polymerica*, *48*: 269.
89. Bai, S. J., Dotrong, M., Soloski, E. J., and Evers, R. C. (1991). *J. Polym. Sci., Part B., Polym Phys.*, *29*: 119.
90. Ueda, M., Abe, T., and Oda, M. (1992). *J. Polym. Sci., Part A, Polym. Chem.*, *30*: 1993.
91. Finocchiaro, P., Montaudo, G., Mertioli, P., Puglisi, P., and Samperi, F. (1996). *Macromol. Chem. Phys.*, *197*: 1007.
92. Saegusa, Y., Kojima, A., and Nakamura, S. (1993). *Makromol. Chem.*, *194*: 777.
93. Tunca, U., and Hizal, G., (1998). *J. Polym. Sci., Part A, Polym. Chem.*, *36*: 1227.
94. Risse, W., Soga, D. Y., and Boettcher, F. P. (1991). *Makromol. Chem. Macromol. Symp.*, *44*: 185.
95. Colquhoun, H. M., and Lewis, D. F. (1988). *Polymer*, *29*: 1902.
96. Colquhoun, H. M., Daniels, J. A., Stephenson, I. R., and Wade, K. (1991). *Polymer Commun.*, *32*: 272.
97. Tunca, U. (1997). *Angew. Makromol. Chem.*, *253*: 89.

98. Douglas, J. E., and Wang, Z. Y. (1995). *Macromolecules, 28*: 5970.
99. Percec, V., Clough, R. S., Rinaldi, P. L., and Litman, V. E. (1991). *Macromolecules, 24*: 5889.
100. Mani, R. S., Zimmermann, B., Bhatnagar, A., and Mohanty, D. K. (1993). *Polymer, 34*: 171.
101. Bhatnagar, A., Mani, R. S., King, B., and Mohanty, D. K. (1994). *Polymer, 35*: 1111.
102. Bhatnagar, A., Mani, R. S., King, B., and Mohanty, D. K. (1996). *Macromol. Chem. Phys., 197*: 315.
103. Dukes, K. E., Forbes, M. D. E., Jeevarajan, A. S., Belu, A. M., DeSimone, J. M., Linton, R. W., and Sheares, V. V. (1996). *Macromolecules, 29*: 3081.
104. Hoffmann, U., Klapper, M., Müller, K. (1993). *Polym. Bull., 30*: 481.
105. Hoffmann, U., Helmer-Metzmann, F., Klaper, J., and Müllen, K. (1994). *Macromolecules, 27*: 3575.
106. Carlier, V., Devaux, J., Legras, R., Bunn, A., and McGrail, P. T. (1994). *(Polymer), 35*: 415.
107. Carlier, V., Devaux, J., Legras, R., and McGrail, P. T. (1994). *Polymer, 35*: 423.
108. Fukawa, I., and Tanabe, T. (1993). *J. Polym. Sci., Part A, Polym. Chem., 31*: 535.
109. Strukeli, M., Hedrick, J. C., Hedrick, J. L. and Twieg, R. I. (1994). *Macromolecules, 27*: 6277.
110. Labadie, J. W., Carter, K. P., Hedrick, J. L., Jonsson, H., Kim, S. Y., and Twieg, R. J. (1993). *Polym. Bull., 30*: 25.
111. Rao, M. R., Rao, V. K., Radakristinan, T. S., and Ramachandram, S. (1992). *Polymer, 33*: 2834.
112. Rao, V. L., and Sivadasan, P. (1994). *Eur. Polym. J., 30*: 1381.
113. Ruan, R., Jiang, Z., Xu, W., Li, G., Wu, Z., Shibata, M., and Yosomiya, R. (1999). *Angew. Makromol. Chem., 270*: 33.
114. Mathias, L. J., Lewis, C. M., and Wiegel, K. N. (1997). *Macromolecules, 30*: 5970.
115. Kim, W. G., and Hay, A. S. (1992). *J.M.S.–Pure Appl. Chem., A29*: 1141.
116. Venkatasubramanian, N., Dean, D. R., Price, G. E., and Arnold, F. E. (1997). *High Perform. Polym., 9*: 291.
117. Keitoku, F., Kakimoto, M., and Imai, Y. (1994). *J. Polym. Sci., Part A, Polym. Chem., 32*: 317.
118. Kim, W. G., and Hay, A. S. (1992). *Makromol. Chem. Macromol. Symp. 54/55*: 331.
119. Akutsu, F., Takahashi, K., Kasashima, Y., Inoki, M., and Naruchi, K. (1995). *Macromol. Rapid Commun., 16*: 495.
120. Matsuo, S., Yakoh, N., Chino, S., Mitani, M., and Tagami, S. (1994). *J. Polym. Sci. Part. Polym. Chem., 32*: 1071.
121. Wang, Z. Y., and Hay, A. S. (1991). *J. Polym. Sci., Part A, Polym. Chem., 29*: 1045.
122. Meng, Y., Hlil, A. R., and Hay, A. S. (1999). *J. Polym. Sci. Part. A., Polym. Chem., 37*: 1781.
123. Liggat, J. J. and Stamland, P. A. (1991). *Polymer Commun., 32*: 450.
124. Singh, R., and Hay, A. S. (1991). *Macromolecules, 24*: 2639.
125. Singh, R., and Hay A. S (1992). *Macromolecules, 25*: 1017.
126. Taguchi, Y., Uyama, H., and Kobayashi, S. (1996). *J. Polym. Sci., Part A., 34*: 561.
127. Ueda, M., Nakayama, T., and Mitsukashi, T. (1997). *J. Polym. Sci., Part A, Polym. Chem., 35*: 371.
128. Ueda, M., Toyoda, H., Nakayama, T., and Abe, T. (1996). *J. Polym. Sci., Part A, Polym. Chem., 34*: 109.
129. Mani, R. S., Weeks, Br., and Mohanty, D. K. (1993). *Macromol. Chem. Phys., 194*: 1935.
130. Percec V., Grigoras, M., Clough, R. S., and Fanjul, J. (1995). *J. Polym. Sci., Part A, Polym. Chem., 33*: 331.
131. Zhang, C., and Wang. Z. Y. (1993). *Macromolecules, 26*: 3324.
132. Bottino, F. A., Di Pasquale, G., Leonardi, N., and Pollicino, A. (1998). *Polymer, 39*: 3199.
133. Bottino, F. A., Di Pasquale, G., Leonardi, N., and Pollicino A. (1995). *J.M.S.–Pure Appl. Chem. A32*: 1947.
134. Yoshida, S., and Hay, A. S. (1997). *J.M.S.–Pure Appl. Chem., A34*: 1299.
135. Ohno, M., Takata, T., and Endo, T. (1994). *Macromolecules, 27*: 3447.
136. Ohno, M., Takata, T., and Endo, T. (1995). *J. Polym. Sci., Part A, Polym. Chem., 33*: 2647.

137. Wang, Z. Y., and Le Guen, A. (1995). *Macromolecules, 28*: 3728.
138. Ritter, H., and Torwirth, R. (1993). *Makromol. Chem., 194*: 1469.
139. Elce, E., and Hay, A. S. (1995). *J.M.S.–Pure Appl. Chem. A32*: 1709.
140. Meier, G., Yang, D., and Nuyken, O. (1993). *Macromol. Chem., 194*: 1901.
141. Meier, G., and Wolf, M. (1997). *Macromol. Chem. Phys., 198*: 2421.
142. Staniland, P. A., Wilde, C. J., Bottino, F. A., Di Pasquale, G., Pollicino, A., and Recca, A. (1992). *Polymer, 33*: 1976.
143. Han, Y.-K., Chi, S. D., Kim, Y. H., Park, B. K., and Jin, J.-I. (1995). *Macromolecules, 28*: 916.
144. Fitch, J. W., Reddy, V. S., Youngman, P. W., Wohlfahrt, G. A., and Cassidy, P. E. (2000). *Polymer, 41*: 2301.
145. De Simone, J. M., and Sheares, V. V. (1992). *Macromolecules, 25*: 4235.
146. De Simone, J. M., Stompel, S., Samulski, E. T., Wang, Y. G., and Brennan, A. B. (1992). *25*: 2546.
147. Cornier, L., Lucotte, G., and Delfort, B. (1991). *Polym. Bull., 26*: 395
148. Gibson, H. W., and Guiliani, B. (1991). *Polymer Commun., 32*: 324.
149. Mercer, F. W., Fone, M. M., Reddy, V. N., and Goodwin, A. A. (1997). *Polymer., 38*: 1989.
150. Tullos, G. L., Cassidy, P. E., and St. Clair, A. K. (1991). *Macromolecules, 24*: 6059.
151. Mercer, F. W., Fone, M. M., and McKenzie, M. T. (1997). *J. Polym. Sci., Part A, Polym. Chem., 35*: 521.
152. Clask, J. H., and Deness, J. E. (1994). *Polymer, 35*: 5124.
153. Park, S. K., and Kim, S. Y. (1998). *Macromolecules, 31*: 3385.
154. Elce, E., and Hay, A. S. (1995). *J. Polym. Sci., Part A, Polym. Chem., 33*: 1143.
155. Zhang, S., Zheng, Y., Wu, Z. W., Tian, M., Yang, D., and Yosomiya, R. (1997). *Polym. Bull., 38*: 621.
156. Zhang, S. J., Cheng, S. Z. D., Zheng, Y. B., Wu, Z. W., Yosomiya, R. (1998). *Acta Polymerica, 49*: 198.
157. Zhang, S. Zheng, Y., Wu, Z. W., Tian, M. W., Yang, D., and Yosomiya, R. (1997). *Macromol. Rapid Commun., 18*: 729.
158. Howe, L. A., and Jaycox, G. D. (1998). *J. Polym. Sci., Part A, Polym. Chem., 36*: 2827.
159. Jaycox, G. D. (1998) *Polymer, 39*: 2598.
160. Corfield, G. C., Wheatley, G. W., and Parker, D. G. (1992). *J. Polym. Sci., Part A, Polym. Chem., 30*: 645.
161. Hedrick, J. L., Volksen, W., and Mohanty, D. K. (1992). *J. Polym. Sci., Part A, Polym. Chem., 30*: 2085.
162. Bourgois, Y., Devaux, J., Legras, R., Charlier, Y., and Hedrick, J. L. (1995). *J. Polym. Sci., Part A, Polym. Chem., 33*: 779.
163. Bennett, G. S., Farris, R. J., and Thompson, S. A. (1991). *Polymer, 32*: 1633.
164. Bennett, G. S., and Farris, R. J. (1994). *J. Polym. Sci., Part A, Polym. Chem., 32*: 73.
165. Lyle, G. D., Senger, D. S., Chen, D. H., Kilic, S., Wu, D. S., Hohanty, D. K., and McGrath, J. E. (1989). *Polymer, 30*: 978.
166. Hedrick, J. L., Yang, A. C. M., Scott, J. C., Economy, J. E., and McGrath, J. E. (1992). *Polymer, 33*: 5094.
167. Taguchi, Y., Ugama, H., and Kobayashi, S. (1995). *Macromol. Rapid Commun., 16*: 183.
168. Lucotte, G., Corrnier, L., and Delfort, B. (1990). *Polym. Bull., 24*: 577.
169. Lee, H.-J., Lee, E.-M., Lee, M.-H., Oh, M.-Ch., Ahn, J.-H., Han, S. G., and Kim, H. G. (1998) *J. Polym. Sci., Part A, Polym. Chem., 36*: 2881.
170. Delfort, B., Lucotte, G., and Corrnier, L. (1990). *J. Polym. Sci., Part A, Polym. Chem., 28*: 2451.
171. Martinez-Nuñez, F., de Abajo, J., Mercier, R., and Sillion, B. (1992). *Polymer, 33*: 3286.
172. Taguchi, Y., Uyama, H., and Kobayashi, S. (1997). *J. Polym. Sci., Polym. Chem., 35*: 271.
173. Clendinning, R. A., Keisey, D. R., Botkin, J. H., Winslow, P. A., Youssefi, M., Cotter, R. J., Matzner, M., and Kwiatkowski, G. T. (1993). *Macromolecules, 26*: 2361.

174. Harris, J. E., Winslow, P. A., Botkin, J. H., Maresca, L. M., Clendinning, R. A., Cotter, R. J., Matzner, M., and Kwiatkowski, G. T. (1993). *Macromolecules, 26*: 2366.

175. Botkin, J. H., Cottner, R. J., Matzner, M., and Kwiatkowski, G. T. (1993). *Macromolecules, 26*: 2372.

176. Kricheldorf, H. R., Chen, X., and Al Masri, M. (1995). *Macromolecules, 28*: 2112.

177. Cao, J., Su, W.-C., Chen, Y., Wu, Z.-W., Shibata, M., and Yosomiya, R. (1996). *Angew. Makromol. Chem., 238*: 191.

178. Cao, J., Su, W.-C., Wang, J., Wu, Z.-W., Shibata, M., and Yosomiya, R. (1997). *Anew Makromol Chem., 250*: 93.

179. Corfield, G. C., and Wheatly, G. W. (1990). *J. Polym. Sci., Part A, Polym. Chem., 28*: 2821.

180. Kricheldorf, H. R., and Stukenbrock, T. (1998). *J. Polym. Sci., Part A, Polym. Chem., 36*: 31.

181. Chu, F., and Hawker, C. J. (1993). *Polym. Bull., 30*: 266.

182. Hawker, C. J., and Chu, F. (1996). *Macromolecules, 29*: 4370.

183. Shu, C.-F., and Leu, C.-M. (1999). *Macromolecules, 32*: 100.

184. Shu, C.-F., Leu, C.-M., and Huang, F.-Y. (1999). *Polymer, 40*: 6591.

185. Martinez, C. A., and Hay, A. S. (1998). *J.M.S.–Pure. Appl. Chem. A35*: 57.

186. Martinez, C. A., and Hay, A. S. (1997). *J. Polym. Sci., Part A, Polym. Chem., 35*: 2015.

187. Fukawa, I., Tanabe, T., and Dozono, T. (1991). *Macromolecules, 24*: 3838.

188. Fukawa, I., and Tanabe, T. (1992). *J. Polym. Sci., Part A, Polym. Chem., 30*: 1977.

189. Ueda, M., and Ichikawa, F. (1990). *Macromolecules, 23*: 926.

190. Ueda, M., Seino, Y., Hamede, Y., Yoneda, M., and Sugiyama, J. (1994). *J. Polym. Sci., Part A, Polym. Chem., 32*: 675.

191. Pandya, A., Yang, J., and Gibson, H. W. (1994). *Macromolecules, 27*: 1367.

192. Yang, J., and Gibson, H. W. (1997). *Macromolecules, 30*: 5629.

193. Gibson, H. W., and Dotson, D. L. (1998). *Polymer, 39*: 6483.

194. Yang, J., and Gibson, H. W. (1999). *Macromolecules, 32*: 8740.

195. Moore, J. S. (1992). *Mokromol. Chem. Rapid commun., 13*: 91.

196. Colquhoun, H. M., Dudman, C. D., Thomas, M., O'Mahoney, C. A., and Williams, D. J. (1999). *J. Chem. Soc. Chem. Commun.* 336.

197. Qui, Y., Chen, T., and Xu, J. (1999). *Polym. Bull., 42*: 245.

198. Chan, K. P., Wang, Y.-F., and Hay, A. S. (1995). *Macromolecules, 28*: 653.

199. Gao, C.-P., and Hay, A. S. (1995). *Polymer, 36*: 4141.

200. Chan, K. P., Wang, Y.-F., Hay, A. S., Hrowowski, X. L., and Cotter, J. (1995) *Macromolecules, 28*: 6705.

201. Wang, Y.-F., Paventi, M., Chan, K. P., and Hay, A. S. (1996). *J. Polym. Sci., Part A, Polym. Chem., 34*: 2135.

202. Ding, Y., and Hay, A. S. (1996). *Macromolecules, 29*: 3090.

203. Wang, Y. F., Paventi, M., and Hay, A. S. (1997). *Polymer, 38*: 469.

204. Shibata, M., Yosomiya, R., Chen, C., Zhou, H., Wang, J., and Wu, Z. (1999). *Eur. Polym. J., 35*: 1967.

205. Wang, Y. F., Chan, K. P., and Hay, A. S. (1996). *Macromolecules, 29*: 3717.

206. Wang, Y. F., Chan, K. P., and Hay, A. S. (1996). *J. Polym. Sci., Part A, Polym. Chem., 34*: 375.

207. Teasley, M. F., and Hsiao, B. S. (1996). *Macromolecules, 29*: 6432.

208. Teasley, M. F., Wu, D. Q., and Harlew, R. L. (1998). *Macromolecules, 31*: 2064.

209. Chen, M. F., Fronczek, F., and Gibson, H. W. (1996). *Macromol. Chem. Phys., 197*: 4069.

210. Chen, M. F., and Gibson, H. W. (1996). *Macromolecules, 29*: 5502.

211. Wang, J., Chen, C., Xun, X., Wang, S., and Wu, Z. (1999). *J. Polym. Sci., Part A, Polym. Chem., 37*: 1957.

212. Jiang, H. Y., Qi, Y.-H., Chen, T. L., Xing, Y., Lin, Y.-H., and Xu, J. P. (1997). *J. Polym. Sci., Part A, Polym. Chem., 35*: 1753.

213. Colquhoun, H. M., Sestina, L. C., Zolotukhin, M. G., and Williams, D. J. (2000). *Macromolecules, 33*: 8907.

214. Risse, W., and Sogah, D. Y. (1990). *Macromolecules*, *23*: 4029.
215. Zolotukhin, M. G., de Abajo, J., Alvarez, J. C., de la Campa, J. G., and Rueda, D. R., (1998). *J. Polym. Sci., Part A, Polym. Chem.*, *36*: 1251.
216. Roovers, J., Cooney, J. D., and Toporowski, P. M. (1990) *Macromolecules*, *23*: 1611.
217. Lindfors, B. E., Mani, R. S., McGrath, J. E., and Mokanty, D. K. (1991). *Makromol. Chem. Rapid Commun.*, *12*: 337.
218. Brink, A. E., Gutzeit, S., Lin, T., Marand, H., Lyon, K., Hua, T., Davis, R., and Riffle, J. S. (1993). *Polymer*, *34*: 825.
219. Bourgeois, Y., Devaux, J., Legras, R., and Parsons, I. W. (1996). *Polymer*, *37*: 3171.
220. Wang, F., Chen, T., and Xu, J. (1998). *Macromol. Chem. Phys.*, *199*: 1421.
221. Wang, F., Li, J., Chen, T., and Xu, J. (1999). *Polymer*, *40*: 795.
222. Parthiban, A., Le Guen, A., Yansheng, Y., Hoffmann, U., Klapper, M., and Müllen, K. (1997). *Macromolecules*, *30*: 2238.
223. Henneuse-Boxus, C., Boxus, T., Duliere, E., Pringalle, C., Tesolin, L., Adriaensen, Y., and Marchand-Brynaert, J. (1998). *Polymer*, *39*: 5359.
224. Franchina, N. L., and McCarthy, T. J. (1991). *Macromolecules*, *24*: 3048.
225. Beasy, M., Wang, F., and Roovers, J. (1994). *Polym. Bull.*, *32*: 281.
226. Wang, F., and Roovers, J. (1994). *J. Polym. Sci., Part A, Polym. Chem.*, *32*: 2413.
227. Wang, F., and Roovers, J. (1993). *Macromolecules*, *26*: 5295.
228. Park, S. K., and Kim, S. Y. (1998). *Macromol. Chem. Phys.*, *199*: 2117.
229. Koch, T., and Ritter, H. (1995). *Macromolecules*, *28*: 4806.
230. Walker, K. A., Markoski, L. J., and Moore, J. S. (1993). *Macromolecules*, *26*: 3713.
231. Jensen, B. J. and Hergenrother, P. M. (1993). *J.M.S–Pure, Appl. Chem. A30*: 449.
232. Klapper, M., Wehrmeister, T., and Müllen, K. (1996). *Macromolecules*, *29*: 5805.
233. Viersen, F. J., Challa, G., and Reedijk, J. (1990). *Polymer*, *31*: 1361.
234. Viersen, F. J., Renkema, J., Challa, G., and Reedijk, J. (1992). *J. Polym. Sci., Part A, Polym. Chem.*, *30*: 901.
235. Viersen, F. J., Challa, G., and Reedijk, J. (1989). *Rec. Trav. Chim. Rays. Bas. 108*: 167 and *108*: 247.
236. Viersen, F. J., Challa, G., and Reedijk, J. (1990). *Polymer*, *31*: 1368.
237. Chen, W., and Challa, G. (1990). *Polymer*, *31*: 2171.
238. Chen, W., Challa, G., and Reedijk, J. (1991). *Polymer Commun.*, *32*: 518.
239. Kricheldorf, H. R. (1991). *Handbook of Polymer Syntheses* (Kricheldorf, H. R. ed.), Marcel Dekker, N.Y., Chapter 9, pp. 554–557.
240. Percec V., and Wang, J. H. (1991). *J. Polym. Sci., Part A, Polym. Chem.*, *29*: 63.
241. Hay, A. S., and Dana, D. E. (1989). *J. Polym. Sci., Part A, Polym. Chem.*, *27*: 873.
242. Dai, D.-R., Marks, T. J., Yang, J., Landquist, P. M., and Wong, G. K. (1990). *Macromolecules*, *23*: 1894.
243. Chao, H.S.-T., Hovatter, T. W., Johnson, B. C., and Rice, S. T. (1989). *J. Polym. Sci., Part A*, *27*: 3371.
244. Unal, H. I., Sanli, O., and Kisakürek, D. (1989). *Polymer*, *30*: 344.
245. Yigit, S., Kisakürek, D., Türker, L., Toppare, L., and Akbulut, U. (1989). *Polymer*, *30*: 348.
246. Sacak, M., Akbulut, U., Kisakürek, D., and Toppare, L. (1989). *Polymer*, *30*: 928.
247. Sacak, M., Akbulut, U., Kisakürek, D., Türker, L., and Toppare, L., (1989). *J. Polym. Sci., Part A, Polym. Chem.*, *27*: 1599.
248. Maier, G., Hecht, R., Nuyken, O., Burger, K., and Helmreich, B. (1993). *Macromolecules*, *26*: 2583.
249. Maier, G., and Hecht, R. (1995). *Macromolecules*, *28*: 7558.
250. Maier, G., and Schneider, J. M. (1998). *Macromolecules*, *31*: 1798.
251. Hedrich, J. L., and Twieg, R. (1992). *Macromolecules*, *25*: 2021.
252. Hedrich, J. L. (1992). *Polymer*, *33*: 3375.
253. Connell, J. W., Hergenrother, P. M., and Wolf, P. (1992). *Polymer*, *33*: 3507.
254. Carter, K. P., Miller, R. D., and Hedrick, J. L. (1993). *Macromolecules*, *26*: 2209.

255. Yu, L., and Hay, A. S. (1998). *J. Polym. Sci., Part A, Polym. Chem., 36*: 1107.
256. Connell, J. W., and Hergenrother, P. M. (1991). *J. Polym. Sci., Part A, Polym. Chem., 29*: 1667.
257. Ding, Y., Hay A. S. (1998). *J. Polym. Chem. Sci., Part A, Polym. Chem., 36*: 519.
258. Hilborn, J. G., Labadie, J. W., and Hedrick, J. L. (1990). *Macromolecules, 23*: 2854.
259. Smith, J. G., Connell, J. W., Hergenrother, P. M. (1992). *Polymer, 33*: 1742.
260. Hedrick, J. L., Hillborn, J., Palmer, T. D., and Labadie, J. W. (1990). *J. Polym. Sci., Part A, Polym. Chem., 28*: 2255.
261. Hedrick, J. L. (1992). *Polymer, 33*: 1399.
262. Hedrick, J. L. (1991). *Macromolecules, 24*: 6361.
263. Labadie, J. W., and Hedrick, J. L. (1992). *Makromol. Chem. Macromol. Symp. 54/55*: 313.
264. Hedrick, J. L., and Labadie, J. W. (1990). *Macromolecules, 23*: 1561.
265. Hedrick, J. L., Twieg, R., Matray, T., and Carter, K. (1993). *Macromolecules, 26*: 4833.
266. Labadie, J. W., Hedrick, J. L., Boyer, S. K. (1992). *J. Polym. Sci., Part A, Polym. Chem., 30*: 519.
267. Connell, J. W., and Hergenrother, P. M. (1992). *Polymer, 33*: 3739.
268. Hedrick, J. L., and Labadie, J. W. (1992). *J. Polym. Sci., Part A, Polym. Chem., 30*: 105.
269. Hedrick, J. L., Labadie, J. W., and Russell, T. P. (1991) *Macromolecules, 24*: 4559.
270. Takekoshi, T., and Terry, J. M. (1997). *J. Polym. Sci., Part A, Polym. Chem., 35*: 759.
271. Singh, R., Hay R. S. (1991). *Macromolecules, 24*: 2643.
272. Singh, R., Hay A. S. (1991). *Macromolecules, 24*: 2640.
273. Singh, R., Hay A. S. (1992). *Macromolecules, 25*: 1033.
274. Konno, K., Ueda, M., Cassidy, P. E., and Fitch, J. W. (1997). *J.M.S.–Pure Appl. Chem., A34*: 929.
275. Konno, K., Deguchi, N., Yonetake, K., Ueda, M., Cassidy, P. E., and Fitch, J. W. (1997). *J. Polym. Sci., Part A, Polym. Chem. Ed., 35*: 605.
276. Konno, K., Kobayashi, T., Yonetake, T., Ueda, M., Fitch, J. W., and Cassidy, P. E. (1998). *Polymer, 39*: 719.
277. Suh, D. H., Chung, E. Y., Hong, Y.-T., Choi, K.-Y. (1998). *Angew. Makromol. Chem., 254*: 33.
278. Strukelj, M., Paventi, M., and Hay, A. S. (1993). *Macromolecules, 26*: 1727.
279. Hedrick, J. L. (1991). *Mecromolecules, 24*: 812.
280. Kricheldorf, H. R., Schmidt, B., and Delius, U. (1990). *Eur. Polym. J., 26*: 791.
281. Ganderon, R., Plummer, C. J. G., Hillborn, J. G., and Knauss, D. M. (1998). *Macromolecules, 31*: 501.
282. Kricheldorf, H. R. (1996). *Silicon in Polymer Syntheses* (Kricheldorf, H. R. ed.), Springer Publ., Berlin, N.Y., Chapter 5.
283. Kricheldorf, H. R., and Jahnke, P. (1990). *Makromol. Chem., 191*: 2027.
284. Matsuo, S., Murakami, T., and Takasawa, R. (1993). *J. Polym. Sci., Part A, Polym. Chem., 31*: 3439.
285. Mercer, F. W., McKenzie, M. T., Easteal, A., and Moses, S. J. (1994). *Polymer, 35*: 5355.
286. Mercer, F. W., Easteal, A., and Bruma, M. (1997). *Polymer, 38*: 707.
287. Kricheldorf, H. R., and Berghahn, M. (1991). *Makromol. Chem. Rapid Commun., 12*: 529.
288. Kricheldorf, H. R., and Jürgens, C. (1993). *Eur. Polym. J., 29*: 903.
289. Dutta, P. K. (1995). *Macromolecular Reports A 32 (Suppl. 4)*, 467.
290. Kricheldorf, H. R., Schwarz, G., and Erxleben, J. (1988). *Makromol. Chem., 18*: 2255.
291. Kricheldorf, H. R., Jahnke, P., and Scharnagl, N. (1992). *Macromolecules, 25*: 1382.
292. Kricheldorf, H. R., Jahnke, P. (1993). *J. Polym. Sci., Part A, Polym. Chem., 30*: 1299.
293. Irvin, J. A., Neef, C. J., Kane, K. M., Cassidy, P. E., Tallos, G., St. Clair, A. K. (1992). *J. Polym. Sci., Part A, Polym. Chem., 30*: 1675.
294. Mercer, F., Goodman, T., Woitowicz, J., and Duff, D. (1992). *J. Polym. Sci., Part A, Polym. Chem., 30*: 1767.
295. Labadie, J. W., and Hedrick, J. L. (1990). *Macromolecules, 23*: 5371.

296. Bochmann, M., Kelly, K., and Lu, J. (1992). *J. Polym. Sci., Part A, Polym. Chem.*, *30*: 2511.
297. Lee, J. I., Kwon, L. Y., Kim, J. H., Choi, K. Y., and Suh, D. H. (1998). *Angew. Makromol. Chem.*, *254*: 27.
298. Percec, V., Wang, J. H., and Okita, S. (1991). *J. Polym. Sci., Part A, Polym. Chem.*, *29*: 1789.
299. Percec, V., Okita, S., and Wang, J. H. (1992). *Macromolecules*, *25*: 64.
300. Percec, V., Wang, J. H., and Okita, S. (1992). *J. Polym. Sci., Part A, Polym. Chem.*, *30*: 1992.
301. Percec, V., Wang, J. H., and Oishi, Y. (1992). *J. Polym. Sci., Part A, Polym. Chem.*, *30*: 439.
302. Wang, Z. Y., and Zhang, C. (1992). *Macromolecules*, *25*: 5851.
303. Zhang, C., and Wang, Z. Y. (1993). *Macromolecules*, *26*: 3330.
304. Wang, Z. Y., and Douglas, J. E. (1997). *Macromolecules*, *30*: 8091.
305. Peason, A. J., and Sun, Lei (1997). *J. Polym. Sci., Part A, Polym. Chem.*, *35*: 447.
306. Saegusa, Y., Kato, T., Oshimui, H., and Nakamura, S. (1992). *J. Polym. Sci., Part A, Polym. Chem.*, *30*: 1401.
307. Srinivasan, S. A., and McGrath, J. E. (1993). *High perform. Polym.*, *5*: 259.
308. Srinivasan, S. A., and McGrath, J. E. (1997). *J. Appl. Polym. Sci.*, *65*: 167.
309. Srinivasan, S. A., Joardar, S. S., Kranbuehl, D., Ward, T. C., and McGrath, J. E. (1997). *J. Appl. Polym. Sci.*, *64*: 179.
310. Abed, J. C., Mercier, R., and McGrath, J. E. (1997). *J. Appl. Polym. Sci.*, *64*: 977.
311. Srinivasan, S. A., and McGrath, J. E. (1998). *Polymer*, *39*: 2415.
312. Wang, B., Wilkes, G. L., Smith, C. D., and McGrath, J. E. (1991). *Polym. Commun.*, *32*: 400
313. Podkościelny, W., Dethloff, J., Dethloff, M., and Brunn, J. (1991). *Angew.Makromol. Chem.* *188*: 143.
314. Podkościelny, W., Dethloff, J., and Dethloff, M. (1991). *Angew. Makromol. Chem.*, *188*: 155.
315. Iizawa, T., Kudo, H., and Nishikubo, T. (1991). *J. Polym. Sci., Part A, Polym. Chem.*, *29*: 1875.
316. Papava, G. Sh., Borisov, G. B., Varbanov, S., Vinogradova, S. U., Oirshak, V. V., Tsiskarishvili, R. P., Sarishvili, Z. M., and Razmadze, G. B. (1988). *Acta Polym.*, *39*: 419.
317. Shaffer, T. (1989). *J. Polym. Sci., Part C, Polym. Letters*, *27*: 457.
318. Kricheldorf, H. R., and Al-Masri, M. (1996). *J. Polym. Sci., Part A, Polym. Chem. 34*: 2037.
319. Kricheldorf, H. R., and Al-Masri, M. (1995). *J. Polym. Sci., Part A, Polym. Chem.*, *33*: 2667.
320. Barton, S. J., Ghortra, J. S., Matthews, A. E., and Pritchard, G. (1989). *Polymer*, *30*: 1546.
321. Hawker, C. J., and Fréchet, J. M. J. (1992). *Polymer*, *33*: 1507.
322. Uhrich, K. E., Hawker, C. J., Fréchet, J. M. J., and Turner, S. R. (1992). *Macromolecules*, *25*: 4583.
323. Cimecioglu, A. L., and Weiss, R. A. (1992). *J. Polym. Sci., Part A, Polym. Chem.*, *30*: 1355.
324. Mates, T. E., and Ober, C. K. (1992). *J. Polym. Sci., Part A, Polym. Chem.*, *30*: 2541.
325. Johnsson, H., Werner, P. E., Gedde, U. W., and Hult, A. (1989). *Macromolecules*, *22*: 1683
326. Kricheldorf, H. R., and Linzer, V. (1995). *Polymer*, *36*: 1893.
327. Percec, V., and Yourd, R. (1989). *Macromolecules*, *22*: 524.
328. Percec, V., and Yourd, R. (1989). *Macromolecules*, *22*: 3229.
329. Percec, V., and Tsuda, Y. (1990). *Macromolecules*, *23*: 5.
330. Percec, V., and Tsuda, Y. (1990). *Macromolecules*, *23*: 3509.
331. Percec, V., and Tsuda, Y. (1991). *Polymer*, *32*: 661.
332. Percec, V., and Tsuda, Y. (1991). *Polymer*, *32*: 673.
333. Ungar, G., Feijoo, J. L., Percec, V., and Yourd, R. (1991). *Macromolecules*, *24*: 1168.
334. Ungar, G., Feijoo, J. L., Percec, V., and Yourd, R. (1991). *Macromolecules*, *24*: 953.
335. Percec, V., and Kawasumi, M. (1991). *Macromolecules*, *24*: 6518.
336. Percec, V., and Zuber, M. (1992). *J. Polym. Sci., Part A, Polym. Chem.*, *30*: 997.
337. Ungar, G., Percec, V., and Zuber, M. (1992). *Macromolecules*, *25*: 75
338. Percec, V., Kawasumi, M., Rinaldi, P. L., and Litman, V. E. (1992). *Macromolecules*, *25*: 3851.
339. Percec, V., and Rodenhouse, R. (1989). *Macromolecules*, *22*: 2043.
340. Percec, V., and Kawasumi, M. (1992). *Macromolecules*, *25*: 3843.
341. Percec, V. Cho, C. G., Pugh, C., and Tomazos, D. (1992). *Macromolecules*, *25*: 1164.

342. Fahey, D. R., and Ash, C. E. (1991). *Macromolecules, 24*: 4242.
343. Durvasula, V. R., Stuber, F. A., and Bhattachavjee, D. (1989). *J. Polym. Sci., Part A., Polym. Chem., 27*: 661.
344. Senn, D. R. (1994). *J. Polym. Sci., Part A, Polym. Chem., 32*: 1175.
345. Hara, A., Oishi, Y., Kakimoto, M., and Imai, Y. (1991). *J. Polym. Sci., Part A, Polym. Chem., 29*: 1933.
346. Kricheldorf, H. R., and Jahnke, P. (1991). *Polym. Bull., 27*: 135.
347. Kricheldorf, H. R., and Jahnke, P. (1992). *Polym. Bull., 28*: 411.
348. Wang, Z. Y., and Hay, A. S. (1992). *Polymer, 33*: 1778.
349. Ding, Y., and Hay, A. S. (1998). *Macromolecules, 31*: 2690.
350. Ding, Y., Hlil, A. R., and Hay, A. S. (1998). *J. Polym. Sci., Part A, Polym. Chem., 36*: 1201.
351. Tsuchida, E., Yamamoto, K., Nishide, H., Yoshida, S., and Jikei, M. (1990). *Macromolecules, 23*: 2102.
352. Tsuchida, E., Yamamoto, K., Jikei, N., and Nishide, H. (1990). *Macromolecules, 23*: 930.
353. Yamamoto, K., Jikei, M., Oi, K., Nishide, H., and Tsuchida, E. (1991). *J. Polym. Sci., Part A, Polym. Chem., 29*: 1359.
354. Tsuchida, E., Yamamoto, K., Jikei, M., and Nishide, H. (1989). *Macromolecules, 22*: 4138.
355. Yamamoto, K., Jikei, M., Kato, J., Nishide, H., and Tsuchida, E. (1992). *Macromolecules, 25*: 2689.
356. Wang, Z. Y., and Hay, A. S. (1991). *Macromolecules, 24*: 333.
357. Babu, J. R., Brink, A. E., Konas, M., and Riffle, J. S. (1994). *Polymer, 23*: 4949.
358. Montoneri, E. (1989). *J. Polym. Sci., Part A, Polym. Chem., 27*: 3043.
359. Wang, Y.-F., and Hay, A. S. (1996). *Macromolecules, 29*: 5050.
360. Wang, Y.-F., Chan, K. P., and Hay, A. S. (1995). *Macromolecules, 28*: 6371.
361. Wang, Y.-F., and Hay, A. S. (1997). *Macromolecules, 30*: 182.
362. Ding, Y., and Hay, A. S. (1998). *Macromolecules, 29*: 6386.
363. Ding, Y., and Hay, A. S. (1997). *Polymer, 38*: 2239.
364. Ding, Y., and Hay, A. S. (1997). *Marcomolecules, 30*: 2527.
365. Ding, Y., and Hay, A. S. (1997). *Macromolecules, 35*: 2961.

8
Polyurethanes

Zoran S. Petrović
Pittsburg State University, Kansas Polymer Research Center, Pittsburg, Kansas

I. INTRODUCTION

The history of polyurethanes started in the 1930s in Germany when Otto Bayer proposed using diisocyanates and diols for preparation of macromolecules. The first commercial polyurethane, based on hexamethylene diisocyanate and butanediol, had similar properties to polyamides and is still used to make fibers for brushes. However, fast growth of the production and expanded application range started in the 1950s with the building of toluene diisocyanate (TDI) and polyester polyol plants for flexible foams in Germany. However, the real jump in applications came with the introduction of polyether polyols in foam formulations. Further development and application of polyurethanes shifted from Europe to the USA and Japan. Today, polyurethanes are about the sixth largest polymer by consumption, right behind high volume thermoplastics, with about 6% of the market. The largest part of the urethane application is in the field of flexible foams (about 44%), rigid foams (about 28%), while 28% are coatings, adhesives, sealants and elastomers (CASE) applications. These data are taken at a certain moment in time (1996) and vary from year to year and region to region, but they illustrate the relative consumption in different categories. Consumption of polyurethanes in different industries is the following: about 40% of PU is used in the furniture industry, 16% in transport, 13% in construction, 7% in refrigeration and about the same in coatings, 6% in the textile industry, 4% in the footwear industry and 8% for other applications. Table 1 illustrates the consumption of urethanes in the United States in 1996.

Polyurethanes are a broad class of very different polymers, which have only one thing in common – the presence of the urethane group:

$$-NH-\overset{\overset{\displaystyle O}{\|}}{C}-O- \tag{1}$$

urethane group

The number of these groups in a polymer can be relatively small compared with other groups in the chain (for example ester or ether groups in elastomers), but the polymer will still belong to the polyurethane group. Varying the structure of polyurethanes, one can vary the properties in a wide range. Polyurethanes are formed by reaction of polyisocyanates with hydroxyl-containing compounds, most frequently

Table 1   US polyurethane 1996 market.[a]

	Consumption, Million of lb
Flexible foam slab	1593
Rigid foam	1268
Molded flexible foam	451
Coatings	309
Binders and fillers	271
Adhesives	183
Cast elastomers	158
Molded thermoplastics	114
Automotive RIM	113
Sealants	70
Spandex fibers	45
Nonautomotive RIM	33
Total	4609

[a]*Chem. Eng. News*, August 4, 22 (1997).

during processing. By selecting the type of isocyanate and polyols, or combination of isocyanates and combination of polyols, one can tailor the structure to obtain desired properties. For this, however, it is necessary to know the relationship between the structure and properties. The flexibility to tailor the structure during processing is one of the main advantages of polyurethanes over other types of polymers. Urethane groups form strong hydrogen bonds among themselves and with different substrates. Strong intermolecular bonds make them useful for diverse applications in adhesives and coatings, but also in elastomers and foams. One of the great advantages of polyurethanes arises from the high reactivity of isocyanates, which can react with a number of substances having different functional groups. This allows polymerization at relatively low temperatures and in short times (several minutes). One group of polymers, which is conditionally treated as urethanes, is polyurea, because urea is often formed during urethane production. Urea is formed in the reaction between isocyanates and amines. The urea group is similar to the urethane group, except that it has two –NH– groups, and can form more hydrogen bonds than the urethane group:

$$
\begin{array}{c}
\quad\;\; O \\
\quad\;\; \| \\
-NH-C-NH- \\
\text{urea group}
\end{array}
\tag{2}
$$

II. ISOCYANATE CHEMISTRY [1–8]

A. Basic Reactions of Isocyanates

The exceptionally high reactivity of the isocyanate group originates from its electronic structure, which can be represented by the following resonance structures:

$$
-\ddot{N}=C=\ddot{O} \longleftrightarrow -\ddot{N}=\overset{+}{C}-\ddot{O}\!:^{-} \longleftrightarrow {}^{-}\ddot{N}-\overset{+}{C}=\ddot{O}\!:
\tag{3}
$$

It follows that the highest electron density is an oxygen (electronegative) and the least on the carbon (electropositive), while nitrogen is somewhat less electronegative than oxygen. Thus, NCO easily reacts with proton donors:

$$R-N=C=O+H-R' \longrightarrow R-N=C=O \rightleftharpoons R-N-C=O \tag{4}$$

Isocyanates are susceptible, however, to nucleophilic as well as electrophilic attacks. Typical nucleophilic reactions of isocyanates are urethane (carbamate) formation with alcohols:

$$R-N=C=O+HOR' \rightleftharpoons R-[NH-\overset{\displaystyle O}{\overset{\|}{C}}-O]-R' \tag{5}$$

and formation of urea (carbamide) with amines:

$$R-N=C=O+H-\underset{\underset{H}{|}}{N}-R' \rightleftharpoons R-[NH-\overset{\displaystyle O}{\overset{\|}{C}}-NH]-R' \tag{6}$$

The reaction of isocyanate with alcohols is strongly exothermic (170–190 kJ/mol). One of the basic reactions in the urethane foam technology is the reaction of isocyanate with water with evolution of carbon dioxide and amine formation:

$$R-N=C=O+H-O-H \rightleftharpoons R-NH-\overset{\displaystyle O}{\overset{\|}{C}}-OH \longrightarrow RNH_2+CO_2 \tag{7}$$

Since the urethane group itself contains active hydrogen, it could react with isocyanate to produce allophanate:

$$R-N=C=O+-NH-\overset{\displaystyle O}{\overset{\|}{C}}-O- \rightleftharpoons \left[\begin{array}{c} -N-\overset{\displaystyle O}{\overset{\|}{C}}-O- \\ R-NHC=O \end{array}\right] \tag{8}$$

allophanate

This reaction proceeds to a significant degree at about 120–140 °C but it could occur also at lower temperatures at high excess of isocyanates. Similar is the reaction of biuret formation from isocyanate and urea groups:

$$R-N=C=O+-NH-CO-NH- \rightleftharpoons \left[\begin{array}{c} -N-\overset{\displaystyle O}{\overset{\|}{C}}-NH- \\ R-NHC=O \end{array}\right] \tag{9}$$

biuret

Biuret formation reaction proceeds to a considerable measure above 100 °C. Both reactions (8) and (9) are utilized to introduce crosslinks with the excess of isocyanate. The previously given reactions are the most frequent ones in the polyurethanes chemistry. There are other important reactions such as the reaction of isocyanate with itself, which may occur during storage or are intentionally carried out to obtain new products. Isocyanates (particularly the reactive aromatic ones) easily form dimers (uretdiones):

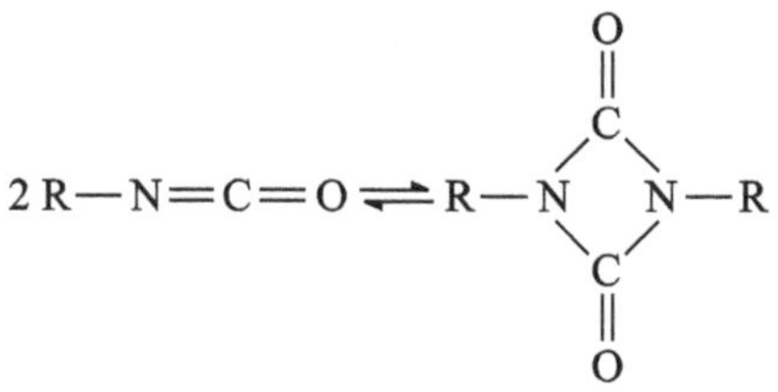

$$2\,R\!-\!N\!=\!C\!=\!O \rightleftharpoons R\!-\!N \underset{\substack{\displaystyle C \\ \parallel \\ O}}{\overset{\substack{O \\ \parallel \\ C}}{\diamondsuit}} N\!-\!R \tag{10}$$

Dimers are formed in presence of mild based such as pyridine or isocyanates themselves. Dimerization can be prevented by adding acids or acid chlorides (e.g., benzoyl chloride). Dimers are thermally unstable, and upon heating they dissociate into starting components. Thus, they are sometimes used to form so called blocked isocyanates, which are quite stable at room temperature but react at elevated temperatures. Strong bases, however, favor the trimerization of isocyanate to form isocyanurate:

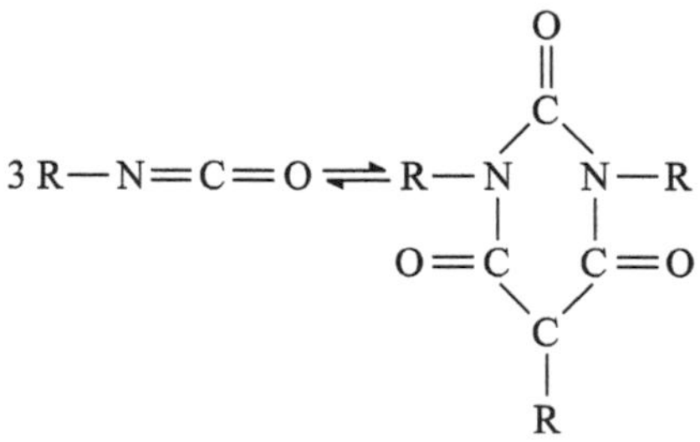

$$3\,R\!-\!N\!=\!C\!=\!O \rightleftharpoons \tag{11}$$

Triisocyanurates possess exceptional thermal stability. The reaction (11) is used in industry to prepare thermally stable foams.

Polymerization of isocyanate to polyisocyanates (polyamide 2) proceeds in presence of anionic polymerization catalysts, such as NaCN, triethylphosphine, butyllithium and strong bases, according to the following scheme:

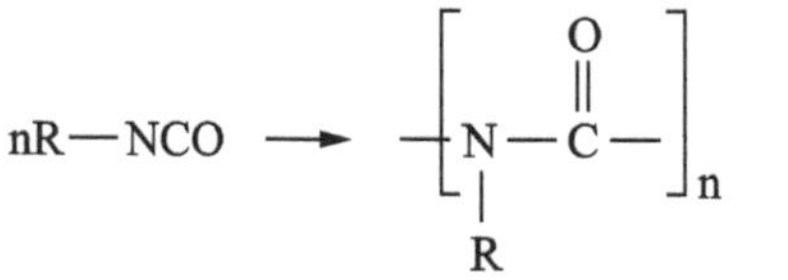

$$nR\!-\!NCO \longrightarrow \left[\begin{array}{c} O \\ \parallel \\ N\!-\!C\!- \\ | \\ R \end{array} \right]_n \tag{12}$$

Polyisocyanates have no commercial application, and the conditions for their formation should be avoided when planning other urethane chemical reactions. An important chemical reaction of isocyanates, which proceeds at high temperature without catalysts, is carbodiimide formation. CO_2 is generated in the process:

$$R\!-\!NCO + OCN\!-\!R \longrightarrow RN\!=\!C\!=\!NR + CO_2 \tag{13}$$
$$\text{(carbodiimide)}$$

This reaction proceeds also at room temperature in the presence of special catalysts (e.g., 1-ethyl-3 methyl-3-phospholin-1-oxide). Carbodiimides are used as stabilizers against hydrolysis of polyester urethanes, since they react with acids produced by hydrolysis and thus slow down the process. Acids are catalysts for hydrolysis of polyesters. The carbodiimide reaction is used to modify isocyanates (e.g., Isonate 143 L from Dow Chemical is carbodiimide modified MDI).

A number of self-reactions of isocyanates create a problem during storage. Acid inhibitors do not really slow down the isocyanate reactions but primarily react with bases, which are accelerators of these processes.

B. Other Isocyanate Reactions

Isocyanates react with organic acids forming unstable intermediaries, which decompose into an amide and carbon dioxide:

$$R-NCO + R'-COOH \longrightarrow R-NH-CO-R' + CO_2 \tag{14}$$

Isocyanate reacts with HCl to form an adduct which decomposes at higher temperatures to starting components:

$$R-NCO + HCl \rightleftharpoons R-NH-CO-Cl \tag{15}$$

To avoid high sensitivity of isocyanates towards moisture and to increase their stability, blocked isocyanates are often used. They are obtained in reactions with some blocking agents, and decompose to isocyanates under certain conditions, most frequently at elevated temperatures. Isocyanates can react with activated methylene groups in the presence of sodium or sodium alcoholate to produce a blocked isocyanate, as in the case of a diester of malonic acid:

$$
R-N{=}C{=}O + H\underset{\underset{\displaystyle COOR'}{|}}{\overset{\overset{\displaystyle COOR'}{|}}{C}}H \rightleftharpoons R-NHCO-\underset{\underset{\displaystyle COOR'}{|}}{\overset{\overset{\displaystyle COOR'}{|}}{C}}H \tag{16}
$$

A frequently used blocking agent is phenol:

$$R-NCO + HO-\bigcirc \rightleftharpoons R-NHCO-O-\bigcirc \tag{17}$$

which produces an adduct that decomposes to the starting components at 160–180 °C or at lower temperatures in the presence of catalysts. Isocyanates react with oximes to produce blocked (masked) isocyanates, which decompose at elevated temperatures to starting components:

$$
R-NCO + HON{=}C\overset{\displaystyle R_1}{\underset{\displaystyle R_2}{\big<}} \rightleftharpoons R-NH-\overset{\overset{\displaystyle O}{\|}}{C}-O-N{=}C\overset{\displaystyle R_1}{\underset{\displaystyle R_2}{\big<}} \tag{18}
$$

Isocyanates react with aromatic and aliphatic anhydrides to give imides:

$$\text{(structure)} + \text{OCN} - \text{R} \longrightarrow \text{(structure)} \text{N} - \text{R} + CO_2 \qquad (19)$$

This reaction can be used to prepare polyimides (from dianhydrides and diisocyanates). Aldehydes and ketones may react with isocyanates to produce unstable cyclic compounds, which decompose according to the scheme:

$$\text{R} - \text{NCO} + \underset{R'}{\overset{R'}{>}}C=O \longrightarrow \underset{R'}{\overset{R'}{>}}\overset{O-C=O}{\underset{}{C-N-R}} \longrightarrow \underset{R'}{\overset{R'}{>}}C=N-R + CO_2 \qquad (20)$$

Isocyanates may undergo addition to olefins (enamines, ketenketales) in the following way:

$$A=B + RNCO \longrightarrow \begin{matrix} A-C=O \\ | \quad | \\ B-N-R \end{matrix} \qquad (21)$$

Isocyanates also react with epoxides to produce cyclic compounds – oxazolidones:

$$H_2C - CH_2 + R - NCO \longrightarrow \begin{matrix} H_2C - N - R \\ | \qquad | \\ H_2C \quad C=O \\ \backslash O / \end{matrix} \qquad (22)$$

III. BASIC COMPONENTS IN URETHANE TECHNOLOGY

A. Isocyanates

Polyurethanes are formed in the reaction of isocyanates with polyols. The most important commercial aromatic isocyanates are toluenediisocyanate (TDI), diphenylmethane diisocyanate (MDI) and naphthalene diisocyanate (NDI), while the important aliphatic isocyanate is hexamethylene diisocyanate (HDI). Cycloaliphatic isocyanates of industrial importance are isophorone diisocyanate (IPDI) and hydrogenated MDI (HMDI). A number of triisocyanates, such as triphenylmethane triisocyanate, are used in coatings and adhesives.

Chemistry and technology of a wide range of isocyanates is given in several books [9,10]. Toluene diisocyanate is usually supplied as the mixture of two isomers: 2,4-TDI and

2,6-TDI with a ratio 80:20 (called TDI 80) or 65:35 (TDI 65).

$$\text{2,4-TDI} \qquad \text{2,6-TDI} \tag{23}$$

TDI is a liquid at room temperature, having density $1.22\,g/cm^3$, boiling point $120\,°C$ at 1333.22 Pa (1 atm) and melting point $13.6\,°C$ (TDI 80) or $5\,°C$ (TDI 65). It is used primarily for flexible foams and different adducts-intermediaries for coatings.

Pure MDI is a solid at room temperature, having melting point $39.5\,°C$ and density $1.18\,g/cm^3$ at $40\,°C$.

$$\text{MDI} \tag{24}$$

In the manufacture of distilled (pure) MDI, a residue is obtained, which contains a mixture of isomers, trimers and isocyanates with a higher degree of polymerization. Such a mixture is a dark brown liquid at room temperature and is called crude MDI or polymeric MDI (PAPI). The dominating species is a triisocyanate with the approximate structure:

$$\tag{25}$$

Pure MDI is used mainly for preparation of thermoplastic elastomers, while crude MDI is used for rigid and partly for flexible foams.

Paraphenylene diisocyanate is another important isocyanate. It produces excellent elastomers but its use is limited due to a very high price.

$$\text{paraphenylene diisocyanate} \tag{26}$$

Aromatic diisocyanates are not suitable for products that are exposed to irradiation and external influences (such as coatings) because of yellowing. Those applications require aliphatic or cycloaliphatic isocyanates. One popular cycloaliphatic isocyanate is isophorone diisocyanate, a liquid at room temperature (melting point is $-60\,^\circ$C) having density $1.06\,\mathrm{g/cm^3}$, molecular weight 222 and boiling point $158\,^\circ$C at 1333.22 Pa:

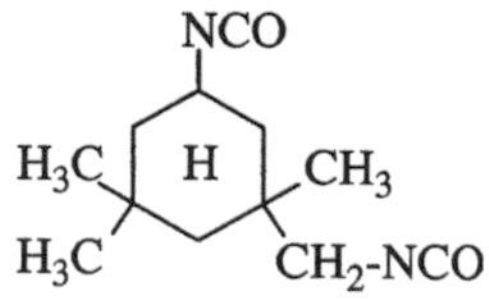

Isophorone diisocyanate

(27)

The reactivity of an isocyanate group depends on the radical to which it is attached, as well as the position in the molecule. In principle aromatic isocyanates are more reactive than the aliphatic ones. The reactivity of an isocyanate group in symmetric diisocyanates decreases after the first group has reacted, which should be taken into account [4]. Reactivity also depends on temperature, and sometimes the difference in reactivity of two isocyanate groups may diminish with increasing temperature. This effect is stronger in the cases with higher activation energies. Table 2 displays rate constants and activation energies for several diisocyanates in the reaction with hydroxyl groups from polyethyleneadipate diol. The constants and their relative ratios are different in reactions with alcohols, amines or water.

The comparison of the reactivity of two groups in various diisocyanates is shown in Table 3. Rate constants k_1 and k_2 show the relative rates for the first and second group (compared with a standard rate). The constant k_2 is obtained after the first group is reacted, and it should be half of k_1 if the reactivity is the same.

Table 2 Rate constants, k, and activation energy, E, in the reaction of isocyanates with polyethyleneglycol adipate diol at $100\,^\circ$C.

Diisocyanate	$k \times 10^4$, $\mathrm{L\,mol\,s}$	E, kJ/mol
p-Phenylene	36.0	46
2,4-TDI	21.0	33.1
2,6-TDI	7.4	41.9
1,5-NDI	4.0	50.2
1,6-HDI	8.3	46.0

Table 3 Relative rate constants of isocyanate groups with a hydroxyl group.

Isocyanate	k_1	k_2
MDI	16	8.6
2,4-TDI	42.5	2
2,6-TDI	5	2
HDI	0.2	–

It follows from Table 3 that the first group in 2,4 TDI is much more reactive than the second one. The difference however, decreases with increasing temperature or in the presence of catalysts.

Reactions of isocyanates can be accelerated either by increasing temperature or adding catalyst. Slowing down the reaction cannot be done by additives if the concentration of isocyanate and polyol is kept constant. Lowering the temperature or diluting the mixture polyol–isocyanate by adding a solvent or neutral diluents would, however, slow down the reaction. Activation energies of the reactions of isocyanates with polyols, as a rule, do not exceed 20–40 kJ/mol. The reaction rates increase with increasing polarity of the medium (e.g., solvent). The reactivity of different groups, proton donors, with isocyanates decreases in the order: aliphatic NH_2 > aromatic NH_2 > primary OH > water > secondary OH > tertiary OH > COOH. Urea group in R-NH-CO-NH-R is more reactive than amide group, R'CONHR, and amide is more reactive than the urethane group, R-NHCOO-R'. This sequence can be changed if the groups with different steric hindrances are attached.

B. Polyols

Second to isocyanate in the technology of polyurethane preparation is polyol. Polyether polyols (polypropylene glycols and triols) having molecular weights between 400 and 10,000 dominate in the foam technology. Foams are usually made with triols, which form crosslinked products with diisocyanates, whereas diols dominate in the elastomer technology. Polyether polyols have higher hydrolytic stability than the polyester polyols, but they are more sensitive to different kinds of irradiation and oxidation at elevated temperatures. Polypropylene oxide (PPO) polyols, also called polypropylene glycols (PPG), are cheaper than other polyols. PPG structure can be represented by the formula:

$$H \left[O-HC-H_2C \right]_n -O-R-O- \left[CH_2-CH-O \right]_n -H \qquad (28)$$
$$\quad\quad\quad\;\; \underset{CH_3}{|} \qquad\qquad\qquad\qquad\quad \underset{CH_3}{|}$$

Group R comes from the starter diol such as ethylene glycol ($R = -CH_2-CH_2-$). If multifunctional starters, such as glycerin, trimethylol propane or sugars are used, the resulting polypropyleneoxide polyol would have the functionality of the starter component.

Due to the weak intermolecular attractive forces (low polarity) and non-crystallizing nature, PPG polyols are liquid at room temperature even at very high molecular weight, unlike polyester polyols, which are often crystalline greases. Weaker interactions on the other hand cause lower strengths of the PPG based urethanes. Viscosity of polyether polyols is a function of the hydroxyl content (due to hydrogen bonding) and molecular weight. PPO diols have viscosities from 110 mPa s (cP) at 20 °C for the molecular weight of 425 to 1720 mPa s for MW = 4000. Glycerin for example has viscosity above 1000 mPa s at 20 °C but when propoxylated to MW = 1000 gives a triol with viscosity of about 400 mPa s.

Polyether polyols based on polytetramethylene oxide (PTMO), sometimes called polytetrahydrofurane (PTHF), have better strengths than PPG polyols, mainly due to their ability to crystallize under stress. Their structure is represented by structural formula (29):

$$HO\text{-}[CH_2\text{-}CH_2\text{-}CH_2\text{-}CH_2\text{-}O]_n\, H \qquad\qquad (29)$$

Polyester polyols are an important class of urethane raw materials, with applications in elastomers, adhesives, etc. They are usually made from adipic acid and ethylene glycols (polyethylene adipate):

$$HO-[(CH_2)_2-O-\overset{\overset{\displaystyle O}{\|}}{C}-(CH)_4\overset{\overset{\displaystyle O}{\|}}{C}O]_n-CH-CH_2-OH \qquad\qquad (30)$$

or butane diol and adipic acid (polybutylene adipate). Both would crystallize above room temperature. In order to reduce their glass transition and destroy crystallinity, copolyesters are prepared from the mixture of ethylene glycol and butane diol with adipic acid. Polycaprolactone diol is another crystallizable polyester diol:

$$HO-[(CH_2)_5-\overset{\overset{\displaystyle O}{\|}}{C}-O-]_nH \qquad\qquad (31)$$

Polyols for coatings, rigid foams, and adhesives may contain aromatic rings in the structure in order to increase rigidity. These polyols may also crystallize, which is important in some applications, e.g., adhesives. Special class of polyols are 'polymer polyols' containing usually copolymers of acrylonitrile and styrene or methylmetacrylate attached to the chains of polyether polyols, forming a dispersion. They are used for high modulus products such as froth and integral skin foams, RIM, shoe soles and one-shot elastomers.

An important but less frequently used group of polyols, polybutadiene diols, are mainly used for elastomers:

$$HO\text{-}[CH_2\text{-}CH=CH\text{-}CH_2]_n\text{-}OH \qquad\qquad (32)$$

Structural formula (32) shows poly-1,4-butadiene (BD), but 1,2-poly BD and the mixture of the two are also produced.

Castor oil is a natural triol with a typical OH number 160 mg KOH/g (functionality = 2.7). Although it has three ester groups, it is not considered a polyester type polyol.

$$\qquad\qquad (33)$$

castor oil

A new class of polyols from vegetable oils could become a significant player in rigid foam technology. An example are soybean oil based polyols [11,12] having the structure:

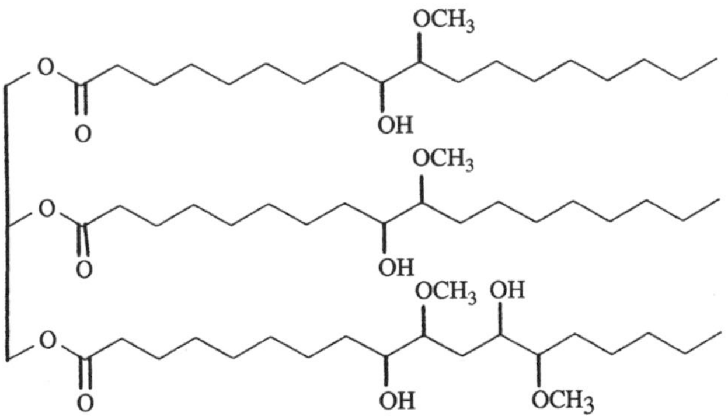

(34)

soy-polyol

The advantage of these polyols is their compatibility with hydrocarbon blowing agents, higher hydrophobicity and improved hydrolytic properties of resulting polyurethanes. They have also better oxidative stability than PPO based polyurethanes, but their viscosity is typically between 2–12 Pa s (2000–12,000 cP). Molecular weight of these polyols is about 1000 and functionality may vary from 2 to 8, but high hydroxyl numbers cause high viscosity. These molecular weights are not sufficient for flexible foams and copolymerization with propyleneoxide and ethylene oxide is necessary to obtain polyols for these applications. Alternative ways of making polyols from triglycerides is by hydrolysis to fatty acids and introduction of OH groups. Although the price of vegetable oils is very competitive with petrochemicals, the number of chemical steps should be minimal in order to have polyols at competitive prices.

C. Catalysts [1,3–8,13,14]

Rapid growth of urethane technology can be attributed to the development of catalysts. Catalysts for the isocyanate–alcohol reaction can be nucleophilic (e.g., bases such as tertiary amines, salts and weak acids) or electrophilic (e.g., organometallic compounds). In the traditional applications of polyurethanes (cast elastomers, block foams, etc.) the usual catalysts are trialkylamines, peralkylated aliphatic amines, triethylenediamine or diazobiscyclooctane (known as DABCO), N-alkyl morpholin, tindioctoate, dibutyltindioctoate, dibutyltindilaurate etc.

Usually a combination of catalysts is required to achieve proper structure and properties, especially in applications such as integral skin foams or reaction injection molding (RIM). The mechanism of the catalysis of isocyanate-alochol reaction in presence of amines is assumed to proceed through an activated complex between amine and isocyanate [15,16]

$$R_3N + R'NCO^- \longrightarrow R'\bar{N}\!=\!\underset{\underset{^+NR_3}{|}}{C}\!-\!O^- \rightleftharpoons R'\bar{N}\!-\!\underset{\underset{^+NR_3}{|}}{C}\!=\!O$$

(35)

Table 4 Relative activity of different catalysts in a model isocyanate-hydroxyl reaction.

Catalyst	Concentration, %	Relative activity
Uncatalyzed	0	1
TMBDA	0.1	56
DABCO	0.1	130
TMBDA	0.5	160
DBTDL	0.1	210
DABCO	0.3	330
Sn-octoate	0.1	540
DBTDL + DABCO	0.1 + 0.2	1000
Sn-octoate	0.3	3500
Sn-octoate + DABCO	0.3 + 0.3	4250

Designations: TMBDA, tetramethylbutane diamine; DBTDL, dibutyltin dilaurate.

The complex then reacts with the alcohol to form an intermediary product, which decomposes to give urethane and regenerate the catalyst:

$$R'\bar{N}-C=O+R''OH \longrightarrow R'-N=C-O^- \longrightarrow R'NH-C=O+NR_3 \qquad (36)$$

In hydroxyl-containing compounds with higher acidity, a transfer of proton from alcohol to amine is possible.

Tin (Sn) catalysts are considerably stronger than amine catalysts, but their mixtures are even more powerful. The reaction rates depend also on the amount of catalyst, which usually is not more than 0.3% in the mixture. Table 4 illustrates relative reactivities (rates) of isocyanates with an alcohol in the presence of different concentrations of the catalysts [17]. The mechanism of metal catalysis is multifaceted and it always involves metal complexes with reacting species, but true nature of the transition states is open to debate [18]. Organometalic catalysts could be lead, zinc, copper, calcium and magnesium salts of fatty acids, such as octanoates or naphthenates. Especially good for application in elastomers are mercury catalysts, since they strongly promote isocyanate–alcohol reaction but are fairly insensitive towards isocyanate–water reaction. Also, they may give long processing (gel) time but once the reaction starts, curing is finished quickly, as required in flooring applications. Gel time can be easily adjusted with catalyst concentration. Unfortunately mercury is undesirable in many applications.

IV. ANALYSIS OF RAW MATERIALS

A. Analysis of Isocyanates

The most important characteristic of polyisocyanates is NCO content. It is determined according to ASTM D1638-74 by dissolving isocyanate in the mixture of toluene and dibutylamine (DBA). DBA reacts with isocyanate and the excess is titrated with HCl solution. The NCO content is calculated from the expression:

$$\%NCO = [(B - S)N \times 4.202]/W \tag{37}$$

where B is the number of ml of HCl used for titration of the blank, S is the number of ml of HCl used for titration of the sample, N is the molarity of HCl solution, and W is the weight of the sample in grams.

Other characteristics of isocyanates that are analyzed are total chlorine content, the content of hydrolyzable chlorine, acid content, freezing point, density and color.

B. Analysis of Polyols

The principal property measured in polyols is hydroxyl content. According to ASTM D4274-88, hydroxyl group content is determined by acetylation and the excess of acid is back titrated with a base. The acetylating agent is usually a solution of acetic anhydride in pyridine. Acetylation is carried out at $100\,°C$. Unreacted anhydride is then converted with water into acid and titrated with $1\,N$ NaOH. Hydroxyl content is usually expressed as hydroxyl number (OH number), which is defined as the number of milligrams of KOH ($M_{KOH} = 56.11$) used for titration of one gram of the sample.

$$\text{OH number (mg KOH/g)} = 56.1(B - A)N/W \tag{38}$$

where A is the number of mL NaOH, B is number of mL NaOH used for titration of blank, N is molarity of the NaOH solution, and W is the weight of the sample in grams.

Hydroxyl content in percent can be calculated from the proportion which takes into account that OH number of 56.1 corresponds to 1.7% OH groups. Thus, the content of OH groups, $X(\%)$, is equal to:

$$X(\%) = \frac{1.7Y}{56.1} \tag{39}$$

where Y is OH number expressed in mg KOH/g. In polyester polyols an important characteristic is acid number, also expressed in mg KOH/g. Other important characteristics are unsaturation, water content (determined by the Karl–Fisher method), Na and K content, density, viscosity, color and the content of suspended matter.

C. Calculation of Equivalent Ratios

If the hydroxyl number of the polyol and the content of NCO in the isocyanate are known, we can easily calculate the stoichiometric amounts of two components. Usually we need to find how much isocyanate (a-grams), having x percent of NCO groups (%NCO), we need to react at the stoichiometric molar ratio (1:1) with b-grams of the polyol component having y (%OH), or vice versa. This relationship is given by the expression:

$$b = a \frac{x}{y} \frac{17}{42} = 0.40476 ax/y \tag{40}$$

Alternatively, we may wish to work with equivalent weights, since it is easy to calculate the stoichiometric ratios. The equivalent weight of an isocyanate component should match the equivalent weight of the polyol component at the equivalent ratio 1:1. Weight equivalent refers to the weight of material that has 1 mol of functional groups, and is obtained by dividing number average molecular weight M_n, by the functionality of the component, f:

$$E = \frac{\overline{M}_n}{f} \qquad (41)$$

If we know the functionality of a component and the content of the groups (%NCO or %OH) we could calculate the number average molecular weight. For polyols the expression would be:

$$\overline{M}_n = \frac{fM_{OH} \times 100}{\%OH} = \frac{f \times 1700}{\%OH} \qquad (42)$$

and for isocyanates:

$$\overline{M}_n = \frac{fM_{NCO} \times 100}{\%NCO} = \frac{f \times 4200}{\%NCO} \qquad (43)$$

Thus, the diol of $M_n = 1000$ would have weight equivalent $E = 500$ [g/equiv] and triol of $M_n = 3600$ would have $E = 1200$ [g/equiv] and MDI equivalent weight would be 125 [g/equiv]. Water is a specific compound, behaving as a two-functional reactant, having $E = 9$ [g/equiv]. Equivalent weight of hexamethylenediamine is $116/2 = 58$. From the above, 125 g of MDI should react with 500 g of the diol with $M_n = 1000$, or 9 g of water, or 58 g of hexamethylenediamine, if 1:1 molar ratio is desired. Equivalent weight of polyols can be calculated from the known OH number:

$$E = \frac{56,100}{OH\#} \qquad (44)$$

D. Infrared Spectra of Polyurethanes

Infrared spectroscopy is a powerful method in analyzing raw materials and finished PU products. Polyols are characterized by the hydrogen bonded OH stretching absorption band at about $3300 \, cm^{-1}$ (3 μm). The difference between ester and ether polyols should be observed at $1280–1150 \, cm^{-1}$ (ester C–O stretching) and 1150–1060 (ether CH_2–O–CH_2). Isocyanate group has very strong absorption at about $2275–2240 \, cm^{-1}$. Assignment of absorption bands in IR spectrum of MDI/butane diol/polyether (PTHF) urethane elastomers is given in Table 5 [19]. Relative intensities refer to the sample with approximately 85% soft segment concentration.

Table 5 IR absorption bands of polyether urethanes.

Wavelength, μm	Frequency, cm^{-1}	Relative intensity	Phase	Urea, urethane	PTHF	Benzene ring
3.06	3268	m		U, UT – ν(N – H)		
3.40	2941	vs			$\nu_a(CH_2)$	
3.50	2857	vs			$\nu_s(CH_2)$	
3.58	2793	m			$\nu_s(CH_2)$	
5.79	1727	m		UT: amide I		
6.12	1634	m		U: amide I		
6.28	1592	m				ν(C–C)
6.35	1575	w		U: amide II		
6.53	1531	s		UT: amide II		
6.61	1513	sh				ν(C–C)
6.71	1490	m	C		$\sigma(CH_2)$	
6.91	1447	m	A		$\sigma(CH_2)$	
7.08	1412	m				ν(C–C)
7.30	1370	s			$W(CH_2)$	
7.63	1311	m				β(C–H)
8.12	1232	sh		UT: amide III		
8.22	1216	s		UT: amide III, (COOC)		
8.28	1208	sh			$t(CH_2)$	
8.99	1112	vs		UT: ν_s(CO–O–C)	ν_a(C–O–C)	
12.94	773			UT: γ(O=C–O)		

s, strong: m, medium; w, weak; v, very; sh, shoulder; A, amorphous; C, crystalline, ν, stretching; ν_a, antisymmetric stretching; ν_s, symmetric stretching; δ, bending; W, wagging; t, twisting; r, rocking; β, in plane bending; γ, out-of-plane bending; U, urea; UT, urethane.

V. POLYURETHANE FOAMS [2,6,8,13,20,21]

Polyurethane foams are the largest group of urethane products, covering about 80% of the total urethane production. Polyurethane foams can be categorized as rigid and flexible. Rigid foams are used primarily for heat insulation in refrigeration and construction, and partly in automobile industry. Flexible foams find their application in furniture, the automobile industry, for packaging, etc. A variety of rigidity grades of flexible foams are manufactured, with grades having rigidity between soft and rigid foams being called semi-rigid. Semi-rigid foams are used for automobile seats and components for interior and exterior safety. Two basic reactions of isocyanates are used in foam production:

$$\text{Isocyanate} + \text{polyol} \longrightarrow \text{polymer}$$
$$\text{Isocyanate} + \text{water} \longrightarrow CO_2 \text{ for foaming}$$

(45)

The correct foaming process requires that these two reactions take place at the same rate. If the polymerization (the first reaction) is faster, the polymer formed will have final strength before foaming and the result will be a high density foam (low degree of foaming). If the second reaction is much faster, the evolved gas will blow the foam. Due to the low 'green' strength and viscosity of the polymerizing mixture, the gas will leave the mixture, and the foam will collapse to a high density foam, as in the first case. In the balanced process, the polymerization should proceed fast enough to give high viscosity and melt strength of the mixture, which will trap fast evolving gas and finish the polymerization at the end of foam growth.

A. The Mechanism of Foam Formation [21,22]

The initial polyol and isocyanate mixture is a low molecular weight, low viscosity fluid, which is reflected in the low strength of the bubble wall formed during foaming. The wall of such a bubble breaks easily and gas escapes. Therefore, it is necessary to increase the strength and elastic properties of the bubble wall (gel strength), which is achieved by increasing the molecular weight of the polymer. The mechanism of the bubble formation is a science 'per se', and it is essential to understand the basics of the process. This process is similar to bubble generation during boiling of a liquid. Gas which is formed in the chemical reaction, or by evaporation of the added low boiling foaming agent, is partially soluble in the polymer mass. When the limit of solubility is reached, i.e., when enough gas is generated to exceed the solubility limit (saturation), the excess separates in the form of bubble. First stage of bubble formation is called nucleation. The number of bubbles will depend on the number of nuclei (seeds) present in the system. Nucleation can be homogeneous (in the absence of foreign particles, nucleants) or heterogeneous (in the presence of nucleants). The bubble nucleus is usually a small amount of air caught in the crevasses or in the roughness on the surface of the solid or liquid particle, in case of heterogeneous nucleation. The beginning of foam formation is characterized by formation of large number of nuclei. Their creation causes refraction of light on the walls of nuclei, which is manifested as whitening of the mass (cream formation) without significant volume increase. The next stage is bubble growth from the nucleus due to the incoming evolved gas, and the volume increase of the foaming mixture. This stage is observed as the foam rise. Stability of a growing bubble depends on the surface tension. If the surface tension is too large and there is no nucleation, a small number of large bubbles will grow, and the shape should be elongated in the direction of rise. Such foams are usually not desirable since they show anisotropy in their mechanical properties. Regulation of bubble growth is achieved by the addition of surfactants (usually silicone copolymers). They lower the surface tension and enable bubble division into smaller, more regularly shaped bubbles. This process is helped by vigorous mixing. Foam rise (due to gas diffusion into the bubbles) is completed when the polymerization has passed the gel point, and the infinite network of the polymer, spanning from one to the other end of the sample, is formed. Gas concentration in the urethane mass varies with time. Figure 1 illustrates three characteristic regions which coincide with the three stages of foam formation; zone I, nucleation (the reaction mass whitens but does not rise which characterizes the cream time), zones II and III coincide with the foam rise.

Figure 1 can be interpreted the following way: gas generated during the foaming process is dissolved in the polymer until it reaches the saturation limit S. The nucleation rate $V_n = 0$. Nucleation does not proceed at low supersaturation ($V_n \to 0$) but will begin at somewhat higher supersaturation and will accelerate to reach the maximum rate ($V_n \to \infty$). When nucleation is practically finished, the concentration of gas in the polymer will decrease due to the diffusion into growing bubbles. Gas concentration in the polymer will decrease with time until reaching the saturation limit, S.

Technological parameters used to characterize the foaming process are cream time, rise time and gel time. Cream time may vary between 0.001 s and 30 s, and rise time is

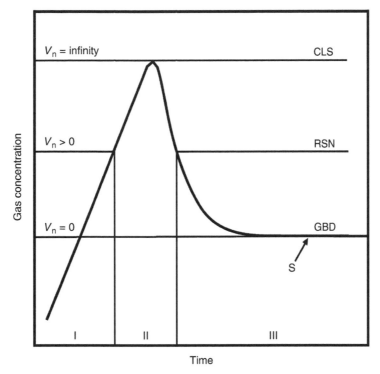

Figure 1 Change of gas concentration in the reaction mixture during foaming and its effect on bubble nucleation rate. V_n, nucleation rate; RSN, rapid self-nucleation with partial release of saturation; GBD, growth by diffusion; S, saturation; CLS, critical limiting supersaturation.

typically between 20 s and 120 s. Gel time is measured by touching the foaming mass with the glass rod. Before gelation polymer is sticky and can be drawn into long fibers.

B. The Role of Components in the Foam Mixture

Typical urethane foam composition is the following:

- isocyanate
- polyol
- water
- physical blowing agent
- amine catalyst
- metal catalyst
- surfactant (usually silicone block copolymer).

Today, foams are almost exclusively made by the one step process called 'one shot' process. This was made possible with the development of new catalysts, which could adjust the two reaction rates: isocyanate with polyol, and isocyanate with water. In this process all components are mixed simultaneously and the mixture is converted into the final product. The alternative process is a two stage 'prepolymer process', which was used earlier before the advent of catalysts, and is still used in special cases and in preparation of elastomers. In this case, the polyol component is reacted with excess of isocyanate to obtain isocyanate terminated prepolymer. The prepolymer is then reacted with a short polyol, water or polyamine, called 'chain extender' or curing agent, to obtain the final product. Polyol and isocyanate determine the physical and mechanical properties of the product. Water is used to produce CO_2 gas, to blow the foam. The resulting amine will react with isocyanate to produce urea groups, which give higher mechanical strength and rigidity than urethane groups. If urea groups are to be avoided, and softer foams are desired, the foaming can be achieved by the addition of physical blowing agents, low boiling liquids such as fluorocarbons, hydrocarbons or carbon dioxide. Most of fluorocarbons are generally banned for industrial used because of their negative effect on the ozone layer. New blowing agents are pentanes, CO_2, or other gases, but the search for the good replacement of fluorcarbons is very active. Physical blowing agents are essential in rigid foams where little or no water is used.

Amine catalysts are primarily used to catalyze the isocyanate–water reaction ('blowing catalyst'), while tin or other metal catalysts are used to regulate the rate of the isocyanate–polyol reaction ('gelling catalyst'). Surfactants are used up to 2 pph (parts per hundred) to regulate the cell size. Higher amounts of the surfactant produce thinner cell walls and smaller cells. An excessive amount would cause collapse of the foam as the walls and ribs of the foam cells could not support the pressure of the gas.

C. Technology of the Flexible Foam Preparation

Flexible foams have their flexibility (low modulus) because of the long, low T_g polyol chains and thus low degree of crosslinking. The flexibility of the foam depends on the molecular weight of the polyol, molar ratio isocyanate/hydroxyl (called also index when multiplied by 100), and the selection of isocyanate (TDI gives higher flexibility than crude MDI). Isocyanate index 100 indicates a 1:1 ratio of the isocyanate and hydroxyl groups, while index 105 shows 5% excess of isocyanate above the stoichiometric ratio. Catalyst selection is crucial for regulating the foaming process and properties. Usually a system of catalysts is used, consisting of one or several amine catalysts and metal organic catalysts. The latter are hydrolytically unstable and should not be added to the polyol component long in advance if water is present in the composition. Preparation of foams requires rapid and efficient mixing of the polyol and isocyanate component during processing. All other components are added to the polyol component or sometimes fed to the mixing machine as the third component (e.g. catalyst). The typical composition for the flexible foam given in Table 6 illustrates the amount of each component in the mixture.

Table 6 Typical formulation for flexible foam.

Component	Parts per hundred (pph)
PPO triol ($M = 3000$), partially terminated with ethylene oxide	100
Water	3.5
Fluorocarbon blowing agent	10
DABCO	0.45
N-ethylmorpholine	0.60
Tin octoate	0.15
Silicone surfactant	1–2
TDI (80/20)	45

We see that the two amine catalysts, DABCO and N-ethylmorpholine, are added in the amount of about 1% based on the polyol component, while the amount of Sn-octoate is 0.15%. While DABCO is a balanced catalyst, which promotes both gelation and foaming, N-ethylmorpholine favors open cell formation. The surfactant has multiple role, to lower surface tension and facilitate division of cells, and since it is a separate phase, to act as a nucleant. Increasing the amount of surfactant gives finer cells with thinner walls until the limit is reached above which it causes foam collapse. Density of flexible foams is usually between 30 and 80 kg/m^3. Density of the polyurethane itself is about 1100 kg/m^3.

Foams may have open or closed cells. Open cells are obtained by crushing the foam after gelation, but the amount of open cells is regulated by the selection of catalysts. Foams used in the furniture industry contain open cells while those used for thermal insulation (rigid foams) are required to have closed cells, since they contain a gas of low thermal conductivity. Polyester urethane flexible foams have better strength and oxidative stability but lower hydrolytic stability than polyether urethane foams. They also show higher hysteresis in the stress–strain cycling test. Polyester urethane foams are more resistant to chemicals, particularly those used for chemical cleaning, but are also more expensive than PPG based foams.

The manufacture of flexible block foams is carried out in a continuous process. The components (polyol), isocyanate and eventually catalysts) are mixed in the head of the mixing machine and poured in the transverse direction of the moving conveyer belt. The liquid mixture starts foaming to form a large foamed bun, which is then sliced into squares of the desired thickness. Such products could be used directly for mattresses, for example.

D. Integral Skin Foams

When foams are made either by free foaming or in a mold, a skin is formed on the foam surface. This fact is utilized to prepare foamed products with a controlled thickness of the skin. The formulation for integral foams generally does not contain water but it has physical blowing agents. The objects are made in closed molds. Density of the skin can be regulated by the mold temperature, amount of the mixture poured in the mold (larger amount exerts higher pressure) and mold release agents (usually silicones). As a rule lower temperature favor thicker skin. Higher pressure and release agents, which act as antifoaming agents in contact with the skin, also favor thicker skin.

E. Microcellular Foams

Microcellular foams (elastomers) differ from classical foams, because of their cell structure, higher density of the foams, which is typically 200 kg/m^3, and the structure of the matrix. Microcellular foams are foamed segmented elastomers with smaller number of round cells, unlike polygonal cells with ribs in standard foams. Because of their superior mechanical properties they are used for shoe soles, car bumpers, etc. They are formed by adding water and excess isocyanate in the elastomer formulation, which liberates CO_2.

F. Rigid Foams

Rigid foam compositions differ from those of flexible foams as they use short triols or higher functionality polyols, typically with $M_n = 400$. They are made with crude MDI, and main part of foaming is done with physical blowing agents. Due to the high concentration

Table 7 Typical formulation of a rigid foam.

Component	Amount, pph
PPG triol	100
Crude MDI	Stoichiometric + 5%
Blowing agent	50
Triethylene diamine (DABCO)	0.5
Surfactant (silicone block copolymer	1.0
Crosslinker (glycerin)	10

of crosslinks the foams are rigid (the glass transition, T_g, of the PU matrix is above room temperature). Part of the rigidity comes from the higher weight ratio of aromatic isocyanates as well as from higher isocyanate index (usually 105 or higher). Higher rigidity can be obtained by using sugar (sorbitol)-based polyols, which have higher functionality ($f = 6$). Due to the large concentration of isocyanate and hydroxyl groups, the reaction is more exothermic than in the case of flexible foams, requiring less powerful catalysts. A typical rigid foam formulation is given in Table 7.

This formulation uses triethylenediamine, which moderately catalyses the polyol/ isocyanate reaction. Crosslinking density is increased by adding low molecular crosslinker, glycerin, and foaming is achieved exclusively with the physical blowing agent. However, water may be added as co-blowing agent to increase mechanical properties. Rigid foams are foamed usually in molds or cavities, as in refrigeration, laminates and packaging. Standard foaming machines are used to mix the components and pour-in-place. Alternatively reaction injection molding (RIM) machines are utilized. Rigid foams can be also applied by spraying.

G. Processing of Polyurethanes [7,23]

Polyurethane foams and cast elastomers are made from liquid compositions, while thermoplastic polyurethanes are processed using standard processing techniques used for thermoplastics, such as injection molding and extrusion. Liquid systems are handled differently since several components have to be mixed and poured in a mold or in open space. Thus the essential part of urethane processing is mixing equipment, which consists of the reservoirs for storage of each component, a metering unit, and a mixing head. The scheme is given in Figure 2.

The process of molding foams or elastomers consists of pumping components at a given ratio (metering) to the mixing head, where the components are mixed to a homogeneous mixture, and pouring in the mold or on a conveyer belt as in the case of flexible foams. The liquids to be pumped, primarily polyols, may have viscosities up to 20,000 mPa s (cP). Low speed gear pumps are used to transport the fluids. The heart of the system is the mixing head. Basically two types are available: low pressure and high pressure mixing heads. Components in low pressure mixing heads are mixed using pressure up to 4 MPa (40 bar) and mechanical stirring. The advantage of low pressure mixers is their lower cost. They can handle low throughput (less than 35 g/s), small part casting (less than 15 g) and they allow processing of the wide range of viscosities. At the end of the process, the head is cleaned with solvents. If low viscosity polyols are used (viscosity not higher than 2000 mPa s) then high pressure machines with 'impingement mixing' can be

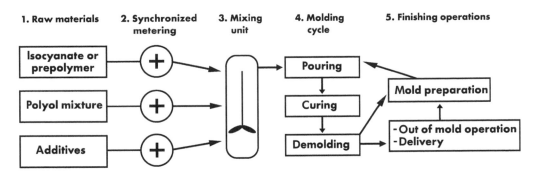

Figure 2 Schematic representation of the polyurethane casting process.

utilized. Viscosity can be reduced by heating the polyol component. Here the two or more component streams are injected into the mixing chamber with high velocity, where they collide and mix by turbulent flow. The advantage of high pressure machines is that they allow exact metering, processing of very fast systems, minimize waste and may not require cleaning between shots (self-cleaning heads). High pressure machines dominate the market.

The molding process could be continuous as in the case of slabstock foam, Figure 2, or discontinuous when molding in the mold is carried out. In the continuous process the mixing head traverses from one side of the conveyer to the other in the perpendicular direction to the direction of conveyer movement and pours the liquid urethane mixture to the paper base on the conveyer. The liquid quickly forms a cream and then rises to form a bun, which is cut to the desired size with razor blades.

H. Reaction Injection Molding (RIM) [24,25]

Reaction injection molding is a variation of the standard high pressure molding with impingement mixing. A very low viscosity mixture is injected into the mold to produce quickly the final part. RIM differs from regular molding in that the formulation of the polyurethane system has to be very fast. This is achieved by replacing the diol crosslinker with diamine crosslinker to obtain polyurea. This technique can be used to produce 'structural foams' (high density rigid foams with a skin) for auto body parts, dashboards and bumpers and also to obtain elastomers and microcellular foams. Components are injected in the mixing chamber of the mixing head under high pressure and mixed by impingement. The piston then injects the accumulated mass into the mold and cleans the chamber for the new shot. When the piston is in the down position the polyol and isocyanate components are recycled. Because of the low viscosity and low pressures RIM technology can be used to mold large parts with metal inserts. The molds for RIM can be made from steel, aluminum or zinc alloys. They are cheaper than the molds for injection molding of thermoplastics. Total consumption of energy is lower than in the competing techniques, and the investment in equipment is lower.

If glass fibers are added to get reinforcement, the method is known as RRIM (Reinforced Reaction Injection Molding). Structural RIM (SRIM) is the process whereby the reinforcement fabric or mat (glass, carbon) are placed in the mold and the resin is injected to impregnate the reinforcement.

VI. ELASTOMERS [4,8,14,26]

Two structural features characterize every useful elastomer: high chain flexibility (i.e., glass transition below the application range) and existence of either chemical or physical crosslinks. Flexibility of chains allows high deformation (uncoiling of the coiled chains) while crosslinks prevent chain slip, which produces plastic (irreversible) deformation. Polyether, polyester or polybutadiene chains having molecular weight above 1000 satisfy the first condition. The glass transition temperature of these materials is usually between $-40\,°C$ and $-80\,°C$. Polyurethane elastomers can be single or two-phase systems. One-phase systems are homogeneous chemically crosslinked polymers. Two-phase systems are block copolymers consisting of a hard and soft phase, separated by an interface. The blocks are called segments. Due to the difference in structure of the blocks, they do not mix but separate into 'domains'. Schematic representation of the segmented polyurethanes is given in Figure 3.

Hard domains are usually prepared from aromatic isocyanates and short glycols or diamines, called chain extenders. Neighboring hard segments are held together by Van der Vaals forces and hydrogen bonds, forming domains, which act as physical crosslinks.

Segmented polyurethanes are usually prepared by a prepolymer process (reaction (46)) and subsequent chain extension (reaction (47)). A prepolymer is prepared by reacting excess of isocyanate with a polyol (diol), typically of the molecular weight 2000.

OCN -R- NCO + HO ᴡᴡᴡᴡᴡᴡᴡᴡᴡ OH + OCN - R - NCO ⟶

⟶ OCN -R- NHCOO ᴡᴡᴡᴡᴡᴡᴡᴡᴡ OOCHNR -NCO   (46)

prepolymer

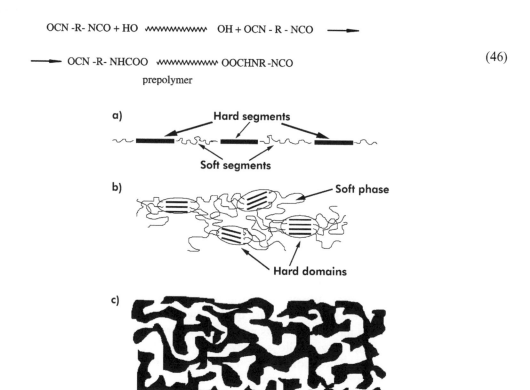

Figure 3 Schematic representation of the structure of the segmented polyurethane chain (a), association of hard segments into domains of globular morphology (b) and co-continuous soft and hard phase morphology (c).

The most frequent chain extenders are butanediol or diamines as in the case of elastomeric fibers. A typical procedure involves mixing MDI and a polyol at 80 °C for several hours under an inert gas blanket. Then the chain extender is added and stirred until the temperature starts rising. The material is then poured into the mold and the temperature increased to 110–130 °C for several hours to promote curing. Post-curing is then carried out for 24 h at 110 °C to complete chemical reaction. Preparation of the elastomer from the prepolymer and chain extender proceeds according to the scheme:

$$OCN\text{-}\text{wwwww}\text{-}NCO + HO\text{-}R'\text{-}OH \longrightarrow$$

prepolymer   chain extender

$$\text{(47)}$$

$$\text{wwwww } NHCOOR'OOCNH \text{wwwww } NHCOOR'OOCNH \text{wwwww}$$

polymer

This method produces polyurethanes with a controlled composition. Alternatively, the polymer can be produced by the one step ('one shot') process, where all components are mixed together at the same time. The resulting polymer has statistical composition (random distribution of polyol and chain extender units in the chain), which depends on the relative reactivity of different diol components. The properties would differ somewhat from those of the polymers made by the prepolymer process. Soft segment concentration is controlled by the chain extender/polyol ratio. The following formula (48) can be used to calculate chain extender (CE)/polyol (POL) molar ratio (r) for the desired soft segment concentration, SSC:

$$r = n_{CE}/n_{Pol}$$
$$r = [100(M_{po} - 34) - SSC(M_{pol} + M_{ISO})]/SSC(M_{CE} + M_{ISO}) \quad \text{(48)}$$

Here, M_{pol}, M_{ISO} and M_{CE} are molecular weights of the polyol, isocyanate and chain extender, respectively.

Setting the number of moles of the polyol, n_{pol}, to be 1, r becomes the number of moles of the chain extender. Number of moles of the diisocyanate, n_{ISO}, at the stoichiometric ratio of NCO/OH groups is the sum of the moles of the polyol and the chain extender, i.e., $n_{ISO} = r + 1$. Thus, a prepolymer for a given SSC should be prepared from one mole of the polyol and $(r + 1)$ moles of diisocyanate and extended with r moles of the chain extender. Number average molecular weight of the soft segment is determined by the selection of the polyol molecular weight. The hard segment molecular weight, M_{nhs}, is determined by the soft segment molecular weight and soft segment concentration, according to the expression:

$$M_{nhs} = (100 - SSC)(M_{pol} - 34)/SSC \quad \text{(49)}$$

At 50% SSC, the number average molecular weights of the hard and the soft segment must be equal.

When stress is applied, soft segments uncoil to give large deformation, while hard domains preventing slippage of the chain past each other, restrain plastic deformation. Properties of a polyurethane elastomer depend on the selection of a diisocyanate, chain extender and polyol but also on the length and concentration of the soft and hard segments. At low concentrations of hard segments (below 30 wt%) the hard domains have globular shapes and are dispersed in the (continuous) matrix of the soft phase. By increasing hard segment concentration, the globules become ellipsoidal and more elongated until they reach rod-like shape. At about 50% of each phase, the most likely morphology is lamellar, i.e., the sample structure consists of alternating layers of the hard phase and soft phase. Both phases are continuous, i.e., they span from one to the other end of the sample. At still higher hard segment concentration, phase inversion occurs and the soft phase becomes discontinuous, dispersed in the hard phase. Thus, by increasing soft segment concentration (SSC) from zero to the maximum value, two phase inversions are observed, the first occurring when soft phase becomes continuous (typically at about 35% SSC) and the second when the hard segment becomes discontinuous (typically at about 65% SSC). Polyurethanes at low SSC are tough, nylon-like polymers, and at high SSC are soft (low durometer hardness) elastomers. At intermediate concentrations they are hard elastomers. Phase separation primarily occurs because of the immiscibility of the hard and soft segments. Degree of phase separation (or phase mixing) affects the properties of the polymers, and it depends on the structure of the soft and hard segments and temperature. Usually the hard phase is crystalline. For example, the melting point of the hard segment consisting of MDI and butane diol is between 180 and 220 °C, and it depends on the molecular weight of the hard segment. The glass transition temperature of the amorphous part of the high molecular weight hard segment is around 80 °C. Phase mixing above the melting point is considerable, being higher in polyester polyurethanes than in polyether urethanes. Polybutadiene soft segments and especially silicone based soft segments have almost complete phase separation even in the melt. By quenching the melt, one can preserve partially mixed phase structure, which, however, will not be stable and will change with time. Slow cooling of the melt or preparation of films from the solution gives maximum phase separation.

The most frequently used diisocyanate in elastomer technology is MDI, although the first elastomers from Bayer Corp. (Vulkolans) were based on NDI. TDI in principle, does not give high quality elastomers unless aromatic diamine chain extenders are used.

Polyester soft segments impart better thermal and oxidative stability, oil and solvent resistance, higher abrasion resistance and strength to elastomers, expecially if they crystallize under stress, but they have lower hydrolytic, acid/base and fungus resistance than polyether urethanes. Polyether urethanes have generally lower T_g and are better suited for low temperatures than the polyester urethanes. Polypropylene oxide-based soft segments are the least expensive, do not crystallize under any conditions and have excellent flexibility. PTMO-based polyurethanes have superior characteristics and an excellent balance of properties.

The most frequently-used chain extender is butanediol, but when higher modulus or strengths are required, then aliphatic–aromatic chain extenders, such as *p*-bis(hydroxy-ethoxy)benzene or aliphatic and aromatic diamines can be used. Primary amines are too fast and unsuitable for work except in special cases. Retardation of the reaction can be achieved by introducing steric hindrances, such as by introduction of chlorine atom as in 3,3'-dichloro-4,4'-diamino phenylmethane (MOCA):

(50)

MOCA is a strong carcinogen and should be used with good safety protection.

One way of influencing properties of urethane elastomers is to use excess of isocyanates. If the NCO/OH ratio is higher than one, the resulting polyurethanes will have higher hardness and strength. Optimal excess of NCO is about 2–5%.

Polyol (soft segment) molecular weight affects the modulus, E, of an elastomer. The theory of rubber elasticity predicts that Young's modulus of an elastomer is inversely proportional to the molecular weight of network chains, M_c:

$$E = \frac{3\rho RT}{\overline{M}_c} \tag{51}$$

This means that longer polyols produce softer polyurethane elastomers. The T_g of the soft phase is also related to the M_c:

$$T_g = T_{g\infty} + \frac{K}{\overline{M}_c} \tag{52}$$

where $T_{g\infty}$ is the glass transition temperature of the linear long polymer, and K is a constant for the given system. Glass transition temperature of the soft phase of an elastomer based on polytetramethyleneoxide is $-43\,°C$ when molecular weight of the polyol is 650, $T_g = -60\,°C$ for the polyol with $M_c = 1000$, or $-86\,°C$ for the $M_c = 2000$. Thus, for semi-rigid elastomers and foams, polyol molecular weight should be below 1000. Modulus of polyurethane elastomers can be elevated by adding fillers. To summarize, the factors determining properties of a polyurethane elastomer are:

1. structure of the polyol
2. type of diisocyanate
3. type of the chain extender
4. molar ratio NCO/OH
5. soft segment concentration
6. molecular weight of the polyol
7. filler.

In all cases above it is understood that the chains are linear and crosslinking was achieved by physical bonds and hard domain formation. Such polymers display typical thermoplastic behavior, i.e., they flow when they are melted and harden by cooling. Domains are destroyed above the melting point of the hard phase but are reformed upon cooling, displaying reversible crosslinking. These materials are called 'thermoplastic urethanes' (TPU). Properties of thermoplastic urethanes are very temperature dependent and their strength decrease dramatically above the glass transition of the hard segments (above $100\,°C$). They also display a large permanent set (irreversible deformation) after being held under stress for a long period, especially at elevated temperatures.

There is another group of polyurethanes that is chemically crosslinked with a crosslinker, either triol or polyamine or polyisocyanate. They are single-phase elastomers, and they display lower strengths than the thermoplastic urethanes. However, their properties are less temperature sensitive, and elastic recovery is generally considerably better (permanent set is smaller) than in TPUs. Their strength can be improved by adding proper fillers. Such systems are called 'cast systems' since they are processed by casting liquid components into a mold. Again, the hardness of these elastomers is governed by the molecular weight of the polyol and its functionality.

A. Processing of Polyurethane Elastomers

Polyurethane elastomers can be processed by casting, milling or calendaring as in the rubber industry or by standard techniques for thermoplastics (injection molding, extrusion).

1. Cast Polyurethane Systems

Both single phase and two phase systems can be processed by casting. Two phase elastomers are prepared from the low molecular weight components (prepolymer and chain extender or polyol, isocyanate and chain extender) with or without catalysts. The composition can be mixed by hand or with mixing equipment as shown earlier and poured into the molds. RIM technology can be utilized to speed the process and to obtain large parts. The advantage of casting segmented polyurethanes is that the structure and thus properties can be tailored according to the processor's desires. Also, chemical crosslinks can be introduced by using components with higher functionality than 3. The equipment cost is modest and the molds may be inexpensive.

2. Vulcanizing Polyurethanes

The rubber industry uses specific processing equipment, and the transition to the standard urethane technology would be costly. Therefore, a family of urethanes was developed that can be processed on standard rubber equipment. High molecular weight polyurethanes are crosslinked using sulfur, peroxides or polyisocyanates. Crosslinking of high molecular weight polymers is called vulcanization. Vulcanization by sulfur and peroxides require polymers with double bonds, while isocyanate crosslinking is carried out through the active OH groups or urea (–NHCONH–), urethane (–NHCO–O–) or amide (–NHCO–) groups.

3. Processing of Thermoplastic Polyurethane Elastomers [23]

TPUs are segmented elastomers with strong physical crosslinks. Polymerization is completed in the manufacturer's plant and the user buys the granulated resin.

Polymerization is usually carried out in the reactor and the melt poured on the conveyer belt to be put through the oven to complete polymerization. The polymer is then ground, extruded, palletized and packaged. Alternatively, liquid components are fed into an extruder, which acts as a reactor. The residence time in the extruder should be long enough to obtain a polymer, which is then extruded, pelletized and packaged.

Thermoplastic urethanes are generally soluble in strongly polar solvents such as dimethylformamide (DMF), dimethylacetamide (DMA) or dimethylsulfoxide (DMSO). Tensile strength of the MDI/polyester diol/butanediol polyurethane may reach 40 MPa, and the hardness varies typically from 60 Shore A to 75 Shore D, depending on hard segment content. Polyurethanes absorb moisture readily. Thus, the material should be stored in a cool, dry area, and must be thoroughly dried before injection molding or extrusion. Typically, injection molding machines have injection screw design with three zones: feed, compression and metering. The recommended range of length to diameter ratio of the screw (L/D) should be between 16/1 and 20/1. The compression ratio is usually between 2:1 and 3:1. Polyurethanes are sensitive to high shear stresses, and the check valves should be such as to minimize the risk. Shot size should be between 25% and 75% of the barrel capacity and clamping pressure about $0.45–0.60 \, tons/cm^2$ ($3–4 \, tons/sq.in$) because of the high viscosity of the polyurethane melt.

Extrusion of TPUs requires extruders with higher torque or hors power drives compared with other thermoplastics. Recommended screw compression ratio is 2.5:1 and minimum L/D ratio of 24:1 for most polyurethanes.

TPUs can be blow molded, and the requirements for the screw design are the same as for extrusion. Both injection molded and blow molded parts should be heated at $100 \, °C$ for 24 hours to reach the equilibrium structure and minimize creep and compression set. The main field of application of thermoplastic urethanes is for various casters, rollers, wheels, flexible clutches, seals and gaskets for hydraulic machines, shoe soles, printing rolls and machine parts.

VII. ELASTOMERIC POLYURETHANE ('SPANDEX') FIBERS

'Spandex' fibers are elastomeric polyurethane fibers used in the manufacturing of high stretch garments, such as swimsuits, sport apparel, etc. Elastomeric fibers are essentially thermoplastic elastomers spun into fibers. Usually diamine chain extenders are used instead of diols to produce polyureas. Polyureas have high melting points (above the onset of degradation) and cannot be spun from the melt. They are spun from the solution. Only non-urea polyurethanes can be melt processed. There are four basic processes for making fibers:

1. wet spinning
2. dry spinning
3. reaction spinning
4. melt spinning.

Wet spinning or solution spinning is a process where the elastomer is dissolved in a solvent, for example DMF or DMA, and forced through the spinneret into the coagulating bath, which contains a nonsolvent for polyurethane miscible with the solvent. Water is a good coagulating agent and is miscible with DMF or DMA. The solution coming out of the spinneret comes in contact with the coagulating nonsolvent, immediately forming the fiber. Initially the solvent is in the fiber, but it diffuses into the bath causing additional coagulation. Thus the residence time in the coagulating bath must be sufficiently long to allow most of the solvent to leave the fiber. This time is controlled by the length of the bath and speed of drawing the fiber. Usually the spinneret contains a large number of holes, producing a number of fibers which are bound together in a multifilament yarn.

The fiber may be stretched by pulling to obtain orientation and improve packing of hard domains. The scheme of the wet spinning process is depicted in Figure 4. The process is relatively slow and requires treatment of the wastewater from the baths.

Dry spinning starts from the solution containing pigments and additives like in wet spinning, but instead of extruding the solution into the coagulating bath, the fibers are extruded in the chamber heated with hot air. The solvent evaporates, and the resulting fiber is wound on a take-up roll. More than 80% of spandex fibers are produced this way since it is faster and more economical than the other techniques.

Reaction spinning involves simultaneous chain extension reaction and spinning. The isocyanate-terminated prepolymer is extruded into the bath containing diamine or

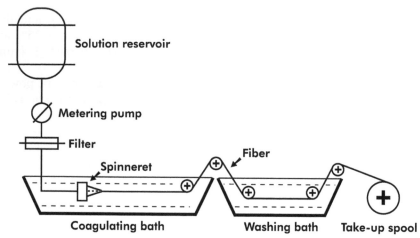

Figure 4 Schematic representation of the wet spinning process. a, reservoir for the solution; b, metering pump; c, filter; d, coagulating bath; e, spinneret; f, washing bath; g, finishing; h, take-up roller.

polyamine. Isocyanate-amine reaction is almost instantaneous. The fiber has fairly rigid skin on the surface, which facilitates wind-up. The advantage of this process is that chemically crosslinked polymers can be obtained. The application of the method is, however, limited.

Melt spinning is essentially an extrusion process applied to segmented polyurethanes.

VIII.   COATINGS [5]

A coating consists of binder (polymer resin), solvent, pigments and filler. Polyurethane resins have a special place among the natural and synthetic binders in the coating industry due to their excellent adhesion to various substrates. Polyurethane paints and varnishes can be classified in several groups:

1.   Two component systems, where one component is polyisocyanate and the second is a polyol with additives. These systems are manufactured with or without solvents.
2.   One component systems that cure with the moisture from the surrounding air. These systems can also be with or without solvents.
3.   One component systems containing a mixture of a polyol and a blocked isocyanate. At elevated temperatures the polyisocyanate is deblocked and reacts with the polyol. Powder coatings also belong to this group.
4.   Non-reactive urethane systems containing a polyurethane dissolved in a solvent. The system dries upon evaporation of the solvent.
5.   Urethane oils or urethane alkyds.
6.   Water based dispersions.

A.   Two-Component Coatings

The two components of this system are the isocyanate and the polyol with all additives. The two components are mixed and applied by some of the standard application techniques (brush, spraying, roller, dipping etc.). Since a requirement for the isocyanate is low vapor pressure, instead of using pure isocyanates, their adducts, polymeric isocyanates, isocyanurates, or prepolymers are preferred. MDI is, however, used in its monomeric form because of its low vapor pressure. An example of an adduct for the coating industry is Bayer's Desmodur L, based on trimethylol propane and TDI:

$$
\begin{aligned}
&\text{CH}_2\text{-O-OC-NH-}\langle\bigcirc\rangle\text{-NCO} \quad (\text{CH}_3) \\
&\text{CH}_3\text{-CH}_2\text{-C-CH}_2\text{-O-OC-NH-}\langle\bigcirc\rangle\text{-NCO} \quad (\text{CH}_3) \\
&\text{CH}_2\text{-O-OC-NH-}\langle\bigcirc\rangle\text{-NCO} \quad (\text{CH}_3)
\end{aligned}
\tag{53}
$$

Aromatic isocyanates are not desirable in coatings that will be exposed to sunlight (exterior applications) because of yellowing. Such systems are preferably based on aliphatic (HDI) or cycloaliphatic isocyanates (isophorone diisocyanate). Polyisocyanates are usually supplied in the form of a concentrated solution (50–80%) to reduce viscosity. Polyols for two component coatings can be polyester, polyether, acrylic resins, or urethane resins containing hydroxyl groups. Other components with hydroxyl groups include epoxy resins, coal tar, cellulose esters, etc. Higher hydroxyl content in the polyol translates into higher crosslinking density, higher film strength and higher resistance to chemicals. Lower hydroxyl contents give better film elasticity. The isocyanate/hydroxyl molar ratio may be off-stoichiometric and is determined by trial and error. At low NCO/OH ratios (less than 1) the coating displays higher elasticity but lower solvent and chemical resistance. It should be emphasized that in two-component coatings, not all isocyanate groups react with polyol, and almost one third reacts with moisture from the surrounding air. This fact should be taken into account when adjusting NCO/OH ratio.

B. One-Component Systems

One-component urethane coatings are usually based on MDI, TDI or HDI terminated prepolymers containing free isocyanate groups. Pigmenting such systems is delicate due to moisture in the pigment, which can cause premature gelation. Therefore, these systems must contain additives for moisture removal such as zeolites (alumosilicates). Since the chain extension in such systems is carried out with water from air, one of the products of the reaction is CO_2, which diffuses from the film without foaming.

If the film is not too thick and the isocyanate content high, the curing process is under control and no bubbles are formed. The prepolymers may be dissolved in a solvent to reduce viscosity. The initial phase of drying consists of solvent evaporation followed by chemical reaction. Reaction rate in one-component coatings depends on the relative humidity. Film thickness with two-component and one-component coatings can be considerable (0.5 to 10 mm). The advantage of these systems is the absence of the unpleasant solvent smell, lower fire hazard and lower price. The disadvantage is poorer wetting of the surface, difficulty in obtaining mat surfaces and possibility of bubble formation. One-component PU solventless systems are used for the preparation of synthetic mortars, floors (PU resin with sand at a ratio 1:15), or elastic athletic tracks (filled with ground rubber).

C. Blocked Isocyanates

Prepolymers and adducts with free NCO groups are sensitive to moisture and have limited shelf life. If the isocyanate groups are blocked (for example with phenol, cresol, ethyl acetoacetate, dimethyl malonate or butane oxime) one can obtain stable systems at room temperature. Heating blocked isocyanates to 160–180 °C (or about 30 °C less in the presence of a catalyst) causes deblocking to occur, and isocyanate can react with the present co-reactant (polyol), i.e., the system becomes the two-component coating. The polyol can be also phenolic, urea or melamine resins wit free OH groups. These systems have to be 'baked' at elevated temperatures. The main application field for these systems is in electroinsulation, such as wire enamels, impregnation varnishes, and for substrate coating. Another important field of application of blocked isocyanate systems includes powder coating. Here the polyol component may be a polyester from terephthalic acid or polyacrylate with free OH groups. The isocyanate component is TDI or IPDI blocked with caprolactam. After heating at 160–200 °C for 10–35 min, the powder melts (on the electrostatically coated metal), and the deblocked isocyanate reacts with the polyol.

D. Non-reactive PU Systems

These coatings are formed by physical drying, i.e., high molecular weight linear polyurethane forms strong secondary bonds with substrate after solvent evaporation. These varnishes have, however, low solvent and chemical resistance. They are used for coating flexible substrates such as leather or flexible parts made of PU foams with integral skin and as the modifiers for printing inks (paints), etc.

E. Urethane Oils or Urethane Alkyds

Urethane oils are solvent-borne, air-drying systems, prepared by reacting partially hydrolyzed drying oils. The oils are triglycerides of fatty acids, which after heating with glycerin and a catalyst produce a mixture of mono- and di-glycerides containing free OH groups. The isocyanates used for reacting with oils are almost exclusively TDI and IPDI. Formulation of the coatings is the same as with classical alkydes, i.e., they must contain metal catalysts (for example cobalt naphthenate), which promote oxidative drying through double bonds in oils. Thus, drying oils with high content of unsaturation, such as linseed, soybean, sunflower and safflower oil, must be used as the base. Molecular weight of drying oils before crosslinking is low in order to have low viscosity. In spite of that, they are diluted with solvents to further reduce viscosity. Urethane alkyds are used as printing inks, wood varnishes, floor coatings, etc.

F. Polyurethane Dispersions

These systems are based on ionomers, i.e., polymers containing ionic groups, mainly anions such as sulfonic or carboxylic groups. They are neutralized with bases to form salts, which contribute to hydrophilicity (formation of bonds with water), and formation of stable dispersions of polymers in water. After drying, a film on the applied surface is formed. Orientation of molecules, hydrogen bonding, as well as coulombic forces, act as a kind of crosslinks. These films are tough and have good oil and water resistance but they are sensitive to polar solvents.

IX. POLYURETHANE ADHESIVES [6,8,27]

Polyurethane adhesives are an important group of materials thanks to the very polar groups in their structure as well as the ability of isocyanates to form chemical bonds with the substrate. According to the method of application adhesives are classified in the following groups:

1. Reactive two-component adhesives based on polyisocyanates and low molecular weight polyols, which form the polymer when mixed together.
2. One-component reactive prepolymer that reacts with moisture from the air to give a polymer.
3. Solution adhesive consisting of the high molecular weight linear polyurethane polymer dissolved in a suitable solvent.
4. Solution of adhesive as in (3) which has polyisocyanate as a crosslinker.
5. Solution of a non-urethane polymer, e.g., polychloroprene, with isocyanate as the crosslinker.
6. Dispersion adhesive containing high molecular weight polyurethane with ionic groups dispersed in water.

One can easily observe the similarity between coatings and adhesives. Polyisocyanates used for adhesives should have high molecular weight and thus, low vapor pressure at the reaction temperature. Adducts such as Desmodur L, or crude MDI are often used, as are triisocyanates such as triphenyl methane-4,4′,4″-triisocyanate (a) or tris(p-isocyanatophenyl) ester of thiophosphoric acid (b) dissolved in methylene chloride or ethyl acetate:

(54)

Since isocyanates are used as adhesion promoters, they can be added in excess up to 50% above stoichiometric ratio in the two-component adhesives. Excess of hydroxyl groups, up to 10%, is used if bonding of flexible materials is carried out. The chains with unreacted OH groups then act as plasticizers. Polyols for adhesives are usually crystallizable components. Polyether polyols are used in spite of the lower adhesivity because of the lower viscosity.

One-component adhesives are either prepolymers or adducts with terminal isocyanate groups, which could react with moisture from air to give a polymer. Excellent adhesion is obtained with high molecular weight polymers.

X. OTHER APPLICATIONS OF POLYURETHANES

A. Sealants [28]

Sealants are very important materials in construction but also in the automotive industry. They are basically low modulus polymers (elastomers) with good adhesion properties, containing high concentration of filler. They are obtained by mixing isocyanate and polyol components or by reacting prepolymer with moisture. Most frequently used polyol components are polypropylene oxide-based polyols because of their lower price and good hydrolytic stability, but other types, such as castor oil and polybutadiene polyols, are used as well. The chemistry of hardening is the same as in other systems with isocyanate and hydroxyl groups.

B. Polyurethane Casting Resins

Casting resins are used in the electrical industry for preparation of insulators, embedding transformers, cable joints, encapsulating electrical components, etc. Although the dominating resins are epoxies, polyurethanes are finding their way too, because of their lower price and better processability at low temperatures. One of the crucial requirements in high voltage electrical insulation is the absence of voids and other forms of trapped gas, which ionizes and causes dielectric breakdown. Thus, any form of bubble formation resulting from the isocyanate–water reaction must be prevented. This is achieved by adding zeolites to the polyol. Zeolites bind water faster than isocyanates, allowing casting in the open air. Standard rigid casting resin consists typically of the PPO-based triol of low molecular weight (about 450), 200 pph of filler (silica, calcium carbonate), zeolite paste (5–10 pph) and crude MDI. Such a compound has properties similar to the epoxy compound but lower price and viscosity. Casting under vacuum assures good quality insulation for medium voltages, up to 80 kV. Cycloaliphatic isocyanates such as IPDI are used for resins exposed to exterior conditions. Increasing molecular weight of the polyols decreases rigidity of the resin, and rubbery insulation can be molded if necessary. Good elastic properties are obtained by using castor oil as the polyol component.

XI. ENVIRONMENTAL STABILITY OF POLYURETHANES

Resistance to various environmental factors, such as heat, light and humidity, is one of the most important properties of materials. Stability of polyurethanes under the influence of these factors varies with the structure of the material. Thus we will discuss just the stability of the urethane group and a few typical chemical groups in urethane materials, such as ether or ester.

A. Thermal Stability of Polyurethanes [29–32]

Thermal stability of urethane group is relatively low. It was already stated that it depends on the groups to which it is attached, being the highest for aliphatic isocyanate–aliphatic alcohol (approx. dissociation temperature 250 °C), then aliphatic alcohol–aromatic isocyanate (200 °C), aliphatic isocyanate-aryl alcohol (180 °C) and the lowest for the aromatic isocyanate–aromatic alcohol (120 °C) [33]. The upper limit temperature for polyurethanes in long term continuous use is set at 120 °C. Thermal degradation of urethanes proceeds in one of the three basic reactions [31]:

1. dissociation of urethane group to initial components (the same is valid for urea group):

$$R\text{–}NH\text{–}COO\text{–}R' \xrightleftharpoons{\Delta} R\text{–}NCO + HOR' \qquad (55)$$

2. formation of primary amine and olefin:

$$R\text{–}NH\text{–}COOCH_2\text{–}CH_2\text{–}R' \rightarrow RNH_2 + CO_2 + R'CH{=}CH_2 \qquad (56)$$

3. decomposition resulting in the formation of a secondary amine and CO_2:

$$R\text{—}NH\text{—}COOR' \longrightarrow \underset{\underset{H}{|}}{R\text{—}N}\text{—}R' + CO_2 \qquad (57)$$

Degradation products can further react generating a number of products. The reactions above take place both in oxidative and inert atmosphere at the same rate (temperature), except that further course of reaction varies. Therefore, burning of polyurethanes can generate toxic isocyanates, which should be taken into account when these products are used for building insulation. Reaction (55) can be slowed down or the temperature of urethane group decomposition increased by replacing hydrogen in the urethane group by methyl group [34,35]. Thermogravimetric analysis of thermoplastic urethanes shows that the onset temperature of degradation is the same in both inert and oxidative atmosphere since it starts in the urethane group, followed by degradation of the soft segments.

Degradation of polyols occurs faster in air than in nitrogen. Polyether urethanes have lower thermal stability than the corresponding polyester urethanes, particularly in presence of oxygen, because of the increased sensitivity of the alpha C-atom in ethers towards oxidation. The sensitivity to oxidation increases in polyesters with decreasing number of CH_2 groups between two ester groups, or if tertiary carbon is present in the chain, as in the case of polypropylene glycols.

Thermal stability of polyurethanes can be judged by loss of weight or loss of properties and the results may not coincide. For example, polyurethanes with continuous hard segment concentration would initially loose weight faster than those low hard segment concentration, but the crystalline structure of the former may retain the properties better than the latter. Thermal stability is affected not only by chemical composition, but also by the shape of the product. In principle, thermal stability is higher if the ratio of surface to volume of the body is lower. Thus, fibers are very sensitive to oxidation because the degradation products can diffuse out easily and have no time to undergo recombination into a more stable product. The order of thermal stabilities of different isocyanate products is given as follows [33]: isocyanurate (>270 °C) > urea (≥ 180 °C) > urethane (≥ 150 °C) > biuret (≥ 120 °C) > allophanate (≥ 120 °C) > uretdion (≥ 120 °C).

B. Resistance to UV-Light of Polyurethanes [36–38]

According to Nevskii et al. [36] polyurethane decomposes in the presence of UV light following several possible routes:

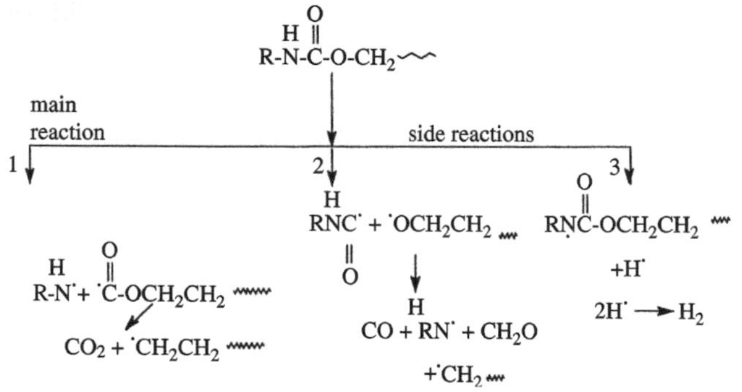

$$
\begin{array}{c}
\overset{\text{O}}{\underset{\text{H}}{\text{R-N-C-O-CH}_2}} \\
\end{array}
\tag{58}
$$

Photolysis processes (decomposition in presence of light) are very complex and depend on a number of conditions such as chemical structure of the urethane group environment, presence of moisture and other agents. It is known that aromatic isocyanates give urethanes that yellow and then become dark brown in presence of UV light. This comes from an extended system of conjugated bonds from aromatic rings and urethane bond, formed during UV irradiation:

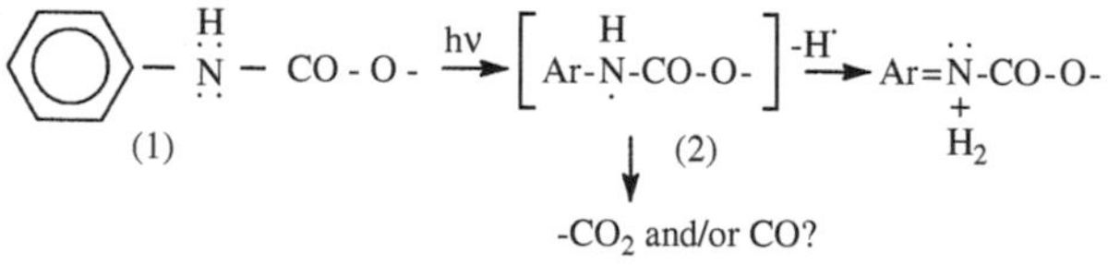

$$
\tag{59}
$$

It was found that isocyanate part in polyesterurethanes based on MDI changes according to Scheme (60). Complexity of the processes occurring under UV irradiation is evident. While thermal stability of urethanes cannot be enhanced (degradation slowed down), they can be fairly efficiently protected against UV irradiation with additives, such as pigments, UV absorbers, inhibitors, etc. As with thermal stability, UV stability of polyetherurethanes is lower than that of polyesterurethanes, especially in the presence of oxygen.

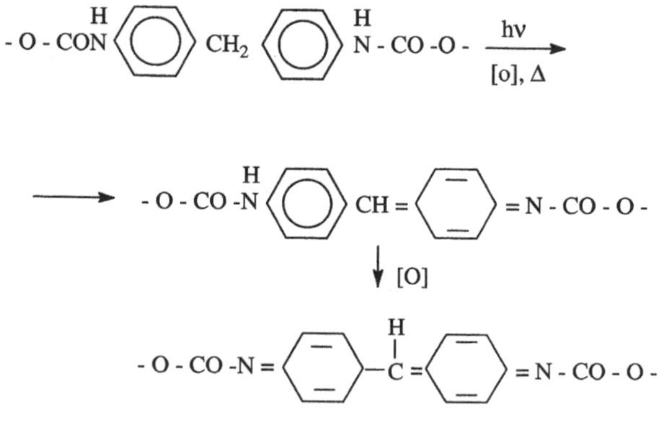

$$
\tag{60}
$$

C. Hydrolytic Stability of Polyurethanes [39]

Hydrolytic stability of polyurethanes depends on the composition and concentration of the weakest groups. Polyesterurethanes have considerably lower resistance to hydrolysis than polyether or polybutadiene based polyurethanes. Hydrolysis of ester groups is catalyzed by acid groups formed as the product of hydrolysis. Therefore, an efficient way of slowing down the process is to block the acid formed, which is usually carried out by adding carbodiimides. Such solutions are temporary since after the consumption of all carbodiimide the process accelerates again.

The urethane group can be considered a combination of the amide and ester groups and is sensitive to moisture. It is highly hydrophilic, and it dissociates in the presence of moisture to give an alcohol, amine and CO_2, according to the following scheme:

$$R{-}NH{-}COOR' + HOH \longrightarrow RNHCOOH + R'OH \tag{61}$$

$$R{-}NH{-}COOH \longrightarrow RNH_2 + CO_2 \tag{62}$$

These reactions proceed at a significant rate at temperatures of 170–190 °C when they are used in recycling urethane products.

XII. SAFETY CONSIDERATIONS WHEN WORKING WITH POLYURETHANE RAW MATERIALS [3,8]

Raw materials for polyurethane preparation are isocyanates, polyols (sometimes polyamines) and catalysts. Polyols are fairly harmless substances, but the polyol component may contain hazardous additives and catalysts. Amine chain extenders and catalysts present usual hazard like all amines, i.e., they are often carcinogenic. This is especially true of MOCA. Metal catalysts are also very toxic and careful handling is necessary.

A major hazard comes from the isocyanate component because of its high reactivity and high concentration. Isocyanates may alter proteins, deactivate enzymes and destroy tissue cells. The main danger comes from inhalation of vapors. Somewhat less harmful is skin contact. Oral intake of isocyanates may also occur. Isocyanates in contact with mucous tissue cause irritation at low concentration. Effect on skin is felt, however, only at high concentrations and prolonged contact. HDI although less reactive, has a stronger irritating effect on skin than TDI. TDI is used in large quantities. Due to its high volatility, TDI has strong biological effects on health, which also depends on the concentration. Table 8 shows the effect of concentration on humans after exposure of several minutes.

Table 8 Acute biological effects of TDI (8).

Concentration of TDI, mg/kg	Effect
< 0.02	Possible asthmatic reactions with hypersensitive humans
0.1–1.5	Small nose and throat irritation
1.3–10	Coughing, reversible bronchitis (lasts several hours)
10–50	Irreversible bronchitis, pulmonary edema
Above 50	Life threatening

Maximal allowable concentration (MAK value) for TDI is 0.02 mg/kg (during eight hour exposure). TDI smell appears above 0.05 mg/kg, which is already above the MAK value and indicates a sufficiently dangerous concentration.

When isocyanate is spilled over the skin, it should be immediately wiped off, and the contact area should be washed with a large amount of water and soap. If isocyanate has reached the eyes, it should be immediately washed with water for several minutes.

Additional washing with a buffer solution for eyes is recommended. If a person swallows isocyanate, he should drink a large amount of water and empty the content of the stomach by vomiting. Protection against isocyanates and safe handling for each individual isocyanate is given in the safety data sheet delivered with the isocyanate.

REFERENCES

1. Saunders, J. H., and Frisch, K. C. (1983). *Polyurethanes, Chemistry and Technology*, Part I. *Chemistry*, Krieger Publishing Co., Malabar, FL.
2. Saunders, J. H., and Frisch, K. C. (1964). *Polyurethanes: Chemistry and Technology*, Part II, Interscience Publishers, New York.
3. Buist, J. M., and Gudgeon, H. (1969). *Advances in Polyurethane Technology*, MacLaren and Son, London.
4. Wright, P., and Cumming, A. (1969). *Solid Polyurethane Elastomers*, MacLaren and Sons, London.
5. Wiegel, K. (1966). *Polyurethane*, Lacke Holz-Verlag GmbH, Mering, Germany.
6. Vieweg, R., and Hochtlen, A. (1966). Polyurethane. In *Kunstoff–Handbuch*, Carl Hanser Verlag, Munchen, Germany.
7. Wirpsza, Z. (1993). *Polyurethanes, Chemistry, Technology and Application*, Ellis Horwood, New York.
8. Oertel, G. (1985). *Polyurethane Handbook*, Hanser Publishers, Munich, Germany.
9. Gorbatenko, V. I., Zhuravlev, E. Z., and Samaray, L. I. (1987). *Izocianati-Metodi sinteza i fziko-hemicheskie svojstva alkil-, aril-i geterilizocianatov*, Naukova Dumka, Kiev.
10. Ulrich, H. (1966). *Chemistry and Technology of Isocyanates*, John Wiley and Sons, New York.
11. Guo, A., Javni, I., and Petrovic, Z. (2000). Rigid polyurethane foams based on soybean oil. *J. Appl. Polym. Sci.*, 77: 467–473.
12. Guo, A., Cho, Y.-J., and Petrovic, Z. S. (2000). Structure and properties of halogenated and non-halogenated soy-based polyols. *J. Polym. Sci. Part A: Polym. Chem.*, 38: 3900–3910.
13. Buist, J. M. (1978). *Development in Polyurethanes – 1*, Applied Science Publishers, London.
14. Hepburn, C. (1973). *Polyurethane Elastomers*, Applied Science Publishers, London.
15. Berlin, A. A., and Shutov, F. A. (2000). Penopolimeri na osnove reakcionosposobni oligomerov, Himiya, Moskva.
16. Baker, J. W., and Gaunt, J. (1949). The mechanism of the reaction of aryl isocyanates with alcohols. Part III. The 'spontaneous' reaction of phenyl isocyanate with various alcohols. Further evidence relating to the anomalous effect of dialkyanilines in the base-catalysed reaction. *J. Chem. Soc.*, 9: 19.
17. Reegan, S. L., and Frisch, K. C. (1971). Catalysis in isocyanate reactions. In *Advances in Urethane Science and Technology* (Frisch, K. C., and Reegan, S. L., eds.), Technomic Publishing Co., Inc., Westport, CT.
18. Huynh-Ba, G., and Jerome, R. (1981). Catalysis of isocyanate reactions with protonic substrates: a new concept for the catalysis of polyurethane formation via tertiary amines and organometallic compounds. In *Urethane Chemistry and Applications* (Edwards, D. N., ed.), ACS, Washington D.C.

19. Nakayama, K., Ino, T., and Matsubara, I. (1969). Infrared spectra and structure of polyurethane elastomers from polytetrahydrofurane, diphenylmethane-4,4′-diisocyanate, and ethylenediamine. *J. Macromol. Sci. Chem.*, *3*: 1005–1020.

20. Pigott, K. A. (1969). Polyurethans. In *Encyclopedia of Polymer Science and Technology*, Vol. 11 (Mark, H., Gaylord, N. G., and Bikales, N. M., eds.), Interscience Publishers, New York, pp. 506–563.

21. Frisch, K. C., and Saunders, J. H. (1972). *Plastics Foams*, Part I, Marcel Dekker Inc., New York.

22. Saunders, J. H. (1960). The formation of urethane foam. *Rubber Chemistry and Technology*, *33*: 1293–1322.

23. BASF. *Elastolan Design and Processing Guide*, BASF Technical Publication 9/93. BASF Corporation, Wyandotte, MI.

24. Becker, W. E. (1979). *Reaction Injection Molding*, Van Nostrand Reinhold Co., New York.

25. Sweeney, F. M. (1987). *Reaction Injection Molding Machinery and Processes*, Marcel Dekker, Inc., New York.

26. Petrovic, Z. S., and Ferguson, J. (1991). Polyurethane elastomers. *Progress in Polymer Science*, *16*: 695–836.

27. *Polyurethan Klebestoffe aus Baycoll, Desmocoll und Desmodur*, Technical Publication from Bayer Co., Leverkusen, Germany.

28. Evans, R. M. (1993). *Polyurethane Sealants, Technology and Applications*, Technomic Publishing Co. Inc., Lancaster, PA.

29. Petrovic, Z. S., Zavargo, Z., Flynn, J. H., and MacKnight, W. J. (1994). Thermal degradation of segmented polyurethanes. *J. Appl. Polym. Sci.*, *51*: 1087–1095.

30. Javni, I., Petrovic, Z. S., Guo, A., and Fuller, R. (2000). Thermal stability of polyurethanes based on vegetable oils. *J. Appl. Polym. Sci.*, *77*: 1723.

31. Saunders, J. R. (1959). The reactions of isocyanates and isocyanate derivatives at elevated temperatures. *Rubb. Chem. Technol.*, *32*: 337.

32. Thimm, T. (1982). Derzeitige Erkentnisse uber physikalische und chemische Vorgange bei der thermischen und thermo-oxidativen Beanschpruchung von Polyurethanelastomeren. Teil 1. Der chemische und morphologische Aufbau von Polyurethanelastomeren und dessen physicallische Veranderung bei thermische Beanschpruchung. *Kauchuk und Gummi, Kunststoffe*, *35*: 568–584.

33. Thimm, T. (1983). Derzeitige Erkentnisse uber physikalische und chemische Vorgange bei der thermischen und thermo-oxidativen Beanschpruchung von Polyurethanelastomeren. Teil 2. Chemische Vorgange I) Allgemeine Ubersicht und Betrachungen Uber thermolitische bzw. Pyrolitische Processe. *Kauchuk, Gummi, Kunststoffe*, *36*: 257–268.

34. Foti, S., Maravigna, P., and Montaudo, G. (1982). Effects of N-methyl substitution on the thermal stability of polyurethanes and polyureas. *Polymer Degradation and Stability*, *4*: 287–292.

35. Flynn, J. H., and Petrovic, Z. (1994). Thermal stability enhancement of polyurethane by surface treatment. *J. Thermal Analysis*, *41*: 549–561.

36. Nevskii, L. V., Tarakanov, O. G., and Belyakov, V. K. (1967). Deagradation of polyurethanes under action of ultraviolet radiation. *Sov. Plastics*, 47–49.

37. Thimm, T. (1984). Derzeitige Erkentnisse uber physikalische und chemische Vorgange bei der thermischen und thermo-oxidativen Beanschpruchung von Polyurethanelastomeren. Teil 2: Chemische Vorgange III) Photooxidation. *Kautchuk, Gummi, Kunststoffe*, *37*: 1021.

38. Hoyle, C. E., and Kim, L.-J. (1987). Effect of crystallinity and flexibility on the photodegradation of polyurethanes. *J. Polym. Sci., Part A: Polym. Chem.*, *25*: 2631–2642.

39. Thimm, T. (1984). Derzeitige Erkentnisse uber physikalische und chemische Vorgange bei der thermischen und thermo-oxidativen Beanschpruchung von Polyurethanelastomeren. Teil 2. Chemische Vorgange II) Thermooxidative und solvolitische Processe. *Kauchuk, Gummi, Kunststoffe*, *37*: 933–944.

9
Polyimides

Javier de Abajo and José G. de la Campa
Institute of Polymer Science and Technology, Madrid, Spain

I. INTRODUCTION

Polyimides are polymers incorporating the imide group in their repeating unit, either as an open chain or as closed rings. However, only cyclic imides are actually of interest concerning polymer chemistry. Thus, under the generic name polyimides, we will exclusively refer to cyclic polyimides in this chapter.

The first reference to a polyimide was dated at the beginning of the 20th century [1], but the actual emergence of polyimides as a polymer class took place in 1955 with a patent of Edwards and Robinson on polymers from pyromellitic acid (1,2,4,5-tetracarboxy-benzene) and aliphatic diamines [2]. Since then, growing interest in polyimides has brought about a big expansion of the science and technology of this family of special polymers, which are characterised by excellent mechanical and electrical properties along with outstanding thermal stability. Among the wide list of reported heat-resistant condensation polymers [3–5], polyimides have gained a prominent position due to their good properties–price–processability balance. And from the production figures, it can be inferred that polyimides stand virtually alone with respect to providing useful, available, technological materials.

Furthermore, while at the beginning polyimides found application in a rather restricted variety of technologies, mainly on the form of films and varnishes for the aerospace and electrical industries, the discovering of addition polyimides, and, more recently, of thermoplastic, processable aromatic polyimides has widened the range of properties and application possibilities to a great extent. Presently, they should be considered as versatile polymers with an almost unlimited spectrum of applications as specialty polymers for advanced technologies [6–12].

In a list of applications of polyimides, the following should be included:

- Insulating films, coatings and laminates
- Molded parts
- Structural adhesives
- Insulating foams
- High-modulus fibers
- High-temperature composites
- Permselective membranes

From the beginning, the major proportion of research effort on polyimides was directed to the development of wholly aromatic species, seeking for high thermal stability. In this respect, wholly aromatic polyimides are materials that can retain their properties almost unchanged for long periods at 250–300 °C. But it was soon realized that the application of aromatic polyimides, and in general aromatic polyheterocycles, was not possible from the melt and, furthermore, their extreme structural rigidity and high density of cohesive energy made them insoluble in any organic media. Given the excellent properties of the aromatic polyimides, structural modifications were soon outlined in order to overcome these limitations, and as a consequence of the many research efforts made in this direction, the chemistry of polyimides has greatly enriched thanks to the many improvements achieved in the last thirty years [9–11,13–15].

II. CONDENSATION POLYIMIDES

A. Polyimides via Poly(amic acid) from Dianhydrides and Diamines. Reaction Conditions and Monomers Reactivity

The polycondensation of an organic dianhydride and a diamine is the traditional method employed in the synthesis of polyimides (Scheme 1).

$$\text{(1)}$$

This general scheme is valid for both aliphatic and aromatic polyimides. Since this is the route preferably used for aromatic, aliphatic and cycloaliphatic polyimides of technical importance, it has been the subject of numerous studies, and the main aspects of the mechanisms and kinetics are fairly well known [16]. It is a two-step reaction. In the first step the nucleophilic attack of the amine groups to the carbonyl groups of the dianhydride gives rise to the opening of the rings yielding an intermediate poly(amic acid) (Scheme 2).

Poly(amic acid)

$$\text{(2)}$$

The symmetrical and unsymmetrical poly(amic acid)s are intended, since both are possible.

The poly(amic acid) is converted, in the second step, to the corresponding polyimide through a cyclodehydration reaction (Scheme 3).

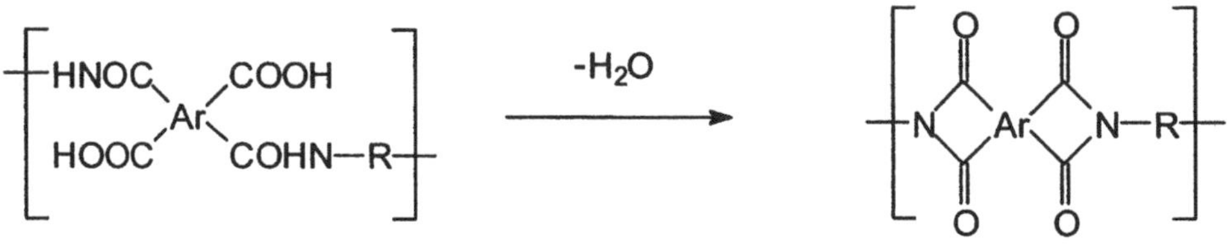

(3)

This simplified scheme may be envisioned in a more complete form by using monofunctional species (Scheme 4).

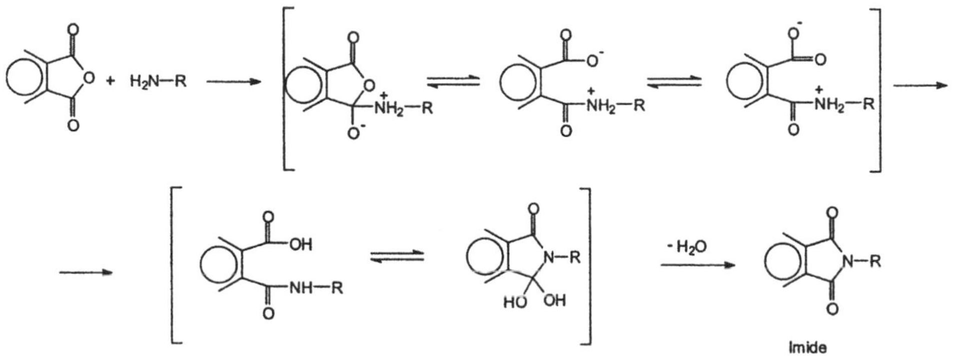

(4)

The first step is crucial to attain high molecular weight, and the second has a great influence in the final nature of the polyimide since a quantitative conversion in the cyclodehydration process is needed to have a pure, fully cyclized polyimide. Highly polar solvents are suitable media to dissolve monomers and poly(amic acid)s. N,N-dimethyl-acetamide (DMA), N,N-dimethylformamide (DMF), dimethylsulfoxide (DMSO), and N-methyl-2-pyrrolidinone (NMP) are the most adequate. Purity of solvents and reactants, and strict stoichiometric balance are requirements of polycondensation reactions that fully fit polyimides synthesis, where a careful control of the reaction variables is essential to achieve high molecular weight [17–19]. For instance, rigorous exclusion of water is a key condition, as well as a moderate polymerization temperature (about 0 °C or less) in poly(amic acid) formation in order to limit the competition of side reactions and a premature release of imidation water.

A comparative study of the influence of side reactions has been made by Kolegov *et al.* [20], who have considered the following sequence of possible reactions (Scheme 5).

The concurrence of these reactions can obviously alter the progress of the main reactions 1 and 2 and may prevent a high molecular weight. Experimental data of polycondensations of diamines and dianhydrides can generally be treated as second order reversible reactions, but the comparatively great magnitude of K_1 allows the calculation of rate constants according to an irrversible reaction. In fact K_1 is greater than K_2, K_4 and

K_5 and K_5 by approximately seven orders of magnitude and over fifteen times greater than K_3 [21].

(5)

The reactants concentration also plays a determinant role. It has been stated that on plotting the inherent viscosity of poly(amic acid) against the initial concentration of monomers, a curve with a maximum can be attained. This maximum is presumably different for each monomers combination and solvent, but from the available data it is accepted that for high molecular weight to be obtained 0.4 to 0.8 mol/L monomer concentration is to be used [22–24]. The figures correlate well with data reported for the synthesis of aromatic polyamides from aromatic diamines and aromatic diacid chlorides [25].

In order to carry out a successful polymerization, a fixed mode of monomers addition has been suggested. Traditionally, the addition of the dianhydride (preferably as a solid) on the diamine solution has been considered as the right mode of addition, and that because the anhydride is sensitive to solvent impurities (water, amines), and even to solvent reaction, in much greater degree than the diamine, so that with the diamine in large excess the main reaction will be favoured [22,26,27]. Furthermore, unlike aromatic diamines, aromatic dianhydrides are not easily dissolved at low temperature.

Nevertheless, the same results can be obtained regardless the order of monomers addition in the synthesis of poly(amic acid)s from pyromellitic dianhydride and oxydianiline if the reaction conditions are stretched in terms of dryness, stoichiometry, and solvent and monomers purity [28]. This indicates that the classical order of monomers addition has been imposed by the sensitivity of dianhydrides to water and solvent impurities more than by reactivity or solubility concerns.

The progress of the polycondensation reaction largely depends also on the nature of the monomers, and particularly on the monomers reactivity. As a rule, electron deficient diamines will react more slowly than electron rich diamines. At this respect, some studies have been made on the reactivity of diamines by conventional methods. A reliable approach to quantify the reactivity of diamines and dianhydrides, is the calculation of molecular parameters by means of the modern methods of Computational Chemistry. The reactivity of diamines against acylating monomers like acid chlorides have been reported [29,30]. Likewise, theoretical calculations can be made to estimate the relative reactivity of diamine and dianhydride monomers.

Quantum semiempirical methods are reliable tools for the determination of parameters involved in the reactivity of organic reactants [31]. In fact, some partial studies were performed by Russian researchers more than twenty years ago to relate electronic parameters with reactivity of polyimide monomers [16]. However, the methods they used to calculate these parameters have been nowadays overcome, and consequently, it seems interesting to obtain new theoretical data that could be correlated with experimental results. Thus, the method AM1 [32] included in the MOPAC package, version 6.0 [33] has been used for the calculations that follow.

In spite of the commercial importance of polyimides and of the huge number of new monomers synthesized in the last twenty years, the amount of kinetic data for the acylation reaction of diamines and dianhydrides is very scarce, and we have only been able to find data for a few diamines and an even shorter number of dianhydrides [34]. As commented before, the acylation reaction between a diamine and a dianhydride takes place by the attack of the lone pair of the nitrogen of the amine to the centre of low electronic density located in the carbonylic carbon of the anhydride. Therefore, the reaction will be controlled by the interaction between the occupied orbitals of the diamine and the unoccupied orbitals of the dianhydride. The reactivity of the amines will be affected by both the electronic density on the nitrogen and by the energy of the Highest Occupied Molecular Orbital (HOMO) [29,30]. In dianhydrides, the reactivity will be determined by the electronic deficiency on the carbonylic carbon and by the energy of the Lowest Unoccupied Molecular Orbital (LUMO).

As the reactivity will be higher when the difference between both orbitals will be lower, higher values of E_{HOMO} and lower values of E_{LUMO} will indicate the more reactive diamines and dianhydrides respectively. Tables 1 and 2 show the main parameters calculated for several diamines and dianhydrides, from which kinetic data could be found in the literature. The calculated values correspond, in all cases, to the more stable conformation. In both cases, diamines and dianhydrides, the differences of charge, either on the amino nitrogen or on the carbonylic carbon, are very scarce and, furthermore, in the case of diamines, because of the fact that the C_{Ar}–N bond is out of the plane of the aromatic ring, the charge transfer from the amine to the ring is difficult. Therefore, the presence of electronwithdrawing groups does not cause a decrease of the charge on the nitrogen but an increase on the polarizability of the N–H bonds.

The values of E_{HOMO} in the diamines are controlled by the character of the groups present in the structure, being higher (higher reactivity) in the case of electron donating groups. In that way, the higher reactivity should correspond to p-phenylene diamine, where the second amino group acts as activating of the first one. The lowest reactivity corresponds to the sulfonyldianiline (DDSO), because of the strong electron withdrawing character of the sulfone group. These values of E_{HOMO} can be related with the

Table 1 Electronic parameters and kinetic data for several diamines and their corresponding monobenzamides.

Diamine	$Q_N{}^a$	E_{HOMO}	$Q_N{}^b$ amide	E_{HOMO} amide	$\log K$ acylation
H₂N—⟨C₆H₄⟩—NH₂	−0.314	−7.92	−0.319	−8.06	2.48
H₂N—⟨C₆H₄⟩—NH₂ (meta)	−0.329	−8.26	−0.330	−8.40	0.00
H₂N—⟨C₆H₄⟩—⟨C₆H₄⟩—NH₂	−0.327	−7.94	−0.328	−8.06	0.37
H₂N—⟨C₆H₄⟩—O—⟨C₆H₄⟩—NH₂	−0.323	−8.11	−0.323	−8.25	0.79
H₂N—⟨C₆H₄⟩—CO—⟨C₆H₄⟩—NH₂	−0.337	−8.65	−0.338	−8.73	−2.17
H₂N—⟨C₆H₄⟩—CH₂—⟨C₆H₄⟩—NH₂	−0.326	−8.29	−0.326	−8.39	0.56
H₂N—⟨C₆H₄⟩—SO₂—⟨C₆H₄⟩—NH₂	−0.354	−8.89	−0.356	−8.99	−2.66
H₂N—⟨C₆H₄⟩—O—⟨C₆H₄⟩—⟨C₆H₄⟩—O—⟨C₆H₄⟩—NH₂	−0.330	−8.32	−0.330	−8.36	0.15

[a]Charge on the nitrogen of any of the amino groups in the diamine.
[b]Charge on the remaining amino group after the formation of the benzamide on the other side.

experimental values of acylation constants shown in Table 1 as it can be seen in Figure 1. A very good linear relationship can be observed, thus confirming the influence of the electronic parameters of the diamines in the determination of reactivity.

The reaction of the first amino group, that is converted to amide, causes a decrease of the reactivity of the second amino group, as it could be expected, which is reflected by a decrease of E_{HOMO} (Table 1). However, contrarily to the expected, a small increase of the electronic density in the amine nitrogen is observed. This effect is probably related with the out of plane situation of the C_{Ar}–N bond, that has been commented above. The decrease in E_{HOMO} is very small in all cases, even for p-phenylene diamine and practically no influence of the structure of the diamine can be observed.

In Table 2 are shown the electronic characteristics of the dianhydrides (E_{LUMO} and charge on the carbonylic carbon) and their acylation constants. In this case, the presence of electronwithdrawing groups causes a decrease of E_{LUMO}. Thus, the most reactive compound is the pyromellitic dianhydride, because of the strong activation produced by the presence of the second anhydride group. Next in reactivity is the dianhydride with the sulfonyl group, and the lower reactivity corresponds to monomers with a long separation between both anhydrides, and with electron donating ether groups. However, in this case, the correlation between theoretical and experimental data is not as good as in the case of diamines, mainly because of the strong deviation of the linear behaviour observed in the case of the pyromellitic dianhydride.

Table 2 Electronic parameters and kinetic data for several dianhydrides and their corresponding monoamides.

Dianhydrides	$Q_{C(C=O)}$[a]	E_{LUMO}	$Q_{C(C=O)}$[b] amide	E_{LUMO} amide	$\log K$ acylation
	0.344	−2.86	0.349	−2.18	0.79
	0.349(m)[c] 0.348(p)	−2.20	0.350(m) 0.349(p)	−1.85	0.13
	0.349(m) 0.351(p)	−2.03	0.349(m) 0.353(p)	−1.68	−0.006
	0.350(m) 0.346(p)	−2.30	0.350(m) 0.346(p)	−2.02	−0.66
	0.351(m) 0.341(p)	−2.45	0.352(m) 0.342(p)	−2.15	1.04
	0.349(m) 0.354(p)	−1.67	0.349(m) 0.354(p)	−1.59	−0.32
	0.349(m) 0.355(p)	−1.53	0.349(m) 0.354(p)	−1.46	−0.80
	0.351(m) 0.345(p)	−2.09	0.349(m) 0.355(p)	−2.01	0.326

[a]Charge on any of the carbonyl groups in the dianhydride.
[b]Charge on the remaining carbonyl groups after the formation of amide on the other side.
[c]m and p refer to the carbonyl in *meta* or *para* position to the substituent.

This must be attributed to the effect produced on the reactivity of the second anhydride group by the formation of the amide in the first one. Also in this case, the occurrence of the first reaction causes a decrease in the reactivity of the second anhydride (an increase of E_{LUMO}), but a very small change of the charge on the carbonylic carbon. However, in this case, the change in the orbitalic energy is significantly higher than for diamines and it depends very much on the structure of the dianhydride (as most of the

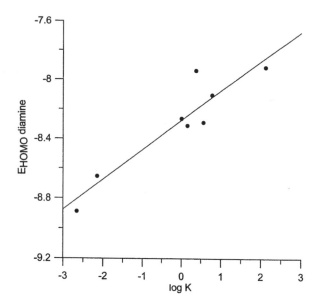

Figure 1   Correlation between E_{HOMO} of the diamines and $\log K$.

dianhydrides are not symmetrical there are two possibilities of ring opening, one with the amide group in *meta* to the substituent and one with the amide in *para*). Although there are small differences between both, they are not significant and consequently the values of E_{LUMO} shown in Table 2 are the mean of both possibilities). The reaction of one group in pyromellitic dianhydride increases E_{LUMO} in 0.68 eV (maximum change for diamines was 0.14 eV), but in the case of the dianhydride with the aliphatic chain and the ether groups between both rings, only an increase of 0.07 eV is observed.

    This means that the reactivity for the global acylation does not depend on the reactivity of the dianhydride but on the reactivity of the less reactive molecule, that is, the monoreacted anhydride. Consequently, it can be confirmed that the reactivity of these species is controlled by the energy of the LUMO. A representation of E_{LUMO} (monoamide) versus $\log K$ is shown in Figure 2. The correlation in this case is very good, thus confirming the usefulness of the electronic parameters to predict the reactivity, even in a semiquantitative way.

    Thus, the value of E_{LUMO} can be used to predict the reactivity of dianhydrides, when no kinetic data are available. In Table 3 are shown the E_{LUMO} values of several important dianhydrides, for which kinetic data are not available.

    All these dianhydrides should have a very high reactivity, because of the lower values of E_{LUMO} for both the dianhydride and the monoamide. In fact, hexafluoroisopropyliden 4,4'-diphthalic anhydride should be only slightly less reactive than benzophenone tetracarboxylic dianhydride, and 2,3,6,7-naphthalene tetracarboxylic dianhydride should be very similar to biphenyl dianhydride. But if the reaction is controlled by the monoamide, as we have postulated, the most reactive dianhydride should be 1,4,5,8-naphthalene tetracarboxylic dianhydride, because E_{LUMO} is almost the same than for pyromellitic dianhydride, but E_{LUMO} monoamide is lower than E_{LUMO} monoamide of the pyromellitic (-2.33 versus -2.18 eV).

    To conclude, it can be said that the reactivity of diamines and dianhydrides to give polyamic acids, and consequently polyimides, is controlled by the energy of the frontier orbitals of both types of molecules. Although the charges could also play a role in

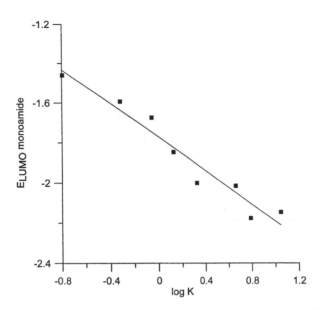

Figure 2 Correlation between E_{LUMO} of the monoreacted dianhydrides and $\log K$.

Table 3 Electronic parameters for dianhydrides from which there are no kinetic data available.

Dianhydride	E_{LUMO}	E_{LUMO} monoamide
	-2.237	-2.03
	-2.15	-1.92
	-2.85	-2.33
	-2.20	-1.905

the control of reactivity, the differences between them are very small and, in addition, in the case of diamines it is very difficult to determine the real value of charge on the nitrogen because the amino group is not in the same plane that the aromatic ring.

For the theoretical study of reactivities, selected diamines and dianhydrides have been chosen along those more frequently used in the preparation of aromatic polyimides. Most of them are commercially available, but some of them have been produced only at laboratory scale.

For some specific applications, particularly for microelectronics, the purification of these monomers is sometimes so critical that the isolation of suitable reactants requires sophisticated purification methods. For instance, miniaturization and tougher processing requirements for advanced microelectronics have forced researchers to attain ultrapure poly(amic acid)s from monomers purified by zone refining, and dianhydrides isolated in solid ingot form [35].

As to the molecular weights of poly(amic acid)s and polyimides, they had been only very seldom measured and reported. Thus, the usual criterion for molecular size in poly(amic acid)s and soluble polyimides had traditionally been the inherent viscosity (η_{inh}) until the size exclusion chromatography techniques (GPC) were refined and implemented in last years. The development of many new soluble thermoplastic polyimides has moved also for a growing interest in knowing the molecular weights, and for an improvement of the analytical technique for the determination of M_n's and M_w's. GPC columns, that can work with aggressive solvents like DMF, DMA or m-cresol at temperatures up to 70–80 °C, are available nowadays and can be used for the analysis of many soluble polyimides [36,37].

Of greatest importance is the cyclodehydration reaction leading from poly(amic acid)s to polyimides. The general approach in the application of insoluble, wholly aromatic polyimides as materials involves the elimination of solvent and water at high temperature. When a poly(amic acid) solution is heated over 200 °C, or at a lower temperature in the presence of a dehydrating agent, such as acetic anhydride/base, the polyimide is attained in few hours. Logically, the first approach received much more attention in the early years, because the research effort was mainly focussed to insoluble polyimides based on pyromellitic dianhydride, although the chemical imidization of poly(amic acid) films in the solid state has been the subject of several studies [38–40]. Thermal imidization associated to classical aromatic polyimides actually needs temperatures of about 300 °C to ensure total rings closing, and that is far from being an optimal approach in many instances because elimination of solvent and water at high temperatures can approach about surface irregularities, microvoids and even polymer degradation. Furthermore, high temperatures help for cross-linking side reactions, for example (Scheme 6):

1. $-NH_2$ end groups with the imide rings of the chains:

2. Thermal imidization by means of non-cyclized ortho-carboxyamides:

(6)

3. Amidation of $-NH_2$ free groups and ortho-carboxyamides:

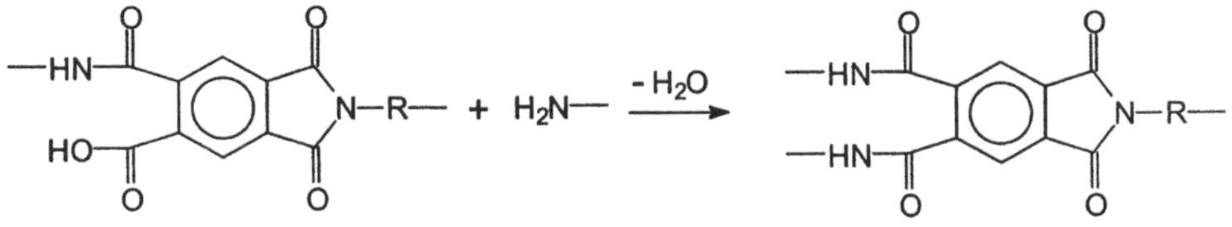

These reactions do help for a faster immobilization of the chain and, consequently, for additional difficulties to get 100% cyclodehydration. The strong interactions between the poly(amic acid) and the solvent also greatly interferes with the intramolecular cyclization, and its presence does not certainly aid a quantitative conversion. It has been demonstrated that solvents and poly(amic acid)s readily give rise to complexes [41], and that solvent rests can remain joined to the polymer even through covalent bonds [42,43].

A novel preparative method of poly(amic acid)s from aromatic diamines and dianhydrides consists of carrying out the polycondensation reaction in a precise mixture of tetrahydrofurane/methanol (9/1 to 6/4 by weight), at room temperature [44]. Average molecular weights (M_w) exceeding 150,000 g/mol have been reported for poly(amic acid)s attained by this method from oxydianiline and pyromellitic anhydride [45]. Moreover, thermal imidization seems to be more easily achievable on replacing classical high boiling amide solvents such as DMA or NMP by the easy to evaporate THF and methanol mixtures [46].

Chemical imidization is normally promoted by acetic anhydride, in combination with organic bases, for instance pyridine or triethylamine, but other dehydrating agents can be used, such as propionic anhydride, trifluoroacetic anhydride, N,N-dicylo-hexylcarbodiimide and the like. Although it can be performed on polyimide films, chemical imidization is mostly carried out in solution, with the final polyimide being collected as a precipitate, but most conveniently remaining dissolved all over the process. A premature precipitation of the polymer does not ensure total imidization at all, as partially imidized species can be insoluble in the organic medium. Temperatures and reaction times amply vary depending on the polymer and the cyclization system. Thus, if the reaction is conducted at room temperature 24 to 48 h are needed for total imidization, while some few hours are enough if the chemical cyclodehydration reaction proceeds at 100 °C.

The imidization process, either thermally or chemically induced, may be followed by a variety of means. It has been traditionally studied on poly(amic acid)s, as well as with molecular models, by IR and NMR spectroscopy [47,48]. But many other analytical methods have been used, for instance: TGA [41,49,50], DSC [42,51], polarizing microscopy [41], gas chromatography [52,53], microdielectrometry [54], or torsional braid analysis [55]. From the numerous contributions on this topic some conclusions can be drawn. Among other features, we remark that a rate reduction of the imidation and the rate constant occurs as the conversion increases, so that it can not be considered as a classical first order reaction. This phenomenon has been explained by considering entropic factors [56]. Since the kinetic data could not be unequivocally assimilated to a determined reaction order, they were interpreted as if the imidization reaction could be divided into rapid and slow first order cyclization steps. The retardation in the ring closure reaction has also been explained by the existence of various amide-acid groups with different reactivities and by the mobility reduction when the linear polymer is converted into a cyclic chain [38,57].

The formation of isoimide as an intermediate step to imide has been confirmed also in many instances (Scheme 7). Isoimides are less stable than imides, so that isoimide formation does not seem to play any significant role when conversion into imide is forced by thermal treatment, but it can affect the imidization process when rings closure is performed by chemical treatment at moderate temperatures. Furthermore, it has been proved that some solvent/anhydride/base combinations clearly favour the formation of isoimide [58], what can in turn, offer some advantage from a practical view point as polyisoimides are much more soluble than polyimides [14,59,60].

Chemical imidization is less attractive for commercial and experimental polyimides that are tested and used in the form of films, but chemical imidation has been the preferred method concerning experimental polyimides that are soluble in organic solvents in the state of full imidation. At this respect it is worthy to remark that 100% conversion in the ring closure step is virtually impossible to achieve, particularly for thermal imidation at high temperature (about 300 °C) in the solid state, due to the complexity of the process and to the inherent molecular regidity of insoluble polyimides. However, for soluble polyimides, solution imidization is possible at mild reaction temperatures, for instance 150–200 °C, with 100% conversion, and avoiding undesirable side reactions which lead to insolubility and infusibility [61].

(7)

By using monomers other than dianhydrides and diamines, a number of methods has been outlined to synthesize polyimides, for instance from tetracarboxylic acids and their half diesters. This method can be successfully applied to the preparation of aliphatic–aromatic polyimides by melt polycondensation of the salt from the diamine and an aromatic tetracarboxylic acid or half-diester (Scheme 8). The reaction achieved some importance when polyimides appeared in the 1950s as an alternative for uses such as fiber-forming and injection-molding polymers [2,62].

$$(8)$$

The method was believed to be valid only for polyimides with melting points low enough to remain molten during the polymerization, and solution methods were considered as not suitable since the 'polyamic salts' are not soluble in aprotic organic solvents, so that only low molecular weight polymeric salts were obtained. The method was not used for many years, mainly because aliphatic polyimides did not show much higher Tm than conventional nylons, and their main advantage, their high T_g value, did not mean any useful improvement as their performances under service conditions were comparable to the semicrystalline aliphatic polyamides, which are in turn much cheaper.

However, polyalkylenimides can be prepared from pyromellitic anhydride and α,ω-diaminoalkanes in solution of NMP. NMP seems to provide a much more convenient medium for these reaction than other organic solvents, and in this way, high molecular weight poly(amic acid)s and polyimides have been attained [63,64].

A revision of this approach has been made in last years, and aliphatic and aromatic salt monomers have been studied as precursors of high molecular weight polyimides. Salt monomers have shown to be actually highly reactive, as they can produce directly polyimides in a very short reaction time, and this feature has recently been observed not only for aliphatic precursors but for aromatic salts as well [65]. Moreover, high molecular weight polyimides can be achieved by combining the salt monomer method with high pressure polycondensation, or with microwave induced polycondensation [65,66]. Other advantages of the salt monomer method is that polycondensations can progress at high conversion in solid state, at temperatures substantially lower than the melting point (T_m) of the polymers, and at a lower temperature than the salt monomer melting point. Thus, imide ring closure takes place simultaneously to water or alcohol elimination, rendering polyimide in an one-step direct reaction without passing through poly(amic acid). In general, better results (higher inherent viscosities) are achieved from half esters than from tetracarboxylic acids, and another feature of this recently revised method is that highly crystalline polyimides, both aliphatic and aromatic, can be attained [66].

Half diesters and their derivatives have been extensively used also in the preparation of aromatic polyimides by the two-step method. The initial step involves the preparation of the modified monomers, which consist of the half esters themselves or of activated derivatives. The activation of the half esters is normally directed to the enhancement of the carboxylic acids reactivity, by converting them into other highly electrophilic groups such as acyl chlorides. The global synthetic route is depicted in Scheme (9). The high reactivity of acid chlorides against diamines, makes the solution method at low temperature not only recommendable but virtually the only possible one if high molecular weights are to be accomplished.

Low to medium molecular sizes can be also obtained from the half esters directly with specific amidation catalysts [67,68].

(9)

Imidization is achieved by thermal treatment of the poly(amic ester) precursor in the usual way, with elimination of alcohol. This method, because of its relative complexity, has not got practical significance for conventional polyimides. However it has been of great importance in the development of photocurable condensation polyimides [8], and to study the behaviour of different isomers as starting materials for model polyimides [67].

B. One-step Polycondensation. Thermoplastic Polyimides

As mentioned before, the first generation of fully aromatic homopolyimides, could be used only in a few application because they had to be applied in the form of soluble polyamic acids, what limited the materials to be transformed almost exclusively into films or coatings. They all had to be synthesized by a two-step method.

Further improvements in the chemistry of polyimides during the last years have been directed towards novel, linear species that are soluble in workable organic solvents or melt-processable while fully imidized. Thus, changes had to be introduced in the chemical structure to adapt the behaviour and performance of these specialty polymers to the demands of the new technologies. As a consequence, a new generation of condensation polyimides has appeared, the so-called thermoplastic polyimides.

The difficulties to process conventional aromatic polyimides are due to their inherent molecular features, what is particularly true for the most popular of them: polypyromellitimides. Molecular stiffness, high polarity and high intermolecular association forces (high density of cohesive energy) make these polymers virtually insoluble in any organic medium, and shift up the transition temperatures well over the decomposition temperatures. Thus, the strategies to novel processable aromatic polyimides have focussed on chemical modifications, mainly by preparing new monomers, that provide less molecular order, torsional mobility and lower intermolecular bonding.

From the various alternatives to design novel processable polyimides some general approaches have been universally adopted:

- Introduction of flexible linkages, which reduces chain stiffness.
- Introduction of side substituents, which helps for separation of polymer chains and hinder molecular packing and crystallization.
- Use of 1,3-substituted instead of 1,4-substituted monomers, and/or asymmetric monomers, which lower regularity and molecular ordering.
- Preparation of co-polyimides from two or more dianhydrides or diamines.

Polyimides with flexible linkages have been known from the advent of high temperature aromatic polyimides. In fact, most of the commercial, fully aromatic polyimides contain ketone or ether linkages in their repeating units [69], and early works in the field soon demonstrated that dianhydrides having two phthalic anhydride moieties joined by bonding groups, gave more tractable polyimides [26,70–73].

Many different linkages have been introduced with these purposes, but the most promising are: –O–, C=O, –S–, –SO$_2$–, –C(CH$_3$)$_2$–, –CH$_2$–, –CHOH–, and –C(CF$_3$)$_2$–. These bonding groups may be located on the dianhydride, on the diamine or on both monomers, or they can even be formed during the polycondensation reaction, when some functional monomers containing preformed phthalimide groups are used [74,75]. The presence of flexible linkages has a dramatic effect on the properties of the final polymers. First, 'kink' linkages between aromatic rings or between phthalic anhydride functions cause a breakdown of the planarity and an increase of the torsional mobility. Furthermore, the additional bonds mean an enlargement of the repeating unit and, consequently, a separation of the imide rings, whose relative density is actually responsible of the polymer tractability. The suppression of the coplanar structure is maximal when bulky groups are introduced in the main chain, for instance sulfonyl or hexafluoro-isopropylydene groups, or when the monomers are enlarged by more than one flexible linkage. Some diamines and dianhydrides with a flexible linkage in their structure have been listed in Tables 4 and 5. The combination of those dianhydrides and diamines, and also the combination of some of them with conventional rigid monomers like benzenediamines, benzidine, pyromellitic dianhydride or biphenyldianhydride, offer a major possibility of different structures with a wide spectrum of properties, particularly concerning solubility and meltability [76–80].

However, very few of the polymers that can be synthesized combining monomers of Tables 3 and 4 have been reported as melt-processable, although many of them are soluble in highly polar organic solvents. All of them show high glass transition temperatures, commonly over 250 °C, and, theoretically, they can develop crystallinity upon a suitable thermal treatment, mainly those containing polar connecting groups. Thus, depending on the nature of X and Y in the general formula (Scheme 10), polyimides can be prepared that show an acceptable degree of solubility in organic solvents.

(10)

Table 6 shows the T_g's and solubility of some selected polyimides among those prepared from monomers of Tables 4 and 5. The combination of non-planar dianhydrides and aromatic diamines containing flexible linkages, provides the structural elements needed for solubility and melt processability. Some aromatic polyimides marketed as thermoplastic materials are based on these statements [69,81–83].

Structural modifications to attain soluble aromatic polyimides have been also carried out by introducing side substituents, alkyl, aryl or heterocyclic rings. One of the first references of this approach described the synthesis of soluble aromatic polyimides containing side phthalimide groups [84,85]. Since then, many attempts have been made to prepare new monomers, diamines and dianhydrides, with pendent groups for novel processable polyimides. Table 7 shows some of these monomers.

Probably, the most promising species are those containing phenyl pendent groups. The phenyl rest does not introduce any relevant weakness regarding thermal stability, and provides a factor of molecular irregularity and separation of chains very beneficial in terms of free volume increasing and lowering of the cohesive energy density [80–91]. Fluorene

Table 4 Diamines for polyimides containing flexible linkages.

Table 5 Dianhydrides for polyimides containing flexible linkages.

Table 6   Properties of selected polyimides from monomers containing flexible bridging groups.

Polymer	T_g (°C)	Solubility	
		NMP	m-cresol
	259		±
	267	++	++
	299		+
	305	++	++
	223		++

diamines and the so-called 'cardo' monomers, can be considered in this section, and they can be seen also as valuable alternatives for the preparation of processable polyimides [92–94].

For the preparation of this new generation of aromatic polyimides, synthetic methods have been outlined which allow to achieve the polymers in their state of full imidization in only one-step. However, the classical sequence poly(amic acid) → polyimide is generally followed somehow, although imidization occurs virtually at the same time that propagation. Amide solvents of high boiling point, as NMP of N-cyclohexylpyrrolidinone (CHP), nitrobenzene, chloroaromatics, phenols, cresols [36,37,61,95–100], and even carboxylic acids as benzoic acid [101], are solvents successfully used for the preparation of processable polyimides. Moreover, the beneficial effect of some basic (isoquinoline, triethylamine, pyridine) and acid (benzoic, hydroxybenzoic, salicylic) catalysts have been observed [80,95]. The reaction proceeds usually at low or moderate temperature in the first stage to favour the formation of a high molecular weight poly(amic acid), while the second part is led at high temperature to promote the cyclodehydration reaction and to force water separation. For the splitting off of water, azeotropic solvents are frequently used too. From a mechanistic point of view, it is to presume that bases help for the nucleophilic attack of the diamine to the anhydride to form amic acid, and acids catalyse the closing of the ring with evolution of water. Nevertheless, the role of acids in the formation of six-membered ring imides, like naphthalimides, needs still an explanation [102].

Table 7 Diamines and dianhydrides used in the preparation of polyimides with bulky side groups.

Diamines	Dianhydrides

(*continued*)

Table 7 Continued.

Diamines	Dianhydrides

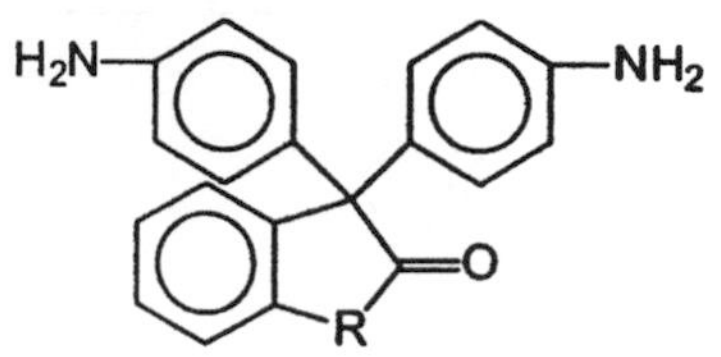

R: O, NH

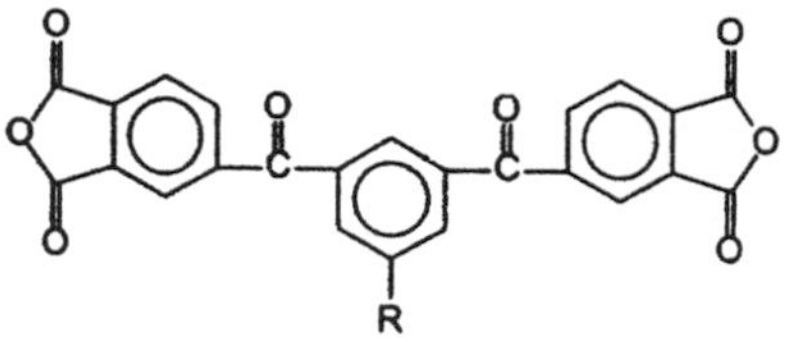

R: *t*-But, Ph, NO₂Ph

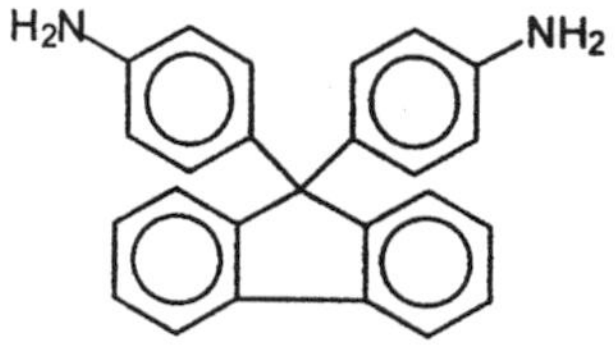

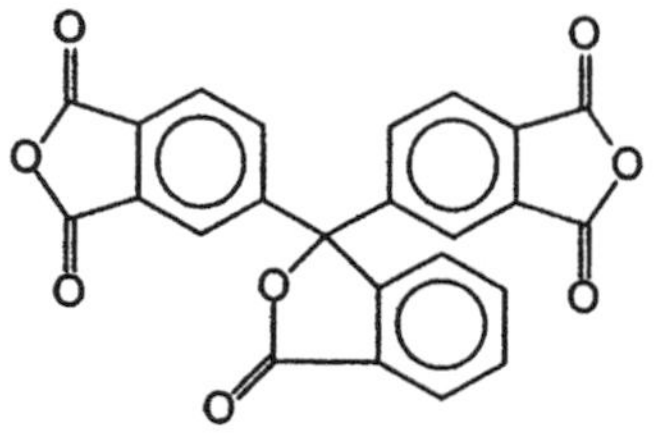

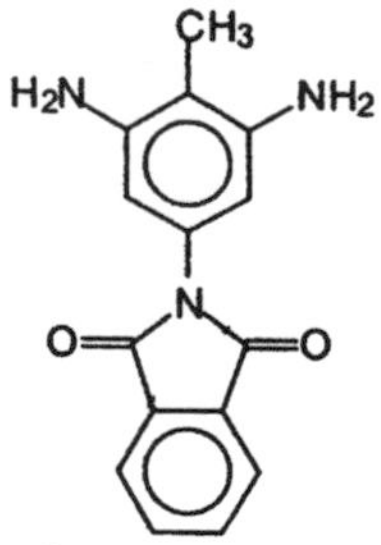

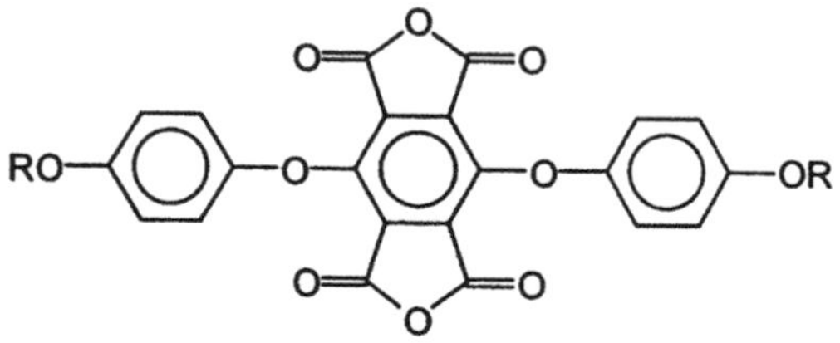

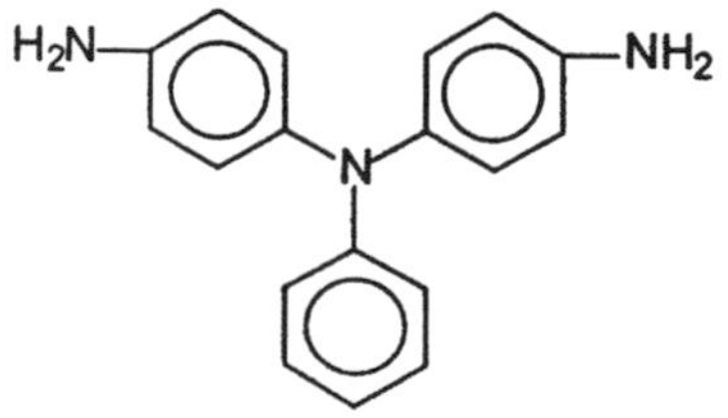

C. Polyimides from Dianhydrides and Diisocyanates

Although the reaction of anhydrides with isocyanates to give imides was very early
reported [103], it was not until about 1970 that the reaction found application in the
synthesis of polyimides and copolyimides [104–106] (Scheme 11).

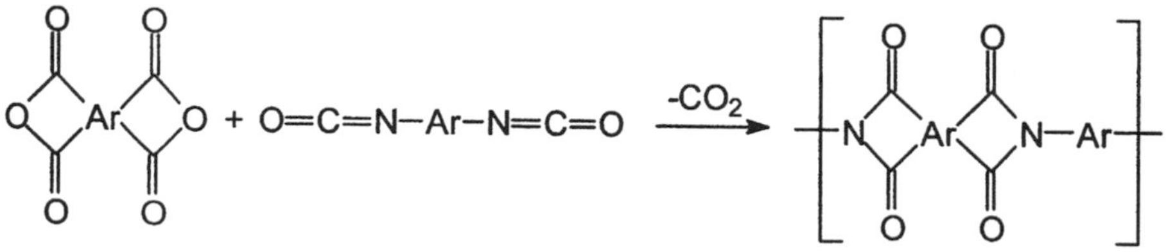

$$-CO_2$$

(11)

The reaction takes different pathways depending on the conditions. In the absence of catalyst the reaction has been claimed to proceed through a seven-membered polycycle intermediate (Scheme 12) that finally gives rise to polyimide with separation of carbon dioxide.

$$\tag{12}$$

Spectroscopic evidence of the seven-membered rings has been found in the preparation of polyimides from pyromellitic dianhydride and methylenediphenyl-diisocyanate (MDI) [105]. The reaction is conducted in solution of aprotic solvents, with reagents addition at low temperature and a maximum reaction temperature of about 130°C. On the other hand, polyimides of very high molecular weight have not been reported by this method. The mechanism is different when the reaction is accelerated by the action of catalysts. Catalytic quantities of water or alcohols facilitate imide formation, and intermediate ureas and carbamates seem to be formed, which then react with anhydrides to yield polyimides [106]. Water as catalyst has been used to exemplify the mechanism of reaction of phthalic anhydride and phenyl isocyanates, with the conclusion that the addition of water, until a molecular equivalent, markedly increases the formation of phthalimide [107] (Scheme 13). The first step is actually the hydrolysis of the isocyanates, and it has been claimed that ureas are present in high concentration during the intermediate steps of the reaction [107]. Other conventional catalysts have been widely used to accelerate this reaction. Thus, tertiary amines, alkali metal alcoholates, metal lactames, and even mercury organic salts have been attempted [108].

$$\tag{13}$$

For the preparation of polyimides, conventional dianhydrides have been combined with aliphatic and aromatic diisocyanates which are well known in the chemistry of polyurethanes. Other more modern diisocyanates have been also studied [109,110].

Diisocyanates containing an aliphatic sequence with phenylisocyanate end groups [111], and diisocyanates containing preformed imide rings [112], have been recently synthesized and used as monomers against aromatic dianhydrides.

Another approach to polyimides from diisocyanates is based on the reaction of isocyanates with half esters. The isocyanate group readily reacts with the carboxy group in solution, without catalyst under mild conditions, to yield amic ester with splitting off of carbon dioxide (Scheme 14).

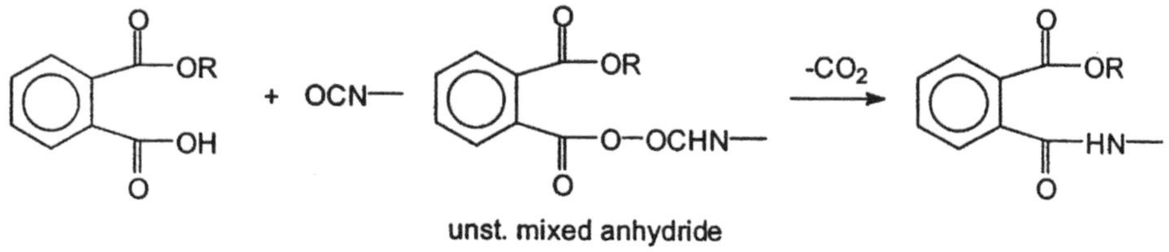

unst. mixed anhydride

(14)

In this way, polyimides have been reported from diisocyanates and half diesters via soluble poly(amic ester)s, which were converted into the final cyclized polyimides with loss of alcohol by the classical imidization method at high temperature [113,114].

A related reaction is the condensation of anhydrides with cyanates to imide carbamates (Scheme 15).

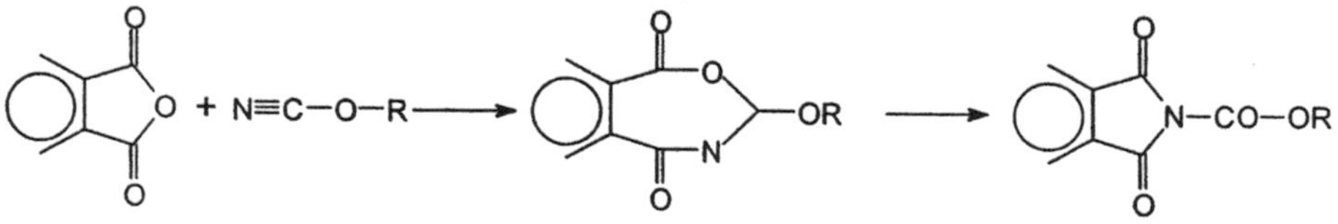

(15)

This reaction has been used to synthesize polyimides (more properly polyimide carbamates) from dianhydrides and dicyanates [23]. The reaction proceeds in nitrobenzene at high temperature, catalysed by triethylamine. The thermal resistance of these polymers is much lower than that of pure aromatic polyimides, and, therefore, the reaction has not found practical application.

D. Other Methods to Condensation Polyimides

1. From Diimides

Diimides of tetracarboxylic acids can be used for the synthesis of polyimides. Several reactions have been used:

(a) Polycondensation of Diimides with Dihalides (Scheme 16).

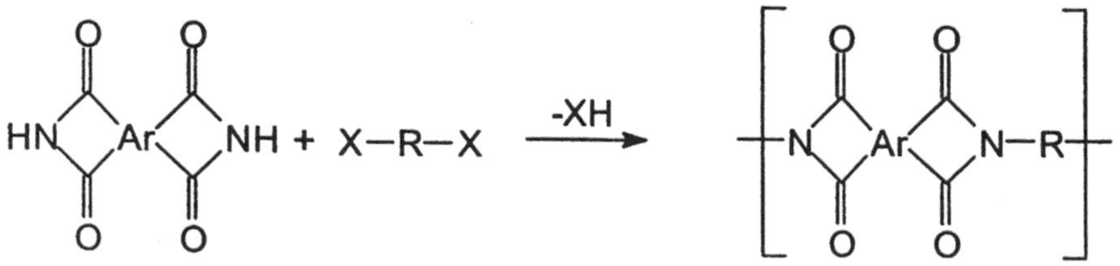

(16)

This reaction is accompanied by the separation of hydrogen halide, so that an acid acceptor is needed to catalize the reaction, which is carried out in polar solvents at high temperature [115]. Aromatic dihalides are not suitable reactants for this reaction, unless activated dihaloaromatic monomers are used [116].

(b) Aminolysis of Diimides by Diamines. Ammonia and amines can readily react with cyclic imides to yield ortho-diamides by nucleophilic attack and subsequent ring opening. On the basis of this old reaction, polyimides have been synthesized from aromatic diimides and diamines. The reaction has been classified as a *migrational polymerization* [117]. It proceeds in solution through a lineal poly(*ortho*-diamide), and this intermediate is converted to the polyimide by heating, with evolution of ammonia, in a similar fashion to the conversion of poly(amic acid)s into polyimides (Scheme 17).

$$(17)$$

(c) Transimidization. Another possibility is the reaction of diamines with *N,N'*-disubstituted diimides, by a synthetic route that can be considered as a transimidization, with evolution of monoamine from the intermediate poly(amic amide). In this exchange reaction, the nature of R plays an important role as the residue R–NH$_2$ has to be eliminated to accomplish ring closing, so that short-length R substituents are, in principle, desirable for this approach (Scheme 18). Nonetheless, the chemistry involved in these reactions has been studied also for the case when R is rather long and constituted by aliphatic aminoacid moieties [118,119].

The global reaction is an equilibrium that moves to the right at relatively high temperature, and it is necessary to have diamine monomers which are more nucleophilic than the monoamine, unless specific catalysts as transition metal salts are employed [120,121]. An alternative method that uses *N,N'*-bis-pyridyl- or *N,N'*-bis-pyrimidyl bisphthalimides as monomers has been developed. By this transimidation approach, 2-aminopyridine or 2-aminopyrimidine are readily displaced by diamines to yield high molecular weight polyimides [122].

$$(18)$$

Polypyromellitimides have been prepared by condensation in solution of NMP from aliphatic diamines and *N,N'*-dialkyloxycarbonyl pyromellitimides. The reaction can be carried out by interfacial polycondensation, as illustrated in Scheme (19) [123,124].

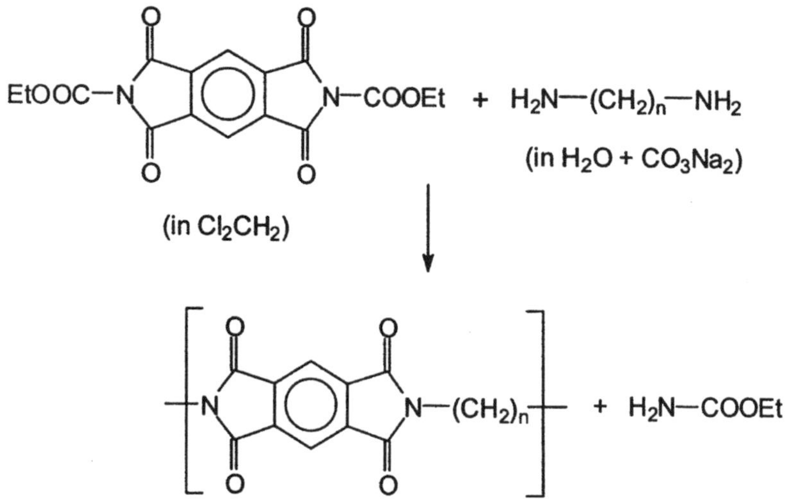

(19)

The same reaction has been studied with aromatic and aliphatic diamines, and the conclusion has been drawn that the procedure is only valid for aliphatic diamines, because the low basicity of aromatic diamines does not allow for polymer formation in mild conditions.

(d) From Diimides it is also Possible to Attain Polyimides by Reaction with Diisocyanates (Scheme 20).

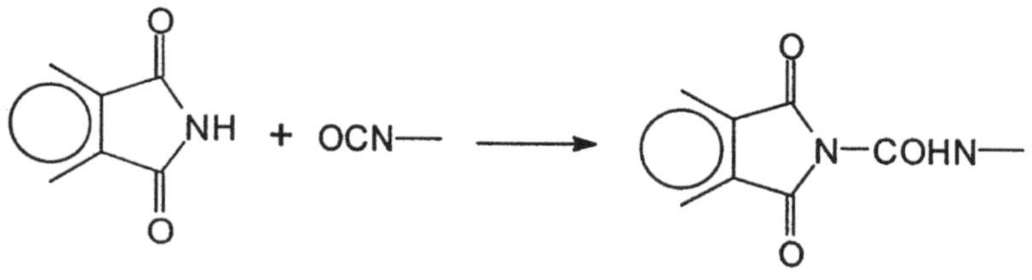

(20)

However, the molecular weights reported for polyimides prepared by this procedure are comparatively low (η_{inh} 0.2–0.3 dL/g) [125,126]. Moreover, the presence of functional grouping consisting of imide plus ureyl linkages makes these polymers thermally unstable. This is also the case of polyimides synthesized from dihydroxyimides and diacyl chlorides [127] or diisocyanates [128] (Scheme 21). The combination in a functional grouping of imide and carbamates also makes these polyimides unstable and easily attackable by nucleophiles.

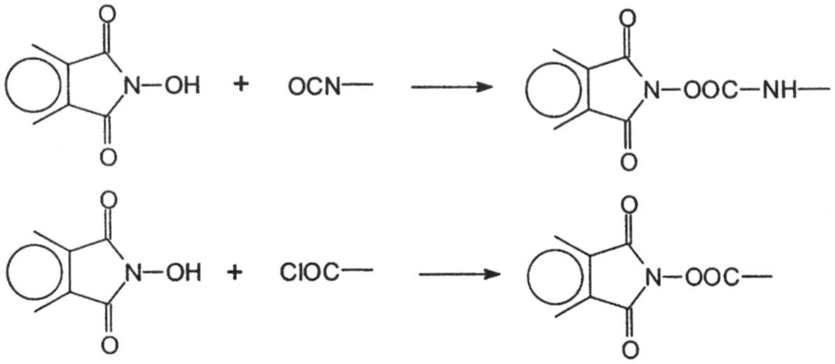

(21)

Another approach is the reaction of diimides with divinyl monomers. Examples of this route, starting from divilylsulfone and pyromellitic, benzophenonetetracarboxylic and cyclopentanetetracarboxylic diimides have been reported (Scheme 22). The polymerizations are carried out in solution in the presence of inhibitors of radical polymerization, and the molecular weights achieved are not very high [129]. By a similar mechanism, polyimides have been prepared from diallylesters and cycloaliphatic [130] and aromatic diimides [131].

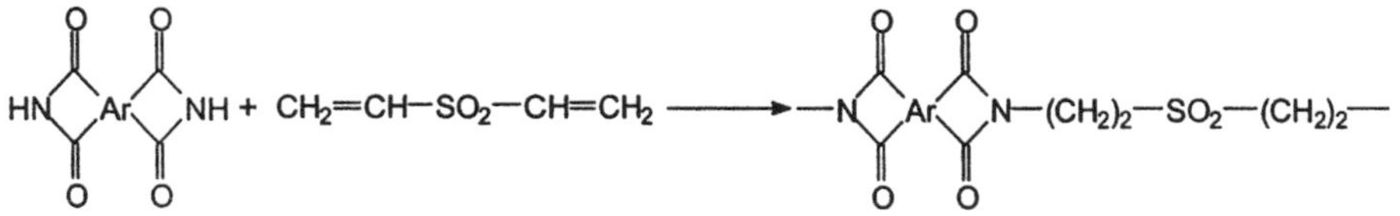

(22)

2. From Silylated Diamines

The *silylation* method has received particular attention during last years for the preparation of a variety of condensation monomers and polymers [132–134]. It has been successfully applied to the synthesis of aromatic polyimides, and can be considered as a recommendable method in some instances, particularly for less reactive diamines because it has been proved that silylated aromatic diamines are more nucleophilic than free diamines [135,136]. By this method, a poly(amic trimethyl silyl ester) is produced in the first step (Scheme 23), which can be converted into polyimide by chemical means [91,93,137].

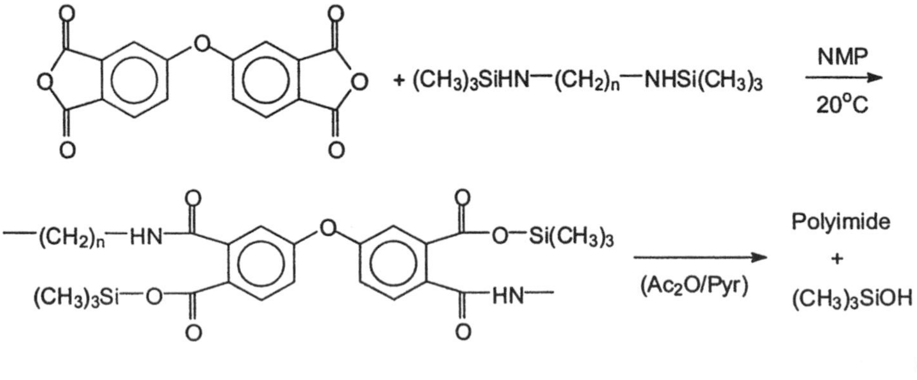

(23)

3. From Dithioanhydrides

The reaction has been described for pyromellitic dithioanhydride and aromatic diamines [138]. It is a two-step reaction that involves the formation of a soluble poly(amic thioacid), which is converted into polyimide by cyclocondensation (Scheme 24). The hydrogen sulphide that separates can be neutralized by an addition reaction with acrylonitrile. This latter works as an effective acceptor against SH_2. Polyimides with η_{inh} higher than 0.3 dL/g have not been attained by this method.

$$(24)$$

4. By a Diels–Alder Reaction

The Diels–Alder reaction of condensation between a diene and a dienophile has been also applied to the preparation of polyimides [139]. Starting from bismaleimides and biscyclopentadienones, a soluble poly(hydrophthalimide) of high molecular weight can be attained by solution polycondensation in refluxing chloroaromatic solvents (Scheme 25) [140,141]. They are readily converted into polyimides by dehydrogenation (boiling nitric acid). However, thermal dehydrogenation results in a decreasing of the molecular weights and, furthermore, full aromatisation of poly(hydrogenated phthalimide)s is difficult to accomplish [142,143].

$$(25)$$

III. CONDENSATION COPOLYIMIDES

Due to the intractability of classical, fully aromatic polyimides, other alternatives were soon envisioned in order to take profit from the intrinsic high thermal stability of these polymers. Copolymerization is an obvious procedure, and thus copolyimides have been developed and have been marketed parallel to aromatic polyimides. Chronologically, poly(ester imide)s and poly(amide imide)s were the first and most important copolyimides, but poly(ether imide)s have got great importance since the first processable poly(ether imide)s appeared in the eighties [74,81,120]. These three families are the most important copolyimides from a practical viewpoint, but other lineal copolyimides have also been described and evaluated, such as poly(anhydride imide)s [144–146].

Many methods of synthesis have been outlined for the preparation of copolyimides, and the most important will be considered here.

A. Poly(ester imide)s

1. From Monomers Containing Ester Groups

They are generally dianhydrides containing ester groups. The polymerization scheme for these monomers to yield polyimides by reaction with diamines is depicted in Scheme (26).

$$+ H_2N-R'-NH_2 \longrightarrow \text{Poly(ester imide)}$$

$$(26)$$

The reaction conditions are similar to those described for polyimides from diamines and aromatic dianhydrides. Here also the aliphatic species (R = alkyl) can be made to react in the melt. The aromatic ones have to react in solution of appropriate organic solvents to yield poly(amic acid ester) in a first step, and poly(ester imide) by cyclo-dehydration in the second step.

A number of dianhydrides containing ester groups have been used for the synthesis of poly(ester imide)s. Most of them are bistrimellitates which are synthesized by condensation of diacetylbisphenols (R = arylene) with two moles of trimellitic anhydride (4-carboxyphthalic anhydride), or by condensation of glycols (R = alkylene) with two moles of alkyltrimellitate by ester interchange in the melt [4,147,148].

The bistrimellitic anhydride esters may also react with diisocyanates to yield poly(ester imide)s in the same way as aromatic dianhydrides and diisocyanates described above.

Bisimides containing ester groups have also been used as monomers against diamines, to prepare lineal poly(ester imide)s by an aminolysis reaction [149].

2. From Imide Containing Monomers

Dianhydrides can react with aminoacids or aminoalcohols to yield diacids or dialcohols containing preformed imide rings of the following general formulae (Scheme 27):

$$\text{HOOC-R-N}{<}\text{Ar}{>}\text{N-R-COOH} \qquad \text{HO-R-N}{<}\text{Ar}{>}\text{N-R-OH}$$

$$(27)$$

Dicarboxyimide   Dihydroxyimide

Likewise, monomers from trimellitic acid anhydride are of the form (Scheme 28):

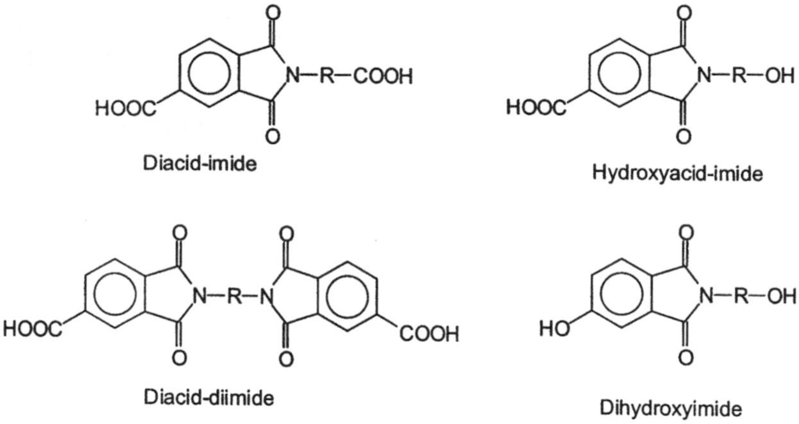

Diacid-imide

Hydroxyacid-imide

Diacid-diimide

Dihydroxyimide

(28)

The condensation of these monomers with dihydroxy or diacid monomers, or self-condensation in the case of hydroxyacids containing imide, renders poly(ester imide)s by polyesterification. Obviously, while diacid-imides and diacid-diimides are suitable monomers for AABB polymers, AB polymers would be attained from hydroxyacid-imides. Depending on the nature of the rest R (aliphatic or aromatic), and the nature of the corresponding comonomer the reaction should be carried out in solution or in the melt. The method, of course, depends also on the reactivity of the functional groups. When acid chlorides are used instead of carboxylic acids, solution or interfacial methods are to be employed [150]. On the other hand, since the transition temperatures of aliphatic–aromatic poly(ester imide)s are low compared to polyimides, melt polycondensation is a general method of synthesis for these copolyimides [151–153]. In fact, ester interchange reactions (through acidolysis or alcoholysis) in the melt have been used in the synthesis of many poly(ester imide)s of varied chemical structure [154–158], including a great deal of novel aliphatic–aromatic and aromatic thermotropic poly(ester imide)s [159].

3. Poly(ester imide) Resins

Most of these copolymers have been described as linear because they are formulated to be soluble in amidic solvents and cresols, however, they become thermosets upon cross-linking at the moment of application. Poly(ester imide) resins are mainly used as electrical insulating materials and they have been for many years the class of copolyimides which have practically deserved most attention, because of their good price-performance balance, good processability and good properties as high temperature insulating varnishes [160]. In fact, the electrical insulators industry undertook a qualitative improvement in the sixties when poly(ester imide) varnishes irrupted in the market, and they have been used and improved from then up to now.

The chemistry involved in curable poly(ester imide) formulations is well established, and it is still based on trimellitic anhydride, methylene dianiline, and low molecular weight polyesters of aromatic dicarboxylic acids and glycols, along with a classical heterocyclic triol, 2,4,6-trishydroxyethyl isocyanurate (THEIC), as it is shown in Scheme (29). The composition is formulated in a way that the final polymers, although linear, contain free − OH groups, both as chain ends and as side reactive groups. At the moment of application the resin is cured through the many free hydroxy groups by the action of diisocyanates, tetraalkoxytitanates, phenolic resins, or a combination of them, upon heating to eliminate solvents at the same time [160].

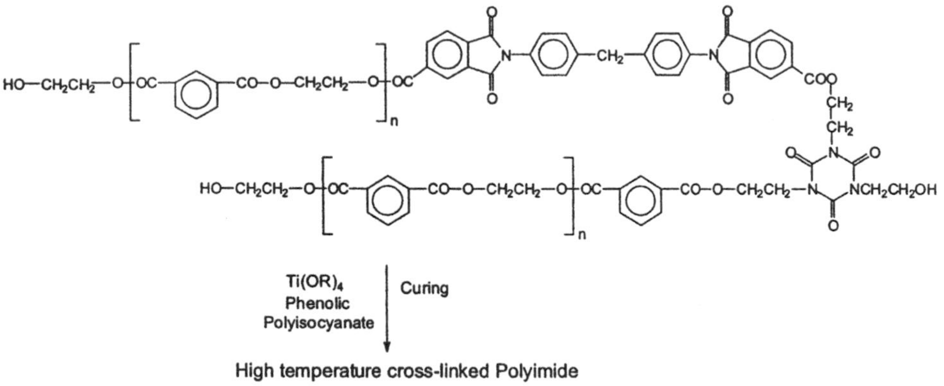

Ti(OR)₄
Phenolic
Polyisocyanate | Curing

High temperature cross-linked Polyimide

(29)

The combination of imide rings and ester linkages in a cured network can be also accomplished by curing of epoxy-imides with reagents (polyamines or polyacids) containing imide rings. Epoxy-imides of very varied composition have been prepared and reported, many of them containing also ester groups [145]. Some generic structures are shown in Scheme (30). The thermal properties of these materials are obviously controlled by the less stable moieties derived from the oxirane ring.

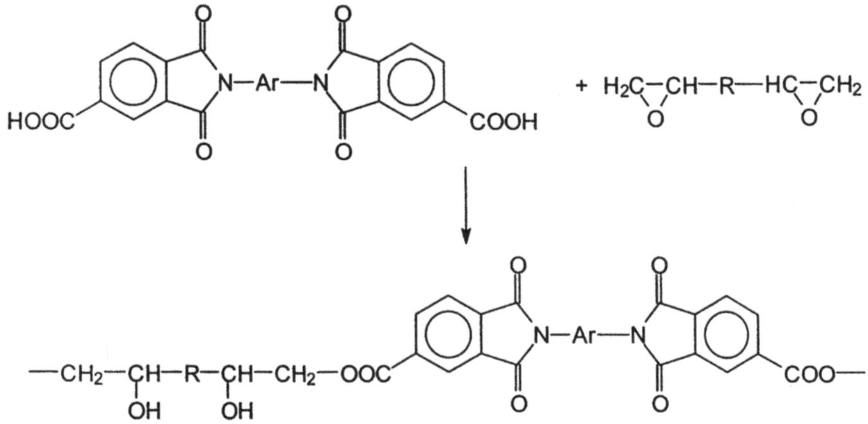

(30)

The combination of properties owing to poly(ester imide)s and epoxy resins in an unique material, has been also attempted by using soluble poly(amic acid)s as effective curing agents of conventional bisglycidyl ethers [161]. The result of these combinations are real interpenetrated polymer networks.

Another approach to poly(imide epoxy) homogeneous materials, is the preparation of diepoxides from imide-containing reactants (Scheme 31). For instance, the synthesis of bis(glycidylester) imides from imide-diacids or bis-imides and epichlorhydrine has been repeatedly reported [162,163]. The combination of them with curing agents (diamines and dianhydrides) leads to cross-linked polymers of rather complex structure and special properties [164–166].

$$(31)$$

B. Poly(anhydride imide)s

These copolyimides were first described in the 1970s [144,167] as thermally resistant polymers that can be prepared in high molecular weight from imide-containing monomers, in the form depicted in Scheme (32). They can be synthesized by melt polycondensation, in solution at high temperature, or even by solid state polymerization if the monomers show very high melting temperature. The monomers are mixed dianhydrides, which can be readily prepared by reaction of imide-containing dicarboxylic acids with acetyl chloride or acetic anhydride. During the polycondensation reaction, acetic anhydride splits off under special conditions of temperature (above 250°C) and low pressure (some mm Hg) thanks to an anhydride interchange reaction. This reaction is the classical pathway for lineal polyanhydrides, described as early as 1932 by Hill and Carothers [168]. Half dianhydrides have the advantage that stoichiometric imbalance is not possible as the reaction consists of the self-polycondensation of a single monomer.

$$(32)$$

Due to their poor resistance to hydrolysis, poly(anhydride imide)s [169,170] could not achieve technical importance, however, they have got a certain importance as biomaterials as they can be designed in a way to contain natural aminoacids [146,171]. They have shown also advantageous features for their processability and biocompatibility [172].

C. Poly(amide imide)s

Like poly(ester imide)s, poly(amide imide)s were reported in the earliest literature on condensation polyimides [4,5,173]. They constitute a polymer class with average properties between aromatic polyimides and aromatic polyamides.

Aromatic poly(amide imide)s are easier to prepare and to process than both aromatic polyimides and aromatic polyamides, because the copolymers are usually soluble in organic solvents. Thanks to that, they have found many practical applications, and the research effort devoted to poly(amide imide)s has been, and is still, considerable.

The general methods to synthesize poly(amide imide)s are rather simple and include those monomers and reactions conducting to polyimides or polyamides, consequently, the number and variety of chemical structures reported are countless [174]. The most important ones will be mentioned here.

1. From Amide-containing Monomers

They belong to reactant bearing the functional groups suitable for the synthesis of polyimides, mainly dianhydrides [147,175] of the form (Scheme 33):

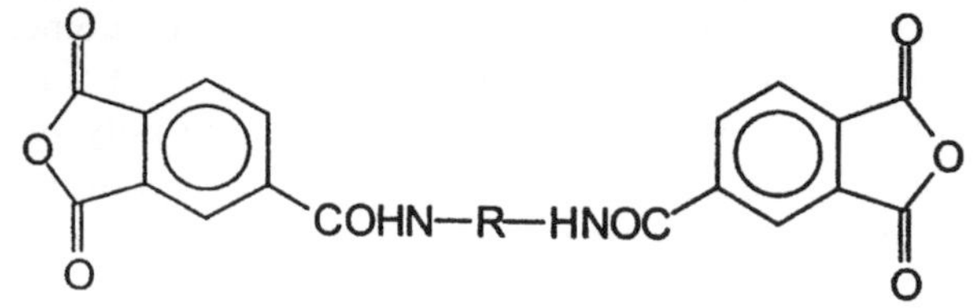

(33)

The reaction of these dianhydrides with diamines yields poly(amide imide)s by the general processes described for polyimides via poly(amic acid) intermediates. Solution polycondensation is the preferred procedure for these reactions. In general, everything mentioned above for the preparation of polyimides from diamines and anhydrides can directly be applied to the synthesis of poly(amide imide)s.

Instead of amide-dianhydrides, amide-diimides may be used in a similar fashion to that used for the preparation of polyimides from bis-phthalimides (Scheme 17). Thus, an alternative method from reactive diamide-diimides and diamines in solution of high-boiling solvents has been used for the preparation of aromatic poly(amide imide)s [149].

Amide-containing diamines can also be used for the synthesis of poly(amide imide)s, by combination with dianhydrides. The use of diamines containing amide groups was actually evaluated at an early date [173,176]. The general synthetic route is exemplified in Scheme (34) with pyromellitic anhydride and 4,4′-diamino-benzanilide [177]:

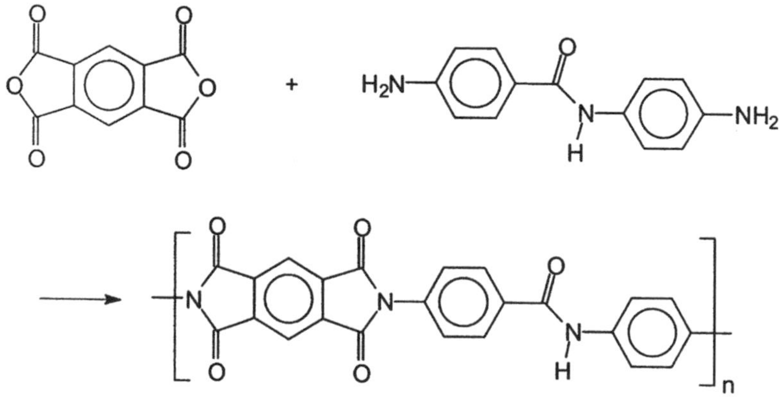

(34)

The possibilities of these approaches can be envisioned as unlimited if one thinks about the numerous possible combinations of different dianhydrides and amine-terminated oligoamides or polyamides. Actually, a lot of poly(amide imide)s have been prepared by this general synthetic route.

2. From Imide-containing Monomers

Virtually any of the traditional methods outlined to prepare polyamides, particularly aromatic polyimides, has readily been used for the synthesis of poly(amide imide)s from diamines or diamine derivatives and imide-containing dicarboxylic acids or acid derivatives (Scheme 35).

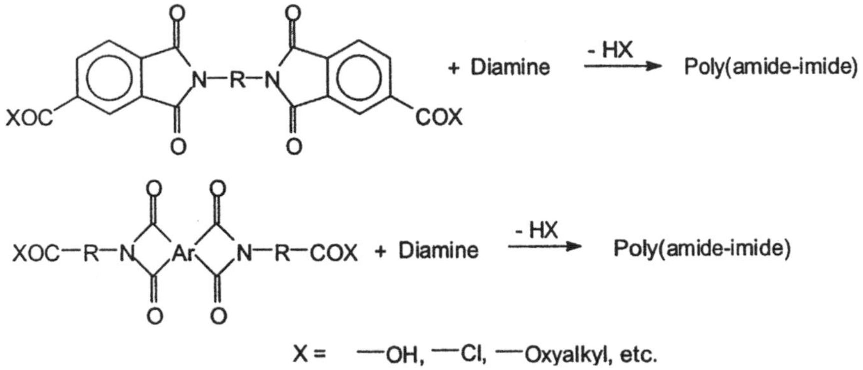

$$X = \quad -OH, \quad -Cl, \quad -Oxyalkyl, \text{ etc.}$$

(35)

Since poly(amide imide)s are generally polymers with high transition temperatures, solution methods at low temperature, or under mild conditions, have mostly served for these reactions, although melt polycondensation has also been used for aromatic–aliphatic species [178]. There are many patents and many examples of such reactions in the scientific literature, which deal mainly with the synthesis and polycondensation of imide-containing diacid chlorides and imide-containing diamines [4,5,11,179–183]. The numerous examples of this route agree with the advantages of the method over the conventional poly(amic acid) route in that the approach of using monomers with preformed imide rings, ensures full imidization, and furthermore, the copolymers do not require as long reaction times as with the two-step classical route.

Monomers derived from trimellitic anhydride, mainly *N*-carboxyphenyltrimel-litimides and *N*-(ω-carboxyalkylene)trimellitimides have been also used many times as starting materials for the synthesis of poly(amide imide)s. These poly(amide imide)s have been traditionally prepared by low temperature solution polycondensation, from diamines and imide-diacid chlorides [182], but they have been also successfully synthesized by the phosphorylation method of direct polyamidation [184], from diamines and imide-diacids [185–188] as depicted in Scheme (36). Trimellitic acid imide (4-carboxyphthal-imide) has also been used for the preparation of poly(amide imide)s, by reaction with aliphatic and aromatic diamines in solution at moderate temperatures [189].

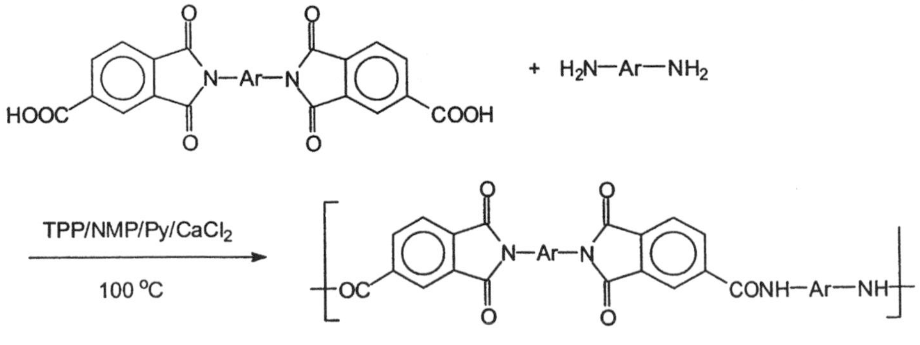

(36)

3. From Acid Anhydrides

An acid anhydride such as trimellitic anhydride can be considered as an ideal monomer since both carboxylic acid and anhydride groups can react with diamines to make up AB poly(amide imide)s. The overall reaction is depicted in Scheme (37).

$$X = \quad -OH, \ -Cl, \ -Oxyalkyl, \text{ etc.}$$

$$(37)$$

All the experience gained from the synthesis of homopolyimides from diamines and dianhydrides, and the knowledge accumulated for many years on the synthesis of polyamides have helped greatly for improving these synthetic routes.

In this regard, the polycondensation of trimellitic anhydride chloride (4-chloro-carbonyl phthalic anhydride) with diamines at low temperature can be considered as a model reaction, and it has been the subject of numerous studies [190–192]. As usual, the first product formed in these reactions is poly(amide amic acid), which is eventually converted into poly(amide imide) by cyclodehydration. Some of these poly(amide imide)s from diamines and trimellitic anhydride have been reported as technical materials, and marketed as high T_g engineering thermoplastics [7].

4. From Diisocyanates

Isocyanates can react with both anhydrides and carboxylic acids to produce high molecular weight polymers. Poly(amide imide)s prepared by this route have actually achieved technical importance as high temperature varnishes and fibers [193,194]. Another approach is to make diisocyanates to react with imide-containing diacids as illustrated in Scheme (38).

$$+ \ OCN-R-NCO \longrightarrow \text{Poly(amide imide) AB} \ + \ CO_2$$

$$+ \ OCN-R-NCO$$

$$(38)$$

Poly(amide imide) AABB $+ \ CO_2$

$$N-R-COOH \ + \ OCN-R'-NCO \longrightarrow \text{Poly(amide imide) AB} \ + \ CO_2$$

Poly(amide imide)s from diisocyanates and trimellitic anhydride or diacid-imides are preferably prepared in solution of NMP or DMA, at relatively high temperature (over 150 °C). Cresols or mixtures cresol-phenol are also suitable solvents. As for homo-polyimides, they are synthesized in a one-stage process, with the polymer, fully imidized, remaining in solution at the end of the reaction. High molecular weight poly(amide imide)s have been attained by using this method [194], although special care has to be taken to limit side reactions conducting to cross-linking and formation of unstable urea linkages [195,196].

D. Poly(ether imide)s

Any polyimide prepared from diamines or dianhydride containing ether linkages should be classified as a poly(ether imide), but only those copolyimides with at least one ether linkage per imide group in the repeating unit will be considered here.

1. From Monomers Containing Ether Linkages

This is the method mostly used for the synthesis of these copolyimides. In fact, many of the countless processable, aromatic polyimides described in last years, have ether linkages in their backbone [74,81,197–199]. The method of synthesis is the general method via poly(amic acid) in most instances, using ether-containing dianhydrides or ether-containing diamines. One of the most representative structure is shown in Scheme (39).

$$(39)$$

Some representative monomers have been shown in Tables 3 and 4. As many of these copolyimides are soluble in organic solvents, the polycondensation reaction can be carried out as a two-step, one pot process, because once the intermediate poly(amic acid)s are formed at low temperature, the imidization step can proceed to quantitative conversion by merely raising the temperature beyond the cyclization temperature, usually 160–200 °C, for a few hours. Water produced by cyclodehydration can be more easily removed if a cosolvent is present that can help for an azeotropic water separation [81,200]. In this way, the polyimide is attained in solution, and it can be isolated by precipitation or by direct transformation from the polymer solution.

Apart form AABB poly(ether imide)s AB polymers have also been described. It has been claimed that the self-condensation of 4-(4-aminophenoxy) phthalic anhydride hydrochloride leads to high molecular weight poly(ether imide)s [201]. Similar compositions have been more recently reported, taking as starting materials 3- and 4-(4-aminophenoxy) phthalic acid hydrochloride, and using the phosphorylation method to attain AB poly(ether imide)s of only moderate molecular weight (η_{inh} 0.15–0.25 dL/g) [202].

2. From Biphenols via Nucleophilic Displacement

The presence of two carbonyl groups in the phthalimide ring greatly activates the free positions for nucleophilic reactions. Thanks to that, biphenolates have been successfully used as monomers for poly(ether imide)s in combination with 3- and 4-substituted bisphthalimides, as depicted in Scheme (40) [203]. The substitution of activated nitro groups has proven to be specially suitable for the attainment of poly(ether imide)s in high yield and high molecular weight [74,81,203]. Nevertheless, the method has not got practical significance, and the conventional route from dianhydrides and diamines containing ether linkages is the preferred alternative to synthesize these copolyimides.

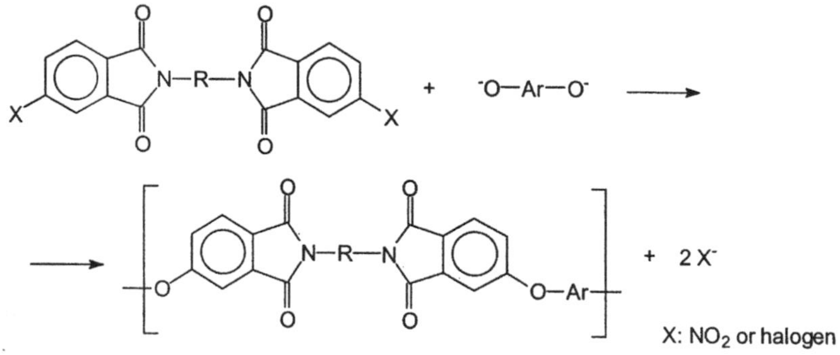

(40)

X: NO$_2$ or halogen

E. Segmented Polyimides

This kind of polyimides consist of a sequential alternance of imide groupings and flexible segments, in a fashion conducting to special copolyimides with specific properties.

Poly(urethane imide)s are a classical example of segmented polyimides. Although there are old examples of poly(urethane imides) synthesized from diisocyanates and dihydroxyimides, the ultimate objective of the reactions outlined to prepare poly(urethane imide)s has commonly been to attain imide-modified polyurethanes [204–206]. They have been prepared from glycols, biphenols and polyols (polyesterdiols, polyetherdiols) and isocyanates-terminated prepolymers obtained from a dianhydride and excess diisocyanate, as illustrated in Scheme (41) with pyromellitic dianhydride and toluylenediisocyanate.

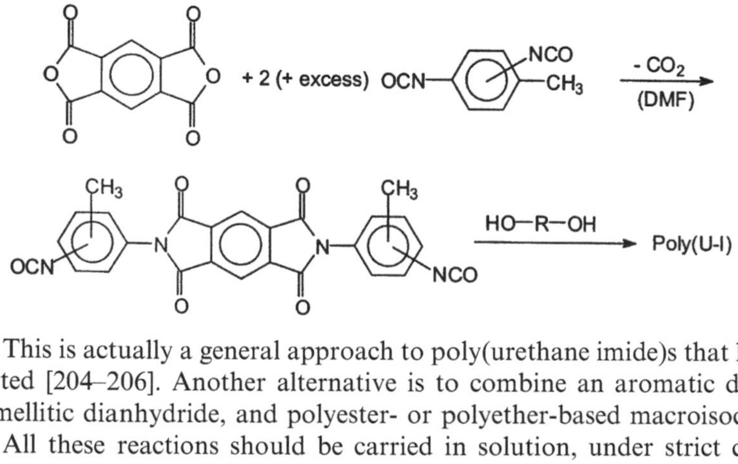

(41)

This is actually a general approach to poly(urethane imide)s that has been repeatedly reported [204–206]. Another alternative is to combine an aromatic dianhydride, mainly pyromellitic dianhydride, and polyester- or polyether-based macroisocyanates [207–209].

All these reactions should be carried in solution, under strict conditions imposed by the sensitiveness of isocyanates and anhydrides to moisture and impurities. The copolymers of this kind should be considered as real segmented polyimides, which contain soft phase and hard phase, analogous to conventional elastomers or thermoplastic elastomers.

There are also examples of segmented polyimides containing short and long sequences of polysiloxane or polyalkyleneoxide. They are mosta frequently prepared by the reaction of dianhydrides with amine-terminated polysiloxanes, or polyalkylene oxides of medium-low molecular weight, usually not greater than 4000 g/mol [210–213], although α,ω-bis(aminophenyl) polydimethylsiloxanes [214,215] and α,ω-bis(amino-propyl) polydimethylsiloxanes of over 10,000 g/mol have been reported [216,217]. The synthetic routes for all these segmented copolyimides involve the conventional pathway in one or two steps, using dianhydrides, particularly PMA, BTDA, or 6F, and amine-terminated flexible oligomers of general formula depicted in Scheme (42). The same method can be applied to the reaction of conventional diamines with anhydride-terminated oligomers (Scheme 42) [213]. The ring closure from poly(amic acid)s to polyimides is commonly performed by thermal treatment, either in solution or in the melt.

These segmented polyimides may show two glass transition temperatures, as it corresponds to microphase-separated structures, and one melting temperature [216]. Their behaviour as materials is similar to that of classical thermoplastic elastomers.

On applying the same concept to other systems, elastomeric polyimides based on amine-terminated butadiene-acrylonitrile oligomers have also been prepared [218] (Scheme 42).

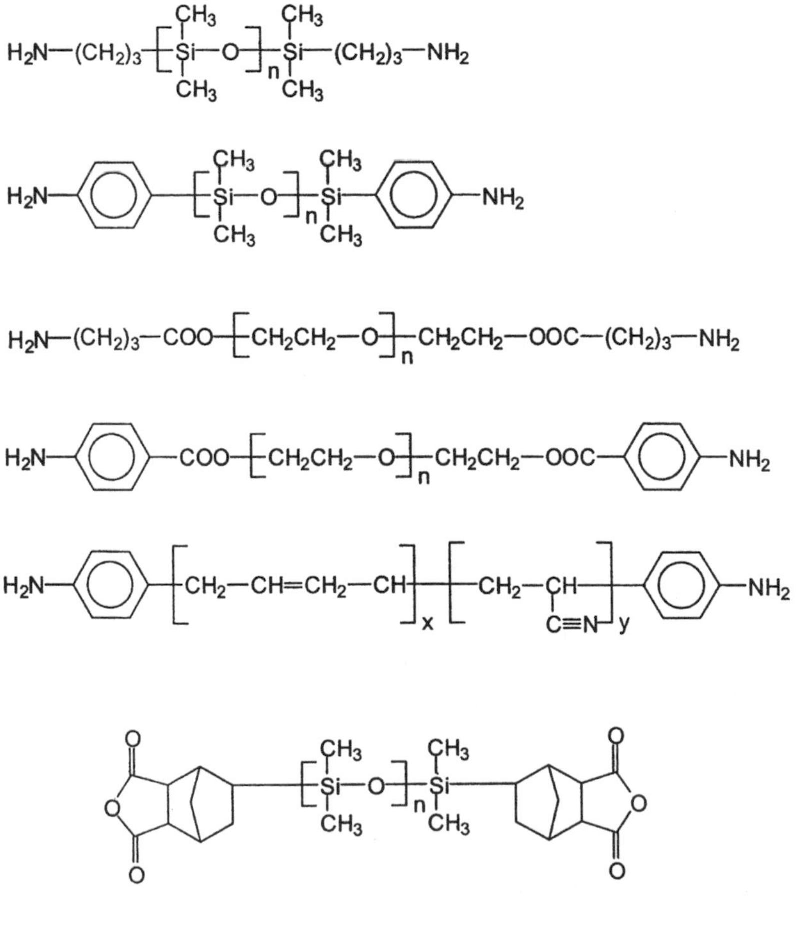

(42)

F. Polyimides Containing Other Heterocycles

Much work has been done on this interesting group of condensation polyimides [219]. Trying to combine the good thermal stability of other heterorings and the availability of polyimide monomers, a considerable number of mixed polyheterocycles has been synthesized. Among other heterocyclic copolyimides, poly(imidazole imide)s [220,221], poly(oxadiazole imide)s [222,223], poly(benzazole imide)s [224,225], poly(quinoxaline-imide)s [226], and poly(quinazolinedione imide)s [227] have been synthesized.

Diamines and dianhydrides containing preformed heterocycles are monomers usually employed. They are normally made to react in the conditions of the two-step traditional method to yield heteroring-modified polyimides.

Monomers containing imide rings may also be used. In that case the monomers have to bear the functions to form the other heterocycle. Thus, benzoxazole-imide copolymers have been prepared from imide-diacids and dihydroxybenzidine [228], and poly(quinoxa-line imide)s have been prepared from tetraamines and imide-containing bisbenzyls by one-step process in m-cresol [229]. As an example, the synthesis of generic poly(oxadiazole imide)s is shown in Scheme (43).

(43)

When poly(hydrazide amic acid) was heated, the conversion of amic acid to imide was essentially complete before the hydrazide linkages were converted into oxadiazole in any substantial extension. In the last step, conversion from polyhydrazide to polyoxadiazole was accompanied by decomposition, so that, for these copolyimides a better method would be through the oxadiazole-containing monomers [221].

Solubility problems comparable to those with homopolyimides appear also with the heterocycle-modified ones: soluble polymeric precursors can be attained, but the final materials are infusible and insoluble in workable solvents. Furthermore, apparent improvements in the thermal stability of polyimides have not been achieved by these approaches. An exception to this rule are the phenylated poly(quinoxaline imide)s which are soluble and show improved thermal stability [229,230].

G. Polymers with Imide Pendant Groups

Early works on this approach concerned the preparation of photocurable polymers based on the photodimerization observed in maleimides and mono- and disubstituted maleimides [231–233], as depicted in Scheme (44).

$$(44)$$

For instance, polymers containing dimethylmaleimide pendant groups can be cross-linked by UV light with the formation of cyclobutadiene bridges, through the known $2 + 2$ photodimerization mechanism. These polymers have nothing to do with heat-resistant polymers; they have been developed and studied because of their potential as photoresists.

On the other hand, a large number of curable and non-curable polymers containing pendant imide rings have been synthesized and studied with the objective of improving the properties of classical addition polymers [234–237]. The synthesis of these polymers fits the general rules to polymerize unsaturated monomers via a chain growth process. Monomers suitable for these purposes are shown in Scheme (45):

$$(45)$$

Condensation polymers have also been modified by incorporating imide rings as pendant substituents. It has been proved that phthalimide group, for instance, is a heat-resistant substituents that can provide better solubility without much damage in thermal resistance. Thus, aromatic polyamides [238–242], and polyimides [85,243,244] with phthalimide or naphthylimide pendant groups, have been reported. Polyimides like those shown in Scheme (46) are prepared by conventional means, and they are soluble in a variety of solvents, but they become totally insoluble upon heating over 200 °C. That is explained by the possibility of a number of intermolecular reactions [85,243].

$$(46)$$

Soluble poly(p-phenylene) with side imide groups has been prepared by a C–C coupling reaction from N,N'-diphenyl-3,6-dibromopyromellitimide. The electron-withdrawing effect of the side imide groups seems to improve the reactivity of the monomer, and facilitates the formation of high molecular weight polymer by Ni(0) promoted dehalogenation reaction. Whereas conventional poly(p-phenylene) is insoluble, the presence of side phthalimide rings makes the modified polymer soluble in polar organic solvents [245].

The approach has been extended to aromatic polymers modified by thermally cross-linkable side imide groups. Thus, trimellitimide groups can be hanged on the main chain of aromatic polymers, such as polystyrene, polycarbonates or poly(phenylene oxide), by grafting reactions through a Friedel–Crafts mechanism [246] as illustrated in Scheme (47). The free carboxy groups of the trimellitimide rests provide real possibilities of ulterior curing or modification.

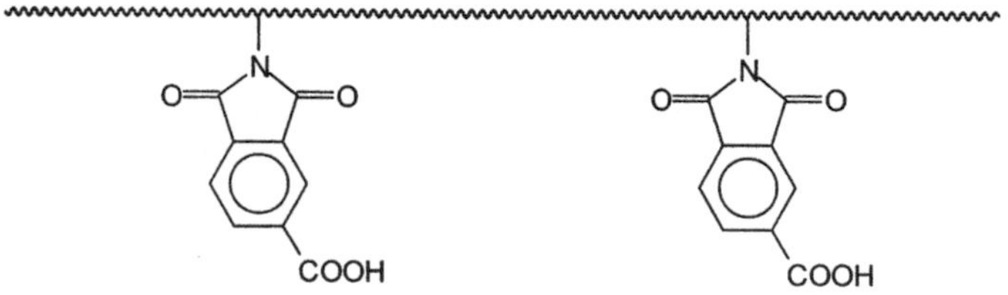

(47)

A number of polymers containing unsaturated imide side groups susceptible to cure by a thermal treatment (maleimide, nadimide, tetrahydrophthalimide), have been described [247,248]. The reactive side groups are generally introduced in this case by means of modified condensation monomers bearing reactive imide groups (Scheme 48). This approach has been applied for the preparation of cross-linkable aromatic polyamides and polyesters [248,249].

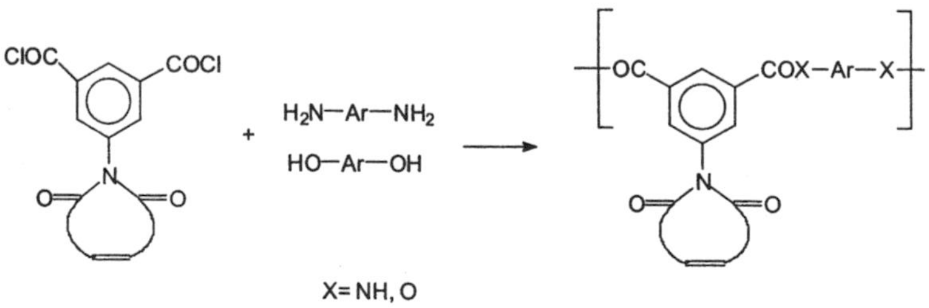

(48)

X= NH, O

Polyethers with pendent nadimide groups have also been synthesized, by ring-opening polymerization of nadimide-oxyranes, as depicted in Scheme (49). The polymers could be cured to cross-linked poly(ether imide)s [250].

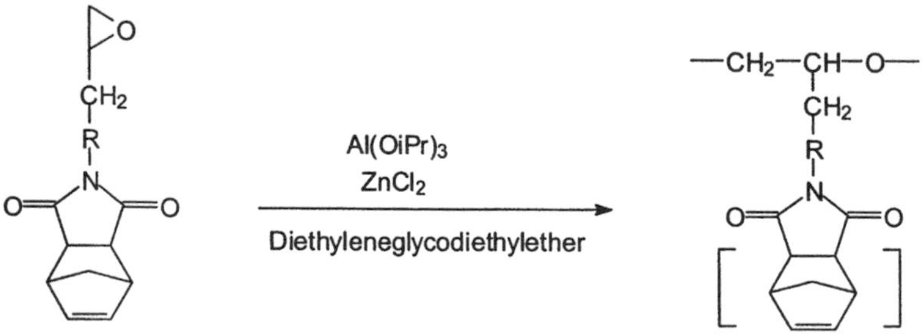

(49)

4. Organic–Inorganic Hybrid Polyimides

The continuous demand for new materials have moved to the investigation of polyimides containing metal atoms. Although many polyimide formulations have been reported containing inorganic fillers, of instance silica, it seems that not more than 10% of silica can be incorporated to the composite material at molecular level. Therefore, methods have been outlined to get hybrid polyimides by the sol–gel process. It implies the combination of polyimide with classical precursors of silica employed in the sol–gel methods, particularly tetraethoxysilane and tetramethoxysilane, which lead to silicon oxide through hydrolysis and polycondensation.

The general synthetic route consists of co-dissolving a poly(amic acid) with a polyalkoxysilane in an organic solvent, as DMA or NMP, and the needed amount of water and catalyst to promote the initial hydrolysis of alkoxysilane. Then, the progress of hydrolysis and condensation reactions conducting to an inorganic network, is achieved by applying a strict heating schedule. The simultaneous conversion of the poly(amic acid) into polyimide leads to a hybrid material of special properties [251–254]. During the process, interchange reactions between the organic and the inorganic matrices help for the formation of an organic–inorganic composite, but a compatible material is only observed for small concentration of the silicon component. The use of poly(amic acid)s containing pendant alkoxysilyl groups, seems to improve the method concerning compatibilization, in such a way that silica-polyimide hybrids can be attained with a homogeneous dense structure at the nanometer scale [255,256].

The presence of silica domains affects all the properties, with clear changes in the density, thermal expansion coefficient, thermal transitions, surface adhesion, water uptake, and mechanical resistance depending on the process and on the inorganic proportion incorporated [251–258].

In another approach, N,N'-bis-(3-triethoxysilylporpyl) diimides are first prepared by the transimidation method from N,N'-disubstituted diimides and aminopropyl-triethoxysilane (Scheme 50). The final hybrid material is attained by thermal treatment in the conditions of the sol–gel method. Transparent xerogels can be obtained by this procedure from a variety of bis-(triethoxysilyl) diimides based on classical dianhydrides [259].

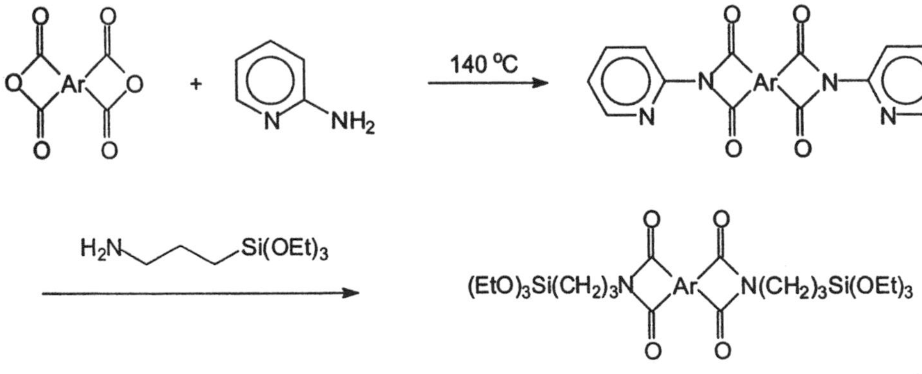

(50)

Not only have homopolyimides been combined with silica, poly(amide imide)/TiO$_2$ nano-composites have also been reported. These hybrids have been prepared with different optical appearance (transparent, translucent or opaque) depending on the ratio poly(amic acid amide) to tetraalkyltitanate used in the preparation. As for SiO$_2$ hybrids, the sol–gel method provides a convenient route to these polyimide-TiO$_2$ hybrids [260].

IV. ADDITION POLYIMIDES

Cyclic imides with carbon–carbon double bonds are susceptible to polymerization by a radical mechanism or by several other mechanisms, such as photocondensation, Diels–Alder addition, or nucleophilic substitution. Although these latter reactions yield step-growth polymers, they will be considered in this part because traditionally they are considered addition polyimides. Furthermore, in many cases they are indeed cross-linkable through polymerizable imide double bonds.

A. Linear Addition Polyimides

The radical polymerization or copolymerization of N-substituted maleimides has been the subject of many studies (Scheme 51) [261–264].

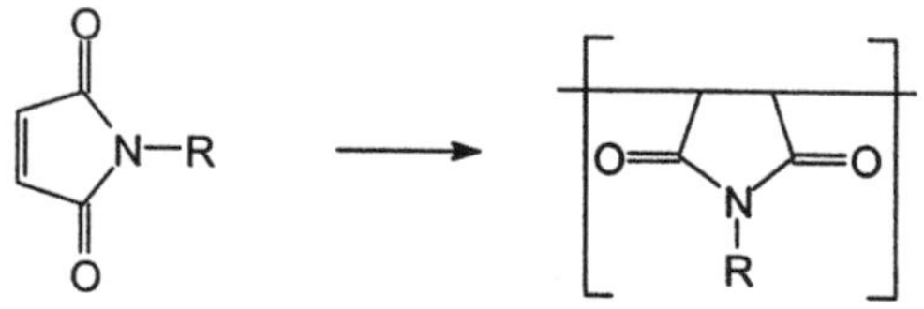

(51)

Although the investigation of polymaleimides and copolymaleimides has served to solve several theoretical aspects on radical polymerization of these unsaturated rings [265,266] and to give a great deal of data on maleimide polymers, they have not yet found end-use applications. The same is true for polyimides synthesized by the so-called photocondensation method, that consists of step-growth polymerization of bisimides with benzene or alkylbenzenes [267–269]. The reaction is induced by UV radiation (Scheme 52).

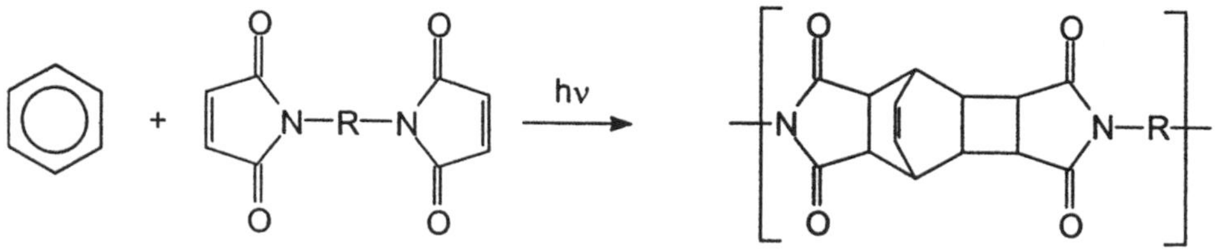

(52)

The aromatic ring works as a difunctional dienophile, and the overall reaction may be considered as extension of the photolytic coupling reactions that are well known for small molecules [270].

Photopolymerization of bisimides by a photocycloaddition mechanism has also been extensively studied [271,272]. The reaction is a real stepwise condensation process since every chain propagation step involves the absorption of a photon [273,274].

Photocondensation is a procedure limited to monomers that do not easily polymerize by a photoinitiated radical mechanism. Otherwise the UV radiation would induce the formation of chain growth homopolymers.

B. Thermosetting Polyimides

Thermosetting polyimides are low molecular weight systems with imide functions in their backbone and reactive terminations capable to react by an addition reaction to give a cross-linked system. These materials were developed in the 1970s in order to fulfil the requirements of the aerospace industry in the domain of high performance adhesives and matrices based on glass, carbon and aramide fibers.

From a practical viewpoint, the compounds most commonly used in the fabrication of addition polyimides are bismaleimides, bisnadimides and ethynyl terminated oligoimides [275]. Poly(bismaleimides) have actually achieved special importance as technical polymer materials [276]. Due to the aliphatic-type linkages, which appear as a consequence of polyaddition, polymers of this type are not as thermally stable as aromatic polyimides are. However, their good processability, the polymerization without volatiles release and the relative low cost of raw materials have helped them to become a real alternative for long term uses up to 200–250 °C.

Typically, bismaleimides are synthesized by reacting a diamine with maleic anhydride in two steps [277,278]. In the first one a bismaleamic acid is formed in a fast, exothermic reaction, that is carried out at room temperature. The second step consist of the imidization of the maleamic acid, usually by chemical means, with acetic anhydride in the presence of a small amount of basic sodium acetate. This step is carried out at moderate temperature to avoid premature polymerization of the double bonds. The use or tertiary amines as catalyst is also possible but the obtained product is less pure than in the case of sodium acetate. In fact, this effect has been used to prepare bismaleimides with a lower melting point, and consequently with a wider processing window [279].

Bismaleimides are readily polymerized to cross-linked materials simply by heating (Scheme 53). The reaction produces no volatile by-products and yields void-free thermosets. The cross-linking density directly depends on the length of the diamine used in the synthesis.

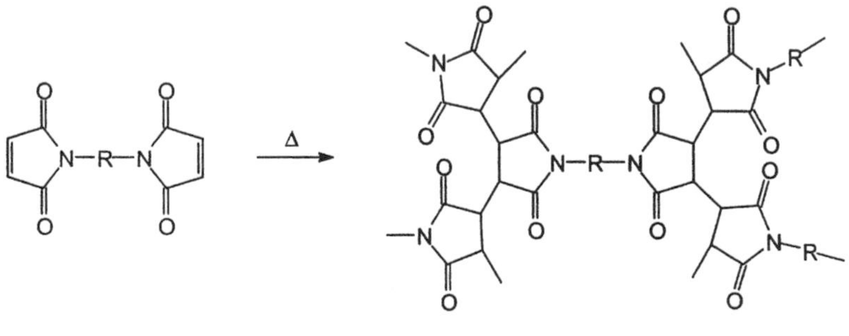

(53)

The structure and thermal characteristics of some low molecular weight bismaleimides are listed in Table 8.

Although bismaleimides have received most of the research effort in this field, other similar bisimides have been also synthesized and evaluated. Bisitaconimides have been

Table 8 Bismaleimides.

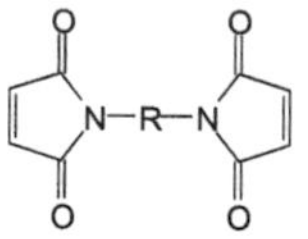

—R—	Melting point (°C)	Exotherm peak (°C)	Ref.
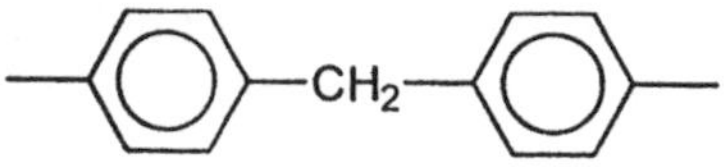	157–158	235	[280,281]
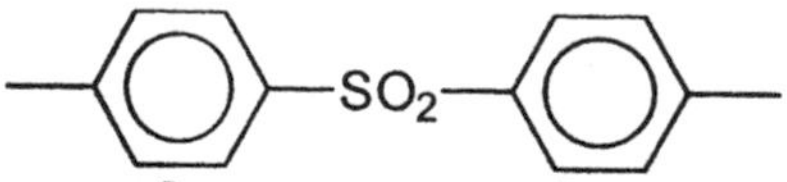	252–255	264	[281]
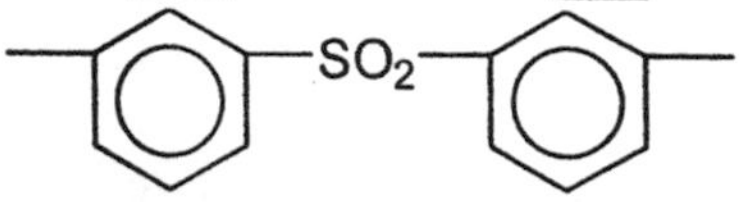	210–211	217	[281]
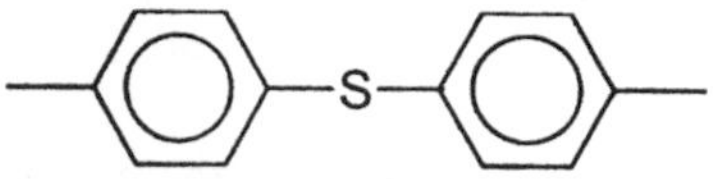	181–182	300	[282]
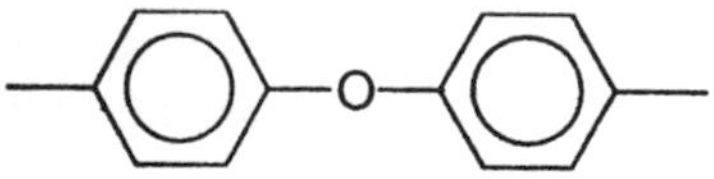	175–180	286	[281,283]
	363(d)		[280,281]
	202–204		[280,281]

prepared by several authors [284–287] in a similar fashion to bismaleimides via an intermediate bis(itaconamic acid) (Scheme 54).

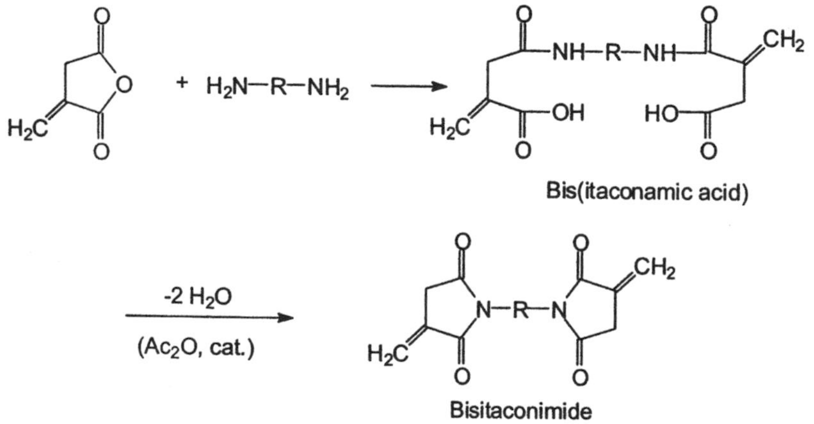

Bis(itaconamic acid)

(54)

Bisitaconimide

The thermal stability of cured bisitaconimides is comparable to that of poly(bis-maleimides). It has been observed that itaconimide groups undergo an isomerization to be converted to citraconimides upon heating (Scheme 55) [284,288–290].

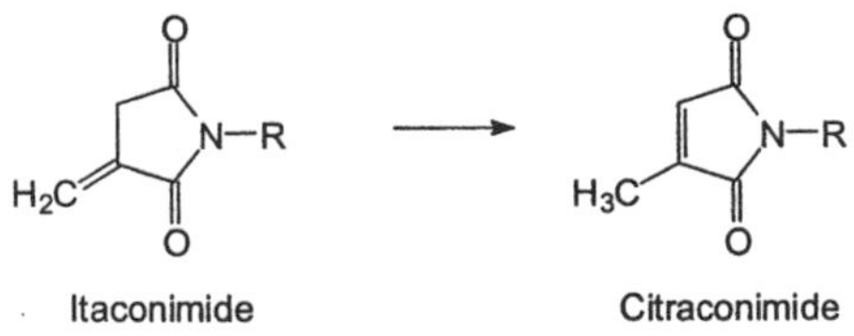

Itaconimide   Citraconimide

(55)

Biscitraconimides have been investigated as cross-linkable systems, in the same way that bismaleimides [291,292]. Their reactivity has been compared and some discrepancies have been found. Although some authors claim that the reactivity of biscitraconimides is higher than that of bismaleimides [293,294] it seems that the reactivity is controlled by the amount of impurities in the system. Carefully purified bismaleimides are more reactive than biscitraconimides, while the converse is true when the purification is not so rigorous [295,296].

Biscitraconimides have been studied in recent years as antireversion agents for S-vulcanized rubbers [297–299]. Reversion causes the loss of cross-linking density and properties. It can be brought about by overcure and/or by high temperature applications. The presence of biscitraconimides minimizes this effect by Diels–Alder reaction with the conjugated polyenes which are formed as a result of reversion [300].

From the beginning one of the critical points in the chemistry of addition polyimides was to adjust the formulations in order to prevent brittleness of the final curing material. The first approach for improving the mechanical behaviour of poly(bismaleimides) was the incorporation of moieties which provide a separation of the two active maleimide groups. Diamines and dithiols have worked as very suitable spacers, capable to react with the double bonds of bismaleimides [301–304]. The reaction takes place by nucleophilic addition (Michael addition) on the electron-deficient double bond, which is activated by the two adjacent carbonyl groups (Scheme 56). The reaction is usually carried out in acidic solvents (*m*-cresol or DMF/acetic acid mixtures) to avoid cross-linking that occurs by reaction of the anionic intermediate with maleimide double bonds [305,306]. The mechanism for the reaction of thiols and maleimides is depicted in Scheme (56). A similar mechanism applies to amines [307–309].

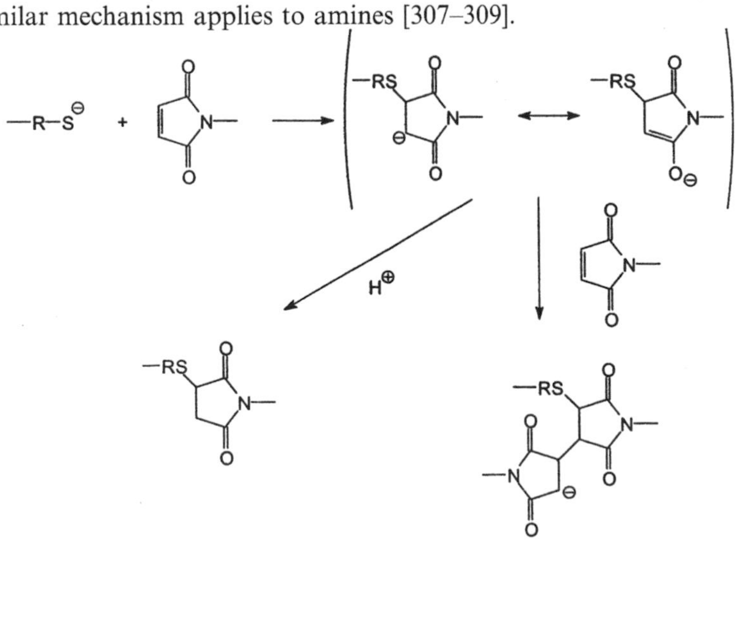

(56)

Apart from these two main reactions, an aminolysis reaction has been detected, giving place to the structure shown in Scheme (57) [307,309].

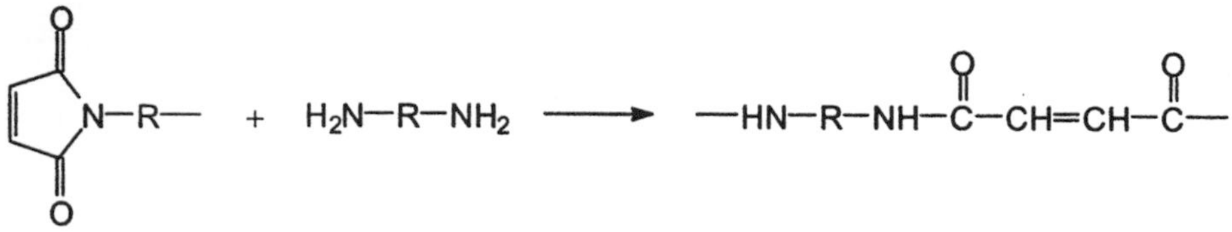

(57)

The reaction with aromatic amines can lead to high molecular weight linear polymers (polyaspartimides). By controlling the stoichiometric ratio of the reactants, the reaction can be used to obtain low molecular weight polyaspartimides end-capped with maleimides (Scheme 58).

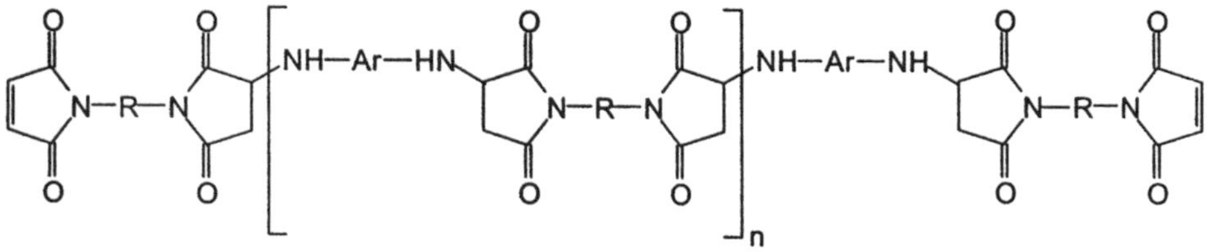

(58)

The same approach was used by making to react aminobenzhydrazide with a bismaleimide (Scheme 59) giving an extended bismaleimide [310,311].

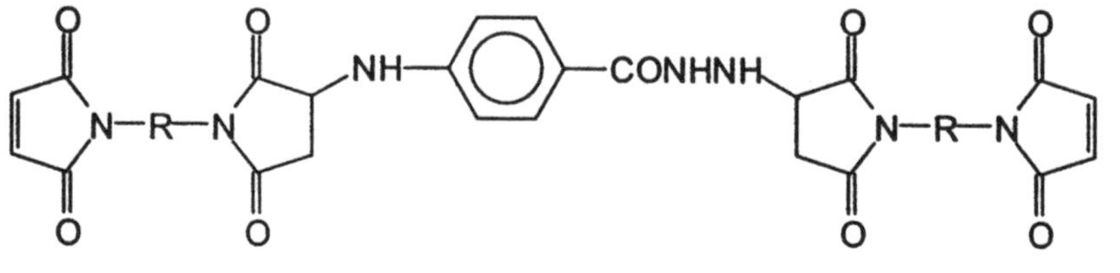

(59)

These products show solubility in polar aprotic solvents, and their low viscosity permits their application in hot melt formulations.

Besides low molecular weight bismaleimides building blocks, long chain maleimide terminated oligomers have been synthesized. The method involves the preparation of an amine terminated oligomer, which is used to prepare the bismaleimides. In that way, maleimide terminated poly(ether-sulfones) [312], phenoxy resins [313], polyamides [314] and polyesters [315–317] have been prepared.

C. Diels–Alder Polymerization

The electron-deficient double bond of maleimides is very reactive towards electron-rich dienes, by means of a Diels–Alder cycloaddition. Consequently, bismaleimides have been used to prepare linear polyimides by reaction with several types of dienes. Furan terminated oligomers have been made to react with bismaleimides [318–320] as shown in Scheme (60).

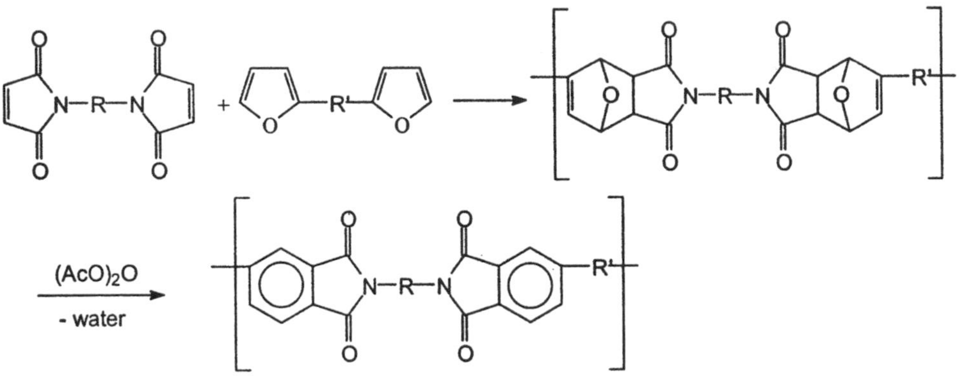

$$(60)$$

The use of silicone linkages in the structure of these adducts causes an increase of solubility (the oligomers are soluble in acetone and THF) while maintaining good thermal properties [321,322].

The same principle has been applied to the polymerization of bismaleimides with bis(benzocyclobutenes). It is known that under appropriate thermal conditions, the strained four-membered ring of benzocyclobutene opens up to give a benzodiene by a concerted process [323,324]. The diene readily reacts with a bismaleimide (Scheme 61).

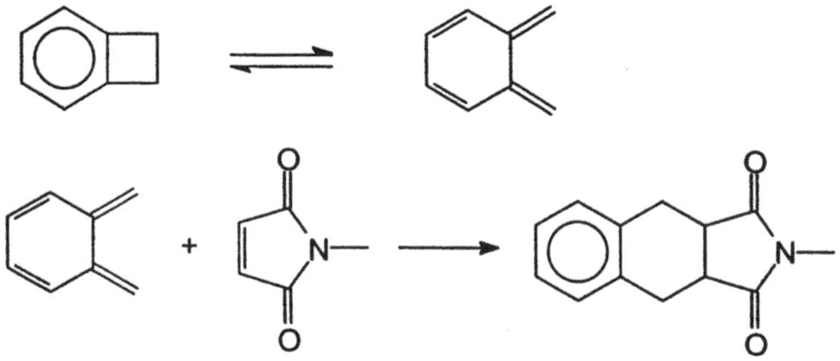

$$(61)$$

Thus, mixtures of bis(benzocyclobutene)s and bismaleimides give rise to thermosetting polyimides with favourable processing conditions and excellent thermal stability [325,326].

More recently, it has been described the synthesis of new Diels–Alder adducts by reaction of bis(2-pyrone) with bismaleimides [327,328]. The reaction is a double Diels–Alder reaction with formation of structures such as those presented in Scheme (62). Tractable polymers were formed only for bismaleimides possessing flexible spacers, while rigid spacers led to cross-linked systems.

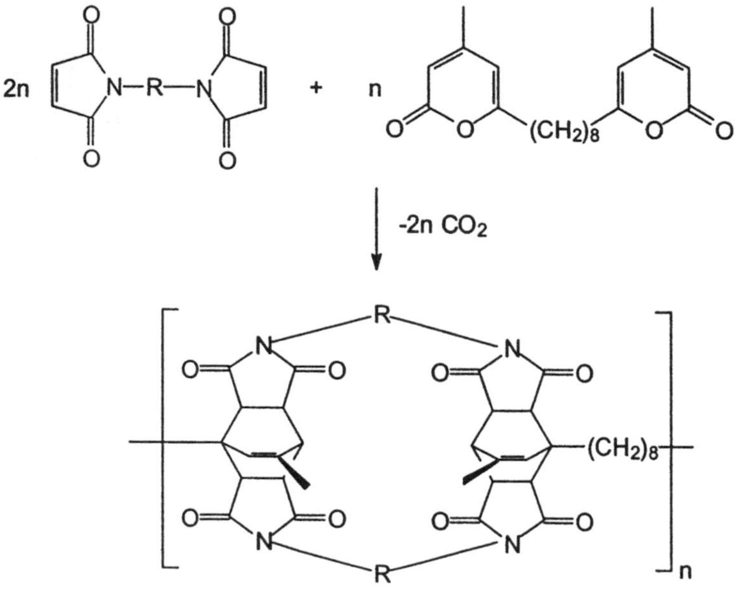

$$-2n \ CO_2$$

(62)

D. Bisnadimides

Bisnadimides are obtained usually in a two-step process. The monomers are dissolved in a polar solvent (NMP, DMF, dyglime, etc.) and made to react at moderate temperature (80 °C). In the second step, the imidization is performed by raising the temperature to around 160 °C. This is the main difference with bismaleimides, where thermal imidization is not possible because of their low-onset curing temperatures. The synthesis of bisnadimides is shown in Scheme (63).

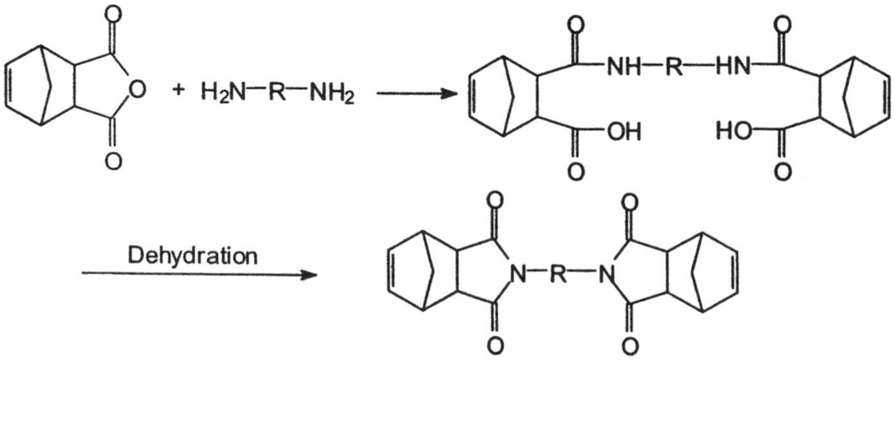

Dehydration

(63)

In order to avoid the problems associated to high cross-linking density, that affect adversely the properties of the final materials, bisnadimides are usually obtained as imide oligomers terminated on nadimide groups. In general, a nadimide terminated oligoimide is prepared by the classical polyimide synthetic method by reacting a dianhydride with an aromatic diamine in the presence of nadic anhydride (Scheme 64) [329].

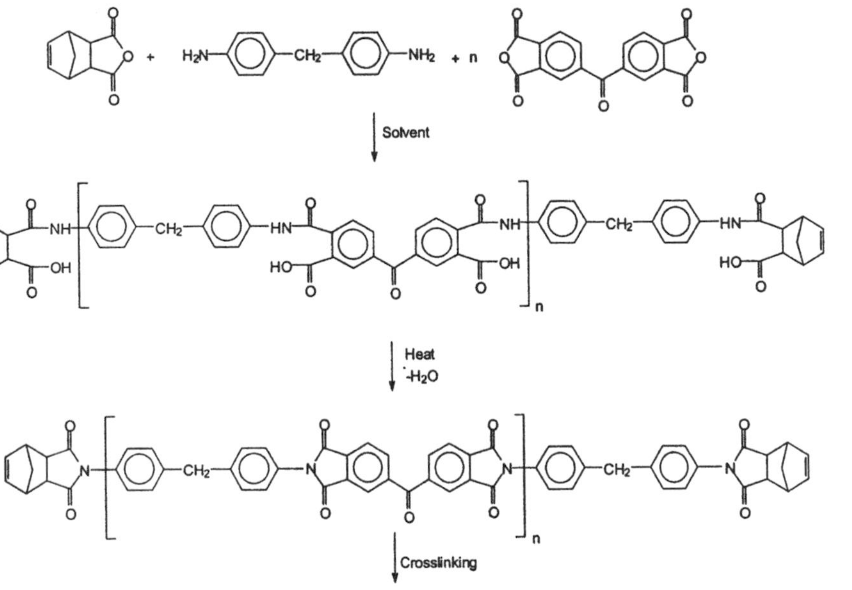

(64)

The first nadimide oligomers were obtained in NMP solution with an equivalent ratio calculated to get a molecular weight of 1300 g/mol [330]. The final step is, as in the case of poly(bismaleimide)s, the cross-linking (thermally induced) of the oligomeric bisimides, and it is always accomplished at the moment of application.

A drawback of these materials is the very high softening temperature of the fully cyclized species, which leads to a narrow processing window. To improve the processability, a new method of obtaining thermosetting polyimides from nadimides has been developed: the *in situ* polymerization of monomeric reactants (PMR), first reported by Serafini *et al.* [331,332] and developed also by other groups [333,334]. In this method a bis-ortho ester replaces the dianhydride and the monomethyl ester of the nadic acid replaces the nadic anhydride. A low boiling point solvent helps for conveniently mixing and applying these formulations. The PMR approach may be considered as the top step achieved so far in the research of cross-linkable polyimides. Typical monomers of PMR polyimides are shown in Scheme (65).

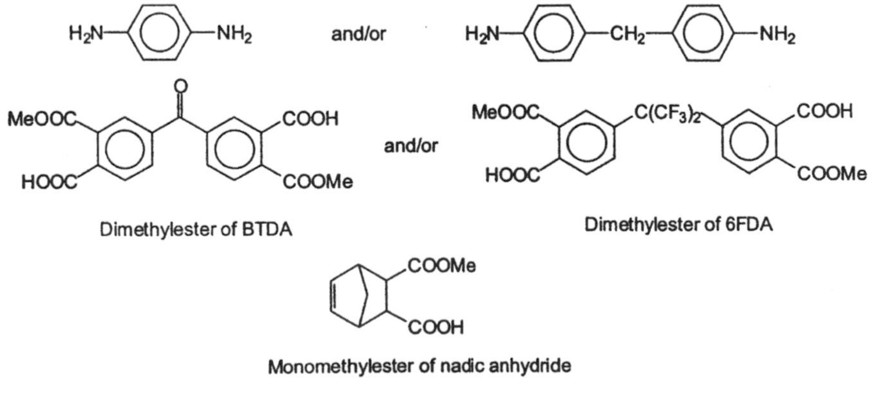

Dimethylester of BTDA

Dimethylester of 6FDA

Monomethylester of nadic anhydride

(65)

The big advantage of PMR, related to other methods of synthesis, is the use of common, easily eliminable, organic solvents. The mixture of monomers, in methanol, is calculated in such a way to obtain a specifically formulated molecular weight. After the solvent removal, at about 120 °C, first the monomers mixture melts, then it condenses to an amic-ester oligomer, which is imidized at arond 170–200 °C, and finally the curing is accomplished by heating to 300–340 °C, generally under pressure to minimize void formation [335,336].

The curing mechanism is rather complex [337–339]. From 200 to 300 °C, isomerization of the as-prepared endo norbornenyl isomer takes place, then a retro Diels–Alder process occurs with evolution of cyclopentadiene, and then polymerization and copolymerization of a number of unsaturated species (cyclopentadiene, maleimide, nadimide and nadimide-cyclopentadiene adducts) proceeds, to give a final material of highly complex composition. The possible structures present in the cross-linked polymer are shown in Scheme (66).

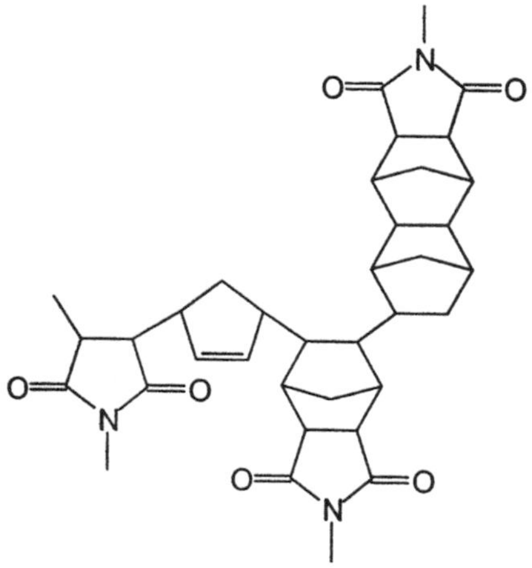

(66)

E. Ethynyl Terminated Oligoimides

Acetylene-terminated oligoimides provide excellent thermal resistance, high service temperature, low moisture absorption and high strength, that is, the best performance of the end-capped imides. However, the main difficulty for the commercial use of these systems is the difficulty in their preparation [340,341].

Usually the acetylene is introduced by means of *m*-ethynlaniline, which is synthesized [342] as summarized in Scheme (67).

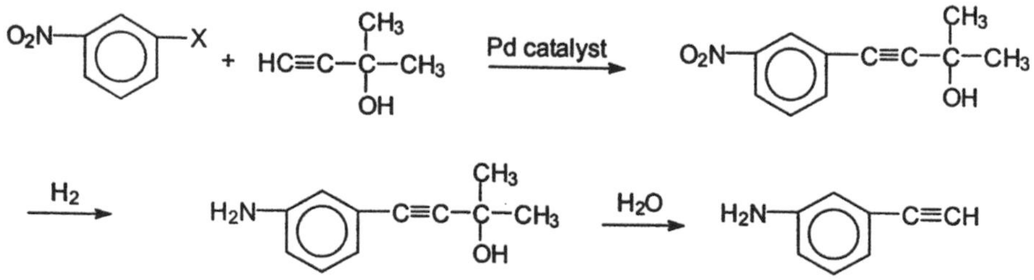

(67)

The ethynyl monomer is reacted with an oligomeric imide terminated in anhydride groups to yield the ethynyl terminated oligomers as it is shown in Scheme (68).

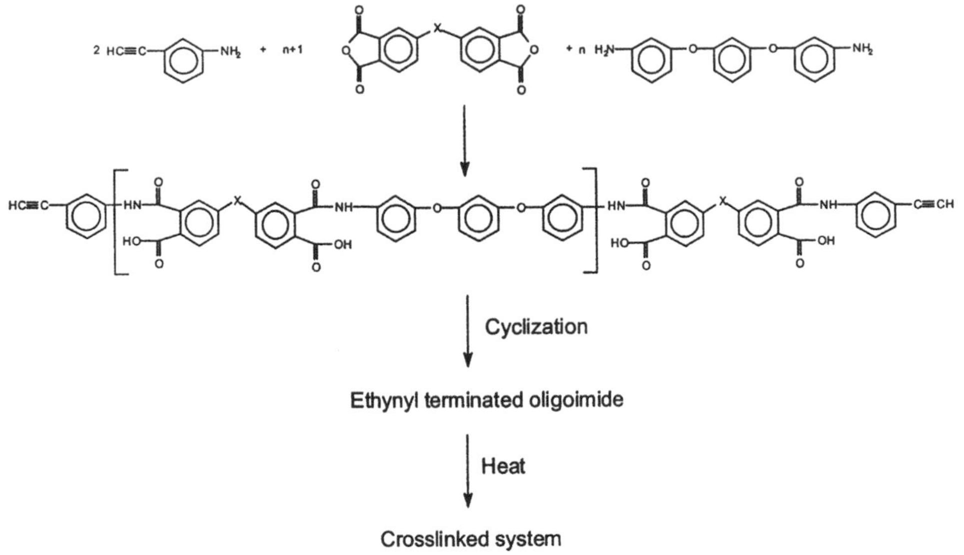

Cyclization

Ethynyl terminated oligoimide

Heat

Crosslinked system

(68)

This type of oligomers suffer from two main problems, the high melting point, that reduces significantly the processing window, and the high rate of cure, as a consequence of the high melting point.

To solve these problems, oligomers containing isoimide units were developed (Scheme 69) [343–345]. They are obtained in the same way than the imide units, but the cyclization of the ethynyl-terminated amic acid oligomers is carried out chemically with dicyclohexyl carbodiimide. The isoimide form has a much lower melting temperature (160 versus 200 °C) and good solubility in low boiling solvents such as THF or dioxane. Because of the lower melting temperature, the isoimide can be cured at a lower temperature, and then, the ethynyl groups polymerize more slowly, what gives longer get times and facilitates processing. The isoimide is transformed into the imide during the cure, thus yielding the same final structure, with the same properties.

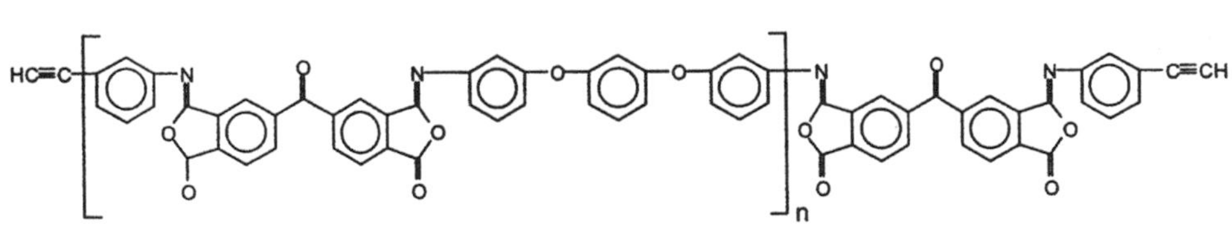

(69)

The 3,3′,4,4′-benzophenonetetracarboxylic anhydride has been replaced by 4,4′-hexafluoroisopropylidenebis(phthalic anhydride) for the preparation of oligomers. This change increases substantially the solubility of the oligomers and improves the processing window.

The reaction mechanism of the acetylene groups is rather complex, and several possible reactions have been described (Scheme 70) [346].

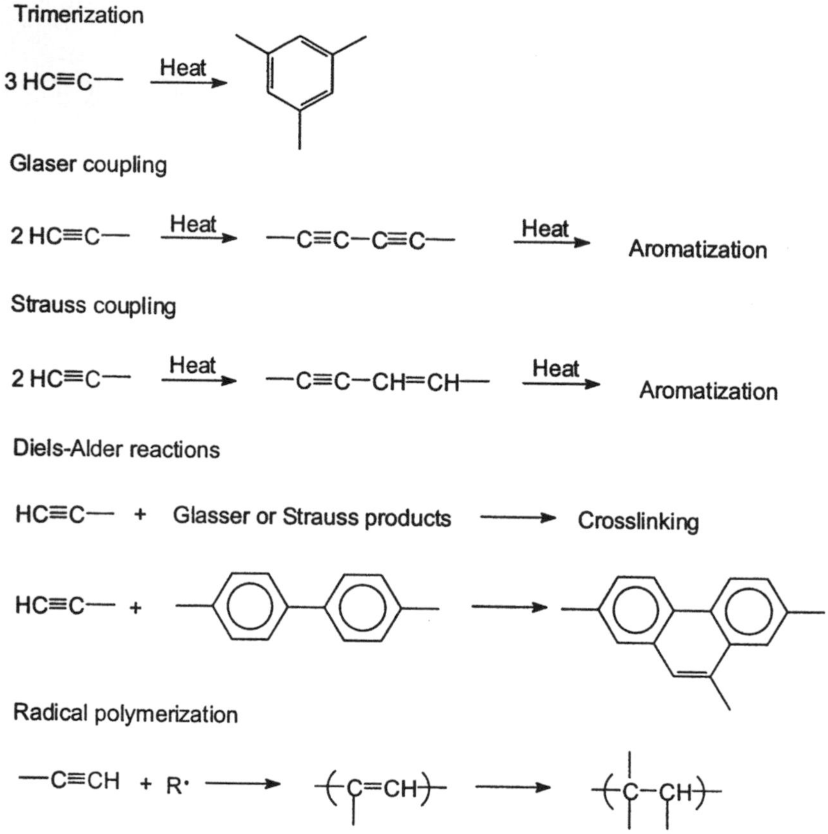
Trimerization
3 HC≡C—
Heat
Glaser coupling
2 HC≡C—
Heat
—C≡C—C≡C—
Heat
Aromatization
Strauss coupling
2 HC≡C—
Heat
—C≡C—CH=CH—
Heat
Aromatization
Diels-Alder reactions
HC≡C— + Glasser or Strauss products
Crosslinking
HC≡C— +
Radical polymerization
—C≡CH + R·

(70)

The last reaction (polymerization) is considered the major curing reaction.

To broaden the processing window and to improve processability, the synthesis of phenylethynyl terminated oligoimides has been also studied (Scheme 71). Phenylethynyl terminated oligomers normally have reduced reactivities compared to the ethynyl compounds and, consequently, provide a wider processing window to the end-capped oligomers [347–349].

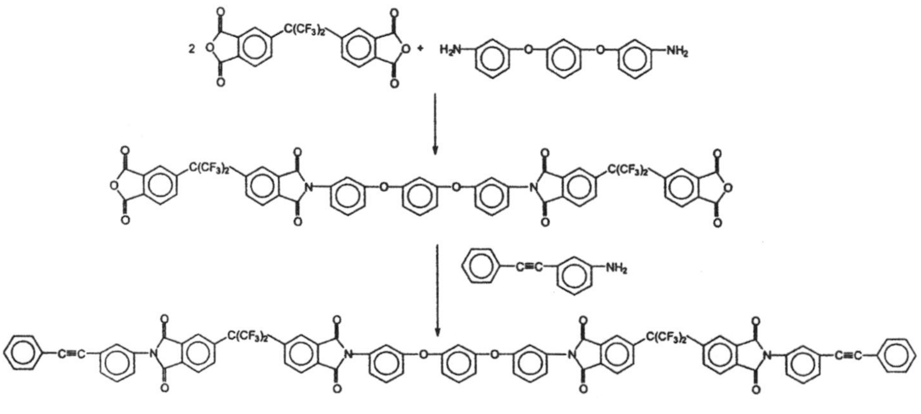

(71)

F. Other End-Capping Groups

Apart from the three reactive end-capping groups we have commented up to now, which are the most frequently used, other chain ends have been tested, capable to give cross-linking by means of an addition reaction. They are summarized in Table 9.

Table 9 Possible end-capping groups for addition polyimides.

Maleimide	
Nadimide	
Allyl-nadimide	$CH_2-CH=CH_2$
Acetylene	$C\equiv CH$
Propargyl ether	$O\text{-}CH_2-C\equiv CH$
Cyanate	$O-C\equiv N$
Cyanamide	$NH-C\equiv N$
Phthalonitrile	$C\equiv N$ $C\equiv N$
Benzocyclobutene	
Biphenylene	
p-Cyclophane	CH_2-CH_2 CH_2-CH_2

REFERENCES

1. Bogert, M. T., and Renshaw, R. R. (1908). *J. Am. Chem. Soc.*, *30*: 1135.
2. Edwards, M. W., and Robinson, I. M. (1955). U.S. Pat. 2710853 to E. I. DuPont de Nemours.
3. Hirsh, S. S., and Lilyquist, M. R. (1967). *J. Appl. Polym. Sci.*, *11*: 305.
4. Lee, H., Stofey, D., and Neville, K. (1967). *New Linear Polymers*, McGraw-Hill, New York.
5. Frazer, A. H. (1968). *High Temperature Resistant Polymers*, Interscience, New York.
6. Mittal, K. L., ed. (1984). *Polyimides. Synthesis, Characterization and Application*, Plenum, New York.
7. Margolis, J. M., ed. (1985). *Engineering Thermoplastics. Properties and Applications*, Dekker, New York.
8. Soane, D. S., and Martynenko, Z. (1989). *Polymers in Microelectronics. Fundamentals and Application*, Elsevier, Amsterdam.
9. Wilson, D., Stenzemberger, H. D., and Hergenrother, P. M. (1990). *Polyimides*, Blackie, New York.
10. Abadie, M. J. M., and Sillion, B. (1991). *Polyimides and Other High-Temperature Polymers*, Elsevier, Amsterdam.
11. Ghosh, M. K., and Mittal, K. L. (1996). *Polyimides. Fundamentals and Applications*, Dekker, New York.
12. Rabilloud, G. (1999). *High Performance Polymers*, Technip, Paris.
13. de Abajo, J. (1992). Polyimides. In *Handbook of Polymer Synthesis* (Kricheldorf, R., ed.), Dekker, New York.
14. Volksen, W. (1994). *Adv. Polym. Sci.*, *117*: 111.
15. de Abajo, J., and de la Campa, J. G. (1998). *Adv. Polym. Sci.*, *140*: 23.
16. Bessonov, M. T., Koton, M. M., Kudryatsev, V. V., and Laius, L. A. (1987). *Polyimides. Thermally Stable Polymers*, Consultants Bureau, New York.
17. Wallach, M. L. (1967). *J. Polym. Sci.*, *C*, *16*: 1191.
18. Sroog, C. E. (1967). *J. Polym. Sci.*, *A-2*: 653.
19. Yang, C.-P., and Hsiao, S.-H. (1985). *J. Appl. Polym. Sci.*, *30*: 2883.
20. Kolegov, V. I., and Frenkel, S. Y. (1976). *Polymer Sci. USSR*, *18*: 1919; Kolegov, V. I. (1976). *Polymer Sci. USSR*, *18*: 1929.
21. Kolegov, V. I. (1977). *Dokl. Akad. Nauk SSSR*, *232*: 848; C. A. (1977), *86*: 140531z.
22. Johnston, N. J. (1972). *J. Polym. Sci.*, *A-1,10*: 2727.
23. Pankratov, V. A., Mairova, A. A., Korshak, V. V., and Vinogradova, S. V. (1975). *Polymer Sci. USSR*, *17*: 2189.
24. Korshak, V. V., Vinogradova, S. V., and Vygodskij, J. S. (1977). *Faserforsch. U. Textiltech.*, *28*: 439.
25. Preston, J. (1975). *Polymer Eng. Sci.*, *15*: 199.
26. Dine-Hart, R. A., and Wright, W. W. (1967). *J. Appl. Polym. Sci.*, *11*: 609.
27. Kumar, D. (1981). *J. Polym. Sci. Polym. Chem. Ed.*, *19*: 795.
28. Volksen, W., and Cotts, P. M. (1984). In Ref. 6, p. 163.
29. Lozano, A. E., de Abajo, J., and de la Campa, J. G. (1993). *Makromol. Chem. Theory Simul.*, *2*: 815.
30. Lozano, A. E., de Abajo, J., and de la Campa, J. G. (1997). *Macromol. Symp.*, *122*: 197.
31. Stewart, J. J. P. (1990). *Semiempirical Molecular Orbital Methods in Reviews in Computational Chemistry* (Lipkowitz, K. B., and Boyd, D. B., eds.), VCH Publishers, New York.
32. Dewar, M. J. S., Zoebish, E. G., Healey, E. F., and Stewart, J. J. P. (1985). *J. Am. Chem. Soc.*, *107*: 3902.
33. MOPAC version 6.0. *Quant. Chem. Prog. Exch.*, 1999.
34. Koton, M. M., Kudriavtsev, V. V., and Svetlichny, V. M. (1984). In Ref. 6, p. 171.
35. Duran, J., and Viswanathan, N. S. (1984). *Polymers in Electronics* (Davison, T., ed.), p. 239, ACS Symposium Series 242, American Chemical Society, Washington D.C.
36. Gao, J. P., and Wang, Z. Y. (1995). *J. Polym. Sci. Part A, Polym. Chem.*, *33*: 1627.

37. Yi, M. H., Huang, W., Jin, M. Y., and Choi, K.-Y. (1997). *Macromolecules*, *30*: 5606.
38. Tsapovetski, M. I., and Laius, L. A. (1982). *Polym. Sci. USSR*, *24*: 1103.
39. Koton, M. M., Meleshko, T. K., Kudryatsev, V. V., Nechayev, P. P., Kamzolkina, Y. V., and Bogorad, N. N. (1982). *Polym. Sci. USSR*, *24*: 715.
40. Koton, M. M., Meleshko, T. K., Kudryatsev, V. V., Gofman, I. V., Kuznetzov, N. P., Dergacheva, Y. N., Bessonov, M. I., Leonov, Y. I., and Gorokhov, A. G. (1985). *Polym. Sci. USSR*, *27*: 905.
41. Brekner, M.-J., and Feger, C. J. (1987). *J. Polym. Sci. Polym. Chem. Ed.*, *25*: 2005; *id. 25*: 2479.
42. Sazanov, Y. N., Krailnikova, L. K., and Schervakova, I. M. (1975). *Eur. Polym. J.*, *11*: 801.
43. Hodgkin, J. H. (1975). *J. Appl. Polym. Sci.*, *20*: 2339.
44. Echigo, Y., Iwaya, Y., Tomioka, I., Furukawa, M., and Okamoto, S. (1995). *Macromolecules*, *28*: 3000.
45. Echigo, Y., Iwaya, Y., Saito, M., and Tomioka, I. (1995). *Macromolecules*, *28*: 6684.
46. Echigo, Y., Iwaya, Y., Tomioka, I., and Yamada, H. (1995). *Macromolecules*, *28*: 4861.
47. Schnider, P., Schmidt, P., Marek, J., Straka, J., Bednár, B., and Králicek, M. (1990). *Eur. Polym. J.*, *26*: 941.
48. Smith, C. D., Mercier, R., Waton, H., and Sillion, B. (1993). *Polymer*, *34*: 4852.
49. Sacher, E., and Sedor, D. G. (1974). *J. Polym. Sci. Polym. Phys. Ed.*, *12*: 629.
50. Numara, S., Fujisake, K., and Kinjo, N. (1984). In Ref. 6, p. 259.
51. Kotera, M., Nishino, T., and Nakamae, K. (2000). *Polymer*, *41*: 3615.
52. Sykes, G. F., and Young, P. R. (1977). *J. Appl. Polym.*, *21*: 2393.
53. Korshak, V. V., Paulova, S. S., Grobkova, P. N., Vlasova, I. V., Vygodskii, Y. S., and Vinogradova, S. V. (1981). *Polym. Sci. USSR*, *23*: 1732.
54. Day, D. R., and Senturia, S. D. (1984). In Ref. 6, p. 249.
55. Palmese, G. R., and Gillham, J. K. (1987). *J. Appl. Polym. Sci.*, *34*: 1925.
56. Kreuz, J. A., Endrey, A. L., Gay, F. P., and Sroog, C. E. (1966). *J. Polym. Sci. A-1*, *4*: 2697.
57. Koton, M. M. (1971). *Polym. Sci. USSR*, *13*: 1513.
58. Roderick, W. R., and Bhatia, P. M. (1963). *J. Org. Chem.*, *28*: 2018.
59. Kurita, K., Suzuki, Y., Enari, T., Kiruchi, M., Nishimura, S.-I., and Ishi, S. (1994). *J. Polym. Sci. Part A, Polym. Chem.*, *32*: 393.
60. Kim, Y. J., and Park, H. P. (1999). *Polym. Int.*, *49*: 8.
61. Kim, Y. J., Glass, T. E., Lyle, G. D., and McGrath, J. E. (1993). *Macromolecules*, *26*: 1344.
62. Edwards, W. M., and Robinson, I. M. (1959). U.S. Pat. 2880230 to E. I. DuPont de Nemours.
63. Preston, J., and Tropsha, Y. (1994). *Polymer Eng. Sci.*, *34*: 305.
64. Konning, C. E., Teuwen, L., Meuer, E. W., and Mooney, J. (1994). *Polymer*, *35*: 4889.
65. Imai, Y. (1996). *Proc. Int. Symp. on Polycondensation*, Paris, p. 24.
66. Imai, Y. (1999). *Adv. Polym. Sci.*, *140*: 1.
67. Houlihan, F. M., Bachman, B. J., Wilkins, C. W., Jr., and Pryde, C. A. (1989). *Macromolecules*, *22*: 4477.
68. Ueda, M. (1993). *Makromol. Chem.*, *194*: 511.
69. Sato, M. (1997). Polyimides. In *Handbook of Thermoplastics* (Olabisi, O., ed.), Dekker, New York.
70. Adrova, N., Bessonov, M., Laius, L. A., and Rudakov, A. P. (1969). *Polyimides: A New Class of Heat-resistant Polymers*, IPST Press, Jerusalem.
71. Dine-Hart, R. A., and Wright, W. W. (1972). *MaKromol. Chem.*, *155*: 237.
72. Gibbs, H. H., and Breder, C. V. (1974). *Polym. Prep.*, *15(1)*: 775.
73. St. Clair, T. L., St. Clair, A. K., and Smith, E. N. (1976). *Polym. Prep.*, *17(2)*: 359.
74. Takekoshi, T., Wirth, J. G., Heath, D. R., Kochanowski, J. E., Manello, J. S., and Weber, M. J. (1980). *J. Polym. Sci. Polym. Chem. Ed.*, *18*: 3069.
75. Davies, M., Hay, J. N., and Woodfine, B. (1992). *High Perform. Polym.*, *5*: 37.
76. Bell, V. I., Stump, B. I., and Gager, H. (1976). *J. Polym. Sci. Polym. Chem. Ed.*, *14*: 2275.
77. Gerber, M. K., Pratt, J. R., and St. Clair, T. L. (1988). *Proc. 3rd Int. Symp. on Polyimides*, Ellenville, p. 101.
78. Malinge, J., Garapon, J., and Sillion, B. (1988). *Brit. Polym. J.*, *20*: 431.
79. Huang, W., Tong, I. T., Xu, J., and Ding, M. (1977). *J. Polym. Sci. A, Polym. Chem.*, *35*: 143.

80. Tamai, S., Yamaguchi, A., and Ohta, M. (1996). *Polymer*, *37*: 3683.
81. Takekoshi, T. (1990). *Adv. Polym. Sci.*, *94*: 1.
82. Abadie, M. J. M., and Sillion, B., eds. (1991). *Polyimides and Other High Temperature Polymers*, Elsevier, Amsterdam.
83. Strong, D. B. (1993). *High Performance and Engineering Thermoplastic Composites*, Technomic Pub., Lancaster, PA.
84. Korshak, V. V., and Rusanov, A. L. (1968). *Izv. Akad. Nauk SSSR. Ser. Khim.*, *10*: 2418; C. A. (1969), *70*: 2054s.
85. Korshak, V. V., Rusanov, A. L., Katsarava, R. D., Niyazi, F. F., and Batirov, I. (1974). *Polym. Sci. USSR*, *16*: 831.
86. Harris, F. W., and Hsu, S.-C. (1989). *High Perform. Polym.*, *1*: 3.
87. Giesa, R., Keller, U., Eiselt, P., and Schmidt, H. W. (1993). *J. Polym. Sci. A, Polym. Chem.*, *31*: 141.
88. Imai,Y., Maldar, N. N., and Kakimoto, M.-A. (1984). *J. Polym. Sci. Polym. Chem. Ed.*, *22*: 2189.
89. Akutsu, F., Kataoka, T., Shimizu, H., Naruchi, K., and Miura, M. (1994). *Macromol. Chem. Rapid Commun.*, *15*: 411.
90. Spiliopoulus, I. K., and Mikroyannidis, J. A. (1996). *Macromolecules*, *29*: 5313.
91. Ayala, D., Lozano, A. E., de Abajo, J., and de la Campa, J. G. (1999). *J. Polym. Sci. A, Polym. Chem.*, *37*: 805.
92. Korshak, V. V., Vinogradova, S. V., and Vygodskii, Y. S. (1974). *J. Macromol. Sci. Rev. Macromol. Chem.*, *C11*: 45.
93. Korshak, V. V., Vynogardova, S. V., Vygodskii, Y. S., Nagiev, Z. M., Urman, Y. G., Alekseeva, S. J., and Slonium, I. Y. (1983). *Makromol. Chem.*, *193*: 445.
94. Yang, C.-P., Ceng, J.-M., and Hsiao, S.-H. (1992). *Makromol. Chem.*, *193*: 445.
95. Kaneda, T., Katsura, T., Nakagawa, K., and Makino, H. (1986). *J. Appl. Polym. Sci.*, *32*: 3133.
96. Sek, D., Pijet, P., and Wanic, A. (1992). *Polymer*, *33*: 190.
97. Shevelev, S. A., Dutov, M. D., Korolev, M. A., and Sapozhnikov, O. Y. (1999). *Polym. Sci. Ser. B*, *41*: 238.
98. Ahn, S. K., Kim, Y. H., Shin, D. C., and Kwon, S. K. (2000). *Bull. Korean Chem. Soc.*, *21*: 377.
99. Wang, C.-S., and Leu, T.-S. (2000). *Polymer*, *41*: 3581.
100. Rusanov, A. L., Komarova, L. G., Prigozhina, M. P., Askadskii, A. A., Kazantseva, V. V., Genies, C., Mericer, R., Sillion, B., Cornet, N., Gebel, G., and Pineri, M. (2001). *Polymer*, *42*: 359.
101. Kuznetzov, A. A., Tsegelskaya, A. Y., Belov, M. Y., Berendyaev, V. I., Lavrov, S. V., Semenova, G. K., Izyumnikov, A. L., Kozlova, N. V., and Kotov, B. V. (1998). *Macromol. Symp.*, *128*: 203.
102. Sek, D., Wanic, A., and Schab-Balcerzak, E. (1995). *J. Polym. Sci. A, Polym. Chem.*, *33*: 547.
103. Wurtz, A. (1854). *Ann. Chim. Phys.*, *42*: 54.
104. Frey, H. E. (1967). U.S. Pat. 3300420, to Standard Oil.
105. Meyers, R. A. (1969). *J. Polym. Sci. A-1*, 7: 2757.
106. Farrisey, W. J., Rose, J. M., and Carleton, P. S. (1970). *J. Appl. Polym. Sci.*, *14*: 1093.
107. Carleton, P. S., Farrisey, W. J., and Rose, J. S. (1972). *J. Appl. Polym. Sci.*, *16*: 2983.
108. Carleton, P. S., Farrisey, W. J., and Rose, J. S. (1970). Ger. Pat. 2001914, to Upjhon.
109. Kilic, S., Mohanty, D. K., Yilgor, I., and McGrath, J. E. (1986). *Polym. Prep.*, *27(1)*: 318.
110. Kakimoto, M. A., Akiyama, R., Megi, Y. S., and Imai, Y. (1988). *J. Polym. Sci. A, Polym. Chem.*, *26*: 99.
111. Avadhani, C. V., Wadgaonkar, P. P., and Vernekar, S. P. (1990). *J. Polym. Sci. A, Polym. Chem.*, *28*: 1681.
112. Barikani, M., Yeganeh, H., and Ataei, S. M. (1999). *Polym. Int.*, *48*: 1264.
113. Alvino, W. M., and Edelman, I. E. (1975). *J. Appl. Polym. Sci.*, *19*: 2961.
114. Alvino, W. M., and Edelman, I. E. (1978). *J. Appl. Polym. Sci.*, *22*: 1983.

115. Nishizaki, S., and Fukami, A. (1965). *Kogyo Kagaku Zasshi, 68*: 383; C. A. (1965), *63*: 3057b.
116. Sideridou-Karayannidou, I., and Karayannidis, G. P. (1987). *J. Macromol. Sci. Chem.*, *A24*: 689.
117. Saidenova, S. B., Zhuvanov, B. A., Arkhipova, I. A., and Rafikov, S. R. (1975). *Polym. Sci. USSR, 17*: 1653.
118. Gagliani, J., and Long, J. V. (1983). U.S. Pat. 4394464.
119. Scola, D. A., and Brunete, C. M. (1987). *Polym. Prep., 28(1)*: 86.
120. Takekoshi, T. (1974). U.S. Pat. 3847870 to General Electric.
121. Takekoshi, T., and Kochanowski, E. J. (1974).U.S. Pat. 3850885 to General Electric.
122. Rogers, M. E., Glass, T. E., Mecham, S. J., Rodrigues, D., Wilkes, G. L., and McGrath, J. E. (1994). *Polymer, 32*: 2663.
123. Imai, Y. (1970). *Polym. Lett., 8*: 555.
124. Taguchi, K. (1980). *Polym. Lett., 18*: 525.
125. Yoda, N. (1968). Jap. Pat. 68, 13076 to Toyo Rayon Ltd.; C. A. (1969), *70*: 29602d.
126. de Abajo, J., Brunet, A. G., Babé, S. G., and Fontán, J. (1974). *An. Quim., 70* 908.
127. Kurita, K., Imajo, H., and Iwakura, Y. (1979). *J. Polym. Sci. Polym. Chem. Ed., 17*: 1619.
128. Imajo, J., Kurita, K., and Iwakura, Y. (1980). *J. Polym. Sci. Polym. Chem. Ed., 18*: 2189.
129. Russo, S., and Mortillaro, L. (1969). *J. Polym. Sci. A-1, 7*: 3337.
130. Yoda, N. (1968). Jap. Pat. 68, 00,629 to Toyo Rayon Ltd.; C. A. (1968), *69*: 11075d.
131. Yoda, N. (1968). Jap. Pat. 68, 25,984 to Toyo Rayon Ltd.; C. A. (1969), *70*: 68913a.
132. Oishi, Y., Kakimoto, M., and Imai, Y. (1987), *Macromolecules, 20*: 703; *ib.* (1988). *Macromolecules, 21*: 547.
133. Kricheldorf, H. R., ed. (1996). *Silicon in Polymer Synthesis*, Springer, Heidelberg.
134. Lozano, A. E., de Abajo, J., and de la Campa, J. G. (1997). *Macromolecules, 30*: 2507.
135. Kricheldorf, H. R. (1970). *Chem. Ber., 103*: 3353.
136. Lozano, A. E., de Abajo, J., and de la Campa, J. G. (1998). *Macromo. Theory Simul., 7*: 41.
137. Becker, K. H., and Schmidt, H. W. (1992). *Macromolecules, 25*: 6784.
138. Imai, Y., and Kojima, K. (1972). *J. Polym. Sci. A-1, 10*: 209.
139. Lenz, R. W. (1967). *Organic Chemistry of Synthetic High Polymers*, Interscience, New York.
140. Stille, J. K., and Anyos, T. (1964). *J. Polym. Sci., A-2*: 1487.
141. Stille, J. K., and Morgan, R. A. (1965). *J. Polym. Sci., A-3*: 2397.
142. Harris, F. W., and Norris, S. O. (1973). *J. Polym. Sci. Polym. Chem. Ed., 11*: 2143.
143. Harris, F. W., and Lainer, L. H. (1975). *Appl. Polym. Symp., 26*: 421.
144. de Abajo, J., Babé, S. G., and Fontán, J. (1971). *Angew. Makromol. Chem., 19*: 121.
145. Fontán, J., Babé, S. G., and de Abajo, J. (1975). *Proc. Int. Symp. on Macromolecules*, Rio do Janheiro, Elsevier, Amsterdam, p. 305.
146. Staubli, A., Ron, E., and Langer, R. (1990). *J. Am. Chem. Soc., 112*: 4419.
147. Loncrini, D. F. (1965). U.S. Pat. 3182074 to General Electric.
148. Loncrini, D. F. (1966). Fr. Pat. 1422925 to Thomson-Houston.
149. Babé, S. G., de Abajo, J., and Fontán, J. (1972). *Angew. Makromol. Chem., 21*: 65.
150. Gabarda, J. P., de Abajo, J., and Fontán, J. (1976). *Rev. Plast. Mod., 31*: 63.
151. Maros, C. L., and de Abajo, J. (1976). *Angew. Makromol. Chem., 55*: 73.
152. Maiti, S., and Das, S. (1981). *J. Appl. Polym. Sci., 26*: 957.
153. Kurita, K., and Matsuda, S. (1983). *Makromol. Chem., 184*: 1223.
154. Kricheldorf, H. R., and Pakull, R. (1988). *Macromolecules, 21*: 551.
155. de Abajo, J., de la Campa, J. G., Kricheldorf, H. R., and Schwarz, G. (1990). *Makromol. Chem., 191*: 537.
156. Kricheldorf, H. R., Schwarz, G., de Abajo, J., and de la Campa, J. G. (1991). *Polymer, 32*: 942.
157. de Abajo, J., de la Campa, J. G., Kricheldorf, H. R., and Schwarz, G. (1994). *Polymer, 35*: 5577.
158. Kricheldorf, H. R., Schwarz, G., Berghahn, M., de Abajo, J., and de la Campa, J. G. (1994). *Macromolecules, 27*: 2540.
159. Kricheldorf, H. R. (1999). *Adv. Polym. Sci., 141*: 83.

160. Lienert, K.-W. (1999). *Adv. Polym. Sci.*, *141*: 45.
161. Gaw, K. O., and Kakimoto, M. (1999). *Adv. Polym. Sci.*, *141*: 107.
162. Mantecón, A., Cádiz, V., Serra, A., and Martínez, P. A. (1987). *Angew. Makromol. Chem.*, *148*: 149.
163. Galiá, M., Serra, A., Mantecón, A., and Cádiz, V. (1995). *J. Appl. Polym. Sci.*, *56*: 193.
164. Martínez, P. A., Cádiz, C., Serra, A., and Mantecón, A. (1985). *Angew. Makromol. Chem.*, *136*: 159.
165. de la Campa, J. G., de Abajo, J., Mantecón, A., and Cádiz, V. (1987). *Eur. Polym. J.*, *23*: 961.
166. Castell, P., Derra, A., Cádiz, V., and Mantecón, A. (1999). *J. Appl. Polym. Sci.*, *72*: 537.
167. Babé, S. G., de Abajo, J., and Fontán, J. (1972). *Anal. Quim.*, *68*: 1259.
168. Hill J. W., and Carothers, W. H. (1932). *J. Am. Chem. Soc.*, *54*: 1569.
169. González, J. I., de Abajo, J., Babé, S. G., and Fontán, J. (1972). *Rev. Plást. Mod.*, *23*: 674.
170. González, J. I., de Abajo, J., Babé, S. G., and Fontán, J. (1976). *Angew. Makromol. Chem.*, *55*: 85.
171. Staubli, A., Mathiowitz, E., and Langer, R. (1991). *Macromolecules*, *24*: 2291.
172. Hanes, J., Chiba, M., and Langer, R. (1998). *Biomaterials*, *19*: 163.
173. Bower, G. M., and Frost, L. W. (1963). *J. Polym. Sci.*, *A-1*: 55.
174. Imai, Y. (1996). *Synthesis of Polyamide-imides*, in Ref. 11, p. 49.
175. Jonquières, A., Dole, C., Clèment, R., and Lochon, P. (2000). *J. Polym. Sci. Part A, Polym. Chem.*, *38*: 614.
176. Freeman, J. H., Traynor, E. J., Miglarese, I., and Lunn, R. H. (1962). *SPE Trans.*, *2*: 216.
177. Dezern, J. F. (1988). *J. Polym. Sci. A, Polym. Chem.*, *26*: 2157.
178. Keske, R. G., Stephens, J. R., and Dunlap, R. W. (1984). *Polym. Prep.*, *25(2)*: 12.
179. Wrasidlo, W. L., and Augl, J. M. (1969). *J. Polym. Sci.*, *A-1*, *7*: 321, 1589.
180. Strul, M., Neamtu, G., Mantaluta, E., and Zugravescu, I. (1971). *Rev. Roum. Chim.*, *16*: 941.
181. Fedyna, V. N., and Suberlyak, O. V. (1974). *Polym. Sci. USSR*, *15*: 1722.
182. de Abajo, J., Gabarda, J. P., and Fontán, J. (1978). *Angew. Makromol. Chem.*, *71*: 143.
183. Maiti, S., and Ray, A. (1983). *J. Appl. Polym. Sci.*, *28*: 225.
184. Yamazaki, N., Matsumoto, M., and Higashi, F. (1975). *J. Polym. Sci. Polym. Chem. Ed.*, *13*: 1373.
185. Yang, C.-P., and Hsiao, S.-H. (1989). *Makromol. Chem.*, *190*: 2119.
186. Hsiao, S.-H., and Yang, C.-P. (1991). *J. Polym. Sci. Part A, Polym. Chem.*, *29*: 447.
187. Yang, C.-P., and Chen, W.-Y. (1993). *J. Polym. Sci. Part A, Polym. Chem.*, *31*: 3081.
188. Yang, C.-P., Chen, R.-S., and Chen, H.-D. (1999). *Polym. J.*, *31*: 1253.
189. de Abajo, J., Babé, S. G., and Fontán, J. (1970). *Rev. Plást. Mod.*, *21*: 177.
190. Alvino, W. M., and Frost, L. W. (1971). *J. Polym. Sci. A-1*, *9*: 2209.
191. Alvino, W. M., and Ray, R. (1980). *J. Polym. Sci. Polym. Chem. Ed.*, *19*: 2551.
192. Lwdneva, O. A., Pariiskii, G. B., Trezvov, V. V., and Toptygin, D. Y. (1984). *Polym. Sci. USSR*, *26*: 707.
193. Terney, S., Keating, J., and Zielinsli, J. (1970). *J. Polym. Sci. A-1*, *8*: 683.
194. Pigeon, R., and Allard, P. (1974). *Angew. Makromol. Chem.*, *40/41*: 139.
195. Nieto, J. L., de la Campa, J. G., and de Abajo, J. (1982). *Makromol. Chem.*, *183*: 557.
196. de la Campa, J. G., and de Abajo, J. (1983). *Eur. Polym. J.*, *19*: 667.
197. Jones, R. J., and Silversman, E. M. (1989). *SAMPE J.*, *25*: 41.
198. Sasuga, T. (1991). *Polymer*, *32*: 1539.
199. Muellerleile, J. T., Brian, G. R., Rodrigues, D. E., and Garth, L. W. (1993). *Polymer*, *34*: 789.
200. Takekoshi, T., Kochanowski, J. E., Manello, J. S., and Webber, M. J. (1986). *J. Polym. Sci. Part C, Polym. Symp.*, *7*: 93.
201. Anonymous (1969). Fr. Pat. 1,575,176 to Farbenfbriken Bayer A.G.; C.A. (1970), *72*: 79671.
202. Im, J. K., and Jung, J. C. (2000). *J. Polym. Sci. A, Polym. Chem.*, *38*: 402.
203. Takekoshi, T. (1987). *Polym. J.*, *19*: 191.
204. Petit, B., and Marechal, E. (1974). *Bull. Soc. Chim.*, *7*: 1597.

205. Salary, J., and Smith, C. H. (1975). *J. Cell. Plast.*, *11*: 262.
206. Matsuda, H. (1975). *Makromol. Chem.*, *176*: 573.
207. Jiang, B., Hao, J., Wang, W., Jiang, L., and Cai, X. (2001). *J. Appl. Polym. Sci.*, *81*: 773.
208. Maisulanis, B., Hrouz, J., Baldrian, J., Ilavský, M., and Dusek, K. (1987). *J. Appl. Polym. Sci.*, *34*: 1941.
209. de Visser, A., Driesen, A. A., and Wolke, J. G. C. (1980). *Makromol. Chem. Rapid. Commun.*, *1*: 177.
210. de Visser, A., Gregonis, D. E., and Wolke, J. G. C. (1978). *Makromol. Chem.*, *179*: 1855.
211. Kawakami, Y., Yu, S.-P., and Abe, T. (1992). *Polym. J.*, *24*: 1129.
212. Jwo, S.-L., Whang, W.-T., and Liaw, W.-C. (1999). *J. Appl. Polym. Sci.*, *74*: 2832.
213. McGrath, J. E., Dunson, D. L., Mecham, S. J., and Hedrick, J. (1999). *Adv. Polym. Sci.*, *141*: 61.
214. Hang, J. H., Cho, K., and Park, C. E. (2000). *Polymer*, *42*: 2513.
215. Rogers, M. E., Glass, T. E., Mecham, S. J., Rodrigues, D., Wilkes, G. L., and McGrath, J. E. (1994). *Polymer*, *32*: 2663.
216. Okamoto, K.-I., Fuji, M., Okamyo, S., Suzuki, H., Tanaka, K., and Kita, H. (1995). *Macromolecules*, *28*: 6950.
217. Berger, A. (1984). In Ref. 6, p. 67.
218. Ezzell, S. A., St. Clair, A. K., and Hinkley, J. K. (1987). *Polymer*, *28*: 1779.
219. Preston, J. (1971). *Encyclopedia of Chemical Technology*, 2nd ed. (Kirk, R. E. and Othmer, D. F., ed.), Wiley, New York, sup. vol., p. 746.
220. Preston, J., and Black, W. B. (1967). *J. Polym. Sci. A-1*, *5*: 2429.
221. Frost, L. W., Bower, G. M., Freeman, J. H., Burgman, H. A., Traynor, E. J., and Ruffing, C. R. (1968). *J. Polym. Sci. A-1*, *6*: 215.
222. Preston, J. (1972). *J. Polym. Sci. Polym. Chem. Ed.*, *10*: 3373.
223. Thaemlitz, C. J., Weikel, W. J., and Cassidy, P. E. (1992). *Polymer*, *33*: 3278.
224. Preston, J., Dewinter, W. F., and Black, W. B. (1969). *J. Polym. Sci. A-1*, *7*: 283.
225. Sakaguchi, Y., and Kato, Y. (1993). *J. Polym. Sci. A, Polym. Chem.*, *31*: 1029.
226. Bruma, M., Sava, I., Hamciuc, E., Belomoina, N. M., and Krongauz, E. S. (1991). *Angew. Makromol. Chem.*, *193*: 113.
227. Saiki, A., Mukai, K., Harada, S., and Miyadera, Y. (1980). *ACS Org. Coat. Plast. Prepr.*, *43*: 459.
228. Preston, J., Dewinter, W. F., and Black, W. B. (1972). *J. Polym. Sci. Polym. Chem. Ed.*, *10*: 1377.
229. Augl, J. M. (1970). *J. Polym. Sci. A-1*, *8*: 3145.
230. Augl, J. M., and Booth, H. J. (1973). *J. Polym. Sci., Polym. Chem. Ed.*, *11*: 2195.
231. Oster, G., and Yang, N. L. (1969). *Chem. Rev.*, *68*: 125.
232. Smets, G. (1974). *Proc. Int. Symp. on Macromolecules*. Rio do Jahneiro (Mano, E. B., ed.), Elsevier, Amsterdam, p. 41.
233. Green, G. E., Stark, B. P., and Zahir, S. A. (1981–82). *J. Macrom. Sci. Rev. Macromol. Chem.*, *C21*: 187.
234. Stevens, M. P., and Jenkins, A. D. (1979). *J. Polym. Sci. Polym. Chem. Ed.*, *17*: 3675.
235. Ichimura, K., Watanabe, S., and Ochi, H. (1976). *Polym. Lett.*, *14*: 207.
236. Majundar, R. N., Yang, S., and Harwood, H. J. (1983). *J. Polym. Sci. Polym. Chem. Ed.*, *21*: 1717.
237. de Abajo, J., Madruga, E. L., San Román, J., de la Campa, J. G., and Guzmán, J. (1992). *Polymer*, *33*: 1090.
238. Yang, C.-P., Iwakura, Y., Uno, K., and Toda, F. (1976). *Makromol. Chem.*, *177*: 3495.
239. de Abajo, J., and de Santos, E. (1983). *Angew. Makromol. Chem.*, *111*: 17.
240. de Abajo, J., de la Campa, J. G., Lozano, A. E., and Alvarez, J. C. (1995). *Adv. Mat.*, *7*: 148.
241. Diakoumakos, C. D., and Mikroyannidis, J. A. (1994). *Polymer*, *35*: 1986.
242. Hamciuc, C., Hamciuc, E., Diaconu, I., Merder, F. W., and Bruma, M. (1997). *J. Macromol. Sci. Pure Appl. Chem.*, *A34*: 143.

243. Korshak, V. V., Rusanov, A. L., Batirov, I., Katsarava, R. D., and Niyazi, F. F. (1978). *Faserforsch. U. Textiltech.*, *29*: 649.
244. Meyer, G. W., Pak, S. J., Lee, Y. J., and McGrath, J. E. (1995). In *High-Temperature Properties and Applications of Polymeric Materials*, ACS Symposium Series, 603 (Tant, M. R., Connell, J. W., and McManus, H. L. N., eds.), Washington.
245. Rhee, T. H., Choi, T., Chung, E. Y., and Suh, D. H. (2001). *Macromol. Chem. Phys.*, *202*: 906.
246. Klebe, J. A., Wroblewsky, H. A., and Gilbert, A. R. (1973). U.S. Pat. 1616832 to General Electric.
247. Serna, F. J., de Abajo, J., and de la Campa, J. G. (1985). *J. Appl. Polym. Sci.*, *30*: 61.
248. Serna, F. J., de la Campa, J. G., and de Abajo, J. (1986). *Angew. Makromol. Chem.*, *139*: 113.
249. Serna, F. J., de Abajo, J., and de la Campa, J. G. (1987). *Brit. Polym. J.*, *19*: 453.
250. Galiá, M., Cádiz, V., Mantecón, A., and Serra, A. (1992). *J. Polym. Sci. A, Polym. Chem.*, *30*: 2379.
251. Morikawa, A., Iyoku, Y., Kakimoto, M., and Imai, Y. (1992). *Polym. J.*, *24*: 107.
252. Iyoku, Y., Kakimoto, M.-A., and Imai, Y. (1994). *High Perform. Polym.*, *6*: 53.
253. Kim, Y., Lee, W. K., Cho, W. J., and Ha, C. S. (1997). *Polym. Int.*, *43*: 129.
254. Ha, C.-S., and Cho, W.-J. (1999). *Macromol. Symp.*, *142*: 205.
255. Smaishi, M., Schotter, J.-C., Lesimple, C., Prevost, I., and Guizard, C. (1999). *J. Membrane Sci.*, *161*: 157.
256. Chen, Y., and Iroh, J. O. (1999). *Chem. Mater.*, *11*: 1218.
257. Hsiue, G.-H., Chen, J.-K., and Liu, Y.-L. (2000). *J. Appl. Polym. Sci.*, *76*: 1609.
258. Liu, J., Gao, Y., Wang, F., and Wu, M. (2000). *J. Appl. Polym. Sci.*, *75*: 384.
259. Hobson, S. T., and Shea, K. J. (1997). *Chem. Mater.*, *9*: 616.
260. Hu, Q., Marand, E., Dhingra, S., Fritsch, D., Wen, J., and Wilkes, G. (1997). *J. Membrane Sci.*, *135*: 65.
261. Matsumoto, A. and Kimura, T. (1998). *J. Appl. Polym. Sci.*, *68*: 1703.
262. Caulfield, M. J., and Solomon, D. H. (1999). *Polymer*, *40*: 1251.
263. Sato, T., Masaki, K., Kondo, K., Seno, M., and Tanaka, H. (1995). *Polym. Bull.*, *35*: 345.
264. Wang, J., Chern, Y., and Chung, M. (2000). *J. Polym. Sci. Part A: Polym. Chem.*, *34*: 3345.
265. Matsumoto, A., Kubota, T., and Otsu, T. (1990). *Polym. Bull.*, *24*: 459.
266. Barrales-Rienda, J. M., de la Campa, J. G., and González, J. (1977). *J. Macromol. Sci.*, *A1, 1*: 267.
267. Kardush, N., and Stevens, M. P. (1972). *J. Polym. Sci. A-1*, *10*: 1093.
268. Zhubanov, B. A., and Akkulova, Z. G. (1974). *Int. Polym. Sci. Tech.*, *1*: 20.
269. Dumas, P. (1974). *Izv. Akad. Nauk. Kaz. S.S.R. Ser. Khim.*, *24*: 78; C. A. (1974), *81*: 78281.
270. Kinstle, J. F., and Sivils, G. (1974). *J. Rad. Cur.*, *1*: 11.
271. De Schryver, F. C., Feast, W. J., and Smets, G (1974). *J. Polym. Sci. A-1*, *8*: 1939.
272. De Schryver, F. C. (1973). *Pure Appl. Chem. 34(2)*: 213.
273. De Schryver, F. C., Boens, N., and Smets, G. (1974). *Macromolecules*, *7*: 399.
274. Dilling, W. L. (1983). *Chem. Rev.*, *83(1)*: 1.
275. Mison, P., and Sillion, B. (1999). *Adv. Polym. Sci.*, *140*: 137.
276. Lin, S.-C., and Pearce, E. M. (1993). *High Performance Thermosets. Chemistry, Properties, Applications*. Hanser Publishers, Munich.
277. Searle, N. E. (1948). U.S. Pat 2444536 to E. I. DuPont de Nemours & Co.
278. Stenzenberger, H. D. (1990). In *Polymides* (Wilson, D. *et al.*, eds.), Blackie, Glasgow.
279. Winter, H., Loontjens, J. A., Mostert, H., and Tholen, M. G. W. (1989). Chemistry and properties of new bismaleimides designed for improved processability. In *Polyimides: Materials, Chemistry and Characterization* (Feger, C., Khojasteh, M. M., and McGrath, J. E., eds.), Elsevier, Amsterdam, p. 229.
280. Crivello, J. V. (1973). *J. Polym. Sci. Polym. Chem. Ed.*, *19*: 451.
281. Stenzenberger, H. D. (1986). *Structural Adhesives* (Kinloch, A. J., ed.), Elsevier, Essex, p. 80.
282. Sergeyev, V. A., Medel'kin, V. I., Yuferov, Y. A., Yerzh, B. V., Komarova, L. I., Bakhmutov, V. I., and Tysiryapkin, V. A. (1984). *Polym. Sci. USSR*, *26*: 2166.

283. Varma, I. K., Gupta, A. K., Sangita, and Varma, D. S. (1983). *J. Appl. Polym. Sci.*, *28*: 191.

284. Hartford, S. L., Subramanian, S., and Parker, J. A. (1978). *J. Polym. Sci. Polym. Chem. Ed.*, *16*: 137.

285. Krishnan, K., and Ninan, K. N. (1991). *Thermochim. Acta*, 189: 241.

286. Ninan, K. N., George, R., Krishnan, K., and Rao, K. V. C. (1989). *J. Appl. Polym. Sci.*, *37*: 127.

287. Vijayan, Bisth, T. M., and Singh, M. M. (1991). *Front. Polym. Res.*, 1st ed. (Prasad, P. N., and Nigam, J. K., eds.), Plenum, New York, p. 559.

288. Galanti, A. V., and Scola, D. A. (1981). *J. Polym. Sci. Polym. Chem. Ed.*, *19*: 451.

289. Galanti, A. V., Keen, B. T., Paterand, R. H., and Scola, D. A. (1981). *J. Polym. Sci. Polym. Chem. Ed.*, *19*: 2243.

290. Galanti, A. V., Liotta, F., Keen, B. T., and Scola, D. A. (1981). *J. Polym. Sci. Polym. Chem. Ed.*, *20*: 233.

291. Liu, F. J., Munukutia, S., Levon, K., and Tesoro, G. (1992). *J. Polym. Sci. Part A, Polym. Chem.*, *30*: 157.

292. Sen, S. R., and Chakravorty, S. (1996). *J. Polym. Sci. Part A, Polym. Chem.*, *34*: 25.

293. Varma, I. K., Fohlen, G. M., and Parker, J. A. (1982). *J. Polym. Sci. Polym. Chem. Ed.*, *20*: 283.

294. Mikroyannidis, J. A. (1984). *J. Polym. Sci. Polym. Chem. Ed.*, *22*: 1717.

295. Barton, J. M., Hamerton, I., Rose, J. B., and Warner, D. (1991). *Polyimides and Other High-Temperature Polymers* (Abadie, M. J. M., and Sillion, B., eds.), Elsevier, Amsterdam.

296. Barton, J. M., Hamerton, I., Rose, J. B., and Warner, D. (1992). *Polymer*, *33*: 3664.

297. Datta, R. N., Talma, A. G., and Wagenmakers, J. C. (1997). *Kautsch. Gummi Kunstst.*, *50*: 274.

298. Datta, R. N., and Ingham, F. A. A. (1997). *Kautsch. Gummi Kunstst.*, *52*: 758.

299. Beers, R. N., Benko, D. A., and Wolski, T. P. (2000). Eur. Pat. Appl. 988999 to Goodyear Tire & Rubber Co., USA.

300. Schotman, A. H. M., van Haeren, P. J. C., Weber, A. J. M., van Wijk, F. G. H., Hofstraat, J. W., Talma, A. G., Steenbergen, A., and Datta, R. N. (1996). *Rubber Chem. Technol.*, *69*: 727.

301. Crivello, J. V., and Juliano, P. C. (1975). *J. Polym. Sci. Polym. Chem. Ed.*, *13*: 1819.

302. White, J. E., Snider, D. A., and Scala, M. D. (1984). *J. Polym. Sci. Polym. Chem. Ed.*, *22*: 589.

303. Stenzenberger, H. D. (1973). *Appl. Polym. Symp.*, *22*: 77.

304. Wang, C.-S., and Hwang, H.-J. (1996). *Polymer*, *37*: 499.

305. Wu, W., Wang, D., and Ye, C. (1998). *J. Appl. Polym. Sci.*, *70*: 2471.

306. Barton, J. M., Hamerton, I., Rose, J. B., and Warner, D. (1994). *High Perform. Polym.*, *6*: 21.

307. White, J. E. (1986). *Ind. Eng. Chem. Prod. Res. Dev.*, *25*: 395.

308. Crivello, J. V. (1973). *J. Polym. Sci. Polym. Chem. Ed.*, *11*: 1185.

309. Grenier-Loustalot, M. F., Gouarderes, F., Joubert, F., and Grenier, P. (1993). *Polymer*, *34*: 3848.

310. Stenzenberger, H. D. (1980). U.S. Patent 4211861.

311. Stenzenberger, H. D. (1986). *Structural Adhesives* (Kinloch, A. J., ed.), Elsevier, Essex, p. 80.

312. Kwiatkowski, G. T., and Brode, G. L. (1974). U.S. Pat. 4.276.344.

313. Rao, B. S. (1988). *J. Polym. Sci. Part C, Polym. Lett.*, *26*: 3.

314. Holub, F. F. and Evans, M. L. (1971). *Ger. Offen.* 2.031.574.

315. Acevedo, M., de la Campa, J. G., and de Abajo, J. (1989). *J. Appl. Polym. Sci.*, *38*: 1745.

316. Acevedo, M., de la Campa, J. G., and de Abajo, J. (1990). *J. Appl. Polym. Sci.*, *41*: 1990.

317. Acevedo, M., de Abajo, J., and de la Campa, J. G. (1990). Polymer, *31*: 1955.

318. Laita, H., Boufi, S., and Gandini, A. (1997). *Eur. Polym. J.*, *33*: 1203.

319. Harwthorne, D. G., Hodgkin, J. H., Jackson, M. B., and Morton, T. C. (1994). *High Perform. Polym.*, *6*: 249.

320. Glousse, C., and Gandini, A. (1999). *Polym. Int.*, *48*: 723.

321. Tesoro, G. C., and Sastri, V. R. (1986). *Ind. Eng. Chem. Prod. Res. Dev.*, *25*: 444.

322. Hao, J., Wang, W., Jiang, B., Cai, X., and Jiang, L. (1999). *Polym. Int.*, *48*: 235.

323. Oppolzer, W. (1978). *Synthesis, 11*: 793.
324. Tan, L. S., Soloski, E. J., and Arnold, F. E. (1986). *Polym. Prepr., 28(2)*: 280.
325. Chuah, H. H., Tan, L. S., and Arnold, F. E. (1989). *Polym. Eng. Sci., 29(2)*: 107.
326. Tan, L. S., and Arnold, F. E. (1990). U.S. Patent 4916235; C.A. 113.173382.
327. Alhakimi, G., Klemm, E., and Goerls, H. (1995). *J. Polym. Sci. Part A, Polym. Chem., 33*: 1133.
328. Kootner, N., and Klemm, E. (1998). *Polym. Bull., 41*: 153.
329. Lubowitz, H. R. (1970). U.S. Pat. 3528950.
330. Lubowitz, H. R. (1971). *Preprints Am. Chem. Soc. Div. Org. Coat. Plast. Chem., 31(1)*: 561.
331. Serafini, T. T., Delvigs, P., and Lithsey, G. R. (1972). *J. Appl. Polym. Sci., 16*: 905.
332. Serafini, T. T. (1973). U.S. Patent 3745149.
333. Russell, J. D., and Kardos, J. L. (1996). *41st Int. SAMPE Symp.*, 120.
334. Hu, A. J., Hao, J. M., He, T., and Yang, S. Y. (1999). *Macromolecules, 32*: 8046.
335. Vannucci, R. D. (1987). *SAMPE Quart., 19(1)*: 31.
336. Hilaire, B., and Verdu, J. (1991). Polyimides and other high temperature polymers. In Proceedings of the 2nd European Technical Symposium on Polyimides and High Temperature Polymers, STEPI 2 (Abadie, M. J. M., and Sillion, B., eds.), Elsevier, Amsterdam, p. 309.
337. Serafini, T. T., and Delvigs, P. (1973). *Appl. Polym. Symp., 22*: 89.
338. Wong, A. C., Garroway, A. N., and Ritchey, W. M. (1981). *Macromolecules, 14*: 832.
339. Hay, I. N., Boyle, I. D., Parker, S. F., and Wilson, D. (1989). *Polymer, 30*: 1032.
340. Alam, S., Kandpal, L. D., and Varma, I. K. (1993). *J. Macromol. Sci. Rev. Macromol. Chem. Phys., C33*: 291.
341. See Ref. 276, p. 147.
342. Sabourin, E. T. (1979). *Symposium on Synthesis and Chemistry of Acetylene Compounds. Div. Petro. Chem., Am. Chem. Soc. Jap.*, Honolulu, p. 233.
343. Landis, A. L., and Naselov, A. B. (1982). *Natl. Sampe Tech. Conf., 14*: 236.
344. Landis A. L. (1984). U.S. Patent 4485231.
345. Landis, A. L. (1983). *Report AFWAL-TR-83-4079*, C. A. (1984), *101*: 1.315.605.
346. Kovar, R. F. (1977). *J. Polym. Sci. Polym. Chem. Ed., 15*: 1081.
347. Paul, C. W., Schultz, R. A., and Fenelli, S. P. (1992). U.S. Pat. 5138028; C.A. *118*: 103616.
348. Paul, C. W., Schultz, R. A., and Fenelli, S. P. (1991). Eur. Pat. 443352.
349. Smith, J. G., Connell, J. W., and Hergenrother, P. M. (1998). *43 Int. SAMPE Symp. 93.*

10

Poly(vinyl aldehyde)s, Poly(vinyl ketone)s, and Phosphorus-Containing Vinyl Polymers

Oskar Nuyken
Technische Universität München, Garching, Germany

I. POLY(ACROLEIN)

(This section was prepared by O. Nuyken, T. Pöhlmann, R. Vogel and U. Anders.)

Acrolein (propenal, acrylaldehyde) is the simplest unsaturated aldehyde, a colorless and volatile liquid with high toxicity and lachrymal irritability [1,2]. The first synthesis from glycerol and from fats by pyrolytic decomposition was described by Redtenbacher in 1848 [3]. Among its typical reactions he recognized that upon standing the fluid acrolein is spontaneously converted to a white, solid, infusible, and insoluble product he called *disacryl*. Later this substance has been proven to be the result of a spontaneous polymerization [4–8]. But it was not before the early 1940s that the career of acrolein as a 'key compound' in organic chemistry began [9,10]. It is mainly used in the production of D,L-methionine and acrylic acid. In polymer chemistry, however, none of the acrolein homopolymers has until now achieved technical significance, although the monomer is difunctional and highly reactive, and the polymers are susceptible to modification reactions [11–13].

A. Manufacture of the Monomer

The oldest method for the preparation of acrolein, the acid-catalyzed thermolysis of glycerol (dehydration) at about 190 °C, is still used today to obtain acrolein on a laboratory scale [3]:

$$\text{(1)}$$

By support of $KHSO_4$ the yield can be enhanced up to 50% [14]. Further possibilities are the reaction of gaseous propene with a suspension of $HgSO_4$ in aqueous sulfuric acid [15]:

$$\text{(2)}$$

or the pyrolytic cleavage of 2,3-dihydropyrane [16,17]:

$$\text{(3)}$$

The first efficient and profitable manufacturing process for acrolein was established by Degussa AG, Germany, in 1942 [8,18–20]. It depends on the gas-phase condensation (addition and dehydration) of formaldehyde with acetaldehyde at 300 to 320 °C. In the presence of alkaline silica gel catalysts yields as high as 82% were achieved.

$$\text{(4)}$$

In 1945, at the same time that the Shell Company commercialized the pyrolysis of diallyl ether [21], acrolein production began.

$$\text{(5)}$$

With the supply of large amounts of propene in the 1950s the search began to find a system for its direct oxidation with molecular oxygen to yield acrolein. Attempts with cuprous oxide marked the beginning of the technical development of alkene oxidation in the gas phase by metal oxide catalysts [22]. But this system showed weak points in the conversion (20%) [23,24] and in the selectivity, with the consequence that most of the propene added had to be recycled and many side products had to be removed. The development and introduction of the bismuth molybdate/bismuth phosphomolybdate system (Sohio, 1957) as a catalyst [25–27] and the following application for propene oxidation opened the door to problem control. Specifically, for the system $BiPMo_{12}O_{52}$ catalyst on a SiO_2 support, a reasonable selectivity (maximum 72%) could be observed. However, the propene conversion (57%) was still low. By a further development toward modern multicompound metal oxide catalysts [28] the propene conversion could be raised from 90 to 98% with a maximum yield of 80 to 90%. The main side product (*ca.* 5 to 10%) is acrylic acid, which can be removed by distillation.

Examples of catalysts are:

FeMoBiCoNiP oxide [29] (Nippon Kayaka), FeMoBiCoNiPK oxide [30] (Nippon Kayaka), FeMoBiCoNiPSm oxide [31,32] (Degussa), MoBiFeCoWKSNaLi oxide [33] (Nippon Shokubai), MoBiFeP oxide [34] (Farbenwerke Hoechst).

Common conditions for a good performance are:

300 to 400 °C reaction temperature, 1.5 to 3.5 s contact time, 5 to 8 vol% propene concentration, 150 to 250 kPa inlet pressure, 1:10 to 20:1% molar ratio propene/air/gas passed over a solid catalyst of suitable shape.

B. Radical Polymerization

Acrolein, a member of the family of the polymerizable 2-alkenales and 2-alkenones, is provided with an extraordinary tendency for polymerization. Therefore, it may only be stored in the presence of a stabilizer (e.g., hydrochinone) in the absence of light, air, and moisture because of spontaneous polymerization. Even small amounts of initiator have the ability to force acrolein polymerization radically, anionically, or cationically, partly in an explosive manner. According to the existing reaction conditions and the catalysts used, it is possible to attain polymers of completely different shapes with characteristic features [9,13,35].

Radical polymerization prinicipally proceeds across the vinyl function [1,2-addition; Scheme (6a)], whereas ionic polymerization yields products mainly by an addition at the carbonyl group [3,4-addition; Scheme (6b)]. However, the third possibility, 1,4-addition across the α,β and C,O double bond, is a subordinate process [Scheme (6c)] [12,36,37].

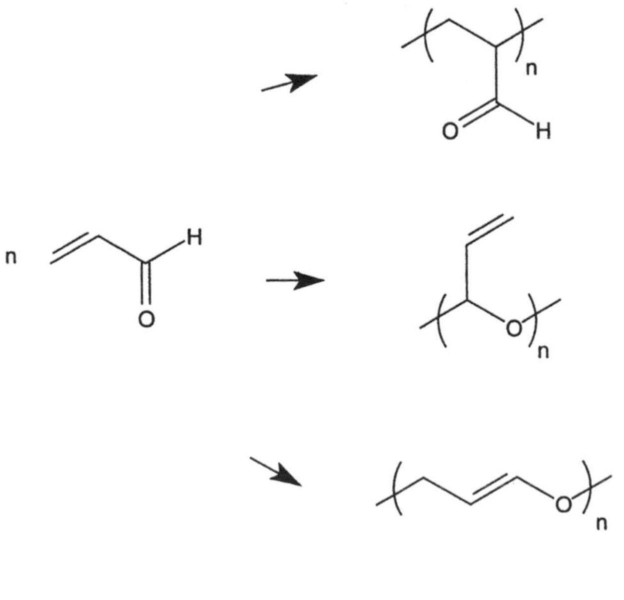

(6a–c)

Because of the polymerization across one of the two double bonds in acrolein polymers, the corresponding function remains pendant at the polymer backbone and is accessible to derivation reactions or for analytical purposes [9,37].

Radical polymerization occurs exclusively across the vinyl function. The remaining pendant formyl groups form hydrates and acetales without effort by intra- and inter-molecular condensation. The following structure elements are able to arise, including the characteristic tetrahydropyrane rings [38–40]:

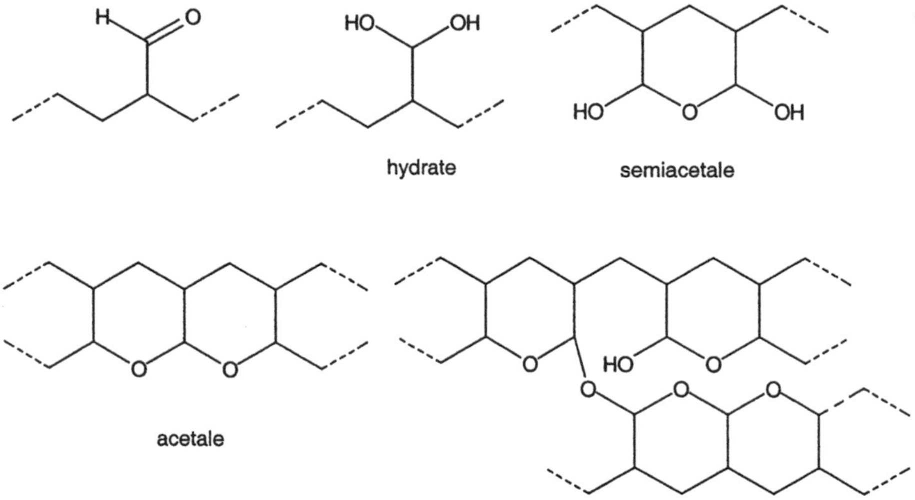

hydrate   semiacetale

acetale

$$\tag{7}$$

Due to numerous chain cross-linkings by actetal groups, radically manufactured acrolein polymers are insoluble in water and in organic solvents. They decompose above 200 °C without fusing. The polymerization itself is carried out in bulk, in aqueous solution, and in organic solvents. The Polymer precipitates from the solution and can be removed by filtration [11]. To start the polymerization the following initiators are used: inorganic peroxides [41], organic peroxides [42,43], azo compounds [42,43], redox initiators [43–45], γ-rays and others [46–49].

1. Polymerization in Bulk

The first spontaneous curing of acrolein observed was also the first polymerization in bulk [3]. Later, this observation was examined more closely [4–6,50–52]. Furthermore, a slow light- or γ-ray-initiated polymerization is possible, yielding highly cross-linked glassy products [46,53,54]. By means of AIBN or peroxides as initiators an explosive course of the reaction is observed that causes problems in the carriage of the reaction heat [42,55]. Therefore, working with only small amounts is recommended.

2. Precipitation Polymerization

The heat problem does not occur during polymerizations in aqueous solution. At 20 °C acrolein is soluble to 21.4% in water, whereas the polymer precipitates from the solution at molecular weights above 50,000 g/mol. The polymerization is started with water-soluble initiators or redox systems. In the case of redox initiators, H_2O_2, $S_2O_8^{2-}$, $P_2O_4^{4-}$, and organic peroxides and hyperoxides serve as oxidizing agents. Typical reducing agents are

Ag(I), Fe(II), and Tl(III) compounds, Na_2SO_3, $NaNO_2$, and polyacrolein hydroxysulfonic acid [13,41,42,56,57]. It is favorable to add the reducing agents to the aqueous solution of the oxidizing agent and acrolein.

3. Polymerization in Emulsion

A very favorable way to obtain acrolein polymers having molar masses of some 100,000 g/mol is by emulsion polymerization [43,58–62]. In oil–water emulsions the water-soluble addition compounds of sulfuric acid (respectively, SO_2) and polyacrolein are used as very suitable emulsifiers to produce stable polymer dispersions. The emulsion polymerization is started by water-soluble redox initiators. The acrolein polymers containing adsorbed or chemical bond SO_2 serve as reducing agents. Together with air in combination with oxygen donors [e.g., $Fe(NO_3)_3 \cdot 9H_2O$, H_2O_2, $K_2S_2O_8$], a powerful redox system is designed [63–65]. Further examples are the systems $K_2S_2O_8/AgNO_3$ [60,61], $K_2S_2O_8/(NH_4)_2SO_4$–Fe(II) compounds, and $K_2S_2O_8/Na_2SO_3$ [55]. Other soluble polymers, such as gelatine, PVA, or methyl cellulose, combined with sulfuric acid or SO_2, also accomplish the double function of emulsifier and reducing agent [63,64]. Polymerization in the inverse emulsion (water–oil) has also been described [66,67]. Aliphatic and aromatic hydrocarbons make up the continuous phase, and acrolein exists in the aqueous phase.

4. Polymerization in Solution

The monomer is soluble in numerous solvents; however, the polymer precipitates from most of these solvents at about 15% conversion during radical polymerization. Molecular weights up to 100,000 g/mol and aldehyde contents above 65% can be achieved when the polymerization is carried out in polar solvents such as DMF, γ-butyrolactone, or pyridine by means of hydroperoxides and nitrous acid derivatives as redox catalysts [68]. Deviations from this behavior are observed if DMF is used as solvent and the polymerization is initiated by AIBN. A microgel is formed here; after 16% conversion the clear reaction solution turns into a transparent gel [69]. Polymerization in the presence of methanol initiated by means of azo compounds or peroxides does yield soluble poly(acrolein), presumably because of the polymer's molecular weight [70].

5. Radiation-Induced Polymerization

Bulk polymerization of acrolein under the influence of γ-rays yields a highly cross-linked glassy polymer, which is completely insoluble in organic solvents and also in aqueous sulfuric acid. Gamma-ray-induced polymerization in solution, especially in water, is much faster than in bulk [46–48,54]. Investigations of radiation-induced polymerizations in bulk or in aqueous solution by means of a ^{60}Co source yielded microspheres of different size containing reactive formyl functions [49,71,72].

6. Solubilization of the Polymers

To solubilize the products of radically induced acrolein polymerization, the following procedures are used.

Disproportionation of the aldehyde and acetale groups pending on the polymer backbone by means of sodium hydoxide solution (Cannizzaro reaction) [73–75]:

$$\xrightarrow{\text{OH}^-} \qquad (8)$$

Formation of water-soluble addition products by the action of sodium bisulfite and aqueous sulfurous acid [76–78]:

$$\xrightarrow[\text{SO}_2]{\text{H}_2\text{O}} \qquad (9)$$

By dialysis of the primary addition product, the following equilibrium can be forced to the right side yielding water-soluble, SO_2-free acrolein hydrate [79]:

$$\rightleftharpoons \qquad + \quad 2\,SO_2 \qquad (10)$$

C. Ionic Polymerization

1. Anionically

In the presence of alkaline metal hydroxides or carbonates, acrolein is converted into oily resinous products [5,6]. This reaction proceeds in a vigorous-to-explosive way by means of strong bases and amines [13]. In the 1950s this techniques was used to produce polymers by anionic polymerization in solution under well-defined conditions. In THF, DMF, toluene, glyme, and other solvents, products with melting and softening points between 90 and 200 °C were obtained which were soluble in organic solvents but insoluble in sulfurous acid [80,81]. Structural analysis of the polymer's repetition units gave rise to the assumption that

chain growth occurs mainly across the carbonyl group (3,4-polymerization,

structure units) [37,81]. Furthermore, there is addition across the vinyl function (1,2-addition) and across both functional groups (1,4-addition) [82,83]. The latter takes place only on a very small scale. Consequently, copolymers are formed that contain the following

structure elements: , partly in block arrangement ($n + m = 1$; $m = 0.7$ to 0.8) [84].

In a water-free medium chain growth polymerization can be initiated by numerous metal-organic or basic compounds, such as trityl sodium [81], butyl lithium [80,81], naphthyl sodium [80,81], benzophenone potassium [81], sodium methoxide [80,81], lithium organocuprates [85] and rhodium(I) complexes [86] or ammonia [87], tert-phosphines [80,88], aliphatic amines [89], cyclic amines [90], and aromatic amines (pyridine [91,92], imidazole [93,94], N-ethylimidazole [95]). The reaction temperatures range from $-60\,°C$ to $+25\,°C$, whereby the reaction rates as well as the properties of the products (composition) are influenced. Higher temperatures lead to products having a higher content of aldehyde side groups and a lower content of vinyl side groups. Weaker bases and solvents with lower polarity also favor the formation of polymers with aldehyde side groups [81]. Acrolein can be polymerized by alkali cyanides in polar solvents such as THF or DMF [96,97]. At reaction temperatures below $-10\,°C$, only 3,4-connected products were obtained.

2. Cationically

Few sources describe the cationic acrolein polymerization in bulk or in homogeneous solution [7,12,42,80,98]. Using trifluoroborane-diethyl ether or triethyloxonium-tetra-fluoroborate as initiators carbonyl and vinyl group containing polymers are obtained at reaction temperatures ranging from $-80\,°C$ to room temperature. The carbonyl content of these polymers varies from 9 to 15 mol%. For this polymerization polar solvents such as nitromethane or nitrobenzene are favorable. When the polymerization is stopped at low conversion soluble products (cf. in 1,4-dioxane, $CHCl_3$, THF, pyridine) are obtained. Adding *tert*-amines during the last step of the polymerizations results in the highest content of carbonyl polymerization [9]. At higher conversions or at prolonged storage the products become cross-linked and insoluble. All these products soften between 80 and $120\,°C$.

D. Copolymerizations

1. Radical Copolymerization

For a list of various characteristics of radical copolymerization, see Table 1.

2. Graft Copolymerization

Acrolein can be grafted onto poly(methyl methacrylate), cellulose, and poly(ethylene) by γ- or electron-beam radiation.

1. A foil of poly(methyl methacrylate) was swollen in aqueous or methanolic acrolein solution and then exposed to γ-radiation of a ^{60}Co source. Graft polymers with aldehyde groups were formed, which show the specific aldehyde-type reactions [46].
2. Cellulose dispersed in an acrolein solution (solvent: water, ethanol, acetone, ether, or benzene) was treated with γ-radiation of a ^{60}Co source at 40 to $43\,°C$. In addition to the formation of a network of cellulose, homopolymerization of acrolein was observed. Homopolymerization of acrolein could be avoided if cellulose was treated with gaseous acrolein at a pressure of 10^{-3} torr before radiation [106].
3. Acrolein was grafted onto poly(ethylene) which was exposed to electron beams. The remaining aldehyde groups could be transformed into hydrazone, oxime, and oxyacid units [107].

3. Oxidative Copolymerization

Acrolein and acrylic acid were copolymerized in aqueous H_2O_2 solution at 60 to $90\,°C$ to form poly(aldehyde carbon acids). The Cannizzaro reaction took place if an aqueous

Table 1 Parameters of the radical copolymerization.

Monomer M_2	r_1	r_2	Temp (°C)	Initiator	Solvent	Refs.
Acrylic acid	0.50 ± 0.30	1.15 ± 0.20	54	AIBN	Water[a]	[99]
	2.40 ± 0.50	0.05 ± 0.05	75	AIBN	Water[b]	[99]
	6.70 ± 3.00	0.00	80	AIBN	Water[c]	[99]
Acryl amide	2.0 ± 0.05	0.76 ± 0.02	20	$K_2S_2O_8 + AgNO_3$	Water	[100]
	1.69 ± 0.1	0.21 ± 0.02	50	AIBN	DMF	[101]
Acryl nitrile	1.09 ± 0.05	0.77 ± 0.1	20	$K_2S_2O_8 + AgNO_3$; $H_2O_2 + NaNO_2$	Water	[100]
	1.60 ± 0.04	0.52 ± 0.02	50	AIBN	DMF	[101]
Butyl acrylate	1.6	0.6	50	$K_2S_2O_8$	Water	[102]
	1.6	0.6	60	AIBN	Dioxane	[103]
	1.2	0.6	60	AIBN		
Ethyl acrylate	1.6	0.6	50	$K_2S_2O_8$	Water	[102]
				AIBN	Dioxane	[103]
Maleic hydrazide	16	0	60	AIBN	DMSO	[104]
Maleimide	3.20	0.12	60	AIBN	DMSO	[104]
Methacryl nitrile	0.72 ± 0.06	1.20 ± 0.08	50	AIBN	Dioxane	[101]
Methyl acrylate	~ 0	7.7 ± 0.2	20	$K_2S_2O_8 + AgNO_3$	Water	[100,101]
	1.6	0.6	50	$K_2S_2O_8$	Water	[102]
	1.2	0.6	60	AIBN	Dioxane	[103]
Methyl methacrylate	0.5	1.0	50	$K_2S_2O_8$	Water	[102]
	0.8	1.2	60	AIBN	Dioxane	[103]
Styrene	0.034	0.32	50	$K_2S_2O_8$	Water	[102]
	0.25	0.25	60	AIBN	Dioxane	[103]
	0.22	0.33	50	AIBN	Dioxane	[105]
Vinyl acetate	3.33 ± 0.1	0.1 ± 0.05	20	$K_2S_2O_8 + AgNO_3$	Water	[100]
2-Vinyl pyridine	~ 4	~ 0	50	AIBN	DMF	[101]

[a] pH 3.
[b] pH 5.
[c] pH 7.

solution or suspension of this polymer material was treated with aqueous NaOH. The aldehyde functions disproportionated into carboxylate and alcohol groups to form poly(hydroxy carboxylates) [108,109].

4. Anionic Copolymerization

Acrolein was anionically copolymerized with acryl amide and methyl vinyl ketone ($r_1 = 2.02$, $r_2 = 0.06$) at 0 °C in THF with imidazole as an initiator [110]. Copolymerizations of acrolein with various aldehydes (e.g., acetaldehyde and benzaldehyde) were carried out in THF at -30 °C with NaCN as initiator [111].

5. Block Copolymers

1. Living oligomers of butadiene were functionalized by the addition of acrolein or ethylene oxide and then treated with acrolein to yield block copolymers. The homopolymerization of acrolein could not be avoided [112,113].
2. Short poly(acrolein) blocks were formed, if α,ω-disodium oligobutadiene (initiated with sodium naphthalene in THF at -40 °C) was treated with

acrolein. Homopolymerization of acrolein did not take place. The acrolein units could be cross-linked after an UV cure to form a poly(acrolein) network that can be used as photo-polymer layers to prepare negative printing plates [114].

6. Graft Copolymerization

Acrolein could be grafted onto imidazole-containing polymers [poly4(5)-vinylimidazole) or copolymers of 4(5)-vinylimidazole with acryl amide, styrene, 1-vinyl-2-pyrrolidone, 4-vinylpyridine, acrylates, and methyl vinyl ketone] in ethanol or an ethanol–water mixture at 0 °C under nitrogen [115–117].

7. Cationic Copolymerization

Cationic copolymerization of acrolein with styrene took place in methylene chloride, toluene, and 1-nitropropane with bortrifluoride-etherate as a catalyst at different temperatures $(-78\,°C$ to $0\,°C)$ [118].

E. Modification Reactions of Poly(acrolein)

1. Radically Polymerized Acrolein (Redox Poly(acrolein))

Redox poly(acrolein) is one of the most reactive polymers and susceptible to a number of modification reactions that lead to high conversions under mild conditions [9,11,37]. Containing one pendant aldehyde function per repetition unit – either free or masked – poly(acrolein) possesses functional groups and can react basically in the following ways [37,39,40]:

(a) As a Polymeric Monoaldehyde (i.e., after the pyran rings' cleavage, the aldehyde functions developed react independent of each other). Examples are oxidations [119] (e.g., with peracetic acid) and reductions [120,121] [e.g., to poly(allylalcohol)] of the C,O group, or reactions with alcohols to acetales [122], amines to imines [39,123], hydroxylamine to oximes [124], or phenylhydrazine to hydrozones [39,123]. The latter serve for the quantitative determination of the aldehyde group content.

(b) In Condensation Reactions. Representative reactions are aldol condensation [125,126] with formaldehyde taking place at the polymers' α-carbons, and Knoevenagel condensation [40,127] with C,H acidic compounds (e.g., malodinitrile).

$$\text{(11)}$$

$$\text{(12)}$$

(c) As a Polymeric Dicarbonyl Compounds. For reasons of their masking in the form of pyran rings, reactions are favored in which two adjacent carbonyl functions are involved. The intramolecular disproportionation reaction by Cannizzaro serves as a well-known example. Under the action of alkali and due to the proximity and reactivity of the aldehyde groups, polymers with pendant hydoxymethyl (CH_2OH) and carboxylate (COO^-) groups are formed [73–75].

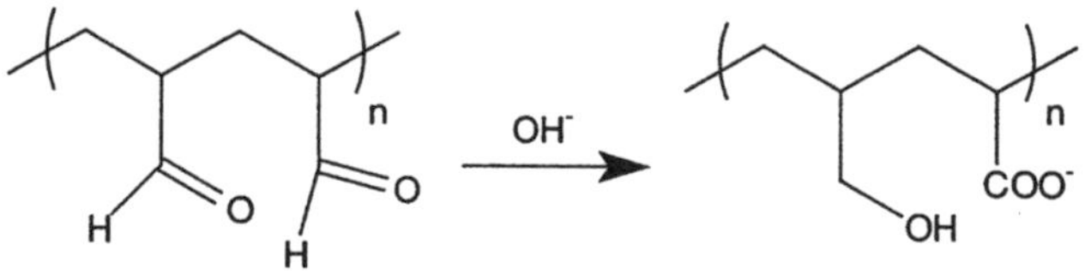

$$\text{(13)}$$

(d) As a Polymeric Semiacetate. The semiacetale hydroxy groups are able to perform characteristic reactions without cleaving the pyran ring structure (e.g., thiol addition) [128].

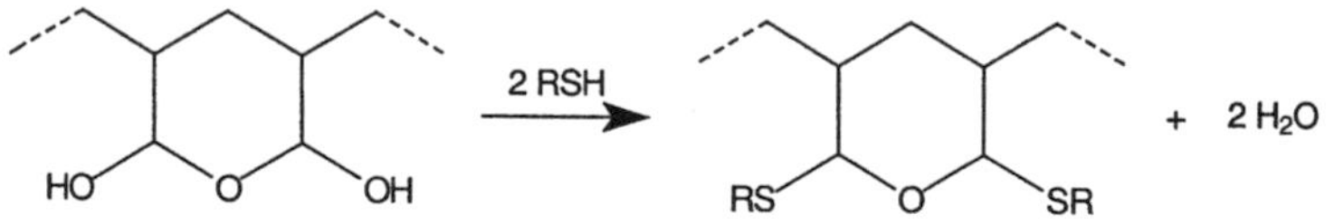

$$\xrightarrow{2\,RSH} \quad + \quad 2\,H_2O \qquad \text{(14)}$$

Because of the insolubility of redox poly(acrolein) [129], modification reactions must always start in heterogeneous systems and lead to soluble products gradually. The already presented water-soluble products of the reaction between poly(acrolein) and Na_2SO_3 or H_2SO_3 [76–78] are still better precursors for modification reactions than is native redox poly(acrolein). They permit a reaction performance in homogeneous media.

Apart from conversions with low-molecular-weight compounds, soluble and insoluble redox poly(acrolein) can react with high-molecular-weight substrates. Connections with the following in vivo and in vitro occurring polymers are good examples of that behavior: poly(vinyl alcohol) [130,131], cellulose [130–132], proteins [130,131,133,134] (e.g., collagen, gelatine [135]), enzymes [130,136,137], lectins [138,139], erythrocytes [140–142] and lymphocytes [140], leukemia cells [140,142], antibodies [133,143,144], and metal complexing agents [145].

2. Anionically Polymerized Acrolein

Due to the high portion of pendant vinyl groups, the following reactions of this polymer material are possible:

1. Co- and graft polymerization with vinyl and acryl monomers in the form of a two-step copolymerization process [146].
2. Autoxidation of the double bond and a subsequent connection with the polymers' remaining aldehyde functions [81].
3. Light-induced cross-linking across the vinyl group [147].

F. Applications and Economic Aspects

The statement that acrolein homopolymers do not find technical applications does not hold for copolymers. The already introduced poly(aldehyde carbon acids) (trade name POC, Degussa, Germany) are strong complexing agents [108,109,148–151]. They are able to form complexes with cations such as Na^+, Mg^{2+}, Ca^{2+}, Fe^{3+}, Mn^{2+}, Cu^{2+} [152] (also reversible), with gaseous ammonia [11], with peroxides for stabilization purposes, or with amino acids. The material is used in water treatment as a water softener in detergents or rinsing agents, and as a supported sequestering agent showing rising complexing activity with increasing aldehyde content. The ability to bind amino acids is utilized in the determination of the C-terminated end in proteins [153].

On a laboratory scale, acrolein homo- and copolymers are tested as polymeric reagents, polymeric complexing agents, and polymeric carriers. Poly(acrolein) microspheres can easily be bound to antibodies, proteins, and drugs containing primary amino groups in a single step under physiological pH [71,72,139–145,154,155]. Aldehyde groups react under mild conditions with primary amino groups forming the corresponding imino (Schiff base) linkage. Reaction with sodium cyanoborhydride as reducing agent forms a stable $-CH_2-NH-$ linkage [134].

(15)

In this way poly(acrolein) particles may play an important role as immunoreagents for biological research.

II. POLYMERS OF CROTONALDEHYDE AND METHACROLEIN

(This section was prepared by O. Nuyken, R. Bayer and J. Bayer.)

A. Crotonaldehyde

1. Properties and Structure

Crotonaldehyde (2-butenal, crotonic aldehyde, β-methacrolein) is a colorless, strong lacrimatory, and toxic liquid. The mutagene potential of crotonaldehyde and its role in cancerogenese has been investigated [156–159]. It has a melting point of $-69\,°C$ and a boiling point of $102.2\,°C$. Crotonaldehyde and water form an azeotrop containing 24.8% water and boiling at $84\,°C$. Other physical properties are given in Refs [160] and [161] and the literature cited therein. Technical crotonaldehyde consists of two isomers, where *trans*-crotonaldehyde has an occurance of more than 95%.

$$\text{(16)}$$

cis   *trans*

The very reactive crotonaldehyde is easily oxidized by contact with air [162–169]. This causes resinifying and darkening. Avoiding the formation of peroxides and iron salts, it can be stored without adding inhibitor. The usual inhibitors are water and hydroquinone. In contact with strong acids, crotonaldehyde forms a dimer that can be separated into *cis*- and *trans*-isomers [170].

2. Synthesis

The general method of producing crotonaldehyde is the aldol condensation of acetaldehyde, followed by dehydration and rectification respectively extraction [160,171].

$$\text{(17)}$$

More details and other synthesis routes are given in Ref. [160].

3. Anionic Polymerization

Anionic polymerization is the best investigated and the most used method to polymerize crotonaldehyde. A great number of publications about the anionic polymerization of crotonaldehyde deals with a method that was first used by Koral [172,173]. The initiation occurs through tertiary phosphines (see Table 2). Koral proposes the following initiation step:

$$\text{(18)}$$

The fact that the rate of polymerization of crotonaldehyde increases with the dielectric constant of the solvent is evidence for an ionic mechanism of the polymerization. An increase in dielectric constant of the medium will favor energetically an increase in the rate of initiation and the stabilization of the zwitterion [172]. It should not influence the rate of propagation and termination.

Radical anions generated from metal–organic compounds are another group of initiators. Based on polarographic investigations and Hückel calculations it was shown that the polymerization of crotonaldehyde with benzophenone radical anions proceeds via formation of a complex radical anion of crotonaldehyde [174]. This complex accepts a

Table 2 Common initiators for anionic polymerization of crotonaldehyde.

Name	Max. molar mass reached	Refs.
2,4-Dimethylbenzophenone		[175]
4-Methylbenzophenone		[175]
1-Benzoylnaphthalene		[175]
Benzophenone		[175]
4-Benzoyl bichloride		[175]
Xanthone		[175]
Potassium diphenylketyle	10,000	[185]
Potassium dihydronaphthalide		[185]
Potassium graphite inclusion compounds		[185]
t-Phosphines (Pr_3P, Bu_3P, $PhEt_2P$, Ph_3P)	3,270	[172,173,176,177]
NaCN	350	[176]
Et_3N	560–830	[178]
Sodium hydroxide	1,000–10,000	[179–183]
Sodium naphthalene, Sodium methanolate		[184]
Various inorganic salts (e.g., K_2CO_3, $NaNH_2$)		[182,183]

second electron from another benzophenone radical anion and a dianion is built that is capable to grow and to build up the polymer chain. This mechanism was corroborated by isolating 2,2′-diphenyl-3-methyl-5-hydroxytetrahydrofurane from the solution [175]. Its formation can be explained by the following mechanism:

$$(19)$$

The dianion is able to grow a polymer chain.

Numerous initiators have been reported to be used in anionic polymerization of crotonaldehyde. Some are shown in Table 2.

Varying the conditions of the polymerization (initiators, temperature, solvent, etc.), polymers with different structures can be prepared. Anionic initiators are leading to polymers containing monomer units bonded together via C–C– or C–O-linkages.

$$(20)$$

In the case of polycrotonaldehyde using sodium dihydronapthalide as initiator, both types of linkages were found [184]. An important change in the structure of the polymer chain was found by Rashkov *et al.* [185]. They compared the polymers started with homogeneous initiators (e.g., potassium ethoxide) with heterogeneous initiators (e.g., graphite inclusion initiators such as C_8K). Homogeneous initiators cause polymerization of the aldehyde groups even at low temperatures (e.g., 10 and 30 °C). Heterogeneous initiators inhibit the side reaction of the aldehyde groups so only the vinyl groups polymerize even at high temperatures and concentrations of the monomer and/or initiator. The authors assume that the propagation of the active chain predominantly proceeds on the surface of the heterogeneous initiator.

Generally it was found that at low temperatures the polymerization of aldehyde groups proceeds to a larger degree [184,185].

A different polymer structure was obtained by using tert-phosphines as initiator. Koral [172,173] found a large amount of free carbonyl groups (conjugated and unconjugated) and ether groups together with a small hydroxyl concentration and some residual unsaturation. This structure results from a vinyl-type polymerization with a simultaneous cyclization of some vicinal, pendant aldehyde groups. The following structure is proposed:

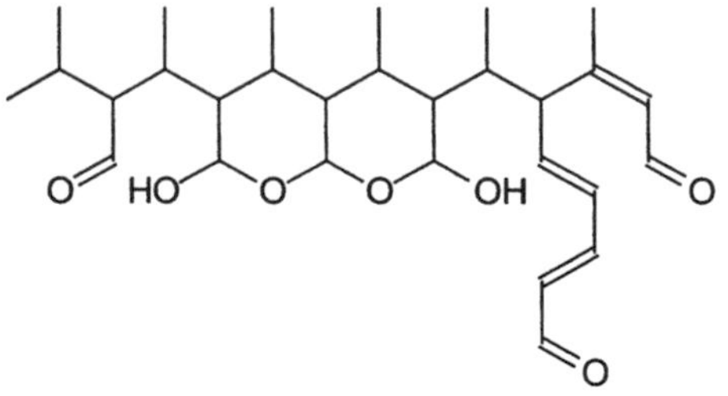

(21)

The anionic polymerization of crotonaldehyde was also carried out under high pressure with Et_3N [186]. It was found that the melting point and molar mass of the polymer increase linearly with rising pressure or temperature.

4. Cationic Polymerization and Field Polymerization

Cationic polymerization of crotonaldehyde is less important than anionic polymerization. With $(EtO)_3Al$ or $(i\text{-}PrO)_3Al$ as initiators, rather unstable polymers were obtained [187]; with H_3PO_4 and PCl_5 only oil was formed [188]. Polymerization of crotonaldehyde can also be induced by high electric fields (several $10^7 \, V/cm$) [189]. Field polymerization results in the growth of organic semiconducting micro needles with side-chain cross-linking and $P_{max} = 3$.

5. Step-growth Polymerization

The polymerization of crotonaldehyde and several amines (butylamine, ethylenediamine, triethylenetetramine, diethylenetriamine, hexamethylenediamine, aniline, melamine, and diaminodiphenylmethane respectively diaminomaleonitrile) proceeds in two steps. In the first step a Schiff-base-reaction between the aldehyde groups and the amino groups take place. In the second step the vinyl groups disappear due to step-growth-polymerization and lead to resins [190–196]. The step-growth-polymerization of crotonaldehyde and alcohols like phenols and glycols leads to resins, too. A review is given in Ref. [160].

6. Copolymerization

Crotonaldehyde acts in copolymerization (with styrene, methacrylic esters, vinyl esters, vinylcaprolactam) as a retarder, and therefore only oligomeric products can be isolated [160,197].

7. Applications

Copolymerization of crotonaldehyde with quinol forms a resin, that chelate divalent cations (e.g., Mg, Co, Fe, Cu, or Cd) [198].

Polycrotonaldehyde bearing specific ligands is used for the removal of drugs and diagnostic substances from blood, which have long half-live times in blood. The microparticles (0.1 up tp 6 µm size) agglomerate and are taken up by the mononuclear phagocytic system) [199]. Cyanoacrylate/crotonaldehyde-copolymers are used for building up dental compositions [200]. Citric acid were produced by hydrolysis of the polymeric product of the lactonization of crotonaldehyde with ketene and water [201–203].

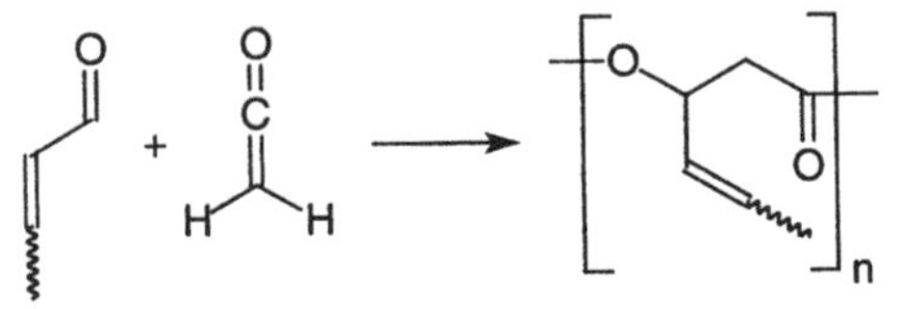

(22)

The copolymerization of pyrrole, crotonaldehyde, and a polymerizable, organic acid leads to water-based resins or coating compounds [204].

B. Methacrolein

1. Properties and Reactions

Methacrolein (2-methylpropenal, methacrylaldehyde, 2-methylacrolein, α-methylacrolein) is a colorless, sharp (stinging) smelling, flammable, highly reactive, and lacrimatory liquid with a melting point of $-81\,°C$ and a boiling point of $68.4\,°C$. Methacrolein and water form an azeotrop containing 6.7% water and boiling at $63.9\,°C$. The solubility in water is 6% at $20\,°C$. Other properties are described in Refs. [205–207].

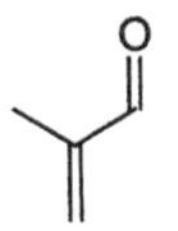

(23)

The reactions of methacrolein are analogous in many respects to those of acrolein. Dimerization of methacrolein occurs similar to the dimerization of acrolein via Diels–Alder addition, where methacrolein reacts both as a diene and a dienophile [205]. By treatment with alkali tri-, tetra-, and pentamers are formed by Michael addition [208,209]. By exposure to air, methacrolein forms peroxides and acids. The peroxide groups can be incorporated into the polymer chain [208]:

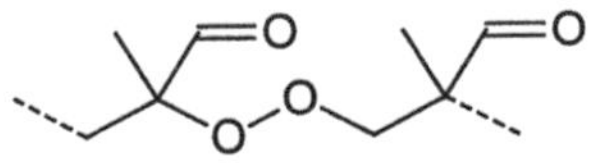

(24)

Avoiding air by storage under nitrogen and avoiding iron salts [210], no inhibitor (hydroquinone) is required.

2. Synthesis

The following methods are used to synthesize methacrolein:

1. Catalytic oxidation of isobutane with oxygen [211–216]:

$$(25)$$

2. Catalytic oxidation of isobutylene with oxygen [206,211,219–226]:

$$(26)$$

3. Catalytic oxidation of tert-butanol with oxygen [206,211,219–227]:

$$(27)$$

4. Cross-condensation of propionaldehyde and formaldehyde with catalysts in the vapor phase followed by dehydration [206,228]:

$$(28)$$

5. Catalytic oxidation of β-methallyl alcohol [229,230]:

$$(29)$$

6. Catalytic oxidation of isobutyraldehyde [231]:

$$(30)$$

For laboratory use methacrolein can also be prepared by heating of Mannich aldehydes [232].

3. Radical Polymerization

Methacrolein polymerization can be carried out in bulk, inorganic solvents, and in water. Heating methacrolein without initiator gives a brittle, yellow polymer which does not contain any free aldehyde groups. Contrary to this free aldehyde groups were found by initiation with peroxo or azo compounds [233]. The polymer was described as clear and glassy. For the radical polymerization of methacrolein in organic solvents, peroxo or azo compounds were used [79]. With ammonium peroxodisulfate in DMF, molar masses of 5,000 up to 21,000 g/mol were reached [234]. The polymerization is aqueous media can be carried out in different ways:

1. Precipitation polymerization with redox systems [235,236] or peroxo compounds [208,237] as initiators. Using peroxo compounds, the resulting molar masses are relatively low (maximally reached 30,000 g/mol) due to the monomer causing chain transfer [208]. This effect is also known from other aldehydes.
2. Suspension polymerization [238].
3. Emulsion polymerization [233,239–241]. In an extended investigation of the redox system $K_2S_2O_8$–$Na_2S_2O_5$, Andreeva et al. [240,242,243] found that polymerization in emulsion failed because the interaction of bisulfite ions with the monomer at the double bond causes a deactivation of the initiator.
    Further, the effect of proton formation during polymerization follows the equation

$$\overset{\bullet}{S}O_4^- + H_2O \longrightarrow SO_4^{2-} + H^+ + \overset{\bullet}{O}H \tag{31}$$

The decrease in pH value that results was investigated with respect to the viscosity of the polymer solution and the structure of the polymer.

4. Anionic Polymerization

Anionic polymerization of methacrolein was investigated by several authors [244–251]. Homogeneous initiators are the anion radicals of naphthalene [242,246], 2,4-dimethylbenzophenone [247], 4-methylbenzophenone [247], 1-benzoylnaphthalene [247], benzophenone [247], 4-benzoylbiphenyl [247], diphenyl ketone [249], dihydronaphthalene [249], and also BuLi [250], NaCN [245], and Bu_3P [245]. Also heterogeneous initiators such as the radical anions of graphite [251] formed from graphite inclusion compounds are used.

It has been suggested that the polymerization of methacrolein initiated by benzophenone or naphthalene anion radicals is accomplished by electron transfer [246,248]. In the case of benzophenone it was shown by polarographic methods and Hückel calculations that the complex radical anion with methacrolein (if it is formed) is unstable and dissociates into benzophenone and the radical anion of methacrolein.

The initiation with heterogeneous inclusion compounds such as C_8K, $C_{16}K$, or $C_{24}K$ is explained as follows [251]: The initiation is preceded by adsorption of the monomer on the initiator surface. The monomer molecule absorbs an electron and converts into an anion radical. The latter remains fixed on the initiator surface due to coulombic interactions with the counterion. After recombination of the radical ends, a dianion is formed that is suitable for propagation. The propagating anion ends probably remain fixed on the initiator surface.

In general four types of linkages between the monomer units are possible and found depending on the reaction conditions [249,250]:

1. Bonded via C–C-linkage:

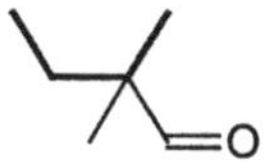

(32)

2. Bonded with C–O-linkage:

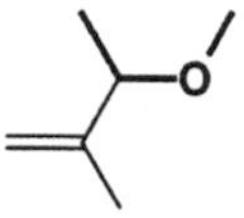

(33)

3. Tetrahydropyrane ring structure:

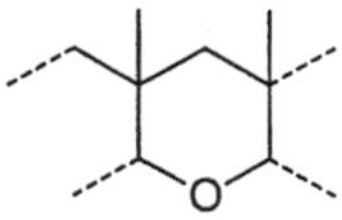

(34)

4. Lactone ring structure:

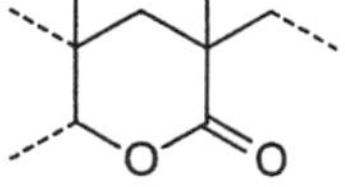

(35)

The relation of the former two is very sensitive to the polymerization conditions. At low temperatures polymerization of the aldehyde groups proceeds to a larger degree. The formation of the dianion (36) was found by Rashkov *et al.* [249]. The authors showed with Hückel calculations that the formation of tetrahydropyrane rings in the propagation reaction of methacrolein is energetically favoured. These calculations lead to the assumption that these rings are formed as a result of the interaction of dianion (36) with methacrolein molecules.

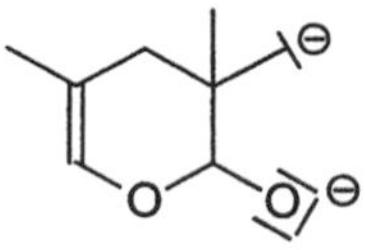

(36)

If the polymerization of methacrolein is initiated by a typical anionic initiator such as BuLi, the polymer obtained does not contain tetrahydropyrane rings [251]. When sodium dihydronaphthalide at higher temperatures and higher monomer concentrations is used the formation of the cyclic lactones (35) is observed. Also side reactions of the aldehyde group take place and the polymers yield is decreased. Koton *et al.* [250] assumed that the side reactions with aldehyde groups are due to the catalytic effect of the propagating anionic ends, consisting of an alcoholate group ...-CH–O–Mt$^+$. Here the same result as in the polymerization of crotonaldehyde is found. By initiation with a typical anionic initiator such as KOEt, side reactions with the aldehyde groups in the polymer proceed to a considerable extend even at higher temperatures. In contrast to that the aldehyde groups were not involved in side reactions if heterogeneous initiators (e.g., graphite inclusion compounds) are used.

Also methacrolein is the base for new types of monomers like 3-methyl-N-(phenylsulfonyl)-1-aza-1,3-butadiene as described in [252], which can be polymerized anionically.

5. Cationic Polymerization

Cationic initiators (BF$_3$-etherate, SnCl$_4$, or AlCl$_3$) are used to form soluble polymers with free aldehyde groups [253]. Tertiary phosphines in the presence of secondary alcohols at low temperatures are used by other authors [233].

Also an unsaturated cyclic acetal (2-isopropenyl-4-methylene-1,3-dioxolane), which is formed from methacrolein and epichlorhydrine, can be polymerized via ring-opening-cationic polymerization [254].

6. Step-growth Polymerization

Methacrolein and the conjugated amine diaminomaleic dinitrile is stepwise polymerized. The first step is a Schiff-base reaction between the aldehyde groups and the amino groups. In the next step the vinyl groups polymerize resulting a resin [195].

7. Methacrolein Copolymers

A general description to prepare copolymers is given in Ref. [207]. Methacrolein copolymers were described with the following comonomers: (1) styrene and vinyl compounds [255–262], (2) vinylidene compounds [263], (3) acrylic acid, derivatives, and substituted acroleins [255,261,264–266], and (4) derivatives of methacrylic acid [255,266].

8. Applications

Crosslinked poly(4-vinylpyridine-*co*-methacrolein) is used as permselective membrane for reverse osmosis [267].

Polymers of methacrolein (10,000 up to 500,000 g/mol) in combination with alkali metal sulfides can coagulate heavy metals from municipals wastewaters [268,269].

Derivatation of methacrolein copolymers with NaHSO$_3$ respectively Na$_2$S$_2$O$_5$ produces copolymers with sulfite groups which leads to water soluble copolymers [270–272].

Copolymers of methacrolein are used as coating material for immuno assays [273].

III. POLY(METHYL VINYL KETONE)

(This section was prepared by O. Nuyken, A. Riederer and S. Büchel.)

Vinyl ketones are an interesting class of monomers because various members of this group polymerize via a radical, anionic, and cationic mechanism. Methyl vinyl ketone (MVK) – also named 3-butene-2-one – is its best examined representative. The physical properties of poly(methyl vinyl ketone) (PMVK) depend on the polymerization conditions and the degree of polymerization. PMVK ranges from a viscous oil to a hard plastic or rubbery mass. Polymers obtained with free radical initiators are amorphous materials with low softening points (about 40 to 80 °C) and poor thermal and chemical stability [274,275]. The molecular weights are relatively low because of the lability of the protons in the α-position to the carbonyl groups.

The polymers are soluble in the monomer and in numerous organic solvents, such as acetone, methyl ethyl ketone, tetrahydrofurane, dioxane, pyridine, or chloroform. They are insoluble in aliphatic and aromatic hydrocarbons, carbon tetrachloride, ethyl ether, and water. The reactivity of the carbonyl groups in homo- and copolymers obtained from MVK allows many modification reactions. PMVK itself has not found commercial applications because of its instability, but great efforts have been made in synthesizing copolymers with a wide range of physical properties: for example, the preparation of oil- and solvent-resistant rubbers with butadiene to replace styrene-butadiene rubbers or the preparation of crosslinked resins by treating MVK-butyl acrylate copolymer with hydrazine or using a divinyl compound as comonomer [275]. Crystalline products are obtained by anionic polymerization with some organometallic compounds. They are soluble in formic acid and show melting points of 140 to 160 °C [276].

A. Monomer Synthesis

MVK, or systematically 3-butene-2-one, was first synthesized in 1906 by heating β-chloroethylketone with diethylaniline [277]:

$$CH_3\!-\!\underset{\underset{O}{\|}}{C}\!-\!CH\!=\!\!CH_2 \tag{37}$$

Alternative synthetic routes are described below.

1. Hydration of vinylacetylene in the presence of mercury salts [278]:

$$HC\!\equiv\!C\!-\!CH\!=\!\!CH_2 \;\; + \;\; H_2O \;\; \xrightarrow{\;Hg(II)\;} \;\; H_3C\!-\!\underset{\underset{O}{\|}}{C}\!-\!CH\!=\!\!CH_2 \tag{38}$$

2. Oxidation of 1-butene (formation of an olefin–mercury–salt complex and its decomposition with acid [278]:

$$H_2C\!=\!CH\!-\!CH_2\!-\!CH_3 \;\; + \;\; O_2 \;\; \xrightarrow[\text{2.H}^+]{\text{1.Hg(II)}} \;\; H_3C\!-\!\underset{\underset{O}{\|}}{C}\!-\!CH\!=\!\!CH_2 \;\; + \;\; H_2O$$

$$\tag{39}$$

3. Thermal dehydration of β-ketoalcohol catalysed by weak acids [279]:

$$HO—CH_2—CH_2—\underset{\underset{O}{\|}}{C}—CH_3 \xrightarrow{[H^+]} H_3C—\underset{\underset{O}{\|}}{C}—CH=\!\!=CH_2 \; + \; H_2O \quad (40)$$

4. Reaction of acetone with formaldehyde in the gas phase passing over lead zeolite or alkali metal hydroxide-impregnated silica gel at 200 to 300 °C [280]:

$$H_3C—\underset{\underset{O}{\|}}{C}—CH_3 + HCHO \longrightarrow HO—CH_2—CH_2—\underset{\underset{O}{\|}}{C}—CH_3 \xrightarrow{[OH]} H_3C—\underset{\underset{O}{\|}}{C}—CH=\!\!=CH_2$$

$$(41)$$

5. Mannich reaction of acetone, formaldehyde, and diethylamine hydrochloride followed by pyrolysis at 150 to 210 °C under reduced pressure [281]:

$$H_3C—\underset{\underset{O}{\|}}{C}—CH_3 \; + \; HCHO + (C_2H_5)_2NH \bullet HCl \longrightarrow H_3C—\underset{\underset{O}{\|}}{C}—CH_2—CH_2—N(C_2H_5)_2 \bullet HCl$$

$$\xrightarrow[\text{80–100 torr}]{\text{150–200°C}} H_2C=\!\!=CH—\underset{\underset{O}{\|}}{C}—CH_3$$

$$(42)$$

The Mannich reaction is generally the method of choice. Some physical properties of MVK are summarized in Table 3.

B. Radical Polymerization

The radical polymerization of MVK is initiated by almost any common free-radical initiator in bulk, solution, emulsion, or suspension. Marvel and Levesque [283] have polymerized MVK in bulk with 0.5% benzoyl peroxide as intiator at 50 °C and found a 1,5-diketone structure, indicating a head-to-tail arrangement of the soluble polymer.

$$\left[\!\!\begin{array}{c} CH_2—\underset{\underset{\underset{CH_3}{|}}{\underset{\|}{C=O}}}{CH}—CH_2—\underset{\underset{\underset{CH_3}{|}}{\underset{\|}{C=O}}}{CH} \end{array}\!\!\right]_n \quad (43)$$

For producing polymers with good color stability, azobisisobutyronitrile (AIBN) is favored. All other catalysts leave residues or degrade the polymer during the

Table 3 Physical properties of MVK.

Property	Value	Refs.
Molecular weight (g/mol)	70.09	
Boiling point (°C)	81.4^{760T}	[275]
	32^{120T}	[282]
	$81.4^{1013\,mbar}$	[278]
	81.4^{750T}	[274]
Refractive index		
n_D^{20}	1.4086	[275]
	1.4084	[274]
n_D^{25}	1.408	[278]
Density, d_4^{20} (g/mL)	0.8636	[275]
	0.8393	[274]
	0.842	[278]
Solubility	Water, organic solvents	[275,278]
Vapor pressure (mbar)	130	[274]
Point of ignition (°C)	-13 to -7	[274]
Ignition temperature (°C)	370	[278]
Inhibitor	Hydrochinone, acetic acid	[278]
Colorless, flammable, toxic lacrimatory, liquid		[274,275,278]

polymerization [275]. UV- or γ-irradiation can start the bulk polymerization [284]. Therefore, quinone is added to the monomer for long-term storage. As already mentioned, bulk polymerization is feasible, but better results are obtained in solvents such as cyclohexane or petroleum ether, which dissolve the monomer but not the polymer (precipitation polymerization). These polymers show not only higher rates of polymerization but also higher molecular weights, which has been attributed to the reduction of termination relative to the propagation rate [285].

An interesting initiator for MVK is N,N-dimethylaniline or N,N-diethylamine [286]. As weak bases they do not polymerize, for example, methyl methacrylate or methyl acrylate.

It seems to be a fact that the α,β-unsaturated carbonyl group is building up the initiating species, which is proposed to be an electron transfer complex of the following type:

(44)

Table 4 Radical polymerization of MVK.

Initiator	Solvent	c(MVK) in mol/L	Temp in °C	Time in h	Yield in %	Ref.
BPO[a]	None	12.33	26	40	40	[288]
	Benzene	3.5	25	100	57	[279]
	Acetone	3.5	25	100	35	[279]
AIBN	EtOH/H$_2$O (7:3)	6.5	50	1.5	62	[274]
K$_2$S$_2$O$_8$	H$_2$O/AgNO$_3$	1.43	30	3	91	[287]

[a]BPO: benzoyl peroxide.

Since MVK is completely miscible with water, an emulsion-type polymerization without additional emulsifiers is possible. For the oxodisulfate-silver nitrate initiator, the following mechanism is suggested:

$$S_2O_8^{2-} + Ag^+ \longrightarrow 2\,SO_4^{2-} + Ag^{3+}$$

$$Ag^{3+} + 2\,H_2O \longrightarrow Ag^+ + 2\,OH^{\bullet} + 2\,H^+ \tag{45}$$

$$OH^{\bullet} + MVK \longrightarrow HO\!-\!MVK^{\bullet}$$

This method allows polymerization at $-15\,°C$. Molecular weights on the order of 4×10^5 g/mol were observed [287]. Higher-molecular-weight PMVK can be obtained by decreasing the solubility of MVK in water by adding sodium chloride and an emulsifier such as potassium caproate [288] (Table 4).

C. Ionic and Group Transfer Polymerization

1. Anionic Polymerization

A wide variety of typical anionic initiators is described for the polymerization of PMVK. Grignard reagents are used as well as complexes formed of alkylaluminum or alkylzinc compounds with alkali metal alkyls (so called '-ate complexes'). Alkali metal initiators and alkoxides are also described. Some examples are given in Table 5.

The ethyl derivatives of aluminum, cadmium, magnesium, and zinc yield highly crystalline polymers. Organometallic complexes such as magnesium diethyl cobalt chloride, which coordinate strongly with the polymer anion and the monomer, produce white crystalline PMVK. PMVK obtained by alkoxides, sodium naphthalene, and *n*-butyllithium are intensively colored because of a partially occurring aldol condensation [292]. The mechanism of these reactions has been studied intensively. It is assumed that

Table 5 Examples for anionic initiators.

Initiator	Solvent	Temp in °C	Time in h	Conversion in %	Comments	Ref.
BuLi	Toluene	−70	72	52	Amorphous	[275]
	Toluene	−78	24	19	Colored	[276]
AlEt$_3$	Toluene	0	168	71	Partially crystalline	[276]
MgEt$_2$	Bulk	−78	24	*50*	Amorphous	[289]
	THF	−78	3	3		[289]
EtMgBr	Bulk	−78	17	36	Crystalline	[289]
PhMgBr	Toluene	−70	24	Trace		[276]
Na/naphthalene	Toluene	0	168	95	Noncrystalline	[276]
CaZnEt$_4$	Toluene	0	96	72	Partially crystalline	[276,290]
LiZnEt$_2$Bu	Toluene	0	168	70	Amorphous	[276]
NaOEt	*n*-Hexane			47	Crystalline	[290,291]
	THF			60	Crystalline	[290,291]

n-BuLi reacts in three ways with MVK:

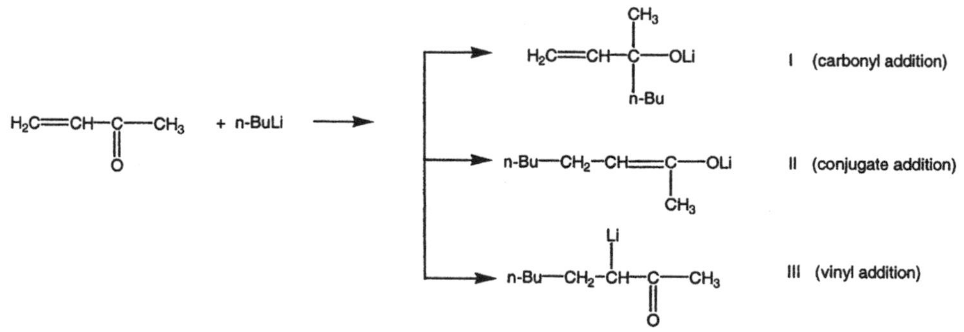

(46)

III acts as the propagating species in the polymerization reaction.

Compounds such as AlEt$_3$ form '-ate complexes', which react to a 'conjugate addition' product (II). In this case propagation occurs via type II intermediates.

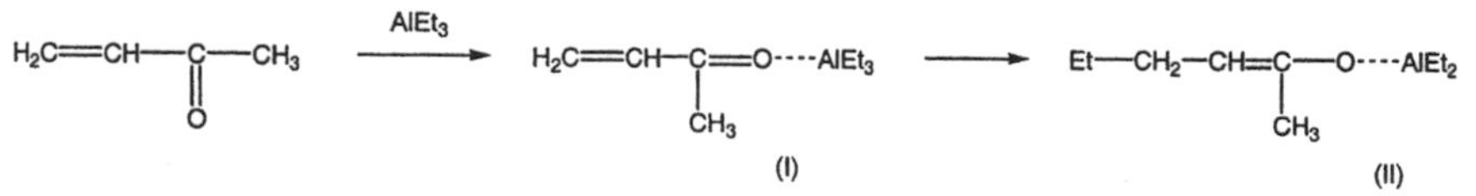

(47)

Other possible products are [294]:

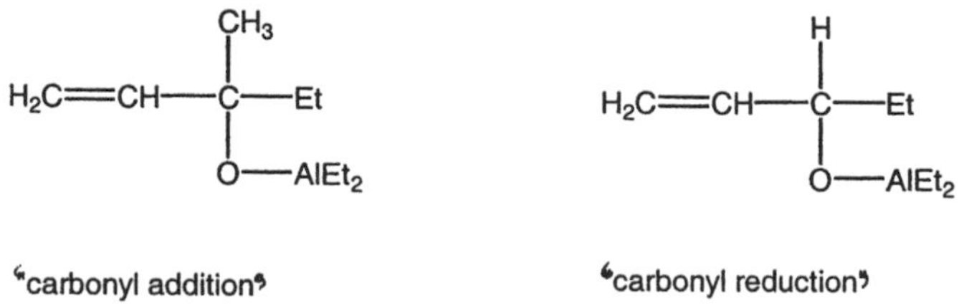

(48)

More details about the polymerization mechanism by '-ate complexes' are given in the section on α,β-unsaturated ketones.

Grignard reagents such as n-BuMgBr react as follows [295]:

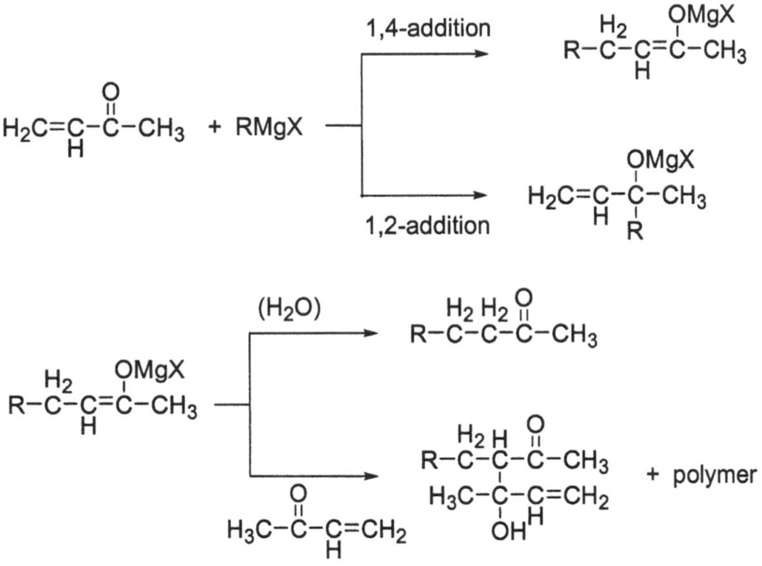

$$
\text{H}_2\text{C}=\underset{\text{H}}{\text{C}}-\overset{\text{O}}{\overset{\|}{\text{C}}}-\text{CH}_3 \;+\; \text{RMgX}
\quad
\begin{cases}
\xrightarrow{\text{1,4-addition}} & \text{R}-\overset{\text{H}_2}{\text{C}}-\underset{\text{H}}{\text{C}}=\overset{\text{OMgX}}{\text{C}}-\text{CH}_3 \\[2em]
\xrightarrow{\text{1,2-addition}} & \text{H}_2\text{C}=\underset{\text{H}}{\text{C}}-\underset{\text{R}}{\overset{\text{OMgX}}{\text{C}}}-\text{CH}_3
\end{cases}
$$

$$
\text{R}-\overset{\text{H}_2}{\text{C}}-\underset{\text{H}}{\text{C}}=\overset{\text{OMgX}}{\text{C}}-\text{CH}_3
\quad
\begin{cases}
\xrightarrow{(\text{H}_2\text{O})} & \text{R}-\overset{\text{H}_2}{\text{C}}-\overset{\text{H}_2}{\text{C}}-\overset{\text{O}}{\overset{\|}{\text{C}}}-\text{CH}_3 \\[2em]
\xrightarrow{\underset{\text{H}}{\overset{\text{O}}{\text{H}_3\text{C}-\overset{\|}{\text{C}}-\text{C}=\text{CH}_2}}} &
\begin{array}{l}
\text{R}-\overset{\text{H}_2}{\text{C}}-\underset{\text{OH}}{\overset{\text{H}}{\text{C}}}-\overset{\text{O}}{\overset{\|}{\text{C}}}-\text{CH}_3 \\
\text{H}_3\text{C}-\text{C}-\underset{\text{H}}{\text{C}}=\text{CH}_2
\end{array}
\;+\; \text{polymer}
\end{cases}
$$

(49)

Alkoxides such as sodium tert-butoxides cause a hydrogen transfer, and therefore the following polymer structure (50) is observed instead of the 'normal' 1,2 addition [296]:

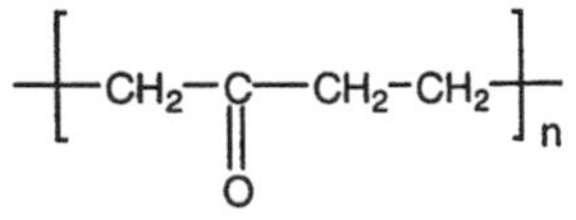

$$
\begin{array}{c}
\text{—}\!\!\left[\text{CH}_2\text{—}\underset{\overset{\|}{\text{O}}}{\text{C}}\text{—CH}_2\text{—CH}_2\right]_{\!n}\!\!\text{—}
\end{array}
$$

(50)

Another interesting initiator for MVK is the system pyridine-water. An initial addition product, β-ketobutanol, is formed, which in the presence of a base, yields a 1,2 addition polymer [297].

2. Cationic Polymerization

Cationic polymerization of MVK is certainly not the method of choice. However, if boron trifluoride etherate was added to a monomer-carbon dioxide mixture in petroleum ether polymerization was observed [298]. Acid-catalyzed polarography of MVK in methanol is also considered to be a cationic polymerization. For the polymer an alternating ketone-ether copolymer structure was suggested [299,300]. The following reaction mechanism is proposed (Structure (51)):

Initiation:

$$H_2C=\underset{H}{C}-\underset{O}{\overset{||}{C}}-CH_3 \ + \ H^+ \ \rightleftharpoons \ H_2C=\overset{+}{\underset{H}{C}}=\underset{CH_3}{\overset{OH}{C}}$$

Propagation:

$$\underset{\underset{CH_3}{|}}{\overset{+}{H\overset{||}{C}}=CH_2} \ + \ \underset{HC=CH_2}{O=\overset{CH_3}{\overset{|}{C}}} \ \rightleftharpoons \ \underset{\underset{CH_3}{|}}{\overset{H_2}{HC}-\overset{|}{C}-O-\overset{+}{\underset{||}{C}}}\qquad\qquad (51)$$

dimer

$$n \text{ dimer} \longrightarrow \overset{H_2}{HC}-\overset{CH_3}{\overset{|}{C}}-O-\overset{CH_3}{\overset{|}{C}}\left[\overset{H_2}{CH}-\overset{CH_3}{\overset{|}{C}}-O-\overset{CH_3}{\overset{|}{C}}\right]_n\overset{H_2}{CH}-\overset{CH_3}{\overset{|}{C}}-O-\overset{CH_3}{\overset{+}{C}}$$

Termination:

$$\overset{CH_3}{\underset{HC}{\overset{|}{\underset{||}{\overset{+}{C}}}}}\ +\ HO-CH_3 \longrightarrow \overset{CH_3}{\underset{HC}{\overset{|}{\underset{||}{C}}}}-O-CH_3 \ + \ H^+$$

3. Group Transfer Polymerization

A nonionic way of polymerizing MVK is the group transfer polymerization (GTP) with dimethylketene methyl trimethylsilyl acetal as intiator and the Me_3SiF_2 anion delivered from tris(dimethylamino)sulfonium difluorotrimethylsilicate ($TASF_2SiMe_3$)

$[(Me_2N)_3S-F_2SiMe_3]$ as catalyst [301]:

$$(52)$$

Metallocene-catalysts were successfully applied as initiators for the GTP [302, 318,319]. Especially adducts of group 4 metallocene-enolates and tris(pentafluoro-phenyl)boranes lead to a rapid polymerization by means of group transfer polymerization of MVK [320]:

$$M = Ti, Zr, Hf \qquad (53)$$

D. Copolymerization

1. Radical Copolymerization

As already mentioned, PMVK has poor mechanical properties. Therefore, numerous copolymers were synthesized with a large number of vinyl monomers and dienes. MVK is

Table 6   Reactivity ratios of some comonomers.

M_1	r_1	M_2	r_2	Temp in °C	Comments	Ref.
MVK	1.78 ± 0.22	Acrylonitrile	0.61 ± 0.04	60		[303]
	1.6 ± 0.1	Butyl acrylate	0.65 ± 0.07	50		[304]
	0.35 ± 0.02	Styrene	0.29 ± 0.04	60		[303]
	7.00	Vinyl acetate	0.05	70		[305]
	8.3	Vinyl chloride	0.10	70		[306]
	1.8	Vinyliden chloride	0.55	70		[306]
	0.29	Methyl acryl amide	3.05	60	In dioxane	[307]
	3.37	Methyl acryl amide	2.04	60	In ethanol	[307]

similar to styrene in its copolymerizability, as indicated by its parameters, $e = 0.7$ and $Q = 1.0$.

The Q-value is identical to that of styrene and e has the opposite polarity [302]. Therefore, the expected equivalent incorporation of the two monomers in the copolymer was found. Some examples of comonomers with the corresponding r_1–r_2 value pairs are given in Table 6.

The initiators used in copolymerization are the same as those in homopolymerization. Benzoyl peroxide is also used in grafting MVK onto poly(cis-1,4-isoprene) to give surface coating materials [308]. The copolymers are also able to undergo some polymer-analogous reactions such as the cross-linking of a n-butyl acrylate/MVK copolymer with sulfur/zinc oxide [294] resulting in disulfide cross-linkages.

2.   Copolymerization in the Presence of Lewis Acids

Despite the results of pure radical copolymerization, it is more difficult to produce copolymers of MVK with styrene under ionic conditions. Only a small amount (about 2%) of styrene is incorporated in the polymer if catalysts such as Et_3Al, Et_2Zn, and Et_2Cd are used [276]. It was more attractive to copolymerize MVK with styrene under catalysis of Lewis acids such as $AlCl_3$, $EtAlCl_2$, or $ZnCl_2$. The products obtained are 1 : 1 copolymers. Although these reactions run without any radical initiator, shown by the addition of hydroquinone, the yield of copolymer can be increased in the presence of traces of benzoyl peroxide [309]. The copolymerization behavior of MVK can be changed by complexation of the monomer with Lewis acids [310]. The 2 : 1 complex $(MVK)_2ZnCl_2$ can be copolymerized with allyl benzene, which is not possible without $ZnCl_2$ [311].

E.   Recent Developments

In recent years, conductive poly(methyl vinyl ketone) homo- and copolymers were prepared [321,322]. Conductivity was achieved by reacting PMVK with a dopant solution containing $POCl_3$. During the reaction double bonds are formed, namely the PMVK is partly converted into poly(acetyl-acetylene). Conductivities from 10^{-7} to $10^{-9}\,S\,cm^{-1}$ could be achieved [323], which varied drastically with the time of treatment with the dopant solution.

$$
\begin{array}{c}
\left[\text{CH}_2\text{--}\underset{\underset{\text{CH}_3}{\overset{|}{\text{C}=\text{O}}}}{\overset{|}{\text{CH}}}\text{--CH}_2\text{--}\underset{\underset{\text{CH}_3}{\overset{|}{\text{C}=\text{O}}}}{\overset{|}{\text{CH}}}\right]_n
\xrightarrow{\text{POCl}_3}
\left[\text{CH}=\underset{\underset{\text{CH}_3}{\overset{|}{\text{C}=\text{O}}}}{\text{C}}\right]_m
\left[\text{CH}_2\text{--}\underset{\underset{\text{CH}_3}{\overset{|}{\text{C}=\text{O}}}}{\text{CH}}\right]_{n-m}
\end{array}
\tag{54}
$$

F. Physical Properties

PMVK easily undergoes degradation processes by heating, irradiating, and treating with bases. At temperatures above 250 °C, PMVK loses water and yields glossy, red, non-cross-linked products.

Cyclization reactions resulting in cyclohexenone structures by intramolecular aldol condensation of neighbored methyl vinyl ketone units are held responsible for the red color [313,314]. This view is supported by UV [315] and IR spectroscopy [316].

In contrast to thermal degradation, amine- or alkali-catalyzed degradation seems to be a chain process. The reaction of discoloration is accelerated and the product turns black and is crosslinked, indicating long conjugated sequences [316]. Photolytic degradation – caused by UV and γ-irradiation – is a more complicated process. The result from irradiating films [317] is a rapid reduction of the molecular weight, followed by the formation of acetaldehyde, carbon monoxide, and methane. These results are interpreted by assuming a concomitant occurrence of Norrish type I (α-scission) and type II (ketone cleavage) reactions:

Type I:

Type II:

$$\tag{55}$$

IV. POLYMERS OF α,β-UNSATURATED KETONES

(This section was prepared by O. Nuyken, K. Losert and V. Knopfova.)

Polymers derived from vinyl ketones ($H_2C = CR'–CO–R''$) have been known since 1903 [324], but no significant commercial application has been found so far. The use of vinyl ketone polymers as materials for billard balls [325] was thwarted by their poor thermal and photochemical stability. Commercial applications will depend on the development of stabilizers that will inhibit the rapid decolorization and degradation of the ketone polymers upon exposure to heat or light. Both vinyl ($R' = H$) and isopropenyl ($R' = CH_3$) ketones are extremely reactive monomers that polymerize spontaneously upon exposure to heat or sunlight. Vinyl ketones are generally toxic, lacrimatory compounds. The normal precautions for handling toxic, flammable liquids should be observed. The polymerization of α,β-unsaturated ketones can be initiated by free-radical, cationic, or anionic catalysts. Monomer reactivity toward the initiator species increases with increased stabilization of the active center produced. The fact that the majority of work on the ionic polymerization of vinyl ketones has been concerned with anionic initiation reflects the ready stabilization of the carbanionic active center by the conjugative effect of the carbonyl group.

Copolymerization data have been tabulated by Greenley [326].

Alkyl vinyl ketones were synthesized in 1906 by heating β-chloroethyl ketones with diethylaniline [327]. A large number of syntheses have been developed since that time, but only three or four have general applicability. Most syntheses are carried out at a low pH value to minimize the base-catalyzed condensation of the vinyl ketones. However, a rather elegant synthetic route for alkyl vinyl ketones involves a base catalyzed condensation of formaldehyde with methyl or ethyl ketones, respectively. Thermal dehydration of the β-ketoalcohol intermediates in the presence of weak acid catalysts produced α,β-unsaturated ketones in 50 to 60% yields. Several variations of this procedure have been reported [328].

Methyl vinyl ketone is synthesized industrially by the hydration of vinylacetylene. The reaction is catalyzed by acetates, formates, or sulfates of mercury, silver, cadmium, copper, or zinc in the presence of acids [329,330]. The oxidation of 1-butene to methyl vinyl ketone in 72% yield by the formation of olefin–mercuric salt complexes followed by the decomposition of these complexes with acid may become commercially feasible [331].

Similar oxidation procedures using cupric salts have also been reported, but only a 40% yield of vinyl ketone was obtained [332].

Preparation of vinyl ketones via a Mannich reaction overcomes many of the drawbacks of the procedures described above. The α,β-unsaturated ketones are obtained in high yields under mild conditions from readily available starting materials (Table 7). Thus this is the best technique for laboratory preparation [333]. The Mannich base, which is formed by heating equimolar quantities of ketone, formaline, and diethylamine hydrochloride for 1 h at 95 °C, is isolated and pyrolyzed at 150 to 210 °C under reduced pressure. The α,β-unsaturated ketone is distilled from the reaction mixture; the

Table 7 Physical properties of polymerizable α,β-unsaturated ketones.

Name and structure[a]	Mol. wt. (g/mol)	bp (°C/mmHg)	d_4^{20} (g/cm^3)	n_D^{20}	Yield (%)	Refs.
2-Methyl-3-oxobutene-1 (methyl isopropenyl ketone) $H_2C-C\overset{CH_3}{\underset{CO-CH_3}{}}$	84.1	98/760	0.8410	1.4220	94.5	[334,335]
3-Phenyl-3-oxopropene-1 (phenyl vinyl ketone) $H_2C=C\overset{H}{\underset{CO-C_6H_5}{}}$	132.2	108–110/13 58–60/0.2	1.060	1.5520 1.5580	82.2	[334,336]
4,4-Dimethyl-3-oxopentene-1 $H_2C=C\overset{H}{\underset{CO-C(CH_3)_3}{}}$	112.2	59–60/103 48–50/50		1.4222 (15 °C)	61.7	[337]
3-(2-Furyl)-3-oxopropene-1 (2-furyl vinyl ketone) $H_2C-C\overset{H}{\underset{CO}{}}$	122.2	76–77/6		1.4219 (14 °C)		[338]

[a]Names in parentheses are those used in this chapter.

diethylamine hydrochloride can be recycled.

$$R-\underset{O}{\overset{\|}{C}}-CH_3 + CH_2O + (H_5C_2)_2NH\cdot HCl \longrightarrow R-\underset{O}{\overset{\|}{C}}-CH_2-CH_2-\underset{\underset{HCl}{\overset{\bullet}{}}}{N(CH_2CH_3)_2}$$

$$R-\underset{O}{\overset{\|}{C}}-CH_2-CH_2-\underset{\underset{HCl}{\overset{\bullet}{}}}{N(CH_2CH_3)_2} \xrightarrow[\text{80-100 mmHg}]{150\text{-}200°C} R-\underset{O}{\overset{\|}{C}}-CH=CH_2 + (H_5C_2)_2NH\cdot HCl$$

(56)

R = Alkyl, Phenyl, etc.

A. Radical Polymerization

Most of the common free-radical systems are effective in initiating vinyl ketone polymerization. Azobisisobutyronitrile (AIBN) is considered the best initiator for producing polymers with good color stability. All other catalysts leave acid residues or degrade the free polymer during the polymerization. Although methyl isopropenyl ketone [339,340], phenyl vinyl ketone [341], and higher alkyl vinyl ketones [342] polymerize readily in bulk at room temperature, better results are obtained in solvents such as cyclohexane or other organic solvents, which dissolve the monomer but not the polymer (precipitation polymerization). In precipitation polymerization, higher polymerization rates and higher molecular weights are observed than in homogeneous solutions under comparable conditions. This phenomenon has been attributed to the reduction of the rate of termination relative to the propagation rate [343].

Feng [344] has described the dibenzoyl peroxide-initiated polymerization of t-butyl vinyl ketone to an amorphous polymer in organic solvents, while several alkyl vinyl ketones have been polymerized in aqueous solutions to low-melting polymers using a potassium persulfate/sodium metabisulfite initiation [345]. Phenyl vinyl ketone was polymerized in an emulsion containing 7.5% soap and 0.2% potassium peroxodisulfate [346].

Chaudhuri [347] reported a thermally initiated polymerization of methyl iso-propenyl ketone in bulk and solution. The reaction order with respect to monomer was less than 2 in homogeneous and greater than 2 in heterogeneous systems. Chain transfer [348] increased in the order benzene < toluene < ethylbenzene as solvents.

Both UV- and γ-irradiation have been applied successfully for the initiation of methyl isopropenyl ketone [349–351] and phenyl vinyl ketone polymerization [341]. Since polymerization initiated by γ-irradiation was inhibited by chinone, a radical mechanism was proposed.

Aliphatic vinyl ketones have been reported to polymerize similarly [349,352]. Although, the introduction of the furan ring (2-furyl vinyl ketone as monomer) does not alter the mode of chain growth with radical initiation. The regularity of the macromolecular structure is, however, accompanied by a serious drawback in terms of yield. Even with very high initiator concentration, the formation of polymer stopped at about 20% conversion. This is due to the retarding effect produced by the attack of primary radical onto the furan ring rather than onto the vinylic function.

The behavior of 2-furyl vinyl ketone is similar to that of 2-vinylfuran, in which the monomer is activated in the 'normal' fashion by the primary radical (addition onto the vinylic bond), but the formed polymer chains act as radical traps through their pendant furan ring. Thus at a critical polymer concentration practically all primary radicals are quenched to form stable furyl radicals, and normal initiation cannot take place. This phenomenon of self-retarding is also responsible for the low molecular weight [353].

Several detailed kinetic studies of the polymerization of alkyl vinyl ketones have been reported. Smets and Oosterbosch conducted a study of both bulk and solution polymerization. They observed that the rate law was one-half order to the initiator in bulk and first order in monomer [354]. The energy of activation was calculated to be about 5 kcal/mol in the temperature range −78 to 20 °C and is comparable with a value of 4.8 kcal/mol for the ^{60}CO-initiated polymerization of methyl vinyl ketone [355].

B. Anionic Polymerization

A wide variety of anionic initiators [352] can also affect the polymerization of vinyl ketones. Crystalline poly(alkyl vinyl ketones) were prepared by precipitation polymerization using metallic lithium or alkyl lithium catalysts. Thomas [342] reported that lithium dust initiation at −25 °C produced two types of poly(isopropyl vinyl ketone). The ether-soluble crystalline fraction was unstable. The highest crystalline samples melted to a colorless liquid at about 220 °C. The polymer degraded during thermal treatment [342]. Meanwhile the amorphous material remained unchanged.

The instability of the crystalline form is said to be caused by the incorporation of a β-ether linkage into the polymer chain, produced by carbonyl addition, which breaks down the ether oxygen at elevated temperatures [342,352].

High polymer yields were obtained [344,356] from the polymerization of t-butyl vinyl ketone at room temperature initiated by potassium, sodium and lithium. Potassium was the most active metal and gave a conversion of 90% after 20 h reaction. Lithium, sodium and potassium gave rise to crystalline, possibly isotactic polymers during heterogeneous polymerization in *n*-heptane, benzene, or toluene and amorphous polymers during homogeneous polymerization in tetrahydrofuran. The softening points of the crystalline polymers were about 240 °C [342]. Atactic polymer of *t*-butyl vinyl ketone was prepared by radical bulk, polymerization with AIBN as initiator at 60 °C. Isotactic polymer of *t*-butyl vinyl ketone was anionically obtained with butyl lithium, Al(*i*-Bu)$_3$, in toluene at 0 °C [357].

The lithium-initiated polymerization of phenyl vinyl ketone was carried out in bulk and in tetrahydrofuran at room temperature [341] and in liquid ammonia at − 78 °C [346]. No difference in the reactivity of sodium and lithium toward this monomer was recognized.

Alkali alkoxides or *n*-butyl lithium-initiated polymerization of 2-furyl vinyl ketone gave high yield; however, concentrations of initiator had to be low to avoid crosslinking [353].

Sodium hydride, lithium aluminum hydride, and lithium borohydride have also been used as initiators for the polymerization of alkyl vinyl ketones [342,344,356]. The noncrystalline products were frequently colored, due to aldol condensation.

$$\text{(57)}$$

R = phenyl

Lyons and Catterall reported on the mechanism of *n*-butyl lithium-initiated polymerization of methyl isopropenyl ketone in benzene at 0 °C [352,358]. Relatively rapid initial consumption of monomer gave rise to a bimodal molecular weight distribution of low M_w which was maintained throughout the entire reaction.

The higher molecular-weight polymer contained some intramolecular cyclized units. The process of cyclization produced water in the reaction mixture. This retarded the polymerization and limited the molecular weight of the polymer [359]. A pseudotermination step was proposed to explain the retention of the bimodal molecular weight distribution throughout the whole polymerization. The following overall reaction scheme is proposed [352,358]:

$$
\text{n-BuLi} \ + \ \text{H}_2\text{C}=\overset{\overset{\displaystyle CH_3}{|}}{\underset{\underset{\displaystyle CH_2}{|}}{C}}-\overset{}{\underset{}{C}}-\text{CH}_3
\longrightarrow
\begin{cases}
\text{n-Bu}-\overset{H_2}{C}-\overset{CH_3}{C}=\overset{CH_3}{C}-\text{OLi} & \text{(conjugate addition)} \\[2mm]
\text{n-Bu}-\overset{H_2}{C}-\underset{\underset{\displaystyle Li}{|}}{\overset{\overset{\displaystyle CH_3}{|}}{C}}-\underset{\underset{\displaystyle O}{\|}}{C}-\text{CH}_3 & \text{(vinyl addition)} \\[2mm]
\text{H}_2\text{C}=\overset{CH_3}{C}-\underset{\underset{\displaystyle \text{n-Bu}}{|}}{\overset{\overset{\displaystyle OLi}{|}}{C}}-\text{CH}_3 & \text{(carbonyl addition)}
\end{cases}
\tag{58}
$$

Several studies have been made of the interaction between initiator compounds and α,β-unsaturated carbonyl compounds. Lithium alkyls were known to react with α,β-unsaturated ketones by either carbonyl, conjugate, or vinyl addition [360]. Under polymerization conditions it was shown for methyl acrylate as the carbonyl compound [361,362] that carbonyl addition takes place only to a negligible extent. The conjugate and vinyl adducts are mesomeric.

Propagation:

$$
\text{n-Bu}-\overset{H_2}{C}-\underset{\underset{\underset{\displaystyle CH_3}{|}}{\underset{\displaystyle C=O}{}}}{\overset{CH_3}{C}}{}^{\ominus}\ \text{Li}^{\oplus}
\xrightarrow{\text{monomer}}
\text{n-Bu}-\overset{H_2}{C}-\underset{\underset{\underset{\displaystyle CH_3}{|}}{\underset{\displaystyle C=O}{}}}{\overset{CH_3}{C}}-\text{CH}_2-\underset{\underset{\underset{\displaystyle CH_3}{|}}{\underset{\displaystyle C=O}{}}}{\overset{CH_2}{C}}{}^{\ominus}\ \text{Li}^{\oplus}
\tag{59}
$$

Termination:

$$
\xrightarrow{\text{H}_2\text{O}}
\text{n-Bu}\left[\overset{H_2}{C}-\underset{\underset{\underset{\displaystyle CH_3}{|}}{\underset{\displaystyle C=O}{}}}{\overset{CH_3}{C}}\right]_n\!\!\!\!H \ + \ \text{LiOH}
\tag{60}
$$

Side reactions limit the molecular weights:

Cyclization:

$$
\text{wwCH}_2-\underset{\underset{\underset{\displaystyle CH_3}{|}}{\underset{\displaystyle C=O}{}}}{\overset{\overset{\displaystyle CH_3}{|}}{C}}-\text{CH}_2-\underset{\underset{\underset{\displaystyle CH_3}{|}}{\underset{\displaystyle C=O}{}}}{\overset{\displaystyle C}{}}\text{w}
\xrightarrow{\text{base}} \cdots \xrightarrow{-\text{H}_2\text{O}} \cdots
\tag{61}
$$

Pseudotermination:

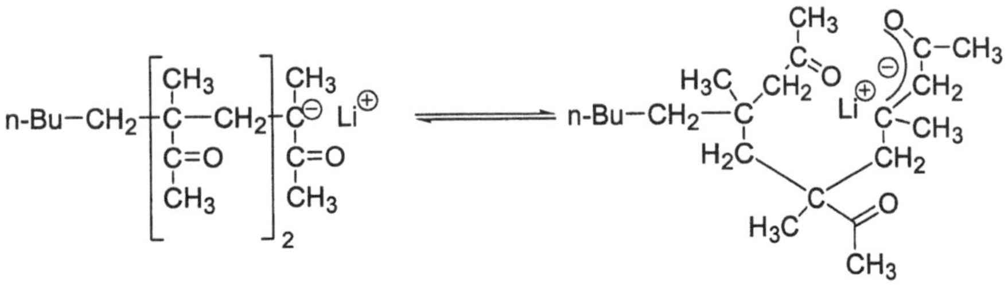

(62)

C. Coordinated and Cationic Polymerization

Polymerization of methyl isopropenyl ketone in toluene and tetrahydrofuran initiated by triethylaluminium at 0 °C is described by Lyons and Catterell [358,363]. The reaction between the metal alkyl and the vinyl ketone gave rise to a yellow '-ate complex' which rearranges to an initiating species. Rapid production of a linear trimer is followed by slow polymerization, giving rise to a bimodal molecular weight distribution. In contrast to this report, Tsuskima and Tsumuta [364] described a unimodal molecular weight distribution by the polymerization of phenyl vinyl ketone initiated with diethyl zinc. The latter was explained in terms of the reversible cyclization of the linear trimer to a pseudoterminated intermediate, which was inactive toward polymerization. When the linear molecular chain was extended by one or two additional units, the low polymer was precipitated and pseudotermination was no longer the favored reaction. For the same reason the reaction proceeded more rapidly in toluene under equivalent conditions than in tetrahydrofuran, where the polymeric products were soluble. The reaction was shown to follow a first-order dependence on both monomer and initial triethylaluminium concentration [365]. The higher-molecular-weight fraction was a white powder with a softening range of 145 to 165 °C. The oligomers were an almost colorless viscous liquid. No crystallinity could be detected in the high molecular weight polymer [363].

A coordinate mechanism for the polymerization is proposed [363]:

Initiation:

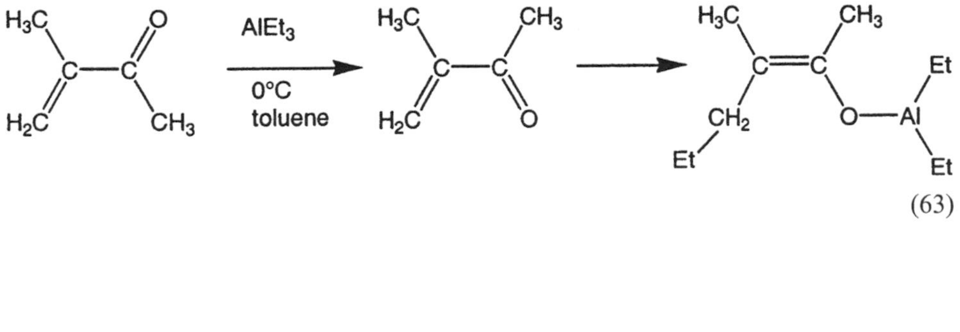

(63)

Propagation:

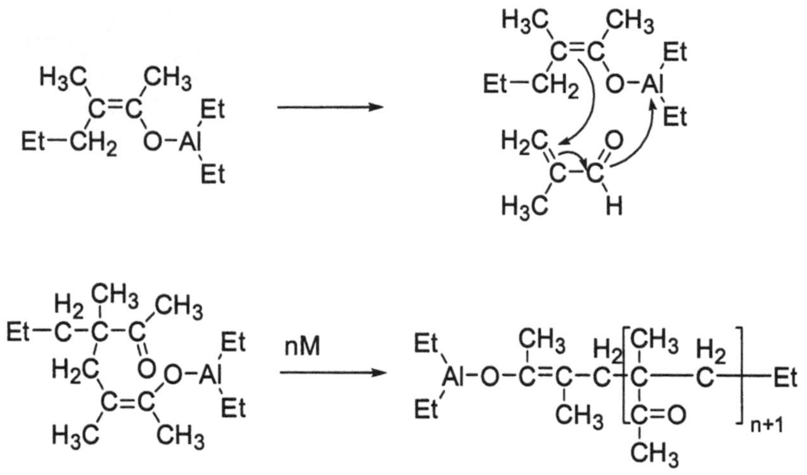

(64)

The reactions between organoaluminum compounds and vinyl ketones have been studied by several authors [366–368].

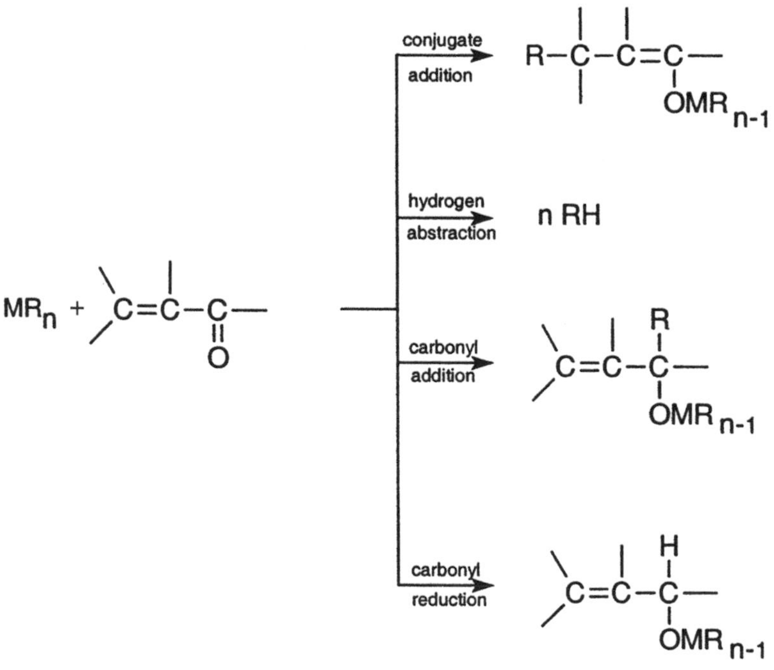

(65)

Conjugate addition was almost quantitative for the reaction between di-*n*-butyl zinc or tri-*n*-butyl aluminum and both methyl isopropenyl ketone and phenyl vinyl ketone. With 2-furyl vinyl ketone an increase in initiator concentration produced crosslinking of the material [353].

Only a few papers have been published regarding the polymerization of α,β-unsaturated ketones begun with cationic initiators. This reflects the fact that stabilization of the active center by the conjugative effect of the carbonyl group is not as simple as in anionic-initiated polymerization. Coleman [369] and Schildknecht [370] describe the homopolymerization of some perfluoralkyl propenyl ketones with boron trifluoride at − 80 °C in bulk and in solution. However, the resulting products have not been characterized adequately.

D. Copolymerization

1. Radical-Initiated Copolymerization

The copolymerization of α,β-unsaturated ketones has been studied extensively in order to improve the poor chemical and thermal stability exhibited by the homopolymers. The vinyl ketones have been copolymerized with most of the common vinyl and diene monomers. The data are given in Ref. [326]. For initiation, the same reagents could be used as for free-radical homopolymerization. Copolymerization was carried out in bulk [371] and in emulsion systems [372]. In copolymerization with methyl methacrylate, vinyl acetate [373], and styrene [371] it was concluded that the relative reactivities of the vinyl ketones increase with the increasing electron-withdrawing nature of the vinyl ketone substituent. Polar and steric effects are not observed. Most of the work has been directed toward the preparation of oil- and solvent-resistant rubbers to replace styrene-butadiene rubber. Emulsion copolymerization of butadiene with methyl isopropenyl ketone yielded rubbers with good solvent resistance and low temperature flexibility, but the products tended to harden on storage and were not compatible with natural rubber [374]. The reactive carbonyl function caused sensitivity to alkine reagents. Copolymers of butylacrylate and methyl vinyl ketone, for example, can be crosslinked by treatment with hydrazine [375].

2. Ionic-Initiated Copolymerization

Nearly all the reported attempts at ionic copolymerization of vinyl ketones led to polymers containing very high ketone content, even when the comonomer was known to homopolymerize under the conditions. Copolymerization of phenyl vinyl ketone and styrene in bulk or in tetrahydrofuran initiated with *n*-butyllithium produced only poly(phenyl vinyl ketone) [341]. The non-incorporation of styrene in the anionic copolymerization was due to the phenyl vinyl ketone enolate anion being sufficiently nucleophilic to add the phenyl vinyl ketone monomer but not the styrene.

E. Physical Properties of the Polymers

The low chemical and thermal stability of poly(vinyl ketones) leads to a sensitivity to degradation reactions. Poly(methyl isopropenyl ketone) lost water at about 250 °C, to yield glassy, red, non-crosslinked products. It was proposed that an intramolecular aldol condensation was responsible for this degradation, and it was shown that 15 to 21% of the oxygen remained in the polymer [349,376].

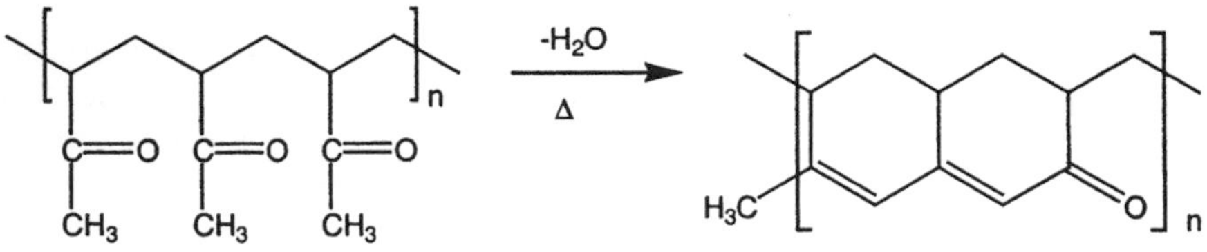

(66)

UV and γ-radiation of poly(methyl isopropenyl ketone) produced random chain scission at 23 °C. The presence of air increases unexpectedly the main chain scission of the polymer under γ-radiation [377]. In a series of publications [378] the radiolysis and photolysis of poly(phenyl vinyl ketone), poly(vinyl benzophenone), and poly(*t*-butyl vinyl ketone) [357] were described. The authors stated that photodegradation of poly(phenyl vinyl ketone) occurred by the abstraction of a hydrogen in the γ-position to a carbonyl group, followed by chain scission by a Norrish type II photoelimination mechanism.

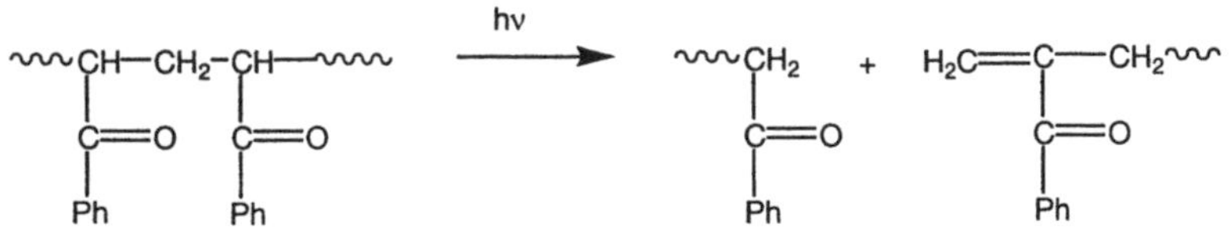

(67)

No crosslinking was observed [378].

V. PHOSPHORUS-CONTAINING VINYL POLYMERS

(This section was prepared by O. Nuyken, A. W. Förtig and T. Komenda.)

The characteristic properties obtained by the incorporation of phosphorus into an organic polymer are reduced flammability, increased adhesion, increased thermal stability, and increased solubility in inorganic solvents. These properties – specific to phosphorus – depend on the number of phosphorus units incorporated into the polymer rather than on the polymer structure [379–384]. The scope of this section is the description of the homo- and copolymerization of phosphorus compounds having one or more olefinic groups. The polymerizability of those compounds depends strongly on the type of bonding of olefinic groups to the phosphorus group. High molar masses were observed for phosphorus monomers, having several other atoms between P and olefin. Only oligomers are formed if the olefin is directly bonded to the phosphorus or connected by an oxygen bridge [379].

A. 1-Alkenylphosphonic Acid

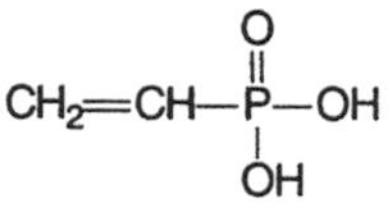

(68)

1

1-Alkenylphosphonic acid (**1**) can be synthesized according to the following scheme [385–388]:

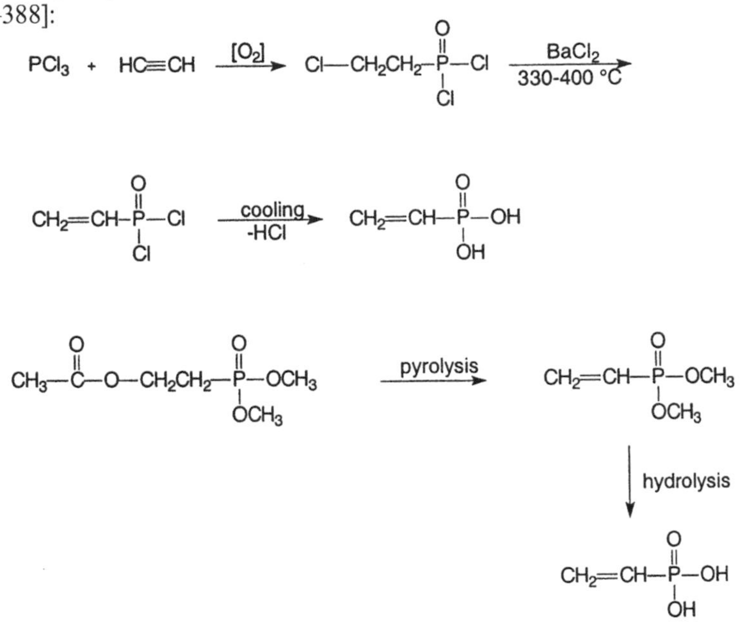

$$PCl_3 \;+\; HC\!\equiv\!CH \xrightarrow{\;[O_2]\;} Cl-CH_2CH_2-\overset{\displaystyle O}{\underset{\displaystyle Cl}{\overset{\displaystyle \|}{P}}}-Cl \xrightarrow[330\text{-}400\,°C]{BaCl_2}$$

$$CH_2\!=\!CH-\overset{\displaystyle O}{\underset{\displaystyle Cl}{\overset{\displaystyle \|}{P}}}-Cl \xrightarrow[\text{-HCl}]{\text{cooling}} CH_2\!=\!CH-\overset{\displaystyle O}{\underset{\displaystyle OH}{\overset{\displaystyle \|}{P}}}-OH$$

(69)

$$CH_3-\overset{\displaystyle O}{\overset{\displaystyle \|}{C}}-O-CH_2CH_2-\overset{\displaystyle O}{\underset{\displaystyle OCH_3}{\overset{\displaystyle \|}{P}}}-OCH_3 \xrightarrow{\text{pyrolysis}} CH_2\!=\!CH-\overset{\displaystyle O}{\underset{\displaystyle OCH_3}{\overset{\displaystyle \|}{P}}}-OCH_3$$

$$\downarrow \text{hydrolysis}$$

$$CH_2\!=\!CH-\overset{\displaystyle O}{\underset{\displaystyle OH}{\overset{\displaystyle \|}{P}}}-OH$$

The polymerization of **1** can be started thermally, with radicals, or by light [385,390,391]. However, since only oligomers were observed, those homopolymerizations are of academic interest only. **1** has been copolymerized with vinyl chloride and vinyl acetate [392], initiated by redox initiators in emulsion. Copolymers of this monomer are also available by hydrolysis of copolymers containing derivatives of 1-alkenylphosphonic acid, such as dichlorides [392–394] or diesters [395]. Copolymers are also described with acrylonitrile, acrylic amide, N-vinylacetamide, and N-vinylpyrrolidone; they are particularly interesting for textile dying, tanning techniques and water separating membranes [396–399].

B. Derivatives of Ethenylphosphonic Acid

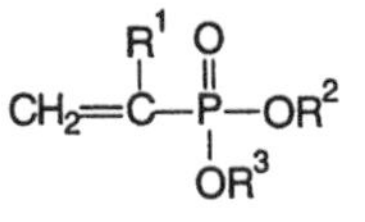

$$CH_2\!=\!\overset{\displaystyle R^1}{\underset{\displaystyle OR^3}{\overset{\displaystyle |}{C}}}-\overset{\displaystyle O}{\overset{\displaystyle \|}{P}}-OR^2$$

(70)

Compounds of type (70) have been synthesized according to the following scheme [389]:

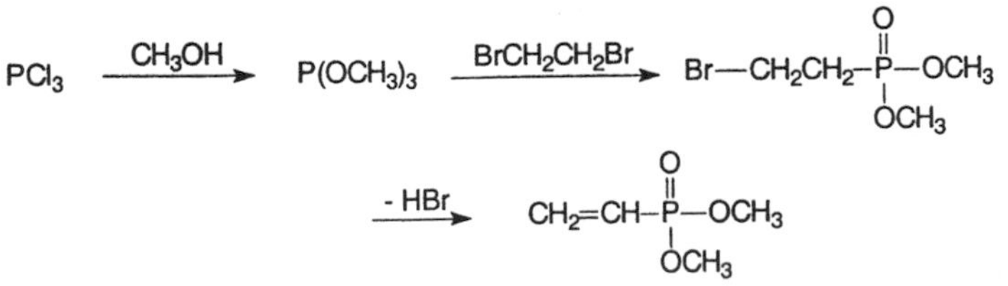

$$PCl_3 \xrightarrow{CH_3OH} P(OCH_3)_3 \xrightarrow{BrCH_2CH_2Br} Br-CH_2CH_2-\overset{\displaystyle O}{\underset{\displaystyle OCH_3}{\overset{\displaystyle \|}{P}}}-OCH_3$$

(71)

$$\xrightarrow{\text{- HBr}} CH_2\!=\!CH-\overset{\displaystyle O}{\underset{\displaystyle OCH_3}{\overset{\displaystyle \|}{P}}}-OCH_3$$

Table 8 summarizes structure and polymerization characteristics of selected derivatives of ethenyl phosphoric acid. The radical homo- and copolymerization of the derivatives of ethenylic phosphonic acid do not yield high molecular weight. However, by

Table 8 Polymerization and copolymerization of compounds of the general structure $CH_2=C(R_1)-P(O)(OR_2)_2$.

Monomer	Comonomer	Reaction conditions	Remarks	Refs.
$R^1 = H$	Styrene			[400]
$R^2 = CH_3$	Glycidyl methacrylate	AIBN	$M = 1{,}170\,g/mol$	[401]
$R^1 = H$		Grignard, Na-naphthalene, $-70\,°C$, THF	$[\eta] = 1.78$ in THF	[402]
$R^2 = C_2H_5$		$(C_2H_5)_3Al$, $98\,°C$, heptane	$[\eta] = 0.78$	[403,404]
	Styrene	BuLi: $-70\,°C$, THF		[394]
		BPO: $120\,°C$, bulk		[405]
	Vinyl acetate	H_2O_2, $80\,°C$		[405]
	Methyl methacrylate	BPO		[405]
$R^1 = H$	Styrene	AIBN, $70\,°C$, THF		[394]
$R^2 = n\text{-}C_3H_7$		BPO		[406]
	Vinyl acetate	BPO, $70\,°C$, suspension		[405]
$R^1 = H$	Vinyl chloride	AIBN, $45\,°C$, emulsion		[405]
$R^2 = n\text{-}C_3H_7$	Vinyl acetate	H_2O_2, $100\,°C$, emulsion		[405]
$R^1 = H$	Styrene	AIBN		[407]
$R^2 = CH_2CH_2Cl$	Acrylonitrile	$K_2SO_3/NaHSO_3$, $50\,°C$, emulsion		[408]
$R^1 = H$	Acrylonitrile	$K_2SO_3/NaHSO_3$, $70\,°C$, emulsion		[399]
$R^2 = H$	Methyl methacrylate	$K_2SO_5/NaHSO_3$, $40\,°C$, emulsion		[408]
$R^1 = H$	Styrene	BPO, $120\,°C$, bulk		[394]
$R^2 = Si(CH_3)_3$		BuLi, $-70\,°C$, THF		
$R^1 = CH_3$		$Al(C_2H_5)_3$, $70\,°C$, bulk		[403]
$R^2 = C_2H_5$				
$R^1 = CH_3$		$Al(C_2H_5)_3$, $70\,°C$, bulk		[403]
$R^2 = N(CH_3)_2$				

application of an ionic initiator such as Grignard reagent [402], sodium naphthaline [402], and trialkylaluminum [403,404], high molar masses are possible.

C. Diene-type Monomers

$$CH_2=CH-\underset{\underset{R^2}{|}}{\overset{R^1}{\underset{|}{C}}}=CH-\overset{O}{\underset{\underset{R^2}{|}}{\overset{\|}{P}}}-R^2$$

$$\tag{72}$$

$$R^1 = H, CH_3$$
$$R^2 = Cl, F, OR, NR_2$$

Dienes (72) can be synthesized according to the following reaction scheme [409]:

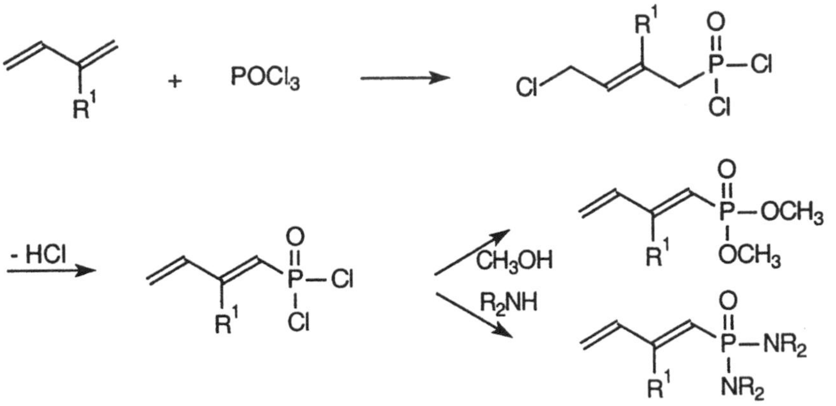

(73)

Compounds of type (72) can be polymerized with radical and anionic initiators. Molecular weights of approximately 5×10^5 g/mol have been observed for the homopolymer of Scheme (72) ($R^1 = CH_3$, $R^2 = OCH_3$) [410]. This polymer contained 60 to 69% 3.4 structure units and 31 to 40% 1.4 structure units [411], which were determined from ^{1}H-NMR data. Anionic polymerization can be initiated by butyllithium, butyl magnesium bromide, and other typical anionic initiators in bulk and in solution [412].

D. Vinyl Esters and Divinyl Esters

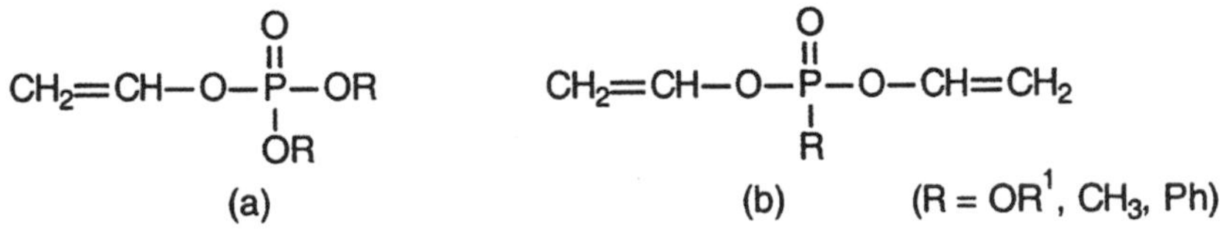

(74)

Compounds of type (74a) are available via the following reaction route:

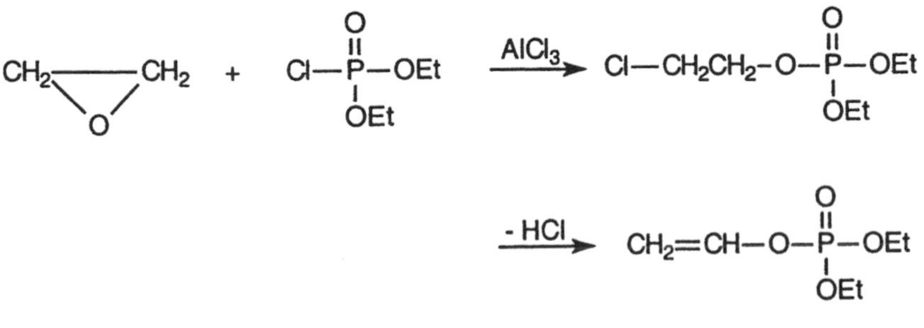

(75)

For the synthesis of compounds of type (74b) the following pathway is described [413]:

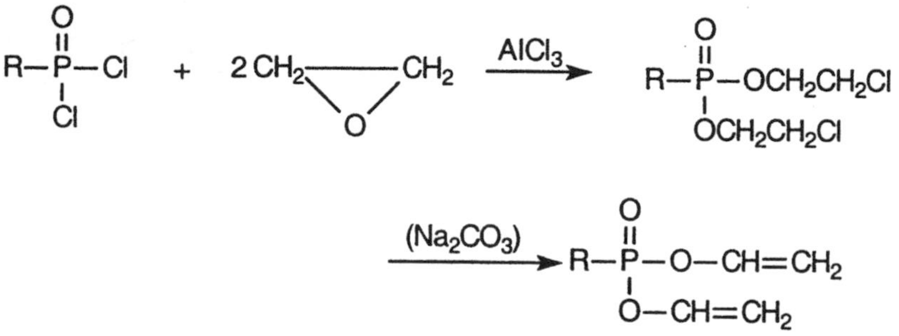

$$\underset{\overset{|}{Cl}}{\overset{\overset{O}{\|}}{R-P-Cl}} \quad + \quad 2\,CH_2\overset{\textstyle\diagdown\!\!\!\diagup}{\underset{O}{}}CH_2 \quad \xrightarrow{AlCl_3} \quad \underset{\overset{|}{OCH_2CH_2Cl}}{\overset{\overset{O}{\|}}{R-P-OCH_2CH_2Cl}}$$

(76)

$$\xrightarrow{(Na_2CO_3)} \underset{\overset{|}{O-CH=CH_2}}{\overset{\overset{O}{\|}}{R-P-O-CH=CH_2}}$$

Radical homopolymerization of (74a) does not give molecular weights higher than 10^3 to 10^4 g/mol [414,415]. Compounds of type (74b) monomers with $R=OR$ form five-membered rings during polymerization. For $R=CH_3$ the formation of six-membered rings is favored; whereas both six- and five-membered rings are formed when $R=Ph$.

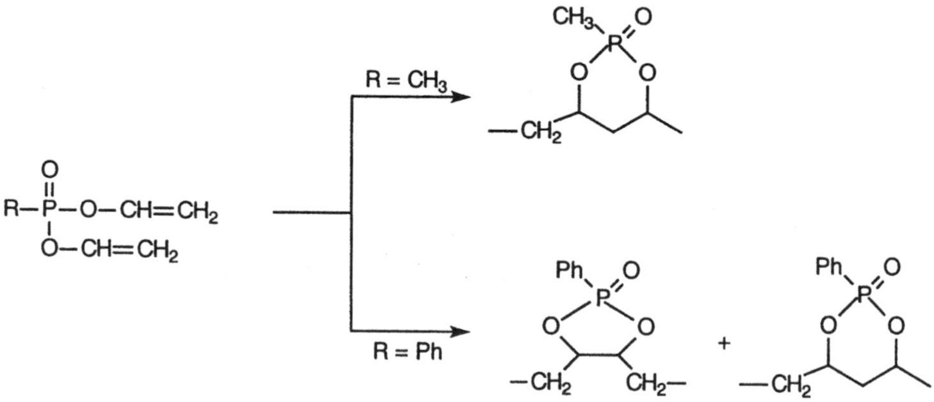

(77)

Hydrolysis of polymers made from (74a)-type monomers by alkali ethanol yields polymers containing the following structure units [414]:

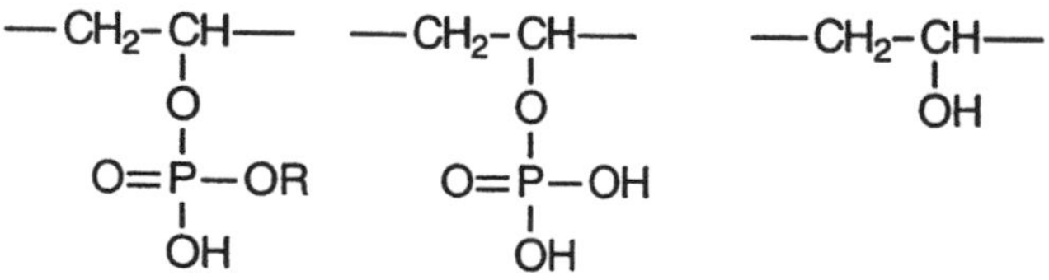

(78)

Monomers of type (74a) can be copolymerized with vinyl chloride, vinyl acetate, and acrylonitrile [384,414–417].

E. Dimethyl Perfluoro(3-vinyloxypropyl)phosphonate

$$F_2C = CFO(CF_2)_3P(O)(OCH_3)_2 \qquad (79)$$

The perfluorenated monomer (Scheme 79) has been synthesized as shown below.

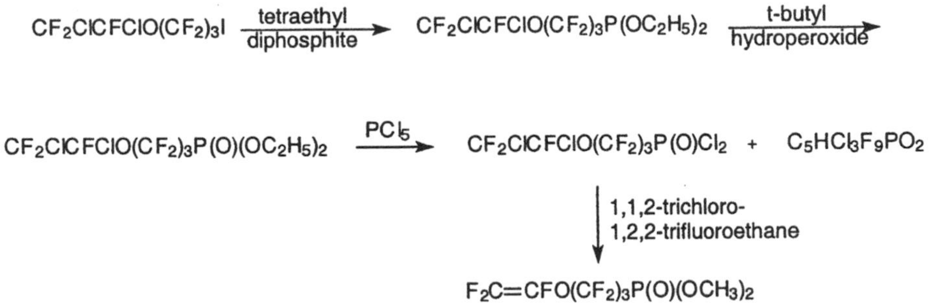

$$CF_2ClCFClO(CF_2)_3I \xrightarrow[\text{diphosphite}]{\text{tetraethyl}} CF_2ClCFClO(CF_2)_3P(OC_2H_5)_2 \xrightarrow[\text{hydroperoxide}]{\text{t-butyl}}$$

$$CF_2ClCFClO(CF_2)_3P(O)(OC_2H_5)_2 \xrightarrow{PCl_5} CF_2ClCFClO(CF_2)_3P(O)Cl_2 \ + \ C_5HCl_3F_9PO_2$$

$$\downarrow \begin{array}{l}\text{1,1,2-trichloro-}\\\text{1,2,2-trifluoroethane}\end{array}$$

$$F_2C{=}CFO(CF_2)_3P(O)(OCH_3)_2$$

$$(80)$$

This monomer can be polymerized radicalically initiated with AIBN. Its copolymerization with tetrafluoroethylene is also described [418].

F. Acrylic Esters, Acrylic Amide, and Styrene-Containing Phosphorus

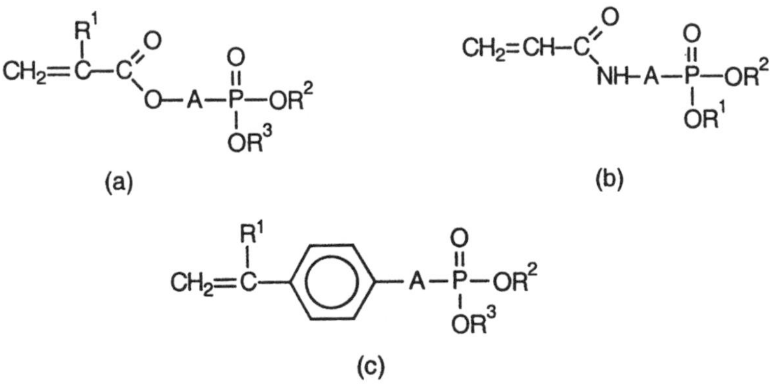

$$(81)$$

Scheme (82) is representative for other synthetic routes for Scheme (81a)-type monomers described in the literature [419–425].

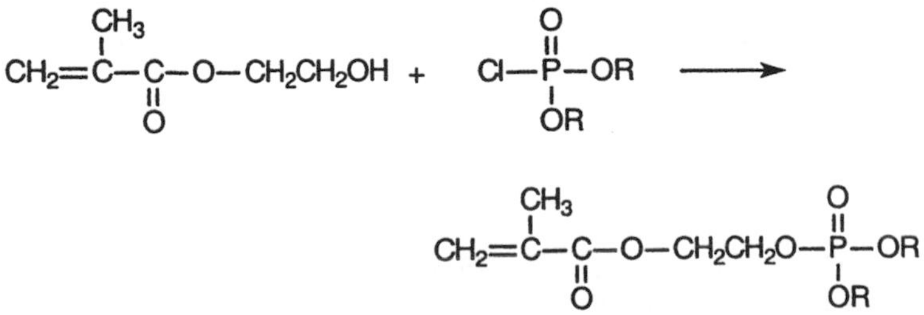

$$(82)$$

Table 9 Structure variations of structures (81a)-, (81b)- and (81c)-type monomers.

R^1	R^2	R^3	A	Ref.
CH_3	H	CH_2Ph	$+CH_2+_2$	[421]
		$(CH_2)_2-NH_2$	$+CH_2+_2$	[430]
			$-\bigcirc-COO+CH_2+_2$	[431]
		$(CH_2)_2-\overset{\oplus}{N}(CH_3)_3$	$+CH_2+_2$	[432]
		$(CH_2)_2-N$ phthalimide		[433]
CH_3	C_2H_5	C_2H_5	$+CH_2+_2 NH-\overset{O}{\overset{\|}{C}}-$	[434]
CH_3	C_2H_5	C_2H_5	$-\overset{CH_3}{\underset{CH_3}{\overset{\|}{\underset{\|}{C}}}}-NH-\overset{O}{\overset{\|}{C}}-$	[434]
	CH_2Ph	CH_2Ph	$+CH_2+_{10}$	[435]

[a]Names in parentheses are those used in this chapter.

The polymerization behavior of Schemes (81a), (81b) and (81c) is similar to that of unsubstituted monomers [384]. Copolymers of Schemes (81a), (81b) and (81c) with 1,3-butadiene [427], acrylonitrile [428], acryl ester [426], and styrene [426] are of technical interest, due to their fire-retarding properties. Detailed investigations are available on monomers of type (81a), which are particular interesting as models for phospholipide-analogous biological membranes [429]. Table 9 shows selected structure variations of those monomers.

Scheme (81c)-type compounds ($A = CH_2CH_2$) can be synthesized according to the following pathway [436,437]:

$$(83)$$

REFERENCES

1. Clayton, G. D., and Clayton, F. E. *Patty's Industrial Hygiene and Toxicology*, 3rd rev. ed., Vol. 2A, Wiley-Interscience, New York, p. 2649.

2. *Registry of Toxic Effects of Chemical Substances 1980*, National Institute for Occupational Safety and Health, Cincinnati, Ohio, 1980.
3. Redtenbacher, J. (1843). *J. Liebig Ann. Chem.*, *47*: 113.
4. Moureu, C., and Dufraisse, C. (1919). *Compt. Rend.*, *169*: 621.
5. Moureu, C., and Dufraisse, C. (1922). *Compt. Rend.*, *175*: 127
6. Moureu, C., and Dufraisse, C. (1923). *Compt. Rend.*, *176*: 624, 797.
7. Dufraisse, C., and Horclois, R. (1930). *Compt. Rend.*, *191*: 1126.
8. Schulz, H., and Wagner, H. (1950). *Angew. Chem.*, *62*: 105.
9. Weigert, M. N., and Haschke, H. (1974). *Chemiker Ztg.*, *98*: 61.
10. Ohara, T., Sato, T., Shimizu, N., Prescher, G., and Schwind, H. (1985). *Ullmann's Encyclopedia of Industrial Chemistry*, 5th ed., Vol. A1 (Bartholomé, E., Biekert, E., Helimann, H., Lei, H., Weigert, W. M., and Wiese, E., eds.), Verlag Chemie, Weinheim, p. 149.
11. Schulz, R. C. (1964). *Encyclopedia of Polymer Science and Technology*, Vol. 1 (Mark, H. F., Gaylord, N. G., and Bikales, N. M., eds.), Wiley, New York, p. 160.
12. Schulz, R. C. (1985). *Encyclopedia of Polymer Science and Engineering*, 2nd ed., Vol. 1 (Mark, H. F., ed.), Wiley, New York, p. 160.
13. Morlock, G. (1987). *Houben-Weyl: Methoden der organischen Chemie*, 4th ed., Vol. E20, Part 2 (Bartl, H., and Falbe, J., eds.), Georg Thieme, Stuttgart, p. 1127.
14. Adkins, H., and Hartung, W. H. (1941). *Organic Syntheses*, Coll. Vol. I, 2nd ed. (Adkins, H., Hartung, W. H., Whitmore, F. C., and Wolfrom, M. L., eds.), Wiley, New York, p. 15.
15. Bayer, O. (1954). *Houben-Weyl: Methoden der organischen Chemie*, 4th ed., Vol. 7, Part 1 (Bartl, H., and Falbe, J., eds.), Georg Thieme, Stuttgart, p. 157.
16. Bayer, O. (1954). *Houben-Weyl: Methoden der organischen Chemie*, 4th ed., Vol. 7, Part 1 (Bartl, H., and Falbe, J., eds.), Georg Thieme, Stuttgart, p. 253.
17. Bremner, J. G. M., Jones, D. G., and Beaumont, S. (1946). *J. Chem. Soc.*, *1018*.
18. Walter, H., and Schulz, H. (1942). U.S. 2,288,335 to Degussa; C.A. (1943), *37*: 143(8).
19. Schulz, H., and Walter, H. (1941). Ger. 707,021 to Degussa; C.A. (1942), *36*: 1955(7).
20. Belg. 435,516 (1939) to Degussa; *Chem. Zentralbl. (1940)*, *111*: 3177.
21. Watson, F. G. (1947). *Chem. Eng.*, *54*: 106; C.A. (1942), *42*: 2756d.
22. Hearne, G. W., and Adams, M. L. (1948). U.S. Pat. 2,451,485 to Shell Development Co.; C. A. (1949), *43*: 2222f.
23. Idol, J. D. (1959). U.S. 2,905,580 to Standard Oil Co., Ohio; C.A. (1960), *54*: 5470f.
24. Hatch, L. F., and Matar, S. (1978). *Hydrocarbon Process*, *57*: 149; C.A. (1978), *89*: 146322r.
25. Callahan, J. L., Foreman, R. W., and Veatch, F. (1960). U.S. 2,941,007 to Standard Oil Co., Ohio; C.A. (1960), *54*: 19487f.
26. Adams, R. C. (1970). *Chem. Ind.*, *52*: 1644; C.A. (1971), *74*: 52939k.
27. Callahan, J. L., Grasselli, R. K., Milberger, E. C., and Strecker, H. A. (1970). *Ind. Eng. Chem. Prod. Res. Develop.*, *9*: 134; C.A. (1970), *73*: 18916v.
28. Offermanns, H., and Prescher, G. (1983). *Houben-Weyl: Methoden der organischen Chemie*, 4th ed., Vol. E3 (Bartl, H., and Falbe, J., eds.), Georg Thieme, Stuttgart, p. 231.
29. Takenaka, S., and Yamaguchi, G. (1974). Jpn. 7,403,510 to Nippon Kayaku Co.; C.A. (1974), *81*: 151543c.
30. Takenaka, S., Kido, Y., Shimabara, T., and Ogawa, M. (1971). Ger. Offen. 2,038,749 to Nippon Kayaku Co.; C.A. (1971), *74*: 99459u.
31. Arntz, D., Prescher, G., and Heilos, J. (1983). Ger. Offen. 3,125,061 to Degussa AG; C.A. (1983), *99*: 70207r.
32. Arntz, D., Prescher, G., Burkhardt, W., Manner, R., and Heilos, J. (1983). Ger. Offen. 3,125,062 to Degussa AG; C.A. (1983), *98*: 143972c.
33. Ohara, T., Ninomiya, M., Yanagizawa, I., Ueshima, M., and Takeda, M. (1972). Jpn. 7,242,242 to Japan Catalytic Chemical Industry Co.; C.A. (1973), *78*: 57754p.
34. Belg. 613,157 (1962) to Farbwerke Hoechst AG; C.A. (1963), *58*: 1352g.

35. Weigert, W. (1974). *Ullmanns Encyclopedie der technischen Chemie*, 4th ed., Vol. 7 (Bartholomé, E., Biekert, E., Hellmann, H., Lei, H., Weigert, W. M., and Wiese, E., eds.), Verlag Chemie, Weinheim, p. 74.

36. Schulz, R. C. (1967). *Vinyl Polymerization*, Part 1 (Ham, G. E., ed.), Marcel Dekker, New York, p. 403.

37. Schulz, R. C. (1964). *Angew. Chem.*, 76: 357; *Angew. Chem. Intern. Ed.*, 3: 416.

38. Schulz, R. C., and Kern, W. (1956). *Makromol. Chem.*, 18/19: 4.

39. Schulz, R. C., Meyersen, K., and Kern, W. (1962). *Makromol. Chem.*, 54: 156.

40. Schulz, R. C., Meyersen, K., and Kern, W. (1962). *Makromol. Chem.*, 59: 123.

41. Sarasvathy, S., and Venkatarao, K. (1982). *Polymer*, 23: 1999.

42. Schulz, R. C. (1955/1956). *Makromol. Chem.*, 17: 62

43. Schulz, R. C., and Cherdron, H. (1969). *Macromolecular Syntheses*, Vol. 3 (Gaylord, N. G., ed.), Wiley, New York, p. 16.

44. Schulz, R. C., Cherdron, H., and Kern, W. (1957). *Makromol. Chem.*, 24: 141.

45. Ryder, E. E., and Pezzaglia, P. (1965). *J. Polym. Sci. Part A*, 3: 3459.

46. Henglein, A., Schnabel, W., and Schulz, R. C. (1959). *Makromol. Chem.*, 31: 181.

47. Toi, Y., and Hachihama, Y. (1959). *Kogyo Kagaku Zasshi*, 62: 1924; C.A. (1962), 57: 12707f.

48. Toi, Y., Fujii, T., and Hachihama, Y. (1964). *Bull. Chem. Soc. Jpn.*, 37: 307.

49. Kumakura, M., and Kaetsu, I. (1985). *J. Polym. Sci. Polym. Chem. Ed.*, 23: 131.

50. Wöhlk, A. (1900). *J. Prakt. Chem.*, 61: 200.

51. Nef, J. U. (1904). *J. Liebig Ann. Chem.*, 355: 221.

52. Lockemann, G., and Liesche, O. (1905). *J. Prakt. Chem.*, 71: 483.

53. Blacet, F. E., Fielding, G. H., and Roof, J. C. (1937). *J. Am. Chem. Soc.*, 59: 2375.

54. Chang, M., Rembaum, A., and McDonald, C. J. (1984). *J. Polym. Sci. Polym. Lett. Ed.*, 22: 279.

55. Joshi, R. M. (1962). *Makromol. Chem.*, 55: 35.

56. Schulz, R. C., Cherdron, H., and Kern, W. (1961). *Houben-Weyl: Methoden der organischen Chemie*, 4th ed., Vol. 14, Part 1 (Bartl, H., and Falbe, J., eds.), Georg Thieme, Stuttgart, p. 1080.

57. Kern, W., Schulz, R. C., and Cherdron, H. (1960). Ger. 1,082,054 to Degussa; C.A. (1961), 55: 13911i.

58. Kern, W., Schulz, R. C., and Cherdron, H. (1959). Ger. 1,062,937 to Degussa; C.A. (1961), 55: 6932d.

59. Kern, W., Schulz, R. C., and Cherdron, H. (1960). Ger. 1,076,372 to Degussa; C.A. (1961), 55: 18178f.

60. Cherdron, H., Schulz, R. C., and Kern, W. (1959). *Makromol. Chem.*, 32: 197.

61. Cherdron, H. (1960). *Kunststoffe*, 50: 568.

62. Andreeva, I. V., Artemyeva, V. N., Nesterov, V. V., and Kukarkina, N. V. (1979). *J. Polym. Sci. Polym. Chem. Ed.*, 17: 3415.

63. Bäder, E., Rink, K.-H., and Trautwein, H. (1966). *Macromol. Chem.*, 92: 198.

64. Rink, K.-H., and Schweitzer, O. (1962). Ger. 1,138,546 to Degussa; C.A. (1963), 58: 4661b.

65. Kekish, G. T. (1970). Ger. Offen. 1,912,811 to Nalco Chem. Co.; C.A. (1971), 74: 23217t.

66. Kekish, G. T. (1969). U.S. 3,457,230 to Nalco Chem. Co.; C.A. (1969), 71: 81900j.

67. Kekish, G. T., James, W. G., and Simons, D. J. (1972). U.S. 3,640,933 to Nalco Chem. Co.; C.A. (1972), 76: 129121p.

68. Kern, W., Schulz, R. C., and Cherdron, H. (1960). Ger. 1,082,055 to Degussa; C.A. (1962), 57: 11384a.

69. Schulz, R. C., Suzuki, S., Cherdron, H., and Kern, W. (1962). *Makromol. Chem.*, 53: 145.

70. Miller, H. C., and Rothrock, H. S. (1953). U.S. 2,657,192 to E. I. du Pont de Nemours & Co.; C.A. (1954), 48: 3723e.

71. Margel, S., and Wiesel, E. (1984). *J. Polym. Sci. Polym. Chem. Ed.*, 22: 145.

72. Kumakura, M., Suzuki, M., and Kaetsu, I. (1984). *J. Colloid. Interface Sci.*, 97: 157.

73. Schulz, R. C., Müller, E., and Kern, W. (1958). *Naturwissenschaften*, *45*: 440.
74. Thiele, H., and Jentsch, F. (1963). *Kolloid Z. Z. Polym.*, *190*: 99.
75. Bergman, E., Tsatsos, W. T., and Fischer, R. F. (1965). *J. Polym. Sci. Part A*, *3*: 3485.
76. Dawson, T. L., and Welch, F. J. (1964). *J. Am. Chem. Soc.*, *86*: 4791.
77. Brit. Pat. 797,459 (1958) to Degussa; C.A. (1959), *53*: 769f.
78. Schweitzer, O., Kern, W., Schulz, R., and Holländer, R. (1957). Ger. 1,019,825 to Degussa; C.A. (1960), *54*: 11584g.
79. Schulz, R. C., Löflund, I., and Kern, W. (1959). *Makromol. Chem.*, *32*: 209.
80. Schulz, R. C. (1965). *Chimia*, *19*: 143.
81. Schulz, R. C., and Passmann, W. (1963). *Makromol. Chem.*, *60*: 139.
82. Thivollet, P., and Golé, J. (1973). *J. Polym. Sci. Polym. Chem. Ed.*, *11*: 1615.
83. Gulino, D., Pascault, J. P., and Quang, T. P. (1981). *Makromol. Chem.*, *182*: 2321.
84. Calvayrac, H., Thivollet, P., and Golé, J. (1973). *J. Polym. Sci. Polym. Chem. Ed.*, *11*: 1631.
85. Han, Y. K., Park, J. M., and Choi, S. K. (1982). *J. Polym. Sci. Polym. Chem. Ed.*, *20*: 1549.
86. Bum Kim and Chong Shik Chin (1984). *Polyhedron*, *3*: 1151.
87. Hank, R., and Schilling, H. (1964). *Makromol. Chem.*, *76*: 134.
88. Horner, L., Jurgeleit, W., and Klüpfel, K. (1955). *J. Liebig Ann. Chem.*, *591*: 108.
89. Yamashita, N., Tadokoro, A., Ozo, E., Maeshima, T., Baianu, I. C., and Lun-Shin Wei (1987). *J. Macromol. Sci. Chem. Part A*, *24*: 1223.
90. Rink, K.-H. (1959). Ger. 1,059,661 to Degussa; C.A. (1961), *55*: 7920i.
91. Yamashita, N., Yoshihara, M., and Maeshima, T. (1972). *J. Polym. Sci. Polym. Lett. Ed.*, *10*: 643.
92. Yamashita, N., Inoue, H., and Maeshima, T. (1979). *J. Polym. Sci. Polym. Chem. Ed.*, *17*: 2739.
93. Yamashita, N., Morita, S., and Maeshima, T. (1978). *J. Macromol. Sci. Chem. Part A*, *12*: 1261.
94. Horiba, M., Yamashita, N., and Maeshima, T. (1986). *J. Macromol. Sci. Chem. Part A*, *23*: 1117.
95. Yamashita, N., Morita, S., Yoneyama, H., and Maeshima, T. (1983). *J. Polym. Sci. Polym. Lett. Ed.*, *21*: 13.
96. Schulz, R. C., Wegner, G., and Kern, W. (1967). *J. Polym. Sci. Part C*, *16*: 989.
97. Schulz, R. C., Wegner, G., and Kern, W. (1967). *Makromol. Chem.*, *100*: 208.
98. Toi, Y., and Hachihama, Y. (1964). *Bull. Chem. Soc. Jpn.*, *37*: 302.
99. D'Alelio, G. F., and Hummer, T. F. (1967). *J. Polym. Sci. Part A-1*, *5*: 77.
100. Schulz, R. C., Cherdron, H., and Kern, W. (1958). *Makromol. Chem.*, *28*: 197.
101. Schulz, R. C., Kaiser, E., and Kern, W. (1962). *Makromol. Chem.*, *58*: 160.
102. Kinoshita, Y., Kobayashi, J., Ide, F., and Nakatsuka, K. (1971). *Kobunshi Kagaku*, *28*: 430; C.A. (1971), *75*: 152181w.
103. Kinoshita, Y., Kobayashi, J., Ide, F., and Nakatsuka, K. (1971). *Kobunshi Kagaku*, *27*, 496; C.A. (1970), *73*: 110180y.
104. Matsubara, Y., Asakura, J., Yamashita, N., Sumitomo, H., and Maeshimo, T. (1969). *Kogyo Kagaku Zasshi*, *72*: 2658; C.A. (1970), *72*: 121983t.
105. Ouchi, T., and Oiwa, M. (1969). *Kogyo Kagaku Zasshi*, *72*: 1587; C.A. (1969), *71*: 113282c.
106. Ishanov, M. M., Azizov, U. A., Nigmankhodzhayeva, M. S., and Usmanov, Kh. U. (1971). *J. Polym. Sci. Part A-1*, *9*: 1013.
107. Omichi, H., Kataki, A., and Okamoto, J. (1989). *J. Appl. Polym. Sci.*, *37*: 2429.
108. Haschke, H. (1972). *Monstsh. Chem.*, *103*: 525.
109. Haschke, H., Morlock, G., and Kunzel, P. (1972). *Chemiker Ztg.*, *96*: 199.
110. Morita, S., Ikezawa, K., Inoue, H., Yamashita, N., and Maeshima, T. (1982). *J. Macromol. Sci. Chem. Part A*, *17*: 1495.
111. Mateo, J. L., and Sastre, R. (1972). *Makromol. Chem.*, *157*: 141.
112. Gulino, D., Golé, J., and Pascault, J. P. (1980). *Polym. Prepr. Am. Chem. Soc. Div. Polym. Chem.*, *21*: 67; C.A. (1981), *95*: 187768h.

113. Gulino, D., Pham, Q. T., and Golé, J. (1981). *ACS Symp. Ser.*, *166*: 307, C.A. (1982), *96*: 20515g.
114. Gulino, D., Golé, J., and Pascault, J. P. (1981). *Polym. Bull.*, *4*: 641.
115. Morita, S., Yamashita, N., and Maeshima, T. (1980). *J. Polym. Sci. Polym. Chem. Ed.*, *18*: 1599.
116. Yamashita, N., Morita, S., and Maeshima, T. (1982). *J. Polym. Sci. Polym. Chem. Ed.*, *20*: 327.
117. Yamashita, N., Morita, S., Kanzaki, K., and Maeshima, T. (1983). *J. Polym. Sci. Polym. Chem. Ed.*, *21*: 191.
118. Kobayashi, K., Sumitomo, H., and Furuya, K. (1977). *J. Polym. Sci. Polym. Chem. Ed.*, *15*: 1503.
119. Schulz, R. C., Löflund, I., and Kern, W. (1958). *Makromol. Chem.*, *28*: 58.
120. Schulz, R. C., and Elzer, P. (1958). *Makromol. Chem.*, *42*: 205.
121. Schulz, R. C., Kovacs, J., and Kern, W. (1962). *Makromol. Chem.*, *54*: 146.
122. Schulz, R. C., Fauth, H., and Kem, W. (1956). *Makromol. Chem.*, *21*: 227.
123. Schulz, R. C., Holländer, R., and Kern, W. (1960). *Makromol. Chem.*, *40*: 16.
124. Schulz, R. C., Fauth, H., and Kern, W. (1956). *Makromol. Chem.*, *20*: 161.
125. Schulz, R. C., Kovacs, J., and Kern, W. (1963). *Makromol. Chem.*, *67*: 187.
126. Bier, G., Hartel, H., and Nebel, I. U. (1966). *Makromol. Chem.*, *92*: 240.
127. Schulz, R. C., Meyersen, K., and Kern, W. (1962). *Makromol. Chem.*, *53*: 58.
128. Schulz, R. C., Müller, E., and Kern, W. (1959). *Makromol. Chem.*, *30*: 39.
129. Hank, R. (1962). *Makromol. Chem.*, *52*: 108.
130. Schulz, R. C., and Löflund, I. (1960). *Angew. Chem.*, *72*: 771.
131. Schulz, R. C. (1962). *Kolloid Z. Z. Polym.*, *182*: 99.
132. Kern, W., Schulz, R. C., and Löflund, I. (1960). Ger. 1,083,051 to Degussa; C.A. (1961), *55*: 14993f.
133. Rembaum, A. (1983). U.S. 4,413,070; C.A. (1984), *100*: 206178f.
134. Slomkowski, S. (1998). *Prog. Polym. Sci.*, *23*(5): 815.
135. Himmelmann, W., Ulrich, H., and Meckl, H. (1963). Ger. 1,156,649 to Agfa AG; C.A. (1964), *60*: 7609d.
136. Manecke, G., and Pohl, R. (1978). *Makromol. Chem.*, *179*: 2361.
137. Brown, E., and Racois, A. (1978). *Makromol. Chem.*, *179*: 2887.
138. Schulz, R. C., and Ziegler, P. (1983). *Makromol. Chem. Rapid Commun.*, *4*: 629.
139. Margel, S. (1982). *FEBS Lett.*, *145*: 341; C.A. (1982), *97*: 158992n.
140. Margel, S. (1982). *Ind. Eng. Chem. Prod. Res. Develop.*, *21*: 343; C.A. (1982), *97*: 87966u.
141. Margel, S., Beitler, U., and Ofarim, M. (1981). *Immunol. Commun.*, *10*: 567; C.A. (1982), *96*: 158743m.
142. Margel, S., Beitler, U., and Ofarim, M. (1982). *J. Cell Sci.*, *56*: 157; C.A. (1983), *98*: 14023v.
143. Rembaum, A., Yen, R. C. K., Kempner, D. H., and Ugelstad, J. (1982). *J. Immunol. Methods*, *52*: 341; C.A. (1982), *97*: 211767w.
144. Margel, S., and Ofarim, M. (1983). *Anal. Biochem.*, *128*: 342.
145. Ramirez, R. S., and Andrade, J. D. (1973). *J. Macromol. Sci. Chem. Part A*, *7*: 1035.
146. Schulz, R. C., and Passmann, W. (1964). *Makromol. Chem.*, *72*: 198.
147. Golé, J., and Calvayrac, H. (1968). *J. Polym. Sci. Part C*, *16*: 3765.
148. Haschke, H., and Morlock, G. (1974). *Tenside Deterg.*, *11*: 57.
149. Haschke, H. (1973). *Papier*, *27*: 490; C.A. (1974), *80*: 55376j.
150. Haschke, H., and Baeder, E. (1970). Ger. Offen. 1,904,940 to Degussa; C.A. (1970), *73*: 88393w.
151. Haschke, H., and Baeder, E. (1971). Ger. Offen. 1,942,566 to Degussa; C.A. (1971), *75*: 6823r.
152. Koton, M. M., Andreeva, I. V., Andreev, P. V., Danilov, L. G., and Rogozina, E. M. (1962). *Dokl. Akad. Nauk SSSR*, *146*: 608; C.A. (1963), *58*: 12697f.
153. Kaufmann, T., and Boettcher, F.-P. (1959). *Ann. Chem.*, *625*: 123.

154. Arshady, R., Margel, S., Pichot, C., and Delair, T. (1999). *Microspheres, Microcapsules & Liposomes, 1*: 165.

155. Margel, S., Sturchak, S., Ben-Bassat, E., Reznikov, A., Nitzan, B., Kransniker, R., Melamed, O., Sadeh, M., Gura, S., Mandel, E., Michael, E., and Burdygine, I. (1999). *Microspheres, Microcapsules & Liposomes, 2*: 11.

156. Chung, F. L., Zhang, L., Ocando, J. E., and Nath, R. G. (1999). *IARC Sci. Publ., 150*: 45.

157. Wu, R., Hernandez, G., Dunlap, R. B., Odom, J. D., Martinez, R. A., and Silks, L. A. P. (1998). *Trends Org. Chem., 7*: 105.

158. Bartsch, H. (1999). *ARC Sci. Publ., 150*: 1.

159. Feron, V. J., Til, H. P., De Vriejer, F., Woutersen, R. A., Cassee, F. R., and Van Bladeren, P. J. (1991). *Mutat. Res., 259*: 363.

160. Baxter, W. F., Jr. (1979). *Kirk-Othmer Encyclopedia of Chemical Technology*, 3rd ed., Vol. 7 (Mark, H. F., Othmer, D. F., Overberger, C. G., and Seaborg, G. T., eds.), Wiley Interscience, New York, p. 207.

161. Blau, W., Baltes, H., and Mayer, D. (1987). *Ullmann's Encyclopedia of Industrial Chemistry*, 5th ed., Vol. A8 (Gerhartz, W., ed.), Verlag Chemie, Weinheim, p. 83.

162. Freure, B. T. (1945). U.S. 2,378,996 to Carbide and Carbon Chemicals Corp.; C.A. (1946), *40*: 981.

163. Kennedy, D. J. (1946). U.S. 2,413,235 to Shawinigan Chem. Inc.; C.A. (1947), *41*: 1696d.

164. Kennedy, D. J. (1946). U.S. 2,413,235 to Shawinigan Chem. Inc.; C.A. (1947), *41*: 1696d.

165. Brit. 595,170 (1947) to Shawinigan Chem. Ltd; C.A. (1948), *42*: 2985c.

166. Trieschmann, H. G. (1951). Ger. 803,296 to Badische Anilin und Soda-Fabrik; C.A. (1951), *45*: 8550d.

167. Herrmann, W. O., and Heohnek, W. (1953). Ger. 870,846 to Consortium für Elektrochemische Industrie GmbH; C.A. (1956), *50*: 4199d.

168. Young, W. G. (1932). *J. Am. Chem. Soc., 54*: 2500.

169. Owen, L. N. (1943). *J. Chem. Soc.*, 463.

170. Spaeth, E., Lorenz, R., and Freund, E. (1947). *Monatsh. Chem., 76*: 297.

171. Shan, Y., Wu, G., Li, W., and Huang, W. (1998). *Jiangsu Shiyou Huagong Xueyuan Xuebao, 10*: 810; C.A. (1999), *131*: 75216g.

172. Koral, J. N. (1963). *Makromol. Chem., 62*: 148.

173. Koral, J. N. (1962). *J. Polym. Sci., 61*: S37.

174. Rashkov, I. B., and Panayotov, I. M. (1972). *Makromol. Chem., 151*: 275.

175. Panayotov, I. M., and Rashkov, I. B. (1972). *Makromol. Chem., 154*: 129.

176. Schulz, R. C., Wegner, G., and Kern, W. (1967). *J. Polym. Sci. Part C, 16*: 989.

177. Koral, J. N. (1964). U.S. 3,163,622; C.A. (1965), *62*: 6594b.

178. Degering, E. D. F., and Stuodt, T. (1951). *J. Polym. Sci., 7*: 653.

179. Annenkova, V. M., Shilyaeva, N. P., and Annenkova, V. Z. (1995). *Vysokomol. Soedin., Ser. A Ser. B, 37*: 1051; C.A. (1995), *123*: 144798x.

180. Ryabova, I. N. (1997). *Kazakhstan. Izv. Minist. Nauki-Akad. Nauk Resp. Kaz., Ser. Khim., 2*: 40; C.A. (1997), *127*: 346694d.

181. Zigon, M., Sebenik, A., and Osredkar, U. (1992). *J. Appl. Polym. Sci., 45*: 597.

182. Hsu, H., Chang, C., Wang, H., and Hsu, A. (1963). *K'o Hsueh Ch'u Pan She*, 1963: 198; C.A. (1965), *63*: 14983h.

183. Sobue, H., and Saito, Y. (1962). *Kogyo Kagaku Zasshi, 65*: 1630; C.A. (1963), *59*: 4046a.

184. Amerik, V. V., Krentsel, B. A., and Shiskina, M. V. (1965). *Vysokomol. Soedin., 7*: 1713; C.A. (1966), *64*: 3698h.

185. Rashkov, I. B., Spassov, S. L., and Panayotov, I. M. (1973). *Makromol. Chem., 170*: 39.

186. Imoto, T., Ota, T., and Matsubara, T. (1961). *Nippon Kagaku Zasshi, 82*: 378; C.A. (1962), *56*: 8922f.

187. Ota, T. (1965). *Nippon Kagaku Zasshi, 86*: 850; C.A. (1966), *65*: 15523b.

188. Miyakawa, T., Yamamoto, J., and Yaku, F. (1965). *Jpn. 10*: 431; C.A. (1966), *64*: 5233a.

189. Migahed, M. D., and Beckey, H. D. (1971). *Kolloid Z. Z. Polym., 246*: 679.

190. Sebenik, A. (1998). *J. Coat. Technol., 70*: 95.
191. Sebenik, A. (1997). *Polym. Eng. Sci., 37*: 421.
192. Sebenik, A. (1994). *J. Appl. Polym. Sci., 54*: 1013.
193. Sebenik, A., Osredkar, U., and Zidar, R. (1990). *Br. Polym. J., 23*: 145.
194. Sebenik, A., Osredkar, U., and Lesar, M. (1990). *Polymer, 31*: 130.
195. Rasmussen, P. G., Reybuck, S. E., Jang, T., and Lawton, R. G. (1998). U.S. 5,712,408 to University of Michigan; C.A. (1998), *128*: 128380c.
196. Reybuck, S., Rasmussen, P. G., Jang, T., and Lawton, R. G. (1997). *Polym. Prepr. (Am. Chem. Soc., Div. Polym. Chem.)*, *38*: 151.
197. Mirzaev, U. M., and Shadrin, I. F. (1990). *Uzb. Khim. Zh., 3*: 46.
198. Sahoo, S. C., Nayak, P. L., and Lenka, S. (1994). *J. Appl. Polym. Sci., 54*: 1185.
199. Weitschies, W., Heldmann, D., Bunte, T., and Speck, U. (1996). Ger. 4,428,056 to Schering A.-G., C.A. (1996), *124*: 242319g.
200. Klee, J. E., Walz, U., and Mülhaupt, R. (2000). WO 2,000,033,793 to Dentsply International Inc.; C.A. (2000), *133*: 48969q.
201. Yasui, K., Noguchi, Y., Arai, K., and Ajiri, M. (1999). Jp. 11,029,519 to Nippon Synthetic Chemical Industry Co., Ltd.; C.A. (1999), *130*: 139078c.
202. Fukuta, Y., Katsuura, A., and Kakimoto, T. (1998). Jp. 10,087,795 to Nippon Synthetic Chemical Industry Co., Ltd.; C.A. (1998), *128*: 308912m.
203. Kakimoto, T., Kawase, Y., and Yoshida, Y. (1990), Jp. 02,091,039 to Nippon Synthetic Chemical Industry Co., Ltd.; C.A. (1990), *113*: 114661y.
204. Guthrie, J. T., Morris, R. A., He, W. D., Pashley, R. M., Ninham, B., and Senden, T. J. (1996). WO 9,601,282 to Australian National University; C.A. (1996), *124*; 205193r.
205. Guest, H. R., Kiff, B. W., and Stansburg, H. A., Jr. (1963). *Kirk-Othmer Encyclopedia of Chemical Technology*, 2nd ed., Vol. 1 (Standen, A., ed.), Wiley Interscience, New York, p. 255.
206. Ohara, T., Sato, Shimizu, N., Prescher, G., Schwind, H., and Weiberg, O. (1985). *Ullmann's Encyclopedia of Industrial Chemistry*, 5th ed., Vol. 1 (Bartholomé, E., Biekert, E., Hellmann, H., Lei, H., Weigert, W. M., and Wiese, E., eds.), Verlag Chemie, Weinheim, p. 149; Vol. 16, p. 609.
207. Schulz, R. C., Cherdron, H., and Kern, W. (1961). *Houben-Weyl: Methoden der organischen Chemie*, Vol. 14, Part 1 (Bartl, H., and Falbe, J., eds.), Georg Thieme, Stuttgart, p. 1087.
208. Kern, W., and Schulz, R. C. (1957). *Angew. Chem., 69*: 153.
209. Gilbert, E. E., and Donleavy, J. J. (1938). *J. Am. Chem. Soc., 60*: 1737.
210. Howlett, J., and Arche, H. R. (1957). U.S. 2,800,434 to Distillers Co. Ltd.; C.A. (1957), *51*: 18396c.
211. Inoue, T., Oyama, S. T., Imoto, H., Asakura, K., and Iwasawa, Y. (2000). *Appl. Catal., A, 191*: 131.
212. Inumaru, K., Ono, A., Kubo, H., and Misono, M. (1998). *J. Chem. Soc., Faraday Trans., 94*: 1765.
213. Matsura, I., and Aoki, Y. (1993). Jp. 05,331,085 to Nippon Catalytic Chem. Ind. Co.; C.A. (1994), *121*: 107989y.
214. Kuroda, T., and Okita, M. (1992). Jp. 04,128,247 to Mitsubishi Rayon Co., Ltd.; C.A. (1992), *117*: 192537j.
215. Kuroda, T., and Okita, M. (1992). Jp. 04,059,739 to Mitsubishi Rayon K. K.; C.A. (1992), *117*: 8674c.
216. Kuroda, T., and Okita, M. (1992). Jp. 04,059,738 to Mitsubishi Rayon K. K.; C.A. (1992), *117*: 8673b.
217. Brit. 964,552 (1962) to ICI; C.A. (1964), *61*: 1759b.
218. Watanabe, S., and Okita, M. (2000). Jp. 2,000,237,592 to Mitsubishi Rayon Co., Ltd.; C.A. (2000), *133*: 193614z.
219. Watanabe, T., and Eino, O. (1997). Jp. 09,323,950 to Asahi Chemical Industry Co., Ltd.; C.A. (1997), *128*: 61263w.

220. Seto, T., Sakai, Y., and Kinoshita, H. (1995). Jp. 07,082,204 to Mitsubishi Kagaku Kk; C.A. (1995), *123*: 169237b.
221. Sakai, Y., Kinoshita, H., Seto, T., and Ishii, Y. (1995). Jp. 07,069,958 to Mitsubishi Kagaku Kk; C.A. (1995), *123*: 143288u.
222. Sakai, Y., and Kinoshita, H. (1993). Jp. 05,286,886 to Mitsubishi Petrochemical Co; C.A. (1994), *120*: 244115d.
223. Watanabe, S., and Ookita, M. (1993). Jp. 05,253,480 to Mitsubishi Rayon Co., Ltd.; C.A. (1994), *120*: 332309a.
224. Matsuura, I. (1992). Eur. 501,794 to Mitsui Toatsu Chemicals, Inc.; C.A. (1993), *118*: 148212w.
225. Sato, K., Harada, H., Hamachi, H., Asakawa, T., and Yamada, I. (1991). Jp. 03,240,748 to Shindaikyo Sekiyu Kagaku K. K.; C.A. (1992), *116*: 84384c.
226. Okita, M. (1990). Jp. 02,227,140 to Mitsubishi Rayon Co., Ltd.; C.A. (1991), *114*: 102996u.
227. Maksudur, K. M., Zhiznevs'kii, V. M., Bazhan, L. V., and Mokrii, E. M. (1997), *Dopov. Nats. Akad. Nauk. Ukr.*, *8*: 149: C.A. (1999), *130*: 138999s.
228. Gresham, W. F. (1951). U.S. 2,549,457 to E. I. du Pont de Nemours & Co.; C.A. (1951), *45*: 8549d.
229. Bayer, O. (1954). *Houben-Weyl: Methoden der organischen Chemie*, Vol. 7, Part 1 (Bartl, H., and Falbe, J., eds.), Georg Thieme, Stuttgart, p. 170.
230. Yoshida, K., and Okita, T. (1998). Jp. 10,072,387 to Mitsubishi Rayon Co., Ltd.; C.A. (1998), *128*: 243744m.
231. So, M., and Oodan, K. (1995). Jp. 07,053,435 to Ube Industries; C.A. (1995), *123*: 260356a.
232. Weigert, M. N., and Haschke, H. (1974). *Chemiker Ztg.*, *98*: 94.
233. Eifert, R. L., and Marks, B. M. (1960). Ger. 1,081,231 to E. I. du Pont de Nemours & Co.; C.A. (1961), *55*: 21673d.
234. Schulz, R. C., Suzuki, S., Cherdron, H., and Kern, W. (1962). *Makromol. Chem.*, *53*: 145.
235. Kern, W., Schulz, R. C., and Cherdron, H. (1959). Ger. 1,082,054 to Degussa; C.A. (1961), *55*: 13911.
236. Brit. 829,601 (1960) to E. I. du Pont de Nemours & Co.; C.A. (1960), *54*: 13743c.
237. Kropa, E. L. (1944). U.S. 2,356,767 to American Cyanamid Co.; C.A. (1945), *39*: 225(6).
238. Marks, B. M. (1958). Ger. 1,040,244 (Am. Prior., 1956) to E. I. du Pont de Nemours & Co.; C.A. (1960), *54*: 26006a.
239. Kern, W., Schulz, R. C., and Cherdron, H. (1959). Ger. 1,062,937 to Degussa; C.A. (1961), *55*: 6932d.
240. Andreeva, I. V., Madorskaya, L. Ya., Sidorovich, A. V., and Koton, M. M. (1972). *J. Polym. Sci. Part A-1*, *10*: 1467.
241. Witby, G. S., Gross, M. D., Miller, J. R., and Costanza, A. J. (1955). *J. Polym. Sci.*, *16*: 549.
242. Andreeva, I. V., Koton, M. M., Getmanchuk, Y. P., Madorskaya, L. Y., Pokrovskii, E. I., and Kol'tsov, A. I. (1967). *J. Polym. Sci. Polym. Symp.*, *16(3)*: 1409.
243. Madorskaya, L. Y., Andreeva, I. V., and Sidorovich, A. V. (1969). *Int. Symp. Macromol. Chem. Prepr.*, *3*: 231.
244. Bell, E. R., Capanile, V. A., and Bergman, E. (1963). U.S. 3,105,801; C.A. (1963), *58*: 14133a.
245. Schulz, R. C., Wegner, G., and Kern, W. (1967). *J. Polym. Sci. Part C*, *16*: 989.
246. Schulz, R. C., and Passmann, W. (1963). *Makromol. Chem.*, *60*: 139.
247. Panayotov, I. M., and Rashkov, I. B. (1972). *Makromol. Chem.*, *154*: 129.
248. Rashkov, I. B., and Panayotov, I. M. (1972). *Makromol. Chem.*, *151*: 275.
249. Rashkov, I. B., Spassov, S. L., and Panayotov, I. M. (1973). *Makromol. Chem.*, *170*: 39.
250. Koton, M. M., Andreeva, I. V., Getmanschuk, Yu. P., Madarskaya, L. Ya., Pokroskii, E. J., Koltsov, A. I., and Filatova, V. E. (1965). *Vysokomol. Soedin.*, *7*: 2039. C.A. (1966), *64*: 11331d.
251. Panayotov, I. M., and Rashkov, I. B. (1973). *J. Polym. Sci. Polym. Chem. Ed.*, *11*: 2615.
252. Bonner, B., Padias, A. B., and Hall, H. K., Jr. (1992). *Polym. Bull.*, *28*: 517.

253. Nagai, Y., and Nakajima, T. (1963). *Kogyo Kagaku Zasshi*, *66*: 1905; C.A. (1964), *61*: 8181e.
254. Park, J. K., Choi, H. K., and Endo, T. (1995). *Korea Polym. J.*, *3*: 106.
255. Ambroz, L., Caskova, J., Jelinek, M., and Majer, J. (1976). *Chem. Prum.*, *26(1)*: 27.
256. Ambroz, L., Caskova, J., Jelinek, M., Majer, J., and Pidrova, V. (1976). *Chem. Prum.*, *26(2)*: 82.
257. Nutting, H. S., and Petrie, P. S. (1942). U.S. 2,256,152 to Dow Chemical Co.; C.A. (1942), *36*: 192(2).
258. Chapin, E. C., and Longley, R. I. (1959). U.S. 2,889,311 to Monsanto Chemical Co.; C.A. (1959), *53*: 16601i.
259. Reinhard, R. H. (1959). U.S. 2,833,743 to Monsanto Chemical Co.; C.A. (1958), *52*: 13323e.
260. Seyeek, O., Bednar, B., Houska, M., and Kaial, J. (1982). *Collect. Czech. Chem. Commun.*, *47*: 785.
261. Kopeikin, V. V., Panarin, E. F., Milewskaya, J. S., and Redi, N. S. (1977). *Vysokomol. Soedin. Ser. A.*, *19*: 861; C.A. (1977), *86*: 171949w.
262. Allen, C. (1942). Can. 409,376 to Shell; C.A. (1943), *37*: 1531(7).
263. Chapin, E. C., and Longley, R. J., Jr. (1959). U.S. 2,893,979 to Monsanto Chem. Co.; C.A. (1959), *53*: 17535h.
264. Belyaev, V. I., and Saprygina, V. M. (1981). *Ref. Zh. Khim.* (1980), Abstr. 19S227; C.A. (1979), *91*: 158212j.
265. Nehar, H. T., and Woodward, C. F. (1947). U.S. 2,416,536 to Rohm & Hass Co.; C.A. (1947), *41*: 4006i.
266. Schauer, J., Hauska, M., and Kalal, J. (1980). *Makromol. Chem.*, *181*: 367.
267. Oikawa, E., and Makino, H. (1989). *Sep. Sci. Technol.*, *24*: 659.
268. Josa, J., Takagi, H., Kuwamoto, T., and Iriguchi, J. (1996). Jp. 08,218,056 to Nippon Catalytic Chem Ind; C.A. (1996), *125*: 283948a.
269. Iriguchi, J., Takagi, H., Kuwamoto, T., and Josa, J. (1996). Jp. 08,208,757 to Nippon Catalytic Chem Ind; C.A. (1996), *125*: 276931m.
270. Chosa, J., Kuwamoto, T., Iriguchi, J., and Matsuda, T. (1995). Jp. 07,010,944 to Nippon Catalytic Chem Ind; C.A. (1995), *122*: 291857y.
271. Iriguchi, J., Chosa, J., Kuwamoto, T., and Matsuda, T. (1994). Jp. 06,329,740 to Nippon Catalytic Chem Ind; C.A. (1995), *122*: 188530n.
272. Iriguchi, J., Chosa, J., Kuwamoto, T., and Matsuda, T. (1994). Jp. 06,329,723 to Nippon Catalytic Chem Ind; C.A. (1995), *122*: 188549a.
273. Kurosawa, S., Kamo, N., Minoura, N., and Muratsugu, M. (1997). *Mater. Sci. Eng., C.*, *C4*: 291.
274. Morlock, G. (1987). *Houben-Weyl: Methoden der organischen Chemie*, 4th ed., Vol. E20 (Bartl, H., and Falbe, J., eds.), Georg Thieme, Stuttgart, p. 1138.
275. Daly, W. H. (1971). *Encyclopedia of Polymer Science and Technology*, Vol. 14 (Mark, H. F., ed.), Wiley, New York, p. 617.
276. Tsuruta, T., Fujio, R., and Furukawa, J. (1964). *Makromol. Chem.*, *80*: 172.
277. Blaise, E. E., and Maire, M. (1906). *Compt. Rend.*, *142*: 215.
278. Bartholomé, E., Biekert, E., Helltnann, H., Lei, H., Weigert, W. M., and Wiese, E., eds. (1977). *Ullmanns Enzyklopädie der Technischen Chemie*, 4th ed., Vol. 14, Verlag Chemie, Weinheim, p. 207.
279. White, T., and Haward, R. N. (1943). *J. Chem. Soc.*, 25.
280. McManon, E. M., Roper, J. H., Utermohler, W. P., Haser, R. H., Harris, R. C., and Brant, J. H. (1948). *J. Am. Chem. Soc.*, *70*: 2971.
281. Farberov, M. J., and Mironov, G. S. (1963). *Dokl. Akad. Nauk SSSR*, *148*: 1095; C.A. (1963), *59*: 5062a.
282. Fleischer, D. (1975). *Polymer Handbook*, 2nd ed. (Brandrup, T., and Immergut, E. H., eds.), Wiley, New York, p. VII/11.
283. Marvel, C. S., and Levesque, Ch. L. (1938). *J. Am. Chem. Soc.*, *60*: 280.

284. Matsuda, T., Yamakita, H., and Fuji, S. (1964). *Kobunshi Kagaku, 21*: 415; C.A. (1964), *61*: 12093b.
285. Haward, R. N. (1948). *J. Polym. Sci., 3*: 10.
286. Ishida, T., Kondo, S., and Tsuda, K. (1977). *Makromol. Chem., 178*: 3221.
287. Whitby, G. S., Gross, M. D., Miller, J. R., and Constanza, A. J. (1955). *J. Polym. Sci., 16*: 549.
288. Marvel, C. S., and Casey, D. S. (1959). *J. Org. Chem., 24*: 957.
289. Lyons, A. R. (1972). *Macromol. Rev., 6*: 251.
290. Tsuruta, T., and Yasuda, Y. (1968). *J. Macromol. Sci. Chem. Part A, 2*: 943.
291. Feng, H. T., Chiu, N. F., and Chiang, T. C. (1963). *K'o Hsueh T'ung Pao, 5*: 51; C.A. (1964), *60*: 4262e.
292. Feng, H. T., Chiu, N. F., and Chiang, T. C. (1965). *K'o Fen Tzu Tung Hsun, 7*: 48; C.A. (1965), *63*: 16478c.
293. Eicher, T. (1966). *The Chemistry of the Carbonyl Group* (Patai, S., ed.), Interscience, London.
294. Caterall, E., and Lyons, A. R. (1971). *Eur. Polym. J., 7*: 849.
295. Kurihara, Y., Higa, T., Sato, K., and Abe, S. (1965). *Bull. Chem. Soc. Jpn., 38*: 29.
296. Iwatsuki, Sh., Yamashita, Y., and Ishii, Y. (1963). *J. Polym. Sci. Polym. Lett., 1*: 545.
297. Yamashita, N., Inoue, H., and Maeshirna, T. (1979). *J. Polym. Sci. Polym. Chem. Ed., 17*: 2739.
298. Schildknecht, C. E., Zoss, A. O., and Grossner, F. (1949). *Ind. Eng. Chem., 41*: 2891.
299. Holleck, L., and Mahapatra, S. (1969). *Monatsh. Chem., 100*: 1928.
300. Paspaleev, E., Kantschev, K., and Batzalova, K. (1981). *Monatsh. Chem., 112*: 287.
301. Webster, W., Hartier, W. R., Sogah, D. Y., Farnham, W. B., and Rajan Babu, T. V. (1983). *J. Am. Chem. Soc., 105*: 5706.
302. Reetz, M. T., Knauf, T., Minet, U., and Bingel, C. (1988). *Angew. Chem., 100*: 1422.
303. Lewis, F. M., Walling, C., Cummings, W., Briggs, E. R., and Wenison, W. J. (1948). *J. Am. Chem. Soc., 70*: 1527.
304. Cooper, W., and Catterall, E. (1956). *Can. J. Chem., 34*: 387.
305. Haas, H. C., and Simon, M. S. (1952). *J. Polym. Sci., 9*: 309.
306. Young, L. J. (1961). *J. Polym. Sci., 54*: 411.
307. Yamashita, N., Ikezawa, K., Ayukawa, Sh., and Maeshirna, T. (1984). *J. Macromol. Sci. Chem. Part A, 21*: 615.
308. Egboh, S. H. O., and Fagbule, M. O. (1988). *Eur. Polym. J., 11*: 1041.
309. Kuran, W., Pasynkiewicz, S. T., Florjanezyk, Z., and Lyszkowa, D. (1978). *J. Polym. Sci. Polym. Chem. Ed., 16*: 867.
310. Pasynkiewicz, S., Diem, T., and Korol, A. (1970). *Makromol. Chem., 37*: 61.
311. Masuda, S., Omochi, T., and Ota, T. (1985). *J. Polym. Sci. Chem. Ed., 23*: 2081.
312. Saegusa, T., Kobayashi, S., and Kimura, Y. (1977). *Macromolecules, 10*: 64.
313. McNeill, I. C., and Neil, B. (1971). *Eur. Polym. J., 7*: 115.
314. Grassie, N., and Hay, J. N. (1963). *Makromol. Chem., 64*: 82.
315. Hay, J. N. (1963). *Makromol. Chem., 67*: 31.
316. Cooper, W., and Catterall, E. (1954). *Chem. Ind.*, 1514.
317. Wissbrun, K. F. (1959). *J. Am. Chem. Soc., 81*: 58.
318. Wulff, G., Birnbrich, P., and Hansen, A. (1988). *Angew. Chem. Int. Ed. Engl., 100*: 1197.
319. Kuroki, M., Watanabe, T., Aida, T., and Inoue, S. (1991). *J. Am. Chem. Soc., 113*: 5903.
320. Spaether, W., Klaß, K., Erker, E., Zippel, F., and Fröhlich, R. (1998). *Chem. Eur. J., 4*: 1411.
321. Ma, J. H., Tauber, J. D., and Ramelow, U. S. (1997). *Turk. J. Chem., 21(4)*: 313.
322. Ramelow, U. S., Ma, J. H., and Tauber, J. D. (1998). *Polym. Prepr. (Am. Chem. Soc. Div. Polym. Chem.), 39(1)*: 179.
323. Kim, Y. M., Ha, C. S., Cho, S. J., Park, D. K., and Cho, W. J. (1992). *Pollimo, 16(5)*: 563.
324. Van Marle, C. M., and Tollens, B. (1903). *Chem. Ber., 36*: 1351.

325. Merling, G., and Koehler, H. (1911). U.S. Pat. 981.668 to Farbenfabrik vorm. Friedr. Bayer & Co.; C.A. (1911), *5*: 1192.
326. Greenley, R. Z. (1980). *J. Macromol. Sci. Chem., Part A, 14(4)*: 445.
327. Blaise, E. E., and Maire, M. (1906). *Compt. Rend., 142*: 215.
328. MeManon, E. M., Roper, J. H., Utermohler, W. P., Haser, R. H., Harris, R. C., and Brant, J. H. (1948). *J. Am. Chem. Soc., 70*: 2971.
329. Carter, A. S. (1933). U.S. Pat. 1,896,161 to E. I. du Pont de Nemours & Co.; C.A. (1933), *27*: 2458.
330. Nicodemus, O., and Weibezahn, W. (1934). Ger. Pat. 590,237 to E. I. du Pont de Nemours & Co.; C.A. (1934), *28*: 2014(3).
331. Belg. Pat. 660,006 (1965) to Imperial Chemical Industries, Ltd.; C.A. (1965), *64*: 598c.
332. Popova, N. I., Kabakova, B. V., Milman, F. A., and Vermel, E. E. (1964). *Dokl. Akad. Nauk SSSR, 155*: 149; C.A. (1964), *60*: 13070e.
333. Farberov, M. I., and Mironov, G. S. (1963). *Dokl. Akad. Nauk SSSR, 148*: 1095; C.A. (1963), *59*: 5062a.
334. Schildknecht, C. E. (1952). *Vinyl and Related Polymers*, Wiley, New York, p. 682ff.
335. Tsuruta, T., Fujio, R., and Furukawa, J. (1964). *Makromol. Chem., 80*: 172.
336. Overberger, C. G., and Schille, A. M. (1963). *J. Polym. Sci., Part C, 1*: 325; C.A. (1961), *54*: 530.
337. Barnes, C. E. (1943). U.S. Pat. 2,309,727 to E. I. du Pont de Nemours & Co.; C.A. (1943), *37*: 3767(8).
338. Sam, J., and Mozingo, J. R. (1969). *J. Pharm. Sci., 58*: 1030.
339. Staudinger, H., and Ritzenthaler, B. (1934). *Ber. dtsch. Chem. Ges., 67*: 1773.
340. McNeill, I. C., and Neil, D. (1971). *Eur. Polym. J., 7*: 115.
341. Mulvaney, J. E., and Dillon, J. G. (1968). *J. Polym. Sci. Part A-1, 6*: 1849.
342. Thomas, P. R., Tyler, G. J., Edwards, T. W., Radeliffe, A. T., and Cubbon, R. C. P. (1964). *Polymer, 5*: 525.
343. Isler, O., Huber, W., Ronco, A., and Korler, M. (1947). *Helv. Chem. Acta, 30*: 1911.
344. Feng, H. T., Chin, N. F., and Chiang, T. C. (1963). *K'o Hsueh, T'ung Pao., 5*: 51; C.A. (1964), *60*: 4262e.
345. Thomas, P. R., and Tyler, G. J. (1962). Brit. Pat. 898,866 to Brit. Nylon Spinners Ltd.; C.A. (1962), *57*: 7472f.
346. Marvel, C. S., and Casey, D. I. (1959). *J. Org. Chem., 24*: 957.
347. Chaudhuri, A. K., and Basu, S. (1959). *Makromol. Chem., 29*: 48.
348. Chaudhuri, A. K. (1959). *Makromol. Chem., 31*: 214.
349. Marvel, C. S., Riddle, E. H., and Corner, J. O. (1942). *J. Am. Chem. Soc., 64*: 92.
350. Matsuzaki, K., and Lay, T. C. (1967). *Makromol. Chem., 110*: 185.
351. Matsuzaki, K., Yoshitnura, M., and Sobue, H. (1964). *Kogyo Kagaku Zasshi, 67*: 944.
352. Lyons, A. R. (1972). *Macromol. Rev., 6*: 251.
353. Labidi, A., Salon, M.-C., Gandini, A., and Cheradame, H. (1985). *Polym. Bull., 14*: 271.
354. Smets, G., and Oosterbosch, L. (1952). *Bull. Soc. Chim. Belg., 61*: 139.
355. Goto, Y., Tabata, Y., and Sobue H. (1964). *Kogyo Kagaku Zasshi, 67*: 1276.
356. Feng, H. T., Chin, N. F., and Chiang, T. C. (1965). *K'o Fen Tzu, T'ung Hsun, 7*: 48; C.A. (1965), *63*: 16478c.
357. Tanaka, H., and Otsu, T. (1977). *J. Polym. Sci. Polym. Chem. Ed., 15*: 2613.
358. Lyons, A. R., and Catterall, E. (1971). *Eur. Polym. J., 7*: 839.
359. Wiles, D. M. (1969). *Structure and Mechanism in Vinyl Polymerization* (Tsuruta, T., and O'Driscoll, K. F., eds.), Marcel Dekker, New York, Chap. 8, p. 260.
360. Eicher, T. (1966). *The Chemistry of the Carbonyl Group* (Patai, S., ed.), Interscience, London, p. 621ff.
361. Kawabata, N., and Tsuruta, T. (1965). *Makromol. Chem., 86*: 231.
362. Kawabata, N., and Tsuruta, T. (1965). *Kogyo Kagaku Zasshi, 68*: 339.
363. Lyons, A. R., and Catterall, E. (1971). *Eur. Polym. J., 7*: 339.

364. Tsushima, R., and Tsuruta, T. (1974). *J. Polym. Sci. Polym. Chem. Ed.*, *12*: 183.
365. Lyons, A. R., and Catterall, E. (1971). *J. Polym. Sci. Part A*, *9*: 1335.
366. Allen, P. E. M., and Casey, B. A. (1970). *Eur. Polym. J.*, *6*: 793.
367. Baba, Y. (1968). *Bull. Chem. Soc. Jpn.*, *41*: 928.
368. Lyons, A. R., and Catterall, E. (1970). *J. Organometal. Chem.*, *25*: 351.
369. Coleman, L. E., and Meinhardt, N. A. (1959). *Fortschr. Hochpolym. Forsch.*, *1*: 159.
370. Schildknecht, C. E., Zoss, A. O., and Grosser, F. (1949). *Ind. Eng. Chem.*, *41*: 2891.
371. Otsu, T., and Tanaka, H. (1975). *J. Polym. Sci. Polym. Chem. Ed.*, *13*: 2605.
372. Marvel, C. S., McCorkle, J. E., Fukuto, T. R., and Wright, J. C. (1951). *J. Polym. Sci.*, *6*: 776.
373. Sastre, R., Acosta, J. L., Garrido, R., and Fontan, J. (1976). *Angew. Makromol. Chem.*, *62(1)*: 85.
374. Whitby, G. S. (1954). *Synthetic Rubber*, Wiley, New York, pp. 697, 726ff.
375. Cooper, W., and Catterall, E. (1956). *Can. J. Chem.*, *34*: 387.
376. Marvel, C. S., and Levesque, C. L., (1938). *J. Amer. Chem. Soc.*, *60*: 280.
377. Schutte, A. R., (1960). *J. Polym. Sci.*, *47*: 267.
378. David, C., Demarteau, W., and Geuskens, G. (1967). *Polymer*, *8*: 49.
379. Schroer, W. D. (1986). *Houben-Weyl: Methoden der Organischen Chemie*, 4th ed., Vol. 20 (Bartl, H., and Falbe, I., eds.), Georg Thieme, Stuttgart, p. 1300ff.
380. Sander, M. (1969). *Encyclopedia of Polymer Science and Technology*, Vol. 10 (Mark, F., Gaylord, N. G., and Bikales, N. M., eds.), Wiley, New York, p. 123.
381. Weil, E. D. (1988). *Encylopedia of Polymer Science and Engineering*, 2nd ed., Vol. 11 (Mark, H. F., ed.), Wiley, New York, p. 96.
382. Sander, M., and Steininger, E. (1967/68). *J. Macromol. Sci. Macromol. Rev.*, Part C, *2*: 1.
383. Banks, M., Ebdon, J. R., and Johnson, M. (1994). *Polymer, 35, 16*: 3470.
384. Shulyndin, S. V., Levin, Y. A., and Ivanov, B. E. (1981). *Russ. Chem. Rev.*, *50*: 865.
385. Rochlitz, F., and Vilsczek, H. (1962). *Angew. Chem.*, *74*: 970.
386. Kabachnik, M. I., and Medved, T. Y. (1959). *Izv. Akad. Nauk SSR Khim.*, *1959*: 2141; C.A. (1960), *54*: 10834e.
387. Kleiner, H. J. (1982). Ger. 3,110,976 to Hoechst AG; C.A. (1983), *98*: 89660j.
388. Kleiner, H. J. (1982). Ger. 3,120,437 to Hoechst AG; C.A. (1983), *98*: 143643w.
389. Zenftman, H., and Calder, D. (1959). Brit. 812,983 to ICI Ltd.; C.A. (1959), *53*: 15647f.
390. Herbst, W., Rochlitz, F., and Vilscsek, H. (1961). Ger. 1,106,963 to Hoechst AG; C.A. (1961), *55*: 26543a.
391. Dürsch, W., Herwig, W., and Engelhardt, F. (1982). Ger. 3,248,491 A1 to Hoechst; C.A. (1984), *101*: 171962y.
392. Messwarb, G., Denk, W., and Scherer, H. (1958). Ger. 1,027,874 to Hoechst AG; C.A. (1960), *54*: 12665i.
393. Krämer, H., Messwarb, G., and Denk, W. (1958). Ger. 1,032,537 to Hoechst AG; C.A. (1960), *54*: 16010h.
394. Hartmann, M., Hipler, U. C., and Carlsohn, H. (1980). *Acta Polym.*, *31*: 165.
395. Schneider, P. (1962). *Houben-Weyl: Methoden der organischen Chemie*, Vol. 14, Part 2 (Müller, E., Bayer, O., Meerwein, H., and Ziegler, K., eds.), Georg Thieme, Stuttgart, p. 704.
396. Engelhardt, F., and Greiner, U. (1982). DOS 3,248,019 to Cassella AG; C.A. (1984), *101*: 171960w.
397. Engelhardt, F., Kühlein, K., Balzer, J., Dürsch, W., and Kleiner, H. J. (1982). DOS 3,248,031 to Cassella AG; C.A. (1984), *101*: 231191h.
398. Engelhardt, F., Kühlein, K., Balzer, J., Dürsch, W., and Kleiner, H. J. (1982). DOS 3,245,541 to Cassella AG; C.A. (1984), *101*: 192675x.
399. Park, C. H., Nam, S. Y., and Lee, Y. M. (1999). *J. Appl. Polym. Sci.*, *74*: 83.
400. Kolesnikov, G. S., Rodionova, E. F., and Safaralieva, I. G. (1963). *Izv. Akad. Nauk SSR Ser. Khim.*, *1964*: 2028; C.A. (1964), *60*: 6933f.
401. Konter, W., and Vehleefld, W. (1976). DOS 2,646,106 to Bayer AG; C.A. (1978), *89*: 11134sy.

402. Tsuda, T., and Yamashita, Y. (1962). *Kogyo Kagaku Zasshi*, 65: 811; C.A. (1962), 57: 15344f.
403. Coover, H. W., and McCall, M. A. (1962). U.S. 3,043,821 to Eastman Kodak; C.A. (1962), 57: 12731f.
404. Imoto, M., Sakae, M., and Ouchi, T. (1979). *Makromol. Chem.*, 180: 2819.
405. Koeh, G., Lederer, M., and Rochlitz, F. (1959). Ger. 1,125,658 to Hoechst AG; C.A. (1962), 57: 2435h.
406. Kolesnikov, G. S., Rodionova, E. F., Fedorova, L. S., and Gavrikova, L. A. (1960). *Vysokomol. Soedin.*, 2: 1432.
407. Konya, S., and Yokoyama, M. (1965). *Kogyo Kagaku Zasshi*, 68: 1080; C.A. (1966), 64: 11326a.
408. Shashoua, V. E. (1959). U.S. 2,888,434 to E. I. du Pont de Nemours & Co.; C.A. (1959), 53: 16554e.
409. Mashlyakovskii, L. N., Berezina, G. G., Zakharov, V. I., and Jonin, B. I. (1979). *Zh. Obshch. Khim.*, 49: 54; C.A. (1979), 90: 187046t.
410. Mashlyakovskii, L. N., Bodrov, S. G., and Okhrimenko, I. S. (1975). *Vysokomol. Soedin. Ser. B*, 17: 715; C.A. (1976), 84: 5433n.
411. Mashyakovskii, L. N., Berezina, G. G., Dogadina, A. V., Ionin, B. I., and Smirnov, S. A. (1976). *Vysokomol. Soedin. Ser. A*, 18: 308; C.A. (1976), 84: 165296y.
412. Kuzina, N. G., Mashlyakovskii, L. N., Podoiskii, A. F., Okhritnenko, I. S. (1980). *Vysokomol. Soedin. Ser A*, 22: 1270; C.A. (1980), 93: 150710r.
413. Upson, R. W. (1953). *J. Am. Chem. Soc.*, 75: 1763.
414. Hayashi, K. (1978). *Makromol. Chem.*, 179: 1753.
415. Corfield, G. C., and Monks, H. H. (1975). *J. Macromol. Sci. Chem.*, 9: 1113.
416. Jin, J. I., and Byun, H. S. (1981). *J. Macromol. Sci. Chem.*, 16: 953.
417. Pyun, C. H., Jin, J. I., and Sohn, Y. S., (1981). *J. Fire Retard. Chem.*, 8: 135.
418. Yamabe, M., Akiyama, K., Akatsuka, Y., and Kato, M. (2000). *Europ. Polym. J.*, 36: 1035.
419. Doiuchi, T., Nakaya, T., and Imoto, M. (1974). *Makromol. Chem.*, 175: 43.
420. Pudovik, A. N., Kashevarova, E. J., and Golovenkina, L. I. (1965). *Vysokomol. Soedin.*, 7: 1248; C.A. (1965), 63: 13420g.
421. Kimura, T., Nakaya, T., and Imoto, M. (1975). *Makromol. Chem.*, 176: 1945.
422. Kimura, T., Nakaya, T., and Imoto, M. (1976). *Makromol. Chem.*, 177: 1235.
423. Ringsdorf, H., Schlarb, B., and Venzmer, J. (1988). *Angew. Chem.*, 100: 117.
424. Ringsdorf, H., and Schlarb, B. (1988). *Makromol. Chem.*, 189: 299.
425. Ringsdorf, H., Schlarb, B., Tyminski, P. N., and O'Brien, D. F. (1988). *Macromolecules*, 21: 671.
426. Khardin, A. P., Kargin, Y. N., and Kryukov, N. V. (1977). *Vysokomol. Soedin. Ser. B*, 19: 105; C.A. (1977), 86: 156055h.
427. Bauer, R. G., Karchok, R. W., and O'Connor, J. M. (1976). *J. Fire Retard. Chem.*, 3: 99.
428. Finke, M., and Rupp, W. (1982). Ger. 3,210,775 to Hoechst AG; C.A. (1984), 100: 68909s.
429. Nakaya, T., and Li, Y. J. (1999). *Prog. Polym. Sci.*, 24: 143.
430. Nakai, S., Nakaya, T., and Imoto, M. (1977). *Makromol. Chem.*, 178: 2963.
431. Nakai, S., Nakaya, T., and Imoto, M. (1978). *Makromol. Chem.*, 179: 2349.
432. Kadoma, Y., Nakabayashi, N., Masuhara, E., and Yamauchi, J. (1979). *Kobunski Ronbunshu*, 35: 423; C.A. (1978), 89: 186031b.
433. Nakai, S., Nakaya, T., and Imoto, M. (1976). *J. Macromol. Sci. Chem.*, 10: 1547.
434. Ignatious, F., Sein, A., Cabasso, I., and Smid, J. (1993). *J. Polym. Sci. Part A: Polym. Chem.*, 31: 239.
435. Inaishi, K., Nakaya, T., and Imoto, M. (1975). *Makromol. Chem.*, 176: 2473.
436. Hartmann, M., Hipler, U., and Huebig, K. (1976). *Z. Chem.*, 16: 487.
437. Tays, P. (1970). *Chem. Ber.*, 103: 2428.

11
Metal-Containing Macromolecules

Dieter Wöhrle
University of Bremen, Bremen, Germany

I. FUNDAMENTALS ABOUT METAL-CONTAINING MACROMOLECULES

A. Classification

In metal-containing macromolecules or macromolecular metal complexes (MMC) (article in the previous edition of the Handbook see [1]) suitable compounds are combined to materials with new unusual properties: organic or inorganic macromolecules with metal ions, complexes, chelates or also metal clusters. These combinations result in new materials with high activities and specific selectivities in different functions. This article concentrates on synthetic aspects of artificial metal-containing macromolecules. Properties are shortly mentioned, and one has to look for more details in the cited literatures. In order to understand what kind of properties are realized in metal-containing macromolecules, in a first view functions of comparable natural systems (a short overview is given below) has to be considered:

- metallo-enzymes for catalysis,
- hemoglobin, myoglobin for gas transport,
- cofactors for electron-interaction,
- apparatus of photosynthesis for energy conversion,
- metallo-proteins and related systems for various functions.

For metal-containing polymers it is important to understand also their molecular arrangements: primary structure (composition of a MMC); secondary structure (steric orientation of a MMC unit); tertiary structure (orientation of the whole MMC); quarternary structure (interaction of different MMCs). The more detailed knowledge about biological macromolecular metal complexes led in the recent years to an intensified research. The activities in this field are parts of IUPAC conferences on Macromolecule-Metal Complexes (MMC I–VII [2]), and are summarized in some monographs and several reviews [3–45].

Various combinations of macromolecules and metal components such as metal ions, metal complexes and metal chelates exist. The side of the macromolecule considers mainly organic polymers, for example, based on polystyrene, polyethyleneimine, polymethacrylic acid, polyvinylpyridines, polyvinylimidazoles and others. The main chain of these polymers can be linear or crosslinked. In several cases a metal is part of the polymer chain leading to new structural units. Inorganic macromolecules like silica, different kinds of sol-gel materials or molecular sieves can be included also if these macromolecules are modified in such a way to carry as active part one metal component in a specific kind of interaction with the carrier. A classification of metal-containing macromolecules is as follows.

Type I: A metal ion, a metal complex or metal chelate is connected with a linear or crosslinked macromolecule by covalent, coordinative, chelate, ionic or π-type bonds (Figure 1). This **type I** is realized by binding of the metal part at a linear, crosslinked polymer or at the outer or interior surface of an inorganic support. Another possibility uses the polymerization or copolymerization of metal containing monomers.

Type II: The ligand of a metal complex or metal chelate is part of a linear or crosslinked macromolecule (Figure 2). Either a multifunctional ligand/metal complex or a multifunctional ligand metal complex precursor are converted in polyreactions to **type II** macromolecular metal complexes.

Type III: The metal is part of a polymer chain or network. This type considers homochain or heterochain polymers with covalent bonds to the metal, coordinative bonds between metal ions and a polyfunctional ligand (coordination polymers), π-complexes in the main chain with a metal, cofacially stacked polymer metal complexes and different types (polycatenanes, polyrotaxanes, dendrimers with metals) (Figure 3).

Type IV: This type is concerned with the physical incorporation of different kinds of metal complexes or metal chelates in linear or crosslinked organic or inorganic macromolecules. The formation and stabilization of metal and semiconductor cluster will be not considered in this review (Figure 4).

Because in most cases no clear IUPAC nomenclature exists for metal-containing macromolecules or macromolecular metal complexes, it is not possible to obtain by a Chemical Abstract literature search a detailed information on them. One has to look for each individual metal, metal ion, metal complex, metal chelate, ligand or also polymer. For **type I** usually rational nomenclature is used (for example: cobalt(II) complex with/ of poly(4-vinylpyridine) or 2,9,16,23-tetrakis(4-hydroxyphenyl)phthalocyanine zinc(II)

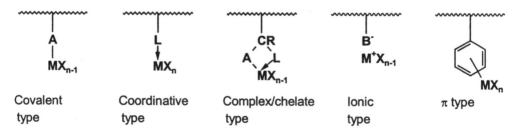

Figure 1 **Type I:** Metal ions, complexes, chelates at macromolecules.

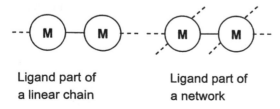

Figure 2 **Type II:** Ligand of metal complexes, chelates as part of linear or crosslinked macromolecules.

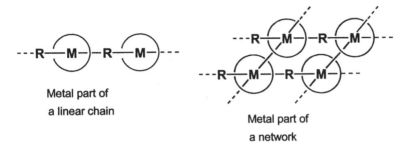

Metal part of
a linear chain

Metal part of
a network

Figure 3   Type III: Metals as part of a linear chain or network.

Physical incorporation

Figure 4   Type IV: Physical incorporation of metal complexes, chelates.

complex covalently bound at poly(methacrylic acid). In the case of **type II** often the metal complex in combination with the term poly is used, e.g., poly(metal phthalocyanines) from 1,2,4,5-tetracyanbenzene. IUPAC nomenclature of **type III** are described as 'regular single-strand' and 'quasi single strand' inorganic and coordination polymers in [46]. The detailed name of the metal complex in polymers or inorganic macromolecules are a common description for **type IV**.

B.   Kinetical, Thermodynamical, and Analytical Aspects of Macromolecular Metal Complex Formation

As in low molecular weight metal complexes, the process of complex formation of metal ion binding in macromolecular metal complexes is accompanied by numerous complicated factors like ion exchange equilibrium, ligand conformational changes, influence in the change of the electrostatic potential, etc. Kind and strength of the formed bonds between metal and ligand depend on the ionisation potential of the metal ion, its electron affinity and the donor properties of the ligand groups. For macromolecular metal complexes either in solution or in the solid state various secondary binding forces are of importance and determine, besides the covalent and ionic bonds, secondary, tertiary and quaternary structures. In addition, specific polymer parameters like degree of crosslinking, distribution of ligands, and, in the case of insoluble polymers, the topography of a macroligand or protecting high molecular weight surrounding must be considered. Many unsolved problems exist in the field of physical chemistry of complex formation, secondary binding forces, composition and reactivity of metal-containing polymers due to their manifold structures. The present situation is best described in [3].

Different models were used to describe the interaction of metal ions with macroligands of **type I** and some **type II** complexes. In one considered model for linear macroligands the polymer ligand L is the central particle, and the metal ion/complex is added in a stepwise manner. In this case the equilibrium constant will not depend on the molecular weight of the macroligand. A second model based on the metal ion M as central particle is described by the Flory concept of infinitely large chains with the reactivity of binding centers independent on their position in the polymer [3]. Another approach calculated the sequence equilibrium which means equilibrium constans for the metal ion binding at different positions at the macroligand ([47,48] and literature cited therein). The equilibrium is usually described by the equilibrium constant $\bar{K}$ of a macroligand L-containing metal ion (M^+) as complexed repeating units [equations (1) and (2)] [3,47–51] (Cp and Cs: initial concentrations of polymer (expressed in repeating units) and metal salt; α: fraction of metal ion/complex not complexed by the polymer).

$$\cdots\text{L}-\text{L}-\text{L}_{n-1}-\text{L}-\cdots \ + \ \text{M}^+ \ \overset{\bar{K}}{\rightleftharpoons} \ \cdots\text{L}-\text{L}-\overset{\overset{\text{M}^+}{\wedge}}{\text{L}}_{n-1}-\text{L}-\cdots \tag{1}$$

$$\bar{K} = \frac{[\text{M}(-\text{L}_n\text{-})]}{[\text{M}^+][(-\text{L}_n\text{-})]} = \frac{1-\alpha}{\alpha(\text{Cp}/n\text{-Cs}(1-\alpha))} \tag{2}$$

The right side of equation (3) is not totally correct because the equilibrium concentration of the macroligand $[-\text{L}_n-] \neq (\text{Cp}/n\text{-Cs}(1-\alpha))$ [47]. The reason is that a sequence of $n+1$ vacant repeating units can consist of different but overlapping neighbor sequences of polymer units. Different length between not complexed sequences exists which influence each other and results in different equilibrium constants $k_1, k_2, k_3 \ldots k_x$ [equations (3)–(5)].

$$\cdots\text{L}-\text{L}-\text{L}-\text{L}-\text{L}-\text{L}-\text{L}-\cdots \ + \ \text{M}^+ \ \overset{k_1}{\rightleftharpoons} \ \cdots\text{L}-\text{L}-\overset{\overset{\text{M}^+}{\wedge}}{\text{L}}-\text{L}-\text{L}-\text{L}-\text{L}-\cdots \tag{3}$$

$$\cdots\text{L}-\text{L}-\overset{\overset{\text{M}^+}{\wedge}}{\text{L}}-\text{L}-\text{L}-\text{L}-\text{L}-\cdots \ + \ \text{M}^+ \ \overset{k_2}{\rightleftharpoons} \ \cdots\text{L}-\text{L}-\overset{\overset{\text{M}^+}{\wedge}}{\text{L}}-\text{L}-\overset{\overset{\text{M}^+}{\wedge}}{\text{L}}-\text{L}-\text{L}-\cdots \tag{4}$$

$$\cdots\text{L}-\text{L}-\overset{\overset{\text{M}^+}{\wedge}}{\text{L}}-\text{L}-\overset{\overset{\text{M}^+}{\wedge}}{\text{L}}-\text{L}-\text{L}-\cdots \ + \ \text{M}^+ \ \overset{k_3}{\rightleftharpoons} \ \cdots\text{L}-\text{L}-\overset{\overset{\text{M}^+}{\wedge}}{\text{L}}-\text{L}-\overset{\overset{\text{M}^+}{\wedge}}{\text{L}}-\text{L}-\overset{\overset{\text{M}^+}{\wedge}}{\text{L}}-\text{L}-\cdots \tag{5}$$

A theoretical model allows to determine k_1 and k_2 on the basis of a numeral fit $\alpha = f(n, [\text{L}]_0, [\text{M}^+]_0, k_1, k_2)$ with $[\text{L}]_0, [\text{M}^+]_0$. The validity of the model was tested for the interaction of Na^+ (as $\text{Na}^+\text{B}[\text{C}_6\text{H}_5]_4^-$) with poly(oxyethylene) in methanol. The best fit between measured and calculated values are found for $n=1$ with $k_1 = 1.9 \, \text{mol}^{-1} \text{L}$ and $k_1/k_2 = 3.5$. Cooperative effects with changes in polymer chain conformation under complex formation must be considered in addition [3]. Bending of a polymer chain by coordination of different ligand groups of one polymer chain leads to an increase of macroligand reactivity (increase of formation constant in comparison to separated, e.g., low molecular weight ligand enters). This was discussed for metal binding at poly(oxyethylene) and poly(4-vinylpyridine) [52,53].

In the case of crosslinked macroligands electrostatic factors significantly influence the composition, structure and stability of a metal complex. Metal ion/complex binding can be described as mentioned before. In heterogeneous systems, when the ligand groups are mainly arranged on a surface with zero concentration in solution, diffusion and topological restrictions must be considered. At low binding center concentrations a Langmuir equation is valid for binding of a metal ion/complex [equation (6)] [3,53] (f: maximum binding of metal ion/complex by a macroligand).

$$\frac{[M^+]}{[M]_{bound}} = \frac{1}{\bar{K}} + \frac{[M]}{f_{max}} \tag{6}$$

One example is the binding of Cu^{2+} by a crosslinked polymer containing bis (carboxymethyl)amino ligand groups with $\bar{K} = 3.5 \times 10^{-3}$ L/mol and $f_{max} = 0.075$ mmol/g [54]. For a non-porous solid matrix containing ligands grafted on a surface the stability of the complex is independent of the degree of surface coverage as shown for $CuCl_2$ or $PtCl_2$ on Aerosil from acetonitrile [55].

The formation of **type II** metal-containing macromolecules obtained by the reaction of bi/multifunctional low molecular weight metal complexes with another bi/multi-functional ligand can be evaluated by usual rate constants, equilibrium and kinetics as known for polycondensation or polyaddition reactions in macromolecular chemistry. Increasing insolubility results easily in chain termination and formation oligomers.

The thermal polycondensation of dihydroxy(metallo)phthalocyanines to cofacially stacked polymer in the solid state as example of a **type III** polymers [equation (7)] is topotactic and under topochemical control, which means that well-defined intermolecular distances and interactions in the lattice control the reaction [56]. Following a kinetic study the fraction of unreacted $-OH$ end groups X over time does not obey a first order kinetics ($X = \exp(-k_2 t^2)$, M = Si, Ge, Sn; $n = 50$–200).

$$n HO–M(Pc)–OH \longrightarrow HO–(M(Pc)–O–)_n H + n-1\, H_2O \tag{7}$$

Besides the kinetic also the thermodynamic during the formation of MMCs is complicated. Changes of the conformation of macromolecules, for example, the chain flexibility, the electrical charges and others influence the thermodynamic parameters such as ΔS in the formation of different types of metal-containing macromolecules [3,57]. The general expression for the reaction is shown in equation (8).

$$\Delta G = -RT \ln \bar{K} = \Delta H - T\Delta S \tag{8}$$

For the formation of a low molecular weight chelate the so-called chelated effect in dependence on kind of solvent interaction is in the order of -5 to -20 kJ/mol mainly determined by entropic terms). The polymer chelate effect for **type I** polymers is more complicated and includes besides the above-mentioned parameters also local, molecular and supramolecular organizations of macromolecules [6,58]. With a low degree of chelation ΔH for macroligands and low molecular weight ligands in the interaction with metal ions are comparable, but ΔS is different (polymer chelate effect) as it was shown for the reaction of amines with Cu^{2+} [59]. For concentrated solutions as well as suspensions, interactions such as intermolecular or supramolecular organizations must be considered and are determined by entropic terms. A more detailed discussion are included in [3].

By intermolecular interactions between the macroligand and the metal ion/complex the temperature of ligand $\leftrightarrow$ gel formation, T_{tr}, is influenced by the ultimate polymer concentration L^{ul} [60]. Above L^{ul} the T_{tr} is independent on the concentration of the polymer and its molecular weight. In the case of Fe^{3+}-polyhydroxamine acid, infinite networks are formed when the probability of intermolecular metal binding is above 50% [6,61].

Type II and III metal-containing macromolecules often form insoluble, more or less crystalline products. Therefore entropic terms going from solution to a crystalline or amorphous precipitate must be considered. Entropic terms are also important for the stabilization of metal clusters or metal complexes/chelates in a high molecular weight surrounding (**type IV** compounds).

During formation of MMCs various thermodynamic side effects driven by a thermodynamically favoured terms can occur. This includes conformational changes, modification of functional groups and also macrochain breakage. Examples of conformational changes are: chain transformation in poly(oxyethylene)-transition metal complexes [61,62], double helix model of poly(oxyethylene)-alkali metal ion complexes [63], conformational modifications of poly(2-vinylpyridine) or poly(amidoamines) during complex formation [64,65], and others. Important to mention here is that chain destruction can occur in **type I** polymers during their formation [3,66,67].

A detailed analysis is the fundamental prerequisite to correlate structure and properties of the new materials. After preparation and isolation of a metal-containing macromolecule at first one has to analyze on the composition of the new material (primary structure). Well-known analytical methods can be used. For soluble compounds usual methods of molecular weight determination can be applied. Microcalorimetric studies allow to measure the enthalpy of formation of a metal-containing macromolecule. In some cases by potentiometric or conductometric measurements complex formation constants can be determined [3,6]. More complicated are the investigation of the secondary, tertiary and quaternary structure of metal-containing macromolecules either in solution or in the solid state. Each method (IR, UV/VIS/NIR, Raman, acoustic, dielectric loss, several methods of x-ray and Mössbauer, ESCA, XAFS, various magnetochemical, ESR techniques, solution/solid NMR, etc.) provides some information on **type I–IV** compounds.

In nearly every case some special analytical investigations must be carried out. This is demonstrated for polyphthalocyanines of **type II** structure. These polymers are obtained by two-dimensional layer growth from various tetracarbonitriles as bifunctional monomers. A polymeric phthalocyanine has in an ideal case a regular planar structure which can be treated in a two-dimensional Cartesian coordinate system allowing positive integers (propagation directions of the polymers are denoted by the letters x and y) [68]. A model describing the structural features such as degree of polymerization, size and shape of polymeric phthalocyanines has been discussed. Equation (9) correlates now the number of macrocycles n (degree of polymerization) with the number of bridged monomers b and the number of end monomers e.

$$n = b/2 + e/4 \tag{9}$$

Evaluation of some data (determination of number of nitrile end groups and groups of bridged monomers by quantitative IR spectra) leads, in dependence on the kind of tetracarbonitrile and reaction conditions, to values of $x = 4$–∞ and $y = 1$–∞. In addition it was shown that the unique structure of polymeric phthalocyanines exhibits fractal properties. They have a regular structure and four fractal dimensions for every size/shape/dilation combination [68]. This important mathematical model can serve as polymer model for discussing basic fractals. Cofacial stacked polymeric phthalocyanines containing four substituents and their possible isomers in such a stacking were also treated mathematically [69].

II. METAL-CONTAINING MACROMOLECULES IN BIOLOGICAL SYSTEMS

A. Metal Complexes in Living Systems

The range of metals used by biological systems is very large, reaching from the alkaline to the transition metals [14–19]. They play an essential role in living systems, both in growth and metabolism. Some metals such as Na, K, Ca, Mg, Fe, Zn are necessary in g quantities. Other trace elements such as Cu, Mn, Mo, Co, V, W, Ni are essential beneficial nutrients at low levels but metabolic poisons at high levels. Some metal ions such as Pb, Cd are called 'detrimental metal ions' because they are toxic and impair the regular course at life functions at all concentrations.

Metal ions such as Ca, Mg, Na, K, Mn exhibit more ionic or coordinative interactions whereas Pt, Hg, Cd, Pb are going more for the covalent bonds, and Ni, Cu, Zn have to be considered as intermediates. In biological systems metal ions can coordinate to a variety of biomolecules such as (Table 1):

- proteins at the (C=O)- or (N–H)-bonds and especially, at N, O, S-donor atoms of side chains;

Table 1 Important bioligand groups and their coordination to metals in natural systems (after Reedijk in [3])

Ligand group	Metal	Substance in which detect or proven
=O	Fe	P-450 enzymes
–OH	Fe, Zn	Carbonic anhydrase
H_2O	Fe, Zn, Ca	Many proteins; additional ligands
O_2/O_2^{2-}	Fe, Cu	Hemoglobin, hemocyanin, hemerythrin
$O_2^{-\bullet}$	Cu, Fe	Superoxide dismutase
$-OOH^{-\bullet}$	Fe	Haemerythrin
Tyrosine	Fe	Oxidases
Glutamase (and Asp)	Fe	Hemerythrin, ribonucleotide reductase
OPO_2R	Ca, Mg	Nucleic acids; ATP
NO_3^-, SO_3^{2-}	Mo	Several reductases
$-Cl^-$	Mn	Mn cluster in photosynthesis
$-S^{2-}$	Fe, Mo	Ferredoxin; nitrogenase
$-SR^-$ (cysteine)	Fe, Cu	Ferrodoxin, plastocyanin, P-450, azurins
Me–S–R (methionine)	Cu, Fe	Plastocyanin, cytochromes, azurins
Imidazole	Cu, Zn, Fe, Mn	Plastocyanin, insulin
Benzimidazole	Co	Vitamin B-12
$(N<)^-$ (peptide)	Cu	Albumin
Tetrapyrroles	Fe, Co, Ni, Mg	Prosthetic groups, hemoglobin
CO	Fe	Toxic for myoglobin; cytochrome oxidase
$(CH_2-R)^-$	Co	Vitamin B-12

- nucleic acids at basic N-donor atoms or at phosphate groups;
- carbohydrates and lipids at (C–O)- and (P–O)-groups;
- in solid bones, teeth, kidney stones.

Metal or metal compound clusters are found, for example, in the respiratory chain (Fe–S clusters) or in the photosynthesis apparatus (Mn clusters).

B. Metal Complexes at Natural Polymer

A bridge between natural and artificial macromolecular metal complexes is the interaction of metal ions/complexes with peptides/proteins [70], nucleic acids/DNA [71,72], enzymes [73], steroids [74], carbohydrates [75]. Biometal-organic chemistry concentrates on such complexes [15]. The reason for the increasing interest in this field lays in medical applications of metal complexes [16,76] (cancer, photodynamic therapy of cancer – immuno-assays, fluorescence markers, enantioselective catalysis, template orientated synthesis of peptides) as exemplarily shown below.

Stable metal complexes can be employed as markers for biochemical and biological systems in immuno-assays, radiographic and electron microscopic investigations of active centers and use as radio pharmaceuticals. Essential is a covalent stable linkage. One simple possibility is the functionalization of peptides and proteins by acylation of, e.g., lysine side chains using succinimyl esters [70]. Modification of this reactive unit with transition metal complexes such as cyclopentadienyl complexes, sandwich complexes or alkinyl clusters leads to the activated carboxylic acid derivatives which can be isolated and reacted with the free amino group of lysine units in peptides and proteins. Fourier-transform-infrared spectroscopy (FT-IR) at $1900-2100 \, cm^{-1}$ allows the detection of the bonded carbonyl complexes down to a dection limit of the picomol region. The carbonyl-metallo-immunoassay (CMIA) has the advantage that no radioactive compounds are necessary and by use of different metal organic markers several immuno assays can be carried out simultaneously. Other possibilities are reviewed in [70].

The chemotherapy of cancer with cytotoxic drugs is one of the major approaches. Most cytotoxic anticancer drugs are only antiproliferative which means that the process of cell division is interrupted. *cis*-Diaminedichloroplatinum(II) (nicknamed cisplatin) is used today routinely against testicular and ovarian cancer. In order to develop new more selective and active anticancer drugs based on platinum, the interaction of the active model compound cisplatin with DNA is important. Structural data have shown that the binding of cisplatin to DNA occurs preferentially at the N7 position of adjacent guanines [72,75,77]. This binding leads to local denaturation of DNA, inhibits the replication process and kills the tumor cells. Because cisplatin possesses two reactive Cl-groups, intrastrand and interstrand crosslinking can occur.

Several ruthenium complexes were investigated in the interaction with proteins, cytochromes and nucleic acids [78]. The reason is to use these Ru-complexes as luminescence sensors (e.g. optical O_2 sensor), to trigger electron transfer and photo-induced electron transfer in proteins and DNA. For example, electrogenerated chemoluminescence (ECL) of $Ru(phen)_3^{2+}$ (phen: 1,10-phenanthroline) can be used to detect the presence of double-stranded DNA (details see [78, p. 642]). $Ru(phen)_3^{2+}$ binds strongly to double-stranded DNA, and minimal binding is observed in the presence of single-stranded DNA. If a given single-stranded DNA sequence is immobilized on an electrode, treatment with a suitable target DNA may generate double-stranded DNA which allows the binding of the Ru-complex and by electrode reactions the detection of ECL.

III. TYPE I: BINDING OF METALS TO MACROMOLECULAR CARRIERS

Several possibilities, as shown in Figure 1, exist for the binding of metal ions/complexes/ chelates to a variety of macromolecules. Methods for the preparation can be subdivided into two main routes:

- Reaction of a macromolecule bearing suitable ligands or reactive substituents for metal ion/complex/chelate binding [equation (10)]
- Homo- or copolymerisation of a vinyl monomer (or other polymerizable groups) bearing a metal complex/chelate or a ligand as a metal complex/chelate precursor [equation (11)].

Along both routes linear or crosslinked materials can be used or obtained. **Type I** compounds with a linear backbone are soluble and can be coated to thin film devices. Crosslinked materials possess in dependence on the amount of crosslinking and procedure of copolymerization pores of different type and size with more or less uniform cross-linked density [79]. One example is amorphous polystyrene crosslinked with divinylbenzene. Non-porous examples are partially crystalline polymers like polyethylene and some inorganic carriers like silica gel. Ligand/metal ion/complex/chelate groups can be distributed on the whole polymer volume or localized only on the carrier surface and connection to the carrier is possible via a direct bond or spacer. All possibilities result in different relativities (properties) of the materials [80,81].

$$
\begin{array}{ccc}
\text{---} \!\!\!\! \underset{L \quad L \quad L}{\overline{\rule{3cm}{0pt}}} \!\!\!\! \text{---} & \xrightarrow[-aX]{+ MX_n} & \text{---} \!\!\!\! \underset{\substack{L \quad L \quad L \\ \text{\scriptsize MX}_{n\text{-}a}}}{\overline{\rule{3cm}{0pt}}} \!\!\!\! \text{---}
\end{array}
\tag{10}
$$

$$
n \; \underset{\substack{L \\ MX_m}}{\square} \longrightarrow \left[\; \underset{\substack{L \\ MX_m}}{\square} \;\right]_n
\tag{11}
$$

A. Anchoring of Metal Complexes or their Ligands at an Organic Macromolecule

1. General Considerations

A macromolecular ligand principally can interact with a metal compound MX_n by covalent, coordinative, ionic, charge-transfer or chelate bindings. The interactions with an organic polymer ligand may occur either through monodentate binding (a) (when MX_n possesses only one coordination vacancy or group for interaction with the polymer ligand) and polydentate binding either intra-(b) or intermolecular (c) [equation (12)]. In the case of linear or branched organic polymers the macromolecular complexes (a) as a rule, are soluble in organic solvents and their structure is identified rather easy. The solubility of the bridged macrocomplexes (b) decreases; they are more stable and have a less-defined structure. The complexes (c) with the intermolecular bridge bonds are insoluble and difficult to characterize. Exemplarily, it was shown for hydroxyamic acid copolymers that infinite networks are formed when the probability of intermolecular binding of metal ions exceeds 50% [82].

$$
\text{(diagram)} \qquad \text{a} \quad \text{b} \quad \text{c}
$$

(12)

The complex formation on the surface of inorganic carriers preferably occurs by the intramolecular types (a) and (b).

The interaction of a polymer ligand with metal ions in aqueous solutions is explained in more detail. Figure 5 shows the dependence of the changes of the hydrogel swelling coefficient of poly(ethylenimine) (PEI) and polyallylamine (PAA·HCl) hydrochloride hydrogel and reduced viscosity of its linear polymer on the concentration of copper sulfate ($C(CuSO_4)$) in aqueous solution (curves a and b) and the ratio of polymer functional groups (C_p) to metal ions concentration (curves c and d) [83]. Characteristic of both investigated systems is the strong compression of hydrogel volume with increasing amount of the metal ion. Attention must be paid to the fact of the influence of the degree of macroligand ionization on the character of the conformational change of the linear segments of the gel. It is seen that under high pH the swelling coefficient of the PEI gel passes through a maximum in the gel-metal ion systems at a concentration of $CuSO_4$ equal to 8×10^{-3} mol/L for Cu^{2+} (molar ratio of Cu^{2+}: PEI = 0.25). The increase of hydrogels swelling degree under complexation with metal ions at high pH can be explained in terms of additional charges in the slightly-charged gel by bivalent metal ions coordinated with the amino groups of PEI. The latter increases the electrostatic energy of the system resulting in an increased swelling coefficient. For the PAA-HCl hydrogel a decrease of the swelling coefficient caused by intramolecular chelation between metal ions and polyligands is observed. This results in additional cross-linking in the network due to the donor–acceptor bonds and compactization of linear parts of polymers between covalently cross-linked points. At low values of pH the complexation proceeds by substitution mechanism of protons of the protonated nitrogen atoms of the gel by metal ions avoiding the stage when the polymer chain acquires charge as it was observed at high pH values

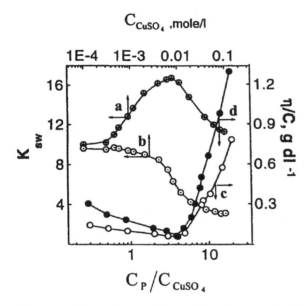

Figure 5 Dependence of the swelling degree and reduced viscosity for PEI gels (a, b), PAA · HCl gel (c) and its linear polymer (d) on the CuSO$_4$ concentration at pH 8.3 (a, c, d) and 6.5 (b).

(see Figure 5(a), gel PEI-Cu^{2+}, curves a and b also). An appropriate correlation between changing of K_{sw} and reduced viscosity of the gel and linear polymer is observed (Figure 5, curves c and d).

In most cases the structure of the local chelated unit in macromolecular metal complexes is the same as in the low molecular weight analogues. But the polymer chain may provide a significant influence. For salicylaldimine ligand the structure of the complex units are different in low molecular weight and macromolecular ones: planar as low molecular weight complex and distored tetrahedral as macromolecular complex [84]. The influence of ring size on the mechanism of the binding of metal ions by polymers can be illustrated in relation to the formation of complexes between TiCl$_4$ and the copolymers of styrene and the diallyl esters of dicarboxylic acids [85]. For $n = 1$ or 2, a mixture of complexes with the *cis*-disposition of **1** and *trans*-disposition of **2** of the carbonyl groups is formed. Increase in the size of the separating bridge ($n > 3$) precludes the formation of type 1 complexes.

The complex formation can be influenced also by the nature of the connecting bridge between the complexing unit and the polymer chain. For example, the transfer of Cu^{2+} from the aqueous to the organic phase (chloroform, toluene) for the formation of a complex with a hydrophobic low molecular weight ligand (compound **3a**) occurs readily. In contrast, complexation by the polymeric analogue **3b** is ineffective. Only the replacement of the short and hydrophobic methylene bridge in compound **3b** by the long hydrophilic ethylenediamine (compound **3c**) or methylamine (compound **3d**) unit leads to appreciable hydrophilicity and spatial mobility of the complexing unit. This results in the diffusion of ions in the polymeric medium and allows the ligands bound to the polymer to be more mobile [86]. By steric hindrance of the macromolecular chain the formation of a multidentate complex often cannot occur. In polystyrene being substituted by bipyridyl groups the formation of a monodentate complex **4** and not of the expected trisbipyridyl complex is observed [87].

3a: R = CH$_2$, R$_1$ = CH$_3$
3b: R = CH$_2$, R$_1$ = polystyrene
3c: R = -NHCH$_2$CH$_2$NH-, R$_1$ = polystyrene
3d: R = -CH$_2$NH-, R$_2$ = polystyrene

4

The closed packing in a polymer chain may lead to uncoordinated ligand groups. Poly(4-vinylpyridine) dissolved in an ethanol/water mixture results with Co-acetylacetonate in a degree of complexation of ~ 0.7. The rate of formation of the Co(II)-complex in with R partly quaternized poly(4-vinylpyridine) decreases due to steric reasons as follows: R = $-CH_3 > -CH_2-C_6H_5$ [88].

Another important point of stereochemical recognition with metal ions called 'template' or 'memory' effect is mentioned. A template effect is observed during the formation of the complexes of corresponding ions with some copolymers followed by cross-linking of the chains [89–92]. The structure of the macrocomplex formed during interaction of the metal ion with the ligand is strictly determined by their nature. If then the metal ion is removed and simultaneously the formed stereostructure of the polymer is preserved, the remained polymer ligand has 'pocket' fitted to the same metal ion (templates) which were removed from the polymeric matrix [equation (13)]. Selectivity and the value of the template effect depend on the spatial organization, on the nature of the complexing ligands and the stabilities of the formed complexes. Examples are complexes of poly(4-vinylpyridine) crosslinked with 1,4-dibromobutane or complexes of polyethyleneimine crosslinked with N,N'-methylenediacrylamide [92].

For the crosslinked polyethyleneimine the distribution coefficients of the non-prearranged polymer between Cu^{2+} and Ni^{2+} is ~ 7.8, whereas for the Cu(II)-prearranged polymer the value is ~ 6.25, and the Ni(II)-prearranged polymer the value is ~ 0.9 which shows different selectivity in metal ion uptake. Catalytic activities for oxidation reactions were investigated.

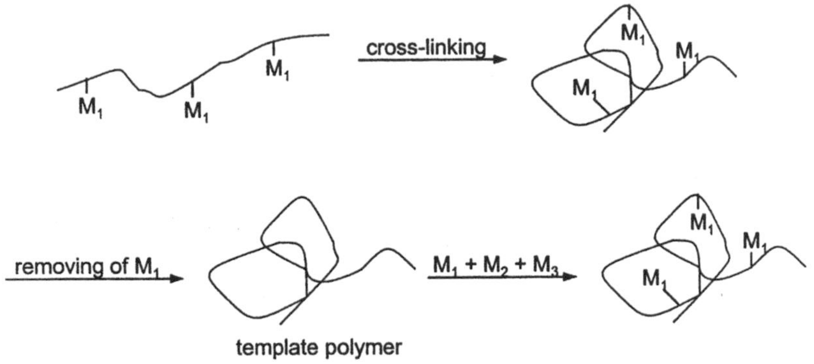

$$(13)$$

template polymer

Another possibility for realizing a template effect used the copolymerization of metal complex vinyl monomers. Copolymerization of Ni(II), Co(II) or Cr(III) complexes of bis[di-4-vinylphenyl)]dithiophosphinates with styrene and ethyleneglycoldimethacrylate yields crosslinked polymers which exhibit after removing of the metal ion in some degree the selectivity of the 'own' metal ion [92,93]. Copolymerization of the Zn(II)-complex of 1,4,7-triazacyclononane with divinylbenzene (molar ratio $\sim 1:3$) results in a macroporous copolymer containing sandwich complexes **5** of the Zn(II) complex [94]. After removal of Zn(II) the prearranged copolymers show now a selectivity of $Cu^{2+}:Zn^{2+}$ up to $157:1$. This means that the thermodynamic stability of the new complex formation dominates in this case over the template effect. But the template effect of Zn^{2+} for Cu^{2+} results in a high selectivity of Cu^{2+} against other transition metal ions such as Fe^{3+}. Altogether the prearrangement effects are difficult to predict and further research is necessary.

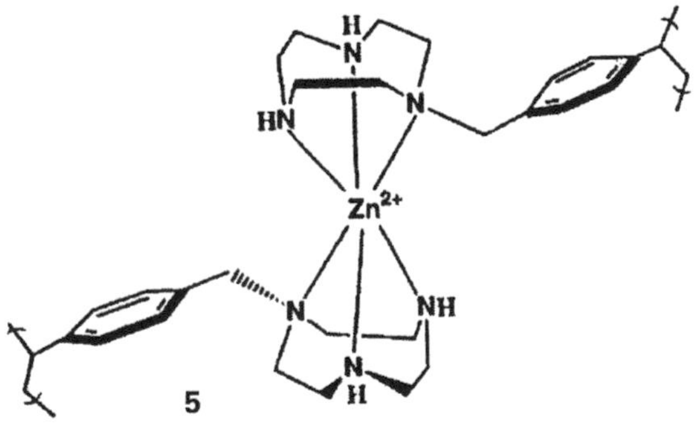

5

B.   Binding of Metal Ions or Complexes at Organic Polymers

Different polymer analogous reactions are applied for the functionalization of polymers by ligands or metal ion/complexes/chelates. The most employed method uses the immobilization of a ligand capable of metal ion complex binding in a second step [3,6,41]. Immobilized lignad groups contain, for example, oxygen, nitrogen, sulfur, phosphorus and arsenic donors. Beside open chain ethers and amines also cyclic ethers and amines are used. Other examples of chelating groups are pyridine-2-aldehyde, iminodiacetic acid, 8-hydroxyquinoline, hydroxylamine, bipyridyl, Schiff bases, Mannich bases, porphyrin-type macrocycles. Often intensively chloromethylated polystyrenes – either linear or with different degrees of crosslinking – are employed as starting material. Water soluble polymers with chelate properties are formed by derivatization of linear polymers such as polyethyleneimine, polyvinylamine, methacrylic acid, polyarylic acid, N-vinylpyrrolidone [3,6,92,95–97]. Other typical ligands are derived from phosphorus compounds like phosphines or phosphates at modified polystyrene for transition metal ion binding [3,6,96]. One example is binding of $PdP(C_6H_5)_3Cl_2$ or $Rh(H)P(C_6H_5)_3(CO)$ at diphenylphosphinated polyethylene $Bu–(CH_2–CH_2–)_n–P(C_6H_5)_2$ obtained by polymerization of ethylene with BuLi and quenching with $(C_6H_5)_2PCl$ [97]. Crosslinked polymers bearing phosphorylic, carboxylic, pyridine, amine and imine functions were used for the binding of Cu^{2+}, Ni^{2+}, Co^{2+} and other transition metal ions. For the well-known metal ion binding at polycarboxylic acids, polyalcohols, polyamines, polyvinylpyridines see [3,6]. In the following only some examples are given.

1. Ethers

Poly(oxyethylene)–metal salt complexes are of interest as solid polymer electrolytes after complex formation with Li^+, Na^+, K^+, Mg^{2+}, Ba^{2+} (see [3,6,41,98] and literature cited therein). The synthesis is carried out by direct interaction of the ligand and metal ions in solution or, if crosslinked poly(oxyethylene) is employed, by immersing the polymer ligand into a solution of the metal salt. As polymer ligands also poly(oxypropylene), crosslinked phosphate esters and ethers were used [99,100]. Polymer cathode materials based on organosulfur compounds are developed for lithium rechargeable batteries with high energy density. A 2,5-dimercapto-1,3,4-thiadiazol-polyaniline composed with Li-counter ions on a copper cathode current collector show high discharge capacity [101].

Crown ether moieties at crosslinked polystyrene are prepared by the reaction of crosslinked chloromethylated polystyrene with hydroxy-substituted crown ethers (in THF in the presence of NaH) [102]. Binding of alkali ions were investigated. Crown ether moieties containing cinnamoyl groups **6** which can be crosslinked by UV-irradiation, are prepared by polymerization of the corresponding vinyl monomer with cinnamoyl and crown ether groups [103].

6

2. Ketones, Carboxylic Acids and Nitriles

Metal acetylacetonates are covalently bound by the reaction of crosslinked chloromethyl-ated polystyrene (DMF, 100 °C) under formation of **7** [104]. Rare earth Eu(III)-complexes of 1-carboxy-8-naphthoyl bound covalently at polystyrene **8** are obtained by Friedel–Crafts acylation of the corresponding naphthalenetetracarboxylic acid anhydride with the polymer followed by reaction with Eu^{3+} [105]. The luminescence properties of lanthanide ions with polycarboxylates were investigated in detail [106]. The effects of the conformation of polymer chains on electron transfer and luminescence behaviour of Co(II)-, Co(III)-ethylenediamine complexes at polycarboxylates were studied [107].

When water-soluble polymers having pendant carboxylic acid residues and powdered metal oxides containing leachable Ca^{2+}, Al^{3+}, etc., ions in the presence of controlled amounts of water, metal cation carboxylate anion salt-bridges are generated which bring about curing or hardening of the formulation [108]. These so-called glass-ionomers are applied as dental biomaterials. An example is a terpolymer based on acrylic acid, itaconic acid and methacrolylglutamic acid **9** hardened with Ca^{2+} or Al^{3+}.

Water soluble macromolecular Pd^{2+} complexes with phase transfer ability employed for the Wacker oxidation of higher alkenes were prepared from ligands such as monobutyl ether of poly(ethylene glycol) functionalized with β,β'-iminodipropionitrile and aceto-nitrile [109]. One example is the polymer ligand **10** complexed with $PdCl_2$. Also other examples are described in [109].

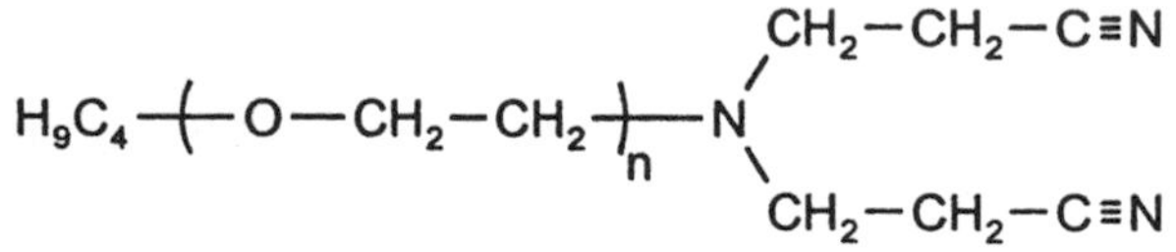

$$H_9C_4 \left(O-CH_2-CH_2 \right)_n N \begin{array}{c} CH_2-CH_2-C\equiv N \\ \\ CH_2-CH_2-C\equiv N \end{array}$$

10

3. Amines, Amido-Oximes and Hydroxamic Acids

Open chain and cyclic amines can coordinate with various metal ions. Poly(ethyleneimine) from 2-methyloxazoline by ring opening polymerization was investigated for Na^+ binding [110]. Various open chain amines and amides, cyclic amines **11** and amides were synthesized starting from crosslinked chloromethylated polystyrene [111]. The modified polymers contain up to 2.7 mmol/g amine or amide groups. They were investigated for the reversible binding of CO^{2+}, Ni^{2+}, Cu^{2+}. Solutions of undoped polyaniline in 1-methyl-2-pyrrolidinone were treated with Cu, Fe and Pd salts [112]. A bathochromic shift of the absorption of polyaniline at $\lambda \sim 640\,nm$ is attributable to charge transfer from the benzoid to the chinoid form of the polymer. The complexes **12** are effective in dehydrogenative oxidation reactions of, e.g., cinnamoyl alcohol.

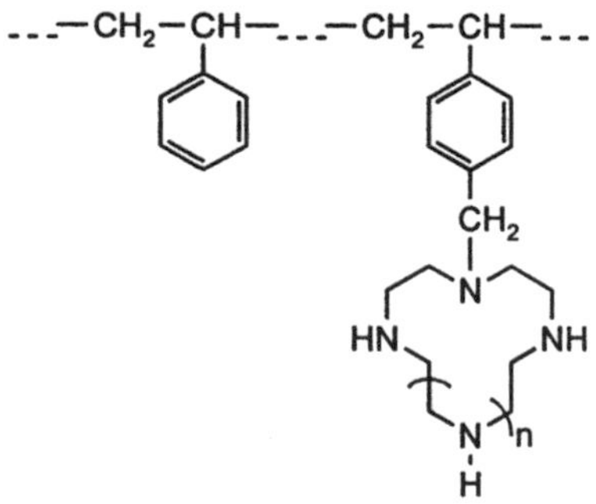

11

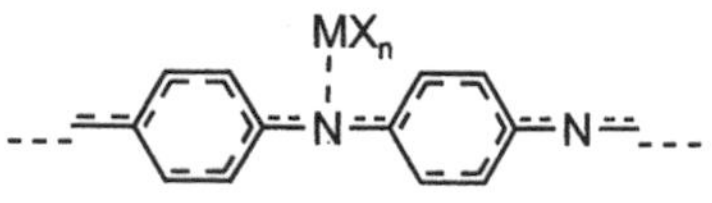

12

Water soluble cetylpyridinium chloride modified poly(ethyleneimine) **13** were investigated for the removal of several cations (Cu^{2+}, Zn^{2+}, Cd^{2+}, Pb^{2+}, etc.) and anions (PO_4^{3-} CrO_4^{2-}) from water [113]. The polymer can form interaction products with negative ions due to electrostatic bonds and also with metal ions due to complex formation. Other basic polymers such as poly(vinylamine), neutral polymers such as polyalcohols and acidic polymers such as poly(acrylic acid) were investigated using the method of 'Liquid-Phase Polymer-Based Retention' for the separation of metal ions from aqueous solution [114].

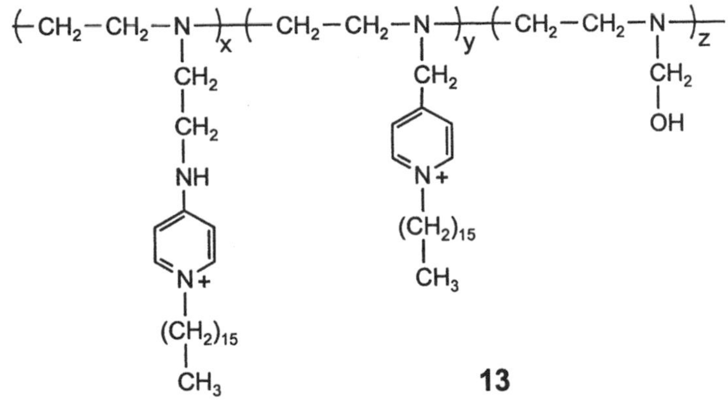

13

A N-isopropylacrylamide-bound hydroxamic acid copolymer **14** was prepared by the reaction of poly(N-isopropyl acrylamide)-co-(2-acryloxysuccinimide) with 6-amino-hexanhydroxamic acid [equation (14)] [115]. This water soluble copolymer after Fe^{3+} uptake quantitatively separates from aqueous solution by heating. By Fe^{3+} uptake of the copolymer the amount of Fe^{3+} in an aqueous solution is reduced from 15.5 ppm to 116 ppb.

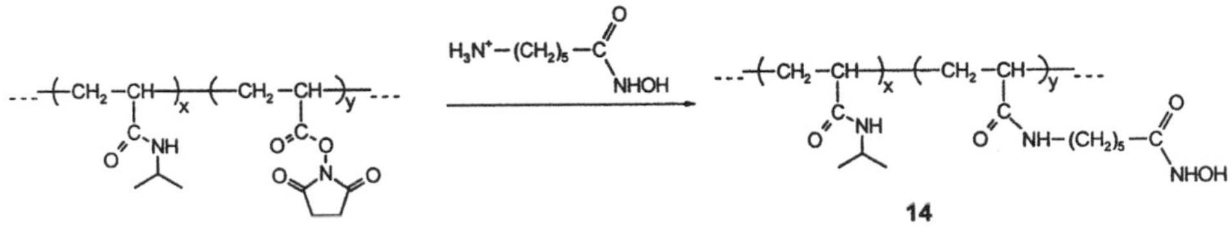

14

$$(14)$$

A crosslinked polystyrene with 2-amido-oxime groups **15** was prepared from crosslinked chloromethylated polystyrene by cyanoethylation and reaction with hydroxyl-amine. This polymer ligand shows a good selectivity for the separation of UO_2^{2+} from sea water [116]. Amideoxime polymers (and their interaction with Cu^{2+}) were also prepared from macroporous acrylonitrile-divinylbenzene co-polymers by reaction with NH_2OH (around 2 mmol/g amideoxime groups in the polymer) [117].

4. Schiff Bases

The reaction of crosslinked polystyrene with 5-chloromethyl-2-hydroxybenzaldehyde followed by interaction with the Co(II) chelate of the Schiff base from 2-hydroxybenzal-dehyde with diaminomaleonitrile yields the polymer chelate **16** (content 0.2 mmol/g chelate centers) [118]. This polymer complex was investigated as catalysts for the conversion of quadricyclane to norbornadiene. Crosslinked chloromethylated polystyrene was reacted with N_2O_3-Schiff base ligands. The resulting macroligands were investigated for the binding of Co^{2+}, Mn^{2+}, Fe^{2+} (formula **17**) [119]. Also cyclic Schiff base chelates were synthesized [120]. Gel-type and macroporous versions of a chiral Mn(III)-salen complex **18** were prepared by the reaction of poly[4-(4-vinylbenzyloxy)salicylaldehyd] at first with a chiral 1,2-diaminocyclohexane to **18a** and then with salicylaldehyde derivatives and a Mn salt to **18b** as shown in equation (15) [121]. These polymers are very active catalysts in the asymmetric epoxidation of alkenes.

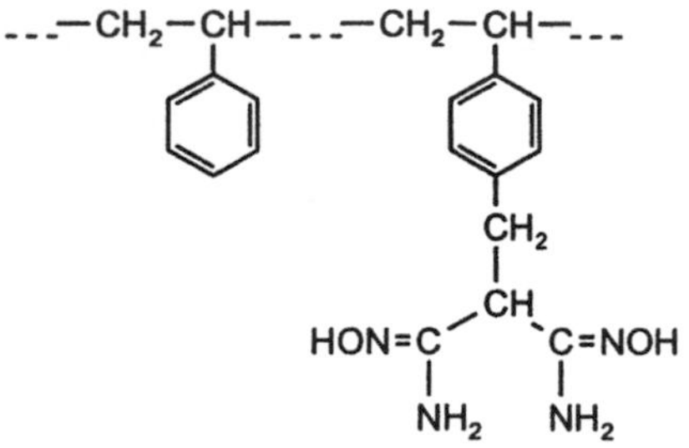

15

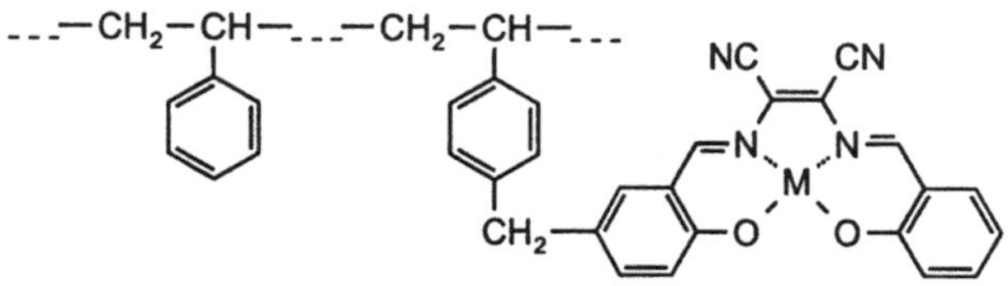

16   (M = Co^{2+})

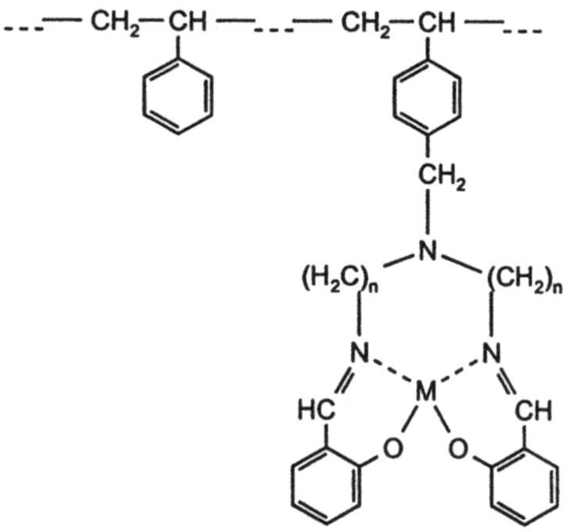

17

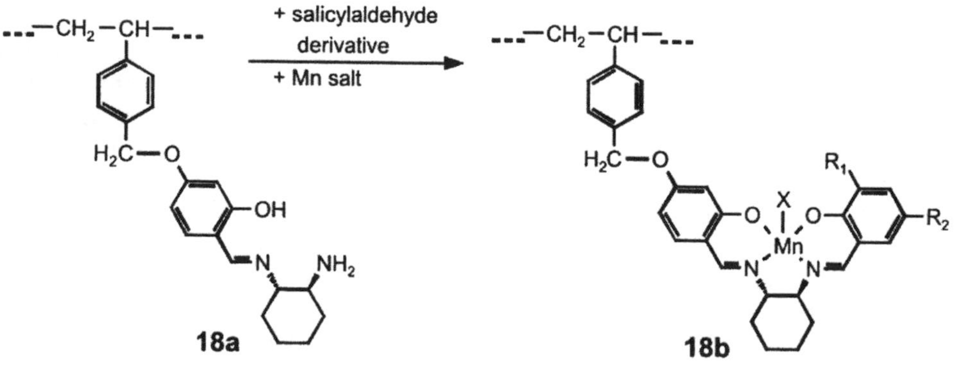

+ salicylaldehyde
derivative

+ Mn salt

18a   **18b**

5. Pyridyl, Bipyridyl and Other Heterocycles

The excellent complexing ability of the pyridine group led to several investigations on the coordination of polymers bearing pyridyl or bipyridyl groups with metal ions like Ru^{2+}, Re^{2+}, Co^{2+} and others [3,6,41,122–124]. Polymers and copolymers of vinylpyridine or N-vinylimidazole can easily interact by coordinative bonds in solution with a variety of transition metal salts, metal complexes and macrocyclic metal chelates such as Schiff base chelates of Co(salen) type, Co(dimethylglyoxim) or porphyrins like 5,10,15,20-tetra-phenylporphyrin [3,5,125–129]. After film casting, binding of oxygen and its separation in membranes were investigated. For the coordinative interaction in analogy to coordinative binding in low molecular weight complexes, the polymer must have groups with σ-donor or π-acceptor properties. In contrast to monoaxial coordination of low molecular weight donors with Co-complexes, polymer donors can interact biaxially with the result of crosslinking, change of polymer conformation and therefore different properties.

Polymer metal complex formation of different polyvinylpyridines in solution, in hydrogels and at interfaces were investigated [83]. In aqueous solution linear or crosslinked polyvinylpyridines in the interaction with H_2PtCl_6 results in reduced viscosities and reduces swelling coefficients, respectively. Complexation leads to molecular bridges and folding of the polymer. Film formation was observed at the interface of poly(2-vinylpyridine) dissolved in benzene and metal salts dissolved in water.

Ru(II), Cu(II), Cr(III) complexes at 2,2′-bipyridyl and poly(4-vinylpyridine) (PVP) are reviewed in [3,6,41]. cis-Ru(II)(2,2′-bipyridyl)$_2^{2+}$(Ru(bpy)$_2^{2+}$) reacts in methanol with PVP to (Ru(bpy)$_2$(PVP)$_2$]$^{2+}$ and with PVP in the presence of pyridine (py) to [Ru(bpy)$_2$(PVP)(py)]$^{2+}$ [130].

A polymer complex containing Ru(bpy)$_3^{2+}$ pendant groups was obtained by the reaction of a lithium substituted polystyrene with 2,2′-bipyridyl followed by interaction with cis-Ru(bpy)$_2^{2+}$ [131]. Another example is binding of 4,4′-dicarboxy-2,2′-bipyridyl at a copolymer of p-aminostyrene followed by reaction with cis-Ru(bpy)$_2^{2+}$ (structure **19**) [132]. Other copolymers with pendant Ru(bpy)$_3^{2+}$ bound via a spacer or containing additional bound 4,4′-bipyridyl are also prepared. These materials are interesting as sensitizers for visible light energy conversion.

(m=0.14, n=0.86, x=0.096)

19

Different polybenzimidazoles bearing cyanomethyl ligands were coordinated with PdCl$_2$ partly with CuCl$_2$ as cocomponent, and investigated for their activity catalyst [133].

Several catalytically active Pd^0-heteroarylene complexes were prepared by the interaction of the polyheterocycles, with $PdCl_2$ followed by reduction with $NaBH_4$ to Pd^0 [134].

6. Porphyrins and Phthalocyanines

A general route that allows binding of different porphyrins at linear polymers was described [135,136]. The substituted porphyrines **20** (R = $-O-C_6H_4-NH_2$), **21** (R = $-NH_2$) and **22** (R= $-NH_2$) contain nucleophilic amino groups of similar reactivities. Therefore, an identical synthetic procedure can be applied to conduct the covalent binding to a polymer with reactive sites. Beside the binding of one porphyrin, the addition of different porphyrins to the reaction mixture allows the fixation of two or three porphyrins at one polymer system in a one-step procedure. Mainly a method was selected where a diluted solution of the polymer was added dropwise to a diluted solution to the porphyrins. If the reaction of poly(4-chloromethylstyrene) is carried out in the presence of an excess of triethylamine, the covalent binding of the porphyrin and a quarternization reaction occur simultaneously. Positively charged polymers **23** soluble in water were obtained. In addition to a porphyrin also viologen as electron relay were covalently bonded at positively charged polystyrene [137].

Negatively charged polymers **24** containing porphyrin moieties are easily synthesized by the reaction of poly(methacrylic acid) (activation of the carboxylic acid group by carbodiimides or triphenylphosphine/CCl_4) with the porphyrins [135,136]. Uncharged water-soluble polymers **25** containing the porphyrin moieties are obtained by the reaction of poly(N-vinylpyrrolidone-co-methacrylic acid) with the low-molecular-weight substituted porphyrins in the presence of the same activating agents for the carboxylic acid groups. Residual carboxylic acid groups were converted to methyl esters. The employed porphyrins **20**–**22** contain four reactive functional groups. Therefore inter- and intramolecular crosslinking may occur in the reaction with the polymers employed. Intermolecular crosslinking could be avoided up to an amount of 2 mol% of applied porphyrins corresponding to one unit of the polymers. Higher amounts of porphyrins result in the formation of gels due to intermolecular crosslinking. Viscosity measurements indicate intramolecular crosslinking (micro-gel formation) in some cases. The porphyrin moieties in the polymers can act as antenna for reactions for electron and photoelectron transfer reactions. By studying these reactions, information concerning the polymer environment can be obtained [136,137].

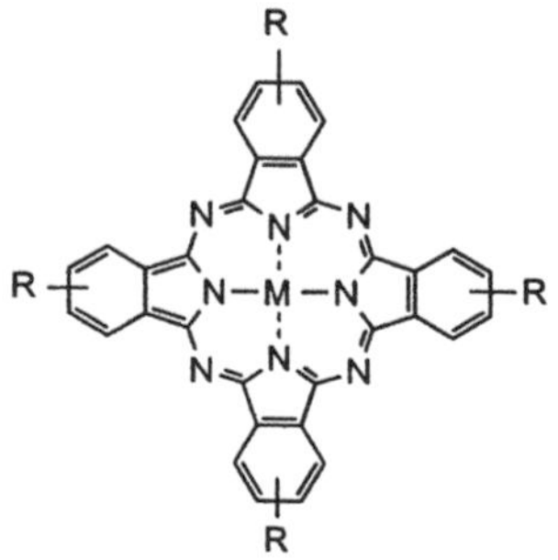

20

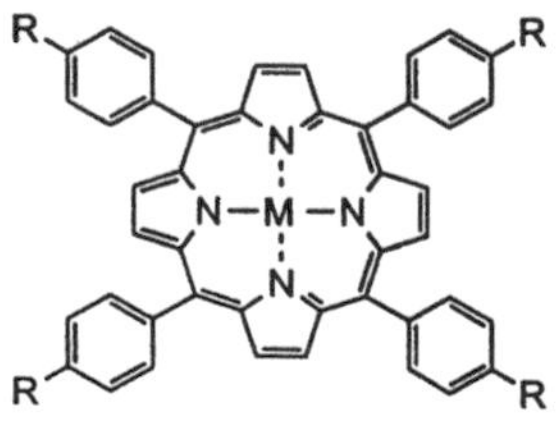

21

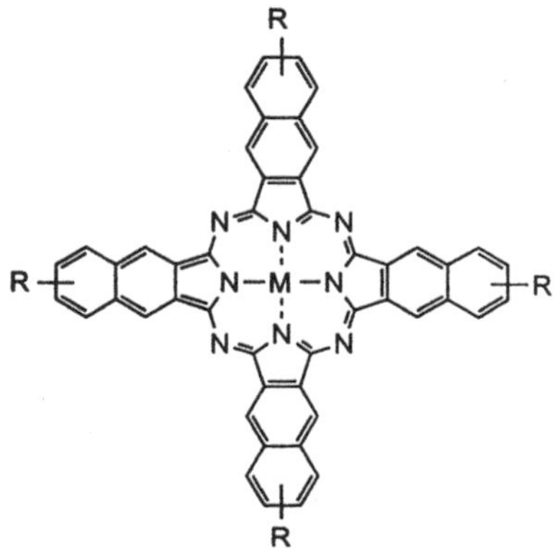

22

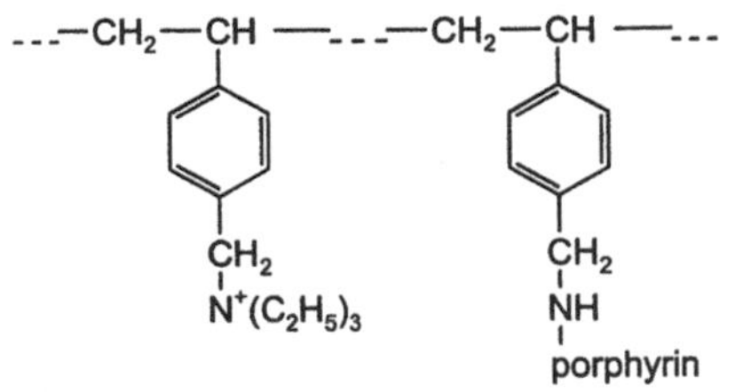

23

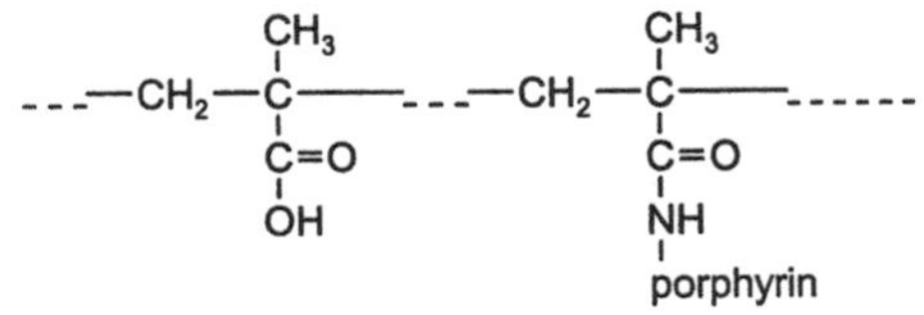

24

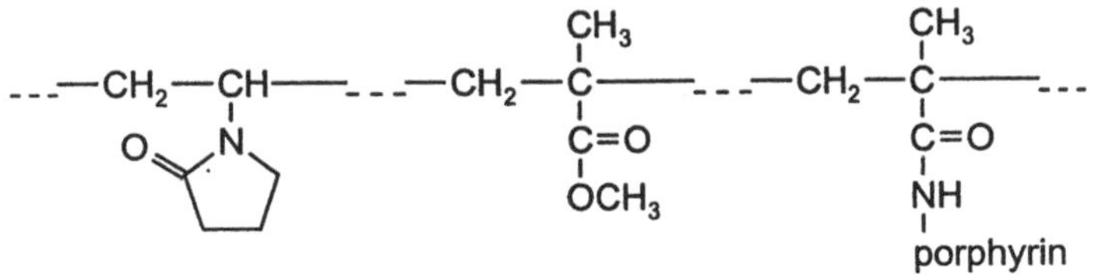

25

Some other reports describe the binding of tetracarboxyphthalocyanines at linear polystyrene [138] or macroporous polystyrene grafted with polyvinylamine [139,140], of chlorosulfonated phthalocyanines at macroreticular polystyrene [141] and of tetra-chlorocarboxyphthalocyanines at poly(γ-benzyl-L-glutamate) [142].

The donor properties of suitable nitrogen containing macromolecular ligands are used in a Lewis base/Lewis acid interaction with cobalt or iron in the core of porphyrin-type compounds to achieve a coordinative binding. Some years ago the coordinative binding of cobalt phthalocyanines **20** with R = –COOH or R = –SO$_3$H was examined taking polymer ligands such a poly(ethyleneimine) [143–145], poly(vinylamine) [143–148], amino group-modified poly(acrylamide) or modified silica gel [146]. For **20** (R = –COOH, M = Co(II)) conclusive evidence of axial coordination was obtained by ESR showing a 5-coordinative complex structure [146]. Increasing concentration of poly(vinylamine) shifted the equilibrium between monomer and aggregated such as dimer form to the monomeric phthalocyanine. A high concentration of polymer ligands separates the chelate molecules in the polymer coil (shielding effect). The materials were investigated as catalysts in oxidation reactions.

Recently, the electrochemical properties of cobalt phthalocyanines included by coordinative binding in membranes of poly(4-vinylpyridine) [149] or poly(4-vinylpyridine-co-styrene) were investigated [150]. The membranes were prepared by dissolving **20** (R = –H, M = Co(II)) in DMF in the presence of poly(4-vinylpyridine). The coordinative interaction of the metal complex to the pyridyl group strongly enhances the solubility of the phthalocyanine in DMF. The film formed on a carrier after casting and evaporation of the solvent is homogenously blue. The pattern in the UV/Vis spectra of the films are comparable to the Co-phthalocyanine dissolved in pyridine showing homogenous monomeric distribution of the metal complex in the polymer. In contrast, the film of the cobalt complex casted from pyridine solution shows a strong resonance broadening of the long wave length band, indicating its crystallinity.

An electrostatic binding occurs easily by ionic interactions of oppositely charged macromolecular carriers and phthalocyanines. Positively charged polymers such as ionenes $[-N^+(CH_3)_2-(CH_2)_x-N^+(CH_3)_2-(CH_3)_2-(CH_2)_y-]_n$ form stoichiometric complexes in the interaction with tetrasulfonated **20** (R = $-SO_3^-$, M = CO(II)) in the composition $N^+/CoPc(SO_3^-)_4$ of 4:1 [146,151,152]. The tendency of aggregation of phthalocyanines in water strongly depends on the hydrophilic character of the kind of latexes based on copolymers of styrene, quarternized *p*-aminomethylstyrene and divinylbenzene [153]. Increasing content of quarternized comonomers enhances the content of non-aggregated **20** (R = $-SO_3^-$, M = Zn(II) absorbing at $\lambda \sim 685$ nm compared to aggregated ones absorbing at $\lambda \sim 640$ nm due to a shielding effect for the positively-charged phthalocyanine.

An ionic binding at charged crosslinked polymers can easily be realized by treating with a solution of an oppositely charged metal complex. Shaking of a positively charged ion exchanger, for example Amberlite, with the negatively charged **20** (R = $-SO_3^-$ M = Zn(II), Al(III)(OH), Si(IV)(OH)$_2$), results in blue-colored polymeric complexes **26** containing monomeric distribution of the MPc [154]. These compounds are very effective photosensitizers for the photooxidation of several substrates by irradiation with visible artificial or solar light.

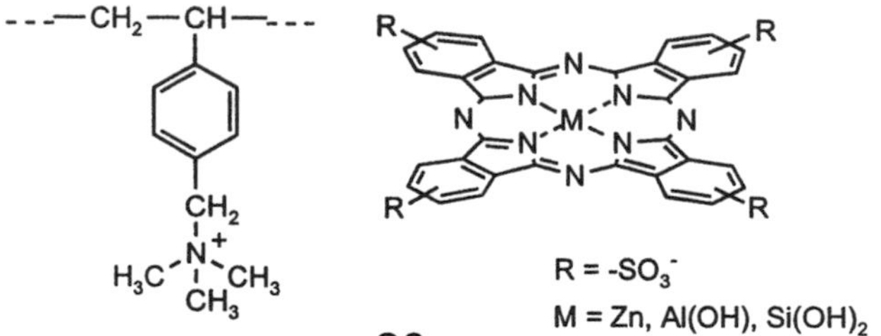

R = -SO$_3^-$

M = Zn, Al(OH), Si(OH)$_2$

26

C. Binding of Metal Complexes on the Surface of Macromolecular Carriers

For different properties such as catalysis it is favourable to create reactive sites on the surface of an organic polymer or an macromolecular inorganic carrier. Anchoring of metal complexes exhibit the advantage of higher reaction rates for reactions at the metal complex centers and the easiness of the separation from the reaction for reuse. Covalent anchorage can be realized by polymerization of different monomers bearing ligand groups **L** for metal complex formation (for example, by mechanical, chemical or irradiated-chemical treatment of the carriers [equation (16)] [3,6]. Gas phase grafting is achieved by polymerization initiated by irradiation (γ-irradiation) accelerated electrons, low-pressure gas discharge [3,6].

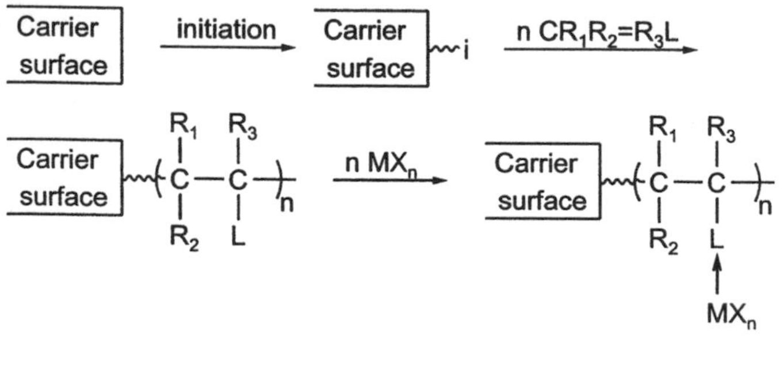

(16)

Some papers describe the grafting on polymers containing bond metal complexes on the surface of organic polymers: polyethylene-graft-poly(methylvinylketone)/Schiff base with 2-aminophenol **27** [6,37,155] or salicylaldehyde hydrazide [156], polyethylene-graft-poly(vinyl-1,3-diketone) [157], polytetrafluorethylene-graft-poly(acrylate)-complexes with 2,2′-bipyridyl or 1,10-phenanthroline [157].

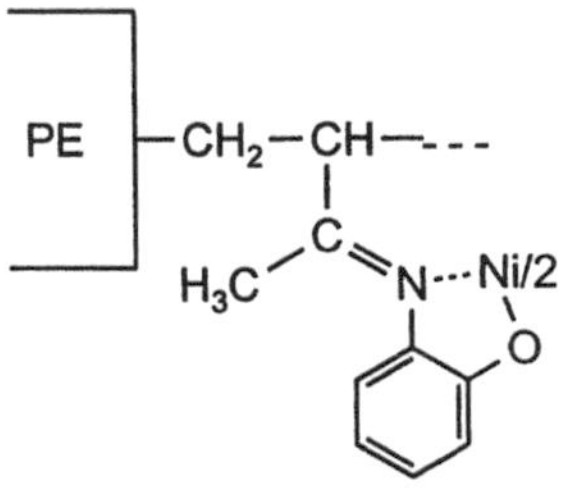

27

More intensively the immobilization of metal complexes on inorganic macromolecules was investigated. The covalent binding was described and reviewed in [3,158]. Some examples are the reactions of $Cp_2Zr(CH_3)_2$ (Cp = cyclopentadienyl) or diorgano-$ZrCl_2$ with silica gel and alumina (after treatment with $AlMe_3$ as catalysts for the olefin polymerization), dichlorotitanium pirocathecolate with silica gel, binding of a nickel P/O chelate at silica gel modified with tetrabenzyltitanium followed by binding of a nickel P/O chelate, and preparation of alumina-supported bis(arene)-Ti and tetra(neopentyl)-Zr [159]. The interest in this work is related to obtain heterogeneous catalysts for the olefin polymerization.

Ligands for transition metal ion interaction at silica gel were obtained by covalent connection of trialkoxysilanes containing a ligand group such as N,N-dimethylamino [160] or ethylenediphenylphosphine [3,6,161,162] silica-grafted 3,3′,5,5′-tetra-tert-butylbiphenyl-2,2′-diylphenylphosphite [96] and trimethylenephosphine covalently linked to silica [163]. Different tridendate bis(2-pyridylalkylamines) have been couple to 3-(glycidyloxypropyl)trimethoxysilane and subsequently grafted onto silica [as an example see **28** in equation (17)] [164]. The ligand concentration varied between 0.29–0.63 mmol g^{-1}. Most ligands selectively absorb Cu^{2+} from aqueous solution containing a mixture of different metal ions. Silica was modified by 3-chloropropyltrimethoxysilane and afterwards reacted with 2-(phenylazo)pyridine which is a good lignad for Ru^{3+} [165]. This macromolecular Ru-complex is a good catalyst for the epoxidation of *trans*-stilbene.

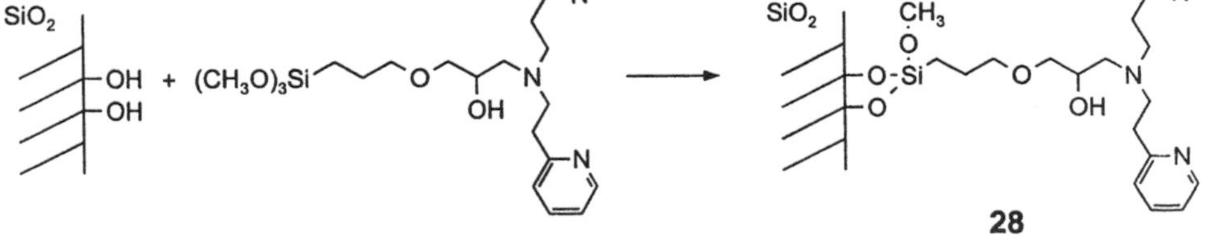

28

(17)

The immobilization of phthalocyanines by covalent binding to inorganic macromolecular carriers such as silica is a prospective approach to achieve heterogenous catalysts and photocatalysts in which the carrier is stable against several chemicals including oxygen. With loadings of $\sim 10^{-5}$–10^{-6} mol per g carrier monomolecular dispersion of the phthalocyanine are achieved [166–168]. Different silica such as macroporous Lichrosorb (surface area $\sim 300\,m^2\,g^{-1}$), macroporous Lichrosphere (surface area $\sim 40\,m^2\,g^{-1}$), Fractosil (surface area $\sim 8\,m^2\,g^{-1}$) and monosphere silica (surface areas between 24 and $1.7\,m^2\,g^{-1}$) – all silica from Merck AG – are employed. In the first step the silica surfaces were modified to obtain chemically active positions for the attachment of substituted phthalocyanines. Functionalization was achieved by reaction with 3-aminopropylsilyl groups for binding of **20** with R = –COCl to synthesize **29** or with 3-chloropropylsilyl groups for the binding of **20** with R = –NH$_2$ to synthesize **30** [equations (18) and (19)]. The loadings are with substituted silyl groups between 10^{-3} and $10^{-4}\,mol\,g^{-1}$ and with phthalocyanines between 10^{-5} and $10^{-6}\,mol\,g^{-1}$. Comparable covalent binding can be carried out also on the surface of titanium dioxide [169].

$$(18)$$

$$(19)$$

For the coordinative binding of phthalocyanines at inorganic carriers, the surface has to be modified. In a one-step-procedure for the preparation of silica modified on the surface with imidazoyl-groups, different silica materials as mentioned before were treated with a mixture of 3-chloropropyltriethoxysilane and an excess of imidazole in *m*-xylene [equation (20)]. Following treatment with different kind of substituted cobalt phthalocyanines, naphthalocyanines and porphyrins **20–22** in DMF led to the modified silica as exemplarily shown with **20** (R = –H) for **31**. The silica contains ~ 0.8–$12\,\mu mol\,g^{-1}$ metal complex moieties [167,170].

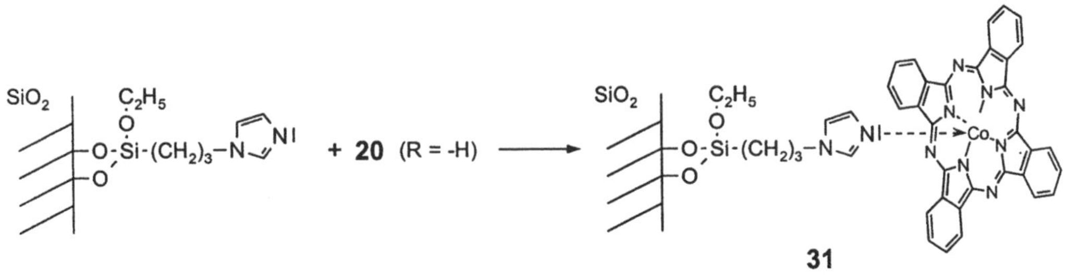

(20)

D. Polymerization of Metal Containing Monomers

Vinyl and related unsaturated groups being substituted by different kind of metals can be employed in polymerization or copolymerization reactions. If no side reactions occur by metals, uniformly substituted chains are obtained. A classification of the monomers is based on the type of bond between the metal and the organic part as shown in Figure 6 [4,12]. Covalent-type compounds contain real organometallic 'metal–carbon' or 'metal–oxygen' bonds. Monomers of the coordinative type are often formed in the interaction of heteroatoms with unshared pairs of electrons and transition metal compounds. Characteristic for π-bound compounds are transition metals of the groups VI A, VII A and VIII of the periodic table. Non-transition metals are more characteristic for the ionic type. True organometallic compounds with metal–carbon bonds are only rarely described. Monomers of the complex/chelate type contain a vinyl group at the complexing ligand for binding of various metal ions. Either the ligand with subsequent metallation or the complex/chelate can be employed in polymerization reactions.

Due to the reactivity of the metal for itself or of the kind of binding to the unsaturated monomer part, several side reactions can occur during the radicalic, anionic, cationic or Ziegler–Natta-type polymerizations. Some aspects of side reactions are [171]:

- By elimination of metal or metal containing groups during the polymerization formation of non-uniform units in the polymer chain.
- Formation of different oxidation states of metals in units in the polymer chain.
- Irregularity in the polymer chain by formation of new chemical bonds between the monomer and the metal containing group.

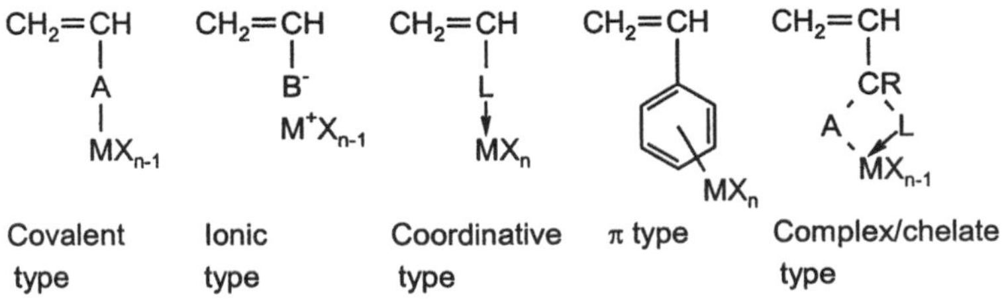

Figure 6 Classification of metal containing monomers for polymerization.

- Formation of new coordination numbers around the metal (mono-, bi-, bridged etc. coordinations) including a changed geometry.
- Side reactions such as hydrolysis, etherisation, salt formation etc.
- Formation of polynuclear, cluster or nano-sized particles during the polymerization.
- Stereoregularities caused by the metal during the polymerization.
- Chain crosslinking, chain transfer.
- Formation of cycles during the polymerization.

Few characteristic examples are given in the following subchapters (for reviews see [4,171]).

1. Covalent-type Monomers

Vinyl magnesium compounds are prepared as conventionally known by the reaction of vinyl halides with magnesium in THF. In the case of vinylmetal halides the low stability of these monomers easily leads to a splitting of the metal from the vinyl group. Styrylmagnesium chloride in polar solvents such as hexamethylphosphoramide results due to the polar C–Mg bond to branched or crosslinked polymers containing the different structural element **32** [172,173]. A convenient method for the preparation of unsaturated monomers with other metals consists of the reaction of vinylmagnesium halide derivatives with, for example, $ClPb(C_6H_5)_3$ or $ClSn(C_2H_5)_3$ for the synthesis of **33** and **34** [174–177]. Vinyl- and styryl organometallic compounds can show in the radicalic polymerization a lower or higher activity compared to styrene. $R_3M–C_6H_4–CH=CH_2$ with $M = Sn(IV)$ or $Pb(IV)$ exhibit a higher reactivity compared to styrene [174,176]. The copolymerization parameters in the copolymerization of styrene (M_1) with **33** (M_2) are $r_1 = 0.98$, $r_2 = 1.22$ and with **34** (M_2, C_6H_5 instead of C_2H_5) are $r_1 = 0.83$, $r_2 = 2.86$, respectively [178]. The medium values of the molecular weight are in general less then 10^4. In the case of the copolymerization of $trans$-$Pd[P(C_4H_9)_3]_2(C_6H_4CH=CH_2)Cl$ (M_2), the copolymerization parameters $r_1 = 1.49$ and $r_2 = 0.45$ show a lower reactivity of the organometallic compound [179]. For high molecular weight polymers it is more suitable to prepare the organometallic polymer by polymer analogous reactions at reactive polymers.

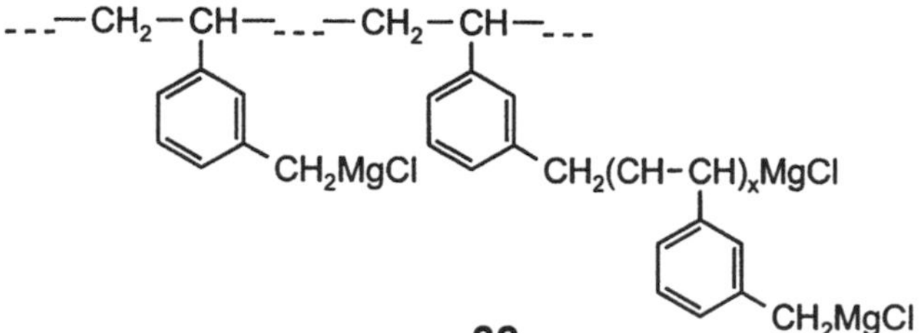

32

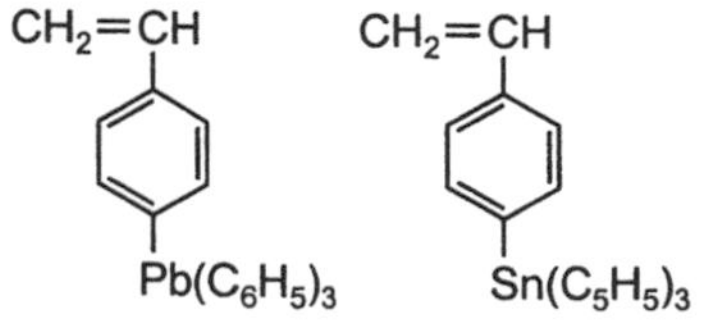

33   **34**

Only a few papers describe the polymerization of unsaturated monomers with a covalent M–O bond. Ziegler–Natta copolymerization of the diisobutylaluminium-alkoxy-isopren derivative 35 with butadiene occurs by a neodynium catalyst in a hydrocarbon solvent [180]. Mainly the monomer 35 in 1,4-*cis* configuration is found in the copolymer. A chiral monomer based on ethyleneglycolmonomethacrylat being substituted by alkoxy derivatives of Ti(IV) and different chiralic substituents was polymerized [181]. Such polymers are interesting as chiralic catalysts.

$$CH_2=C-CH=CH_2$$
$$|$$
$$CH_2-CH_2-OAl(iso-C_4H_9)_2$$

35

2. Ionic-type Monomers

Salts of unsaturated carboxylic acids such as acrylates or methacrylates are most important among the ionic-type monomers. As a general method of synthesis, the reaction of (hydro)oxides, (hydro)carbonates of metals as well as their alkyl(aryl) derivatives with unsaturated carboxylic acids is carried out [4]. Co(II), Ni(II), Zn(II) and Cu(II) acrylates exhibit the diacrylate structure 36 with one or two additionally coordinated solvent molecules such as water [182,183].

$$CH_2=CH-C\overset{O}{\underset{O}{\diagdown}}\overset{L}{\underset{L}{M}}\overset{O}{\underset{O}{\diagup}}C-CH=CH_2$$

36

Different transition metal salts of acrylate polymerize at 60 °C with AIBN, e.g., in ethanol under dissociation-excluding conditions [183,184]. The resulting metal-containing polymers are as expected insoluble in organic solvents but they are converted to soluble polyacrylic acid in a methanol–HCl mixture. The reactivity of the metal-acrylates in the homopolymerization decreases as follows: Co(II) > Ni(II) > Fe(III) > Cu(II).

3. Coordinative-type Monomers

Coordinative-type bonds are formed by various unsaturated donor ligands containing single electron pairs carring N-, O-, S- or P-atoms [4]. N-vinyl monomers are based on differently substituted vinylpyridines, imidazoles, benzimidazoles, unsaturated nitriles (acrylonitrile, methacrylonitrile), amides (acrylamide) and cyclic amines (ethylene imine) interacting from solution with various transition metals such as Cu(II), Co(II), Pd(II), Ru(II), Os(II), Pt(IV). Characteristic compositions of vinylpyridine (VP) complexes are: Cu(2-VP)Cl$_2$ [185], Pd(4-VP)Cl$_2$ [186], Co(4-VP)$_4$(NCS)$_2$ [187]. The crystal structure of the complex Co(1-vinylbenzimidazole)$_2$Cl$_2$ is shown in Figure 7 [188]. The N(3)-atoms of the two ligands and the chlorine atoms are located at the apexes of a distorted tetrahedron.

After analysis of IR spectra of acrylonitrile complexes with Al(III), Zn(II), Ni(II), Tl(IV), Pd(II) the probable structure is described by electron density transfer to the metal: $R-C\equiv N^+-M^-Cl_n$ [189–194]. Styryl phosphine complexes 37 of Co(II), Ni(II) or Pd(II)

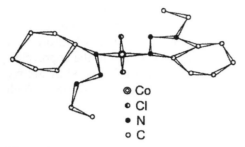

◎ Co
○ Cl
● N
○ C

Figure 7   Structure of Co(1-vinylbenzimidazole)$_2$Cl$_2$.

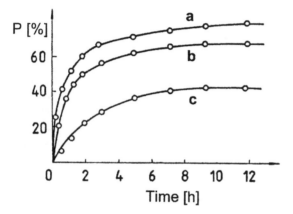

Figure 8   Time dependence of the polymer yield for the polymerization of 1-vinylimidazole (VIA, a), Mn(VIA)$_4$Cl$_2$ (b), Ni(VIA)$_4$Cl$_2$ (c) in ethanol with 4 mol% azobisisobutyronitrile at 70°C.

are prepared by direct interaction of the styryl phosphine with metal halides or ligand exchange with complexes containing substitutable low molecular weight ligands (phosphines, nitriles, acetylacetonates) [195].

CH$_2$=CH   CH=CH$_2$

R$_2$P   PR$_2$

MX$_n$

37

Polymerizations and copolymerizations of various coordinative-type monomers were intensively investigated in solution or the bulk [4]. A great influence on the kind of ligand, metal ion and also solvent on the probability of the polymerization under radicalic initiation was found. Due to side reactions often the polymer yield of the coordinative-type monomers are lower compared to the polymer yield of the free monomer ligand (Figure 8) [196].

In the copolymerization of styrene (M_1) with Zn(II)-complexes of N-vinylbenz-imidazole ($Zn(VBI)_2Cl_2$) (M_2) the reactivity of the coordinative-type monomer is lower ($r_1 = 4.0$, $r_2 = 0.24$) also in comparison to the copolymerization of styrene and N-vinylbenzimidazole ($r_1 = 2.8$, $r_2 = 0.36$) [197].

4. π-Type Monomers

Various strategies for the synthesis of metallocene monomers were described in [198]. Vinylmetallocenes **38** like vinylferrocene or η^5-(vinylcyclopentadienyl)-dicarbonylnitro-sylmanganese **39** are prepared several decades ago by synthesizing the vinyl group in the metallo-derivatives [199]. Other π-type compounds such as η^6-(styrene)tricarbonyl-chromium **40** are obtained by reaction of styrene with triamine-tricarbonylchromium [200].

Polymerization of vinylferrocene, for example, is carried out by different initiations including the Ziegler–Natta type one [3,201]. The transition metal ion such as Fe(II) can participate during the radicalic polymerization by transfer of an electron from a Fe atom to the terminal chain radical. Therefore higher values of constant of chain transfer are determined: $k_{ct}/k_p = 8 \times 10^{-3}$ at 60 °C; for styrene $k_{ct}/k_p = 6 \times 10^{-5}$.

By Heck-type coupling liquid crystalline rigid-rod polymers containing [1,3-(diethynyl)cyclobutadiene] cyclopentadienyl moieties were prepared [202]. One example is the reaction of the diethynyl derivative **41** with a 2,5-diodothiophene **42** to the polymer **43** [equation (21)] which show lyotropic nematic phases.

$$(21)$$

5. Complex/Chelate-type Monomers

Monomers of metal complexes/chelates suitable for polymerizations, and also poly-condensations or polyaddition reactions, can be employed successfully for the preparation of metal containing polymers. In most cases polymerizations are carried out in the presence of a comonomer to obtain polymers with sufficient solubility. For polymeriza-tions either a vinyl group containing ligand is polymerized followed by introduction of a metal ion in the macroligand, or the vinyl group containing metal complex/chelate is directly converted into the MMC. In some cases when chain transfer due to a transition metal ion in the core of a ligand occurs, the polymerization of the ligand is preferable. The literature on this subject is reviewed in [3,6,29,39].

Azacrown ether metal ion complexes were synthesized by copolymerization of vinyl azacrown ethers with styrene, acrylic acid, methacrylic acid and N-vinylpyrrolidone [203]. The linear copolymers are soluble in some organic solvents or water. The coordination properties with transition metal ions were studied by UV/VIS and ESR. Some of the ligands exhibit high selectivity for Au-ion binding. One example is shown in **44**.

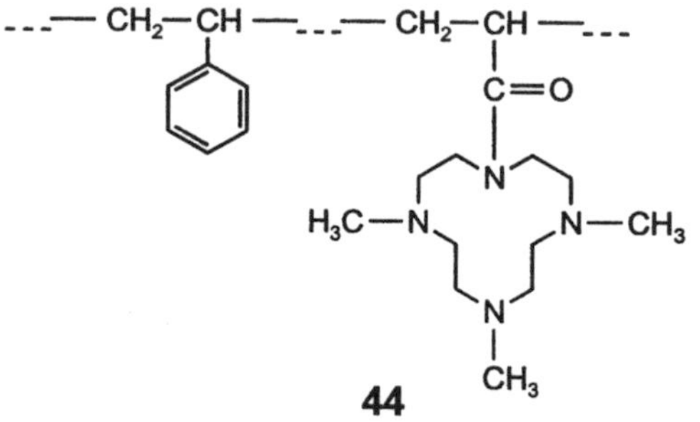

$$\text{--- CH}_2\text{--CH ---...--- CH}_2\text{--CH ---...}$$

44

4-Methyl-4'-vinyl-2,2'-bipyridyl was copolymerized with styrene and then treated with *cis*-Ru(bpy)$_2$Cl$_2$ to form pendant Ru(bpy)$_3^{2+}$ **45** (see [3,6,37,204,205] and literature cited therein). In order to study ionic domains around the Ru complex also copolymers with acrylic acid were synthesized. In solution or as thin films photophysical properties and photo-induced electron transfer were investigated. Photoluminescence properties were also studied for polysiloxane pendant Ru(bpy)$_3^{2+}$ prepared from the corresponding substituted dihydroxysilanes [206]. Reductive electropolymerization of Fe(II), Ru(II) complexes containing 2,2'-bipyridyl and others in acetonitrile in the presence of Et$_4$NClO$_4$ (as electrolyte) on Pt results in films of <1 µm thickness [207].

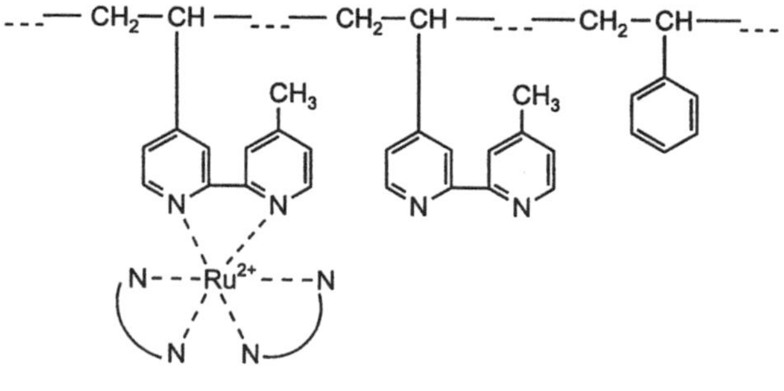

45

Polymer cobalt (II) Schiff base chelates based on the radical copolymerization of 4,4'-divinylsalens with styrene or 4-vinylpyridine are described [208–210]. One example of a polymer employing 2-hydroxy-5-vinylbenzaldehyde and 1,2-diaminoethane is shown in **46**. The Co(II) chelates were investigated for their activity in the binding of oxygen and its activation for the oxidation of 2,6-di-tert-butylphenol.

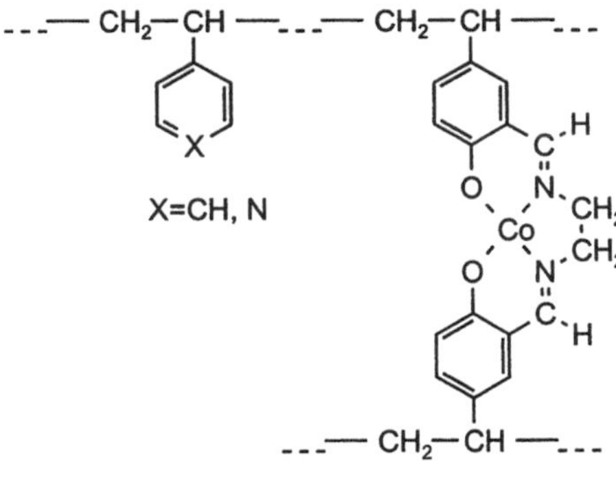

46

The polymerization and copolymerization of vinyl group containing porphyrins and related compounds was reviewed in [29,31,39,211] and more recently described in detail in [212–224]. Either the vinyl-substituted ligands or their metal complexes are employed mainly in copolymerizations. In order to avoid crosslinking, mainly monosubstituted acryloyloxy, methacryloyloxy and acrylamido-substituted porphyrins containing various, metal ions were studied for polymerization reactions in solution [214–217,221–223] (see, e.g., **47–49** from [224]). In contrast, the thermal bulk polymerization of octasubstituted phthalocyanines such as their octaacryloyloxy and octamethacryloyloxy derivatives in the presence of 2,2′-azoisobutyronitrile (AIBN) resulted in insoluble products by crosslinking [225]. The radicalic coplymerization of analogously tetrasubstituted phthalocyanines with a great excess of N-vinylcarbazole led to soluble non-crosslinked coplymers [226].

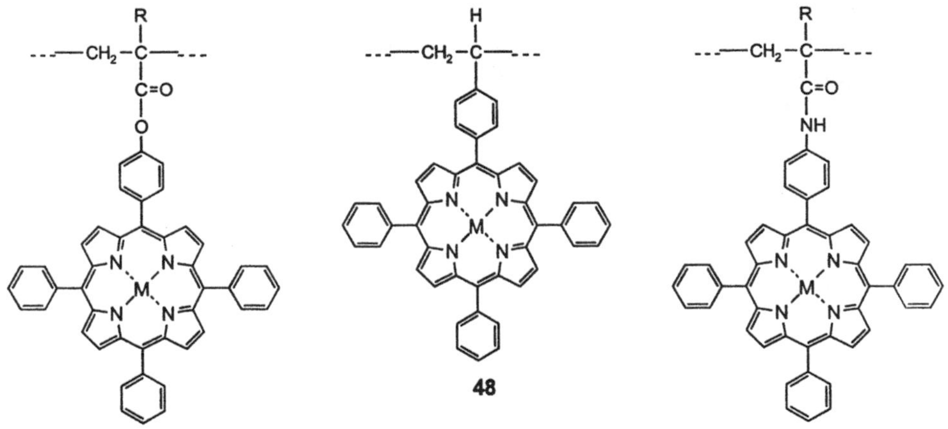

47   **48**   **49**

The homo- and copolymerization of porphyrins and phthalocyanines being substituted by four methacryloyloxy or 2,4-hexadienoyloxy groups, such as the zinc complexes of 2,9,16,23-tetrakis(4-(methacryloyloxy)- and 2,9,16,23-tetrakis(4-(11-methacryloyloxyundecyloxy)phenoxy)phthalocyanines **50** and **51** were investigated [227]. The copolymerizations were carried out with styrene in DMF initiated by AIBN at 60 °C. In order to avoid crosslinking and to obtain DMF soluble copolymers, only a small amount of the phthalocyanine in relation to the comonomer styrene was employed, and the copolymerizations were stopped at yields of around 10%. The copolymers contain around 0.06–0.003 mmol phthalocyanine units per g polymer (~ 8–0.6 wt%). The molecular weights of the copolymers after GPC is in the order of $\overline{M}_n \sim 2 \times 10^4$ and $\overline{M}_w \sim 4 \times 10^4$.

50: R= —O—⟨⟩—O-C-C=CH₂ (O, CH₃)

51: R= —O—⟨⟩—O-(CH₂)₁₁—O-C-C=CH₂ (O, CH₃)

2,3,7,8,12,13,17,18-Octakis[[(4-methoxycinnamoyl)oxyalkylthio]-tetraazaporphyrins **52** with different methylene chain lengths were investigated by UV irradiation in solution and as spin-coated films in the solid state [228,229]. In THF or toluene solution the $Z \to E$ isomerization of the double bond up to a photostationary state is the dominating process. The spin-coated films are optically transparent, do not scatter light nor are birefringent, suggesting that they are amorphous or microcrystalline. Irradiation was carried out at the absorption of the cinnamoyl group at 313 nm with $3\,\mathrm{mW\,cm^{-2}}$, and a decrease of this absorption occurs. After $\sim 1200\,\mathrm{s}$ of irradiation time the films became completely insoluble in organic solvents by intermolecular crosslinking of the cinnamoyl groups through [2 + 2]-photocycloaddition and partially also poymerization. The highest conversion rate of the cinnamoyl groups with $\sim 75\%$ were observed for **52** with eleven methylene groups. Totally glassy films absorbing at $\sim 670\,\mathrm{nm}$, which may be interesting for optical applications, were obtained.

R = —C(O)=CH—⟨⟩—OCH₃   n = 3,6,8,11   M = 2H, Mg, Zn, Cu, Ni

52

The electrochemical polymerization of π-electron-rich aromatics, such as aniline, pyrrole and thiophene, to obtain electrically conducting polymers is well-known. Some reports describe the polymerization of amino-, pyrrolyl- and hydroxy-substituted tetraphenylporphyrins and suitable substituted phthalocyanines (for reviews see [230,231]) (anodic electropolymerization of 2,9,16,23-tetraaminophthalocyanine (M = Co(II), Ni(II)) [231,232] and 2,9,16,23-tetra(1-pyrrolylalkyleneoxy)phthalocyanines (M = 2H, Zn(II), Co(II) [232])) under formation of polymers **53** and **54** shown as idealized structures. Depending on the reaction conditions the film thicknesses are between around 50 nm and several μm. The films remain electroactive at the electrochemical potential so that oxidation or reduction current envelope grows with each successive potential cycle. Electrochromism, redox mediation and electrocatalysis of the electrically conducting films are summarized in [230,231].

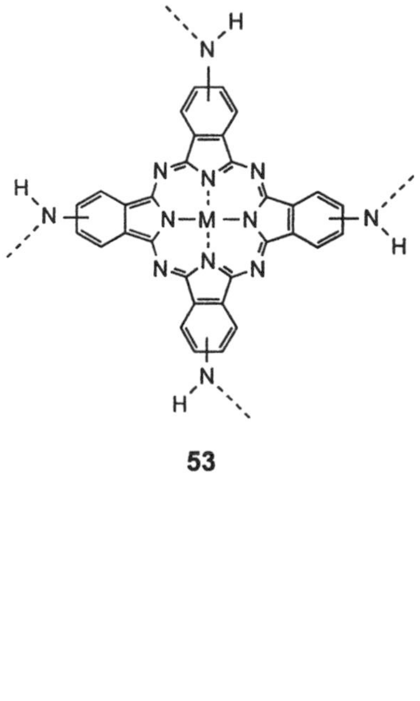

53

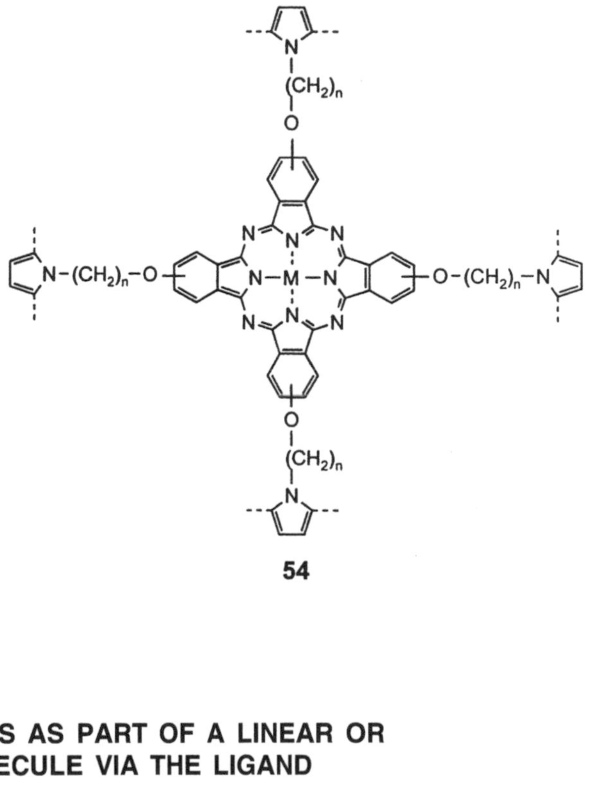

54

IV. TYPE II: METAL COMPLEXES AS PART OF A LINEAR OR CROSSLINKED MACROMOLECULE VIA THE LIGAND

As pointed out in Figure 2, metal complexes/chelates can be via their ligands part of a polymer chain or network. The ligands can be of non-cyclic or cyclic type. Two general routes are used to synthesize such polymer metal complexes:

- A bifunctional/higher-functional ligand (followed by metallation afterwards) or metal complex/chelate are reacting by self-condensation or with another bifunctional/higher-functional comonomer.
- A bifunctional/higher-functional ligand in the presence of a metal or a metal complex precursor is reacting under polymer metal complex formation.

The work in this field was reviewed in [1,3]. In elder work less care was taken to characterize the structural uniformity and molecular weight which is due to the fact that these polymers are often not soluble or fusible. But present techniques of instrumental analysis allows a more or less detailed analysis of the solids.

A. Non-cyclic Organic Ligand Type

Polymeric Schiff base chelates of Co(salen)-type are prepared for example by the reaction of a dihydroxy substituted Schiff base ligand with a biscarboxylic acid derivative or by the reaction of a bis(o-hydroxyaldehyde) with a diamine (for example **55** [233–235]). A series of Schiff-base Cu(II)-complexes were synthesized by a transesterification reaction of random liquid crystalline polymers with a functionalized tetradendate low molecular weight Cu(II)-complex [236]. The organometallic unit was incorporated between 5 and 20 mol% without disrupting the liquid crystallinity. The aim of the research was to obtain new magneto-active organic systems which combine the anisotropic paramagnetic susceptibility of metal entities and the cooperative re-orientation of liquid crystals in external fields (for a review see [43]). Disadvantageous is the not sufficient analytical characterization of the polymers.

$M = Co^{2+}, Cu^{2+}, Zn^{2+}, Ni^{2+}, Mn^{3+}$

55

Thermotropic liquid crystalline polymers containing β-diketonato groups (–R–CO–CH$_2$–CO–R) capable of metal ion binding (Cu^{2+}, Ni^{2+}) in the main chain of polyesters or in the side chain of polyarylates are described [237]. Binding of Pt^{4+} at poly(ethyleneimine) and other polymers containing 1,2-diimino parts in the polymer chain were prepared for the use in chemotherapy of cancer [238].

B. Cyclic Organic Ligand Type

Porphyrins, phthalocyanines, hemiporphyrazines and tetraazaannulenes were intensively investigated as cyclic ligands for polymer metal complexes [31,32]. The polymeric chelates are generally obtained as insoluble brown-to-black powders. In some cases for device construction film formation during the preparation process was achieved.

Low molecular weight phthalocyanines are prepared by cyclotetramerization of unsubstituted or substituted phthalic acid derivatives such as phthalonitrile in the presence of suitable metals. If bifunctional tetracarbonitriles are employed, polymeric phthalocyanines **56** and **57** are obtained [equations (22) and (23)]: 1,2,4,5-tetracyano-benzene, oxyalkyleneoxy-, oxyaryleneoxy- [239–242], tetrathio- or dioxadithia-bridged [243], bisphthalonitriles and tetracarboxylic acid derivatives such as 1,2,4,5-benzenetetra-carboxylic acid dianhydride [244–246]. The reactions of the carbonitriles were carried out in bulk in the presence of a metal at $T = 200–400\ °C$. Dark-blue till black-coloured polymers insoluble in any organic solvent were obtained. Recently the synthesis of the polymers by the reaction of tetracarbonitriles under mild conditions in boiling pentanol-1 in the presence of Li-pentanolate was described [247].

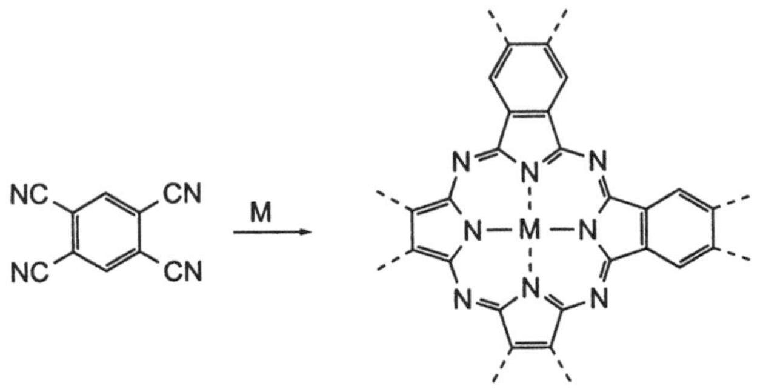

56

(22)

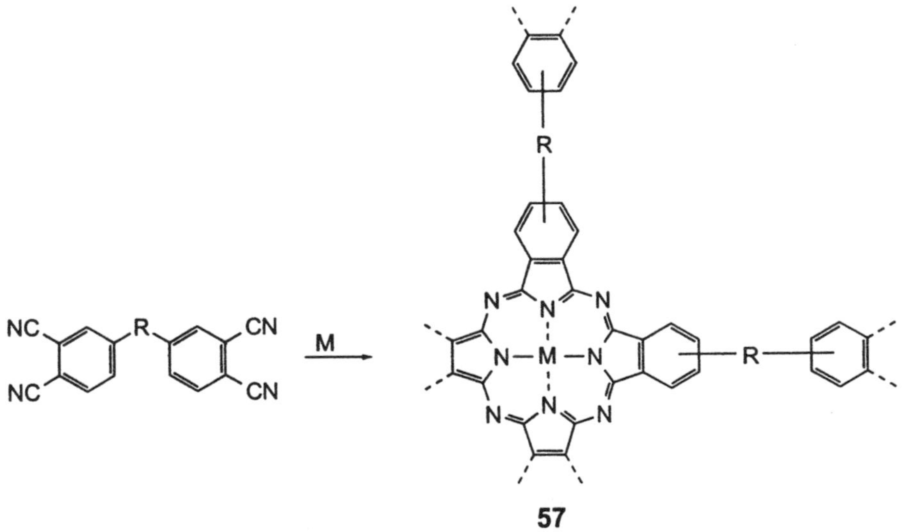

57

(23)

For complete characterization of the polymers the following points must be considered: structural uniformity, nature of end groups, metal content and degree of polymerization (molecular weight). Only in few reports sufficient statements regarding these points were made [239,241,247], and the preparation of pure polymeric phthalocyanines needs a lot of experience. UV/Vis spectroscopy (in conc. sulfuric acid or in reflexion as solid) and IR spectroscopy (in KBr) are suitable methods to evaluate in first steps the structural purity. During the polycyclotetramerization of tetracarbonitriles the formation of the by-products polyisoindolenines and polytriazines can occur. They are covalently incorporated as co-units into the polymers and cannot be separated from phthalocyanines structural elements. The bulk reactions of tetracaronitriles with Cu or CuCl$_2$ result in structural uniform polymers whereas with Mg, Al, V, Cr, Mn, Fe, Co, Ni, Zn some impurities are included. The pentanol-1 method allows introduction of different metals in structurally pure polymers [247].

The bulk reactions lead to cyano end-groups which can be hydrolyzed under drastic conditions (KOH in triethylene glycol in the presence of a small amount of water at 160 °C) to carboxylic end groups [239].

The degree of polymerization (number of connected phthalocyanine rings) were carried out by exact end group determination using the IR spectroscopy method.

This method was practically applied in a few cases [239,241,244] and theoretically evaluated [248,249]. After quantitative nitrile group determination the molecular weight of polymeric Cu-phthalocyanines are increasing with increasing flexibility between the two reactive sides of a tetracarbonitrile or the bridging group X: for polymers 56 from 1,2,4,5-tetracyanobenzene >4000, for polymers 57 (with oxyaryleneoxy-bridges) >6000, for 57 (with oxyalkyleneoxy-bridges) up to infinite.

For several investigations like electrical, photoelectrical, catalytic and photocatalytic properties thin films on flat surface (e.g., glass, Ti, ITO, KCl) or coatings on particles (e.g., SiO_2, TiO_2, Al_2O_3) are necessary. Because polymeric phthalocyanines are insoluble and not vaporizable, special techniques must be employed. They include the reactions of gaseous tetracarbonitriles with films or coatings of metals or metal salts on flat surfaces [240,250,251] or inorganic powdered particles [252,253].

The mechanism of film growth of 56 was discussed in [240,250]. After formation of the first few layers of polymeric phthalocyanines, copper atoms diffuse from the copper film to the growing polymer film surface in order to react with 1,2,4,5-tetracyanobenzene at first to octacyanophthalocyanine and then to oligomeric and polymeric phthalocyanines. By ESCA spectra ~0.7% of free Cu in the polymeric films were found. In dependence of the deposited Cu-film thicknesses of ~1.5 till 20 nm adhering films of the polymers 56 with thicknesses of ~46 till 230 nm were obtained For the ratio of the thickness of the polymer film to the copper film in every case an average value of ~25 was determined. The films exhibit good electrical conductivities.

For the preparation of coatings of phthalocyanines of SiO_2 or TiO_2 two routes were used [equation (24)] [253]: route a after adsorption of a metal carbonyl at T_1 (40–60 °C) their decomposition to metals at T_2 (130–320 °C) and subsequent reaction with the nitrile at T_3 (200–350 °C); route b direct reaction of the adsorbed metal carbonyl at T_4 (180–250 °C with the nitrile). The amount of loading on quartz particles with polymeric phthalocyanines of ≈ 2 wt% were calculated from the amount of employed metal carbonyl and by parallel experiments from the reaction of phthalonitrile with $Co_2(CO)_8$.

$$M_x(CO)_y \ + \ SiO_2 \ \text{or} \ TiO_2$$

(24)

The investigation of properties of polymeric phthalocyanines concentrates on thermal stability [239,244,245,254], electrical conductivity and redox behaviour of thin films [30,240,250,251,255], catalytic activity [253], electrocatalytic activity for the O₂ reduction [30] and photochemical properties [253].

The preparation of polymeric hemiphorphyrazine **58** (polyhexazocyclanes) [256,257] and polymeric tetraaza [14]-annulenes **59** [258,259] were described several years ago, and they were not so well structurally characterized.

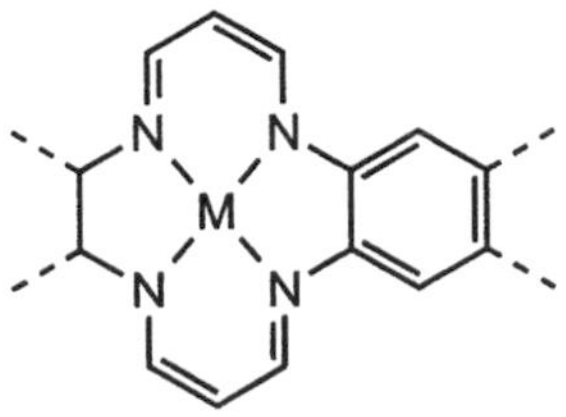

58 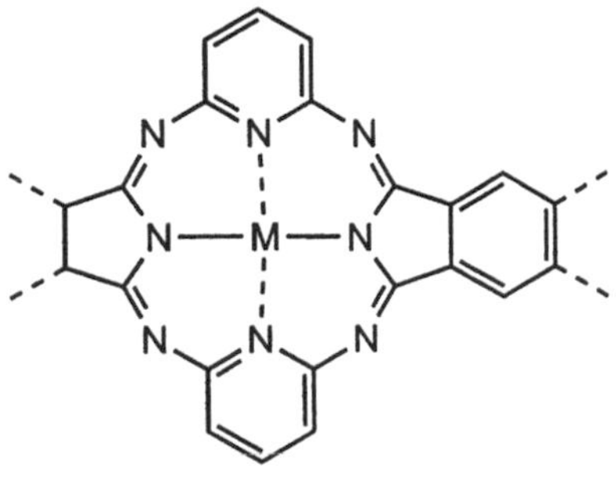   **59**

In few cases linear chain structured polymeric metal complexes were prepared. A linear polymeric phthalocyanine **60** was obtained as film by the electrochemical polymerization of the corresponding monomer [260]. The synthesis of structural uniform ladder polymers **61** based on the hemiporphyrazine structures was achieved by a repetitive Diels–Alder reaction [261,262]. Recently, linear oligomeric porphyrines covalently connected via meso-meso-positions up to 128 units were synthesized [263].

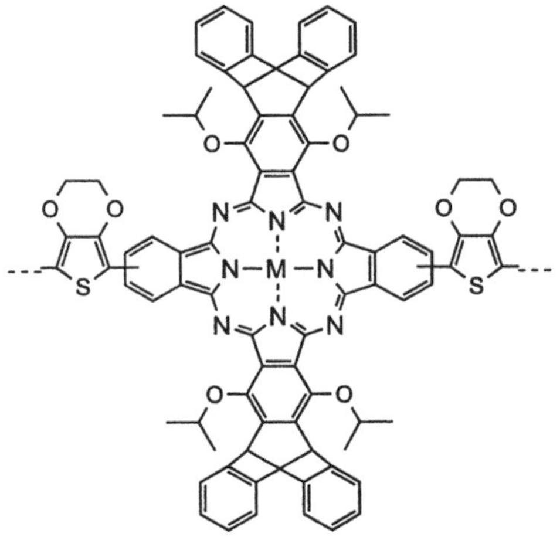

60 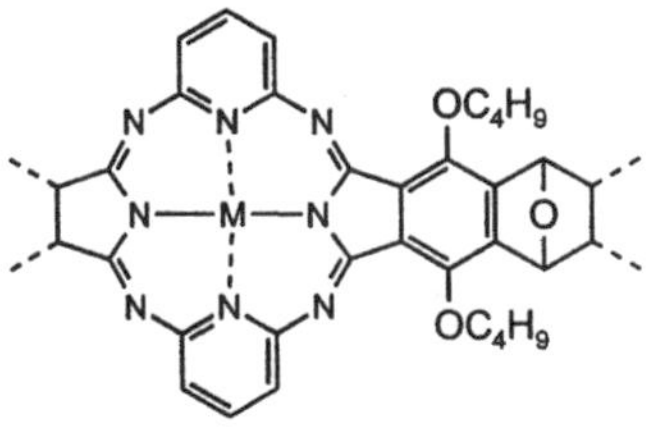   **61**

Low molecular weight higher functional substituted macrocyclic metal complexes (M = Co, Ni, Cu, Zn) were converted with other bifunctional compounds to polymers. By the reaction of tetraaminophthalocyanine in the presence of another diamine with benzenetetracarboxylic acid dianhydride, at first in dimethylsulfoxide (DMSO) soluble amide-carbocylic acid copolymers were obtained and after film casting and heating to ~325 °C converted into films of insoluble poly(metal phthalocyanine)imide copolymers [264]. A high thermal stability of these colored polyimides were found.

Several of functional Fe(III)- and Co(II)-phthalocyanines and their polymers as models for catalase, peroxidase, oxidase and oxygenase enzymes were synthesized ([265] and references cited therein). Copolyesters **62** containing Fe(III)- and Cu(II) phthalocyanines were obtained by polycondensation of phthalocyanine dicarboxylic acid dichlorides with terephthalic acid dichloride and aliphatic diols. Green or blue colored fibres could be obtained by melt spinning of the copolyesters containing below 1 mol% of the metal complex [265]. The polymers were investigated as catalysts for the thiol oxidation.

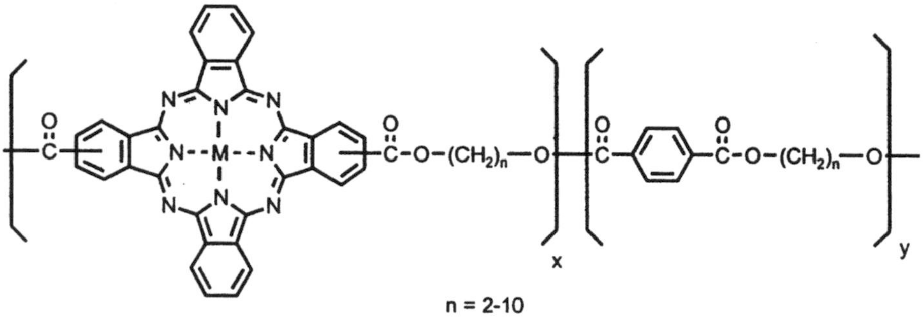

n = 2-10

62

By the Heck coupling reaction soluble and processible polymeric porphyrins **63** containing phenylene vinylene units were prepared [equation (25)] [266]. After GPC polymers with reasonable molecular weights $\overline{M_w} \sim 10^4 \, \mathrm{mol \, g^{-1}}$ were obtained. Good photoconductivities and good quantum yields for photochrage generation (e.g., 2.8%) were observed (applied field ~620 kV cm^{-1}). Due to steric hindrance the porphyrin and phenylene groups are out of plane and every porphyrin behave comparable to a monomeric porphyrin.

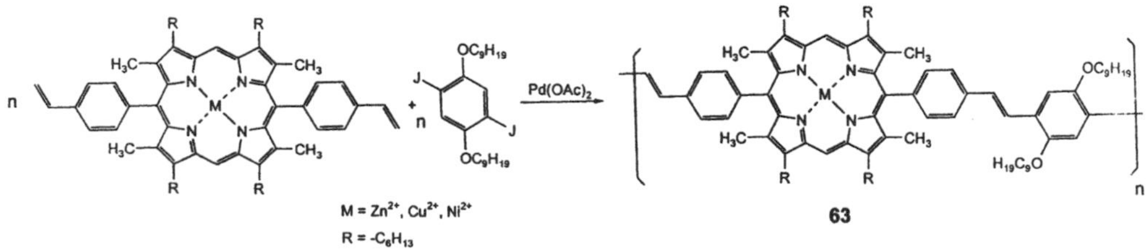

$$ \tag{25} $$

Conjugated polymers **64** containing zinc-porphyrin units linked by acetylene units were obtained by the Glas–Hay coupling of meso-diethynyl zinc-porphyrins [267]. Some results on third-order non-linear optical phenomenon were observed. A porphyrin-polyimide system was designed for photorefractive polymers [268] with a high temporal stability in dipole orientation without significant decay in the nonlinearity at higher temperatures.

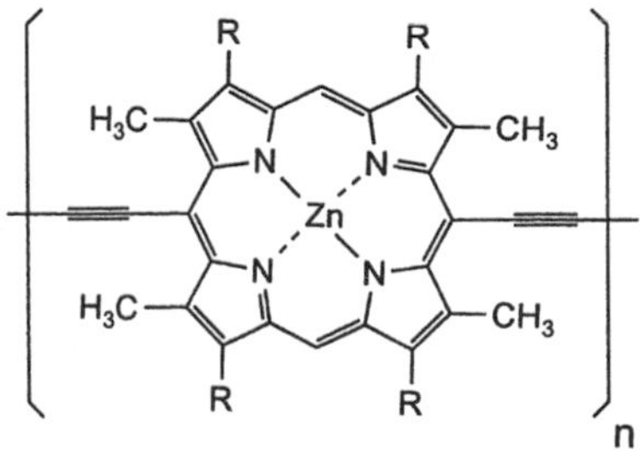

64

Interesting self-assembled multi-layer thin films **65** of covalently bonded porphyrins were utilized starting from chloromethylphenylsilylated oxide surfaces in the reaction with tetrapyridyl-substituted porphyrins and dichloro-*p*-xylene [269]. The growth of the film was studied by monitoring the absorption intensity of the Soret-band. It is said that a highly ordered and closely packed film was obtained.

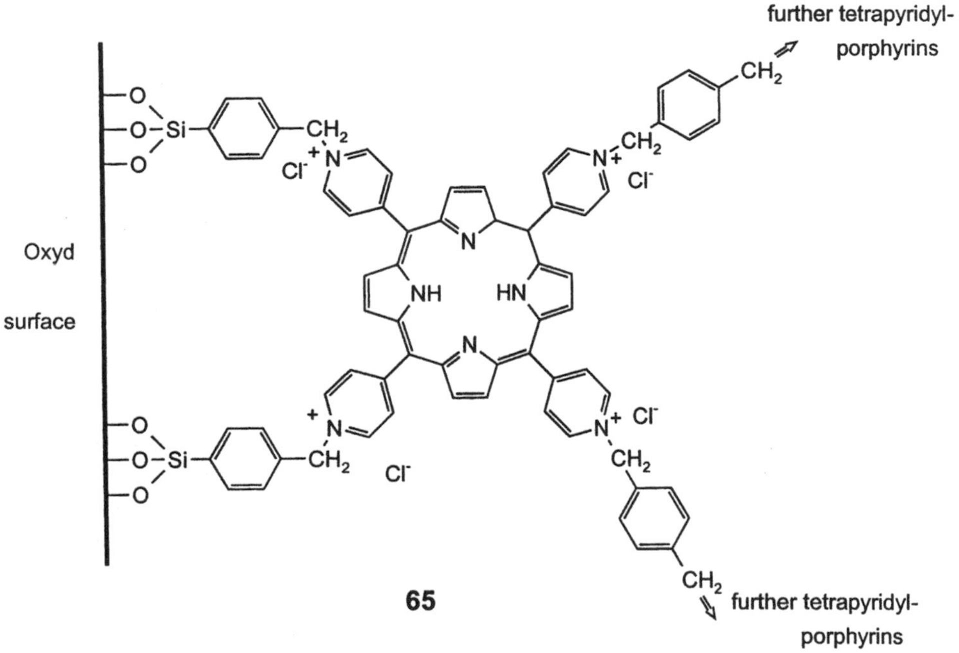

65

Also the interfacial polycondensation technique, in which reactive comonomers are dissolved in separate immiscible solutions and thereby constrained to react only at the interface between two solutions, was employed for the synthesis of chemically asymmetrical polymeric porphyrins containing different metals [270]. Tetrakis(4-amino-phenyl)-, tetrakis(4-hydroxyphenyl)porphyrins or aliphatic diamines in one solvent were reacted with tetrakis(4-chlorocarboxyphenyl)porphyrin or aliphatic diacylchlorides, respectively, in the other solvent. Typical film thicknesses are in the range of 0.1–10 μm. The unique chemical asymmetry is shown by distinctive difference in the concentration and type of functional groups present. Photoactivities of the polymeric porphyrin films were measured in dry sandwich cells.

V. TYPE III: METAL COMPLEXES OR METALS AS PART OF A LINEAR OR CROSSLINKED MACROMOLECULE VIA THE METAL

A great variety of possibilities were realized to have the metal as part of a linear or crosslinked macromolecule (Figure 3). One possibility is the covalent incorporation of a metal into homochain or heterochain polymers. Coordinative bonds between a metal and another element can occur in various combinations. Recently, the supramolecular organization under formation of metal containing coordination polymers were described. Different bonds were realized in stacking of metal chelates. In addition, metal containing catenanes and dendrimers are mentioned in this subchapter.

A Homochain Polymers with Covalent Metal–Metal Bonds

Examples of polymers of the type $-(M-M-)_n$ containing a bond in the main chain are known with metals or semimetals like B, Si, Ge, Sn, As, Sb, Th and Po [1]. Examples are shown in **66**. Well-described are polydiorganosilicones (polysilanes) which are investigated for use as precursors of ß-silicon carbide, as photoconductors, in nonlinear optics and microlithography. Preparations and properties are reviewed in [1,42,271,272].

$$
\cdots-\underset{\underset{R}{|}}{\overset{\overset{R}{|}}{Si}}-\cdots \quad
\cdots-\underset{\underset{R}{|}}{\overset{\overset{R}{|}}{Ge}}-\cdots \quad
\cdots-\underset{\underset{R}{|}}{\overset{\overset{R}{|}}{Sn}}-\cdots \quad
\cdots-\underset{}{\overset{\overset{R}{|}}{As}}-\cdots \quad
\cdots-\underset{}{\overset{\overset{R}{|}}{Sb}}-\cdots
$$

66

One general route of synthesis is the reaction between dihalides in the presence of sodium [equation (26)]. Polygermyne $(GeH)_n$ films are obtained by treating a $CaGe_2$ film with conc. aqueous HCl [273]. Employing trifunctional halides $(RMCl_3)$, cross-linked polymers such as polysilines **67a** and polygermines **67b** can be synthesized. Such polymers absorb up to $\lambda = 800$ nm and are therefore sensitive to visible light. One-dimensional metals of d^8-complexes with a quadric-planar surrounding are good electrical conductors. Partial doping of tetracyano-platinate (Kroogmann-salt) with columnar stacks results in metallic like conductivity [274,275]. A linear chain polymer containing Rh–Rh bonds are obtained by galvanostatic reduction of $[Rh_2(CH_3CN)_{10}](BF_4)_4$. The polymer $\{[Rh(CH_3CN)_4](BF_4)_{1.5}\}_n$ exhibits Rh–Rh distances of 0.28442 and 0.29277 nm with Rh in the oxidation state 1.6 [276]. Calculation on the band structure were reported.

$$
n\,halM(R)_x hal + 2n\,A \rightarrow -[-M(R)_x-]_n- + 2n\,Ahal
$$
$$
R = alkyl, aryl\ (x = 1, 2); A = alkali\,metals
$$

(26)

$$
\left(-\underset{\underset{}{}}{\overset{\overset{R}{|}}{Si}}\right)_n \qquad \left(-\underset{}{\overset{\overset{R}{|}}{Ge}}\right)_n
$$

67a   **67b**

B. Heterochain Polymers with Covalent Bonds Between Metals and Another Element

Heterochain polymers of the type $-(M-X-)n$ contain polar M–X bonds (for reviews see [1,9–11,42,43]). Such polymers are often prepared by polycondensation of a bifunctional metal halides (M = B, Si, Ge, Sn, Pb, Sb, Ni, Pd, Pt, Ti, Hf) with a bifunctional Lewis base such as a diol, diamine, dihydrazine, dihydrazide, dioxime, diamideoxime, dithiol, diacetylene [equation (27)]. Another possibility is the polyaddition of a bifunctional metal hydride to bifunctional alkenes [equation (28)].

$$n \, halM(R)_x hal + n \, H-L-R'-L-H \rightarrow -[-M(R)_x-L-R'-L]_n- + 2n \, HCl \qquad (27)$$

$$n \, HM(R)_x H + n \, H_2C=CH-R'-CH=CH_2$$
$$\rightarrow -[-M(R)_x-CH_2-CH_2-R'-CH_2-CH_2-]_n- \qquad (28)$$

Group 10 metal-poly(yne)s exhibit even in solution a unique linear rigid rod-like structure in which the *trans* positions of square planar group metals such as Pt or Pd are linked by conjugated diacetylenes like butadiyne [277–279]. These polymers form lyotropic liquid crystals which exhibit a response to an external electrical field [277,280]. The polymers are prepared by polycondensation, for example, of $PtCl_2$ or $PdCl_2$ and 1,4-diethynylbenzene in the presence of a copper halide in amines [equation (29)]. The soluble polymers have a molecular weight of more than 10^5. Analogously polymers **68** with 2,3-diphenylthieno[3,4]pyrazino building blocks were prepared. These polymers show a good photocurrent as sandwich-diodes [281]. Chiral poly(yne)s **69** containing 1,1'-bi-2-naphthol exist in a helical conformation [282]. Also Pt acetylide dendrimers were prepared [283]. Other examples are polyarylene cobalt-cyclopentadienylenes prepared by the reaction of diacetylenes with $CpCo(PPh_3)_2$ [284].

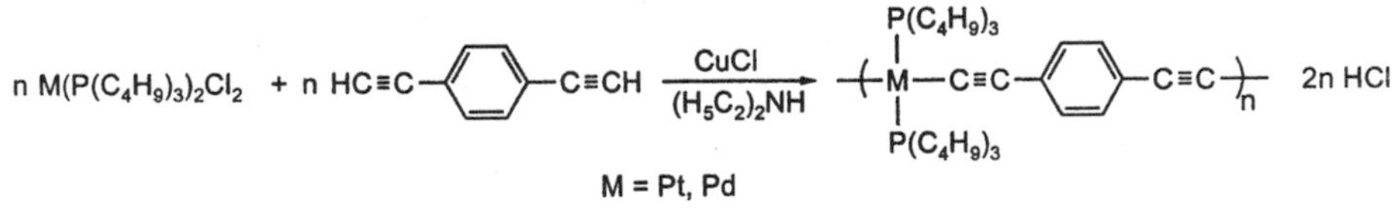

$$(29)$$

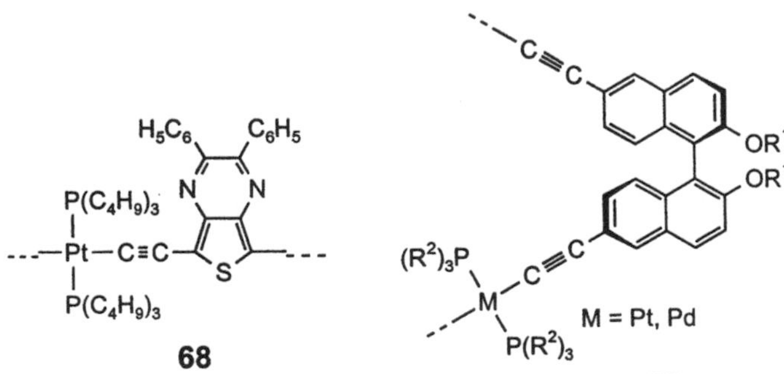

68

69

One-dimensional chains of heterometallic polymers with covalent metal–metal bonds are interesting as molecular conductors (see. ref. cited in [285]). A complex system of four different elements surrounded by insulting organic materials with this aim is described in [285]. An alloy $K_6Ag_2Sn_2Te_9$ was treated with 1,2-diaminoethane, and to the resulting solution a saturated aqueous solution of tetraethylammonium iodide was added. One-dimensional chains of the composition $(Et_4N)_4[Au(Ag_{1-x}Au_xSn_2Te_9]$ $(x = 0.32)$ **70** were obtained. Band structures are discussed and a band gap of 0.45 eV was found.

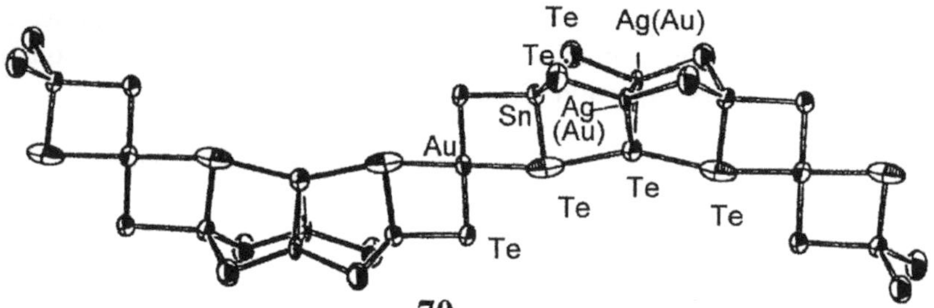

70

A golden-colored polymeric organometallic oxide of the formula $\{H_{0.5}[CH_3)_{0.92}ReO_3]\}_\infty$ **71** was prepared by heating methylthioxorhenium in water [286]. The structure is described in terms of double layers with corner-sharing CH_3ReO_5 octahedra (A,A′) containing intercalated water molecules (B) in a AA′BB′... layer sequence. Hints for other inorganic macromolecules are given: LiB_x (linear unbranched borynide chains in a lithium matrix) [287], two-dimensional layers of self-organized 1,2-bis(chloromercurio)tetrafluorobenzene [288], framework-structured Sb(III)-phosphate [289], graphite-structured $[(Me_3Sn)_3O]Cl$ [290], three-dimensional structure $RbCuSb_2Sl_4$ [291], layer-structured $[Cu_4(OH)_4][Re_4(Te)_4(CN)_{12}]$ [292], linear polymeric Mo/Ag/S-complexes [293] (all these references contain additional literature to comparable macromolecules).

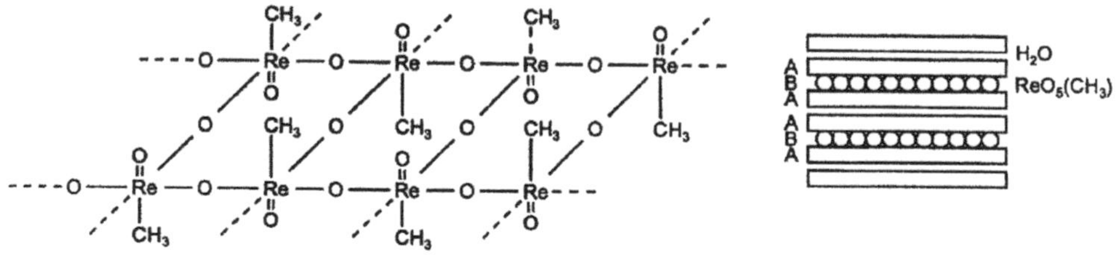

71

C. Heterochain Polymers with Coordinative Bonds Between Metals and Another Element

Coordination polymers are prepared by the reaction of a bifunctional or higher-functional electron donor/Lewis base groups containing single electron pairs ($=O$, $=S$, $=NR$, $-NR_2$, $-O^-$, $-S^-$, $-NR^-$) with a metal ion of Lewis acid properties [equation (30)].

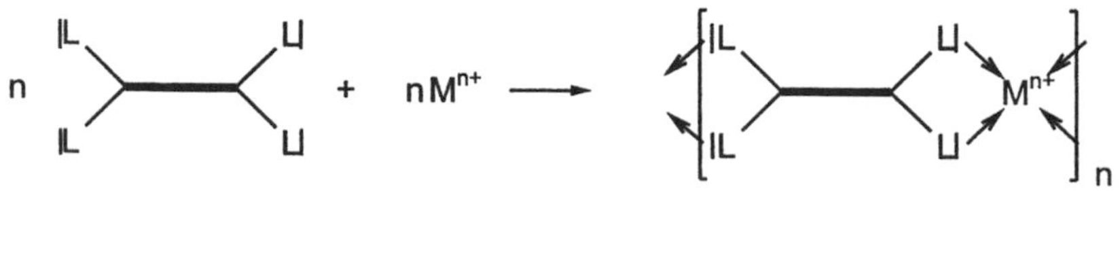

$$(30)$$

1.   Chain Forming Coordination Polymers

Some decades ago described coordination polymers were reviewed in [1,3,294]. They were obtained in general as insoluble dark-colored powders and are therefore difficult to characterize. As compressed powders these polymers often exhibit a high electronic conductivity. Some examples of these polymers are given:

- Reactions of dicarboxylic acids with $SnCl_2$ or uranyl salts [295,296].
- Polymers from transition metal ions with aromatic bis(o-hydroxy acids) [297], dihydroxyquinones [298–300], dihydroxyquinoxalines or -quinolines [301–304].
- Coordination polymers of transition metal ions with different tetrathiolates such as tetrathiooxalate [305], tetrathiosquarate [306], tetrathiofulvalene tetrathiolate [307,308], benzene or naphthalenetetrathiolates [309,310]. Such polymers exhibit electrical conductivities up to $30\,S\,cm^{-1}$ (for the investigation of the electronic structures [311]).

Thermodynamically very stable coordination polymers of Cu^I and Ag^I **72** are obtained by ligand exchange reaction of acetonitrile complexes of, e.g., Cu(I) against phenanthroline monomers in a solvent mixture of acetonitrile and tetrachloroethane [equation (31)] [312]. n-Hexyl side chains are necessary for good solubility of the polymer in less polar solvents. Because the coordination sphere of each metal center is fully saturated by exactly two chelating moieties of the ligand monomers and all carbon atom are *ortho* to the phenanthroline-N-atoms substituted by bulky groups (requiring pseudotetrahydral coordination of Cu(I)) branches and crosslinking is avoided. Intramolecular ring closure is also avoided by an intrinsically rigid p-terphenylene bridge. Excellent soluble Ru^{II} coordination polymers were prepared by the reaction of the metal monomer [Ru(R$_2$bpy) Cl$_3$]$_x$ with a tetrapyridophenazine [313]. The polymers are considered as ribbon-like polyelectrolytes with a coiled shape. In the structure comparable liquid crystalline coordination polymers based on oligopyridines were recently shortly reviewed [314].

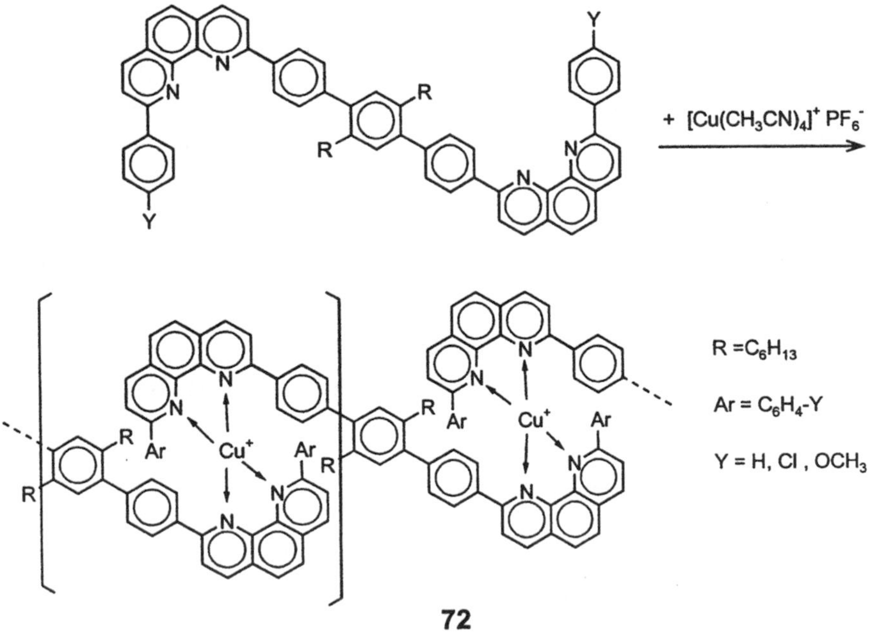

$$R = C_6H_{13}$$

$$Ar = C_6H_4\text{-}Y$$

$$Y = H,\ Cl\ ,\ OCH_3$$

(31)

72

Organo-copper compounds are important for the (C–C)-connection. In the solid state such copper compounds can exist as linear polymer. An example is the cyanocuprat [2-(Me$_2$NCH$_2$)(C$_6$H$_4$)$_2$CuLi$_2$(CN)(THF)$_4$]$_x$ structured as polymeric chain with alternating cuprat-anions (Ar$_2$Cu)$^-$ and cations (LiCNLi)$^{+\cdot}$ [315,316]. 2,4,6-Tris[4-pyridyl]-methyl-sulfanyl]-1,3,5-triazine forms with Ag$^+$ a one-dimensional chain polymer with nanometer tubes **73a** (M = Ag(II)) [317]. These tubes are connected via Ag–N and Ag–S bonds to linear chains **73b**. Other comparable polymers are mentioned in [317].

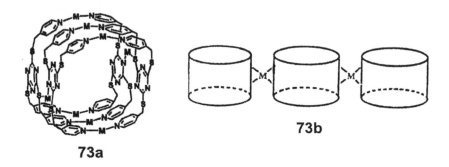

73a

73b

D. Supramolecular Organization of Coordination Polymers

Numerous papers were recently published about the crystal engineering of supramolecular solids. The below mentioned literature contains exemplarily selected results. Further references are given therein. Tetrahedral, trigonal or octahedral acting metal ion centers (e.g. Zn^2, Cd^{2+}, Ag$^+$, Cu$^+$) coordinate with ligands containing N-donor (amines, N-heterocycles), *O*-donor (carboxylic acids), cyano or thioether ligands. The interaction can be seen also as a connection of Lewis-acid metal centers with polyfunctional Lewis bases. Beside the synthesis of new microporous materials especially the reversible inclusion of guest molecules and the use as catalysts or magnetic materials are interesting.

The supramolecular organization of coordination polymers is classified into not-penetrating and penetrating networks. The coordination numbers and geometries of the metal centers and the functionality of ligands allow in principle to pre-determine the lattice structure. But it is difficult to predict not-penetrating or penetrating network formation. Often guest molecules like solvents are included in the lattice. The thermal stabilities are low compared to inorganic networks such as zeolites. An excess of a strong mono-functional Lewis-base destroys by their coordination to the metal ion the structure of the coordination polymers.

1. Non-interpenetrating Coordination Polymers

Several compounds with the simple formulae, e.g., AuJ, PdCl$_2$, MoJ$_3$, AuCN are in fact crosslinked owing to coordination of halide or pseudohalide bridges [318]. Cyano-bridged one- to three-dimensional coordination polymers based on [M(CN)$_6$]$^{3-}$ (M = Fe, Cr, Mn, etc.) have attracted great attention because of rich structures and magnetic behaviour [319]. Prussian blue Fe$_4$[Fe(CN)$_6$]$_3 \cdot$ H$_2$O consists of a three-dimensional network and is obtained by mixing dilute equimolar solutions of K$_4$[Fe(CN)$_6$] and FeCl$_3$ as colloid with an average diameter of 23 nm and a molecular weight of $\sim 7 \times 10^6$ mol g^{-1} [320]. This polymer is interesting as photosensitizing device, rechargeable battery material, memory device and for electrochromic displays [321]. The three-dimensional polymers

[SmFe(CN)$_6$]·4H$_2$O and [TbCr(CN)$_6$]·4H$_2$O exhibit long-range ferromagnetic ordering below 3.5 K and highest known Curie temperature ($T_C = 11.7$ K) for 4f–3d molecule based magnets [322]. Unusual magnetic properties were measured for two-dimensional nets prepared from K$_3$[Fe(CN)$_6$] and 2,2′-bipyrimidine with Nd(NO$_3$)$_3$ [323]. Alternating fused rows or rhombus-like Fe$_2$Nd$_2$(CN)$_4$ rings and six-sided Fe$_4$ND$_4$(CN)$_4$ rings forming the net, and the bipyrimidine coordinates to the Nd ions in a chelating fashion. Mesoscopic layers of a cyanobridged Cu/Ni-coordination compound were obtained coating at first quartz- or membrane filters with an aqueous layer of the detergent **74** [324]. This film was dipped in an aqueous solution of K$_2$[Ni(CN)$_4$] and then in an aqueous solution of Cu(NO$_3$)$_2$. Thus a two-dimensional layer of Cu[Ni(CN)$_4$] is formed between the layers of **74** (Figure 9).

Based on bifunctional space ligands several building blocks as shown in Figure 10 were realized recently [325,326]: (a) diamond analogues [327], (b) honeycomb analogues [328], (c) square lattice analogues [329], (d) ladder and stair analogues [325,330], (e) brick analogues [331], (f) octahedral structures [332,333], (g) helical structures [334] (for other examples see [335] and literature cited therein). The coordination polymers are hold together by 'coordinative covalent' bonds. The preparation is often relatively easy. In a one-pot synthesis a metal salt is mixed in solution with the bifunctional ligand and under

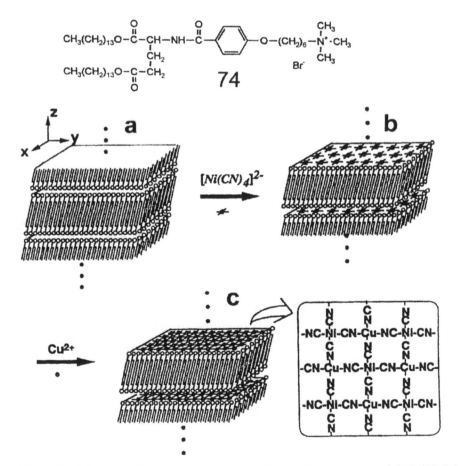

Figure 9 Schematic view on the step-wise synthesis of an aggregate of Cu[Ni(CN)$_4$]. (a) Layer of **74**. (b) Film with **74** with K$_2$[Ni(CN)$_4$]. (c) Film of **74** with Cu[Ni(CN)$_4$].

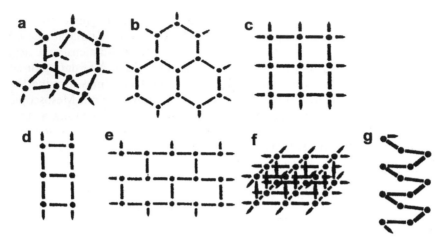

Figure 10 Schematic representation of simple network structures with metal units as points and ligand units as small rods.

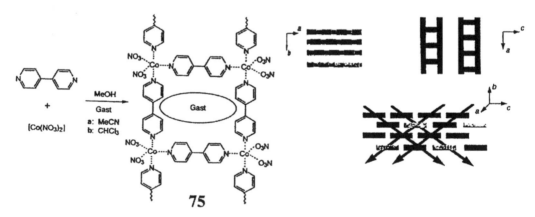

75

Figure 11 Synthesis and molecular structure of the coordination polymer **75** prepared from $CoNO_3$ and 4,4′-bipyridine.

slow crystallization the coordination polymer is obtained. Very important is the molar ratio of the two compounds. The host lattice often includes solvent molecules. Few examples are given below.

Mixing of a solution of $Co(NO_3)_2 \cdot 6H_2O$ with 4,4′-bipyridine (molar ratio 1:1.5) results after few hours in red crystals of **75** [325]. In air the crystals are stable only for few hours. After X-ray analysis **75** consists of a molecular ladder structure (Figure 11). By self-organization the 4,4′-bipyridine ligands coordinate to Co^{2+} and forming the side part and the rung of a ladder in the direction of the a-axis (Figure 11). The infinite ladders which include solvent molecules are shifted by a rung distance against each other and stacked like a stair in the direction of the b-axis. A two-dimensional square lattice of **76** is obtained from Cd^{2+} and 4,4′-bipyridine [329]. Aromatic guest molecules can be incorporated, and the polymer catalyze the cyanosilylation of aldehydes. The structure of [Ag(4,4′-bipyridine)NO_3] consists of linear silver-ligand chains which are crosslinked by an Ag–Ag interaction [336]. Infinite channels (2.3×0.6 nm) are formed which can reversibly incorporate PF_6^-, MoO_4^{2-}, BF_4^- and SO_4^{2-} ions.

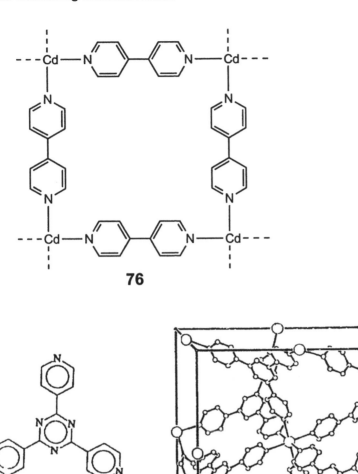

76

77

78

Figure 12 Unit cell of the coordination polymer **78** prepared from $HgClO_4$ and the ligand **77** (Hg as great circles, C- and N-atoms as small circles).

Inserting in the ligand 4,4'-bipyridine between the two bipyridyl a biphenylene unit (0.85 nm longer than **75**) extraordinarily big square-grid coordination polymers with dimensions of about 2×2 nm were obtained [337]. The guest o-xylene (used as solvent was included and occupies 58% of the crystal volume. The big square cavities are packed and create big rectangular channels. Reaction of 2,4,6-tri(4-pyridyl)-1,3,5-triazine **77** with $HgClO_4$ in tetrachloroethane yields three-dimensional structures **78** containing solvent molecules (Figure 12) [338]. The Hg centers are forming a slightly distored octahedral geometry. An infinite network **80** was obtained from hexakis(imidazol-1-ylmethyl)benzene **79** as ligand in the interaction with CdF_2 (Figure 13) [339]. The dithia-ligand **81** interacts with $AgBF_4$ in CH_3CN under formation of a two-dimensional layer compound **82** (with BF_4^- between coplanar layers) (Figure 14) [340].

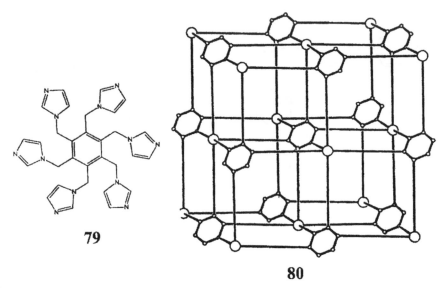

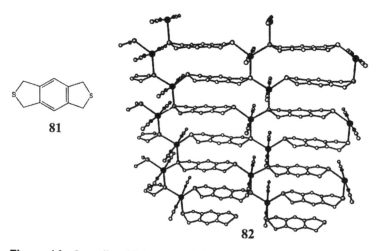

Figure 13 Infinite three-dimensional network of the coordination polymer **80** prepared from CdF_2 and **79** (Cd as great circles, imidazole-bridge as lines).

Figure 14 Lamella of **82** prepared from $AgBF_4$ and the ligand **81**.

Helical coordination polymers containing great chiral cavities or channels are interesting for stereospecific synthesis, separation of enantiomers and stereoselective catalysis. Crystals of **83** are obtained from a solution of nickel(II) acetate and benzoic acid in methanol which is covered with a solution of 4,4′-bipyridine in the presence of benzene or nitrobenzene [334]. Each helix winding contains three complex units and consists in the chain of Ni^{2+} (with binding of benzoate) and the bipyridine. Because each helix is shifted against each other, great cavities are formed containing, e.g., benzene or nitrobenzene.

A column layer structured coordination polymer **84** (Figure 15) was obtained from Cu^{2+} in the reaction with pyrazine-2,3-dicarboxylate (pzdc) and pyrazine [341]. The polymer consists of a two-dimensional layer of Cu-pyrazinedicarboxylate which are connected by columns of pyrazine. After desorption of water at 100 °C, methane could be adsorbed reversibly. Single-, double- and three-dimensional structures of rare-earth metal coordination polymers are formed by hydrothermal synthesis through the reaction of a rare-earth metal(III) nitrate (M = La, Ce, Eu) with 3,5-pyrazoledicarboxylic acid in water for 3 days at 150 °C in a Teflon autoclave [342]. The three-dimensional framework contains nine-coordinated lanthanide metal centers. Hexagonal layered networks based on $[M_2(OOCCF_4)_4]$ (M = Ru, Rh) as donors and tetracyanoquinodimethane as acceptor are described in [343], and three-dimensional salts consisting of nitrile ligands in $[\{Cu\{C[C(CN)_2]_3\}(H_2O)_2\}_n]$ are described in [344]. Two-dimensional magnetic materials based on nitroxides are prepared from nitronyl-nitroxide ligands in the interaction with Mn^{2+} [345].

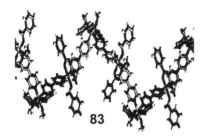

The reaction of the Ni(II)-complex **85** of bis(2-pyridylcarbonyl)amine with Fe(II) perchlorate in methanol results in dark purple hexagonal crysatal of **86** [346]. Figure 16 shows the stepwise growth to a graphite-like polymer coordination complex with large cavities. Two- and three-dimensional polyrotaxane coordination polymers were described recently [347]. The first step results in the pseudorotaxane **89** by threading the cucurbituric **87** with the bipyridyl derivative **88** (Figure 17). In the second step the polyrotaxane **90** is obtained by reaction of **89** with $Cu(NO_3)_2$ in the presence of oxalate ions. In **90** the compound **87** is threaded on a two-dimensional polymer network. The stacking of the layers in a distance of 1.29 nm results in one-dimensional channels.

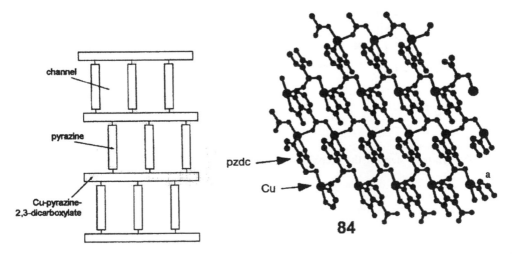

Figure 15 Schematic representation of the column-layer structure of the coordination polymer **84** prepared from Cu^{2+} and pyrazine-2,3-dicarboxylate (pzdc) and pyrazine (right) and the molecular structure (left).

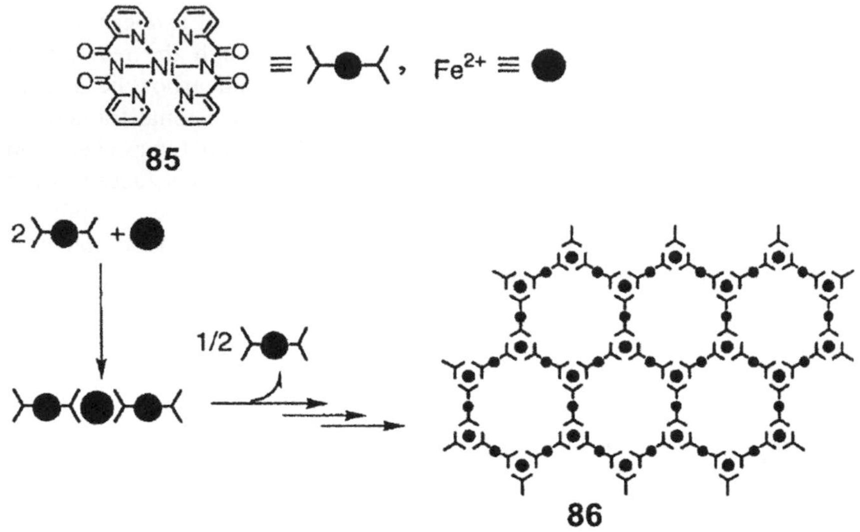

85

86

Figure 16 Schematic representation of the stepwise growth of the Ni-complex **85** with Fe(II) perchlorate to the coordination polymer **86**.

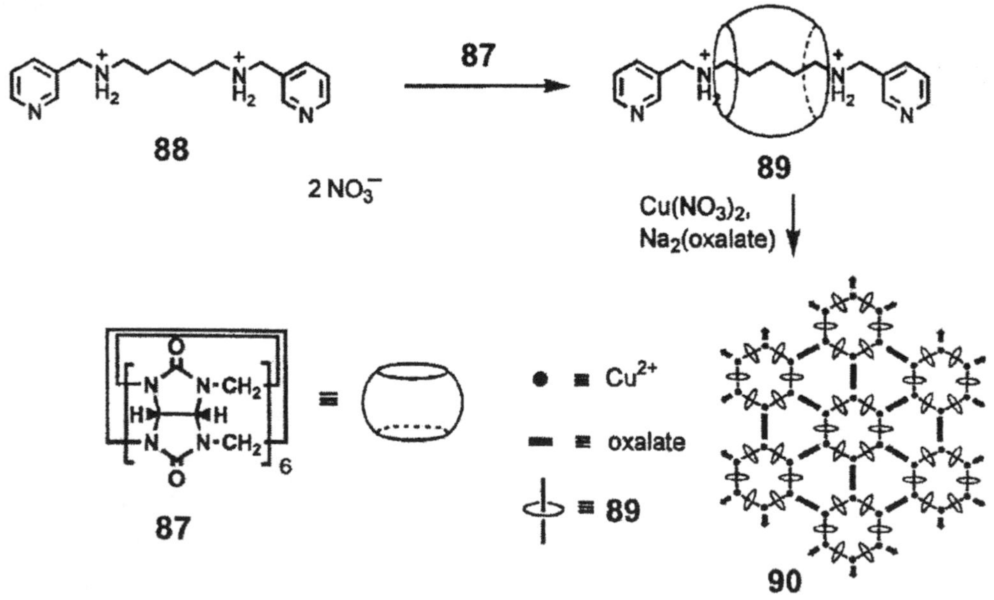

88

2 NO$_3^-$

89

Cu(NO$_3$)$_2$, Na$_2$(oxalate)

87

● ≡ Cu^{2+}

▬ ≡ oxalate

≡ **89**

90

Figure 17 Synthetic step for the preparation of the two-dimensional polyrotaxane **90**.

Another concept is based on neutral, polyfunctional Lewis acid spacers for the construction of stair and ladder structures [330]. Ortho-phenylene(indium-bromide) **91** exists as a THF stabilized dimmer. The *p*-orbitals at indium are orientated perpendicular to the diindacyclus. Reaction of **91** with pyrazine in a molar ratio of 1:1 or 1:2 in THF results in the stair structured molecule **92** whereas in a molar ratio of 1:4 the ladder structured **93** (both contains additional THF) is obtained (Figure 18). The polymers are

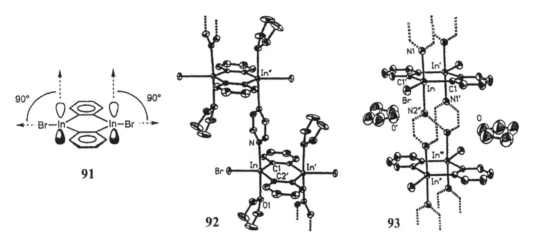

Figure 18   Molecular structure of ortho-phenylene/indium-bromide **91** and its coordination polymers **92** and **93** by reaction with pyrazine.

insoluble in unpolar solvents but can be dissolved in coordination solvents destroying then the polymer structure.

Porphyrins are participating in nature in electron transfer, energy transfer and redox-catalysis. Therefore these macrocycles are also interesting as building blocks in coordination polymers. Tetrasubstituted porphyrins being substituted by groups capable of coordination to metal ions are employed for the preparation of porphyrin coordination polymers. Mixing of $Hg(ClO_4)_2$ in methanol with 5,10,15,20-tetrapyridylporphyrin (**21** with pyridyl groups instead of $-C_6H_4-R$) in 1,1,2,2-tetrachloroethane results in crystals of a coordination polymer [348]. The Hg-centers form by coordinating to the pyridyl-groups an infinite net. In the cavities tetrachloroethane and ClO_4^- are encapsulated. Different networks with great cavities are obtained when 5,10,15,20-tetrapyridylporphyrin is reacted with Co^{2+} or Mn^{2+} in aqueous solution under conditions of hydrothermal synthesis in a Teflon autoclave at 200 °C for 2 days [349].

As another ligand the 5,10,15,20-tetracarboxytetraphenylporphyrin **21** (R = –COOH) was employed [350]. The Zn-complex of this porphyrin was treated with 4,4'-bipyridine in a mixture of methanol and ethyl benzoate to give crystals of **94**. Figure 19 shows interlinked arrays of the porphyrin and bipyridyl. The structure represents an open three-dimensional network in which the individual metallo porphyrin units are cross-linked both axially as well as equatorially by ion-impairing interactions. Partially pyridyl-substituted porphyrins are reacted with $PdCl_2$ to definite oligomers [351].

2.   Interpenetrating Coordination Polymers

As mentioned before, coordination polymers can form porous structures (with channels, cavities) where solvent molecules are included. In several cases penetrating structures are obtained in which cavities belonging to one lattice frame-work is occupied by one or more independent lattice frame-works. These penetrating structures can be separated only by splitting of bonds. Some relations to catenanes, rotaxanes and molecular knots can be taken into account. A classification of interpenetrating coordination polymers with a catenated structure is given in Figure 20: (a, b) linear structures, (c) interpenetration of ladders, (d) inclined interpenetration of ladders, (e) interpenetration of undulating layers,

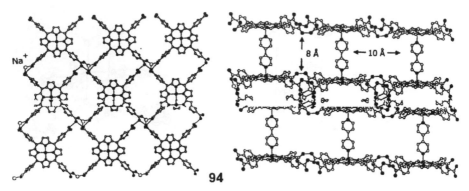

Figure 19 Different views on the crystal structure of the coordination polymer **94** prepared from **21** (R = –COOH, M = Zn(II) and 4,4′-bipyridyl.

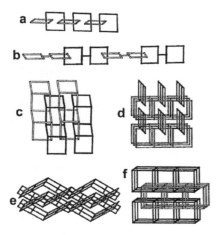

Figure 20 Classification of interpenetrating coordination polymers. (a and b) One-dimensional linear structures; (c and d) two-dimensional interpenetration of ladders; (e and f) three-dimensional interpenetration of andulating and multiple layers.

(f) interpenetration of multiple layers [352]. A recent review describe in detail penetrating networks [353]. Only very few examples are given below.

The ligand 1,4-bis(imidazol-1-ylmethyl)benzene **95** leads in the interaction with Ag^+ to a polymer **96** of the composition $[Ag_2(\mathbf{95})_3(NO_3)_2]_\infty$ [353]. Two one-dimensional chains penetrate in a polycatenane analogous structure (Figure 21). In contrast now, the reaction of **95** with Zn^{2+} yields a polymer **97** of the composition $[Zn(\mathbf{95})_2](NO_3)_2 \cdot 4.5\,H_2O$ [354]. This polymer consists of two independent parallel lying two-dimensional networks. Zinc is tetracoordinated (Figure 22).

An example for an three-dimensional penetrating network is the solvated $[(ZnCl_2)_3(\mathbf{77})_2]$ **98** [353]. The triazine molecules are the part of threefold connecting knots (Figure 23). The zinc atoms consists of a nearly tetrahedral coordination geometry with two N-atoms of **77** and two chlorine ligands. A distorted network is obtained. The self-assembly of $CuSO_4$ in water and 1,3-bis(4-pyridyl)propane in ethanol results in **99** which presents a three-dimensional architecture sustained by two different types of coordination polymers: one-dimensional ribbons of rings and two-dimensional layers (Figure 24) [352].

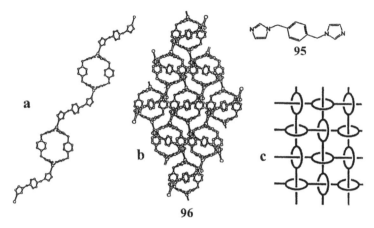

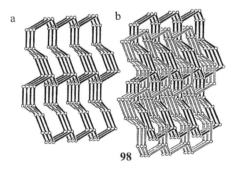

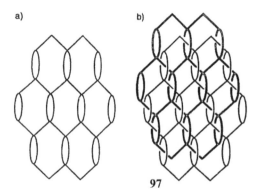

Figure 21 Coordination polymer **96** prepared from the ligand **95** and AgNO₃. (a) One single chain; (b) polyrotoxane analogous layer of different chains; (c) schematic representation of the structure.

Figure 22 Schematic representation of the coordination polymer **97** prepared from the ligand **95** and Zn(NO₃)₂. Zn are the centers of four-fold knots; the ligand **95** represents the connections between them. (a) Part of the network; (b) interpenetrating network.

Figure 23 Schematic representation of the coordination polymer **98** prepared from the ligand **77** and ZnCl₂. (a) Part of the network; (b) interpenetrating network.

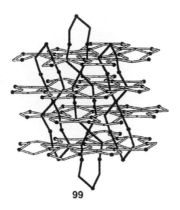

99

Figure 24 Schematic representation of the coordination polymer **99** (simplified by showing only the central methylene atoms).

E. Metallocenes as Part of a Polymer Chain

The arrangement of bifunctional π-electron-rich charged aromatics connected by π-bonds in the main chain as shown in the general formula **100** has been described in some reviews [42,355–358]. Polymer containing directly linked metallocenes are obtained by, for example, step growth polycondensation of 1,1-dithioferrocenes with 1,1-diodoferrocenes (M < 10 000) [359]. Also poly(1,1-ruthenocenylenes) were described [359]. Face to face polydecker sandwich complexes via a naphthalene spacer **101** are prepared by a monomer dianion through reaction with a transition metal ion [357,360]. Polydecker sandwich complexes containing 2,3-dihydro-1,3-diborolo ligands and Rh^{2+} are described in [358]. The Barat complex $[Cp_3Ba]^-$ consists in the solid state on a linear coordination polymer **102** in which Ba^{2+} ions are surrounded in a tetraedric coordination by four Cp anions (Figure 25) [361]. The polymer **102** is prepared by reaction of a Wittig-reagent with CpH under formation of $[Cp]^-[R_4P]^+$ which is reacting *in situ* to $[Cp_3Ba]^-[Bu_4P]^+$ [equation (32)]:

$$Bu_3P = CHCH_2CH_2CH_3 + CpH + Cp_2Ba \rightarrow [Cp_3Ba]^-[Bu_4P]^+ \tag{32}$$

Several papers report on polyester, polyamides, polyurethanes containing metallocenylenes [359,362] and polyferrocenylenes-bridged by $-(CH_2)_n-$, $(SiR_2)-$, $-(GeR_2)-$ [359,363]. The synthesis of poly(1,1'-ferrocenylene-*p*-oligophenylenes) such as **103** with a degree of polycondensation of ~55 was realized by a Pd-catalyzed polycondensation of 1,1'-bis(p-bromophenyl)ferrocene with 2,5-dialkylbenzene-1,4-diboronic acids [364].

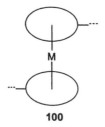

100

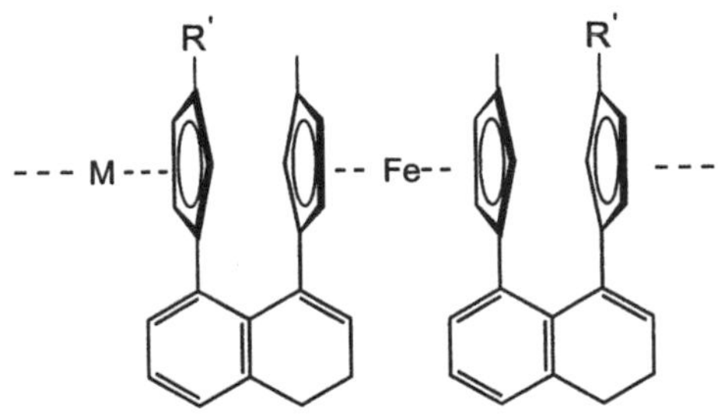

101

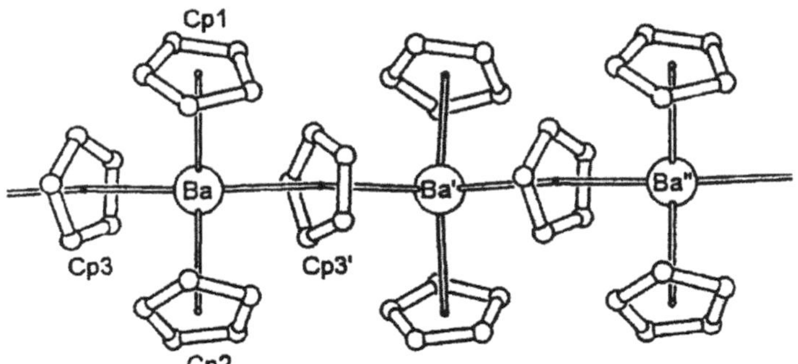

102

Figure 25 Part of the structure of the linear $[Cp_3Ba]^-$ chain of the polymer **102**.

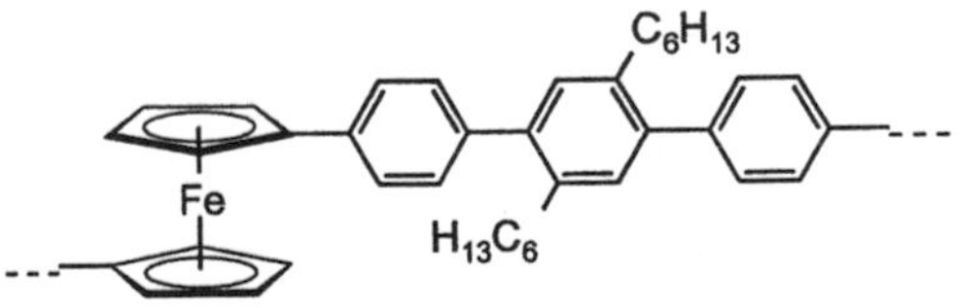

103

Interesting polymers are obtained by the ring-opening polymerisation of various metallocenophanes to high molecular weight poly(metallocenes) **104** ($M_w = 10^5$–10^6, $M_n > 10^5$ [equation (33)] [355,356,363,365–372]. The polymers contain between the metallocenes the following bridging blocks X in **104**: $-(CR_2-CR_2)-$ [42,372], $-(SiR_2)-$ [42,366], $-(GeR_2)-$ [42], $-(PR)-$ [42], $-(SnR_2)-$ [367], $-(S)-$ [42], $-[B(N(SiMe_3)_2]-$ [368]. The ring opening can be achieved thermally, via anionic or cationic initiation or by the use of transition-metal catalysts. Different copolymers were also prepared [366,369,370,373]. Two examples of a di-block- and tri-block-copolymer are shown in **105** and **106**. These poly(metallocenes) are interesting materials for coating of dielectrics [371] and precursors for magnetic ceramics [374]. The block-copolymers have been found to undergo microphase separation and in a suitable solvent to form interesting morphologies based on isolated or collections of spherical and cyclindrical micelles/nanotubes [369,370, 373,374]. It is pointed out in [374] that such superlattices may be used for the fabrication of magnetic multilayers for data storage, patterned redox-active thin-film electronic devices and designer ceramic architectures.

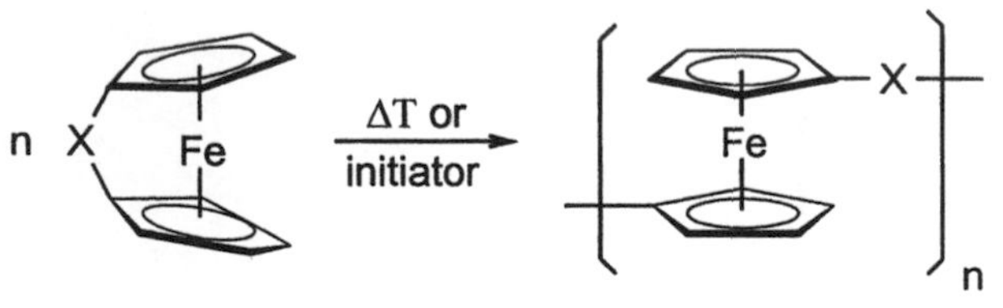

(33)

104

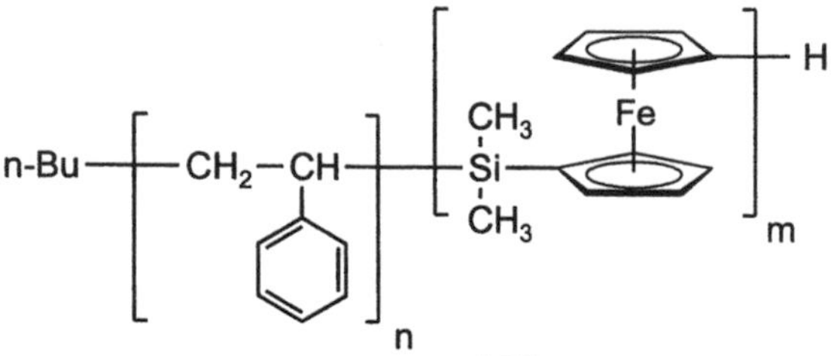

105

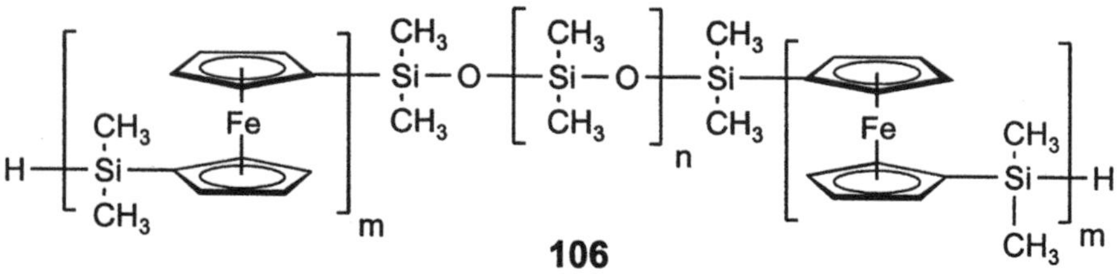

106

F. Cofacial Stacked Polymeric Metal Complexes

A metal ion M as part of a polymer chain is surrounded by a multivalent ligand. As a result, stacked arrangements **107** are realized with a face-to-face orientation of the ligands. Porphyrins, phthalocyanines, naphthalocyanines, etioporphyrin, tetraaza[14]annulene, hemiporphyrazines, and related macrocycles as ligands have been most intensively investigated. Because several reviews summarize preparations and properties (conductivity, photoconductivity, NLO, electroluminescent detectors) of these materials, only a short overview on these polymers is given [1,3,34,40,375–377]. Several years ago, unsubstituted macrocycles were employed as ligands. These polymers are much less soluble. Recently, the use of substituted macrocycles has led to soluble polymers which are analytically easier to characterize.

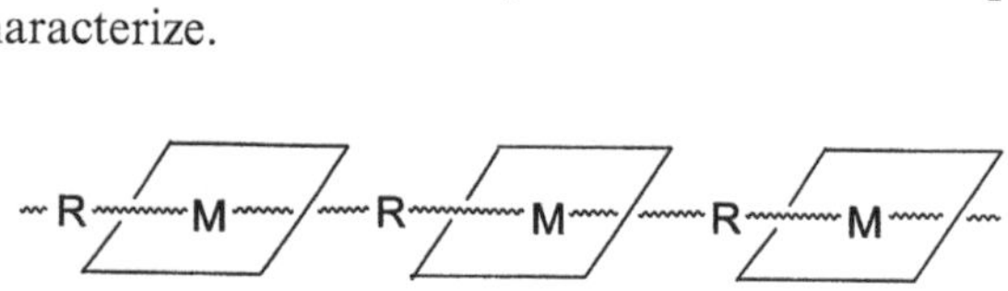

107

1. Covalent/Covalent Bonds Between the Central Metal Ions

The structural arrangement is shown in **108**. Dihydroxy-substituted macrocycles with Me = Si(IV), Ge(IV), or Sn(IV) are converted by heating in bulk or in a high-boiling solvent to the μ-oxo-bridged polymers. The topotactic polymerization mechanism is described in [378]. Depending on the reaction conditions, the degree of polymerization is <150. Use of octasubstituted phthalocyanines (oxyalkyl, methyleneoxyalkyl, crown ether, or ester groups) in $Si(OH)_2PcR_8$, partially with trifluoroacetyl as leaving group ($Si(OCOCF_3)_2PcR_8$), enhance the solubility of the polymers and favor the polycondensation [379–383]. High molecular weight polyenes were obtained and the rotational dynamic behavior of the polymers was investigated [380]. Langmuir–Blodgett films were prepared and liquid crystalline behavior of the polymer was observed [380–383]. Polymers with different bridges between the stacked macrocycles have been described [31,34,376].

M = Si(IV), Ge(IV), Sn(IV)
108

2. Covalent/Coordinative Bonds Between the Central Metal Ions

As shown in **109** these stacked macrocycles contain Al(III), Ga(III) or Cr(III) with fluoro as the bridging atom or Co(III), Fe(III), Mn(III), Cr(III) or Rh(III) with $-C\equiv N$, $-S-C\equiv N$, or $-N_3$ as the binding groups. Details for the preparations have been described [31,34,376].

M = Al(III), Ga(III), Cr(III)   R = -F⁻
M = Fe(III), Co(III), Rh(III), Mn(III), Cr(III)   R = -CN⁻, -S-CN⁻, -N₃⁻
109

3. Coordinative/Coordinative Bonds Between the Central Metal Ions

In this case, macrocycles with transition metal ions in the oxidation state +2 and capable of hexacoordination and neutral organic donor containing two groups or heteroatoms for coordination are used in stoichiometric amounts as starting materials (arrangement **110**) [31,34,376]. The degree of polymerization depends on the reaction conditions and is between 20 and 50.

M = Fe(II), Co(II), Ru(II); Os(II)   R = pyrazine, tetrazine, triazine,
p-diisocyanobenzene, p-phenylenediamine, fumaronitril
110

4. Self-organization to Cofacial Arrangements

Phthalocyanines substituted in the peripheric positions by long chain substituents can form liquid crystalline discotic phases [384]. Polymerization of liquid crystalline phthalocyanines with polymerizable substituents leads to supramolecular structures with fixed columnar orientation [385].

The cofacial stacked polymers are characterized by different chemical stabilities. Whereas polymers **108** may be unaffected by aqueous HF at $100\,^\circ$C, aqueous 2 M NaOH at reflux and concentrated H_2SO_4, polymers **110** are not stable and are reacting with an excess of a monofunctional and even a bifunctional donor ligand under heating. The thermal stability in general increases in the sequence: **110 < 109 < 108**. These polymers can exhibit very high electrical conductivities as compressed powders of $\sim 10^2\,\mathrm{S\,cm^{-1}}$ [34,376].

G. Metallodendrimers

Metallodendrimers are an interestingly represented class of molecules in the area of dendrimer chemistry. They combine dendritic structures with the specific activity of metal complex centers. Metal coordination has facilitated the synthesis of a number of dendritic, supramolecuar structures. Metals have been incorporated in all of the topologically different parts of dendrimers in the repeat or branching unit, in the molecular core, in the peripheral units of dendrimers. Because this field of metallodendrimers were reviewed recently [386,387], only few examples are given below. Other supramolecular organizations such as catenanes and rotaxanes were mentioned before in this chapter.

The synthesis of a dendrimer **111** of the fourth generation with a substituted tetraphenylporphyrin in the cores is shown in equation (34) [388]. The covergent strategy compared to the divergent one is more usefully in order to avoid several preparative steps. Structurally comparable dendrimer porphyrins bearing negatively or positively-charged groups at the peripheral units and therefore being water soluble are described [389]. A silicon(IV)-phthalocyanine **112** substituted at the axial positions with dendritic group shows in the solid state no aggregation by phthalocyanine–phthalocyanine interaction [390,391].

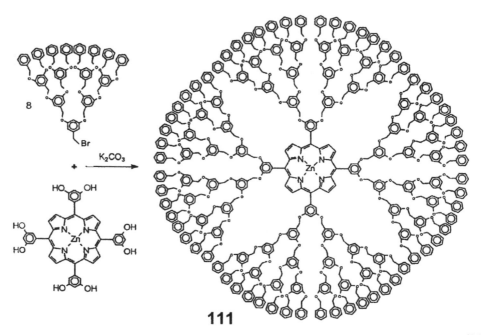

111

(34)

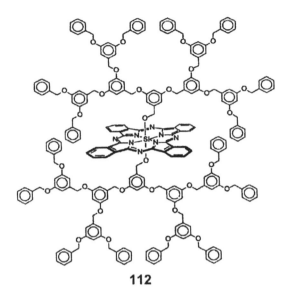

112

One example for a dendrimer substituted at the peripheral units with a metal complex is shown in equation (35). Structure **113** is obtained by treatment of the fourth generation of poly(propylene imine) dendrimer (DAB-dendr-$(NH_2)_x$) ($x = 32$) with chlorocarbonylferrocene and the PF_6^- salt of chlorocarbonylcobaltocenium [392]. A static substitution by the two different metallocenes are observed.

$$\text{(35)}$$

113

Different types of metallodendrimers with metal compounds at each repeating unit throughout the dendrimer are reviewed in [386]. They can be considered due to a high concentration of metal centers as nanoparticle equivalents. One example is platinum acetylide dendrimer **114** by a stoichiometric transmetallation to form acetylides [393]. This metallodendrimer has a precedent earlier syntheses of linear Pt acetylide coordination polymers.

114

The properties of metallodendrimers in the direction of potential applications are reviewed in [386,387]: sensors, binding of small molecules, catalysis, reactive centers, molecular antenna by visible light irradiation.

VI. METALS INCORPORATED PHYSICALLY INTO MACROMOLECULES

This part concentrates on metal complexes dissolved or dispersed in macromolecules. The stabilization of metal or semiconductor clusters in organic or inorganic macromolecules is not part this review.

In principle, every metal complex (or metal salt) can be incorporated into an organic polymer monomolecular or aggregated with the result of a solid composite material. Because of several hundred papers that have been published in this field, the following survey considers only few examples. The combination of organic polymers with included metal complexes offers the advantage of preparing more flexible films with higher mechanical stability. In addition, polar polymers impress the activity in devices with metal complexes as active parts.

Film preparation from a solution containing the polymer and metal complex homogeneously dissolved or the dissolved polymer and dispersed metal complex on a suitable carrier [Pt, Au, carbon (in the form of graphite or glassy carbon) and inorganic semiconductors (ITO, SnO_2, Si, GaAs)] were carried out. The layer thicknesses varied a great deal, extending from approximately, 50 nm to a few μm. The number of active centers of a metal complex in the films coated on a carrier is on the order of 10^{-10}–10^{-6} mol/cm^2. The apparent concentration of a metal complex in a polymer is often as high as 0.1–5 M. The different methods as reviewed in [30] are:

1. *Casting from solution.*   The solution is spread onto a carrier followed by careful evaporation of the solvent.

2. *Spin coating.* A small amount of solution is dropped onto a spinning carrier. The thickness of a film depends on the rotation speed, the evaporation rate of the solvent and the initial viscosity of the solution.

3. *Dip coating.* A carrier is dipped into a solution and then dried. Higher concentration and longer soaking time yield thicker coatings.

4. *Coating and adsorption process.* First, a layer of the polymer is coated onto a carrier from solution. Then, in a second step, this coated material is placed in contact with an aqueous solution containing the metal complex for a few minutes or several hours.

In order to obtain a homogeneous distribution of the metal complex and a smooth film formation, the experimental conditions must be optimized carefully. Low concentrations of a metal complex (<1–10 wt%) in a polymer can result in mononuclear distribution. The polymer appears in this case to function as a solvent that minimizes complex–complex interaction by separation. Depending on the solubility of the metal complex in the solvent used for casting, higher concentrations (>1–10 wt%) can result in aggregation or microcrystal formation. Some examples for metal complexes in polymers are given.

Tris(2,2′-bipyridine)ruthenium(II) ($Ru(bpy)_3^{2+}$) is dispersed in a Nafion membrane coated on an ITO electrode. By cyclic voltammetry oxidation and re-reduction of $Ru(bpy)_3^{2+}/Ru(bpy)_3^{3+}$ was studied. A critical distance of the Ru-complexes for efficient charge hopping between redox centers of approximately 1.3 nm is found and calculated. The maximum of a Poisson distribution of distances is ≥ 0.1 M of $Ru(bpy)_3^{2+}$ [394,395].

Photocurrent has been successfully obtained with a $Ru(bpy)_3^{2+}/MV^{2+}$ (methylviologen, 1,1′-dimethyl-4,4′-bipyridinium) photochemical reaction couple by coating a polymer-pendant $Ru(bpy)_3^{2+}$ film on an electrode [396–399], for which the photochemical reaction products of monomeric $Ru(bpy)_3^{2+}$ and $MV^{+\bullet}$ are too short-lived to induce photocurrent from solution. The acceptor, MV^{2+}, can either be present in solution in contact with the Ru-comple-coated electrode or as a second polymer layer on top of the polymer-Ru-complex film. Also O_2 can work as electron acceptor in such systems [400,401]. The presence of a MV^{2+} in second polymer layer in the $Ru(bpy)_3^{2+}/O_2$ system enhanced the photocurrent for which MV^{2+} works as an efficient electron transport mediator from $Ru(bpy)_3^{2+}$ to O_2 [402].

Intermolecular interactions of metal complexes can be seen by photophysical and photochemical measurements. Some metal complexes, such as $Ru(bpy)_3^{2+}$ or $Ru(II)$-tris(4,7-diphenyl-1,10-phenanthroline)$^{2+}$, exhibit a high quantum yield and long life time of excited states, which is quenched by acceptors such as oxygen (energy transfer under formation of singlet oxygen) or MV^{2+} (electron transfer and formation of methylviologen cation radical) [403–405]. Inclusion in polymers can enhance the life times because of the microenvironmental effect of the surrounding polymer and reduced bimolecular annihilation processes [3,404].

Thin films consisting of a mixture of Zn(II)-phthalocyanine **20** (R $=$ –H) with polymers were prepared by drop coating or spin coating on conducting ITO or gold [406–409]. Polymers of different polarity and coordination ability such as polystyrene, poly(acrylonitrile), poly(1-vinylcarbazole), poly(4-vinylpyridine) or poly(vinylidenfluoride) were used. In order to achieve conducting pathways between the porphyrin type compounds in the film around equal weight mixtures of the polymer and the phthalocyanine were dissolved in a polar solvent such as N,N-dimethylacetamide and drop coated on a carrier. Homogeneously colored, pin hole free films were obtained.

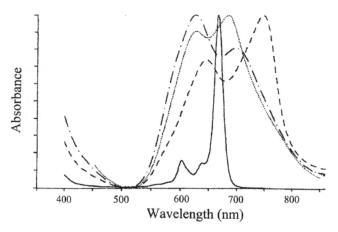

Figure 26 Visible absorption spectra of different films of **20** (R = –H, M = Zn(II). — · — = α-modification and - - - - = β-modification by vapor deposition. · · · · · · = mixture (1 : 1) with poly(vinylidene fluoride). —— = dissolved in *N,N*-dimethylacetamide.

The films with the Zn(II)-phthalocyanine show significant differences in the electronic absorption spectra compared to those obtained in solution and prepared by vapor deposition (Figure 26). Whereas dissolved phthalocyanines exhibits the Q-band transition at ~670 nm, the layers with different polycrystalline structure lead to different electronic spectra. The α- an β-modification are characterized by a slip stack orientation of adjacent phthalocyanine rings giving distinguishable bands in the visible region [410]. Exemplarily, the spectrum of **20** (R = –H, M = Zn(II)) in Figure 26 shows that in polymers embedded particles exhibit a different polymorphic arrangement of molecules. The amount of splitting indicates the strength of intermolecular electronic interaction thus to be most intense in β-crystals, intermediate in α-crystals and weak of the zinc complex dispersed in poly(vinylidene fluoride).

The phthalocyanine containing polymer films were electrochemically investigated for their electrochromic reductions and reoxidations [406,411]. Under irradiation the reduction of O_2 to water was studied in photoelectrochemical cells [407,409,412]. Especially Zn(II)-phthalocyanine in poly(vinylidene fluoride) shows high cathodic photocurrents. Also the electrochemical carbox dioxide and proton reduction by Co(II)-phthalocyanines in a low concentration monomolecular in a polyvinylpyridine matrix were investigated as part of a photoenergy systems [413,414]. As an active catalyst for proton reduction also a bipyridyl platinum complex in a polymer Nafion membrane was found [415]. In order to construct such a photochemical energy conversion system, the research in this field was extended for the electrocatalytic water oxidation to O_2 [416–419]. The Ru-complexes *cis*-[Ru(bpy)$_2$Cl$_2$] and especially Ru-red ([(NH$_3$)$_5$–Ru–O–Ru(NH$_3$)$_4$–O–Ru(NH$_3$)$_5$]$^{6+}$) are active as electrocatalysts.

Anionic metal complexes, such as tetrasulfophthalocyanine or disulfoferrocene, are incorporated from solution during the oxidative electrochemical polymerisation of pyrrole [420–423]. After re-reduction of the polypyrrole film the phthalocyanine derivative remains in the film, and the observed electrochemical conductivity is explained by ion transport of small electrolyte cations [422].

Co(II) and Fe(II) Schiff bases and porphyrin complexes were included by casting in various copolymers containing vinylpyridine or *N*-vinylimidazole units. Details on preparation of thin films and their use in O_2-transporting membranes are given in the reviews [3,5].

The excitation and emission spectra of lanthanide metal ions, such as Tb^{3+} or Eu^{3+} in polycarboxylates (ionic interaction), were investigated in detail [424,425]. The interest is to study the structure of ionomers in solution and as solid. Because of a linear increase of the luminescence intensity with increasing concentrations of lanthanides in a polymer film, a mononuclear distribution at lower concentrations is shown. In a copolymer of styrene and acrylic acid a decrease of the luminescence intensity at >4–6 wt%, and in poly(acrylic acid) or poly(styrene-co-maleic acid) at >15 wt% shows formation of ionic aggregates [426].

Luminescence behavior and electron transfer of various transition metal complexes with 4,4′-bipyridyl, EDTA or 1,2-diaminoethane in the interaction with polyelectrolytes were studied in solution in detail [427,428]. The systems are prepared by mixing solutions of the polymers (in excess) and the metal complexes. The measurements show monomolecular distribution of the metal complexes and allow study depending on the pH-value-dependent conformational transition of the polymer chains as shown in Figure 27.

Transition metal salts are stabilized by interaction with part of a polymer chain. Examples are poly(styrene) $AlCuCl_4$-Co complexes [429,430], CuCl complexes at poly(styrene) modified with amino groups [429], $PdCl_2$ and RhCl complexes at poly(styrene-co-butadiene) [431,432], $PdCl_2$ at poly(acrylonitrile) [433] and $RhH(CO)(PPh_3)_3$, $PtCl_2$–$SnCl_2$, $RuCl_3$–$CoCl_2$, $Rh_2(CO)_4Cl_2$ at different polymer phosphine ligands [434], $PdCl_2$ at different organic polymers (poly(benzimidazole), cyanomethylated cross-linked poly(styrene), cross-linked poly(acrylonitrile) [435] and carboxylated Co(II)- and Fe(III)-phthalocyanines in rayon fibres [436]. These complexes are investigated mainly as catalysts in different reactions. Polymer (e.g., poly(ethylene oxide))/inorganic salt (e.g., alkali salts) complexes have been actively investigated for solid-state ionic conductivity to develop materials for commercial applications (battery, electrochromic devices, moisture or gas sensors) [3,437–440]. Films are prepared easily by casting a solution of the polymer and metal salt followed by drying.

Various metal complexes such as metal-phthalocyanines, metal-salenes or Ru-pyridyl complexes were incorporated in molecular sieves such as cavity-structured zeolites (faujasites, supercages with 1.3 nm diameter), channel-structured aluminum phosphates

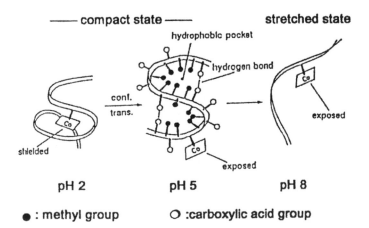

● : methyl group   O :carboxylic acid group

Figure 27 Schematic representation of conformational transitions of Co(III)-(ethylenediamine)$_2$ at poly(methacrylic acid) at different pH.

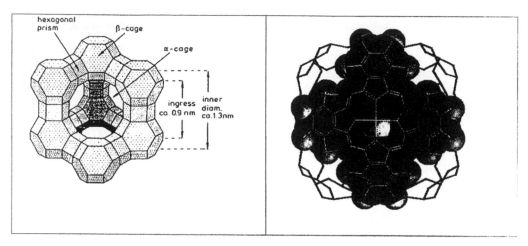

Figure 28 Molecular sieve zeolite faujasite. Left: Structure of the zeolite with supercages of 1.3 nm diameter. Right: Model for the incorporated metal phthalocyanine **20** (R= –H) with diameter 1.2 nm.

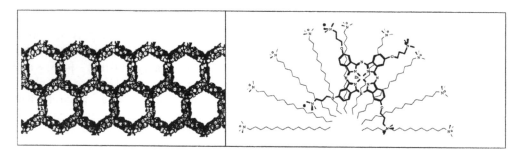

Figure 29 Molecular sieve silicate MCM-41. Left: Structure of hexagonal pores with 2.4 nm diameter. Right: Model for the incorporation of **20** (R = –O–CH$_2$–CH$_2$–N$^+$ (CH$_3$)$_3$. M = Zn(II) in the columnar orientated detergent cetyltrimethylammonium chloride surrounded by the MCM-41 channels.

(AlPO$_4$-5, channel diameter 0.73 nm) and channel-structured silicates MCM-41 (channel diameter 3.2 nm) [441,442]. Different strategies for the inclusion, as exemplarily mentioned, for phthalocyanines were applied. Whereas the zeolite encaged phthalocyanines (**20** R = –H, M = Co(II), Ru(II), etc.) are, for example, synthesized by the reaction of a transition metal ion-exchanged zeolites with phthalonitrile in a closed bomb vessel [443], substituted derivatives of phthalocyanines were added to the mixture of the hydrothermal synthesis of the molecular sieve in the cases of AlPO$_4$-5 and MCM-41 [444,445]. The cavity size of the faujasite zeolite agrees well with the diameter of the phthalocyanine (Figure 28) [443]. Phthalocyanines included in MCM-41 can be found in the columnar-orientated detergent (used as templates for synthesis) of the channels (Figure 29) [445]. Loadings of $\sim 10^{-5}$ per g molecular sieve were achieved. The materials are intensively green-colored and contain the phthalocyanines in the monomeric state as shown in the UV/Vis reflectance spectra by the Q-band transition at ~ 680 nm. One example for the inclusion of the phthalocyanine (**20** R = –O–CH$_2$–CH$_2$–N$^+$(CH$_3$)$_3$, M = Zn(II)) is shown in Figure 29. The molecular sieve-encapsulated phthalocyanines were investigated as catalysts [443] or in the photophysical hole burning optical information storage [446].

REFERENCES

1. Wöhrle, D. (1992). Polymers with metals in the backbone. In *Handbook of Polymer Synthesis* (Kricheldorf, H. R., ed.), Marcel Dekker, New York, Part B: 1133–1195.
2. MMC II (1988). *J. Macromol. Sci.-Chem., A25*, Vol. 10, and 11. MMC III (1990), *J. Macromol. Sci.-Chem., A26*, Vol. 2, and 3; Vol. 9–11. MMC IV (1992). *Makromol. Chem. Macromol. Symp. 59.* MMC V (1994). *Macromol. Symp. 80.* MMC VI (1996). *Macromol. Symp. 105.* MMC VII (1998). *Macromol. Symp. 131.*
3. Ciardelli, F., Tsuchida, E., and Wöhrle, D., eds. (1996). *Macromolecule Metal Complexes*, Springer-Publishers, Heidelberg.
4. Pomogailo, A. D., and Savost'yanov, V. S. (1994). *Synthesis and Polymerization of Metal Containing Monomers*, CRC Press, Boca Raton.
5. Tsuchida, E., ed. (1991). *Macromolecular Complexes, Dynamic Interactions and Electronic Processes*, VCH Publishers, New York.
6. Pomogailo, A. D., and Uflyand, I. E. (1991). *Macromolecular Metal Chelates*, Nauka, Moscow.
7. Ray, N. H. (1978). *Inorganic Polymers*, Academic Press, New York.
8. Stone, F. G. A., and Graham, W. A. G., eds. (1962). *Inorganic Polymers*, Academic Press, New York.
9. Carraher, C. E., Sheats, J. E., and Pittman, C. U., eds. (1978). *Organometallic Polymers*, Academic Press, New York.
10. Carraher, C. E., Sheats, J. E., and Pittman, C. U., eds. (1982). *Advances in Organometallic and Inorganic Polymer Science*, Marcel Dekker, New York.
11. Sheats, J. E., Carraher, C. E., and Pittmann, C. U., eds. (1985). *Metal-Containing Polymeric Systems*, Plenum Press, New York.
12. Pittmann, C. U. et al. (1996). *Metal Containing Polymeric Materials*, Plenum Press, New York.
13. Scientific Journal: *J. Inorg. Organomet. Polym.*, Plenum Press, New York and London.
14. Reedijk, J., ed. (1993). *Bioinorganic Catalysis*, Marcel Dekker, New York.
15. Sigel, H., ed. (1985). *Metal Ions in Biological Systems* (Over 25 Volumes), Marcel Dekker, New York. Spiro, T. G., ed. (1985). *Metal Ions in Biology*, (7 Volumes), Wiley and Sons, New York. Sigel, A., and Sigel, H., eds. (1996). *Metal Ions for Biological Systems*, Marcel Dekker Inc., New York.
16. *Metal-Based Drugs, An International Journal*, Vol. 5, Freund Publishing House, Tel Aviv, Israel (1998).
17. Deutsche Forschungsgemeinschaft (1997). *Bioinorganic Chemistry, Transition Metals in Biology and their Coordination Chemistry*, Wiley-VCH-Verlag, Weinheim.
18. Williams, R. J. P., and Frausto da Silva, J. R. R. (1995). *The Natural Selection of the Chemical Elements*, Oxford University Press.
19. Berthon, G., ed. (1995). In *Handbook of Metal-Ligand, Interactions in Biological Fluids*, Marcel Dekker Inc., New York.
20. Kepler, B. K. (1993). *Metal Complexes in Cancer Chemotherapy*, VCH Publishers, New York.
21. Pomogailo, A. D. (1988). *Polymeric Immobilized Metallocomplex Catalysts*, Nauka Moscow.
22. Hartley, F. R. (1985). *Supported Metal Complexes, A New Generation of Catalysts*, Reidel Pub. Co, Dordrecht.
23. Yermakov, Y. I., Kuznetsov, B. N., and Zakharov, V. A. (1981). *Catalysis by Supported Complexes*, Elsevier, Amsterdam.
24. Yermakov, Y. I., and Likholobov (1987). *Homogeneous and Heterogeneous Catalysis*, VNU Science Press, Utrecht.
25. Bekturov, E. A., and Kudaibergenov, S. E. (1996). *Catalysis by Polymers*, Hüthig und Wepf-Verlag, Heidelberg.
26. Jacobs, P. A., Jaeger, N. I., Kubelkova, L., and Wichterlova, B., eds. (1991). *Zeolite Chemistry and Catalysis, Studies in Surface Science and Catalysis*, Vol. 69, Elsevier Science Pub., Amsterdam.

27. Karge, H., and Weitkamp, J., eds. (1989). *Zeolites as Catalysts, Sorbents and Detergent Builders, Studies in Surface Science and Catalysis*, Vol. 46, Elsevier Science Pub., Amsterdam.
28. Ramamurthy, V., ed. (1991). *Photochemistry in Organized and Constrained Media*, VCH Publishers, New York.
29. Wöhrle, D. (1983). *Adv. Polym. Sci.*, *50*: 45.
30. Kaneko, M., and Wöhrle, D. (1988). *Adv. Polym. Sci.*, *84*: 141.
31. Wöhrle, D. (1989). Phthalocyanines in polymer phases. In *Phthalocyanines, Properties, and Applications*, Vol. 1, (Leznoff, C. C., and Lever, A. B. P., eds.), VCH Publishers, New York, p. 55. Wöhrle, D. (2001). *Macromol. Chem. Rapid. Commun.*, *22*: 68.
32. Wöhrle, D., and Pomogailo, A. (2001). Metal-containing macromolecules. In *Advanced Functional Molecules and Polymers*, Vol. 1, (Nalwa, H. S., ed.), Gordon and Breach, Amsterdam, pp. 87–161.
33. Korshak, V. V., and Kozyreva, N. M. (1983). *Russ. Chem. Rev.*, *54*: 1091.
34. Hanack, M., and Lang, M. (1994). *Ad Mater.*, *6*: 819; ibid (1975). *Chemtracts-Org. Chem.*, *8*: 131.
35. Ozin, G. A., and Gil, C. (1989). *Chem. Rev.*, *89*: 1749.
36. Sherrington, D. C. (1988). *Pure Appl. Chem.*, *60*: 401.
37. Pomogailo, A. D. (1992). *Russ. Chem. Rev.*, *61*: 133; Pomogailo, A. D., and Uflyand, I. E. (1990). *Adv. Polym. Sci.*, *97*: 61.
38. Rehahn, M. (1998). *Acta Polym.*, *49*: 201.
39. Tsuchida, E., and Nishide, H. (1977). *Adv. Polym. Sci.*, *24*: 1.
40. Hanack, M., Deger, S., and Lange, A. (1988). *Coord. Chem. Rev.*, *83*: 115.
41. Biswas, M., and Mukherjee, A. (1994). *Adv. Polym. Sci.*, *115*: 89.
42. Manners, I. (1996). *Angew. Chem.*, *108*: 1713.
43. Oriol, L., and Serrano, J. L. (1995). *Adv. Mater.*, *7*: 348.
44. Batten, S. R., and Robson, R. (1998). *Angew. Chem.*, *110*: 1558.
45. Pomogailo, A. D. (1997). *Russ. Chem. Rev.*, *66*: 679.
46. Nomenclature for single-strand and quasi-single-strand inorganic and coordination polymers (1985). *Pure Appl. Chem.*, *57*: 151.
47. Szymanski, R. (1991). *Makromol. Chem.*, *192*: 757; Szymanski, R. (1987). *Makromol. Chem.*, *188*: 2605.
48. Ono, K., Konami, H., and Murakami, K. (1979). *J. Phys. Chem.*, *83*: 2665.
49. Awano, H., Ono, K., and Murakami, K. (1989). *J. Macromol. Sci. Chem.*, *A26*: 567.
50. Awano, H., Ono, K., and Murakami, K. (1982). *Bull. Chem. Soc. Jpn.*, *55*: 2530.
51. Arkhipovitch, G. N., Dubrovski, S. A., Kazanskii, K. S., and Shupik, A. N. (1981). *Vysokomol. Soedin.*, *A23*: 1653.
52. Polinskii, A. S., Pshezhetskii, V. S., and Kabanov, V. A. (1981). *Dokl. Akad. Nauk. SSSR*, *256*: 129; Polinskii, A. S., Pshezhetskii, V. S., and Kabanov, V. A. (1982). *Vysokomol. Soedin.*, *A25*: 72; Polinskii, A. S., Pshezhetskii, V. S., and Kabanov, V. A. (1985). *Vysokomol. Soedin.*, *A27*: 2295.
53. Wong, L.-H., and Smid, J. (1977). *J. Am. Chem. Soc.*, *99*: 5637.
54. Arai, K., and Ogiwara, Y. (1986). *J. Polym. Sci.*, *24*: 2027.
55. Filippov, A. P. (1983). *Eksp. Khim.*, *19*: 463.
56. Beltsios, K. G., and Carr, S. H. (1989). *J. Polym. Sci., Part C, Polym. Lett.*, *27*: 355.
57. Pomogailo, A. D., Sokol'skii, D. V., and Baishiganov, E. (1974). *Dokl. Akad. Nauk, SSSR*, *218*: 1111.
58. Uflyand, I. E., Vainshtein, E. F., and Pomogailo, A. D. (1991). *Zh. Obshch. Khim.*, *61*: 1790.
59. Barbucci, R., Campbell, M. J. M., Casolaro, M., Nocentini, M., Reginato, G., and Ferruti, P. (1986). *J. Chem. Soc. Dalton. Trans.* 2325.
60. Graessley, W. W. (1974). *Adv. Polym. Sci.*, *16*: 1.
61. Takadoro, H., Chatani, Y., Yoshihara, T., Tahara, S., and Murahashi, S. (1964). *Makromol. Chem.*, *73*: 109.
62. Wetton, R. E., Jamess, D. B., and Whiting, W. (1976). *J. Polym. Sci. Polym. Lett. Ed.*, *14*: 577.

63. Parker, J. M., Wright, P. V., and Lee, C. C. (1981). *Polymer*, *22*: 1305.
64. Higuchi, N., Hiraoki, T., and Hikichi, K. (1979). *Polymer J.*, *11*: 139.
65. Barbucci, R., and Ferruti, P. (1979). *Polymer*, *20*: 1061.
66. Kuzaev, A. I., Pomogailo, A. D., and Mambetov, U. A. (1981). *Vysokomol. Soedin Ser.*, *A23*: 213.
67. Roman, E., Valenzuela, G., Gargallo, L., and Radic, D. (1983). *J. Polym. Sci. Polym. Chem. Ed.*, *21*: 2057.
68. Knothe, G., and Wöhrle, D. (1989). *Makromol. Chem.*, *90*: 1573; Knothe, G. (1992). *Makromol. Chem., Theory Simul.*, *1*: 187.
69. Knothe, G. (1993). *Makromol. Chem., Theory Simul.*, *2*: 917.
70. Severin, K., Bergs, R., and Beck, W. (1998). *Angew. Chem.*, *110*: 1722.
71. Chen, H., Ogo, S., and Fish, R. H. (1996). *J. Am. Chem. Soc.*, *118*: 4993.
72. Bloemink, M. J., and Reedijk, J., (in Ref. [62], p. 641).
73. Ryabov, A. D. (1991). *Angew. Chem.*, *103*: 945.
74. Jaouen, G., Vessieres, A., and Butler, I. S. (1993). *Acc. Chem. Res.*, *26*: 361.
75. Krawielitzki, S., and Beck, W. (1997). *Chem. Ber.*, *130*: 1659.
76. Guo, Z., and Sadler, P. J. (1999). *Angew. Chem.*, *111*: 1610.
77. Reedijk, J. (1996). *Chem. Commun.*, 801.
78. Several articles (1997). In *J. Chem. Educ.*, *74*: 633–651.
79. Guyot, A., and Bartholin, M. (1982). *Progr. Polym. Sci.*, *8*: 277.
80. Hodge, P., and Sherrington, D. C., eds. (1981). *Polymer Supported Reactions in Organic Synthesis*, John Wiley and Sons, New York.
81. Ise, N., and Tabuski, I., eds. (1980). *Speciality Polymers*, Iwanami, Shoton Publ. Tokyo.
82. Miroshnik, L. V., Dubina, A. M., and Tolmachev, V. N. (1980). *Koord. Khim.*, *6*: 870; Rosthauser, J. W., and Winston, A. (1981). *Macromolecules*, *14*: 538.
83. Bekturov, E. A. (2000). *Macromol. Symp.*, *156*: 231.
84. Kalalova, E., Populova, O., Stokrova, S., and Stopka, P. (1983). *Collect. Czech. Chem. Commun.*, *48*: 2021.
85. Bol'shov, A. A., Pomogailo, A. D., and Leonov, I. D. (1972). In *Kompleksnaya Pererabotka Mangyshlakskoi Nefti (Complex Processing of Mangyshlak Petroleum)*, Vol. 4, Alma-Ata, Nauka, p. 110.
86. Warshawsky, A., Deshe, A., Rossey, G., and Patchornik, A. (1984). *React. Polym.*, *2*: 301.
87. Drago, P. S., Nyberg, E. D., and El A'mma, A. G. (1981). *Inorg. Chem.*, *20*: 2461.
88. Tsuchida, E., Karino, Y., Nishide, H., and Kurimura, Y. (1974). *Makromol. Chem.*, *175*: 161.
89. Wulff, G., Vesper, W., Crobe-Einsler, R., and Sarhan, A. (1977). *Makromol. Chem.*, *178*: 2799; Wulff, G. (1995). *Angew. Chem.*, *107*: 1958.
90. Mosbach, K. (1994). *Trends in Biochem. Sci.*, *19*: 19; Whitcombe, M. J., Rodriguez, M. E., Villar, P., and Vulfson, E. N. (1995). *J. Am. Chem. Soc.*, *117*: 7105.
91. Efendiev, A. A., Orudgev, D. D., and Kabanov, V. A. (1980). *Dokl. Akad. Nauk SSSR*, *255*: 1393.
92. Efendiev, A. A. (1994). *Macromol. Symp.*, *80*: 289; Efendiev, A. A. (1998). *Macromol. Symp.*, *131*: 29.
93. Braun, U., and Kuchen, W. (1984). *Chem.-Ztg.*, *107*: 255.
94. Chen, H., Olmstead, M. M., Albright, R. L., Devenyi, J., and Fish, R. H. (1997). *Angew. Chem.*, *109*: 624.
95. Geckeler, K., Lange, G., Eberhart, H., and Bayer, E. (1980). *Pure Appl. Chem.*, *52*: 1883.
96. Leeuwen, P. W. N. M., Jongsma, T., and Challa, G. (1994). *Macromol. Symp.*, *80*: 241.
97. Karakhanov, E. A., Runova, E. A., Berezkin, G. V., and Neimerovets, E. B. (1994). *Macromol. Symp.*, *80*: 231.
98. Takeoka, S., Horiuchi, K., Yamagata, S., and Tsuchida, E. (1991). *Macromolecules*, *24*: 2003.
99. Chiang, C. K., Davies, G. T., Harding, C. A., and Takahashi, T. (1985). *Macromolecules*, *18*: 825.
100. Giles, J. M. R., and Greenhall, M. P. (1986). *Polym. Commun.*, *27*: 360.

101. Oyama, N., and Hatozaki, O. (2000). *Macromol. Symp.*, *156*: 171.
102. Sinta, R., Lamb, B., and Smid, J. (1983). *Macromolecules*, *16*: 1382.
103. Shirai, M., Veda, A., and Tanaka, M. (1985). *Makromol. Chem.*, *186*: 2519.
104. Fei, C. P., and Chan, T. H. (1982). *Synthesis*, *6*: 467.
105. Ueba, Y., Zhu, K. J., Banks, E., and Okamoto, Y. (1982). *J. Polym. Sci., Polym. Chem. Ed.*, *20*: 1272; Joshi, R. H., and Patel, M. N. (1983). *Macromol. Sci.*, *19A*: 919.
106. Okamoto, Y. (1992). *Macromol. Chem., Macromol. Symp.*, *59*: 83; Okamoto, Y., and Kido, J. (1991). In *Macromolecular Complexes* (Tsuchida, E., ed.), VCH Publishers, New York, p. 143.
107. Kurimura, Y., Sairenchi, Y., and Nakayama, S. (1992). *Macromol. Chem., Macromol. Symp.*, *59*: 199; Kurimura, Y. (1991). In *Macromolecular Complexes* (Tsuchida, E., ed.), VCH Publishers, New York, p. 93.
108. Culbertson, B. M., Xie, D., and Thakur, A. (1998). *Macromol. Symp.*, *131*: 11.
109. Karakhanov, E., Filippova, T., Maximov, A., Predeina, V., and Restakyan, F. (1998). *Macromol. Symp.*, *131*: 87; ibid (2000). *Macromol. Symp.*, *156*: 137.
110. Harris, C. S., Shriver, D. F., and Ratner, M. A. (1986). *Macromolecules*, *19*: 987.
111. Nicolaus, V., and Wöhrle, D. (1992). *Angew. Makromol. Chem.*, *198*: 179.
112. Hirao, T., Higuchi, M., and Yamaguchi, S. (1998). *Macromol. Symp.*, *131*: 59.
113. Palmer, V., Zhou, R., and Geckeler, K. E. (1994). *Angew. Markomol. Chem.*, *215*: 175.
114. Geckeler, K. E. (2000). *Macromol. Symp.*, *156*: 29.
115. Bergbreiter, D. E., Koshti, N., Franchina, J. G., and Frels, J. D. (2000). *Angew. Chem.*, *112*: 1082.
116. Kise, H., and Sato, H. (1985). *Makromol. Chem.*, *186*: 2449.
117. Boudakgi, A., Jezierska, J., and Kolarz, B. N. (1992). *Makromol. Chem., Macromol. Symp.*, *59*: 343.
118. Wöhrle, D., and Buttner, P. (1985). *Polym. Bull.*, *13*: 57.
119. Aeissen, H., and Wöhrle, D. (1981). *Makromol. Chem.*, *182*: 2961.
120. Paredes, R. S., Valera, N. S., and Lindoy, L. F. (1986). *Austr. J. Chem.*, *39*: 1081.
121. Sherrington, D. C., Karjalainen, J. K., Canali, L., Deleuze, H., and Hormi, O. E. O. (2000). *Macromol. Symp.*, *156*: 125.
122. Kaneko, M., and Yamada, A. (1984). *Adv. Polym. Sci.*, *55*: 1
123. Kaneko, M., and Yamada, A. (1985). In *Metal Containing Polymeric Systems* (Sheats, J. E., Carraher, C. E., and Pittmann, C. U., eds.), Plenum Publishing Corporation, New York, p. 249.
124. Sasaki, T., and Matsunaga, F. (1968). *Bull. Chem. Soc. Jpn.*, *41*: 2440.
125. Challa, G., Chen, W., and Reedijk, J. (1992). *Makromol. Chem. Macromol. Symp.*, *59*: 59.
126. Wöhrle, D., Bohlen, H., and Blum, J. K. (1986). *Makromol. Chem.*, *187*: 2081; Wöhrle, D., Bohlen, H., Aringer, C., and Pohl, D. (1984). *Makromol. Chem.*, *185*: 669.
127. Nishide, H., Kuwahara, M., Ohyanagi, M., Funada, Y., Kawakami, H., and Tsuchida, E. (1986). *Chem. Lett.*, 43.
128. Tsuchida, E., Nishide, H., Ohyanagi, M., and Okada, O. (1988). *J. Phys. Chem.*, *92*: 6461; Nishide, H., Yuasa, M., Hasegawa, E., and Tsuchida, E. (1987). *Macromolecules*, *20*: 1913.
129. Sugie, K. (1988). *Polym. Mater. Sci. Eng.*, *59*: 133.
130. Shimidzu, T., Izaki, K., Akai, Y., and Iyoda, T. (1981). *Polym. J.*, *13*: 889.
131. Kaneko, M., Nemoto, S., Yamada, A., and Kurimura, Y. (1980). *Inorg. Chim. Acta*, *44*: L289.
132. Kurimura, Y., Shinozaki, N., Ito, F., Uratani, Y., Shigehara, K., Tsuchida, E., Kaneko, M., and Yamada, A. (1982). *Bull. Chem. Soc. Jpn.*, *55*: 380.
133. Sherrington, D. C., and Tang, H.-G. (1994). *Macromol. Symp.*, *80*: 193.
134. Rusanov, A. L., Vol'pin, M. E., Beliy, A. A., and Chigladze, L. G. (1994). *Macromol. Symp.*, *80*: 215.
135. Wöhrle, D., and Krawczyk, G. (1986). *Makromol. Chem.*, *187*: 2535; Wöhrle, D., Krawczyk, G., and Paliuras, M. (1988). *Makromol. Chem.*, *189*: 1001; Wöhrle, D., Krawczyk, G., and Paliuras, M. (1988). *Makromol. Chem.*, *189*: 1013.

136. Wöhrle, D., Gitzel, J., Krawczyk, G., Tsuchida, E., Ohno, H., Okura, I., and Nishisaka, T. (1988). *J. Macromol. Sci., A25*: 1227.
137. Wöhrle, D., Paliuras, M., and Okura, I. (1991). *Makromol. Chem., 192*: 819.
138. Yamaguchi, H., Fujiwara, R., and Kusuda, K. (1986). *Makromol. Chem., Rapid Commun., 7*: 225.
139. Shutten, J. H., and Zwart, J. (1979). *J. Mol. Catal., 5*: 109.
140. Shutten, J. H. (1980). *Angew. Makromol. Chem., 89*: 201.
141. Gebler, M. (1981). *J. Inorg. Nucl. Chem., 43*: 2759.
142. Hanabusa, K., Kobayashi, C., Koyama, T., Masuda, E., Shirai, H., Kondo, Y., Takemoto, K., Izuka, E., and Hojo, N. (1986). *Makromol. Chem., 187*: 753.
143. Zwart, J., van der Weide, H. C., Bröker, N., Rummens, C., Schuit, G. C. A., and German, A. L. (1977/78). *J. Catal., 3*: 151.
144. Brouwer, W. M., Piet, P., and German, A. L. (1985). *J. Mol. Catal., 31*: 169.
145. Shirai, H., Maruyama, A., Konishi, M., and Hojo, N. (1980). *Makromol. Chem., 181*: 1003.
146. Shutten, J. H., and Zwart, J. (1979). *J. Mol. Catal., 5*: 109.
147. Brouwer, W. M., Piet, P., and German, A. L. (1985). *J. Mol. Catal., 25*: 335.
148. Shutten, J. H., Piet, P., and German, A. L. (1979). *Makromol. Chem., 180*: 2341.
149. Yoshida, T., Kamato, K., Tsukamoto, M., Iida, T., Schlettwein, D., Wöhrle, D., and Kaneko, M. (1995). *J. Electroanal. Chem., 385*: 209.
150. Zhao, F., Zhang, J., Abe, T., Wöhrle, D., and Kaneko, M. (1999). *J. Mol. Catal., A: Chem., 145*: 245.
151. Buck, T., Bohlen, H., Wöhrle, D., Schulz-Ekloff, G., and Andreev, A. (1993). *J. Mol. Catal., 80*: 253.
152. van Herk, A. M., Tullemans, A. H. J., van Welzen, J., and German, A. L. (1988). *J. Mol. Catal., 44*: 269; Hassanein, M., and Ford, W. T. (1988). *Macromolecules, 21*: 526; Przywarska-Boniecka, H., Trynda, L., and Antonini, E. (1975). *Eur. J. Biochem., 52*: 567.
153. Spiller, W., Wöhrle, D., Schulz-Ekloff, G., Ford, W. T., Schneider, G., and Stark, J. (1996). *J. Photochem. Photobiol. A. Chem., 95*: 161.
154. Bartels, O., Wöhrle, D., Gerdes, R., Schneider, G., and Schulz-Ekloff, G. (2000). *J. Inform. Record., 25*: 251.
155. Uflyand, I. E., Pomogailo, A. D., Golubeva, N. D., and Sheinker, V. N. (1985). *Proc. XVII Euop. Congr. Mol. Spectros*, Madrid, p. 152.
156. Pomogailo, A. D., Uflyand, I. E., and Golubeva, N. D. (1985). *Kinetics and Catalysis, 26*: 1104.
157. Uflyand, I. E., Pomogailo, A. D., Gorbumova, M. O., Starikov, A. G., and Sheinker, V. N. (1987). *Kinetics and Catalysis, 28*: 613. Uflyand, I. E., Kuzharov, A. S., Gorbumova, M. O., Sheinker, V. N., and Pomogailo, A. D. (1989). *React. Polymer, 11*: 221.
158. Ciardelli, F., Altomare, A., Conti, G., Arribas, G., Mendez, B., and Ismayel, A. (1994). *Macromol. Symp., 80*: 29.
159. Ittel, S. D. (1990). *J. Macromol. Sci.-Chem., A27*: 1133.
160. Kurusu, Y. (1992). *Makromol. Chem., Macromol. Symp., 59*: 313.
161. Steele, M. R., Macdonald, P. M., and Ozin, G. A. (1993). *J. Am. Chem. Soc., 115*: 7285.
162. Steigerwald, M. C., and Brus, L. E. (1990). *Acc. Chem. Res., 23*: 183; Wang, Y. (1991). *Acc. Chem. Res., 24*: 133.
163. Pan, C., and Zong, H. (1994). *Macromol. Symp., 80*: 265.
164. Hoorn, H. J., de Joode, P., Driessen, W. L., and Reedijk, J. (1996). *Recl. Trav. Chim. Pays-Bas, 115*: 191.
165. Jorna, A. M. J., Boelrijk, E. M., Hoorn, H. J., and Reedijk, J. (1996). *React. Funct. Polym., 29*: 101.
166. Buck, T., Wöhrle, D., Schulz-Ekloff, G., and Andreev, A. (1991). *J. Mol. Catal., 70*: 259.
167. Fischer, H., Schulz-Ekloff, G., Buck, T., and Wöhrle, D. (1994). *Erdöl, Erdgas, Kohle, 110*: 128.

168. Schneider, G., Wöhrle, D., Spiller, W., Stark, J., and Schulz-Ekloff, G. (1994). *Photochem. Photobiol.*, *60*: 333.
169. Gerdes, R. (2000). Doctor Thesis, Universität Bremen.
170. Buck, T., Bohlen, H., Wöhrle, D., and Schulz-Ekloff, G. (1993). *J. Mol. Catal.*, *80*: 253.
171. Pomogailo, A. D., and Dzardimalieva, G. I. (1998). *Russ. Chem. Bull.*, *47*: 2319.
172. Hatada, K., Nakanishi, H., Ute, K., and Kitayama, T. (1986). *Polym. J.*, *18*: 581.
173. Bochkin, A. M., and Pomogailo, A. D. (in press). *Izv. Akad. Nauk SSSR Khim.*
174. Krongauz, E. S. (1966). *Advances in the Field of Organoelement Polymer Synthesis*, (Korshak, V. V., ed.), Nauka, Moscow, p. 129.
175. Braun, D. (1961). *Angew. Chem.*, *73*: 197.
176. Koton, M. M., Kiseleva, T. M., and Florinsky, F. S. (1959). *Izv. Akad. Nauk SSSR, OKhN*, 948.
177. Korshak, V. V., Polyakova, A. M., and Tambovtseva, E. S. (1959). *Izv. Akad. Nauk SSSR, OKhN*, 742.
178. Koton, M. M., and Dokukina, L. F. (1964). *Vysokomol. Soedin.*, *6*: 1791.
179. Fujuita, N., and Sonogashira, K. (1974). *J. Polym. Sci.*, *12*: 2845.
180. Marina, N. G., Duvakina, N. V., Monakov, Yu, B., Kuchin, A. V., and Tolstikov, G. A. (1984). *Dokl. Akad. Nauk SSSR*, *276*: 635.
181. Pomogailo, A. D., and Dzhardimalieva, G. I. (1998). *Russian Chem. Bull.*, *47*: N12.
182. Besecke, S., Schroeder, G., Ude, W., and Baumgartner, E. (1984). German Patent 3.224.927; *Chem. Abstr.*, *100*: 104028q.
183. Wojtczak, Z., and Gronowski, A. (1984). *Polimery*, *27*: 471.
184. Dzardimaliyeva, G. I., Pomogailo, A. D., Davtyan, S. P., and Ponomarev, V. I. (1988). *Izv. Akad. Nauk SSSR Khim.*, 1531.
185. Agnew, N. H., and Brown, M. E. (1971). *J. Polym. Sci. Part A-1*, *9*: 2561.
186. Karklin, L. N., Klyuyev, M. V., and Pomogailo, A. D. (1983). *Kinet. Katal.*, *24*: 408.
187. Agnew, N. H., and Brown, M. E. (1974). *J. Polym. Sci., Polym. Chem.*, *12*: 1493.
188. Sokol, V. I., Nikolaev, V. P., Butman, L. A., Poraj-Koshits, M. A., Domnina, E. S., Skvortsova, G. G., and Makhno, L. P. (1979). *Zh. Neorg. Khim.*, *24*: 727.
189. Ikegami, T., and Hirai, H. (1970). *J. Polym. Sci. A-1*, *8*: 195.
190. Matkovsky, P. E., Leonov, I. D., Beikhold, G. A., Pomogailo, A. D., Kissin, Yu.V., and Chirkov, N. M. (1970). *Izv. Akad. Nauk SSSR Khim.*, 1311.
191. Matkovsky, P. E., Leonov, I. D., Kissin, Yu.V., Chirkov, N. M., Pomogailo, A. D., and Beikhold, G. A. (1968). *Izv. Akad. Nauk SSSR Khim.*, 930.
192. Kuran, W., Pasinkievicz, S., Florjanczcyk, Z., and Lisztyk, E. (1976). *Makromol. Chem.*, *177*: 2627.
193. Firsov, V. A., Zaitsev, Yu.S., Kucher, R. V., and Kisel, N. G. (1976). *Physico-Chemical Properties and Synthesis of High-Molecular Compounds*, Naukova Dumka, Kiev, p. 45.
194. Schrauzer, G. V. (1959). *J. Am. Chem. Soc.*, *81*: 5310.
195. Ragg, P. L. (1974). U.S.S.R. Patent, 445.203; *Bull. Izobr.* (1971). *36*: 146.
196. Skushnikova, A. I., Domnina, E. S., and Skvortsova, G. G. (1982). *Vysokomol. Soedin.*, *Ser. B*, *24*: 11.
197. Danilovtseva, E. N., Skushnikova, A. I., Domnina, E. S., and Afonin, A. V. (1989). *Vysokomol. Soedin.*, *Ser. B*, *31*: 777.
198. Miller, E. J., Naughton, M. J., Weigelt, C. A., Bradt, J. E., Serth, J. A., Ofslager, C. L., and O'Brien, W. J. (1992). *Makromol. Chem., Macromol. Symp.*, *59*: 135.
199. Korshak, V. V., and Kozyreva, N. M. (1985). *Usp. Khim.*, *54*: 1841.
200. Pittman, C. U. Jr., Grube, P. L., and Ayers, O. E. (1972). *J. Polym. Sci. A-1*, *10*: 379.
201. Sheats, J. E. (1981). *J. Macromol. Sci.-Chem. A1*, *15*: 1173.
202. Altmann, M., Enkelmann, V., Lieser, G., and Bunz, H. F. (1995). *Adv. Mater.*, *7*: 726.
203. Psheshetskii, V. S., Lishinsky, V. L., Kokorin, A. I., Tsarkova, L. A., Rakhnianskaya, A. A., and Pertsov, N. V. (1992). *Makromol. Chem., Macromol. Symp.*, *59*: 163.
204. Kaneko, M., and Yamada, A. (1984). *Adv. Polym.-Sci.*, *55*: 1.

205. Kaneko, M., and Yamada, A. (1985). *Metal Containing Polymeric Systems* (Sheats, J. E., Carraher, C. E., and Pittman, C. U., eds.), Plenum, New York, p. 249.

206. Nagai, K., Nemoto, N., Ueno, Y., Ikeda, K., Takamiya, and Kaneko, M. (1992). *Makromol. Chem., Macromol. Symp.*, *59*: 257.

207. Leidner, C. R., and Murray, R. W. (1984). *J. Am. Chem. Soc.*, *106*: 1606.

208. Challa, G., Chen, W., and Reedijk, J. (1992). *Macromol. Chem., Makromol. Symp.*, *59*: 59.

209. Wöhrle, D., Bohlen, H., and Meyer, G. (1984). *Polym. Bull.*, *11*: 145, Wöhrle, D., Bohlen, H., and Meyer, G. (1984). *Polym. Bull.*, *11*: 151.

210. Fujii, Y., Kikuchi, K., Matsutani, K., Ota, K., Adashi, M., Syoji, M., Haneishi, I., and Kuwana, Y. (1984). *Chem. Lett.*, 1487.

211. Tsuchida, E. (1979). *J. Macromol. Sci., Chem.*, *A13*: 545.

212. Tsuchida, E., Nishide, H., Yuasa, M., Hasegawa, E., Eshima, K., and Matsushita, Y. (1989). *Macromolecules*, *22*: 2103.

213. Nishide, H., Yuasa, M., Hasegawa, E., and Tsuchida, E. (1987). *Macromolecules*, *20*: 1913.

214. Tsuchida, E. (1989). *J. Macromol. Sci., Chem.*, *A13*: 545.

215. Kajiwara, A., and Kamachi, M. (1989). *Polym. J.*, *21*: 593.

216. Kajiwara, A., Aramata, K., Nomura, S., Morishima, A., and Kamachi, M. (1992). *Chem. Lett.*, 95.

217. Kamachi, M., Cheng, X. S., Kida, T., Kajiwara, A., Shibasaka, M., and Nagata, S. (1987). *Macromolecules*, *20*: 2665.

218. Hiroyoshi, H., Nakata, T., and Komatsu, S. (1991). *Bull. Chem. Soc. Jpn.*, *64*: 2300.

219. Van der Pol, J. F., Neeleman, E., Nolte, R. J. M., Zwikker, J. W., and Drenth, W. (1989). *Makromol. Chem.*, *190*: 2727.

220. Watanabe, N., Cheng, X. S., Harada, A., Kamachi, M., Nakayama, H., Mori, W., and Kishita, M. (1989). *Polym. J.*, *21*: 8, 633.

221. Pomogailo, A. D., Razumov, V. F., and Voloshanovskii, I. S. (2000). *J. Porphyrins Phthalocyanines*, *4*: 45.

222. Kitsenko, N. A., Ishkov, Y. V., Voloshanovskii, I. S., Aliev, Z. G., and Pomogailo, A. D. (1995). *Russ. Chem. Bull.*, *44*: 1758.

223. Pomogailo, A. D., and Bravaya, N. M. (1996). *Russ. Chem. Bull.*, *45*: 2773.

224. Kamachi, M. (1998). *Macromol. Symp.*, *131*: 69.

225. Watanabe, N., Cheng, X. S., Harada, A., Kamachi, M., Nakayama, H., Mori, W., and Kishita, M. (1989). *Polym. J.*, *21*: 1109.

226. Itoh, H., Kondo, S., Matsuda, E., Hanabusa, K., Shirai, H., and Hojo, N. (1986). *Makromol. Chem., Rapid Commun.*, *7*: 585.

227. Eichhorn, H., Sturm, M., and Wöhrle, D. (1995). *Macromol. Chem. Phys.*, *196*: 115.

228. Eichhorn, H., Rutloh, M., Wöhrle, D., and Stumpe, J. (1996). *J. Chem. Soc., Perkin Trans.*, *2*: 1801.

229. Eichhorn, H., Bruce, D. W., and Wöhrle, D. (1998). *Adv. Mater.*, *10*: 419.

230. Bedioui, F., and Devynk, J. (1995). *Acc. Chem. Res.*, *28*: 30.

231. Guarr, T. F. (1997). *Handbook of Organic Conducting Molecules and Polymers*, Vol. 2 (Nalwa, H. S., ed.), John Wiley & Sons, Chichester, p. 461.

232. Trombach, N. (2000). Doctor Thesis, Universität Bremen. Hild, O. (1999). Master Thesis, Universität Bremen.

233. Marcu, M., Lazarescu, S., and Grigorin, G. E. (1986). *Polymer Bull.*, *16*: 103.

234. Sawodny, W., Grünes, R., and Reitzle, H. (1982). *Angew. Chem., Int. Ed. Engl.*, *21*: 775.

235. Manecke, G., Wille, W. E, and Kossmehl, G. (1972). *Makromol. Chem.*, *160*: 111.

236. Moore, J. S., and Stupp, S. I. (1988). *Polymer Bull.*, *19*: 251; ibid (1988). *Macromolecules*, *21*: 1217, 1222, 1228.

237. Hanabusa, K., Suzuki, T., Koyama, T., and Shirai, H. (1991). *Makromol. Chem.*, *192*: 233; ibid. *Makromol. Chem.*, *193*: 2149.

238. Neuse, E. W. (1994). *Macromol. Symp.*, *80*: 111.

239. Wöhrle, D., Marose, U., and Knoop, R. (1985). *Makromol. Chem.*, *186*: 2209.

240. Wöhrle, D., Schmidt, V., Schumann, B., Yamada, A., and Shigehara, K. (1987). *Ber. Bunsenges. Phys. Chem.*, *97*: 975.
241. Wöhrle, D., and Schulte, B. (1988). *Makromol. Chem.*, *189*: 1167, 1229.
242. Djurado, D., Tadlaoui, S., Hamwi, A., and Cousseins, J. C. (1991). *Synth. Met.*, *41–43*: 2595, Liao, M. S., and Kuo, K. T. (1993). *Polymer J.*, *25*: 947.
243. Gürek, A. G., and Bekaroglu, Ö. (1997). *J. Porphyrins Phthalocyanines*, *1*: 67, 227.
244. Wöhrle, D., and Preussner, E. (1985). *Makromol. Chem.*, *186*: 2189.
245. Wöhrle, D., and Hündorf, U. (1985). *Makromol. Chem.*, *186*: 2177.
246. Achar, B. H., Fohlen, G. M., and Parker, J. A. (1982). *J. Polym. Sci., Polym. Chem. Ed.*, *20*: 1785.
247. Wöhrle, D., Benters, R., Suvorova, O., Schnurpfeil, G., Trombach, N., and Bogdahn-Rai, T. (2000). *J. Porphyrins Phthalocyanines*, *4*: 491.
248. Knothe, G., and Wöhrle, D. (1989). *Makromol. Chem.*, *190*: 1573.
249. Knothe, G. (1992). *Makromol. Chem., Theory Simul.*, *1*: 187; Knothe, G. (1994). *J. Inorg. Organomet. Polym.*, *4*: 325.
250. Wöhrle, D., Schumann, B., Schmidt, V., and Jaeger, N. (1987). *Makromol. Chem. Macromol. Symp.*, *8*: 195.
251. Wöhrle, D., Bannehr, R., Jaeger, N., and Schumann, B. (1983). *J. Mol. Catal.*, *21*: 255.
252. Wöhrle, D., Buck, T., Hündof, U., Schulz-Ekloff, G., and Andreev, A. (1989). *Makromol. Chem.*, *190*: 961.
253. Wöhrle, D., Suvorova, O., Trombach, N., Schupak, E. A., Gerdes, R., Semenov, M. N., Bartels, O., Zakurazhov, A. A., and Wendt, A. (2001). *J. Porphyrins Phthalocyanines*, *5*: 381.
254. Wöhrle, D., and Schulte, B. (1985). *Makromol. Chem.*, *186*: 2229.
255. Wöhrle, D., Bannehr, R., Jaeger, N., and Schumann, B. (1983). *Angew. Makromol. Chem.*, *117*: 103.
256. Manecke, G., and Wöhrle, D. (1968). *Makromol. Chem.*, *120*: 192.
257. Korshak, V. V., and Vinogradova, S. V. (1975). *Faserf. Textiltechnik*, *26*: 318.
258. Wöhrle, D., and Müller, R. (1976). *Makromol. Chem.*, *177*: 2241.
259. Müller, R., and Wöhrle, D. (1978). *Makromol. Chem.*, *179*: 2161.
260. Kingsborough, R. P., and Swager, T. M. (2000). *Angew. Chem.*, *112*: 3019.
261. Rack, M., and Hanack, M. (1994). *Angew. Chem.*, *106*: 1712.
262. Hanack, M., and Stihler, P. (1998). *Macromol. Symp.*, *131*: 49.
263. Aratani, N., Osuka, A., Kim, Y. H., Jeong, D. H., and Kim, D. (2000). *Angew. Chem.*, *112*: 1517.
264. Achar, B N., Fohlen, G. M., and Parker, J. A. (1985). *J. Polym. Sci., Polym. Chem. Ed.*, *23*: 801.
265. Shirai, H. et al. (1998). *J. Porphyrins Phthalocyanines*, *2*: 31.
266. Bao, Z. N., Chen, Y. M., and Yu, L. P. (1994). *Macromolecules*, *27*: 4629. ibid (1993). *Macromolecules*, *26*: 5281.
267. Anderson, H. L., Martin, S. J., and Bradley, D. D. C. (1994). *Angew. Chem. Int. Ed. Engl.*, *33*: 655.
268. Peng, Z. H., Bao, Z. N., and Yu, L. P. (1994). *J. Am. Chem. Soc.*, *116*: 6003.
269. Li, D., Buscher, C. T., and Swanson, B. I. (1994). *Chem. Mater.*, *6*: 803.
270. Wamser, C. C., Bard, R. R., and Senthilathipan, V. (1989). *J. Chem. Soc.*, *111*: 8485; Wamser, C. C. (1991). *Mol. Cryst. Liq. Crist.*, *194*: 65; Wamser, C. C., Senthilathipan, V., and Li, W. (1991). *SPIE*, *1436*: 114.
271. Miller, R. D. (1989). *Angew. Chem. Adv. Mater.*, *101*: 1773.
272. West, R. (1986). *J. Organomet. Chem.*, *300*: 327.
273. Vogg, G., Martin, S., and Stutzmann, M. (2000). *Adv. Mater.*, *12*: 1278.
274. Krogman, K. (1969). *Angew. Chem.*, *81*: 10.
275. Miller, J. S., ed. (1982). *Extended Linear Chain Compounds*, Vol. 1 (Plenum Press, New York).
276. Finnis, G. M., Canadell, E., Campana, C., and Dunbar, K. R. (1996). *Angew. Chem.*, *108*: 2946.

277. Hagihara, N., Sonogashira, K., and Takahashi, S. (1981). *Adv. Polym. Sci.*, *41*: 149.
278. Davidson, P. J., Lappert, M. F., and Pearce, R. (1976). *Chem. Rev.*, *2*: 219. Schrock, R. R., and Parshall, G. W. (1976). *Chem. Rev.*, *2*: 243.
279. Lang, H. (1994). *Angew. Chem.*, *106*: 569.
280. Abe, A., Tabaa, S., and Kimur, N. (1991). *Polymer J.*, *23*: 69; Dray, A. (1991). *Macromolecules*, *25*: 3473.
281. Younus, M., Köhler, A., and Cron, S. (1998). *Angew. Chem.*, *110*: 1998.
282. Onitsuka, K., Harada, Y., Takei, F., and Takahashi, S. (1998). *Chem. Commun.*, 643.
283. Oshiro, N., Takei, F., Onitsuka, K., and Takahashi, S. (1998). *J. Organomet. Chem.*, *569*: 195.
284. Nishihara, H., Shimura, T., Ohkubo, A., Matsuda, N., and Aramaki, K. (1993). *Adv. Mater.*, *5*: 752.
285. Dhingra, S. S., Kowach, G. R., and Kremer, R. K. (1997). *Angew. Chem.*, *109*: 1127.
286. Mattner, M. R., Herrmann, W. A., Berger, R., Gerber, C., and Gimzewski, J. K. (1996). *Adv. Mater.*, *8*: 654.
287. Wörle, M., and Nesper, R. (2000). *Angew. Chem.*, *112*: 2439.
288. Tschinkl, M., Schier, A., Riede, J., and Gabbai, F. P. (1999). *Angew. Chem.*, *111*: 3769.
289. Adair, B. A., de Delgado, G. D., and Cheetham, A. K. (2000). *Angew. Chem.*, *112*: 761.
290. Räke, B., Müller, P., Roesky, H. W., and Uson, I. (1999). *Angew. Chem.*, *111*: 2069.
291. Hanko, J. A., and Kanatzidis, M. G. (1998). *Angew. Chem.*, *110*: 354.
292. Mironov, Y. V., Virovets, A. V., Artmekina, S. B., and Feodorov, V. E. (1998). *Angew. Chem.*, *100*: 2656.
293. Heng, Y., Zhang, W., Wu, X., Sheng, T., Wang, Q., and Lin, P. (1998). *Angew. Chem.*, *110*: 2662.
294. Dey, A. K. (1986). *J. Indian Chem. Soc.*, *63*: 557.
295. Ibidapo, T. A. (1990). *Adv. Polym. Sci.*, *30*: 1151.
296. Carraher, C. E., and Schroeder, J. A. (1975). *J. Polym. Sci., Polym. Lett. Ed.*, *13*: 215.
297. Srivastava, P. C., Pandeya, K. B., and Nigram, H. I. (1973). *J. Inorg. Nucl. Chem.*, *35*: 3613.
298. Wrobleski, J. T., and Brown, D. B. (1979). *Inorg. Chem.*, *18*: 498, 2738.
299. Theocharis, C. R. (1987). *J. Chem. Soc., Chem. Commun.*, 80.
300. Rao, T. R., Rao, P. R., Lingaiah, P., and Deshmukh, L. S. (1991). *Angew. Makromol. Chem.*, *191*: 177.
301. Poddar, S. N., and Saha, N. (1970). *Indian J. Appl. Chem.*, *33*: 244.
302. Banerjie, V. et al. (1979). *Polymer Bull.*, *1*: 685.
303. Banerjie, V. et al. (1980). *Makromol. Chem. Rapid Commun.*, *1*: 41.
304. Kumari, V. D., Khave, M., and Munishi, K. N. (1985). *Indian J. Chem.*, *24A*: 72.
305. Reynolds, J. R., Lillya, C. P., and Chien, J. C. W. (1987). *Macromolecules*, *20*: 1184.
306. Schumater, R. R., and Engler, E. M. (1977). *J. Am. Chem. Soc.*, *99*: 5521.
307. Rivera, N. M., Engler, E. M., and Schumater, R. R. (1979). *J. Chem. Soc., Chem. Commun.*, 184.
308. Ribas, J., and Cassoux, P. C. (1981). *R. Seances Acad. Sci.*, *293*: 665.
309. Dirk, C. W., Mintz, E. A., Schoch, K. F., and Marks, T. J. (1986). *Advances in Organometallic and Inorganic Polymer Science* (Carraher, C. E., Sheats, J. E., and Pittmann, C. U., eds.), Marcel Dekker, New York, p. 275.
310. Teo, B. K., Wudl, F., Hauser, J. J., and Krüger, A. (1977). *J. Am. Chem. Soc.*, *99*: 4862.
311. Böhm, M. C. (1984). *Phys. Stat. Sol. (B)*, *121*: 255.
312. Velton, U., Lahn, B., and Rehahn, M. (1997). *Macromol. Chem. Phys.*, *198*: 2789.
313. Kelch, S., and Rehahn, M. (1997). *Macromolecules*, *36*: 6185.
314. Tschierske, C. (2000). *Angew. Chem.*, *112*: 2547.
315. Krause, N. (1999). *Angew. Chem.*, *111*: 83.
316. Kronenberg, C. M. P., Jastrzebski, J. T. B. M., Spek, A. L., and van Koten, G. (1998). *J. Am. Chem. Soc.*, *120*: 9688.
317. Hong, M., and Zhao, Y. (2000). *Angew. Chem.*, *112*: 2586.

318. Ray, N. H. (1978). Academic Press, New York. *Inorganic Polymers*. Rheingold, A. (1987). Kroschwitz et al. eds., (Wiley Interscience, New York), Encyclopedia of Polymer Science and Engineering, *Inorganic Polymers*. Vol. 8, p. 138. Stone, F. G. A., and Graham, W. A. G., eds. (1962). Academic Press, New York. *Inorganic Polymers*.

319. See for example: Ohba, M., Usuki, N. Fukita, N., and Okawa, H. (1999). *Angew. Chem.*, *111*: 1909. Langenberg, K. V., Batten, S. R., Berry, K. J., Hockless, D. C. R., Moubaraki, B., and Murray, K. S. (1997). *Inorg. Chem.*, *36*: 5006. Holmes, S. M., and Girolami, G. S. (1999). *J. Am. Chem. Soc.*, *121*: 5593.

320. Rosthauser, J. W., and Winston, A. (1981). *Macromolecules*, *14*: 538.

321. Buser, H. J. et al. (1997). *Inorg. Chem.*, *16*: 2704. Kaneko, M. Hou, X.-H., and Yamada, A. (1984). *Bull. Chem. Soc. Jpn.*, *57*: 156. Kaneko, M. (1987). *J. Macromol. Sci. Chem.*, *A24*: 357.

322. Hulliger, F., Ladolt, M., and Vetsch, H. (1976). *J. Solid State Chem.*, *18*: 283.

323. Ma, B. Q., Gao, S., Su. G., and Xu, G. X. (2001). *Angew. Chem.*, *113*: 448.

324. Kimizuka, N., Handa, T., Ichinose, I., and Kunitake, T. (1994). *Angew. Chem.*, *106*: 2576.

325. Losier, P., and Zaworotko, M. J. (1996). *Angew. Chem.*, *108*: 2957.

326. Zaworotko, M. J. (1998). *Angew. Chem.*, *110*: 1269.

327. MacGillivray, L. R., Subramanian, S., and Zaworotko, M. J. (1994). *J. Chem. Soc., Chem. Commun.*, 1325. Zaworotko, M. J. (1994). *Chem. Soc. Rev.*, 283.

328. Gardner, G. B., Venkataraman, D., Moore, J. S., and Lee, S. (1993). *Nature, 374*; 792.

329. Fujita, M., Kown, Y. J., Washizu, S., and Ogura, K. (1994). *J. Am. Chem. Soc.*, *116*: 1151.

330. Gabbai, F. P., Schier, A., and Riede, J. (1998). *Angew. Chem.*, *110*: 646.

331. Fujita, M. et al. (1995). *J. Am. Chem. Soc.*, *117*: 7287.

332. Subramanian, S., and Zaworotko, M. J. (1995). *Angew. Chem.*, *107*: 2295.

333. Mattner, M. R., Herrmann, W. A. et al. (1996). *Adv. Mater.*, *8*: 654.

334. Kumar, B., Seward, C., and Zaworotko, M. J. (1999). *Angew. Chem.*, *111*: 584.

335. Janiak, C. (1997). *Angew. Chem.*, *109*: 1499.

336. Hoskins, B. F., and Robson, R. (1990). *J. Am. Chem. Soc.*, *112*: 1546.

337. Biradha, K., Hongo, Y., and Fujita, M. (2000). *Angew. Chem.*, *112*: 4001.

338. Battern, S. R., Hoskins, B. F., and Robson, R. (1995). *Angew. Chem.*, *107*: 884.

339. Hoskins, B. F., Robson, R., and Slizys, D. A. (1997). *Angew. Chem.*, *109*: 2861.

340. Shimizu, G. K. H. (1998). *Angew. Chem.*, *110*: 1510.

341. Kondo, M., Okubo, T., and Kitagawa, S. (1999). *Angew. Chem.*, *111*: 190.

342. Pan, L., Huang, X., Wu, Y., and Zheng, N. (2000). *Angew. Chem.*, *112*; 537.

343. Miyasaka, H., Campos-Fernandez, C. S., Clerac, R., and Dunbar, K. R. (2000). *Angew. Chem.*, *112*: 3989.

344. Triki, S., Pala, J. S., Decoster, M., Molinie, P., and Toupet, L. (1999). *Angew. Chem.*, *111*: 155.

345. Fegy, K., Lueneau, D., Ohm, T., Paulsen, C., and Rey, P. (1998). *Angew. Chem.*, *110*: 1331.

346. Kamiyama, A., Nogushi, T., Kajiwara, T., and Ito, T. (2000). *Angew. Chem.*, *112*: 3260.

347. Lee, E., Heo, J., and Kim, K. (2000). *Angew. Chem.*, *112*: 2811; Lee, E., Kim, J., Heo, J., Whang, D., and Kim, K. (2001). *Angew. Chem.*, *113*: 413.

348. Batten, S. R., Hoskins, B., and Robson, R. (1995). *Angew. Chem.*, *107*: 884.

349. Lin, K.-J. (1999). *Angew. Chem.*, *111*: 2894.

350. Diskin-Posner, Y., Dahal, S., and Goldberg, I. (2000). *Angew. Chem.*, *112*: 1344.

351. Drain, C. M., Nifiatis, F., Vasenko, A., and Batteas, J. D. (1998). *Angew. Chem*, *110*: 2478.

352. Carlucci, L., Ciani, G., Moret, M., Proserpio, D. M., and Rizatto, S. (2000). *Angew. Chem.*, *112*: 1566.

353. Batten, S. R., and Robson, R. (1998). *Angew. Chem.*, *110*: 1558.

354. Hoskins, B. F., Robson, R., and Slizys, D. A. (1997). *J. Am. Chem. Soc.*, *119*: 2952; ibid. (1997). *Angew. Chem.*, *109*: 2430.

355. Manners, I. (1994). *Adv. Mater.*, *6*: 68.

356. Manners, I. (1998). *Can. J., Chem.*, *76*: 371.

357. Rosenblum, M. (1994). *Adv. Mater.*, *6*: 159.

358. Siebert, W. (1991). *Russ. Chem. Rev.*, *60*: 784.
359. Neuse, E. W., and Rosenberg, H. (1981). In *Metallocene Polymers*, (Marcel Dekker, New York, 1970). Neuse, E. W. (1981). *J. Macromol. Sci.-Chem.*, *A16*: 3.
360. Nugent, H. M., Rosenblum, M., and Klemarczyk, P. (1993). *J. Am. Chem. Soc.*, *115*: 3848.
361. Harder, S. (1998). *Angew. Chem.*, *110*: 1357.
362. Nuyken, O., Burkhardt, V., Pöhlmann, T., and Heberhold, M. (1991). *Makromol. Chem.*, *Macromol. Symp.*, *44*: 195; Miller, R. D. (1989). *Angew. Chem. Adv. Mater.*, *101*: 1773.
363. Manners, I. (1994). *Adv. Mater.*, *6*: 68; Hnyene, M., Yassar, A., Escorne, M., Percheron-Guegan, A., and Garnio, F. (1994). *Adv. Mater.*, *6*: 564.
364. Knapp, R., Velten, U., and Rehahn, M. (1998). *Polym.*, *23*: 5827.
365. Manners, I. (1996). *Polyhedron*, *15*: 4311.
366. Ni, Y., Rulkens, R., and Manners, I. (1996). *J. Am. Chem. Soc.*, *118*: 4102, 12683.
367. Rulkens, R., Lough, L., and Manners, I. (1996). *Angew. Chem. Int. Ed. Engl.*, *35*: 1805.
368. Braunschweig, H., Dirk, R., Müller, M., Nguven, P., Resendes, R., Gates, D. P., and Manners, I. (1997). *Angew. Chem.*, *109*: 2433.
369. Resendes, R., Massay, J. A., Dorn, H., Power, K. N., Winnik, M. A., and Manners, I. (1999). *Angew. Chem.*, *111*: 2738.
370. Raez, J., Barjovanu, R., Massay, J. A., Winnik, M. A. and Manners, I. (2000). *Angew. Chem.*, *112*: 4020.
371. Resendes, R., and Manners, I. (2000). *Adv. Mater.*, *12*: 327.
372. Nelson, J. M., and Manners, I. (1997). *Chem. Eur. J.*, *3*: 573.
373. Massay, J. A., Power, K. N., Mitchell, A., Winnik, M. A., and Manners, I. (1998). *Adv. Mater.*, *10*: 1559.
374. MacLachlan, M. J., Manners, I., and Ozin, G. A. (2000). *Adv. Mater.*, *12*: 675.
375. Dirk, C. W., Mintz, E. A., Schoch, K. F., and Marks, T. J. (1986). In *Advances in Organometallic and Inorganic Polymer Science* (Carraher, C. E., Sheat, J. E., and Pittman, C. U., eds.), Marcel Dekker, New York, p. 275.
376. Schultz, H., Lehman, H, Rein, M., and Hanack, M. (1991). *Structure and Bonding*, *74*: 41.
377. Marks, T. J. (1990). *Angew. Chem.*, *102*: 886.
378. Beltsios, K. G., and Carr, S. H. (1989). *J. Polym. Sci., Part C, Polym. Lett.*, *27*: 355.
379. Schwiegk, S., Fischer, H., Xu, Y., Kremer, F., and Wegner, G. (1991). *Makromol. Chem.*, *Macromol. Symp.*, *46*: 211.
380. Dulog, L., Gittinger, A., Roth, S., and Wagner, T. (1993). *Makromol. Chem.*, *194*: 493.
381. Crockett, G. M., Campbell, A. J., and Ahmed, F. R. (1990). *Polymer*, *31*: 602.
382. Kentgens, A. P. M., Markies, B. A., van der Pol, J. F., and Nolte, R. J. M. (1990). *J. Am. Chem. Soc.*, *112*: 8800.
383. Ferenz, A., Ries, R., and Wegner, G. (1993). *Angew. Chem., Int. Ed. Engl.*, *32*: 1184.
384. Snow, A. N., and Barger, W. R. (1989). In *Phthalocyanines, Properties and Applications* (Leznoff, C. C., and Lever, A. B. P., eds.), VCH Publishers, New York, p. 341; Nolte, R. J. M., and Drenth, W. (1992). In *Inorg. and Organomet. Polymers with Special Properties* (Laine, R. M., ed.), Kluwer Academic Publishers, The Netherlands, p. 223; Espinet, P., Esternelas, M. A., Oro, L. A., Serrano, J. L., and Sola, W. (1992). *Coord. Chem. Rev.*, *117*: 215; Engel, M. K., Bassoul, P., Bosio, L., Lehmann, H., and Hanack, M. (1993). *Liq. Cryst.*, *15*: 704. Chandrasekhar, S., and Ranganath, G. S. (1990). *Rep. Prog. Phys.*, *53*: 570; Sielcken, O. E., van Lindert, H. C. A., Drenth, W., Schooman, J., Schram, J., and Nolte, R. J. M. (1989). *Ber. Bunsenges. Phys. Chem.*, *93*: 702.
385. Van der Pol, J. F., Neeleman, E., Nolte, R. J. M., and Zwikker, J. W. (1989). *Makromol. Chem.*, *190*: 2727. Van Nostrum, C. F., Nolte, R. J. M., Devillers, M. A. C., Oestergetel, G. T., Teerenstra, M. N., and Schouten, A. U. (1993). *Macromolecules*, *26*: 3306.
386. Gorman, C. (1998). *Adv. Mater.*, *10*: 295.
387. Hecht, S., and Frechet, J. M. J. (2001). *Angew. Chem.*, *113*: 77.
388. Pollak, K. W., Sanford, E. M., and Frechet, J. M. J. (1998). *J. Mater. Chem.*, *8*: 519.
389. Tomioka, N., Takasu, D., Takahashi, T., and Aida, T. (1998). *Angew. Chem.*, *110*: 1611.

390. Brewis, M., Clarkson, G. J., Goddard, V., Helliwell, M., Holder, A. M., and McKeown, N. B. (1998). *Angew. Chem., Int. Ed.*, *37*: 1092; McKeown, N. B. (1999). *Adv. Mater.*, *11*: 67.
391. Kimura, M., Sugihara, Y., Muto, T., Hanabusa, K., Shirai, H., and Kobayashi, N. (1999). *Chem. Eur. J.*, *5*: 3495.
392. Casado, C. M., Gonzales, B., Cuadrado, I., Alonso, B., Moran, M., and Losada, J. (2000). *Angew. Chem.*, *112*: 2219.
393. Ohshiro, N., Takei, F., Onitsuka, K., and Takahashi, S. (1996). *Chem. Lett.*, 871.
394. Lin, R.-J., Onikubo, T., and Kaneko, M. (1993). *J. Electroanal. Chem.*, *348*: 189.
395. Zhang, J., Yagi, M., and Kaneko, M. (1996). *Macromol. Symp.*, *105*: 59; ibid (1996). *J. Electroanal. Chem.*, *412*: 159.
396. Kaneko, M., Imai, Y., and Tsuchida, E. (1991). *J. Chem. Soc., Faraday Trans.*, 187: 83.
397. Kaneko, M., Ochiai, M., and Yamada, A. (1982). *Macromol. Chem. Rapid Commun.*, *3*: 299.
398. Kaneko, M., Yamada, A., Oyama, N., and Yamaguchi, S. (1982). *Macromol. Chem. Rapid Commun.*, *3*: 769.
399. Kaneko, M., Moriya, S., Yamada, A, Yamamoto, H., and Oyama, N. (1984). *Electrochim. Acta*, *29*: 115.
400. Kaneko, M., and Yamada, A. (1986). *Electrochim. Acta.*, *31*: 273.
401. Kaneko, M. (1987). *J. Macromol. Sci. Chem.*, *A24*: 357.
402. Ueno, Y., Yamada, K., Yokata, T., Ikeda, K., Takamiya, N., and Kaneko, M. (1993). *Electrochim. Acta.*, *38*: 129.
403. Demas, J. N., and DeGraff, B. A. (1992). *Makromol. Chem., Macromol. Symp.*, *59*: 35; *J. Macromol. Sci. Chem.*, *A25*: 1189.
404. Kaneko, M., and Yamada, A. (1984). *Adv. Polym. Sci.*, *55*: 1; (1985). In *Metal Containing Polymer Systems* (Sheats, J. E., Carraher, C. E., and Pittman, C. U., eds.), Plenum Press, New York, p. 249.
405. Kaneko, M., and Hayakawa, S. (1988). *J. Macromol. Sci. Chem.*, *A25*: 1255.
406. Wöhrle, D., Kaune, H., Schuman, B., and Jaeger, N. I. (1986). *Makromol. Chem.*, *187*: 2947.
407. Kaneko, M., Wöhrle, D., Schlettwein, D., and Schmidt, V. (1988). *Makromol. Chem.*, *189*: 2419.
408. Schlettwein, D., Kaneko, M., Yamada, A., Wöhrle, D., and Jaeger, N. I. (1991). *J. Phys. Chem.*, *95*: 1748.
409. Schlettwein, D., Jaeger, N. I., and Wöhrle, D. (1992). *Makromol. Chem., Macromol. Symp.*, *59*: 267.
410. Stillman, M. J., and Nyokong, T. (1989). *Phthalocyanines – Properties and Applications* (Leznoff, C. C., and Lever, A.B. P., eds.), VCH Publishers, New York, p. 133.
411. Wöhrle, D., Schlettwein, D., Kirschenmann, M., Kaneko, M., and Yamada, A. (1990). *J. Macromol. Sci.-Chem.*, *A27*: 1239.
412. Schlettwein, D., Wöhrle, D., and Jaeger, N. I. (1991). *Ber. Bunsenges. Phys. Chem.*, *95*: 1526.
413. Yoshida T., Kamato, K., Tsukamoto, M., Iida, T., Schlettwein, D., Wöhrle, D., and Kaneko, M. (1995). *J. Electroanal. Chem.*, *385*: 209; ibid (1996). *J. Electroanal. Chem.*, *412*: 125.
414. Zhao, F., Zhang, J., Abe, T., Wöhrle, D., and Kaneko, M. (1999). *J. Mol. Catal. A: Chem.*, *145*: 245; Zhao, F., Zhang, J., Abe, T., Wöhrle, D., and Kaneko, M. (2000). *J. Porphyrins Phthalocyanines*, *4*: 31.
415. Abe, T., Takahashi, K., Shiraishi, Y., Toshima, N., and Kaneko, M. (2000). *Makromol. Chem. Phys.*, *201*: 102.
416. Yagi, M., Kinoshita, K., and Kaneko, M. (1996). *J. Phys. Chem.*, *100*: 11099.
417. Yagi, M., Tokita, S., Nagoshi, K., Ogino, I., and Kaneko, M. (1996). *J. Chem. Soc., Faraday Trans.*, *92*: 2457.
418. Kinoshita, K., Yagi, M., and Kaneoko, M. (1999). *J. Mol. Catal. A: Chem.*, *142*: 1.
419. Yagi, M., Sukegawa, N., Kasamatsu, M., and Kaneko, M. (1999). *J. Phys. Chem. B*, *103*: 2151.
420. Skotheim, T., Velazquez, M., and Linhous, C. A. (1985). *J. Chem. Soc., Chem. Commun.*, 612.
421. Bull, R. A., Fan, F. R., and Bard, A. J. (1984). *J. Electrochem. Soc.*, *131*: 687.

422. Choi, C. S., and Tachikawa, H. (1990). *NTIS Chem.*, *90*: 20; AD-A 217 677.
423. Walton, D. J., and Hall, C. E. (1991). *Synth. Met.*, *45*: 363.
424. Okamoto, Y. (1992). *Makromol. Chem., Macromol. Symp.*, *59*: 83.
425. Okamoto, Y., and Kido, J. (1991). In *Macromolecular Complexes* (Tsuchida, E., ed.), VCH Publishers, New York, p. 143.
426. Okamoto, Y., Kido, J., Brittain, H. G., and Paoletti, S. (1988). *J. Macromol. Sci. Chem.*, *A25*: 1383.
427. Kurimura, Y., Sairenchi, Y., and Nakayama, S. (1992). *Makromol. Chem., Macromol. Symp.*, *59*: 199.
428. Kurimura, Y. (1991). In *Macromolecular Complexes* (Tsuchida, E., ed.), VCH Publishers, New York, p. 93.
429. Hirai, H. (1990). *J. Macromol. Sci., Chem.*, *A27*: 1293.
430. Toshima, N., Kanaka, K., Komiyama, M., and Hirai, H. (1988). *J. Macromol. Sci. Chem.*, *A25*: 1349.
431. Bronshtein, L. M., Larikova, I. E., and Valetsky, P. M. (1987). *Polym. Sci. USSR*, *29*: 2653.
432. Mirzoeva, E. S., Bronshtein, L. M., and Valetsky P. M. (1989). *Polym. Sci. USSR*, *31*: 2898.
433. Jia, C.-G., Jin, F.-Y., Pan, H.-Q., Hung, M.-Y., and Jiang, Y.-Y., (1984). *Macromol. Chem. Phys.*, *195*: 751.
434. Pan, C., and Zong, H. (1994). *Macromol. Symp., 80*: 265.
435. Sherrington, D. C., and Tang, H.-G. (1994) *Macromol. Symp.*, *80*: 193.
436. Shirai, H., Hanabusa, K., Koyama, T., Tsuiki, H., and Masuda, E. (1992). *Makromol. Chem., Macromol. Symp.*, *59*: 155.
437. Ono, K. (1985). *High Polym. Jpn.*, *34*: 766.
438. Orihara, K., and Yonekura, H. (1990). *J. Macromol. Sci. Chem.*, *A27*: 1217.
439. Hegenmüller, P., and Gool, W. V., eds. (1980). *Solid. Electrolytes*, Plenum Press, New York.
440. Bogdanov, B., Uzov, C. and Michaelov, M. (1992). *Acta Polym.*, *43*: 202.
441. Albert, G., and Bein, T., eds. (1996). Comprehensive supramolecular chemistry, Vol. 7, *Solid State Supramolecular Chemistry: Two- and three-dimensional Inorganic Networks*, Pergamon, Oxford.
442. Wöhrle, D., and Schulz-Ekloff, G. (1994). *Adv. Mater.*, *6*: 875.
443. Balku, K. J. (1996). *Phthalocyanines – Properties and Applications*, Vol. 4 (Leznoff, C. C., and Lever, A. B. P., eds.), VCH Publishers, New York, p. 287.
444. Wöhrle, D., Sobbi, A. K., Franke, O., and Schulz-Ekloff, G. (1995). *Zeolites*, *15*: 450.
445. Ganschow, M., Wöhrle, D., and Schulz-Ekloff, G. (1999). *J. Porphyrins Phthalocyanines*, *3*: 299.
446. Ehrl, M., Deeg, F. W., Bräuchle, C., Franke, O., Sobbi, A., Schulz-Ekloff, G., and Wöhrle, D. (1994). *J. Phys. Chem.*, *98*: 47.

12

Conducting Polymers

Herbert Naarmann
(emerit) BASF AG Ludwigshafen

I. INTRODUCTION

Since the fascinating field of electrically conducting polymers was discovered more than 30 years ago [1], it has been the object of intense research. However, it took a long time to learn that the benefits of these polymers lie less in providing substitutes for conventional metals than in opening up new fields of application.

The Royal Swedish Academy of Sciences decided to award the Nobel Prize in Chemistry for 2000 to three scientists:

- Professor Alan J. Heeger at the University of California at Santa Barbara, USA,
- Professor Alan G. MacDiarmid at the University of Pennsylvania, USA, and
- Professor Hideki Shirakawa at the University of Tsukuba, Japan,

They are rewarded 'for the discovery and development of electrically conductive polymers' [2].

Electrically conducting polymers are materials with an extended system of C=C conjugated bonds. They are obtained by reduction or oxidation reactions (called doping), giving materials with electrical conductivities up to 10^5 S/cm. These materials differ from polymers filled with carbon black or metals because the latter are only conductive if the individual conductive particles are mutually in contact and form a coherent phase.

This review concerns the synthesis routes, polymerization techniques, doping, orientation, and development of well-defined, highly conducting polymeric materials.

Electrically conducting materials are complied, their specific properties and potential applications are described.

Numerous attempts have been made to synthesize 'conductive organic materials'. The first was the synthesis of poly(aniline) by F. Goppelsroeder in 1891 [3]. After decades interest grew in organic polymers as insulators, but not as electrical conductors.

In the late 1950s organic semiconductors became the focus of investigations. Preliminary studies in this field up until the mid-1960s are reviewed in [4]. The semiconducting polymers were termed 'covalent organic polymers', 'charge-transfer complexes', and 'mixed polymers'. Highest conductivity values reached about 10^{-3} S/cm.

Systematic work on this field began in the 1960s. Oxidative coupling was systematically extended and became established as the general structural method for synthesizing poly(aromatic)s and poly(heterocycle)s [5] with conductivities of 10^{-1} S/cm. At that time, the results aroused great astonishment, because it seemed a paradox that insulators well known as organic compounds should suddenly become conductive.

Not only was this the highest value yet obtained for a polymer, but these were the first polymers capable of conducting electricity.

The polymers also displayed photovoltaic and thermoelectric properties. After the great surprise and no less incredulity as to how polymeric organic materials can suddenly conduct electricity had subsided, the serious business of elucidating the structure, type of charge, mechanism, etc. was pursued relentlessly and with some success.

As early as 1969, it was pointed out that complex formation between electron acceptors and electron donors increase the conductivity by several orders of magnitude [6].

Analogous effects can be achieved by:

- Increasing the degree of polymerization
- Increasing the pressure
- Raising the temperature
- Irradiation.

A crucial task was the search for defined structures with conjugated π systems, starting from characterized prepolymers, e.g., poly(vinylmethylketone) to poly(cyclohexenone) or heterobridged or substituted poly(arylenes), e.g., by condensation of p,p'-dialkinylbenzene with reactive intermediates (pyrones, coumarins, cyclopentadienones).

In the search of easy-to-manufacture, highly-stable compounds with a known number of double bonds, perylene derivatives of the imide-type and imidazole-type were studied for their electrical photo- and dark conductivities.

Interesting differences in the conductivity were found to be a function of the substituent and the crystallinity of the samples. The formation of charge-transfer complexes with tetracyanoquinone dimethane (TCNQ), tetracyanoethylene (TCNE) and iodine (I_2) increased the conductivity by a factor of 1000, thereby allowing a conductivity similar to that of graphite (10^1 S/cm) to be attained in some cases.

Translating the system to polymeric charge-transfer complexes of the type polymer with donor + acceptor monomer, polymer with donor + polymer with acceptor, or polymer with acceptor + donor monomer led to a new class of compounds [6] that have electrical conductivities of up to 10^2 S/cm.

The idea of inserting electron acceptors and donor groups alternatively in one molecule was realized in the synthesis of substituted ladder-like poly(quinones) with –S–, –NH groups [7].

A large number of potential applications suggested themselves, i.e., thermostable polymers, coatings, organic electrical contacts, photoelectric devices, photocells as well as pigments with outstanding light-fastness and thermal stability.

Other potential applications are resistance thermometers, thermistors, photoconductors, photodiodes, photoelements, solar batteries, electrical reproduction of information, electroluminescence, electrostatic storage batteries, image storage, and catalysis in chemical and biochemical systems [5].

In 1964 Little theoretically evaluated the possibility of superconductivity in polymers and suggested a model, consisting of a polyene chain with cyanine, dyelike substituents [8].

The work on CT complex radical cations by Heeger et al. [9] was another important milestone.

Interest heightened and became acute from 1975 when **IBM** scientists showed that crystalline poly(sulfurnitride), (SN)$_n$, was superconductive [10] and MacDiarmid's group [11] reported the doping of poly(acetylene) films prepared by Shirakawa [12] reaching conductivity values of 0.4 S/cm (bromine doped) and 38 S/cm (iodine doped), later Heeger reported a conductivity of 3000 S/cm (also iodine doped) [2].

Some kind of breakthrough was reached and led to new consideration because the Shirakawa poly(acetylene) was dopable, but not the polyene called cuprene or niprene. This material was in large quantities available as a film with metallic lustre (deposited on the vessel walks synthesizing cyclooctatetraene).

The important point is that science must blaze the trail for technology, and the efforts made and the successes scored in this direction are evident from scientific seminars and publications. This applies to the chemists, in the synthesis of polymers with good mechanical properties and defined structures; to the physicists, in clarifying the relationships between charge carriers, mobility, and polymer structure; and to the engineer, in opening up virgin territory in finding applications for the new materials.

II. PRINCIPLES OF ELECTRICAL CONDUCTION

The electrical properties of materials are determined by their electronic structure (Figure 1). The band theory accounts for the different behaviours of metals, semiconductors, and insulators.

The band gap is the energy spacing between the highest occupied energy level (valence band) and the lowest unoccupied energy level (conduction band). Metals have a zero band gap which means that they have a high electron mobility, i.e., conductivity. Semiconductors have a narrow band gap (ca. 2.5–1.5 eV), conductivity only occurs on excitation of electrons from the valence band to the conduction band (e.g., by heating). If the band gap is larger (3 eV), electron excitation is difficult; electrons are unable to cross the gap and the material is an insulator.

Electrically conducting organic materials such as poly(phenylene), poly(acetylene) or poly(pyrrole) are, however, peculiar in that the band theory cannot explain why the charge-carrying species (electrons or holes) are spinless. Conduction by polarons and bipolarons is now thought to be the dominant mechanism of charge transport in organic materials. This concept also explains the drastic deepening of color changes produced by doping. A polaron (a term used in solid-state physics) is a radical cation that is partially

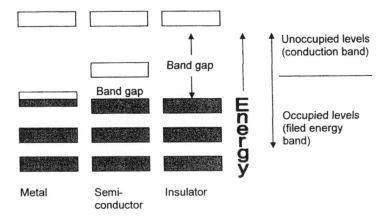

Figure 1 Model of band structure.

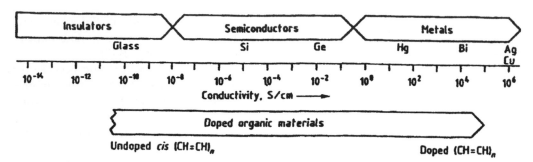

Figure 2 Comparison of the electrical conductivity (300 K) of organic and inorganic materials and the effect of doping [2,13–15].

delocalized over several monomer units (e.g., in a polymer segment). The bipolaron is a diradical dication. Low doping levels gives rise to polarons, whereas higher doping levels produce bipolarons. Both polarons and bipolarons are mobile and can move along the polymer chain [13–16].

III. DOPING

The process that transforms insulating polymers (e.g., poly(acetylene), conductivity 0.1 S/cm) to excellent conductors (Figure 2) is the formation of charge-transfer complexes by electron donors such as sodium or potassium (*n* doping, reduction) or by electron acceptors such as I_2, AsF_5, or $FeCl_3$ (*p* doping, oxidation). The doped polymer backbone becomes negatively or positively charged with the dopant forming oppositely charged ions (Na^+, K^+, I_3^-, I_5^-, AsF_6^-, $FeCl_4^-$). The polymer can be switched between the doped, conductive state and the undoped, insulating state by applying an electric potential that makes the counterions move in and out. This switching corresponds to charging and discharging when these materials are used as electrodes in rechargeable batteries [2,13–15].

The chemistry of doping and the distribution of doping in poly(acetylene) has been treated in detail also by Pekker and Janossy [16].

IV. MEASUREMENT

Electrical conductivity is a measure of the flow of current through a material for a given applied voltage.

The electrical conductivity σ is reciprocal ohms or siemens per centimetre ($\Omega^{-1} cm^{-1}$ or S/cm) [16].

V. TYPES OF ELECTRICALLY CONDUCTING ORGANIC MATERIALS

Of the plethora of systems containing conjugated double bonds, poly(acetylene)s, poly(heterocycle)s, and poly(aminoaromatic) compounds are undoubtedly the most popular both in regard to their electrical conductivity and their stability and ease of preparation. Poly(acetylene), poly(pyrrole), and poly(aniline) are the most intensively studied polymers.

VI. POLY(ACETYLENE)

Poly(acetylene) (PAC) exists in various isomeric forms:

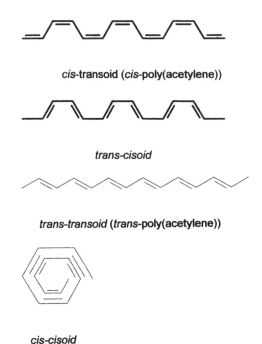

cis-transoid (cis-poly(acetylene))

trans-cisoid

trans-transoid (trans-poly(acetylene))

cis-cisoid

The *cis-cisoid* PAC has not yet been prepared in pure form. Model reactions, however, have shown that cyclic and helical structures are flexible [17–19].

Cis-poly(acetylene) is relatively unstable and reverts to the thermodynamically stable *trans*-poly(acetylene) via the metastable *trans-cisoid* form.

VII. VARIOUS TYPES AND SYNTHESIS OF POLY(ACETYLENE)

A historical overview was given in the first edition of this book [20]. But two publications should be mentioned:

In 1948 Reppe [21] prepared Cuprene film with a metallic luster. In 1961 Hatano reported the polymerization of acetylene with a AlEt$_3$/Ti(OBu)$_4$ catalyst to give polymers with conductivities up to 0.001 S/cm [22]. Since then intensive work has been carried out on the various polymer types, reviews are given in [13–15,20].

Later in 1974 Japanese scientists published the polymerization of acetylene [12] on the surface of a high concentrated solution of Ziegler–Natta catalyst, receiving also poly(acetylene) films with a metallic lustre. These small film pieces—inspite of their impurities (O $\sim$ 1.0%, Ti + Al $\sim$ 0.5%)—had one remarkable property they were 'dopable' reaching values of up to 2500 S/cm. What was the reason for that unusual behaviour in case of other known poly(acetylene)s, e.g., the cuprene film with a lustre like copper or nickel and was produced in large quantities and sizes?

This question was the starting point for extensive studies [23]. These showed that poly(acetylene) with lowest degree of crosslinking have the greatest crystallinity and electrical conductivity, and that such highly crystalline polymers have the lowest capacity for absorbing oxygen. Furthermore, oxygen absorption considerably reduces the crystallinity. These results motivated researchers to make better polymers.

VIII. CONSEQUENCES: NEW TYPES AND METHODS

The search for easy-to-manufacture, highly stable compounds with a known number of double bonds also focused on perylene derivatives. Further investigation led to the concept of ribbon-like polymers (e.g., by repetitive Diels–Alder addition [24] and ladder-like self-dopant systems [25]).

(a) An interesting method is the polymerization of butenyne:

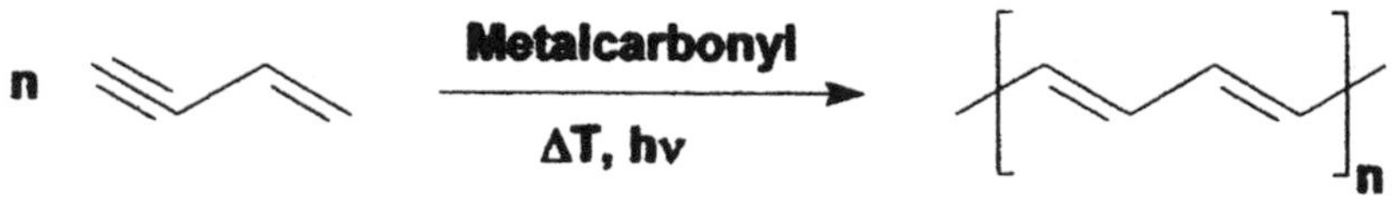

(b) The Feast method [26] for producing 'Durham PAC' proceeds according to the following scheme: 7,8-Bis(trifluoromethyl)tricyclo(4,2,2,0)-deca-3,7,9-triene polymerizes by undergoing ring opening and yields poly (acetylene) through elimination of 1,2-bis(trifluoromethyl)benzene:

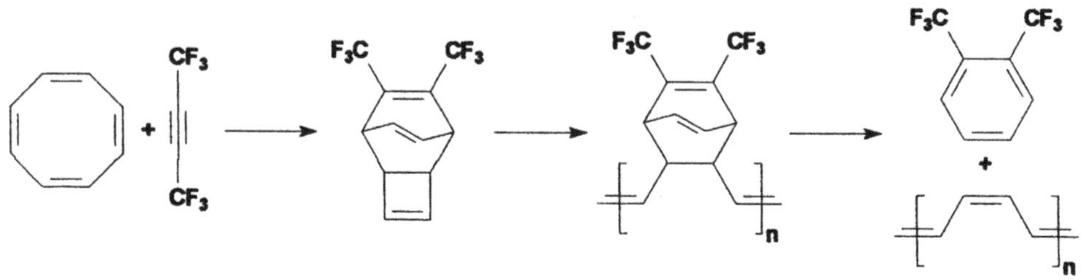

(c) In the Grubbs method [27] poly(benzvalene) is isomerized in the presence of $HgCl_2$ to PAC:

Both of these methods start off with certain monomers that are converted to soluble prepolymers that then yield insoluble perconjugated polymers after thermal treatment.

(d)   Elimination reactions [28,29]:

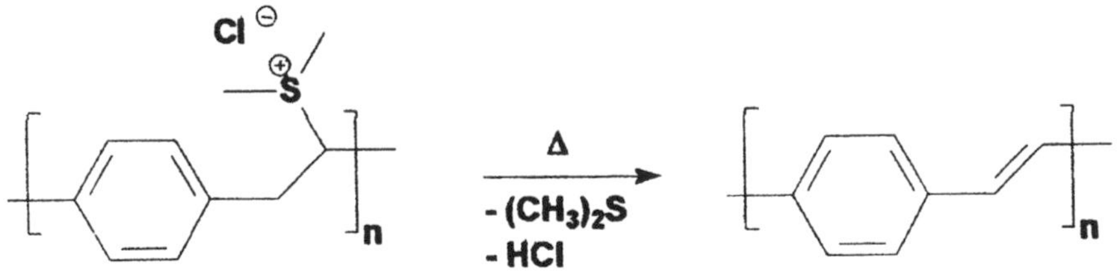

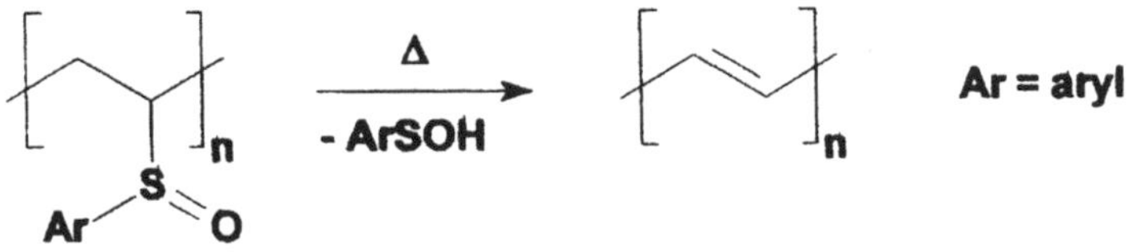

(e)   Cyclooctatetraene is also polymerized to give soluble polyenes [30]:

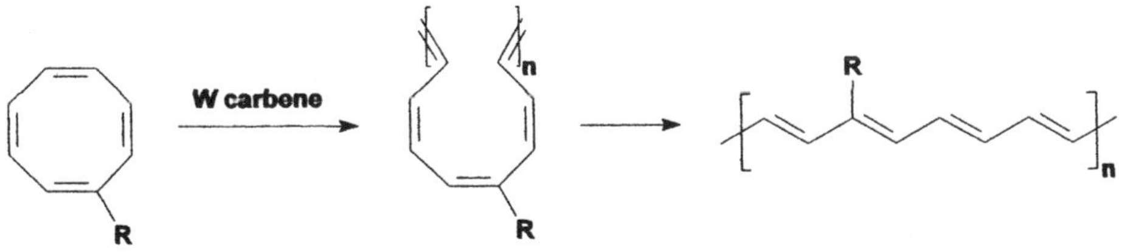

A.   Modification of Poly(acetylene)

1.   Cycloaddition

A variety of chemical modifications result from radical addition or cyclo-additions to the $(CH)_x$ backbone, e.g., with chlorosulfonyl isocyanate. The ring of the adduct thus formed can be opened by alkalis. The reaction scheme for cyclo-addition of chlorosulfonyl isocyanate and ring opening to substituted hydrophilic poly(acetylene) is as follows:

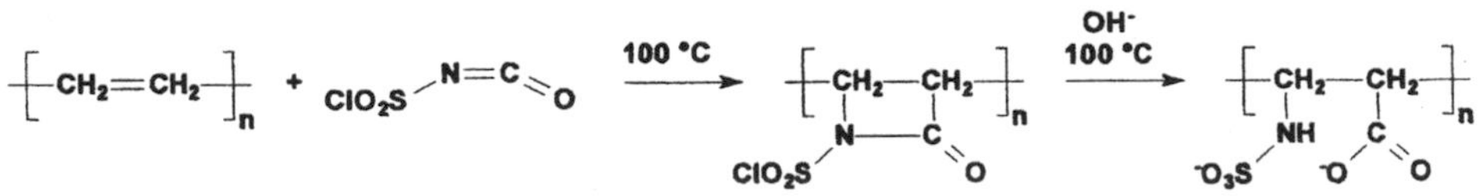

With 3-chloroperbenzoic acid, the dominant reaction is the formation of oxirane structures, which can react further. Metal carbonyls, e.g., $Fe_3(CO)_{12}$, react only with *cisoid* units. Otherwise the metal atoms combine with two different units of the poly(acetylene)

or isomerization occurs, resulting in *cis* configurations. All these types of reactions have been confirmed by IR spectroscopy. *CO* insertion can also be observed with molybdenum carbonyls. Cyclo-addition of maleic anhydride (MA) and 3,4-dichloromaleic anhydride (DCMA) leads to adducts like that shown below. The adduct formed by DCMA is worth mentioning because it gives rise to fusible poly(acetylene) (165–80°C) [31].

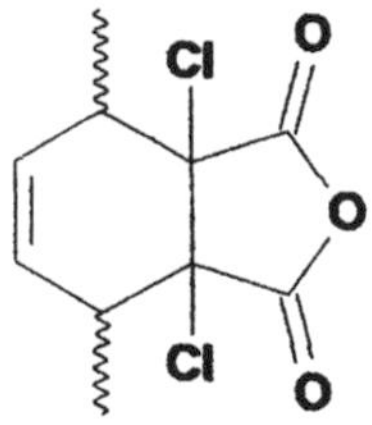

DCMA modified (CH)ₓ

2. Modification of Polymerization Conditions

An important progress (concerning $(CH)_x$ properties) occurred by a comparison of the various types of poly(acetylene) [23] and revealed some astonishing correlations: conductivity was directly proportional to crystallinity and inversely proportional to the number of sp^3 orbitals. This discovery was the key to the production of new poly(acetylene) types with fewer defects and greater stability. Another important advance was the modification of the polymerization conditions, e.g., using silicone oil or other viscous media. For instance, $(CH)_x$ can be polymerized at room temperature to yield a new $(CH)_x$ poly(acetylene) of at least the same quality as the standard $(CH)_x$ obtained at $-78\,°C$ by Shirakawa and co-workers [22]. Ageing of the standard catalyst brings about another surprising improvement in the $(CH)_x$ properties. The resulting reduction in the number of sp^3 orbitals, i.e., the production of a defect-free system, is of great benefit—you can stretch this $(CH)_x$ [17].

Special techniques were applied to orient the $(CH)_x$ in order to attain high conductivities (i.e., values up to $100\,000\,\mathrm{S\,cm^{-1}}$ [32] and parallel fibrils. Similarly, it is possible to make transparent $(CH)_x$ films with a conductivity of over $5000\,\mathrm{S\,cm^{-1}}$. The poly(acetylene) is produced on a plastic film and stretched together with the supporting material. Later it is complexed, e.g., with iodine, under standard conditions.

The standard Shirakawa type is crosslinked and contains an sp^3 fraction of approximately 2%.

The new BASF technique involves polymerization at room temperature (instead of $-78\,°C$) and the use of a tempered catalyst. The stretched poly(acetylene) product has parallel fibrils. It is linear (no sp^3 fractions), is highly orientable (can be stretched by up to 660%), and has a conductivity exceeding $10^5\,\mathrm{S/cm^{-1}}$. A convincing demonstration of the high anisotropy (1 : 100) in the stretched polymer are laid across each other, polarized light (sunlight) is extinguished in the region of overlap [33] in a manner similar to the effect of crossed Nicol prisms.

Figure 3 shows the equipment for the new BASF technique and process. As seen polymerization doesn't occur in a shaken or stirred vessel but on an even polymerization desk. This process was also developed as a continuous one.

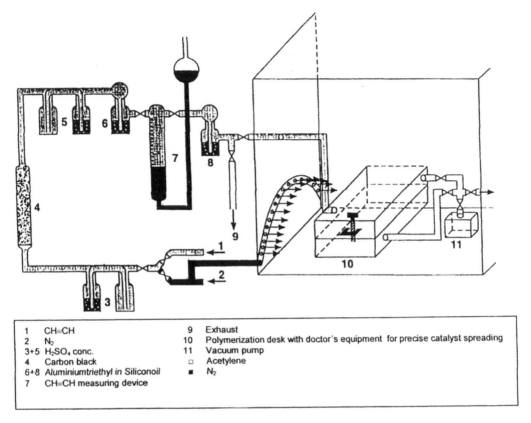

1	$CH{\equiv}CH$	9	Exhaust
2	N_2	10	Polymerization desk with doctor's equipment for precise catalyst spreading
3+5	H_2SO_4 conc.	11	Vacuum pump
4	Carbon black	▫	Acetylene
6+8	Aluminiumtriethyl in Siliconoil	■	N_2
7	$CH{\equiv}CH$ measuring device		

Figure 3 Glove-box—pilot plant.

Details are given under [17], also the preparation procedure of various poly-(acetylene) types.

IX. CATALYST

A crucial point mainly in acetylene polymerization is the catalyst influence of impurities, preparation of the catalyst system, changes in the catalyst according to the preparation temperature, examination by IR or NMR annealing of the catalyst and modifications, including preparation of the catalyst, details under [17].

X. ORIENTATION PROCESSES

Orientation processes are powerful methods that are used to improve conductivity and other material properties (e.g., transparency, anisotropy). Orientation can be achieved in several ways, including stretching.

Mechanical stretching can be performed after polymerization, e.g., in noncross-linked polymers. In the case of poly(acetylene)s prepared with aged Ziegler–Natta catalysts [34] stretching increases conductivity from 2500 S/cm to values as high as 10^5 S/cm.

Continuous electrochemical polymerization (e.g., of poly(pyrrole)) on the surface of a rotating drum permits simultaneous peeling off, mechanical stretching, and orientation $\sigma \leq 200\,\text{S/cm}$. Greater stretching rates and therefore greater conductivities are reported in poly(pyrrole perchlorate) films (σ up to $10^3\,\text{S/cm}$) [35]. Biaxially stretched films yielded conductivities of $800\,\text{S/cm}$ parallel to the stretching direction and $290\,\text{S/cm}$ in the cross direction.

Stretched poly(phenyl vinylenes) and poly(thienyl vinylenes) yielded conductivities of *ca.* $10^3\,\text{S/cm}$ [36].

Orientation can also occur during polymerization or by performing polymerization in an oriented matrix consisting of liquid crystals and using magnetic fields [37]. Variants are the use of liquid crystal matrices during the electromechanical synthesis of poly-(heterocycle)s [38] and the synthesis of polymers (e.g., substituted thiophenes) with liquid crystal side chains that contain sulfonate groups [39]. The sulfonate groups act as 'self dopants' and the liquid crystal side chains are responsible for orientation.

Polymerization of extremely thin poly(acetylene) films ($<1\,\mu\text{m}$) on crystal surfaces by epitaxial growth (e.g., on frozen benzene) [40] also induced orientation in the deposited polymer layer. Substituted poly(pyrrole) films with a high anisotropy can be produced by the Langmuir–Blodgett technique [41–43].

XI. WHAT ABOUT STABILITY?

The importance of stability was recognized early and in particular, oxygen absorption and storage stability were investigated. Stability is a relative term, being generally understood to mean the constancy of material properties. In practice it means that the properties of the materials used to make a product should undergo no changes during normal use (including storage), at least for the duration of their life cycle. The life cycle is extremely short for disposable articles, such as those used for personal hygiene or in medicine, but it may be several years for products such as domestic appliances, tools, machines and cars, and even decades for construction materials for bridges, buildings, etc.

The stability, particularly the susceptibility to autooxidation, is the Achilles' heel of the new materials as well as of organic polymers in general. The problem of oxidative damage has therefore been the object of intensive research. Poly(acetylene) $(CH)_x$, manufactured with Ziegler–Natta, Luttinger or other catalysts were used as model compounds.

All organic polymers degrade on exposure to oxygen, particularly in the presence of sunlight, but the extent of degradation varies markedly with the structure of the polymer. Normal $(CH)_x$ is particularly susceptible to reaction with oxygen (Figure 4).

The stability of $(CH)_x$ synthesized with different catalysts increases in the order:

Luttinger type $L-(CH)_x$ < Shirakawa type $S-(CH)_x$ < new type $N-(CH)_x$

The BASF type $-N-(CH)_x$, is the optimum material. Due to its special method of synthesis, it has a minimal sp^3 fraction, a high *cis* content (80% *cis* isomer synthesized at RT), a high density, very thin fibrils, and a high conductivity after doping with iodine. Both $N-(CH)_x$ and highly stretched $(CH)_x$ has greater stability than the usual systems (such as those of Shirakawa and Luttinger), probably due to the higher density and very low defect rate of the former [17] and less impurities (only ppm amounts of O, Ti, Al and Si).

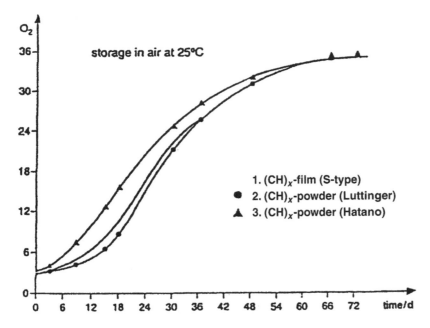

Figure 4   Autooxidation of poly(acetylene) [20].

The mechanism of polymer degradation usually entails the absorption of energy (thermal or UV), leading to the formation of active free radicals that partake in chain scission and crosslinking. Pristine S–(CH)$_x$ appears to undergo some spontaneous or thermal degradation at all practical temperatures. Even at $-78\,°C$, *cis–trans* isomerization occurs.

The *cis* content (%) of N–(CH)$_x$ film versus storage time in months at $22\,°C$ under nitrogen decreases very slowly compared with that of S–(CH)$_x$ to reach 78% after 3 months. This higher stability of N–(CH)$_x$ at $22\,°C$ is also confirmed by thermal isomerization studies, which show a higher energy barrier ($25\,kcal\,mol^{-1}$) for N–(CH)$_x$ compared with $17\,kcal\,mol^{-1}$ for typical S–(CH)$_x$. Also, we observed that N–(CH)$_x$ film that has been stored for 3 months under nitrogen gives, after doping with iodine, conductivities as high as $2000\,S/cm$, whereas 1 month-old S–(CH)$_x$ only $150\,S/cm^{-1}$. The infrared spectrum of the 3 month-old N–(CH)$_x$ sample is similar to that of the initial one [17].

More recently, Wegner et al. [44] studied the properties of oxidized poly(acetylene) by EFTEM (energy-filtering transmission electron microscopy). They demonstrated that oxidation proceeds homogeneously and the anion distribution is uniform without any sign of nucleation.

XII. THERMAL POLYMERIZATION OF ALKYNES TO (CH)$_X$ [45]

Since doped poly(acetylene)s were shown to possess metallic conductivity, this class of organic polymers has been studied intensively. Regardless of the catalyst system used, it is an inherent disadvantage of the washing process which is required to remove the catalyst or residues from the desired polymer preparation. Therefore a method was developed that avoids this pitfall and provides polyene films that form a solid layer on glass surfaces, ceramic plates, tubes, etc.

Table 1 High-temperature polymerization of alkynes and diynes [17].

Monomer	Electrical conductivity of polymerfilm [S/cm]	
Propyne	10^{-8}	10^{-1}
1-Hexyne	10^{-9}	10^{-2}
Butadiyne	10^{-8}	10
1,5-Hexadiyne	10^{-6}	60

Allene (propadiene) is condensed under high vacuum into a 2-litre round bottomed flask and, after several freeze-and-pump cycles to remove residual oxygen and volatile impurities, the reaction vessel is sealed (internal pressure at room temperature approximately 1000 mbar).

After 20 hours at 640 °C (a high-temperature drying oven) the inner surface of the flask is covered completely by a deep-black, shiny film which can be removed in patches and whose electrical conductivity is 10^{-8} S/cm^{-1}. When this film is doped by treating it for 30 min with a saturated solution of iodine in carbon tetrachloride its appearance changes to a shiny gold. After solvent removal, the conductivity is found to have increased by a factor of 109. Several other alkynes were polymerized under above conditions and the results of these experiments are summarized in Table 1.

XIII. WHAT IS THERE BESIDES POLY(ACETYLENE)

Poly(pyrrole) and poly(thiophene), both first described in 1963 as electrically conducting materials [5], experienced a renaissance when Diaz and Street gave a new attention to the electrochemical oxidation of pyrrole and Garnier to the poly(thiophene) field transistor. Poly(phenylenevinylene), poly(aniline), poly(phenylenesulfide), poly(carbazole), poly(indole), poly(pyrene) and polyene fulvene are just a few of the large number of electrically conducting polymers with specific properties and interest [20].

XIV. SUBSTITUTED POLY(ACETYLENE)S [20,45]

The synthesis of polymers from substituted acetylene monomers is directed toward the preparation of substituted, conjugated chains which ameliorate the negative properties of poly(acetylene)s (e.g., sensitivity to air, insolubility, and infusibility) while maintaining the desired electrical properties of acetylene's conjugated backbone alkyl- and aryl substituted polymers result. They are soluble (e.g., in toluene and cyclohexane) and proccessible, but have low conductivities (<0.1 S/cm) compared with the unsubstituted poly(acetylene).

A new development are polymers from phenylacetylene substituted Schiff's base monomers with conductivities of 10^{-2} S/cm but high environmental stability up to 300 °C and above [43,46].

XV. POLY(DIACETYLENE)S

In contrast to poly(acetylene), poly(diacetylene)s have limited electrical conductivity (≤0.1 S/cm), but can be obtained as large, single crystals.

The polymerization of diacetylene is an example of a topochemical polymerization in which 1,4-addition of 1,3-diyne units takes place in the crystalline state. The reaction does not require a catalysts and is performed by irradiation of the diacetylene crystals with visible or UV light, X-rays, γ-rays, or by annealing the crystals below their melting point. The unreacted monomer is then extracted with a suitable solvent (e.g., hexane, toluene), leaving a single, dark red crystal of poly(diacetylene). This unique polymerization process was already observed in 1882 [47] and has now been intensively studied [48]. A correct interpretation of the phenomena was first provided in the early 1970s [13–15,49].

More details concerning mono or multilayer applications and the synthesis of diacetylenes are summarized in [20].

XVI. POLY(PYRROLE)—ANOTHER STEP FORWARD

Poly(pyrrole) is a polysalt that can be produced in the form of powders, coatings, or films; it is intrinsically conductive, and exceptionally stable and can be quite easily produced also continuously, e.g., by electrochemical techniques (Figure 5).

The preparation of poly(pyrrole) (pyrrole red and pyrrole black) by oxidation of pyrrole dates back to 1888 [50] and by electrochemical polymerization to 1957 [51]. A fairly long period elapsed before this organic π-system attracted general interest and was found to be electrically conductive [5] in 1963.

Conductive poly(pyrrole) films are obtained directly by anodic polymerization of pyrrole in aqueous or organic electrolytes [52]. They are black and under suitable reaction conditions, can be detached from the anode in the form of self-supporting films. The conducting salt use in the electrolyte solution is incorporated in the film as a counterion.

In contrast to poly(acetylene), poly(pyrrole) has a high mechanical and chemical stability and can be produced continuously as flexible film (thickness 80 μm; trade name: Lutamer, BASF) by electrochemical techniques [53].

Other electrochemical polymerizable heterocycles are thiophene, furan and their substituted and oligomeric derivatives.

The polymerization starts initial by an oxidation step, followed by a radical cation formation, coupling reaction, deprotonation, and a one-electron oxidation in order to regenerate the aromatic system [54] scheme [20].

Also important is the reaction introducing oxygen—forming a labile –CO–NH– group in the pyrrole ring system.

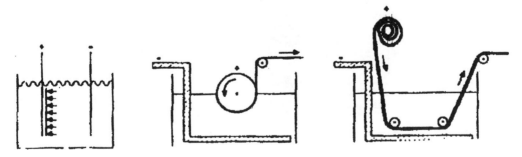

Figure 5 Electrochemical poly(pyrrole) synthesis. Left discontinuous process [52] center and right BASF methods, center-continuous by rotating-drum electrode (US Patent 4468 291, June 27, 1993) and right-rotating band electrode.

The quality of the polymers is greatly influenced by many factors, e.g., impurities, electrode material, pressure, concentrations, temperature and comonomers. The most decisive, however, are the current density and the electrolyte, particularly the conduting anion X^- because it is incorporated into the polymer as a counterion.

The properties of the counterion (e.g., its size, geometry, charge) influence the properties of the polymer.

In general, one anion is incorporated for every three pyrrole units. Exceptions are pyrrole- or thiophenesulfonic acids where the counterion is coupled directly to the monomer (self doping) [55]. Some typical conducting anions are fluoroborate, perchlorate, aromatic sulfonic acids, penicillin, n-dodecyl sulfate phthalocyanine sulfonic acid, poly(styrenesulfonicacid), styrene sulfonic acid, and heparin [13].

Interesting conductive salts that affect optical activity in poly(pyrrole) are (+) or (−)-camphor sulfonic acid (used for racemate separations) [20].

The anion X^- can also be released, e.g., by applying a negative potential. Release can be specifically controlled, offering interesting possibilities for active counterions of medical interest (e.g., heparin and monobactam) that are incorporated into poly(pyrrole) [56].

Poly(pyrrole) functionalized with peptides are reported in Refs. [57–59].

By changing reaction conditions, polymers with different surface morphologies (e.g., an open porous structure) can be obtained (Figure 6).

Figure 6 Poly(pyrrole) with defined holes and counterion [60].

Table 2 Properties of conductive polymers

Conductivity (S/cm)	10^{-1} to 10^2
Film gauge (µm)	
(a) Self-supporting films	*ca.* 30 upward
(b) Coatings	0.01 upward
Specific surface	
Nitrogen surface (m^2/g)	5–50
X-ray crystallinity (%)	15
Oxygen absorption [% (w/w)/day]	0.1/30

Preparation procedures of films and powders are given under [20], e.g. electrochemical production of poly(pyrrole)films—discontinuously or continuously:

- Preparation of Finely Divided Poly(pyrrole)
- Optimization of Synthesis Conditions of Poly(pyrrole) from Aqueous Solution.
- Coating Surfaces with Poly(pyrrole) (e.g. Poly(ester), Poly(amide)).
- Chemical Modification of Surfaces (e.g. ceramics, glass).

Properties and potential uses are shown in Table 2 and Figure 7.

The conductivity of poly(pyrrole) film suggests applications such as flexible conductive paths in printed circuits, heating films, and film keyboards.

Poly(pyrrole) films show good electromagnetic shielding effects of about 40 dB over a wide range of frequencies (0–1500 MHz) [20].

Although the new systems appear to be promising, stability problems may be encountered.

Poly(pyrrole) is sensitive to moisture because this leads to leaching of the counterion and thus to a decrease in conductivity. This can be avoided by use of appropriate hydrophobic or polymeric counterions (e.g., camphor sulfonic acid or poly(styrenesulfonicacid) or by incorporating hydrophilic compounds.

Poly(pyrrole) in which the counterion is 4-hydroxyphenyl sulfonic acid do not undergo any change in conductivity if they are exposed to nitrogen for two month at 140 °C. No change in conductivity was observed when these polymers were stored in the laboratory for three years at room temperature and 55% R.H.

Poly(pyrrole)s obtained by synthesis in aqueous electrolytes maintain their conductivity at a level of about 20 S/cm. Poly(pyrrole)s with perchlorate as a counterion are unstable under atmospheric conditions but can be used as electrodes in rechargeable batteries [13].

Poly(pyrrole) is a suitable electrode material for rechargeable electrochemical cells.

Storage of poly(pyrrole) films (80 µm thickness) over 19 years (1982–2001) in a plastic wrapper showed a loss of conductivity <5% and no change in flexibility [14]. The influence of oxygen on the properties is described in [15].

XVII. POLY(PYRROLE)-BATTERIES [54,60]

The advantage of polymer electrodes is that they can be easily shaped, allowing novel battery design. Polymer cells with poly(pyrrole) and lithium electrodes have been developed [13]. In the flat cell, the poly(pyrrole) and lithium films are sandwiched together, in the cylindrical cell, the two films are wound concentrically. Their energy per unit mass

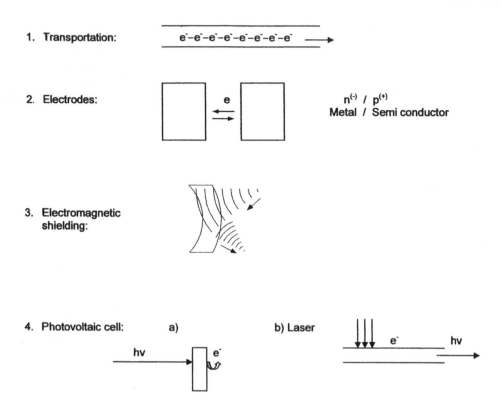

1. Transportation:

2. Electrodes: $n^{(-)}$ / $p^{(+)}$
                        Metal / Semi conductor

3. Electromagnetic
   shielding:

4. Photovoltaic cell: a)   b) Laser

   hv   e⁻   hv

5. Antistatic equipment, immersion heating systems, sensors, transistors or reversible
   batteries, electronic devices, medical application, drug release [13, 57–59].

- chemical deposition of conducting materials, e.g., pyrrole on ceramic glass, plastic
- materials, glass-fibres, carbon fibres
- piezo ceramic devices
- deposition of structured electrical conducting polymers by irradiation
- deposition of metals (copper) on deposited conducting polymers (printed circuits)
- elastic poly(pyrrole)s (elongation | 100%)
- optical data storage
- separation of DNA on poly(pyrrole) films [cit.ed in 58]
- controlled exchange of actinide and lanthanide cations [59].

Figure 7 Some typical application for conductive polymers (e.g. poly(pyrrole)s).

and their discharge characteristics are similar to those of the nickel-cadmium cells now
on the market. More than 500 charging and recharging cycles have been achieved with
laboratory cells. Applications include dictaphones and pocket radios [13].

XVIII. NEW MOLECULAR ARRANGEMENTS

Under oxidation conditions (electrochem. or. chem.), starting from substituted
benzenes or derivatives, triphenylene structures (benzo[1,2:3,4:5,6]tris[arylenes]) are

formed, e.g.:

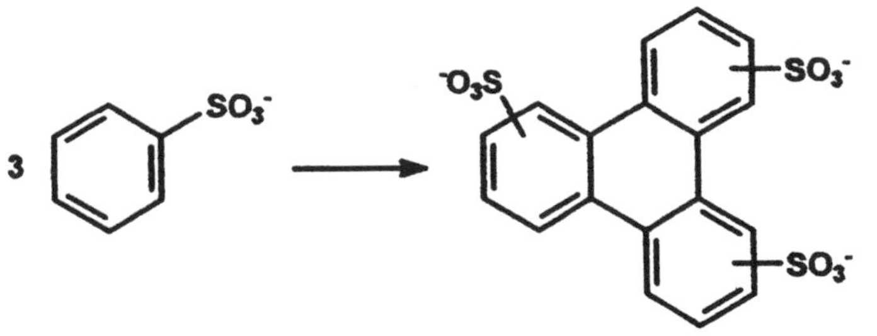

If phenylsulfonic acid is used as a counterion (current density (i) $>10\,\text{mA/cm}^2$, the trisulfonic acid of triphenylene is formed to some extent [17].

Another simple and convincing experiment is the conversion (trimerization) of o-dimethoxybenzene to the analogous triphenylene ether [61].

Condensed ring systems like naphthoquinone are trimerized analogously, and heterocycles can similarly be oxidized to 'tris[hetarylenes]' [62]. Fo example, using pyrrole as a starting monomer under strongly oxidizing conditions (electrochemical oxidation with current densities $>10\,\text{mA/cm}^2$ or redox potentials | 1,5 V), triphenylene structures were formed that could be oxidized to new macrocycle assembilies [43].

1. The resulting poly(pyrrole) showed no C–H signals after pulsing (^{1}H-NMR).
2. When tetradeutero (2,3,4,5)pyrrole was used as the starting monomer, the resulting poly(pyrrole) showed no detectable amounts of D.
3. Poly(pyrrole) degradation (pyrolysis at 600 °C or anodic over-oxidation) gave benzene, indole, carbazole, etc, fragments.
4. When a poly(pyrrole) film was treated with an aq. $CuSO_4$ solution, Cu^{2+} was sequestered (verified by spectroscopy) [63,64].

All these phenomena are in accordance with the proposed macrocyclic structure and not with a linear one.

This condensation seems to be a general approach to the synthesis of new types of disc-like tridentate polymeric structures. The pyrrole units can be imagined to conform to a three-dimensional fullerene type structure in which one carbon from each of the pentagonal units is replaced by a nitrogen (Figures 8 and 9).

It is worth mentioning that indophenins, including both oligomers and polymers, represent a new type of electrically conducting materials.

XIX. POLY(THIOPHENE)

Since the first report of thiophene polymerization in 1883 [65] decades passed until 1963 thiophene lead to an electrically conducting material [5] and was considered to be an attractive monomer for conducting polymers [66]. Interest is still growing because thiophene is easier to handle than pyrrole (less sensitive to oxygen) and allows the simple preparation of substituted monomers leading to soluble polymers. Three different methods have been used to produce poly(thiophene): chemical oxidation coupling with organometallic agents or by the Grignard reaction and electrochemical oxidation. The preparation techniques for films, powders, and coatings are similar to those of pyrroles

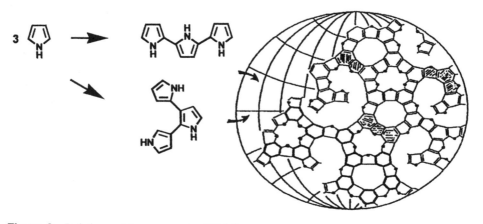

Percorjugated Organic Polymers

Figure 8 Poly(pyrrole) macrocycles [17,64].

but thiophene needs stronger oxidants for its polymerization than pyrrole or other heterocycles. Standard oxidation potentials follow:

- Pyrrole $+0.8$ V   Dipyrrole $+0.6$ V
- Indole $+0.9$ V
- Azulene $+0.9$ V
- Thiophene $+1.6$ V   Dithiphene $+1.2$ V
- Furan $+1.85$ V

The oxidation potential correlates to the ease of polymerization [66].

Figure 10 illustrates the correlation between UV–VIS absorption and the oxidation potential of thiopene oligomers [67]. The oxidation potential of the series $(\text{thiophene})_n = n = 1\text{--}6$ decreases with the increasing number of thiophene units.

XX. SUBSTITUTED POLY(THIOPHENE)S

Substituted poly(thiophene)s are soluble in common organic solvents with conductivities up to 100 S/cm.

Figure 9 Model of poly(pyrrole) macrocycle with counterion [17].

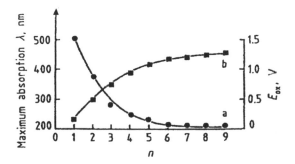

Figure 10 Correlation between oxidation potential (a) and maximum absorption wavelength (b) of thiophene oligomers [67]; Oxidation potential measured in 0.5 mol/L Et_4NBF_4 in acetonitrile versus saturated calomel electrode.

Almost parallel to the polymerization of thiophene, much work has also been done with the electropolymerization of 3-methylthiophene. The structure, morphology, electronic, and electrical properties of poly(3-methylthiophene) have been studied by several authors [20,68].

The 3-alkylsubstituted thiophenes with alkylchain size larger or equal than butyl were found to be soluble in common organic solvent and therefore an important amount of work is being developed since their discovery [69].

Polysubstituted thiophenes with other groups different to n-alkyl chain are also known. For example, alkylsulfonate, alkoxy, amide, poly(ether), and acylgroup were introduced in 3-position in thiophene and their electroobtained polymers were studied. A water soluble poly(alkane)sulfonatederivative of thiophene has also been reported [55] which would be an intrinsically conducting polymer (self-doped). Composites of poly(thiophene)s with poly(methylmethacrylate) and poly(vinylchloride) were prepared as transparent films by electropolymerization of thiophene in the presence of the respective polymers (Lit. cit. under [20]). Poly(hexylthiophene) doped with electrons shows superconductivity [70].

The synthetic methods used to polymerize the 3-alkyl thiophene do not differ substantially from these employed for thiophene. The good choice of the solvent is important to ensure a complete dissolution of the monomer and the electrolyte in the electrosynthesis case. Chemically polymerization is based on the Grignard coupling method used by Yamamoto et al. [71] and later revised by Kobayashi et al. [72]. Some polymerizations carried out with chemical oxidants are also known with poly(3-alkyl-thiophene) [73].

The electropolymerization of alkylthiophene is carried out mostly in a one-compartment cell, using platinum as working electrodes and nitrobenzene as solvent. Lower temperatures, preferentially between 5–15 °C and current densities ranging 1–2 mA/cm^2 are applied in galvanostatic conditions [74]. Dihydroisothiophene and aromatic bridged thiophene see [75].

A new type of a substituted thiophene represents bis(thienyl) coronene, leading to an alternating polymer with bis-thiophene and coronene units [76]. Synthesis and properties of processable poly(thiophene) are described in [77].

XXI. OLIGOTHIOPHENES

The oligothiophenes have reached more prominence in recent years. The progress of these materials is summarized by Bäuerle [78]. Dithienylpolyenes, thienylene polyenylene oligomers and polymers are described by Spangler and He [77].

XXII. CHARACTERIZATION [79–82]

A number of different characterization methods have been performed on poly(thiophene) and poly(alkylderivative)s. NMR of electropolymerized poly(thiophene) films has been studied by Hotta et al. and Osterholm et al. An infrared study about vibrational key band on poly(thiophene) films and FT-IR spectra were also published. Resonance Raman scattering on poly(methylthiophene) and x-ray scattering on poly(thiophene) were performed; x-ray photoelectron spectroscopy has been reported on FeCl$_3$-doped poly(hexylthiophene) and electrochemically obtained poly(thiophene) [20]. Time resolved fluorescence studies on thiophene oligomers are given by Chosrovian et al. [79]. For studies on oligothiophene films see [80,81].

Xanes (x-ray absorption near edge structure) investigations showed that polymer chain variations decrease with increasing size of alkyl-side chains on the thiophene backbone [82].

XXIII. POLY(PHENYLENEVINYLENE) PPV

Poly(p-phenylenevinylene) [83] was first synthesized by the Wittig condensation of terephthaldehyde and p-xylenebis(triphenylphosphonium)chloride. The method which is now mainly employed starts from a soluble poly(sulfonium salt) [28] for example:

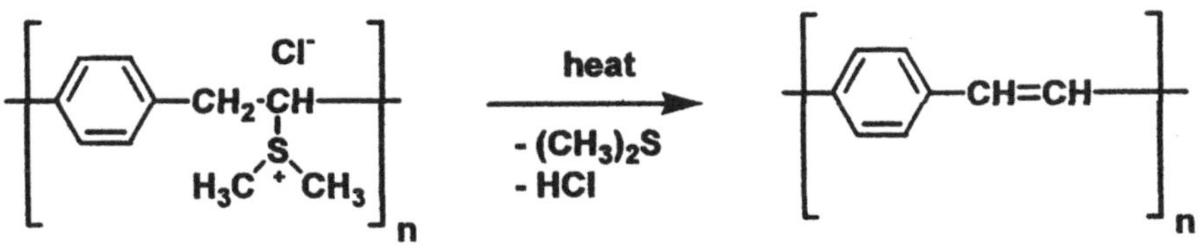

The poly(sulfonium salt) is obtained by the reaction if α,α-dichloro-p-xylene with excess dimethyl sulfide (50 °C, 20 h) and polymerization with sodium hydroxide (0 °C, 1 h).

Instead of the dimethylsulfonium salt often tetrahydrothiophene or tetrahydro-pyrane salts are used.

The starting precursor poly(sulfonium salt)s are prepared by polymerizing the monomer sulfonium salt with an equimolar amount of NaOH in aqueous solution at 0 °C for 1 h. Further details and similar preparation techniques are cited in [20].

A complete different route to PPVs [84] starts from tetrahalogenated xylylen-derivatives and by an dehalogenation reaction yielding aryl substituted PPVs.

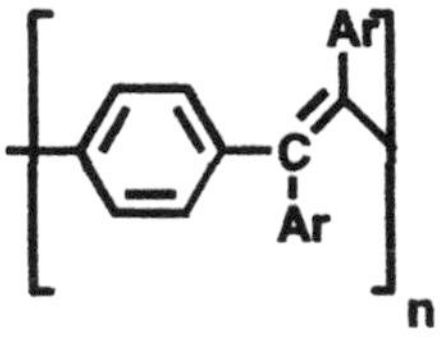

But comparing both PPVs synthesized by the sulfonium route with those prepared by the dehalogenation route, you'll find remarkable differences; PPVs prepared by the sulfoniumroute:

- contain impurities (S, O, Cl up to 1% and more)
- show electrically conductivity (10^{-4} S/cm)
- show electroluminescence

PPVs prepared by the dehalogenation route:

- are very pure (impurities <10 ppm)
- are not electrically conducting ($<10^{-12}$ S/cm)
- show no electroluminescence [85].

The first doped poly(phenylenevinylene) was reported by Karasz using AsF_5 as dopant; its conductivity was 3 S/cm. Stretching the precursor film at high temperature and doping with AsF_5 or SO_3 yielded conductivities up to 2800 S/cm.

Substituted derivatives, copolymers, and blends show better thermal stability. Highly conductive graphite films have been prepared by pyrolysis of poly(phenylenevinylene) (>3000°C), stretched samples doped with SO_3 had conductivities of 10^5 S/cm [20]. Poly(phenylenevinylene) can be used as tunable polymer diodes or luminescent electrodes [83].

XXIV. LED'S LIGHTEMITTING DIODES [86–89]

Recently LED-devices using organic materials (e.g., oligothiophenes or mainly PPVs) are attracting increasing attention because these materials allow the manufacturing of large displays.

Light emission is caused by the recombination of electrons and construction of the device is seen in [20].

An electron is injected into the PPV from the cathode while a hole is injected from the anode. The hole and the electron migrate and when they meet each other, they recombine and visible light is the result.

The frequency of light emitted is approximately equal to the difference of the oxidation and reduction potential of the polymer—the electrochemical band gap.

A challenge in the operating of LEDs is the fact that the hole-mobility is higher than the electron-mobility [89]. Therefore the barrier height (resistance) between the polymer and each of the electrodes must be roughly equal. The LED-efficience is about 4%, the brightness is acceptable and a potential market are displays and, e.g., backlighting.

XXV. POLY(PHENYLENESULFIDE) [17,20]

Poly(p-phenylesulfide), PPS, (white powder, T_g 92 °C, mp 270–290 °C, 65% crystallinity) was the first melt-proccessible polymer to be doped with strong electron acceptors (e.g., AsF_5) to yield highly conductive products [90]. The first laboratory synthesis of PPS was reported by Macallum [91] and involved the melt reaction of 1,4-dichlorobenzene, sulfur and sodium carbonate. A commercially product has been available as powder, film or fiber since 1973 from Phillips Petroleum under the trade name Ryton; it is produced from 1,4-dichlorobenzene and sodium sulfide (high-pressure process) in a polar solvent (N-methylpyrrolidone) [92].

In laboratory scale, obviously low-pressure processes are mostly preferred due to handling of chemicals and safety considerations. Several other synthetic routes have been attempted to prepare PPS.

XXVI. SYNTHESIS OF PPS

The Macallum synthesis of PPS was studied in detail by Lenz and co-workers [93] who developed an improved synthetic route based on the self-condensation of metal-p-halogenothiophenoxide. This method although used less drastic conditions than Macallum's synthesis required extensive washing of the product in order to remove the residual metal contaminates. Port et al. [94,95] obtained PPS from copper(I)-4-bromothiophenoxide in a mixture of 10:1 quinoline:pyridine at 198–235 °C.

The optimal conditions of this reaction, the influence of the solvent in yield and molecular weight have been examined by Lovell [93]. Evidences of disulfide bonding in the PPS were found by IR and ^{1}H-NMR analysis of the toluene soluble fractions. Studies of the homopolymerization of thiophenol in the presence of sulfuric acid have been carried out by Zuk [96] and Viswanathan [97].

The oxidative condensation of thiophenol with thionyl chloride in the presence of Lewis acid was studied by Wejchan-Judek [97].

The high-temperature condensation of aromatic halides (diphenylsulfide, thianthrene, diphenyldisulfide, thiophenol and direct reactions of benzene with sulfur) in the presence of aluminium chloride was examined by Sergeyev et al. [98]. The reaction of 1,4-dichlorobenzene or 1,4-dibromobenzene with sodium sulfide at 195 °C in N-methylpyrrolidone (NMP) has been studied under normal atmospheric pressure, the kinetic of polymerization was found to be of second order. The same reaction in NMP was revised by Russian authors [99]. The polymerization of diphenyldisulfide produced PPS when stoichiometric amounts of diphenyldisulfide and antimoniumpentachloride were used in nitrobenzene at 20 °C. The S–S bond cleavage is catalyzed by the Lewis acid and the authors claim to have obtained an identic product compared with the commercial one with a molecular weight >1000 [100].

Another process involves self-condensation of a halogenated thiophenol resulting in 'head-to-tail' polymerization [93]:

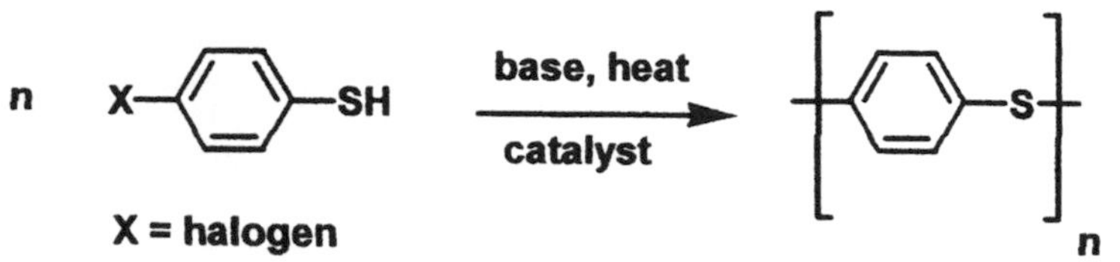

X = halogen

The electrooxidative polymerization of thiopenol in nitromethane at room temperature was studied by Tsuchida et al. [100]. The elctropolymerization was carried out in the presence of 1.5 M trifluoracetic acid. The presence of sulfonic acid or stannic chloride produced also PPS. The white polymer precipitates from the solution during the synthesis and is soluble in NMP with m.p. 180–190 °C and $M_w > 1000$. The variation of solvent and acid on the polymer yield and molecular weight was examined. According to the mechanism of polymerization proposed by the authors, the reaction proceeds at first with the formation of diphenyldisulfide as intermediate, which would be rearranged subsequently to PPS. Recently a new process for the synthesis of PPS was developed. Tsuchida [100] reported that diphenyldisulfide (DPS) mixed with a small amount of dichloromethane and subjected to (air)oxidation polymerization at 20 °C in the presence of vanadylacetylacetonate produced high yield of PPS. The same author succeeded in the electrolytic polymerization of DPS to produce PPS by using tetra-n-butylammonium-tetrafluorborate in dichloromethane with addition to trifluoraceticacid and trifluoracetic-anhydride.

XXVII. DOPING OF PPS [90]

Doping of PPS with AsF_5 produces conductivities up to 2×10^{-2} S/cm [90]. Attempts to make n-doped PPS have been unsuccessful. In addition to AsF_5, doping with $FeCl_3$, H_2SO_4, $HClO_4$, FSO_3H, CF_3SO_3H, $AlCl_3$, TaF_5 has been investigated [101]. Heavily doped PPS undergoes intramolecular cross-linking to form benzothiophene rings:

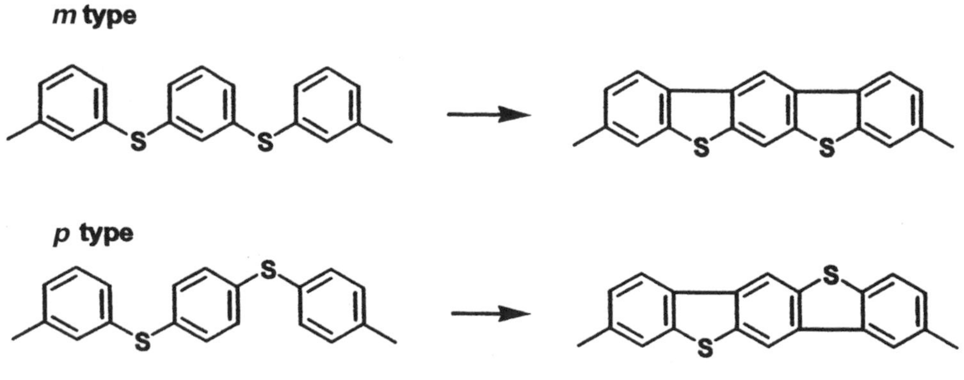

m type

p type

XXVIII. POLY(*m*-PHENYLENESULFIDE)

Poly(*m*-phenylenesulfide) and substituted (methyl-, fluoro-) poly(*p*-phenylenesulfide) are prepared by analogous reactions from appropriate monomers.

By heavily doping ring closure occurs (formula above).

XXIX. OTHER DERIVATIVES

PPS substituted with methoxy, acetyl, and hydroxy groups were analyzed by Laakso et al. [102–104]. The introduction of methoxy- and acetylgroups also increases the molecular weight and lead to higher conductivity. Blends, blocks, and statistic copolymers of 1,4-phenylensulfide with 2-methyl-1,4-phenylensulfide have been synthesized and their thermal behavior was examined by differential scanning calometry (DSC) [96].

Poly(phenylenesulfides) are used where resistance to heat and chemicals is required, for example in the chemical process industry (pump housings, impellers, valves, metering devices, tanks, coil bobbins) [13].

Poly(*p*-phenyleneselenide) and poly(*p*-phenylenetelluride) have been prepared by techniques similar to those used to prepare PPS [102]. Conductivities of AsF_5-doped samples reach 0.01 S/cm.

PPS combining biphenyl or terphenyl moieties in the main chain are reported in the literature [103].

XXX. POLY(ANILINE): HISTORICAL BACKGROUND AND METHODS OF SYNTHESIS [20]

Aniline was variously discovered and designated by Unverdorben as 'Kristallin', by Runge as 'Kyanol', by Fritsche as 'Aniline' (from Spanish anil, indigo), and by Zinin as 'Benzidam'. In 1854 for recovering aniline by reducing nitrobenzene with iron filings in the presence of dilute acids was developed by Perkin into an industrial process in 1857. The dye industry was born. One way of making dyes from aniline is afforded by oxidation. As early as 1860, an industrial process for manufacturing oxidation dyes was presented by Calvert, Clift, and Lowe.

In 1891, the practical use of poly(aniline) was described by Goppelsroeder, The chemical constitution of aniline dyes was mainly elucidated by Willstätter and Green around the turn of the century. The synthesis of poly(aniline) is remarkably simple. Aniline is chemically (1863 Lightfool) or electrochemically (1865 Letheby) oxidized, whereby a quinone diimine is formed by the oxidation and subsequent dehydrogenation of 2 molecules of aniline. Multiple repetition of this process with simultaneous dehydrogenation affords emeraldine and then nigraniline, which is a long-chain molecule consisting of eight benzene rings and paraquinoid groups that are linked in the para position by nitrogen atoms. This converts to pernigraniline and finally to aniline black.

The reaction postulated by Willstätter and Green is shown in Figure 11.

Recent studies of poly(aniline) suggest that the polymer can exist in a wide range of structures, which can be regarded as copolymers of reduced (amine) and oxidized (imine)

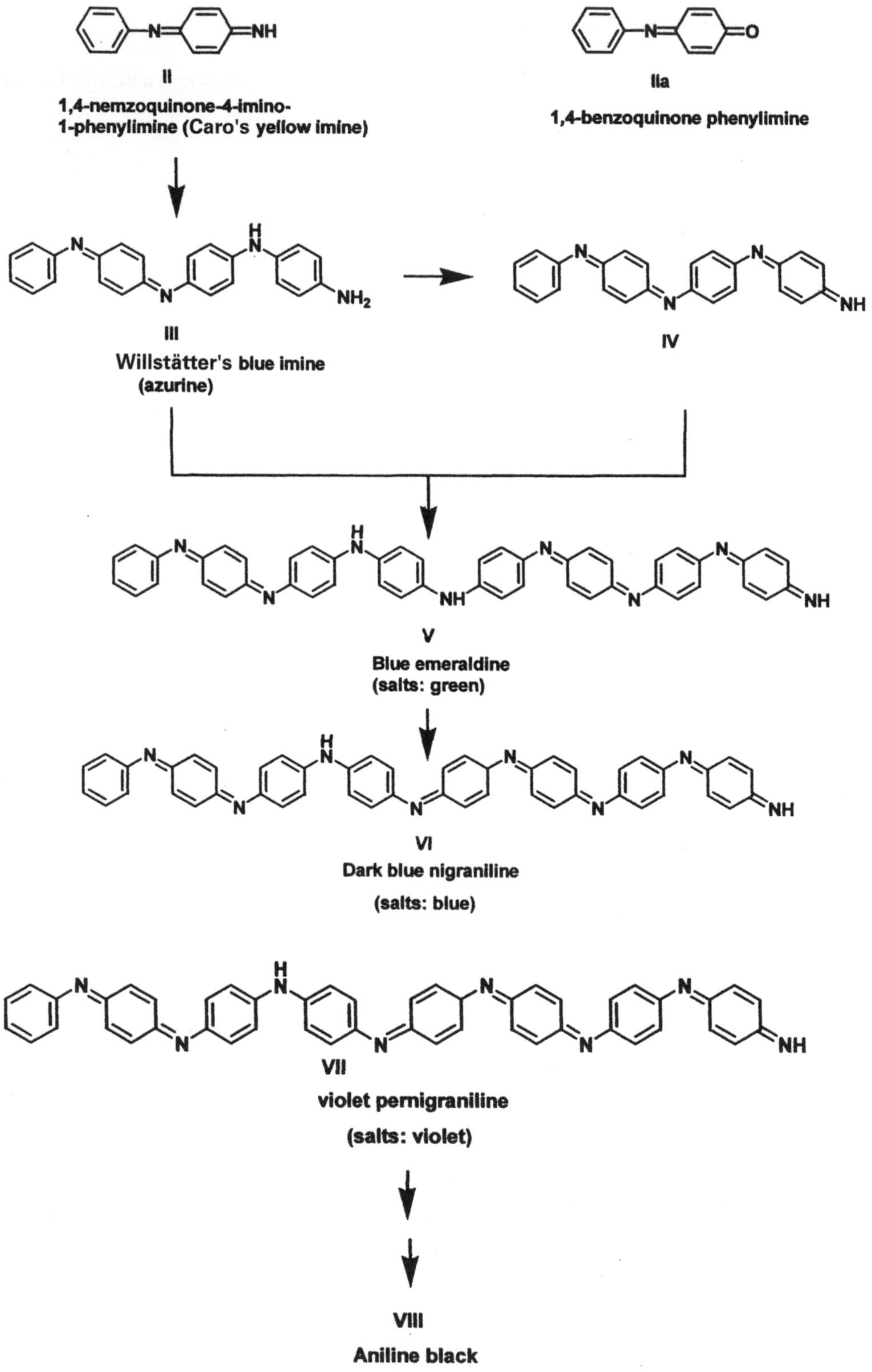

Figure 11 Aniline oxidation steps.

units of the form:

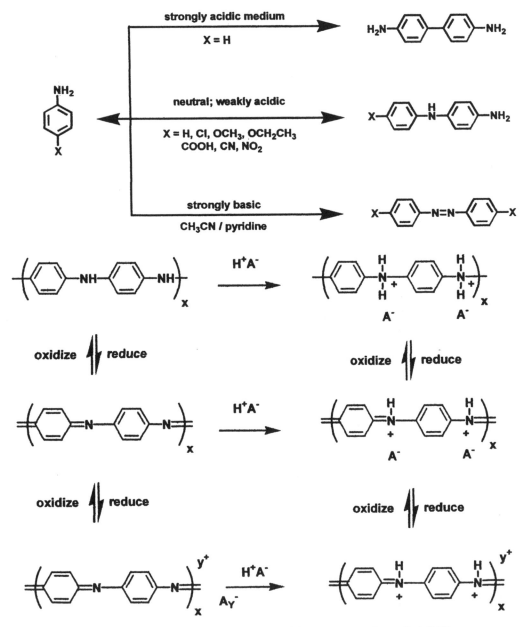

A decisive handicap in the production of poly(aniline) is, however, the appearance of benzidene; see Figure 12 [103].

Figure 12 Redox behavior of polyaniline (as proved by spectroscopic studies) [108].

Honzl [103,105,106] synthesized poly(aniline)s by not using aniline as the starting monomer. In this way, he arrived at a 'poly(aniline)' that is free from benzidene. The Honzl synthesis starts with

ROOC

O=⬡=O   **and**   H₂N—⬡—NH—⬡

COOR

Reaction scheme in [20].

Alternative structures can be obtained by polycondensation of quinones and aromatic amines. They have been described by Hall, Jr. Yields are generally high (>70%). The reaction is usually carried out at 250 °C for several hours. This temperature is crucial for obtaining reasonable molar masses.

XXXI. EXPERIMENTAL DETAILS OF POLY(ANILINE) PREPARATIONS

A. Chemical Oxidation

- Preparation of aniline black
- Electrochemical oxidation
- Electrochemically synthesized poly(aniline) see [20].

The phenomena of aniodic oxidation and of discharge by means of reversed polarity have been known since the time 1891 of Goppelsroeder [3].

A new mechanism by which poly(aniline) conducts electricity has been established by J. P. Travers and M. Nechtschein. The conduction process can be accounted for in terms of electrons hopping between localized states under the assistance of proton transfer, for which the presence of water plays an essential role [107] (Figure 12).

Poly(aniline) is primarily of interest because it can be used as electrode material. It is the preferred choice of all conducting polymers. Its discharge capacity is greater than that of poly(pyrrole)$^{(+)}$/perchlorate$^{(-)}$ (which is theoretically limited to 88 Ah/kg), its self-discharge is better than that of, e.g., poly(thiophene) or Ni/Cd batteery systems. Genies has announced a rechargeable battery of the type poly(aniline/propylene) carbonate–$LiClO_4$/Li–Al.

The polymer is made by aniline oxidation with ammonium persulfate in NH_4, 2.3 HF as solvent. The discharge capacity of the polymer is 100 Ah Kg^{-1} at 25 °C and 140 Ah/kf at 40 °C for current densities of 0.5 mA/cm^2 and for an amount of material giving a capaicty of 10 mAh.

The voltage is open circuit for the fully charged battery is 3.6 V. The average utilizable potential is 2.8–3 V. The energy density for the polymer lies between 280 and 420 Wh/kg. The ratio of the amounts of electricity in discharge and charge is one for several hundred deep cycles. Its behavior with regard to self-discharge and to constant applied voltage (floating life) is excellent [109].

XXXII. PROPERTIES AND USES [13]

The conductivity of poly(aniline) is $10\,\mathrm{S/cm}$ up to pH 4, but decreases to $10^{-10}\,\mathrm{S/cm}$ above pH 4. The polymers are stable up to $250°\mathrm{C}$ (undoped) and up to *ca.* $150°\mathrm{C}$ (doped). Conduction in polyaniline can be accounted for in terms of electrons hopping under the assistance of proton transfer, for which the presence of water plays an essential role [107].

In contrast to other electrically conducting polymers, poly(aniline)s may be doped with protons. Poly(aniline) is available as powder, film or fibrils [109].

The use of poly(aniline) as a battery electrode on account of its redox and proton transfer behavior was described in 1968 [110,111]. In 1986 poly(aniline) was presented as a 'novel conducting polymer'; its preparation and redox behavior were described [112–114]. The phenomena of anodic oxidation and discharge by means of reversed polarity have been known since 1891 [3].

XXXIII. FURTHER APPLICATION [88]

Like poly(pyrrole)s, poly(aniline)s show excellent antistatic behavior and have a high shielding efficiency for electromagnetic interference and are useful as anticorrosive protections [115].

XXXIV. POLY(TOLUIDINE)S, PT [115]

Poly(toluidine)s refer to *o*, *m*, and *p*-methyl substituted aniline polymers which exist like poly(aniline) in different discrete oxidation states. The preparation techniques are similar to PT those of poly(aniline). The similarity of the substituted PT and the unsubstituted aniline are evident from IR and Raman spectroscopy, the processability of PT is better due to the advantage of higher solubility but the conductivity is one to two orders of magnitude lower than that of poly(aniline).

Blending of PT with other polymers leads to excellent mechanical properties.

Other substituted poly(aniline)s (e.g., with alkoxy-groups) are reported, see also Arand, Palaniappan and Suthyanarayana [115].

XXXV. POLY(PHENYLENE) [12,13]

First attempts to prepare poly(phenylene) date back to 1842 [78]. Riese describes a process in which poly(phenylene) ($n = 13$) is synthesized from 1,4-dibromobenzene and sodium. Further methods are the Ullmann reaction, thermal decomposition of diazonium salts, coupling of phenylene dihalogenide – Grignard compounds.

In 1963 the Kovacic method was systematically extended by varying reaction conditions (temperature, catalysts, Lewis acid, and oxidants) and starting materials, leading to electrically conducting polymers [5,115].

Figure 13 shows the correlation between synthesis conditions and yielded poly(phenylenes) using the oxidative coupling reaction.

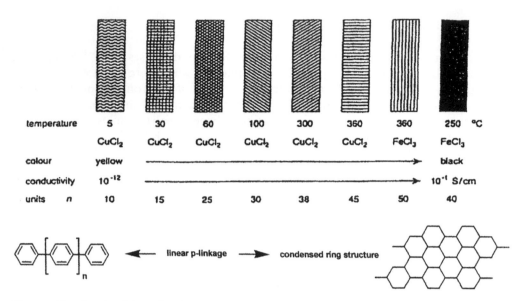

Figure 13 Conductivity of poly(phenylenes) as a function of number of aromatic units *n* and synthesis conditions [5].

XXXVI. POLY(PHENYLENE)S WITH ALTERNATING GROUPS

For examples with thiophene [74,116], with bithiophene [117] or with azomethin-units [115] see the cited literature. Substituted phenylenes with 1–6 thienyl groups were described in [114].

XXXVII. MISCELLANEOUS POLYMERS

There has been a flood of literature concerning new electrically conducting polymers [117].

Bridged macrocyclic complexes are mainly derivatives of tetraazaporphyrin or phthalocyanine:

Tetraazaporphyrin   **Phthalocyanine**

The synthesis and doping of bridged phthalocyanine polymers is reported by Venkatachalam in [115]. The bridging ligand can be linked by two σ bonds, by two coordinate bonds, or by one σ and one coordinate bond.

After doping with iodine, conductivities up to 750 S/cm are obtained. Substituted phthalocyanines are soluble and can be cast as films or handled by the Langmuir–Blodgett technique to give ultrathin, well-defined molecular layers. The insoluble powders can only be processed by press sintering.

Organo-metallic conductive polymers (OCP) can be divided in two groups:

1. OCPs without intrinsic interchain metal–metal electronic interaction, e.g., poly(vinylferrocene)
2. OCPs with intrinsic interchain metal–metal electronic interaction, e.g., poly(1,1'-ferrocenylene).

Data concerning, synthesis, doping and structure are reviewed by Nalwa [118] and cited in [115].

XXXVIII. PHENALENE-*m*-COMPLEXES [119]

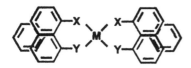

M = Fe, Ni, Pd, Pt, Co, Cu

p-Doped, these complexes reach conductivities up to 10 S/cm, they are used as photovoltaic devices for panels, shielding, etc.

Other electrically conducting systems like bimetallic complexes, thioxolato polymers, polymers with tetrathiafulvalene moieties or poly(phosphazane)s are reviewed in [120].

- Alternative Polymers "pseudo-aromatics", received by polycondensation reaction
- Poly(cyclohexenone): by condensation of vinylmethylketone
- Poly(quinone)s: by condensation of chloanil (tetrachloroquinone).

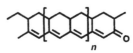

Si or Ge containing conductive polymers, e.g., Si-bridged thiophene macromolecules, α-silylated oligothiophenes, Ge-homopolymers and copolymers. Synthesis routes, structures and application for electroluminescence devices are reviewed in [115]. **Poly(silane)s** become semiconducting upon treatment with AsF_5 [121].

Poly(azine)s and poly(azene)s are nitrogen-containing analogues of poly(acetylene). They are environmentally stable, have also a good thermal stability and react with iodine yielding conductivities about 10^{-2} S/cm.

Synthesis, doping, photochemistry, structure investigations and theoretical consideration are reviewed by W. B. Euler [115]; for a special preparation see [122].

Poly(vinamidine), polymeric malondialdehyde dianils were synthesized by Gompper and coworkers [123].

Ar = aryl

The polymers are air and waterstable and reached, doped with (I_2 or $FeCl_3$), values of 0.2–50 S/cm and represent a novel type of electrically conducting polymers.

Poly(azepine) [124]. Phenyl azide can be photopolymerized in the gas phase to yield films of poly(1,2-azepine) with a conductivity of 10^{-2} S/cm (after doping).

The photolysis reaction is also successful for various substituted azides (trimethyl-, trimethoxyphenyl) yielding substituted polymers.

Poly(indole) [20]. Electrochemical oxidation of indole (counterion ClO_4^- or BF_4^-) yields brittle films with conductivities of *ca.* 0.01 S/cm.

Poly(indole) has found applications as an organic polymer coating. The performance of layered semiconductors has been shown to be improved by the electropolymerization of layers of poly(indole) on the defective sites of the surface. Carbon fibers may be coated with poly(indole) by electropolymerization. More recently, poly(indole) has been employed for the polymer coating for a glucose sensor [125].

Poly(indole) was studied by R. Holze [126] and shows similar properties like poly(aniline).

Poly(carbazole) [20]. Solution of carbazole in acetonitrile may be electrochemically oxidized (counterion ClO_4^- or BF_4^-) at a platinum anode to give electrically conductive films with poor mechanical stability. The polymers obtained by chemical coupling are mores stable. Poly(carbazole) has also been obtained by vacuum evaporation of carbazole and by chemical condensation. Doping with I_2 or $NOBF_4$ leads to conductivities an high as 1 S/cm.

Poly(azulene) [13] is synthesized by electrochemical polymerization with ClO_4^- as counterion (similar to that of poly(pyrrole) yielding amorphous polymer films which can be peeled from the anode. Conductivity is 0.01 S/cm. The films may be electrochemically and reversibly discharged to the nonconducting form.

The simultaneous polymerization and oxidation of azulene with bromine or iodine in acetonitrile have recently been reported. The resultant slightly soluble poly(azulene)-bromine and insoluble poly(azulene)-iodine complexes have lower electrical conductivities than the electrochemically produced polymer, 5×10^{-3} and 10^{-6} S/cm, respectively. Removal of soluble oligomers from the former leads to a slight improvement of the electrical conductivity. [119,127].

Poly(pyrene) [13]. Electrochemical oxidation of pyrene (counterion BF_4^-, ClO_4^- or AsF_6^-) yields insoluble, brittle films with conductivities up to 1 S/cm.

Poly(pyrene) has also been synthesized by Lewis acid oxidative coupling of the monomer. These polymers are contaminated with low-molecular-weight materials and contain extensive polynuclear structures [119].

Alternating pyrene/thiophene polymers were synthesized by Thelakkat [76]. Poly(2,5-furanvinylene) [128].

O or NH

used as electrodes in Li-polymerbatteries

Poly(fulvene)s [13,20]. Poly(fulvene)s are formed by cationic polymerization of 6,6-dimethylfulvene followed by chemical or electrochemical dehydrogenation, conductivity 10^2 S/cm (after I_2 doping).

Poly(indophenine)s [17] are prepared by reaction of isatin with thiophene (Figure 14) and have conductivities up to 10^{-2} S/cm without additional dopant. The polymers are soluble in dimethyl sulfoxide and can be cast as films; the material is thermostable up to 230 °C [129].

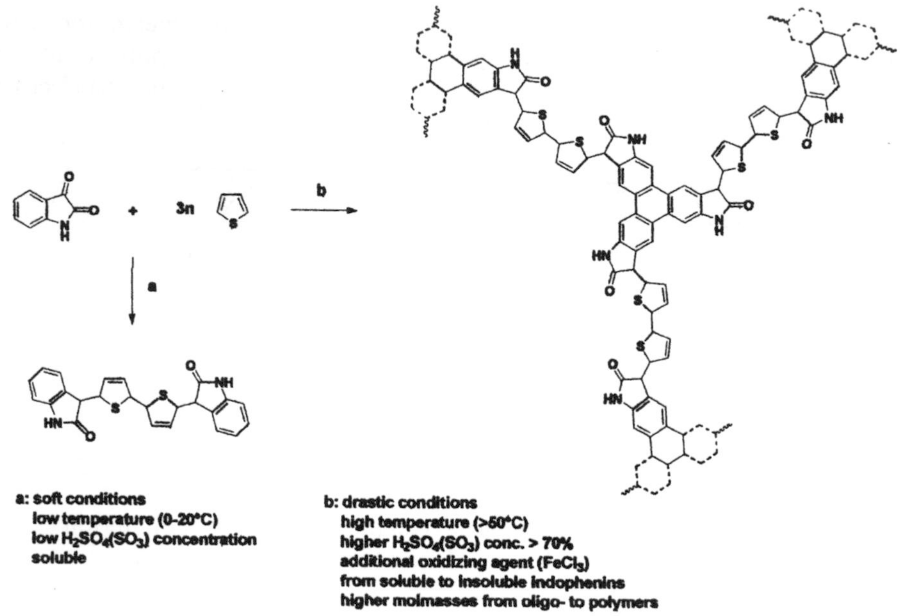
a
b
a: soft conditions
low temperature (0-20°C)
low $H_2SO_4(SO_3)$ concentration
soluble
b: drastic conditions
high temperature (>50°C)
higher $H_2SO_4(SO_3)$ conc. > 70%
additional oxidizing agent ($FeCl_3$)
from soluble to insoluble indophenins
higher molmasses from oligo- to polymers

Figure 14 Isatin/thiophene condensation to indophenins.

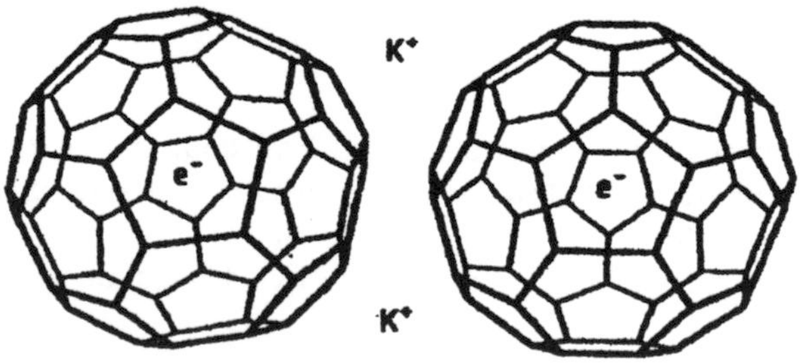
K⁺
e⁻
e⁻
K⁺

Figure 15 Potassium-doped C_{60} molecules.

Types of polycondensation – reactions:

Buckminsterfullerene [130]. Figure 15 is an allotrope of carbon. When doped with potassium, it reaches conductivities up to 500 S/cm; cooling to 18 K makes it superconducting.

In 1985 the fullerenes were discovered by R. F. Carl, H. W. Kroto and R. E. Smalley, and in 1996 their investigations were honored by the nobel prize of chemistry.

Low-band-gap aromatic polymers [131]. An important prerequisite for obtaining low-band-gap polymers is that the ground state of the polymer has a quinonoid contribution. These polymers have the general formula:

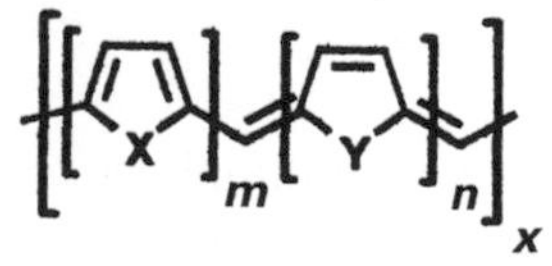

X, Y = SO_2, SO, S, N–R, CH=CH

Benzenoid and anthracenoid precursors are:

or

Low- or narrow-band-gap systems (gap $<0.5\,eV$) already contain a high proportion of thermally excited electrons at room temperature. These materials can be oxidized electrochemically at a low oxidation potential to give conducting polymers. Conductivities up to $10\,S/cm$ are reached (after chemical or electrochemical doping) [13].

Theoretical tailoring of electrically conducting polymers, mentioning thiazole-based polymers and N- and O-containing analogues of poly(isonaphthothiophene) are reported by Bakhshi [132] and earlier [75].

Charge-transfer complexes [133] are combinations of electron donor compounds with electron accepter compounds. They are synthesized, for example, by electrochemical charging of aromatic hydrocarbon with ClO_4^-. The charge-transfer complexes are assembled in defined stacks and have conductivities up to $10^3\,S/cm$, in some cases with superconductivity. Examples are given below:

Bisethylene dithiotetrathiofulvalene – I_3 complex, T_c, 8.1 K [134]

Dicyanodiimine quinone – copper complex, $\sigma = 10^3\,S/cm$ [135].

Finally it should be remembered that all electrically conducting organic materials are due to the various dopant charge-transfer complexes, and coming back to the roots—in the 60s—polymer CT-complexes are found to be electrically conducting materials with up to $10^2\,S/cm$ [6].

XIL. TWO DIMENSIONAL POLYMERS [24]

Various polymerization route with unique starting monomer [64] have been developed, e.g., topochemical poly(diacetylene) synthesis with pyrrole-, thiophene-, –Si-groups; dienophilic addition of maleic imides, Diels–Alder polymers via cyclopentadienone derivatives.

Methods of preparation, properties and characterization are reviewed [129]: 'polymers with extraordinarily electron mobility'.

XL. MAGNETIC ORDER IN CONDUCTING POLYMERS

Poly(1,3-dithienyl-5-phenylverdazyl) is an example of a well characterized, chemically stable organic ferromagnet obtained via a reproducible synthetic route [136].

Ar = aryl

The search for organic polymeric with ferromagnetic properties is currently an area of very intense research activity; the materials would open a complete new field of application namely in combination with electrical conductivity [137].

XLI. CARBON-NANO-TUBES 'THE DERNIER CRI' [138]

CNTs represent a new generation of materials, materials from semiconductors to metals; catalyzed by fullerenes [130] the CNTs are 'nano-wires' for electronic devices, assembled to well defined aggregates they may be used as field transistors.

XLII. APPLICATIONS AND GOALS FOR THE FUTURE [88,121]

During the thirty-five years since their discovery in the 1960s, the understanding of the fundamental chemistry, electrochemistry, physics and processing—also continuously—conducting polymers has already taken place. Early problems associated with insolubility and environmental stability still exist—that's the 'Achilles heel'—but have often been overcome for many applications. We learned to handle instability materials like lithium or the glow incandescent filament in the bulb by using environmental adapted methods.

To the promising groups of applications envisaged back in 1969 [5], viz. sensors, photocells, solar batteries, electrostatic accumulators, photodiodes xergraphy, etc., can be added important new types of piezo ceramic elements, elastic poly(pyrrole) film (elongation > 100%) useful for plasters and tubes, self dopant materials [55], printed circuits, deposition by irradiation [17], laser processing, surface structure modification [139], corrosion protection, and NLO [88].

The class of electrically conducting polymers—a future oriented emerging technology [140,141] represent the basis for a new generation of 'intelligent' materials, not to substitute metals but to open new areas.

ACKNOWLEDGEMENTS

This chapter is a short resumé of my many years of research activity (1960–95 at BASF) and represents a selection from more than 500 patents and 100 scientific publications as well as numerous internal reports. I want to thank my colleagues at BASF for their efficient teamwork, especially Chem. Ing. Hellwig (model substances). Dr. Haberkorn (structural investigations), Dr. Heckmann (morphology), Dr. Denig (analysis), Dr. Simak (IR spectroscopy), Dr. Voelkel (NMR spectroscopy), Dr. Schlag (conductivitiy measurements), Dr. Naegele (electrochemistry), Dr. Penzien/Dr. Köhler (synthesis of poly-(acetylene), poly(pyrrole), etc.) Dr. Theophilou (stretched poly(acetylene)), Dr. Cosmo (modified poly(acetylene)), Dr. Martinez (poly(thiophene)), Dr. Kallitsis (terphenylenes, Dr. Lang/Dr. Hmyene (verdazyles, ferromagnetic materials) and Dr. van Eyk/Dr. Thelakat (substituted pyrroles and thiophenes).

Special thanks are also due to my partners in the BMFT projects, Professors Wegner, Paasch, Rentsch, Hörhold, Hünig, Gompper, Müllen, Hanack, Schwoerer and Dorman and their colleagues, for fruitful co-operation. Thank also to Dr. F. Beck (now Professor in Duisburg), who measured the electrical conductivity of polyconjugated systems at the beginning of the 1960s.

REFERENCES

1. Katon, J. E. (1968). *Organic Semiconducting Polymers*, Marcel Dekker, New York.
2. Conductive polymers, The Nobel Prize in Chemistry, 2000 Kungl. Vetenskapsakademie, Inform. Dept. E-mail: info@kva.se. Web site: www.kva.se.
3. Goppelsroeder, F. (1891). Die International Elektrochemische Ausstellung *18*: 978; *19*: 1047.
4. Pohl, H. A. (1968): Electronic Behavior of Organic Macromolecular Solids, Bolto, B. A.: Semiconducting Organic Polymers Containing Metal Groups, and D. D. Eley: Semiconducting Biological Polymers in *Organic Semiconducting Polymers* (Katon, J. E., ed.), Marcel Dekker, New York.
5. Naarman, H., and Beck, F.: Neuartige Polymerisate aus aromatischen und heterocyclischen Verbindungen und ihre elektrophysikalischen Eigenschaften, GDCh, Meeting, Munich Oct. 12, 1964; Naarmann, H., Beck, F., and Kastning, E. G. (1964). DE 1178529 to BASF AG, C.A. (1964), *62*: 19664; Naarmann, H. (1969). *Naturwissenschaften 56*: 308.
6. Naarmann, H. (1969). *Angew. Chem. Int. Ed. Engl.*, *8*: 915; Willersinn, H., Naarmann, H., and Schneider, K. (1969). DE 1953898 to BASF AG, C. A. (1969), *75*: 37498; Naarmann, H., Conducting Polymers, (1990). *Adv. Mater.*, *2*: 345.
7. Naarmann, H. (1964). DE 1179715 to BASF AG, C.A. (1964), *62*: 3430.
8. Little, W. A. (1964). *Phys. Rev.*, *135*: A1416.
9. Heeger, A. J., Coleman, L. B., Cohen, M. J., Sandmann, D. I., Yamagishi, F. G., and Garito, A. F. (1973). *Solid State Commun.*, *12*: 1125.
10. Greene, R. Street, G. B., and Süter, L. J. (1975). *Phys. Rev. Lett.*, *34*: 577.
11. Chiang, C. K., Fincher, C. R. Jr., Park, Y. W., Heeger, A. J. Shirakawa, H., Louis, E. J., Gau, S. C., and MacDiarmid, A. G. (1977). *Phys. Rev. Lett.*, *39*: 1098.
12. Ito, T., Shirakawa, H., and Ikeda, S. (1974). *J. Polym. Sci., Polym. Chem. Ed.*, *12*: 11.
13. Naarmann, H. (1992). Polymers, Electrically Conducting, In *Ullmann's Encyclopedia of Industrial Chemistry*, Vol. A 21, VCH, Weinheim, pp. 429.
14. Unpublished results, personal comm. by H. Naarmann.
15. Bartl, A., Dünsch, I., Schmeisser, D., Goepel, W., and Naarmann, H. (1995). *Synth. Met.*, *69*: 389.

16. Menke, K., and Roth, S. (1986). *Chemie in unserer Zeit*, *20*: 33; Pekker S., and Janossy, A. (1986). *Handbook of Conducting Polymers* (Skotheim, T. A., ed.), Marcel Dekker, New York, pp. 45.
17. Naarmann, H. (1997). *Handbook of Organic Conductive Molecules and Polymers*, Vol. 3 (Nalwa, H. S., ed.), John Wiley, New York, p. 98.
18. Bozovic, I. (1987). *Mod. Phys. Lett. B*, *1*: 81; Naarmann, H. (1987). *Synth. Met.*, *17*: 225.
19. Minxie, Q. (1984). *Synth. Met.*, *30*: 1.
20. Naarmann, H. (1992). *Handbook of Polymer Synthesis*, Part B (Kricheldorf, H. R. ed.), Marcel Dekker, New York, pp. 1353, 1362, 1363.
21. Reppe, W. (1948). *Justus Liebig Ann. Chem.*, *560*: 140.
22. Hatano, M., Kambera, S., and Shigeharu, O. (1961). *J. Polym. Sci.*, *51*: S26.
23. Haberkorn, H., Naarmann, H., Penzien, K. Schiag, I., and Simak, P. (1987). *Synth. Met.*, *5*: 51.
24. Naarmann, H., Müllen, K., Wegner, G., Hanack, M., Schwoerer, M., and Dormann, E. BMFT Forschungsbericht 03M4019, Polymere mit zweidimensionalen Strukturen, Februar 1991.
25. Naarmann, H. (1964). DE 1179715 to BASF AG, C.A. (1964), *62*: 3430; DE 1197228, C.A. (1965), *63*: 64028.
26. Edwards, H. H., and Feast, W. J. (1980). *Polymer*, *27*: 595; Feast, W. J. (1986). *Handbook of Conducting Polymers* (Skotheim, T. A., ed.), Marcel Dekker, New York, pp. 35.
27. Swager, T. M., Dougherty, D. A., and Grubbs, R. H. (1988). *J. Am. Chem. Soc.*, *110*: 2973; Swager, T. M., Dougherty, D. A., and Grubbs, R. H. (1988). *J. Am. Chem. Soc.*, *777*: 4413.
28. Kanabe, M., and Okawara, M. (1968). *J. Polym. Sci., Polym. Chem. Ed.*, *6*: 1058.
29. Reibel, D. Nuffer, R., and Mathis, C. (1992). *Macromolecules*, *25*: 7090.
30. Ginsburg, E. J., Gorman, C. B., Grubbs, R. H., Klavetter, F. L., Lewis, N. S., Marder, S. R., Perry, I. W., and Sailor, M. J. (1990). *NATO ASI Ser.*, *182*: 65.
31. Haberkorn, H., Heckmann, W., Köhler, G., Naarmann, H., Nickl, J., Schlag, J., and Simak, P. BMFT Forschungsbericht 03C 1340, July 1985, I pp. 1–29, II pp. 1–75 BASF AG.
32. Schimmel, T., Glaeser, D., Schwoerer, M., and Naarmann, H. (1991). *Synth. Met. 41*: 19; Naarmann, H. and Theophilou, N. (1987). *Synth. Met.*, *22*: 1.
33. Naarmann, H. (1990). *Conjugated Polymeric Materials* (Breda, J. L., and Chance, R. R., eds.), Kluwer Academic, Dordrecht, pp. 11.
34. BASF, BMFT-Forschungsbericht: Entwicklung von elektrisch leitfähigen Alternativ-Polymeren, 03 C 134-0, Chap. II, Ludwigshafen, Aug. 1, 1982–July 31, 1985, pp. 1; Naarmann, H. (1987). *Synth, Met.*, *17*: 223; Naarmann, H., and Theophilou, N. (1987). *Synth. Met.*, *22*: 1; Schimmel, T., and Schwoerer, M. (1988). *Solid State Commun.*, *65*: 1311; Tsukamato, J., Takahashi, A., and Kawasaki, K. (1990). *Jpn. J. Appl. Phys.*, *29*: 125.
35. Yamaura, M., Hagiwara, M., Demura, T., and Iwata, K. (1989) *Synth. Met.*, *28*: C157.
36. Gagnon, D., Capistran, J., Karasz, F., and Lenz, R. (1984). *Polym. Prepr. (Am. Chem. Soc., Div. Polym. Chem.)*, *25*: 284; Gangon, D., Karasz, F., Thomas, E., and Lenz, R. (1987). *Synth. Met.*, *20*: 85.
37. Shirakawa, H., Chen, Y.-C., Akagi, K., and Norahara, T. (1986). *Synth. Met.*, *14*: 173, 199; Aldissi, M. (1985). *J. Polym. Sci., Polym. Lett. Ed.*, *23*: 167; Shirakawa, H., Akagi, K., and Katayama, S. (1988). *J. Macromol. Sci. Chem.*, *A25*: 643; Shirakawa, H., Akagi, K., and Suezaki, M. (1989). *Synth. Metals*, *28*: D 1; Moutaner, A. (1989). *Synth. Met.*, *28*: D 19.
38. Naarmann, H., Portugali, M., (1985). BASF AG, DE 3533252, C.A. (1985), *106*: 225801.
39. Naarmann, H. (1991). *Synth. Met.*, *41*: 1.
40. Mac Diarmid, A. G., Woerner, T., Heeger, A. G., and Feldbium, A. (1982). *J. Polym. Sci., Polym. Lett. Ed.*, *20*: 305; (1984). *22*: 119.
41. Shimidzu, T. (1987). *Langmuir*, *3*: 1169.
42. Nakahara, H. (1988). *Thin Solid Films 60, 87*: 153; Watanabe, J., Hong, K., and Rubner, M. F. (1989). *Synth. Met.*, *28* C: 473.
43. Nakomura, T., Tanaka, H., Matsumoto, M., Tachibana, H., Manda, E., and Kawabatoy, Y. (1988). *Synth. Met.*, *27*: B601.

44. Lieser, G., Schmid, S. C., and Wegner, G. (1996). *J. of Microsc.*, *183*: 53.
45. Hopf, H., Kretschmer, O., and Naarmann, H. (1989). *Adv. Mater.*, *1*: 445; H. W. Gibson (1986). Substituted Poly(acetylenes) in *Handbook of Conducting Polymers* (Skotheim, T., ed.), Marcel Dekker, New York, p. 405.
46. Kim, I., and Lee, D. J. (1995). *Synth. Met.*, *69*: 25.
47. Bayer, A., and Landsberg, L. (1882). *Ber. Dtsch. Chem. Ges.*, *15*: 52; Bohlmann, F. (1957). *Angew. Chem.*, *69*: 82.
48. Wegner, G. (1979). *Molecular Metals*, Chap. 4.0, Plenum Publ. (Halfield, E. W., ed.), New York 1979, p. 209.
49. Hädicke, E., Mez, E. C., Krauch, C. H., Wegner, G., and Kaiser, I. (1971). *Angew. Chem. 83*: 253.
50. Dennstedt, M., and Zimmermann, J. (1888). *Ber. Dtsch. Chem., Ges., 21*: 1478; Grossauer, A. (1974). *Die Chemie der Pyrrole*, Springer-Verlag, Berlin, p. 149.
51. Lund, H. (1957). *Acta Chem. Scand.*, *11*: 1323; Stanienda, A. (1967). *Z. Naturforsch.*, *228*: 1107.
52. Kanazawa, K. K., Diaz, A. F., Geiss, R. H., Gill, W. D., Kuak, J. F., Logan, J. A., Rabolt, J. F., and Street, G. B. (1979). *J. Chem. Soc., Chem. Commun., 19*: 854; Diaz, A. F., and Bargon, J. (1986). Electrochemical Synthesis of Conducting Polymers in *Handbook of Conducting Polymers*, Vol. 1 (Skotheim, T. A., ed.), Marcel Dekker, New York, p. 82; Street, G. B. (1986). From Powder to Plastics in *Handbook of Conducting Polymers,* Vol. 1 (Skotheim, T. A., ed.), Marcel Dekker, New York, p. 266.
53. Naarmann, H., Köhler, G., and Schlag, J. (1982). US 4468291 to BASF AG, C.A. (1982), *100*: 93546.
54. Genies, E. M., Bidan, G., and Diaz, A. F. (1983). *J. Electroanal. Chem. Interfacial Soc.*, *149*: 101; (1982). *129*: 1685; Lacroix, I. Ch. et al. (1998). *Chem. Eur.*, *4*: 1667.
55. Naarmann, H., Köhler, G. (1986). BASF, DE 3425511 to BASF AG; C.A. (1986), 104:158211; Patil, A.O., Ikenoue, Y., Wudl, F., and Heeger, A. (1987). *J. Am. Chem. Soc.*, *109*: 1858.
56. Naarmann, H. (1994). *Macromol. Symp.*, *80*: 129; Naarmann, H. (1988). *Angew. Makromol. Chem., 162*: 1; Naarmann, H. (1993). *J. of Polym. Science, Polym. Symp., 75*: 53; Borsdorf, H., and Naarmann, H. (1987). DE 3607302 to BASF AG, C.A. (1987), *108*: 28574: Naarmann, H. (1993). *Intrinsically Conducting Polymers* (Aldissi, M., ed.) Kluver Academic Publ., Netherlands.
57. Garnier, F. (1994). *J. Am. Chem. Soc.*, *119*: 8813; Wang, I. Y. (1994). *Proc. Natl Acad. Sci. USA., 91*: 3201.
58. Investigation with BASF Poly(pyrrole) films by Th. Dandekar, Europ. Molec. Biolog. Laboratory.
59. Jérome, C., Mertens, M., Martinot, L., Jérome, R., Strivay, D., and Weber, G. (1998). *Radiochim. Acta, 80*: 193.
60. Naarmann, H. (1991). *Sience and Applications of Conducting Polymers* (Salaneck, W. R., ed.) Adam Hilger, Bristol, p. 82.
61. Hanack, M., Naarmann, H., and Mattmer, R. (1995). *J. Synth. Org. Chem.*, *5*: 477.
62. Bergmann, J., and Ekklund, N. (1980). *Tetrahedron*, *36*: 14.
63. van Eyk, St. I., Naarmann, H., Nigel, P., and Walker, P. C. (1993). *Synth. Met.*, *58*: 233.
64. Naarmann, H. (1994). *Frontiers of Polymers and Advanced Materials* (Prassad, P. N. ed.), Plenum Press New York, p. 333.
65. Meyer, V. (1883). *Ber. Dtsch. Chem. Ges.*, *16*: 1465.
66. Tourillon, G. (1986). Poly(thiophene) and its Derivatives in *Handbook of Conducting Polymers*, Vol. 1 (Skotheim, T. A., ed.), Marcel Dekker, New York, p. 293.
67. Martinez, F., Voelkel, R., Naegele, D., and Naarmann, H. (1989). *Mol. Cryst. Liq. Cryst.*, *167*: 227.
68. Tourillon, G., and Garnier, F. (1982). *J. Electroanal. Chem.*, *134*: 173; Garnier, F., Tourillon, G., Gazard, M., and Dubois, J. (1983). *J. Electroanal. Chem. Interfacial Electrochem.*, *148*: 299; Tourillon, G. (1986). *Handbook of Conducting Polymers* (Skotheim, T. A., ed.), Marcel Dekker, New York, p. 293.

69. Sato, M., Tanaka, S., and Kaeriyama, K. (1987). *Makromol. Chem., 188*: 1763; Hotta, S., Rughooputh, S., Heeger, A., and Wudi, F. (1987). *Macromolecules, 20*: 212; Nowak, M., Rughooputh, S., Hotta, S., Heeger, A., and Wudi, F. (1987). *J. Polym. Sci., Polym. Phys. Ed., 25*: 1071; Rughooputh, S., Hotta, S., Heeger, A. and Wudi, F. (1987). *J. Polym. Sci., Polym. Phys. Ed., 25*: 1071; Hotta, S., Rughooputh, S., and Heeger, A. (1987). *Synth. Met., 22*: 79; Patil, A., Ikenoue, Y., Wudi, F., and Heeger, A. (1987). *J. Am. Chem. Soc., 109*: 1858.

70. Schön, J. H., Dodabaladur, A., Bao, Z., Kloc, Ch., Schenker, O., and Batlogg, B. (2001). *Nature, 410*: 189.

71. Yamamoto, T., Sanechika, K., and Yamamoto, A. (1980). *J. Polym. Sci. Polym. Lett. Ed., 18*: 9; Lin, J., and Dudeck, P. (1980). *J. Polym. Sci. Polym. Chem. Ed., 18*: 2869; Yamamoto, T., Sanechika, K., and Yamamoto, A. (1980). *J. Polym. Sci. Polym. Lett. Ed., 18*: 9; Yamamoto, T., Sanechika, K., and Yamamoto, A. (1983). *Bull Chem. Soc. Jpn., 56*: 1503.

72. Kobayashi, M., Chung, C., Moraes, F., Heeger, A. and Wudi, F. (1984). *Synth. Met., 9*: 77.

73. Yoshino, K., Nakajima, S., Onoda, M., and Sugimoto, R. (1989). *Synth. Met., 250*: 349.

74. Martinez, F., Retuert, I., and Neculqueo, G., and Naarmann, H. (1995). *Intern. J. Polymeric Mater., 28*: 51.

75. Köhler, G., and Naarmann, H. (1986). DE 3502937, C.A. (1986), *105*: 180519 and DE 3435947 to BASF AG, C.A. (1986), *105*: 53075, (Dithienylbenzene, Dihydroisothio-naphthene).

76. Thelakkat, M., and Naarmann, H. (1995). *Synth. Met., 68*: 153.

77. Kaeriyama, B. (1997). *Handbook of Organic Conductive Molecules and Polymers*, Vol. 2 (Nalwa, H. S. ed.), John Wiley, New York, p. 105.

78. Bäuerie, P. (1998). *Oligothiophenes, Electronic Materials: The Oligomer Approach* (Müllen, K., Wegner, G., eds.), Wiley-VCH, Weinheim, p. 105.

79. Chosrovian, H., Rentsch, S., Dahm, D. U., Birckner, E., and Naarmann, H. (1993). *Synth. Met., 60*: 23.

80. Oeter, D., Ziegler, Ch., Goepel, W., and Naarmann, H. (1993). *Synth. Met., 67*: 267.

81. Egelhaaf, H. J., Bäuerle, P., Rauer, K., Hoffmann, V., and Oelkrug, D. (1993). *J Mol. Struct., 293*: 249.

82. Winter, I., Hormes, J., and Hiller, M. (1995). *Nucl. Instrum. Phys. Res., B 97*: 287.

83. Scherf, V., and Müllen, K. (1993). DE 4331401.5, C.A. (1993), *123*: 170626; Friend, R. (1993). Minisymposium "Polymer LEDs" Eindhoven, 1993 D. D. C. Bradley, Internat. Conf. of Luminescence, Storrs, CT, 1993.

84. Hörhold, H. H., Helbig, M., Raabe, D., Scherf, U., Stockmann, R., and Weiß, D. (1987). *Z. Chem., 27*: 126.

85. Personal information by the author.

86. Schwoerer, M. (1994). *Phys. Bl., 50*: 52.

87. Winstel, G. (1987). Electroluminescent Materials and Devices in *Ullmann's Encyclopedia of Industrial Chemistry*, Vol. 19, VCH, Weinheim, p. 255.

88. Stenger-Smith, I. D. (1998). *Prog. Polym. Sci., 29*: 57.

89. Friend, R. H. and Holmes, A. B. (1994). *Synth. Met., 64*: 3.

90. Shacklette, L. W., Elsenbaumer, R. L., Chance, R., and Eckhardt, H. (1981). *J. Chem. Phys., 75*: 1919.

91. Macallum, A. (1948). *J. Org. Chem., 13*: 154.

92. Edmonds, S., and Hill, H. Jr. (1971). Philips Petroleum, US 3607843, C.A. (1971), *76*: 15187.

93. Lenz, R. W., Handlorits, C. E., and Smith, H. A. (1962). *J. Polym. Sci., 58*: 351.

94. Port, A. B., and Still, R. (1979). *Polym. Degrad. Stab.*, 7: 133.

95. Lovell, P. and Still, R. (1987). *Makromol. Chem., 755*: 1561.

96. Zuk, A., Wejchan-Judek, M., and Rogal, E. (1978). *Polymer, 79*: 438.

97. Viswanathan, P., and Vasudevan, P. (1982). *Angew. Makromol. Chem., 102*: 17; Wejchan-Judek, M., Rogal, E., and Zuk, A. (1981). *Polymer, 22*: 845.

98. Sergeyev, V., and Nedelkin, V. (1986). *J. Polym. Sci., Polym. Chem. Ed., 24*: 3153.

99. Kamkina, M., Annenkova, V., Chaleyllen, A., Abrazova, O., and Voronkov, M. (1988). *Vysokomol. Soedin., Ser. B, 30*: 620.

100. Tsuchida, E., Yamamoto, K., Nishide, M., and Yoshida, S. (1987). *Macromolecules, 20*: 2030; Tsuchida, E., Nishide, H., Yamamoto, K., and Yoshida, S. (1987). *Macromolecules, 20*: 2315.
101. Tsukamoto, I., and Matsumura, K. (1984). *Jpn. J. Appl. Phys., 23*: 584.
102. Elsenbaumer, R. L., and Schacklette, L. W. (1986). *Handbook of Conducting Polymers*, (Skotheim, T. A. ed.), Marcel Dekker, New York, p. 214.
103. Kallitsis, J. K., and Naarmann, H. (1992). *Makromol. Chem., 193*: 2345; Nastopoulos, V., Kallitsis, J. K., Naarmann, H., Dideberg, O., and Dupout, L. (1997), *Acta Cryst., C53*: 248.
104. Genies, E. (1988). *New Journal of Chemistry, 12*: 184; Naarmann, H. (1989). Brite Report RI IB 0109-DB.
105. Honzl, J., and Metalova, M. (1969). *Tetrahedron, 25*: 3641.
106. Everaerts, A., Roberts, S., and Hall, H. K. (1968). *J. Polym. Sci., Polym. Chem. Ed., 24*: 1703.
107. Travers, J. P., and Netschein, M. (1987). *Synth. Met., 21*: 135.
108. Bloor, D., and Monkman, A. (1987). *Synth. Met., 21*: 175.
109. Genies, E. M., and Hany, P., and Santier, Ch. (1989). *Synth. Met., 28C*: 647.
110. Heeger, A. J. (pp. 1–12, 105–115), Genies, E. (pp. 93–104), MacDiarmid, A. G., and Epstein, A. J. (pp. 117–127) (1990). *Science and Application of Conducting Polymers* (Salaneck, W. R., Clark, D. T., and Samuelsen, E. J., eds.), Adam Hilger, New York; Buret, R., Desagher, S., Jozefowicz, M., Perichon, J., and Yu, L. T. (1968). *Electrochim. Acta, 13*: 1441, 1451.
111. Jozefowicz, M., Yu, L. I., Perichon, J., and Buret, R. (1969). *J. Polym. Sci., Part C, 22*: 1187; Buret, R., Desagher. S., Jozefowicz, M., Perichon, J., and Yu, L. T. (1968). *Electrochimica Acta, 13/2*: 1441, 1451; Jozefowicz, M., Perichon, J. H., Tseyu, L., and Buret, U. R. E. (1970). Brit. Pat. 1,216.549, C.A. *73*: 51728; Syed, A. A., Dinesan, M. K. (1992) *React. Polym., 17*: 145.
112. Huang, W.-S., Humphrey, B. D., and MacDiarmid, A. G. (1986). *J. Chem. Soc., Faraday Trans., 1*: 2385; Syed, A. A., Dinesan, M. K. (1990). *Synth. Met., 36*: 209.
113. Kricheldorf, H. R., and Schwarz, G. (1992). *Handbook of Polymer Synthesis*, Part A (Kricheldorf, H. R., ed.), Marcel Dekker, New York.
114. Kallitsis, J. K., and Naarmann, H. (1992). DE 4223810.2 to BASF AG, C.A. (1992), *120*: 217259.
115. Venkatachalam, S. et al. (1997). *Handbook of Organic Conductive Molecules and Polymers*, Vol. 2 (Nalwa, H. S., ed.), John Wiley, New York, p. 741.
116. Ruiz, J. R., and Reynolds, J. R. (1991). *Synth. Met., 41*: 783.
117. Reynolds, J. R., Ruiz, J. P., Child, A. D., Nayak, K., and Marynick, D. S. (1991). *Macromolecules, 24*: 678; Shiroka, Y. (1997). *Functional Monomers and Polymers* (Takemoto, K., Offenbrite, R. M., Kamachi, M., eds.), Marcel Dekker, New York, p. 117.
118. Nalwa, H. S. (1990). *Appl. Organometal. Chem., 4*: 91.
119. Franz, K. D., Münch, V., Penzien, K., and Naarmann, H. (1983). US 4410693, C.A. (1983), *100*: 8860 and US 4468509 to BASF AG, C.A. (1984), *101*: 24124.
120. Naarmann, H., and Theophilou, N. (1988). Electroresponsive Molecular and Polymeric Systems, (Skotheim, T. A., ed.), Marcel Dekker, New York, p. 2.
121. West, R. (1986). *J. Organomet. Chem., 300*: 327.
122. Würthwein, W., Buhmann, K., and Naarmann, H. (1992). DE 4223264.3 to BASF AG, CAN (1992), *20*: 269854.
123. Gompper, R. Müller, Th. I., and Polborn, K. (1998). *J. Mater. Chem., 8*: 2011.
124. Meijer, A. W., Nijhius, S., Van Vroomhoven, F., and Havinga, E. (1989). *Conjugated Polymeric Materials*, Vol. 182 (Bredes, L. J., and Chance, R. R., eds.), Klüver Acad. Publ., Dordrecht, p. 115.
125. Pandey, P. C. (1988). *J. Chem. Soc., Faraday Trans. 1, 84*: 2259.
126. Holze, R., and Lippe, J. (1992). Dechema-Monographie 125, 679.
127. Neoh, K. G., Tang, E. T., and Tan, T. C. (1988). *Polym. Bull., 79*: 325.
128. Nickl, J., Möhwald, H., and Naarmann, H. (1985). DE 3409655 to BASF AG, C.A. (1985), *111*: 118184.

129. Polymers with extraordinarily electron mobility Rentsch, S., Paasch, G., Dormann, E., Schwoerer, M., Hanack, M., Hörhold, H. H., Wegner, G., Müllen, K., and Naarmann, H. BMFT Report 03M40458 Sept. 1994; Köhler, G., and Naarmann, H. (1986). DE 3618838 to BASF AG; C.A. (1986), *106*: 102868.

130. Sleight, A. W. (1991). *Nature, 350*: 557.

131. Hanack, M., Hieber, G., Dewald, G., and Ritter, H. (1990). *Science and Application of Conducting Polymers* (Salaneck, W. R., Clark, D. T., and Samuelsen, E. J., eds.), Adam Hilger, New York, p. 153.

132. Bakhshi, A. K. (1995). *Mat. Sci and Eng., C4*: 249.

133. Billingham, N. C., and Calvert, P. D. (1989). *Adv. Polym. Sci., 90*: 4; Goddings, E. P. (1976). *Endeavour*, 34; 125; (1988). *Synth. Met., 27*: 1.

134. Bechard, K., and Jerome, D. (1982). *Spektrum der Wissenschaft 9*: 38; Jerome, D. (1988). *Synth. Met., 21A*: 183.

135. Hünig, S., Sinzger, K., Jopp, M., Bauer, D., Bietsch, W., von Schütz, J. V., Wolf, H. C., Kremer, R. K., Metzenthin, T., Bau, R., Khan, S. J., Lindbaum, A., Lengauer, C. L., and Tillmanns, E. (1993). *J. Am. Chem. Soc., 115*: 7696.

136. Hmyene, M., Naarmann, H., Winter, H., Pilawa, B., and Dormann, E. (1994). *J. Phys. Condens. Matter, 6*: L511.

137. Cosmo, R., Dormann, E., Gotscha, B., Naarmann, H., and Winter, H. (1991). *Synth. Met., 41*: 369; Dormann, E., and Winter, H. (1993). *Magnetism in Organic Materials, Physica Scripta*, Vol. T 49, p. 731; Dormann, E., Polymere mit besonderen Eigenschaften im Hinblick auf Ferromagnetismus, BMFT Projektnr. 03M4067-6 1/1994.

138. Lieber, Ch., Avouris, Ph., and Remskar, M. (2001). *Science, 292*: 479, 702, 706.

139. Bargon, R., and Baumann, R. (1993). *Microelectronic. Eng., 20*: 55; Phillips, H. M., Smagling, M. C., and Sawerbrey, R. (1993). *Microelectron. Eng., 20*: 73.

140. Naarmann, H. (1993). *J. Polym. Sci., Polym. Symp., 75*: 53.

141. Bakhshi, A. K., and Rattan, P. (1997). *Curr. Sci., 73*: 8, 648.

13
Photoconductive Polymers

P. Strohriegl
Universität Bayreuth, Makromolekulare Chemie I, and Bayreuther Institut für Makromolekülforschung (BIMF), Bayreuth, Germany

J. V. Grazulevicius
Kaunas University of Technology, Kaunas, Lithuania

I. FOREWORD

Since 1992 when the first edition of the *Handbook of Polymer Synthesis* was published a number of new applications for photoconductive polymers or, to put it correct, charge transport materials, have appeared. The most successful development are organic light emitting diodes (OLEDs) which right now enter the market as bright displays for cellular phones and car radios. Other imortant areas are organic field effect transistors, solar cells and lasers.

For this reason the review has been thoroughly updated mainly in the Sections V.B and V.C which deal with conjugated polymers, a very active research area in which A. Heeger, A. McDiarmid and H. Shirakawa received the Nobel Prize in 2000. A large number of new polymers and up-to-date references have been included.

II. INTRODUCTION

Photoconductivity is defined as an increase of electrical conductivity upon irradiation. According to this definition photoconductive polymers are insulators in the dark and become semiconductors if illuminated. In contrast to electrically conductive polymers photoconductors do not have free carriers of charge. In photoconductors these carriers, electrons or holes, are generated by the action of light. The carriers of electricity can also be photogenerated extrinsically in an adjacent charge generation layer, and injected into the polymer which in this case acts as a charge transporting material.

Only polymers capable of both producing charge carriers upon exposure to light and transporting them through the bulk are true photoconductors. Polymers that do not absorb the incident light but accept charges generated in an adjacent material are merely charge transport materials.

The discovery of photoconductivity dates back to 1873 when W. Smith found the effect in selenium. Based on this discovery C. F. Carlson developed the principles of the xerographic process already in 1938. Photoconductivity in polymers was first discovered in 1957 by H. Hoegl [1,2]. He found that poly(N-vinylcarbazole) (PVK) sensitized with suitable electron acceptors showed high enough levels of photoconductivity to be useful in practical applications like electrophotography. As a result of the following activities IBM introduced its Copier I series in 1970, in which an organic photoconductor, the charge transfer complex of PVK with 2,4,7-trinitrofluorenone (TNF), was used for the first time [3]. The photoconductor was a 13 μm single-layer device. It was prepared by casting a tetrahydrofuran solution containing PVK and TNF onto an aluminum substrate [4]. Since then numerous photoconductive polymers have been described in literature and specially in patents. The ongoing interest in photoconducting polymers is connected with an increasing need for low cost, easy to process and easy to form large area materials.

The polymeric photoconductors used in practice are based on two types of systems. The first one are polymers in which the photoconductive moiety is part of the polymer, for example a pendant or in-chain group. The second group involves low molecular weight chromophores imbedded in a polymer matrix. These so called molecularly doped polymers are widely used today. Almost 100% of all xerographic photoreceptors at present are made of organic photoconductors [5]. The main area of application of polymeric photo-conductors is electrophotography [6]. Photoconductive polymers are used in photocopiers, laser printers, electrophotographic printing plates, and electrophotographic microfilming. During the last decade, photoconductive or more precisely charge transporting polymers have been widely used in photorefractive composites [7] and in organic light emitting diodes (OLEDs) [8,9]. An upcoming field for the application of charge-transporting polymers are photovoltaic devices [10,11].

The process of electrophotography is schematically shown in Figure 1. It is a complex process involving at least five steps [12].

1. *Charge.* In the first step the surface of the photoconductor drum is uniformly charged by a corona discharge.
2. *Expose.* Parts of the photoconductor are discharged by light reflected from an image. So the information is transferred into a latent, electrostatic image on the surface of the photoconductor.
3. *Develop.* Electrostatically charged and pigmented polymer particles, the toner, are brought into the vicinity of the oppositely charged latent image transforming it into a real image.
4. *Transfer.* The toner particles are transferred from the surface to a sheet of paper by giving the back side of the paper a charge opposite to the toner particles.
5. *Fuse.* In the last step the image is permanently fixed by melting the toner particles to the paper between two heated rolers. The photoconductor drum is cleaned from any residual toner and is ready for the next copy.

Organic electrophotographic photoreceptors are also widely used in laser printers [13,14]. The principal of these printers is almost the same as in a photocopier except the direct generation of the image by a laser instead of the optical system in a copier. Photoreceptors of the laser printers have to absorb in the near infrared range of spectrum. The third area in which photoconductive polymers or polymer composites are applied are electrophotographic printing plates.

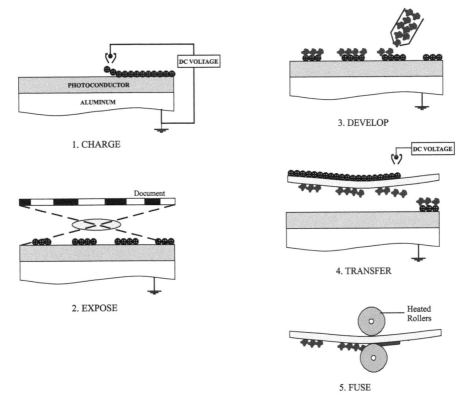

Figure 1 Principles of the xerographic process (for explanations see text).

The first comprehensive reviews on photoconductive polymers were published by Stolka alone [15] and in co-authorship with Pai [16]. Chemical aspects of the topic were later reviewed by several authors [17–19]. In the work of Mylnikov photoconductivity of polymers was reviewed within the framework of semiconductor physics [20], whereas Haarer [21] has concentrated mainly on the transport properties of photoconductive polymers. In their comprehensive book, Borsenberger and Weiss described all aspects of photoconductive materials [6].

Photoconductive polymers can be p-type (hole-transporting), n-type (electron-transporting), or bipolar (capable of transporting both holes and electrons). Typically, bipolarity can be accomplished by adding electron-transporting molecules such as TNF to a donorlike, hole-transporting polymer such as PVK. Most of practical photoconductive polymers are p-type, however recently much attention is paid to electron-transporting and bipolar polymers [22].

III. BASIC PRINCIPLES OF PHOTOCONDUCTIVITY

Since the major goal of this chapter is the description of the different classes of photoconductive polymers, the underlying physical principles will be only briefly discussed. For more detailed reviews dealing with photoconductor physics the reader is referred to the literature [21–24].

The process of photoconduction involves several steps [15].

A. Absorption of Radiation

The first step to a charge carrier generation is the absorption of radiation. Photo-conductive materials are truly photoconductive only in the range of wavelength of absorption. Thus PVK is a photoconductor only in the UV range. To produce carriers by visible light sensitizing dyes or electron acceptors forming coloured charge transfer complexes must be added.

B. Generation of Charge Carriers

By the absorption of light the active groups are excited and form closely bound electron–hole pairs. The key process that determines the overall photogeneration efficiency is the following field induced separation into free charge carriers. This process competes with the geminate recombination of the electron–hole pair. A theoretical description of this process is provided by Onsager's [25] theory for the dissociation of ion pairs in weak electrolytes in the presence of an electric field. The model has been successfully applied to amorphous photoconductors [26]. It was found that the photogeneration efficiency, in other words quantum yield of the process, is a complicated function of several variables such as electric field strength, temperature, and separation distance. The predicted relationship is in good agreement with experimental data for doped polymers like N-isopropylcarbazole in polycarbonate [27], triphenylamine doped polycarbonate [28] and PVK [29,30].

The quantum yields in 'pure' photoconductors absorbing in the UV range are usually low and strongly field dependent. So at room temperature and an excitation wavelength of 345 nm the quantum yield Φ for PVK rises from 0.01% at 10^4 V/cm to about 6% at 10^6 V/cm [28]. Substantially higher values for Φ are obtained in the presence of complexing additives like dimethyl terephthalate [31,32]. The addition of suitable electron acceptors which form colored charge-transfer complexes is a proven way to increase the photogeneration efficiency. 2,4,7-Trinitrofluorenone (TNF) in combination with PVK is so effective that the combination was used in the IBM copier I, the first commercial copier with an organic photoconductor.

C. Injection of Carriers

An injection of carriers only occurs if an extrinsic photogenerator is used together with a charge transporting material. Usually dye particles are dispersed in a polymer matrix or evaporated on top of a conductive substrate and then covered with the charge transporting polymer. The carriers are generated in the visible light-absorbing material and injected into the polymer.

D. Carrier Transport

The photogenerated or injected charge carriers move within the polymer under the influence of the electric field. In this process the photoconductive species, for example carbazole groups in PVK, pass electrons to the electrode in the first step and thereby become cation radicals. The transport of carriers can now be regarded as a thermally activated hopping process [33–37], in which the hole hops from one localized site to another in the general direction of the electric field (Figure 2). The moving cation radical can accept an electron from the neighboring neutral carbazole group which in turn becomes a hole, and so on. Effectively the hole moves within the material while electrons

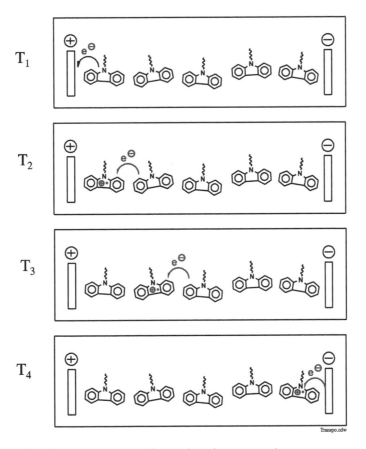

Figure 2 Principles of carrier transport (for explanations see text).

only jump among neighboring species. Hole transport can therefore be described as a series of redox reactions among equivalent groups.

During transit, the carriers do not move with uniform velocity but reside most of the time in localized states (traps) and only occasionally are released from these traps to move in field direction. This trapping process is responsible for the extremely low hole mobilities in photoconductive polymers. For PVK room temperature mobilities from 3×10^{-8} to $10^{-6}\,\mathrm{cm^2/Vs}$ ($E = 10^5\,\mathrm{V/cm}$) have been reported [6]. Since the transport of holes can be described as a series of electron transfer reactions with a certain activation energy it is not surprising that the carrier mobility is temperature- and field-dependent.

IV. EXPERIMENTAL TECHNIQUES

For the characterization of polymeric photoconductors two established methods exist: the Time of Flight (TOF) and the xerographic method. Both methods provide information about the two fundamental parameters that characterize a photoconductive material: carrier mobility μ and quantum yield Φ.

The principle of TOF method is shown in Figure 3. A thin film of photoconductive material is sandwiched between a conductive substrate, for example an aluminized

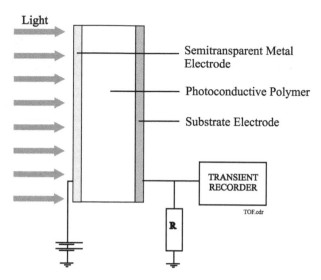

Figure 3 Typical time-of-flight (TOF) setup for measuring hole mobilities in polymers.

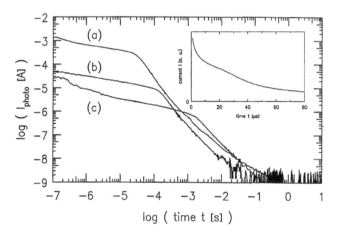

Figure 4 Typical experimental photocurrent of polysiloxane 13 ($m = 3$) at an electric field of 3×10^5 V/cm ($T = 293\,K$). The arrow marks the transit time.

mylar film, and a semitransparent top electrode and connected to a voltage source and a resistor R. Because of the blocking electrodes the source voltage appears across the film. A thin sheet of charge carriers is generated near the top electrode by a short pulse of strongly absorbed light. Due to the influence of the applied field the carriers drift across the sample towards the bottom electrode. The resulting current is measured in the external circuit at the resistor R. A typical experimental photocurrent for the polysiloxane 11c ($m = 3$) with pendant carbazolyl groups is shown in Figure 4 [38].

In the double logarithmic plot of photocurrent versus time the bend at the transit time t_t is clearly detectable. The effective carrier mobility μ is calculated from the transit time according to Equation (1)

$$\mu = d/t_t E \tag{1}$$

where d denotes the sample thickness and E is the electric field strength. With $d = 6.7\,\mu m$, $E = 4.6 \times 10^5\,V/cm$ and a transit time t_t of $2.8 \times 10^{-5}\,\mu s$ an effective carrier mobility of $1 \times 10^{-4}\,cm^2/Vs$ is calculated from Figure 4. Note that for the conjugated trimer (74) with its high mobility the transit time can be seen even in a linear plot of I_{photo} vs. time (inset).

The carrier mobility μ is temperature- and field-dependent. Many theories have been developed to explain the temperature dependence, but no comprehensive model is yet available. It is still not clear whether the charge carrier mobility follows a simple Arrhenius relationship ($\log\mu \cong 1/T$) as predicted by Gill [33] or if the more complex relationship $\log\mu \cong 1/T^2$ proposed by Bässler [39] is valid. The relationship between the mobility μ and the electrical field strength E is equally unclear. Here Gill's model predicts a $\log\mu \cong E^{1/2}$ dependence which is consistent with a Pool–Frenkel formalism, whereas Bässler's calculations lead to a $\log\mu \cong E$ dependence. A detailed description of the different models and results obtained by fitting experimental mobility data to those models is beyond the scope of this chapter. It shall only be pointed out here that the main difficulty is the limited range of temperature and electric field in which carrier mobilities can be measured [38]. Additional experiments are necessary to understand the mechanism of carrier transport in photoconductive polymers in detail.

V. CLASSES OF PHOTOCONDUCTIVE POLYMERS

Several polymer types and classes are known to exhibit photoconductivity. Consequently no preferred method of synthesis exists. The known photoconductive polymers are prepared by almost all common methods like free-radical, cationic, anionic, coordination, and ring-opening polymerization, step-growth polymerization, and polymeranalogous reactions. The only common requirement for all photoconductive materials is that they have to be of extreme purity. It is well known [40–42] that even traces of impurities act as traps and have drastic influence on both quantum yield and carrier mobility.

From the structural point of view the photoconductive polymers described in this chapter can be divided into three groups (Figure 5):

- Polymers with pendant or in-chain electronically isolated photoactive groups with large π-electron systems, for example, aromatic amino groups, like carbazole or condensed aromatic rings, like anthracene
- Polymers with π-conjugated main chain like polyacetylene and poly(1,4-phenylenevinylene)
- Polymers with σ-conjugated backbone, like organopolysilanes

A. Polymers with Pendant or in-Chain Electronically Isolated Photoactive Groups

An aromatic amino group is a common building block of many known photoconductive or charge transporting materials. Many practical systems used in electrophotography belong to this category. The active groups in these materials are either part of the polymer structure or low-molecular dopants imbedded in a polymer matrix. The later group of

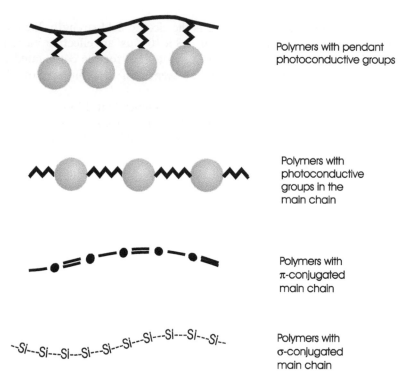

Polymers with pendant photoconductive groups

Polymers with photoconductive groups in the main chain

Polymers with π-conjugated main chain

Polymers with σ-conjugated main chain

Figure 5   Different types of photoconductive polymers.

materials of which numerous examples exist especially in the patent literature will not be discussed here.

1.   Carbazole-Containing Polymers

Since the discovery of photoconductivity in poly(N-vinylcarbazole) (PVK) [1,2] a variety of polymers with carbazole groups have been synthesized and their photophysical properties have been investigated. The main topic of this article is the synthesis of photoconductive polymers, so minor attention is given to their photophysical properties. PVK (2b) can be synthesized by free-radical, cationic, or charge-transfer initiated polymerization of N-vinylcarbazole (2a). A detailed description of the PVK synthesis is given in Chapter 2 of this handbook.

(2)

(a)   (b)

Poly(N-ethyl-2-vinylcarbazole) (Structure 3a) has been prepared by free-radical polymerization, whereas poly(N-ethyl-3-vinylcarbazole) (3b) was synthesized by cationic polymerization with a boron trifluoride initiator [43]. The 2-isomer is reported to exhibit higher carrier mobility than PVK, while that of the 3-isomer is lower [44].

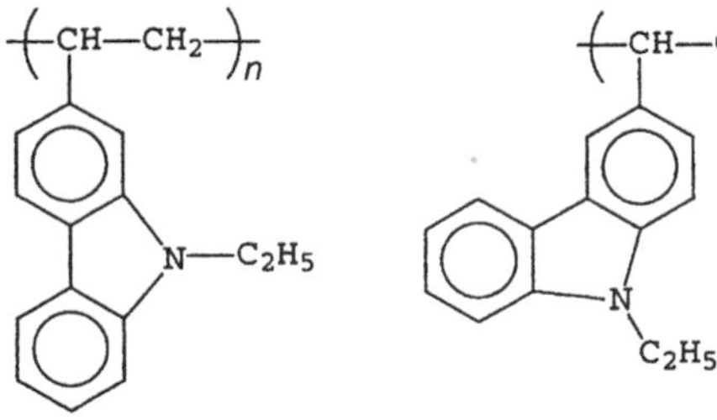

(3)

(a)   (b)

Tazuke and Inoue [45] reported on the synthesis of a polyvinyl derivative having a pendant dimeric carbazole unit, 1,2-trans-bis(9H-carbazol-9-yl)cyclobutane (DCZB). Poly(trans-1-(3-vinyl-)-carbazolyl)-2-(9-carbazolyl)cyclobutane) (4a) was prepared by cationic polymerization of the corresponding monomer with boron trifluoride. The reaction yielded a polymer of relatively high molecular weight ($M_n = 2.5 \times 10^5$, $M_w = 5.8 \times 10^5$). Copolymers of the vinyl derivative of DCZB with N-ethyl-3-vinylcarbazole were also obtained. Fluorescence spectroscopy data have indicated that the polymer (4a) does not form excimers. The photoconductive properties of polymer (4a) as well as of its copolymers have been studied by the xerographic technique, both in the presence and in the absence of the sensitizer TNF [46,47]. The photoconductivity of (4a) is increased compared to PVK when the charge transfer band of the complex is irradiated. Better photoconductive properties of (6a) correlate with its photophysical properties. Excimer formation is sterically hindered by DCZB groups whereas energy migration occurs efficiently in it. Charge transfer interaction with TNF is also stronger for (4a) than for PVK.

Several polyacrylates and polymethacrylates with pendant carbazole groups have been described. Poly(2-(N-carbazolyl)ethyl acrylate) (formula 4b) has been prepared by free radical polymerization of the corresponding monomer [48].

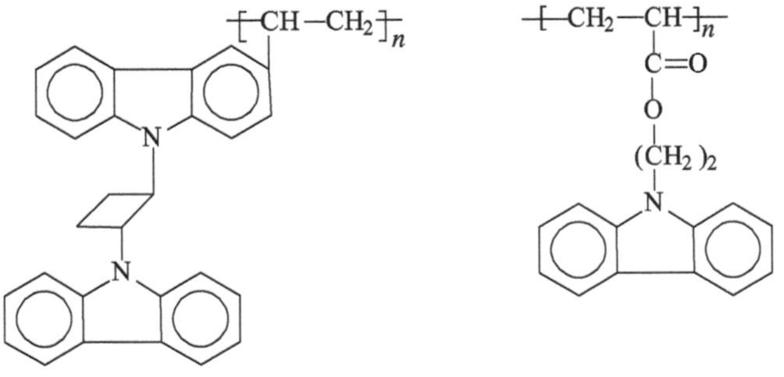

(4)

(a)   (b)

The polymer exhibits a charge carrier mobility of $7 \times 10^{-6}\,cm^2/Vs$ (20 °C, $5 \times 10^5\,V/cm$) which is higher than in PVK. The enhanced carrier mobility in the carbazole containing polyacrylate is apparently due to the lack of excimer-forming sites in it. Polymer (4b) has also been prepared anionically with ethyl magnesium chloride/benzalaceto-phenone as catalyst [49,50] to yield an almost exclusively isotactic product. Due to the insolubility of the polymer in the toluene/diethyl ether mixture in which the polymerization was carried out the molecular weight is low and the product shows a broad molecular weight distribution. Nevertheless time of flight measurements show that the carrier mobility of the isotactic material (1.7×10^{-5} cm^2/Vs, 20 °C, 2×10^5 V/cm) is about six times higher than the mobility of the atactic polymer. The authors concluded that stereoregular structures enhance the hole drift mobility of pendant-type photoconductive polymers. However, the relatively small increase of the measured mobilities should be interpreted with caution because it is well known that even traces of impurities may have a drastic influence on the carrier mobility.

A series of polyacrylates and polymethacrylates (5a) in which the carbazolyl groups are separated from the polymer backbone by alkyl spacers of variable length have been prepared by different methods as shown in the Scheme 5 [51]. The molecular weights of the polymers obtained by free-radical polymerization with AIBN in toluene solution are rather low and all polymers exhibit a broad molecular weight distribution. The reason is the low solubility of the polymers in the polymerization solvent toluene. In more polar solvents like tetrahydrofuran the molecular weight is limited by chain transfer reactions. High-molecular weight poly(meth)acrylates ($M_w = 100{,}000\text{–}150{,}000$, $M_n = 50{,}000\text{–}70{,}000$) were obtained by polymeranalogous reaction of ω-hydroxyalkylcarbazoles with poly(meth)-acryloylchloride. IR and ^{1}H NMR spectroscopy as well as elemental analysis show that the reaction yields poly(meth)acrylates with an almost quantitative degree of substitution.

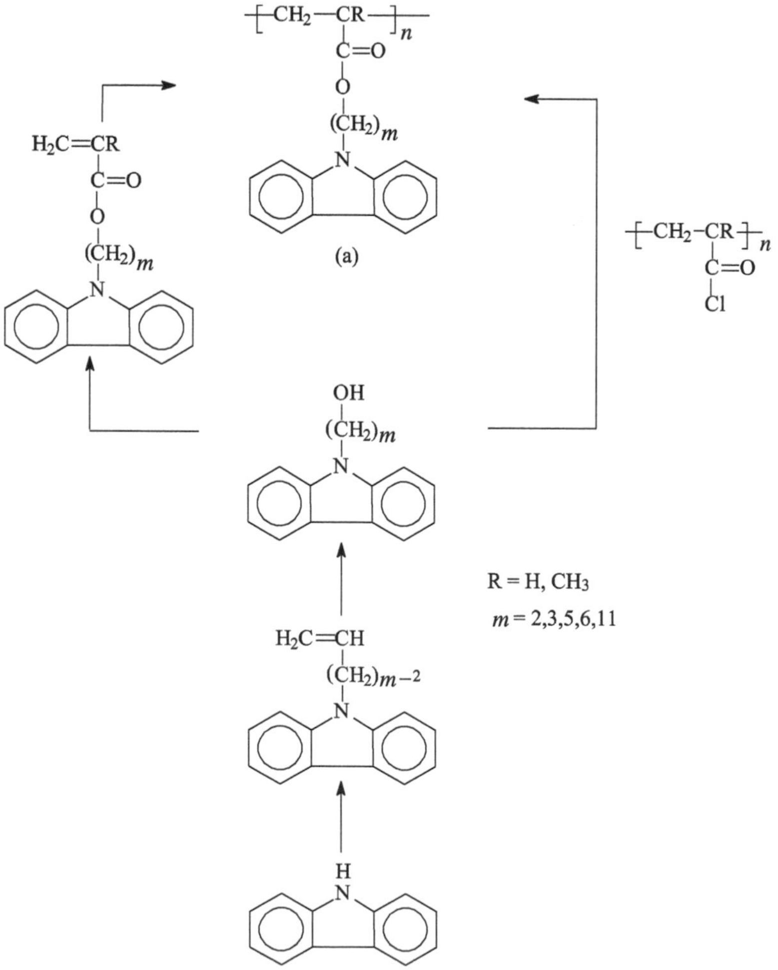

(5)

R = H, CH$_3$

m = 2,3,5,6,11

The polyacrylate (6) with a pendant dimeric carbazole unit, 1,2-trans-bis(9H-carbazol-9-yl)cyclobutane (DCZB), does not show excimer fluorescence and exhibits improved hole drift mobility [52]. It is obtained by free-radical polymerization of the corresponding acrylate [53]. The molecular weight of the polymer (6) established by vapour pressure osmometry is 46,000. The hole drift mobility of polymer (6) is more than ten times higher than that of PVK or poly(9-ethyl-3-vinylcarbazole).

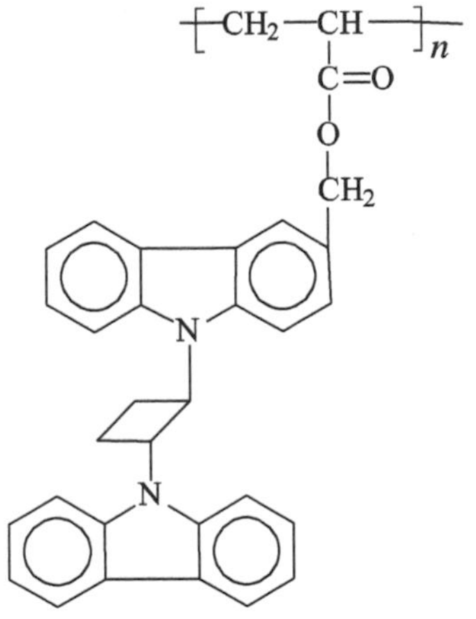

(6)

It was established that the elevated hole drift mobility of DCZB polymers is due to the reduced concentration of trapping sites which are in fact excimer-forming sites. This was confirmed by the temperature and electric field dependencies of the hole mobility. These observations support the idea that charge transport and exciton transport have many features in common [54].

The cationic polymerization of 2-(N-carbazolyl)ethyl vinyl ether with boron trifluoride etherate or ethylaluminum dichloride as initiator has been described by several authors [55–58] (Scheme 7). Low-molar-mass polymers were obtained with both initiators [56]. In the case of boron trifluoride etherate the molecular weight (M_n) was 3160, and the ethyl–aluminum dichloride initiated polymerization yielded poly(2-(N-carbazolyl)ethyl vinyl ether) (11b) with $M_n = 24,500$. At longer reaction times with ethylaluminum dichloride, considerable amounts of insoluble material were formed by cross-linking reactions. The data on the photoconductivity of the polymer (7b) are contradicting. In a steady state measurement Okamoto et al. [55] found that the photocurrent in the polymer (7b) is much lower than that in PVK. However xerographic discharge measurements carried out by Turner and Pai showed that the samples of the polymer (7b) prepared with boron trifluoride as initiator had carrier mobilities only slightly lower than that of PVK [56]. The samples of (7b) prepared with ethylaluminum dichloride showed a high level of charge trapping that stems from impurities in the polymer film.

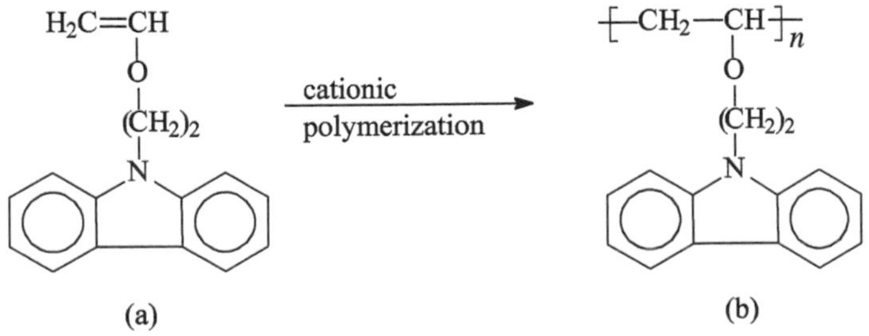

(7)

(a)   (b)

Again, it becomes evident that it is almost impossible to compare the results of photoconductivity measurements from different authors because of the different methods of polymer synthesis, purification, and the varying measurement techniques.

Gaidelis et al. [59] reported that the carrier mobilities of poly-(N-epoxypropyl-carbazole) (PEPK) (8b) are more than an order of magnitude higher than the values reported for PVK. This observation later was confirmed by the data of Wada [60]. Because of this property PEPK can be used as a charge transporting material in xerocopier drums [61,62]. It was also used in electrophotographic microfilming [63]. High-molecular-weight PEPK is prepared by substituting halogen atoms of epihalohydrin polymers with carbazole in organic solvents in the presence of inorganic bases and phenol radical chain inhibitors, like 2,6-di-tert-butyl-p-cresol [64] (Scheme 8).

$$(8)$$

The weight average molecular weight (M_w) of PEPK synthesized by such a method is 440,000.

Oligomeric PEPK was produced industrially according to Scheme (9) [65].

$$(9)$$

Apart from hydroxy end groups PEPK (9b) contains also unsaturated end groups [66]. Propenylcarbazole groups appear in the oligomer during anionic polymerization of the monomer (9a) as the result of a chain transfer reaction [67]. PEPK exhibits the best film forming properties when its molecular weight (M_w) is in the range from 1000 to 1500. The glass transition temperature of an oligomer of such molecular weight is 65–75 °C.

Brominated analogues of PEPK enable to obtain electrophotographic layers of enhanced electrophotographic photosensitivity [68]. The most promising from the point of view of convenience of synthesis and photoactivity among the brominated poly(carbazolyloxiranes) is poly(3,6-dibromo-9-(2,3-epoxypropyl)carbazole) (10a). It is synthesized mainly by cationic ring-opening polymerization of the corresponding oxirane monomer using Lewis acids [69] or triphenylcarbenium salts [70] as initiators. The molecular weight of the oligomers (10a) usually does not exceed 2000. Because of the presence of heavy bromine atoms, the glass transition temperature of these oligomers is higher than that of unbrominated PEPK. Their film-forming properties are usually inferior to those of PEPK. Polymerization via activated monomer mechanism in the presence of diols allows to prepare bifunctional oligomers of 3,6-dibromo-9-(2,3-epoxypropyl)carbazole having hydroxyl end-groups and a flexible oxyalkylene fragment in the main chain [71]. They show high electrophotographic photosensitivity when sensitized and good film-forming properties [72].

Poly((2-(9-carbazolyl)ethoxymethyl)oxirane) (10b) has been synthesized both by cationic polymerization of the corresponding epoxy monomer with Lewis acids [73], triphenylcarbenium salts [74] and by anionic polymerization initiated with KOH [75] or by potassium alkalide, potassium hydride, and potassium tert-butoxide [76]. Since chain transfer reactions to (2-(9-carbazolyl)ethoxymethyl)oxirane are not as intense as in the case of EPK polymerization oligomers (10b) of higher molecular weight can be prepared using both cationic and anionic initiators. Polymerization with potassium hydride yields polymers of a degree of polymerization up to 62. Since the carbazole units in (10b) are removed from the main chain compared to PEPK it has a lower glass transition temperature and exhibits good film-forming properties in a wide range of molecular weights. Xerographic photosensitivity of its layers doped with TNF is lower than that of the corresponding layers of PEPK.

(a)   (b)   (10)

A series of polysiloxanes with pendant carbazolyl groups (11c) have been synthesized by the reaction of poly(hydrogenmethylsiloxane) with various ω-alkenylcarbazoles [77].

(a)   (b) $m = 3,5,6,11$ (c)   (11)

Detailed time of flight measurements [78] have shown that the polysiloxane (11c) with the shortest spacer ($m = 3$) exhibits a carrier mobility which is about one order of magnitude higher than that for PVK. The data of Goldie et al. [79] corroborate this observation. The activation energy for carrier transport derived from the temperature dependence of the carrier mobility is 0.6 eV for all the polysiloxanes and for both PVK and N-isopropylcarbazole in a polycarbonate matrix. The fluorescence spectra [80] of the extremely pure polysiloxanes prepared starting from synthetic carbazole show that these polymers, due to the conformational freedom of the carbazole groups, are free of excimer forming sites.

Thermotropic liquid crystalline side group polymers with carbazolyl groups have been reported by Lux et al. [81]. The idea behind this work was to make a liquid crystalline polymer with a photoconductive mesogenic unit. It should be possible to orient such a polymer by means of an electric or magnetic field at elevated temperatures where it exhibits a mesophase and to freeze this orientation by cooling down below the glass transition temperature. In the polysiloxanes (12) a carbazole group is incorporated into a mesogenic unit. The polymers are prepared by a multistep synthesis the last step of which is the polymer analogous reaction of the mesogenic unit with an alkenyl-terminated spacer and poly(hydrogenmethylsiloxane) [77]. The polymers exhibit broad mesophases, for example polymer (12) with a spacer of three methylene units ($m = 3$) has a glass transition at 69 °C and a smectic mesophase up to the clearing point at 215 °C. Unfortunately, the polysiloxanes show almost no photoconductivity.

(12)

The influence of liquid crystalline media on the hole transport of organic photoconductors has been demonstrated by Ikeda et al. [82]. They have established that DCZB dissolved in polymer liquid crystals showed improved hole drift mobility owing to the orientation of the carrier molecules. The same research group [83] has prepared copolymers of acrylates with side chain mesogens and dimeric carbazoles (13). Incorporation of the DCZB moieties into the copolymers resulted in homogeneous dispersion of carrier groups, but a great extent of destabilization of the liquid crystalline phase was observed. Nevertheless the hole drift mobility was found to be enhanced in copolymer films with more ordered structure of the DCZB moieties, indicating that orientation of the photoconductive groups is favourable for the charge carrier transport.

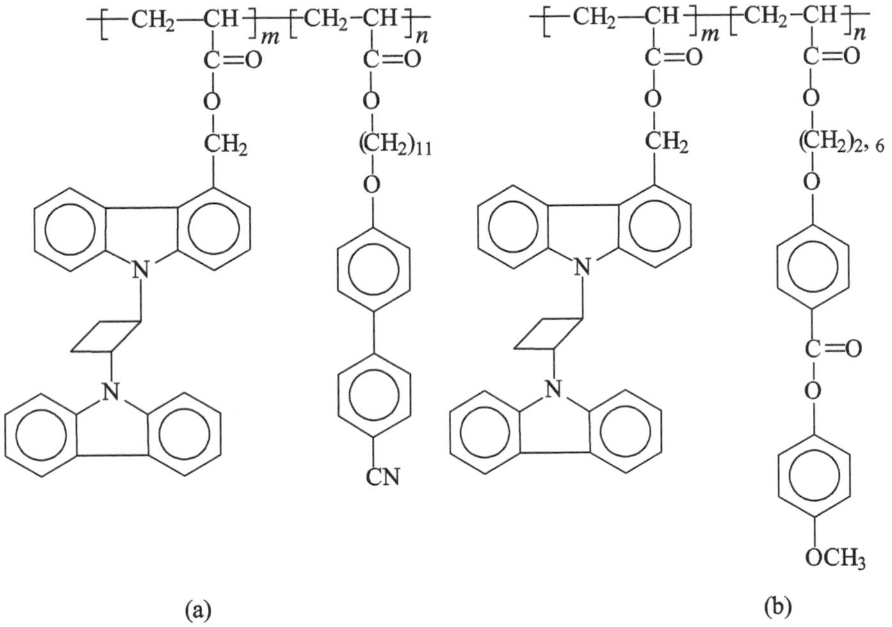

(13)

(a)   (b)

Apart from polymers containing both photoconductive and liquid crystalline side groups a lot of attention has been paid to the synthesis of polymers in which both photoconductive and nonlinear optical chromophores are present. Polymers showing both second-order nonlinear optical and photoconductive properties are photorefractive and have potential application in data storage and image processing as well as in medicine [84]. Carbazolyl-containing photorefractive polymers have been reviewed [85]. An example of such functional polymer is given in the Scheme 14. Tamura et al. [86,87] have synthesized polyacrylates and polymethacrylates having carbazole and tricyanovinylcarbazole side groups. 5-(N-carbazolyl)pentyl methacrylate and acrylate were polymerized using AIBN as an initiator. The resulting polymer was reacted with tetracyanoethylene to tetracyanovinylate ca. 20% of the carbazole units.

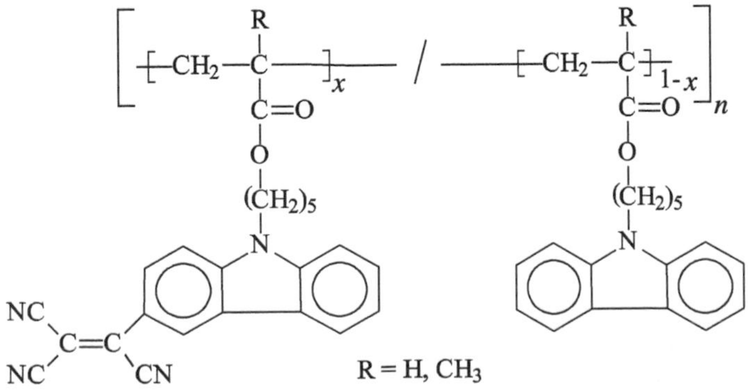

(14)

R = H, CH$_3$

All polymers discussed above have pendant carbazolyl groups. Only few poly-condensates in which the carbazolyl group is part of the main chain have been reported. Tazuke et al. [88–90] have synthesized polyurethanes, poly-Schiff bases and polyamides containing DCZB moieties in the main chain. Polyurethanes containing DCZB moieties in the main chain (15) were prepared by treating trans-1,2-bis(3-hydroxy-methyl-9-carbazolyl)cyclobutane with the corresponding diisocyanate in the presence of dibutyltin dilaurate [88]. The molecular weight of the polymer synthesized using hexamethylene diisocyanate as a linking agent was 2700, and that of the polymer prepared with toluylene diisocyanate was 16,000. Polymers (15) exhibit almost exclusively monomer fluorescence in dilute solution, i.e., they practically have no intramolecular excimer-forming sites. Their complexes with TNF show better photoconductive properties than PVK-TNF.

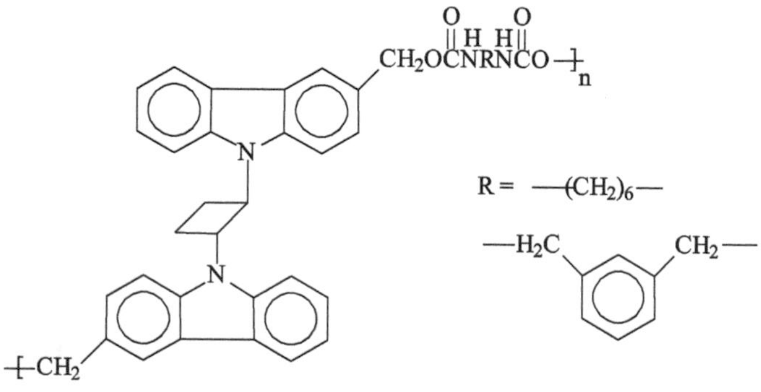

$$R = -(CH_2)_6-$$

(15)

Polyimines (16) containing DCZB moieties and a spacer of variable number of methylene groups have been synthesized by Natansohn et al. [91] from trans-1,2-bis(3-formyl-9-carbazolyl)cyclobutane and the corresponding aliphatic diamine. The charge transfer complexes of the polyamines (16) with tetracyanoethylene and TNF have been analyzed both in solution and in solid state. These polyimines do form charge transfer complexes with both TNF and tetracyanoethylene, but these complexes have a solution like behavior, i.e., the components are relatively free to move around. Charge carrier transport in the polyimines (16) has been studied by the time-of-flight technique [92]. The hole mobility in polyimines (16) is higher than that in PVK.

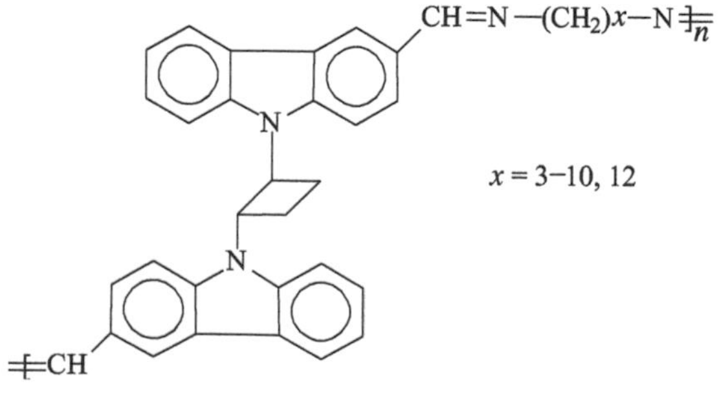

$x = 3–10, 12$

(16)

2. Other Photoconductive Polymers with Non-Conjugated Main Chain

Besides polymers with a carbazole moiety a number of polymers with various pendant aromatic amino groups have been reported. Poly(N-vinyldiphenylamine) (17a) and poly(4-diphenyl-aminostyrene) (17b) have been reported in early patents reviewed by Stolka and Pai [16]. The polymers were claimed to be useful in electrophotography.

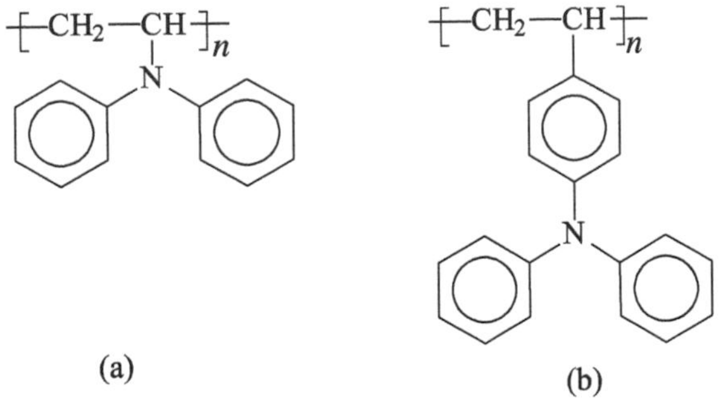

(a) (b)

(17)

A detailed photoconductivity study has been carried out with a number of polymethacrylates with pendant aromatic amino groups [93]. Among seven polymethacrylates that have been synthesized from the corresponding methacrylate monomers by free-radical polymerization, poly(2-(N-ethyl-N-3-tolylamino)ethyl methacrylate (18a) and poly((4-diphenylamino)phenylmethylmethacrylate) (18b) exhibit carrier mobilities that exceed the values of PVK by about one order of magnitude at all electric fields.

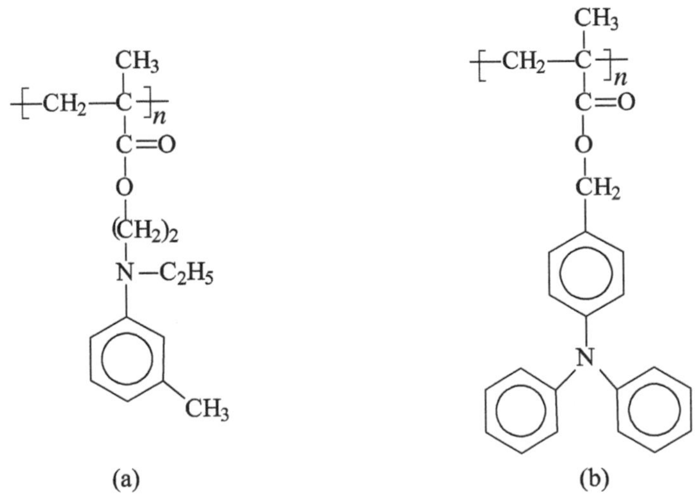

(a) (b)

(18)

Aromatic amino group-containing polymethacrylates alone only exhibit charge carrier generation when irradiated with UV light in the range of absorption. The charge-transfer complex of polymer (18a) with TNF (2:1 mol ratio) displays photoconductivity in visible light. Xero-graphic discharge experiments of these polymers in combination with a thin selenium layer proved aromatic amino group-containing polymethacrylates to be useful for application.

Another series of soluble hole-transporting polymers containing pendant arylamine groups were prepared by anionic polymerisation of newly synthesized vinylarylamines [94]. The general structure of the poly(vinylarylamines) reported is shown in Scheme (19). n-Butyllithium was used for the initiation of the anionic polymerization. The molecular weight of the polymers obtained is not high. M_w varies from 5000 to 15,700. The glass transition temperature is in the range of 130–150 °C. Poly(vinylarylamines) (19) have been used as hole transport materials in two-layer light-emitting diodes with tris(8-quinolinato)aluminium as electron transporting and emitting layer.

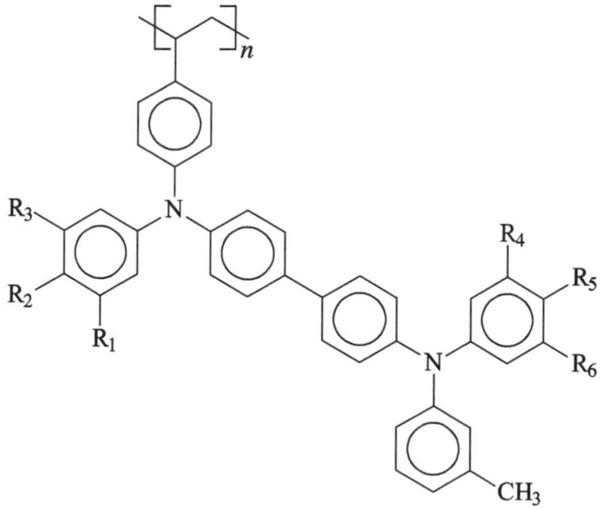

(a) R_1, R_3, R_4, R_6 = H;
R_2, R_5 = OCH_3

(b) R_1, R_2, R_4, R_5, R_6 = H;
R_3 = CH_3

(c) R_1, R_2, R_4, R_5 = H;
R_3, R_6 = F

(d) R_2, R_5 = H;
R_1, R_3, R_4, R_6 = F

(19)

Ulanski et al. [95] have reported that poly((E,E-[6,2]-paracyclophane-1,5-diene) (20b) shows relatively high photoconductivity especially when it is doped with tetracyanoethylene (TCNE). The polymer (20b) is obtained either by free-radical or cationic polymerization of the corresponding monomer (20a) [96]. Cationic polymerization is favored.

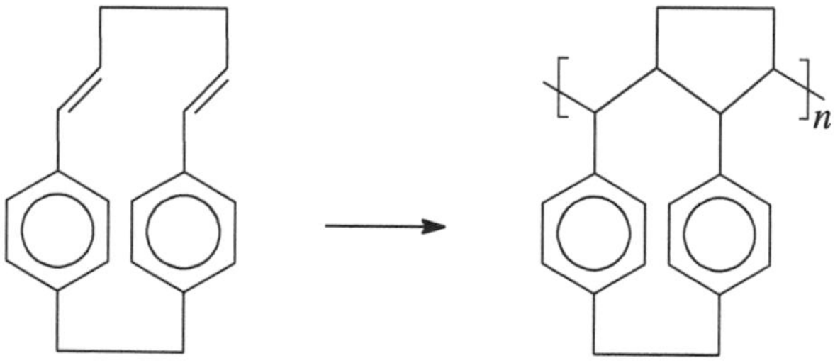

(20)

(a)   (b)

At a field of 4×10^5 V cm^{-1} pure polymer (20b) shows a mobility of 1.2×10^{-6} cm^2 V^{-1} s^{-1} and that doped with 4% of TCNE exhibits a mobility of 3.6×10^{-5} cm^2 V^{-1} cm^{-1} [97].

Polymers containing triphenyldiamine (TPD) moieties in the main chain, obtained by step growth polymerization are of increasing interest both as photoreceptors and for light emitting diodes. A series of TPD-containing condensation polymers is described in the patent [98]. The structure of one such polymer is shown in Scheme (21). The application of the hole-transporting polymers instead of the low-molar-mass compounds for the charge transport layers of photoreceptors prevents penetration of the small molecules from the charge transport to the charge generation layer.

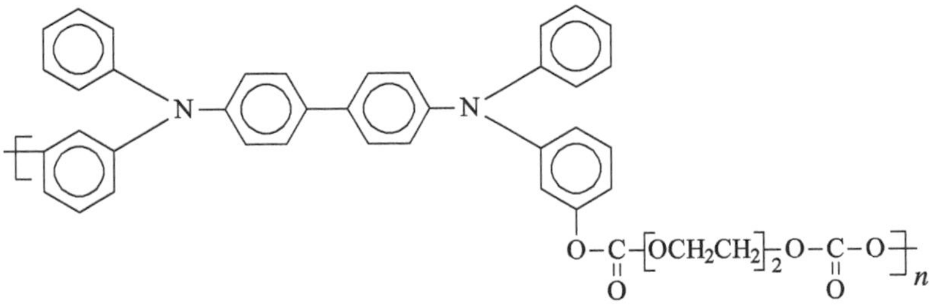

(21)

A poly(arylene ether sulfone) (22) containing TPD moieties was synthesized by the reaction of the corresponding bisphenol with 4,4′-difluorodiphenylsulfone [99]. The weight average molecular weight of the polymer (22) was determined to be 9300. Its thermal properties are excellent for the application in electroluminescent devices as hole transport layer. The glass transition temperature of the polymer (22) is 190 °C.

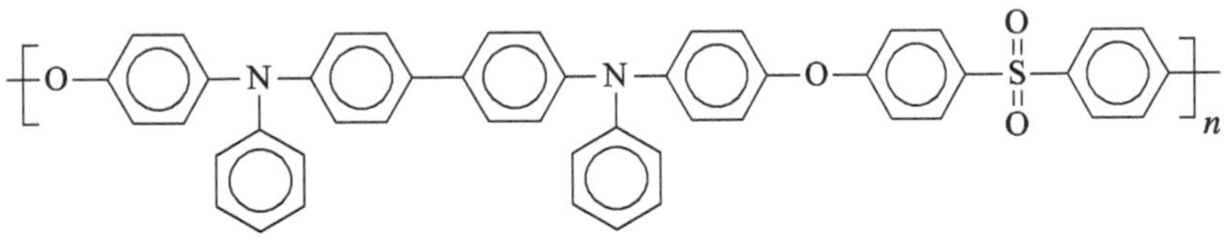

(22)

Polycarbonates [100] and polyethers [101] containing triphenylamine moieties in the polymer backbone have also been synthesized and used as hole transport materials in light-emitting diodes.

Crosslinkable charge transport materials recently attract much attention since they allow to prepare multilayer devices by low cost techniques, i.e., the combination of spin coating and crosslinking. Nuyken et al. [102] have reported the synthesis of photocrosslinkable derivatives of TPD containing oxetane functionalities. One example of a photocrosslinkable TPD is shown in Scheme (23). The photocrosslinking was carried out by a cationic mechanism. The resulting films are resistant against solvents to use in subsequent spin coating. The performance of single- and two-layer electroluminescent devices based on the crosslinked polymers is reported to be greatly enhanced relative to those containing the non-crosslinked compound (23) what is explained by the improved stability of the crosslinked layer.

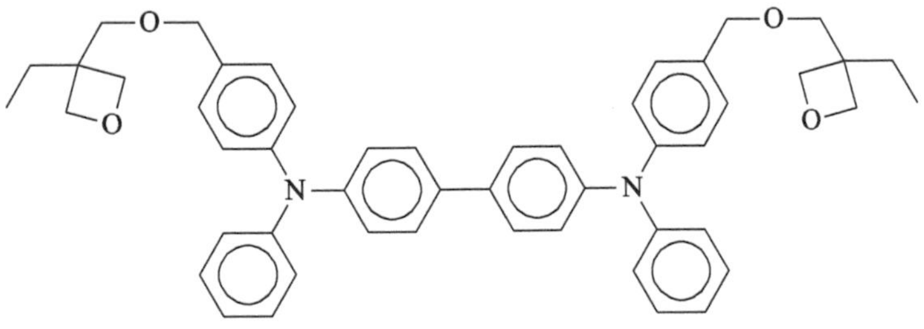

(23)

B. Polymers with π-Conjugated Main Chain

A number of photoconductive polymers and oligomers with conjugated double bonds along the polymer chain have been reported in the literature. Among π-conjugated polymers are polyacetylene and its derivatives, polydiacetylenes, polyarylenes like poly(phenylenevinylene) or poly(phenylenesulfide), polythiophene and poly(3-alkylthiophenes), polybenzothiazoles and others. These polymers are insulators in the dark and exhibit photoconductivity when illuminated. After chemical or electrochemical oxidation or reduction these π-conjugated polymers become conductive. In this section we are going to describe only those π-conjugated polymers which have attracted much attention as photoconductors.

The photoconductivity in polyacetylene, the simplest conjugated polymer, has been the subject of intense investigations [103–106]. Transient photoconductivity measurements on a picosecond time scale have been carried out [107–112]. These ultrafast methods are a powerful tools to investigate the transport properties as well as the recombination kinetics of charged excitations. It was found [107] that the photocurrent in trans-polyacetylene consists of two components: a fast component which relaxes on a picosecond time scale and for which a carrier mobility of about $1\,\text{cm}^2\,\text{V}^{-1}\,\text{s}^{-1}$ was reported [110,111] and a slow component with carrier lifetimes up to seconds.

Some polyacetylene derivatives have also been thoroughly investigated. Kang et al. [113,114] reported on photoconductivity measurements in trans-poly(phenylacetylene) and its charge transfer complexes. Trans-poly(phenylacetylene) was prepared by the polymerization of the corresponding monomer with $W(CO)_6$ in carbon tetrachloride solution under UV irradiation [115]. The reaction yielded a room-temperature-soluble polymer of molecular weight (M_n) 80,000. Poly(2-chloro-1-phenylacetylene) was synthesized by a similar procedure. M_n of the polymer was 400,000. Steady-state and pulsed photoconductivities were explored in amorphous films of poly(phenylacetylene) and of that doped with inorganic and organic electron accepting compounds like iodine and 2,3-dichloro-5,6-dicyano-p-benzoquinone and dyes like pyronin Y and methylene blue [114,115]. It was concluded that the transport mechanism in these systems is significantly different from the hopping transport which occurs in PVK and its charge-transfer complexes. Cis-poly(phenylacetylene) can also be converted to a photoconductive material. It has been done by irradiating with ^{60}Co and electron beam, doping with iodine and ferric chloride and sensitizing with 4-isothiocyanatofluorescein or TNF [116–118]. Cis-poly(phenylacetylene) was prepared by a direct method of polymerization of phenylacetylene into a polymer film with a rare-earth coordination catalyst. The cis-content of poly(phenylacetylene) obtained by this method was more than 90%. The molecular weight (M_n) was about 10^5 as measured by gel permeation chromatography.

Pfleger et al. [119] have studied photoconduction in undoped poly(phenylacetylene) which they prepared by coordination polymerization of phenylacetylene using the methatesis catalyst $WOCl_4/Ph_4Sn$. The polymer thus obtained was predominantly in the cis-transoidal form, as demonstrated by IR spectra, and had a molecular weight of (M_n) 91,000. The photoconduction threshold has been detected at 410 nm, although absorption of the film extends up to 550 nm. It is suggested that the mechanism of photogeneration is intrinsic by nature. The formation of initial charge carrier pairs occurs by an exciton autoionization process [42].

Poly(N-2-propynylcarbazole) (24a) and poly(N-2-propynylphenothiazine) (24b) have been prepared with $Ti(OBu)_4/Et_3Al$ as initiator [120]. Polymer (24a) was only partly soluble in some solvents like tetrahydrofuran, chloroform, nitrobenzene, and p-dichlorobenzene. In contrast to $Ti(OBu)/Et_3Al$ initiation polymerization of 3-(N-carbazolyl)-1-propyne with $MoCl_5$ and WCl_6 based catalysts gave high yields of yellow polymer insoluble in any solvent [121].

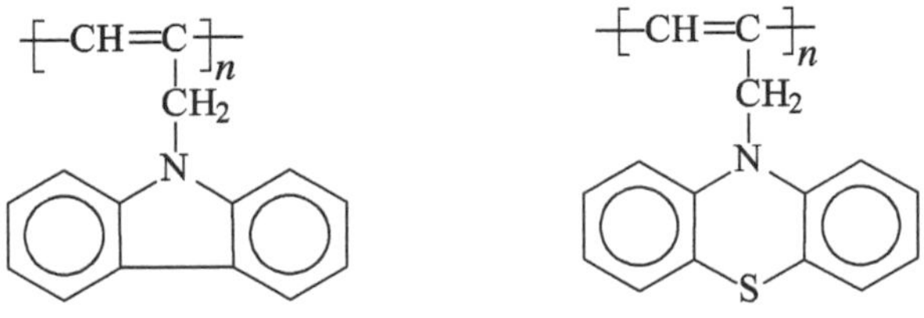

(24)

(a)   (b)

Copolymerization of 3-(N-carbazolyl)-1-propyne with tert-butylacetylene initiated by $MoCl_5/(C_4H_9)_4Sn$ yielded copolymers of high molecular weight ($M_w = 350,000$) completely soluble in toluene and chloroform [121]. Polymers (24a) and (24b) were found to show photoconductivity. Charge carrier photogeneration in these polymers and some related copolymers has been studied in detail [122,123].

Poly(1,6-heptadiyne) derivatives containing a carbazole moiety (25b) were synthesized by metathesis cyclopolymerization of bis(N-carbazolyl)-n-hexyl dipropargylmalonate (25a) [124]. The resulting polymer exhibited good solubility in common organic solvents and could easily be cast on a glass plate to give violet, shiny thin films. The number-average molecular weight values of the polymer were in the range from 3.2×10^4 to 8.9×10^4. Polymer (25b) shows two maximum values of the photocurrent around 350 nm and around 700 nm. The photo- to dark-conductivity ratio without doping was found to be in the range of 30–50 at 10^3–10^4 V/cm.

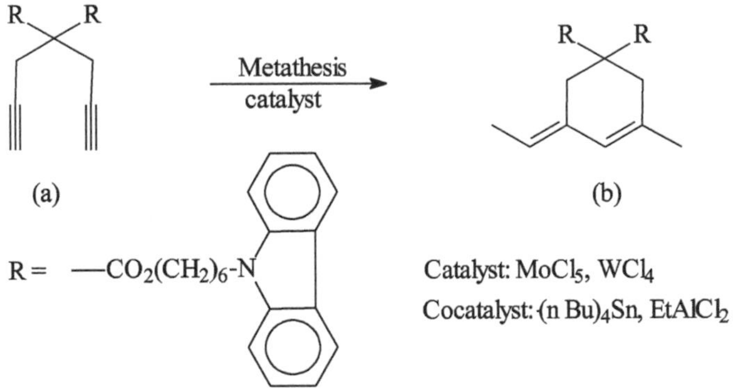

(a)   (b)   (25)

$R = \quad -CO_2(CH_2)_6-N$

Catalyst: $MoCl_5$, WCl_4
Cocatalyst: $(n\,Bu)_4Sn$, $EtAlCl_2$

Polydiacetylenes like poly(2,4-hexadiyne-1,6-diol bis(p-toluenesulfonate)) (26) have been studied by several authors [111,112,125–128].

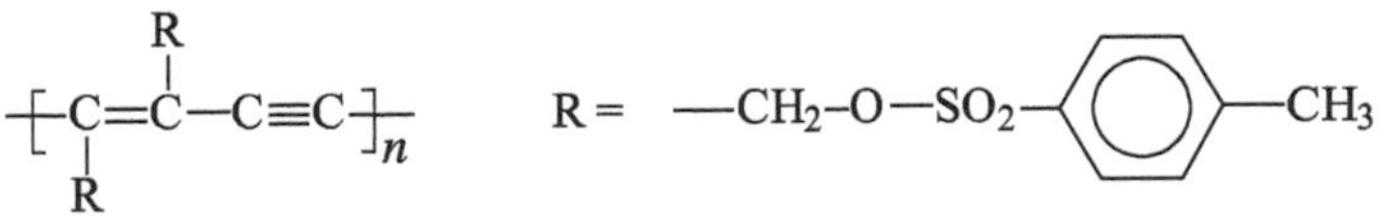

$R = \quad -CH_2-O-SO_2-$⟨benzene⟩$-CH_3$ (26)

They are unique in that that they can be obtained as polymer single crystals and therefore they have found a considerable interest in fundamental studies. A carrier mobility of $5\,cm^2\,V^{-1}\,s^{-1}$ has been reported for polymer (26) [112]. The field and temperature dependencies of the mobility have been investigated in detail [128].

Many years ago photoconductivity has been reported in a number of polyaryl-enes like poly(phenylenevinylene) (PPV) (27a), poly(phenyleneazomethine) (27b) and poly(phenylene sulfurdiimide) (27c) [16].

$$\text{--}\!\!\left[\text{--}\!\!\bigcirc\!\!\text{--CH=CH--}\right]_n \quad \text{--}\!\!\left[\text{--}\!\!\bigcirc\!\!\text{--CH=N--}\right]_n \quad \text{--}\!\!\left[\text{--}\!\!\bigcirc\!\!\text{--N=S=N--}\right]_n \quad (27)$$

(a)   (b)   (c)

These and a number of related polymers like poly(styrylpyrimidines), poly(quina-zones), poly(pyrrones), and poly(benzoxazoles) have already been reviewed by Stolka and Pai in 1978 [16]. They stated that there were some major problems with these polymers: complicated synthesis, in many cases poorly identified structures, and with a few exceptions insolubility and intractability. Large efforts have been made since then to overcome these difficulties. Two major pathways have been established which lead to tractable materials. Proper substitution of a rigid conjugated polymer leads to a soluble and fusible material. A second approach to improve the processability of conjugated polymers is to adopt a two step synthesis. In this case a nonconjugated polymer which can be readily converted to the desired material by heat treatment and which has good stability and processing properties is used as a precursor.

PPV (27a) has been prepared by a number of different methods which were studied in detail by Hörhold and Opfermann [129]. It can be synthesized by bifunctional carbonyl olefination of terephthalaldehyde according to Wittig's reaction and from p-xylylene-bis-(diethyl phosphonate) as well as by dehydrochlorination of p-xylylene dichloride with sodium hydride in N,N-dimethylformamide and with potassium amide in liquid ammonia. Another route to PPV used today is the precursor route, first described by Wessling [130–133] and Kanabe [134], starting from the monomers p-xylylene-bis(dimethylsulfo-nium tetrafluoroborate) [134] or chloride (Scheme 28) [130–133].

$$\text{ClCH}_2\text{--}\bigcirc\text{--CH}_2\text{Cl} + (\text{CH}_3)_2\text{S} \longrightarrow \text{Cl}^{\ominus}\ \overset{\overset{\displaystyle CH_3}{|}}{\underset{\underset{\displaystyle CH_3}{|}}{\overset{\oplus}{S}}}\text{--CH}_2\text{--}\bigcirc\text{--CH}_2\text{--}\overset{\overset{\displaystyle CH_3}{|}}{\underset{\underset{\displaystyle CH_3}{|}}{\overset{\oplus}{S}}}\ \text{Cl}^{\ominus}$$

(a)   (b)   (c)

$$\xrightarrow{\text{OH}^{\ominus}} \text{--}\!\!\left[\text{--}\!\!\bigcirc\!\!\text{--}\underset{\underset{\displaystyle CH_2\text{--CH--}}{}}{\overset{\overset{\displaystyle H_3C\overset{\oplus}{-}S-CH_3}{}}{}}\right]_n \xrightarrow{\triangle} \text{--}\!\!\left[\text{--}\!\!\bigcirc\!\!\text{--CH=CH--}\right]_n$$

(d)   (e)

$$(28)$$

The latter is polymerized to yield a water soluble sulfonium salt polyelectrolyte (28d) which is then purified by dialysis [135]. The precursor polymer is converted to PPV (28e) by the thermal elimination of dimethyl sulfide and HCl. The method has been later developed by Hörhold et al. [136], Lenz et al. [137,138], Murase et al. [139] and Bradley [140]. One of the major improvements was the use of tetrahydrothiophene instead of dimethyl sulphide in the synthesis of the precursor polymer [141]. The use of the cyclic leaving group facilitates the elimination when the precursor polymer is heated at 230–300 °C and leads to PPV with reduced amounts of defect structures in the polymer chain.

The photoconductivity of PPV prepared by the precursor route has been studied by several groups [142–145]. The polymer has a photoconductivity threshold at 506 nm that coincides well with the absorption edge [145]. Measurements of the transient photocurrent indicate a dispersive type of transport. The current is predominantly carried by holes with mobilities in the range from 10^{-3} to $10^{-4}\,\mathrm{cm^2\,V^{-1}\,s^{-1}}$. PPV was the first π-conjugated polymer in which the phenomenon of electroluminescence was demonstrated and from which light-emitting diodes were fabricated [146].

Soluble analogues of PPV with variety of substituents have been synthesized by different methods in Hörholds laboratories [147–154] and in other groups [155–159]. The synthetic routes to PPV have been recently reviewed by Holmes [9].

π-Conjugated polymers [160,161] and copolymers [162–164] of 9,9-dialkylfluorenes now attract strong interest as blue-emitting polymers showing high hole mobilities and having good prospects of commercial application in light-emitting diodes. Poly(2,7-fluorenes) are prepared via Suzuki coupling [160,161,165] and nickel(0) catalyzed reductive coupling [166] while the copolymers are also prepared by Wittig reaction [163] and Heck reaction [164]. The most widely studied among the poly(fluorenes) is poly(9,9′-dioctylfluorene) (29) [167–169]. This polymer forms a well defined thermotropic liquid crystalline state that can be aligned on rubbed substrates and can be either quenched into a glass or crystallized [161]. Polarized absorption and emission spectra of the polymer show a high degree of orientation, indicating strong potential for use in polarized electroluminescent devices. Poly(9,9′-dioctylfluorene) exhibits relatively high hole mobility, which is necessary, since in order to ensure an acceptable power efficiency high brightness of electroluminescent devices should be reached at low bias voltages. The as-spin coated ('isotropic') polymer shows hole mobility of $3 \times 10^{-4}\,\mathrm{cm^2\,V^{-1}\,s^{-1}}$ [168]. In addition, hole transport is nondispersive, which points to a high degree of chemical purity and regularity. Homogeneous nematic alignment of poly(9,9′-dioctylfluorene) films on rubbed polyimide results in more than one order of magnitude increase in Time of Flight hole mobility normal to the alignment direction. A hole mobility of $8.5 \times 10^{-3}\,\mathrm{cm^2\,V^{-1}\,s^{-1}}$ at an electric field of $10^4\,\mathrm{V\,cm^{-1}}$ is reported for the aligned quenched film of poly(9,9′-dioctylfluorene) [169].

(29)

R R

R = C_8H_{17}

Conjugated triphenyldiamine (TPD) based oligomers (30) have been prepared by polycondensation of the corresponding bis(sec-amines) and diodides [170]. The number average molecular weight of the oligomers (30) ranges from 1400 to 1800. Their glass transition temperatures are ca. 130 °C.

(30)

A polymer, incorporating both TPD and phenylenevinylene segments (31) has by recently reported [171]. This polymer possesses excellent film-forming properties, good thermal stability, and high electrochemical reversibility. It was prepared by the Wittig–Horner polycondensation reaction between a TPD-based dialdehyde and 1,4-xylylene diphosphate.

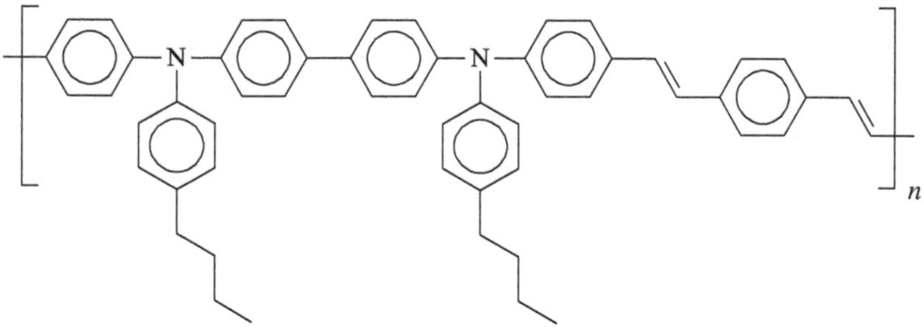

(31)

Poly(9-hexyl-3,6-carbazolyleneethynylene) (32c) has been prepared by palladium catalyzed polycondensation of 3,6-diiodo-9-hexylcarbazole (32a) and 3,6-diethynyl-9-hexyl-carbazole (32b) [172]. The polymer has a number average molecular weight M_n of 3000. By fractionation a polymer with M_n of 6400 has been obtained.

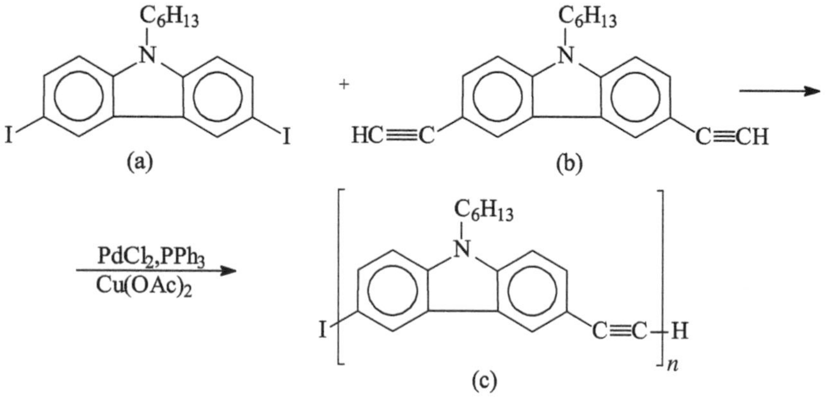

(32)

Polymer (32c) is soluble in common organic solvents. The trimer model compound of the polymer (32c) 3,6-bis((9-hexyl-3-carbazolyl)ethynyl)-9-hexylcarbazole (33) forms a stable glass with a glass transition at 41 °C. The trimer as well as the dimer were synthesized by stepwise reactions of the derivatives of 9-hexyl-carbazole [172]. Time-of-flight experiments have revealed carrier mobilities up to 2×10^{-4} cm^2 V^{-1} s^{-1} at an electric field of 6×10^5 V/cm in the trimer (33).

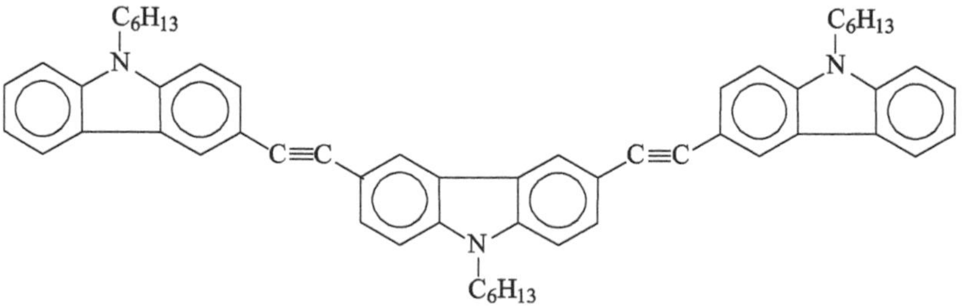

(33)

C. Polymers with σ-Conjugated Main Chain

Polysilylenes (polysilanes) (34b) have received widespread interest. Their electronic properties are associated with σ-electron conjugation in the silicon backbone which allows a significant delocalization of electrons along the chain. In the usual synthesis of polysilylenes, diorganodichlorsilanes (34a) are treated with sodium metal in a hydrocarbon diluent [173]. In order to recreate the surface of the sodium metal permanently ultrasound is used in these reactions [174,175].

$$
\underset{\substack{\mathrm{R_2}}}{\overset{\substack{\mathrm{R_1}}}{\mathrm{Cl-Si-Cl}}} \ + \ 2\,\mathrm{Na} \ \longrightarrow \ \left[\!\!\underset{\substack{\mathrm{R_2}}}{\overset{\substack{\mathrm{R_1}}}{\mathrm{Si}}}\!\!\right]_n
$$

(34)

$$\mathrm{R_1, R_2 = alkyl, \ cycloalkyl, \ aryl}$$

(a)   (b)

Poly(methylphenylsilylene) (PMPS) obtained by this method has a high molecular weight and a narrow molecular-weight distribution ($M_n = 184{,}000$, $M_w/M_n = 1.4$) [175]. PMPS is the most thoroughly studied polysilylene. Photoconductivity measurements of this polymer have been carried out by several groups [175–192]. The quantum yield of the charge carrier generation Φ in PMPS is rather low (3×10^{-3} charges per photon at an electric field of 3×10^5 V/cm) [181], while the hole drift mobility is rather high. Most of the authors report room temperature mobilities of about $10^{-4}\,\mathrm{cm^2\,V^{-1}\,s^{-1}}$ at an electric field of 10^5 V/cm [175,179]. Higher hole mobilities exceeding $10^{-3}\,\mathrm{cm^2\,V^{-1}\,s^{-1}}$ at room temperature have been recently observed in self-organized individual oligomerhomologues of poly(dimethylsilylene) [193]. Since there is no apparent difference in charge carrier mobility in PMPS and poly(dialkylsilylenes) [183] it can be assumed that the charge-carrier transport proceeds predominantly along the σ-delocalized Si backbone. The temperature and field dependencies of the carrier mobility in PMPS have been studied in great detail and were discussed in relation with different theoretical models [179].

In order to increase the quantum yield of charge carrier generation doping of PMPS with different additives has been studied [184–187]. Doping of the polymer with electron scavenging compounds generally resulted in soaring of the Φ values and plummeting of the μ values.

The influence of hole trapping substances, which at the same time are transport-active, on the photoconductivity of PMPS has also been investigated [186,188,189]. Aromatic amines with different ionization potentials have been examined. It turned out that a small amount (1%) of N,N′-diphenyl-N,N′-bis(3-methylphenyl)-(1,1′-biphenyl)-4,4′-diamine (TPD) (36a) did not exert any influence on the carrier mobility, while other amines strongly diminished it. This observation was explained by the fact that the ionization potential of TPD is equal to that of PMPS while the ionization potentials of the other amines studied are lower.

A decrease of the hole drift mobility was also observed in carbazole containing polysilylenes relative to PMPS [194]. The polysilylenes of which the repeat units are shown in Scheme (35) were prepared either by Wurtz polycondensation of the corresponding diorganodichlorosilanes [194–196] or by chemical modification of poly(alkylphenylsilylenes) [194]. For electronic and/or steric reasons it appeared to be impossible to prepare the homopolymer of 9-carbazolylmethyldichlorosilane. In the presence of simple diorganodichlorosilanes like Me_2SiCl_2 or $MePhSiCl_2$ low-molecular weight copolysilylenes have been prepared. Homo- and copolysilylenes have been synthesized by Wurtz polycondensation of (3-(9-carbazolyl)-propyl)methyldichlorosilane and by copolycondensation with Me_2SiCl_2 or $MePhSiCl_2$. The number average molecular weight M_n of the homopolymers did not exceed 3000 while M_n of the copolymers with a low content of carbazole-containing units reached 40,000. In order to prepare polysilylenes containing –CzMeSi– and –(CzPh)MeSi– units (where Cz = carbazolyl) partial dearylation of PMPS with triflic acid was carried out followed by nucleophilic displacement of triflate groups with CzLi or CzPhLi. The products of these poymeranalogous reactions where terpolymers since n-BuLi was added at the end of the reactions to avoid the presence of unreacted triflate groups.

$$R = CH_3, C_2H_5$$

The room temperature charge carrier mobilities of PMPS containing from 5 to 15% units of –(CzPr)MeSi–, –(CzPh)MeSi–, or –(CzPh)EtSi– doped with TNF were in the range from $3–8 \times 10^{-5} cm^2 V^{-1} s^{-1}$. The highest charge photogeneration quantum yield ($\Phi = 0.24$ with $E = 100 V/\mu m$) was observed in the TNF-doped copolymer containing –(CzPh)MeSi– (35c) units [194].

A spectacular effect was observed upon doping of large amounts of the TPD derivative   N,N'-bis(4-methylphenyl)-N,N'-bis(4-ethylphenyl)-(1,1'-(3,3'-dimethyl)biphenyl)-4,4'-diamine (36b) into PMPS. The hole mobility of the composite reached values of the order of $10^{-1} cm^2 V^{-1} s^{-1}$, comparable to those measured in molecular crystals and much in excess of μ values for either undoped polysilylene or pure (36b). A similar effect was observed with some other aromatic amines, including TPD (36a). According to Bässler [190] this observation suggests that the polymer matrix imposes structural constraints on the charge carrying molecules that favor intramolecular charge exchange and minimize disorder effects. The hole mobility approaching $10^{-1} cm^2 V^{-1} s^{-1}$ at $E = 2.5 \times 10^5 V/cm$ and 295 K is the highest reported μ value for disordered organic systems. The hole mobility of pure PMPS is enhanced by almost three orders of magnitude.

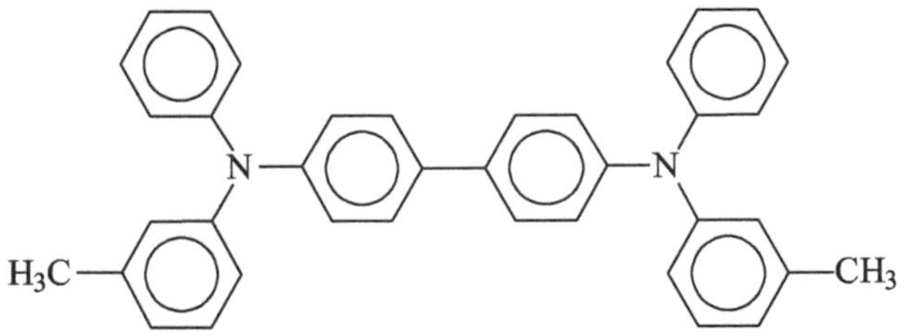

(a)

(36)

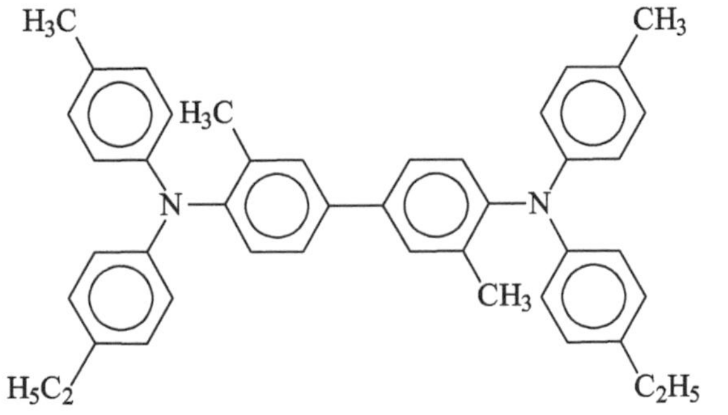

(b)

Unfortunately the good photoconductive properties of PMPS are accompanied by a degradation of the silicon backbone when the material is irradiated at wavelengths corresponding to the absorption of the σ-conjugated system [191]. In spite of its low photochemical stability, PMPS in combination with an effective charge generating material, such as amorphous selenium or phthalocyanine pigments can be applied in high-sensitive photoreceptors [197,198].

REFERENCES

1. Hoegl, H., Süs, O., and Neugebauer, W. (1959). Ger. Offen. 106 8115 to Kalle AG; C.A. (1961), *55*: 20742a.
2. Hoegl, H. (1965). *J. Phys. Chem.*, *69*: 755.
3. Schattuck, M. D., and Vahtra, U. (1969). U.S. Pat. 3 484 327.
4. Schaffert, R. M. (1971). *IBM J. Res. Dev.*, *15*: 75.
5. Law, K. Y. (1993). *Chem. Rev.*, *93*: 449.
6. Borsenberger, P. M., and Weiss, D. S. (1993). *Organic Photoreceptors for Imaging Systems*, Marcel Dekker.
7. Peyghambarian, N., and Kippelen, B. (1988). *Mat. Res. Soc. Symp. Proc.*, *488*: 39.
8. Greiner, A. (1998). *Polym. Adv. Technol.*, *9*: 371.
9. Kraft, A., Grimsdale, A. C., and Holmes, A. (1998). *Angew. Chem.*, *110*: 416.
10. Granström, M., Petritsch, K., Arias, A. C., Lux, A., Andersson, M. R., and Friend, R. (1998). *Nature*, *395*: 257.
11. Chen, L., Godowsky, D., Inganäs, O., Hummelen, R. A., Swensson, M., and Andersson, M. R. (2000). *Adv. Mater.*, *12*: 1367.
12. Burland, D. M., and Schein, L. B. (1986). *Physics Today*, *5*: 46.
13. Lutz, M. (1985). *J. Imaging Techn.*, *11*: 254.
14. Kukuta, A. (1990). In *Infrared Absorbing Dyes* (Matsuoka, M., ed.), Plenum Press, New York, Chapter 12.

15. Stolka, M. (1988). *Encyclopedia of Polymer Science and Engineering*, Vol. 11, Wiley-Interscience, New York, p. 154.
16. Stolka, M., and Pai, D. M. (1978). *Adv. Polym. Sci.*, *29*: 1.
17. Wiedemann, W. (1982). *Chem. Ztg.*, *106*: 275.
18. Biswas, M., and Uryu, T. (1986). *J. Macromol. Sci., Rev. Macromol. Chem. Phys.*, *C26*: 248.
19. Strohriegl, P., and Grazulevicius, J. V. (1997). In *Handbook of Organic Conductive Molecules and Polymers. Vol. 1 Charge-Transfer Salts, Fullerenes and Photoconductors* (Nalwa, H. S., ed.), J. Wiley & Sons, Chichester, p. 553.
20. Mylnikov, V. (1994). *Adv. Polym. Sci.*, *115*: 1.
21. Haarer, D. (1990). *Angew. Makromol. Chem.*, *183*: 197.
22. Gražulevičius, J. V., and Strohriegl, P. (2001). In *Handbook of Advanced Photonic Materials and Devices. Vol. 10 Light-Emitting Diodes and Polymer Devices* (Nalwa, H. S., ed.), Academic Press, San Diego, p. 233.
23. Pearson, J. M., and Stolka, M. (1981). *Polymer Monographs. Poly(N-vinylcarbazole)*, Vol. 61, Gordon & Breach, New York.
24. Mort, J., and Pfister, G. (1982). *Electronic Properties of Polymers* (Mort, J., and Pfister, G., eds.), Wiley, New York, p. 215.
25. Onsager, L. (1938). *Phys. Rev.*, *54*: 554.
26. Pai, D. M., and Eck, R. C. (1975). *Phys. Rev.*, *B11*: 5163.
27. Borsenberger, P. M., and Contois, L. E. (1979). *J. Appl. Phys.*, *50*: 914.
28. Borsenberger, P. M., Contois, L. E., and Hoesterey, D. C. (1978). *J. Chem. Phys.*, *68*: 637.
29. Borsenberger, P. M., and Ateya, A. T. (1978). *J. Appl. Phys.*, *49*: 4035.
30. Kaul, H., and Haarer, D. (1987). *Ber. Bunsenges. Phys. Chem.*, *91*: 845.
31. Hughes, R. C. (1971). *Chem. Phys. Lett.*, *8*: 403.
32. Hughes, R. C. (1971). *J. Chem. Phys.*, *55*: 5422.
33. Gill, W. D. J. (1972). *Appl. Phys.*, *43*: 5033.
34. Pai, D. M. (1970). *J. Chem. Phys.*, *52*: 2285.
35. Scher, H., and Montroll, E. W. (1975). *Phys. Rev.*, *B12*: 2455.
36. Chen, J., and Slovik, J. H. (1975). *Solid State Commun.*, *17*: 783.
37. Pfister, G., and Griffith, C. (1978). *Phys. Rev. Lett.*, *40*: 659.
38. Domes, H., Fischer, R., Haarer, D., and Strohriegl, P. (1989). *Makromol. Chem.*, *190*: 165.
39. Bässler, H., Schönherr, G., Abkowitz, M., and Pai, D. M. (1982). *Phys. Rev.*, *B26*: 3105.
40. Pai, D. M., Yanus, J. F., and Stolka, M. (1984). *J. Phys. Chem.*, *88*: 4714.
41. Hirsch, J. (1979). *J. Phys. Chem.*, *12*: 321.
42. Nespurek, S. (1993). *Synth. Met.*, *61*: 55.
43. Limburg, W. W., and Williams, D. J. (1973). *Macromolecules*, *6*: 787.
44. Williams, D. J., Limburg, W. W., Pearson, J. M., Goedde, A. O., and Yanus, J. F. (1975). *J. Chem. Phys.*, *62*: 1501.
45. Inoue, T., and Tazuke, S. (1981). *J. Polym. Sci., Polym. Chem. Ed.*, *19*: 2861.
46. Tazuke, S., Inoue, T., Saito, S., Hirota, S., and Kokaido, H. (1985). *Polym. Photochem.*, *6*: 221.
47. Tazuke, S., Inoue, T., and Kokaido, H. (1985). *Polym. Photochem.*, *6*: 385.
48. Oshima, R., Uryu, T., and Seno, M. (1985). *Macromolecules*, *18*: 1043.
49. Uryu, T., Okhawa, H., and Oshima, R. (1987). *Macromolecules*, *20*: 712.
50. Uryu, T., Ohahu, K., and Matsuzaki, K. (1974). *J. Polym. Sci., Polym. Chem. Ed.*, *12*: 1723.
51. Strohriegl, P. (1990). *Mol. Cryst. Liq. Cryst.*, *1983*: 261.
52. Sasakawa, T., Ikeda, T., and Tazuke, S. (1989). *Macromolecules*, *22*: 4253.
53. Tazuke, S., Inoue, T., Tanabe, T., Hirota, S., and Saito, S. (1981). *J. Polym. Sci., Polym. Letters Ed.*, *19*: 11.
54. Yokoyama, M., Shimokihara, S., Matsubara, A., and Mikawa, H. (1982). *J. Chem. Phys.*, *76*: 724.
55. Okamoto, K., Itaya, A., and Kusabayashi, S. (1975). *Polym. J.*, *7*: 622.
56. Turner, R. S., and Pai, D. M. (1979). *Macromolecules*, *12*: 1.
57. Haque, S. A., Uryu, T., and Ohkawa, H. (1988). *Makromol. Chem.*, *188*: 2523.

58. Haque, S. A., and Uryu, T. (1988). *Polym. J., 20*: 163.
59. Gaidelis, V., Krisciunas, V., and Montrimas, E. (1976). *Thin Solid Films, 38*: 9.
60. Sasabe, H., and Wada, T. (1989). In *Comprehensive Polymer Science*, Vol. 7 (Allen, G., and Bevigton, J., eds.), Pergamon Press, Oxford, p. 179.
61. Matsui, N., and Takano, S. (1991). EP 462; C.A. *116*: 245244z.
62. Yashigawa, T., and Shimizu, Y. (1992). JP 04 15 656; C.A. *116*: 265595.
63. Vapsinskaite, I., Gaidelis, V., Girdziusas, A., Sidaravicius, J., Skarzinskas, V., Rakauskas, J., and Sarkovas, A. (1982). *Zh. Nauchn. Prikl. Fotogr. Kinematogr. (Russ.), 27*: 255.
64. Shimizu, Y., and Yanagida, M. (1992). JP 04 180 923; C.A. *117*: 213260f.
65. Undzenas, A., Gaidelis, V., Sidaravicius, J., Kavaliunas, R., Zdanavicius, J., and Duobinis, N. (1976). U.S.S.R. Patent 503 200; C.A. *85*: 114788y.
66. Bliumbergas, R., Duobinis, N., Kavaliunas, R., Urbonaviciene, J., Gaidelis, V., Undzenas, A., and Grazulevicius, J. V. (1978). U.S.S.R. Patent 762 394.
67. Buika, G., and Grazulevicius, J. V. (1993). *Eur. Polym. J., 29*: 1489.
68. Grazulevicius, J. V., Kavaliunas, R., Stanisauskaite, A., Kutkevicius, S., and Undzenas, A. (1984). U.S.S.R. Patent 1 254 711.
69. Grazulevicius, J. V., Kublickas, R., and Kavaliunas, R. (1994). *J. Macromol. Sci., Pure Appl. Chem., A31*: 1301.
70. Grazulevicius, J. V., Kublickas, R., and Undzenas, A. (1992). *Eur. Polym. J., 28*: 539.
71. Grazulevicius, J. V., and Kublickas, R. (1991). *Eur. Polym. J., 27*: 1411.
72. Kavaliunas, R., Grazulevicius, J. V., Undzenas, A., and Kublickas, R. (1991). U.S.S.R. Patent 1 672 582.
73. Bruzga, P., Grazulevicius, J. V., and Kavaliunas, R. (1991). *Eur. Polym. J., 27*: 707.
74. Bruzga, P., Grazulevicius, J. V., Kavaliunas, R., Kublickas, R., and Liutviniene, I. (1993). *Polym. Bull., 30*: 509.
75. Kavaliunas, R., Undzenas, A., and Urbonaviciene, J. (1982). *Zh. Nauch. Prikl. Fotogr. Kinematogr. (Russ.), 27*: 291.
76. Buika, G., Grazulevicius, J. V., Stolarzewicz, A., and Grobelny, Z. (1995). *Macromol. Chem. Phys., 196*: 1287.
77. Strohriegl, P. (1986). *Makromol. Chem., Rapid Commun., 7*: 771.
78. Domes, H., Fischer, R., Haarer, D., and Strohriegl, P. (1989). *Makromol. Chem., 190*: 165.
79. Goldie, D. M., Hepburn, A. R., Moud, J. M., and Marshall, J. M. (1993). *Synth. Met., 57*: 5026.
80. Hennecke, M., and Strohriegl, P. (1988). *Makromol. Chem., 198*: 2601.
81. Lux, M., Strohriegl, P., and Höcker, H. (1987). *Makromol. Chem., 188*: 811.
82. Ikeda, T., Mochizuki, H., Hayashi, Y., Sisido, M., and Sasakawa, T. (1991). *J. Appl. Phys., 70*: 3689.
83. Ikeda, T., Mochizuki, H., Hayashi, Y., Sisido, M., and Sasakawa, T. (1991). *J. Appl. Phys., 70*: 3696.
84. Zilker, S. J. (2000). *ChemPhysChem., 1*: 72.
85. Zhang, Y., Wada, T., and Sasabe, H. (1998). *J. Mater. Chem., 8*: 809.
86. Tamura, K., Hall, H. K., Jr., and Peyghambarian, N. (1992). *Appl. Phys. Lett., 60*: 1803.
87. Kippelen, B., Tamura, K., Peyghambarian, N., Padias, A. B., and Hall, H. K., Jr. (1993). *Phys. Rev., B48*: 10710.
88. Tazuke, S., Inoue, T., and Tanabe, T. (1981). *J. Polym. Sci., Poly. Lett. Ed., 19*: 11.
89. Subramaniam, P., Sasakawa, T., Ikeda, I., Tazuke, S., and Srinivasan, M. (1987). *J. Polym. Sci., Polym. Chem. Ed., 25*: 1463.
90. Subramaniam, P., Sasakawa, T., Ikeda, T., and Tazuke, S. (1987). *Makromol. Chem., 188*: 1147.
91. Yang, H., Jin, A., and Natansohn, A. (1992). *J. Polym. Sci., Polym. Chem. Ed., 30*: 1953.
92. Natansohn, A., Yang, H., Murto, D. K., and Popovic, Z. D. (1993). *Chem. Mater., 5*: 1370.
93. Stolka, M., Pai, D. M., Refner, D. S., and Yanus, J. C. (1983). *J. Polym. Sci. Polym. Chem. Ed., 21*: 969.

94. Bellmann, E., Shaheen, S. E., Grubbs, R. H., Marder, S. R., Kippelen, B., and Peyghambarian, N. (1999). *Chem. Mater.*, *11*: 399.
95. Ulanski, J., Kubacki, J., Glowacki, I., Kryszewski, M., and Glatzhofer, D. T. (1992). *J. Appl. Polym. Sci.*, *44*: 2103.
96. Longane, D. T., and Glatzhofer, D. T. (1986). *J. Polym. Sci., Polym. Chem. Ed.*, *24*: 1725.
97. Ulanski, J., Sielski, J., Glatzhofer, D. T., and Kryszewski, M. (1990). *J. Phys.*, *D23*: 75.
98. Yanus, J. F., Pai, D. M., Murti, D. K., Hsiao, C.-K., and Defeo, P. J. (1997). U.S. Patent 5,698,359.
99. Kido, J., Harada, G., and. Nagai, K. (1996). *Polym. Adv. Tecnnol.*, *7*: 31.
100. Hosokawa, C., Kawasaki, N., Sakamoto, S., and Kusumoto, T., (1992). *Appl. Phys. Lett.*, *61*: 2503.
101. Bacher, A., Bleyl, I., Erdelen, C. H., Haarer, D., Paulus, W., and Schmidt, H.-W. (1997). *Adv. Mater.*, *9*: 1031.
102. Bayerl, M. S., Braig, T., Nuyken, O., Müller, D. C., Groß, M., and Meerholz, K. (1999). *Macromol. Rapid Commun.*, *20*: 224.
103. Tani, T., Gill, W. D., Grant, P. M., Clarke, T. C., and Street, G. R. (1980). *Synth. Met.*, *1*: 301.
104. Tani, T., Grant, P. M., Gill, W. D., Street, G. B., and Clarke, T. C. (1980). *Solid State Commun.*, *33*: 499.
105. Miura, M., Kamagami, S., Takezoe, H., Fukuda., A., and Kuze, E. (1983). *Jpn. J. Appl. Phys.*, *22*: 1915.
106. Guilaud, G., Maitrot, M., Mathis, C., Francois, B., and Andre, J. J. (1984). *J. Phys. Paris Lett.*, *45L*: 265.
107. Tubino, R., Dorsinvile, R., Walser, A., Ses., A., and Alfano, R. R. (1989). *Synth. Met.*, *28D*: 175.
108. Townsed, P. D., and Friend, R. H. (1989). *Synth. Met.*, *28D*: 181.
109. Bleier, H., Donovan, K., Friend, R. H., Roth, S., Rothberg, L., Tubino, R., Vardney, Z., and Wilson, G. (1989). *Synth. Met.*, *28D*: 189.
110. Phillips, S. D., Yu, G., and Heeger, A. J. (1989). *Synth. Met.*, *28D*: 669.
111. Moses, D., Sinclair, M., Phillips, S., and Heeger, A. J. (1989). *Synth. Met.*, *28D*: 675.
112. Moses, D., Sinclair, M., Phillips, S., and Heeger, A. J. (1989). *Synth. Met.*, *28D*: 675.
113. Kang, E. T., Ehrlich, P., Bhatt, A. P., and Anderson, W. A. (1984). *Macromolecules*, *17*: 1020.
114. Kang, E. T., Ehrlich, P., and Anderson, W. A. (1984). *Mol. Cryst. Liq. Cryst.*, *106*: 305.
115. Masuda, T., Kuwane, Y., Yamamoto, K., and Highashimura, T. (1980). *Polym. Bull.*, *2*: 823.
116. Yang, M. J., Zhao, J., and Shen, Z. Q. (1989). *J. Polym. Sci., Polym. Chem. Ed.*, *27*: 3829.
117. Yang, M. J., Zhao, J., and Shen, Z. Q. (1991). *Makromol. Chem.*, *192*: 1225.
118. Yang, M. J., Zhao, J., and Shen, Z. Q. (1991). *Polym. J.*, *23*: 963.
119. Pfleger, J., Nespurek, S., and Vohlidal, J. (1989). *Mol. Cryst. Liq. Cryst.*, *166*: 143.
120. Kminek, I., Cimrova, V., Nespurek., S., and Vohlidal, J. (1989). *Makromol. Chem.*, *190*: 1025.
121. Nakano, M., Masuda, T., and Highashimura, T. (1995). *Polym. Bull.*, *34*: 191.
122. Kalvoda, L., Kminek, I., Cimrova, V., Nespurek, S., Schnabel, W., and Sedlacek, J. (1990). *Colloid. Polym. Sci.*, *268*: 1024.
123. Cimrova, V., and Nespurek, S. (1993). *Mol. Cryst. Liq. Cryst.*, *228*: 201.
124. Park, J.-W., Lee, J.-H., Cho, H.-N., and Choi, S.-K. (1993). *Macromolecules*, *26*: 1191.
125. Chance, R. R., and Baugham, R. H. (1976). *J. Chem. Phys.*, *64*: 3889.
126. Donovan, K. J., and Wilson, E. G. (1981). *Phil. Mag. Sect. B*, *44*: 31.
127. Sidiqui, A. S. (1984). *J. Phys. Chem.*, *17*: 683.
128. Blum, T., and Bässler, H. (1988). *Chem. Phys.*, *123*: 431.
129. Hörhold, H.-H., and Opfermann, J. (1970). Makromol. Chem., *131*: 105.
130. Wessling, R., and Zimmermann, R. (1968). U.S. Patent 3 401 152; C.A. *69*: 87735q.
131. Wessling, R., and Zimmermann, R. (1968). U.S. Patent 3 404 132; C.A. *69*: 107665x.
132. Wessling, R., and Zimmermann, R. (1970). U.S. Patent 3 532 643; C.A. *74*: 3994r.
133. Wessling, R., and Zimmermann, R. (1972). U.S. Patent 3 706 677; C.A. *78*: 85306n.

134. Kanabe, M., and Okawara, M. (1968). *J. Polym. Sci., A1*: 1056.
135. Capistran, J. D., Gagnon, D. R., Antoun, S., Lenz, R. W., and Karasz, F. E. (1984). *Polym. Prepr. (Am. Chem. Soc., Div. Polym. Chem.)*, 25: 282
136. Hörhold, H.-H., Palme, H. J., and Bergmann, R. (1978). *Zeitschr. Polymerforsch.*, 29: 299.
137. Gagnon, D. R., Capistran, J. D., Karasz, F. E., and Lenz, R. W. (1984). *Polym. Prepr. (Am. Chem. Soc., Div. Polym. Chem.)*, 25: 284.
138. Antoun, S., Gagnon, D. R., Karasz, F. E., and Lenz, R. W. (1986). *J. Polym. Sci., Polym. Lett. Ed.*, 24: 503.
139. Murase, I., Ohnishi, T., Noguchi, T., and Hirooka, M. (1984). *Polym. Commun.*, 25: 327.
140. Bradley, D. D. C. (1987). *J. Phys. D: Appl. Phys.*, 20: 1389.
141. Lenz, R. W., Han, C. C., Stenger-Smith, J., and Karasz, F. E. (1988). *J. Polym. Sci., Polym. Chem. Ed.*, 26: 3241.
142. Tokito, S., Tsutsui, T., Tanaka, R., and Saito, S. (1986). *Jpn. J. Appl. Phys.*, 25L: 680.
143. Tagiguchi, T., Park, D. H., Ueno, H., Yoshihino, K., and Sugimoto, R. (1987). *Synth. Met.*, 17: 657.
144. Bleier, H., Shen, Y., Bradley, D. D. C., Lindenberger, H., and Roth, S. (1989). *Synth. Met.*, 29E: 73.
145. Obrzut, J., Obrzut, M. J., and Karasz, F. E. (1989). *Synth. Met.*, 29E: 103.
146. Burroughes, J. H., Bradley, D. D. C., Brown, A. R., Marks, R. N., MacKay, K., Friend, R. H., Burn, P. L., and Holmes, A. B. (1990). *Nature*, 347: 539.
147. Hörhold, H.-H., Helbig, M., Raabe, D., Opfermann, J., Scherf, U., Stockmann, R., and Weiß, D. (1987). *Zeitschr. für Chemie*, 27: 126.
148. Hörhold, H.-H., Räthe, H., Helbig, M., and Opfermann, J. (1987). *Makromol. Chem.*, 188: 2083.
149. Hörhold, H.-H., Räthe, H., and Opfermann, J. (1986). *Acta Polym.*, 37: 369.
150. Raabe, D., Hörhold, H.-H., and Scherf, U. (1986). *Makromol. Chem., Rapid Commun.*, 7: 613.
151. Hörhold, H.-H., and Helbig, M. (1987). *Makromol. Chem., Macromol. Symp.*, 12: 229.
152. Hörhold, H.-H., Gottshaldt, J., and Opfermann, J. (1977). *J. Prakt. Chem.*, 39: 611.
153. Hörhold, H.-H., Bergmann, R., Gottschaldt, J., and Drefahl, G. (1974). *Acta Chim. Acad. Sci. Hung.*, 81: 239.
154. Raabe, D., Hörhold, H.-H., Paar, D., and Scherf, U. (1984). Ger. (East) Patent 220 023; C.A. 101: 152554a.
155. Feld, W. A., Ganesen, A., and Nymberg, D. D. (1983). *Polym. Prepr. (Am. Chem. Soc., Div. Polym. Chem.)*, 24: 43.
156. Feast, W. J., and Millichamp, I. S. (1983). *Polym. Commun.*, 24: 102.
157. DeKoninck, L., and Smets, G. (1969). *J. Polym. Sci., Part A-1*, 7: 3313.
158. Oyoshino, K., Yin, X. N., Muro, K., Kiyomatsu, S., Morita, S., Zakhidov, A., Noguchi, T., and Ohnishi, T. (1993). *Jpn. J. Appl. Phys., Part 2, 32(3A)*: 357.
159. Martelock, H., Greiner, A., and Heitz, W. (1991). *Makromol. Chem.*, 192: 967.
160. Grice, A. W., Bradley, D. D. C., Bernius, M. T., Inbasekaran, M., Wu, W. W., and Woo, E. P. (1998). *Appl. Phys. Lett.*, 73: 629.
161. Grell, M., Bradley, D. D. C., Inbasekaran, M., and Woo, E. P. (1997). *Adv. Mater.*, 9: 798.
162. Kim, C. Y., Cho, H. N., Kim, D. Y., Kim, Y. C., Lee, J. Y., and Kim, J. K. (1999). U.S. Patent 5,876,864.
163. Cho, H. N., Kim, D. Y., Kim, Y. C., Lee, J. Y., and Kim, C. Y. (1997). *Adv. Mater.*, 9: 327.
164. Cho, H. N., Kim, J. K., Kim, D. Y., Kim, C. Y., Song, N. W., and Kim, D. (1999). *Macromolecules*, 32: 1476.
165. Ranger, M., Rondeu, D., and Leclerc, M. (1997). *Macromolecules*, 30: 7686.
166. Pei, Q., and Yang, Y. (1996) *J. Am. Chem. Soc.*, 118: 7416.
167. Janietz, S., Bradley, D. D. C., Grell, M., Inbasekaran, M., and Woo, E. P. (1998). *Appl. Phys. Lett.*, 73: 2453.
168. Redecker, M., Bradley, D. D. C., Inbasekaran, M., and Woo, E. P. (1998). *Appl. Phys. Lett.*, 73: 1565.

169. Redecker, M., Bradley, D. D. C., Inbasekaran, M., and Woo, E. P. (1999). *Appl. Phys. Lett.*, *74*: 1400.
170. Thelakkat, M., Fink, R., Haubner, F., and Schmidt, H.-W. (1997). *Macromol. Symp.*, *125*: 157.
171. Liu, Y., Liu, M. S., and Yen, A. K.-Y. (1999). *Acta Polym.*, *50*: 105.
172. Beginn, C., Grazulevicius, J. V., Strohriegl, P., Simmerer, J., and Haarer, D. (1994). *Macromol. Chem. Phys.*, *195*: 2353.
173. West, R. (1986). *J. Organomet. Chem.*, *300*: 327.
174. Matyjaszewski, K., Chen, Y. L., and Kim, H. K. (1988). *ACS Symp. Ser.*, *360*: 78.
175. Strohriegl, P., and Haarer, D. (1991). *Makromol. Chem., Macromol. Symp.*, *44*: 85.
176. Stolka, M., and Abkowitz, M. A. (1987). *J. Noncryst. Solids*, *97–98*: 1111.
177. Kepler, R. G., Zeigler, J. M., Harrah, L. A., and Kurtz, S. R. (1987). *Phys. Rev.*, *B35*: 2818.
178. Fujino, M. (1987). *Chem. Phys. Lett.*, *136*: 451.
179. Abkowitz, M. A., Rice, M. J., and Stolka, M. (1990). *Phil. Mag.*, *B61*: 25.
180. Klingensmith, K. A., Downing, J. W., Miller, R. F., and Michl, J. (1986). *J. Am. Chem. Soc.*, *108*: 7438.
181. Eckhardt, A., Yars, N., Wolny, T. S., Nespurek, S., and Schnabel, W. (1994). *Ber. Bunsenges. Phys. Chem.*, *98*: 853.
182. Hattori, R., Aoki, Y., and Shirafuji, J. (1993). *J. Non-Cryst. Solids*, *164–166*: 1275.
183. Kepler, R. G., Zeigler, J. M., Harran, L. A., and Hurtz, S. R. (1987). *Phys. Rev.*, *B35*: 2818.
184. Lagarde, M., and Dubois, J. C. (1991). *Proc. 7th Int. Symp. Electrets* (Reimund, G.-M., ed.), IEEE, New York, p. 868.
185. Brynda, E., Nespurek, S., and Schnabel, W. (1993). *Chem. Phys.*, *175*: 459.
186. Stolka, M., and Abkowitz, M. A. (1993). *Synth. Met.*, *54*: 417.
187. Eckhard, A., Herden, V., Nespurek, S., and Schnabel, W. (1995). *Philos. Mag.*, *B71*: 239.
188. Yokoyama, Y., and Yokoyama, M. (1990). *Solid State Commun.*, *73*: 199.
189. Abkowitz, M. A. (1992). *Phil. Mag.*, *B65*: 817.
190. Bässler, H. (1993). *Adv. Mater.*, *5*: 662.
191. Miller, R. D., and Michl, J. (1989). *Chem. Rev.*, *89*: 1359.
192. Kminek, I., Brynda, E., and Schnabel, W. (1991). *Eur. Polym. J.*, *227*: 1073.
193. Okumoto, H., Yatabe, T., Shimomura, M., Kaito, A., Minami, N., and Tanabe, Y. (2001). *Adv. Mater.*, *13*: 72.
194. Lemmer, M., Bebin, P., Selphure, M., Marc, N., and Moisan, J.-Y. (1996). *Polimery (Pol)*, *41*: 508.
195. Lemmer, M., Selphure, M., Marc, N., and Moisan, J.-Y. (1997). *Polym. Adv. Technol.*, *8*: 116.
196. Lemmer, M., Bebin, P., Selphure, M., Marc, N., and Moisan, J.-Y. (1997). *Polym. Adv. Technol.*, *8*: 125.
197. Stolka, M., Yuh, H.-J., McGrane, K., and Pai, D. M. (1987). *J. Polym. Sci., Polym. Chem. Ed.*, *25*: 823.
198. Yokoyama, K., and Yokoyama, M. (1989). *Chem. Lett.*, 1005.

14

Polymers for Organic Light Emitting Devices/Diodes (OLEDs)

O. Nuyken, E. Bacher, M. Rojahn, V. Wiederhirn and R. Weberskirch
Technische Universität München, Garching, Germany

K. Meerholz
Universität Zu Köln, Köln, Germany

I. INTRODUCTION

Facing the 21st century, the development of new techniques that are able to display data faster, more detailed and in mobile applications, is one of the prospering scientific fields. One approach for lightweight, flexible, power-efficient full-color displays are organic light emitting diodes (OLEDs). Such devices with their low driving voltage, bright color and high repetition rate (e.g. for video-application) are ideal for usage in miniature displays as well as in large area screen [1–3]. The basic principle of these devices are electroluminescent 'semiconducting' organic materials packed between two electrodes. After charge injection from the electrodes into the organic layer and charge migration within this layer, electrons and deficient electrons (so called 'holes') can recombine to form an excited singlet state. Light emission of the latter is then a result of relaxation processes [4–6]. To achieve high electroluminescence efficiencies, the materials have to fulfill several specific requirements including low injection barriers at the interface between electrodes and organic material, balanced electron- and hole-density and mobility and high luminescence efficiency. Furthermore, the recombination zone should be located away from the metal cathode to prevent annihilation of the exited state. Since no material known to date is able to meet all these criteria, modern OLEDs consist — besides the transparent substrate (e.g., glass, PET), anode (most commonly indium tin oxide, ITO) and metal cathode (e.g., Mg–Ag-alloy) — of several organic layers for charge injection, transport and/or emission [7,8] (the principal set-up is shown in Scheme 1).

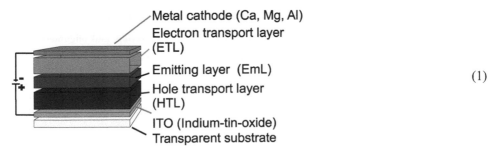

(1)

In such multilayer diodes, each layer can be separately optimized concerning injection barriers, charge mobility and density and quantum efficiency. Much of the motivation for studying organic materials stems from the potential to tailor desirable optoelectronic properties and process characteristics by manipulation of the primary chemical structure. Objecting optimal charge transport, recombination probability and light emission and consequently a maximum external efficiency of the device, various substances have been developed, modified and tested in the last few years. For hole transport/electron blocking layers, triarylamine- and pyrazoline-structures (see Scheme 2) were found to be most promising [9–11].

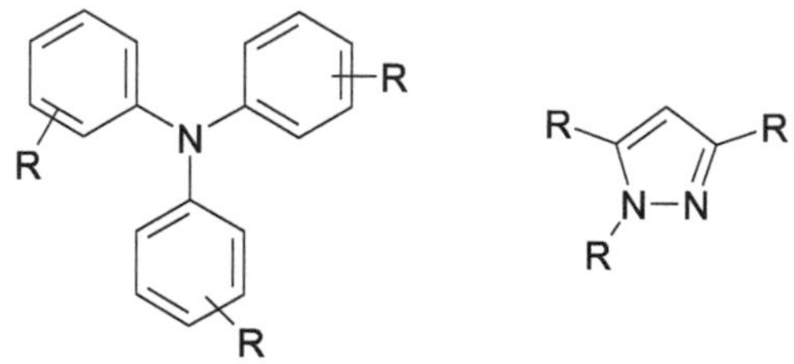

Triarylamines   Pyrazolines

(2)

For electron transport/hole blocking purposes, a wide variety of electron-deficient moieties are well known, e.g., 1,3,4-oxadiazoles [12], 1,2,4-triazoles [13], 1,3-oxazoles, pyridines and quinoxalines [14] (see Scheme 3). Materials with conjugated π-electron system (e.g., styrylarylenes, arylenes, stilbenes, oligo- and poly(thiophene)s — see Scheme 3) are widely used as combined charge transport and luminescence layers as well [12,15].

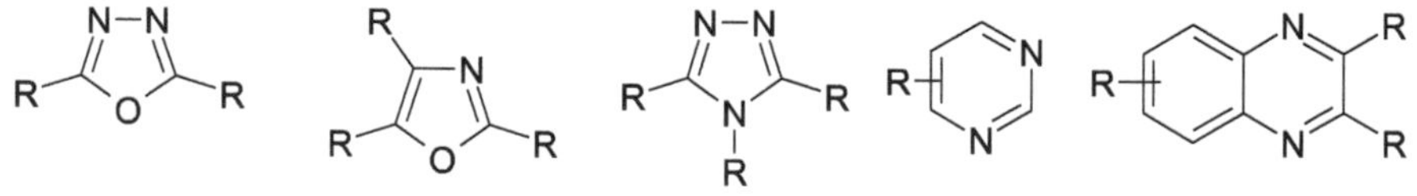

1,3,4-Oxadiazoles   1,3-Oxazoles   1,2,4-Triazoles   Pyrimidines   Quinoxalines

(3)

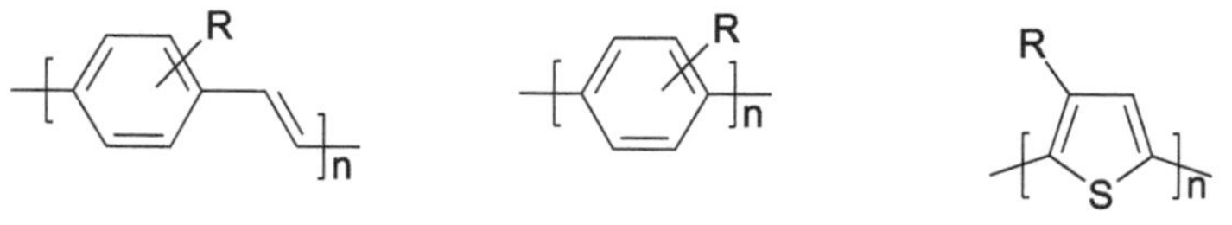

Arylene-vinylenes   Arylenes   Thiophenes

Basic structures of electron transport/hole blocking materials and oligomeric and polymeric materials for charge transport and luminescence

Two basic principles are commonly used for the preparation of OLEDs: the sublimation method, in which the organic layers are prepared by vapor deposition results in well-defined layers of excellent purity but tolerates only low molecular mass molecules with high temperature stability [16]. The less expensive preparation out of solution, requires soluble substances or precursors [17] and is therefore widely used in combination with polymers because of their homogeneity, good layer-building-properties and long-term form stability resulting in a long device lifetime.

The goal of this article is to describe the scope and limitations of synthetic routes that have been used to produce suitable oligomers and polymers for LED application. The polymers in this article will be discussed on the basis of their backbone structure and the synthetic strategy of their formation and are divided into completely π-conjugated polymers, non-conjugated polymers and polymers with defined segmentation (see Structure 4).

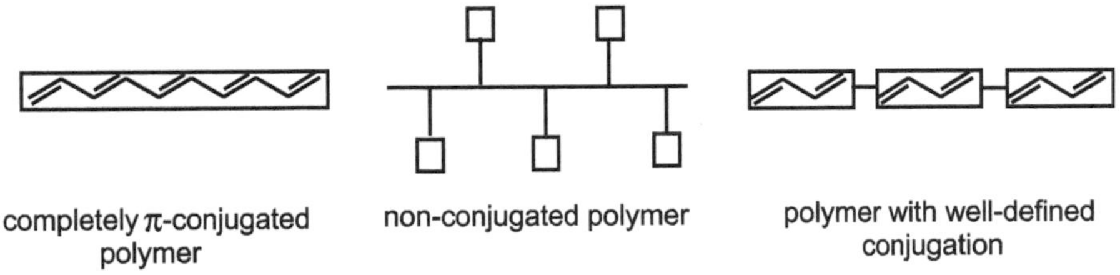

(4)

| completely π-conjugated polymer | non-conjugated polymer | polymer with well-defined conjugation |

II. π-CONJUGATED POLYMERS

Since the discovery of electrically conductive polymers by Heeger, MacDiarmid and Shirakawa et al. in 1977 [18] — resulting in the Nobel Prize in Chemistry 2000 [19] — π-conjugated systems have a major role in the field of so called 'plastic electronics'. Key property of these polymers is the conjugated double bond along the polymeric backbone, allowing charge migration after injection *via* electrodes.

A. Poly(*p*-phenylene-vinylene)s (PPV)

The first polymers used for light emitting diodes — discovered by Friend and Holmes et al. in 1990 [20] — and still the most common ones used in recent devices, are completely π-conjugated poly(*p*-phenylene-vinylene)s. These polymers — which can be used in single layer devices as both charge-transport and green emitting materials — will be discussed on the synthetic strategy of their formation.

1. Precursor Routes

Unsubstituted poly(phenylene-vinylene)s (PPVs) are insoluble in any known solvent. To improve solubility and with that processability unsubstituted PPVs were first synthesized using precursor routes like the so called Wessling- (or sulfonium-) route [21–24]. Accordingly, the condensation is performed with solubilized monomers, and a soluble polymeric intermediate is formed. The latter is converted to PPV in a final reaction step, that is preferentially carried out in the solid state, allowing the formation of homogeneous PPV films or layers. Following this route, a soluble precursor polymer with excellent film forming properties is obtained by base induced polyreaction of *p*-xylylene-α,α'-bisdialkylsulfonium salts. After spin coating, the precursor polymer is converted by polymer analogous heat induced elimination to the corresponding PPVs (Scheme 5).

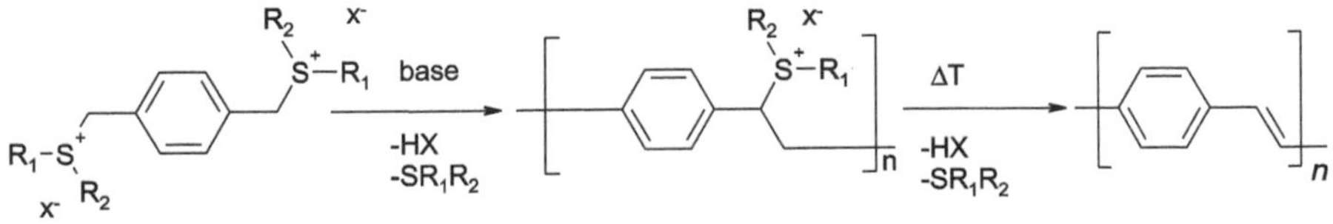

(5)

In general, any functionalized poly(*p*-xylylene) with leaving group in the α-position to the aromatic moieties can be used as precursor, as long as they fulfill the basic requirements of OLED-techniques (i.e., solubility, transparency, excellent film forming properties, good thermal stability after processing, etc.). Commonly used as leaving groups beside the sulfonium group are halogens [25,26] (so called 'Gilch-procedure'), hydrohalogenides [27], alkoxides [28] and alkylsulfinyles (known as 'Vanderzane-procedure') [29].

To avoid unwanted side reactions and damages of other device-layers during thermal conversion (e.g., by oxidation or reaction with volatile corrosive elimination products), organic-solvent soluble PPV derivatives such as poly(2-methoxy-5-(2'-ethylhexyloxy)-*p*-phenylenvinylene (MEH-PPV) or poly 2,5-dihexyloxy-*p*-phenylenevinylene (DH-PPV) (Scheme 6) have been developed. These materials can be spin-coated from solution after the conversion step. Another advantage of these PPV-derivatives is the possibility to modify the electronic properties of the film with different substitution patterns. Therefore all kind of organic substituents have been introduced into the aromatic system to alter the structure of the aromatic building block, including alkoxy-, alkyl-, cholestanoxy and silicium containing groups [30–35] (Scheme 6).

MEH-PPV: $R_1 = OCH_3$, $R_2 = H$, $R_3 = OCH_2CH(C_2H_5)(C_4H_9)$, $R_4 = H$

DH-PPV: $R_1 = OC_6H_{13}$, $R_2 = H$, $R_3 = OC_6H_{13}$, $R_4 = H$

BCHA-PPV: R_1 = Cholestanoxy, $R_2 = H$, R_3 = Cholestanoxy, R_4

DEO-PPV: $R_1 = O-(C_2H_4)_2-C_2H_5$, $R_2 = H$, $R_3 = O-(C_2H_4)_2-C_2H_5$, $R_4 = H$

BuEH-PPV: $R_1 = C_4H_9$, $R_2 = H$, $R_3 = OCH_2CH(C_2H_5)(C_4H_9)$, $R_4 =$

DMOS-PPV: $R_1 = Si(CH_3)_2C_8H_{17}$, $R_2 = H$, $R_3 = H$, $R_4 = H$

$$(6)$$

A precursor route not involving heteroatoms in the precursor polymers has also been developed. It is based on the oxidation of soluble poly(*p*-xylylene)s to corresponding PPVs by using stoichiometrical amounts of 2,3-dichloro-5,6-dicyano-1,4-benzochinone (DDQ) (Scheme 7) but is restricted so far to a-phenyl-substituted poly(*p*-xylylene)s [36].

$$(7)$$

Beside spin-coating-based preparation techniques, the so-called chemical-vapor-deposition-route (CVD) has gained considerable attention as a solvent free preparation process. Following this route, the starting materials are pyrolized after vaporization, followed by CVD and polymerization of the monomers on the substrate. Finally, the halogeno-functionalized poly(p-xylylene) is converted to PPV by polymer-analogous thermoconversion (Scheme 8) [25,37,38].

$$(8)$$

2. Polycondensation and C–C-Coupling Routes

Some drawbacks of the precursor routes mentioned above have been overcome by the use of polycondensation- and C–C-bond-coupling reactions. To produce soluble PPV-, poly(thiophene)-, or poly(pyrrol) derivatives for spin coating preparation, various types of transition metal catalyzed reactions, such as the Heck-, Suzuki-, and Sonogashira-reaction, Wittig- and Wittig–Horner-type coupling reactions, or the McMurry- and Knoevenagel-condensation have been utilized.

A typical example of the Pd catalyzed Heck reaction of 1,4-dibromo-2-phenylbenzol with ethylene to obtain the poly(phenylphenylene vinylene) [39] is depicted in Scheme 9. A common drawback of this reaction-type is the insufficient regioselectivity, resulting in 1,1 diarylation of the product (>1%, depending on the substituents) [40].

$$(9)$$

In order to avoid this problem, the Suzuki coupling is used as well to obtain various substituted PPVs. Therefore an aromatic diboronic acid or ester and dibromoalkylene are reacted in the presence of a Pd catalyst as depicted in Scheme 10 [41].

$$(10)$$

Cyano derivatives of PPV with high oxidation potential are commonly synthesized by Knoevenagel condensation of substituted terephthaldehyde with benzene-1,4-diacetonitriles yielding an alternating copolymer type product (see Scheme 11) [42]

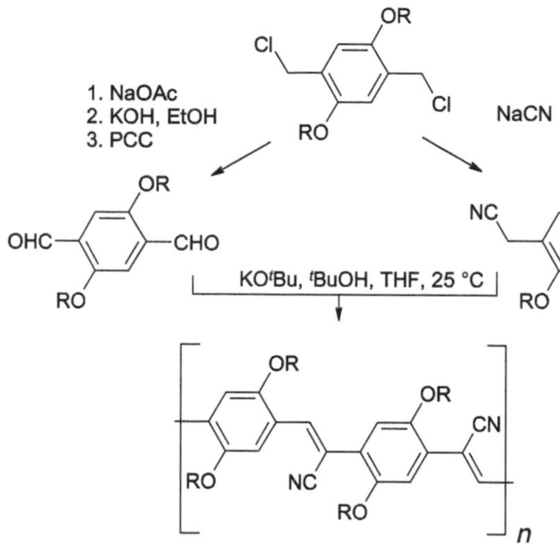

(11)

Schlüter et al. described the synthesis of soluble PPV derivatives from substituted aromatic dialdehydes *via* McMurry-type polycondensation reaction. With this low valent titanium catalyzed reaction (see Scheme 12), the obtained products are characterized by a double bond *cis/trans* ratio of about 0.4 and an average degree of polymerization of about 30 [43].

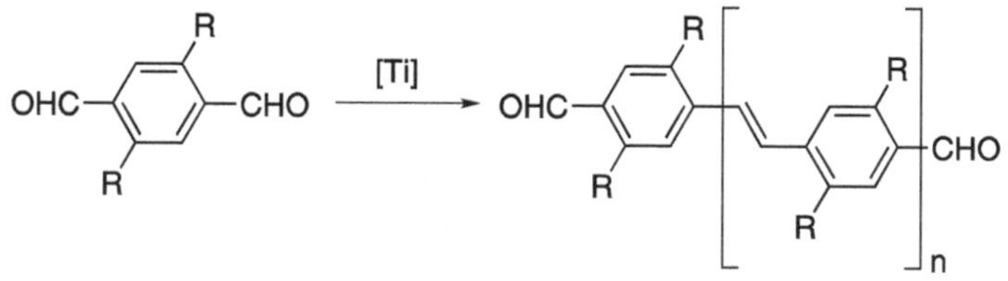

(12)

Phenylic substituents at the vinylene positions — increasing both solubility of the polymer and stability of the double bond — can be achieved by reductive dehalogenation polycondensation of 1,4-bis(phenyldichlormethyl)benzene derivatives with chromium(II)-acetate as reducing agent [44] (see Scheme 13).

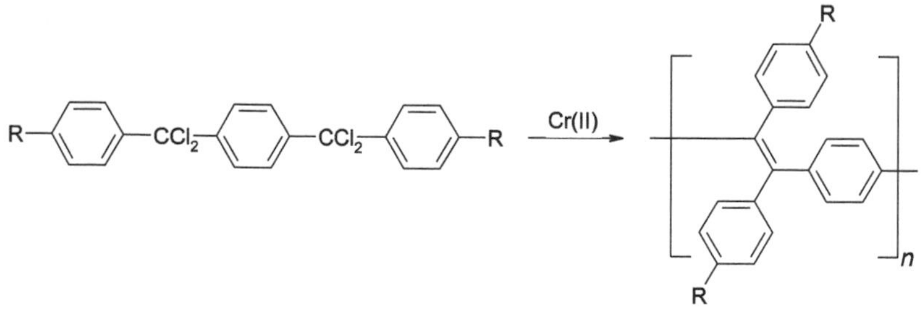

(13)

A further route leading to unsubstituted PPV was published by Grubbs et al. [45], utilizing ring-opening olefin metathesis reaction as shown in Scheme 14. Starting from bicyclic monomers with bicyclo(2.2.2)octadiene skeleton, the ring-opening metathesis polymerization (ROMP) is performed with Schrock-type molybdenum carbene catalysts. The obtained, well defined, nonconjugated soluble precursors, containing carboxylic ester functions, are then thermally converted to the conjugated PPV.

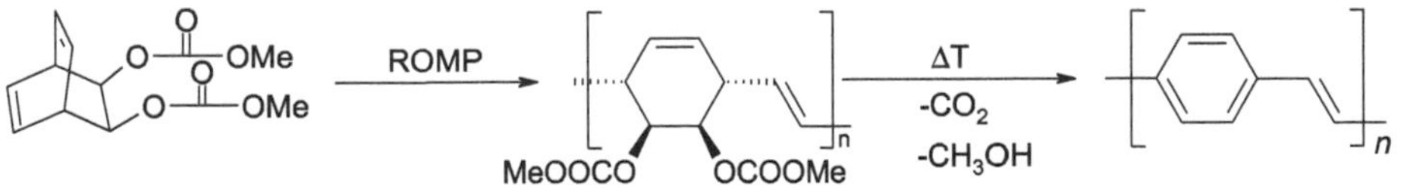

$$(14)$$

The Wittig reaction (see Scheme 15) is also a commonly used method for yielding PPV derivatives from arylene bisphosphonium salts and bisbenzaldehydes. Since only products of moderate molecular weight are obtained, more interest in this reaction is given in the field of spacer segregated poly(*p*-phenylene vinylene)s with defined conjugation length (see III.A) [46]. An improvement concerning the degree of polymerization is obtained by the Horner modification of the Wittig procedure ('Wittig–Horner reaction'). Following this route, the bisphosphonium salt is replaced by bisphosphonates or aromatic bisphosphine oxide monomers [47].

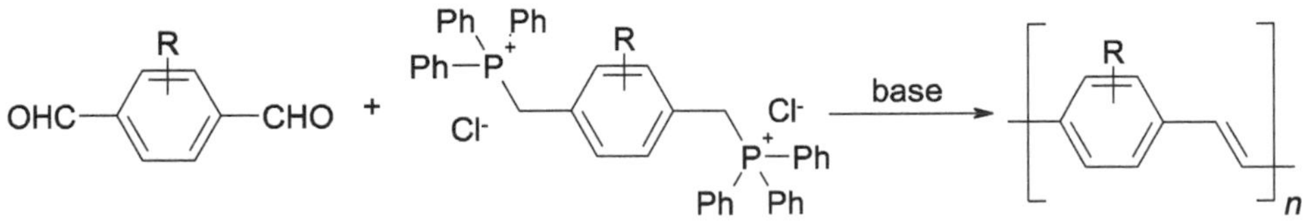

$$(15)$$

Due to the side chain induced twist within the main chain the effective conjugation length is notably effected in soluble PPVs. A strategy to overcome this problem and to develop more rigid conjugated systems has been presented by Davey and co-workers in 1995 who prepared poly(phenylene-ethynylene)-type polymers according to the following scheme (Scheme 16) [48].

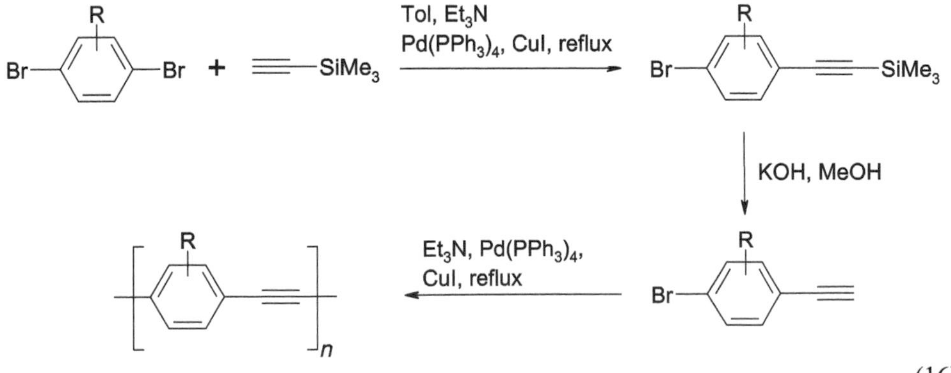

$$(16)$$

3.   Other Poly(phenylene-vinylene)s

Oligo- and poly(*m*-phenylene-vinylene) derivatives are not accessible via the polymeriza-tion approach analogue to the Wessling- or Gilch-route. Accordingly, other methods as the reductive dehalogenation polycondensation or the Wittig-type reaction as shown in II.A.1 and II.A.2. are used for their formation. Despite their increased solubility, the 1,3-phenylene-units within the poly(*m*-phenylene-vinylene)s act as conjugation barriers so that their usage in OLED techniques is very limited.

Oligomers of (*o*-phenylene-vinylene)s can be obtained using various C–C-coupling and polycondensation methods. For higher oligomers and polymers, the Stille-type coupling of 1,2-diiodobenzene or 1,2-bis(2-iodostyryl)benzene with bis(tri-*n*-butylstannyl)-ethylene was introduced by Müllen et al. [49] (see Scheme 17).

$$(17)$$

The *o*-phenylene-vinylene-structure represent an intermediate case between the *p*- and *m*-derivatives, allowing an extended *p*-conjugation and simultaneously disturbing it by the non-planar geometry between the vinylene units. Utilizing three different alkyle chains leads to the PPV-copolymer "Super Yellow" — commercially available from Corion Organic Semiconductors GmbH — which shows the best efficiency and lifetime of PPV-derivatives upto now (see Scheme 18):

$$(18)$$

B.   Heteroaromatic Systems

Heteroaromatic systems, such as the widely used poly(thiophene)s can be obtained by simple oxidative polymerization of the soluble monomers or oligomers either by electrochemical means or oxidizing agent such as $FeCl_3$ [50,51]. This common route is also used to synthesize a variety of mono- and dialkyl-, -alkoxy-, and -alkylsulfonic acid substituted and therefore soluble poly(thiophene)s [52–57] (Scheme 19) and can also be utilized to obtain poly(pyrrole)s. The disadvantage of this polymerization methods however is the regiorandom structure of the polymeric product with non-reproducible properties.

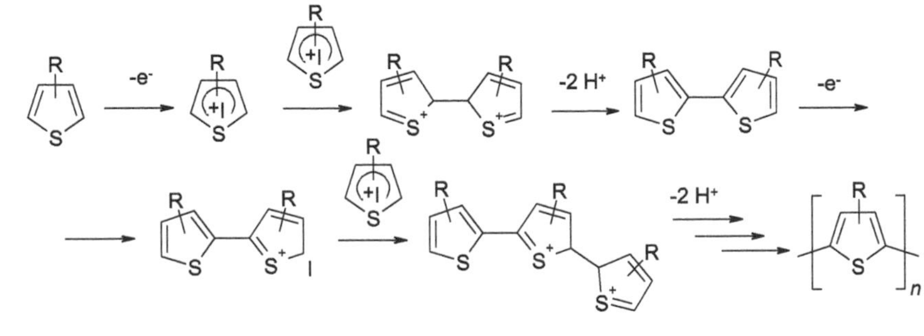

(19)

For better defined poly(thiophene) structures a variety of organometallic mediated synthesis have been introduced. Most widely employed are Grignard-type organo-magnesium compounds in addition to a nickel catalyst. Highly regioregular head-to-tail 3-alkylpoly(thiophene)s are obtained following the synthetic route of McCullough et al. (see Scheme 20).

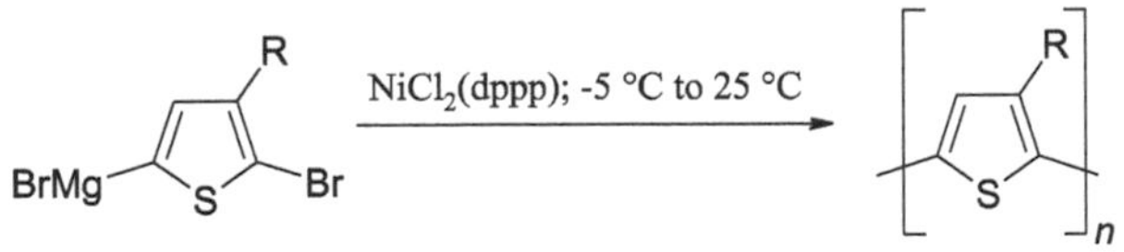

(20)

Polymers — prepared via the polymerization of 2-bromomagnesio-5-bromo-3-alkylthiophenes — exhibit enhanced conductivity and optical properties when compared with regiorandom materials [58,59]. Another approach to regioregular alkylpoly(thio-phene)s is the usage of zinc instead of magnesium in nickel- or palladium catalyzed polymerizations [60,61]. Due to the improvements, these synthetic methods are by far the most valuable synthetic routes to these materials. In contrast, the regioselective synthesis of substituted poly(pyrrole)s was not reported to date.

Heterocyclic, electron deficient conjugated systems like poly(1,3,4-oxadiazole)s, poly(1,3-oxazole)s and poly(1,2,4-triazole)s are applied in organic light emitting diodes as electron transport and hole blocking layers. The synthetic strategies for their formation are as manifold as the structures themselves, reaching from polymerization of functional monomers to polymer analogue formation of the conjugated system (e.g., by ring closure dehydration, dehalogenation, etc.). For further details is referred to the reviews of Schmidt et al. [14] and Feast et al. [62].

C. Light Emitting Polymers (LEPs) Based on Polyfluorenes

A second important class of π-conjugated polymers are polyfluorenes, which were obtained the first time by oxidative polymerization of 9-alkyl- and 9,9-dialkylfluorenes with ferric chloride [63]. These polymers showed low molecular weight and some degree of branching and non-conjugated linkages through positions other than 2 and 7.

A very successful way to improve regiospecificity and to minimize branching was the synthesis through transition-metal-catalyzed reactions of monomeric 2,7-dihalogenated fluorenes. The palladium-catalyzed synthesis of mixed biphenyles from phenylboronic acid and aryl bromide discovered by Suzuki et al. [64] tolerates a large variety of functional groups and the presence of water. This method can also be used to prepare perfectly alternating copolymers.

1.   Polyfluorene-Homopolymers

Polyfluorenes with alkyl substituents at C9 are soluble in conventional organic solvents such as aromatic hydrocarbons, chlorinated hydrocarbons and tetrahydrofuran, which made them useful to prepare thin films for OLEDs. As a consequence many efforts have been undertaken to synthesize a large number of high-molecular-weight, 9-mono-, or disubstituted very pure fluorene-based polymers.

(21)

9,9-Disubstituted 2,7-bis-1,3,2-dioxaborolanylfluorene is allowed to react with a variety of dibromoarenes in the presence of a catalytic amount of (triphenylphosphine) palladium (Scheme 21). The improved process yields high-molecular-weight polymers with a low polydispersity (<2) in less than 24 h reaction time, whereas the conventional Suzuki coupling process can take up to 72 h and more to deliver polymers of modest molecular weights. Optimized LEDs based on these polymers, made by improved Suzuki poly-fluorene chemistry, exhibited light emission exceeding 10,000 cd/m^2 with a peak efficiency of 22 lm/W [3].

2.   Polyfluorene Copolymers

The described synthesis of polyfluoren homopolymers allows also the design of alternating copolymers. Instead of the 2,7-dibromofluorene a variety of dibromoarenes can be used in the Pd-catalyzed C–C coupling polymerization reaction. An important group of comonomers are tertiary aromatic amines, which have been known as excellent hole-transport materials and have found many applications as photoconductors and in LEDs. The resulting alternating copolymers are all blue emitters, excellent film formers and show high hole mobilities [65]. These materials can be used as emitters as well as hole transporters in LED devices.

This alternating copolymer concept has been extended to other conjugated monomers as shown in Scheme (22). All synthesized copolymers [66] are of high molecular weight, are highly photoluminescent and their emissive colours can be qualitatively correlated to the extent of delocalization in the comonomers. For example the thiophene copolymer emits bluish green light, but the bithiophene copolymer emits yellow light.

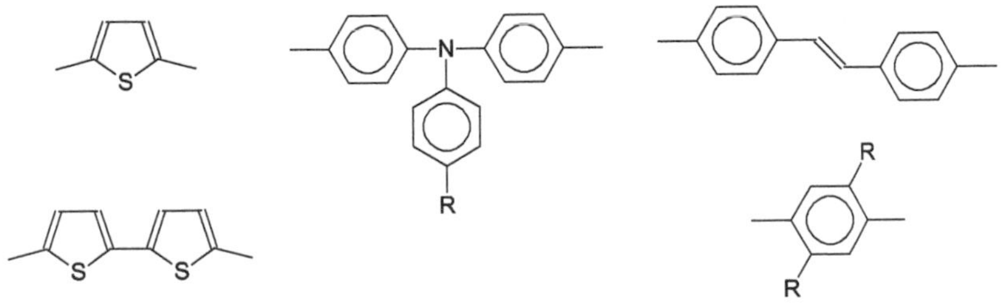

(22)

No other polymer class offers the full range of color with high efficiency, low operating voltage and high lifetime when applied in a device. The polyfluorene-based materials seem to be very viable for commercial applications. A special group within the polyfluorenes are poly-spiro-derivatives. They offer a wide range of accessible colours. Compared to standard polyfluorenes they are morphologically more stable and do not form aggregates as easily (see Scheme 23):

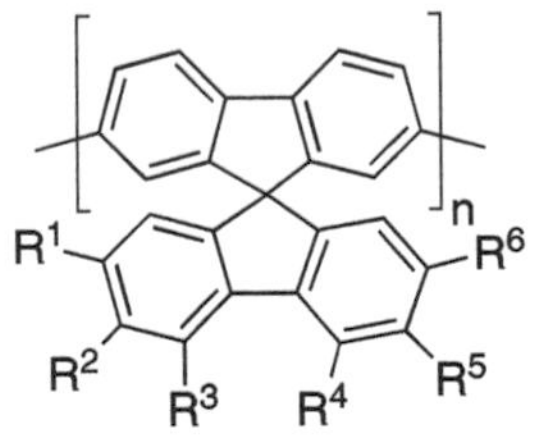

(23)

D. Poly(p-phenylene)s

Poly(p-phenylene) (PPP) represents a wide class of interesting conjugated polymers for PLED applications. To be exact, the formerly described polyfluorenes also belong to this class of polymers. Wide bandgaps are typical for PPPs and allow emission of blue light. Since the design of efficient long-lived blue emitters remains a significant challenge to the field, polymers, such as poly-p-phenylenes, are attractive candidates for consideration. As with the PPVs, most PPPs are characterized by their insolubility and infusibility, properties that were a considerable hindrance towards structural characterization and processing. Thus research activities were directed to form PPP films via soluble thermally converted precursor polymers on the one hand and the development of soluble, substituted PPPs on the other hand.

First attempts to generate poly(p-phenylene) were undertaken by Kovacic et al. in the 1960s [67]. He reported the oxidative treatment of benzene with copper(II)-chloride in the presence of strong Lewis acids (e.g. aluminum trichloride) which led to a condensation of the aromatic rings by forming radical cations as reactive intermediates. The benzene units are preferentially connected in the 1,4-position, but crosslinking and oxidative condensation to highly condensed aromates and a maximum degree of condensation of about 10 make this reaction interesting only for historical aspects.

1. PPPs by Transition-Metal-Catalyzed Condensation Reactions

The availability of newer, more effective methods for aryl–aryl coupling has been an important driving force for the development of new synthetic strategies for PPPs and other polyarylenes. Transition metal catalysis, such as the Pd(0)-catalyzed aryl–aryl coupling developed by Suzuki [63] and nickel(0)-catalyzed or -mediated coupling

according to Yamamoto [68] have been employed most successfully. An example for the
Ni(0)-catalyzed coupling is the coupling of 1,4-dibromo-2-methoxycarbonylbenzene to
poly(2-methoxycarbonyl-1,4-phenylene) as a processable PPP precursor [69]. The aromatic polyester PPP precursor is then converted to carboxylated PPP and thermally
decarboxylated to PPP with copper(II)-oxide catalysts (Scheme 24).

$$\tag{24}$$

A second, very fruitful synthetic principle for structurally homogenous, processable
PPP derivatives involves the preparation of soluble PPPs by the introduction of solubilizing side groups. The pioneering work here was carried out in the late 1980s, when
soluble poly(2,5-dialkyl-1,4-phenylene)s were prepared for the first time [70]. The Suzuki
aryl–aryl cross-coupling method (Scheme 25), adapted to polymers by Schlüter, Wegner
et al., made it possible to synthesize solubilized PPPs with a dramatically increased
molecular weight of up to 100 phenylene units.

R = alkyl, alkoxy

$$\tag{25}$$

Soluble PPPs not only contain alkyl substituents, they were also synthesized with
alkoxy groups and with ionic side groups like carboxy and sulfonic acid functions, which
are able to form PPP polyelectrolytes [71].

It is also possible to synthesize chiral PPPs as Scherf et al. reported [72]. They are
composed of chiral cyclophane subunits, made by a Suzuki-type aryl–aryl cross-coupling
reaction of the corresponding diboronic acid and dibromo derivatives. The monomers
containing cyclic $-O-C_{10}H_{20}-O-$ loops were separated into the pure enantiomers and used
to generate the corresponding stereoregular iso- and syndiotactic PPP-derivatives
(Scheme 26). The isotactic derivative possesses a chirality of its main chain.

$$\tag{26}$$

An important aspect concerning the electronic properties of the PPP is the influence of substituents at the phenylene units. In unsubstituted PPP, there is a twist angle of 23° between adjacent phenylene units [73]. This seems to be significant, but the π-overlap is a function of the cosine of the twist angle, so a fair amount of conjugative interaction remains even at 23°. If substituents are placed along the PPP backbone (e.g., at the 2- and 5-positions), the solubility is enhanced, but the π-overlap is reduced dramatically. The resulting twist angles reach from 60° to 80° depending on the length of the alkyl substituents [74].

The described facts show the synthetic demands for being able to prepare processable, and structurally defined PPPs, in which the π-conjugation remains nearly intact or is even increased compared to the parent PPP system. To realize this principle it is necessary to prepare structures in which the aromatic subunits could be obtained in a planar or only slightly twisted conformation in spite of the introduction of substituents. One of the first examples was the synthesis of polyfluorenes via oxidative coupling of fluorene derivatives as described above [63]. Another possibility to reach this aim is the preparation of 'stepladder' PPPs.

Monomers like the 2,7-dibromo-4,9-dialkyl-4,5,9,10-tetrahydropyrenes (Scheme 27) represent suitable starting monomers for the realization of such 'stepladder' structures. These difunctionalized tetrahydropyrene monomers were first prepared by Müllen et al. [75] and reacted in a Yamamoto-type coupling [76]. Reaction of the dibromide with a stoichiometric amount of a low-valent nickel(0) complex gave a poly(4,9-dialkyl-4,5,9,10-tetrahydropyrene-2,7-diyl) (PTHP) as a new, completely soluble type of PPP derivative, in which each pair of neighboring aromatic rings is doubly bridged with ethano linkages. The solubilizing alkyl substituents are attached at such positions on the periphery of the molecule that they cannot cause twisting of the main chain. The number-average molecular weight was $M_n = 20,000$, corresponding to 46 THP units.

(27)

The luminescence characteristic of PTHP suggests that it is a potential candidate for the active component in OLEDs. Investigations showed the appearance of a quite intense blue-green electroluminescence with a quantum yield of up to 0.15% (single layer construction ITO/PTHP/Ca).

The 'stepladder' concept can be logical continued towards a completely planar ladder polymer to minimize the mutual distorsion of adjacent main chain phenylene units. The complete flattening of the conjugated π-system by bridging all the phenylene subunits should then lead to maximum conjugative interaction. As with the PTHP systems, alkyl or alkoxy side chain should lead to soluble polymers. This idea was realized first in 1991 with the first synthesis of a soluble, conjugated ladder polymer [77]. The preparation is according to a so-called classical route, in which an open-chain, single stranded precursor polymer was closed to give a double stranded ladder polymer (Scheme 28). In the synthesis of this LPPP, the precursor polymer is initially prepared by Suzuki aryl–aryl coupling of an aromatic diboronic acid and an aromatic dibromoketone.

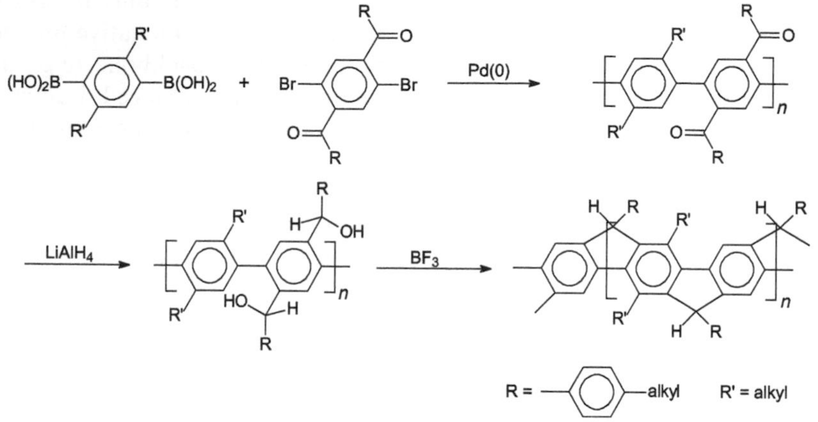

(28)

R = ⟨benzene⟩–alkyl   R' = alkyl

The cyclization to structurally defined, soluble LPPP takes place in a two-step sequence, consisting of a reduction of the keto group followed by ring closure of the secondary alcohol groups in a Friedel–Crafts-type alkylation. The resulting ladder polymer has an average molecular weight of 25,000, corresponding to 65 phenylene units. LPPP is characterized by unusual electronic and optical properties as a consequence of planarization of the chromophore. The absorption maximum undergoes a bathochromic shift to a λ_{max} value of 440–450 nm for the $\pi \rightarrow \pi^*$ transition compared to PPP with $\lambda_{max} = 336$ nm [78]. The photoluminescence of LPPP in solution is a very intensive blue, but the bulk properties are surprising different: Although efficient LEDs can be assembled, the emission of the solid state film is yellow in the case of photoluminescence and electroluminescence. In comparison to the former described PTHPs the quantum yield is with ca. 1% much higher [79].

2. Other Routes to Poly(p-phenylene)s

Recently the most popular synthetic routes to PPPs are the transition-metal-catalyzed condensation reactions discussed above, but several other syntheses were developed to generate PPP and its derivatives.

About 40 years ago, Marvel et al. described [80] the polymerization of 5,6-dibromocyclohexa-1,3-diene to poly(5,6-dibromo-1,4-cyclohex-2-ene), followed by a thermally induced, solid state elimination of HBr with formation of PPP (Scheme 29). The products, however, indicate some structural defects like incomplete cyclization and crosslinking.

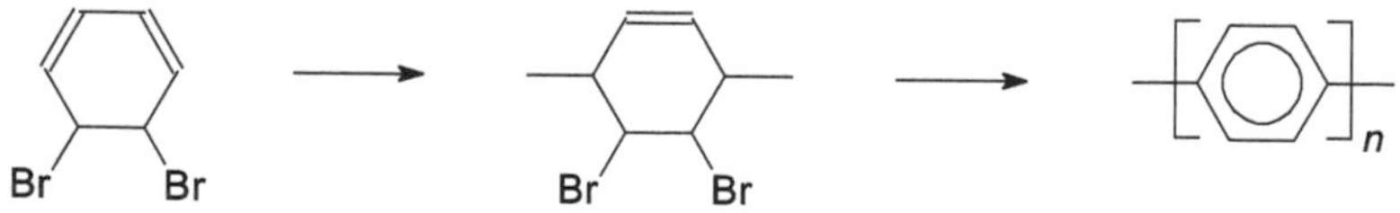

(29)

More than two decades later, Ballard et al. developed an improved precursor route, starting from 5,6-diacetoxycyclohexa-1,3-diene (Scheme 30), the so called ICI route [81,82]. The soluble precursor polymer is then aromatized thermally to PPP via elimination of two molecules of acetic acid per structural unit. The polymerization of the monomer, however, does not proceed as a uniform 1,4-polymerization: beside the regular 1,4-linkages about 10% of 1,2-linkages are formed as a result of a 1,2-polymerization of the monomer.

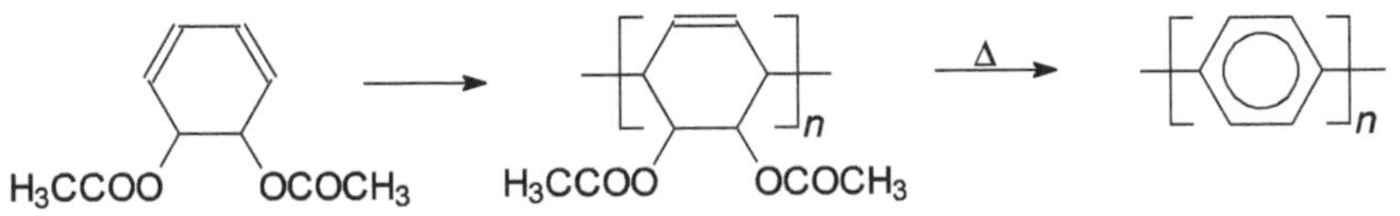

$$(30)$$

An improved precursor route to high molecular weight, structurally regular PPP by transition metal-catalyzed polymerization of a cyclohexa-1,3-diene derivative to a stereoregular precursor polymer was described by Grubbs et al. [83] and MacDiarmid [78] et al (Scheme 31). The final step of the reaction sequence is the thermal, acid-catalyzed elimination of acetic acid, to convert the precursor into PPP. They obtained PPP films of a definite structure, which were unfortunately contaminated with large amounts of polyphosphoric acid, the acidic reagent employed.

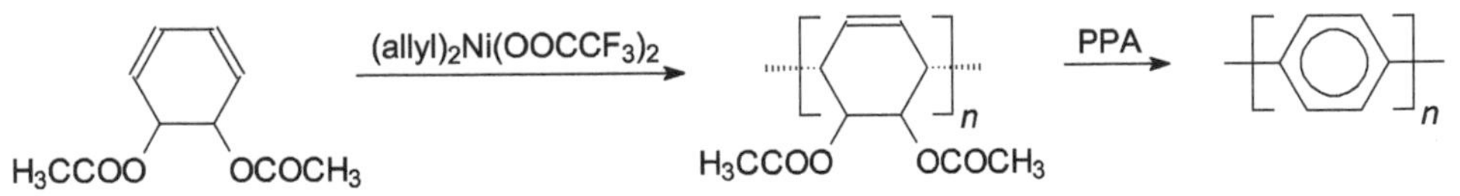

$$(31)$$

Another possibility to receive PPP derivatives is the Bergman cyclization (Scheme 32), starting from substituted enediynes, e.g., 1-phenyl-hex-3-en-1,5-diyne, leading to poly(2-phenyl-1,4-phenylene). It is also possible to synthesize the structurally related poly(2-phenyl-1,4-naphthalene) from 1-phenylethynyl-2-ethynylbenzene [84,85].

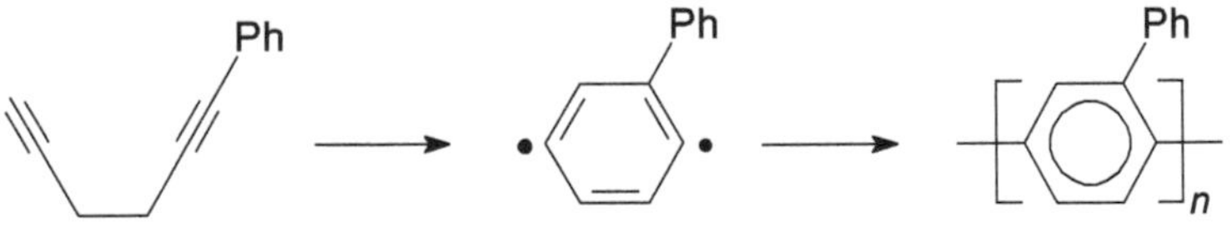

$$(32)$$

Although PPPs and its derivatives reveal extraordinarily high thermal and oxidative stabilities, corresponding single-layer OLEDs exhibit only low electroluminescence efficiencies. Higher efficiencies have been achieved by preparing polymer blends or by virtue of two-layer OLED-constructions. External efficiencies up to 3% were determined for an ITO/PVK/poly(2-decyloxy-1,4-phenylene)/Ca — OLED [86,87].

These recent achievements are the result of new and more efficient synthetic methods, which permit chemo- and regioselective syntheses and allow molecular weights high enough to cast films with good integrity resulting in reasonable efficiencies [88].

III. CONDUCTING POLYMERS WITH ISOLATED CHROMOPHORES

A major challenge in the last few years remained the development of materials that combine the processability of polymers with the defined optical and electrical properties of low molecular weight chromophores. Strategies to optimize electrooptical and mechanical properties in OLEDs have generally implemented the use of chromophores with defined conjugation length either inserted in the polymer backbone or alternatively attached as pendant side groups. By manipulation of the primary chemical structure, processing characteristics as well as optoelectronic properties, i.e. regulating the HOMO and LUMO energy levels allowing fine tuning of charge injection properties can be controlled.

A. Nonconjugated Polymers with Side Chain Chromophores

One approach to obtain electroluminescent material with defined conjugation length is to link the chromophore as side chains to a nonconjugated polymer backbone. This concept was found to be of great advantage for the transformation of crystalline materials to amorphorous ones and combines high loading efficiency without segregation of the fluorophores. Moreover, as the synthesis occurs without any transition metal catalyst, that can be difficult to remove and might act as quencher, the obtained polymers show high purity grade.

Based on this approach, suitable monomers with defined electron or hole transporting units were prepared and polymerized by different methodologies. Following this strategy polymethacrylates, polystyrenes, polynorbornenes, polysiloxanes, polyethers, and polyesters with different side chain chromophores have been synthesized.

Free radical polymerization of functionalized methacrylate monomers containing charge transporting units has been reported by several groups [89–92]. Electron transporting as well as hole transporting material have been synthesized by this method (Scheme 33). A tri-functional copoly(methacrylate) bearing a blue emitting distyrylbenzene chromophore, an aromatic oxadiazole electron transporting unit and a cross-linkable cinnamate showing strong electroluminescence and good solution processability has been synthesized. Although the luminescent properties would be satisfactory for application in electroluminescent devices, one of the major problems related to this application, namely stability could not be achieved with polymethacrylate backbone polymers. It was shown by Cacialli et al. [91] that both the methacrylate backbone and the chromophore groups were susceptible to alteration processes.

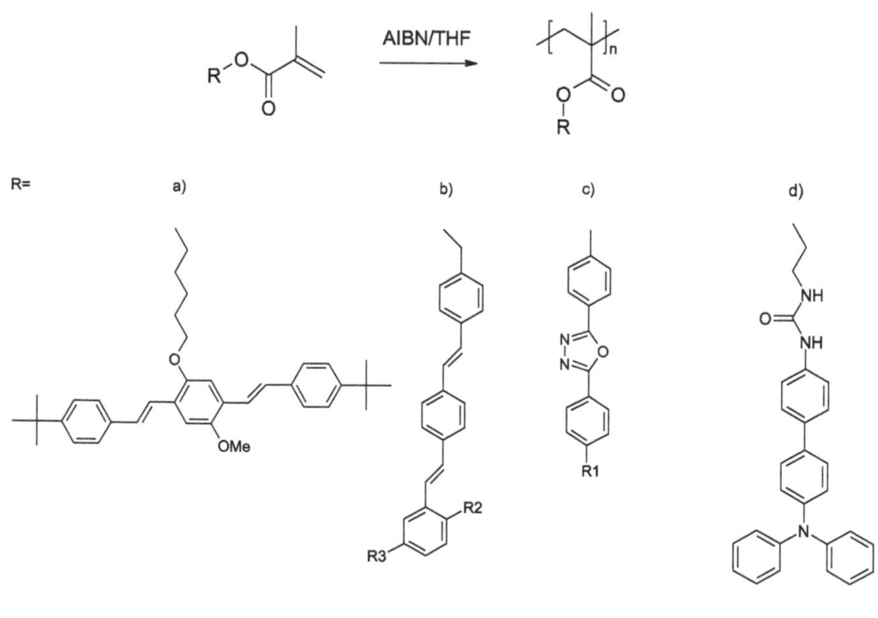

(33)

Styrene derivatives represent a second important class of monomers that can be polymerized by free radical polymerization. By doing so, copolymers with TPD and oxetane side chains were obtained where the oxetane group was used for covalent cross-linking [93] (Scheme 34). The resulting film showed excellent mechanical integrity proving this method to be very useful to improve mechanical properties. Moreover, oligo(PPV)s [94] or electron transporting units such as oxadiazoles [95] were also introduced as pendant groups with a polystyrene backbone.

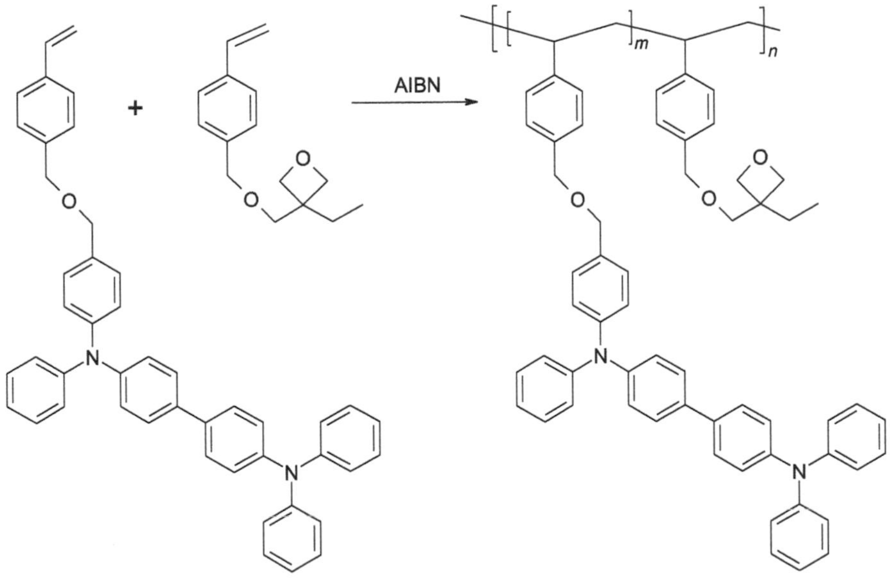

(34)

A second possibility to attach the chromophores to the polymer backbone is the polymer analogue modification with the appropriate electron or hole transporting unit. Complete substitution of poly(p-acetoxystyrene) was achieved through Williamson condensation with chloromethylstilbene resulting in a polymer with $M_n = 22{,}000$ Da, $M_w/M_n = 2.1$ and $T_g = 71\,^\circ$C (Scheme 35 [96]).

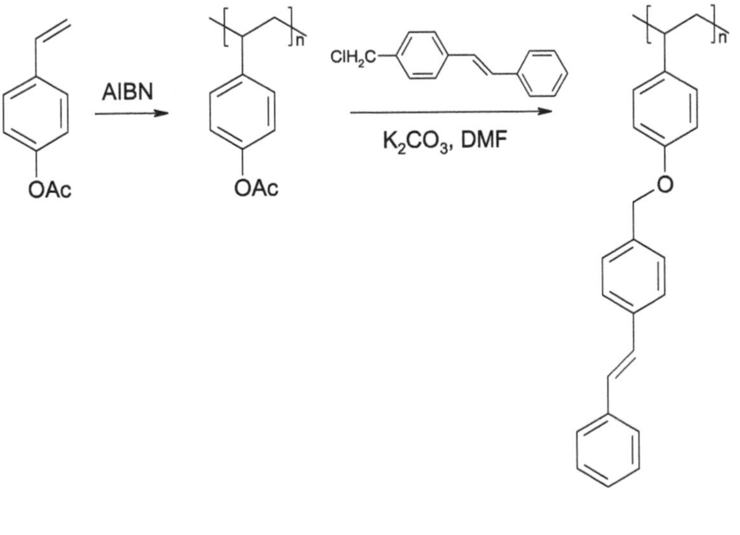

(35)

A similar synthetic concept has been used to attach pyrene emitting groups on poly(methylhydrosiloxane) by hydrosilation in the presence of chloroplatinic acid [89,97]. (Scheme 36). The reaction was complete after one week giving a polymer with $M_n = 6800$ Da with an emission wavelength of 500 nm in a single-layer device, appearing light blue.

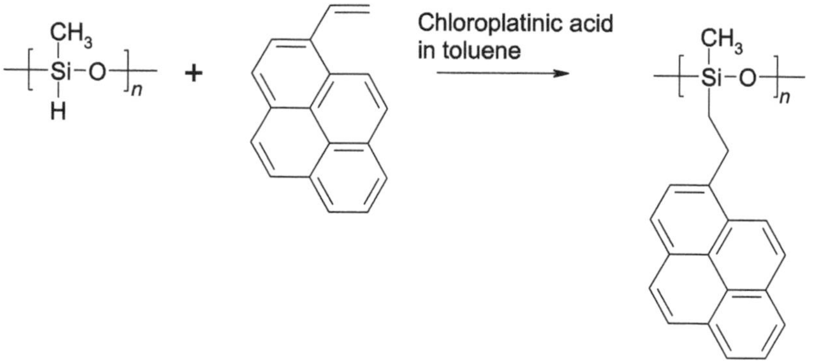

(36)

Moreover, dimethylsilane modified oxadiazole units were used to functionalize polystyrene-block-polyisoprene copolymers via Heck reaction or hydrosilation. The degree of side chain functionalization varied from 4–44% [98].

Anionic polymerization of styrene bearing TPD-like side chains has also been reported [99] (Scheme 37). A wide range of hole transporting polymers were prepared with high T_gs ranging from 132–151 °C to increase thermal and long-term stability of the device.

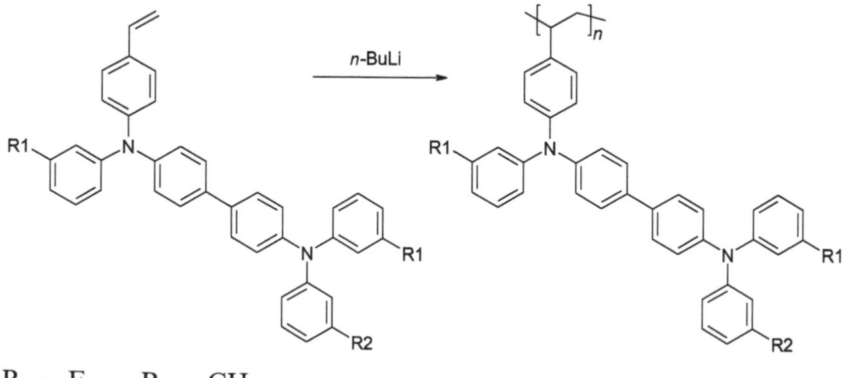

(37)

$R_1 = F$ $R_2 = CH_3$

The ring-opening metathesis polymerization of side chain functionalized norbornene monomers was applied to obtain electroluminescent polymers (Scheme 38) [100,101]. Homo- and copolymers with 25 or 50 repeat units were prepared with M_n ranging from 19,400–53,000 Da and $M_w/M_n = 1.02$–1.04. The color of emission could be fine tuned by varying the monomer ratio.

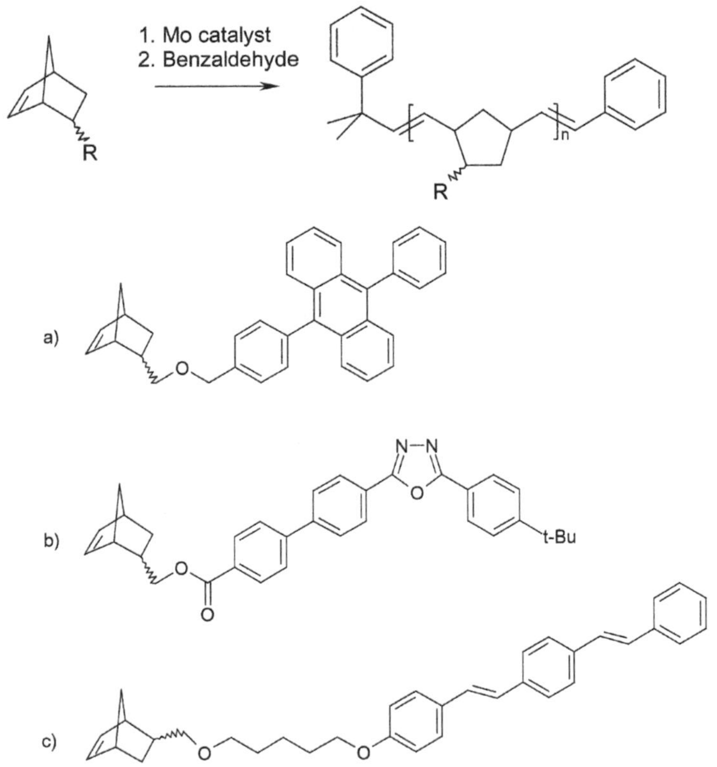

(38)

Poly(norbornenes) containing pendant triarylamines have also been synthesized by ruthenium catalyzed ring-opening metathesis polymerization [102]. By varying the polarity and the length of the linker between polymer backbone and the triarylamine functionality the device characteristics could be tuned (Scheme 39).

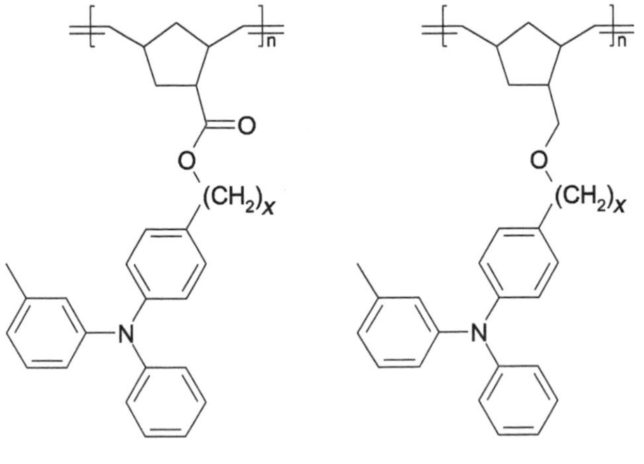

(39)

The substitution of ester groups by less polar ether functionalities enhances thereby external quantum efficiencies, lowers the operating voltage and improves the stability of the device. Further improvement is obtained by reducing the length of the alkyl linker.

A further interesting concept is the synthesis of a polymer having a fully conjugated backbone and pendant side-chain chromophores, combining electron-transport, hole-transport and light emitting properties in a single polymeric material. Hybrid polymers of this type with oxadiazole side chains were obtained by Bao et al. [103] via Heck (Scheme 40) or Stille reaction (Scheme 41). The polymers prepared via the Heck reaction showed a molecular weight $M_n = 28,500$ Da with a polydispersity of 3.65 whereas the Stille coupling resulted in lower molecular weight polymers of 8100 Da with polydispersities of 1.67. The polymer with a PPV-like backbone indicated yellow-orange light emission whereas polymers containing a thiophene group emit red-orange light. Both polymers showed better external quantum efficiencies and better charge injection properties compared to those without any side chain.

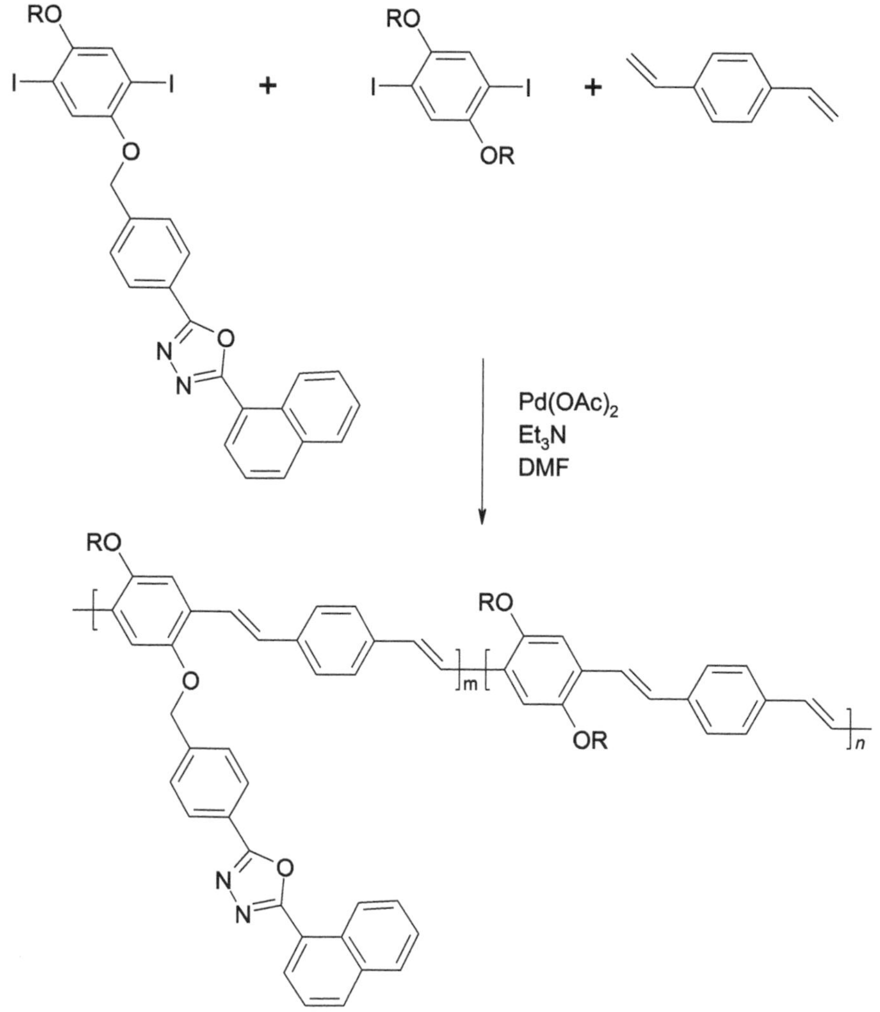

(40)

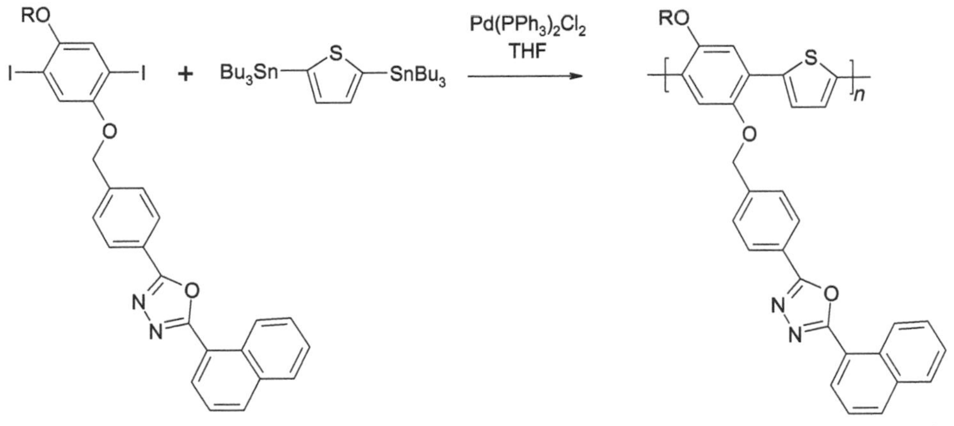

(41)

B. Main Chain Polymers with Defined Segmentation

The effective conjugation length in main chain polymers can be controlled in two ways, either by steric hindrance of π-conjugated segments that are associated in a non-coplanar way or by introducing conjugation interrupters. Both concepts have been used frequently in the past decade to gain better control over the specific device properties, including electroluminescence, photoconductivity and photovoltaic effects.

Segmented conjugated polymers have the advantages over fully conjugated ones that their electronic properties are independent of the degree of polymerization and can be easily tuned by varying the substituents or the conjugation length.

1. Conjugated Main-chain Polymers with Twisted Conformation

Since the π-overlap is a function of the cosine of the twist angle of adjacent aromatic units two approaches have been used in the past to control the effectice conjugation length, either by changes in the polymer geometry or topology. A twisted conformation of the polymer backbone was achieved by introduction of alkyl or alkoxy substituents in 2,5 positions along the PPP backbone causing twist angles of 60–80 ° [104].

A second powerful approach are meta linkages in PPV (Scheme 42) that led also to the interruption of conjugation due to the non-coplanar arrangement of adjacent conjugated repeat units [105,106]. This concept was also applied to prepare chiral polymers using 1,1-binaphthyl units [107,108] (Scheme 41) or conjugated polymers with anthracene units [109–111].

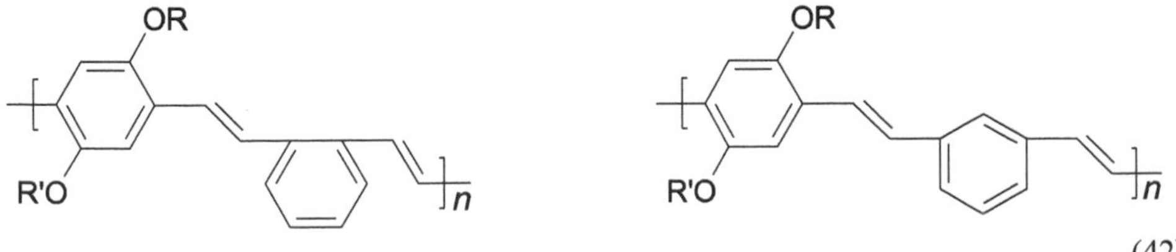

(42)

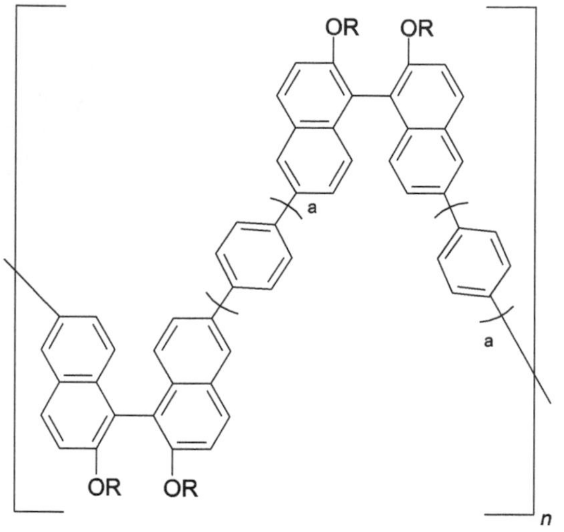

(43)

2. Main Chain Polymers with Non-conjugated Interrupters

More important from a synthetic point of view, however, is the concept of non-conjugative interrupters. In such a polymeric material the optoelectronic properties can be tailored by the proper selection of the chromophore unit whereas the physical properties can be adjusted by the non-chromophoric part. Burn et al. [112] introduced the notion of isolated chromophores in 1992 by selectively eliminating one of two leaving groups of the precursor polymer to give a *conjugated–non-conjugated* polymer.

The design of such polymers with controlled chromophore length is quite a large area, as chromophores and spacers can be combined in an almost infinite way. Chromophores can be hole-transporting triarylamine or electron-transporting ones such as oxadiazoles, with ethers, esters [113], amines, amides, imides [114–116], silanes, fluorenylidene as possible conjugation interrupters [117,118].

Flexible alkyl spacer through ether linkage were introduced by Wittig reaction between the corresponding aldehyde and triphenylphosphonium salt. This has been used for example to synthesize various polymers where the oligomeric PPV segments are separated by polymethylene spacers, which show blue or blue-green emission (Scheme 44) [119].

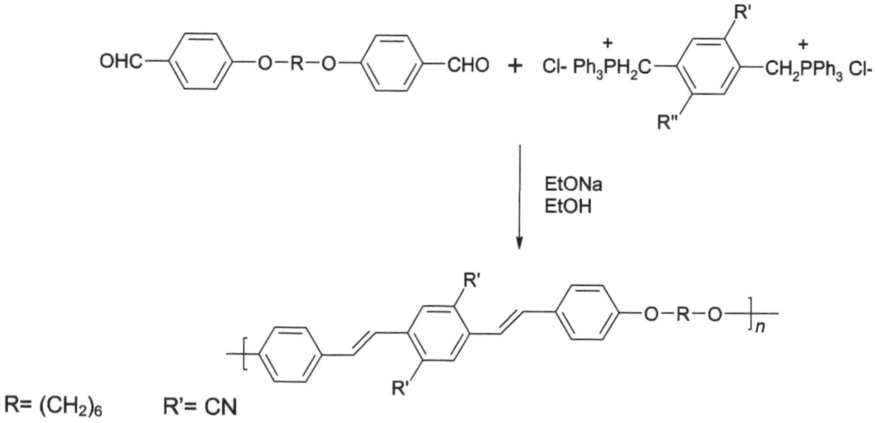

(44)

R= $(CH_2)_6$ R'= CN

Polymers with other substituents [117,120] or similar structures [121,122] were also synthesized by this method.

Hadziioannou et al. introduced a dialkylsilyl spacer between PPV-like segments [123] via Heck reaction (Scheme 45), where R is an alkoxy or an alkyl group. For $a = 1$, the polymer can be obtained via Wittig reaction between the silicon containing dialdehyde and the appropriate triphenylphosphonium halide [124].

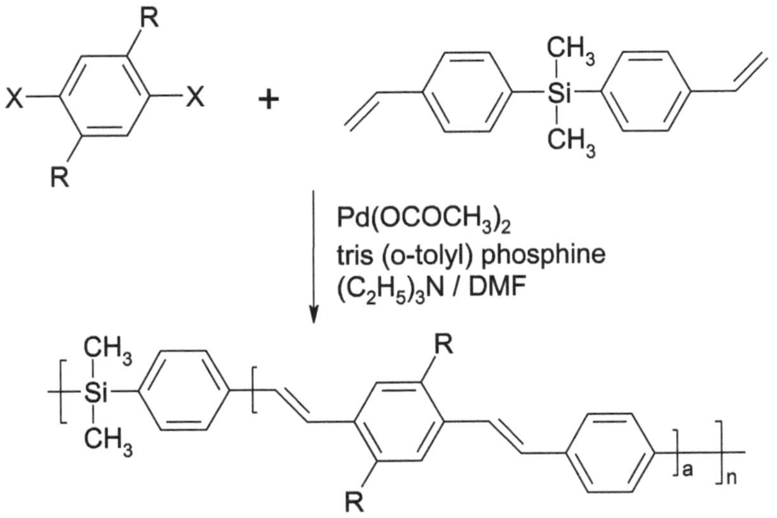

(45)

The same spacer has also been used to interrupt the conjugation between oligothiophenes [125,126]. In this case, the insertion of the dialkylsilyl group occurs via a Wurtz coupling or a palladium-catalyzed polycondensation. The general structure of the obtained polymers is depicted in Scheme (46) where R represents an alkyl chain.

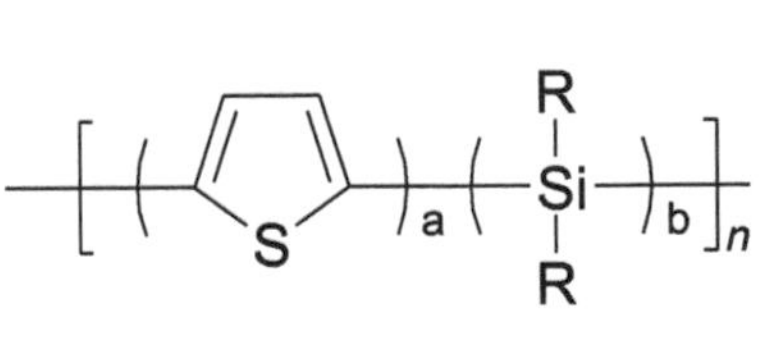

(46)

This class of polymers shows not only excellent environmental stability but can also be easier processed with the addition of long, flexible hydrocarbon side chains in comparison to most of the conjugated polymers. Moreover, the electroluminescence characteristics in single-layer devices indicated a strong dependence on the number of thiophene monomers and were found to be gradually shifted from green to red with an increasing number of thiophene units from three to seven.

Fluorenylidene linkages have been widely used as conjugation interrupters due to their stiff conformation and to the ability to increase solubility of the resulting polymer and were therefore incorporated in several PPP based polymers by the Suzuki polycondensation reaction [127]. Following this concept a blue emitting polymer based on

fluorenylidene segmented oligo(p-phenylene)s was synthetized by Nuyken et al.
[128,129] as described in Scheme (47).

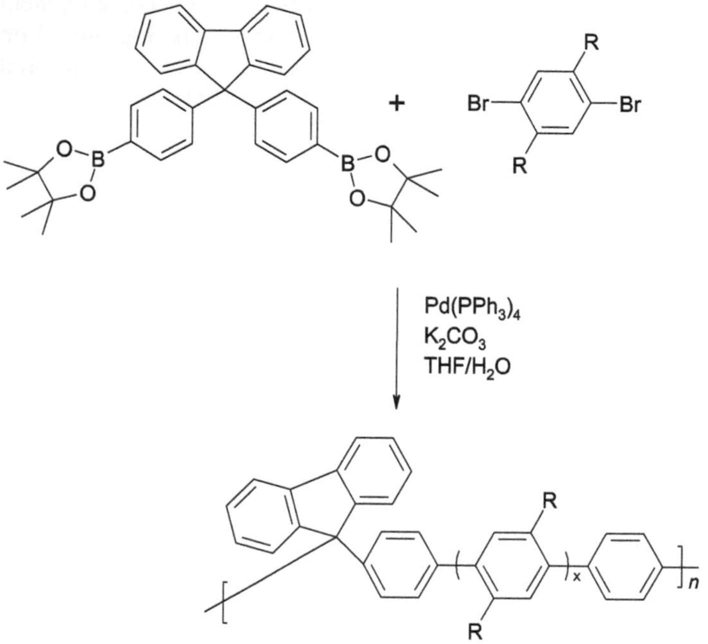

$$(47)$$

The fluorenylidene linker concept was also applied to the synthesis of triarylamine
segmented polymers via a Hartwig–Buchwald reaction [132] (Scheme 48).

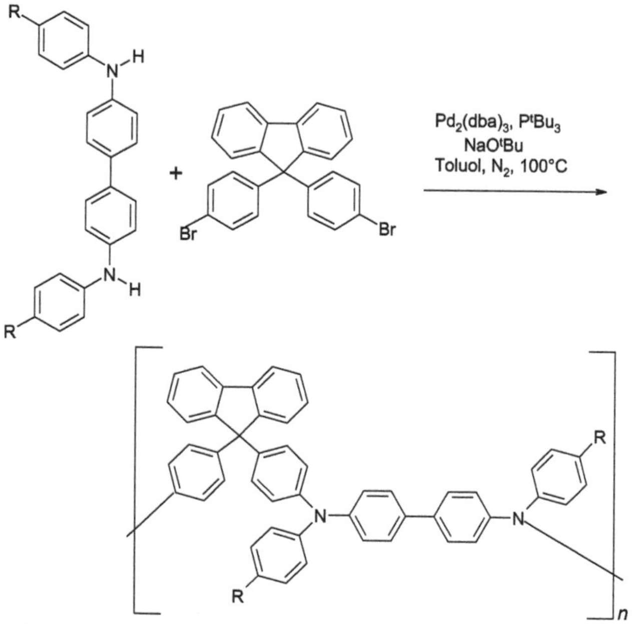

$$(48)$$

Dba = dibenzylidene aceton

Moreover, Miller et al. introduced fluorenylidene linker as well as other conjugation
breakers such as hexafluoropropylidene [127] via the nickel-mediated Yamamoto reaction
[133] as described in Scheme (49).

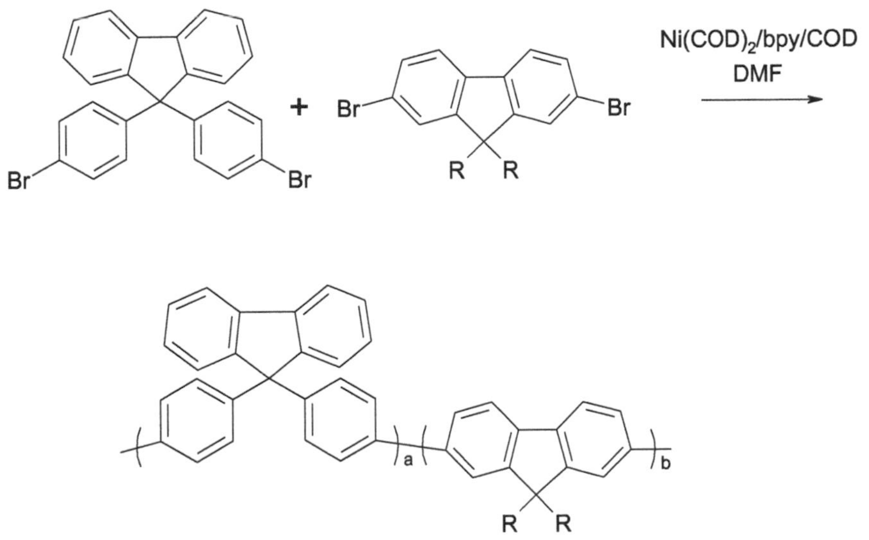

(49)

COD = cyclooctadiene; bpy = bipyridine

Two copolymers were synthesized with different ratios of the fluorene-based monomer and the fluorenylidene linker. Both polymers were obtained in high yields and high molecular weight with $M_w = 55,000$–$89,500$ Da and exhibited excellent thermal stability. Glass transition temperatures ranged from 153 °C to 197 °C with decomposition temperature (5% weight loss measurend by TGA analysis) of 440 to 450 °C.

Wegner et al. developed a class of polymers containing sequences linked by ethylene, vinylene or ethynylene groups [130]. PPV-like polymers containing an adamantane spacer group (Scheme 50) [131] are also accessible via the palladium catalyzed Suzuki reaction. The polymers possessed excellent thermal stability and shows an onset of thermal decomposition temperature at 362 °C and a T_g at 151 °C due to the chain rigidity.

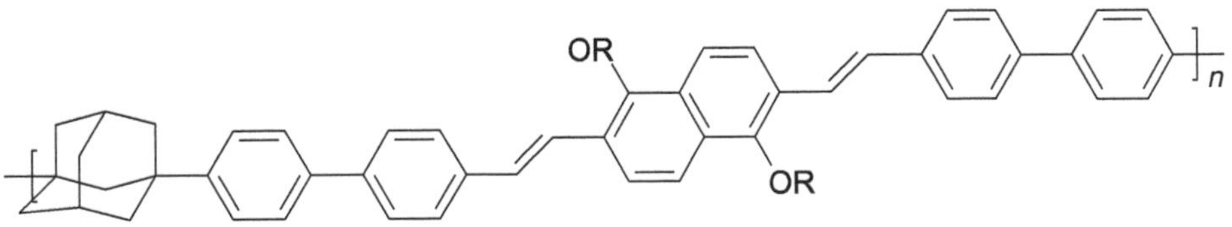

(50)

IV. CONCLUSIONS

In the past decade a great deal of attention has been focused on the preparation and characterization of π-conjugated oligomers and polymers due to their potential application as novel materials for optoelectronics. Initial attempts have been employed oxidative polymerization methods which led very often to low molecular weight compounds with poor defined polymer structure. The availability of newer, more efficient methods for aryl–aryl coupling has been the most important driving force for the development of new synthetic polymerization strategies. In particular, the Suzuki coupling and Yamamoto coupling have been utilized most successfully allowing the regio- and chemoselective synthesis of defined oligomers and polymers with excellent film forming properties and tunable photoluminescence characteristics.

REFERENCES

1. Friend, R. H., Burroughes, J., and Shimoda, T. (1999). *Physics World, 12*: 35.
2. Theis, D. (1996). Status and challenges of flat panel display technologies. In *Inorganic and Organic Electroluminescence* (Mauch, R. H., and Gumlich, H.-E., eds.), Wissenschaft und Technik Verlag, Berlin, S. 3 ff.
3. Friend, R. H., Gymer, R. W., Holmes, A. B., Burroughes, J. H., Marks, R. N., Taliani, C., Bradley, D. D. C., Dos Santos, D. A., Brédas, J. L., Lögdlund, M., and Salaneck, W. R. (1999). *Nature, 397*: 121.
4. Deußen, M., and Bässler, H. (1997). *Chem. Unserer Zeit, 31*: 76.
5. Parker, I. D. (1994). *J. Appl. Phys., 75*: 1656.
6. Vestweber, H., Oberski, J., Greiner, A., Heitz, W., Mahrt, R. F., and Bässler, H. (1993). *Adv. Mater. Opt. Elect., 2*: 197.
7. Tak, Y.-H., Vestweber, H., Bässler, H., Bleyer, A., Stockmann, R., and Hörhold, H.-H. (1996). *Chem. Phys., 212*: 471.
8. Littmann, J., and Martic, P. (1992). *J. Appl. Phys., 72*: 1957.
9. Tamoto, N., Adachi, C., and Nagai, K. (1997). *Chem. Mater., 9*: 1077.
10. Sano, T., Fujii, T., Nishio, Y., Hamada, Y., Shibata, K., and Kuroki, K. (1995). *Jpn. J. Appl. Phys., 34*(6A), Part 1: 3124.
11. Saito, S., Tsutsui, T., Era, M., Takada, N., Adachi, C., Hamada, Y., and Wakimoto, T. (1993). *Proc. SPIE, 1910*: 212.
12. Chen, C. H., Shi, J., and Tang, C. W. (1998). *Macromol. Symp., 125*: 1.
13. Kido, J. (1993). *Jpn. J. Appl. Phys.*, Part 2, *32*: L917.
14. Thelakkat, M., and Schmidt, H.-W. (1998). *Polym. Adv. Technol., 9*: 429.
15. Kalinowski, J. (1999). *J. Phys. D: Appl. Phys., 32*: R179.
16. Weaver, M. S., and Bradley, D. D. C. (1996). *Synth. Met., 83*: 61.
17. Holmes, A. B., Bradley, D. D. C., Brown, A. R., Burn, P. L., Burroughes, J. H., Friend, R. H., Greenham, N. C., Gymer, R. W., Halliday, D. A., Jackson, R. W., Kraft, A., Martens, J. H. F., Pichler, K., and Samuel, I. D. W. (1993). *Synth. Met., 57*: 4031.
18. Shirakawa, H., Louis, E. J., MacDiarmid, A. G., Chiang, C. K., and Heeger, A. J. (1977). *J. Chem. Soc. Chem. Comm.*, 579.
19. The Royal Swedish Academy of Sciences, Information Department, Stockholm, Sweden, *The Nobel Prize in Chemistry, 2000: Conductive Polymers*; Press release: http://www.nobel.se/chemistry/laureates/2000/press.html.
20. Borroughes, J. H., Bradley, D. D. C., Brown, A. R., Marks, R. N., Mackay, K., Friend, R. H., Burns, P. L., and Holmes, A. B. (1990). *Nature, 347*: 539.
21. Wessling, R. A., and Zimmermann, R. G. (Dow Chemical), US-B 3401152, 1968.
22. Wessling, R. A., and Zimmermann, R. G. (1968). *Chem. Abstr., 69*: 87735q.
23. Wessling, R. A. (1985). *J. Polym. Sci. Polym. Symp., 72*: 55.
24. Garay, R. O., Baier, U., Bubeck, C., and Müllen, K. (1993). *Adv. Mater., 5*: 561.
25. Staring, E. G. J., Braun, D., Rikken, G. L. J. A., Demandt, R. J. C. E., Kessener, Y. A. R. R., Bouwmanns, M., and Broer, D. (1994). *Synth. Met., 67*: 71.
26. Gilch, H. G., and Wheelwright, W. L. (1966). *J. Polym. Sci. A1, 4*: 1337.
27. Hörhold, H.-H., Helbig, M., Raabe, D., Opfermann, J., Scherf, U., Stockmann, R., and Weiß, D. Z. (1987). *Chemistry, 27*: 126.
28. Greiner, A., Mang, S., Schäfer, O., and Simon, P. (1997). *Acta Polym., 48*: 1.

29. Louwet, F., Vanderzane, D., Gelan, J., and Mullens, J. (1995). *Macromolecules*, 28: 1330.
30. Gettinger, C. L., Heeger, A. J., Drake, J. H., and Pine, D. J. (1994). *J. Chem. Phys.*, 101: 1673.
31. Schwartz, B. J., Hide, F., Andersson, M. R., and Heeger, A. J. (1997). *Chem. Phys. Lett.*, 265: 327.
32. Staring, E. G. J., Demandt, R. C. J. E., Braun, D., Rikken, G. L. J. A., Kessener, Y. A. R. R., Venhuizen, T. H. J., Wynberg, H., Hoeve, W. T., and Spoelstra, K. J. (1994). *Adv. Mater.*, 6: 934.
33. Wudl, F., Höger, S., Zhang, C., Pakbaz, P., and Heeger, A. J. (1993). *Polym. Prepr. Am. Chem. Soc., Div. Polym. Chem.*, 34: 197.
34. Garay, R. O., Mayer, B., Karasz, F. E., and Lenz, R. W. (1995). *J. Polym. Sci.: A 1*, 33: 525.
35. Höger, S., McNamara, J. J., Schricker, S., and Wudl, F. (1994). *Chem. Mater.*, 6: 171.
36. Schäfer, O., Mang, S., Arici, E., Lüssem, G., Unterlechner, C., Wendorff, J. H., and Greiner, A. (1998). *Macromol. Chem. Phys.*, 199: 807.
37. Iwatsuki, S., Kubo, M., and Kumeuchi, T. (1991). *Chem. Lett.*, 1071.
38. Schäfer, O., Greiner, A., Pommerehne, J., Guss, W., Vestweber, H., Tak, H. Y., Bässler, H., Schmidt, C., Lüssem, G., Schartel, B., Stümpflen, V., Wendorff, J. H., Spiegel, S., Möller, C., and Spiss, H. W. (1996). *Synth. Met.*, 82: 1.
39. Greiner, A., and Heitz, W. (1988). *Macromol. Rapid Commun.*, 9: 581.
40. Martelock, H., Greiner, A., and Heitz, W. (1991). *Macromol. Chem.*, 192: 967.
41. Koch, F., and Heitz, W. (1997). *Macromol. Chem. Phys.*, 198: 1531.
42. Moratti, S. C., Holmes, A. B., Baigent, D. R., Friend, R. H., Greenham, N. C., Grüner, J., and Palmer, P. J. (1995). *Synth. Met.*, 71: 2117.
43. Rehahn, M., and Schlüter, A.-D. (1990). *Macromol. Chem. Rapid Commun.*, 11: 375.
44. Hörhold, H.-H., Gottschaldt, J., and Opfermann, J. (1977). *J. Prakt. Chem.*, 319: 611.
45. Conticello, V. P., Gin, D. L., and Grubbs, R. H. (1992). *J. Am. Chem. Soc.*, 114: 9708.
46. Sokolik, I., Yang, Z., Karasz, R. E., and Morton, D. C. (1993). *J. Appl. Phys.*, 74: 3584.
47. Hörhold, H.-H., and Helbig, M. (1987). *Macromol. Chem. Macromol. Symp.*, 12: 229.
48. Swanson, L. S., Lu, F., Shinar, J., Ding, Y. W., and Barton, T. J. (1993). *Proc. SPIE Int. Soc. Opt. Eng.*, 1910: 101; Weder, C., and Wrighton, M. S. (1996). *Macromolecules*, 29: 5157.
49. Mauermann-Düll, H., Böhm, A., Fiesser, G. and Müllen, K. (1996). *Macromol. Chem. Phys.*, 197: 413.
50. Roncali, J. (1992). *Chem. Rev.*, 92: 711.
51. Sugimoto, R., Takeda, S., Gu, H. B., and Yoshino, K. (1986). *Chem. Express*, 1: 635.
52. Leclerc, M., Diaz, F. M., and Wegner, G. (1989). *Macromol. Chem.*, 190: 3105.
53. Österholm, J.-E., Laakso, J., Nyholm, P., Isotalo, H., Stubb, H., Inganäs, O., and Salaneck, W. R. (1989). *Synth. Met.*, 28: C435.
54. Ikenoue, Y., Saida, Y., Kira, M., Tomozawa, H., Yashima, H., and Kobayashi, M. (1990). *J. Chem. Soc., Chem. Commun.*, 1694.
55. Andersson, M. R., Pei, Q., Hjertberg, T., Inganäs, O., Wennerström, O., and Österholm, J.-E. (1993). *Synth. Met.*, 55/57: 1227.
56. Souto Maior, R. M., Hinkelmann, K., Eckert, H., and Wudl, F. (1990). *Macromolecules*, 23: 1268.
57. Zagorska, M., and Krische, B. (1990). *Polymer*, 31: 1379.
58. McCullough, R. D., Tristram-Nagle, S., Williams, S. P., Lowe, R. D., and Jayaraman, M. J. (1993). *J. Am. Chem. Soc.*, 115: 4910.
59. McCullough, R. D., Williams, S. P., Jayaraman, M., Reddinger, J., Miller, L., and Tristram-Nagle, S. (1994). *Mater. Res. Soc. Symp. Proc.*, 328: 215.
60. Yamamoto, T., Osakada, K., Wakabayashi, T., and Yamamoto, A. (1985). *Macromol. Chem. Rapid Commun.*, 6: 671.
61. Chen, T.-A., O'Brien, R. A., and Riecke, R. D. (1993). *Macromolecules*, 26: 3462.
62. Feast, W. J., Tsibouklis, J., Pouwer, K. L., Groenendaal, L., and Meijer, E. W. (1996). *Polymer*, 37: 5017.

63. Fukuda, M., Sawaka, K., and Yoshino, K. (1989). *Jpn. J. Appl. Phys.*, *28*: 1433.
64. Miyaura, N., Yanagi, T., and Suzuki, A. (1981). *Synth. Commun.*, *11*: 513.
65. Redecker, M., Bradley, D., Inbasekaran, M., Wu, W., and Woo, E. (1999). *Adv. Mater.*, *11*: 241.
66. Bernius, M., Inbasekaran, M., Woo, E., Wu, W., and Wujkowski, L. (2000). *J. Mater. Sci.: Mater. Electron.*, *11*: 111.
67. Kovacic, P., and Jones, M. B. (1987). *Chem. Rev.*, *87*: 357.
68. Kanbara, T., Saito, N., Yamamoto, T., and Kubota, K. (1991). *Macromolecules*, *24*: 5883.
69. Chaturvedi, V., Tanaka, S., and Kaeriyama, K. (1993). *Macromolecules*, *26*: 2607.
70. Rehahn, M., Schlüter, A.-D., Wegner, G., and Feast, W. J. (1989). *Polymer*, *30*: 1054.
71. Vahlenkamp, T., and Wegner, G. (1994). *Macromol. Chem. Phys.*, *195*: 1933.
72. Huber, J., and Scherf, U. (1994). *Macromol. Rapid Commun.*, *15*: 897.
73. Elsenbaumer, R. L., Shacklette, L. W. (1986). In *Handbook of Conducting Polymers*, Vol. 1 (Skotheim, T. A., ed.), Marcel Dekker, New York, Chapter 7.
74. Park, K. C., Dodd, L. R., Levon, K., and Kwei, T. K. (1996). *Macromolecules*, *29*: 7149.
75. Kreyenschmidt, M., Uckert, F., and Müllen, K. (1995). *Macromolecules*, *28*: 4577.
76. Saito, N., Kanbara, T., Sato, T., and Yamamoto, T. (1993). *Polym. Bull.*, *30*: 285.
77. Scherf, U., and Müllen, K. (1991). *Macromol. Chem., Rapid Commun.*, *12*: 489.
78. Gin, D. L., Avlyanov, J. K., and MacDiarmid, A. G. (1994). *Synth. Met.*, *66*: 169.
79. Grüner, J., Wittmann, H. F., Hamer, P. J., Friend, R. H., Huber, J., Scherf, U., Müllen, K., Moratti, S. C., and Holmes, A. B. (1994). *Synth. Met.*, *67*: 181.
80. Marvel, C. S., and Hartzell, G. E. (1959). *J. Am. Chem. Soc.*, *81*: 448.
81. Ballard, D. G. H., Courtis, A., Shirley, I. M., and Taylor, S. C. (1983). *J. Chem. Soc., Chem. Comm.*, 954.
82. Ballard, D. G. H., Curtis, A., Shirley, I. M., and Taylor, S. C. (1987). *Macromolecules*, *21*: 1787.
83. Gin, D. L., Conticello, V. P., and Grubbs, R. H. (1992). *J. Am. Chem. Soc.*, *114*: 3167.
84. Tour, J. M., and John, J. A. (1993). *Polym. Prepr. (J. Am. Chem. Soc. Div. Polym. Chem.)*, *34*(2): 372.
85. John, J. A., and Tour, J. M. (1994). *J. Am. Chem. Soc.*, *116*: 5011.
86. Yang, Y., Pei, Q., and Heeger, A. J. (1996). *Synth. Met.*, *78*: 263.
87. Yang, Y., Pei, Y. Q., and Heeger, A. J. (1996). *J. Appl. Phys.*, *79*: 934.
88. Scherf, U. (1999). *Top. Curr. Chem.*, *201*: 163.
89. Kolb, E. S., Gaudiana, R. A., and Mehta, P. G. (1996). *Macromolecules*, *29*: 2359.
90. Li, X.-C., Yong, T.-M., Gruener, J., Holmes, A. B., Moratti, S. C., Cacialli, F., and Friend, R. H. (1997). *Synth. Met.*, *84*: 437.
91. Cacialli, F., Li, X.-C., Friend, R. H., Moratti, S. C., and Holmes, A. B. (1995). *Synth. Met.*, *75*: 161.
92. Meier, M., Buchwald, E., Karg, S., Poesch, P., Greczmiel, M., Strohriegl, P., and Riess, W. (1996). *Synth. Met.*, *76*: 95.
93. Bayerl, M. S. (2000). Ph.D., TU München.
94. Hochfilzer, C., Tasch, S., Winkler, B., Huslage, J., and Leising, G. (1997). *Synth. Met.*, *85*: 1271.
95. Sato, H., Sakaki, Y., Ogino, K., and Ito, Y. (1997). *Polym. Adv. Technol.*, *8*: 454.
96. Aguiar, M., Hu, B., Karasz, F. E., and Akcelrud, L. (1996). *Macromolecules*, *29*: 3161.
97. Bisberg, J., Cumming, W. J., Gaudiana, R. A., Hutchinson, K. D., Ingwall, R. T., Kolb, E. S., Mehta, P. G., Minns, R. A., and Petersen, C. P. (1995). *Macromolecules*, *28*: 386.
98. Hou, S. J., Gong, X., and Chan, W. K. (1999). *Macromol. Chem. Phys.*, *200*: 100.
99. Bellmann, E., Shaheen, S. E., Grubbs, R. H., Marder, S. R., Kippelen, B., and Peyghambarian, N. (1999). *Chem. Mater.*, *11*: 399.
100. Boyd, T. J., Geerts, Y., Lee, J.-K., Fogg, D. E., Lavoie, G. G., Schrock, R. R., and Rubner, M. F. (1997). *Macromolecules*, *30*: 3553.

101. Lee, J.-K., Schrock, R. R., Baigent, D. R., and Friend, R. H. (1995). *Macromolecules*, *28*: 1966.
102. Bellmann, E., Shaheen, S. E., Thayumanavan, S., Barlow, S., Grubbs, R. H., Marder, S. R., Kippelen, B., and Peyghambarian, N. (1998). *Chem. Mater.*, *10*: 1668.
103. Bao, Z., Peng, Z., Galvin, M. E., and Chandross, E. A. (1998). *Chem. Mater.*, *10*: 1201.
104. Brouwer, H. J., Hilberer, A., Krasnikov, V. V., Werts, M., Wildeman, J., and Hadziioannou, G. (1997). *Synth. Met.*, *84*: 881.
105. Pang, Y., Li, J., Hu, B., and Karasz, F. E. (1999). *Macromolecules*, *32*: 3946.
106. Ahn, T., Jang, M. S., Shim, H.-K., Hwang, D.-H., and Zyung, T. (1999). *Macromolecules*, *32*: 3279.
107. Musick, K. Y., Hu, Q.-S., and Pu, L. (1998). *Macromolecules*, *31*: 2933.
108. Gomez, R., Segura, J. L., and Martin, N. (1999). *Chem. Commun. (Cambridge)*, 619.
109. Baumgarten, M., Caparros, D., Yuksel, T., Karabunarliev, S., and Rettig, W. (2000). *Polym. Prepr. Am. Chem. Soc., Div. Polym. Chem.*, *41*: 776.
110. Klaerner, G., Davey, M. H., Chen, W.-D., Scott, J. C., and Miller, R. D. (1998). *Adv. Mater. (Weinheim, Ger.)*, *10*: 993.
111. Lee, J.-I., Klaerner, G., and Miller, R. D. (1999). *Chem. Mater.*, *11*: 1083.
112. Burn, P. L., Holmes, A. B., Kraft, A., Bradley, D. D. C., Brown, A. R., Friend, R. H., and Gymer, R. W. (1992). *Nature*, *356*: 47.
113. Von Seggern, H., Schmidt-Winkel, P., Zhang, C., and Schmidt, H. W. (1994). *Macromol. Chem. Phys.*, *195*: 2023.
114. Park, H. K., and Ree, M. (2001). *Synth. Met.*, *117*: 197.
115. Pyo, S. M., Kim, S. I., Shin, T. J., Park, H. K., Ree, M., Park, K. H., and Kang, J. S. (1998). *Macromolecules*, *31*: 4777.
116. Pyo, S. M., Kim, S. I., Shin, T. J., Ree, M., Park, K. H., and Kang, J. S. (1999). *Polymer*, *40*: 125.
117. Segura, J. L. (1998). *Acta Polym.*, *49*: 319.
118. Kim, D. Y., Cho, H. N., and Kim, C. Y. (2000). *Prog. Polym. Sci.*, *25*: 1089.
119. Cheng, M., Xiao, Y., Yu, W. L., Chen, Z. K., Lai, Y. H., and Huang, W. (2000). *Thin Solid Films*, *363*: 110.
120. Zyung, T., Hwang, D.-H., Kang, I.-N., Shim, H.-K., Hwang, W.-Y., and Kim, J.-J. (1995). *Chem. Mater.*, *7*: 1499.
121. Hay, M., and Klavetter, F. L. (1995). *J. Am. Chem. Soc.*, *117*: 7112.
122. Kim, D. U., Tsutsui, T., and Saito, S. (1995). *Polymer*, *36*: 2481.
123. Hilberer, A., Van Hutten, P. F., Wildeman, J., and Hadziioannou, G. (1997). *Macromol. Chem. Phys.*, *198*: 2211.
124. Kim, H. K., Ryu, M.-K., and Lee, S.-M. (1997). *Macromolecules*, *30*: 1236.
125. Malliaras, G. G., Herrema, J. K., Wildeman, J., Wieringa, R. H., Gill, R. E., Lampoura, S., and Hadziioannou, G. (1993). *Adv. Mater. (Weinheim)*, *5*: 721.
126. Herrema, J. K., Wildeman, J., Gill, R. E., Wieringa, R. H., van Hutten, P. F., and Hadziioannou, G. (1995). *Macromolecules*, *28*: 8102.
127. Kreyenschmidt, M., Klaerner, G., Fuhrer, T., Ashenhurst, J., Karg, S., Chen, W. D., Lee, V. Y., Scott, J. C., and Miller, R. D. (1998). *Macromolecules*, *31*: 1099.
128. Faber, R., Stasko, A., and Nuyken, O. (2000). *Macromol. Chem. Phys.*, *201*: 2257.
129. Faber, R. (2000). PhD, TU München.
130. Remmers, M., Schulze, M., and Wegner, G. (1996). *Macromol. Rapid Commun.*, *17*: 239.
131. Zheng, S., Shi, J., and Mateu, R. (2000). *Chem. Mater.*, *12*: 1814.
132. Mielke, G. F. (2000) PhD, TU München.
133. Yamamoto, T., Morita, A., Miyazaki, Y., Maruyama, T., Wakayama, H., Zhou, Z. H., Nakamura, Y., Kanbara, T., Sasaki, S., and Kubota, K. (1992). *Macromolecules*, *25*: 1214.

15
Crosslinking and Polymer Networks

Manfred L. Hallensleben
Institut für Makromolekulare Chemie, Universität Hannover, Hannover, Germany

I. INTRODUCTION

In this article chemical crosslinking reactions are dealt with respect to network formation starting from individual monomeric, oligomeric or polymeric molecules. Any type of chemical reaction of functional groups may be used for the purpose of polymer network formation that allows for almost quantitative conversion of these functionalities; if not, considerable difficulties with respect to network stability, long term stability of physical properties of the network, and chemical transformation of unreacted functionalities may arise. Since the chemical and physical properties of a respective polymer network strongly depend on the chemical nature of the monomeric units in the chains and also depend on the crosslink density, a wide variety of physical properties of crosslinked polymeric materials is available and considerable technological input is made to design the network in order to match the demands.

This contribution to chemical crosslinking does not include the use of electron beam or γ-irradiation. These methods have some advantages over the use of chemical crosslinking agents as they do not leave behind toxic, elutable agents. Also it does not include peroxide initiated radical crosslinking of saturated polymers which proceeds randomly by hydrogen abstraction from chain segments and coupling reactions of these radical sites.

This contribution does also not include 'physical' almost reversible crosslinking due to microphase separation of block copolymers, to strong hydrogen bonding or to ionic interactions or to crystallite formation.

II. DEFINITION OF POLYMER NETWORKS

Any formation of a polymer network starts from monomeric, oligomeric or polymeric individual molecules which react in solution, in melt or in the solid state. It is necessary that at least a small fraction of these molecules has a functionality $f \geq 3$ to undergo bond formation with another individual. From each individual molecule may emanate zero to f bonds to neighboring molecules and thus this molecule may participate in the formation of a large cluster of molecules which is called a macromolecule. In the so-called sol–gel transition, an infinitely large macromolecule is formed. This infinitely large macromolecule is called a *gel* whereas a collection of finite clusters is called a *sol* independently from the fact that the gel may be formed by crosslinking the molecules in the solid state. A gel usually coexists with a sol: the finite clusters are then trapped in the interior of the gel. *Gelation* is the phase transition from a state without a gel to a state with a gel, i.e., gelation involves the formation of an infinite network [1,2,5,6–9].

The conversion factor p (see W. H. Carothers, $p =$ extent of reaction) is the fraction of bonds which have been formed between the monomers of the system, i.e., the ratio of the actual number of bonds at the given moment to the maximally possible number of such bonds. Thus, for $p = 0$, no bonds have been formed and all monomers remain isolated 1-clusters. In the other extreme, $p = 1$, all possible bonds between monomers have been formed and thus all monomers in the system have clustered into one infinite network, with no sol phase left. Thus for small p no gel is present whereas for p close to unity one such network exists. The gel is, in fact, considered as one molecule. Therefore, there is in general a sharp phase transition at some intermediate critical point $p = p_c$, where an infinite cluster starts to appear: a gel for p above p_c, a sol for p below p_c. This point $p = p_c$ is the gel point and may be the analog of a liquid–gas critical point: For p below p_c, only a sol is present just as for T above T_c only a supercritical gas exists. But for p above p_c, sol and gel coexist with each other; similarly for T below T_c vapor and liquid coexist at equilibrium on the vapor pressure curve. However, we do not assert that these thermal phase transitions and gelation have the same critical behavior. Also, in gelation there is no phase separation: Whereas the vapor is above the liquid, the sol is within the gel. The liquid–gas transition is a thermodynamic phase transition whereas gelation deals with geometrical connections (i.e., with bonds). At least in simple gelation models the temperature plays only a minor role compared with its dominating influence on the thermodynamic phase transitions. Such simple gelation theories often make the assumption that the conversion p alone determines the behavior of the gelation process, though p may depend on temperature T, concentration c of monomers, and time t.

Early theoretical approaches to the gel-formation [1–4] as the Flory–Stockmayer theory do not take into account several aspects which naturally occur as the individual molecules grow to form the gel, such as cyclic bond formation, excluded volume effects and steric hinderance. The Flory–Stockmayer theory assumes that in the gelation process each bond between two individual monomeric, oligomeric or polymeric molecules is formed randomly. Thus this theory assumes point-like monomers. This apparently is not the case when already existing macromolecules are crosslinked, i.e., in vulcanization reactions as well as in copolymerization reactions of macromolecules with the functionality $f \geq 3$ with bifunctional monomers.

Besides the polymer networks which are generated from homogeneous solution or in bulk either by crosslinking processes of already existing pre-polymers or by crosslinking copolymerization reactions and which are completely insoluble in any solvent, there are also existing network particles in much smaller dimensions which are called *microgels* and which form in very dilute solution in copolymerizing a monofunctional and a difunctional vinyl monomer, e.g., such as styrene and divinylbenzene, or which are formed in emulsion copolymerization of such comonomers. A microgel is an *intramolecularly crosslinked macromolecule* which is dispersed in normal or colloidal solutions, in which, depending on the degree of crosslinking and on the nature of the solvent, it is more or less swollen [10]. The IUPAC Commission on Macromolecular Nomenclature recommended *micronetwork* as a term for microgel [11] and defined it as *a highly ramified macromolecule of colloidal dimensions*. However, 'micro' refers to dimensions of more than one micrometer whereas the dimensions of the so-called microgels are in the range of nanometers.

Probably most network structures obtained by copolymerization reactions of bifunctional monomers and larger fractions of monomers with a higher functionality are inhomogeneous, consisting of more densely crosslinked domains embedded in a less densely crosslinked matrix, often with fluent transitions.

Besides the inhomogeneity due to a non-uniform distribution of crosslinks, other inhomogeneities due to pre-existing orders, network defects (unreacted groups, intramolecular loops and chain entanglements) or inhomogeneities due to phase separation during the crosslinking process may contribute to network structures [7]. It may be concluded therefore that network inhomogeneity is a widespread structural phenomenon of crosslinked polymers.

For any existing polymer network the most important parameters are the crosslink density, the functionality of the crosslinks, that is the number of elastic network chains tied to one given crosslink, the number of dangling chains (with only one end attached to the network), molecular weight and molecular weight distribution of the elastic chains in the network, the number of loops and the number of trapped entanglements.

III. THEORETICAL CONSIDERATIONS

Polymerization reactions comprising monomers of the A–B plus A_f type (with $f > 2$) in the presence of B–B monomers will lead not only to branching but also to a crosslinked polymer structure. Branches from one polymer molecule will be capable of reacting with those of another polymer molecule because of the presence of the B–B reactant. Crosslinking can be pictured as leading to the structure **I** in which two polymer chains have been joined together (crosslinked) by a branch. The branch joining the two chains is referred to as a crosslink.

A crosslink can be formed whenever there are two branches that have different functional groups at their ends, that is, one has an A group and the other a B group. Crosslinking will also occur in other polymerization reactions involving reactants with functionalities f greater than two. These include the polymerizations

$$A - A + B_f \rightarrow$$
$$A - A + B - B + B_f \rightarrow$$
$$A_f + B_f \rightarrow$$

In order to control the crosslinking reaction so that it can be used properly it is important to understand the relationship between gelation and conversion, that is consumption of monomers and/or functional groups, that is also called extent of reaction. Two general approaches have been used to relate the extent of reaction at the gel point to the composition of the polymerization system based on calculating when X_n and X_w, respectively, reach the limit of infinite size.

$$X_n \rightarrow \infty$$

The first one considering the gel point when the number average degree of polymerization X_n becomes infinite $X_n \rightarrow \infty$ in a polycondensation reaction was given by the pioneer W. H. Carothers himself [12]. This approach is based on the simple assumption that the reactive groups in the system only are consumed by chemical reaction; no branching or cyclization events are taken into account. If the average functionality of all functional groups present in the system of two monomers A and B in equimolar amounts is named f_{avg}, the average functionality of a mixture of monomers is the average number of functional groups per monomer molecule and is given by

$$f_{avg} = \sum N_i f_i \bigg/ \sum N_i$$

which of course is the general formula to calculate the average specifics of a great number of individuals. Thus for a system consisting of 2 moles of lycerol (a triol, $f = 3$) and 3 moles of adipic acid (a diacid, $f = 2$), the total number of functional groups is 12 per 5 monomer molecules, and f_{avg} therefore simply is 12/5 or 2.4. For a system consisting of equimolar amounts of glycerol, adipic acid, and acetic acid (a monoacid), the total number of functional groups is 6 per 3 monomer molecules and f_{avg} simply is 6/3 or 2.

In a system containing stiochiometric numbers of A and B groups, the number of monomer molecules present initially is N_0 and the corresponding total number of functional groups is $N_0 f_{avg}$. If N is the number of molecules after reaction has occurred, then $2(N_0 - N)$ is the number of functional groups that have reacted. The extent of reaction p is the fraction of functional groups lost

$$p = 2(N_0 - N)/N_0 f_{avg}$$

while the degree of polymerization is

$$X_n = N_0/N$$

This is the so-called Carothers equation which relates the degree of polymerization to the number of molecules present in the polymerizing system. From combination of both these equations it follows that

$$X_n = 2/2 - p f_{avg}$$

or by rearrangement

$$p = 2/f_{avg} - 2/X_n f_{avg}$$

This equation is equivalent to the Carothers equation, and in this expression it relates to the extent of reaction and degree of polymerization to the average functionality f_{avg} of the system.

At the gel point the number average degree of polymerization X_n becomes infinite and therefore the second term in the previous equation is zero. Thus, the critical extent of reaction p_c at the gel point is given by

$$p_c = 2/f_{avg}$$

This equation allows us to calculate the extent of reaction to which the reaction has to be pushed to reach the onset of gelation in the reaction mixture of reacting monomers from its average functionality.

In the example given above of reacting a dibasic acid, adipic acid, with a trifunctional alcohol, glycerol, which is of the type A^2B^3, we have to take 2 moles of glycerol and 3 of adipic acid, or 5 altogether, containing 12 equivalents and $f_{avg} = 12/5 = 2.4$. Then at $X_n = \infty$, $p = 2/2.4$ and the limit of reaction will be $5/6 = 0.833$. This, in fact, represents the *maximum* amount of reaction that can occur before gelation under any distribution of combinations, provided only, that the reaction is all intermolecular.

$$X_w \rightarrow \infty$$

Flory [1,2] and also Stockmayer [3,4] used a statistical approach to derive an expression for predicting the extent of reaction at the time where gelation will occur by calculating when X_w approaches infinite size. This statistical approach in its simplest form assumes that the reactivity of all functional groups of the same type is the same and independent of molecular size and shape. It is further assumed that there are no intramolecular reactions between functional groups on the same molecule such as cyclization reactions.

For the ease of demonstration how the branching reaction in a step-growth polymerization reaction of $A–A + B–B + A_f$ molecules proceeds, Flory has used a simple picture to sketch the branching procedure which at some critical point finally leads to gelation [13]

$$A-A + B-B + A_f \rightarrow A_{(f-1)}-A(B-BA-A)_nB-BA-A_{(f-1)}$$

The center unit in Figure 1 is given by the segment to the right of the arrow with the two A_f at the end as branching sites. Infinite networks are formed when n number of chains or chain segments give rise to more n chains through branching of some of them. The criterion for gelation in a system containing a reactant of functionality f is that at least

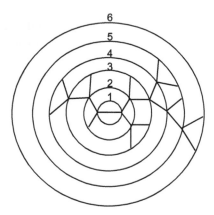

Figure 1 Schematic representation of a trifunctionally branched three-dimensional polymer molecule [13].

one of the $(f-1)$ chain segments radiating from a branch unit will in turn be connected to another branch unit (note: f is not identical to f_{avg} used by Carothers [12]). The probability for this occurring is simply $1/(f-1)$ and the critical branching coefficient α_c for gel formation is

$$\alpha_c = 1(f-1)$$

When $\alpha(f-1)$ equals 1, a chain segment will, on average, be succeeded by $\alpha(f-1)$ chains. Of these $\alpha(f-1)$ chains a portion α will each end in a branch point so that $\alpha^2(f-1)^2$ more chains are created. The branching process continues with the number of succeeding chains becoming progessively greater through each succeeding branching reaction.

If all groups (of the same kind) are equally reactive, regardless of the status of other groups belonging to the same unit, the probability P_A that any particular A group has reacted equals the fraction of the As which have reacted; similarly, P_B is defined. If r is the ratio of all A to all B groups, then

$$P_B = r P_A$$

since the number of reacted A groups equals the number of reacted B groups.

The probability that a given functional group (A) of a branch unit is connected to a sequence of $2n+1$ bifunctional units followed by a branch unit is

$$[P_A P_B (1-\rho)]^n P_A P_B \rho$$

where ρ is the ratio of As belonging to branch units to the total number of As. Then

$$\alpha = \sum_{n=0}^{\infty} [P_A P_B (1-\rho)]^n P_A P_B \rho$$

$$= P_A P_B \rho / [1 - P_A P_B (1-\rho)]$$

$$= r P_A^2 \rho / [1 - r P_A^2 (1-\rho)] = P_B^2 \rho / [r - P_B^2 (1-\rho)]$$

It will depend on the analytical circumstances which of the unreacted groups, A or B, is the one to determine which of the equations will be used.

Combination of $\alpha_c = 1(f-1)$ and $\alpha = rP_A^2\rho / [1 - rP_A^2(1-\rho)] = P_B^2\rho / [r - P_B^2(1-\rho)]$ yields a useful expression for the extent of reaction (of the A functional groups) at the gel point

$$p_c = 1/\{r[1 + \rho(f-2)]\}^{1/2}$$

When the two functional groups are present in equivalent numbers, $r = 1$ and $P_A = P_B = P$, then

$$\alpha = P^2\rho/[1 - P^2(1-\rho)]$$

and

$$p_c = 1/[1 + \rho(f-2)]^{1/2}$$

In the reaction of glycerol, $f = 3$, with equivalent amounts of several diacids, the gel point was observed [14,15] at an extent of reaction of 0.765. The predicted values of p_c are 0.709 and 0.833 calculated from [13] (Flory, statistical) and [12] (Carothers), respectively. Flory [13] studied several systems composed of diethylene glycol ($f = 2$), 1,2,3-propane-tricarboxylic acid ($f = 3$), and either succinic or adipic acid ($f = 2$) with both stoichiometric and nonstoichiometric amounts of hydroxyl and carboxyl groups, see Table 1.

The observed p_c values as in many other similar systems fall approximately midway between the two calculated values. The Carothers equation [12] gives a high value for p_c. The experimental p_c values are close to but always higher than those calculated from the Flory equation [13]. Two reasons can be given for this difference: first the occurence of intramolecular cyclization and second unequal functional group reactivity. Both factors were ignored in the theoretical derivations for p.

Although both the Carothers and statistical approaches are used for the practical prediction of gel points, the statistical approach is the more frequently employed. The statistical method is preferred, since it theoretically gives the gel point for the largest sized molecules in a size distribution.

Some theoretical evaluations of the effect of intramolecular cyclization on gelation have been carried out [6,16,17]. The main conclusion is that, although high reactant concentrations decrease the tendency toward cyclization, there is at least some cyclization occurring even in bulk polymerizations. Thus, even after correcting for unequal reactivity of functional groups, one can expect the actual p_c in a crosslinking system to be larger than a calculated p_c value.

Table 1 Gel point for polymers containing tricarboxylic acid [13].

		Extent of reaction at gel point (p_c)		
$r = [CO_2H]/[OH]$	ρ	Calculated from [12]	Calculated from [13]	Observed
1.000	0.293	0.951	0.879	0.911
1.000	0.194	0.968	0.916	0.939
1.002	0.404	0.933	0.843	0.894
0.800	0.375	1.063	0.955	0.991

IV. CROSSLINKING — CONCEPT

Among all crosslinking strategies which are used to synthesize polymer networks, three different classes are in common application:

1. One-shot crosslinking of multifunctional monomers or copolymerization with difunctional monomers,
2. two-stage crosslinking via prepolymers,
3. crosslinking of high molecular weight polymers.

Into the first category of crosslinking strategies fall the formation of poly(styrene-*co*-divinylbenzene) resins, the methacrylic resins and some others, and among those also a small fraction of the so-called microgels. In general, these resins are formed of monomers which in linear polymerization lead to thermoplastic polymers such as poly(styrene), polyacrylics or methacrylics a.s.o. High glass transition temperature of the linear polymers and high melt viscosity makes it unattractive to process premade linear thermoplastics prior to a second step of crosslinking reaction. Incorporation of pendant C–C– double bonds into the linear chains by copolymerization with small quantities of a difunctional monomer and thereby avoiding early stage crosslinking is difficult to handle and such polymers would be very sensitive to undergo uncontrolled network formation.

One-shot crosslinking of multifunctional monomers and copolymerization therefore is limited to the radical induced copolymerization of styrene and some derivatives with divinylbenzene or of methacrylates with ethyleneglycol dimethacrylate as crosslinker in suspension polymerization to form densely crosslinked polymer beads for applications such as ion exchange resins, Merrifield resins, polymer supports for chemical reagents especially with the aspect of combinatorial syntheses.

Into the second category of crosslinking strategies fall the processes of preparing polymer networks which make use of prepolymers. These are two-stage processes in which in the first stage, overhelmingly in step-growth polymerization reactions, prepolymers are prepared with molecular weight mostly ranging from 1 to 6×10^3 which are soluble in organic solvents, fusible and have low melt viscosity. The second stage curing is achieved either by heat — thermosetting — or, when necessary, by the addition of appropriate curing agents. Most prominent examples are epoxy resins, phenol-formaldehyde resins, unsaturated polyesters, and the polyurethane networks.

Into the third catagory fall the vulcanization reactions of elastomers. These polymers expose C–C double bonds incorporated in the main chain segments which are necessary for the crosslinking process referred to as vulcanization. Natural rubber and the synthetic elastomers have glass transition temperatures far below the temperature range in which the crosslinked rubbers are used. The molecular weight of the applied polymers is in the range of $2–5 \times 10^5$, and natural rubber with an upper molecular weight fraction of $2–4 \times 10^6$ has to be degraded to this molecular weight level by mechanical treatment referred to as mastication. The basis of all processes that come after mastication and before vulcanization are the operations of blending rubber mixtures, mixing with all the vulcanization ingredients, calendering, frictioning, extrusion, moulding and combining with textile fabrics or cords is the flow or viscous deformation of the rubber, more precisely the rheological behavior. Extrusion, calendering and frictioning all involve vigorous mechanical working in large machines and hence enormous energy consumption and heat generation.

A. General Classification of Prepolymers [18]

Curing reactions applied to epoxy prepolymers, unsaturated polyesters, resoles, and novolacs make use of three general classes of prepolymers which are distinguished by the number and location of sites of functional groups available for subsequent crosslinking reactions. These three general classes have been defined as discussed in the following sections.

1. *Random prepolymers.* Random prepolymers are those built up from polyfunctional step-growth monomers which have been reacted randomly and which are capable of forming crosslinked polymers directly. Monomer conversion in the first-stage polymerization reaction for the formation of these prepolymers is stopped short and kept below the critical conversion at which network formation would occur. Crosslinking in the second-stage, step-growth polymerization reaction is achieved simply by heating to carry the original reaction past the critical conversion. For this reason, the term thermoset is applied to these prepolymers, and these are exemplified by the phenol-formaledehyde resole resins and the glycerol polyesters. The term *structoset* has been applied to the other two classes of prepolymers to distinguish them from the thermoset type because in the other two classes the second-stage crosslinking reaction requires the addition of a catalyst or monomer, and generally proceeds by a reaction different from the first-stage reaction.

2. *Structoterminal prepolymers.* Structoterminal prepolymers are those in which the reactive sites are located at the ends of the polymer chains. These first-stage polymers give maximum control of the length and type of chain in the final network polymer. The epoxy prepolymers may be considered examples of this class if the second-stage reaction occurs overwhelmingly through reaction of the terminal epoxide functional groups. If the aliphatic hydroxyl groups along the chain in epoxy prepolymers become significantly involved in the crosslinking reaction, then these polymers are more properly included in the third class of prepolymers.

3. *Structopendant prepolymers.* Structopendant prepolymers are those in which the crosslink sites are distributed in either a regular or random order along the chain. Examples of this class are the unsaturated polyesters and the novolac resins.

V. PHENOL-FORMALDEHYDE RESINS

Phenol-formaldehyde condensates were among the first synthetic polymeric materials on the market. It was Baekeland at the beginning of the 20th century who in 1907 defined the differences between basic or acidic reaction conditions and the different molar ratios on the reaction procedure and the resulting molecular structure. He was able to manufacture a thermosetting resin and made applications for a patent [19] (Bakelite).

Most phenolic resins are heat hardenable or thermosetting. The resin may be delivered to the user ready to be cured or it may be in the temporarily thermoplastic novolac form to which a hardener, commonly hexamethylenetetramine–urotropin, will be added. The major categories of uses for phenolics are

- Molding compounds
- Coatings
- Industrial bonding resins.

The latter includes resins for grinding wheels and coated abrasives, laminating, plywood adhesives, glass wool thermal insulation and bonded organic fiber patting, foundry sand bonding, wood waste bonding, and other miscellaneous applications.

A. Reaction of Phenol and Formaldehyde Under Basic Conditions

The base-catalyzed first-step reaction of phenol ($f = 3$, because reaction can take place in two *ortho* and one *para* position) and formaldehyde ($f = 2$) with an excess of formaldehyde of about 15 mol% closely resembles an aldol addition and yields mixtures of monomolecular methylolphenols and also dimers, trimers and the corresponding polynuclear compounds according to a generalized reaction scheme given in (1b). In commercial processes formaldehyde is added in aqueous solution. Sodium hydroxide, ammonia and hexamethylenetetramine–urotropin, sodium carbonate, calcium-, magnesium-, and barium-hydroxide and tertiary amines are used as catalysts. After the hydroxybenzyl alcohol has been formed in the first step, the condensation steps to form oligomers are likely to be a Michael type of addition to a base-induced dehydration product of the hydroxybenzyl alcohol. Detailed studies have been presented by Martin [20] and Megson [21].

$$\tag{1}$$

Such mixtures, whose exact composition depend on the phenol–formaldehyde ratio and the reaction conditions employed, are termed *resoles* or *resole prepolymers*. The resoles are generally neutralized or made slightly acidic before the second-stage reaction is accomplished by heating. The second-stage polycondensation and crosslinking takes place by the formation of methylene and dibenzyl ether linkages between the benzene rings to yield a network structure of type I. The relative importance of the methylene and ether bridges is not well established, although both are definitely formed. Higher reaction temperatures favor the formation of the methylene bridges.

B. Curing of Resol Prepolymers

Heat curing of resols usually is carried out at temperatures in the range 130–200 °C. Below 150 °C the formation of dibenzyl ether bridges is predominant whereas at higher temperatures methylene bridge formation is favored. This was nicely shown by the investigations of Kämmerer et al. who carried out polycondensation reaction of 2,6-bis(hydroxymethyl-4-methylphenol to the corresponding poly(benzyl ether) [24] with molecular weights ranging from 2500 to 20,000.

Although at lower temperatures only water is liberated but also water and formaldehyde at temperatures above 150 °C [25], the water to formaldehyde ratio is not an exact measure of the ratio of benzyl ether to methylene bridge formation, because it is known that the yield of isolable formaldehyde is considerably less than the theoretical yield [21].

If curing is carried out above 180 °C in the presence of air, some oxidation reaction takes place which gives a reddish color to the final product. Quinone structures are responsible for the color and researchers were able even to isolate quinone methides formed in pyrolysis reactions [26].

C. Reaction of Phenol and Formaldehyde Under Acidic Conditions

The reaction between phenol and formaldehyde under strongly acidic conditions can be regarded as an electrophilic substitution reaction, route (b) in Scheme 1 [28]. The catalysts most frequently used are sulfuric acid, oxalic acids or *p*-toluene sulfonic acid. By the addition of a proton to formaldehyde a hydroxymethylene carbenium ion is formed which undergoes an electrophilic hydroxyalkylation reaction mostly in the o-position of phenol. From this o-methylol phenol compound water is eliminated by reaction of the methylol group with a proton thus yielding a benzylium type carbenium ion which then undergoes very fast alkylation reaction of a second phenol molecule in the o-position with the generation of a new proton [20–22,27]. Continued methylolation and methylene bridge formation by these reactions leads to the formation of polynuclear compounds of considerable complexity. Under strongly acidic conditions, methylol substitution and methylene-bridge formation both occur predominantly at p-positions [29]. The pH most favorable for the formation of the o-products is between 4 and 5.

D. Curing of Novolac Prepolymers

Novolacs require an auxiliary chemical crosslinking agent. The most widely used crosslinker is hexamethylenetetramine, and the products in this curing reaction are influenced by the molar ratio of phenol nuclei to hexamethylenetetramine. At a phenol nucleus to hexamethylenetetramine ratio of 6 : 1, the products turn out to contain little or even no nitrogen, and the reaction appears to an almost entirely one of methylene-bridge formation. At a mole ratio 0.5 : 1 or higher, nitrogen enters into the product, and the nitrogen content of the products can come close to 10% with the amount of ammonia evolved proportionately decreased.

The reaction of curing is not clear. It is known that under controlled conditions phenol and hexamethylenetetramine form a crystalline salt of the stiochiometric composition $C_6H_{12}N_4 \cdot 3C_6H_5OH$ [30] which, when heated, evolves ammonia with the formation of an insoluble, infusible polymer [31]. In the presence of water, hexamethylenetetramine hydrolyzes with the formation of two moles of dimethylolamine DMA, one mole of formaldehyde and two moles of ammonia. Water is ubiquitous in novolacs and therefore under basic reaction conditions in the presence of *tert* and *sec* amines and also ammonia as shown in the chart, methylene bridges are formed by entering formaldehyde into the reaction. With increasing amounts of hexamethylenetetramine, the benzylamine type bridges become predominant.

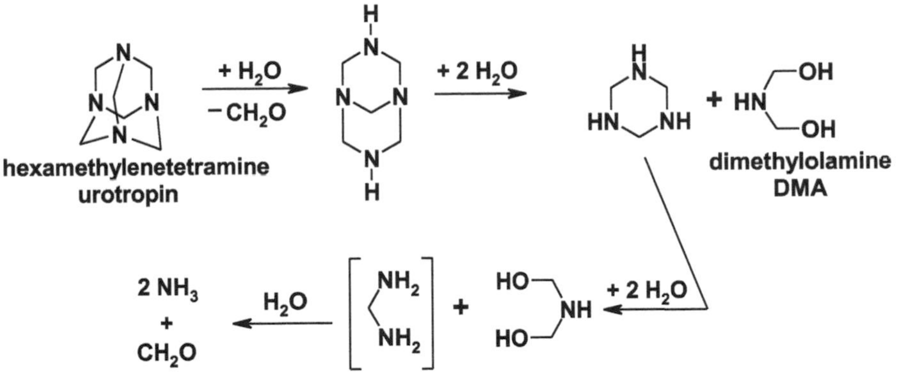

hexamethylenetetramine urotropin

dimethylolamine DMA

Cured novolacs show a more or less slightly yellow color. There is some indication in the literature that the benzylamine type bridges are converted to azomethines by hydrogen elimination under heating conditions applied in the curing reaction [20].

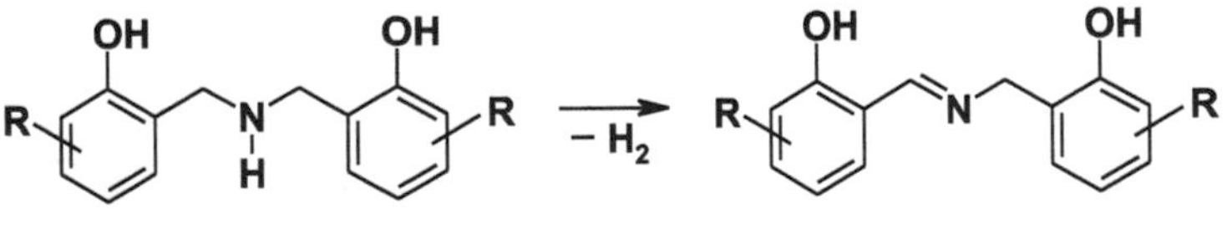

Bender et al. found that the *o,o'*-compounds have a much more rapid cure rate than isomeric 'novolacs' [23]. The gel times for the 2,2', 4,4', and 2,4' isomers at 160 °C have been reported to 60, 175, and 240 sec, respectively.

VI. UREA- AND MELAMIN-FORMALDEHYDE RESINS

Urea **1** ($f = 4$) and melamin **2**, 2,4,6-triamino-1,3,5-triazin ($f = 6$) under basic or acidic conditions react with formaldehyde ($f = 2$) rather similar to the phenol–formaldehyde reaction. The reaction products are called aminoplastics.

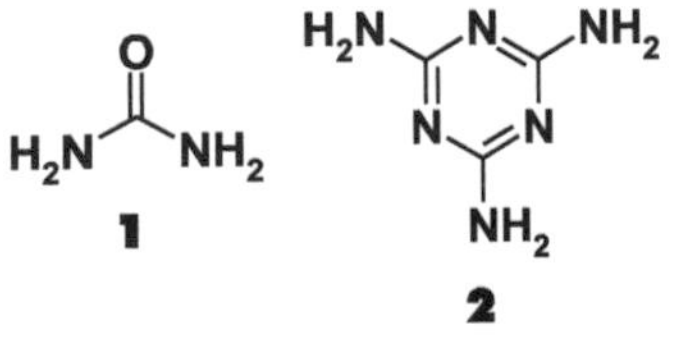

1

2

Polymerization of urea and formaldehyde in a 1.5:1 ratio in the first-stage reaction yields various methylolureas as prepolymers [32–36], which in a second-stage reaction are cured by heat (thermosetting) under neutral or slightly acidic conditions. Control of the extent of reaction is achieved by pH (by the use of buffers) and temperature control. The reaction rate increases with increasing acidity [37,38]. The prepolymer can be made at varying pH levels depending on the reaction temperature. Polymerization is stopped by bringing the pH close to neutral and cooling.

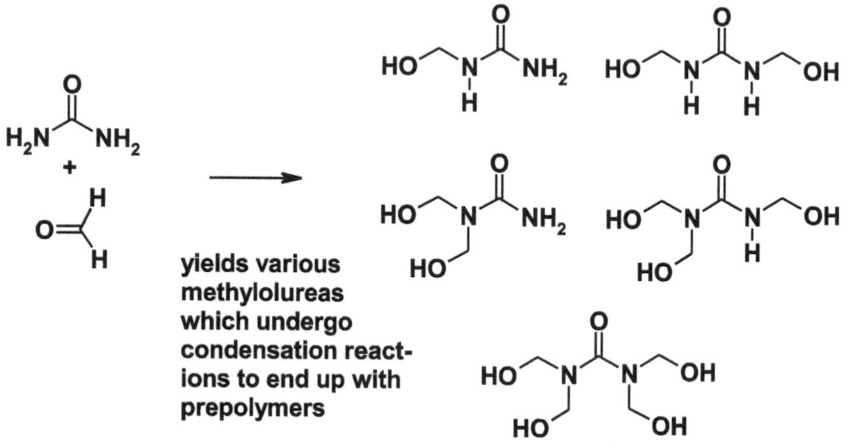

yields various methylolureas which undergo condensation reactions to end up with prepolymers

The second-stage, crosslinking reaction of the prepolymers under acidic conditions causes the formation of a network containing principally a random mixture of linear and branched substituted trimethylenetriamine repeating units and, to some extent, also methylene ether bridges and methylene bridges [35,39]. The latter are exclusively formed under strongly acidic conditions [40].

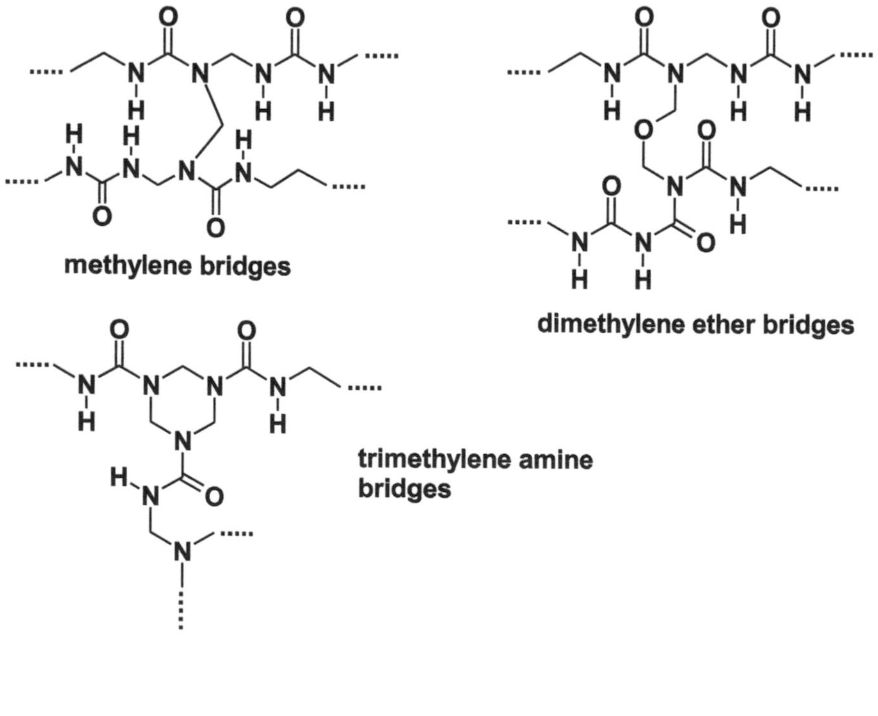

methylene bridges

dimethylene ether bridges

trimethylene amine bridges

The formation and crosslinking of random prepolymers from melamine, 2,4,6-triamino-1,3,5-triazin, and formaldehyde follows in a similar manner [33,34,41–43], but, unlike urea, melamin readily forms polymethylol compounds with two methylol groups on a single nitrogen atom. Paper chromatographic separation of the products of this reaction, in which an excess of formaldehyde greater than 2.1 was used, revealed the presence of all possible methylol compounds from the monosubstituted to the hexasubstituted derivatives [44].

$$R = H \text{ or } CH_2OH$$

VII. EPOXY RESINS

Epoxy resins as a class of crosslinked polymers are prepared by a two-step polymerization sequence. The first step which provides prepolymers, or more exactly: preoligomers, is based on the step-growth polymerization reaction of an alkylene epoxide which contains a functional group to react with a bi- or multifunctional nucleophile by which prepolymers are formed containing two epoxy endgroups. In the second step of the preparation of the resins, these tetrafunctional (at least) prepolymers are cured with appropriate curing agents. Table 2 compiles a representative selection of di- and multi-epoxides both as alkyl and cycloalkyl epoxides and the most widely used curing reagents.

The most widely used pair of monomers to prepare an epoxy prepolymer are 2,2′-bis(4-hydroxyphenyl)propane (referred to as bisphenol-A) and epichlorohydrin, the epoxide of allylchloride. The formation of the prepolymer can be seen to involve two different kinds of reactions. The first one is a base-catalyzed nucleophilic ring-opening reaction of bisphenol-A with excess of epichlorohydrin to yield an intermediate β-chloro alcoholate which readily loses the chlorin anion reforming an oxirane ring. Further nucleophilic ring-opening reaction of bisphenol-A with the terminal epoxy groups leads to oligomers with a degree of polymerization up to 15 or 20, but it is also possible to prepare high molecular weight linear polymers from this reaction by careful control of monomer ratio and reaction conditions [45]. The two ring-opening reactions occur almost exclusively by attack of the nucleophile on the primary carbon atom of the oxirane group [46].

Depending on the conditions of the polymerization reaction, these low molecular weight polymers can contain one or more branches as a result from the reaction of the pendant aliphatic hydroxyl groups with epichlorohydrin monomer. In most cases, however, the chains are generally linear because of the much higher acidity of the phenolic hydroxyl group. At high conversions, when the concentration of phenolic hydroxyl groups drops to a very low level, under the base-catalyzed reaction conditions formation and reaction of alkoxide ions become competitive and polymer chain branching may occur.

Polymers of this type with molecular weight exceeding 8000 are undesirable because of their high viscosity and limited solubility, which make processing in the second-stage, crosslinking-reaction difficult to perform. The oligomers of the diglycidylether of bisphenol-A (DGEBA) are the most commonly epoxy resins, therefore a great deal of

Table 2.

Aliphatic epoxy monomers and
pre-polymers (selection)

Curing agents

prim./sec. Amines

tert. Amines

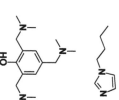

Acid anhydrides

Aliphatic-cycloaliphatic epoxy compounds

Polymerization catalysts

such as amine complexes of Lewis acids [65]
or diaryliodonium salts [66], photocrosslinking [67]

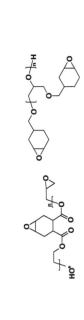

investigations with respect to the processibility behavior before crosslinking is focused on this oligomer [47].

A. Aliphatic-Cycloaliphatic Epoxy Compounds and Prepolymers [48]

Aliphatic-cycloaliphatic epoxy compounds (ACECs) contain different epoxy groups in the molecule: glycidyl, i.e., 2,3-epoxypropyl groups, and cycloaliphatic, i.e., 1,2-epoxycyclopentane or 1,2-epoxycyclohexane rings, for which molecules 3 and 4 are characteristic.

The most important feature of ACECs is the different reactivity of the cycloaliphatic epoxy group and the glycidyl epoxy group with various curing agents. This property affects some important properties of ACECs. Table 2 contains a good selection of ACECs which have been described in the literature. It is possible to consume different epoxy groups consecutively in the course of curing [49,50]. In the early stages of curing, reaction of carboxyl groups with cycloaliphatic epoxy groups prevails, resulting in the formation of a polymer chain with a loose crosslinking. In later stages, the chain extension and dense crosslinking proceeds as a result of the conversion of glycidyl groups and of the remaining cycloaliphatic epoxy groups. Eventually, the network density is achieved. The network density is determined by the ACEC–hardener ratio and by the conditions of the curing process.

The reaction sequence is different if amines are applied as curing agents. In the first stage the glycidyl groups react followed by the cycloaliphatic epoxy groups which then enter into the reaction with the curing agent.

Nevertheless, the sequential entering of different epoxy groups into the reaction, irrespective of the acidic or basic character of the curing agent, is a very important feature of the crosslinking process of ACECs because it conditions the formation of a regular polymer network [51].

B. Curing

The epoxy prepolymers are considered as *structopendant* prepolymers because of the pendant aliphatic hydroxyl groups or as *structoterminal* prepolymers with respect to the terminal epoxy groups [52].

An acid anhydride as curing agent is bifunctional ($f=2$) and crosslinking occurs primarily through the hydroxyl groups. In this reaction, the prepolymer acts as a *structopendant* prepolymer. Maleic anhydride introduces C–C– double bonds into the resin. Mostly phthalic anhydride and pyromellitic anhydride are used.

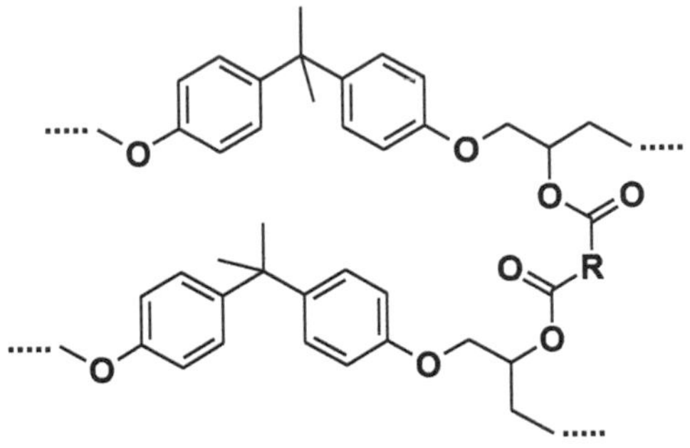

crosslinking with anhydride reacting with pendant aliphatic hydroxyl groups

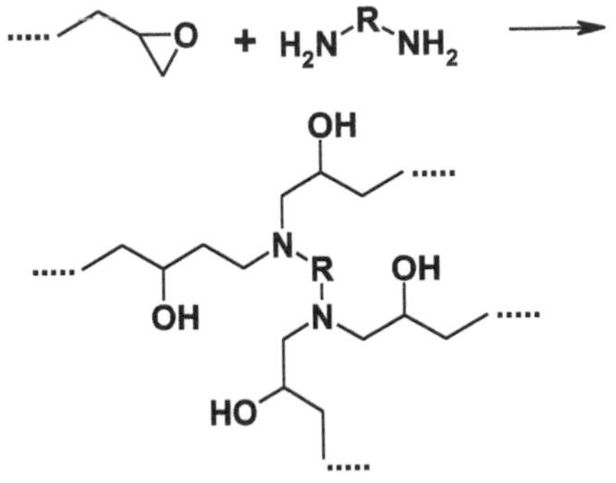

crosslinking with an amine, both primary and secondary amino groups are reactive

Anhydrides react initially with the hydroxyl groups in the prepolymers to form half-esters, and the generated carboxyl groups in this half-ester can condense with another hydroxyl group. Also the reaction of the carboxyl group with an epoxy group is possible [53,54], but these reactions are much slower than the initial alcohol–anhydride reaction and are not shown in the above picture. For these reasons dianhydrides are very effective crosslinking agents, and because of the great number of hydroxyl groups in the prepolymer, curing with dianhydrides can form very densely crosslinked, second-stage polymers if used in relatively high concentrations.

The prepolymer is a *structoterminal* prepolymer when amines are used as crosslinkers. Crosslinking in this case involves the base-catalyzed ring-opening of the oxirane groups. Both primary and secondary amines are used as crosslinking agents [55]. Since each N–H bond is reactive in this process, primary and secondary amine functional groups have a crosslinking functionality f equal to two and one, respectively. A variety of amines such as diethylene triamine ($f=5$), triethylene tetramine ($f=6$), m-phenylene-diamine ($f=4$) and others are used as crosslinking agents. The presence of other reactants is required to foster this ring-opening reaction because the nucleophilic ring-opening reaction of an amine with an oxacyclopropane is not only accelerated by, but, in fact, requires the presence of an active proton-donor [56]. Anhydrous diethylamine and oxacyclopropane do not react, but the reaction proceeds readily in the presence of catalytic amounts of proton-donating agents like water, methanol or ethanol [57]. Similarly, the reaction of epoxybenzylacetophenone with morpholine or with piperidine in benzene or ether is extremely slow, but proceeds smoothly in methanol at room temperature [58]. The reaction of phenyl glycidyl ether with diethylamine in the absence of solvents shows a sigmoidal rate curve, which can be attributed to the autocatalytic effect of the hydroxyl groups in the product [59], while in proton-donating solvents the reaction is greatly accelerated and the sigmoidal form of the rate curve disappears. By protonation of the oxacyclopropane oxygen, an intermediate oxonium ion is formed which facilitates the nucleophilic attack on the carbon atom. In the case of the epoxy end groups of the prepolymers, this nucleophilic attack is exclusively directed to the *sec* carbon atom. Phenol has been found to be a particularly useful proton-donating accelerator. And it has been shown also that the reaction of oxacyclopropane with aniline in the presence of small amounts of water [60] or acids [61] is proportional to the concentration of water or to the strength of the acid. Different mechanisms have been proposed by Smith [56], Tanaka [62], and King et al. [63], but they have not yet been confirmed [64].

VIII. CROSSLINKING–POLYURETHANE NETWORKS

Structoterminal prepolymers with two isocyanate endgroups prepared by reaction of polyethers containing two hydroxyl endgroups with diisocyanates are the basis for the formation of polyurethane networks. They can be made either in melt or in solution, but polyurethanes with melting points much above 200 °C are difficult to prepare in melt because of the thermal instability of the urethane linkage above 220 °C [68]. The molecular weight of the prepolymers generally is in the range of 1–10×10^3.

The fundamental reactions of an isocyanato group which proceed easily at room temperature or slightly above are reaction with (i) an aliphatic or aromatic hydroxyl group in a reactivity order primary > secondary > tertiary OH-group, and with (ii) primary or (iii) secondary amines. With carboxylic acids (iv) an amide is formed and CO_2 is liberated, and with water (v) isocyanates give substituted carbamic acids which decarboxylate with extreme ease to give an amine which is recycled into reaction (ii). Thus, in (i)–(iii) linkages are formed which directly help to build up a polymer chain, and in (iv) and (v) functional groups are created which can further react.

At elevated temperatures ($120-140\,^\circ$C), the structures formed in (i)–(iii) are able to undergo further reaction with isocyanate groups according to (vi)–(viii), which, in this way, can be used for crosslinking.

A. Crosslinking

One-shot crosslinking is a step-growth polymerization of a difunctional alcohol with a diisocyanate in the presence of a small amount of a polyfunctional alcohol. In the presence of small quantities of water, carbon dioxide is liberated from hydrolysis of some isocyanate groups and acts as a foaming agent in polyurethane foam production.

Two-stage crosslinking, in which in the first stage is the synthesis of a prepolymer containing two isocyanato endgroups in the classical way of reaction (a diol either of low or of high molecular weight with an excess of diisocyanate) and the second step to form the network, can be accomplished by

1. addition of multifunctional alcohols, and the resulting bridges are urethane linkages,
2. addition of diamines which extend the linear prepolymer chain via urea linkages, which, in turn, add to other isocyanate endgroups to form biuret branching sites and eventually crosslinks,
3. excess diisocyanate, and the network is formed by the reaction of isocyanate with the preformed urethane linkages according to reaction (vi), and the bridges are of an allophanate structure,

Table 3 Rate constants for reactions of diisocyanate monomers with different substrates [69].

Diisocyanate monomer	Rate constant,[a] $k \times 10^{-4}$, liters mole^{-1} sec^{-1}				
	Hydroxyl	Water	Urea	Amine	Urethane
p-Phenylene	36.0	7.8	13.0	17.0	1.8
2-Chloro-1,4-phenylene	38.0	3.6	13.0	23.0	–
2,4-Tolylene	21.0	5.8	2.2	36.0	0.7
2,6-Tolylene	7.4	4.2	6.3	6.9	–
1,5-Naphthalene	4.0	0.7	8.7	7.1	0.6
Hexamethylene	8.3	0.5	1.1	2.4	2×10^{-5}

[a]For reactions at 100°C, except for 1,5-naphthalene diisocyanate, 130°C.

4. at elevated temperatures at which the isocyanate end groups of the prepolymer react intermolecularly with urethane linkages in the main chain thus also forming allophanate bridges.

Again, if water is present, the network is expanded by the carbon dioxide liberated from hydrolysis reaction of isocyanate groups, and the resulting primary amino groups are recycled into the reaction.

The relative rates of the different types of chain extension and crosslinking reactions will depend in part on the structure of the diisocyanate monomer envolved as indicated by the rate constants for reactions of several diisocyanate monomers with water and with various functional groups which can be found in polyurethanes [69]. It is noteworthy that hexamethylene diisocyanate reacts very slowly with urethane groups and therefore would be a very poor crosslinking agent. In addition, it should be mentioned that the relative rates of the various reaction can be changed significantly by the presence of a catalyst and by the type of the catalyst which, in general, is a base, i.e., an amine, or a metal salt (Table 3).

IX. UNSATURATED POLYESTERS UPs

Unsaturated polyesters have a widespread field of applications. In almost all cases, unsaturation in these materials is introduced by the acid component when the prepolymers are manufactured. These prepolymers can be of either the *structoterminal* or *structopendant* type depending on the location of the unsaturated linkages. The average molecular weight is in the range of $1-5 \times 10^3$.

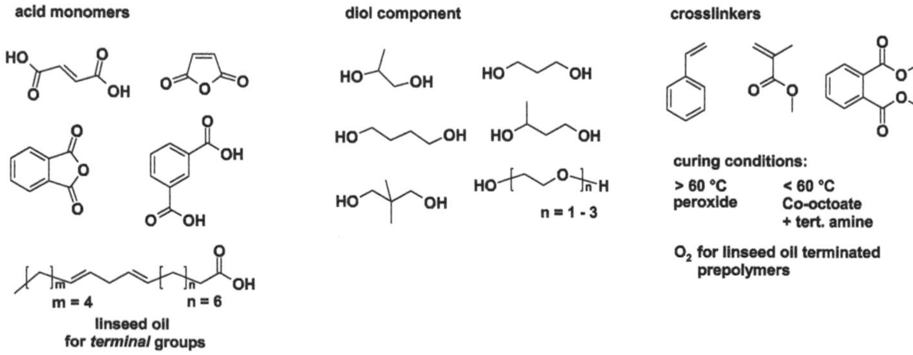

acid monomers

diol component

crosslinkers

m = 4   n = 6
linseed oil
for *terminal* groups

n = 1 - 3

curing conditions:

> 60 °C   < 60 °C
peroxide   Co-octoate
                 + tert. amine

O$_2$ for linseed oil terminated
prepolymers

Structopendant unsaturated polyesters, containing double bonds within the polymer chain, are produced by step-growth polycondensation reaction of an unsaturated diacid or anhydride, such as fumaric acid or maleic anhydride, with a diol. Structural unsymmetry in the diol component lowers the viscosity of the prepolymer. Mostly, crosslinking of the *structopendant* unsaturated polyester is accomplished by copolymerization with alkene monomers such as styrene, methyl methacrylate, or others using radical initiators.

Structoterminal polyesters have terminal C–C–double bonds which are introduced by terminating the step-growth polycondensation reaction by the addition of an unsaturated monocarboxylic acid. The monocarboxylic acid is usually a fatty acid derived from linseed oil, and the polyester is referred to as an *alkyd resin*. Crosslinking is accomplished most simply by oxidation with atmospheric oxygen.

X. SILICON RUBBER

Network formation to build up crosslinked silicones is based on linear polysiloxane precursors. In most cases, poly(dimethylsiloxane) is used, a smaller fraction of products also contains phenyl substituents to silicon. The precursors are all prepared by the usual way of ring-opening polymerization of cyclic tri- or tetrasiloxanes which are previously prepared by cyclocondensation of the corresponding dichlorosilanes [70]. Silicon rubbers are very flexible because of the very low glass transition temperature T_g of $\sim -100\,°C$. Higher stiffness is achieved by the addition of fillers such as silicates which by means of their HO–Si-groups at the surface interact with the silicon Si–O–Si-linkage via hydrogen bonding.

A. Curing

Curing is achieved either by random radical crosslinking of polysiloxanes by heating with peroxides or by room temperature vulcanization techniques making use of reactive end groups of the precursors.

B. Radical Crosslinking

The radical crosslinking method involves heating the polysiloxane with dicumyl peroxide, ditertiary butyl peroxide, benzoyl peroxide, or bis-2,4-dichlorobenzoyl peroxide. The peroxide radical abstracts hydrogen from the polymer chain and creates a radical site on the interior of the chain. Two such sites interact to randomly form the crosslink. The major disadvantage of this technique is its commercial inefficiency. Obviously, vulcanization can only be carried out in a mould to produce the final silicon rubber product.

C. Crosslinking Via Reactive *Structoterminal* Precursors

Platinum catalyzed anti-Markoffnikov addition of hydrosilanes to C–C–double bonds is a widely applied reaction to form Si–C linkages. For this hydrosilylation reaction the platinum based catalyst has to be added only in the ppm scale. Two different polysiloxane components are necessary to achieve network formation by the so-called addition vulcanization of polysiloxanes, one *structoterminal* polysiloxane precursor providing vinyl endgroups and a polysiloxane crosslinker providing hydrosilane groups as chain segments in the main chain.

The advantage of this addition vulcanization is that no revision occurs because no byproducts are formed which might interfere with the network in terms of a reversible network degradation. Furthermore, although this vulcanization reaction is considerably accelerated at elevated temperatures. For a given receipe, the curing characteristics at different temperatures are shown in Table 4.

A second group of room temperature vulcanization techniques have been developed based upon linear polysiloxane chains terminated by hydroxyl groups. Curing can be achieved by two ways which both make use of hydrolyzation reactions of labile Si–O–R bonds. Two-component vulcanization RTV-2.

The so-called *RTV-2* method — room temperature vulcanization of a two component system — adds a crosslinking agent such as tri- or tetraalkoxysilane and a metallic salt catalyst to hydroxyl terminated polysiloxane precursors. The hydroxyl end groups react with the silicic ester, e.g., tetraethyl silicate, in a condensation reaction and ethanol is liberated. This condensation reaction is catalyzed by stannous-based catalysts such as dibutyltindilaurate.

Table 4 Curing times for a typical addition vulcanization reaction of silicones [71], probe thickness 1 cm.

Processing time at room temperature	60 min
Demoulding at room temperature	After 10 hr
Final hardness at room temperature	After 24 hr
Final hardness at 50 °C	After 1 hr
Final hardness at 100 °C	After 10 min
Final hardness at 150 °C	After 5 min

Water is provided by atmospheric moisture. The alcohol liberated from the condensation reaction has to be removed and this is sufficiently achieved by diffusion into the environment. If the vulcanization is carried out in a closed system at elevated temperature, there is the danger of revision.

The so-called *RTV-1* method — room temperature vulcanization of a one component system — is based on the finding that hydroxyl terminated siloxanes do not react with certain crosslinking agents under strictly dry conditions. Technically, this is achieved by the addition of an excess of crosslinker which reacts much faster with water than with the silanol groups and thereby acts as a drying agent. As atmospheric moisture diffuses into the system, crosslinking starts to occur. Therefore, these one-component silicon rubber precursors are stored in one-compartment cartridges and can be applied very easily.

The different reaction steps envolved are demonstrated below for an acetoxy system: in the first step, under dry conditions the hydroxyl terminated polymer reacts with triacetoxymethylsilane to form a diacetoxy-terminated siloxane:

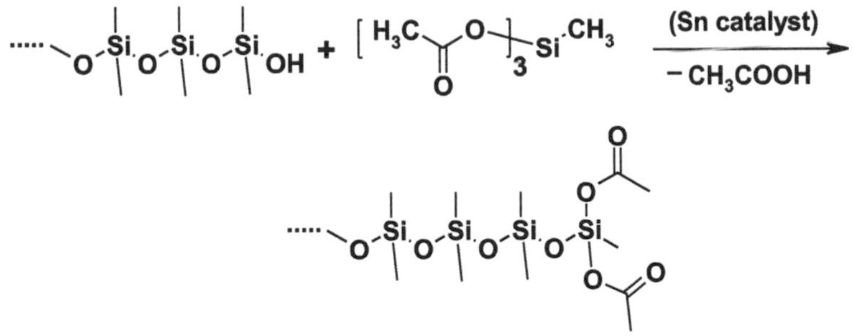

By the addition of water, the silylacetoxy end groups are hydrolyzed and a silanol end group is set free which in the next step of reaction can undergo condensation reaction with an acetoxy group of a second polymer molecule. By consecutive condensation reactions the polysiloxane network is formed.

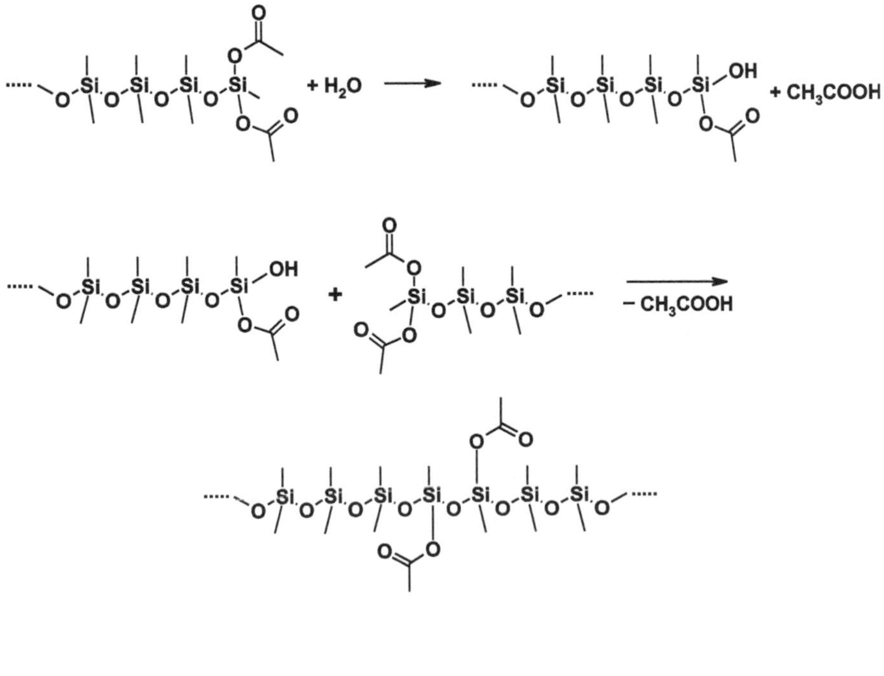

Table 5 Types of active sites in RTV-1 type silicon rubbers [71].

Type	Reactive site	Condensation fragment
Acetoxy		Acetic acid
Oxime		Oxime
Amine		Amine
Amide		Amide
Aminoxy		Hydroxylamine
Isopropeneoxy		Aceton
Alkoxy		Alcohol

R = H , alkyl   R' = alkyl , aryl

On this basis, a number of active sites have been developed for the production of RTV-1 type silicon rubbers (Table 5). The key for this curing behavior is that atmospheric moisture is sufficiently active to start and accelerate the crosslinking reactions.

XI. (METH)ACRYLIC NETWORKS

The monomers which are most widely used in photopolymerization processes to form networks are acrylates. The reason is that they polymerize fast. Methacrylates generally polymerize more slowly but, due to the stiffer main chain, yield harder products. By copolymerization of monoacrylates ($f=2$) with di- ($f=4$) or triacrylates ($f=6$), crosslinked networks are formed. In order to avoid the presence of free monomer in the cured product, monoacrylates are sometimes omitted. The acrylic esters of the lower mono-, di- or trialcohols or the lower ethylene or propylene glycols are liquids of low viscosity and, especially with the lower alcohols, of repellent odor. They are often used in coating formulations as reactive diluents for the more viscous oligomers.

Oligomers serve to reduce the volatility, toxicity, odor, polymerization shrinkage and to improve the properties of the cured material. Frequently used oligomers are as follows.

Epoxy (meth)acrylates, e.g., made by reacting epoxides such as DGEBAs (diglycidyl ethers of bisphenol-A) with (meth)acrylic acid. Although these compounds are no epoxides but have only been derived from epoxides, they are still generally called epoxy (meth)acrylates.

R = H, CH₃

Urethane (meth)acrylates may be obtained by reacting hydroxyalkyl (meth)acrylates, diisocyanates and diols. A typical example of the overall reaction is:

Polyester (meth)acrylates can be made by reacting polyesters with (meth)acrylic acid:

Polyether (meth)acrylates can be made in an analogous way.

Siloxane (meth)acrylates may be obtained in the same way by using siloxanes with terminal hydrosilyl groups which were reacted with either allyl alcohol or allyl glycidyl ether.

In this way, by variation of the length and the composition of the moiety between the (meth)acrylate groups, a large number of linear α,ω diacrylates and dimethacrylates have been synthesized. Crosslinking is achieved by free radical polymerization, by photopolymerization, or by other techniques [72].

(Meth)acrylates with more than three polymerizable vinyl groups ($f \geq 6$) have become interesting materials for low shrinkage network formation, e.g., in the field of dental composites. Branched methacrylates with four or even more methacrylic groups can be prepared easily by a Michael addition of the amino group of diamines or polyamines to the C–C double bond of 2-methacryloyloxyethyl acrylate **5** [73] to yield exclusively the methacrylate terminated products, i.e., tetramethylene diamine reacts to yield **6** almost quantitatively [74]. Also other amines have been reacted such as **7–9** [75]. These highly branched molecules or dendrimers with (meth)acrylate groups in the outer sphere [74] are of interest because they combine low viscosity and low shrinkage behavior.

Also in the field of hydrogels, (meth)acrylates play an important role. Since there is quite a number of hydrophilic methacrylate-based monomers available, radical copolymerization with an appropriate crosslinking monomer leads to network formation. The most widely used monomers are 2-hydroxyethyl methacrylate HEMA, 2-aminoethyl methacrylate and the alkyl derivatives, amethacrylamid and the alkyl derivatives, ethyleneglycol dimethacrylate EGDMA, and polyethyleneglycol dimethacrylate PEGDMA.

XII. MICROGELS

A microgel is an *intramolecularly crosslinked macromolecule* which is dispersed in normal or colloidal solutions, in which, depending on the degree of crosslinking and on the nature of the solvent, it is more or less swollen [76]. The IUPAC Commission on Macromolecular Nomenclature recommended *micronetwork* as a term for a microgel [77] and defined it as *a highly ramified macromolecule of colloidal dimensions*. However, 'micro' refers to dimensions of more than one micrometer whereas the dimensions of the so-called microgels are in the range of nanometers.

Historically, in the early 1930s Staudinger and Husemann were the first who wanted to and really did synthesize a microgel. They polymerized divinylbenzene in very dilute solution at 60 °C for several days and expected that the product should be a colloidal molecule of a globular shape. What they received [78] was a solution of low viscosity and molecular weight osmotically determined was between 2 and 4×10^4. They concluded that *this polymer is a product consisting of strongly branched, three-dimensional molecules*. In natural rubber, microgels were assumed to be present and also in the production of poly(butadiene) [79]. Baker first called attention to microgels as by-products in GR-S polymerizations and briefly described them [80]. In 1958, Shashoua and Beaman prepared microgels by emulsion copolymerization of styrene and methyl acrylate, respectively, with a small amount of divinylbenzene as crosslinker and also acrylonitrile for which N,N'-methylene bisacrylamide proved to be the best crosslinker. They published electron micrographs showing a very narrow size distribution of the microgel particles. They stated that "*each microgel particle is a single macromolecule and that the swelling forces of solvation give rise to dispersion to molecular size*" [81]. Furthermore they stated, that "*the size of a microgel can be varied at will, within the range of 50 to 2000 Å, by merely changing the emulsion polymerization conditions.*" Medalia postulated that solvent-dispersed microgels are thermodynamically true solutions [82], and Cragg and Manson stated that "*these microgel particles belong neither to sol nor to gel, but in a rather paradoxical way, to both*" [83]. Microgels may, under suitable conditions, agglomerate into a gel phase, but the gel so formed can be dispersed again by mechanical agitation.

A. Methods for Preparing Microgels

1. Emulsion (co)polymerization of Monomers

Emulsion polymerization — macroemulsion or microemulsion — is the most efficient synthetic route to prepare microgels. In emulsion polymerization, the dimensions of the micelles as the micro-continuous reactors in which conversion of monomers to polymers is performed, determine the size of the netted particles. Hence, although these tiny particles have the same netted structure as typical gels, they are discrete particles. Among those early recipes to carry out emulsion crosslinking copolymerization to end up with microgels, the recipe given by Shashoua and Beaman is still of actuality and a representative example [81]. In these experiments, the crosslinker concentration was rather low (mole fraction <0.05, and they reported "a tendency for the emulsion polymerization systems to coagulate during the course of polymerization. This is particularly great when high concentrations of crosslinking agent are employed" [81]. In the 1970s, in the emulsion copolymerization of styrene with technical divinylbenzene, Hoffmann increased the crosslinker content up to 17% [84] and he applied an excess amount of emulsifier so that monomer droplets were absent. This article also gives much information about the behavior of the microgels in solution depending on the degree of crosslinking. Some further work should be consulted [85–91], especially the extensive work of Funke et al. on microgels [76].

An even more efficient way to synthesize microgels is microemulsion polymerization. Three characteristic features distinguish micro- from macroemulsion polymerization [92,93]:

1. no monomer droplets exist (see also [84]) but only micelles or microemulsion droplets which are believed to be identical,
2. the initiator stays in the microemulsion droplets only and polymerization occurs only there, provided oil-soluble initiators are used,
3. the reaction mixture is optically transparent and in an equilibrium state.

Antonietti et al. [94] studied microemulsion copolymerization of styrene and 1,3-diisopropenylbenzene using a combination of poly(oxyethylene) and sodium dodecyl sulfate as emulsifier and they received microgels with diameters ranging from 60 to 170 nm.

2. Emulsion (co)polymerization of Prepolymers — Unsaturated Polyesters

From unsaturated polyesters with carboxylic end groups at both chain ends, after neutralization they are efficient emulsifiers for lipophilic monomers [95], and with styrene as comonomer microgels can be prepared with rather uniform diameter [96]. By using lipophilic initiators, such as 2,2′-azobis(isobutyronitrile) (AiBN), in the microemulsion copolymerization, diffusion of monomers is too slow compared with the reaction rate. Therefore, copolymerization is confined to the coherent, lipophilic phase [97,98] and very small microgel particles with a rather uniform size result. Research work by Funke and cited by Funke [76] indicates the usefulness of microemulsion copolymerization to convert unsaturated polyesters into microgels.

B. Solution Polymerization

Microgel formation in free radical solution polymerization can be traced back to 1935 [78]. Even at moderate solution concentrations of a monovinylic/divinylic comonomer pair, intramolecular crosslinking is always observed. At the beginning of a free radical polymerization reaction in solution, the polymer radicals are rather isolated from each other. Hence the local concentration of pendant vinyl groups inside a macroradical coil is much higher than their overall concentration in the reaction mixture. Consequently, the probability of the radical chain end to react with a pendant group of its own chain is strongly favored, and in the early stage of radical crosslinking copolymerization chain cycles are predominately formed thus leading to a decreased size of coils of the same molecular weight [76,99–102].

In ultradilute solution, polymers with pendant vinyl groups have been reported to undergo self-crosslinking. Mecerreyes et al. have prepared copolymers with pendant methacryloyl groups by subsequent chemical modification of poly(styrene-co-hydroxy-ethyl methacrylate) [103]. The extent of crosslinking was in the range from 72 to 92% and the so-formed nanoparticles were very small (3.8–13.1 nm, dynamic light scattering).

'Living' anionic polymerization of divinylbenzene or ethylene glycol dimethacrylate to end up with microgels is only of minor importance, but may be useful for the preparation of microgels with a large content on 'living' carbanionic sites. The reports in the literature with respect to the initiator to monomer ratio for the crosslinking of divinylbenzene are somewhat contradictory [104–106].

C. Other Techniques

Under industrial aspects, crosslinking of natural rubber and also synthetic rubber lattices inside the latex particles might be of interest. In 1954 a patent had already been given to Revertex Ltd. in the UK for curing natural rubber latex with peroxides without appreciable coagulation [107]. A natural rubber latex was radiation induced (with ^{60}Co γ-rays) vulcanized by adding a hydrophobic polyfunctional monomer such as neopentyl glycol diacrylate or 1,3-butylene glycol diacrylate [108]. Also, microparticles of 1.8 μm in average size were formed out of emulsified droplets of liquid poly(butadiene) containing ≤ 2 isocyanate or epoxy groups/mole with amines as curing agents [109]. The use of microgels as fillers in a rubber matrix has been claimed [110–115].

XIII. ELASTOMERS

Polydienes and synthetic rubbers based on dienes, that are butadiene and isoprene, comprise the large bulk of those polymers used as elastomers. They all have in common that the glass transition temperature T_g is far below room temperature. Elastomers therefore are polymers in the molten state and behave like liquids of very high viscosity. Upon prolonged tension, chain slippage occurs and the sample deforms. Crosslinking is an absolute requirement if elastomers are to have their essential property of rapidly and completely recovering from deformations. The term *vulcanization* is used synonymously with crosslinking in elastomer technology.

The characteristics of polydienes and synthetic rubber based on dienes are C–C double bonds as main chain segments. The most widely used examples are:

- natural rubber — poly(*cis*-1,4-isoprene)
- synthetic polyisoprenes
- synthetic polybutadienes, solution or emulsion polymerized
- poly(styrene-*co*-butadiene) random copolymers, solution or emulsion polymerized
- poly(butadiene-*co*-acrylonitrile)
- poly(styrene-*co*-butadiene-*co*-acrylonitrile)
- poly(2-chlorobutadiene)
- poly(isoprene-*co*-isobutene).

A. Unaccelerated Sulfur Vulcanization

Sulfur vulcanization, as first discovered by Goodyear in 1839 and Hancock in 1843, leads to poor rubber qualities and to wastage of sulfur atoms through formation of three types of inefficient or useless structures: long polysulfidic crosslinks **10**, intrachain cyclic monosulfides **11**, and vicinal crosslink pairs **12** which act as a single crosslink [116–118].

10   **11**   **12**

For reasons associated with these useless or labile crosslinks, the unaccelerated sulfur vulcanization is commercially inattractive.

More attractive, however, is the question of what the chemical mechanism of unaccelerated sulfur vulcanization is. A free radical mechanism as first assumed [119–121] had to be abandoned because no evidence was found that free radicals are envolved [116,117]; these sulfur–olefin reactions are insensitive to free-radical initiators and do not respond to free-radical retarders or inhibitors.

Instead, an ionic mechanism is favored which involves the reaction of a highly polarized sulfur–sulfur bond, present in either elemental sulfur or in organic polysulfides, with a carbon–carbon double bond to form an intermediate persulfonium ion **13** which, in turn, can undergo several different types of reactions, particularly proton transfer and hydride ion abstraction reactions [116,117,122].

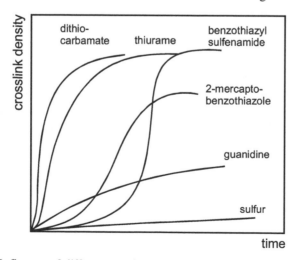

B. Accelerated Sulfur Vulcanization

The so-called accelerated sulfur vulcanization adds accelerators and activators to the rubber/sulfur mixture in order to enhance the rate of crosslink built-up. A schematic drawing of the influence of the different accelerators is given in Figure 2.

Figure 2   Influence of different accelerators.

The different accelerators which are in use can be classified into five groups:

1. Guanidine

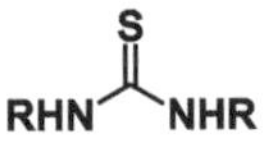

R = phenyl, toluoyl

R = alkyl

2. Dithiocarbamate

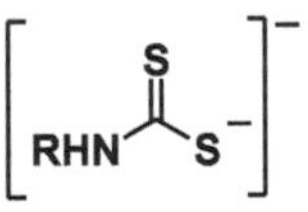

3. Thiurame

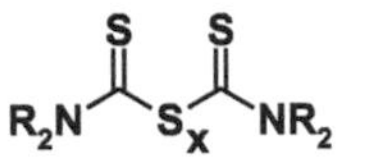

$x = 1, 2, 4$

R = alkyl

4. 2-Mercaptobenzothiazol

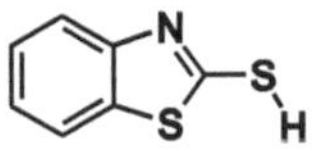

5. Benzothiazyl sulfenamide

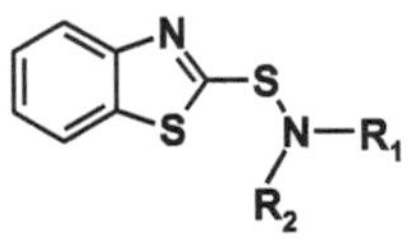

$R_1 = R_2 = H$, alkyl,

cycloalkyl, phenyl

Activators preferentially are ZnO and fatty acids, e.g., stearic acid.

The now generally accepted reaction scheme for the course of accelerated sulfur vulcanization which originally was proposed for natural rubber [116], is an interplay of the active rubber sites, i.e., methylene groups adjacent to main chain carbon–carbon double bonds, sulfur, the accelerator, and, usually, zinc oxide over different stages. Evidence for some rubber-bound intermediates has been obtained [123–125]. However, there is as yet still a debate on the chemical nature of the active sulfurating agent.

It is also accepted — and there is much experimental evidence — that during vulcanization once formed polysulfidic crosslinks are desulfurated by crosslink shortening and finally leading to mono- and disulfidic crosslinks. The sulfur removed from the polysulfidic crosslinks is recycled into the vulcanization process and is able to sulfurate more alkene sites to form additional crosslinks. Zinc complexes derived from the zinc oxide and the accelerator intermediately formed may well be responsible for this desulfuration reaction [126–129].

Vulcanization is performed at rather high temperatures close to 180 °C. Thermal decomposition of polysulfidic crosslinks readily occurs thus forming cyclic mono- and disulfides which are elastically ineffective and also conjugated di- and triene structures in the rubber backbone [129].

Model reactions have contributed much to the understanding of the sulfuration and desulfuration processes [130–134], but they have not solved all problems completely. In a model reaction with cyclohexene representing a poly(butadiene) main chain segment, with the use of dimethyldithiocarbamate and zinc oxide, the zinc dimethyldithioncarbamate **14** was postulated to promote the vulcanization reaction via an ionic mechanism [131]. However, in the same paper evidence was provided that tetramethylthiuram mono- and disulfide **15** react via a free-radical mechanism.

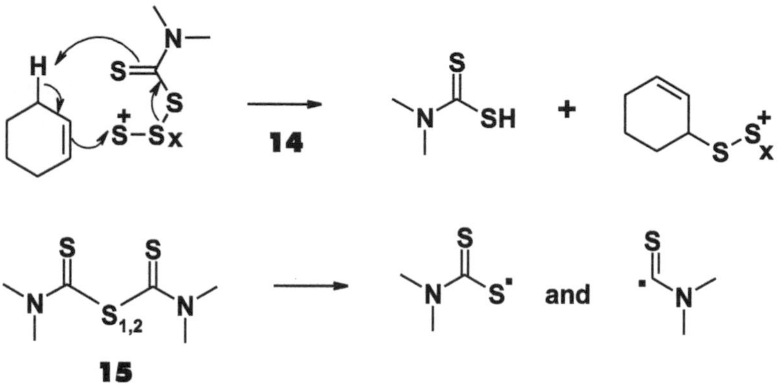

14

15

Pendent sulfur groups are the supposed precursors prior to an interchain crosslink formation. The synthesis of model pendent groups containing benzothiazolyl functions [132,133] has enabled their thermal behavior to be studied directly. Again, the experiments evidenced that the zinc complexes play an important active role in the desulfuration reaction of the pendent polysulfidic groups.

The question of the involvement of zinc in the actual reaction in which carbon-sulfur bonds are formed in the rubber chain remains the major unresolved problem in sulfur vulcanization. Polysulfidic pendent groups are formed initially in the absence of zinc [124,125], probably via accelerator polydisulfides [126,127].

Summarized from all findings reported in the literature, the mechanism of accelerated sulfur vulcanization probably involves initial formation of an accelerator polysulfide [126,135–142], 2,2'-dithiobenzothiazole **16** taken for example. The accelerator polysulfide **17** reacts with a poly(butadiene) chain segment at the allylic hydrogen [118] to form a rubber polysulfide pendent group with one accelerator fragment at the polysulfide group **18**.

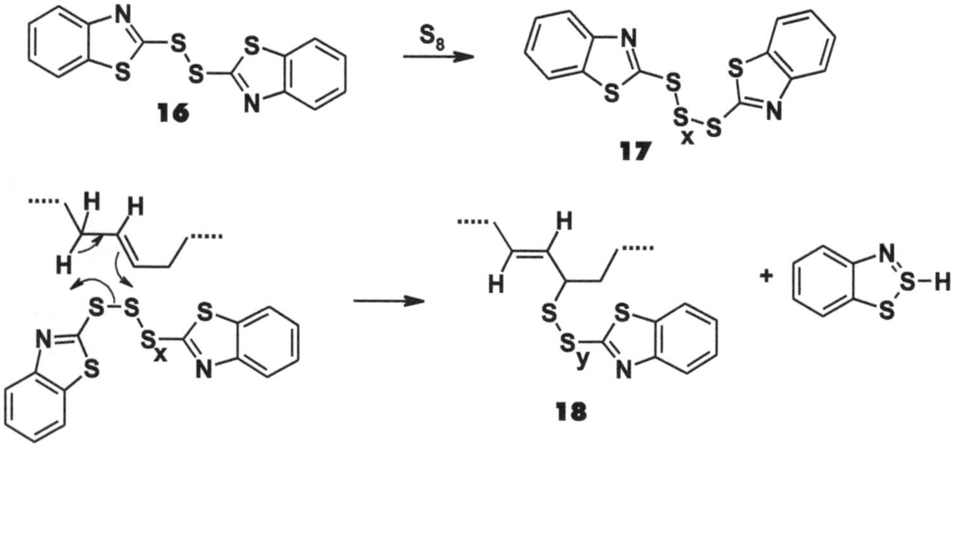

16

17

18

Crosslinking takes place by the corresponding intercatenar action between the accelerator fragment terminated pendent polysulfide chain end and a poly(butadiene) chain segment.

The involvement of zinc in increasing the efficiency of crosslinking is regarded to occur by some chelation of zinc with electron donating sites as ligands L such as amines, carboxylic groups of the activator, or the double bonds of the polymer chain.

XIV. CONCLUDING REMARKS

Crosslinking reactions to form polymer networks has already created new polymeric materials in prehistoric time measured on the time-scale of polymer science. The beginning is Charles Goodyear's discovery of the hot sulfur vulcanization of natural rubber and by this he made natural rubber applicable [143]. This earned him the title of founder of the commercial rubber industry which has since developed immensely and produces a large number of high-tech products. Almost more than half a century later in the early 1900s, Leo Hendrik Baekeland [144] developed the first successful crosslinked or synthetic network polymer obtained by a controlled, step-growth polymerization reaction. This polymer, introduced commercially under the trade name of Bakelite®, was the condensation product of phenol and formaldehyde. Later, other formaldehyde-based network polymers were developed, including the urea-formaldehyde resins and the melamine-formaldehyde resins. In the late 1930s and early 1940s, Otto Bayer discovered that diisocyanates and diols react to form polyurethanes of high molecular weight [145], a very versatile polyaddition reaction which initiated an overwhelming industrial development in the production of polyurethane foams and related materials. Silicone rubber was added to the market in the 1950s and has developed since into many fields of applications. In the early 1980s the liquid silicone rubber technology was established. Crosslinked styrene-divinylbenzene copolymers are the classical basis for ion exchange resins which were first marketed in the 1950s. Crosslinked polymer gels made possible gel permeation chromatography GPC (or SEC) for the determination of molecular weight and molecular weight distribution of soluble macromolecules [146], and they very efficiently served for the development of any kind of absorption chromatography. The first chemical synthesis with the support of a crosslinked polymer carrier was reported in 1963 [147] by Robert Bruce Merrifield and the Nobel Prize awarded to him in 1984 underlines the importance of this first example of a solid phase synthesis on polymer support which was followed by the development of numerous polymeric reagents [148]. Presently, the fast expanding technique of combinatorial organic synthesis traces back to this fruitful strategy. Although the analysis of crosslinked polymer networks is an almost critical task and particular application of crosslinked polymers and elastomers affords a very specific manufacture of the particular product, crosslinked polymers are ubiquitous and have a great potential for further developments.

REFERENCES

1. Flory, P. J. (1941). *J. Am. Chem. Soc.*, *63*: 3083, 3091, 3096.
2. Flory, P. J. (1953). *Principles of Polymer Chemistry*, Cornell University Press, Ithaca, New York.
3. Stockmayer, W. H. (1943). *J. Chem. Phys.*, *11*: 45.
4. Stockmayer, W. H. (1944). *J. Chem. Phys.*, *12*: 125.
5. de Gennes, P. G. (1979). *Scaling Properties in Polymer Physics*, Cornell University Press, Ithaca, New York.
6. Gordon, M., and Ross-Murphy, S. B. (1975). *Pure Appl. Chem.*, *43*: 1.
7. Dušek, K., and Prins, W. (1969). *Adv. Polym. Sci.*, *6*: 1.
8. Dušek, K. (1979). *Makromol. Chem. Suppl.*, *2*: 35.
9. Stauffer, D., Coniglio, A., and Adam, M. (1982). *Adv. Polym. Sci.*, *44*: 103.
10. Funke, W., Okay, O., and Joos-Müller, B. (1998). *Adv. Polym. Sci.*, *136*: 140.
11. IUPAC Macromolecular Division (1995). *Commission on Macromolecular Nomenclature*, Recommendations.
12. Carothers, W. H. (1936). *Trans. Faraday Soc.*, *32*: 39.
13. Flory, P. J. (1941). *J. Am. Chem. Soc.*, *63*: 3083.
14. Kienle, R. H., and Petke, F. E. (1940). *J. Am. Chem. Soc.*, *62*: 1053.
15. Kienle, R. H., and Petke, F. E. (1941). *J. Am. Chem. Soc.*, *63*: 481.
16. Harris, F. H. (1955). *J. Chem. Phys.*, *23*: 1518.
17. Kilb, R. W. (1958). *J. Phys. Chem.*, *62*: 969.
18. Lenz, R. W. (1967). *Organic Chemistry of Synthetic High Polymers*, Interscience Publishers, New York, p. 166.
19. Baekeland, L. H. (1909). U.S. Patent 939.966; U.S. Patent 942.852.
20. Martin, R. W. (1956). *The Chemistry of Phenolic Resins*, John Wiley & Sons, New York.
21. Megson, N. J. L. (1958). *Phenolic Resin Chemistry*, Butterworth, London.
22. Kornblum, N., Smiley, R. A., Blackwood, R. K., and Iffland, D. C. (1955). *J. Am. Chem. Soc.*, *77*: 6269.
23. Bender, H., Farnham, A. G., Guyer, J. W., Apel, F. N., and Gibb, T. B. (1952). *Ind. Eng. Chem.*, *44*: 1619.
24. Kämmerer, H., Kern, W., and Henser, G. (1958). *J. Polym. Sci.*, *28*: 331.
25. Hanus, H., and Fuchs, E. (1939). *J. Prakt. Chem.*, *153*: 327.
26. Cavitt, S. B., Sarrafizadeh, H., and Gardner, P. D. (1962). *J. Org. Chem.*, *27*: 1211.
27. Pepper, D. C. (1941). *Chem. Ind.*, *60*: 866.
28. Baekeland, L. H. (1912). *Ind. Eng. Chem.*, *4*: 739.
29. Rodia, J. S., and Freeman, J. H. (1959). *J. Org. Chem.*, *24*: 21.

30. Smith, L. H., and Welch, K. N. *J. Chem. Soc.*, *1934*: 729.
31. Lebach, H. (1909). *Z. Angew. Chem.*, *22*: 1600.
32. Rider, S. H., and Hardy, E. E. (1962). Prepolymer technology for crosslinked plastics. In *Polymerization and Polycondensation Processes* (Platzer, N. A. J., ed.), American Chemical Society, Van Nostrand Reinhold, New York, Chapter 13.
33. Updegraff, I. H., and Suen, T. J. (1977). Condensations with formaldehyde. In *Polymerization Processes* (Schildknecht, C. E., and Skeist, I., eds.), Wiley-Interscience, New York, Chapter 14.
34. Drumm, M. F., and LeBlanc, J. R. (1972). The reactions of formaldehyde with phenols, melamin, aniline and urea. In *Step-Growth Polymerizations* (Solomon, D. H., ed.), Marcel Dekker, New York, Chapter 5.
35. Ebdon, J. R., and Heaton, P. E. (1977). *Polymer*, *18*: 971.
36. Kumlin, K., and Sominson, R. (1978). *Angew. Makromol. Chem.*, *72*: 67.
37. Crowe, G. A., Jr., and Lynch, C. C. (1948). *J. Am. Chem. Soc.*, *70*: 3795; *71*: 3731 (1949).
38. de Jong, J. I., and de Jong, J. (1952). *Rec. Trav. Chim.*, *71*: 643; *72*: 139 (1953).
39. Tomita, B., and Ono, H. (1979). *J. Polym. Sci., Polym. Chem. Ed.*, *17*: 3205.
40. Zigeuner, G., and Pitter, R. (1955). *Monatsh. Chem.*, *68*: 57.
41. Brydson, J. A. (1970). *Plastic Materials*, 2nd edn., Van Nostrand Reinhold, New York, Chapters 19–23.
42. Tomita, B. (1977). *J. Polym. Sci., Polym. Chem. Ed.*, *15*: 2347.
43. Dawbarn, M., Ebdon, J. R., Hewitt, S. J., Hunt, J. E. B., Williams, I. E., and Westwood, A. R. (1978). *Polymer*, *19*: 1309.
44. Koeda, K. (1954). *J. Chem. Soc. Japan, Pure Chem. Sect.*, *75*: 571.
45. Reinking, N. H., Barnabeo, A. E., and Hale, W. F. (1963). *J. Appl. Polym. Sci.*, *7*: 2135.
46. Chapman, N. B., Isaacs, N. S., and Parker, R. E. (1959). *J. Chem. Soc.*, *1959*: 1925.
47. Koike, T. (1999). *Adv. Polym. Sci.*, *148*: 139.
48. Batog, A. E., Pet'ko, I. P., and Penczek, P. (1999). *Adv. Polym. Sci.*, *144*: 49.
49. Pet'ko, I. P., et al. (1984). *Plast. Massy*, *5*: 47.
50. Nechitaylo, L. G., et al. (1987). *Kinet. Katal.*, *28*: 1322.
51. Andrianov, N. A., and Emel'yanov, V. N. (1976). *Usp. Khim.*, *45*: 1817.
52. Chelnokova, G. I., Rafikov, S. R., and Korshak, V. V. (1949). *Dokl. Akad. Nauk SSSR*, *64*: 353.
53. Tanaka, Y., and Kakiuchi, H. (1963). *J. Appl. Polym. Sci.*, *7*: 1063.
54. Frisch, W., and Hofmann, W. (1961). *Makromol. Chem.*, *44*: 8.
55. Badran, B. H., Yehia, A. A., and Abdel-Bary, E. M. (1977). *Europ. Polym. J.*, *13*: 155.
56. Smith, I. T. (1961). *Polymer*, *2*: 95.
57. Horne, W. H., and Shriner, R. I. (1932). *J. Am. Chem. Soc.*, *54*: 2925.
58. Cromwell, N. H., and Barker, N. G. (1950). *J. Am. Chem. Soc.*, *72*: 4110.
59. Shechter, L., Wynstra, J., and Kurkjy, R. P. (1956). *Ind. Eng. Chem.*, *48*: 94.
60. Lebedev, N. N., and Smirnova, M. M. (1960). *Izvest. Vysshikh Uchebn. Zavedenii, Khim. Tekhnol.*, *3*: 104; *Chem. Abstr.*, *54*: 16149c (1960).
61. Gough, L. J., and Smith, I. T. (1960). *J. Oil Colour Chemists Assoc.*, *43*: 409.
62. Tanaka, Y., and Mika, T. F. (1973). *Epoxy Resin Chemistry and Technology*, Marcel Dekker, New York, p. 135.
63. King, J. J., and Bell, J. P. (1979). *Epoxy Resin Chemistry*, ACS Symposium, Vol. 114, p. 225
64. Rozenberg, B. A. (1986). *Adv. Polym. Sci.*, *75*: 113.
65. Harris, J. J., and Temine, S. C. (1966). *J. Appl. Polym. Sci.*, *10*: 523.
66. Crivello, L. V., and Lam, J. H. W. (1979). *Epoxy Resin Chemistry*, ACS Symposium, Vol. 114, p. 1.
67. Lohse, F., and Zweifel, H. (1986). *Adv. Polym. Sci.*, *78*: 61.
68. Bayer, O. (1947). *Angew. Chem.*, *A59*: 257.
69. Cooper, W., Pearson, R. W., and Drake, S. (1960). *Ind. Chemist*, *36*: 121.
70. Kochs, P. (1987). In *Houben-Weyl, Methoden der organischen Chemie*, Bd. 20, Tl. 3, Hrsg (Bartl, H., and Falbe, J., eds.), Georg Thieme Verlag, Stuttgart, p. 2219.
71. *Silicone, Chemie und Technologie*, Vulkan-Verlag, Essen, 1989.

72. Kloosterboer, J. G. (1988). *Adv. Polym. Sci.*, *84*: 1.
73. Luchtenberg, J., and Ritter, H. (1994). *Macromol. Rapid Commun.*, *15*: 81.
74. Moszner, N., Völkel, T., and Rheinberger, V. (1996). *Macromol. Chem. Phys.*, *197*: 621.
75. Klee, J. E., Neidhart, F., Flammersheim, H.-J., and Mülhaupt, R. (1999). *Macromol. Chem. Phys.*, *200*: 517.
76. Funke, W., Okay, O., and Joos-Müller, B. (1998). *Adv. Polym. Sci.*, *136*: 140.
77. IUPAC Macromolecular Division (1995). *Commission on Macromolecular Nomenclature, Recommendations.*
78. Staudinger, H., and Husemann, E. (1935). *Ber. Dtsch. Chem. Ges.*, *68*: 1618.
79. Elford, W. J. (1930). *Proc. Royal Soc.*, *106B*: 216.
80. Baker, W. O. (1949). *Ind. Eng. Chem.*, *41*: 511.
81. Shashoua, V. E., and Beaman, R. G. (1958). *J. Polym. Sci.*, *33*: 101.
82. Medalia, A. I. (1951). *J. Polym. Sci.*, *6*: 423.
83. Cragg, L. H., and Manson, J. A. (1952). *J. Polym. Sci.*, *9*: 265.
84. Hoffmann, M. (1974). *Makromol. Chem.*, *175*: 613.
85. Price, C., Forget, J. L., and Booth, C. (1977). *Polymer*, *18*: 528.
86. Kunz, D., Thurn, A., and Burchard, W. (1983). *Coll. Polym. Sci.*, *261*: 635.
87. Kunz, D., and Burchard, W. (1986). *Coll. Polym. Sci.*, *264*: 498.
88. Wolfe, M. S., and Scopazzi, C. (1989). *J. Coll. Int. Sci.*, *133*: 265.
89. Ma, G. H., and Fukutomi, T. (1991). *J. Appl. Polym. Sci.*, *43*: 1451.
90. Rodriguez, B. E., Wolfe, M. S., and Fryd, M. (1994). *Macromolecules*, *27*: 6642.
91. Nomura, M., and Fujita, K. (1993). *Polymer Int.*, *30*: 483.
92. Kuo, P.-L., Turro, N. J., Tseng, Ch.-M., El-Aasser, M. S., and Vanderhoff, J. W. (1987). *Macromolecules*, *20*: 1216.
93. Candau, F., and Ottewill, R. H. (1990). *An Introduction to Polymer Colloids*, Kluwer Academic Press, Dordrecht.
94. Antonietti, M., Bremser, W., and Schmidt, M. (1990). *Macromolecules*, *23*: 3796.
95. Yu, Y. Ch., and Funke, W. (1982). *Angew. Makromol. Chem.*, *103*: 187.
96. Funke, W., Kolitz, R., and Straehle, W. (1979). *Makromol. Chem.*, *180*: 2797.
97. Flammer, U., Hirsch, M., and Funke, W. (1994). *Makromol. Chem., Rapid Commun.*, *15*: 519.
98. Liang, L., and Funke, W. (1996). *Macromolecules*, *29*: 8650.
99. Malinsky, J., Klaban, J., and Dušek, K. (1971). *J. Macromol. Sci., Chem.*, *A5*, 1071.
100. Galina, H., and Rupicz, K. (1980). *Polym. Bull.*, *3*: 473.
101. Matsumoto, A., Matsuo, H., and Oiwa, M. (1987). *Makromol. Chem., Rapid Commun.*, *8*: 373.
102. Dušek, K., Matejka, L., Spacek, P., and Winter, H. (1996). *Polymer*, *37*: 2233.
103. Mecerreyes, D., Miller, R. D., Lee, V., Hawker, C. J., Hedrick, J. L., Wursch, A., Volksen, W., Magbitang, T., and Huang, E. (2001). *Adv. Mater.*, *13*: 204.
104. Hiller, J. C., and Funke, W. (1979). *Angew. Makromol. Chem.*, *76/77*: 161.
105. Eschwey, H., Hallensleben, M. L., and Burchard, W. (1973). *Makromol. Chem.*, *73*: 235.
106. Lutz, P., and Rempp, P. (1988). *Makromol. Chem.*, *189*: 1051.
107. Brit. Pat. 738.279 (1954) to: Revertex Ltd., inventor: G. Stott.
108. JP 59124935 (1984) to: Japan Atomic Energy Research Institute.
109. JP 01182326 (1989) to: Toray Ind., Inc. Japan, inventor: K. Oka.
110. DE 4220563, US 5395891 (1992) to: Bayer AG, inventors: W. Obrecht, P. Wendling, R. H. Schuster, A. Bischoff.
111. DE 1972672.9, EP 854170 (1997) to: Bayer AG, inventors: W. Obrecht, T. Scholl, U. Eisele, W. Jeske, P. Wendling.
112. DE 19701487.9 (1997) to: Bayer AG, inventors: W. Obrecht, T. Scholl, U. Eisele, W. Jeske, P. Wendling.
113. DE 19834802, EP 99114743.0 (1998) to: Bayer AG, inventors: W. Obrecht, T. Scholl, P. Wendling, V. Monroy, M. Well.
114. DE 19834803, EP 99114744.8 (1998) to: Bayer AG, inventors: W. Obrecht, T. Scholl, P. Wendling, V. Monroy, M. Well.

115. DE 19834804 (1998) to: Bayer AG, inventors: W. Obrecht, T. Scholl, P. Wendling, V. Monroy, M. Well.
116. Bateman, L., Moore, C. G., Porter, M., and Saville, B. (1963). In *The Chemistry and Physics of Rubber-like Substances* (Bateman, L., ed.), Maclaren, London, Chapters 4 and 15.
117. Bateman, L., Moore, C. G., Porter, M., and Saville, B. (1958). *J. Chem. Soc., 1958*: 2866.
118. Skinner, T. D. (1972). *Rubber Chem. Technol., 45*: 182.
119. Farmer, E. H., and Shipley, F. W. (1946). *J. Polym. Sci., 1*: 293.
120. Farmer, E. H., and Shipley, F. W. *J. Chem. Soc., 1947*: 1519.
121. Farmer, E. H. *J. Soc. Chem. Ind., 66*: 86.
122. Ross, G. W. (1958). *J. Chem. Soc., 1958*: 2856.
123. Campbell, D. S. (1970). *J. Appl. Polym. Sci., 14*: 1409.
124. Parks, C. R., Parker, D. K., Chapman, D. A., and Cox, W. L. (1970). *Rubber Chem. Technol., 43*: 572.
125. Parks, C. R., Parker, D. K., and Chapman, D. A. (1972). *Rubber Chem. Technol., 45*: 467.
126. Coran, A. Y. (1978). In *Science and Technology of Rubber* (Eyrich, F. R., ed.), Academic Press, New York, Chapter 7.
127. Coleman, M. M., Shelton, J. R., and Koenig, J. L. (1974). *Ind. Eng. Chem. Prod. Res. Div., 13*: 154.
128. Porter, M. In *Organic Chemistry of Sulfur* (Oae, S., ed.), Plenum Press, New York, Chapter 3.
129. Morrison, N. J., and Porter, M. (1984). *Rubber Chem. Technol., 57*: 63.
130. Wolfe, J. R., Jr., Pugh, T. L., and Killian, A. S. (1968). *Rubber Chem. Technol., 41*: 1329.
131. Wolfe, R. J., Jr. (1968). *Rubber Chem. Technol., 41*: 1339.
132. Morrison, N. J. (1984). *J. Chem. Soc. Perkin Trans., 1*: 101.
133. Morrison, N. J. (1984). *Rubber Chem. Technol., 57*: 86.
134. Morrison, N. J. (1984). *Rubber Chem. Technol., 57*, 97.
135. Manik, S. P., and Banerjee, S. (1970). *Rubber Chem. Technol., 43*, 1294.
136. Manik, S. P., and Banerjee, S. (1971). *J. Appl. Polym. Sci., 15*: 1341.
137. Duchacek, V. (1972). *Angew. Makromol. Chem., 23*: 21.
138. Duchacek, V. (1974). *J. Appl. Polym. Sci., 18*: 125.
139. Duchacek, V. (1975). *J. Appl. Polym. Sci., 19*: 1617.
140. Duchacek, V. (1976). *J. Appl. Polym. Sci., 20*: 71.
141. Duchacek, V. (1978). *J. Appl. Polym. Sci., 22*: 227.
142. Chapman, D. A. (1978). *J. Elast. Plast., 10*: 129.
143. Goodyear, Ch. (1844). US Patent 3633.
144. Baekeland, L. H. (1913). *Ind. Eng. Chem., 5*: 506.
145. Bayer, O. (1941). *Ann., 549*: 286.
146. Seidl, J., Malinský, J., Dušek, K., and Heitz, W. (1967). *Adv. Polym. Sci., 5*: 113.
147. Merrifield, R. B. (1963). *J. Am. Chem. Soc., 85*: 2149.
148. Kirschning, A., Monenschein, H., and Wittenberg, R. (2001). *Angew. Chem., 113*: 670; *Angew. Chem. Int. Ed., 40*: 650 (2001).

RECENT REVIEWS

Blackley, D. C. (1997). Chemically-modified lattices: 1. Prevulcanized lattices. In *Polymer Lattices*, Chapman & Hall, London, Chapter 13.
Carfagna, C., Amendola, E., and Giamberini, M. (1997). Liquid crystalline epoxy based thermosetting polymers. *Prog. Polym. Sci., 22*: 1607.
Funke, W., Okay, O., and Joos-Müller, B. (1998). *Adv. Polym. Sci., 136*: 140.
Gaw, K. O., and Kakimoto, M. (1999). Polyimide-epoxy composites. *Adv. Polym. Sci., 140*: 107.
Goosey, M., Roth, M., Kainmüller, T., and Seitz, W. (1999). Epoxide resins and their formulation. In *Plast. Electron.*, 2nd edn. (Goosey, M., ed.), Kluwer Academic, Dordrecht.

Grobelny, J. (1999). Structural investigations of unsaturated and cross-linked polyesters by nuclear magnetic resonance spectroscopy. *J. Macromol. Sci., Rev. Macromol. Chem. Phys.*, *C39*: 405.

Guo, Q. (1999). Thermosetting polymer blends: miscibility, crystallization, and related properties. *Plast. Eng. (N.Y.)*, *52*: 155.

Hu, H. S.-W., and James, R. (1999). Epoxy networks from a fluorodiimidediol. In *Fluoropolymers*, Vol. I (Hougham, G., ed.), Kluwer Academic/Plenum Publ., New York, p. 181.

Koike, T. (1999). Viscoelastic behavior of epoxy resins before crosslinking. *Adv. Polym. Sci.*, *148*: 139.

Lopata, V. J., Saunders, C. B., Singh, A., Janke, C. J., Wrenn, G. E., and Havens, S. J. (1999). Electron-beam-curable epoxy resins for the manufacture of high-performance composites. *Radiat. Phys. Chem.*, *56*: 405.

Malik, M., Choudhary, V., and Varma, I. K. (2000). Current status of unsaturated polyester resins. *J. Macromol. Sci., Rev. Macromol. Chem. Phys.*, *C40*: 139.

Mormann, W. (1998). Liquid crystalline thermoset networks by step polymerization of cyanates and epoxides. *Wiley Polym. Networks Group Rev. Ser.*, *1*: 347.

Morrison, N. J., and Porter, M. (1984). *Rubber Chem. Technol.*, *57*: 63.

Ogura, I. (1999). Low dielectric constant epoxy resins. In *Handb. Low High Dielectr. Constant Mater. Their Appl.*, Vol. I (Nalwa, H. S., ed.), Academic Press, San Diego, p. 213.

Sandler, S. R., and Karo, W. (1992). *Polymer Syntheses*, Vol. II, Academic Press, Inc., Boston, Chapter 2.

Sandler, S. R., and Karo, W. (1992). *Polymer Syntheses*, Vol. II, Academic Press, Inc., Boston, Chapter 1.

Sandler, S. R., and Karo, W. (1992). *Polymer Syntheses*, Vol. II, Academic Press, Inc., Boston, Chapter 3.

Sandler, S. R., and Karo, W. (1992). *Polymer Syntheses*, Vol. II, Academic Press, Inc., Boston, Chapter 4.

Vilgis, T. A. (1989). Polymer networks. In *Comprehensive Polymer Science*, Vol. 6 (Allen, G., ed.), Pergamon Press, Oxford, Chapter 8.

White, J. E., Brennan, D. J., Silvis, H. C., and Mang, M. N. (2000). Epoxy-based thermoplastics: new polymers with unusual property profiles. *ACS Symp. Ser.*, *755*: 132.

16

Biodegradable Polymers for Biomedical Applications

Samuel J. Huang
Institute of Materials Science, University of Connecticut, Storrs, Connecticut

I. INTRODUCTION

Traditionally synthetic polymers were designed and manufactured with long term stability, as they were mostly used as coatings, packagings and structures. Since 1970s biodegradable polymers with controllable lifetime have received attentions as biomedical and environmentally compatible consumer products materials [1–7]. Biodegradable polymers are essential in the design, synthesis and applications of biomedical implants and drug delivery system whereas biodegradable polymers prepared from renewable and sustainable resources can be generally disposed through composting. These polymers share many similar structural units but they are different in manner in how they are degraded. Biomedical materials are used and degraded in comparatively narrow range of environments whereas consumer products materials are used and degraded, by contrast, in board environments. This chapter describes synthetic biodegradable polymers for biomedical applications.

II. BIOMEDICAL POLYMERS

The first and most successful use of biodegradable polymers is in the area of degradable and absorbable sutures [8,9]. Biodegradable polymers base sutures, drug release delivery systems [10], scaffolds [11,12], and tissue engineering devices [13–15] are current areas of interests. Hydrophobic polyesters derived from glycolic acid (GA) and lactic acid (LA) represent the most commonly used materials with copolyesters of GA, LA and other cyclic esters and carbonate monomers becoming available recently. Among these poly(lactic acid) (PLA) has become the main polymer as the monomer lactic acid is obtained from fermentation of agricultural and food byproducts [16]. PLA and its copolymers have wide ranges of chemical and physical properties and they represent the most important biodegradable biomedical polymers.

III. POLYESTERS

Aliphatic polymers undergo hydrolysis, both acid or base catalyzed and enzyme catalyzed faster than aromatic polyesters and are generally preferred as biomaterials than aromatic polyesters.

A. Polyesters from Hydroxyacids

These aliphatic polyesters can be obtained by catalyzed dehydration of hydroxyacids and, more efficiently, by ring opening polymerization of the cyclic esters of hydroxyacids (equations (1) and (2)). Catalysts are generally used to facilitate the polymerization. Among the effective catalysts are Lewis acids in the form of metal salts of Sn, Zn, Ti, Al, and rare earth metals [17–23]; alkali metal alkoxides and super-molecular complexes [20,24,25]; and acids [26].

Glycolide   Polyglycolic Acid   (1)

Polymerization of glycolide.

Glycolide   Lactide   Poly(glycolide-co-lactide)
or
PGA / PLA

Copolymerization of glycolide and lactide.

(2)

Trimethylene Carbonate, TMC   *p*-Dioxanone   ε-caprolactone, CL

Thermal dehydration polymerization of hydroxyacids, as shown in (1) for poly(lactic acid) (PLA) is a high energy required process and PLA of low molecular weight (up to thousands) is obtained together with the cyclic dimmer (lactide) [27]. Higher molecular weight PLLA and its copolymers with glycolic acid and ε-hydroxycaproic acid can be obtained in direct thermal polymerization in organic solvents [28]. Ring opening polymerizations of cyclic esters with transition metal catalysts are the most effective methods for obtaining polymers of high molecular weight in good yield in bulk. It is generally accepted that these polymerizations proceed via acyl cleavage with insertion of monomers between the metal–carbon bonds of the active sites [20,29]. Sn (II) esters are commonly used since these are easily obtainable and are approved in food products by USDA. Copolymers of glycolic acid and lactic acid by direct thermal polymerizations have T_ms 135 °C, which is lower than that of copolymers with similar compositions obtained by ring opening polymerization of glycolide and lactide, T_m 145 °C [28].

Copolyesters from direct polymerization of L-lactic acid and ε-hydroxycaproic acid, ring opening polymerization of L-lactide and ε-caprolactone, and sequential polymerization, PLLA with ε-caprolactone (CL) have different properties (Table 1). Different sequencing of the repeating monomeric structural units in these copolymers was suggested as the reason for the difference in property. PLLA–PCL–PLA block copolymers obtained from copolymerization of PCL of various molecular weights (530, 2000, 43,000, and 80,000) and L-lactide have higher T_gs (30–62 °C), T_m (54–58 °C) for PCL blocks, and T_m (153–172 °C) for the PLLA blocks [29], Tab suggesting longer block lengths than those obtained from polymerization of PLLA and CL. When high MW PCLs were copolymerized with lactide the block copolyesters thus obtained had lower MW than expected, indicated substantial ester exchange during the thermal polymerization process.

The physical properties of the copolyesters vary greatly with the composition and sequence. Materials with properties as weak elastomers to hard thermoplastics can be obtained [30]. The tensile modulus and tensile strength were much higher for PLLA, PDLA, and PCL homopolymers than those of the copolymers. Crosslinking with peroxides increases the impact and tensile strength of PCL and PLA copolymers [31–33]. Micromonomers were prepared from ring opening polymerization of cyclic ester with functionalized initiator (3).

Synthesis of polycaprolactone micromonomer.

(3)

Table 1 Copolyesters of L-lactic acid (LLA) and ε-hydroxycaproic acid (HCA).

Method of polymerization		M_w	T_g, °C	T_m, °C
Direct	LLA + HCA	120,000	24	Amorphous
Ring opening	L-Lactide + CL	120,000	34	Amorphous
Sequential	PLLA + CL	130,000	36	127

Data from [28].

B. Poly(ester-co-ether)s

Poly(ester-co-ether)s have been prepared by using polyether with hydroxy-terminal as co-initiator in the ring opening polymerizations of cyclic esters. Among these lactide received the most attention [34–41]. Typically lactide, poly(ethylene glycol) and stannous 2-ethyl hexanoate were heated under nitrogen at 120–150 °C for up to 24 hr. The poly(ester-ether)s thus obtained typically show only one, indicating only one amorphous phase. Only crystalline phase for PLLA is observed with low Mw PEG and both crystalline phases for PEG and PLLA are observed when PEG block size approaches 4000 [41].

C. Hydroxylated Polyesters

Condensation polymerization of glycols with tartaric acid results in poly(alkylene tartrate)s [42–44]. The hydroxylated polyesters from C2 and C4 are hydrophilic and water soluble and those from C6 and higher glycol are water insoluble with increase in hydrophobicity with increasing size of glycols. Crosslinkable unsaturated poly(alkylene tartrate)s are obtained by adding maleic anhydride to the polymerization of glycols with tartaric acid (4) [45]. Poly(tartrate) was obtained from the condensation of tartaric acid ketal with tartaric acid diacetate [46].

Synthesis of unsaturated copolymers based on poly(dodecamthetylene) tartarate.

(4)

Table 2 Typical PLLA–PEG–PLLA triblock copolyesters with PEG M_w 1000.

M_n(NMR)	M_n(GPC)	M_w	MWD	T_g,°C	T_m,°C	ΔH_m, J/g	T_c,°C
10,657	12,293	20,590	1.67	34.5	156	47.1	76.2
18,440	15,636	27,296	1.75	42.9	161	44.5	91
26,511	23,318	40,344	1.73	50.6	167	48.4	107

Data from [41].

D. Carboxylated Polyesters

Poly(β-malic acid) is the most simple carboxylated polyester [47–49]. It is prepared by ring opening polymerization of the mono-benzyl ester β-lactone of malic acid and subsequent debenzylation. It has been explored as drug carrier. Reaction of itaconic anhydride with PCL with hydroxy terminals results in polyesters with carboxylic and C=C double bond functional terminals, suitable for further reactions to form networks and gels, (5) and (6) [50,51].

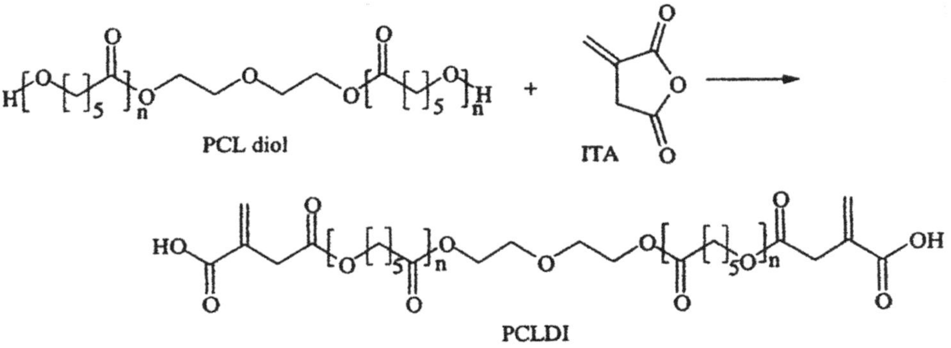

Synthesis of PCL diol end-capped with itaconic anhydride, PCLDI.

(5)

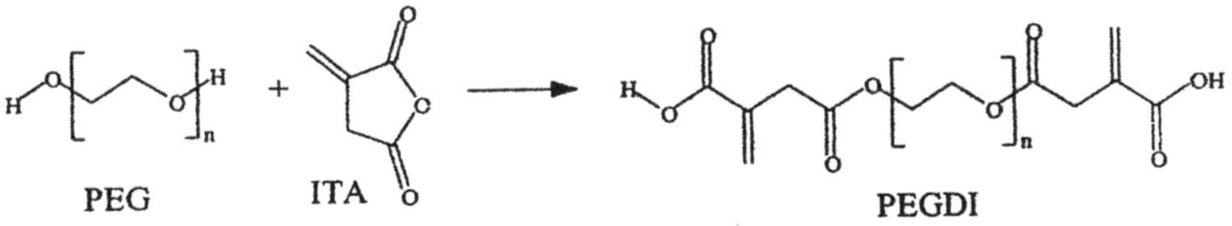

Synthesis of poly(ethylene glycol) end-capped with itaconic anhydride.

(6)

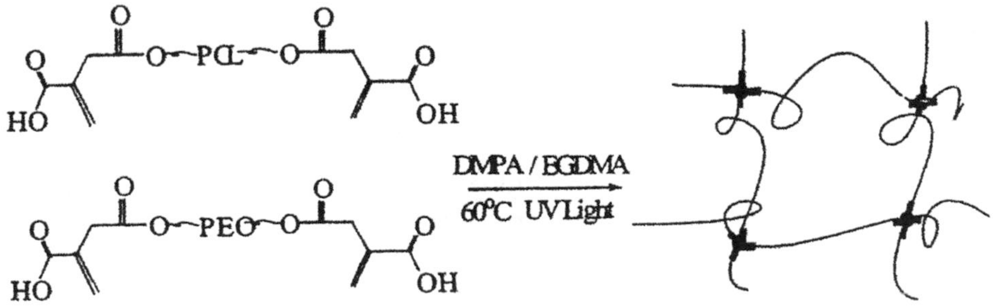

Synthesis of PEG-PEG crosslinked gels.

(7)

E. Polyorthoesters

Transesterification reaction between cyclic orthoesters and glycols gives polyorthoesters. They can also be obtained through the reactions of diketene acetals with glycols [52–56]. These esters are relatively stable in bases and are hydrolyzed slowly in physiological pH and fast in low pH condition. They have been explored as drug delivery systems.

F. Polycarbonates

Trimethylene carbonate, TMC, is a commonly used co-monomer for poly(glycolide-*co*-lactide) base sutures [57,58]. Incorporation of the carbonate structure provides flexibility and toughness to the otherwise rigid and brittle copolyesters.

IV. POLY(AMIDE-ESTER)S

Polydepsipeptides, poly(α-aminoacid-alt-α-hydroxyacid)s can be obtained by ring opening polymerization of morpholine-2,5-dione derivatives which are prepared from α-amonoacids and α-hydroxyacids, (8) and (9) [59]. Those prepared from optically active monomers are partially crystalline whereas those prepared from racemic monomers are amorphous. T_gs of the poly(amide-ester) from valine and lactic acid are between 90–92 °C, 30 degrees higher than that of PLA. Depsipeptides have been explored as sutures [60]. Alternating poly(amide-ester)s have also been prepared from α-aminoacids, ε-aminocaproic acid with β-hydroxyacid [61] and glycolic acid [62]. Alternating amide and ester structure containing poly(amide-ester)s have been prepared from α-amino acids, glycols, and dicarboxylic acids, (10) [63–65]. These poly(amide-ester)s are biodegraded according to known enzyme specificity.

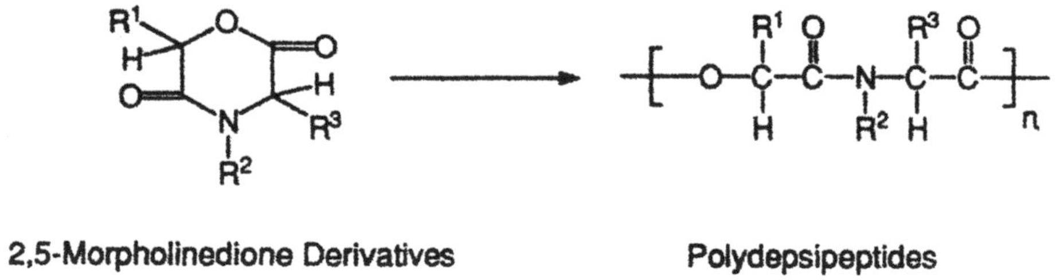

2,5-Morpholinedione Derivatives   **Polydepsipeptides**

Synthesis of polydepsipeptides.

(8)

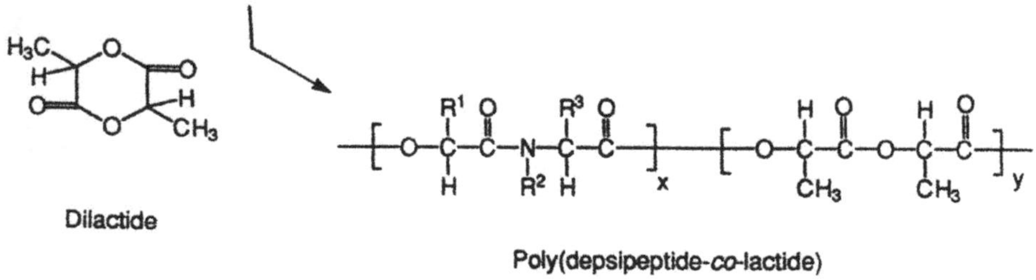

Dilactide

Poly(depsipeptide-*co*-lactide)

Synthesis of poly(depsipeptide-*co*-lactide).

(9)

Synthesis of poly(Z-Tyr-Tyr-Y-imnocarbonate).

(11)

Alternating poly(amide-ester)s have been prepared from aminoalcohols and alkanedicarboxylic acids [67]. These are partially crystalline polymers with various rates of hydrolysis and subtilisin catalyzed degradations.

α,ω-Diaminoalkanes were converted into α,ω-di(hydroxy acetamido)alkanes, which were then polymerized with succinyl chloride into poly(amide-ester)s [67,68]. These polymers are hydrolyzed faster than polyesters but slower than polyamides.

V. POLYAMIDES AND POLYAMINOACIDS

Polyamides are generally degraded slower than polyesters with similar structures [1–4]. As a result their uses as biodegradable materials have not been explored as often as that of polyesters. Interests in degradable polyamides have been mainly directed toward those derived from α-aminoacids [70]. Since polypeptides are out of the scope of this chapter and will be described here, poly(glutamic acid) and its esters have been studied as drug release systems [71]. However their syntheses, generally by ring opening polymerization of N-carboxyanhydrides are tedious, and slower hydrolysis rates limited their potential use. Poly(aspartic acid), PASP, on the other hand, can be easily obtained on large scales by thermal polymerization aspartic acid to give polysuccinimide, PSI, followed by hydrolysis to give PASP [72]. There are only few biomedical applications reported [73].

Synthesis of poly(aspartic acid), PASP.

(12)

VI. POLYENAMINES

Condensation of diamines with diacetoacetyl compounds give poly(enamine-carbonyl)s [74–77]. The enamine-carbonyl system form hydrogen-bonded rings and are stabilized. They are slightly acidic and form metal chelates. They are hydrolyzed faster in acidic than in neutral and basic conditions. The hydrolysis produces diamines (basic) and diacetoacetyl compounds (acidic) providing self-buffering of the pH of the system. This is similar to the hydrolysis of polypeptides.

VII. POLYETHERS

Polyoxyethylenes, poly(ethylene glycol) with M_w less than 20,000 PEG, are obtained from the polymerization of oxirane and are described in another chapter. They received increasing interest as biomedical polymers due to their bio-/blood compatibility in linear, grafted, and crosslinked gel forms [78–88]. PEGs have been functionalized with various terminals for chemical modifications, (7) [51,89].

VIII. POLYPHOSPHAZENES

Polyphosphazenes with hydrolytically sensitizing groups are easily hydrolyzed to give ammonium and phosphate compounds and have been explored as biodegradable biomedical polymers [90,91]. These groups include aminoacid esters, glucosyl, glyceryl, glycolate, lactate, and imidazoyl. Similar to the polyenamines polyphosphazenes hydrolysis produces self-buffering ammonium phosphate systems.

IX. SUTURES AND WOUND REPAIRS

Synthetic sutures are the most successful commercial products of biodegradable polymers [57,58]. Poly(glycolide), PGA, was the first biodegradable synthetic suture [92–94]. Copoly(glycolide-co-lactide), PGLA, usually 90/10 came later. All these partially crystalline polymers are rigid and brittle. Braided fibers are used as sutures. The hydrophobic polyesters cause blood proteins deposition and scar tissues formations which is one draw back of the materials. Dioxanone, trimethylene carbonate, and ε-caprolactone are added to the PGLA polymerization to provide flexibility and toughness. Block copolymers with blocks containing different composition and sequence are the bases for sutures of various properties. PLA and PCL copolymers films have been used for wound dressing [95].

X. IMPLANTS AND SCAFFOLD FOR TISSUE ENGINEERING

The results on initial approaches using biodegradable polyesters as implant materials were mixed. The hydrolysis and enzymatic (and microbial) degradations of hydrophobic aliphatic polyester proceed in selective manners. The amorphous regions of the polymers are degraded prior to the crystalline region forming small crystalline particles in the case of linear polymers [96–98]. In cases of small surface to volume implants the inside part of the implants was found to be degraded faster than the outside due to the self-catalysis of the not yet diffuse oligomeric acid formed during the hydrolysis resulting in harrowing of the implants. All these might lead to complications.

Hydrophilic/hydrophobic systems are closer to bio-systems and provide better diffusion of the degraded products out of the implants in addition to the observed better blood compatibility [99]. Binary systems containing crosslinked poly(2-hydroxyethyl methacrylate), PHEMA, and PCL were found to have higher strength and better biocompatibility than PHEMA, a commonly used biomedical hydrogel [100–103]. A composite artificial tendon scaffold implant constructed with PHEMA/PCL matrix reinforced with PGA fibers was successfully tested in rabbit [104]. Polyether base polymers that can be injected and then transformed into gels chemically or thermally are of great potential as tissue engineering scaffold materials [105].

Increasing the surface areas of implants is essential for the uses of hydrophobic biodegradable polyesters as scaffolds for tissue engineering [14,15,106,107]. These might include, nets, porous foams, membranes, non-wovens, harrow tubes, etc.

XI. RELEASE AND DELIVERY SYSTEMS

Although intense efforts have been directed toward the use of biodegradable polyesters and polyanhydrides for drug release and delivery the results were mixed at best [108,109]. The complicated degradation profiles, production of acids, and poor proteins deposition characteristics contribute to the observed results. The use of microspheres improves the reformance [110]. Hydrophilic–hydrophobic materials are more suitable due to their better biocompatibility and diffusion characteristics. Swellable poly(alkylene tartrate)s, poly-enamine, and oxidized poly(vinyl alcohol) were found to be effective matrix materials for controlled and sustained releases [77,111–113]. Hydrogels have become the most studied release systems [114,115]. PLA/PEG block copolymers are degraded slowly in blood and are thus potentially useful [116]. Poly(γ-glutamic acid) crosslinked with dihaloalkane gels were found to be effective in the release of hormones for 20–30 days [117]. It has been shown that the attachment of PEG to polylysines reduces the toxicity of polylysines in gene delivery applications [118].

XII. SUMMARY

Even with the rapidly increased interest in biodegradable polymers for biomedical applications the advance has been 'material limited' due to the difficulty in balancing the controlled lifetime, property, and biocompatibility.

REFERENCES

1. Huang, S. J. (1995). *J. Macromol. Sci.–Pure Appl. Chem.*, *A32*(4): 593–597.
2. Huang, S. J. (1985). In *Encyclopedia of Polymer Science and Engineering*, Vol. 2 (Kroschwitz, ed.), John Wiley & Sons, pp. 220–243.
3. Huang, S. J. (2002). In *Degradable Polymers — Principles and Applications*, 2nd edn. (Scott, G., ed.), pp. 17–26.
4. Huang, S. J. (1994). In *The Encyclopedia of Advanced Materials* (Bloor, D., Brook, R. J., Flemings, M. C., and Mahajan, S., eds.), Pergamon, Oxford, pp. 338–249.
5. Albertsson, A. C., and Huang, S. J., eds. (1995). *Degradable Polymers, Recycling, and Plastics Waste Management*, Dekker, New York.
6. Doi, Y., and Fukuda, eds. (1994). *Biodegradable Plastics and Polymers*, Elsevier, Amsterdam.
7. Guillet, J., ed. (1973). *Polymers and Ecological Problems*, Plenum Press.
8. Charles, E. L. (1954). US patent 2,668162.
9. Frazza, E. J., and Schmidt, E. E. (1971). *J. Biomed. Mater. Res. Symp.*, *1*: 43–58.
10. Langer, R. S., and Peppas, N. A. (1981). *Biomaterials*, *2*: 201–214.
11. Ambrosio, L. K., Caprino, G., Nicolais, L., Nicodemo, L., Guida, G., Huang, S. J., and Ronica, D. (1988). *Composite Structures*, *4*(2): 2337–2344.
12. Vacanti, C. A., and Vacanti, J. P. (1994). *Otolaryngol. Clin. North Am.*, *27*: 263.
13. Langer, R., and Vacanti, J. (1993). *Science*, *260*: 920.
14. Baldwin, S. P., and Saltzman, W. M. (1996). *Trends in Polym. Sci.*, *4*: 177–182.
15. Kohn, J., ed. (1996). *Tissue Engineering*, Vol. 12, Mater. Research Soc. Bulletin.
16. Brady, J. M., Cutwright, D. E., Miller, R. A., and Battistone, G. (1973). *J. Biomed. Mater. Res.*, *7*: 155.
17. Carother, W. H., and Hill, I. W. (1932). *J. Am. Chem. Soc.*, *54*: 1559.
18. Trofimoff, L., Aida, T., and Inoue, S. (1987). 991.
19. Kohn, F. E., van de Berg, J. W. A., van de Ridder, and Feijen, J. (1984). *J. Appl. Polym. Sci.*, *29*: 4265.
20. Duda, A., and Panzek, S. (1990). *Macromolecules*, *23*: 1636.
21. Dahlman, J., and Rather, G. (1993). *44*: 103.
22. Kricheldorf, H. R., Kreiser-Saunders, I., and Boeticher, C. (1995). *36*: 1253.
23. Degree, P., Dubois, P., Jerome, R., Jacobsen, S., and Fritz, H.-G. (1999). *Macromol. Symp.*, *144*: 289–301.
24. Spassky, N., Simic, V., Huber-Pfalzgrai, L. G., and Mortaudo, M. S. (1999). *Macromol. Symp.*, 257–267.
25. Moon, S. I., and Kimura (2000). *J. Polym. Sci. Part A: Polym. Chem.*, *38*: 1673–1679.
26. Jedlinski, Z., Kurcok, P., and Lenz, R. W. (1995). *J. Macromol. Sci.–Pure Appl. Chem.*, *A32*(4): 797–810.
27. Asakura, S., and Katayama, Y. (1964). *J. Chem. Soc. Japan*, *67*: 956.
28. Ajioka, M., Suizu, H., Higuchi, C., and Kashima, T. (1988). *Polym. Degrad. Stab.*, *59*: 137–143.
29. Lostocco, M., and Huang, S. J. (1997). In *Polymer Modification* (Swift, G., and Carraher, C., Jr., eds.), Plenum Press, New York, pp. 45–57.
30. Hiljanen-Vainio, M., Karjalainen, T., and Seppala, J. (1996). *J. Appl. Polym. Sci.*, *59*: 1281–1288.
31. Nijenhuis, A. J., Grijpma, D. W., and Pennings, A. J. (1996). *Polymer*, *37*: 2783–2791.
32. Han, Y., Edelman, P. G., and Huang, S. J. (1988). *J. Macromol. Sci.–Chem.*, *A25*(5–7): 847–869.
33. Lostocco, M. R., Borzacchiello, A., and Huang, S. J. (1998). *Macromol. Symp.*, *130*: 151–160.
34. Kricheldorf, H. R., and Dunsing, R. (1986). *Makromol. Chem.*, *187*: 1611.
35. Kimura, Y., Matsuzaki, Y., Yamane, H., and Kitao, T. (1989). *Polymer*, *30*: 1342.
36. Deng, X. M., Xiong, C. D., Cheng, L. M., and Xu, R. P. (1990). *J. Polym. Sci. Polym. Lett.*, *28*: 411.
37. Zhu, K. J., Lin, X., and Yang, S. (1990). *J. Appl. Polym. Sci.*, *39*: 1.

38. Kricheldorf, H. R., and Meier-Haack, J. (1993). *Makromol. Chem.*, *194*: 463.
39. Jedlinski, Z., Kurcok, P., Wallach, W., Janeczek, H., and Radecka, I. (1993). *Makromol. Chem.*, *194*: 1681.
40. Cerrai, P., Tricoli, M., Lelli, L., Guerra, R. S., Casone, M. G., and Gusti, P. (1994). *J. Mater. Sci.*, 308–313.
41. Onyari, J., and Huang, S. J. (2003). *Macromol. Symp.*, *193*(1): 143–158.
42. Bitritto, M. M., Bell, J. P., Brencke, G. M., Huang, S. J., and Knox, J. R. (1979). *J. Appl. Polym. Sci.: Appl. Polym. Sym.*, *35*: 415.
43. DiBenedetto, L. J., and Huang, S. J. (1989). *Polym. Prepr.*, *30*(1): 453.
44. Huang, S. J., Kitchen, O., and DiBenedetto, L. J. (1990). *Polym. Mater. Sci. Eng.*, *62*: 804.
45. Borzacchiello, A., Ambrosio, L., Nicolais, L., and Huang, S. J. (2000). *Bioact. Compat. Polym.*, *15*: 61–71.
46. Bengs, H., Bayer, U., Ditzinger, Krone, V., Lill, K., Sandow, J., and Walcj, A. (1996). Abstract of International Symposium on Biodegradable Materials, Hamburg, p. 79.
47. Vert, M., and Lenz, R. W. (1979). *Polym. Prepr.*, *20*(1): 608.
48. Vert, M., and Lenz, R. W. (1981). US Patent 4,265,247.
49. Braud, C., Bunel, C., and Vert, M. (1985). *Polym. Bulletin*, *13*: 293–299.
50. Eschback, F. O., and Huang, S. J. (1991). *Polym. Prepr.*, *34*: 848–849.
51. Ramos, M. S., and Huang, S. J. (2002). In *Functional Condensation Polymers* (Carraher, C. E., Jr., and Swift, G. G., eds.), Kluwer Academic/Plenum Publishers, New York, pp. 185–198.
52. Heller, J. R., Sparer, V., and Zeutner, G. M. (1990). In *Biodegradable Polymers as Drug Delivery Systems*, Vol. 45 (Chasin, M., and Langer, R., eds.), Marcel Dekker, New York, pp. 121–161.
53. Heller, J., Perhale, D. W. H., and Helwing, R. F. (1980). *J. Polym. Sci., Polym. Lett. Ed.*, *18*: 619.
54. Heller, J., Fritzinger, B. K., Ng, S. Y., and Penhale, D. W. H. (1985). *J. Control. Release*, *1*: 225.
55. Fournie, P., Domurado, D., Guerin, P., Braud, C., Vert, M., and Pontikis, R. (1992). *J. Bioact. Compat. Polym.*, *7*: 113.
56. Choi, N. S., and Heller, J. (1979). US Patent 4,180,646.
57. Bencewicz, B. C., and Hopper, P. K. (1991). *J. Bioact. Compat. Polym.*, *6*: 64.
58. Roby, M. S. (1998). In *Biomedical Polymer* (Huang, S. J., and Roby, M. S., eds.), ACS, PMSE Div. Workshop Abstract.
59. Jorres, V., Keul, H., and Hocker, H. (1996). Abstract of International Symposium on Biodegradable Materials, Hamburg, p. 81.
60. Goodman, M., and Kirschenbaum, G. S. (1979). US Patent 30,170.
61. Keul, H., Robertz, B., and Hocker, H. (1999). *Macromol. Symp.*, *144*: 47–61.
62. Ouchi, T., Okamoto, Y., Shiratani, M., Jinno, M., and Ohya, Y. (1994). In *Biodegradable Plastic and Polymers* (Doi, Y., and Fukuda, eds.), Elsevier Science B. V., pp. 528–533.
63. Huang, S. J., and Ho, L. (1999). *Macromol. Symp.*, *144*: 7–32.
64. Endo, T., and Kubota, H. (1990). Abstract of International Symposium on Biodegradable Polymers, Toyko, p. 114.
65. Huang, S. J., Ho, L., Huang, M. T., Koenig, M. F., and Cameron, J. A. (1994). In *Biodegradable Plastics and Polymers* (Doi, Y., and Fukuda, K., eds.), pp. 3–10.
66. James, K., and Kohn, J. (1996). *MRS Bulletin*, 22–26.
67. Huang, S. J., and Roby, M. S. (1992). In *Biodegradable Polymers and Plastics* (Vert, M., Feijen, J., Albertsson, A., Scott, G., and Chiellini, E., eds.), Royal Soc. Chem., pp. 149–155.
68. Barrow, T. H. (1982). US Patent 4,343,931.
69. Barrow, T. H. (1985). US Patent 4,529,793.
70. Gonsalves, K. E., and Mungara, P. M. (1996). *Trims in Polym. Sci.*, *4*(1): 25–31.
71. Kishida, A., Murakami, K., Goto, H., Akashi, M., Kubota, H., and Endo, T. (1998). *J. Bioact. Compat. Polym.*, *13*: 270–278.
72. Roweton, S., Huang, S. J., and Swift, G. (1997). *J. Environ. Polym. Degrad.*, *5*(1): 175–181.
73. Hayashi, T., and Twatsuki, M. (1990). *Biopolymers*, *29*: 549–557.

74. Huang, S. J., Pavlisko, J., and Hong, E. (1978). *Poly. Prepr.*, *19*: 57.
75. Huang, S. J., and Kitchen, O. (1990). *Polym. Prr.*, *31*(2): 207–208.
76. Edelman, P. G., Mathison, R. J., and Huang, S. J. (1985). In *Advances in Polymer Synthesis* (Culbertson, B. M., and McGrath, J. E., eds.), Plenum, New York, pp. 275–290.
77. Huang, S. J., Ho, L., Hong, E., and Kitchen, O. (1994). *Biomaterials*, *15*(15): 1243–1247.
78. Nagaoka, S., Mori, Y., Takiuch, H., Yokota, Y., Tanzawa, H., and Nishiumi, S. (1984). *Polymers as Biomaterials* (Shalaby, S. W., Hoffman, A. S., Ratner, B. D., and Horbett, T. A., eds.), Plenum Press, New York, pp. 36–374.
79. Norman, M. E., William, P., and Illum, L. (1992). *Biomaterials*, *13*(12): 841–849.
80. Chaikof, E. L., and Mirrell, E. W. (1990). *J. Coll. Interface Sci.*, *137*(2): 340–349.
81. Desai, N. P., and Hubbell, J. A. (1991). *Biomaterials*, *12*: 144–153.
82. Llanos, G. R., and Sefton, M. V. (1992). *Biomaterials*, *13*(7): 421–424.
83. Amiji, M., and Park, K. (1992). *13*(10): 682–692.
84. Desai, N. P., Hossainy, S. F. A., and Hubbell, J. A. (1992). *Biomaterials*, *13*(7): 417–420.
85. Dong, C.-M., Qiu, K.-Y., Gu, Z.-W., and Feng, X.-D. (2001). *J. Polym. Sci. Part A: Polym. Chem.*, *40*: 409–415.
86. Boogh, L., Pettersson, B., and Manson, J.-A. E. (1999). *Polymer*, *40*: 2249–2261.
87. Wan, Q., Schricker, S. R., and Culbertson, B. M. (2000). *J. Macromol. Sci. Pure Appl. Chem.*, *A37*: 1301–1315.
88. Wahlstroem, A., and Vamling, L. (2000). *45*: 97–103.
89. Herman, S., Hooftmann, G., and Schacht, E. (1995). *J. Bioac. Compat. Polym.*, *10*: 145–187.
90. Allcocl, H. R. (1999). *Macromol. Symp.*, *144*: 33–46.
91. Lemmouchi, Y., Schacht, E., and Dejardin, S. (1998). *J. Bioact. Compat. Polym.*, *13*: 4–18.
92. Schmitt, E. E., and Polistina, R. A. (1967). US Patent 3,297,033.
93. Schmitt, E. E., Epstein, M., and Polistina, R. A. (1969). US Patent 3,422,871.
94. Frazza, E. J., and Smitt, E. E. (1971). *J. Biomed. Mater. Res. Symp.*, *1*: 43–58.
95. Jugens, Ch., Kreiser-Saunders, I., and Krischeldorf, H. R. (1996). Abstract of International Symposium on Biodegradable Materials, Hamburg, pp. 74–575.
96. Benedict, C. V., Cook, W. J., Jarett, P., Cameron, J. A., Huang, S. J., and Bell, J. P. (1983). *J. Appl. Polym. Sci.*, *28*: 335.
97. Benedict, C. V., Cameron, J. A., and Huang, S. J. (1983). *J. Appl. Polym. Sci.*, *28*: 335.
98. Li, S., and Vert, M. (1995). In *Degradable Polymers* (Scott, G., and Gilead, D., eds.), Chapman and Hall, London, pp. 43–87.
99. Hubbell, J. A. (1994). *Trends in Polym. Sci.*, *2*: 20.
100. Davis, P. A., Nicolais, L., Ambrosio, L., and Huang, S. J. (1988). *J. Bioact. Compat. Polym.*, *3*: 205.
101. Davis, P. A., Huang, S. J., Ambrosio, L., and Nicolais, L. (1989). In *High Performance Biomaterials* (Szycher, ed.), Technomic, Lancaster, pp. 343–368.
102. Eschbach, F. O., and Huang, S. J. (1994). *J. Bioact. Compat. Polym.*, *9*(1): 29–54.
103. Eschbach, F. O., Huang, S. J., and Cameron, J. A. (1994). *J. Bioact. Compat. Polym.*, *9*(2): 210–221.
104. Davis, P. A., Huang, S. J., Ambrosio, L., Bronca, D., and Nicolais, L. (1991). *J. Mater. Sci.: Mater. in Medicine*, *3*: 359.
105. Hubbell, J. A. (1996). *MRS Bulletin*, *21*(11): 33–35.
106. Lu, L., and Mikos, A. G. (1996). *MRS Bulletin*, *21*(11): 28–32.
107. Cima, L., Langer, R., and Vacanti, J. P. (1991). *J. Bioact. Compat. Polym.*, *6*: 232–240.
108. Baker, R. (1987). *Controlled Release of Biologically Active Agents*, Wiley-Interscience, New York.
109. Chasin, M., and Langer, R., eds. (1990). *Biodegradable Polymers as Drug Delivery Systems*, Marcel Dekker, New York.
110. Edlund, U., and Albertsson, A.-C. (2000). *J. Bioact. Biocompat. Polym.*, *15*: 214–229.
111. DiBenedetto, L. J., and Huang, S. J. (1989). *Polym. Prepr.*, *30*(1): 453.
112. Huang, S. J., and Kitechen, O. (1990). *Polym. Prepr.*, *31*(2): 54.

113. DiBenedetto, L. J., and Huang, S. J. (1994). *Polym. Degrad. Stab.*, *45*: 249–257.
114. Ottenbrite, R. M., Huang, S. J., and Park, K., eds. (1996). *Hydrogels and Biodegradable Polymers for Biomedical Applications*, ACS Symp. Series 627.
115. Huang, S. J., Seery, T. A., and Swift, G., eds. (1999). Biomedical applications of water-soluble polymers and hydrogels. *Macromol. Sci.-Pure Appl. Chem.*, *A36*: 7 & 8.
116. Novakova, K., Laznicek, M., Rypacek, F., and Machova, L. (2002). *J. Bioact. Compat. Polym.*, 285–296.
117. Fan, K., Gonzales, D., and Sevoian, M. (1996). *J. Environ. Polym. Degrad.*, *4*(4): 253–260.

17

Controlled/Living Radical Polymerization

Krzysztof Matyjaszewski and James Spanswick
*Center for Macromolecular Engineering, Carnegie Mellon University,
Pittsburgh, Pennsylvania*

I. CHEMISTRY OF CONTROLLED/LIVING RADICAL POLYMERIZATION (CRP)

Conventional free radical polymerization (RP) has many advantages (see Chapters 2–4 discussing specific polymers prepared by radical polymerization). The procedure can be used for the (co)polymerization of a very large range of vinyl monomers under undemanding conditions; requiring the absence of oxygen, but tolerant to water, and can be conducted over a large temperature range (-80 to $250\,°C$) [1]. This is why nearly 50% of all commercial synthetic polymers are prepared using radical chemistry providing a spectrum of materials for a range of markets. Many additional vinyl monomers can be copolymerized via a radical route leading to an infinite number of copolymers with properties dependent on the proportion of incorporated comonomers. The major limitation of RP is poor control over some of the key structural elements that allow the preparation of well defined macromolecular architectures such as molecular weight (MW), polydispersity, end functionality, chain architecture and composition.

Living polymerization was first defined by Szwarc [2] as a chain growth process without chain breaking reactions (transfer and termination). Such a polymerization provides end-group control and enables the synthesis of block copolymers by sequential monomer addition. However, it does not necessarily provide polymers with MW control and narrow molecular weight distribution (MWD). Additional prerequisites to achieve these goals include that the initiator should be consumed at the early stages of polymerization and that the rate of initiation and the rate of exchange between species of various reactivity should be at least as fast as propagation [3–5]. It has been suggested to use the term *controlled* polymerization if these additional criteria are met [6]. This term was proposed for systems, which provide control of MW and MWD but in which chain breaking reactions continue to occur, as in RP. However, the term *controlled* does not specify which features are controlled and which are not controlled. Another option would be to use the term '*living*' polymerization (with quotation marks) or *apparently living* which could indicate a process of preparing well-defined polymers under conditions in which chain breaking reactions undoubtedly occur, as in radical or carbocationic polymerization [7]. The term *controlled/living* could also describe the essence of these systems [6] and will be used in this chapter as we discuss in detail the polymerization procedures that have been developed for control over radical copolymerization of vinyl monomers.

Well-defined polymers with precisely controlled structural parameters are accessible through living ionic polymerization processes, however, ionic living polymerization requires stringent process conditions and the procedures are limited to a relatively small number of monomers [8–10]. Therefore, it remained desirable to prepare, by free radical means which are more practical for industrial manufacturing procedures, new well-defined block and graft copolymers, materials with star, comb and network topology, end-functional polymers and many other materials prepared under mild conditions, from a larger range of monomers, than available for ionic living polymerizations [11].

The concept of living radical polymerization was first discussed by Otsu [12] but did not come to the forefront of scientific scrutiny until after the publishing of the influential work of Georges [13] in 1993 who had built upon the earlier work of Rizzardo [14]. Georges pointed out to the scientific community that controlled radical polymerization was feasible. This is one reason why since 1995 we have witnessed a real explosion of academic and industrial research on controlled/living radical polymerizations (CRP) with over five thousand papers and hundreds of patents devoted to disclosing, and improving the various types of CRP discussed in this chapter, and to developing an understanding of the implications of molecular structure on material properties. In all of the CRP processes developed to date there is a low occurrence of side reactions (e.g., termination or chain transfer) due to creation of a dynamic equilibrium between a dormant species present in large excess and a low concentration of active radical sites. By reducing the instantaneous concentration of active radicals, and hence the number of side reactions, polymerization is able to proceed in a controlled manner. This results in the formation of (co)polymers having predictable MW and controllable MWD with MW increasing as a function of time in a batch polymerization process, all the while maintaining a narrow MWD. In almost all of the references included in this chapter initiation efficiencies are high, and the experimental molecular weight is close to the theoretic molecular weight and MWD is less than 1.3. CRP is also able to produce materials with well-defined block lengths, complex architecture, and functionalized chain ends.

There are several requirements that have to be met for any process that claims to control radical polymerizations, including assuring quantitative initiation and suppressing the contribution of chain breaking reactions. All of the controlled/living radical based processes developed to meet these requirements, along with many other new living polymerization systems, such as carbocationic, ring-opening, group transfer, ligated anionic polymerization of acrylates, etc., depend upon the existence of a dynamic equilibration between an active and a dormant species. In CRP the equilibrium is between growing free radicals and some kind of dormant species [15]. The equilibrium is established via activation (k_a) and deactivation (k_d) steps.

Currently three approaches generally appear to be successful at controlling radical polymerization and the major processes will be discussed in historical order.

1. Thermal homolytic cleavage of a weak bond in a covalent species which reversibly provides a growing radical and a less reactive radical (a persistent or stable free radical) (Scheme 1). There are several examples of persistent radicals but it seems that the most successful are nitroxides [13,14,16], triazolinyl radicals

[17,18], bulky organic radicals, e.g., trityl [19–21] or compounds with photolabile C–S bonds [22] and some organometallic species [23–26].

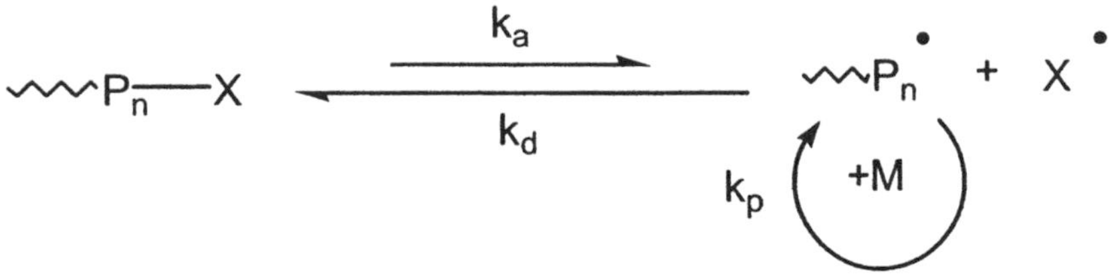

$$\text{$\sim\!\!\sim\!$P}_n\text{—X} \;\underset{k_d}{\overset{k_a}{\rightleftharpoons}}\; \text{$\sim\!\!\sim\!$P}_n\!\!^\bullet \; + \; \text{X}^\bullet$$

$$k_p \;(\; +\text{M}\;)$$

Scheme 1

A subset of this process is the transition metal catalyzed, reversible cleavage of the covalent bond in the dormant species via a redox process (Scheme 2). Since the key step in controlling the polymerization is transfer of an atom (or group) between a dormant chain and a transition metal catalyst in a lower oxidation state forming an active chain end and a transition metal deactivator in a higher oxidation state, this process was named atom transfer radical polymerization (ATRP) [27–32].

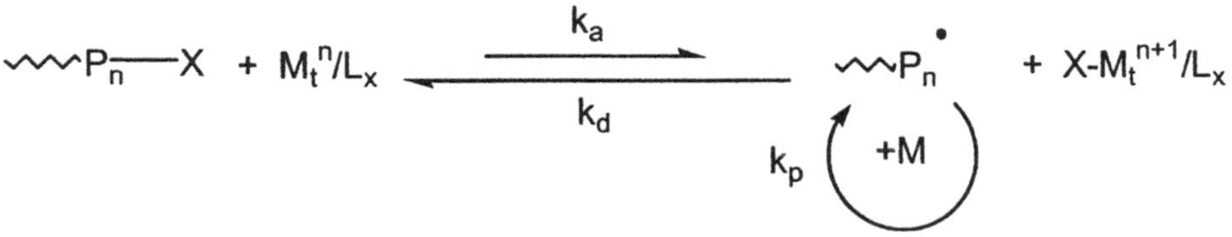

$$\text{$\sim\!\!\sim\!$P}_n\text{—X} \; + \; M_t^n/L_x \;\underset{k_d}{\overset{k_a}{\rightleftharpoons}}\; \text{$\sim\!\!\sim\!$P}_n\!\!^\bullet \; + \; \text{X-}M_t^{n+1}/L_x$$

$$k_p \;(\; +\text{M}\;)$$

Scheme 2

2. The second approach to CRP is based on a thermodynamically neutral exchange process between a growing radical, present at very low concentrations, and dormant species, present at much higher concentrations (generally three to four orders of magnitude) (Scheme 3). This degenerative transfer process can employ alkyl iodides [33,34], unsaturated methacrylate esters [35,36], or thioesters [37,38]. The latter two processes operate via addition-fragmentation chemistry.

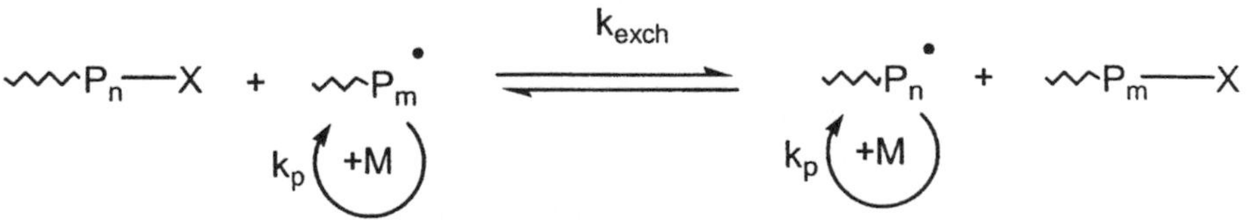

$$\text{$\sim\!\!\sim\!$P}_n\text{—X} \; + \; \text{$\sim\!\!\sim\!$P}_m\!\!^\bullet \;\overset{k_{exch}}{\rightleftharpoons}\; \text{$\sim\!\!\sim\!$P}_n\!\!^\bullet \; + \; \text{$\sim\!\!\sim\!$P}_m\text{—X}$$

$$k_p \;(\; +\text{M}\;) \qquad\qquad k_p \;(\; +\text{M}\;)$$

Scheme 3

3. Finally, there is a third approach that has not yet been as extensively examined as the above systems. This process is the reversible formation of persistent radicals, by reaction of the growing radicals with a species containing an even number of electrons, which do not react with each other or with monomer (Scheme 4). Here, the role of a reversible radical trap may be played by

phosphites [39] or some reactive, but non-polymerizable alkene, such as tetrathiofulvalenes, stilbene or diphenylethylene [40,41].

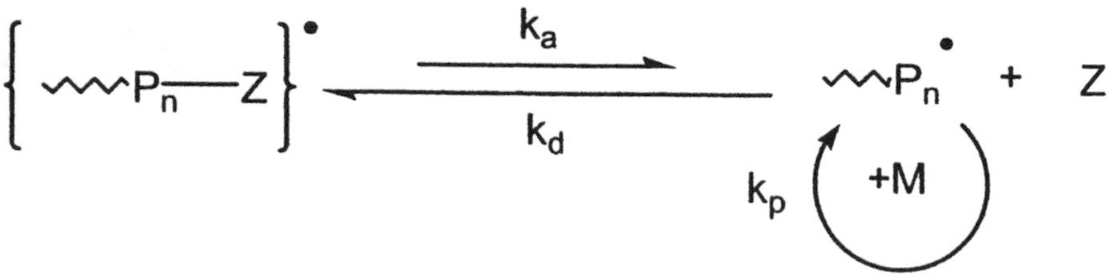

Scheme 4

In the remaining pages of this chapter we will discuss the chemistry of these successful approaches to controlled/living radical polymerization and some examples of new materials prepared by these techniques will be discussed.

Several reviews devoted to CRP have been already been published, and readers may refer to proceedings from ACS Meetings on CRP [42,43], general reviews on CRP [44–48], reviews on ATRP [30,49–54], on macromolecular engineering and materials prepared by ATRP [55], on nitroxide mediated polymerization (NMP) [56–58], on catalytic chain transfer [59,60], and on reversible addition fragmentation transfer polymerization, RAFT [61].

II. CONTROLLED/LIVING RADICAL POLYMERIZATION BASED ON REVERSIBLE THERMAL CLEAVAGE OF WEAK COVALENT BONDS

The homolytic cleavage of weak covalent bonds results in the formation of an active radical capable of propagating the polymerization and a counter radical which, in principle, should only be involved in the reversible capping of the growing chains. The stable counter radicals should *not* react with themselves, with monomer to initiate the growth of new chains, or participate in other side reactions such as the abstraction of β-H atoms. These persistent radicals should be relatively stable, although some recent data indicate that their slow decomposition may help in maintaining appropriate polymerization rates [17].

There are several examples of persistent radicals used in controlled radical polymerization but perhaps the most extensively studied are nitroxides, specifically TEMPO [62,63] (Scheme 5). Hawker showed that the two radicals on the right-hand side of Scheme 5 were not closely associated with each other during the polymerization [64] allowing for a statistical replacement of the nitroxide with functional end groups [65]. Interesting results were also obtained with organometallic species, especially with paramagnetic high spin cobalt (II) compounds [23]. However, often a particular trap acts efficiently only for one class of monomers. For example, Co(II) porphyrine derivatives are excellent for controlling the polymerization of acrylates [66] but poor for styrene, while for methacrylates they act as very efficient transfer reagents (catalytic chain transfer). The nitroxide TEMPO is efficient only for the CRP of styrene and its copolymers, however, some newly developed nitroxides have also been successful for acrylates [67]. Consequently, the range of monomers controllably polymerizable by this procedure is slowly expanding [58].

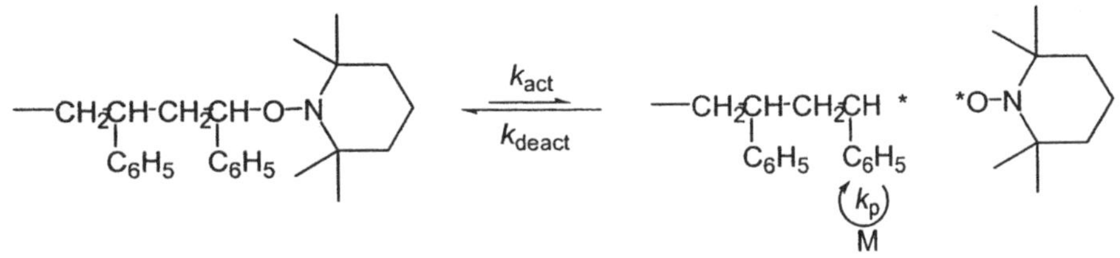

Scheme 5

Nitroxides were originally described in the patent literature as agents in the polymerization of (meth)acrylates [14], but the resulting products were essentially either stable oligo-polymeric alkoxyamines or unsaturated species. It was only after the seminal paper by Georges using TEMPO in styrene polymerizations at elevated temperatures ($>120\,^{\circ}$C), that real advances in controlled radical polymerization were made [13]. Initial results were most encouraging, since they employed very simple reaction conditions (bulk styrene, $[BPO]_o:[TEMPO]_o = 1.3:1$ and simple heating) and obtained the desired outcome ($DP_n = \Delta[Sty]/[TEMPO]_o$ in the range of $M_n = 1000$ to 50,000 and with low polydispersities, $M_w/M_n < 1.3$). The reactions were slow with rates similar to the thermal polymerization of styrene. Under typical conditions, the majority of the chains are present in the form of alkoxyamines, which are the covalent bonded dormant species, while a very small fraction of radicals are continuously generated by thermal initiation and by the thermal cleavage of the alkoxyamines ($[P^*] \approx 10^{-8}$ M) [68]. Chains continuously terminate by coupling/disproportionation and lead to an excess of TEMPO via the persistent radical effect ($[TEMPO] \approx 10^{-5}$ M) [69–71]. The alkoxyamine functional group on the chain ends can also slowly decompose and generate unsaturated structures and a hydroxylamine (Scheme 6) [72] that can be reoxidized to TEMPO in the presence of traces of oxygen.

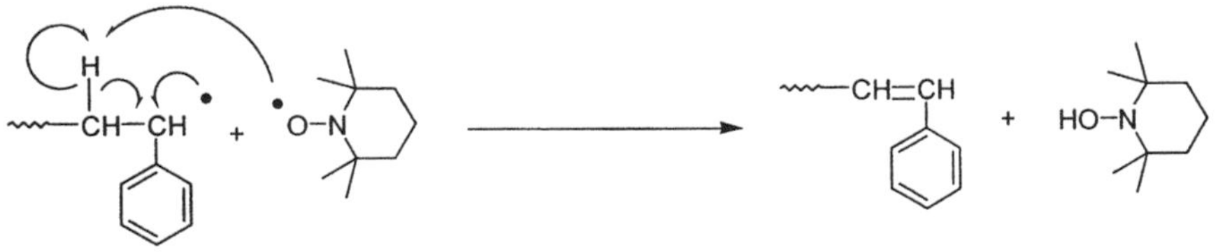

Scheme 6

In the system described by Georges control was initially relatively good but decreased as the reaction progressed and molecular weights exceed $M_n = 20,000$, however, more recent work indicates that molecular weights over 150,000 can be obtained [58,67]. Typically, above 80% of chains are in the form of dormant, potentially active species but this number drops as chain length increases, the remaining ~20% of chains are terminated and not capable of growth. Under appropriate conditions it is possible to conduct chain extensions and therefore prepare block copolymers. Several improvements to the original system have been made; these include the use of different initiators such as AIBN instead of BPO [73], using a simple pure thermal process [74,75], or preformed alkoxyamines, so-called unimolecular initiators [76]. Also di- and multi-functional initiators have been successfully used to make novel materials with chains growing in several directions, or from multiple sites on a backbone polymer [57,77]. The rate of polymerization can be increased over that of TEMPO mediated systems by using new nitroxides, which are sterically bulkier and dissociate easier, thereby providing a larger equilibrium constant. Examples include phosphoric and phosphonic acid cyclic and acyclic nitroxide derivatives [78,79] including N,N-(2-methylpropyl-1)-(1-diethylphosphono-2,2-dimethyl-propyl-1-)-N-oxyl, (SG1) expanding the range of monomers polymerizable by (NMP) (see Scheme 7) [67].

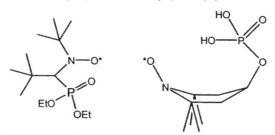

Scheme 7 Phosphorous containing nitroxides.

Rates of propagation for nitroxide mediated systems follow a simple law (Eq. 1) and depend on the concentration of radicals, which are defined by the equilibrium constant (K_{eq}), and the concentration of dormant species [P-SFR] and SFR (Eq. 2), where [SFR] is the concentration of the persistent radical.

$$R_p = -d[M]/dt = k_p[M][P^*] = k_p[M]K_{eq}[\text{P-SFR}]/[\text{SFR}] \quad (1)$$

$$[P^*] = K_{eq}[\text{P-SFR}]/[\text{SFR}] \quad (2)$$

However, when the equilibrium constants are very small the polymerizations are slow, as in the classic case of the TEMPO mediated polymerization of styrene, $K_{eq} \approx 10^{-11}\,\text{M}$ at 130 °C. In that case, the rate can be increased to an acceptable level by increasing the number of radicals either from thermal initiation by the monomer or by adding a second conventional radical initiator, which has an appropriate lifetime at the polymerization temperature, such as dicumyl peroxide [68,80,81]. In that case, the concentration of radicals is defined by the balance between rates of initiation and termination:

$$[P^*] = R_i/R_t \quad (3)$$

A stationary concentration of SFR must therefore self adjust and be reduced to fulfill the equilibrium requirement and obey both equations (2) and (3).

Another approach to increase rates is to reduce the concentration of the SFR, such as TEMPO, by other reactions. The lower thermal stability of 4-oxoTEMPO results in its continuous decomposition, thereby reducing its concentration and resulting in a shift of the equilibrium towards more growing radicals, and finally faster rates. The decomposition/dissociation may also be catalyzed intra- or inter-molecularly by addition of acid derivatives and acetyl compounds (potentially acid generators) [82,83]. The principle of low thermal stability of persistent radicals was also employed in the use of triazolinyl radicals, which decompose at elevated temperatures and spontaneously reduce their concentration [17]. Research is presently being focused on the high throughput synthesis for the design of new alkoxyamine initiators for nitroxide mediated living free radical procedure [84] and Hawker has shown that the rates of polymerization can be significantly enhanced, even when compared to the second generation α-hydrido-based alkoxyamines recently developed. He has demonstrated that intramolecular H-bonding is a powerful tool for increasing the performance of alkoxyamine initiators for nitroxide mediated living free radical polymerizations. Increases in the rate of polymerization (*ca.* 1000%) were observed for polar monomers such as acrylamides and especially acrylates [67], while only moderate improvements were obtained for non-polar monomers, such as styrene and isoprene. In each case, the degree of control during the polymerization was improved, leading to lower polydispersities and a better correlation between experimental and theoretical molecular weights. Nitroxide mediated polymerization has also been conducted in heterogeneous systems including emulsion [85,86], miniemulsion [87], and suspension [88,89], however, as fully discussed below for biphasic ATRP reactions, an understanding of partition coefficients for all components of the system between all phases is critical for a controlled polymerization [90,91].

Probably the most important factor for the future of NMP will be the development of new compounds that allow polymerization and copolymerization of a broader range of monomers under milder reaction conditions; we should however note that nitroxide mediated polymerization has already been applied to styrene [92], acrylates [93], acrylamides [94], acrylonitrile [67], dienes [95], and recently polymerization of ethylene has been claimed to be controlled [96,97]. NMP has also been extended to functional monomers such as sodium styrene sulfonate [98], 2-vinylpyridine [99,100], 3-vinyl pyridine [101,102], and 4-vinylpyridine [103]. However, since a nitroxide residue ends up at the end of each chain, these new compounds should be inexpensive, and introduce no adverse properties (color, poor thermal stability, etc.) to the final material.

III. TRANSITION METAL CATALYZED PROCESSES — ATOM TRANSFER RADICAL POLYMERIZATION

Atom transfer radical polymerization (ATRP) is based on the reversible transfer of halogen atoms, or pseudo-halogens, between a dormant species (P_n–X) and a transition metal catalyst (M_t^n/L) by redox chemistry. The alkyl (pseudo)halides are reduced to active radicals and transition metals are oxidized via an inner sphere electron transfer process [28,50]. In the most studied system, the role of the activator is played by a copper(I) species complexed by two bipyridine ligands and the role of deactivator by the corresponding copper(II) species. Scheme 8, shows such a system with the values of the rate constant for activation (k_a), deactivation (k_d), propagation (k_p) and termination (k_t) for a bulk styrene polymerization at 110°C [32]. The rate coefficients of termination decrease significantly with the progress of the polymerization reaction due to the increase in the chain length and increased viscosity of the system. In fact, the progressive reduction of k_t is one of the most important features of many controlled radical polymerizations [104].

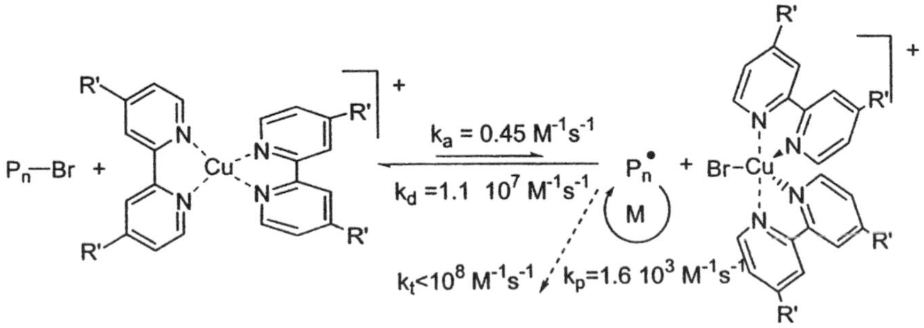

$$k_a = 0.45 \text{ M}^{-1}\text{s}^{-1}$$
$$k_d = 1.1 \; 10^7 \text{ M}^{-1}\text{s}^{-1}$$
$$k_t < 10^8 \text{ M}^{-1}\text{s}^{-1} \quad k_p = 1.6 \; 10^3 \text{ M}^{-1}\text{s}^{-1}$$

Scheme 8

The main difference between nitroxide mediated systems and ATRP is that the latter can be used for a much larger range of monomers, including methacrylates, is practical for a full range of copolymerizations, and it is generally much faster [105]. The rate of propagation for an ATRP (Eq. 4) can be adjusted conveniently, not only by the concentration of deactivator but also by the concentration of activator, since catalysis is at the very nature of ATRP [51]. The activity of the catalyst can be adjusted by selection of the ligand [106,107] and optionally addition of a solvent [108]. The ligand can also be selected for the reaction medium and can encompass hydrophilic or hydrophobic substituents, or in the case of polymerization conducted in supercritical carbon dioxide, fluroalkyl groups [109].

$$R_p = -d[M]/dt = k_p[M][P^*] = k_p[M]\{k_a[\text{P-X}][\text{Cu(I)}]\}/\{k_d[\text{X-Cu(II)}]\} \tag{4}$$

Polydispersities in ATRP, and in other controlled radical reactions, depend on relative rates of propagation and deactivation [5] (Eq. 5):

$$M_w/M_n = 1 + [(k_p[\text{RX}]_o)/(k_d[\text{X-Cu(II)}])](2/p - 1) \tag{5}$$

Thus, polydispersities decrease with conversion, p, with the rate constant of deactivation, k_d, and with the concentration of deactivator, [X-Cu(II)], however, they increase with the propagation rate constant, k_p, and the concentration of initiator, [RX]$_o$. This means that more uniform polymers are obtained at higher conversions, when the concentration of deactivator in solution is high and the concentration of initiator is low. Also, more uniform polymers are formed when the deactivator is very reactive (e.g., copper(II) complexed by bipyridine or triamine) and monomer propagates slowly (e.g., styrene rather than acrylate).

Chain breaking reactions do occur in these controlled radical systems [110], fortunately, at typical reaction temperatures, the contribution of transfer is relatively small. For example, in the polymerization of styrene, less than 10% of chains participate in transfer to monomer before reaching $M_n = 100,000$. However, since the contribution of transfer progressively increases with chain length molecular weights should be limited by the appropriate ratio of monomer to initiator concentrations (for styrene $\Delta[M]/[I]_o < 1000$).

Termination does occur in radical systems and currently cannot be completely avoided. On the other hand, since termination is second order with respect to radical concentration and propagation is first order, the contribution of termination increases with radical concentration, and therefore also with the polymerization rate, consequently, most controlled radical polymerizations are designed to be slower than conventional systems. It is possible to generate relatively fast controlled radical polymerizations, but only for the most reactive monomers, such as acrylates, and/or for relatively short chains. For short chains, the absolute concentration of terminated chains is still high but their percentile in the total number of chains is small enough so as not to affect end functionalities and blocking efficiency. A typical proportion of terminated chains lies between 1 and 10%, with a large fraction of those being very short chains that may not markedly affect the properties of the synthesized polymers and copolymers. It is possible to measure the evolution of concentration of terminated chains by following the copper(II) species by EPR in a system starting from pure copper(I) catalyst. Commercially in a system using a higher cost low molecular weight initiator the addition of copper(II) to the system will increase initiator efficiency by reducing termination reactions between low molecular weight radicals.

The list of monomers polymerized successfully by ATRP is extensive and polymerizations have been investigated with a wide range of transition metals including copper [27], ruthenium [111], iron [112–116], rhodium [117], rhenium [118]. The main requirement for a transition metal catalyst to be suitable for an ATRP is an ability to undergo a one electron redox reaction with an appropriate redox potential selected for the (co)monomers being polymerized. The initial range of monomers, which started with polymerization and copolymerization of styrene, acrylates, and methacrylates [28,119], have been extended to substituted styrenes [120], including 4-acetoxy styrene [121], benzyl ethers [122], and 4-trimethylsilyl derivatives [123]; substituted acrylates include methyl and *n*-butyl [28,124–127], ethyl [128], *t*-butyl [129–132], and isobornyl [133,134]; substituted methyl methacrylates [29,112,135–138], and various other alkyl methacrylates [131,134,139–144], including hydroxyethyl methacrylate [145,146], 2-(N-morpholino)ethyl methacrylate [147], 2-(dimethylamino)ethyl methacrylate [148,149], acrylamides [150,151], including methacrylamides [152–154], and substituted acrylamides, N-*t*-butylacrylamide homopolymer and N-(2-hydroxypropyl)methacrylamide [153], also vinylpyridine [100,155] and dimethylitaconate [156]. In addition, several other monomers have been successfully copolymerized using ATRP and include, for example, isobutylene and vinyl acetate [157].

A big advantage of any radical process, ATRP included, is its tolerance to many functional groups such as amido, amino, ester, ether, hydroxy, siloxy and others. All of them have been incorporated as substituents into (meth)acrylate monomers and successfully polymerized. One current exception is a 'free' carboxylic acid group which potentially complexes with the catalyst and disables ATRP, and therefore, presently, it has to be protected. Recent work has shown that monomers bearing ionic substituents such as sodium 4-vinylbenzoate, sodium 4-vinylbenzylsulfonate and 2-trimethylammonioethyl methacrylate methanesulfonate and triflate, and dimethylaminoethyl methacryate can be polymerized directly [148].

Another advantage of ATRP is a multitude of commercially available initiators. Nearly all compounds with halogen atoms activated by the presence of β-carbonyl, phenyl, vinyl or cyano groups have been used as efficient initiators. Also compounds with a weak halogen–heteroatom bond can be used, such as sulfonyl halides [31]. Small molecule initiators can carry additional functionalities, a few examples are shown in Scheme 9, the functionality is incorporated at the residual chain end.

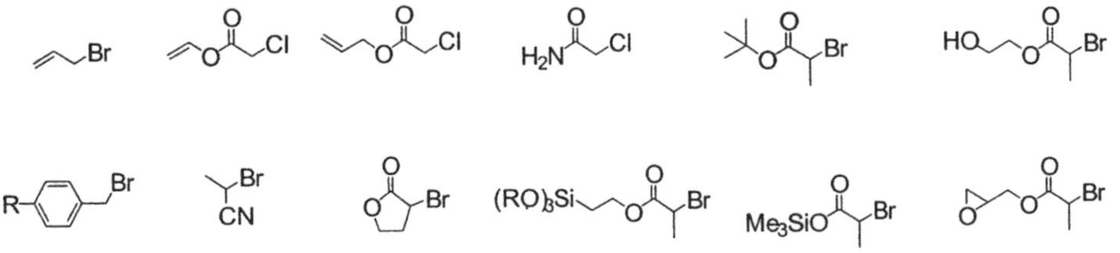

Scheme 9 Some low MW functional ATRP initiators.

Many compounds with multiple active halogen atoms have been used to initiate bi- or multi-directional growth to form ABA block copolymers and star-like polymers and copolymers [158]. Active halogens can be incorporated at the chain ends of polymers prepared by other techniques such as cationic, anionic, ring-opening metathesis and conventional radical processes to form macroinitiators. Such macroinitiators have been successfully chain extended via ATRP to form novel diblock, and triblock copolymers [159–162]. A useful tool that is available for the preparation of block copolymers when the second monomer to be polymerized is a methacrylate is the halogen switch technique [163], which allows one to match the rate of initiation with the rate of propagation.

When the active halogen is incorporated along the backbone of a (co)polymer, graft copolymers are formed. Many commercial polymers including modified poly-butene, polyisobutylene, polyethylene, and polyvinyl chloride have been used as macroinitiators for the preparation of graft copolymers by the 'grafting from' procedure [164–166].

The halogen atoms, at the active chain ends, can be removed either by a reduction process or transformed to other useful functionalities [167], as shown for styrene and acrylate systems (Scheme 10) [168].

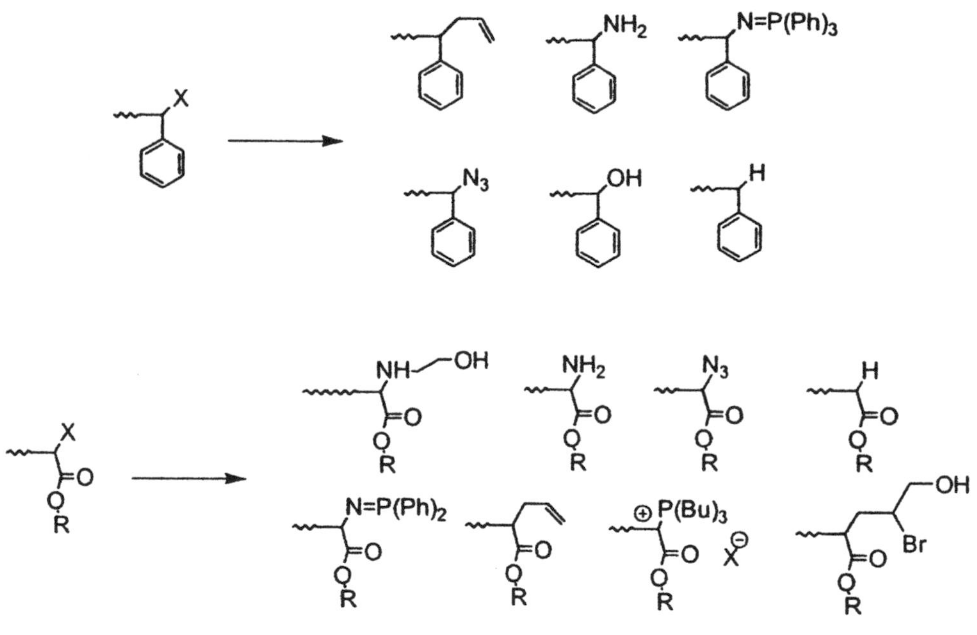

Scheme 10

ATRP has been successfully carried out in bulk, in solution [27,28], as well as in aqueous solution [157], emulsion [169], miniemulsion [170], and suspension [135,171], and in other media (e.g., liquid or supercritical CO_2 [109] or ionic liquids) [172,173]. Typical temperature range for a polymerization is from sub-ambient temperature to $+130\,^\circ$C. Molecular weights for linear and graft copolymers range from $200 < M_n < 500{,}000$ (however the molecular weight of bottle-brush copolymers and particle tethered copoly-mers can reach well into the millions), and polydispersities are low, $1.05 < M_w/M_n < 1.3$, depending on the catalyst used, and also on the relative and absolute catalyst and initiator concentrations.

Copolymerization is facile and many statistical, gradient and block copolymers have been prepared [143,174,175]. The reactivity ratios are nearly identical to conventional radical processes [50,176]. The key feature of ATRP is a transition metal compound, that is made available to participate in a redox cycle with the initiator or growing polymer chain, most often this is accomplished by complexation of the transition metal with a suitable ligand. This ligand should assure solubility of *both* oxidation states of the catalyst, adjust its electronic and steric properties, and should enhance the versatility of atom transfer chemistry when compared to other reactions. The catalyst complex should allow for a dynamic atom transfer by the reversible expansion of the coordination sphere. Successful ATRP polymerizations have been conducted with transition metal complexes based on Cu, Ru, Fe, Ni, Pd, Rh [105,126,135,138,139,116,157]. Ligands are usually mono or polydentate species such as ethers, amines, pyridines, phosphines and the corresponding polyethers, polyamines and polypyridines. The transition metal complex is very often a metal halide but pseudohalides, carboxylates and compounds with non-coordinating triflate [177] and haxafluorophosphate anions [128] have been also used successfully. Transition metal salts comprising an onium counterion [178,179], and solutions of transition metals salts in ionic liquids, have also been used for ATRP [172,173].

IV. DEGENERATIVE TRANSFER

Control by degenerative transfer (DT) involves perhaps the smallest change from a conventional free radical process of all the controlled/living polymerization processes developed to date. A recent review of various methods of telomer synthesis [180] discusses the different types of transfer agents and monomers and the contribution of the techniques of telomerization to CRP (includes discussion of iodine transfer polymerization, RAFT, and macromolecular design through interchange of xanthates (MADIX)) [181,182].

DT relies on a thermodynamically neutral (degenerative) transfer reaction. The key for control is a minimal energy barrier for that reaction. Conventional free radical initiators are used, i.e., peroxides and diazenes, at temperatures typical for radical polymerization and the polymerization is carried out in the presence of a compound with a labile group or atom which can be either reversibly abstracted or added-fragmented by the growing radical. The simplest examples are reactions in the presence of alkyl iodides [33,183–184]; Scheme 11:

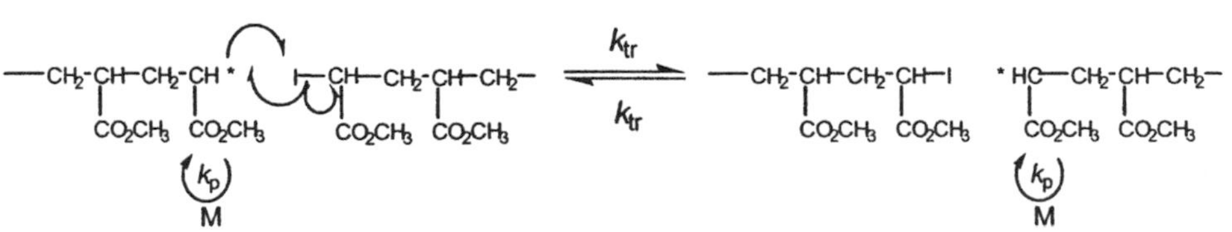

Scheme 11

unsaturated methacrylate esters [36]:

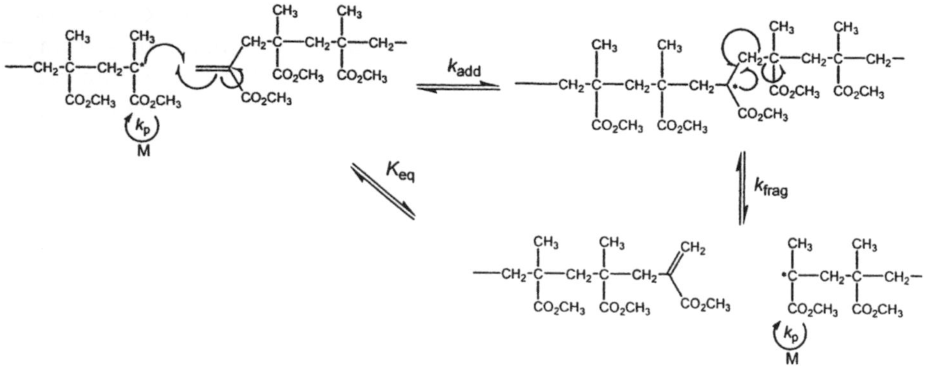

Scheme 12

and dithioesters [37]:

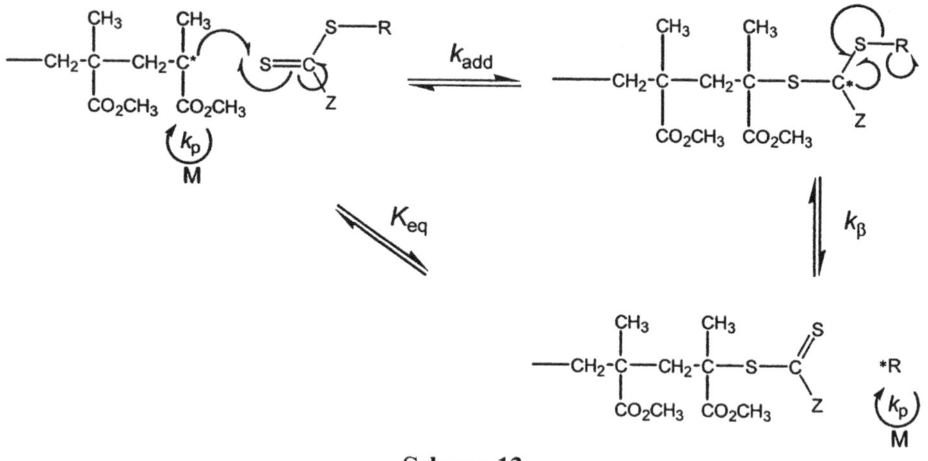

Scheme 13

Polymerization rates in degenerative transfer are typically the same as in a conventional radical polymerization process, however, molecular weights and polydispersities are much lower [183]. The degree of polymerization is roughly defined by the ratio of the concentration of converted monomer to the added transfer agent (more precisely a sum of concentrations of transfer agent and consumed initiator):

$$DP_n = \Delta[M]/([TA] + \Delta[I]) \tag{6}$$

Polydispersities do not depend on the concentration of transfer agent, since it defines both chain length and rate of deactivation:

$$M_w/M_n = 1 + (k_p/k_{tr})(2/p - 1) \tag{7}$$

A key feature for degenerative transfer is the relative rate of transfer (k_{tr}) or of addition (k_{add}), often fragmentation is faster than addition. Three factors determine the overall relative rate of degenerative transfer. One is the structure of the alkyl group in the initial transfer agent, the second is that of transferable atom or group and the third can be the substituent stabilizing the radical. It appears that for degenerative transfer, the only acceptable atom is iodine with the transfer coefficient in polymerization of styrene and acrylates being in the range of $k_{tr}/k_p \approx 2$ to 3. Degenerative transfer with bromine or chlorine was much too slow; the polymerizations behaved the same as without added transfer agent. Transfer coefficients for aryl halcogenides are also relatively slow; rates for aryl sulfides correspond to that for chlorides, aryl selenides to bromides and potentially only tellurides could have sufficient transfer rates, similar to those for iodides (see Curran, D. P. [185]).

The other class of compounds useful for degenerative transfer reactions are those with either $C = C$ or $C = S$ double bonds. Methacrylate derivatives have transfer rates similar to that of the propagation of methacrylates, and are successful only for the polymerization of methacrylates [35,36]. Due to steric effects the intermediate radical shown in Scheme 12 cannot react directly with monomer but only fragment. Unfortunately, mono substituted alkenes such as styrenes and acrylates react with the intermediate radicals and give branched structures, i.e., there is inefficient fragmentation.

Among compounds with $C = S$ double bonds, dithiocarbamates were initially used. This system was used by Otsu in the first studies of controlled radical polymerizations, and he termed them iniferters [12,46]. The main mode of action for these compounds was, however, a photochemical cleavage rather than bimolecular degenerative transfer ($k_{tr}/k_p < 0.1$)). Subsequently replacement of the electron donating group in dithiocarbamates ($-NR_2$) or xanthates, ($-OR$) by an electron neutral ($-Me, -Ph$) group, or electron withdrawing ($-CN$) group increased enormously the relative rates of degenerative transfer to values of $k_{tr}/k_p > 100$ [37]. This new process, called reversible addition fragmentation transfer (RAFT) [186] can be applied to the polymerization of many monomers including styrene, (meth)acrylates and vinyl benzoate [187,188] has been conducted in emulsion systems [189], with functional monomers such as 4-acetoxy styrene [190], and enables the synthesis of new block copolymers. However, the efficiency of the block copolymer synthesis, as well as the consumption of the initial transfer agent depends strongly on the structure of the alkyl precursor. For example, cumyl derivatives have been excellent transfer agents in RAFT but, isobutyrate derivatives were unsuccessful in polymerization of MMA. As described by Moad [61], the choice of CTA is critical in producing near-monodisperse polymers *via* the RAFT process. It was noted that fragmentation efficiency is governed largely by the steric hindrance of the leaving group. However, the stability of the leaving radical cannot be ignored. Another consideration for choosing an appropriate CTA is the ability of the leaving radical to initiate polymerization. Ideally, the leaving group of the CTA would preferentially fragment, yielding a radical that would quickly add to monomer, Scheme 13. Indeed the role of the structure of the chain transfer agent in the polymerization of *N,N*-dimethyl-*s*-thiobenzoylthiopropionamide was examined by Donovan and coworkers [191], and they attributed the success of *N,N*-dimethyl-*s*-thiobenzoylthiopropionamide as the CTA to faster initiation rates of acrylamido radical and the increased steric bulk of the leaving group. The reaction can also be conducted in emulsion systems [188,192].

This would indicate that RAFT is in many ways similar to NMP and ATRP in that the components that contribute to the dynamic fast and reversible equilibrium between dormant and growing species have to be selected for each monomer, if the full benefits of a controlled polymerization are to be optimized, a set of universal reagents, or conditions, do not yet exist for any of these systems.

Work on design and use of molecules suitable as iniferters continues and recently several block copolymers such as poly(vinyl acetate-*b*-styrene-*b*-vinyl acetate), have been prepared utilizing di-Et 2,3-dicyano-2,3-di(*p*-*N*,*N*-diethyldithiocarbamymethyl)phenylsuccinate (DDDCS) as a multi-functional iniferter. Under heating without ultra-violet (UV) light, DDDCS acts as a thermal iniferter by the reversible cleavage of the hexa-substituted C–C bond, while under UV light irradiation at ambient temperature, it serves as a photoiniferter by the reversible cleavage of the two diethyldithiocarbamyl (DC) functional groups [193]. The polymerization proceeds by a CRP mechanism realized by a macro-iniferter technique. The macro-iniferters were designed and synthesized by CRP of vinyl monomers. The polymers bearing alpha- and omega-DC end groups are macro-iniferters and can be used for the preparation of ABA triblock copolymers with different block components [193]. Thioether-thiones have been used for the preparation of several different block copolymers [194].

Another approach that has provided some level of control over radical polymerization has been the use of cobalt complexes as transfer agents and has been employed for polymerization of styrene, acrylates and methacrylates [195].

V. COMPARISON OF VARIOUS METHODS OF CONTROLLING RADICAL POLYMERIZATION

Currently, the three most efficient methods of controlling radical polymerization are NMP, ATRP and degenerative transfer. Each of these methods has advantages over the other processes and also some drawbacks that may direct the choice of process employed for preparation of a particular material. The relative advantages and limitations of each method can be grouped into four categories. They include range of monomers, reaction conditions, active end groups and other required components such as catalysts, accelerators, etc.

Specific nitroxides have to be selected for specific monomers [58]. TEMPO can be successfully applied only to styrene and copolymers due to its relatively small equilibrium constant. Polymerization of acrylics requires the use of either nitroxides with a higher equilibrium constant (phosphate derivatives) or those with a lower thermal stability (4-oxy TEMPO). Homopolymerization of methacrylates still await the development of a suitable nitroxide [196], although methacrylate containing copolymers can be prepared [67]. Together these limitations indicate that nitroxides still have to be developed that will allow for greater freedom in cross-propagation reactions to afford increased capability to prepare copolymers and block copolymers. Typical reactions are carried out in bulk and at high temperatures (>120 °C for TEMPO) because the reactions are inherently slow. Polymerizations in solution, dispersion and emulsion have been reported [85,91,171]. The initiator can be either a combination of conventional initiator and free nitroxide (1.3 : 1 ratio is apparently the best) [13] or a preformed alkoxyamine can be used [76]. End groups in the dormant species are alkoxyamines although some unsaturated species formed by abstraction of β-H atoms or other inactive groups formed by side reactions, e.g., termination can also be present. Alkoxyamines are relatively expensive since they are just beginning to become commercially available on an industrial scale. Nitroxides are generally difficult to remove from the chain end, although chain end functionalization chemistry is being developed [198]. On the more positive side the process typically does not require a catalyst and is carried out at elevated temperatures, optionally in commercially available standard free radical polymerization equipment. The polymerizations are usually slow, although some acceleration was reported in the presence of additional radical initiators [81,199], sugars, acyl compounds [83], and acids [82] and anhydrides [200] that act to control the concentration of the deactivator.

ATRP has been used successfully for the largest range of monomers, although the direct polymerization of vinyl acetate and acrylic acids has not yet been successful. ATRP has been carried out in bulk, solution, dispersion and emulsion at temperatures ranging from $-20\,°C$ to $130\,°C$. Some tolerance to oxygen has been reported in the presence of zero-valent metals [201]. The catalyst complex is based on a transition metal that regulates both polymerization rate and polydispersity furthermore since the catalyst must be available for the reaction to occur both oxidation states should be sufficiently accessible in the reaction medium. The catalyst can be selected to facilitate cross-propagation for the synthesis of difficult block copolymers, and can scavenge some oxygen through *in situ* formation of the deactivator, but in homogeneous systems it should be removed or recycled from the final polymerization product since the concentration of the transition metal complex is generally higher than desired in most products. In some supported or hybrid catalyst systems the concentration of transition metal in the final product may be acceptably low [202–205]. Perhaps the biggest advantage of ATRP is the readily accessible inexpensive initiators whose active end group, normally consists of simple halogens. This is especially important for lower molecular weight polymers due to the high proportion of the end groups. Additionally, there is a multitude of commercially available macro-initiators for ATRP. Moreover, the halogen end groups can be easily displaced with other useful functionalities using S_N2, S_N1, radical or other chemistries [206,207].

Most of the work reported in the open literature has used Schlenk techniques for the polymerizations but this reflects a desire to obtain reproducible kinetics and the use of monomers stored long term under normal laboratory conditions, rather than indicating a need for excessive purification of commercially available materials. It is expected that in commercial scale operations use of standard industrially available radically polymerizable monomers would not require any pretreatment of the reaction medium prior to initiation of the controlled polymerization.

Degenerative transfer can potentially be used for any radically polymerizable monomer. However, reactions of vinyl esters are apparently more difficult and RAFT polymerization of vinyl benzoate requires very high temperatures ($T \sim 150\,°C$). It may be difficult to assure an efficient cross-propagation for some systems [208]. In principle, all classic radical systems can be converted to RAFT, or to another degenerative transfer process, in the presence of efficient transfer reagents. With the current systems the end groups are alkyl iodides, methacrylates or thioesters. The latter are colored and can provide some odor for low molar mass species and require radical chemistry for removal and displacement. Methacrylate oligomers are efficient only for the polymerization of methacrylates. No transition metal catalyst is needed for activation in degenerative transfer since that role is fulfilled by addition of a standard radical initiator however this results in the incorporation of some undesired end groups. The amount of termination is governed by the amount of decomposed initiator. A potential disadvantage of degenerative transfer is that there is always a low molecular weight reactive radical available for termination reactions, in contrast to the ATRP and TEMPO systems where as conversions increase only reactive radicals associated with longer chains exist, and termination reactions occur more slowly.

Thus, the prime advantage of the nitroxide mediated system is the absence of any metal. ATRP may be especially well suited for low molar mass functional polymers due to the low cost of end groups and easier catalyst removal from low viscosity systems. It may be also very suitable for the synthesis of 'difficult' block copolymers and some special hybrids with end functionalities. However, it requires catalyst removal or the use of a supported catalyst. Degenerative transfer, and especially RAFT, should be successful for the polymerization of many less reactive monomers and for the preparation of high molecular weight polymers. It is likely that the search for new efficient transferable groups will continue due to some color and odor limitations of the sulfur containing compounds currently employed.

VI. NEW MATERIALS BY CONTROLLED/LIVING RADICAL POLYMERIZATION

After all this discussion about radical polymerization and new methods to develop processes to obtain better control of the polymerization, the question remains: Why? Why should one use these novel methods to polymerize vinyl monomers? The answer that first comes to mind is supplementation of anionic and cationic polymerization as the primary means of obtaining well-defined (co)polymers, in these cases by radical polymerization processes which are more tolerant of impurities, functional groups and are applicable to a wider range of monomers. This increased level of control over radical polymerization will allow industry to tailor a material to the requirements of a specific application using the most robust polymerization process available, ensuring the polymers have the optimal balance of physical and chemical properties for a given application.

Well-defined (co)polymers are generally recognized as polymers with molecular weights defined by $DP_n = \Delta[M]/[I]_o$, and with low polydispersities, say, $M_w/M_n < 1.3$ (an arbitrary figure). However, such homopolymers are of little interest commercially; in some instances, materials with broad molecular weight distributions are desired for various rheological reasons. What controlled/living polymerizations offer is the ability to prepare entirely new polymers with a myriad of compositions, architectures, and functionalities (Figure 1) with each polymer chain in the bulk material having the same microstructure (composition, architecture and functionality) (Table 1), and not a distribution of composition and properties from chain to chain.

VII. COMPOSITIONS

When two or more monomers are combined and polymerized, statistical copolymers are formed where the relative compositions of the monomers in the polymer chain is a function of the reactivity ratios and the monomer feed ratios at the instant of polymerization. In conventional radical polymerization, high molecular weight polymer is formed early in the reaction and then is irreversibly terminated. As one monomer is generally consumed faster than the other(s), there is a faster depletion of that monomer compared to the other monomers fed to the reactor. At higher conversions, the more reactive monomer will likely be present only in low amounts, while the other(s) will be present in higher amounts, which leads to polymers that contain lower (or zero) amounts of the first monomer when compared to the chains prepared early in the polymerization. This gradient of compositions from chain to chain can be overcome by continuously adding monomer(s) so that the monomer feed remains relatively stable throughout the polymerization. In contrast, for controlled polymerizations, all chains grow at nearly the same rate, with little irreversible termination. The relative rate of monomer consumption

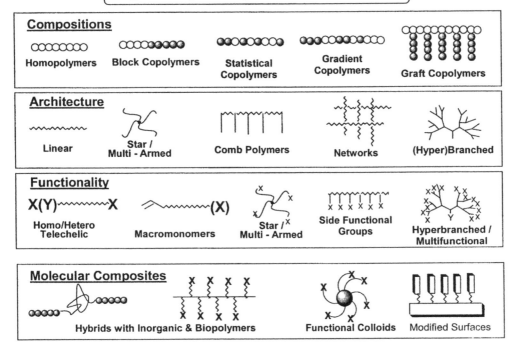

Figure 1 Molecular structures possible with controlled/living polymerizations.

(based on the reactivity ratios) is nearly the same as in a conventional process [50,209]. What is different is that the relative amount of monomer A vs. B in the polymer chains does not vary *from* chain *to* chain, but *along* the chains themselves. This results in the preparation of novel gradient copolymers [210], where composition of the copolymer gradually changes from a higher concentration of one monomer to the other along the length of the chain. Such polymers have been prepared by nitroxide based systems [63,211], by ATRP [157,212], and by RAFT [38,213] (Table 1).

Instead of a gradual change in the composition, an abrupt transition from one monomer to another may be desired as in segmented copolymers, i.e., block and graft copolymers. Block copolymers can be prepared in one of two manners: through the use of macroinitiators or by sequential addition of monomer. Macroinitiators can be prepared by a number of polymerization techniques, including controlled/living radical polymerization. In this case, a monomer is polymerized and the polymer is isolated then dissolved in a second monomer and used to initiate polymerization, in this manner, there is a very clean break between monomer units (blocks). Such a methodology has been used to prepare block copolymers that act as thermoplastic elastomers [175,238] and as amphiphilic copolymers [149,239,240]. The isolated macroinitiator approach has been extended to prepare ABC and ABCBA block copolymers by sequential polymerization of three different monomers [241]. In another approach to block copolymers a second monomer can be added at the end of the polymerization of the first monomer. This sequential addition of monomer may result in a slight taper or gradient of the transition from block A

Table 1 Summary of CRP copolymerizations.

TEMPO derivatives

St/nBMA; St/ClMS; St/MMA [214]; St/AN [215,216]; St/NVC [215,217]; St/VP [218]; St/AcOSt [214]; St/BrSt, St/MSt, St/BuSt, with MOTEMPO [219]; St/CMI [220]; St/BMI [221]; and CMSt/ MVB-TMS [222].

ATRP systems

St/MA [210]; St/MMA; St/nBA [174,223]; St/BuMA [224]; MMA/BA [176,225]; MMA/nBMA [209]; St/MMA [226]; MMA/MA [227,126]; St/AN [212]; MMA/HEMA [138,134]; MMA/MAA [134]; St/EPSt z[228]; MMA/NCMI [229]; St/Mah and St/AEMI or St/PMI [230].

CCT and RAFT

MMA/MA [231]; MMA/nBMA [232]; HEMA/MMA [37].

Gradient copolymers [233]

St/4-acetoxySt by Nitroxide [211], St/ACN [212,218], St/MA [234], St/nBuA [210], by ATRP and St/cMA using iniferters [213], St/Mah [235], using ATRP and RAFT.

Alternating copolymers

St/N-substituted maleimides [221,230], MA/isobutene and MA isobutyl vinyl ether [236], iso-Bu vinyl ether (iBVE) with electron-withdrawing monomers such as maleic anhydride and N-substituted maleimides [237].

---

to block B if monomer A is not completely consumed. Novel materials may be developed by adjusting the length and degree of this taper and this affects the properties of the resulting block copolymer [54,210].

Nitroxide-mediated polymerization has been used to prepare many block copolymers: p(4-CMSt)/St [242–244]; p(BrSt)/St and St/p(BrSt) [219]; St/tBuSt [92]; p(tBOSt)/St [245]; St/PIMS [246]; St/MPCS [247]; p(AcOSt)/MPVB [248]; p(St-*r*-CMI)/St [217]; p(St-*r*-NVC)/St [217]; St/StAN [216]; p(SSt)/DMAM and p(SSt)/SSC [249]; p(SSt)/ VN [250]; p(nBA)/St and p(nBA)/St [251]; p(St)/MA [252]; p(4VP)/St and p(CMSt)/St [253]; p(St)/DMA [94]; p(EBPBB)/St [254]; p(St)/BD and p(St)/IP [95,255]; p(St)/nBA and nBA/St [67]; various isoprene block copolymers [256]; p(St-*alt*-Mah)St [257]; olefins/ acrylates [258]; poly(2,5-dioctyloxy-1,4-phenylenevinylene)/St-co-p(CMSt) [259]. Table 2 lists the block copolymers prepared by ATRP and includes the catalyst complex employed. In reference [241] Davis provides a good review of block copolymers prepared by ATRP.

A benefit of the relatively stable end groups of polymers prepared by controlled/ 'living' polymerizations, is that they can be isolated and stored as macroinitiators with relative ease. Such is not the case for polymers prepared by ionic polymerizations; the active anion or cation will be quenched by advantageous moisture. This also allows one to modify polymers prepared by other methods so that they can become macroinitiators for controlled/'living' radical polymerization. Such 'mechanism transformation' can be used to prepare a wide array of novel polymers; block copolymers of combinations of radically prepared polymers with those synthesized by step-growth polymerizations [160,276], ROMP [159,277], cationic [161,278] and anionic polymerizations [255,279] have been prepared (Table 3).

Some examples of materials prepared from the presently extended range of controllably polymerizable monomers are seen in Table 4 where block copolymers with two disparate ionic blocks have been prepared.

Table 2 Summary of block copolymers prepared using ATRP.

1st Block	2nd Block	Catalyst	Investigator
p(BMA)-Br	MMA	Ni(NCN')Br	Jerome [135]
p(BMA)-Cl	St	CuCl/bpy	Zou [224,260]
p(MMA)-Cl	BMA	RuCl$_2$(PPh$_3$)$_3$/Al(OiPr)$_3$	Sawamoto [139]
p(MMA)-Cl	MA and BA	NiBr$_2$(Pn-Bu$_3$)$_2$/Al(OiPr)$_3$	Sawamoto [126]
p(MMA)-Cl	nBA	CuCl/dNbpy	Matyjaszewski [261]
p(MMA)-Cl	St	CuCl/bpy	Ying
p(MMA)-Cl	St	CuCl/bpy	Qin [262]
p(MA)-Cl, or -Br	MMA	CuCl/dNbpy	Matyjaszewski [261]
p(nBA)-Br[a]	MMA	CuCl/HMTETA[b]	Matyjaszewski [238]
p(St)-Br	MMA and HEMA	CuCl/bpy	Ying
p(St)-Br	NPMA	CuBr/bpy	Liu [142]
p(St)-Br	MMA	Cu(PF$_6$)$_2$/dMbpy[c]	Schubert [128]
Br-p(nBA)-Br	MMA	NiBr$_2$(PPh$_3$)$_2$/Al(OiPr)$_3$	Jerome [175]
Br-p(nBA)-Br	MMA	CuCl/dNbpy	Matyjaszewski [261]
Br-p(nBA)-Br	MA-POSS	CuCl/PMDETA	Matyjaszewski [263]
p(St)-Cl	MA	CuCl/bpy	Matyjaszewski [27]
p(St)-Cl or -Br	nBA	CuCl/bpy	Vairon [223]
p(St)-Br	tBA	CuBr/PMDETA	Matyjaszewski [264,265]
p(tBA)-Br	St	CuBr/PMDETA	Matyjaszewski [264]
Br-p(St)-Br	tBA	CuBr/PMDETA	Matyjaszewski [130]
St	tBMA, MAA	CuCl/bpy	Wang [266]
Br-p(tBA)-Br	St and MA	CuBr/PMDETA	Matyjaszewski [264]
p(tBA-b-St)-Br	MA	CuBr/PMDETA	Matyjaszewski [130]
p(MMA)-Cl	DMAEMA	CuCl/HMTETA	Matyjaszewski [149]
Cl-p(MMA)-Cl	DMAEMA	CuCl/HMTETA	Matyjaszewski [149]
p(MA)-Br	DMAEMA	CuCl/HMTETA	Matyjaszewski [149]
p(MMA)-Cl	HEMA	CuCl/bpy	Matyjaszewski [145]
p(MMA)-Cl	HEMA	CuCl/bpy	Ying [267]
p(MMA)-Cl	VP	CuCl/Me$_6$TREN	Matyjaszewski [268]
p(nBA)-Br	HEA-TMS	CuBr/PMDETA	Matyjaszewski [239]
p(HEA-TMS)	nBA	CuBr/PMDETA	Matyjaszewski [239]
p(MA)-Br	DMA	CuBr/Me$_6$cyclam	Matyjaszewski [152]
p(nBA)-Br	HPMA	CuBr/Me$_6$cyclam	Matyjaszewski [152]
p(St)-Br	MAIpGlc	CuBr/bpy	Fukuda [269]
P(St)-Br	AcGEA		Li [270]
p(OEGMA)	NaVB	CuBr/bpy	Armes [271]
p(FOMA)	MMA and DMEMA	CuCl/dR$_{f6}$bpy	Matyjaszewski [109][d]
p(MMA)	BzMA	CuBr/fluoro triamine	Haddleton [272][e]
p(MMA)	ACN	Cr(OAc)2	Alipour [273]
EbriB[f]	MMA/BA and MMA/nBMA	CuBr/dAbpy[2]	Matyjaszewski [274]
p(nBA)-Br	St	CuBr/dAbpy[2]	Matyjaszewski [274]
p(4-amino styrene)	St	CuBr/bpy	Patten [275]

[a]Sequential monomer addition without isolation of macroinitiator; [b]N,N,N',N'',N''',N'''-hexamethyltriethylene-tetraamine; [c]4,4'-dimethyl-2,2'-bipyridine; [d]conducted in supercritical CO$_2$; [e]conducted in a fluorous biphasic system; [f]ethyl 2-bromoisobutyrate, [2]4,4'-di(5-alkyl)-2,2'-bipyridine.

Table 3 Block copolymers prepared from a combination of ionic and CRP polymerization techniques.

Methods	Monomers
Cationic/ATRP	St/St, MMA, MA [158]
Cationic/ATRP	IB/St, MMA, MA, IA [133]
Cationic/ATRP	IB/MMA(3 arm) [280]
Cationic/ATRP	β-pinene/MMA [281]
Cationic/Nitroxide	Cyclohexeneoxide/St [282]
Cationic ROP/Nitroxide	THF/St [283,284]
Cationic ROP/ATRP	THF/St, MA, MMA [161]
Cationic ROP/ATRP	THF/St(4 armed star) [285]
Cationic ROP/ATRP	THF/St(mikto 4 armed star) [286]
Caionic/Iniferter	THF/MMA [287]
ATRP/Cationic ROP	St/THF [288,289], St/DOP [290]
ATRP/Cationic ROP	St/1,3-Dioxepane(mikto 4 armed star) [291]
Anionic/Nitroxide	Bd/St [255,292]
Anionic/Nitroxide	EO/St [293]
Anionic/Nitroxide	tBuMA/St; bottlebrush copolymers [294]
Anionic/ATRP	St/St, MA, MMA, nBA [279,295]
Anionic/ATRP	IP-St/St [279]
Anionic/ATRP	IP/St [296]
Anionic/ATRP	(meth)acrylate/methacrylate [297]
Anionic ROP/Nitroxide	Caprolactone/St [298]
Anionic ROP/Nitroxide	Ethylene oxide/St [299]
Anionic ROP/ATRP	St then PDMS/nBA, MMA [300]
Anionic ROP/ATRP	PEO the St to form core shell copolymers [301]
ROMP/ATRP	CPD, NB/St, MA [159]
ROMP/ATRP	BD/St [277,302]

Table 4 Ionic block copolymers.

Macroinitiator	2nd Block	Method
MEMA	4-VBK	ATRP [147]
PEG	MAA	ATRP [303]
4VPC16Br[a]	DMAA	Nitroxide [304]
Na St sulfonate	St	Nitroxide [305]
Na 2-Acrylamido-2-methylpropanesulfonate-N,N-dimethylacrylamide-styrene block copolymer [305]		
PAA-poly(benzyl ether) anionic linear-dendritic block amphiphiles [306]		

[a](4VPC16Br): N-hexadecyl-4-vinylpyridinium bromide, (DMAA): N,N-dimethylacrylamide.

Graft copolymers are a special class of segmented copolymer which can be prepared by use of a macroinitiator which contains multiple initiating sites *along* the polymer chain; initiation at these sites allows for the growth of polymer chains from the backbone [307–309]. The degree of branching can vary from a few grafts per chain to a graft site from every monomer unit along the backbone polymer (Table 5) [309].

Table 5 Examples of 'grafting from' using CRP methods.

Backbone	Grafts	Methods
Styrene	Styrene	FRP/Nitroxide [77]
MMS	Styrene	FRP/Nitroxide [310]
α-Olefin	Styrene	Metallocene/Nitroxide [258]
Polypropylene	Styrene	Commercial/Nitroxide [311]
Vinyl Chloride	St, MA, nBA, MMA	Commercial/ATRP [164]
p(IB-co-St)	Styrene	Commercial/ATRP [312,313]
p(SEP)	EMA	Commercial/ATRP [314]
Polyethylene	St, MMA	Commercial/ATRP [165]
HEA, MAOETMACl	St	FRP(latex)/ATRP [315]
St	St, nBA, MMA	Nitroxide/ATRP [308]
St-ClMeSt	OFPA	Nitroxide/ATRP [316]
St, MMA	St	ATRP/Nitroxide [317]
HEMA	nBA	ATRP/ATRP [309]

Table 6 Commercially available macroinitiators transformed into CRP initiators.

Macroinitiator	CRP Process/Reference
Poly(ethylene oxide)	ATRP [320,321], RAFTMoad [38]
Poly(propylene oxide)	ATRP [322,323]
Poly(ethylene adipate)	TEMPO [324]
Poly(β-cyclodextran)	ATRP [325]
Poly(dimethylsulfoxide)	ATRP [326,300]
Poly(ethylene-co-butylene)	ATRP [327], RAFT [328]
Polybutadiene	ATRP [322]
Polysulfone	ATRP [160]
Poly(methylphenylsilylene)	ATRP [329]
Polyphenylenes	ATRP [330]
Poly(p-phenylene vinylene)	Nitroxide + C60 [331]
Radical polymerization	Nitroxide [332,333], ATRP [334,335]
Redox	ATRP [334]
Telomerization	ATRP [336,337]
Electropolymerization	ATRP [338]
Polyether dendrimer	Nitroxide [339,340], ATRP [339,341]
Hyperbranched	ATRP [342]

Backbone macroinitiators can be prepared by any polymerization process and several commercially available polymers (Table 6) have been used as macroinitiators including polyethylene [156,318], polyisobutylene [312,319], and PVC [164,166] for preparation of both block and graft copolymers.

Block, graft, star and surface tethered hybrid copolymers have been prepared by use of inorganic macroinitiators [326,343,344].

Graft copolymers have also been prepared by grafting through techniques. Nitroxide mediated copolymerization has been successful using styrene as comonomer and p(CL), p(LA), or p(EG) [345] as macromonomers, also p(EO) [346]. NVP and NBA have been copolymerized with p(St) [347] and p(MMA) macromonomers [348] and homopolymerized p(IBVE) using ATRP [349]. Polydimethylsiloxane macromonomers have been copolymerized with MMA using ATRP [350] and RAFT [351]. In both systems it was found that the use of a compatible macroinitiator assisted in incorporation of the macromonomer [352].

VIII. ARCHITECTURE

Another area where controlled/living radical polymerizations can make a significant contribution is in the development of polymers with unique architectures. When an initiator site is incorporated into a monomer, branching of the polymer chain can be induced. When such functionalized monomers are homopolymerized, hyperbranched polymers are obtained [307,353,354]. When they are copolymerized with conventional monomers, polymers with a random distribution, or a gradient of branching along the chain can be obtained [307,353]. Homopolymerization of these monomers using techniques that do not consume the initiating sites for the controlled/living radical polymerization results in a polymer with initiating sites at every repeat unit [309]. By using such a polymer as a macroinitiator, graft copolymers with very densely grafted polymer chains have been obtained, including preparation of cylindrical core/shell or amphiphilic bottle brush copolymers [355]. The macromolecules are very large ($M_n = 5,000,000$, $M_w/M_n = 1.2$) and have been called by the trivial name 'bottle brush' copolymers due to their shape. Such macromolecules with styrene and acrylate grafts have been prepared by ATRP from poly(2-(2-bromoisobutyryloxy)ethyl methacrylate [240,309], with attached block copolymers [264,355,356]. The individual macromolecules have been resolved by atomic force microscopy with length in the range of 100 nm and width 10 nm (Figure 2). The AMF image of an unusual non-symmetrical bottle brush copolymer prepared from a backbone gradient copolymer is shown in Figure 3.

An extension of this concept of 'grafting from' is the formation of surface tethered copolymers. TEMPO moieties containing reactive groups that could be used to tether the initiator to silicon surfaces (wafers or gel particles) have been prepared [344,357,358] and this has been extended to ATRP [359–361]. The tethered initiators have been used to initiate CRP forming attached copolymers trivially named 'brush' (co)polymers. One of the major difficulties associated with growing the polymers off the surfaces, which Wirth [150] had not addressed but that Fukuda [343] had considered, is the extremely low

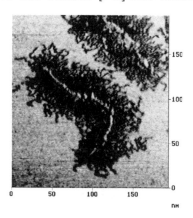

Figure 2 AFM image of poly(butyl acrylate) brushes on mica [263].

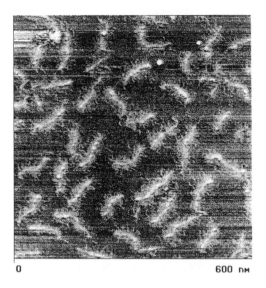

0   600 nm

Figure 3 AFM image of poly(butyl acrylate) brushes on mica [356].

concentration of initiating sites. This leads to a low concentration of radical mediators (i.e., free nitroxide for NMP or Mt $n + 1$ for ATRP) in the contacting solution and leads to an uncontrolled polymerization. Hawker added a small amount of unattached 1-phenylethyl-TEMPO to the system and was able to control the polymer growth from the surface [344], the free polymer chains were separated from those attached to the surface by washing the surface with an appropriate solvent. Later, addition of the persistent radical alone was also shown to be effective at providing controlled polymerization from surfaces [359]. Two groups of workers initially examined functionalization of silica surfaces followed by polymerization of a range of vinyl monomers forming homopolymers and block copolymers [344,359]. Monomers included styrene [360], MMA [140], and acrylamide. Amphiphilic block copolymers were prepared by ATRP [151,362,363], and by RAFT [364]. Tethered PS-*b*-PMMA was prepared by sequential carbocationic polymerization of styrene followed by ATRP of MMA [365,366].

Controlled polymerization from organic, silicon based and carbon particles, and gold surfaces has also been demonstrated [360,367–369]. Bio-active particles were prepared using nitroxide based CRP [370], functional carbon particles were also prepared with nitroxides [369], and ATRP has been used for CRP from silica particles [371–374], and from luminescent particles [375].

Another approach to core shell polymers, or multiarmed star polymers is the arm first approach, where a growing polymer formed by a CRP is copolymerized with a difunctional monomer to form a crosslinked core with the attached first formed arms [131,376,377]. Other surfaces include organic resins and latexes [315].

IX. FUNCTIONALITY

Controlled/'living' radical polymerizations have great potential for the production of polymers of lower molecular weight, but with high degrees of functionality [44]. Precise control of the end groups is readily attained in controlled radical polymerizations, this methodology is ideally suited to preparing telechelic materials [63,168,207,208,378–380].

An example is the preparation of poly(butyl acrylate) with α,ω-hydroxyl groups that can be used as a replacement for poly(ethylene glycol) in polyurethane synthesis [381].

X. APPLICATIONS

Applications discussed in the literature for materials prepared through CRP range from replacement of existing products in existing markets to novel material concepts, creating new applications such as some very novel approaches to drug delivery through the synthesis of well-defined diblock copolymers by ATRP. The block copolymers with a short hydrophobic block $(5 < \text{d.p.} < 9)$ were explored in detail for the development of new colloidal carriers for the delivery of electrostatically charged compounds (e.g., DNA), through the formation of polyion complex micelles [303]. A similar approach has been taken in electronics manufacture [382] where the self-organizing ability of materials prepared by CRP is being exploited.

Some existing markets targeted by materials prepared by CRP are:

Adhesives [383–393]
Sealants [381,394,395]
Emulsifiers [265]
Polymer blend compatibilizers [396,397]
Coatings [398–409]
Toners [410–415]
Dispersants [416–422]
Lubricants [144,423–427]
Curable sealing compositions [428–431]
Elastomeric materials [432,433]
Drug delivery [434,435]
Cosmetics [436–438]
Materials comprising specific bulk physical properties [439–441]

The above references have focused on applications identified by corporate research, in patents and patent applications, but some new applications are also being disclosed by academic workers [442].

Although this has been but a brief review of novel materials prepared using controlled radical polymerizations, one can easily see that, regardless of the type of controlled radical polymerization employed, these methodologies open the door to a wide range of novel polymers with unique properties. Indeed control over polymer sequence distributions continuously expanding and recently multi-block heteropolymer chains with up to 100 blocks in an ordered sequence and controllable block lengths have been reported [443]. Only time will tell, but undoubtedly the question is not if such materials will find commercial uses, but one of when and how.

XI. CONCLUSIONS

Radical polymerizations are widely used in industrial processes, accounting for the synthesis of nearly 50% of all polymeric materials. The widespread use of radical polymerization is due to its unique ability to easily and readily prepare high MW polymers from a variety of monomers, under relatively mild reaction conditions. To extend the usefulness of radical polymerization, various systems have been developed to allow for the 'control' of the polymerization such that termination and transfer processes can be avoided, or at least minimized. Towards this end, three systems have shown some ability to solve this problem; these are the nitroxide mediated polymerization (NMP), atom transfer radical polymerization (ATRP) and radical addition-fragmentation transfer (RAFT). All three have their benefits and deficiencies, but each may be particularly suited for certain applications, i.e., high molecular weight polymers vs. low molecular weight telechelic oligomers, etc.

It has been demonstrated that polymers with novel compositions, architectures, and functionality can be readily prepared by using these methods. Although some terminal functionality of the chains is lost due to unavoidable termination reactions, these materials may provide unique properties that will be good enough, or significant enough, to be used in new applications.

ACKNOWLEDGMENTS

The financial support from the National Science Foundation, the US Environmental Protection Agency and members of ATRP/CRP Consortia: Akzo Nobel, Asahi, Atofina, Bayer, BFGoodrich, BYK, Cabot, Ciba, DSM, Elf, Geon, GIRSA, JSR, Kaneka, Mitsubishi, Mitsui Chemical, Motorola, 3M, Nalco, Nippon Goshei, Nitto Denko, PPG, Rohm & Haas, Rohmax, Sasol, Solvay, Teijin and Zeon at CMU is appreciated.

REFERENCES

1. O'Driscoll, K. F., and Russo, S., eds. (1996). *2nd International Symposium on Free Radical Polymerization: Kinetics and Mechanisms*, held in Santa Margherita Ligure, Genoa, Italy, 26–31 May 1996, 328 pp.
2. Szwarc, M. (1956). 'Living' polymers. *Nature, 178*: 1168–1169.
3. Matyjaszewski, K. (1995). Introduction to living polymerization. Living and/or controlled polymerization. *J. Phys. Org. Chem., 8*(4): 197–207.
4. Matyjaszewski, K., and Lin, C. H. (1991). Exchange reactions in the living cationic polymerization of alkenes. *Makromol. Chem., Macromol. Symp., 47* (*Int. Symp. Mech. Kinet. Polym. React.: Their Use Polym. Synth.*, 1990): 221–237.
5. Litvinenko, G., and Mueller, A. H. E. (1997). General kinetic analysis and comparison of molecular weight distributions for various mechanisms of activity exchange in living polymerizations. *Macromolecules, 30*(5): 1253–1266.
6. Matyjaszewski, K., and Mueller, A. H. E. (1997). Macromolecular nomenclature note no. 12. Naming of controlled, living and 'living' polymerizations. *Polym. Prepr., 38*(1): 6–9.
7. Matyjaszewski, K. (1994). From 'living' carbocationic to 'living' radical polymerization. *J. Macromol. Sci., Pure Appl. Chem., A31*(8): 989–1000.
8. Szwarc, M. (1968). *Carbanions, Living Polymers, and Electron Transfer Processes*, p. 695.
9. Matyjaszewski, K., ed. (1996). *Cationic Polymerizations: Mechanisms, Synthesis, and Applications*, Marcel Dekker, New York, pp. 768.
10. Webster, O. W. (1991). Living polymerization methods. *Science, 251*(4996): 887–893.
11. Smid, J. (2001). Historical perspectives on living anionic polymerization. *Macromol. Symp., 174* (Polymerization Processes and Polymer Materials I): 229.

12. Otsu, T., Yoshida, M., and Tazaki, T. (1982). A model for living radical polymerization. *Makromol. Chem., Rapid Commun.*, *3*(2): 133–140.
13. Georges, M. K., et al. (1993). Narrow molecular weight resins by a free-radical polymerization process. *Macromolecules*, *26*(11): 2987–2988.
14. Solomon, D. H., Rizzardo, E., and Cacioli, P. (1985). Free radical polymerization and the produced polymers. In *Eur. Pat. Appl. 135280*, Commonwealth Scientific and Industrial Research Organization, Australia, 63 pp.
15. Matyjaszewski, K. (2001). Macromolecular engineering by controlled/living ionic and radical polymerizations. *Macromol. Symp.*, *174* (Polymerization Processes and Polymer Materials I): 51–67.
16. Steenbock, M., et al. (1996). Synthesis of block copolymers by nitroxyl-controlled radical polymerization. *Acta Polym.*, *47*(6/7): 276–279.
17. Steenbock, M., et al. (1998). Decomposition of stable free radicals as 'self-regulation' in controlled radical polymerization. *Macromolecules*, *31*(16): 5223–5228.
18. Klapper, M., Brand T., Steenbock, M., and Mullen, K. (2000). Triazolinyl radicals: towards a new mechanism in controlled radical polymerization. *ACS Symp. Ser. 768*: 152–166.
19. Braun, D. (1968). Stable polyradicals. *J. Polym. Sci., Part C, 24*: 7–13.
20. Borsig, E., et al. (1969). Reinitiation reactions of poly(methyl methacrylate) with labile bound fragments of initiator. *Angew. Makromol. Chem., 9*: 89–95.
21. Roussel, J., and Boutevin, B. (2001). Synthesis of methyl methacrylate oligomers with substituted tetraphenylethane initiator. *Polym. Int., 50*(9): 1029–1034.
22. Otsu, T. (2000). Iniferter concept and living radical polymerization. *J. Polym. Sci., Part A: Polym. Chem., 38*(12): 2121–2136.
23. Wayland, B. B., et al. (1994). Living radical polymerization of acrylates by organocobalt porphyrin complexes. *J. Am. Chem. Soc., 116*(17): 7943–7944.
24. Kwon, T. S., et al. (1997). Living radical polymerization of styrene with diphenyl diselenide as a photoiniferter. Synthesis of polystyrene with carbon-carbon double bonds at both chain ends. *J. Macromol. Sci., Pure Appl. Chem., A34*(9): 1553–1567.
25. Takagi, K., et al. (1999). Controlled radical polymerization of styrene utilizing excellent radical capturing ability of diphenyl ditelluride. *Polym. Bull. (Berlin), 43*(2–3): 143–150.
26. Kwon, T. S., et al. (2001). Radical polymerization of methyl methacrylate with diphenyl diselenide under thermal or photoirradiational conditions. *J. Macromol. Sci., Pure Appl. Chem., A38*(5 & 6): 591–604.
27. Wang, J.-S., and Matyjaszewski, K. (1995). Controlled/ 'living' radical polymerization. Atom transfer radical polymerization in the presence of transition-metal complexes. *J. Am. Chem. Soc., 117*(20): 5614–5615.
28. Wang, J.-S., and Matyjaszewski, K. (1995). Controlled/'living' radical polymerization. Halogen atom transfer radical polymerization promoted by a Cu(I)/Cu(II) redox process. *Macromolecules, 28*(23): 7901–7910.
29. Kato, M., et al. (1995). Polymerization of methyl methacrylate with the carbon tetrachloride/ dichlorotris-(triphenylphosphine)ruthenium(II)/methylalumin um bis(2,6-di-tert-butylphen-oxide) initiating system: possibility of living radical polymerization. *Macromolecules, 28*(5): 1721–1723.
30. Sawamoto, M., and Kamigaito, M. (1996). Living radical polymerizations based on transition metal complexes. *Trends Polym. Sci., 4*(11): 371–377.
31. Percec, V., and Barboiu, B. (1995). 'living' radical polymerization of styrene initiated by arenesulfonyl chlorides and CuI(bpy)nCl. *Macromolecules, 28*(23): 7970–7972.
32. Matyjaszewski, K., Patten, T. E., and Xia, J. (1997). Controlled/'living' radical polymerization. Kinetics of the homogeneous atom transfer radical polymerization of styrene. *J. Am. Chem. Soc., 119*(4): 674–680.
33. Matyjaszewski, K., Gaynor, S., and Wang, J.-S. (1995). Controlled radical polymerizations: the use of alkyl iodides in degenerative transfer. *Macromolecules, 28*(6): 2093–2095.

34. Lansalot, M., Farcet, C., Charleux, B., Vaison, J.-P., and Piri, R. (1999). Controlled free-radicle miniemulsion polymerization of styrene using degenerative transfer. *Macromolecules*, *32*: 7354–7360.

35. Krstina, J., et al. (1995). Narrow polydispersity block copolymers by free-radical polymerization in the presence of macromonomers. *Macromolecules*, *28*(15): 5381–5385.

36. Moad, C. L., et al. (1996). Chain transfer activity of omega-unsaturated methyl methacrylate oligomers. *Macromolecules*, *29*(24): 7717–7726.

37. Chiefari, J., et al. (1998). Living free-radical polymerization by reversible addition-fragmentation chain transfer: the RAFT process. *Macromolecules*, *31*(16): 5559–5562.

38. Chong, Y. K., et al. (1999). A more versatile route to block copolymers and other polymers of complex architecture by living radical polymerization: the RAFT process. *Macromolecules*, *32*(6): 2071–2074.

39. Greszta, D., Mardare, D., and Matyjaszewski, K. (1994). Radical polymerization of vinyl acetate in the presence of trialkyl phosphites. *Polym. Prepr.*, *35*(1): 466–467.

40. Harwood, H. J., et al. (1996). Investigation of statistical block, and graft copolymerizations using NMR-sensitive initiators and macroinitiators. *Macromol. Symp.*, *111*: 25–35.

41. Wieland, P. C., Raether, B., and Nuyken, O. (2001). A new additive for controlled radical polymerization. *Macromol. Rapid Commun.*, *22*(9): 700–703.

42. Matyjaszewski, K., ed. (1998). Controlled radical polymerization. In *Proceedings of a Symposium at the 213th National Meeting of the American Chemical Society, held 13–17 April 1997, in San Francisco, California*, ACS Symp. Ser. 685.

43. Matyjaszewski, K., ed. (2000). Controlled/living radical polymerization. Progress in ATRP, NMP, and RAFT. In *Proceedings of a Symposium on Controlled Radical Pcolymerization held on 22–24 August 1999, in New Orleans*, ACS Symp. Ser. 768.

44. Matyjaszewski, K. (1996). Controlled radical polymerization. *Curr. Opin. Solid State Mater Sci.*, *1*(6): 769–776.

45. Colombani, D. (1997). Chain-growth control in free radical polymerization. *Prog. Polym. Sci.*, *22*(8): 1649–1720.

46. Otsu, T., and Matsumoto, A. (1998). Controlled synthesis of polymers using the iniferter technique: developments in living radical polymerization. *Adv. Polym. Sci.*, *136*(Microencapsulation, Microgels, Iniferters): 75–137.

47. Korolev, G. V., and Marchenko, A. P. (2000). 'Living'-chain radical polymerization. *Russ. Chem. Rev.*, *69*(5): 409–434.

48. Fukuda, T., and Goto, A. (2000). Kinetics of living radical polymerization. *ACS Symp. Ser.*, *768*(Controlled/Living Radical Polymerization): 27–38.

49. Matyjaszewski, K., and Gaynor, S. G. (2000). Free radical polymerization. *Appl. Polym. Sci.*, 929–977.

50. Matyjaszewski, K. (1998). Radical nature of Cu-catalyzed controlled radical polymerizations (atom transfer radical polymerization). *Macromolecules*, *31*(15): 4710–4717.

51. Patten, T. E., and Matyjaszewski, K. (1999). Copper(I)-catalyzed atom transfer radical polymerization. *Acc. Chem. Res.*, *32*(10): 895–903.

52. Sawamoto, M., and Kamigaito, M. (1999). Transition-metal-catalyzed living-radical polymerization. *Chemtech.*, *29*(6): 30–38.

53. Kamigaito, M., Ando, T., and Sawamoto, M. (2001). Metal-catalyzed living radical polymerization. *Chem. Rev. (Washington, D C)*, *101*(12): 3689–3745.

54. Matyjaszewski, K., and Xia, J. (2001). Atom transfer radical polymerization. *Chem. Rev. (Washington, D C)*, *101*(9): 2921–2990.

55. Patten, T. E., and Matyjaszewski, K. (1998). Atom-transfer radical polymerization and the synthesis of polymeric materials. *Adv. Mater.*, *10*(12): 901–915.

56. Hawker, C. J. (1996). Advances in 'living' free-radical polymerization: architectural and structural control. *Trends Polym. Sci.*, *4*(6): 183–188.

57. Malmstroem, E. E., and Hawker, C. J. (1998). Macromolecular engineering via 'living' free-radical polymerizations. *Macromol. Chem. Phys.*, *199*(6): 923–935.

58. Hawker, C. J., Bosman, A. W., and Harth, E. (2001). New polymer synthesis by nitroxide mediated living radical polymerizations. *Chemical Reviews*, *101*(12): 3661–3688.

59. Gridnev, A. (2000). The 25th anniversary of catalytic chain transfer. *J. Polym. Sci., Part A*: *Polym. Chem.*, *38*(10): 1753–1766.

60. Gridnev, A., and Ittel, S.D. (2001). Catalytic chain transfer in free-radicle polymerizations. *Chem. Rev.*, *101*: 3611–3659.

61. Rizzardo, E., et al. (2000). Synthesis of defined polymers by reversible addition-fragmentation chain transfer: the RAFT process. *ACS Symp. Ser.*, *768*(Controlled/Living Radical Polymerization): 278–296.

62. George, M. K., et al. (1994). Taming the free-radical polymerization process. *Trends Polym. Sci. (Cambridge, U K)*, *2*(2): 66–72.

63. Hawker, C. J., et al. (1996). Well-defined random copolymers by a 'living' free-radical polymerization process. *Macromolecules*, *29*(7): 2686–2688.

64. Hawker, C. J., Barclay, G. G., and Dao, J. (1996). Radical crossover in nitroxide mediated 'living' free radical polymerizations. *J. Am. Chem. Soc.*, *118*(46): 11467–11471.

65. Turro, N. J., Lem, G., and Zavarine, I. S. (2000). A living free radical exchange reaction for the preparation of photoactive end-labeled monodisperse polymers. *Macromolecules*, *33*(26): 9782–9785.

66. Davis, T. P., et al. (1995). Cobalt-mediated free-radical polymerization of acrylic monomers. *Trends Polym. Sci. (Cambridge, U K)*, *3*(11): 365–373.

67. Benoit, D., et al. (1999). Development of a universal alkoxyamine for 'living' free radical polymerizations. *J. Am. Chem. Soc.*, *121*(16): 3904–3920.

68. Greszta, D., and Matyjaszewski, K. (1996). Mechanism of controlled/'living' radical polymerization of styrene in the presence of nitroxyl radicals. Kinetics and simulations. *Macromolecules*, *29*(24): 7661–7670.

69. Fischer, H. (1997). The persistent radical effect in 'living' radical polymerization. *Macromolecules*, *30*(19): 5666–5672.

70. Fischer, H. (1999). The persistent radical effect in controlled radical polymerizations. *J. Polym. Sci., Part A*: Polym. Chem., *37*(13): 1885–1901.

71. Fischer, H., and Souaille, M. (2001). The persistent radical effect in living radical polymerization — borderline cases and side-reactions. *Macromol. Symp.*, *174*(Polymerization Processes and Polymer Materials I): 231–240.

72. Ananchenko, G. S., and Fischer, H. (2001). Decomposition of model alkoxyamines in simple and polymerizing systems. I. 2,2,6,6-tetramethylpiperidinyl-N-oxyl-based compounds. *J. Polym. Sci., Part A: Polym. Chem.*, *39*(20): 3604–3621.

73. Matyjaszewski, K., et al. (1995). Synthesis of well defined polymers by controlled radical polymerization. *Macromol. Symp.*, *98* (35th IUPAC International Symposium on Macromolecules, 1995): 73–89.

74. Mardare, D., and Matyjaszewski, K. (1994). Thermal polymerization of styrene in the presence of stable radicals and inhibitors. *Polym. Prepr.*, *35*(1): 778–779.

75. Devonport, W., et al. (1997). 'Living' free-radical polymerizations in the absence of initiators: controlled autopolymerization. *Macromolecules*, *30*(7): 1929–1934.

76. Hawker, C. J. (1994). Molecular weight control by a 'living' free-radical polymerization process. *J. Am. Chem. Soc.*, *116*(24): 11185–11186.

77. Hawker, C. J. (1995). Architectural control in 'living' free radical polymerizations: preparation of star and graft polymers. *Angew. Chem., Int. Ed. Engl.*, *34*(13/14): 1456–1459.

78. Shigemoto, T., and Matyjaszewski, K. (1996). Controlled radical polymerization of styrene in the presence of nitronyl nitroxides. *Macromol. Rapid Commun.*, *17*(5): 347–351.

79. Benoit, D., et al. (2000). Kinetics and mechanism of controlled free-radical polymerization of styrene and n-butyl acrylate in the presence of an acyclic beta-phosphorylated nitroxide. *J. Am. Chem. Soc.*, *122*(25): 5929–5939.

80. Fukuda, T., et al. (1996). Mechanisms and kinetics of nitroxide-controlled free radical polymerization. *Macromolecules*, *29*(20): 6393–6398.

81. Greszta, D., and Matyjaszewski, K. (1997). TEMPO-mediated polymerization of styrene: rate enhancement with dicumyl peroxide. *J. Polym. Sci., Part A: Polym. Chem., 35*(9): 1857–1861.

82. Georges, M. K., et al. (1994). Narrow polydispersity polystyrene by a free-radical polymerization process-rate enhancement. *Macromolecules, 27*(24): 7228–7229.

83. Malmstrom, E., Miller, R. D., and Hawker, C. J. (1997). Development of a new class of rate-accelerating additives for nitroxide-mediated 'living' free radical polymerization. *Tetrahedron, 53*(45): 15225–15236.

84. Bosman, A.W., et al. (2001). High-throughput synthesis of nanoscale materials: Structural optimization of functionalized one-step star polymers. *J. Am. Chem. Soc. 123*: 6461–6462.

85. Marestin, C., et al. (1998). Nitroxide mediated living radical polymerization of styrene in emulsion. *Macromolecules, 31*(12): 4041–4044.

86. Charleux, B., et al. (2000). Emulsion polymerization in the presence of a stable free radical. In *Eur. Pat. Appl. 970973*, Elf Atochem S.A., Fr., EP. 21 pp.

87. MacLeod, P. J., et al. (1999). Stable free radical miniemulsion polymerization. *Polym. Mater. Sci. Eng., 80*: 539–540.

88. Schmidt-Naake, G., Drache, M., and Taube, C. (1999). TEMPO-controlled free radical suspension polymerization. *Angew. Makromol. Chem., 265*: 62–68.

89. Taube, C., and Schmidt-Naake, G. (2001). TEMPO-controlled radical suspension polymerization in an oil/water system in an autoclave. *Chem. Eng. Technol., 24*(10): 1013–1017.

90. Ma, J. W., et al. (2001). Nitroxide partitioning between styrene and water. *J. Polym. Sci., Part A: Polym. Chem., 39*(7): 1081–1089.

91. Qiu, J., Charleux, B., and Matyjaszewski, K. (2001). Controlled/living radical polymerization in aqueous media: homogeneous and heterogeneous systems. *Prog. Polym. Sci., 26*(10): 2083–2134.

92. Jousset, S., Hammouch, S. O., and Catala, J. M. (1997). Kinetic studies of the polymerization of p-tert-butylstyrene and its block copolymerization with styrene through living radical polymerization mediated by a nitroxide compound. *Macromolecules, 30*(21): 6685–6687.

93. Georges, M. K., et al. (1995). Homoacrylate polymerization processes with oxonitroxides. In *US 5412047*, Xerox Corp., USA, US. 16 pp.

94. Li, D., and Brittain, W. J. (1998). Synthesis of Poly(N,N-dimethylacrylamide) via nitroxide-mediated radical polymerization. *Macromolecules, 31*(12): 3852–3855.

95. Georges, M. K., Hamer, G. K., and Listigovers, N. A. (1998). Block copolymer synthesis by a nitroxide-mediated living free radical polymerization process. *Macromolecules, 31*(25): 9087–9089.

96. Minaux, E., et al. (2001). Method for controlled free radical polymerization or copolymerization of ethylene under high pressure in the presence of indoline nitroxide radicals. In *Eur. Pat. Appl. 1120430*, Atofina, Fr., EP. 24 pp.

97. Moffat, K. A., et al. (1995). Stable free radical polymerization process and thermoplastic materials produced therefrom. In *US 5449724*, Xerox Corp., USA, 8 pp.

98. Keoshkerian, B., and Georges, M. K. (1995). Water soluble polymers of narrow molecular weight distribution. In *PCT Int. Appl. WO 9526987*, Xerox Corp., USA, 35 pp.

99. Wohlrab, S., and Kuckling, D. (2001). Multisensitive polymers based on 2-vinylpyridine and N-isopropylacrylamide. *J. Polym. Sci., Part A: Polym. Chem., 39*(21): 3797–3804.

100. Chalari, I., Pispas, S., and Hadjichristidis, N. (2001). Controlled free-radical polymerization of 2-vinylpyridine in the presence of nitroxides. *J. Polym. Sci., Part A: Polym. Chem., 39*(17): 2889–2895.

101. Ding, X. Z., et al. (2000). Behavior of 3-vinylpyridine in nitroxide-mediated radical polymerization: the influence of nitroxide concentration, solvent, and temperature. *J. Polym. Sci., Part A: Polym. Chem., 38*(17): 3067–3073.

102. Ding, X. Z., et al. (2001). Solvent effects on TEMPO-mediated radical polymerizations: behaviour of 3-vinylpyridine in a protic solvent. *Eur. Polym. J., 37*(8): 1561–1569.

103. Fischer, A., Brembilla, A., and Lochon, P. (1999). Nitroxide-mediated radical polymerization of 4-vinylpyridine: study of the pseudo-living character of the reaction and influence of temperature and nitroxide concentration. *Macromolecules*, *32*(19): 6069–6072.

104. Shipp, D. A., and Matyjaszewski, K. (1999). Kinetic analysis of controlled/'living' radical polymerizations by simulations. 1. The importance of diffusion-controlled reactions. *Macromolecules*, *32*(9): 2948–2955.

105. Matyjaszewski, K., and Wang, J.-S. (1996). Atom- or group-transfer radical polymerization and polymers produced by the process. In *PCT Int. Appl.*, WO 9630421, Carnegie Mellon University, 1996, 129 pp.

106. Xia, J., and Matyjaszewski, K. (1997). Controlled/'living' radical polymerization. Atom transfer radical polymerization using multidentate amine ligands. *Macromolecules*, *30*(25): 7697–7700.

107. Xia, J., Gaynor, S. G., and Matyjaszewski, K. (1998). Controlled/'living' radical polymerization. Atom transfer radical polymerization of acrylates at ambient temperature. *Macromolecules*, *31*(17): 5958–5959.

108. Matyjaszewski, K., Nakagawa, Y., and Jasieczek, C. B. (1998). Polymerization of n-butyl acrylate by atom transfer radical polymerization. Remarkable effect of ethylene carbonate and other solvents. *Macromolecules*, *31*(5): 1535–1541.

109. Xia, J., et al. (1999). Atom transfer radical polymerization in supercritical carbon dioxide. *Macromolecules*, *32*(15): 4802–4805.

110. Matyjaszewski, K., et al. (1997). Observation and analysis of a slow termination process in the atom transfer radical polymerization of styrene. *Tetrahedron*, *53*(45): 15321–15329.

111. Kotani, Y., Kamigaito, M., and Sawamoto, M. (1999). Transition-metal-mediated living-radical polymerization of styrene: design of initiating systems. Book of Abstracts, 218th ACS National Meeting, New Orleans, Aug 22–26, 1999, p. OLY-520.

112. Matyjaszewski, K., et al. (1997). Controlled/'living' radical polymerization of styrene and methyl methacrylate catalyzed by iron complexes. *Macromolecules*, *30*(26): 8161–8164.

113. Kotani, Y., Kamigaito, M., and Sawamoto, M. (1999). FeCp(CO)2I: a phosphine-free half-metallocene-type iron(II) catalyst for living radical polymerization of styrene. *Macromolecules*, *32*(20): 6877–6880.

114. Louie, J., and Grubbs, R. H. (2000). Highly active iron imidazolylidene catalysts for atom transfer radical polymerization. *Chem. Commun. (Cambridge)*, (16): 1479–1480.

115. Ando, T., Kamigaito, M., and Sawamoto, M. (1997). Iron(II) chloride complex for living radical polymerization of methyl methacrylate. *Macromolecules*, *30*: 4507–4510.

116. Gibson, V.C., et al. (2002). Four-coordinate iron complexes bearing α-diimine ligands: efficient catalysts for Atom Transfer Radical Polymerization (ATRP). *Chem. Commun. (Cambridge)*: 1850–1851.

117. Percec, V., et al. (1996). Metal-catalyzed 'living' radical polymerization of styrene initiated with arenesulfonyl chlorides. From heterogeneous to homogeneous catalysis. *Macromolecules*, *29*(10): 3665–3668.

118. Kotani, Y., Kamigaito, M., and Sawamoto, M. (1999). Re(V)-mediated living radical polymerization of styrene: 1 ReO2I(PPh3)2/R-I initiating systems. *Macromolecules*, *32*(8): 2420–2424.

119. Watanabe, Y., et al. (2001). Ru(Cp)Cl(PPh3)2: a versatile catalyst for living radical polymerization of methacrylates, acrylates, and styrene. *Macromolecules*, *34*(13): 4370–4374.

120. Qiu, J., and Matyjaszewski, K. (1997). Polymerization of substituted styrenes by atom transfer radical polymerization. *Macromolecules*, *30*(19): 5643–5648.

121. Gao, B., et al. (1997). Living atom transfer radical polymerization of 4-acetoxystyrene. *Macromol. Rapid Commun.*, *18*(12): 1095–1100.

122. Doerffler, E. M., and Patten, T. E. (2000). Benzyl acetate and benzyl ether groups as latent initiator sites for atom transfer radical polymerization. *Macromolecules*, *33*(24): 8911–8914.

123. McQuillan, B. W., and Paguio, S. (2000). Living radical polymerization of trimethylsilyl-styrene. *Fusion Technol.*, *38*(1): 108–109.

124. Matyjaszewski, K., et al. (2001). Tridentate nitrogen-based ligands in Cu-based ATRP: a structure-activity study. *Macromolecules*, *34*(3): 430–440.

125. Patten, T. E., et al. (1996). Polymers with very low polydispersities from atom transfer radical polymerization. *Science (Washington, DC)*, *272*(5263): 866–868.

126. Uegaki, H., et al. (1998). NiBr2(Pn-Bu3)2-mediated living radical polymerization of methacrylates and acrylates and their block or random copolymerizations. *Macromolecules*, *31*(20): 6756–6761.

127. Percec, V., Barboiu, B., and Kim, H. J. (1998). Arenesulfonyl halides: a universal class of functional initiators for metal-catalyzed 'living' radical polymerization of styrene(s), methacrylates, and acrylates. *J. Am. Chem. Soc.*, *120*(2): 305–316.

128. Schubert, U., et al. (1999). Controlled polymerization of methyl methacrylate and ethyl acrylate using tris(4,4'-dimethyl-2,2'-bipyridine)copper(II) hexafluorophosphate complexes and aluminum isopropoxide. *Polym. Bull.*, *43*(4–5): 319–326.

129. Bednarek, M., Biedron, T., and Kubisa, P. (2000). Studies of atom-transfer radical polymerization (ATRP) of acrylates by MALDI TOF mass spectrometry. *Macromol. Chem. Phys.*, *201*(1): 58–66.

130. Davis, K. A., and Matyjaszewski, K. (2000). Atom transfer radical polymerization of tert-butyl acrylate and preparation of block copolymers. *Macromolecules*, *33*(11): 4039–4047.

131. Zhang, X., Xia, J., and Matyjaszewski, K. (2000). End-functional poly(tert-butyl acrylate) star polymers by controlled radical polymerization. *Macromolecules*, *33*(7): 2340–2345.

132. Ma, Q., and Wooley, K. L. (2000). The preparation of t-butyl acrylate, methyl acrylate, and styrene block copolymers by atom transfer radical polymerization: precursors to amphiphilic and hydrophilic block copolymers and conversion to complex nanostructured materials. *J. Polym. Sci., Part A: Polym. Chem.*, *38*(Suppl.): 4805–4820.

133. Coca, S., and Matyjaszewski, K. (1997). Block copolymers by transformation of 'living' carbocationic into 'living' radical polymerization. II. ABA-type block copolymers comprising rubbery polyisobutene middle segment. *J. Polym. Sci., Part A: Polym. Chem.*, *35*(16): 3595–3601.

134. Simal, F., Demonceau, A., and Noels, A. F. (1999). Highly efficient ruthenium-based catalytic systems for the controlled free-radical polymerization of vinyl monomers. *Angew. Chem., Int. Ed.*, *38*(4): 538–540.

135. Granel, C., et al. (1996). Controlled radical polymerization of methacrylic monomers in the presence of a bis(ortho-chelated) arylnickel(II) complex and different activated alkyl halides. *Macromolecules*, *29*(27): 8576–8582.

136. Haddleton, D. M., et al. (1997). Atom transfer radical polymerization of methyl methacrylate initiated by alkyl bromide and 2-pyridinecarbaldehyde imine copper(I) complexes. *Macromolecules*, *30*(7): 2190–2193.

137. Liu, B., and Hu, C. P. (2001). The reverse atom transfer radical polymerization of methyl methacrylate in the presence of some polar solvents. *Eur. Polym. J.*, *37*(10): 2025–2030.

138. Moineau, G., et al. (1998). Controlled radical polymerization of methyl methacrylate initiated by an alkyl halide in the presence of the Wilkinson catalyst. *Macromolecules*, *31*(2): 542–544.

139. Kotani, Y., et al. (1996). Living radical polymerization of alkyl methacrylates with ruthenium complex and synthesis of their block copolymers. *Macromolecules*, *29*(22): 6979–6982.

140. Haddleton, D. M., et al. (1999). 3-Aminopropyl silica supported living radical polymerization of methyl methacrylate: dichlorotris(triphenylphosphine)ruthenium(II) mediated atom transfer polymerization. *Macromolecules*, *32*(15): 4769–4775.

141. Acar, A. E., Yagci, M. B., and Mathias, L. J. (2000). Adventitious effect of air in atom transfer radical polymerization: air-induced (reverse) atom transfer radical polymerization of methacrylates in the absence of an added initiator. *Macromolecules*, *33*(21): 7700–7706.

142. Liu, Y., Wang, L., and Pan, C. (1999). Synthesis of block copoly(styrene-b-p-nitrophenyl methacrylate) and its derivatives by atom transfer radical polymerization. *Macromolecules*, *32*(25): 8301–8305.

143. Moineau, C., et al. (1999). Synthesis and characterization of poly(methyl methacrylate)-block-poly(n-butyl acrylate)-block-poly(methyl methacrylate) copolymers by two-step controlled radical polymerization (ATRP) catalyzed by NiBr2(PPh3)2, 1. *Macromolecules*, *32*(25): 8277–8282.

144. Roos, S., et al. (2001). Atom transfer polymerization for manufacture of long-chain alkyl poly(meth)acrylates as lubricating oil additives. In *PCT Int. Appl.*, Rohmax Additives GmbH, Germany, WO. 54 pp.

145. Beers, K. L., et al. (1999). Atom transfer radical polymerization of 2-hydroxyethyl methacrylate. *Macromolecules*, *32*(18): 5772–5776.

146. Robinson, K. L., et al. (2001). Controlled polymerization of 2-hydroxyethyl methacrylate by ATRP at ambient temperature. *Macromolecules*, *34*(10): 3155–3158.

147. Malet, F. L. G., Billingham, N. C., and Armes, S. P. (2000). Controlled/'living' polymerization of 2-(N-morpholino)ethyl methacrylate by atom transfer radical polymerization in aqueous solution at 20.degree.C. *Polym. Prepr.*, *41*(2): 1811–1812.

148. Zhang, X., Xia, J., and Matyjaszewski, K. (1998). Controlled/'living' radical polymerization of 2-(dimethylamino)ethyl methacrylate. *Macromolecules*, *31*(15): 5167–5169.

149. Zhang, X., and Matyjaszewski, K. (1999). Synthesis of well-defined amphiphilic block copolymers with 2-(dimethylamino)ethyl methacrylate by controlled radical polymerization. *Macromolecules*, *32*(6): 1763–1766.

150. Huang, X., Doneski, L. J., and Wirth, M. J. (1998). Surface-confined living radical polymerization for coatings in capillary electrophoresis. *Anal. Chem.*, *70*(19): 4023–4029.

151. Huang, X., and Wirth, M. J. (1999). Surface initiation of living radical polymerization for growth of tethered chains of low polydispersity. *Macromolecules*, *32*(5): 1694–1696.

152. Teodorescu, M., and Matyjaszewski, K. (1999). Atom transfer radical polymerization of (meth)acrylamides. *Macromolecules*, *32*(15): 4826–4831.

153. Teodorescu, M., and Matyjaszewski, K. (2000). Controlled polymerization of (meth)acrylamides by atom transfer radical polymerization. *Macromol. Rapid Commun.*, *21*(4): 190–194.

154. Rademacher, J. T., et al. (2000). Atom transfer radical polymerization of N,N-dimethylacrylamide. *Macromolecules*, *33*(2): 284–288.

155. Ramakrishnan, A., and Dhamodharan, R. (2000). A novel and simple method of preparation of poly(styrene-b-2-vinylpyridine) block copolymer of narrow molecular weight distribution: living anionic polymerization followed by mechanism transfer to controlled/'livingc' radical polymerization (ATRP). *J. Macromol. Sci., Pure Appl. Chem.*, *A37*(6): 621–631.

156. Fernandez-Garcia, M., et al. (2001). Atom-transfer radical polymerization of dimethyl itaconate. *Macromol. Chem. Phys.*, *202*(7): 1213–1218.

157. Matyjaszewski, K., et al. (1997). Improved processes based on atom (or group) transfer radical polymerization and novel (co)polymers having useful structures and properties. In *PCT Int. Appl.* WO 9718247, 1997, Carnegie Mellon University, USA, WO. 182 pp.

158. Coca, S., and Matyjaszewski, K. (1997). Block copolymers by transformation of living carbocationic into living radical polymerization. *Polym. Prepr.*, *38*(1): 693–694.

159. Coca, S., Paik, H.-J., and Matyjaszewski, K. (1997). Block copolymers by transformation of 'living' ring-opening metathesis polymerization into controlled/'living' atom transfer radical polymerization. *Macromolecules*, *30*(21): 6513–6516.

160. Gaynor, S. G., and Matyjaszewski, K. (1997). Step-growth polymers as macroinitiators for 'living' radical polymerization: synthesis of ABA block copolymers. *Macromolecules*, *30*(14): 4241–4243.

161. Kajiwara, A., and Matyjaszewski, K. (1998). Formation of block copolymers by transformation of cationic ring-opening polymerization to atom transfer radical polymerization (ATRP). *Macromolecules*, *31*(11): 3489–3493.

162. Matyjaszewski, K., et al. (1998). Atom or group transfer polymerization and manufacture of polymers using the same. In *PCT Int. Appl. WO 9801480*, Carnegie-Mellon University, USA, 105 pp.

163. Matyjaszewski, K., et al. (1998). Utilizing halide exchange to improve control of atom transfer radical polymerization. *Macromolecules*, *31*(20): 6836–6840.

164. Paik, H. J., Gaynor, S. G., and Matyjaszewski, K. (1998). Synthesis and characterization of graft copolymers of poly(vinyl chloride) with styrene and (meth)acrylates by atom transfer radical polymerization. *Macromol. Rapid. Commun.*, *19*(1): 47–52.

165. Matyjaszewski, K., et al. (2000). Graft copolymers of polyethylene by atom transfer radical polymerization. *J. Polym. Sci., Part A: Polym. Chem.*, *38*(13): 2440–2448.

166. Matyjaszewski, K., Gaynor, S. G., and Coca, S. (1998). Controlled atom or group-transfer radical polymerization, coupling of molecules, multifunctional polymerization initiators, and formation of telechelic functional material. In *PCT Int. Appl. WO 9840415*, Carnegie Mellon University, USA, 230 pp.

167. Bon, S. A. F., Steward, A. G., and Haddleton, D. M. (2000). Modification of the .omega.-bromo end group of poly(methacrylate)s prepared by copper(I)-mediated living radical polymerization. *J. Polym. Sci., Part A: Polym. Chem.*, *38*(15): 2678–2686.

168. Matyjaszewski, K., et al. (1998). Synthesis of functional polymers by atom transfer radical polymerization. *ACS Symp. Ser.*, *704*(Functional Polymers): 16–27.

169. Gaynor, S. G., Qiu, J., and Matyjaszewski, K. (1998). Controlled/'living' radical polymerization applied to water-borne systems. *Macromolecules*, *31*(17): 5951–5954.

170. Matyjaszewski, K., et al. (2000). Atom transfer radical polymerization of n-butyl methacrylate in an aqueous dispersed system: a miniemulsion approach. *J. Polym. Sci., Part A: Polym. Chem.*, *38*(Suppl.): 4724–4734.

171. Qiu, J., Charleux, B., and Matyjaszewski, K. (2001). Progress in controlled/living polymerizations (CLP) in aqueous media. Part II. Conventional polymerization in aqueous media. *Polimery (Warsaw, Pol)*, *46*(9): 575–581.

172. Carmichael, A. J., et al. (2000). Copper(I) mediated living radical polymerisation in an ionic liquid. *Chem. Commun.*, *14*: 1237–1238.

173. Sarbu, T., and Matyjaszewski, K. (2001). ATRP of methyl methacrylate in the presence of ionic liquids with ferrous and cuprous anions. *Macromolecular Chemistry and Physics*, *202*(17): 3379–3391.

174. Arehart, S. V., and Matyjaszewski, K. (1999). Atom transfer radical copolymerization of styrene and n-butyl acrylate. *Macromolecules*, *32*(7): 2221–2231.

175. Moineau, G., et al. (2000). Synthesis of fully acrylic thermoplastic elastomers by atom transfer radical polymerization (ATRP). 2. Effect of the catalyst on the molecular control and the rheological properties of the triblock copolymers. *Macromol. Chem. Phys.*, *201*(11): 1108–1114.

176. Roos, S. G., Mueller, A. H. E., and Matyjaszewski, K. (1999). Copolymerization of n-butyl acrylate with methyl methacrylate and PMMA macromonomers: comparison of reactivity ratios in conventional and atom transfer radical copolymerization. *Macromolecules*, *32*(25): 8331–8335.

177. Woodworth, B. E., Metzner, Z., and Matyjaszewski, K. (1998). Copper triflate as a catalyst in atom transfer radical polymerization of styrene and methyl acrylate. *Macromolecules*, *31*(23): 7999–8004.

178. Teodorescu, M., Gaynor, S. G., and Matyjaszewski, K. (2000). Halide anions as ligands in iron-mediated atom transfer radical polymerization. *Macromolecules*, *33*(7): 2335–2339.

179. Matyjaszewski, K., et al. (2000). Catalytic processes for the controlled polymerization of free radically (co)polymerizable monomers and functional polymeric systems prepared thereby. In *PCT Int. Appl. WO 0056795*, Carnegie Mellon University, USA, 200 pp.

180. Boutevin, B. (2000). From telomerization to living radical polymerization. *J. Polym. Sci., Part A: Polym. Chem.*, *38*(18): 3235–3243.

181. Charmot, D., et al. (2000). Controlled radical polymerization in dispersed media. *Macromol. Symp.*, *150*(Polymers in Dispersed Media): 23–32.
182. Destarac, M., et al. (2001). Synthesis method for polymers by controlled radical polymerization with xanthates. In *PCT Int. Appl. WO 0142312*, Rhodia Chimie, Fr., 46 pp.
183. Gaynor, S. G., Wang, J.-S., and Matyjaszewski, K. (1995). Controlled radical polymerization by degenerative transfer: effect of the structure of the transfer agent. *Macromolecules*, *28*(24): 8051–8056.
184. Farcet, C., et al. (2000). Polystyrene-block-poly(butyl acrylate) and polystyrene-block-poly[(butyl acrylate)-co-styrene] block copolymers prepared via controlled free-radical miniemulsion polymerization using degenerative iodine transfer. *Macromol. Rapid Commun.*, *21*(13): 921–926.
185. Curran (1992). In Vol. 4 of Trost, B. M., and Fleming, I., eds. (1992). *Comprehensive Organic Synthesis: Selectivity, Strategy and Efficiency in Modern Organic Chemistry*, 9-volume set, 10400 pp.
186. Moad, G., et al. (2000). Living free radical polymerization with reversible addition-fragmentation chain transfer (the life of RAFT). *Polym. Int.*, *49*(9): 993–1001.
187. Tsavalas, J. G., et al. (2001). Living radical polymerization by reversible addition-fragmentation chain transfer in ionically stabilized miniemulsions. *Macromolecules*, *34*(12): 3938–3946.
188. Butte, A., Storti, G., and Morbidelli, M. (2001). Miniemulsion living free radical polymerization by RAFT. *Macromolecules*, *34*(17): 5885–5896.
189. Monteiro, M. J., and de Barbeyrac, J. (2001). Free-radical polymerization of styrene in emulsion using a reversible addition-fragmentation chain transfer agent with a low transfer constant: effect on rate, particle size, and molecular weight. *Macromolecules*, *34*(13): 4416–4423.
190. Kanagasabapathy, S., Sudalai, A., and Benicewicz, B. C. (2001). Reversible addition-fragmentation chain-transfer polymerization for the synthesis of poly(4-acetoxystyrene) and poly(4-acetoxystyrene)-block-polystyrene by bulk, solution and emulsion techniques. *Macromol. Rapid. Commun.*, *22*(13): 1076–1080.
191. Donovan, M. S., Stanford, T. A., Lowe, A. B., Sumerlin, B. S., Mitsukami, Y. and McCormick, C. L. (2002). RAFT polymerization of N,N-dimethylacrylamide in water. *Macromolecules 35*: 4570–4572.
192. de Brouwer, H., et al. (2000). Living radical polymerization in miniemulsion using reversible addition-fragmentation chain transfer. *Macromolecules*, *33*(25): 9239–9246.
193. Qin, S.-H., and Qiu, K.-Y. (2001). A facile method for synthesis of ABA triblock copolymers with macro-iniferter technique. *Polymer*, *42*(7): 3033–3042.
194. Greiner, A., et al. (2000). Method for synthesis of block polymers by controlled radical polymerization with the aid of thioether-thiones. In *Fr Demande 2794464*, Rhodia Chimie, Fr., 27 pp.
195. Moad, G., et al. (1996). Polymerization in aqueous media in the presence of cobalt complexes. In *PCT Int. Appl. WO 9615158*, E.I. Du Pont De Nemours and Company, USA; Commonwealth Scientific and Industrial Research Organization, 26 pp.
196. Cresidio, S. P., et al. (2001). Alkoxyamine-mediated 'living' radical polymerization: MS investigation of the early stages of styrene polymerization initiated by cumyl-TEISO. *J. Polym. Sci., Part A: Polym. Chem.*, *39*(8): 1232–1241.
197. Charleux, B. (2000). Theoretical aspects of controlled radical polymerization in a dispersed medium. *Macromolecules*, *33*(15): 5358–5365.
198. Harth, E., et al. (2001). Chain end function in nitroxide-mediated 'living' free radical polymerizations. *Macromolecules*, *34*(12): 3856–3862.
199. Matyjaszewski, K., and Greszta, D. (1998). Rate enhancement of nitroxyl radical-mediated polymerization with controlled growth steps for producing homo- and copolymer including block and graft copolymer. In *PCT Int. Appl. 9807758*, Carnegie-Mellon University, USA, 44 pp.

200. Baumann, M., and Scmidt-Naake, G. (2001). Acetic anhydride — accelerating agent for nitroxide-controlled free-radical copolymerization of styrene and acrylonitrile. *Macromolecular Chemistry and Physics, 202*(13): 2727–2731.

201. Matyjaszewski, K., et al. (1998). Controlled radical polymerization in the presence of oxygen. *Macromolecules, 31*(17): 5967–5969.

202. Kickelbick, G., Paik, H.-J., and Matyjaszewski, K. (1999). Immobilization of the copper catalyst in atom transfer radical polymerization. *Macromolecules, 32*(9): 2941–2947.

203. Haddleton, D. M., et al. (2000). Atom transfer polymerization mediated by solid-supported catalysts. *ACS Symp. Ser., 760*.

204. Hong, S. C., Paik, H.-J., and Matyjaszewski, K. (2001). An immobilized/soluble hybrid catalyst system for atom transfer radical polymerization. *Macromolecules, 34*(15): 5099–5102.

205. Shen, Y., Zhu, S., and Pelton, R. (2000). Packed column reactor for continuous atom transfer radical polymerization: methyl methacrylate polymerization using silica gel supported catalyst. *Macromol. Rapid Commun., 21*(14): 956–959.

206. Matyjaszewski, K., et al. (1997). Functional polymers by atom transfer radical polymerization. *Polym. Mater. Sci. Eng., 76*: 147–148.

207. Coessens, V., Pintauer, T., and Matyjaszewski, K. (2001). Functional polymers by atom transfer radical polymerization. *Prog. Polym. Sci., 26*(3): 337–377.

208. Destarac, M., et al. (2000). Dithiocarbamates as universal reversible addition-fragmentation chain transfer agents. *Macromol. Rapid Commun., 21*(15): 1035–1039.

209. Haddleton, D. M., et al. (1997). Identifying the nature of the active species in the polymerization of methacrylates: inhibition of methyl methacrylate homopolymerizations and reactivity ratios for copolymerization of methyl methacrylate/n-butyl methacrylate in classical anionic, alkyllithium/trialkylaluminum-initiated, group transfer polymerization, atom transfer radical polymerization, catalytic chain transfer, and classical free radical polymerization. *Macromolecules, 30*(14): 3992–3998.

210. Matyjaszewski, K., et al. (2000). Gradient copolymers by atom transfer radical copolymerization. *J. Phys. Org. Chem., 13*(12): 775–786.

211. Gray, M. K., et al. (2001). Gradient copolymerization of styrene and 4-acetoxystyrene via nitroxide-mediated controlled radical polymerization. *Polym. Prepr. (Am. Chem. Soc., Div. Polym. Chem.), 42*(2): 337–338.

212. Greszta, D., Matyjaszewski, K., and Pakula, T. (1997). Gradient copolymers of styrene and acrylonitrile via atom transfer radical polymerization. *Polym. Prepr., 38*(1): 709–710.

213. Zaremskii, M. Y., et al. (1997). Synthesis of compositionally homogeneous gradient copolymers of styrene with methyl acrylate by 'quasi-living' radical polymerization. *Vysokomol. Soedin, Ser. A Ser. B, 39*(8): 1286–1291.

214. Barclay, G. G., et al. (1998). The preparation and investigation of macromolecular architectures for microlithography by 'living' free radical polymerization. *ACS Symp. Ser., 706*(Micro- and Nanopatterning Polymers): 144–160.

215. Fukuda, T., et al. (1996). Well-defined block copolymers comprising styrene-acrylonitrile random copolymer sequences synthesized by 'living' radical polymerization. *Macromolecules, 29*(8): 3050–3052.

216. Baumert, M., and Mulhaupt, R. (1997). Carboxy-terminated homo- and copolymers of styrene using dicarboxylic acid-functional azo initiator and 2,2,6,6-tetramethyl-1-piperidyloxyl (TEMPO). *Macromol. Rapid Commun., 18*(9): 787–794.

217. Baethge, H., Butz, S., and Schmidt-Naake, G. (1997). Living free-radical copolymerization of styrene and N-vinylcarbazole. *Macromol. Rapid Commun., 18*(10): 911–916.

218. Pozzo, J.-L., et al. (1997). 'Living' free-radical polymerization process: a new approach towards well-defined photochromic (co)polymers. *Mol. Cryst. Liq. Cryst. Sci. Technol., Sect. A, 298*: 437–443.

219. Yoshida, E., (1996). Synthesis of a well-defined polybromostyrene by living radical polymerization with a nitroxyl radical. *J. Polym. Sci., Part A: Polym. Chem., 34*(14): 2937–2943.

220. Butz, S., Baethge, H., and Schmidt-Naake, G. (2000). N-oxyl mediated free radical donor-acceptor co- and terpolymerization of styrene, cyclic maleimide monomers and n-butyl methacrylate. *Macromol. Chem. Phys.*, *201*(16): 2143–2151.

221. Lokaj, J., Vlcek, P., and Kriz, J. (1999). Poly(styrene-co-N-butylmaleimide) macroinitiators by controlled autopolymerization and related block copolymers. *J. Appl. Polym. Sci.*, *74*(10): 2378–2385.

222. Bignozzi, M. C., et al. (1999). Lithographic results of electron beam photoresists prepared by living free radical polymerization. *Polym. Bull.*, *43*(1): 93–100.

223. Cassebras, M., et al. (1999). Synthesis of di- and triblock copolymers of styrene and butyl acrylate by controlled atom transfer radical polymerization. *Macromol. Rapid Commun.*, *20*(5): 261–264.

224. Zou, Y.-S., et al. (1998). Atom transfer radical copolymerization of styrene and butyl methacrylate. *Hecheng Huaxue*, *6*(1): 1–3, 7.

225. de la Fuente, J. L., et al. (2001). Sequence distribution and stereoregularity of methyl methacrylate and butyl acrylate statistical copolymers synthesized by atom transfer radical polymerization. *Macromolecules*, *34*(17): 5833–5837.

226. Kotani, Y., Kamigaito, M., and Sawamoto, M. (1998). Living random copolymerization of styrene and methyl methacrylate with a Ru(II) complex and synthesis of ABC-type 'block-random' copolymers. *Macromolecules*, *31*(17): 5582–5587.

227. Matyjaszewski, K., et al. (1995). 'Living' and controlled radical polymerization. *J. Phys. Org. Chem.*, *8*(4): 306–315.

228. Jones, R. G., Yoon, S., and Nagasaki, Y. (1999). Facile synthesis of epoxystyrene and its copolymerizations with styrene by living free radical and atom transfer radical strategies. *Polymer*, *40*(9): 2411–2418.

229. Jiang, X., et al. (2000). Atom transfer radical copolymerization of methyl methacrylate with N-cyclohexylmaleimide. *Polym. Int.*, *49*(8): 893–897.

230. Chen, G.-Q., et al. (2000). Synthesis of alternating copolymers of N-substituted maleimides with styrene via atom transfer radical polymerization. *Macromolecules*, *33*(2): 232–234.

231. Pierik, B., Masclee, D., and Van Herk, A. (2001). Catalytic chain transfer copolymerization of methyl methacrylate and methyl acrylate. *Macromol. Symp.*, *165*(Developments in Polymer Synthesis and Characterization): 19–27.

232. Suddaby, K. G., Hunt, K. H., and Haddleton, D. M. (1996). MALDI-TOF mass spectrometry in the study of statistical copolymerizations and its application in examining the free radical copolymerization of methyl methacrylate and n-butyl methacrylate. *Macromolecules*, *29*(27): 8642–8649.

233. Greszta, D. (1997). *Synthesis of novel gradient copolymers via atom transfer radical polymerization.* PhD thesis CMU, 276 pp.

234. Matyjaszewski, K., Greszta, D., and Pakula, T. (1997). Thermal properties of gradient copolymers and their compatibilizing ability. *Polym. Prepr.*, *38*(1): 707–708.

235. Zhu, M., et al. (2001). Radical reverse addition-fragmentation chain transfer polymerization of maleic anhydride with styrene and new functional block copolymer prepared therefrom. *Gaofenzi Xuebao*, (3): 415–417.

236. Coca, S., and Matyjaszewski, K. (1996). Alternating copolymers of methyl acrylate with isobutene and isobutyl vinyl ether using ATRP. Book of Abstracts, 211th ACS National Meeting, New Orleans, LA, March 24–28, p. OLY-087.

237. Zhu, M., et al. (2001). Controlled radical copolymerization of iso-butyl vinyl ether with electron-withdrawing monomers. *Gaofenzi Xuebao*, (3): 418–421.

238. Matyjaszewski, K., et al. (2000). Simple and effective one-pot synthesis of (meth)acrylic block copolymers through atom transfer radical polymerization. *J. Polym. Sci., Part A: Polym. Chem.*, *38*(11): 2023–2031.

239. Muehlebach, A., Gaynor, S. G., and Matyjaszewski, K. (1998). Synthesis of amphiphilic block copolymers by atom transfer radical polymerization (ATRP). *Macromolecules*, *31*(18): 6046–6052.

240. Muller, A. H. E., et al. (2001). Unimolecular amphipolar nanocylinders via a 'grafting from' process using ATRP. *Polym. Mat. Sci. Eng., 84*: 91–92.

241. Davis, K. A., and Matyjaszewski, K. (2001). ABC triblock copolymers prepared using atom transfer radical polymerization techniques. *Macromolecules*, *34*(7): 2101–2107.

242. Bertin, D., and Boutevin, B. (1996). Controlled radical polymerization. Synthesis of chloromethylstyrene/styrene block copolymers. *Polym. Bull.*, *37*(3): 337–344.

243. Lacroix-Desmazes, P., et al. (2000). Synthesis of poly(chloromethylstyrene-b-styrene) block copolymers by controlled free-radical polymerization. *J. Polym. Sci., Part A: Polym. Chem.*, *38*(21): 3845–3854.

244. Yoshida, E., and Fuji, T. (1997). Synthesis of well-defined polychlorostyrenes by living radical polymerization with 4-methoxy-2,2,6,6-tetramethylpiperidine-1-oxyl. *J. Polym. Sci., Part A: Polym. Chem.*, *35*(12): 2371–2378.

245. Ohno, K., et al. (1998). Nitroxide-controlled free-radical polymerization of para-tert-butoxystyrene. Kinetics and applications. *Macromol. Chem. Phys.*, *199*(2): 291–297.

246. Mariani, M., et al. (1999). Diblock and triblock functional copolymers by controlled radical polymerization. *J. Polym. Sci., Part A: Polym. Chem.*, *37*(9): 1237–1244.

247. Wan, X., et al. (1998). 'Living' free radical synthesis of novel rodcoil diblock copolymers with polystyrene and mesogen-jacketed liquid crystal polymer segments. *Chin. J. Polym. Sci.*, *16*(4): 377–380.

248. Bignozzi, M. C., Ober, C. K., and Laus, M. (1999). Liquid-crystalline side chain-coil diblock copolymers by living free radical polymerization. *Macromol. Rapid Commun.*, *20*(12): 622–627.

249. Gabaston, L. I., et al. (1999). Direct synthesis of novel acidic and zwitterionic block copolymers via TEMPO-mediated living free-radical polymerization. *Polymer*, *40*(16): 4505–4514.

250. Nowakowska, M., Zapotoczny, S., and Karewicz, A. (2000). Synthesis of poly(sodium styrenesulfonate-block-vinylnaphthalene) by nitroxide-mediated free radical polymerization. *Macromolecules*, *33*(20): 7345–7348.

251. Listigovers, N. A., et al. (1996) Narrow-polydispersity diblock and triblock copolymers of alkyl acrylates by a 'living' stable free radical polymerization. *Macromolecules*, *29*(27): 8992–8993.

252. Zaremski, M. Y., et al. (1999). Utilization of nitroxide-mediated polymerization for synthesis of block copolymers. *Russ. Polym. News*, *4*(1): 17–21.

253. Bohrisch, J., Wendler, U., and Jaeger, W. (1997). Controlled radical polymerization of 4-vinylpyridine. *Macromol. Rapid Commun.*, *18*(11): 975–982.

254. Barbosa, C. A., and Gomes, A. S. (1998). Living tandem free-radical polymerization of a liquid-crystalline monomer. *Polym. Bull.*, *41*(1): 15–20.

255. Kobatake, S., et al. (1998). Block copolymer synthesis by styrene polymerization initiated with nitroxy-functionalized polybutadiene. *Macromolecules*, *31*(11): 3735–3739.

256. Benoit, D., et al. (2000). Accurate structural control and block formation in the living polymerization of 1,3-dienes by nitroxide-mediated procedures. *Macromolecules*, *33*(2): 363–370.

257. Benoit, D., et al. (2000). One-step formation of functionalized block copolymers. *Macromolecules*, *33*(5): 1505–1507.

258. Stehling, U. M., et al. (1998). Synthesis of poly(olefin) graft copolymers by a combination of metallocene and 'living' free radical polymerization techniques. *Macromolecules*, *31*(13): 4396–4398.

259. Stalmach, U., et al. (2000). Semiconducting diblock copolymers synthesized by means of controlled radical polymerization techniques. *J. Am. Chem. Soc.*, *122*(23): 5464–5472.

260. Zou, Y., et al. (1998). Controlled sequential and random radical copolymerization of styrene and butyl methacrylate by atom transfer radical polymerization. *Chin. J. React. Polym.*, *7*(2): 75–79.

261. Shipp, D. A., Wang, J.-L., and Matyjaszewski, K. (1998). Synthesis of acrylate and methacrylate block copolymers using atom transfer radical polymerization. *Macromolecules*, *31*(23): 8005–8008.

262. Qin, D.-Q., Qin, S.-H., and Qiu, K.-Y. (2000). A reverse ATRP process with a hexasubstituted cethane thermal iniferter diethyl 2,3-dicyano-2,3-di(p-tolyl)succinate as the initiator. *Macromolecules*, *33*(19): 6987–6992.

263. Pyun, J., and Matyjaszewski, K. (2000). Synthesis of hybrid polymers using atom transfer radical polymerization: homopolymers and Block Copolymers from polyhedral oligomeric silsesquioxane monomers. *Macromolecules*, *33*(1): 217–220.

264. Davis, K. A., Charleux, B., and Matyjaszewski, K. (2000). Preparation of block copolymers of polystyrene and poly(t-butyl acrylate) of various molecular weights and architectures by atom transfer radical polymerization. *J. Polym. Sci., Part A: Polym. Chem.*, *38*(12): 2274–2283.

265. Burguiere, C., et al. (2001). Block copolymers of poly(styrene) and poly(acrylic acid) of various molar masses, topologies, and compositions prepared via controlled/living radical polymerization. Application as stabilizers in emulsion polymerization. *Macromolecules*, *34*(13): 4439–4450.

266. Wang, G., and Yan, D. (2001). Preparation of amphiphilic PS-b-PMAA diblock copolymer by means of atom transfer radical polymerization. *J. Appl. Polym. Sci.*, *82*(10): 2381–2386.

267. Wang, X.-S., Luo, N., and Ying, S.-K. (1999). Controlled radical polymerization of methacrylates at ambient temperature and the synthesis of block copolymers containing methacrylates. *Polymer*, *40*(14): 4157–4161.

268. Xia, J., Zhang, X., and Matyjaszewski, K. (1999). Atom transfer radical polymerization of 4-vinylpyridine. *Macromolecules*, *32*(10): 3531–3533.

269. Ohno, K., Tsujii, Y., and Fukuda, T. (1998). Synthesis of a well-defined glycopolymer by atom transfer radical polymerization. *J. Polym. Sci., Part A: Polym. Chem.*, *36*(14): 2473–2481.

270. Li, Z.-C., et al. (2000). Synthesis of amphiphilic block copolymers with well-defined glycopolymer segment by atom transfer radical polymerization. *Macromol. Rapid Commun.*, *21*(7): 375–380.

271. Furlong, S. A., and Armes, S. P. (2000). Facile synthesis of hydrophilic-hydrophilic copolymers via atom transfer radical polymerization. *Polym. Prepr. (Am. Chem. Soc., Div. Polym. Chem.)*, *41*(1): 450–451.

272. Haddleton, D. M., Jackson, S. G., and Bon, S. A. F. (2000). Copper(I)-mediated living radical polymerization under fluorous biphasic conditions. *J. Am. Chem. Soc.*, *122*(7): 1542–1543.

273. Alipour, M., et al. (2001). Living radical polymerization of methylmethacrylate, methylacrylate and their block copolymers with acrylonitrile by atom transfer radical polymerization. *Iran Polym. J.*, *10*(2): 99–106.

274. Matyjaszewski, K., et al. (2000). Water-borne block and statistical copolymers synthesized using atom transfer radical polymerization. *Macromolecules*, *33*(7): 2296–2298.

275. Gravano, S., et al. (2002). Poly(4-(aminomethyl)styrene)-b-polystyrene: Synthesis and unilamellar vesicle formation. *Langmuir 18*: 1938–1941.

276. Nakagawa, Y., Miller, P. J., and Matyjaszewski, K. (1998). Development of novel attachable initiators for atom transfer radical polymerization. Synthesis of block and graft copolymers from poly(dimethylsiloxane) macroinitiators. *Polymer*, *39*(21): 5163–5170.

277. Bielawski, C. W., Morita, T., and Grubbs, R. H. (2000). Synthesis of ABA triblock copolymers via a tandem ring-opening metathesis polymerization: atom transfer radical polymerization approach. *Macromolecules*, *33*(3): 678–680.

278. Coca, S., and Matyjaszewski, K. (1997). Block copolymers by transformation of 'living' carbocationic into 'living' radical polymerization. *Polym. Prepr. (Am. Chem. Soc., Div. Polym. Chem.) 38*(1): 693–694.

279. Acar, M. H., and Matyjaszewski, K. (1999). Block copolymers by transformation of living anionic polymerization into controlled/'living' atom transfer radical polymerization. *Macromol. Chem. Phys., 200*(5): 1094–1100.

280. Keszler, B., Fenyvesi, G., and Kennedy, J. P. (2000). Novel star-block polymers: three polyisobutylene-b-poly(methyl methacrylate) arms radiating from an aromatic core. *J. Polym. Sci., Part A: Polym. Chem., 38*(4): 706–714.

281. Lu, J., et al. (2001). Synthesis of benzyl chloride capped beta-pinene macroinitiator and block copolymers. *Gaofenzi Xuebao*, (3): 357–360.

282. Yildirim, T. G., et al. (1999). Synthesis of block copolymers by transformation of photosensitized cationic polymerization to stable free radical polymerization. *Polymer, 40*(13): 3885–3890.

283. Yoshida, E., and Sugita, A. (1998). Synthesis of poly(styrene-b-tetrahydrofuran-b-styrene) triblock copolymers by transformation from living cationic into living radical polymerization using 4-hydroxy-2,2,6,6-tetramethylpiperidine-1-oxyl as a transforming agent. *J. Polym. Sci., Part A: Polym. Chem., 36*(12): 2059–2068.

284. Yagci, Y., Duez, A. B., and Oenen, A. (1997). Controlled radical polymerization initiated by stable radical terminated polytetrahydrofuran. *Polymer, 38*(11): 2861–2863.

285. Guo, Y.-M., Pan, C.-Y., and Wang, J. (2001). Block and star block copolymers by mechanism transformation. VI. Synthesis and characterization of A4B4 miktoarm star copolymers consisting of polystyrene and polytetrahydrofuran prepared by cationic ring-opening polymerization and atom transfer radical polymerization. *J. Polym. Sci., Part A: Polym. Chem., 39*(13): 2134–2142.

286. Guo, Y.-M., and Pan, C.-Y. (2001). Block and star block copolymers by mechanism transformation. Part V. Syntheses of polystyrene/polytetrahydrofuran A2B2 miktoarm star copolymers by transformation of CROP into ATRP. *Polymer, 42*(7): 2863–2869.

287. Acar, M. H., et al. (2000). Synthesis of block copolymer by combination of living cationic and iniferter polymerization systems. *Polymer, 41*(18): 6709–6713.

288. Liu, Y., Ying, S., and Wan, X. (1999). Synthesis of block copolymers via transformation of living free radical polymerization into living cationic ring-opening polymerization. *Polym. Prepr, 40*(2): 1053–1054.

289. Xu, Y., and Pan, C. (2000). Block and star block copolymers by mechanism transformation. I. Synthesis of PTHF-PSt-PTHF by the transformation of ATRP into CROP. *J. Polym. Sci., Part A: Polym. Chem., 38*(2): 337–344.

290. Xu, Y., Pan, C., and Tao, L. (2000). Block and star block copolymers by mechanism transformation. II. Synthesis of poly(DOP-b-St) by combination of ATRP and CROP. *J. Polym. Sci., Part A: Polym. Chem., 38*(3): 436–443.

291. Guo, Y.-M., Xu, J., and Pan, C.-Y. (2001). Block and star block copolymers by mechanism transformation. IV. Synthesis of S-(PSt)2(PDOP)2 miktoarm star copolymers by combination of ATRP and CROP. *J. Polym. Sci., Part A: Polym. Chem., 39*(3): 437–445.

292. Miura, Y., et al. (1999). High-yield synthesis of functionalized alkoxyamine initiators and approach to well-controlled block copolymers using them. *Macromolecules, 32*(25): 8356–8362.

293. Hua, F. J., and Yang, Y. L. (2000). Synthesis of block copolymer by 'living' radical polymerization of styrene with nitroxyl-functionalized poly(ethylene oxide). *Polymer, 42*(4): 1361–1368.

294. Cheng, C., and Yang, N.-L. (2001). Advances of well-defined alkoxyamine-based polyfunctional macroinitiators for structural control of graft polymers. *Polym. Mater. Sci. Eng., 85*: 512–513.

295. Liu, F., et al. (1999). The synthesis of block copolymer through the combination of living anionic polymerization and controlled radical polymerization. *Polym. Prepr.,* *40*(2): 1032–1033.

296. Tong, J.-D., Ni, S., and Winnik, M. A. (2000). Synthesis of polyisoprene-b-polystyrene block copolymers bearing a fluorescent dye at the junction by the combination of living anionic polymerization and atom transfer radical polymerization. *Macromolecules, 33*(5): 1482–1486.

297. Tong, J.-D., et al. (2001). Synthesis of meth(acrylate) diblock copolymers bearing a fluorescent dye at the junction using a hydroxyl-protected initiator and the combination of anionic polymerization and controlled radical polymerization. *Macromolecules, 34*(4): 696–705.

298. Yoshida, E., and Osagawa, Y. (1998). Synthesis of poly(ε-caprolactone) with a stable nitroxyl radical as an end-functional group and its application to a counter radical for living radical polymerization. *Macromolecules, 31*(5): 1446–1453.

299. Wang, Y., Chen, S., and Huang, J. (1999). Synthesis and characterization of a novel macroinitiator of poly(ethylene oxide) with a 4-hydroxy-2,2,6,6-tetramethylpiperidinyloxy end group: initiation of the polymerization of styrene by a 'living' radical mechanism. *Macromolecules, 32*(8): 2480–2483.

300. Miller, P. J., and Matyjaszewski, K. (1999). Atom transfer radical polymerization of (meth)acrylates from poly(dimethylsiloxane) macroinitiators. *Macromolecules, 32*(26): 8760–8767.

301. Gnanou, Y., and Taton, D. (2001). Stars and dendrimer-like architectures by the divergent method using controlled radical polymerization. *Macromol. Symp., 174*(Polymerization Processes and Polymer Materials I): 333–341.

302. Grubbs, R. H., and Bielawski, C. Triblock and diblock copolymers prepared from telechelic ROMP polymers. *WO 2000055218,* California Institute of Technology, USA.

303. Ranger, M., et al. (2001). From well-defined diblock copolymers prepared by a versatile atom transfer radical polymerization method to supramolecular assemblies. *J. Polym Sci., Part A: Polym. Chem., 39*(22): 3861–3874.

304. Fischer, A., Brembilla, A., and Lochon, P. (2000). Synthesis of new amphiphilic cationic block copolymers and study of their behavior in aqueous medium as regards hydrophobic microdomain formation. *Polymer, 42*(4): 1441–1448.

305. Visger, D. C., et al. (1998). Block copolymers prepared by stabilized free-radical polymerization. In *Eur. Pat. Appl. 887362,* The Lubrizol Corporation, USA, 12 pp.

306. Zhu, L., et al. (2000). Synthesis and solution properties of anionic linear-dendritic block amphiphiles. *J. Polym. Sci., Part A: Polym. Chem., 38*(23): 4282–4288.

307. Hawker, C. J., et al. (1995). Preparation of hyperbranched and star polymers by a 'living', self-condensing free radical polymerization. *J. Am. Chem. Soc., 117*(43): 10763–10764.

308. Grubbs, R. B., et al. (1997). A tandem approach to graft and dendritic graft copolymers based on 'living' free radical polymerizations. *Angew. Chem., Int. Ed. Engl., 36*(3): 270–272.

309. Beers, K. L., et al. (1998). The synthesis of densely grafted copolymers by atom transfer radical polymerization. *Macromolecules, 31*(26): 9413–9415.

310. Sun, Y., Wan, D., and Huang, J. (2001). Preparation of a copolymer of methyl methacrylate and 2-(dimethylamino)ethyl methacrylate with pendant 4-benzyloxy-2,2,6,6-tetramethyl-1-piperidinyloxy and its initiation of the graft polymerization of styrene by a controlled radical mechanism. *J. Polym. Sci., Part A: Polym. Chem., 39*(5): 604–612.

311. Miwa, Y., et al. (1999). Living radical graft polymerization of styrene to polypropylene with 2,2,6,6-tetramethylpiperidinyl-1-oxy. *Macromolecules, 32*(24): 8234–8236.

312. Fonagy, T., Ivan, B., and Szesztay, M. (1998). Polyisobutylene-graft-polystyrene by quasi-living atom-transfer radical polymerization of styrene from poly(isobutylene-co-p-methyl-styrene-co-p-bromomethylstyrene). *Macromol. Rapid Commun., 19*(9): 479–483.

313. Fonagy, T., Ivan, B., and Szesztay, M. (1998). Poly(isobutylene-g-styrene) graft copolymers by quasiliving atom transfer radical grafting of styrene. *Polym. Mater. Sci. Eng., 79*: 3–4.

314. Pan, Q., et al. (1999). Synthesis and characterization of block-graft copolymers composed of poly(styrene-b-ethylene-co-propylene) and poly(ethyl methacrylate) by atom transfer radical polymerization. *J. Polym. Sci., Part A: Polym. Chem.*, 37(15): 2699–2702.

315. Guerrini, M. M., Charleux, B., and Vairon, J.-P. (2000). Functionalized latexes as substrates for atom transfer radical polymerization. *Macromol. Rapid Commun.*, 21(10): 669–674.

316. Liu, B., Yuan, C. G., and Hu, C. P. (2001). Synthesis and characterization of a well-defined graft copolymer containing fluorine grafts by 'living'/controlled radical polymerization. *Macromol. Chem. Phys.*, 202(12): 2504–2508.

317. Liu, B., et al. (1999). Well-defined multi-branch copolymer synthesis by 'living' free radical polymerization. *Hecheng Xiangjiao Gongye*, 22(6): 373.

318. Liu, S., and Sen, A. (2001). Syntheses of polyethylene-based graft copolymers by atom transfer radical polymerization. *Macromolecules*, 34: 1529–1532.

319. Hong, S. C., Pakula, T., and Matyjaszewski, K. (2001). Preparation of polyisobutylene graft copolymers with different compositions and side chain architectures through ATRP. *Polym. Sci. Eng.*, 84: 767–768.

320. Jankova, K., et al. (1998). Synthesis of amphiphilic PS-b-PEG-b-PS by atom transfer radical polymerization. *Macromolecules*, 31(2): 538–541.

321. Jankova, K., et al. (1999). Controlled/'living' atom transfer radical polymerization of styrene in the synthesis of amphiphilic diblock copolymers from a poly(ethylene glycol) macro-initiator. *Polym. Bull.*, 42(2): 153–158.

322. Wang, X.-S., et al. (1999). The synthesis of ABA block copolymers by means of 'living'/ controlled radical polymerization using hydroxyl-terminated oligomers as precursor. *Eur. Polym. J.*, 36(1): 149–156.

323. Huang, C., Wan, X., and Ying, S. (2000). Synthesis of poly(propylene glycol-g-styrene) from atom transfer radical polymerization. *Gaofenzi Xuebao*, 4: 467–471.

324. Yoshida, E., and Nakamura, M. (1998). Synthesis of poly(ethylene adipate) with a stable nitroxyl radical at both chain ends, and applications to a counter radical for living radical polymerization. *Polym. J.*, 30(11): 915–920.

325. Haddleton, D. M., and Ohno, K. (2000). Well-defined oligosaccharide-terminated polymers from living radical polymerization. *Biomacromolecules*, 1(2): 152–156.

326. Matyjaszewski, K., et al. (1998). Synthesis of block, graft and star polymers from inorganic macroinitiators. *Appl. Organomet. Chem.*, 12(10/11): 667–673.

327. Jankova, K., et al. (1999). Synthesis by ATRP of poly(ethylene-co-butylene)-block-polystyrene, poly(ethylene-co-butylene)-block-poly(4-acetoxystyrene), and its hydrolysis product poly(ethylene-co-butylene)-block-poly(hydroxystyrene). *Macromol. Rapid Commun.*, 20(4): 219–223.

328. De Brouwer, H., et al. (2000). Controlled radical copolymerization of styrene and maleic anhydride and the synthesis of novel polyolefin-based block copolymers by reversible addition-fragmentation chain-transfer (RAFT) polymerization. *J. Polym. Sci., Part A: Polym. Chem.*, 38(19): 3596–3603.

329. Lutsen, L., et al. (1998). Poly(methylphenylsilylene)-block-polystyrene copolymer prepared by the use of a chloromethylphenyl end-capped poly(methylphenylsilylene) as a macromolecular initiator in an atom transfer radical polymerization of styrene. *Eur. Polym. J.*, 34(12): 1829–1837.

330. Tsolakis, P. K., Koulouri, E. G., and Kallitsis, J. K. (1999). Synthesis of rigid-flexible triblock copolymers using atom transfer radical polymerization. *Macromolecules*, 32(26): 9054–9058.

331. de Boer, B., et al. (2001). Supramolecular self-assembly and opto-electronic properties of semiconducting block copolymers. *Polymer*, 42(21): 9097–9109.

332. Li, I. Q., et al. (1997). Block copolymer preparation using sequential normal/living radical polymerization techniques. *Macromolecules*, 30(18): 5195–5199.

333. Yoshida, E., and Tanimoto, S. (1997). Living radical polymerization of styrene by a stable nitroxyl radical and macroazoinitiator. *Macromolecules*, 30(14): 4018–4023.

334. Paik, H.-J., et al. (1999). Block copolymerizations of vinyl acetate by combination of conventional and atom transfer radical polymerization. *Macromolecules, 32*(21): 7023–7031.

335. Destarac, M., and Boutevin, B. (1999). Use of a trichloromethyl-terminated azo initiator to synthesize block copolymers by consecutive conventional radical polymerization and ATRP. *Macromol. Rapid Commun., 20*(12): 641–645.

336. Destarac, M., Pees, B., and Boutevin, B. (2000). Radical telomerization of vinyl acetate with chloroform. Application to the synthesis of poly(vinyl acetate)-block-polystyrene copolymers by consecutive telomerization and atom transfer radical polymerization. *Macromol. Chem. Phys., 201*(11): 1189–1199.

337. Zhang, Z., Ying, S., and Shi, Z. (1999). Synthesis of fluorine-containing block copolymers via ATRP. 1. Synthesis and characterization of PSt-PVDF-PSt triblock copolymers. *Polymer, 40*(5): 1341–1345.

338. Alkan, S., et al. (1999). Block copolymers of thiophene-capped poly(methyl methacrylate) with pyrrole. *J. Polym. Sci., Part A: Polym. Chem., 37*(22): 4218–4225.

339. Leduc, M. R., et al. (1996). Dendritic initiators for 'living' radical polymerizations: a versatile approach to the synthesis of dendritic-linear block copolymers. *J. Am. Chem. Soc., 118*(45): 11111–11118.

340. Emrick, T., Hayes, W., and Frechet, J. M. J. (1999). A TEMPO-mediated 'living' free-radical approach to ABA triblock dendritic linear hybrid copolymers. *J. Polym. Sci., Part A: Polym. Chem., 37*(20): 3748–3755.

341. Hovestad, N. J., et al. (2000). Copper(I) bromide/N-(n-octyl)-2-pyridylmethanimine-mediated living-radical polymerization of methyl methacrylate using carbosilane dendritic initiators. *Macromolecules, 33*(11): 4048–4052.

342. Maier, S., et al. (2000). Synthesis of poly(glycerol)-block-poly(methyl acrylate) multi-arm star polymers. *Macromol. Rapid Commun., 21*(5): 226–230.

343. Ejaz, M., et al. (1998). Controlled graft polymerization of methyl methacrylate on silicon substrate by the combined use of the Langmuir-Blodgett and atom transfer radical polymerization techniques. *Macromolecules, 31*(17): 5934–5936.

344. Husseman, M., et al. (1999). Controlled synthesis of polymer brushes by 'living' free radical polymerization techniques. *Macromolecules, 32*(5): 1424–1431.

345. Hawker, C. J., et al. (1997). Living free radical polymerization of macromonomers. Preparation of well defined graft copolymers. *Macromol. Chem. Phys., 198*(1): 155–166.

346. Wang, Y., and Huang, J. (1998). Controlled radical copolymerization of styrene and the macromonomer of PEO with a methacryloyl end group. *Macromolecules, 31*(13): 4057–4060.

347. Matyjaszewski, K., et al. (1998). Hydrogels by atom transfer radical polymerization. I. poly(N-vinylpyrrolidinone-g-styrene) via the macromonomer method. *J. Polym. Sci., Part A: Polym. Chem., 36*(5): 823–830.

348. Roos, S. G., Muller, A. H. E., and Matyjaszewski, K. (2000). Copolymerization of n-butyl acrylate with methyl methacrylate and PMMA macromonomers by conventional and atom transfer radical copolymerization. *ACS Symp. Ser., 768*: 361–371.

349. Yamada, K., et al. (1999). Atom transfer radical polymerization of poly(vinyl ether) macromonomers. *Macromolecules, 32*(2): 290–293.

350. Shinoda, H., Miller, P. J., and Matyjaszewski, K. (2001). Improving the structural control of graft copolymers by combining ATRP with the macromonomer method. *Macromolecules, 34*(10): 3186–3194.

351. Shinoda, H., and Matyjaszewski, K. (2001). Improving the structural control of graft copolymers. Copolymerization of poly(dimethylsiloxane) macromonomer with methyl methacrylate using RAFT polymerization. *Macromolecular Rapid Communications, 22*(14): 1176–1181.

352. Shinoda, H., Matyjaszewski, K., Okrasa, L., Mierzwa, M., and Pakula, T. (2003). Structural control of poly(methyl methacrylate)-g-poly(dimethylsiloxane) copolymers using controlled radical polymerization: effect of the molecular structure on morphology and mechanical properties. *Macromolecules, 36*: 4772–4778.

353. Gaynor, S. G., Edelman, S., and Matyjaszewski, K. (1996). Synthesis of branched and hyperbranched polystyrenes. *Macromolecules, 29*(3): 1079–1081.

354. Ishizu, K., and Mori, A. (2000). Synthesis of hyperbranched polymers by self-addition free radical vinyl polymerization of photo functional styrene. *Macromol. Rapid Commun., 21*(10): 665–668.

355. Cheng, G., et al. (2001). Amphiphilic cylindrical core-shell brushes via a 'grafting from' process using ATRP. *Macromolecules, 34*(20): 6883–6888.

356. Boerner, H. G., et al. (2001). Synthesis of molecular brushes with block copolymer side chains using atom transfer radical polymerization. *Macromolecules, 34*(13): 4375–4383.

357. Husemann, M., et al. (2000). Manipulation of surface properties by patterning of covalently bound polymer brushes. *J. Am. Chem. Soc., 122*: 1844–1845.

358. Lindsley, C. W., et al. (2000). Rasta silanes: new silyl resins with novel macromolecular architecture via living free radical polymerization. *J. Comb. Chem., 2*(5): 550–559.

359. Matyjaszewski, K., et al. (1999). Polymers at interfaces: using atom transfer radical polymerization in the controlled growth of homopolymers and block copolymers from silicon surfaces in the absence of untethered sacrificial initiator. *Macromolecules, 32*(26): 8716–8724.

360. Bottcher, H., et al. (2000). ATRP grafting from silica surface to create first and second generation of grafts. *Polym. Bull., 44*(2): 223–229.

361. Kim, J.-B., et al. (2003). Kinetics of surface-initiated atom transfer radical polymerization. *J. Polym. Sci., Part A: Polym. Chem. 41*: 386–394.

362. Kong, X., et al. (2001). Amphiphilic polymer brushes grown from the silicon surface by atom transfer radical polymerization. *Macromolecules, 34*(6): 1837–1844.

363. Zhao, B., and Brittain, W. J. (2000). Synthesis, characterization, and properties of tethered polystyrene-b-polyacrylate brushes on flat silicate substrates. *Macromolecules, 33*(23): 8813–8820.

364. Baum, M., and Brittain, W. J. (2002). Synthesis of polymer brushes on silicate substrate via reversible addition fragmentation chain transfer technique. *Macromolecules, 35*: 610–615.

365. Zhao, B., et al. (2000). Nanopattern formation from tethered PS-b-PMMA brushes upon treatment with selective solvents. *J. Am. Chem. Soc., 122*(10): 2407–2408.

366. McGall, G., (2001). Macromolecular arrays on polymeric brushes and methods for preparing the same. In *Eur. Pat. Appl. 1081163*, Affymetrix, Inc., USA, 32 pp.

367. Kim, J.B., Bruening, G.L., and Baker, G.L. (2001). Synthesis of triblock copolymers on gold by surface-initiated atom transfer radical polymerization. *Polym. Mat. Sci. Eng., 84*: 700.

368. Matyjaszewski, K., et al. (2000). Organic-inorganic hybrid polymers from atom transfer radical polymerization and poly(dimethylsiloxane). *ACS Symp. Ser., 729* (Silicones and Silicone-Modified Materials).

369. Whitehouse, R. S., et al. (1999). Particles having an attached stable free radical, polymerized modified particles, and their manufacture. In *PCT Int. Appl.*, Cabot Corporation, USA, 58 pp.

370. Becker, M. L., Liu, J., and Wooley, K. L. (2001). Preparation of tritrpticin block copolymer bio-conjugates by fluorinated nitroxide mediated radical polymerization on solid support. Abstracts of Papers, 222nd ACS National Meeting, Chicago, IL, United States, August 26–30, 2001, p. OLY-017.

371. Pyun, J., and Matyjaszewski, K. (2001). Synthesis of nanocomposite organic/inorganic hybrid materials using controlled/'living' radical polymerization. *Chem. Mater., 13*(10): 3436–3448.

372. Von Werne, T., and Patten, T. E. (1999). Preparation of structurally well-defined polymer-nanoparticle hybrids with controlled/living radical polymerizations. *J. Am. Chem. Soc., 121*(32): 7409–7410.

373. von Werne, T., and Patten, T. E. (2001) Atom transfer radical polymerization from nanoparticles: a tool for the preparation of well-defined hybrid nanostructures and for understanding the chemistry of controlled/'living' radical polymerizations from surfaces. *Journal of the American Chemical Society, 123*(31): 7497–7505.

374. Carrot, G., et al. (2001). Atom transfer radical polymerization of n-butyl acrylate from silica nanoparticles. *J. Polym. Sci., Part A: Polym. Chem.*, *39*(24): 4294–4301.

375. Farmer, S. C., and Patten, T. E. (2000). Synthesis of luminescent organic/inorganic polymer nanocomposites. *Polym. Mater. Sci. Eng.*, *82*: 237–238.

376. Xia, J., Zhang, X., and Matyjaszewski, K. (1999). Synthesis of star-shaped polystyrene by atom transfer radical polymerization using an 'arm first' approach. *Macromolecules*, *32*(13): 4482–4484.

377. Baek, K.-Y., Kamigaito, M., and Sawamoto, M. (2001). Core-functionalized star polymers by transition metal-catalyzed living radical polymerization. 1. Synthesis and characterization of star polymers with PMMA arms and amide cores. *Macromolecules*, *34*(22): 7629–7635.

378. Zhu, Y., et al. (1998). Nitroxide-mediated radical polymerization: end-group analysis. *ACS Symp. Ser.*, *685*(*Controlled Radical Polymerization*): 214–224.

379. Gridnev, A. A., Simonsick, W. J., Jr., and Ittel, S. D. (2000). Synthesis of telechelic polymers initiated with selected functional groups by catalytic chain transfer. *J. Polym. Sci., Part A: Polym. Chem.*, *38*(10): 1911–1918.

380. Kusakabe, M., and Kitano, K. (1997). Preparing (meth)acrylic polymers having functional groups at the chain ends. In *Eur. Pat. Appl. EP 789036*, Kaneka Corporation, Japan, 24 pp.

381. Nakagawa, Y., et al. (2000). Hydroxy-terminated vinyl polymer compositions and their uses. In *Jpn Kokai Tokkyo Koho 2000053723*, Kanegafuchi Chemical Industry Co., Ltd., Japan, 27 pp.

382. Carter, K. R., et al. (2000). Process for manufacture of integrated circuit device using organosilicate insulative matrices. In *US 6143643*, International Business Machines Corp., USA, 7 pp.

383. Fujita, M., Hasegawa, N., and Nakagawa, Y. (2000). Alkenyl-terminated vinyl polymer adhesive compositions with improved adhesion. In *Jpn Kokai Tokkyo Koho 2000154347*, Kanegafuchi Chemical Industry Co., Ltd., Japan, 16 pp.

384. Hasegawa, N., et al. (2000). Adhesive composition with heat, weather, wearing, and water resistance. In *Jpn Kokai Tokkyo Koho 2000086999*, Kanegafuchi Chemical Industry Co., Ltd., Japan, 12 pp.

385. Yamamoto, M., et al. (2001). Non-crosslinking type adhesive compositions, their manufacture and room temperature-bondable adhesive sheets containing them. In *Jpn. Kokai Tokkyo Koho 2001288442*, Nitto Denko Corp., Japan, 9 pp.

386. Yamamoto, M., et al. (2001). Adhesives containing (meth)acrylate type block copolymers and adhesive sheets and their manufacture. In *Jpn. Kokai Tokkyo Koho 2001234147*, Nitto Denko Corp., Japan, 15 pp.

387. Yamamoto, M., et al. (2001). Acrylic block polymer-based adhesive compositions with good adhesive properties, adhesive sheets, and their manufacture. In *Jpn. Kokai Tokkyo Koho 2001234146*, Nitto Denko Corp., Japan, 15 pp.

388. Yamamoto, M., et al. (2000). Pressure-sensitive adhesive composition containing styrene-acrylic block polymers. In *Eur. Pat. Appl. 1008640*, Nitto Denko Corporation, Japan, 32 pp.

389. Fujita, M., Kusakabe, M., and Kitano, K. (1999). High-solids curable siliconized vinyl adhesive compositions. In *PCT Int. Appl. WO 9905215*, Kaneka Corporation, Japan, 61 pp.

390. Nakano, F., et al. (2001). UV-curable adhesive compositions, manufacturing method thereof, and adhesive sheets therefrom. In *Jpn. Kokai Tokkyo Koho 2001303010*, Nitto Denko Corp., Japan, 13 pp.

391. Doi, T., et al. (2001). UV-crosslinkable pressure-sensitive adhesive compositions containing styrene-acrylic block copolymers and their preparation. In *Eur. Pat. Appl. 1127934*, Nitto Denko Corporation, Japan, 19 pp.

392. Lesko, P. M., and Blankenship, R. M. (2001). Pressure-sensitive adhesive compositions having improved peel strength and tack and method for coating substrates with the same. In *Eur. Pat. Appl. 1078935.*, Rohm and Haas Company, USA, 8 pp.

393. Kamifuji, F., et al. (2001). Block copolymer-based adhesive compositions and their production. In *Jpn. Kokai Tokkyo Koho 2001207148*, Nitto Denko Corp., Japan, 12 pp.
394. Nakagawa, Y., et al. (2000). Resin compositions, star polymers and process for their production. In *PCT Int. Appl. WO 0011056.*, Kaneka Corporation, Japan, 57 pp.
395. Nakagawa, Y., et al. (2000). Manufacture of alkenyl-terminated vinyl polymers, their silyl-terminated crosslinkable derivatives, and curable compositions containing them. In *Jpn. Kokai Tokkyo Koho 2000038404*, Kanegafuchi Chemical Industry Co., Ltd., Japan, 14 pp.
396. Bertin, D., Boutevin, B., and Robin, J.-J. (1999). Graft copolymer obtained by free-radical polymerization with stable free radicals, its preparation and uses. In *Eur. Pat. Appl. 906937*, Elf Atochem S.A., Fr., 17 pp.
397. Cardi, N., et al. (1999). Thermoplastic compositions containing compatibilizers. In *Eur. Pat. Appl. 906909*, Enichem S.p.A., Italy, 5 pp.
398. Anderson, L. G., et al. (2000). Thermosetting compositions containing carbamate-functional polymers prepared by atom-transfer radical polymerization and coatings therefrom. In *PCT Int. Appl. WO 0012566*, PPG Industries Ohio, Inc., USA, 59 pp.
399. Barkac, K. A., et al. (2000). Thermosetting compositions containing hydroxy-functional polymers prepared by atom-transfer radical polymerization and powder coatings therefrom. In *PCT Int. Appl. WO 0012579*, Ppg Industries Ohio, Inc., USA, 62 pp.
400. Barkac, K. A., et al. (2000). Thermosetting compositions containing epoxy-functional polymers prepared by atom-transfer radical polymerization and coatings therefrom. In *PCT Int. Appl. WO 0012581*, PPG Industries Ohio, Inc., USA, 63 pp.
401. Barkac, K. A., et al. (2000). Thermosetting compositions containing carboxylic acid functional polymers and epoxy functional polymers prepared by atom transfer radical polymerization. In *PCT Int. Appl. WO 0012583*, Ppg Industries Ohio, Inc., USA, 72 pp.
402. Barkac, K. A., et al. (2000). Thermosetting compositions containing carboxylic acid functional polymers prepared by atom transfer radical polymerization. In *PCT Int. Appl. WO 0012625*, Ppg Industries Ohio, Inc., USA, 60 pp.
403. Coca, S., et al. (2000). Thermosetting compositions containing epoxy-functional polymers prepared by atom-transfer radical polymerization and coatings therefrom. In *PCT Int. Appl. WO 0012580*, PPG Industries Ohio, Inc., USA, 67 pp.
404. Coca, S., et al. (2001). Pigment dispersions containing dispersants having pendent hydrophilic polymeric segments prepared by controlled radical polymerization. In *PCT Int. Appl. WO 0144388*, Ppg Industries Ohio, Inc., USA, 46 pp.
405. McCollum, G. J., Anderson, L. K., and Coca, S. (2000). Electrodepositable coating compositions comprising onium salt group-containing polymers prepared by atom transfer radical polymerization. In *PCT Int. Appl. WO 0012636*, Ppg Industries Ohio, Inc., USA, 53 pp.
406. McCollum, G. J., Anderson, L. K., and Coca, S. (2000). Electrodepositable coating compositions comprising amine salt group-containing polymers prepared by atom transfer radical polymerization. In *PCT Int. Appl. 0012637*, Ppg Industries Ohio, Inc., USA, 52 pp.
407. Fujii, T., Ogawa, T., and Yoshihara, I. (2001). Powder paint compositions containing vinyl polymers. In *Jpn. Kokai Tokkyo Koho 2001064573*, Kansai Paint Co., Ltd., Japan, 13 pp.
408. Andruzzi, L., et al. (2000). Block copolymers with perfluorinated side-chain blocks and preparation thereof for marine antifouling materials. In *PCT Int. Appl. WO 0042084*, Jotun A/S, Norway, 35 pp.
409. Masuda, T. (2000). Resin compositions for powder paints. In *Jpn. Kokai Tokkyo Koho 20000144017*, Kanegafuchi Chemical Industry Co., Ltd., Japan, 11 pp.
410. Kazmaier, P. M., et al. (1997). Processes for preparing telechelic, branched and star thermoplastic resins. In *Eur. Pat. Appl. 814097*, Xerox Corporation, USA, 17 pp.
411. Kazmaier, P. M., et al. (1996). Process for preparation of branched polymer for electrophotographic toners. In *Eur. Pat. Appl. 735064*, Xerox Corp., USA, 31 pp.
412. Keoshkerian, B., Drappel, S., and Georges, M. K. (1996). Ink-jet toner compositions and their preparation. In *Eur. Pat. Appl. 707018*, Xerox Corp., USA, 19 pp.

413. Keoshkerian, B., et al. (1997). Polymerization processes in the presence of free radical initiators, stable free radicals and electron acceptors. In *Eur. Pat. Appl. 773232*, Xerox Corp., USA, 22 pp.

414. Keoshkerian, B., et al. (1998). Polymerization and crosslinking processes and resin particles having narrow particle size distribution. In *Eur. Pat. Appl. 826697*, Xerox Corp., USA, 10 pp.

415. Keoshkerian, B., et al. (1999). Free-radical addition polymerization of monomer giving thermoplastics having a narrow polydispersity and useful in toners. In *Eur. Pat. Appl. 897930*, Xerox Corporation, USA, 16 pp.

416. Auschra, C., Muhlebach, A., and Eckstein, E. (2000). Pigment compositions containing polymers manufactured by atom-transfer-radical polymerization. In *PCT Int. Appl. WO 0040630*, Ciba Specialty Chemicals Holding Inc., Switz, 43 pp.

417. White, D., Coca, S., and O'Dwyer, J. B. (2001). Pigment dispersions containing dispersants prepared by controlled radical polymerization having hydrophilic and hydrophobic segments. In *PCT Int. Appl. WO 0144389*, Ppg Industries Ohio, Inc., USA, 40 pp.

418. Woodworth, B. E., et al. (2001). Pigment dispersions containing dispersants having pendent hydrophobic polymeric segments prepared by controlled radical polymerization. In *US 6306209*, PPG Industries Ohio, Inc., USA, 14 pp.

419. Olson, K. C., et al. (2001). Pigment dispersions containing dispersants prepared by controlled radical polymerization. In *US 6326420*, Ppg Industries Ohio, Inc., USA, 11 pp.

420. Fischer, M., et al. (1998). Process for preparing polymers in the presence of triazolyl radicals, block copolymer dispersants obtained thereby, and their use. In *PCT Int. Appl. WO 9811143*, BASF A.-G., Germany; Max-Planck-Gesellschaft zur Forderung der Wissenschaften e.V., Berlin; Fischer, Michael; Koch, Jurgen; Paulus, Wolfgang; Mullen, Klaus; Klapper, Markus; Steenbock, Marco; Colombani, Daniel, M., 44 pp.

421. Klier, J., Tucker, C. J., and Ladika, M. (2000). Preparation of a water-soluble polymeric dispersant having narrow polydispersity index. In *PCT Int. Appl. WO 0017242*, The Dow Chemical Company, USA, 17 pp.

422. Coca, S., and O'Dwyer, J. B. (2002). Pigment dispersions containing dispersants having core and arm star architecture prepared by controlled radical polymerization. In *US 6336966*, PPG Industries Ohio, Inc., USA, 18 pp.

423. Roos, S., Eisenberg, B., and Mueller, M. (2001). Atom transfer radical polymerization for manufacture of polyacrylates and polymethacrylates as lubricating oil additives. In *PCT Int. Appl. WO 0140317*, Rohmax Additives G.m.b.H., Germany, 43 pp.

424. Visger, D. C., and Lange, R. M. (1999). Vinyl aromatic-(vinyl aromatic-co-acrylic) block copolymers prepared by stabilized free radical polymerization. In *Eur. Pat. Appl. 945474*, The Lubrizol Corporation, USA, 22 pp.

425. Visger, D. C., and Lange, R. M. (2000). Radial polymers prepared by stabilized free radical polymerization and lubricating oil compositions. In *PCT Int. Appl. WO 0024795*, The Lubrizol Corporation, USA, 43 pp.

426. Scherer, M., and Souchik, J. (2001). Synthesis of long-chain polymethacrylates by atom transfer radical polymerization for manufacture of lubricating oil additives. In *PCT Int. Appl. WO 0140334*, Rohmax Additives G.m.b.H., Germany, 48 pp.

427. Scherer, M., Souchik, J., and Bollinger, J. M. (2001). Gradient atom transfer radical polymerization of long-chain alkyl methacrylates for manufacture of lubricating oil additives. In *PCT Int. Appl. WO 0140333*, Rohmax Additives G.m.b.H., Germany, 48 pp.

428. Fujita, T., Nakagawa, Y., and Hasegawa, N. (2001). Curable rubber compositions containing polyethers and vinyl polymers with good compatibility. In *Jpn. Kokai Tokkyo Koho 2001329065*, Kanegafuchi Chemical Industry Co., Ltd., Japan, 34 pp.

429. Hasegawa, N., and Nakagawa, Y. (2001). Sealing compositions comprising living-polymerized vinyl polymers for siding boards. In *Jpn. Kokai Tokkyo Koho 2001271055*, Kanegafuchi Chemical Industry Co., Ltd., Japan, 34 pp.

430. Hasegawa, N., et al. (2001). Curable compositions comprising crosslinkable silyl group-containing vinyl polymers and siloxane bond-containing compounds. In *Jpn. Kokai Tokkyo Koho 2001011321*, Kanegafuchi Chemical Industry Co., Ltd., Japan, 22 pp.

431. Hasegawa, N., and Nakagawa, Y. (2001). Fireproof sealing compositions containing crosslinkable vinyl polymers and blowing agents. In *Jpn. Kokai Tokkyo Koho 2001354830*, Kanegafuchi Chemical Industry Co., Ltd., Japan, 34 pp.

432. Brinkmann-rengel, S., et al. (2000). Elastomeric block copolymers, their production and their use. In *PCT Int. Appl. WO 0063267*, Basf A.-G., Germany, 52 pp.

433. Mulhaupt, R., et al. (2000). Method for producing graft copolymers containing an elastic backbone. In *PCT Int. Appl. WO 0077060*, BASF A.-G., Germany, 43 pp.

434. Brocchini, S. J., and Godwin, A. (2001). Preparation of uniform molecular weight polymers by controlled radical polymerization. In *PCT Int. Appl. WO 0118080*, School of Pharmacy, University of London, UK, 65 pp.

435. Won, C.-Y., Zhang, Y., and Chu, C.-C. (2000). Hydrogel-forming system with hydrophobic and hydrophilic components. In *PCT Int. Appl. WO 0060956*, Cornell Research Foundation, Inc., USA, 38 pp.

436. Adams, G., et al. (2000). Polysiloxane block copolymers in topical cosmetic and personal care compositions. In *PCT Int. Appl. WO 0071607*, Unilever PLC, UK; Unilever NV; Hindustan Lever Limited, 31 pp.

437. Adams, G., et al. (2000). Polysiloxane block copolymers in topical cosmetic and personal care compositions. In *PCT Int. Appl. WO 0071606*, Unilever PLC, UK; Unilever NV; Hindustan Lever Limited, 40 pp.

438. Midha, S., and Nijakowski, T. R. (1998). Personal care compositions containing graft polymers. In *PCT Int. Appl. WO 9851261*, The Procter & Gamble Company, USA, 43 pp.

439. Kitano, K., and Nakagawa, Y. (2000). Resin compositions for making toner-transfer rollers of electrophotographic copiers or printers and the rollers. In *PCT Int. Appl. WO 0011082*, Kaneka Corporation, Japan, 59 pp.

440. Schimmel, K. F., et al. (2001). Block copolymers of (meth)acrylate monomers as flow control agents for coatings. In *US 6288173*, Ppg Industries Ohio, Inc., USA, 13 pp.

441. Schimmel, K. F., et al. (2001). Thermosetting coating compositions containing flow modifiers prepared by controlled radical polymerization. In *US 6197883*, PPG Industries Ohio, Inc., USA, 18 pp.

442. Stenzel-Rosenbaum, M. H., et al. (2001). Porous polymer films and honeycomb structures made by the self-organization of well-defined macromolecular structures created by living radical polymerization techniques. *Angew. Chem., Int. Ed.*, 40(18): 3428–3432.

443. Wu, C., et al. (2001). Novel synthesis of long multi-block heteropolymer chains with an ordered sequence and controllable block lengths. *Chin. J. Polym. Sci.*, 19(5): 451–454.

Index

Milton Keynes UK
Ingram Content Group UK Ltd.
UKHW051857071024
449327UK00025B/1995